Clinical Radiology
of Exotic Companion Mammals

Clinical Radiology
of Exotic Companion Mammals

Vittorio Capello, DVM
Clinica Veterinaria S.Siro
Clinica Veterinaria Gran Sasso
Milano, Italy

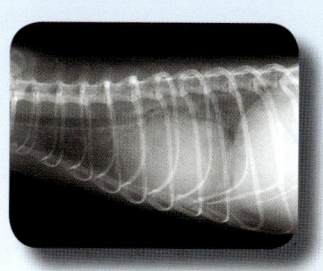

Angela M. Lennox, DVM, Dipl ABVP (Avian)
Avian & Exotic Animal Clinic
Indianapolis, IN, USA

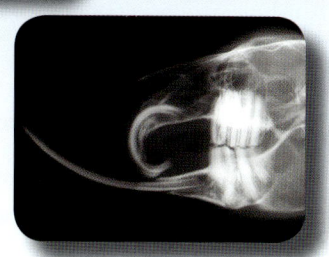

With:

William R. Widmer, DVM, MS, Dipl ACVR
Professor of Diagnostic Imaging
Department of Veterinary Clinical Sciences
School of Veterinary Medicine
Purdue University
Lafayette, IN, USA

WILEY-BLACKWELL
A John Wiley & Sons, Ltd., Publication

Edition first published 2008
© 2008 Vittorio Capello and Angela Lennox

Blackwell Publishing was acquired by John Wiley & Sons in February 2007. Blackwell's publishing program has been merged with Wiley's global Scientific, Technical, and Medical business to form Wiley-Blackwell.

Editorial Office
2121 State Avenue, Ames, Iowa 50014-8300, USA

For details of our global editorial offices, for customer services, and for information about how to apply for permission to reuse the copyright material in this book, please see our website at www.wiley.com/wiley-blackwell.

Authorization to photocopy items for internal or personal use, or the internal or personal use of specific clients, is granted by Blackwell Publishing, provided that the base fee is paid directly to the Copyright Clearance Center, 222 Rosewood Drive, Danvers, MA 01923. For those organizations that have been granted a photocopy license by CCC, a separate system of payments has been arranged. The fee codes for users of the Transactional Reporting Service are ISBN-13: 978-0-8138-1049-2/2008.

Designations used by companies to distinguish their products are often claimed as trademarks. All brand names and product names used in this book are trade names, service marks, trademarks or registered trademarks of their respective owners. The publisher is not associated with any product or vendor mentioned in this book. This publication is designed to provide accurate and authoritative information in regard to the subject matter covered. It is sold on the understanding that the publisher is not engaged in rendering professional services. If professional advice or other expert assistance is required, the services of a competent professional should be sought.

Library of Congress Cataloguing-in-Publication Data
Capello, Vittorio.
  Clinical radiology of exotic companion mammals / Vittorio Capello, Angela M. Lennox ; with William R. Widmer.
    p. ; cm.
  Includes bibliographical references and index.
  ISBN-13: 978-0-8138-1049-2 (alk. paper)
  ISBN-10: 0-8138-1049-3 (alk. paper)
  1. Veterinary radiography. I. Lennox, Angela M. II. Widmer, William R. III. Title.
  [DNLM: 1. Radiography--veterinary. 2. Animals, Domestic. SF 757.8 C238c 2008]
  SF757.8.C26 2008
  636.089'60757--dc22
                      2008020121

A catalogue record for this book is available from the U.S. Library of Congress.

For Wiley-Blackwell:
Graphic Designer: **Vittorio Capello, DVM**

Set in Baskerville by Vittorio Capello and Angela Lennox
Printed in Singapore by Markono Print Media Pte. Ltd.

Disclaimer

  The contents of this work are intended to further general scientific research, understanding, and discussion only and are not intended and should not be relied upon as recommending or promoting a specific method, diagnosis, or treatment by practitioners for any particular patient. The publisher and the author make no representations or warranties with respect to the accuracy or completeness of the contents of this work and specifically disclaim all warranties, including without limitation any implied warranties of fitness for a particular purpose. In view of ongoing research, equipment modifications, changes in governmental regulations, and the constant flow of information relating to the use of medicines, equipment, and devices, the reader is urged to review and evaluate the information provided in the package insert or instructions for each medicine, equipment, or device for, among other things, any changes in the instructions or indication of usage and for added warnings and precautions. Readers should consult with a specialist where appropriate. The fact that an organization or Website is referred to in this work as a citation and/or a potential source of further information does not mean that the author or the publisher endorses the information the organization or Website may provide or recommendations it may make. Further, readers should be aware that Internet Websites listed in this work may have changed or disappeared between when this work was written and when it is read. No warranty may be created or extended by any promotional statements for this work. Neither the publisher nor the author shall be liable for any damages arising herefrom.

1 2008

Angela M. Lennox, DVM,
Dipl ABVP (Avian)
Avian & Exotic Animal Clinic
Indianapolis, IN USA

Vittorio Capello, DVM
Clinica Veterinaria S.Siro
Clinica Veterinaria Gran Sasso
Milano, Italy

Dr. Angela Lennox graduated in 1989 from Purdue University School of Veterinary Medicine, and has practiced avian and exotic animal medicine exclusively since 1991. She is the owner of the Avian & Exotic Animal Clinic of Indianapolis.

Dr. Lennox was awarded board certification in avian medicine by the American Board of Veterinary Practitioners in 2003, and is an adjunct professor at Purdue University where she teaches courses in exotic pet medicine to both veterinary and veterinary technician students.

She is Past President of the Association of Exotic Mammal Veterinarians (2004-2007).

Dr. Lennox has lectured extensively in the US and internationally, and was voted Exotic Animal Speaker of the Year at the North American Veterinary Conference in 2007.

She is editor of *Rabbit and Rodent Dentistry Handbook* (Blackwell; formerly Zoological Education Network) and has both written for and served as guest editor for numerous publications, including *Journal of Avian Medicine and Surgery, Seminars in Exotic Pet Medicine, Veterinary Clinics of North America, Exotic DVM* magazine and the *Journal of Exotic Mammal Medicine and Surgery.* She is a member of the advisory board of *Exotic DVM* magazine and Small Mammal Program Chair, for the North American Veterinary Conference (NAVC).

Dr. Lennox was awarded Exotic DVM of the Year at the International Conference on Exotics (ICE) in 2005.

Dr. Vittorio Capello graduated in 1989 from the School of Veterinary Medicine of the University of Milano, Italy. He has practiced exotic animal medicine exclusively since 1993, and provides professional services for two veterinary clinics in Milano. Dr. Capello's focus has been exotic companion mammal medicine and surgery.

Dr. Capello has lectured, published, and taught exotic animal courses and practical labs throughout Italy, other parts of Europe, and the United States. He has been a guest lecturer for four years at the International Conference on Exotics, where he was voted Most Appreciated Speaker in 2003 and 2004, and received the Exotic DVM Paper of the Year Award in 2005. He has also been a speaker at the European Congress of Veterinary Dentistry, the North American Veterinary Conference, the British Small Animal Veterinary Conference, and a speaker and teacher for the Association of Exotic Mammal Veterinarian wet labs.

He was an associate professor for the year 2005/06 at the School of Veterinary Medicine of Milano, Italy.

Dr. Capello is the author of *Rabbit and Rodent Dentistry Handbook* (Blackwell; formerly Zoological Education Network) and has written 30 articles for *Exotic DVM* magazine, *Journal of Exotic Mammal Medicine and Surgery,* and for the *Journal of Exotic Pet Medicine.* Other works include the Small Rodent Surgeries section in *The Exotic Guidebook* (ZEN).

He is a member of the advisory board of *Exotic DVM* magazine, and member of the Association of Exotic Mammal Veterinarians.

William R. Widmer, DVM, MS, Dipl ACVR
Professor of Diagnostic Imaging
Department of Veterinary Clinical Sciences
School of Veterinary Medicine
Purdue University
Lafayette, IN USA

William Widmer is a Professor of Diagnostic Imaging at the Purdue University School of Veterinary Medicine, where he has been a faculty member since 1998. Upon graduation from Purdue in 1969, he was engaged in private practice for 15 years before returning to Purdue for graduate studies and a residency in radiology. He earned a Master of Science degree in 1986, a residency certificate in 1987 and was awarded Diplomate status by the American College of Veterinary Radiology in 1988. Dr. Widmer was granted sabbatical leave in 1996 and 2003 to study magnetic resonance imaging and helical computed tomography at the Indiana University School of Medicine. His résumé includes collaborative research on various forms of arthritis, intervertebral disk disease, and urinary bladder cancer in animals, and authorship of numerous scientific articles proceedings and abstracts. Dr. Widmer has contributed to numerous textbooks, including the *Textbook of Veterinary Diagnostic Radiology,* edited by DE Thrall. Teaching endeavors at Purdue University include courses taught at the Schools of Veterinary Medicine and Pharmacy. He is a frequent lecturer at state, national, and international venues and has been active in the American College of Veterinary Radiology. Dr. Widmer is a reviewer for scientific journals and acts as a consultant for state and local veterinarians.

ACKNOWLEDGMENTS

Circle City Veterinary Clinic, Indianapolis, IN

Sarah Dehn, RVT
Avian & Exotic Animal Clinic, Indianapolis, IN

Amy Durand
Avian & Exotic Animal Clinic, Indianapolis, IN

Kelly Edgley, DVM
Avian & Exotic Animal Clinic, Indianapolis, IN

Amanda Laney, RVT
Veterinary Specialty Center, Indianapolis, IN

Susan E. Orosz, PhD, DVM, Dipl ABVP (Avian), Dipl ECAMS
Bird and Exotic Pet Wellness Center, Toledo, OH

Germana Scerbanenco, DVM
Milano, Italy

Pat Thorne
Avian & Exotic Animal Clinic, Indianapolis, IN

CONTRIBUTORS

Claudio Bussadori, DVM, Dipl ECVIM
Clinica Veterinaria Gran Sasso, Milano, Italy

Alberto Cauduro, DVM
Milano, Italy

Stefania Gianni, DVM
Clinica Veterinaria S.Siro, Milano, Italy

Margherita Gracis, DVM, Dipl AVDC, Dipl EVDC
Clinica Veterinaria Gran Sasso, Milano, Italy

Cathy Johnson-Delaney, DVM, Dipl ABVP (Avian)
Avian & Exotic Animal Medical Center
Kirkland, WA

Jöerg Mayer, DVM, MSc, MRCVS
Clinical Assistant Professor
Tufts University
Cummings School of Veterinary Medicine
North Grafton, MA

Yasutsugu Miwa, DVM
Veterinary Medical Center
University of Tokyo, Japan

Purdue University School of Veterinary Medicine, Large Animal Clinic
West Lafayette, IN

Giorgio Romanelli, DVM, Dipl ECVS
Clinica Veterinaria Nerviano, Nerviano, Milano, Italy

FOREWORD

Radiology is the extension of the clinician's eyes and fingers for rapidly determining a working diagnosis. Normal radiographic standards for anatomy and common clinical conditions have been determined in small and large animal medicine. Not so in small mammal medicine. Normals and standards have not been established but rather assumed by clinicians as they practice in their own vacuums-- their own clinics. With a greater understanding of normal radiographic anatomy, small mammal practitioners are able to visualize more subtle abnormalities. This greatly enhances the quality of care for their small mammal patients.

For the first time, Drs. Capello and Lennox have succeeded in bringing together the radiographic features of normal anatomy and those for common clinical conditions. The quality of their expansive work is evident in just a flip of several pages. The anatomy for a wide range of species is presented clearly and concisely, with descriptions to enhance understanding of the radiographic images. Clinicians will find the text user-friendly, allowing them to make more accurate diagnoses more quickly.

Radiography is the most common imaging modality used in practice, but CT and MRI are allowing us to visualize anatomy like never before. This is an exciting time for extending the limits of anatomic and hence pathologic knowledge by the use of these newer technologies. While the bulk of the text describes radiographic anatomy and common pathologies encountered in practice, it includes CT images as well. These images show us the level of detail of the future, making anatomy an even broader pillar for clinical medicine. From an anatomist's and exotic animal clinician's perspective, I was truly excited to see this awesome book. It kept me turning the pages to drink in the images much longer than my staff wanted me to! Not only should it find a home with every clinician who does small mammal medicine, it will become the standard for every one of us. This is truly great stuff!

Susan E. Orosz
PhD, DVM, Dipl ABVP (Avian), Dipl ECAMS

PREFACE

Radiology is a common diagnostic procedure in all fields of clinical veterinary practice, and exotic animal medicine is no exception. In addition to hematology and biochemistry analysis, radiology is likely the most commonly performed diagnostic test. With this in mind, Clinical Radiology of Exotic Companion Mammals may not appear to be a presentation of the "cutting edge" of exotic animal medicine. In actuality, there is very little reported in the scientific literature on this subject, with the exception of the very fine Silverman/Tell *Atlas of Exotic Mammal Radiology* which gives an outstanding presentation of normal radiographic anatomy of the most commonly encountered exotic pet mammals. This book intends to take that basic knowledge a step further, and includes:

- A wider range of exotic mammal species beyond rabbits, rodents and ferrets, including species such as marsupials, hedgehogs, and pot belly pigs. As few experts agree on an official list of "appropriate exotic companion mammals" our text has chosen to include most of the species (with a few additions) included for coverage in the American Board of Veterinary Practitioners upcoming Exotic Companion Mammal specialty boards. Due to space considerations (and lack of personal clinical material), a number of animals occasionally encountered by exotic mammal practitioners were not included, specifically primates, foxes, and the kinkajou.

- An extensive review of radiographic pathologic patterns. This represents the most important section of this book, and abnormalities are presented chapter by chapter with normal radiographic patterns and related gross and clinical images. Of course the scope of potential radiographic abnormalities that might be faced in clinical practice is endless, but this book attempts to aid the practitioner in familiarization with a wide range of common and uncommon lesions.

- An extensive introductory chapter, which includes a review of the basic principles of radiology, instruments, materials, and radiologic techniques. This section makes no claim to replace traditional radiology textbooks in this regard, but attempts to focus on those aspects which can "make or break" radiographic quality when applied to exotic companion mammals.

- An extensive visual section, dedicated to proper positioning of the patient, which is absolutely critical and often problematic, especially in smaller-sized patients. We hope this section will be useful for the exotic mammal technician as well.

- A review of anesthetic protocols useful for diagnostic radiology when appropriate.

- A brief discussion of more advanced radiologic techniques including CT scan and High Definition Digital Radiology, and their application to exotic companion mammals.

The visual impact and graphics have been prepared and arranged to allow quick, easy reference in the clinical setting. Text and captions have been limited to essential content, and notes and legends incorporated onto the image to prevent the difficulty of matching numbers/letters to the actual image.

We hope this book will be useful both for practitioners who make exotic companion mammals their practice focus, and for those willing to include these special pets into their regular caseload. In the end the goal can be best summed up as this: we hope this reference can eliminate the number of times we've had to examine a radiograph, and be faced with the fact we might know what we were seeing "if only this were a very small dog or cat…"

Vittorio Capello

Angela M. Lennox

CONTENTS

Chapter 1
The BASICS of RADIOLOGY

**Introduction to Radiology
of the Exotic Companion Mammal**   **2**
Principles of X-ray Production and Image Formation   2
X-ray Production   2
Control of Radiographic Exposure
   and Adjusting Settings on the X-ray Machine   2

Recording the X-ray image   **3**
Cassette and Films   3
Factors Affecting Radiographic Quality   4

**Radiographic Equipment
for the Small Exotic Mammal Patient**   **4**
Obtaining the Radiograph   6

Digital Radiography   **7**

X-ray Projections   **8**
Standard Nomenclature   8

Patient Positioning   **9**
Manual Restraint vs. Pharmacologic Restraint   9
Sedation and Anesthetic Considerations   9
Radiation Safety   9

Patient Positioning Techniques   **10**
Total Body   10
Whole Body Skeleton   14
Head   15
Thorax; Cervical and Thoracic Vertebral Column   26
Abdomen and Lumbar Vertebral Column   28
Thoracic Limb   30
Pelvis and Pelvic Limb   33

Common Errors in Radiographic Technique   **37**

Processing X-ray Film   **39**
Storage of Radiographs   39

Contrast Radiography   **40**
Myelography   40

References   43

**Introduction to Computed Tomography
in Exotic Companion Mammals**   **44**
CT of Exotic Companion Mammals   44
Basic Operation of a CT scanner   **46**
Two-Dimensional Reconstruction   **47**

Three-Dimensional Reconstructions   **48**
Volume Rendering   48
Surface Rendering   49

References   49

Radiation Safety   **50**

References   51

Chapter 2
RABBIT

The Normal Head   **54**
Lateral Projection   54
Oblique Projection   55
Ventrodorsal Projection   56
Rostrocaudal Projection   57
Intraoral Projections   58
The Nasolacrimal Duct   59

Abnormalities of the Head   **60**
Diseases of Incisor Teeth   60
Diseases of Cheek Teeth   62
Periapical Infections and Osteomyelitis   67
Miscellaneous   73

Computed Tomography of the Head   **76**
The Normal Head   76
Acquired Dental Disease,
Periapical Infection, and Osteomyelitis   80

The Normal Total Body   **88**
Lateral Projection   88
Ventrodorsal Projection   89

The Normal Thorax
The Cervical and Thoracic Vertebral Column   **90**
Lateral Projection   90
Ventrodorsal Projection   92

Abnormalities of the Thorax   **94**
Diseases of the Lungs   94
Diseases of the Mediastinum   97

The Normal Abdomen
The Lumbar Vertebral Column   **100**
Lateral Projection   100
Ventrodorsal Projection   102
Miscellaneous   104

Abnormalities of the Abdomen — 106
Diseases of the Stomach — 106
Diseases of the Intestine — 110
Diseases of the Liver — 116
Diseases of the Kidneys and Ureters — 118
Diseases of the Urinary Bladder and Urethra — 122
Diseases of the Uterus and Vagina — 126

Abnormalities of the Vertebral Column — 130
Diseases of the Thoracic Vertebral Column — 130
Diseases of the Lumbosacral and Caudal Vertebral Column — 131

Myelography of the Normal Vertebral Column — 134
Lateral Projection — 134
Ventrodorsal Projection — 135

Myelography of the Abnormal Vertebral Column — 136
Lateral Projection — 136
Ventrodorsal Projection — 137

The Normal Thoracic Limb — 138
Lateral Projection — 138
Caudocranial Projection of the Proximal Thoracic Limb — 140
Craniocaudal Projection of the Distal Thoracic Limb — 141

Abnormalities of the Thoracic Limb — 142
Diseases of the Humerus — 142
Diseases of the Radius, Ulna, and Elbow joint — 143
Lesions of the Carpus, Metacarpus, and Phalanges — 147

The Normal Pelvic Limb — 148
Lateral Projection — 148
Ventrodorsal Projection of the Pelvis
Craniocaudal Projection of the Proximal Pelvic Limb — 150
Craniocaudal Projection of the Distal Pelvic Limb — 151

Abnormalities of the Pelvic Limb — 152
Diseases of the Pelvis — 152
Diseases of the Femur — 155
Diseases of the Tibia and Fibula — 157
Diseases of the Tarsus, Metatarsus, and Phalanges — 163

References — 164

Chapter 3
GUINEA PIG

The Normal Head — 168
Lateral Projection — 168
Oblique Projection — 169
Ventrodorsal Projection — 170
Rostrocaudal Projection — 171

Abnormalities of the Head — 172
Diseases of Incisor Teeth — 172
Diseases of Cheek Teeth — 174
Periapical Infections and Osteomyelitis — 176

Computed Tomography of the Head — 178
The Normal Head — 178
Acquired Dental Disease, Periapical Infection and Osteomyelitis — 182

The Normal Total Body — 184
Lateral Projection — 184
Ventrodorsal Projection — 185

The Normal Thorax
The Cervical and Thoracic Vertebral Column — 186
Lateral Projection — 186
Ventrodorsal Projection — 188

Abnormalities of the Thorax — 190
Diseases of the Lungs — 190
Diseases of the Chest — 191

The Normal Abdomen
The Lumbar Vertebral Column — 192
Lateral Projection — 192
Ventrodorsal Projection — 194
Miscellaneous — 196

Abnormalities of the Abdomen — 198
Diseases of the Stomach — 198
Diseases of the Intestine — 201
Diseases of the Kidneys and Ureters — 202
Diseases of the Urinary Bladder and Urethra — 204
Diseases of the Female Genital Tract — 206
Miscellaneous — 209

The Normal Thoracic Limb — 210
Lateral Projection — 210
Caudocranial Projection of the Proximal Thoracic Limb — 211
Craniocaudal Projection of the Distal Thoracic Limb — 212

Abnormalities of the Thoracic Limb — 213

The Normal Pelvic Limb — 214
Lateral Projection — 214
Ventrodorsal Projection of the Pelvis
Craniocaudal Projection of the Proximal Pelvic Limb — 216
Craniocaudal Projection of the Distal Pelvic Limb — 217

Abnormalities of the Pelvic Limb — 218
Diseases of the Femur — 218
Diseases of the Stifle Joint — 219
Diseases of the Tibia and Fibula — 220
Diseases of the Tarsus, Metatarsus, and Phalanges — 221

References — 221

Chapter 4
CHINCHILLA

The Normal Head — 224
Lateral Projection — 224
Oblique Projection — 225
Ventrodorsal Projection — 226
Rostrocaudal Projection — 227

Abnormalities of the Head — 228
Diseases of Incisor Teeth — 228
Diseases of Cheek Teeth — 229

Computed Tomography of the Head — 232
The Normal Head — 232
Acquired Dental Disease — 236

The Normal Total Body — 238
Lateral Projection — 238
Ventrodorsal Pojection — 239

The Normal Thorax
The Cervical and Thoracic Vertebral Column — 240
Lateral Projection — 240
Ventrodorsal Projection — 241

The Normal Abdomen
The Lumbar Vertebral Column — 242
Lateral Projection — 242
Ventrodorsal Projection — 244
Miscellaneous — 246

Abnormalities of the Abdomen — 248
Diseases of the Gastrointestinal Tract — 248
Diseases of the Female Genital Tract — 252

Abnormalities of the Vertebral Column — 253

The Normal Thoracic Limb — 254
Lateral Projection — 254
Craniocaudal Projection — 255

The Normal Pelvic Limb — 256
Lateral Projection — 256
Ventrodorsal Projection of the Pelvis
Craniocaudal Projection of the Proximal Pelvic Limb — 258
Craniocaudal Projection of the Distal Pelvic Limb — 259

Abnormalities of the Pelvic Limb — 260
Diseases of the Pelvis — 260
Diseases of the Tibia and Fibula — 261

References — 262

Chapter 5
DEGU

The Normal Head — 266
Lateral Projection — 266
Oblique Projection — 267
Ventrodorsal Projection — 268
Rostrocaudal Projection — 268

Abnormalities of the Head — 269
Diseases of Cheek teeth — 269

The Normal Whole Body Skeleton — 270
Lateral Projection — 270
Ventrodorsal Projection — 271
Miscellaneous — 272

References — 273

Chapter 6
RAT

The Normal Head — 276
Lateral Projection — 276
Oblique Projection — 277
Ventrodorsal Projection — 278
Rostrocaudal Projection — 278

Abnormalities of the Head — 279
Diseases of Cheek Teeth — 279

The Normal Total Body — 280
Lateral Projection — 280
Ventrodorsal Projection — 282

Miscellaneous Abnormalities — 284
Diseases of the Thorax — 284
Diseases of the Abdomen — 285
Diseases of the Limbs — 287

References — 288

Chapter 7
MOUSE

The Normal Whole Body Skeleton — 292
Lateral Projection — 292
Ventrodorsal Projection — 293

Miscellaneous Abnormalities — 294
Diseases of the Abdomen — 294

References — 295

Chapter 8
HAMSTER

The Normal Head — 298
Lateral Projection — 298
Oblique Projection — 299
Ventrodorsal Projection — 300
Rostrocaudal Projection — 301

Abnormalities of the Head — 302
Diseases of Incisor Teeth — 302
Diseases of Cheek Teeth — 303

The Normal Whole Body Skeleton — 304
The Golden Hamster — 304
Lateral Projection — 304
Ventrodorsal Projection — 306
Miscellaneous — 309

The Russian Hamster — 310
Lateral Projection — 310
Ventrodorsal Projection — 311

Abnormalities of the Abdomen — 312
Diseases of the Abdominal Cavity — 312
Diseases of the Gastrointestinal Tract — 312
Diseases of the Urogenital Tract — 314

Abnormalities of the Vertebral Column — 315

Abnormalities of the Thoracic Limb — 316

Abnormalities of the Pelvic Limb — 319

References — 323

Chapter 9
PRAIRIE DOG
and other SQUIRREL-LIKE RODENTS

The Normal Head — 326
Lateral Projection — 326
Oblique Projection — 327
Ventrodorsal Projection — 328
Rostrocaudal Projection — 328
Intraoral Projections — 329

Abnormalities of the Head — 330
Diseases of Incisor Teeth — 330
Diseases of Cheek Teeth — 333

The Normal Total Body — 334
Lateral Projection — 334
Ventrodorsal Projection — 335

The Normal Thorax
The Cervical and Thoracic Vertebral Column — 336
Lateral Projection — 336
Ventrodorsal Projection — 338

The Normal Abdomen
The Lumbar Vertebral Column — 340
Lateral Projection — 340
Ventrodorsal Projection — 342

Abnormalities of the Abdomen — 344
Diseases of the Intestine — 344

The Normal Thoracic Limb — 346
Lateral Projection — 346
Caudocranial Projection — 347

The Normal Pelvic Limb — 348
Lateral Projection — 348
Craniocaudal Projection — 349

The Chipmunk — 350
The Normal Whole Body Skeleton — 350
Lateral Projection — 350
Ventrodorsal Projection — 351

Abnormalities of the Head of other Squirrel-like Rodents — 352

Miscellaneous Abnormalities of other Squirrel-like Rodents — 354

References — 355

Chapter 10
FERRET

The Normal Head — 358
Lateral Projection — 358
Oblique Projection — 359
Ventrodorsal Projection — 360
Rostrocaudal Projection — 361

Abnormalities of the Head — 362

The Normal Total Body — 364
Lateral Projection — 364
Ventrodorsal Projection — 365

The Normal Thorax
The Cervical and Thoracic Vertebral Column — 366
Lateral Projection — 366
Ventrodorsal Projection — 367

Abnormalities of the Thorax — 368
Diseases of the Lungs — 368
Diseases of the Heart — 369
Diseases of the Mediastinum — 370
Diseases of the Ribs — 373

The Normal Abdomen
The Lumbar Vertebral Column — 374
Lateral Projection — 374
Ventrodorsal Projection — 376

Abnormalities of the Abdomen — 378
Diseases of the Stomach — 378
Diseases of the Intestine — 380
Diseases of the Liver — 384
Diseases of the Spleen — 385
Diseases of the Urogenital Tract and the Adrenal Glands — 386
Miscellaneous — 389

The Normal Thoracic Limb — 390
Lateral Projection — 390
Craniocaudal Projection — 391

Abnormalities of the Thoracic Limb — 392
Diseases of the Scapula — 392
Diseases of the Radius, Ulna, and Elbow Joint — 394

The Normal Pelvic Limb — 398
Lateral Projection — 398
Ventrodorsal Projection of the Pelvis
Craniocaudal Projection of the Proximal Pelvic Limb — 400
Craniocaudal Projection of the Distal Pelvic limb — 401

Abnormalities of the Pelvic limb — 402
Diseases of the Pelvis — 402
Diseases of the Femur — 403
Diseases of the Tibia and Fibula — 407
Diseases of the Tarsus, Metatarsus, and Phalanges — 408

References — 409

Chapter 11
SKUNK

The Normal Head — 412
Lateral Projection — 412

The Normal Thorax
The Cervical and Thoracic Vertebral Column — 413
Lateral Projection — 413

Abnormalities of the Thorax — 414
Diseases of the Mediastinum — 414

The Normal Abdomen
The Lumbar Vertebral Column — 418
Lateral Projection — 418

The Normal Thoracic Limb — 420
Lateral Projection — 420

Abnormalities of the Thoracic Limb — 421

The Normal Pelvic Limb — 422
Lateral Projection — 422
Ventrodorsal Projection of the Pelvis — 423
Craniocaudal Projection of the Distal Pelvic Limb — 424

Abnormalities of the Pelvic Limb — 425
Diseases of the Femur — 425
Diseases of the Tibia and Fibula — 426
Miscellaneous — 427

References — 427

Chapter 12
SUGAR GLIDER

The Normal Whole Body Skeleton — 430
Lateral Projection — 430
Ventrodorsal Projection — 432

References — 435

Chapter 13
VIRGINIA OPOSSUM

The Normal Head — 438
Lateral Projection — 438
Ventrodorsal Projection — 439

The Normal Thorax
The Cervical and Thoracic Vertebral Column — 440
Lateral Projection — 440
Ventrodorsal Projection — 441

The Normal Abdomen
The Lumbar Vertebral Column — 442
Lateral Projection — 442
Ventrodorsal Projection — 443

The Normal Thoracic Limb — 444
Lateral Projection — 444
Caudocranial Projection of the Proximal Thoracic Limb — 448
Craniocaudal Projection of the Distal Thoracic Limb — 449

The Normal Pelvic Limb — 450
Ventrodorsal Projection of the Pelvis
Craniocaudal Projection of the Proximal Pelvic Limb — 450
Craniocaudal Projection of the Distal Pelvic Limb — 451
Lateral Projection of the Distal Pelvic Limb — 452

Miscellaneous Abnormalities — 453
Diseases of the Teeth — 453
Diseases of the Skeleton — 454

References — 455

Chapter 14
POTBELLIED PIG

The Normal Head — 458
Lateral Projection — 458
Oblique Projection — 460

The Normal Thorax
The Cervical and Thoracic Vertebral Column — 462
Lateral Projection — 462
Ventrodorsal Projection — 464

The Normal Abdomen
The Lumbar Vertebral Column — 466
Lateral Projection — 466
Ventrodorsal Projection — 468

The Normal Thoracic Limb — 470
Lateral Projection of the Distal Thoracic Limb — 470
Craniodaudal Projection of the Distal Thoracic Limb — 471

The Normal Pelvic Limb — 472
Lateral Projection of the Pelvis
and of the Proximal Pelvic Limb — 472
Ventrodorsal Projection of the Pelvis
Craniocaudal Projection of the Proximal Pelvic Limb — 474
Lateral Projecton of the Distal Pelvic Limb — 476
Craniocaudal Projection of the Distal Pelvic Limb — 477

Abnormalities of the Pelvic Limb — 478

References — 479

Chapter 15
AFRICAN PYGMY HEDGEHOG

The Normal Whole Body Skeleton — 482
Lateral Projection — 482
Ventrodorsal Projection — 483

The Normal Head — 484

Miscellaneous Abnormalities — 485
Diseases of the Thorax — 485
Diseases of the Abdomen — 486

References — 487

Final Cut — 489

References — 491

Index — 495

The BASICS of RADIOLOGY

V. Capello, A. Lennox, W.R. Widmer

BASICS of RADIOLOGY

Introduction to Radiology of the Exotic Companion Mammal

This chapter presents the basic principles of physics regarding radiography of exotic companion mammals, that are most useful to the practitioner. While general principles do not differ among species, there are unique considerations for radiography of small species typically encountered in an exotic mammal practice.

Additional information is available elsewhere (see references).

Principles of X-ray Production and Image Formation

X-rays are a type of ionizing electromagnetic radiation. Non-ionizing forms of electromagnetic radiation include light, radio and radar waves. Electromagnetic radiation is an energy form that travels through space as a combination of electric and magnetic fields. Electromagnetic radiation does not require a medium for transmission and travels at the speed of light, about 3.0×10^8 m/s.

X-rays have both wave and particle characteristics. They have short wavelengths (between 0.1 and 1.0 Ångströms (1Å = 10^{-10} m., i.e. 1/10 of one millionth of a mm) allowing penetration of materials that will absorb or reflect visible light. X-rays also behave as if they were small "packets" of energy called *quanta* or photons. It is sometimes easier to understand the actions of x-rays when they are considered as photons of energy, especially for mathematical calculations. When x-rays interact with living tissues, biologic molecules (DNA) may be damaged by ionization or excitation. Fluorescence occurs when x-rays strike certain inorganic materials, causing a brief flash of light. This property is used to help record the radiographic image with conventional screen-film systems.

Radiography is the use of x-rays to make a film record of the internal structure of the body. Some x-rays are absorbed, but those exiting the patient are recorded on a photographic film, making a 2-dimensional image of a 3-dimensional patient. The process of making a radiographic image is similar to making a black and white photographic image, except x-rays are used instead of light. Radiology is the mainstay of veterinary diagnostic imaging because of its cost effectiveness and user-friendly nature.

X-ray Production

A large diode called an x-ray tube is used to generate x-rays (Figure 1.1). The cathode has a small filament that is heated in a fashion similar to the filament of an incandescent light bulb. A weak electrical current in the filament causes electrons to "boil off" of the hot filament and hover in the focusing cup of the cathode. The anode is copper and contains a tungsten target that is directly opposed to the cathode. X-rays are produced when electrons near the cathode are accelerated across the tube and strike the tungsten target of the anode. The interaction of the high-speed electrons with the atoms of the target results in x-ray emission. These events are rapid and are characterized by a single pulse of x-ray energy. A significant amount of heat is generated during an x-ray exposure and over 90% of the kinetic energy of the electrons is converted to heat with the remainder given off as x-ray energy.

Control of Radiographic Exposure and Adjusting Settings on the X-ray Machine

The quantity and quality of x-rays produced are manipulated by three factors: milliamperage (mA), exposure time in seconds (s) and kilovoltage peak (kVp), and are described in more detail below:

MILLIAMPERAGE (mA) - Milliampéres control the number of electrons that are produced by the filament. Increasing the mA will increase the number of electrons that are available to collide with the target. In this way, mA increases the number of x-rays produced. Doubling the mA will double the amount (**quantity**) of x rays produced.

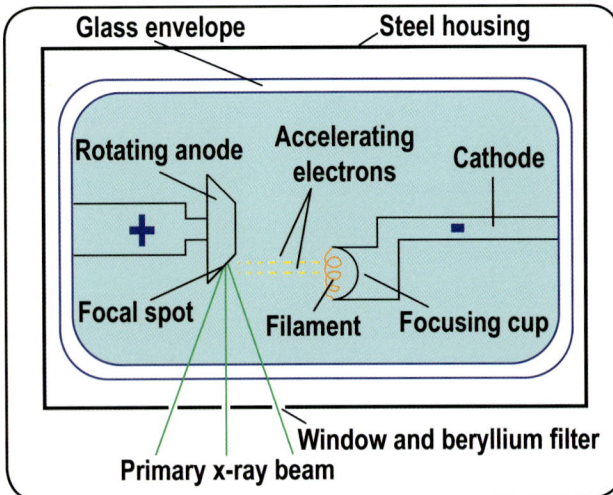

Figure 1.1. Diagram of an X-ray tube.
Glass Envelope - A vacuum chamber is produced by evacuating the glass envelope.
Electrons - Electrons travel from the filament to the target when a large positive electrical potential is applied to the anode.

Filament - The filament is a tungsten coil that emits ("boils off") electrons when heated by passing an electrical current.
Anode - The anode is the positive electrode, is usually made of copper and contains the tungsten target. It is either stationary or rotary, depending on the tube capacity.
Cathode - The cathode is the negative electrode and contains the tungsten filament.
Beryllium Window - A thin window formed in the glass envelope acts as an exit portal for the x-ray beam. The window absorbs a small number of x-rays.
Focusing Cup - The focusing cup is a small depression in the cathode that contains the filament. An electrical potential is applied to the focusing cup to restrict the electron beam diameter, which should equal the size of the target.
Primary X-ray Beam - X-rays, produced by collision of high-speed electrons with the target, pass directly through the window of the x-ray tube.
Secondary Radiation - Includes radiation scattered from the patient (scatter radiation) and off-focus radiation, which is not emitted from the target. Scatter radiation is the most important form of secondary radiation and can degrade radiographic quality, "fogging" the radiographic image.
(Courtesy of William R. Widmer, DVM)

EXPOSURE TIME (s) - In diagnostic radiography the exposure time is usually a fraction of a second. By increasing the period of x-ray production (exposure time), the number of x-rays generated will be increased. Doubling the exposure time will also double the amount of x-rays produced. Traditionally milliamperage (mA) and time (s) are expressed as a multiple (mAs).

KILOVOLTAGE PEAK (kVp) - Kilovoltage controls the speed at which the electrons strike the target. By increasing the kVp, a larger electrical potential is placed between the anode and cathode, accelerating the electrons across the tube. When high speed, energetic electrons strike the anode, shorter wavelength, more penetrating x-rays are generated. Therefore, kVp controls penetration or x-ray **quality**. The term kVp refers to the maximum kilovoltage or "peak" that is produced during an exposure.

Recording the X-ray Image

Cassette and Films

When x-rays strike the patient, there is differential absorption depending on tissue composition and thickness. X-ray/tissue interaction is a complex process that is beyond the scope of this discussion (further information can be found in the references at the end of this chapter.

X-rays that are not absorbed by the patient pass through and are recorded by a film-screen combination held within a light-proof cassette. The cassette contains one or two intensifying screens that convert most of the incident x-rays into visible light, which exposes the x-ray film. A few x-rays escape conversion to light and help expose the film. Exposed film is processed in a dark room to produce a finished radiograph.

X-ray film consists of a clear polyester base and has a gelatin-based coating of photosensitive silver halide crystals on both surfaces. The silver halide is sensitive to various forms of radiation, including x-rays, visible light and gamma rays. Unfortunately x-ray film is also sensitive to pressure, heat, and certain chemicals. When silver halide crystals are exposed, they undergo a physical change rendering them sensitive to silver precipitation during the development process. This results in a black area on the finished radiograph. During development, unexposed crystals are washed free of the emulsion leaving a clear area that appears white when the finished radiograph is placed on a light box. The silver halide emulsion is more sensitive to visible light than x-rays, hence, intensifying screens are used to convert oncoming x-ray photons to visible light within the x-ray cassette. This increased efficiency greatly reduces the amount of x-rays needed to obtain a diagnostic radiograph. This results in shorter exposure time and less x-ray exposure to the patient and personnel. A disadvantage of using cassette-film systems is loss of detail or image sharpness. When x-rays are converted to light, image sharpness is less than what is obtained with direct x-ray exposure of the radiograph.

Screens are available in high speed or "fast," regular speed or "par," and low speed or "detail." Speed of x-ray cassettes (and film) is akin to film speed in conventional photography. High-speed screens allow less exposure time for adequate film density (blackness), but provide poor detail and produce a grainy image. Regular-speed screens require intermediate exposure for proper film density, provide intermediate detail, and represent the best compromise for general use. Because of small patient size, high and regular speed systems are of limited value in exotic practice. Low-speed screens require more exposure but provide good detail and a sharper, non-grainy image. Many different types of screens and film combinations are available.

Like intensifying screens, x-ray film is available in different speeds, e.g., fast, par, and slow. Par speed film is widely used because it usually gives acceptable image sharpness and has a relatively wide latitude for exposure. Various types of x-ray film have differing sensitivity to the visible light spectrum. Thus, film choice is limited by the type of screen that is used. For instance, green light-emitting screens require green-sensitive x-ray film. In general, it is best to select one type of film that can be used with various screens.

Mammography film is a single-emulsion, ultra-slow speed film that can be used with special low speed cassettes or, for maximum detail, without cassettes. When cassettes are not used, exposure is by direct interaction of x-rays with the emulsion of the film, which is protected in a light-proof envelope. For this reason, mammography film is especially advantageous in very small patients that are frequently encountered in exotic animal practice. However, exposure of slow speed film requires more exposure than with conventional film-screen systems. This can be increased by raising mA or time, or both. In machines with lower, fixed mA, this necessitates general anesthesia to avoid motion artifact due to longer exposure time. Various screen and film combinations require different radiographic machine settings (mA, s and kVp). Because of additional detail needed for the exotic patient, most techniques call for slow-speed cassettes and films.

Small dental films (#2, 31mm x 41mm or #0, 22mm x 35 mm) are often useful for dental radiography (Figure 1.7). Due to the inherent difficulty of use of these films with a standard radiography machine, a dental radiographic unit is preferred (Figure 1.8).

Ideal equipment and combinations for small exotic patients are discussed below.

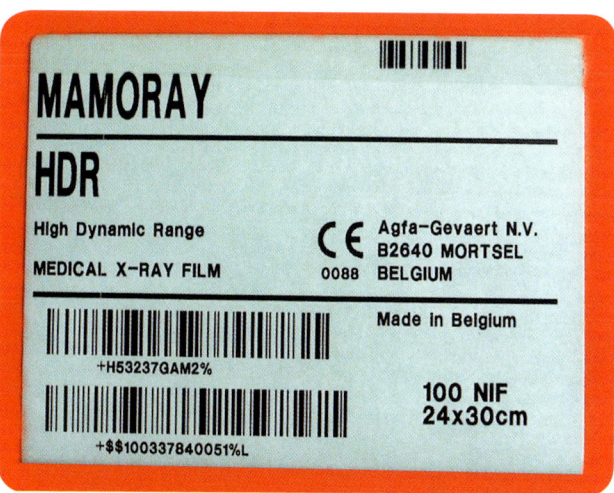

Figure 1.2. High-resolution mammography x-ray films are particularly advantageous.

Factors Affecting Radiographic Quality

Density

The finished radiograph must have adequate precipitation of neutral silver to be of diagnostic value (similar to a black and white photograph). Film blackness or radiographic density is affected by several factors, including amount of exposure (kVp, mAs) and the distance between the x-ray tube and recording plane. Radiographic density should not be confused with physical density (grams/cm^3) which relates to mass per unit volume of a substance. Therefore, the term "radiopacity" is used to define the various radiographic densities (shades of gray) seen on a radiograph. Substances like bone absorb a relatively large amount of incident radiation, produce white areas on the radiograph, and are said to be radiopaque. Air, on the other hand, absorbs few x-rays producing dark areas on a radiograph and is radiolucent. Fat, water, and soft tissues have intermediate opacities. The five basic radiographic opacities from least to most opaque are air, fat, water (soft tissue), bone and metal. These opacities can be distinguished by the naked eye because of their unique ability to absorb incident x-rays. Unfortunately, all soft tissues (given equal thickness) cannot be distinguished from each other on a conventional radiograph.

Detail

Detail is actually spatial resolution and our ability to resolve closely spaced line pairs projected on a test radiograph. For example mammography film can resolve two lines spaced 0.1 mm apart while high speed film might only resolve lines 1.2 mm apart. The most important factors influencing detail are patient motion, object to film distance, size of the focal spot, focal-film distance, and the speed of intensifying screens and film.

Distortion

Distortion occurs when the object being imaged is misrepresented, i.e. the object is magnified, unequally magnified, foreshortened or elongated. This is caused by natural divergence of the x-ray beam or a relatively large object-to-film distance (Figure 1.9).

Contrast

Contrast is the difference in film radiodensity of various parts of the image. If many shades of gray can be seen, then the radiograph is said to have a long scale of contrast. If few shades of gray can be seen, the scale is short. A long scale of contrast (many gray shades) is preferred in diagnostic imaging. Type of film, kVp and mAs, and intensifying screens influence contrast, as well as both natural and artificial contrast materials. Contrast radiography is discussed in more detail later in this chapter.

Radiographic Equipment for the Small Exotic Mammal Patient

Most standard radiography equipment is useful for exotic mammals. However, certain features are helpful for obtaining optimal radiographs. As these patients are often small, techniques usually employ high detail or low-speed screens, requiring increased exposure. As mentioned previously, mammography may call for 25 times the mAs of par or high-speed techniques. Many x-ray machines have a fixed lower end mA, often as low as 20, which is suitable for general veterinary use. However, these low powered, fixed mA generators require longer exposure time (s) in order to achieve a mAs setting acceptable for low-speed screens. For example, at fixed mA 20, the operator would need to select a time of 0.25 s to achieve a mAs of 5.0. The machine should allow the operator to select a higher mA to avoid increasing exposure time. A disadvantage of longer exposure times is increased risk of motion artifact, in particular from normal higher rate respiratory motions of small exotic mammals.

X-ray machines have adjustable lead plates called collimators, which define the area irradiated by the x-ray beam (Figure 1.5).

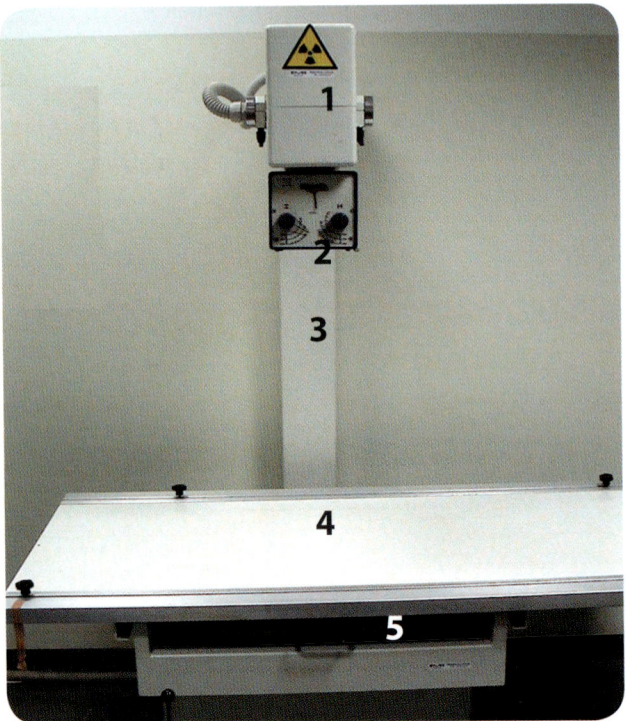

Figure 1.3. Radiology unit.
1. X-ray tube and housing. 2. Collimator of the x-ray beam. 3. Sliding stand for adjusting focal-film distance (FFD). 4. Table including the scatter reducing grid. 5. Drawer where the cassette is placed when the scatter grid is used.

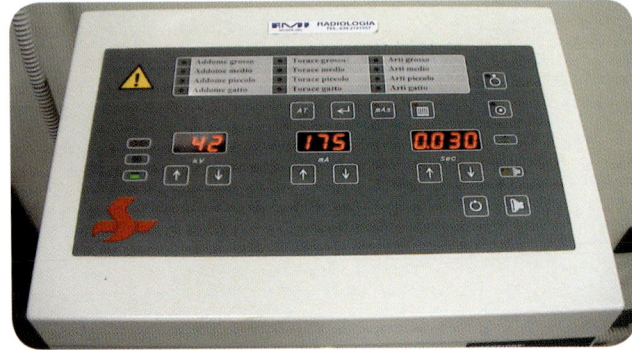

Figure 1.4. Control consol of an x-ray machine. For exotic mammal practice, an x-ray machine should have kVp adjusted in increments of one unit; mA in increments of 25 units; and time in increments of 0.001 seconds.

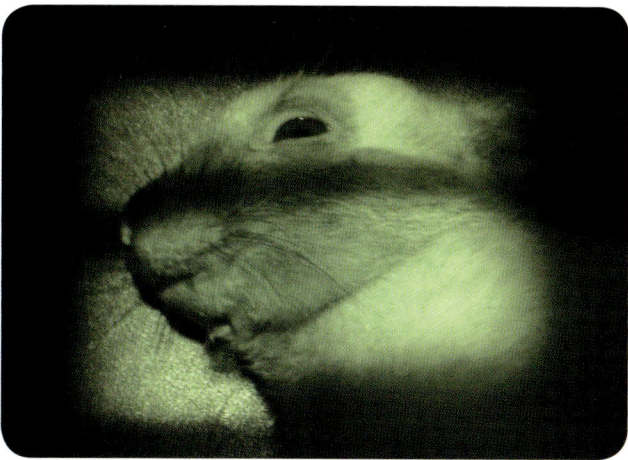

Figure 1.5. Standard radiographic machines allow collimation of the x-ray beam over the area of interest.

Adantages of proper collimation include:
- The same cassette and film can be used for more than one radiographic projection of the same patient (Figure 1.6.).
- During manual restraint, the hands of the operator can be safely excluded from the area to be irradiated.
- Scatter is reduced, limiting personnel exposure and improving image quality.

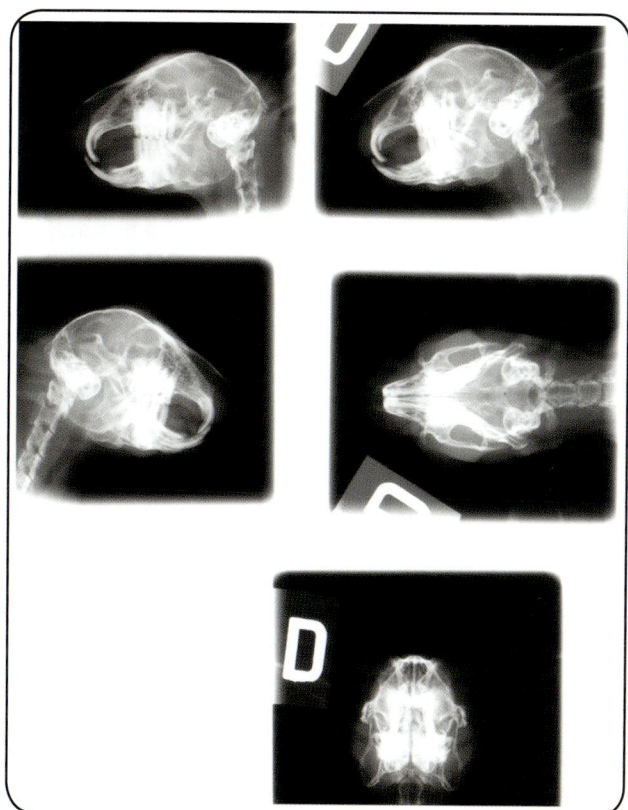

Figure 1.6. Depending on patient size, four to six skull images can be made on a single standard 24 x 30 cm film.

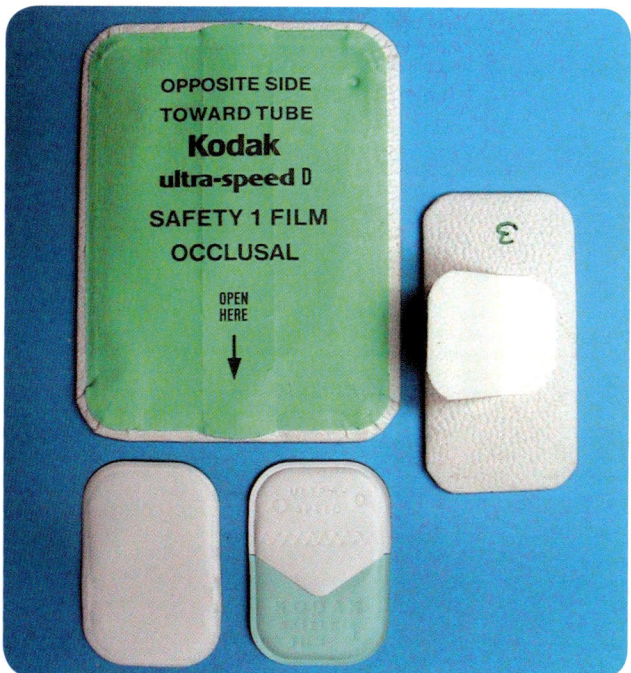

Figure 1.7. Images of the skull, single teeth, groups of teeth, other intraoral views or anatomical regions of small exotic mammals can be obtained using 57x76 mm occlusal size dental films or small dental periapical sized film. These small films are ideally exposed with a dental radiographic unit. The use of a dental radiographic unit makes patient positioning easier when using dental films.
(Courtesy of Margherita Gracis, DVM).

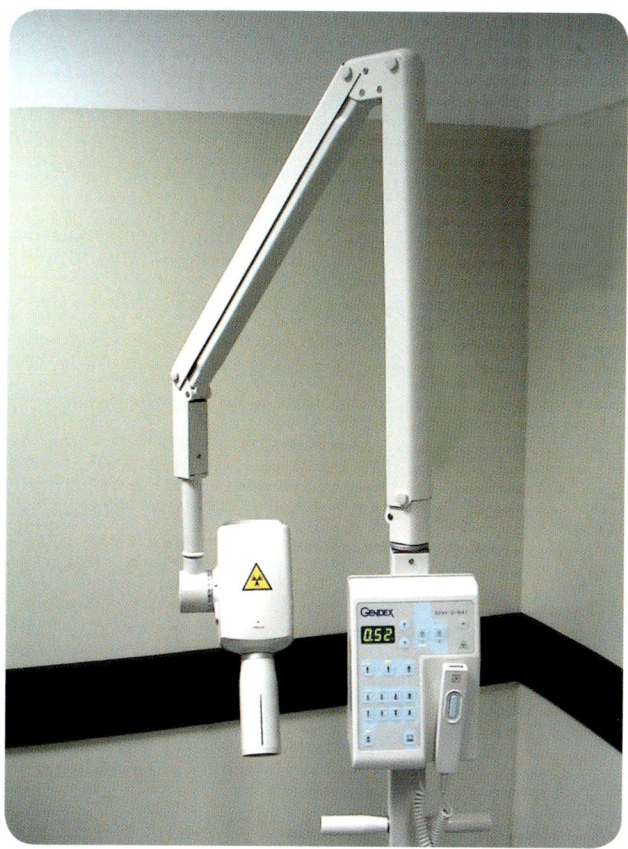

Figure 1.8. Dental radiographic unit.
(Ccourtesy of Margherita Gracis, DVM).

Obtaining the Radiograph

With the exception of larger species such as the potbellied pig, most exotic mammal patients are placed directly on the cassette (table top technique) and the Bucky tray is not used. With the patient positioned on the cassette, the thickness of the area to be radiographed is measured. Measurements in cm are used to determine the kVp setting. Milliampére/second settings (mAs) relate to the body region being imaged, e.g., thorax, abdomen, extremity, etc. Some standard technique charts suggest kVp settings depending on thickness of the tissue to be irradiated, and mAs (or s in fixed mA machines) depending on body region, for example thorax vs. extremities for each exposure. The manufacturer of the radiographic machine often provides a technique chart for specific recommended film/cassette combinations based on measurement in cm and anatomic region (bone vs. abdomen vs. thorax). However, in many cases the manufacturer cannot provide a technique chart for unusual or specialized combinations, for example when utilizing mammography film. Therefore, the practitioner must develop a species-specific technique chart. While a technique chart can be based to a large extent on trial and error, a standard calibration method utilizing a test animal is preferred. It should be kept in mind that a technique chart based simply on measurement and area of interest is often inadequate for the exotic mammal practitioner, as the tissues of herbivorous mammals are often very different from that of carnivores. The best solution is to develop a technique chart based on species, in addition to measurement and area of interest.

Correct placement of the patient and cassette with respect to the center of the x-ray beam is important because of the divergent nature of the x-ray beam (Figure 1.9). X-rays at the periphery of the primary beam are not parallel and will falsely project the image of an object at the periphery of the recording plane. Distortion also occurs when an object is not kept in close contact with the recording plane (Figure 1.9). Magnification can be deliberately utilized to improve visualization of small structures (Figure 1.9). There are two options: with the "tabletop technique," the patient is raised above the cassette by placing it on a block of radiolucent foam; with the "under table technique," the cassette is positioned below the x-ray table, while the patient is positioned normally on the table top.

These techniques increase the object-film distance, which results in magnification of the projected image and decreases the focal-object distance. A trade off with most magnification procedures is loss of detail because of increased *penumbra*

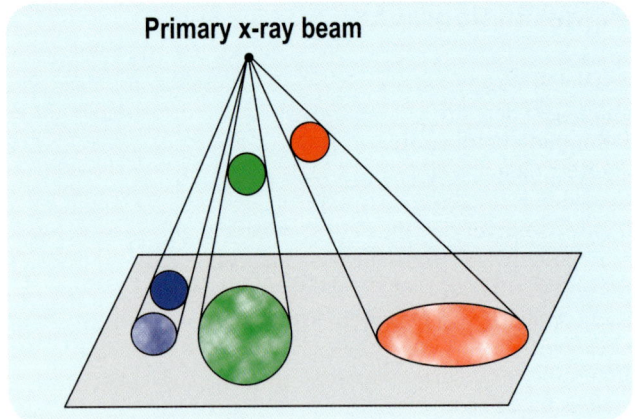

Figure 1.9. Diagram of the magnification and distortion artifacts.

effect, which is discussed below. One way to overcome *penumbra* is to increase FFD (focal-film distance), which necessitates increasing the exposure.

The focal-film distance (FFD) is decreased by lowering the x-ray tube closer to the patient and the cassette. An objective-film distance of 12 inches (31 cm) and a focal-film distance of 20 inches (51 cm) have been reported to provide a magnification factor of approximately 2.0. It should be kept in mind that intensity of x-rays decreases in relation to the square of the FFD. Increasing FFD will decrease the *penumbra* effect but x-ray beam intensity must increase at a rate equal to the square of the increase of FFD: four times if FFD is doubled, and nine times if FFD is tripled. The opposite must be considered if FFD is decreased.

Scattered Radiation

When x-rays penetrate tissues, secondary scattered x-rays are produced which travel in mostly oblique directions. This radiation can create degradation of the image with loss of details. This primarily occurs when the thickness of the patient exceeds 10 cm. In order to prevent this, the table of the radiographic machine includes an anti-diffusion device, called the "scatter-reducing grid" which absorbs secondary radiation, thus improving image quality. The drawback of the use of the grid is that a portion of x-rays for production of the image is reduced as well. Since most exotic mammal patients do not exceed 10 cm, the use of the grid is not usually necessary.

Digital Radiography

Digital radiography is the newest advance in veterinary imaging. There are two basic digital imaging systems:
1) computed radiography (CR)
2) direct digital radiography (DDR).
Computed radiographic detectors are similar to a large floppy computer diskette. These flexible detectors are placed in a standard x-ray cassette sans intensifying screens. A two-step process is used for recording the CR image. First, the latent image is captured or "trapped" with x-ray sensitive phosphors of the image plate (this is similar to recording the latent image on conventional x-ray film). Second, the image plate is processed by a reader that changes the latent image of the phosphor to light photons that are converted to an electrical signal. The reader is a stand-alone unit similar to an x-ray processor. An analog-to-digital converter transforms image data to a digital format and sends it to a computer.

Direct digital systems are so named because unlike CR they send digital information directly to the image computer without a reading step. Types of DDR detectors include:
1) flat panel detectors
2) charge coupled device (CCD) detectors.
Flat panel detectors are rigid plates that look like a conventional x-ray cassette. They have several layers of semiconductors for image capture, transistors and microscopic circuits. Flat panel detectors convert x-ray energy into electrical signals, which are digitized and sent directly to a computer. Charge coupled device detectors consist of a small chip that is similar to those used for digital cameras. A scintillator converts incoming x-rays to light photons. The light signal is minimized by an optical system before reaching the chip where light is converted to electric charges. A digital signal is sent from the chip directly to the imaging computer.

With digital imaging, no x-ray film, dark room, or film storage are needed. Additional advantages include improved subject contrast compared to conventional x-ray systems, flexibility of a digital format, and ability to manipulate and potentially improve images by adjusting parameters such as brightness and contrast. Disadvantages of digital imaging include significant start up costs, limited image resolution and the need for multiple image displays throughout the hospital and archiving (storage) hardware. Veterinarians should be aware that only DICOM (Digital Imaging and Communications in Medicine) file format (.dcm) are legally acceptable images. Image files with tags such as .jpg and .tiff can be altered and are not considered original images. This is akin to an actual hard copy radiograph compared to a copy film or a digitized radiograph that has been acquired by a scanner (Figure 1.90) or a digital camera.

Computed radiographic systems are presently the most cost-effective system for veterinary medicine. A serious concern for use of any digital system in small exotic species is resolution. Digital systems lack the inherent resolution indigenous to conventional analog systems. This should be overcome by larger matrices and software manipulation as digital technology continues to be developed.

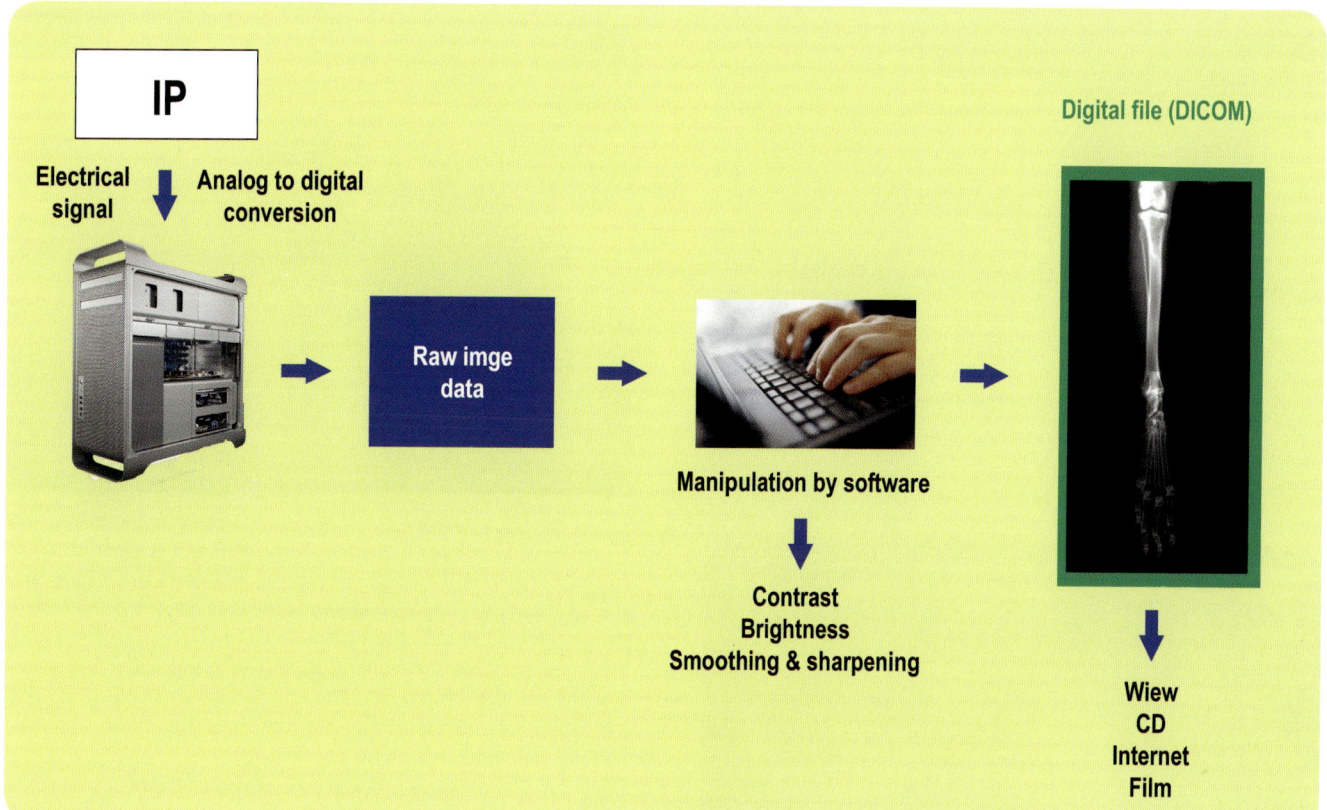

Figure 1.10. Basic concept of digital radiography. An image plate (IP) is used to capture the radiographic exposure. With both computed and direct digital radiography the information from the IP is sent to a computer for initial processing into raw data. Further processing involves manipulation of the data for display and interpretation in a DICOM format.
(Courtesy of William R. Widmer, DVM)

BASICS of RADIOLOGY   8

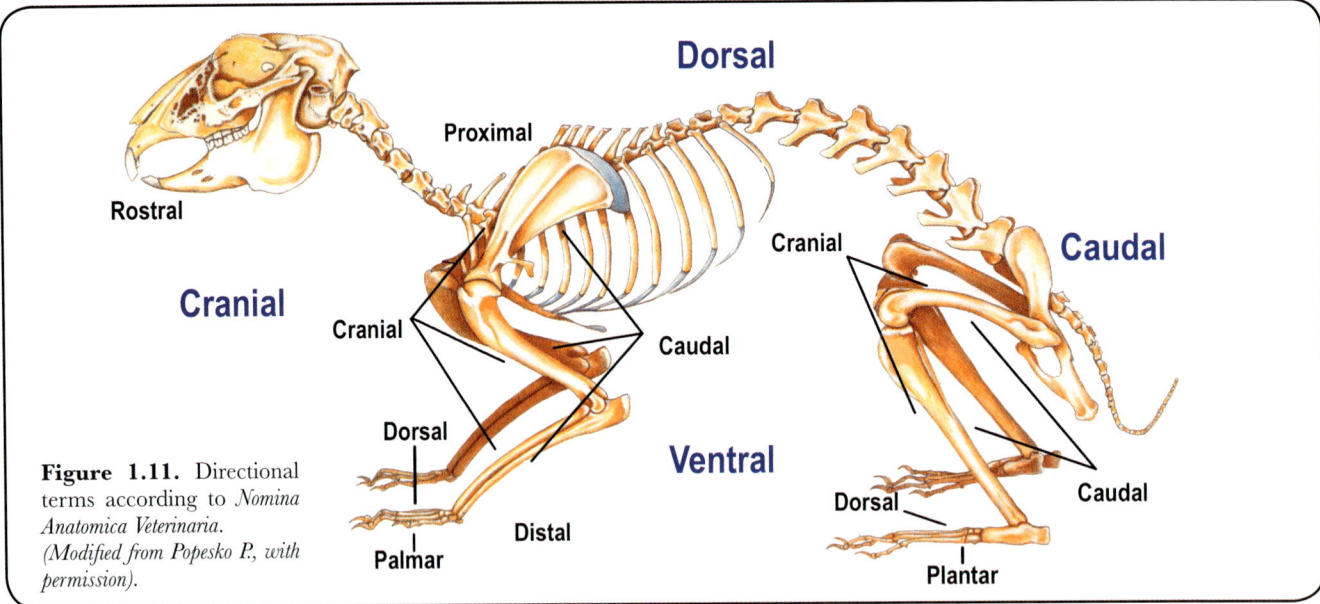

Figure 1.11. Directional terms according to *Nomina Anatomica Veterinaria*. (Modified from Popesko P., with permission).

X-ray Projections

Standard Nomenclature

X-ray projections or "views" are named according to beam entry and beam exit.

The *Nomina Anatomica Veterinaria* provides the correct directional terms that are used for describing radiographic projections (Figure 1.11). For instance a thoracic radiograph of a rabbit in dorsal recumbency with the x-ray tube overhead and the cassette beneath the patient is a ventrodorsal projection. Oblique projections are more complex, but are named by the same method, beam entry to beam exit. The angle of obliquity is often designated by placing the degrees of obliquity between directional terms. (Figure 1.12). By using beam entry to beam exit nomenclature the exact location of structures can be determined on the finished radiograph.

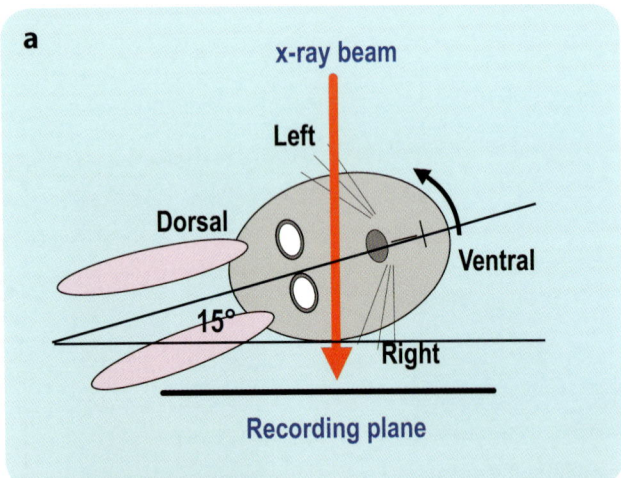

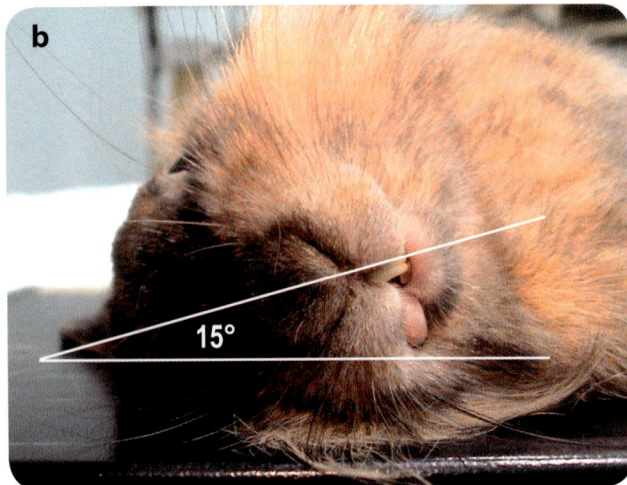

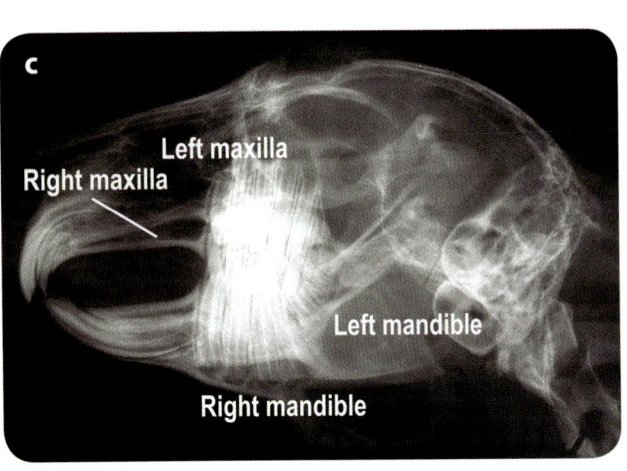

Figure 1.12a,b,c. Left 15° Ventral-Right Dorsal Projection (Lt 15° V-Rt D) of the head of a rabbit (a, b). Notice that the obliquity is 15 degrees ventral from a true right lateral projection. This tells the operator to find the point of beam entry by rotating the ventral aspect of the head 15 degrees toward the tube from a strict right lateral projection. In this example, structures on the right side of the head are projected ventral to those on the left side (c).
(Figure 1.12a. courtesy of William R. Widmer, DVM)

Patient Positioning

Correct positioning of the patient is critical in order to obtain diagnostic quality radiographs. Evaluation of poor radiographs can lead to diagnostic errors and should be avoided. General anesthesia facilitates superior positioning, is often safer and less stressful for the patient, and can reduce the need for repeated x-ray exposure due to inopportune patient movement.

Manual Restraint vs. Pharmacologic Restraint

Due to small patient size and resistance to restraint, sedation and general anesthesia may be necessary to obtain diagnostic quality radiographs. However, the practitioner will face situations where anesthesia is not feasible or judged to be unsafe, especially with the critical patient. As overuse of manual restraint may also produce stress, the need for pharmacologic versus manual restraint must be carefully evaluated for each individual patient.

The following points are important when considering manual restraint for radiography:
1. The owner must be informed in advance about different options;
2. Owners must never be allowed to assist with manual restraint for radiography for reasons related to radioprotection and owner and patient safety;
3. Consider simple sedation to reduce the stress of manual restraint, unless contraindicated (see below);
4. Manual restraint must be performed by the veterinarian or an experienced assistant who can perform restraint quickly and in a coordinated manner, and can recognize signs of patient distress.
5. The person performing manual restraint must employ safe radioprotection techniques, including the use of lead shielding and avoidance of the primary x-ray beam.
6. The procedure should be discontinued immediately should the patient display signs of stress secondary to manual restraint, or complications from anesthesia.

Sedation and Anesthetic Considerations

Protocols for sedation and anesthesia in exotic mammal species have been extensively reported in the literature, and this discussion is beyond the scope of this chapter. However, an overview of basic principles is in order.

Protocols should be selected based on overall patient condition, patient size, and anticipated length of the imaging procedure. In many cases, simple sedation alone can facilitate radiography, especially when attempting manual restraint.

Anesthesia for radiography should follow the same guidelines as that for any other procedure, and should include careful monitoring and attention to details such as maintenance of normothermia and normovolemia.

Induction of general anesthesia can be performed with injectable or inhalant drugs, and each method has both advantages and disadvantages. Induction with injectable agents eliminates stress and apnea that often occur with inhalant agents alone, and allows brief discontinuation of maintenance gas to facilitate positioning for radiography. Time to induction is often shorter with injectable agents, and restraint is limited to that required for injection of the drug. Disadvantages are often inherent in the drug selected, and can include cardiac and pulmonary depression and hypovolemia; however, newer injectable agents with wider margins of safety are being used successfully in small exotic patients. A disadvantage of injectable agents is a longer recovery time.

Induction with inhalant agents necessitates use of a single agent, and allows chamber induction in cases where handling and restraint is dangerous or not feasible. All inhalant agents used for induction or maintenance can cause significant respiratory and cardiac depression, and are not always safer than injectable agents. Time to recovery is faster with inhalant agents. The presence of a face mask and anesthetic circuit can interfere with proper radiographic positioning. This is most important in certain procedures such as radiographs of the skull. Radiographs for diagnosing dental disease in rabbits and rodents must be of excellent quality, and patient positioning must be accurate. For this and other similar procedures, injectable anesthesia may provide an important advantage in allowing time for accurate positioning in between views. When inhalant agents are used alone, temporary discontinuation of gas delivery to facilitate positioning often results in an uneven anesthetic plane with potentially dangerous multiple partial recoveries and re-inductions throughout the procedure, which should be avoided.

Radiation Safety

A complete review of principles of radiation safety is included at the end of this chapter.

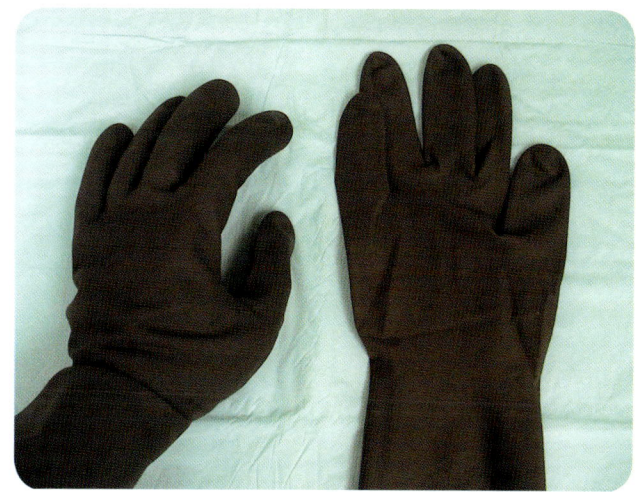

Figure 1.13. Regular full lead gloves greatly hamper manual restraint of small exotic animals. All gloves, regardless of thickness, are designed to reduce scatter radiation, and are not intended for use under the primary x-ray beam. These single use lead gloves are much thinner and allow increased tactile perception, but provide less protection depending on manufacturer and composition. While these thin gloves may be adequate for sheilding scatter radiation and are regularly used by one of the authors outside the US (V. Capello), it should be noted they do not meet the recommendations of the National Council on Radiation Protection (NCRP).

Basics of Radiology

Patient Positioning Techniques

Symmetry is an extremely important factor in patient positioning, with the exception of oblique or other special projections. Contoured pieces of foam are useful to obtain proper positioning. Heavier devices like sandbags are rarely used in exotic mammal patients. Tape is useful for securing patients directly onto the cassette or the radiographic table. Foam and tape are not radiographically apparent, therefore will not interfere with the image of the patient.

For the following descriptions of patient positioning, the cassette and the film were placed in a horizontal position, and the x-ray beam was directed vertically, perpendicular to the film.

In this series of images, the blue dot represents the center of the x-ray beam, and the rectangular overlays represent the area of collimation.

Most of the patients pictured in this series were induced with injectable anesthetic drugs and maintained with isoflurane and oxygen by face mask. The facemask was usually removed during x-ray exposure for demonstration purposes. Some of the smaller patients were induced and maintained with inhalant agents alone.

Cassette sizes were 40 x 30 cm, and 30 x 24 cm.

TOTAL BODY

In larger exotic mammal species, the whole body projection generally includes all structures except for the tail and distal part of the limbs. In some instances, the perineal area is included. Evaluation of the head and skull requires a specific study confined to the head only. In larger species, the whole body overview should never be used when a specific radiographic study of the thorax, abdomen, or limbs is indicated. The x-ray beam will not be properly centered on the area of interest, and technique (in particular kVp) will not be ideal for all sections within the area of exposure.

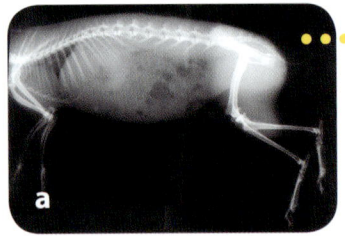
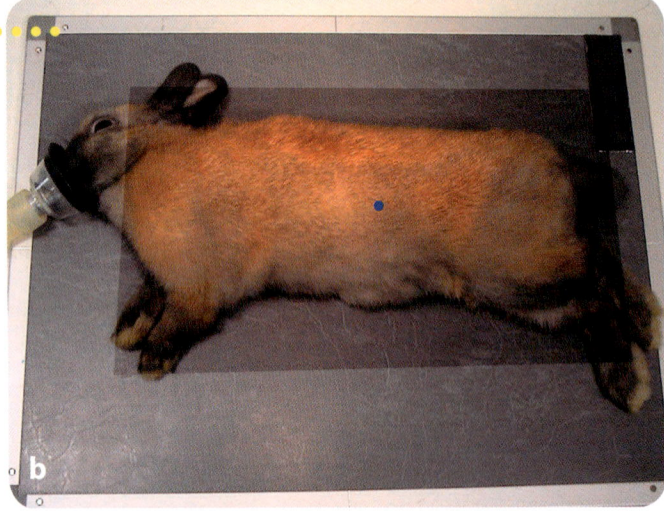

Figure 1.14a,b.
Lateral projection.
The patient is placed in right or left lateral recumbency.
Right lateral projection is shown. When the goal is an overview of the total body, the limbs can be left in a neutral position.

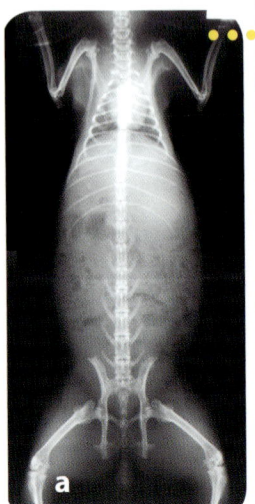
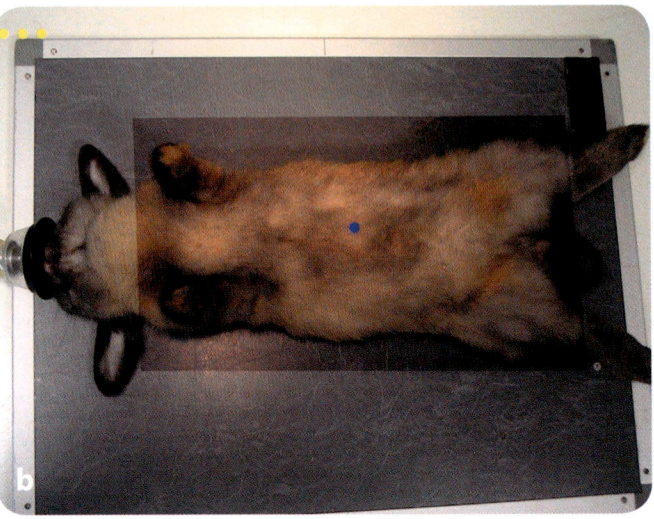

Figure 1.15a,b.
Ventrodorsal projection. The patient is placed in dorsal recumbency, with the head and limbs in neutral position. Dorsoventral projection is also possible, with the patient placed in ventral recumbency. In the authors' experience, it is easier to obtain a symmetrical projection in the ventrodorsal position, rather than the dorsoventral position.

POSITIONING

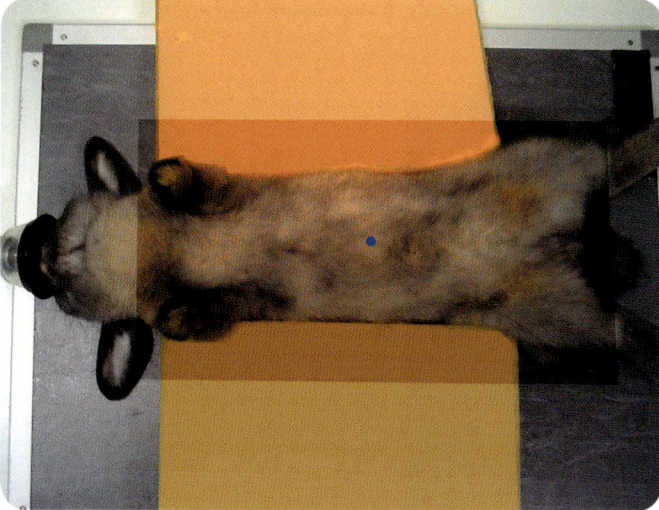

Figure 1.16. Ventrodorsal projection. In very thin patients, blocks of radiolucent foam can be used to maintain the body in proper dorsal recumbency.

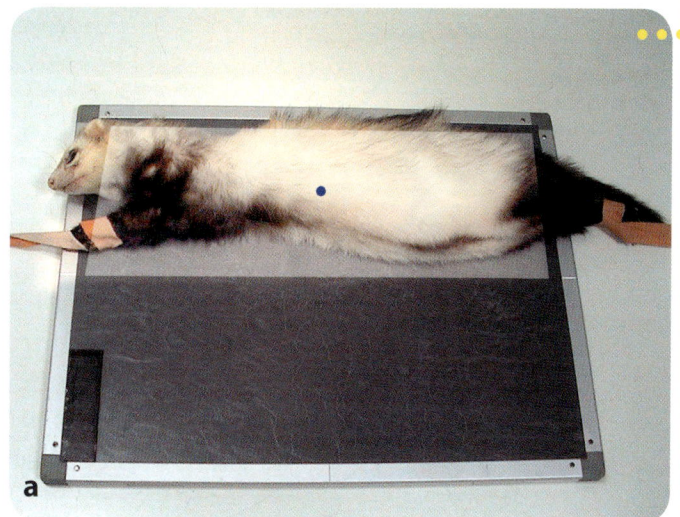

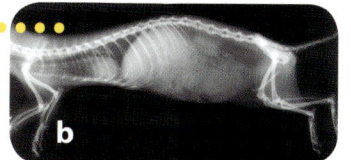

Figure 1.17a,b. Lateral projection. The patient is placed in right or left lateral recumbency. The right lateral projection is shown. The whole body projection should never be used for a specific radiographic study of the thorax or the abdomen in a large mammal like the ferret, because the x-ray beam is not properly centered for these anatomical structures. When the goal is a survey of the total body, the limbs can be left in a neutral position. However, it is better to hyperextend the thoracic limbs cranially and the pelvic limbs caudally and secure them with radiolucent tape.
A single larger cassette can be divided in two areas by collimating the x-ray beam. The same film can then be used both for the lateral and ventrodorsal projections.

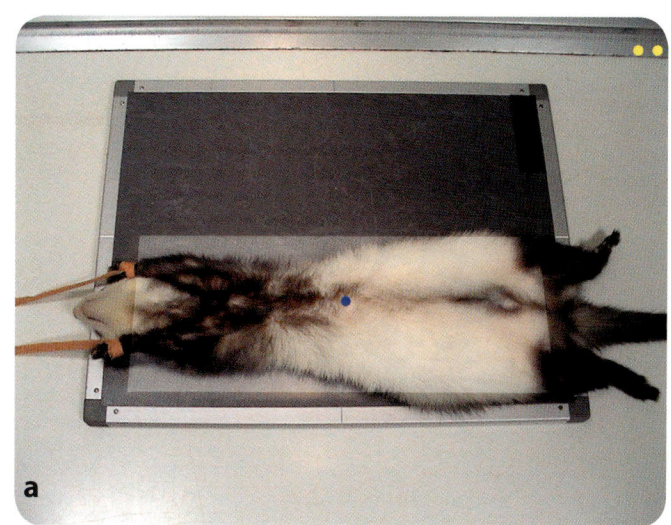
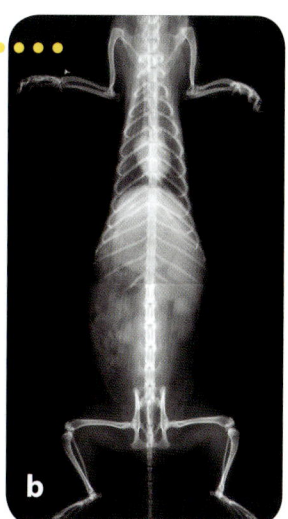

Figure 1.18a,b. Ventrodorsal projection. The patient is placed in dorsal recumbency, with the head and limbs in neutral position or hyperextended.

A number of projections can be obtained with manual restraint in many species, with good results.

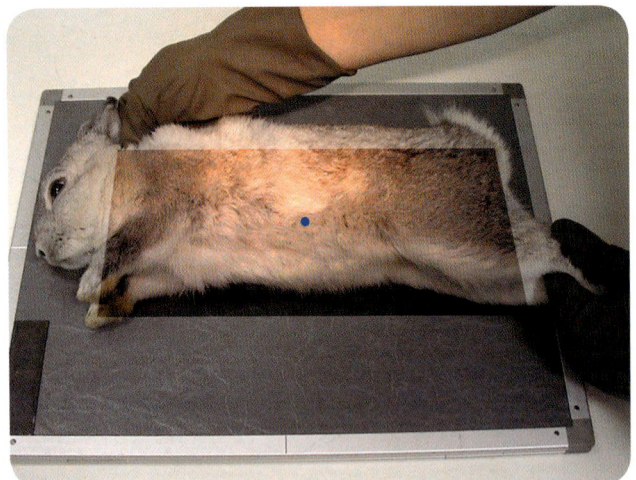

Figure 1.19. Lateral projection, conscious rabbit. The rabbit's skin is held at the nape of neck by the thumb and the first two fingers. The rabbit is then gently positioned in right or left lateral recumbency and allowed to adapt to this position for a few seconds. Most rabbits will accept this form of restraint. The operator should use the dominant hand to scruff.

Once the patient accepts lateral recumbency, the operator grasps the feet and gently extends them caudally. At no time is the patient actually restrained by the pelvic limbs.

During restraint, the operator will be able to detect if the patient is relaxed or ready to struggle, in which case the limbs are released immediately while the scruff is maintained, and the rabbit repositioned starting once again from standing position. This maneuver is critical to avoid serious injury, including damage to the lumbar vertebral column due to kicking out with powerful pelvic limbs.

An assistant collimates the x-ray beam and obtains the exposure, avoiding the gloved hands of the operator.

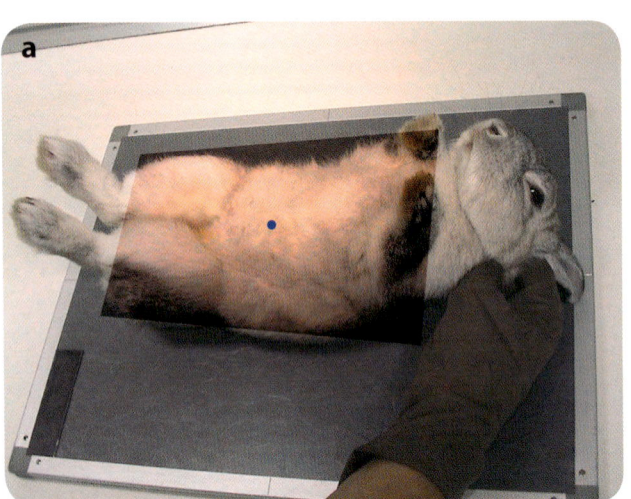
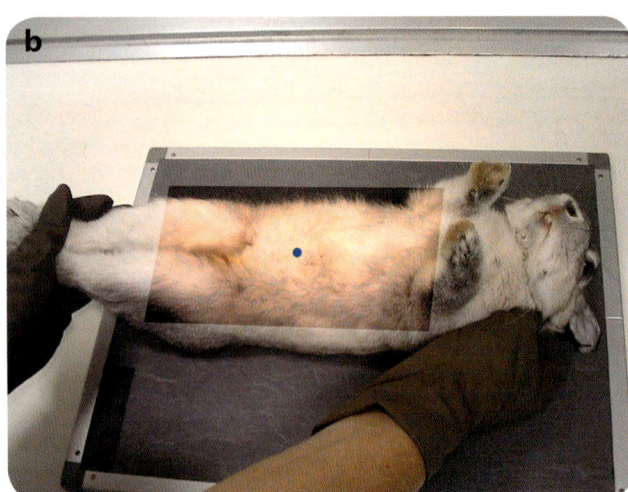

Figure 1.20a,b. Ventrodorsal projection, conscious rabbit. The rabbit is held as described above, and gently placed in dorsal recumbency. The pelvic limbs can then be extended caudally. Rabbits often accept dorsal recumbency better than lateral recumbency.

POSITIONING

Figure 1.21a,b. Lateral projection, conscious chinchilla. The chinchilla is restrained as described for the rabbit, using just two fingers to grasp skin and not fur, which could result in coat damage. The pelvic limbs are grasped and gently hyperextended caudally.

The same method of restraint is adequate to obtain a total body projection. However, the thoracic limbs are in neutral position and will be superimposed over the thorax.

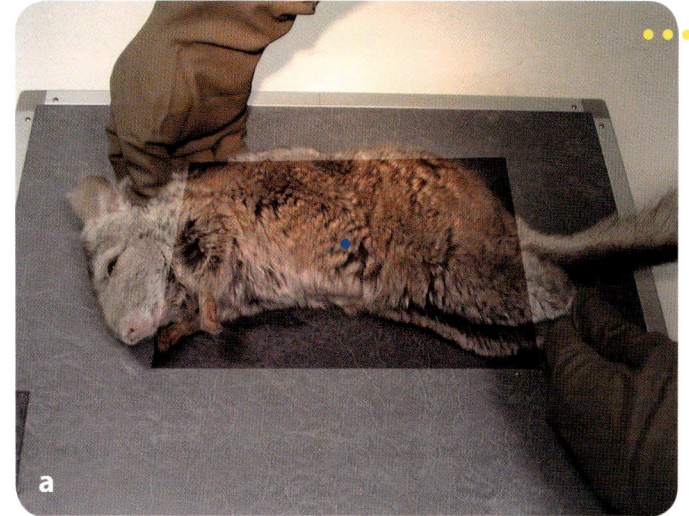
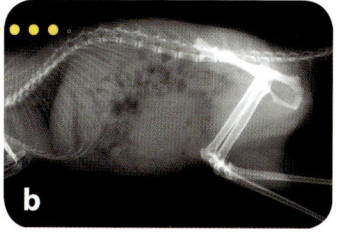

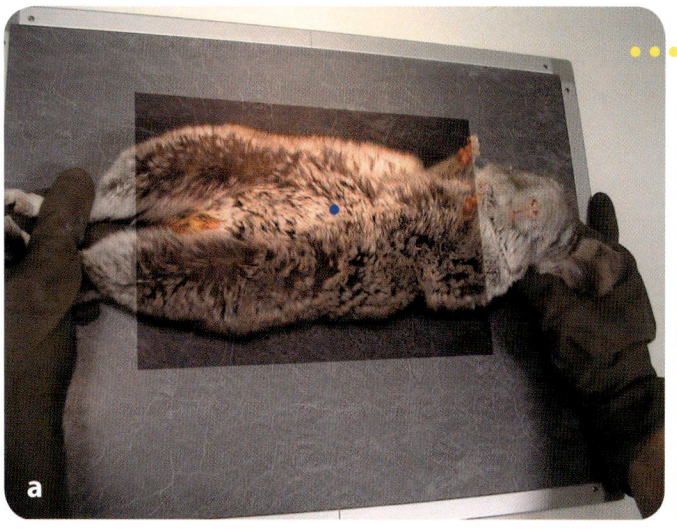
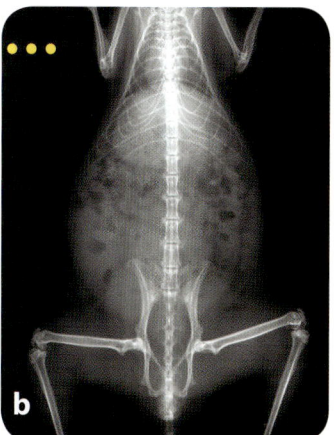

Figure 1.22a,b. Ventrodorsal projection, conscious chinchilla. Restraint is similar to that described above for the rabbit.

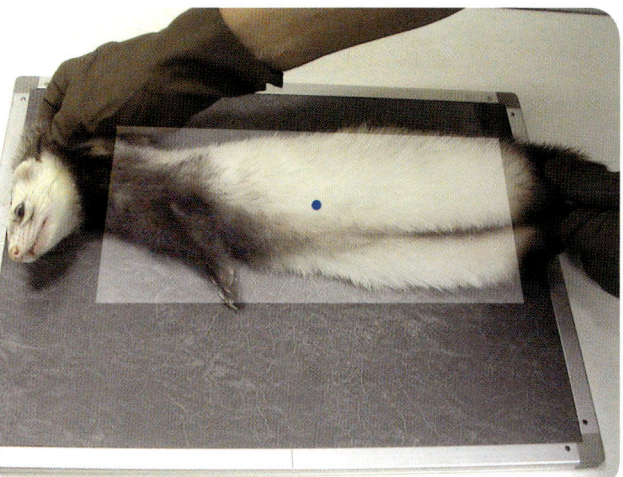
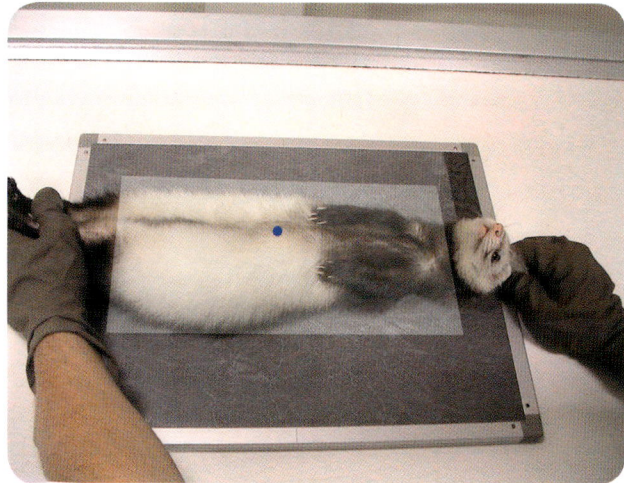

Figure 1.23. Lateral projection, conscious ferret. The ferret is held by the skin of the neck, using the thumb and the first two fingers and moved into lateral recumbency. Ferrets usually accept this form of restraint well. The operator then grasps the feet and gently extends them caudally. An assistant collimates the x-ray beam and collects the image, taking care to avoid the gloved hands of the operator.

Figure 1.24. Ventrodorsal position, conscious ferret. The ferret is held by the skin of the neck as described above, and gently placed in dorsal recumbency. The ferret usually accepts dorsal recumbency better than lateral recumbency.

WHOLE BODY SKELETON

In smaller exotic mammal species, the skull, extremities and tail are often included in a survey radiograph of the whole body skeleton. When inhalant anesthesia is used, the face mask is removed for radiography.

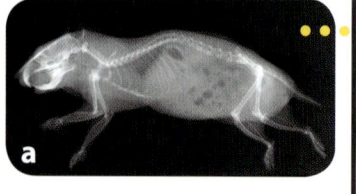
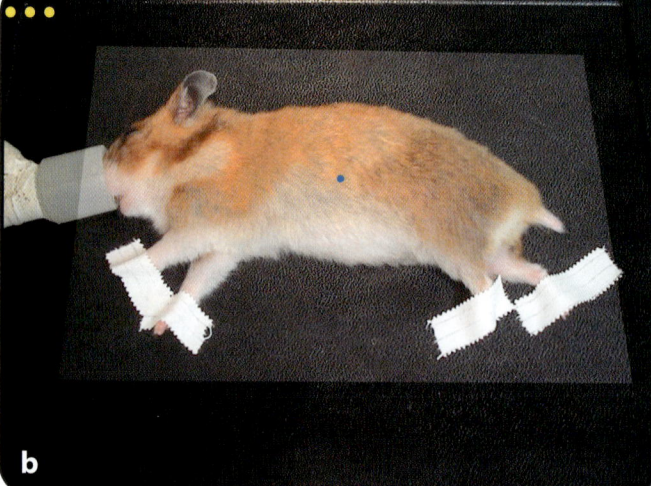

Figure 1.25a,b. Lateral projection. The patient is placed in right or left lateral recumbency. The right lateral projection is shown. Thoracic limbs are hyperextended cranially; the pelvic limbs are hyperextended caudally, securing the extremities with radiolucent tape.

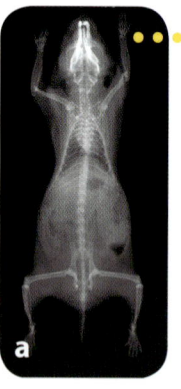
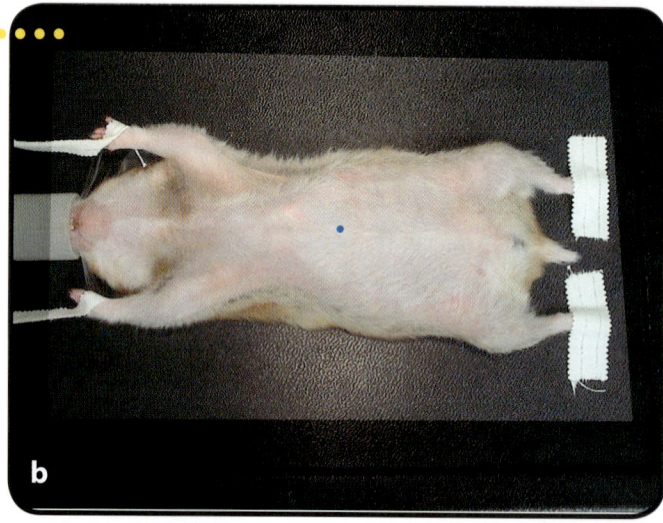

Figure 1.26a,b. Ventrodorsal projection. The patient is placed in dorsal recumbency, with the thoracic limbs hyperextended cranially and the pelvic limbs hyperextended caudally and secured with radiolucent tape.

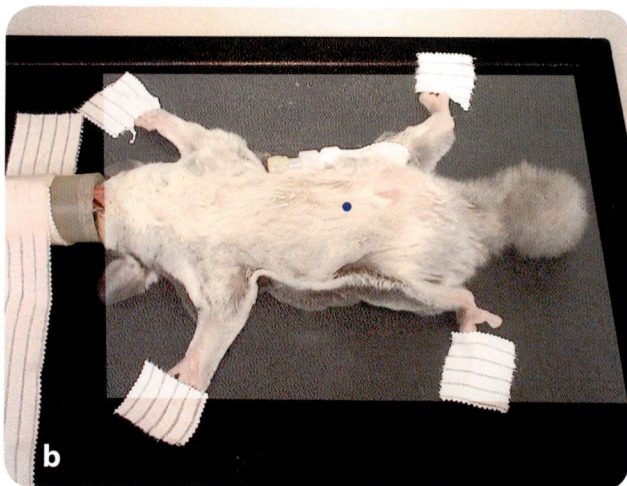

Figure 1.27a,b. The same guidelines described above for the golden hamster can be applied to other small species, for example, this sugar glider. The limbs are gently secured with tape. An intraosseous catheter was placed in this patient for therapeutic purposes.

POSITIONING

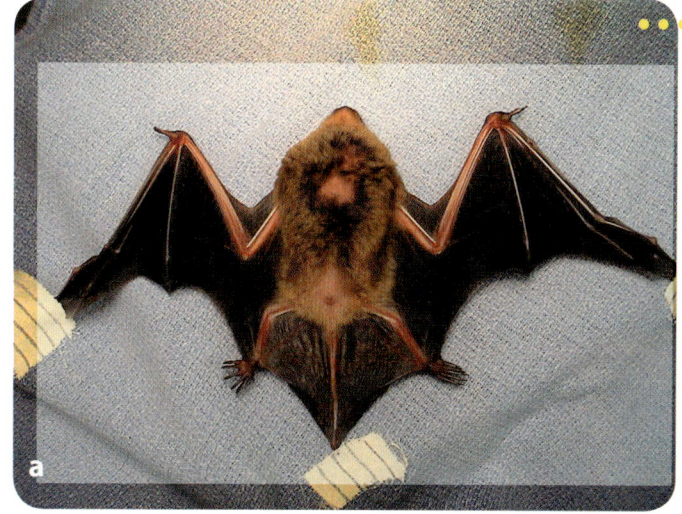
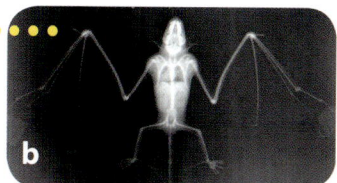

Figure 1.28a,b. Ventrodorsal projection. The patagium is secured with small pieces of radiolucent tape. The thoracic limbs are slightly extended taking care to protect the delicate phalanxes, especially in smaller patients. The tail, the caudal patagium and the pelvic limbs are fully extended.

HEAD

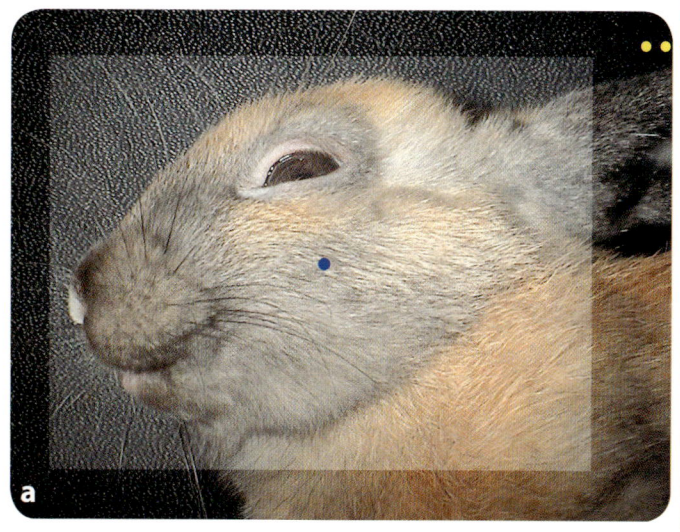
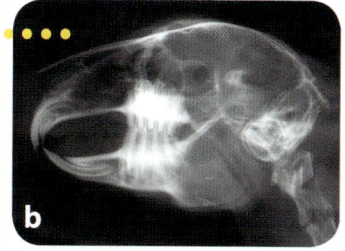

Figure 1.29a,b,c,d. Lateral projection. The patient is placed in right or left lateral recumbency. Right lateral projection is shown. Viewed from above, the head must appear flat and horizontal (a). Proper positioning is confirmed by checking the frontal view. (c) The philtrum should appear parallel to the cassette for an optimal lateral view in most rabbits. In larger rabbits, the shape of the head often requires the nose to be raised slightly for optimal positioning. Contoured pieces of radiolucent foam are positioned under the rostral head (d). Dwarf breeds have a short and rounded head, which often does not require adjustment for a true lateral projection.

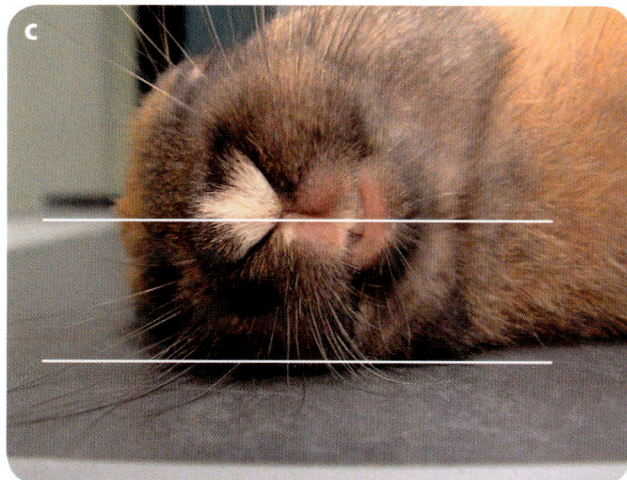
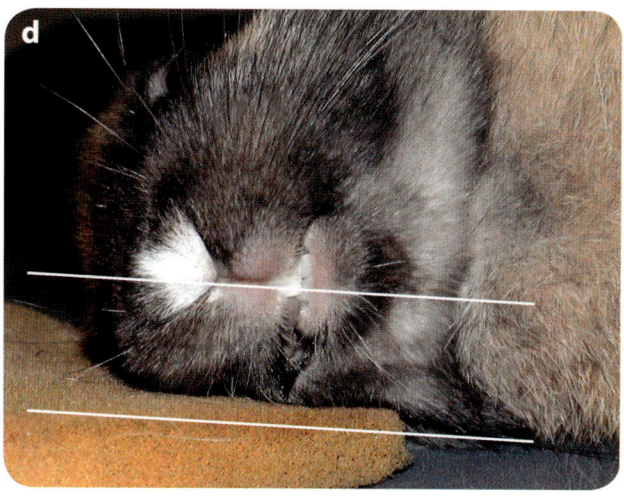

16 BASICS of RADIOLOGY

Veterinarians traditionally use several systems to describe various oblique positions, which are sometimes contradictory and confusing. Human and veterinary radiologists have standardized the naming of oblique positions by describing the position of the beam in relation to the object of interest, from beam entry to beam exit. (Figure 1.11a,b,c).

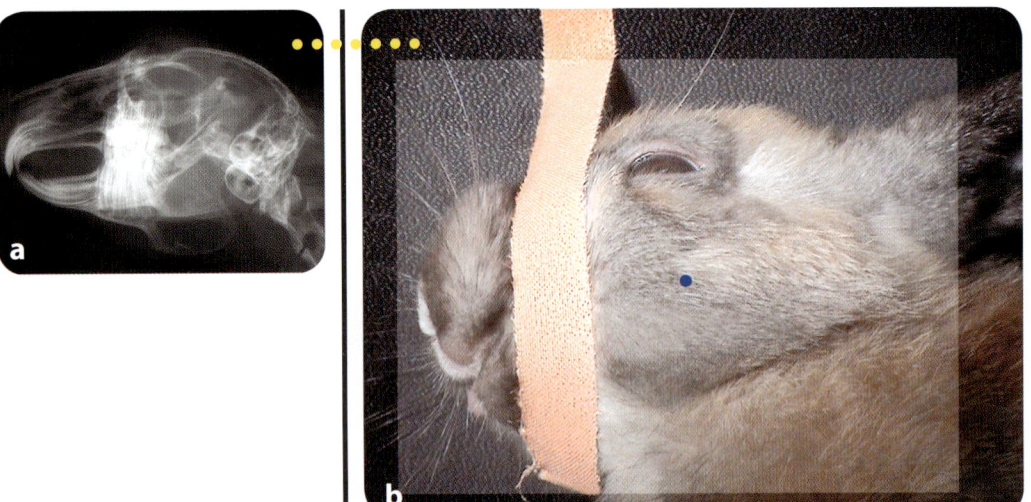

Figure 1.30a,b,c,d. This right lateral oblique projection is properly defined as Left 15° Ventral-Right Dorsal Oblique (Lt 15° V-Rt D Oblique). The patient is placed in right lateral recumbency, as for the right lateral projection. The head is slightly rotated counter clockwise using radiolucent tape or a contoured piece of foam. When viewed from above, the head must appear slightly oblique (b).
Proper right oblique position is checked from the frontal view, where the nose of the patient and the philtrum appear slightly oblique in relationship to the cassette (c). A useful aid for obtaining correct oblique projection in the rabbit is to place the pinna of the inferior ear under the head (c). The superimposition of the cartilage does not create radiographic artifact. An ideal oblique projection for the rabbit's head is approximately 10°-20° rotation from true lateral.

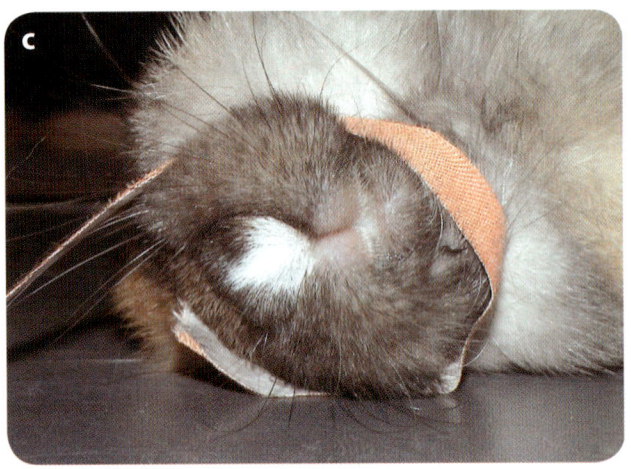

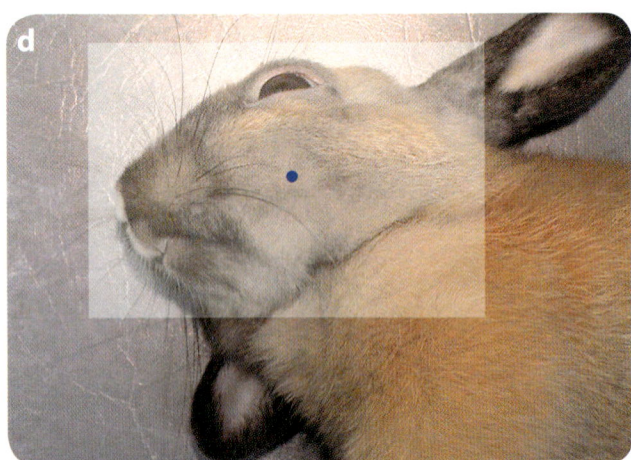

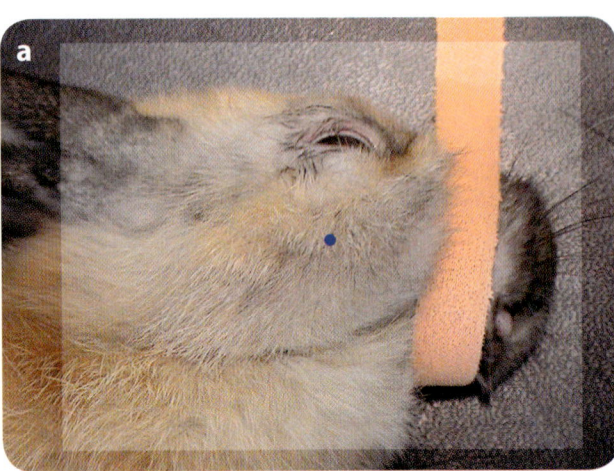

Figure 1.31a,b. Left lateral oblique projection is properly defined as Right 15° Ventral-Left Dorsal Oblique (Rt 15° V-Lt D Oblique). The patient is placed in left lateral recumbency. The same procedure is repeated as for the right lateral oblique projection (a), but the head is rotated slightly clockwise (b).

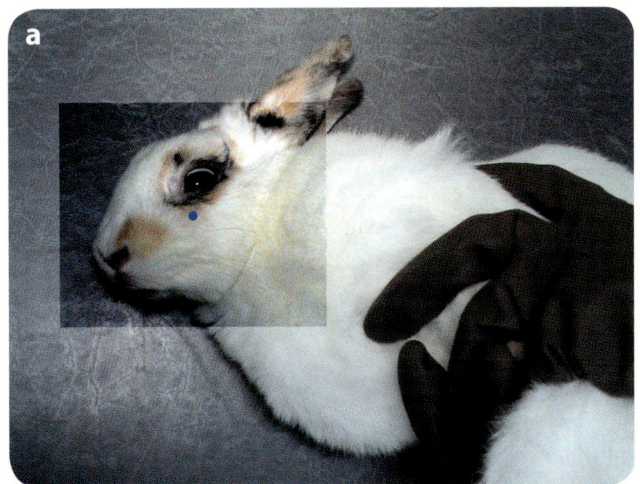

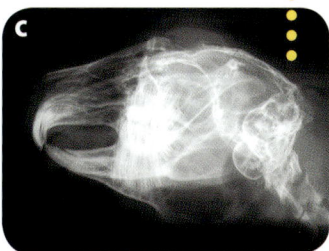

Figure 1.32a,b,c. Oblique radiographs in the conscious patient. While it is possible to obtain a preliminary oblique radiograph of the head without anesthesia, results are usually described as less than optimal, and would only be considered as a test radiograph or in cases of extreme anesthetic risk. This technique should not be considered when attempting to evaluate dental disease.

The rabbit is carefully suspended with the limbs hyperextended caudally. The head is gently placed on the cassette in lateral recumbency. The rabbit will attempt to flex the head laterally, producing an oblique position.

This form of manual restraint naturally produces a 30° oblique projection, which is considered excessive for ideal evaluation of the apexes of cheek teeth. To compensate, the rabbit can be rotated obliquely in the opposite direction.

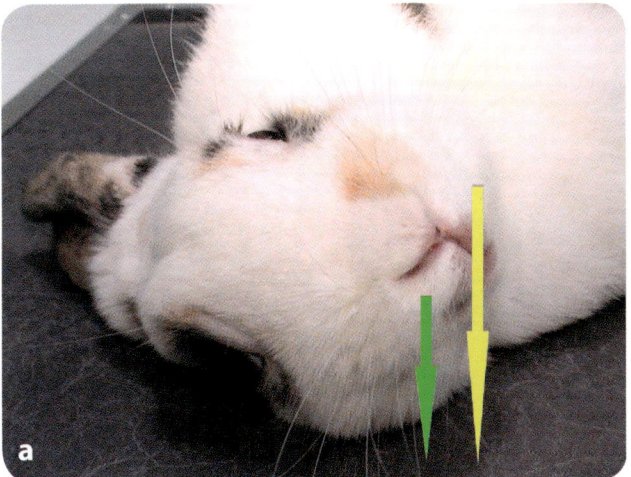
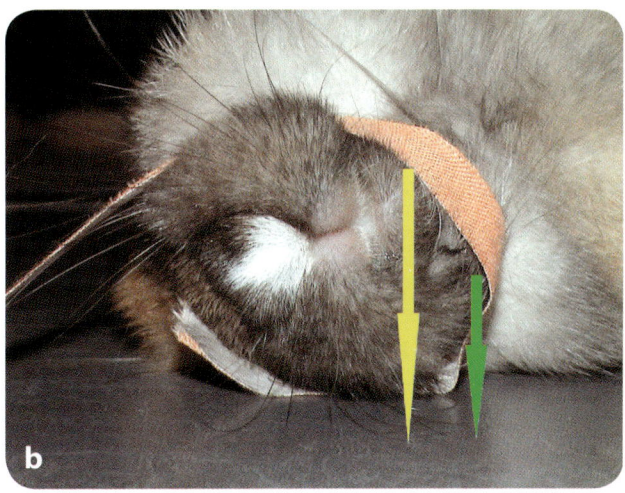

Figure 1.33a,b. Note that when the conscious rabbit is suspended in a right lateral position (a), the radiographic projection of the head will be Left 30° Dorsal-Right Ventral Oblique (yellow arrow: left mandible; green arrow: right mandible), due to the slight lateral voluntary hyperflexion of the head. The opposite is noted in figure (b): this rabbit under general anesthesia has been placed in right lateral recumbency and positioned Left 15° Ventral-Right Dorsal Oblique (yellow arrow: left mandible; green arrow: right mandible). A conscious rabbit suspended in the left lateral position will be Right 30° Dorsal-Left Ventral Oblique.

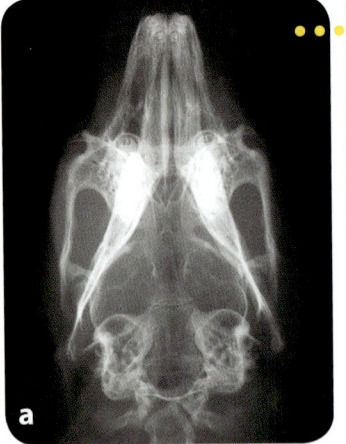
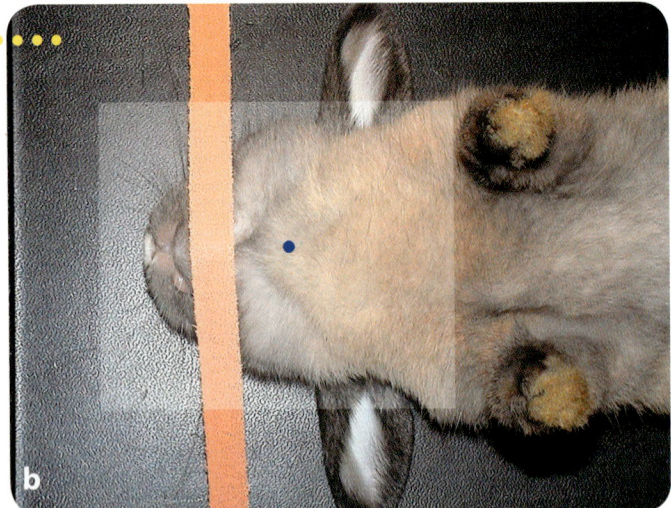

Figure 1.34a,b,c. Ventrodorsal projection. The patient is placed in dorsal recumbency. The neck is hyperextended and a strip of tape used to secure the head parallel to the cassette. Viewed from above, the head must appear flat and horizontal (b). The proper ventrodorsal position is confirmed from the frontal view, where the head should appear symmetrical and the incisor teeth perpendicular to the cassette (c). Respiration may be impaired if the patient is maintained in this position for an extended time. Therefore, the image should be obtained as quickly as possible.

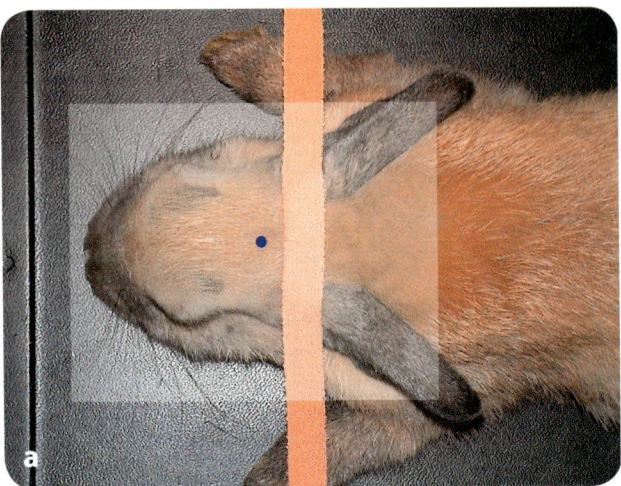

Figure 1.35a,b. Dorsoventral projection. The patient is placed in ventral recumbency. The neck is hyperextended and secured with tape to assure the head is parallel to the cassette.
In the authors' experience, it is easier to obtain a precise projection from the ventrodorsal position than from the dorsoventral position. Considerations related both to positioning and respiration are the same as for the ventrodorsal projection.

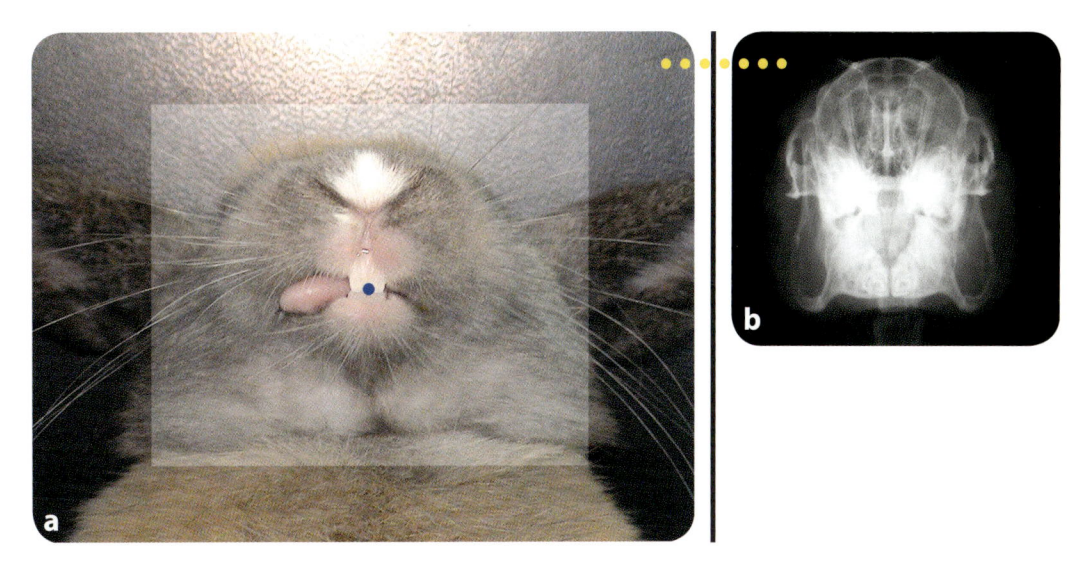

Figure 1.36a,b,c,d,e. Rostrocaudal projection. The patient is placed in dorsal recumbency, with the head hyperflexed so that the palate is perpendicular to the cassette (a). Proper vertical position is confirmed from the frontal (c) and the lateral view (d), where the head must appear symmetrical and not deviated laterally. Radiolucent foam can be used to maintain the head in position (d) and to prevent projections that are not truly perpendicular, as shown in (e). If the patient is not intubated, respiration may be impaired when the neck is hyperflexed. Therefore, this position should be maintained only for enough time to obtain the radiograph.

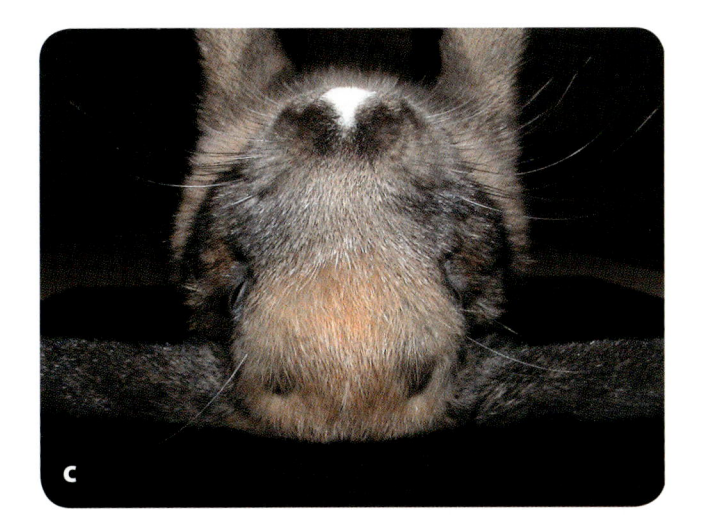

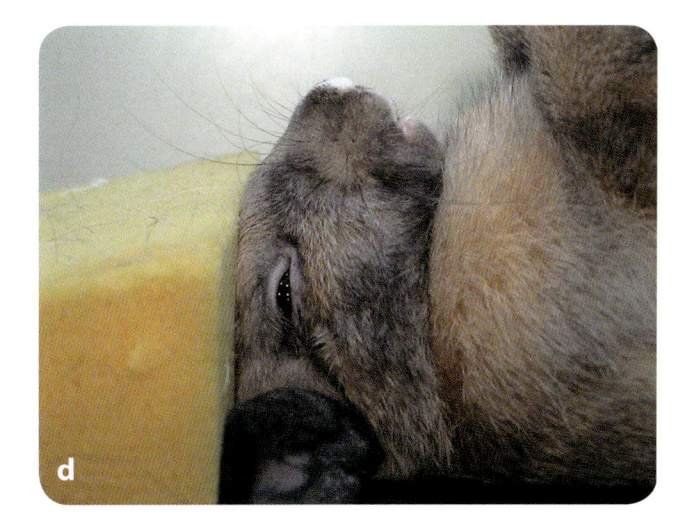

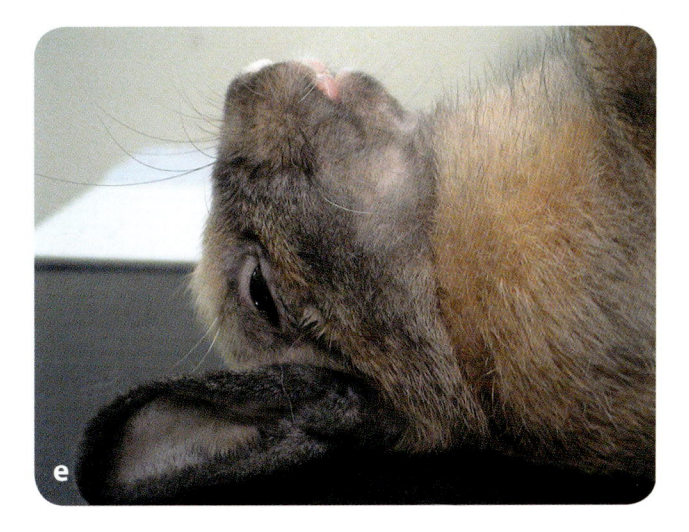

20 BASICS of RADIOLOGY

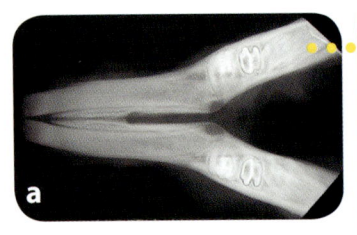

(Courtesy of Margherita Gracis, DVM).

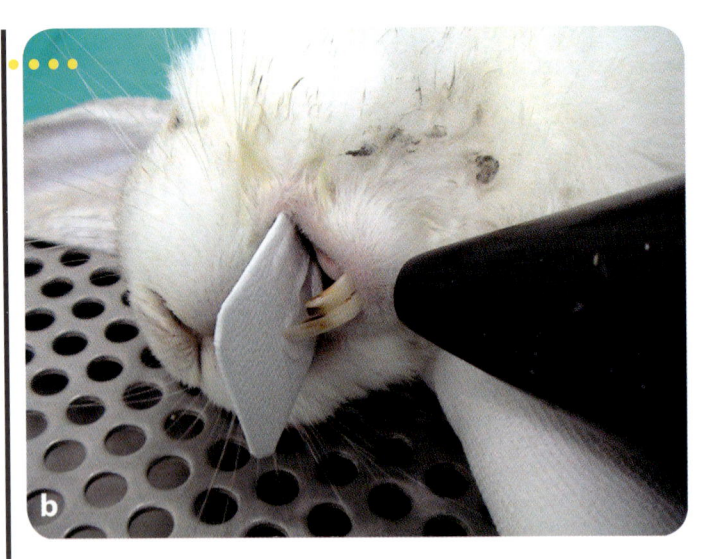

(Courtesy of Margherita Gracis, DVM).

Figure 1.37a,b.
Intraoral placement of a non-screen (bitewing) dental film for radiographs of the mandibular incisor teeth. The film is gently introduced between the mandibular and maxillary dental arcades as far caudally as possible, and the beam is directed perpendicular to the bisecting angle between the long axis of the film and the long axis of the teeth.
(Courtesy of Margherita Gracis, DVM).

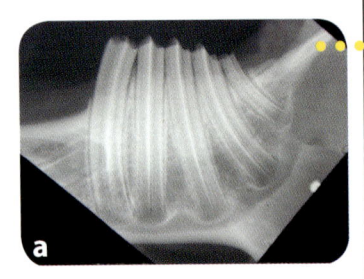

Figure 1.38a,b.
Lateral projection using an occlusal dental film and a dental radiographic unit.
(Courtesy of Margherita Gracis, DVM).

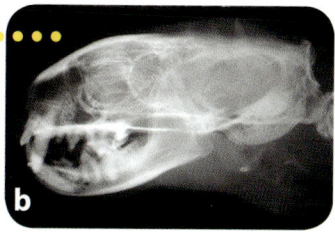

Figure 1.39a,b,c.
Lateral projection. The patient is placed in right or left lateral recumbency. The right lateral projection is shown. Viewed from above the head should appear flat and horizontal (a). Proper horizontal position is confirmed from the frontal view. As the head of the ferret is slightly conical, it may be necessary in some patients to raise the nose slightly to obtain a true lateral projection (c).

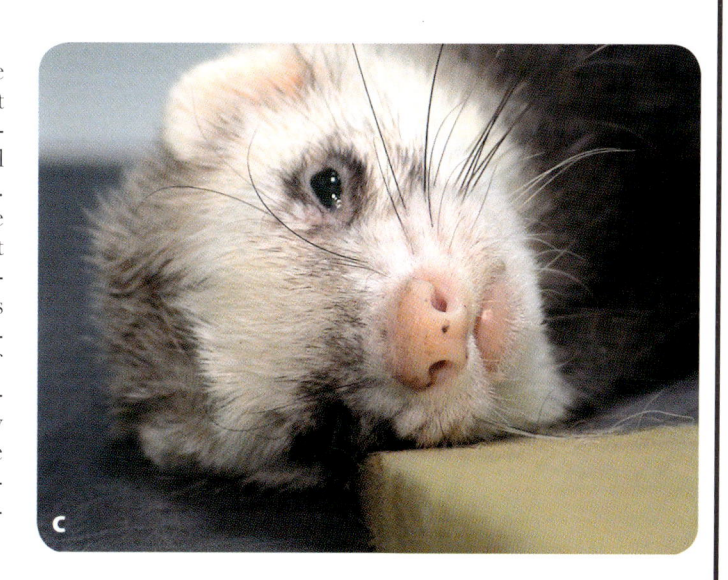

Figure 1.40a,b.
Lateral projection, open mouth. This technique helps reduce superimposition and allows better radiographic study both of the maxilla and the mandible. In this patient, a modified syringe barrel is used as a radiolucent speculum.

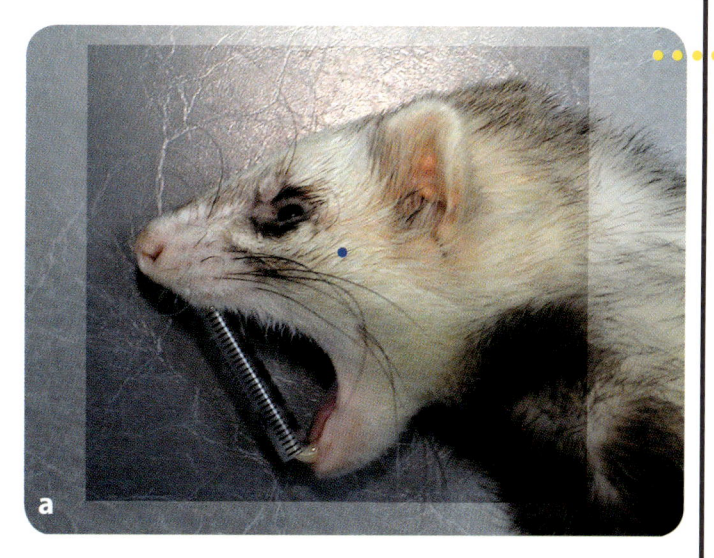

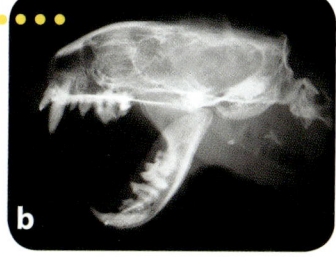

22 BASICS of RADIOLOGY

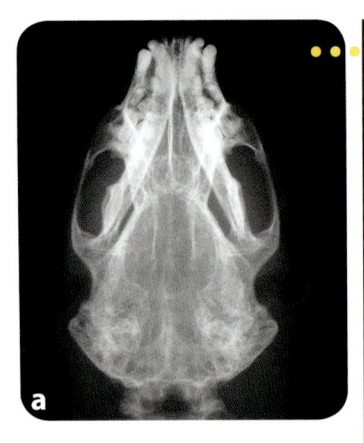

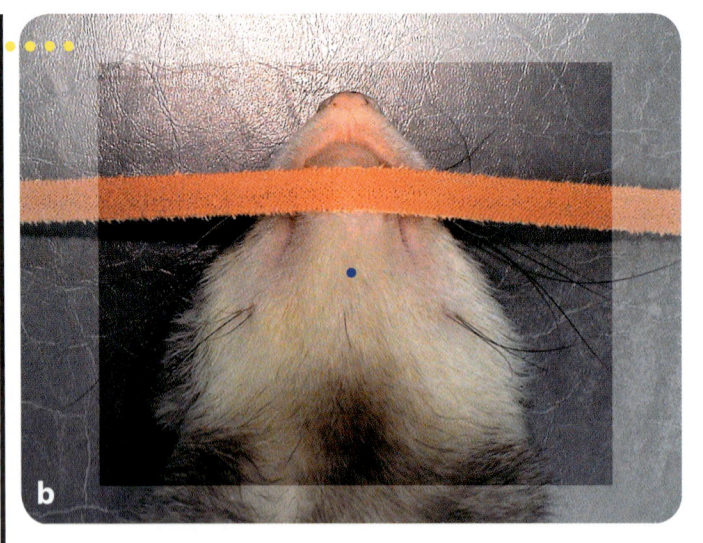

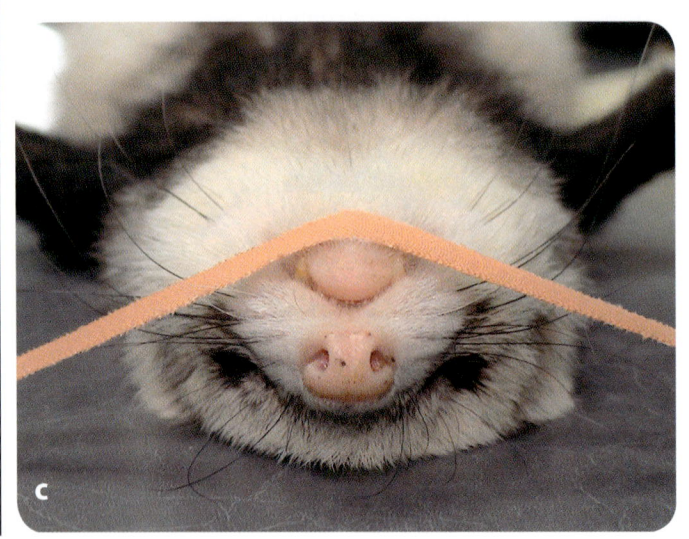

Figure 1.41a,b,c. Ventrodorsal projection. The patient is placed in dorsal recumbency, and the neck is hyperextended and secured with tape. Viewed from above the head should appear flat and horizontal (b). Proper ventrodorsal position is confirmed from the frontal view, and the head must appear symmetrical and not deviate laterally (c).

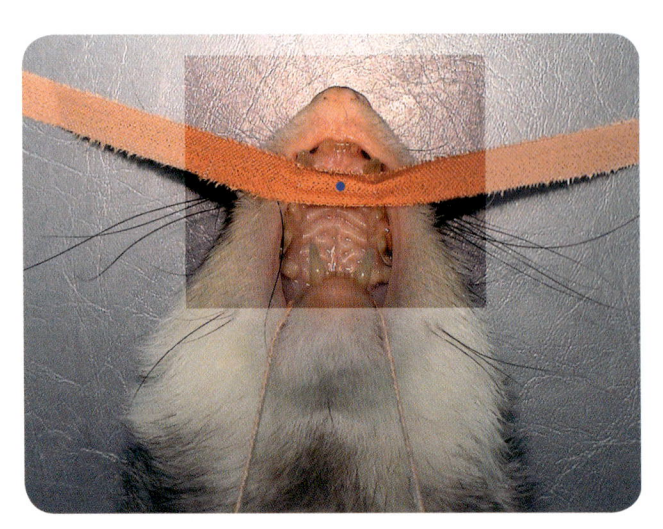

Figure 1.42. Ventrodorsal projection of the maxilla. The patient is placed in dorsal recumbency, with the maxilla secured to the cassette with radiolucent tape positioned over the palate, behind the maxillary canine teeth. The mouth is maintained in open position with another strip of tape secured behind the mandibular canine teeth.

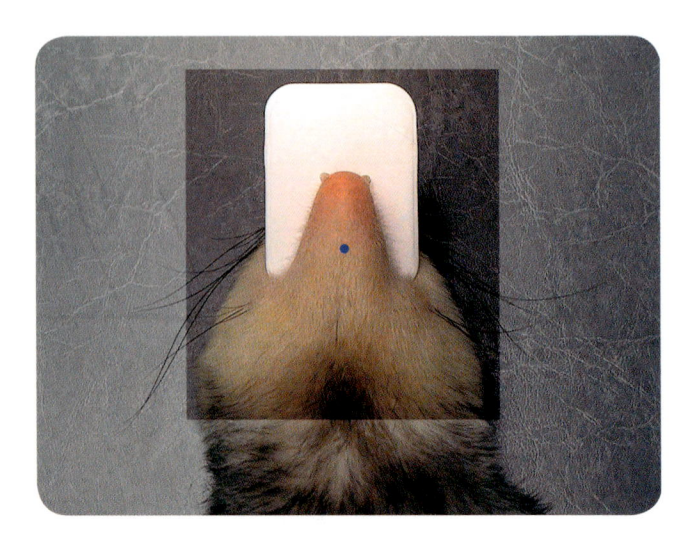

Figure 1.43. Ventrodorsal projection of the mandible obtained with dental bitewing film. The film is inserted into the mouth as far caudally as possible with the ferret in dorsal recumbency.

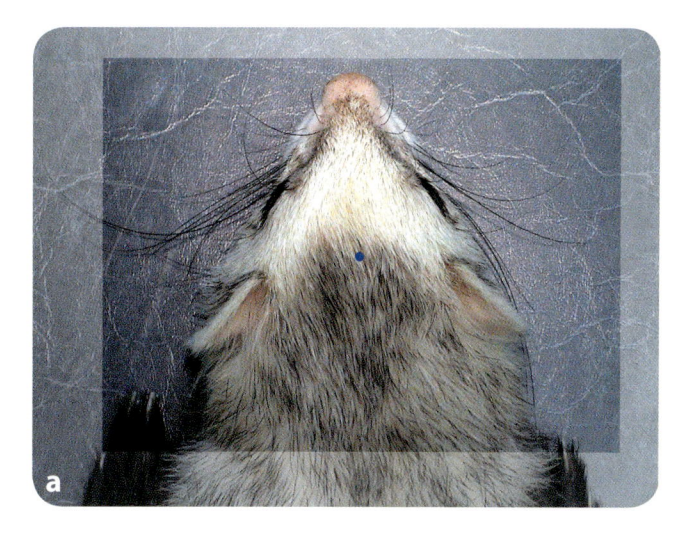

Figure 1.44a,b. Dorsoventral projection. The patient is placed in ventral recumbency. Viewed from above, the head must appear flat and horizontal (a). Proper ventrodorsal position is confirmed by viewing the patient from the front, and observing that the head is symmetrical without lateral deviation (b).

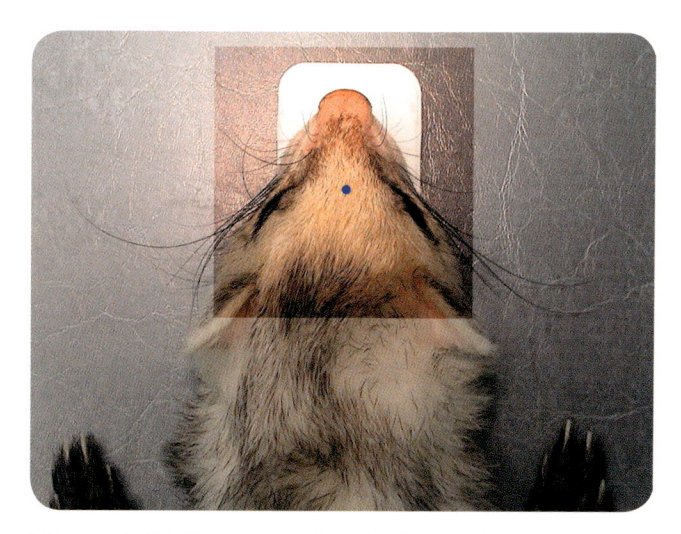

Figure 1.45. Dorsoventral projection of the maxilla using dental bitewing film. The film is inserted as far caudally as possible into the mouth with the ferret in ventral recumbency.

24 BASICS of RADIOLOGY

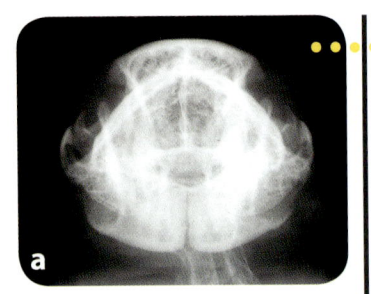

Figure 1.46a,b,c,d. Rostrocaudal projection. The patient is placed in dorsal recumbency, with the head flexed to position the palate perpendicular to the cassette (a). Proper vertical position is confirmed by viewing the patient from above, and observing that the head is symmetrical and not deviated laterally. Radiolucent foam is useful to maintain proper position (d).

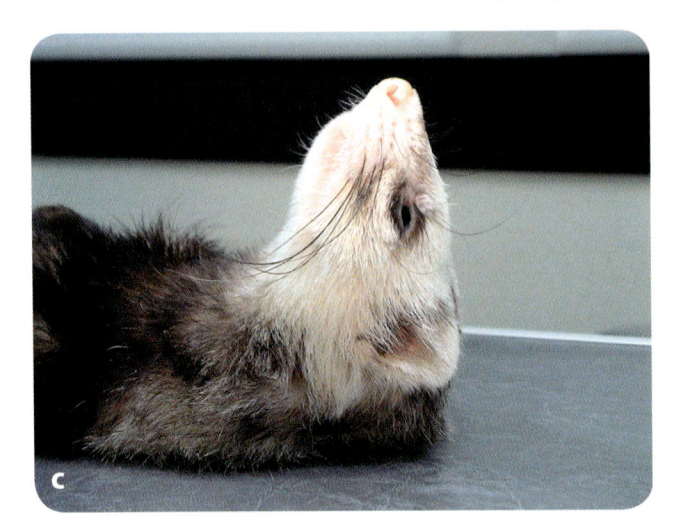

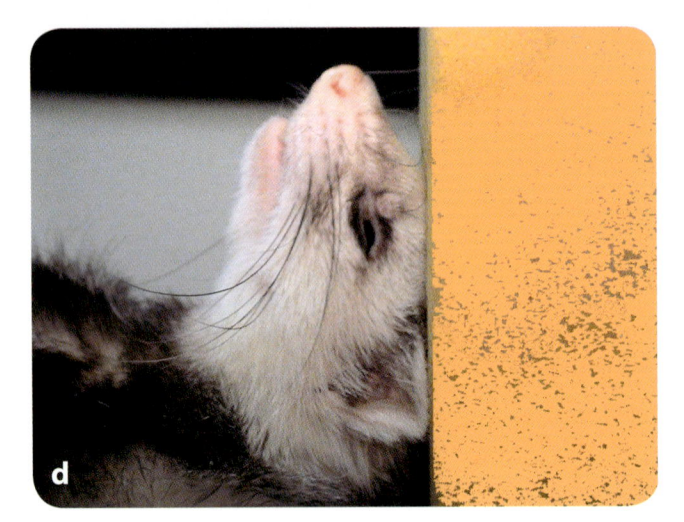

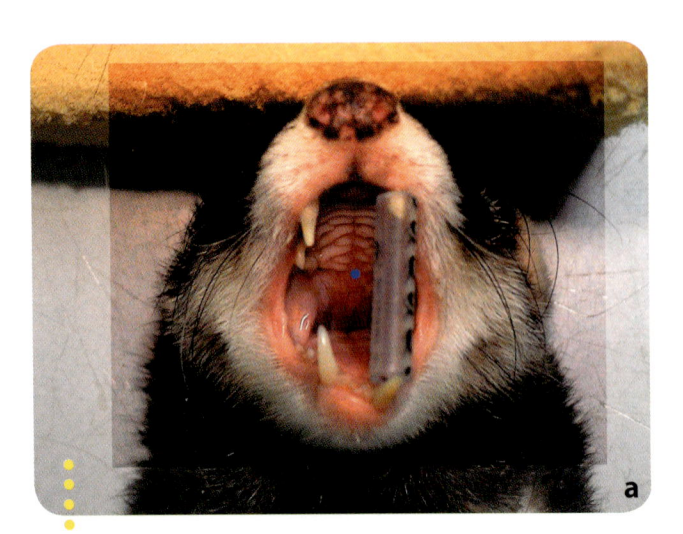

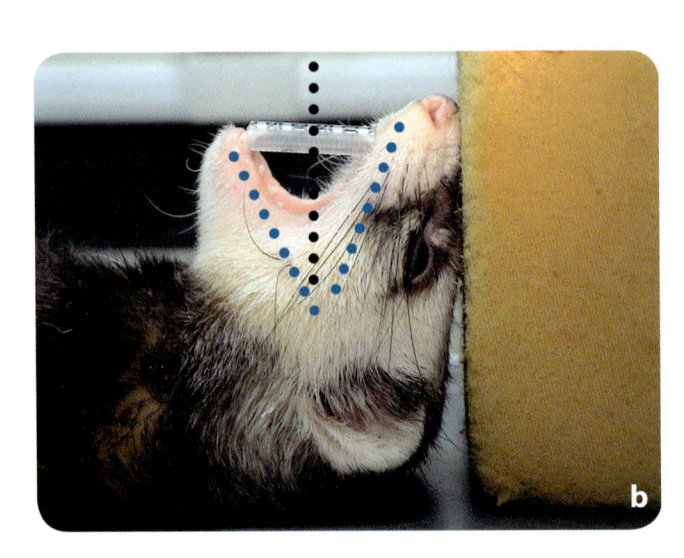

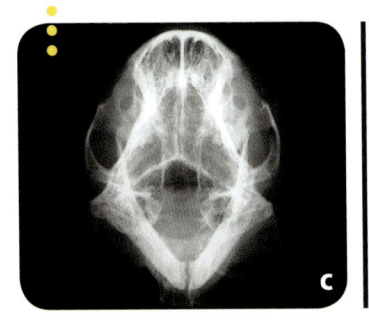

Figure 1.47a,b,c. Rostrocaudal projection, open mouth. The patient is placed in dorsal recumbency with the mouth maintained in an open position (a). The head is flexed slightly so that the radiographic beam will bisect the angle between the long axis of the maxilla and the mandible (b). Proper vertical position is checked from above, by observing that the head is symmetrical and not deviated laterally. Radiolucent foam is often necessary to maintain the head in the proper vertical position (b).

Figure 1.48a,b,c. Oblique projections. The right lateral oblique projection is properly defined as Left 30° Ventral-Right Dorsal Oblique. The patient is placed in right lateral recumbency. The head is positioned using radiolucent tape or a contoured piece of foam. When viewed from above, the head must appear slightly oblique (b). Proper position is confirmed from the frontal view (c).
An ideal oblique projection for the ferret is approximately 30° rotation.
For the left lateral oblique projection (Rt 30° Ventral-Lt Dorsal Oblique) the patient is placed in left lateral recumbency, and the procedure repeated as above, with the head rotated slightly clockwise.

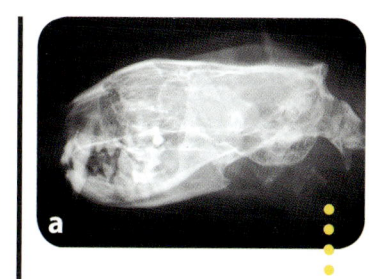

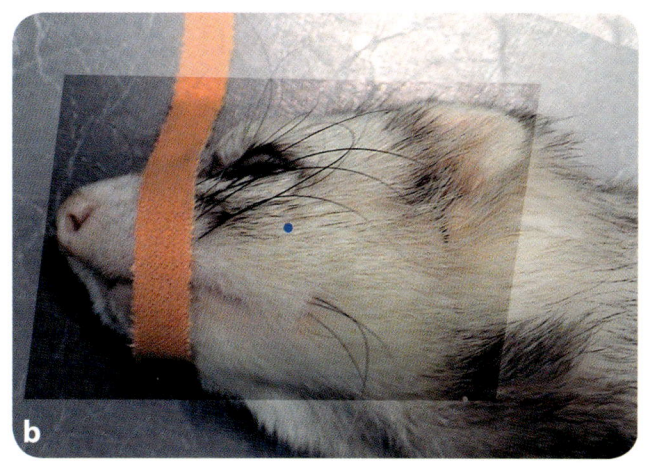

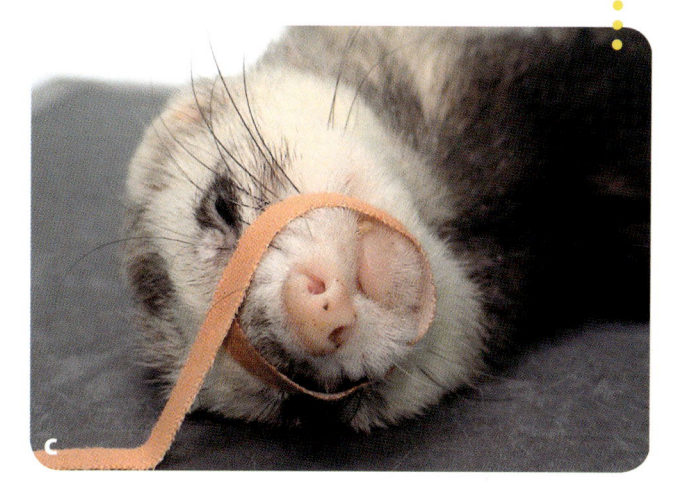

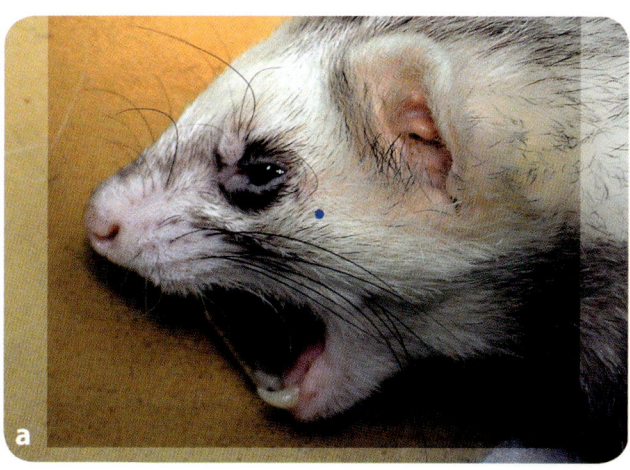

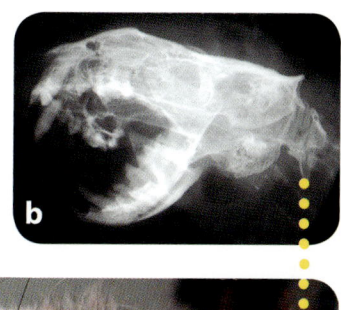

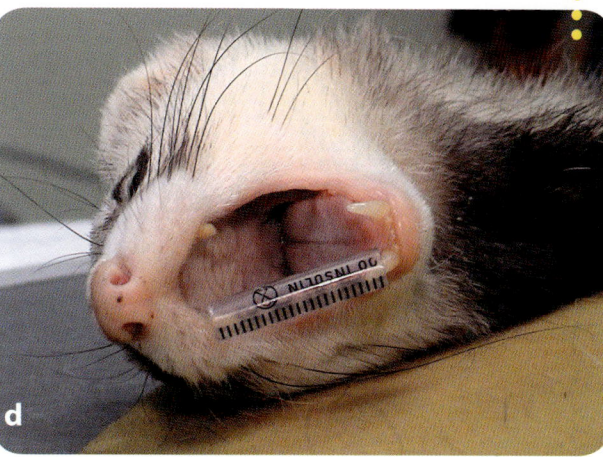

Figure 1.49a,b,c,d. Left 30° Dorsal-Right Ventral Oblique projection, open mouth. The patient is placed in right lateral recumbency. Radiolucent foam is used to position the head at a 30° angle dorsoventrally (a). Proper position is confirmed by observing the frontal view (c). This technique allows oblique projection of both the maxilla and the mandible, but it is particularly useful for the mandible. The opposite oblique projection (Left 30° Ventral-Right Dorsal Oblique projection), can be obtained with the head positioned at the opposite 30° angle. Proper position is also confirmed from the frontal view (d). This technique allows oblique projection of both the maxilla and the mandible, but it is particularly useful for the maxilla.

BASICS of RADIOLOGY

THORAX;
CERVICAL and THORACIC VERTEBRAL COLUMN

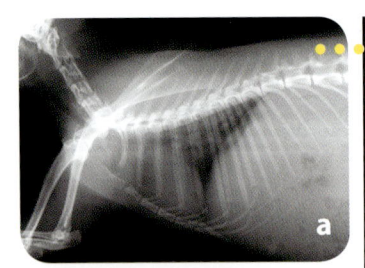

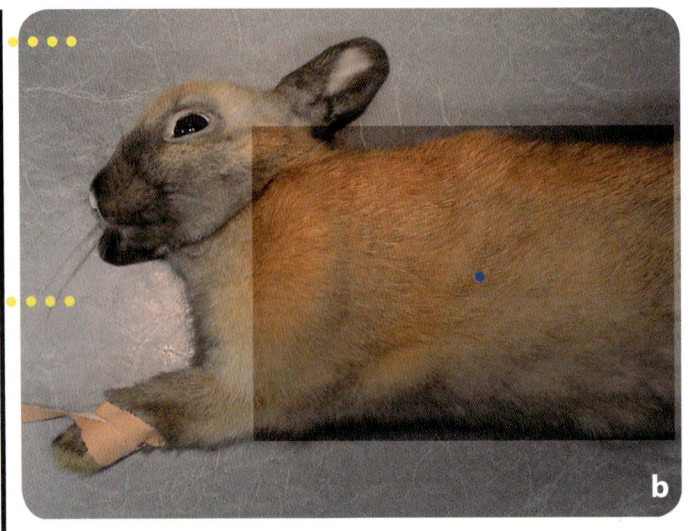

Figure 1.50a,b,c.
Lateral projection.
The patient is placed in right (b) or left (c) lateral recumbency. The thoracic limbs are extended cranially in order to prevent superimposition of the proximal limb over the thorax.
In some cases it may be useful to make radiographs in both lateral views, for example in cases of lung or mediastinal masses.
Respiration may be impaired when the limbs are hyperextended in lateral position. Therefore, this position should be maintained only for enough time to obtain the radiograph.

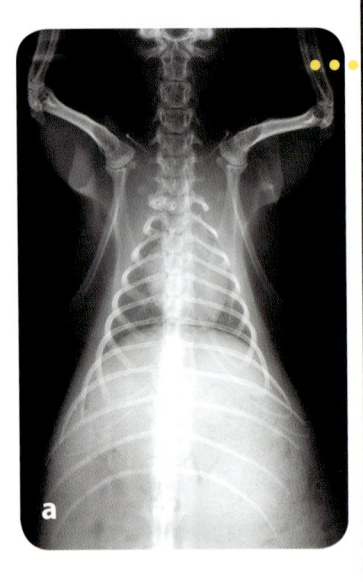

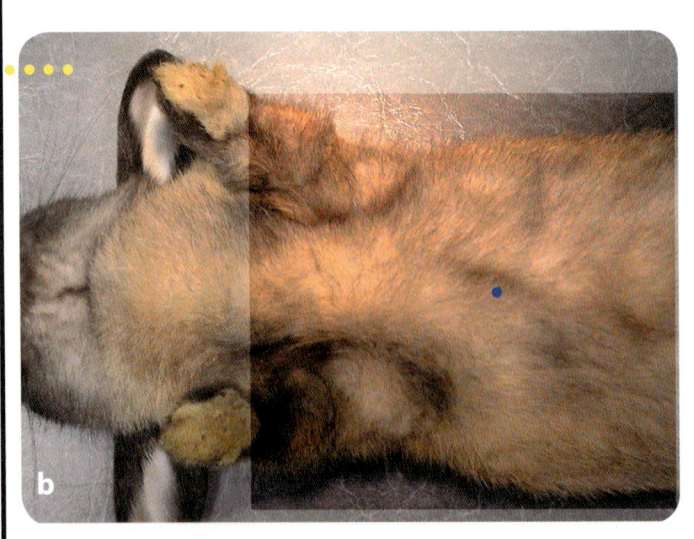

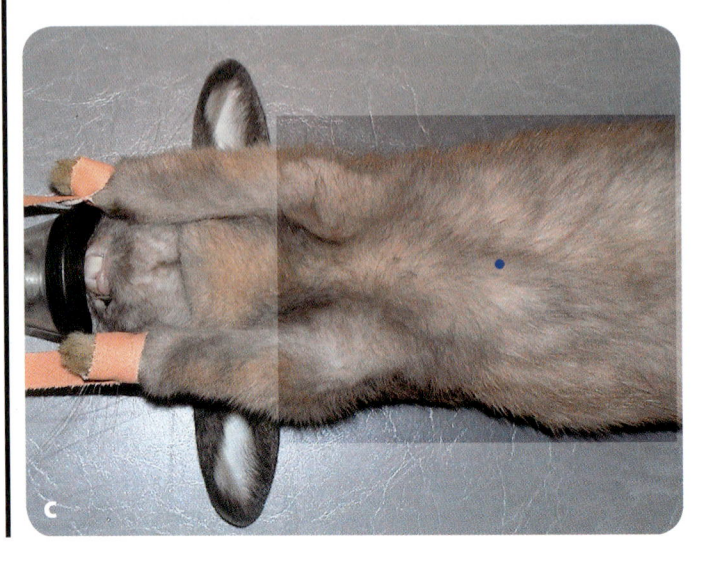

Figure 1.51a,b,c.
Ventrodorsal projection. The patient is placed in dorsal recumbency. The thoracic limbs may be left in neutral position (b) or extended cranially (c). In neutral position, there will be mild superimposition of the scapulae over the lung fields. When thoracic limbs are extended, respiration can be severely impaired, especially in obese or dyspneic patients, even in the presence of an endotracheal tube. This position should be maintained only long enough to obtain the radiograph.

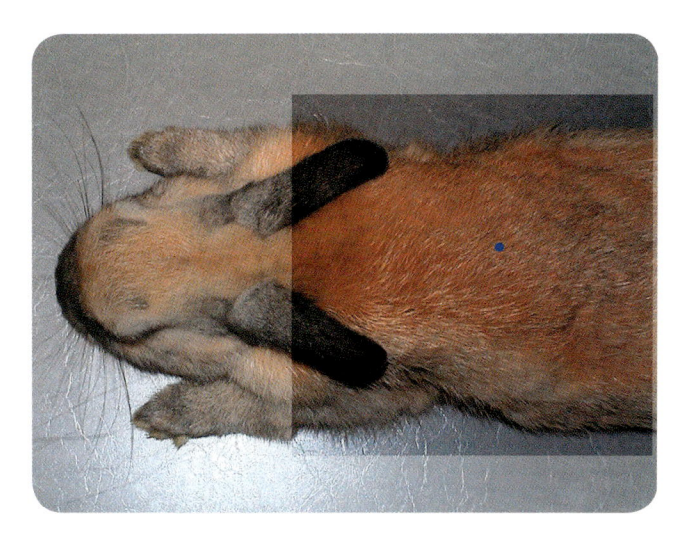

Figure 1.52. Dorsoventral position. The patient is placed in sternal recumbency. This position is considerably less stressful, especially in patients with dyspnea. However, in the authors' experience it is easier to obtain a symmetrical projection from the ventrodorsal position, rather than from this position. Mild superimposition of the scapulae on the lung fields can be present. The ears are positioned to minimize superimposition over the thorax.

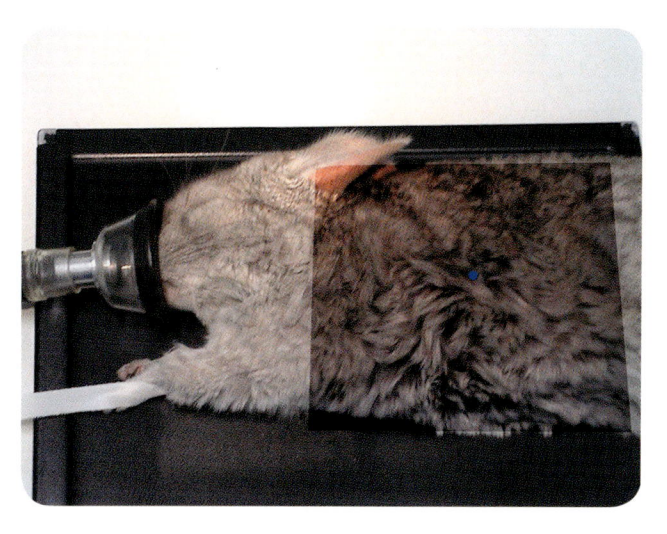

Figure 1.53. Positioning for a lateral radiograph in a chinchilla.

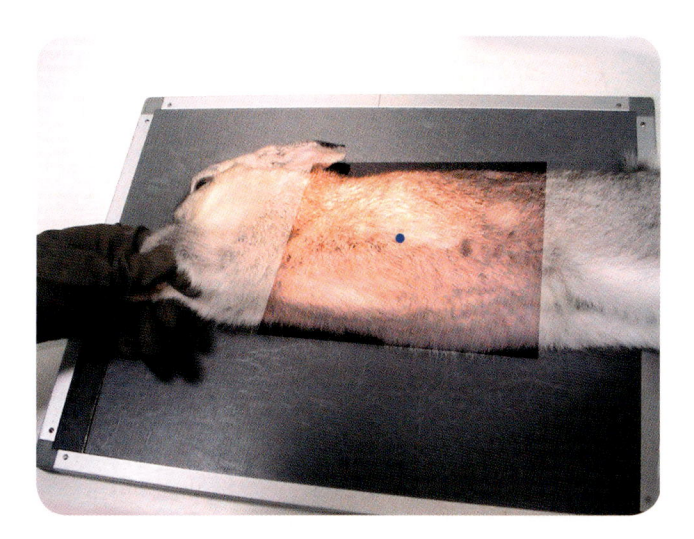

Figure 1.54. Lateral projection in the conscious patient. This technique is possible with careful restraint, and is performed as described for radiographs of the whole body in the conscious patient with an additional step: the thoracic limbs must be extended cranially to avoid superimposition of limbs over the thorax. An assistant maintains the scruff while the operator extends thoracic and pelvic limbs. If the patient tolerates this position, the assistant may be able to release the scruff while the operator maintains limb extension.

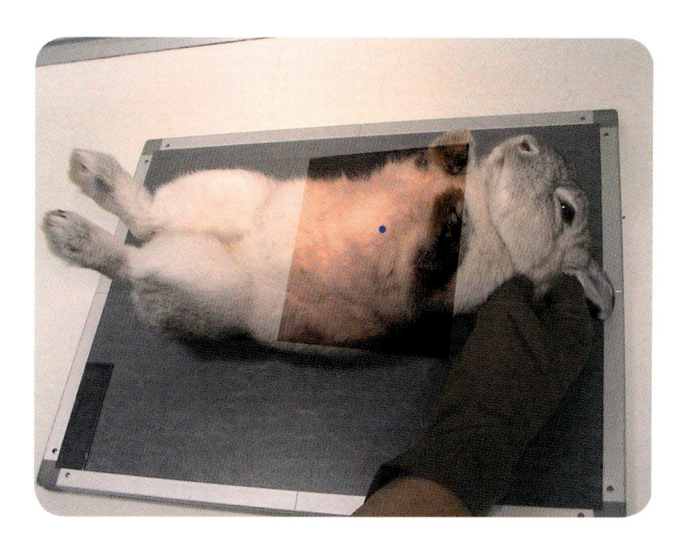

Figure 1.55. Ventrodorsal projection in the conscious patient. Usually, this position can be maintained by simply scruffing the rabbit with the limbs in neutral position as described above for ventrodorsal projection of the total body. Very calm rabbits may tolerate dorsal recumbency by simply extending the thoracic limbs and the pelvic limbs.

ABDOMEN and LUMBAR VERTEBRAL COLUMN

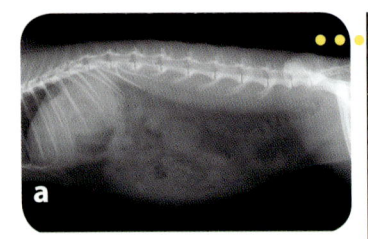

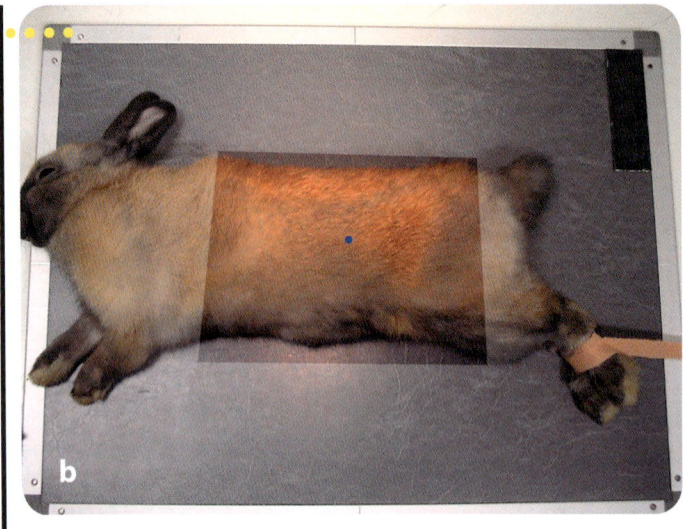

Figure 1.56a,b.
Lateral projection.
The patient is placed in right or left lateral recumbency. Right lateral projection is shown.
The pelvic limbs are extended caudally in order to prevent superimposition of the femurs over the caudal abdominal area.
For animals with suspected uroliths, the image must include the entire urinary system, including the distal urethra, penis or vulva.

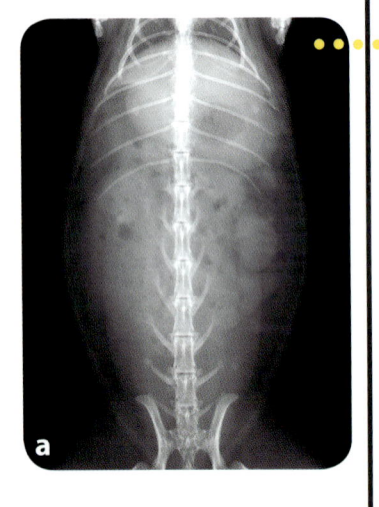

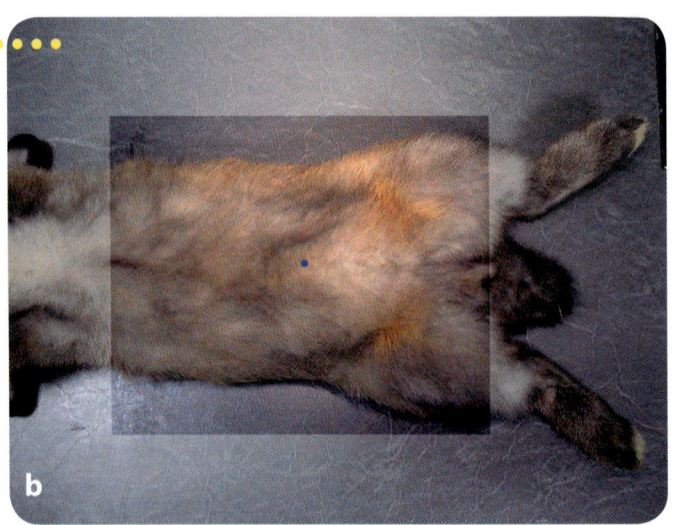

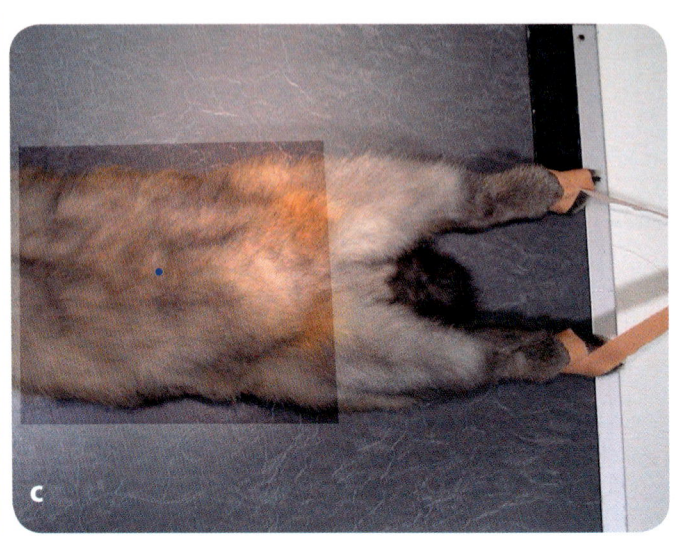

Figure 1.57a,b,c.
Ventrodorsal projection. The patient is placed in dorsal recumbency. The pelvic limbs may be left in a neutral, but slightly abducted position (b) or extended caudally (c).
Dorsoventral projection is also possible, with the patient placed in ventral recumbency. In the authors' experience, it is easier to obtain a symmetrical projection from the ventrodorsal position.
Lateral and ventrodorsal projections with manual restraint are obtained using techniques described above.

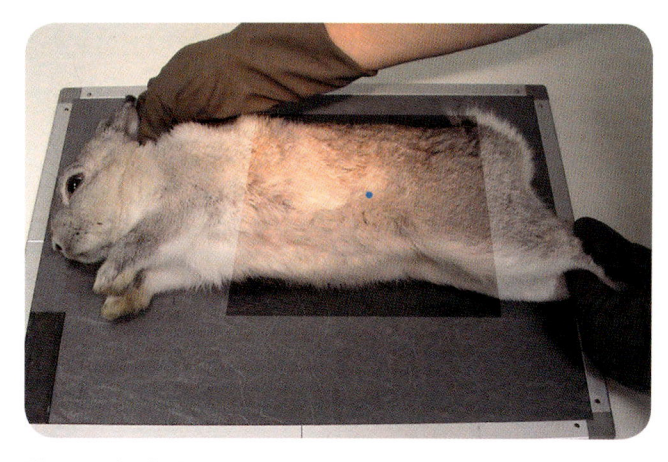

Figure 1.58. Lateral projection of the abdomen in the conscious patient. Restraint is the same as described for lateral projection of the total body, and the x-ray beam is collimated over the area of interest.

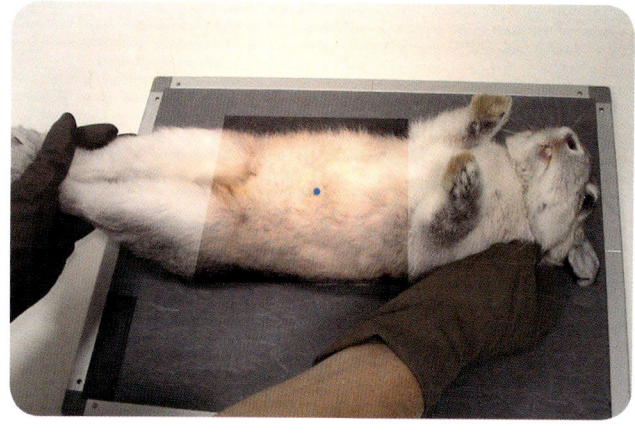

Figure 1.59. Ventrodorsal projection of the abdomen in the conscious patient. Restraint is the same as described for ventrodorsal projection of the total body. Pelvic limbs can be held in neutral or extended position.

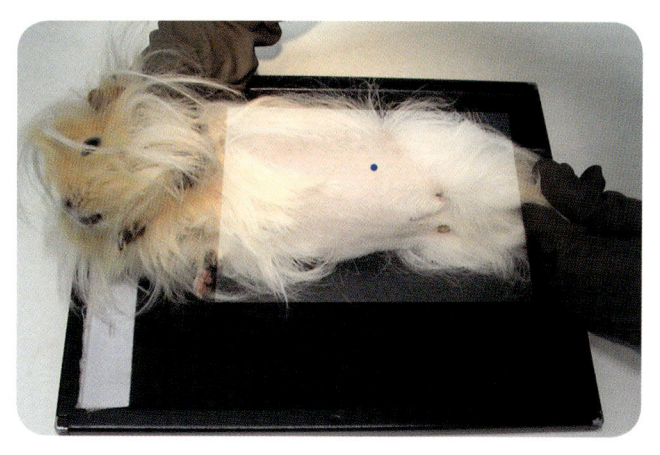

Figure 1.60. Lateral projection of the abdomen in the conscious patient. Guinea pigs generally do not tolerate scruffing. Lateral recumbency is obtained by gently grasping the head and neck. The pelvic limbs can be extended with the other hand, or left in a neutral position.

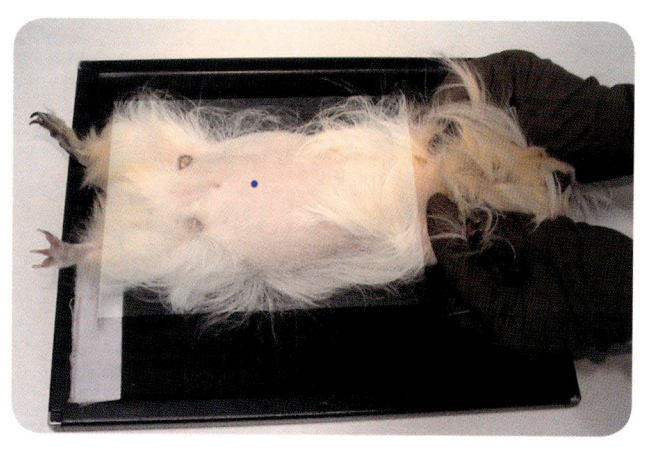

Figure 1.61. Ventrodorsal projection. Guinea pigs usually tolerate ventrodorsal recumbency better than lateral. Position is maintained by simply grasping the thoracic limbs.

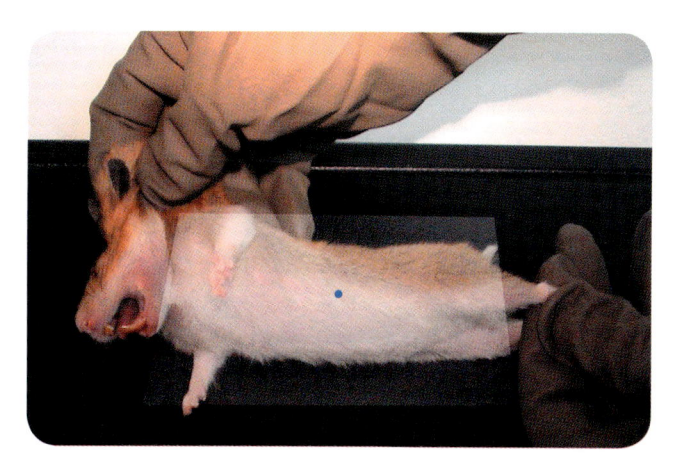

Figure 1.62. Lateral projection in the conscious hamster. The hamster is held as described for the rabbit, using just two fingers. The pelvic limbs are grasped and gently extended caudally. This restraint allows a complete total body projection with thoracic limbs in a neutral projection, which will be slightly superimposed over the thorax. The assistant accurately collimates the x-ray beam in order to prevent inclusion of the operator's fingers.

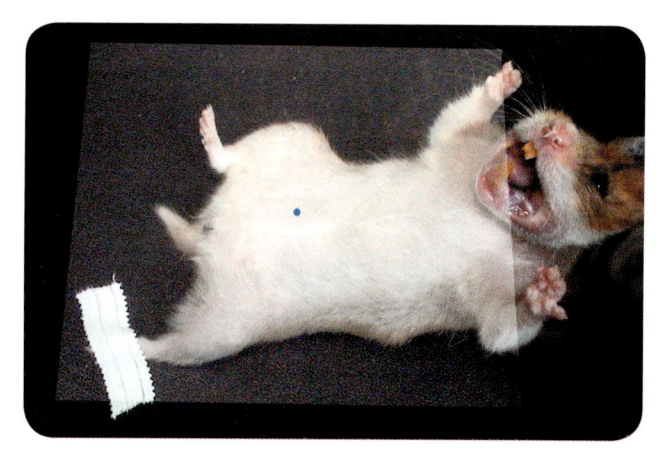

Figure 1.63. Ventrodorsal projection in the conscious hamster. Restraint is repeated the same way holding the hamster in dorsal recumbency. To avoid the operator's fingers, the x-ray is collimated more caudally, and includes just the abdomen. In calmer patients, an assistant can secure the pelvic limbs with tape. Manual restraint should be used only when absolutely necessary, as positioning will likely not be ideal.

THORACIC LIMB

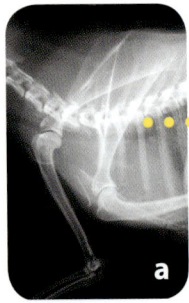

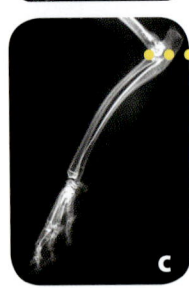

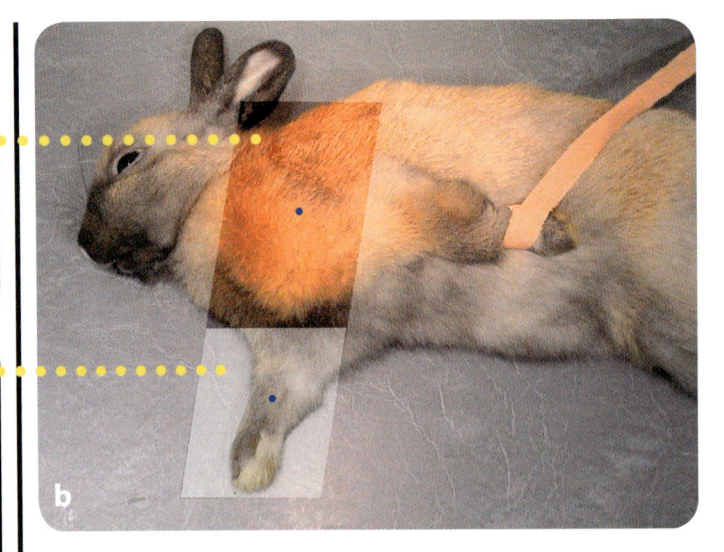

Figure 1.64a,b,c.
Lateral projection.
The rabbit is placed in right or left lateral recumbency, with the limb of interest closest to the cassette. The contralateral limb is extended caudally to prevent superimposition. In larger exotic mammal species, kVp must be adjusted from proximal (thicker) to distal (thinner) portion of the limb.

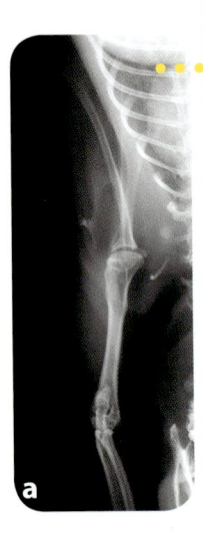

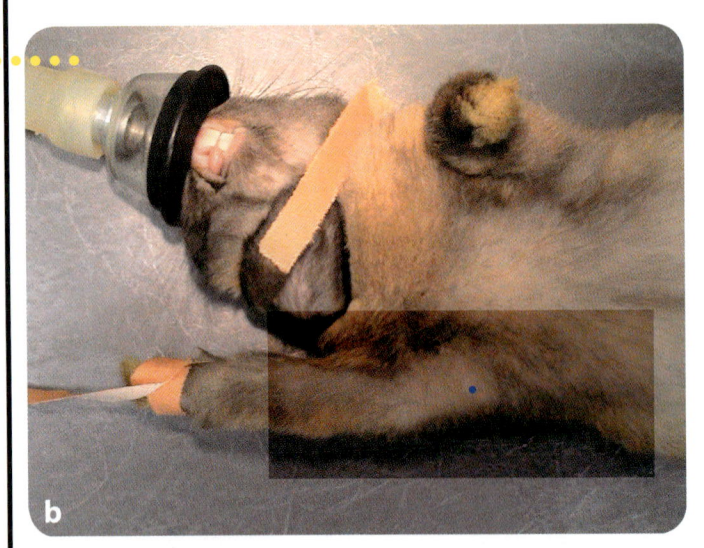

Figure 1.65a,b.
Caudocranial/Palmardorsal. The rabbit is placed in dorsal recumbency, and the thoracic limb is extended cranially. The head and ears are deflected laterally to prevent superimposition. The limb must be well secured to prevent rotation which will result in an oblique projection. Minor adjustment will result in an oblique projection of the elbow, if desired.

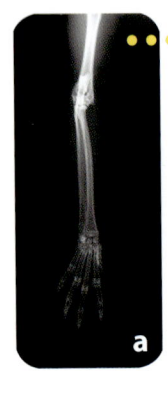

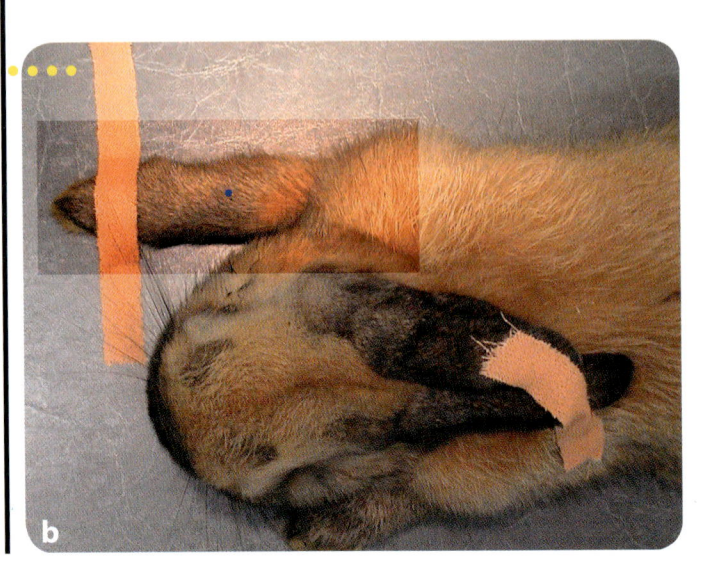

Figure 1.66a,b.
Craniocaudal/Dorsopalmar projection. The rabbit is placed in sternal recumbency, with the thoracic limb extended cranially. This projection produces an excellent projection of the distal segments of the thoracic limb (radius, ulna, carpus, metacarpus, and phalanges), superior to that provided by a palmardorsal projection. Head and ears are deflected laterally to prevent superimposition.

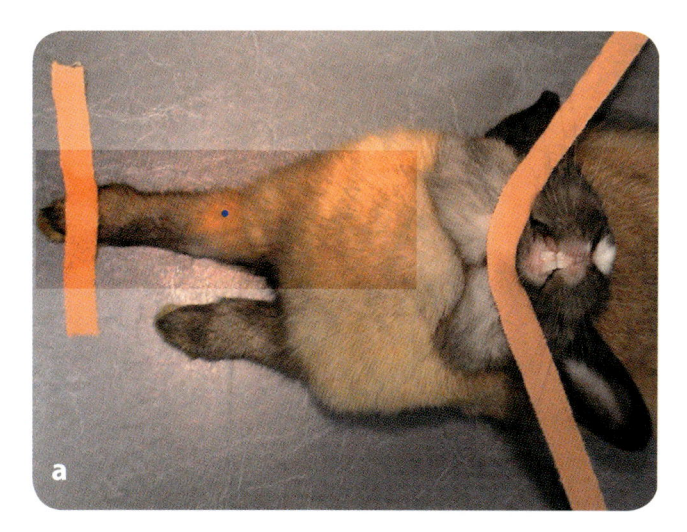

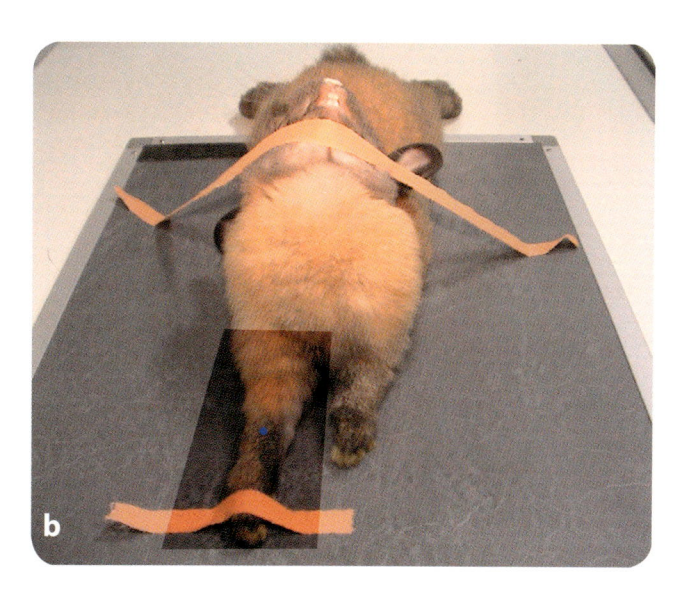

Figure 1.67a,b. Craniocaudal/Dorsopalmar projection. The rabbit is placed in sternal recumbency, with the thoracic limb fully extended cranially. In order to include the proximal segments of the thoracic limb, the head and the neck of the patient must be extended caudally. Even in full extension, this technique does not provide a true craniocaudal projection of the proximal segments of the thoracic limb (scapula and humerus). This position can impair respiration if the patient is not intubated, and should therefore be maintained only for the time needed to take the radiograph.
This position usually reduces the risk of inadvertently obtaining an oblique projection of the distal thoracic limb.

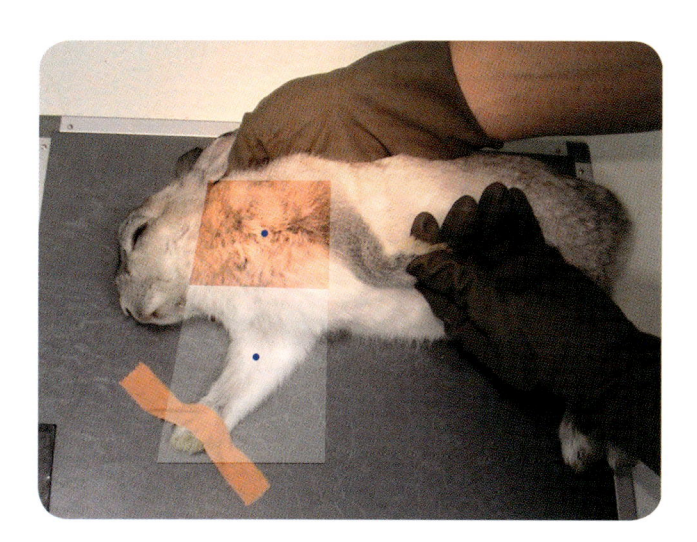

Figure 1.68. Lateral projection in the conscious rabbit. The operator restrains the rabbit in lateral recumbency by scruffing with one hand, while an assistant secures the thoracic limb in slight extension. The operator deflects the contralateral thoracic limb caudally to prevent superimposition.

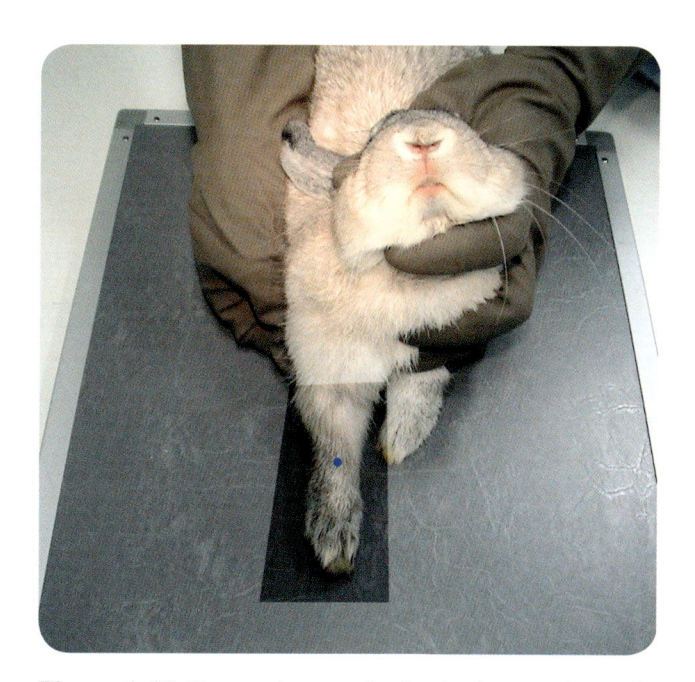

Figure 1.69. Dorsopalmar projection in the conscious rabbit. The distal portion of the thoracic limb can be radiographed by placing the rabbit in sternal recumbency and extending the humerus forward from behind the elbow, while extending the head caudally. An assistant carefully collimates the x-ray beam to avoid including the hands of the operator in the primary beam.

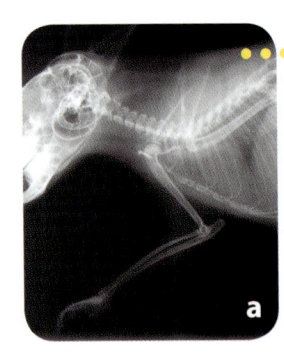

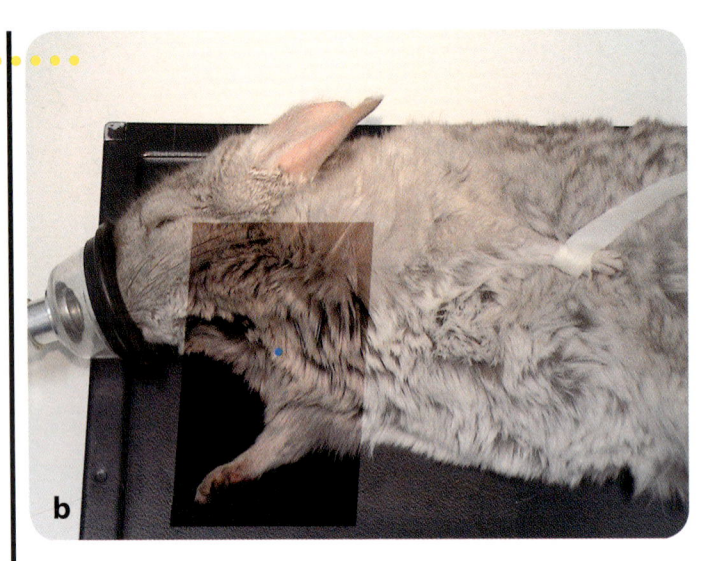

Figure 1.70a,b. The same positioning techniques demonstrated in the anesthetized rabbit can be utilized in the chinchilla and other small exotic mammals as well.

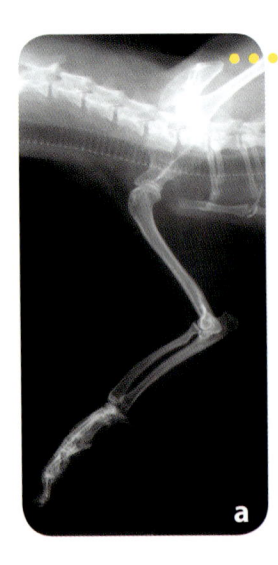

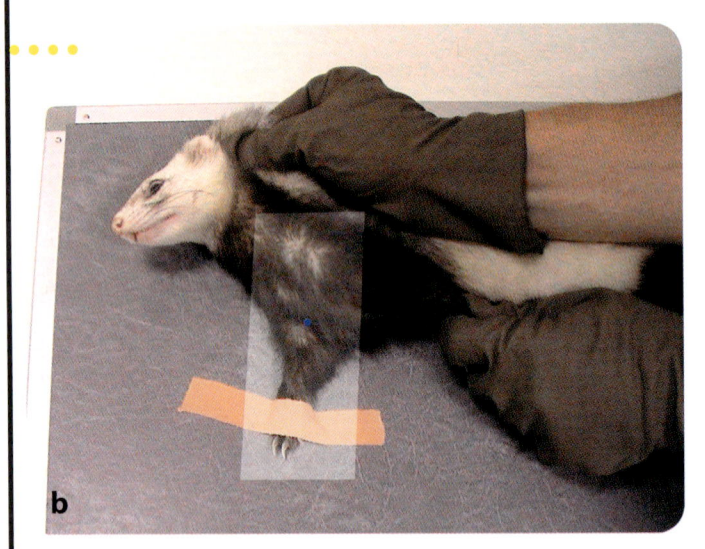

Figure 1.71a,b. Some of the same techniques demonstrated in the conscious rabbit can used in other exotic mammal species as well. Lateral projection. The operator restrains the ferret in lateral recumbency scruffing with one hand, while an assistant secures the thoracic limb in slight extension. The operator deflects the contralateral limb caudally to prevent superimposition.

PELVIS and PELVIC LIMB

Figure 1.72a,b.
Lateral projection of the pelvis. The rabbit is placed in lateral recumbency, with the pelvic limbs hyperextended caudally and parallel. A perfect lateral projection of the pelvis is obtained only when pelvic limbs are parallel to each other, which unfortunately produces superimposition of the femurs. In reality, the true lateral view of the pelvis is less useful than dorsoventral or slightly oblique views.

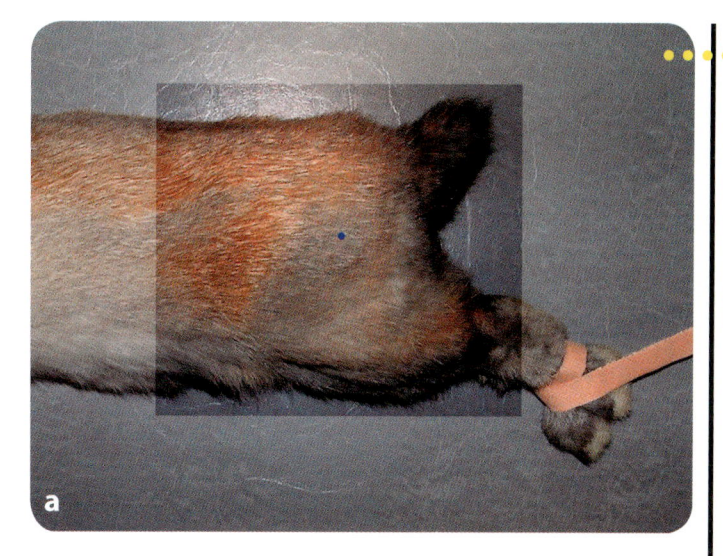

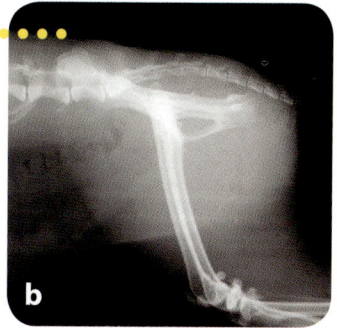

Figure 1.73a,b,c.
Oblique projection of the pelvis and lateral projection of the femur. The rabbit is placed in lateral recumbency. The contralateral limb is extended caudally. This prevents superimposition of the femurs and produces a slightly oblique view of the pelvis.
This position is also useful for lateral projection of the distal limb (tibia, fibula, tarsus, metatarsus and phalanges). Proper kVp settings are lower for the distal limb than for the pelvis because of less tissue thickness.

Figure 1.74a,b.
An oblique projection of the pelvis and lateral projection of the femur can also be obtained with caudal hyperextension and abduction of the contralateral pelvic limb. This produces a 45° oblique projection of the pelvis (Lt 45° Ventral-Rt Dorsal Oblique is shown) and true lateral view of the proximal femur, including the hip joint.

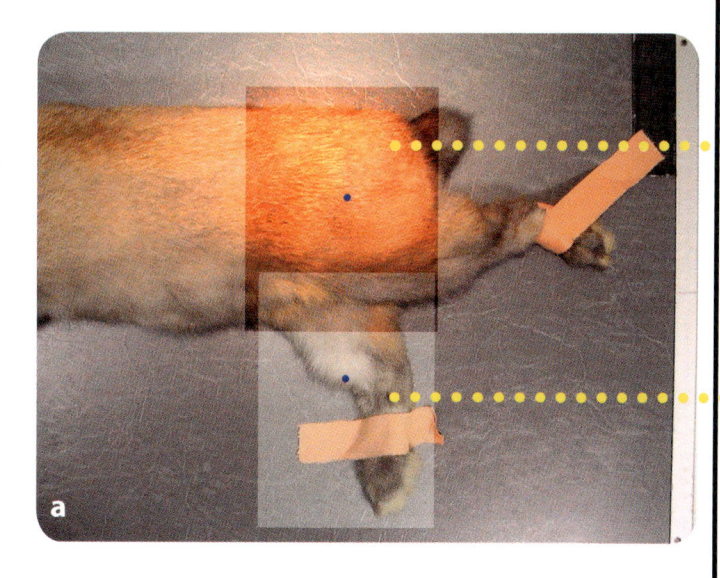

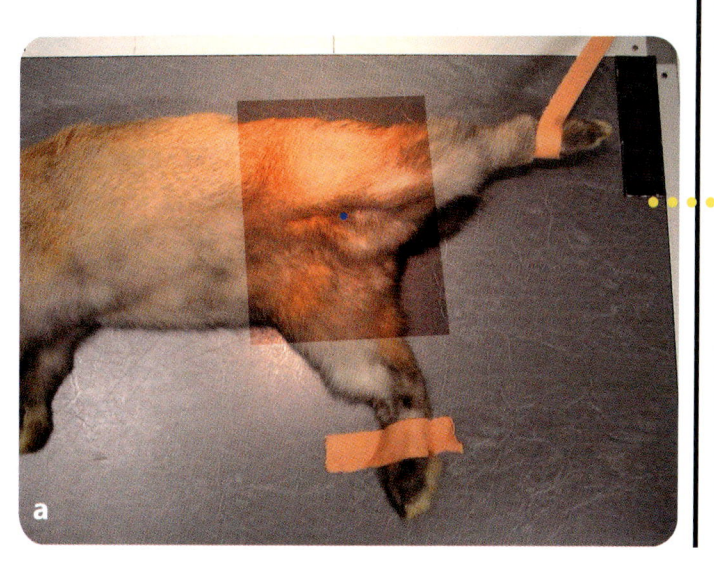

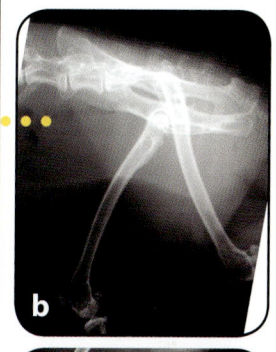

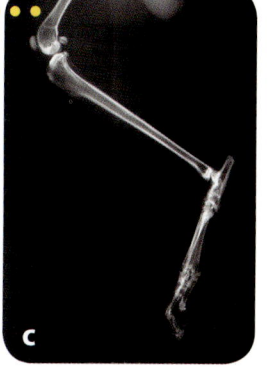

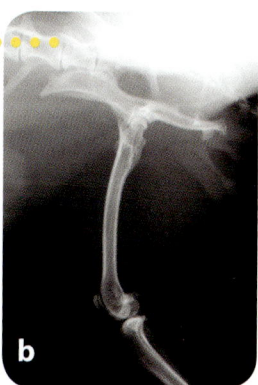

34　BASICS of RADIOLOGY

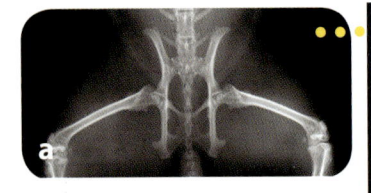

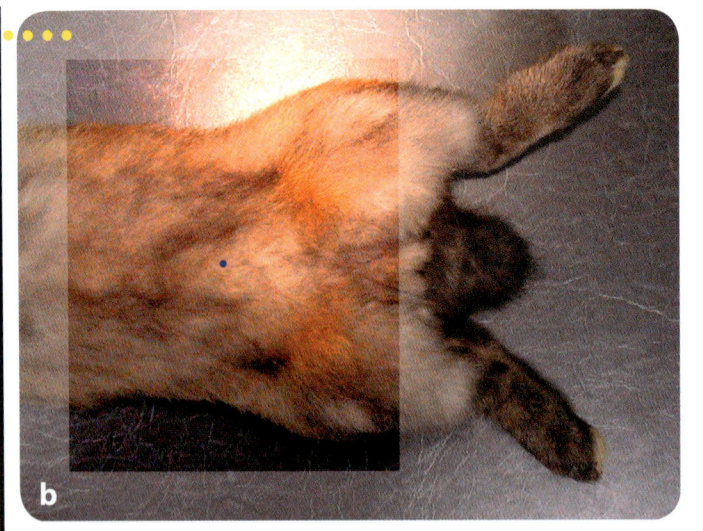

Figure 1.75a,b.
Ventrodorsal projection of the pelvis and craniocaudal projection of the femurs. The rabbit is placed in dorsal recumbency. The pelvic limbs can be left in a slightly abducted neutral position, which results in a "frog position" view of the hip joint, and less optimal view of the femurs.

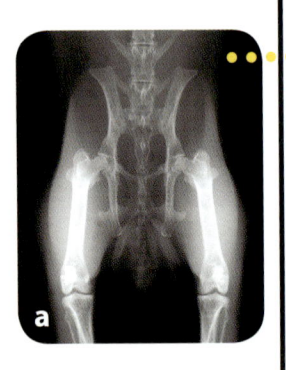

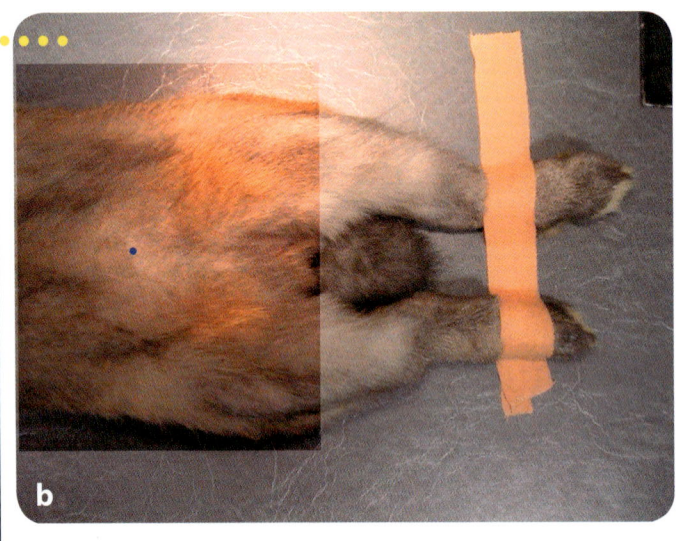

Figure 1.76a,b.
Alternatively, the pelvic limbs can be extended caudally and secured with tape. This technique produces a more standard view of the hip joints and the heads of the femurs.

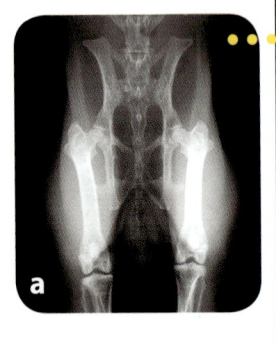

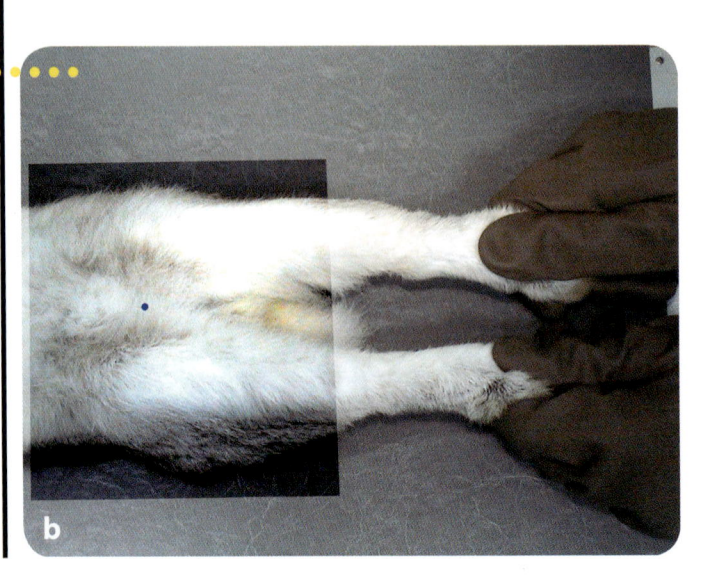

Figure 1.77a,b.
An ideal radiographic study of the pelvis is made with internal rotation of the distal femurs, while keeping the femoral diaphyses parallel, similar to the technique used for an OFA study of hip dysplasia in dogs.

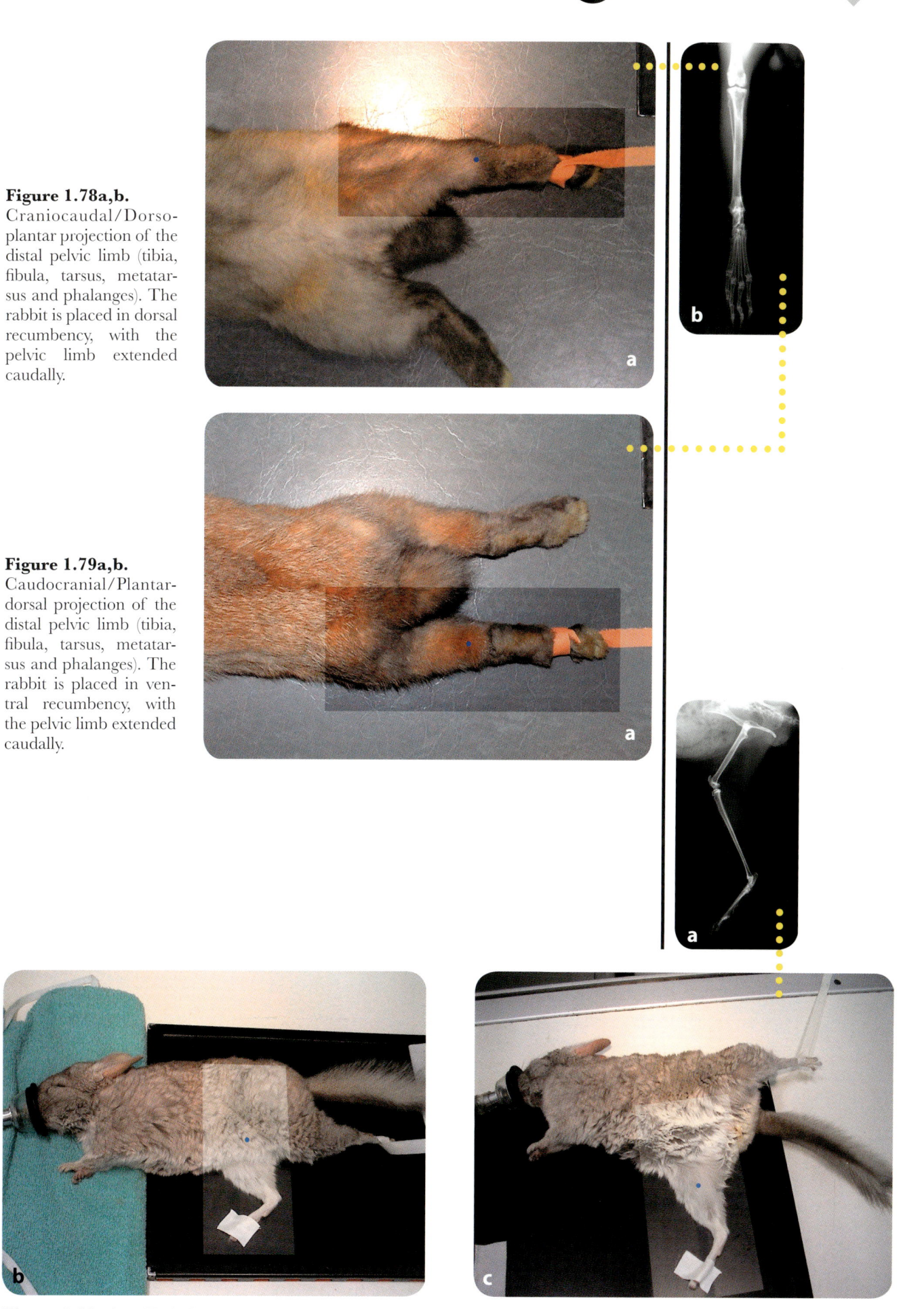

Figure 1.78a,b. Craniocaudal/Dorsoplantar projection of the distal pelvic limb (tibia, fibula, tarsus, metatarsus and phalanges). The rabbit is placed in dorsal recumbency, with the pelvic limb extended caudally.

Figure 1.79a,b. Caudocranial/Plantardorsal projection of the distal pelvic limb (tibia, fibula, tarsus, metatarsus and phalanges). The rabbit is placed in ventral recumbency, with the pelvic limb extended caudally.

Figure 1.80a,b,c. Techniques demonstrated in the anesthetized rabbit can be utilized in other exotic mammal species as well.

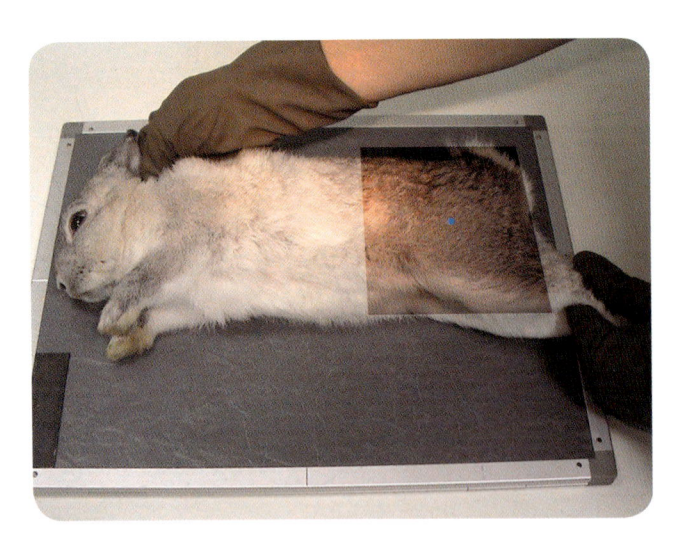

Figure 1.81. Lateral projection of the pelvis in the conscious patient. Restraint is the same as described for lateral projection of the whole body survey and the abdomen, and the x-ray beam is collimated over the area of interest.

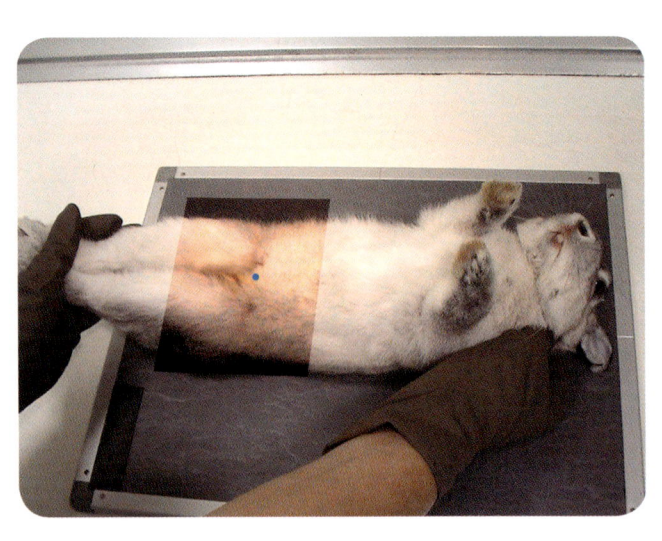

Figure 1.82. Ventrodorsal projection in the conscious patient. Restraint is the same as described for ventrodorsal projection of the whole body survey and the abdomen. Pelvic limbs can be held in neutral or extended position.

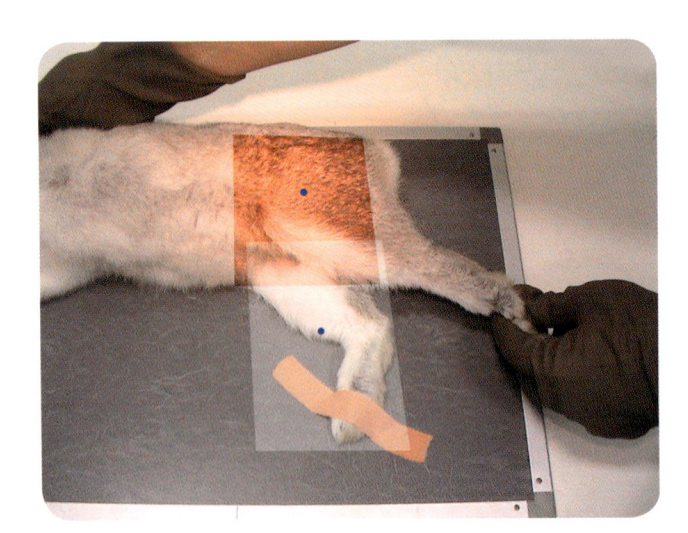

Figure 1.83. Lateral projection of the distal pelvic limb (tibia, fibula, tarsus, metatarsus and phalanges). The rabbit is restrained in lateral recumbency with the scruffing technique. The contralateral pelvic limb is deflected caudally with the other hand while an assistant secures the limb of interest in slight extension with tape. An oblique projection of the pelvis and lateral projection of the femur can be obtained by extending and simultaneously abducting the contralateral pelvic limb.

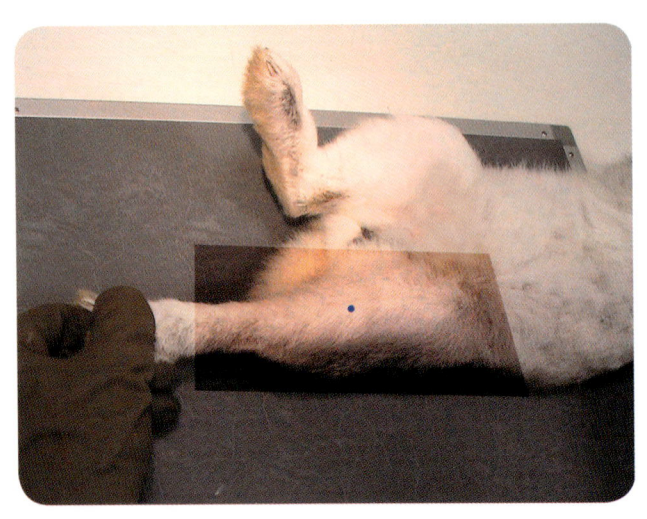

Figure 1.84. Craniocaudal projection of the pelvic limb of a rabbit. This projection (with the exception of the digits) can be obtained by scruffing the rabbit in dorsal recumbency, and grasping the digits to extend the pelvic limb.

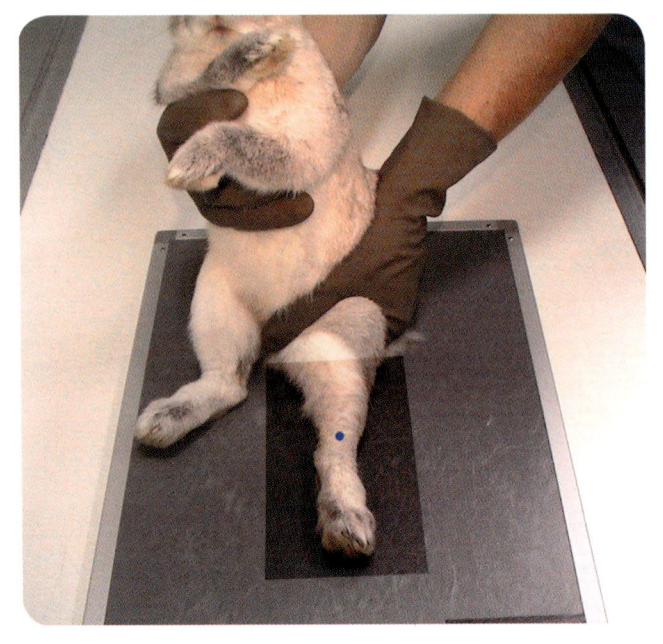

Figure 1.85. Craniocaudal/Dorsoplantar projection of the pelvic limb of a rabbit. The rabbit is restrained in a "sitting" position with one hand, and the femur positioned by extending the stifle joint with the other hand.

Common Errors in Radiographic Technique

Errors in radiographic technique may result in an image with limited diagnostic information. The most common errors include poor density (too dark or too light); narrow gray scale; motion artifact due to improper restraint or excessive exposure time; and improper positioning. Omissions also include failure to include the target area of interest (Figure 1.86a,b), or failure to obtain optimal views essential for diagnosis (Figure 1.87a,b).

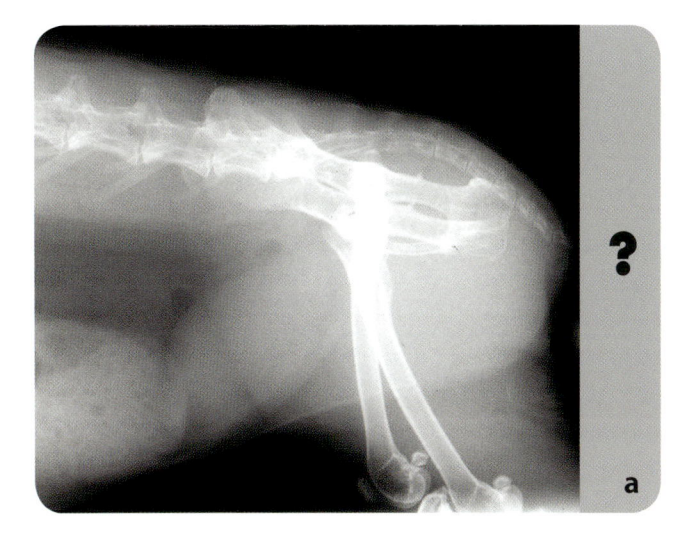

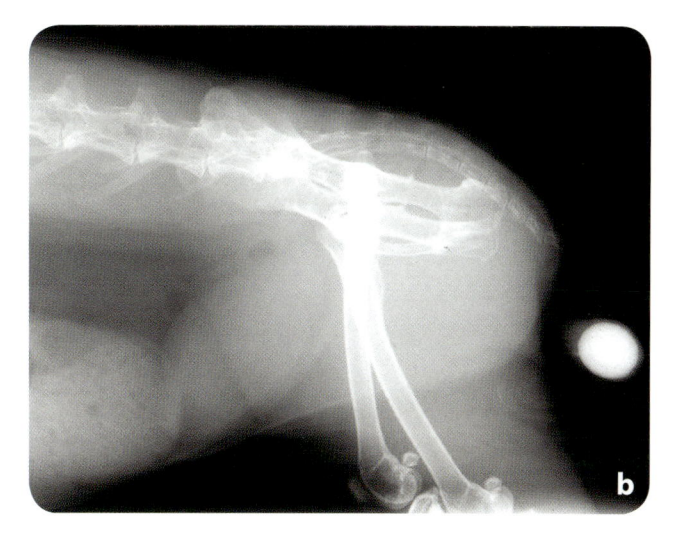

Figure 1.86a,b. Improper centering of the patient within the radiographic beam or improper collimation of the radiographic beam can lead to misdiagnosis. History and physical exam in this male rabbit included dysuria and distended urinary bladder. The beam is properly centered over the bladder; however improper collimation led to exclusion of the urolith in the distal urethra.

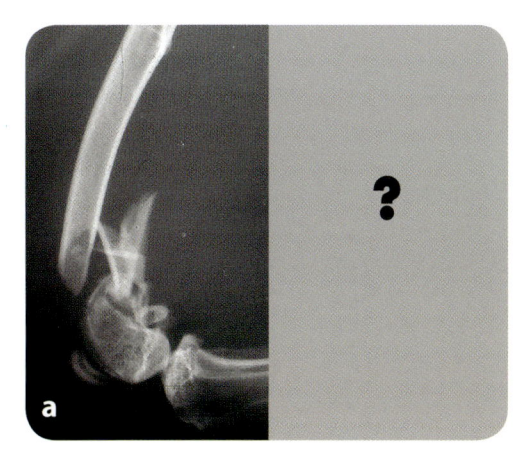

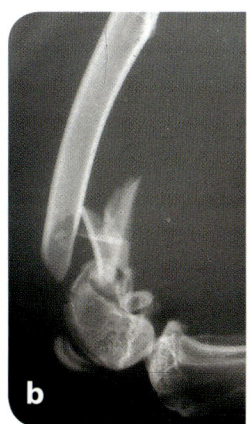

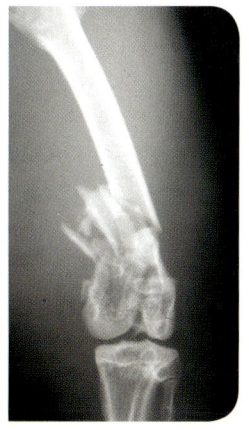

Figure 1.87a,b. The lateral projection alone is not adequate for proper evaluation of this severe comminuted fracture of the femur (a). The craniocaudal projection shows that an intercondylar fracture is also present. This additional information affects case management and prognosis.

Besides magnification artifact discussed earlier, one of the most common problems is distortion artifact due to beam divergence. Tissues and organs far from the central x-ray beam project a shadow that may not represent their true shape, as only a small part of the patient is under the central part of the x-ray beam. This can be overcome by multi-beam centering, where multiple projections are taken of various portions of the patient.

While important, it should be kept in mind that this artifact is much more common in radiography of larger species than it is in small exotic mammal patients. It should also be noted that a so-called lateral projection of certain anatomical structures may in reality be an oblique projection; for example, the lateral projection of the pelvis, the verberal bodies (Figure 1.88), or the pairs of ribs (Figure 1.89).

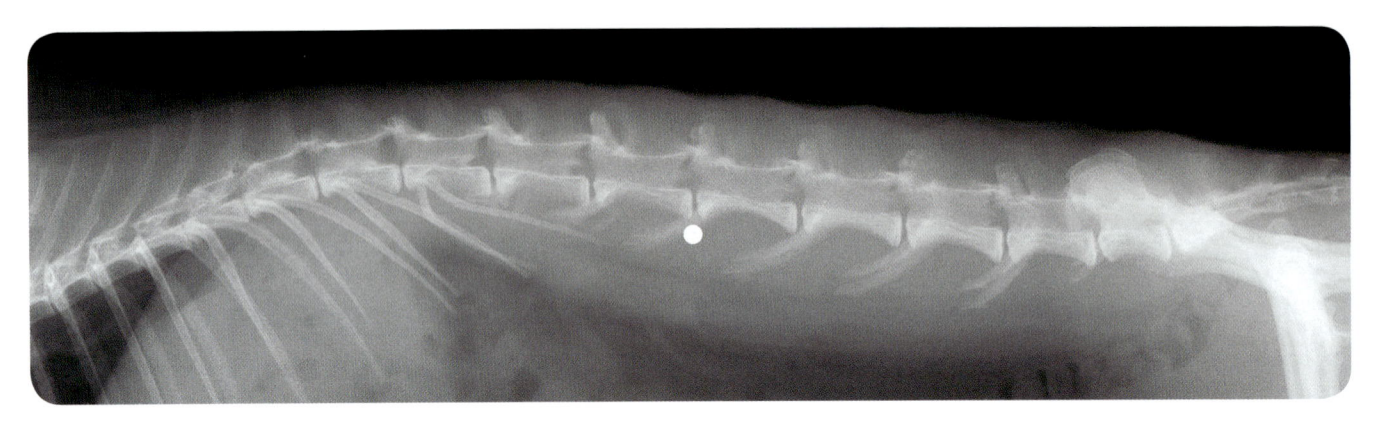

Figure 1.88. Example of the distortion artifact. The radiograph of this vertebral column has been centered (white dot) on the L2-L3 intervertebral space. Only the body of L2 and L3 and L2-L3 intervertebral space are perfectly lateral. Vertebrae more distant from the center are more subject to distortion artifact. For example, the thoracic intervertebral spaces appear narrower than they are, and this projection is not useful to evaluate them.

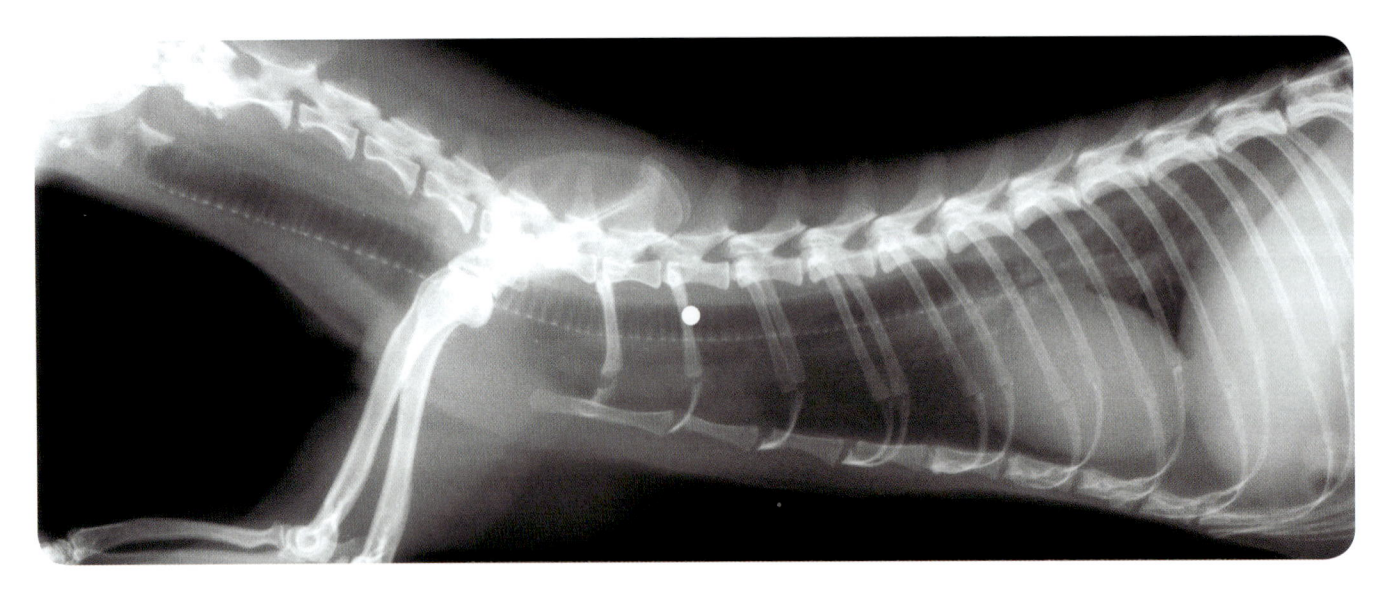

Figure 1.89. Example of the distortion artifact. This radiographic study of the thoracic cavity and the trachea (lateral projection) has been centered on the second rib. Due to distortion artifact, only the first and second pairs of ribs are perfectly lateral. More caudal ribs are not superimposed due to divergence of the x-ray beam.

Processing X-ray Film

Manual and automatic processing allows the latent image on the x-ray film to become visible to the human eye. A detailed description of the fundamentals of processing of x-ray film is beyond the scope of this chapter, but this technology is covered in several references. Briefly, the latent image on an exposed x-ray film is converted to a radiograph in three main steps:

1) Development. Silver halide crystals representing the latent image are precipitated to elemental silver, which is black.
2) Fixation. Unprecipitated silver halide crystals are removed, producing clear areas on the finished radiograph. These appear white when the radiograph is placed on a light box.
3) Wash. The x-ray film is rinsed with water to remove residual chemicals. After drying, the radiograph is ready for interpretation. Automatic processors produce a dry, ready to read radiograph in 90 seconds to two minutes while manual processing requires about an hour for completion. Manually processed radiographs must be dried before viewing while automatically processed radiographs are dried during the processing cycle.

The use of an automatic processor ensures consistency and reduces manual labor. Processing time is decreased, which is important in a busy practice, especially when the patient is anesthetized, and/or when a study requires a series of radiographs.

Radiographic quality depends on good processing practices. These include:
1) changing processor chemicals before they are exhausted;
2) maintaining correct temperature and processing time;
3) regularly scheduled maintenance
4) provision of a dryer for manually processed radiographs.

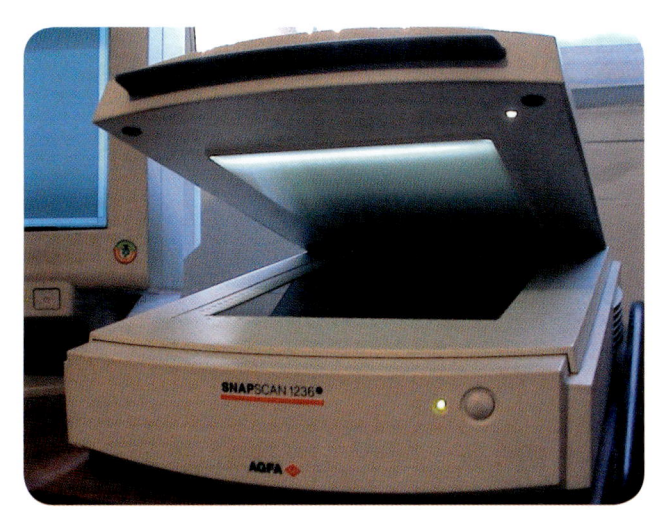

Figure 1.90. Scanner for transparencies. The ideal digital scanner has the capacity to scan 21 cm by 27 cm x-ray films. This size is adequate for typical small exotic radiographs but may be too small for radiographs of larger patients. Scanners are available at consumer electronics stores, connect to a computer via a USB port, and come with user-friendly, easily installed software. The scanner allows the user to select resolution when scanning, from lower resolution often up to 1000 p.p.i or more. In the authors' experience, scanning radiographs takes time but results in a high quality digital image.

Storage of Radiographs

There are many advantages to saving radiographs in digital form, including reduction of physical storage needs, protection from image deterioration, manipulation (particularly enlargement) of the image, and the ability to transmit and share radiographic images via the internet. Conventional radiographs can be saved in digital forms by scanning the image with a digital scanner (Figure 1.90). Many veterinary software packages allow digital images of any kind to be attached to client records. Some common image software programs allow manipulation of brightness and contrast and for some devices hue and saturation in order to maximize quality of the final image. Digital storage of images must be protected by appropriate backup to prevent inadvertent loss of data. Unfortunately, a conventional radiograph (analog) will lose some degree of information when it is converted to a digital format. In addition, digitization will not improve the quality of a suboptimal conventional radiograph.

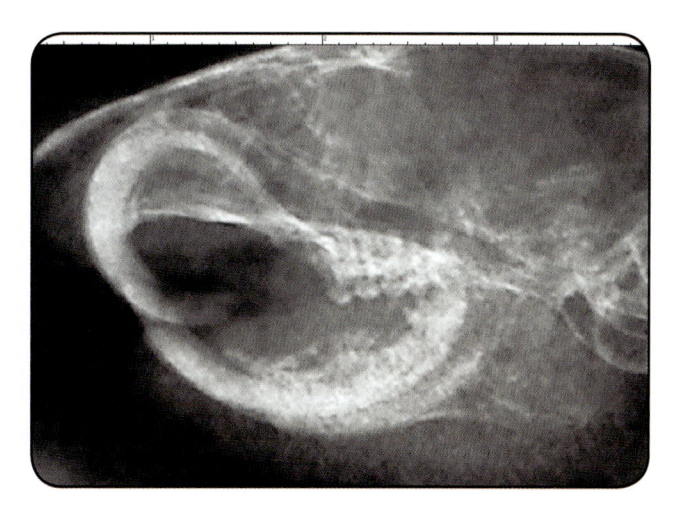

Figure 1.91. Radiograph of the cheek teeth of a chipmunk, photographed with a digital camera with a resolution capability of 300 p.p.i. The actual width of this picture is about 3.5 cm. When magnified to 8.5 cm (the actual size of this photo) the image displays poor resolution with visible pixels.

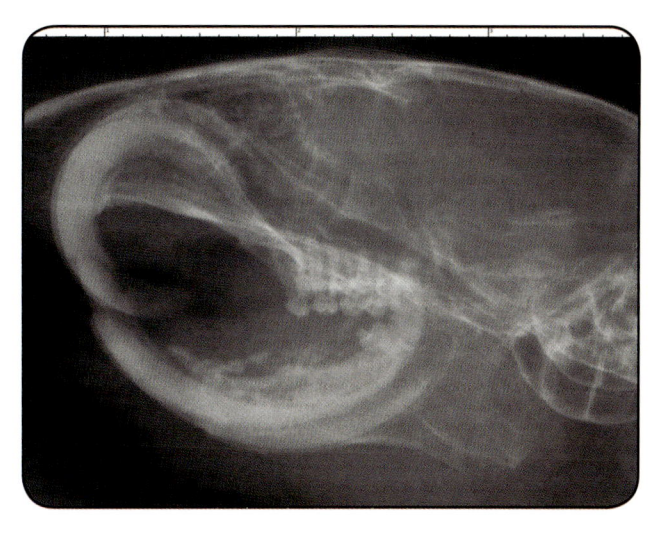

Figure 1.92. Same radiograph, scanned at 1000 p.p.i. with a commercial digital scanner. As long as the quality of the original radiograph is good, the scanned image will often be high quality as well.

Some practitioners advocate photographing radiographs with a digital camera. However, even under ideal circumstances, radiographs captured via digital photography and manipulated digitally are less than optimal and far inferior to the use of a film scanner (Figures 1.91, 1.92).

It should be kept in mind that there may be legal issues involved in the use of digital images, and original radiographs may be required in certain legal proceedings.

Contrast Radiography

Contrast radiography alters natural subject contrast by utilizing artificial media that are either radiopaque (positive contrast) or radiolucent (negative contrast). Commonly used positive contrast materials include barium sulfate and organic iodine compounds, which are triiodinated benzoic acid derivatives. Ionic and non-ionic media are available and although more expensive, non-ionic media are the safest. Negative contrast materials include gases such as air, oxygen and CO_2. Contrast media are used in special procedures such as myelography, portography, angiography, celiography, fistulography, cystography, and excretory urography.

Barium compounds are preferred for the radiographic study of the gastrointestinal tract unless perforation is suspected. The primary use of negative media is for cystography of the urinary bladder. The advent of diagnostic ultrasound has largely replaced the use of contrast procedures in dogs and cats, as well as exotic species. However, gastrointestinal and urinary studies remain useful.

Use of contrast media in exotic species is largely extrapolated from experiences in dogs and cats. Because of species differences in physiology and anatomic structure, caution should be exercised when utilizing these techniques in exotic animal medicine. The normal appearance of the gastrointestinal barium series in the ferret, and use of GI contrast in other species have been described.

Myelography

Myelography allows indirect visualization of the spinal cord utilizing a radiopaque contrast medium injected into the subarachnoid space (Figure 1.93). It may be indicated in cases of severe neurologic deficits following trauma and suspected focal, multifocal or diffuse spinal cord lesions, with or without vertebral lesions visible on survey radiography. The contrast medium can be injected into the cerebellomedullary cistern *(cisterna magna)* or directly into the ventral subarachnoid space at the most caudal lumbar vertebra, where the spinal cord terminates and continues as the cauda equina.

Contrast medium mixed with cerebrospinal fluid (CSF) surrounds the spinal cord, demonstrating extradural, intradural and medullary (cord) lesions. Extradural lesions are seen as a filling defect of the contrast columns surrounding the cord (example: prolapsed intervertebral disk). Extradural lesions may also cause compression of the spinal cord. Intradural lesions also cause a focal filling defect in the contrast column, but occur within the subarachnoid space. Nerve root tumors are a cause of intradural lesions. Medullary lesions are associated with cord swelling and tend to obstruct the subarachnoid space in a symmetrical fashion, also attenuating the contrast columns. While iohexol (Omnipaque®, GE Health Care) is a safe agent, myelography is not entirely risk-free. Convulsions and death are the most commonly reported complications, and in other species, are related to the iodine component of the molecule and direct neurotoxicity of the negatively charged iodine moiety of the side chains, anesthesia and sometimes effects of the puncture itself. Administration of intravenous fluids during myelography is always used to enhance excretion in dogs, cats, and horses. Normal myelographic appearance has been reported in the rabbit and guinea pig.

Diseases of the vertebral column such as protrusion of the intervertebral disk have not been reported in rabbits and

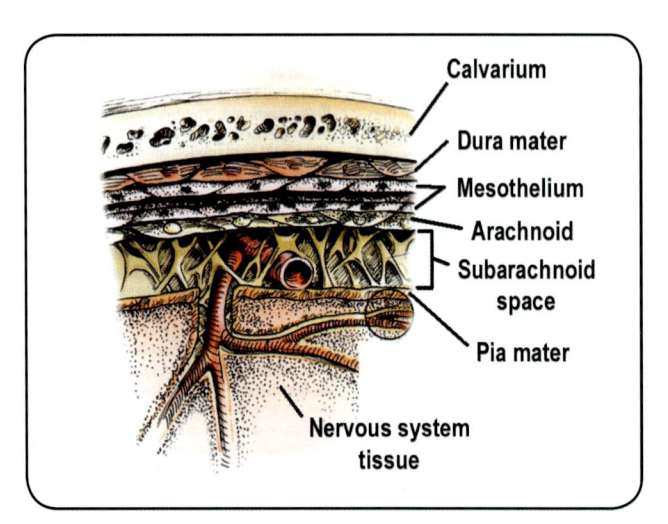

Figure 1.93. Diagram of the relationship between the nervous tissue of the spinal cord and the meninges *(modified from J.F. Zachary, College of Veterinary Medicine, University of Illinois. Reprinted from: McGavin MD, Zachary JF: Pathologic Basis of Veterinary Disease, 4th edition. Mosby Elsevier, 2007; with permission).*

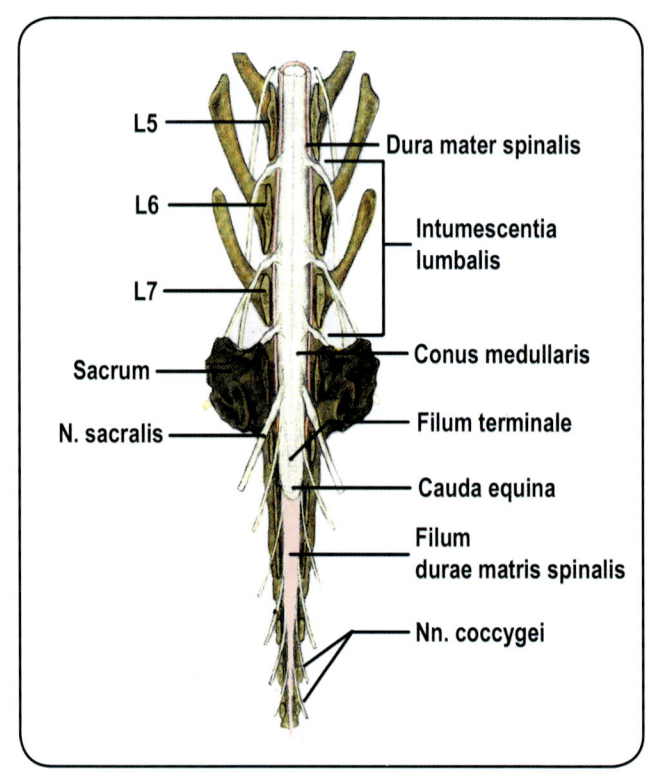

Figure 1.94. Diagram of the lumbar tract of the spinal cord and the *cauda equina* in the rabbit *(modified from Barone R. et al. Atlas d'Anatomie du Lapin, Masson et Cie, 1973; with permission).*

MYELOGRAPHY

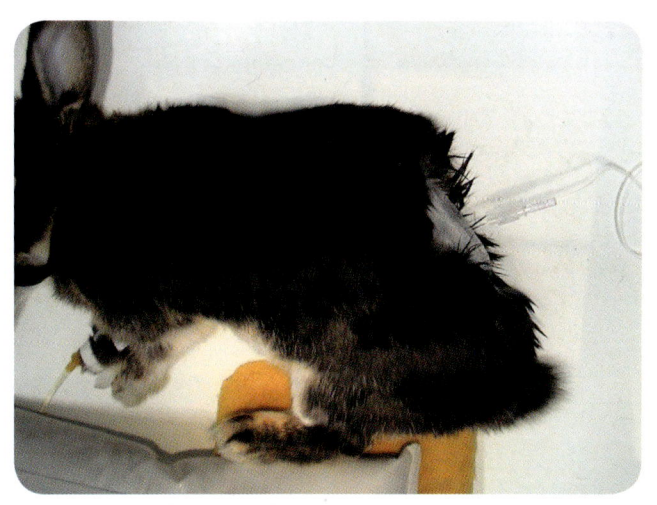

Figure 1.95. Position of the rabbit in preparation for insertion of the spinal needle.

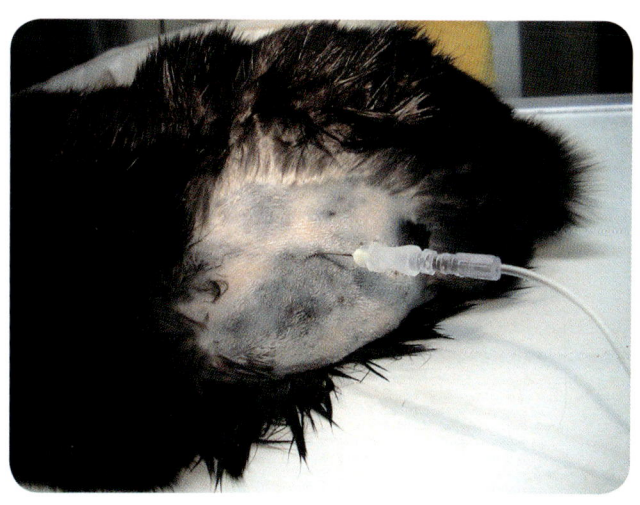

Figure 1.96. The needle is inserted slightly lateral to the spinal process of the vertebra.

ferrets. Therefore, the most common indication in these species is most likely acute lesions following a traumatic injury, or in the case of ferrets, neoplasia.

Myelography is performed with the patient anesthetized and intubated, and with intravenous catheter placement. Some authors recommend intravenous administration of anticonvulsive drugs along with injection of the contrast medium, while some advocate having these drugs on hand to use if indicated. Methylprednisolone sodium succinate 30 mg/kg by slow IV infusion is administered before the injection of the iohexol in order to prevent anaphylactic reactions. Anecdotal reports of cerebellomedullary cistern injection of contrast medium of the rabbit have resulted in the death of the patients, suggesting an apparent anatomic and/or physiologic contraindication to this procedure.

The authors are unaware of any successful attempts to perform this type of myelography in the rabbit. Injection for lumbar myelography should be made caudally to avoid densely packed neural tissue while still providing access to the subarachnoid space. In traditional pet medicine, the preferred site is L5-6. The spinal cord is still present in rabbits at this location, indicating a more caudal site (L6-7 or L7-S1) may be ideal (Figure 1.94). However, successful myelography of rabbits with injection of contrast between the fifth and sixth lumbar vertebrae (L5-L6) using a 23 to 27 gauge needle has been performed without apparent complication (perso-

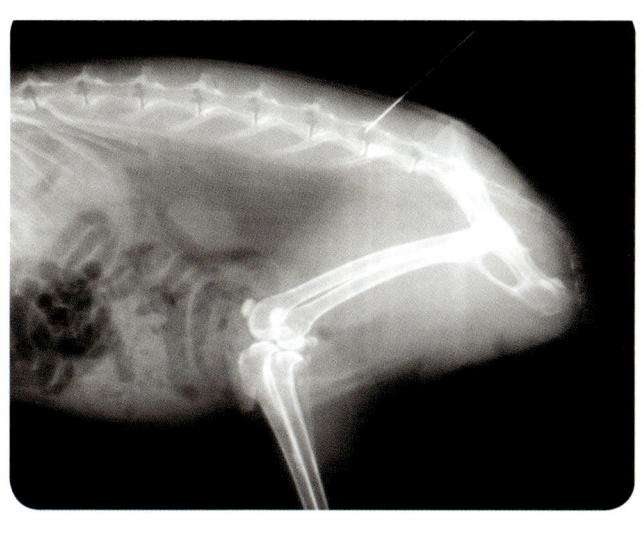

Figure 1.97. Technique for lumbar myelography. The needle is advanced into the L5-L6 interspace. This lateral projection shows initial placement of the spinal needle. The needle should be advanced to the floor of the vertebral canal before injection of contrast medium (see Figure 1.98.) Prior to injection of contrast medium, proper needle position is confirmed radiographically.
(Courtesy of Stefania Gianni, DVM)

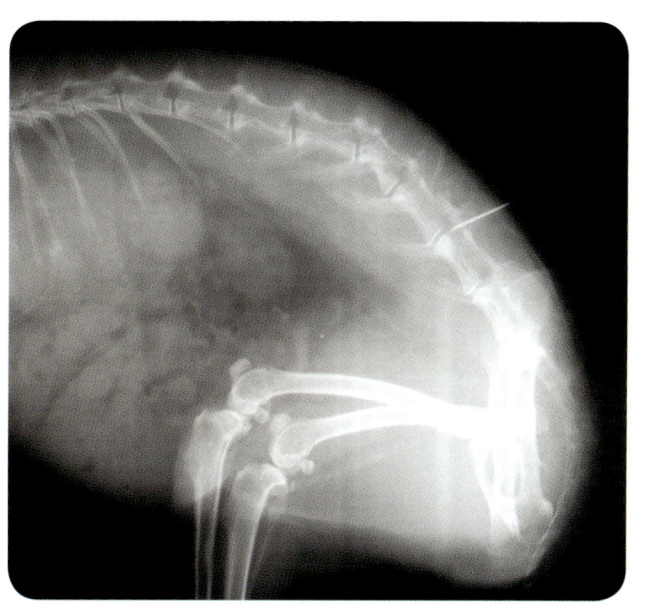

Figure 1.98. Positioning of the patient must be done with care. Due to the extent of the injury, this patient was scheduled for humane euthanasia immediately after the procedure, which was performed for investigative purposes. This radiograph demonstrates how aggressive hyperflexion can worsen an existing fracture and potentially create more damage to the spinal cord. Myelography can still be performed in these patients, but must be done with care, avoiding or minimizing hyperflexion of the spine.
(Courtesy of Stefania Gianni, DVM)

nal communication, Marguerite Knipe, UC Davis, October 2007).

The rabbit is placed in lateral recumbency with the pelvic limbs hyperflexed to enhance flexion and access to the intervertebral space (Figures 1.95, 1.96). Hyperflexion must be performed very gently or avoided in cases of visible or suspected vertebral fractures to prevent further damage to the spinal cord (Figure 1.98). The needle is inserted laterally to the spinal process into the intervertebral space. Pelvic limb twitch usually indicates that the needle has been inserted properly through the *cauda equina*. The needle is retracted slightly until a drop of CSF fluid appears in the needle hub. It should be noted that fluid is not always visible, even with proper needle placement. The insertion of the needle may be performed under guided fluoroscopy, or proper placement confirmed radiographically (Figure 1.97). Injection of a very small volume of iohexol may help confirm proper needle placement. Once placement is confirmed, the entire volume can be injected. A commonly reported dosage of iohexol is 0.45 ml/kg. Immediately after injection, the patient is positioned for lateral, dorsoventral, and right and left oblique projections, which are collected as rapidly as possible before dispersion of the contrast medium. If only a single cassette is available, film can be changed quickly in the dark room and saved in another cassette for later processing.

Putting It All Together for Optimal Images

Production of the highest quality radiographs requires attention to the following details:

- The patient must be properly anesthetized or safely restrained manually to limit patient motion.
- Optimal screens and films must be selected based on patient size and diagnostic goal.
- The object of interest must be centered in the image.
- The patient must be properly positioned according to requirements of the desired projection.
- All projections required for that radiographic study must be acquired rapidly and efficiently.
- Patient identification and directional markers must be clearly labeled and incorporated into the radiographic exposure, not handwritten on the finished radiograph.
- Oblique projections must be labeled according to beam entry and beam exit, to allow clear understanding of patient positioning and aid in identification of lesions.
- Identification must occur in conjunction with the x-ray exposure. Handwritten identification made on the finished radiograph will lead to errors and will not satisfy legal requirements.
- The identification label should not obscure the area of interest.
- Apparatus associated with the patient such as tubes, monitoring equipment, etc. must not obscure the area of interest.
- Other external objects such as collars, or the hands of the operator must not appear in the image.
- In the authors' experience, the best images of smaller exotic patients are made with low speed cassettes and mammography film, with a radiographic machine capable of higher mA settings.

References:

Barone R, Pavaux C, Blin PC, Cuq P: Atlas d'Anatomie du Lapin, Masson et Cie; 1973.

Brenner SZG, Hawkins MG, Tell LA, Hornof WJ, Plopper CG, Verstraete FJM: Clinical anatomy, radiography, and computed tomography of the chinchilla skull. Comp. Cont. Ed. 2005;27:933-944.

Capello V, Gracis M: Radiology of the skull and teeth. In: Lennox A, ed. Rabbit and Rodent Dentistry Handbook. Ames, Iowa: Blackwell Publishing, (Formerly Zoological Education Network, Lake Worth, FL); 2005;65-99.

Capello V. Rabbit and Rodent Dentistry Wet Lab. Proc Assoc Ex Mam Vet. 2006:1-8.

Crossley DA: Rodent and rabbit radiology. In: DeForge DH, Colmery BH III,eds. An Atlas of Veterinary Dental Radiology. Ames: Iowa State University Press; 2000:247-260.

Crossley DA, Jackson A, Yates J, et al. Use of computed tomography to investigate cheek tooth abnormalities in chinchillas *(Chinchilla laniger)*. J Sm Anim Pract.1998;39:385-389.

Girling SJ: Mammalian imaging and anatomy. In: Meredith A, Redrobe S, eds. BSAVA Manual of Exotic Pets, 4th ed. Quedgeley, UK: British Small Animal Veterinary Association; 2005:1-12.

Kalendar, WA: Computed Tomography. Fundamentals, System Technology, Image Quality, Applications. Publicis MCD Verlag: Werbeagentur GmbH, Munich, 2000.

McGavin MD, Zachary JF: Pathologic Basis of Veterinary Disease, 4th ed. Philadelphia, PA: Mosby Elsevier;2007.

Morgan JP, Silverman S: Techniques of Veterinary Radiography, 4th ed. Davis, CA: Veterinary Radiology Associates;1984.

Nemetz LP: Principles of High Definition Digital Radiology for the Avian Patient. Proc Assoc Av Vet. 2006:39-45.

Popesko P, Rijtovà V, Horàk J: A Colour Atlas of Anatomy of Small Laboratory Animals. Vol. I: Rabbit, Guinea Pig. Bratislava: Príroda Publishing House; 1990.

Silverman S, Tell LA: Radiology equipment and positioning techniques. In: Radiology of Rodents, Rabbits and Ferrets. An Atlas of Normal Anatomy and Positioning. Philadelphia, PA: Elsevier Saunders; 2005:1-8.

Tell LA, Silverman S, Wisner E: Imaging techniques for evaluating the head of birds, reptiles and small exotic mammals. Exotic DVM. 2003; 5(2):31-37.

Tell LA, Silverman S, Wisner E: Imaging techniques for evaluating the respiratory system of birds, reptiles and small exotic mammals: Exotic DVM. 2003; 5(2):38-44.

Thrall D: Veterinary Diagnostic Radiology, 4th ed Phildelphia, PA:W.B. Saunders; 2002:4.

Ticer JW: Radiographic Technique in Veterinary Practice, 2nd ed. Philadelphia, PA:W.B. Saunders;1984.

Widmer WR, Thrall DE, Shaw SM: Effects of low-level exposure to ionizing radiation: current concepts and concerns for veterinary workers. Vet Radiol and Ultrasound. 1996;37:227-239.

The Fundamentals of Radiography, 12th ed. Rochester, NY: Eastman Kodak, Co;1980.

Introduction to Computed Tomography in Exotic Companion Mammals

V. Capello, A. Cauduro, W.R. Widmer*

*Authors listed in alphabetical order

Computed tomography (CT) uses a rotating x-ray tube and a computer to obtain cross sectional image slices of the tissues of a patient. The patient moves along the gantry on a couch as exposures are made (Figure 1.99). The name: "tomography" comes from *"tomos"* = to cut, and: *"gramma"* = letter; i.e: image.

The concept of "slice" imaging originated from the need to overcome superimposition of adjacent anatomic structures that is indigenous to conventional radiography. Computed tomography was developed by Godfrey Hounsfield and A. J. Cormack in the early 1970s, and the first scanner was used to image the human head in 1972. Seven years later Hounsfield and Cormack were awarded the Nobel Prize for Medicine. In 1979, a second-generation scanner was introduced which had the capability to image the entire body. Third and fourth-generation whole body scanners were developed in the late '70s and '80s, improving scan speed and image resolution. However, scan time was still limited by scanner configuration because the x-ray tube had to be reversed after each 360 degree rotation around the patient (otherwise the electrical cables supplying the tube would become entangled). In 1987, the first continuously rotating scanner or spiral (helical) scanner was produced which utilized slip rings rather than electrical cables to supply the x-ray tube. This dramatically reduced scan time because image data could be continuously acquired compared to the start-stop motion of single slice scanners.

CT of Exotic Companion Mammals

The use of spiral CT for exotic mammals is a new discipline, and few references are available. Imaging exotic patients is a challenge because of their small size and diverse anatomical structure. Because spiral scanners offer very thin slices (less than 1 mm) and large image matrices (512 x 512), resolution is superior to that obtained with most single slice scanners. High resolution CT images can be magnified 1.5 to 2.0 times with computer software, allowing better detection of subtle anatomic changes. With very thin slices obtained with spiral CT scanners, each voxel of the image slice is nearly cube-shaped allowing accurate reconstruction in almost any image plane. Viewing slices of the patient in axial and dorsal planes as well as the standard transverse plane offers a tremendous advantage over conventional radiography. Contrast resolution, e.g., the ability to see various soft tissues, is also superior with spiral CT scanners. By adjusting image contrast (window level and width) to the anatomic structure of interest, i.e., soft tissue, bone, lung, etc., the problem of low soft to hard tissue ratio in small mammals can be minimized.

The patient is usually positioned in ventral recumbency, with the head elevated slightly.

Preliminary studies by the authors presented here focus on the anatomy of the normal and pathologic skull and teeth of rabbits, guinea pigs, and chinchillas.

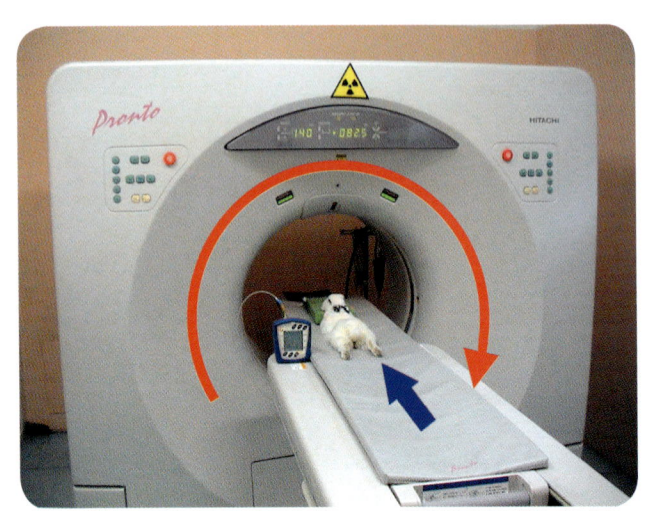

Figure 1.99. Spiral CT scanners acquire a contiguous series of slices of the patient through continuous rotation of the x-ray tube while the patient moves through the tunnel.

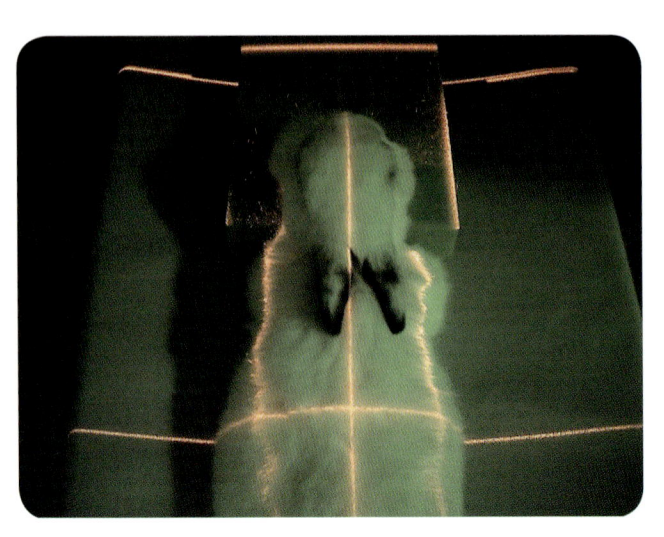

Figure 1.100. Rabbit under anesthesia with injectable drugs positioned in ventral recumbency with the head elevated by a foam cube. Symmetry is critical for generation of slices. A laser beam is used to confirm proper positioning.

Before scanning, a scout radiographic view is obtained in both dorsoventral and lateral projections (Figure 1.101). Scout projections are used to select the anatomic region to be scanned and can later be used as a reference for location of individual slices. The dorsoventral projection is useful for evaluating bilateral symmetry and the lateral projection is useful for the selection of the angle of the scan plane. A provisional transverse scan through the tympanic bullae can also be made to check for proper position of the head if desired. Slice thickness, slice interval (overlap), extent of the scan, matrix, and field of view are selected prior to scanning.

Scanning plane angles for the skull of the rabbit, guinea pig, and chinchilla shown in this book are perpendicular to the palatine bone (Figure 1.102). Further investigative studies may suggest additional advantageous angles, in particular for the study of the mandible and cheek teeth. Note that this particular scanning angle produces slices that are not parallel to the axis of cheek teeth, therefore each slice, depending on thickness, may intersect more than one tooth (Figure 1.102b, arrows). The scan plane is easily adjusted to the desired angle. For example, in b, the plane could be angled such that the cheek teeth are within the image plane.

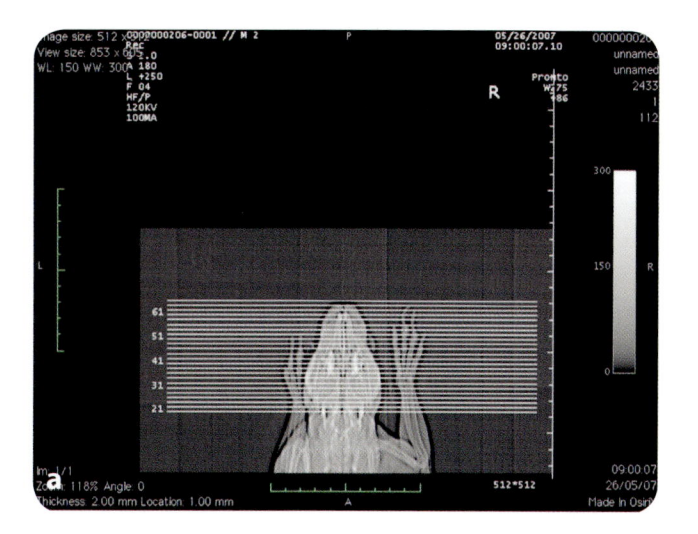

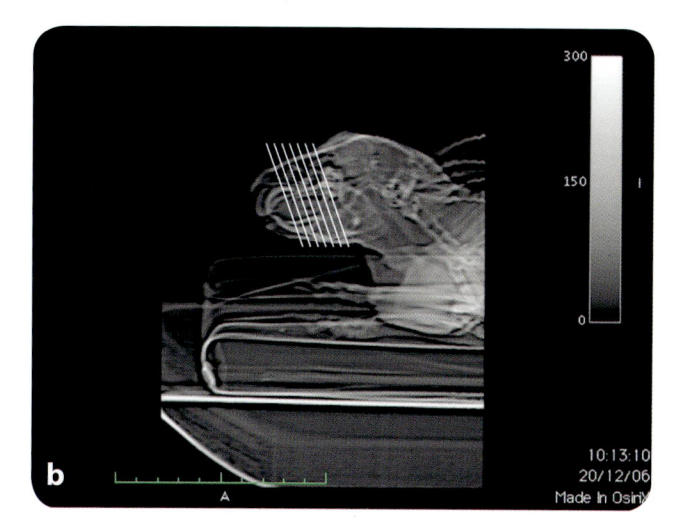

Figure 1.101a,b. Dorsoventral (a) and lateral (b) "scout" projections of a rabbit.

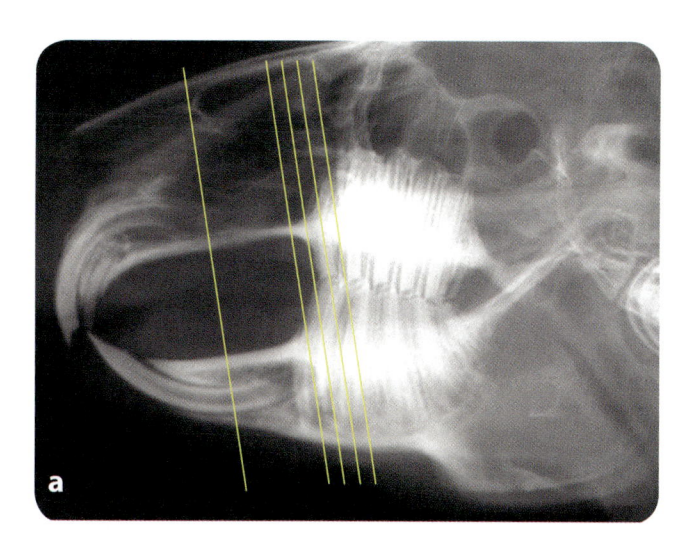

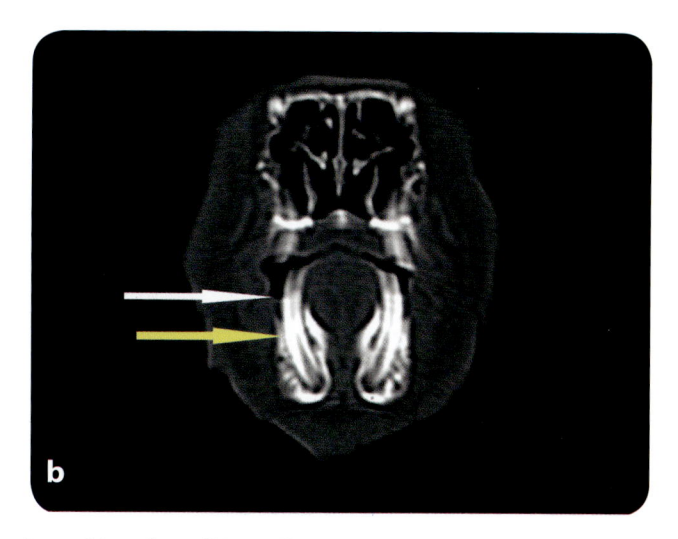

Figure 1.102a,b. Scanning planes perpendicular to the palatine bone (a) and possible artifact showing two mandibular cheek teeth in the same slice (b).

Basic Operation of a CT Scanner
W.R. Widmer

A narrow cone shaped x-ray beam is used for CT acquisition and traverses a very small volume or "slice" of tissue as it moves through an arc of 360 degrees. An array of electronic x-ray detectors records the multiple exposures generated by the x-ray tube. Because the x-ray beam is narrow, slice thickness is generally 1-5 mm (less than 1 mm with new spiral scanners). X-rays exiting the animal excite the detectors, producing an electrical signal that is proportional to x-ray intensity. A computer processes, digitizes, and stores the electrical signal. The ability of tissues within the image slice to attenuate incident x-rays determines the intensity of the x-rays reaching the detectors. Tissues that are thick or have a high physical density or atomic number attenuate more of the x-ray beam than tissues that are thin or have a low physical density or atomic number.

After each exposure, a computer calculates the attenuation for each narrow band of tissue that is interrogated by the x-ray beam. Because each part of the tissue slice is "seen" many times from various directions, it is possible to divide the three-dimensional tissue slice into a grid of tiny blocks. Each tiny block of the tissue grid is called a volume element or voxel (Figure 1.103). Voxels have length, width, and depth. The end (length and width) of a voxel is square and represented as a picture element or a pixel on the two-dimensional hard copy image. Depth of a voxel is determined by the thickness of the tissue slice. With single slice scanners, voxels are rectangular shaped, but with spiral scanners, cube-shaped voxels can be obtained, improving manipulation of the image (see next section). Through a complex reconstruction process, the computer analyzes the mean attenuation of each voxel and assigns a CT number. Each CT number indicates the relative attenuation of the tissue contained within each voxel. Water is the reference standard and is assigned a CT number of 0; cortical bone is about +1000 and air is about -1000. Fluids, soft tissues, and fat have intermediate CT numbers.

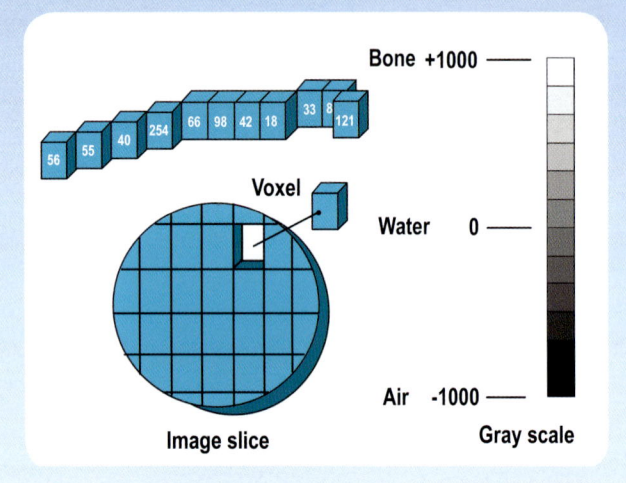

Figure 1.103. Each voxel is assigned a number that represents relative x-ray attenuation and is displayed in gray scale on the monitor.
(Courtesy of William R. Widmer, DVM)

Figure 1.104. Diagram of the window of CT numbers
(Courtesy of William R. Widmer, DVM)

A print-out of an image slice displaying a matrix of CT numbers would be impossible to decipher; therefore, they are converted to various shades of gray on the final image. Generally 64 shades of gray are used to make the final image and each shade of gray represents a group of several CT numbers (If each CT number were assigned a gray shade, the difference between successive numbers could not be distinguished by the human eye). The window of CT numbers representing the gray scale can be adjusted to maximize the tissue of interest (Figure 1.104). The window width determines the range of CT numbers over which the available gray scale is spread. The window level is the median CT number in the window. For instance a soft tissue study might have a window width of 40 with a window level or median gray of 0-100. For bone imaging the window width is set at 300-400 with a window level of 1000. For pulmonary imaging, a window width of 1200 and a window level of -700 to-900 is commonly used. Choice of window settings affects image contrast. Narrow windows provide high contrast and better differentiation of soft tissues because gray scale is spread over a relative few CT numbers. Narrow window settings are commonly used for abdominal imaging. On the other hand, wide window settings reduce the contrast because available gray shades are spread over a wide range of CT numbers. Wide window settings are preferred for imaging bone because soft tissue structures lack contrast and are suppressed while bony structures are well-visualized.

Data is acquired in the transverse plane (axial slices, see Figure 1.101 and Figure 2.80a-l) but can be reformatted by the computer and displayed in sagittal, dorsal, and transverse planes (Figure 1.105). This capability is called multiplanar reformation and has been exploited by the advent of spiral CT (Figure 1.105). It must be remembered that while the actual image slice has three dimensions (x, y, and z), printed images and images seen on a monitor have only two dimensions. However, with interactive monitor viewing, the actual image volume can be viewed as a series of slices by moving the computer mouse. This maneuver can give the interpreter a three-dimensional sense of spatial orientation of the various structures in the image volume. Dedicated imaging software programs allow various reconstruction techniques, including volume rendering (VR) techniques (Figure 1.106, 1.107), and shaded surface displays (SSD) (Figure 1.108). Shaded surface displays present a contoured surface map of the entire image volume. The volume can be rotated on the monitor to allow the observer to visualize any surface. Shaded surface displays have limited usefulness because deep structures are masked, but they are helpful for evaluating fractures and other abnormalities of the skull. So-called volume rendering formats show the entire image volume with certain CT numbers suppressed, e.g., only numbers below or above a threshold are chosen. For instance only a map of the pulmonary vasculature can be displayed, with the remainder of the lung suppressed. When the image volume or "slab" is presented in this fashion, there may be a perception of depth and 3D effect, but the image is actually two-dimensional. While reformations give the interpreter a different perspective, transverse slices provide the most information and are always interpreted before MPR or volume rendering is performed on the image data.

The multiplanar reformtion and the volume rendering images in this book were produced with OsiriX software. OsiriX is an open source program written by Antoine Rosset, MD (Department of Radiology, University of Los Angeles) for DICOM medical imaging and image processing.

Free download and documentation can be found at: **www.osirix-viewer.com/**

Similar software programs are available for PC users, and vary in functionality and price.

TWO-DIMENSIONAL RECONSTRUCTION
Multiplanar Reformation

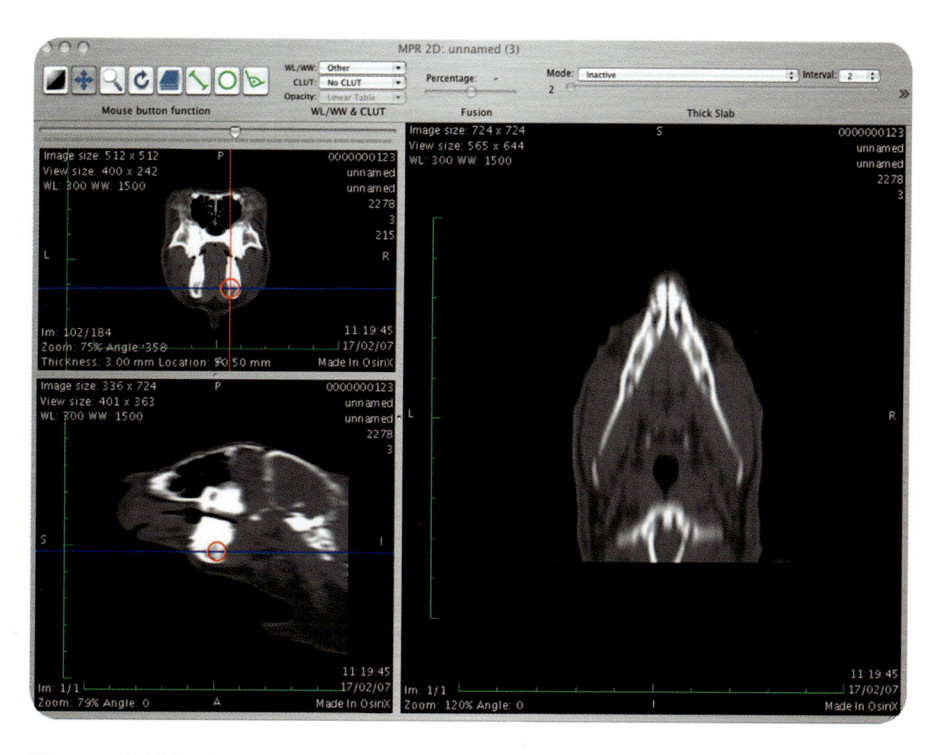

Figure 1.105. A multiplanar reformation (MPR) presents the three basic planes (axial, lateral, coronal) all together. By moving the "target" (the small red circle) the section planes are selected and automatically reconstructed.

THREE-DIMENSIONAL RECONSTRUCTIONS
Volume Rendering

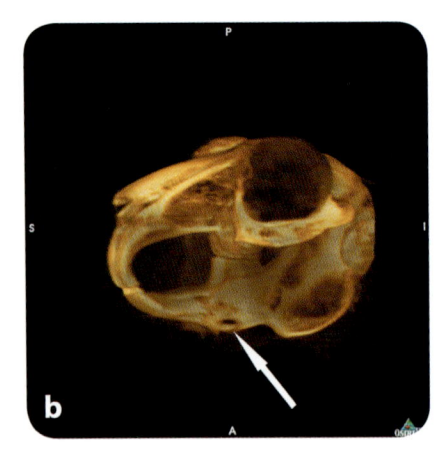

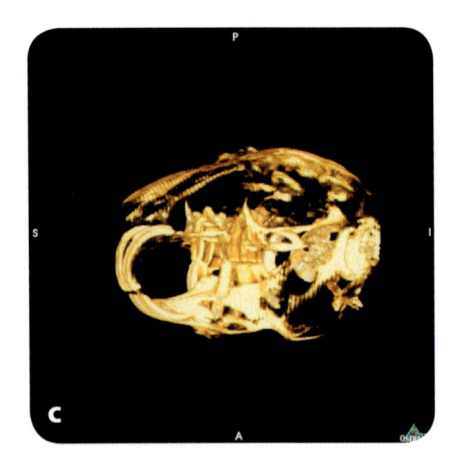

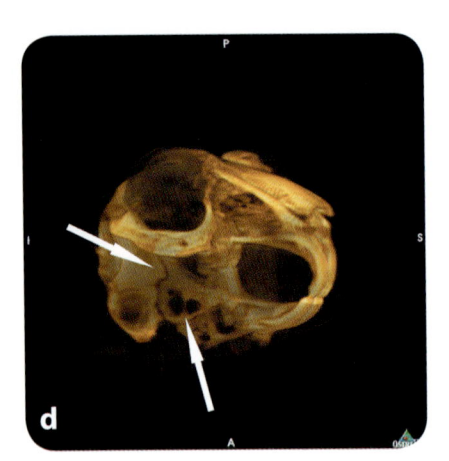

Figure 1.106a,b,c,d. 3D volume rendering format allows additional analysis of the CT data set. A rabbit head is shown in this series. Adjustment of gray scale pixel values allows 3D rendering from the most superficial tissues (a) to deepest bone tissues (b,c). Teeth and the thickest areas of the bone are emphasized in (c). A volume rendering can be rotated in any position in space (d). In this example, the right mandible is affected by extensive changes suggesting osteomyelitis, and a longitudinal fracture between the body of the mandible and the masseteric fossa is visible (arrows).

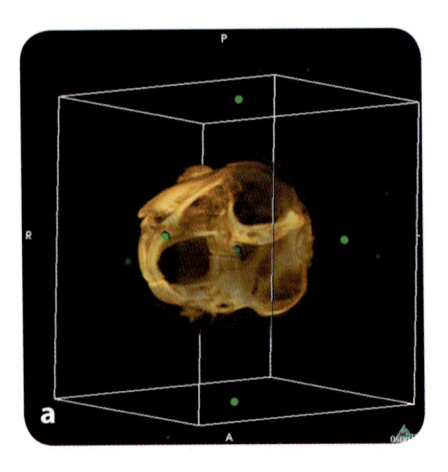

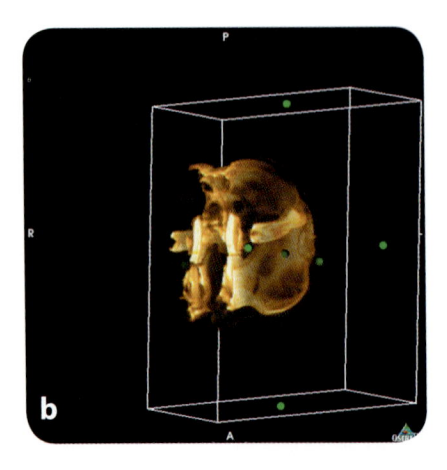

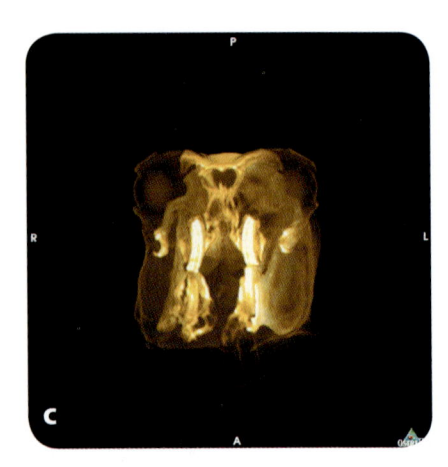

Figure 1.107a,b,c. The 3D volume rendering format is used to visualize an entire block of tissue or body part, such as the head. Volume rendering generates a 3D model with a complex computer graphics program. All voxels that represent a specific tissue or organ volume are assigned a similar gray scale value or, in some cases, a specific color. To make volume rendering more representative of actual organ structure, the tissues of non-interest are deleted, enhancing the tissues of interest. High speed computers allow the image to be rotated in any direction. Volume rendering is most effective when only a few tissues or structures are represented, for example the pulmonary vasculature or the bronchial tree. Volume renderings are presently the most useful reconstructive method in CT, however they should only be used in conjunction with transverse gray scale images because they are only a representation of the actual data set (a). Movement of a single side effectively crops away that aspect of the 3D image, allowing visualization of inner anatomical structures (b). A volume rendering cropped at the level of the cheek teeth shows the axial section of a severely osteomyelitic area of the right mandible (c).

THREE-DIMENSIONAL RECONSTRUCTIONS
Surface Rendering

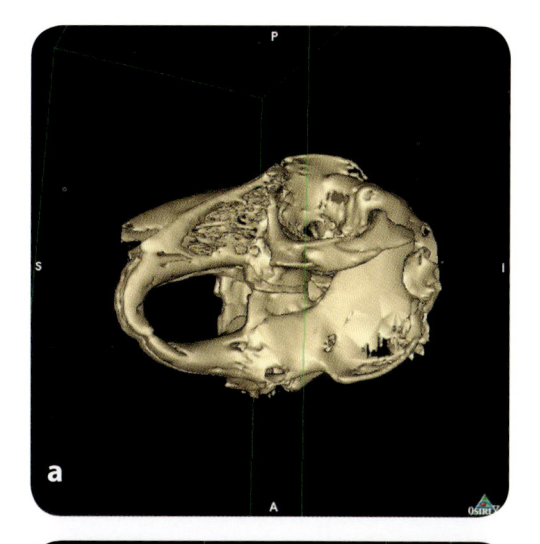

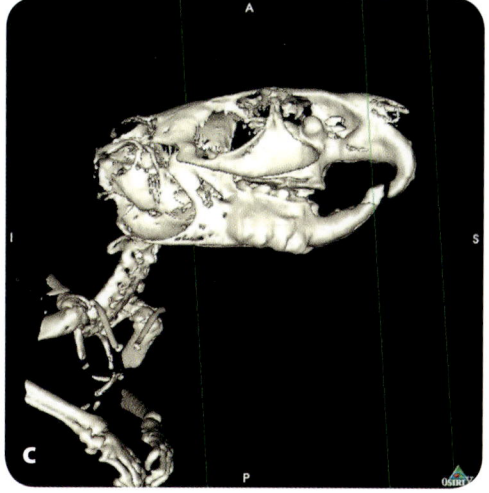

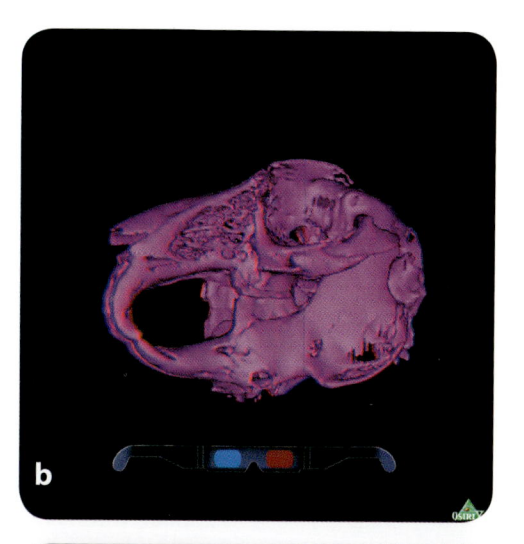

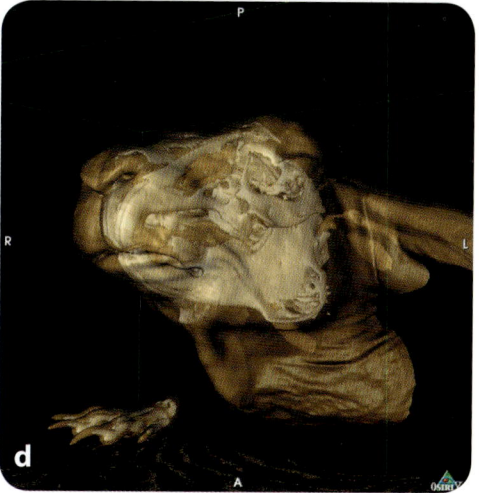

Figure 1.108a,b,c,d. 3D surface rendering (Shaded Surface Display - SSD) permits virtual reconstruction of the actual surface of tissues, allowing the analysis of complex structures. The primary use of SSD is for hard tissues, like bones and teeth (a). The 3D surface rendering can be rotated in any position in space, aiding in diagnosis and treatment planning. Computer software programs like OsiriX® allow the creation of 3D stereo images, which can give a true 3D perception (if the observer wears blue/red glasses). Details of the skull of a chinchilla are shown in (c). A double surface reconstruction (soft tissues/hard tissues) is also possible (d). This can be useful for the analysis of relationship between different organs and tissues.

References:

Brenner SZG, Hawkins MG, Tell LA, Hornof WJ, Plopper CG, Verstraete FJM: Clinical anatomy, radiography, and computed tomography of the chinchilla skull. Comp Cont Ed. 2005; 27: 933-944.

Chesney CJ: CT scanning in chinchillas. J Small Anim Pract. 1998; 39(11): 550.

Crossley DA, Jackson A, Yates J, Boydell IP: Use of computed tomography to investigate cheek tooth abnormalities in chinchillas (*Chinchilla laniger*). J Small Anim Pract. 1998; 39(8): 385-389.

Dal Pozzo G: Compendio di Tomografia Computerizzata e TC multistrato. UTET, Torino, 2006.

De Rycke LM, Gielen IM, Van Meervenne SA, Simoens PJ, Van Bree HJ: Computed tomography and cross-sectional anatomy of the brain in clinically normal dogs. Am J Vet Res. 2005; 66(10):1743-1756.

De Voe RS, Pack L, Greenacre CB: Radiographic and CT imaging of a skull associated osteoma in a ferret. Veterinary Radiology & Ultrasound 2002; 43(4): 346–348.

Fike JR, LeCouter RA, Cann CE: Anatomy of the canine orbital region. Multiplanar imaging by CT. Veterinary Radiology & Ultrasound 1984; 25(1): 32-36.

Garland MR, Lawler LP, Whitaker BR, et al, Modern CT applications in veterinary medicine. Radiographics 2002; 22(1):55-62.

Kalendar WA: Computed Tomography. Fundamentals, System Technology, Image Quality, Applications. Publicis MCD Verlag: Werbeagentur GmbH, Munich, 2000.

Porat-Mosenco Y, Schwarz T, Kass PH: Thick-section reformatting of thinly collimated computed tomography for reduction of skull-base-related artifacts in dogs and horses. Veterinary Radiology & Ultrasound 2004; 45(2): 131–135.

Prokop M, Galanski M: Tomografia computerizzata spirale e multistrato. Masson, 2006.

Silverman S, Tell LA: Radiology of Rodents, Rabbits and Ferrets. An Atlas of Normal Anatomy and Positioning. Philadelphia, PA: Elsevier Saunders; 2005: 1-8.

Tell LA, Silverman S, Wisner E: Imaging techniques for evaluating the head of birds, reptiles and small exotic mammals: Exotic DVM. 2003; 5(2), 31-37.

Radiation Safety
W.R. Widmer

Living tissues are subject to the ionizing properties of x-rays and exposure may result in deleterious effects. Ionizing radiation can damage many intracellular components, including DNA, RNA, enzymes, and intracellular organelles. However, damage to DNA is critical because it is central to all cellular functions, including reproduction. Therefore, DNA damage can lead to such harmful effects as cancer induction, cataractogenesis, organ fibrosis, teratogenesis, and mutation. These are considered late effects of chronic low-level exposure to ionizing radiation and should not be confused with acute effects from single, large dose exposures. Acute effects from massive radiation doses are often lethal and include damage to the gastrointestinal tract and bone marrow depletion. Workers in the field of radiology are not at risk for acute effects, but are subject to chronic low dose exposures.

Cancer induction is the most important harmful effect of ionizing radiation faced by radiology workers. Risk of cancer induction is a stochastic effect based on probability of occurrence. There is no threshold dose before cancer induction occurs and any dose of radiation, however small, carries some risk. The probability of cancer induction increases with increasing radiation dose. The actual risk for workers sustaining low doses of ionizing radiation over a working lifetime is unknown. Large epidemiologic studies with a sample population of millions are needed to determine risk and have not yet been performed. However, smaller studies have shown that a single dose of 0.1 Sv (10 rems) would have a 0.8% risk of cancer induction. This means that if 100,000 persons received a single whole body dose of 0.1 Sv, there would be 800 extra cancer deaths in that population's lifetime. Mutation is considered a stochastic effect while cataractogenesis is not. The latter is also dose dependent, but has a threshold below which no damage to the lens occurs. Effects like cataractogenesis are known as non-stochastic or deterministic effects. Teratogenesis is caused by exposure of the embryo or fetus to ionizing radiation. This is a complex area and a detailed description of embryonic and fetal effects of ionizing radiation is beyond the scope of this chapter. The mammalian embryo is extremely sensitive to ionizing radiation and subject to lethal and teratogenic effects. Therefore, pregnant or potentially pregnant women must avoid potential x-ray exposure. Table 1 provides information on acceptable radiation doses for occupational workers and a comparison of doses received from various sources of ionizing radiation.

Sources of radiation exposure for x-ray workers include the primary beam and radiation scattered from the patient. Factors of radiation safety are aimed at limiting exposure to scattered radiation. These include common sense, use of shielding, distance, and time. Good radiation practices stipulate that operators of x-ray equipment keep out of the path of the primary beam. Hence, most radiation exposure is from x-rays scattered when the primary beam interacts with the patient. "Scatter" radiation has low energy and is easily attenuated and absorbed by the human body. Use of protective lead shielding (0.5 mm thickness), including aprons, gloves, and thyroid shields will nearly eliminate the absorption of scattered x-rays. However, these shielding devices will not attenuate high-energy photons in the primary beam and operators should keep them out of the primary x-ray beam. Protective gloves with less than 0.5 mm Pb thickness have increased dexterity, but are unsafe and should not be used. Chemical restraint, tape, sandbags, and other positioning aids will help overcome obstacles created by use of shielding devices. State laws and/or federal regulations mandate the use of protective shielding and monthly monitoring of exposures to personnel. Lead impregnated eyeglasses are effective for reducing exposure to the lens and even ordinary prescription glasses are beneficial. Laboratory glasses are an acceptable alternative to expensive lead-based eyeglasses.

An important factor that reduces exposures from scattered radiation is distance. Because x-rays diverge as they travel, their intensity falls off rapidly with respect to distance from their source. This phenomenon is expressed by the formula: Intensity = k x $1/d^2$, where I is the intensity of the x-rays at a given distance from their source, k is a constant and d is the distance from the point of origin of the x-rays. Therefore, if a worker were to increase his distance by a factor of two, intensity (exposure) would be decreased by 1/4. Obviously, maximizing distance from a patient during an exposure is an important radiation safety practice. A final factor in radiation safety is time. Using the lowest possible exposure setting to obtain a given mAs will reduce exposure to personnel. For instance, 0.5 s x 10 mA and 0.1 s x 50 mAs both provide five mAs and will result in the same radiographic image. However, the second setting has a shorter exposure time and will produce less scatter radiation.

The previous discussion emphasized radiation safety practices that will minimize whole body and local exposures to x-ray workers. The most common breach in radiation safety is exposure to the skin when positioning the animal. Although the skin is less radiosensitive than internal organs, chronic exposure can lead to dermal neoplasms such as basal cell carcinoma and squamous cell carcinoma. Like other neoplasms, there is no threshold for induction by radiation, thus radiation workers must never compromise good safety principles.

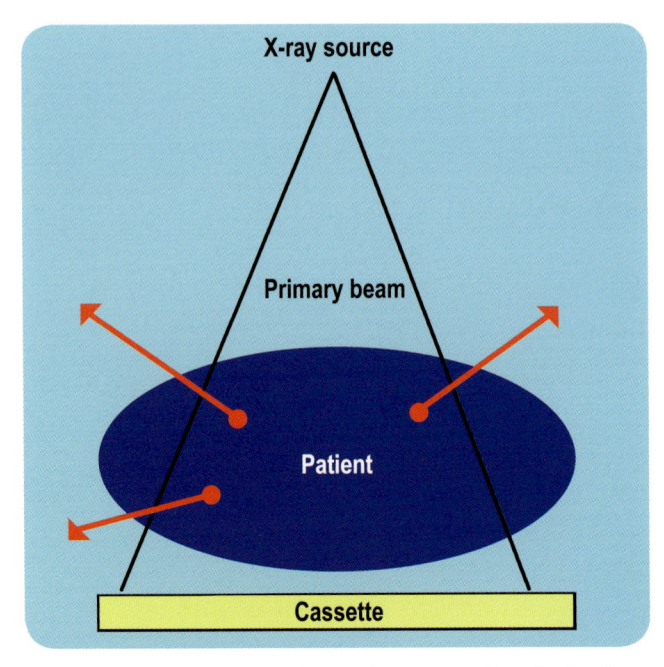

Figure 1.109. Scatter is the main source of radiation hazard. *(Courtesy of William R. Widmer, DVM)*

MAXIMUM PERMISSIBLE DOSE (MPD) RECOMMENDATIONS[1]

Body Part	Average per Week mrem	Max per Quarter rem	Max per Year rem
Occupational (for workers)			
Whole body, gonads, blood-forming organs & lens of eye	1	3	5
Skin of whole body	-	10	30
Hands, forearms, head, neck, feet & ankles	1.5	25	75
Fertile women (with respect to fetus - entire gestation period)	-	-	0.5
Non-occupational or Occasional Exposure (for general public)			
Whole body	0.1		0.5
Students[2]	-		0.1
Population[3]			0.17

[1.] Exposure of patients to radiation for medical and dental purpose is not included in the MPD equivalent.

[2] The dose limit for students is a guideline for non-occupational personnel who are occasionally exposed to ionizing radiation for training or educational purposes. This does not imply that all students and/or persons under 18 years of age should routinely engage in this activity. However, it does suggest that in these instances, risk of low-level radiation exposure is acceptable in light of the benefit of being educated.

[3] Population dose limits have been set to define the risk of genetic and somatic effects to the population at large. The population dose limit is one-third of the non-occupational dose.

Table 1. Maximum permissible dose (MPD) recommendations.

References:
Thrall D: Veterinary Diagnostic Radiology, 4th ed. Phildelphia, PA: W.B. Saunders; 2002:4.
Ticer JW: Radiographic Technique in Veterinary Practice, 2nd ed. Phildelphia, PA: W.B. Saunders; 1984.
Widmer WR, Thrall DE, Shaw SM: Effects of low level exposure to ionizing radiation: current concepts and concerns for veterinary workers. Vet Radiol and Ultrasound 1996; 37:227-239.
The Fundamentals of Radiography, 12th ed. Rochester, NY: Eastman Kodak, Co.; 1980.

RABBIT

The NORMAL HEAD
Lateral Projection

Figure 2.1. Proper lateral projection is obtained by slightly raising the nose of the patient. The vertical margins of the lips must be parallel to the cassette.

Figure 2.2a,b. 160% of actual size. The most common indication for skull radiographs in this species is for evaluation of dental disease. The lateral projection is useful for evaluation of the occlusal plane of the cheek teeth, occlusion of the maxillary and mandibular incisor teeth, and for detection of overgrowth of incisor teeth. In a well-positioned lateral view, there is exact superimposition of the right and left tympanic bullae, rostral margins of the mandibular rami, and rostral orbital margins. In addition, the hard palate in the diasatema between incisor and cheek teeth appears as a single radiopaque line. The radiopaque lines of the maxillary and mandibular diastemata should normally be slightly oblique, converging rostral to the incisor teeth. The normal appearance of the cheek teeth occlusal plane is zigzag (blue line), with a slight rostrodorsal to caudoventral slope in relation to the dorsal aspect of the nasal bones. The apex of the mandibular incisor teeth (red circle) normally extends to the level of the ipsilateral first cheek tooth.

Maxillary incisor teeth are more curved than mandibular incisors, and their apex is normally located half the length of the diastema, at some distance from the corresponding radiopaque hard palate (yellow circle).

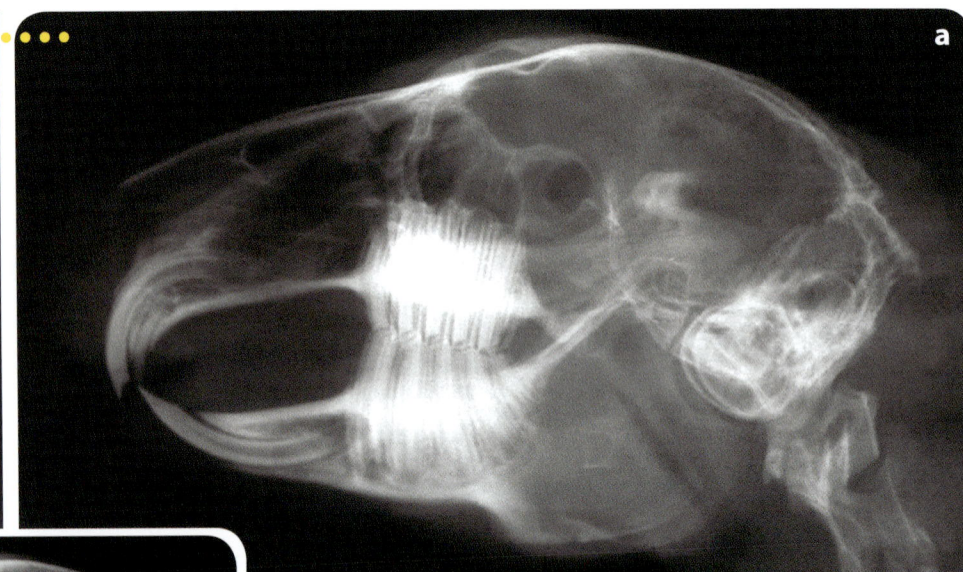

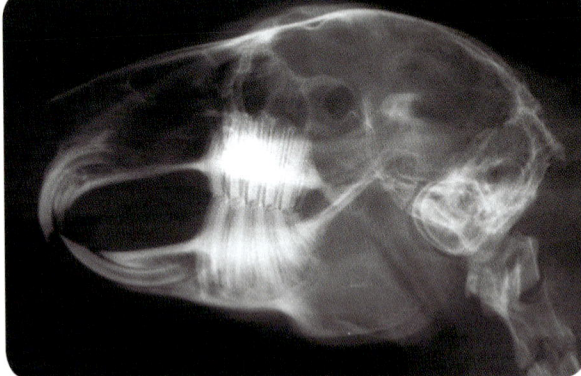

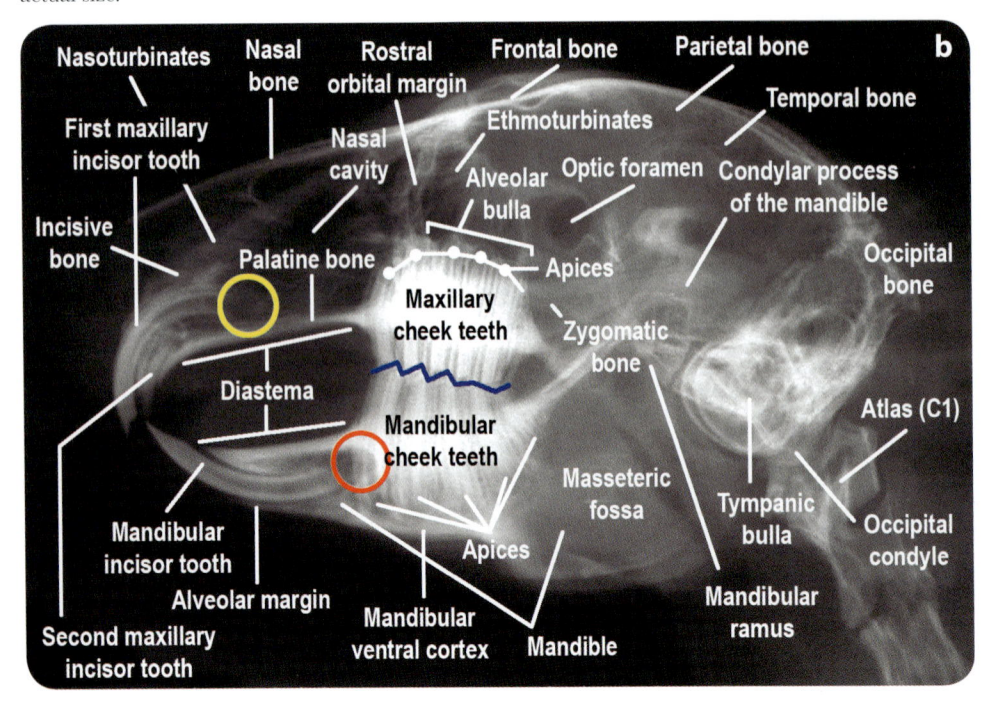

Figure 2.3. Lateral radiograph of the skull of a 1.5 Kg dwarf rabbit, actual size.

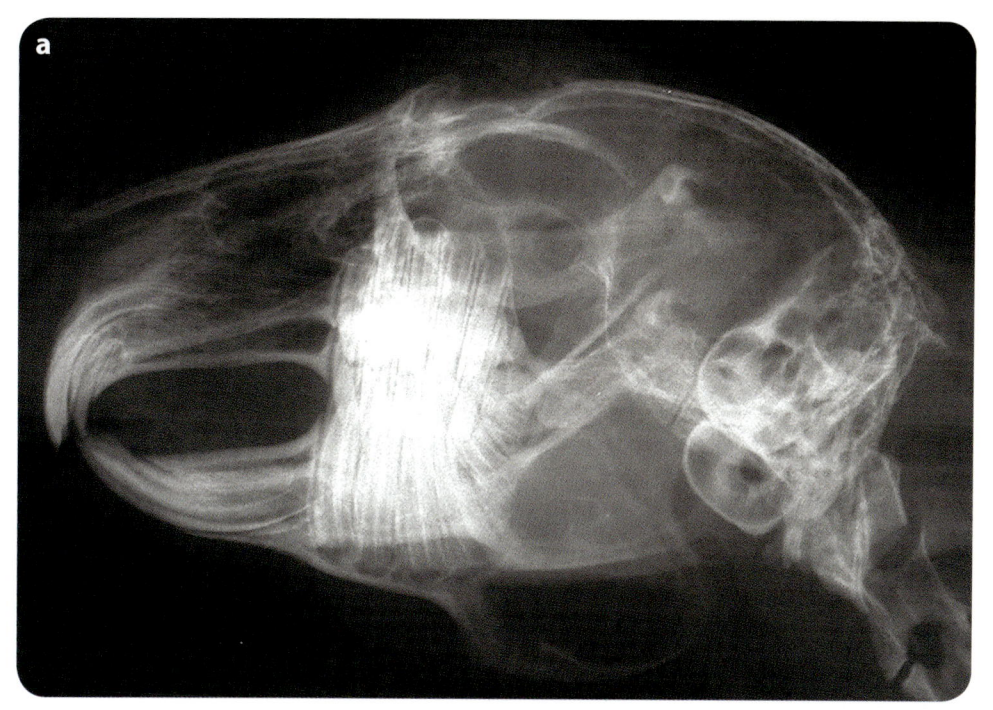

Oblique Projection

Figure 2.4a,b.
Oblique radiographs of the skull of a rabbit. To obtain a proper oblique projection, the head is rotated slightly (10°-20°). An ideal projection demonstrates minimal rotation allowing clear visualization of the apex of the mandibular cheek teeth on one side and the maxillary teeth on the opposite side. Right and left tympanic bullae and condylar processes should appear just dorsal and ventral to each other. The hard palate will appear as a double radiopaque line.

Each root appears completely surrounded by a thin radiolucent periodontal space and an adjacent radiopaque lamina dura (wall of the alveolus). The periodontal space is normally wider at the apex, where dental germinal tissues are located.

The root of the maxillary third cheek tooth is usually longer than the adjacent cheek teeth. The first and especially the last maxillary cheek teeth are much smaller and shorter than the other maxillary teeth. On the lower jaw, the last mandibular cheek tooth is smaller and shorter than the other mandibular teeth. The long axis of the first, second, and third mandibular cheek teeth is almost perpendicular to the long axis of the mandible, while the fourth and fifth cheek teeth are slightly curved distally and positioned obliquely. The contralateral view should always be obtained for comparison. Each projection is useful for evaluation of the roots of the mandibular teeth of one side and the maxillary teeth of the opposite arcade.

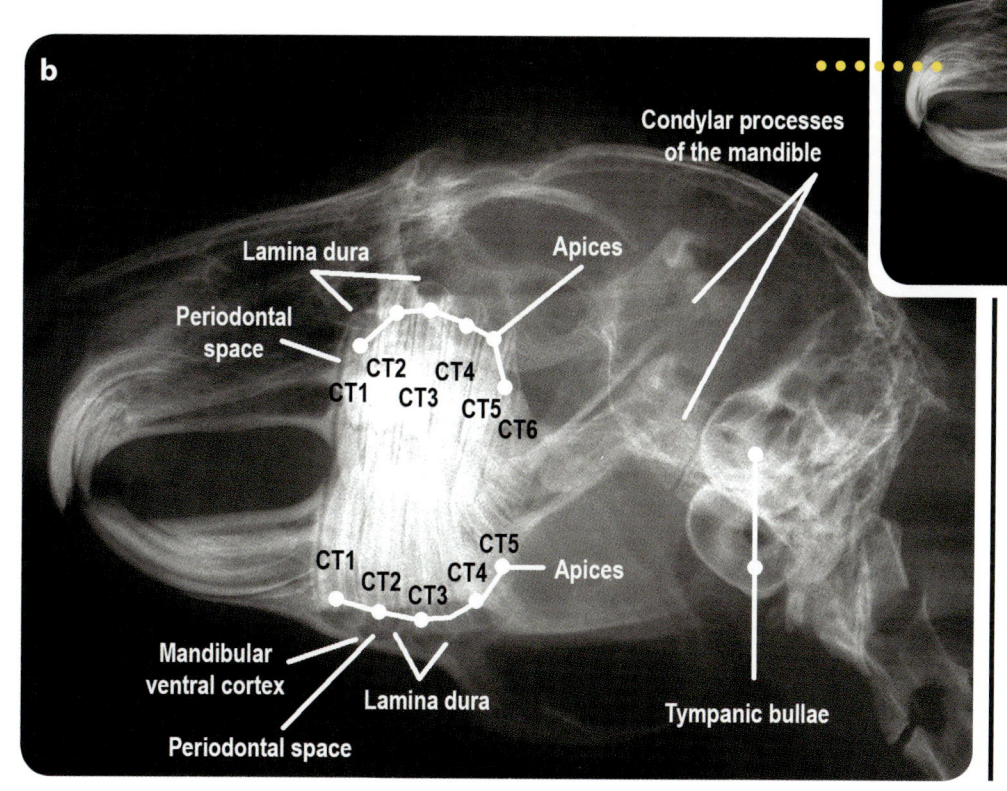

Figure 2.5. Rotation beyond 30° makes distinction between artifact and abnormalities more difficult. Evaluation of the apexes is affected by distorsion due to the projection.

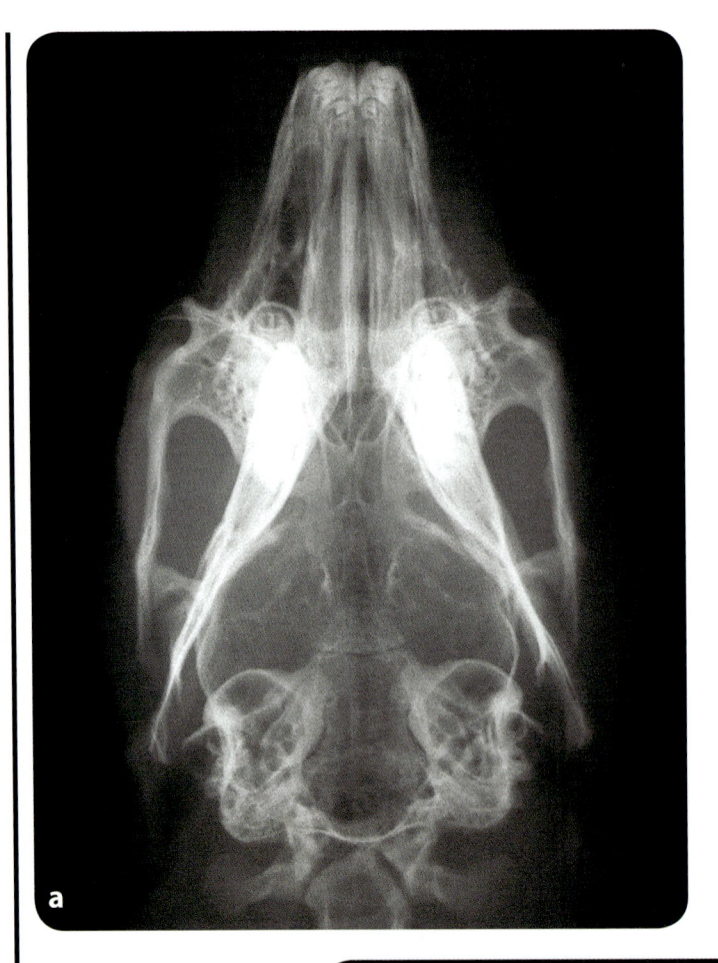

Ventrodorsal Projection

Figure 2.6a,b. The patient is placed in ventrodorsal or dorsoventral recumbency. Excellent positioning is essential for meaningful evaluation of the ventrodorsal (or dorsoventral) projection. The zygomatic arches and tympanic bullae must appear perfectly symmetrical. This view may be useful to evaluate overgrowth of cheek teeth, despite the fact that that normal superimposition of mandibular and maxillary structures complicates radiographic interpretation. The bony profile of the mandible and the maxilla may also be evaluated.

The ventrodorsal projection is the most important view for detection of abnormalities of the tympanic bullae.

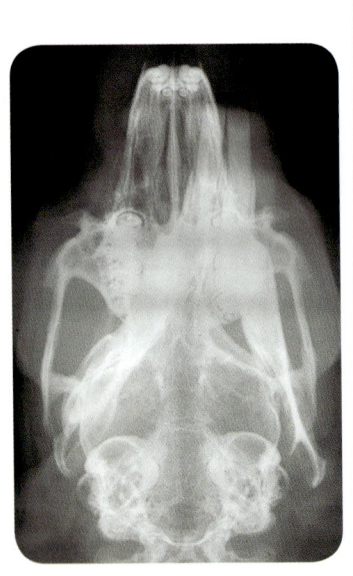

Labels (Figure 2.6b):
- First maxillary incisor tooth
- Second maxillary incisor tooth
- Maxillary CT1
- Incisive bone
- Facial tuber of maxilla
- Maxilla
- Zygomatic bone
- Mandible
- Angular process of the mandible
- Pterygoid bone
- Ear canal
- Occipital condyles
- Tympanic bulla
- Atlas (C1)
- Foramen magnum

Figure 2.7. Superimposition of the mandible impairs evaluation of the maxillary cheek teeth. Shifting the tip of the mandible laterally permits visualization.

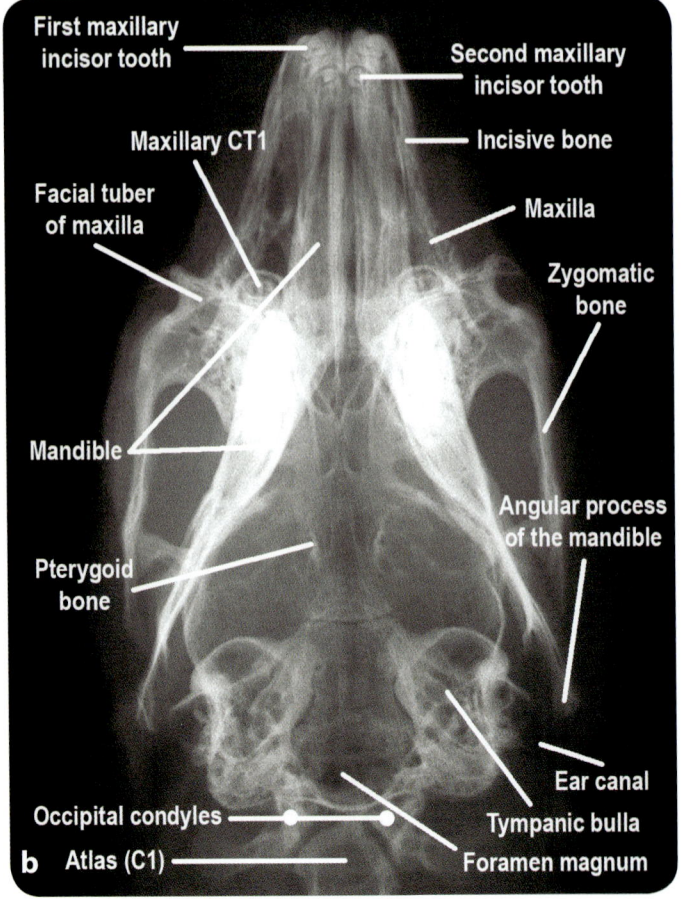

Rostrocaudal Projection

Figure 2.8a,b. Rostrocaudal radiograph of the skull of a rabbit. A perfectly symmetrical projection is essential for meaningful evaluation. Ideal positioning and beam angle allow evaluation of the occlusal plane of the cheek teeth. Slight flexion or extension of the head may compromise interpretation. This view is occasionally useful for evaluation of the calvaria and zygomatic arches, especially in cases of trauma or neoplasia.

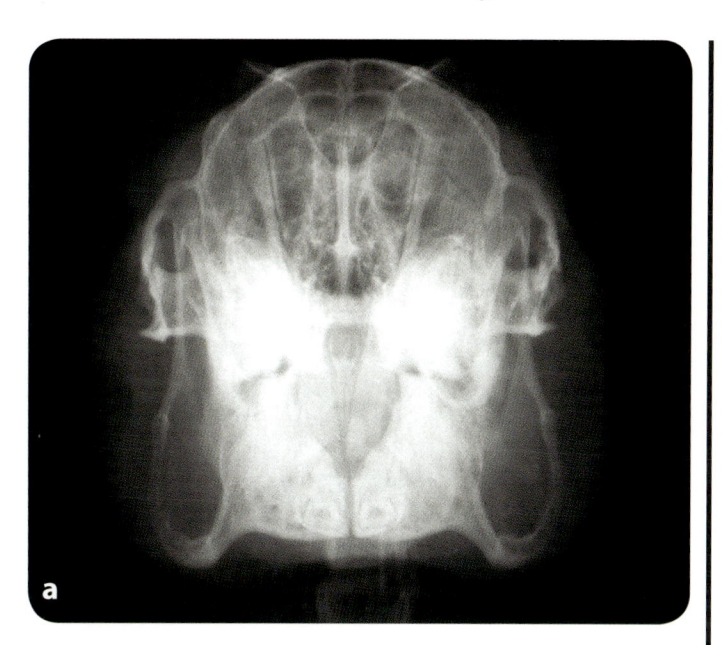

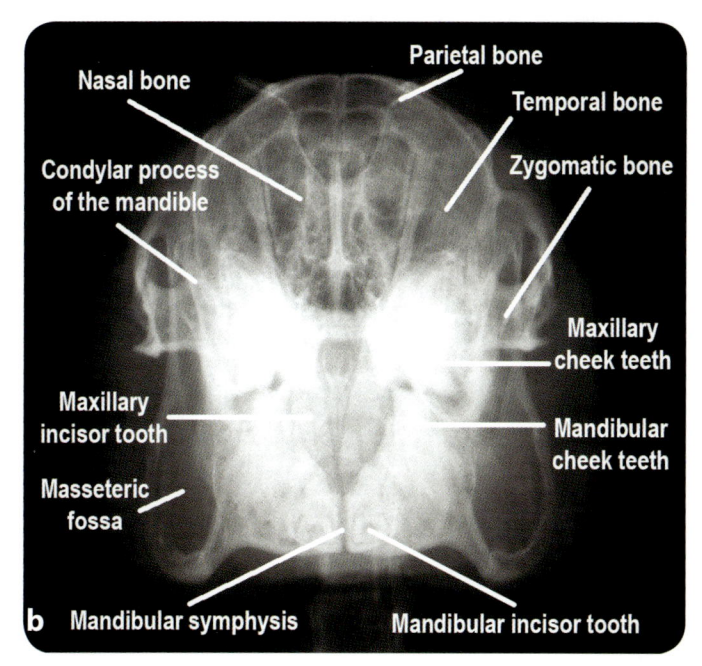

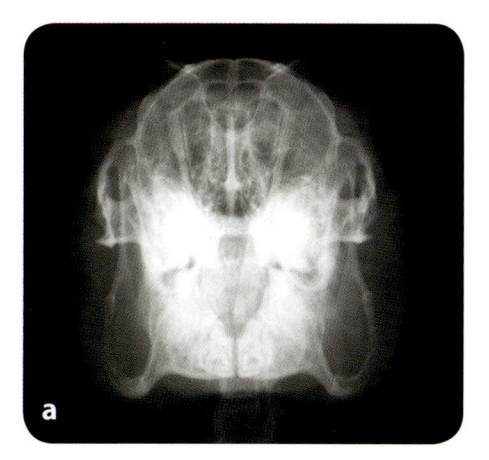

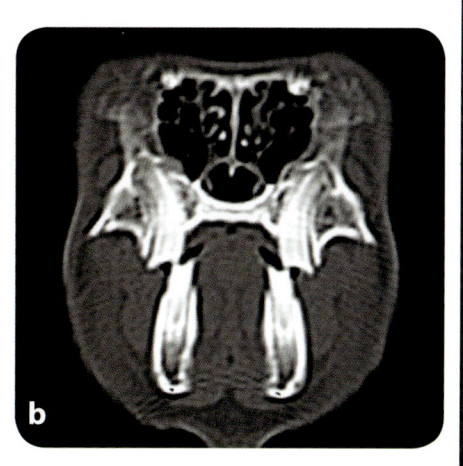

Figure 2.9a,b. Comparison between standard radiographic and CT rostrocaudal projections. CT is superior for evaluation of the shape and orientation of the teeth and the tooth roots.

Intraoral Projections

Figure 2.10. Intraoral placement of a bitewing dental film. The film should be gently introduced between the mandibular and maxillary dental arcades as caudal as possible. Size of the patient determines the practicality of these types of views.
(Courtesy of Margherita Gracis, DVM)

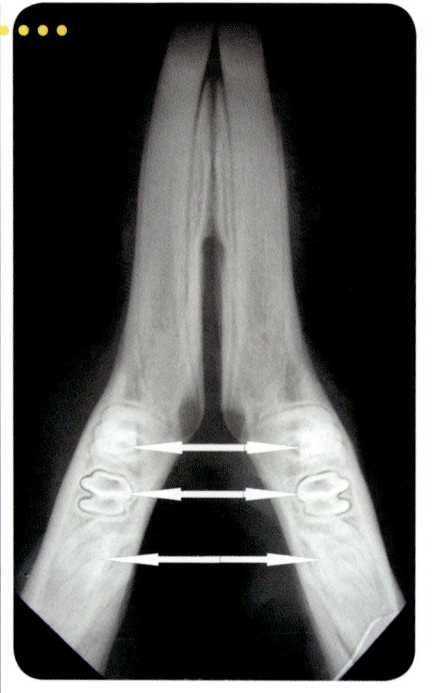

Figure 2.11. Intraoral view of the mandibular incisor teeth. Arrows: first, second and third mandibular cheek teeth.
(Courtesy of Margherita Gracis, DVM)

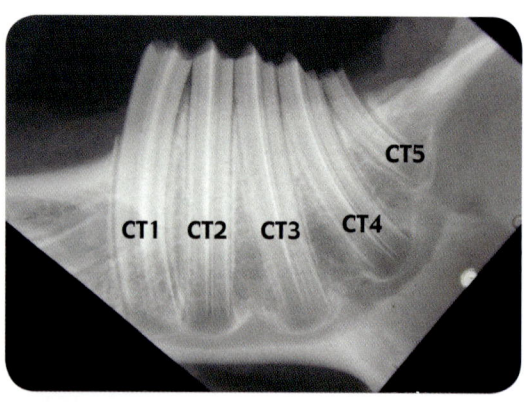

Figure 2.12. Intraoral view of the mandibular left cheek teeth.
(Courtesy of Margherita Gracis, DVM)

The Nasolacrimal Duct

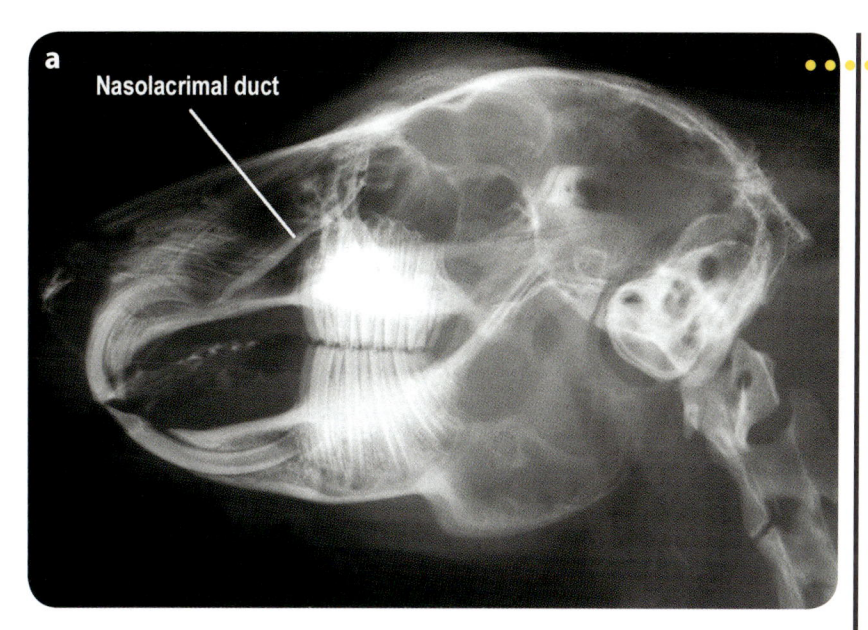

Figure 2.13a,b. Lateral (a) and ventrodorsal (b) contrast radiographs of the left nasolacrimal duct (dacryocystorhinogram). The duct begins at the nasolacrimal punctum, courses within the maxilla to the first incisor root apex, bends dorsally and medially, eventually opening within the nasal mucosa next to the nostril. Dacryocystorhinography can be performed using 0.3-0.5 ml of sterile, water-soluble, organic iodide contrast medium injected into the duct through the nasolacrimal punctum through a 24-22-gauge cannula. Note the flattened occlusal plane of the cheek teeth and obvious incisor teeth malocclusion in this patient.

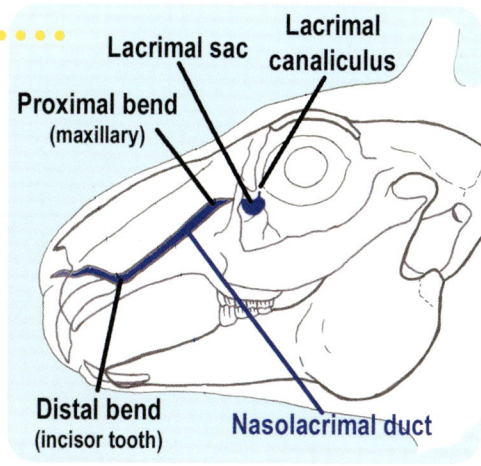

Figure 2.14. Diagram of the head of a rabbit illustrating the nasolacrimal duct. The duct originates at the single nasolacrimal punctum located in the ventral eyelid. It courses ventromedially then curves dorsally and medially to the apex of the ipsilateral first maxillary incisor tooth. The distal opening of the nasolacrimal duct is located in the dorsolateral margin of the nostril.

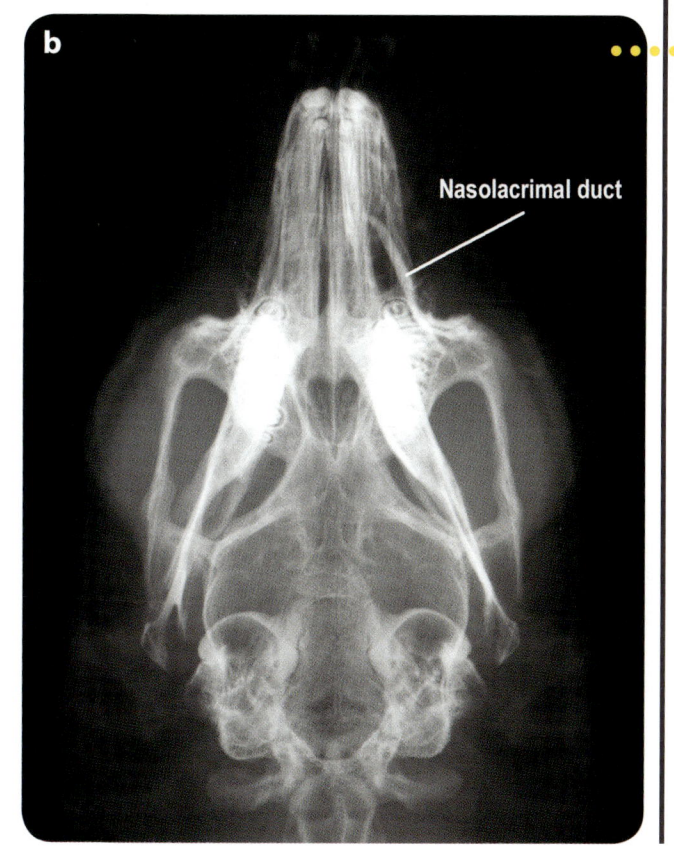

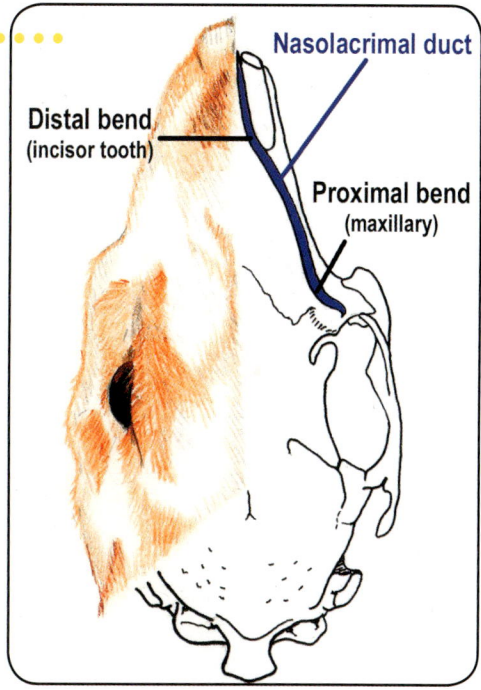

Figure 2.15. Diagram modified from Quesenberry KE, Carpenter JW: Ferrets, Rabbits and Rodents: Clinical Medicine and Surgery 2nd ed., 2004, p. 422.

ABNORMALITIES of the HEAD
Diseases of Incisor Teeth

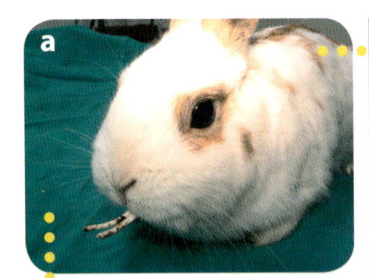

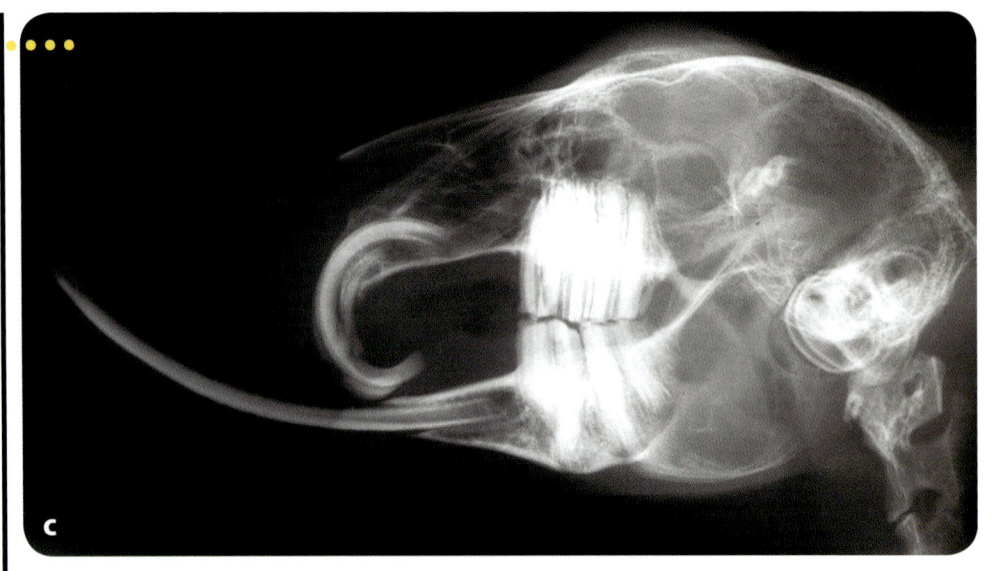

Figure 2.16a,b,c. Incisors of a rabbit (a,b) and lateral radiograph of the skull (c) demonstrating severe malocclusion of maxillary incisor teeth (c) and extraordinary overgrowth of mandibular incisors. Abnormalities of the cheek teeth include uneven occlusal plane.

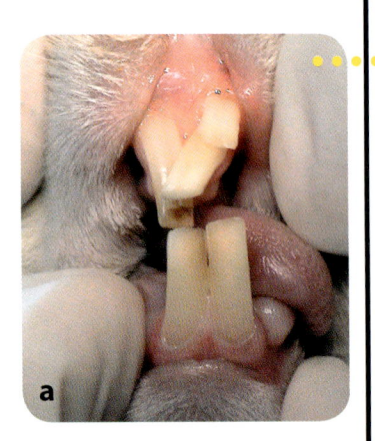

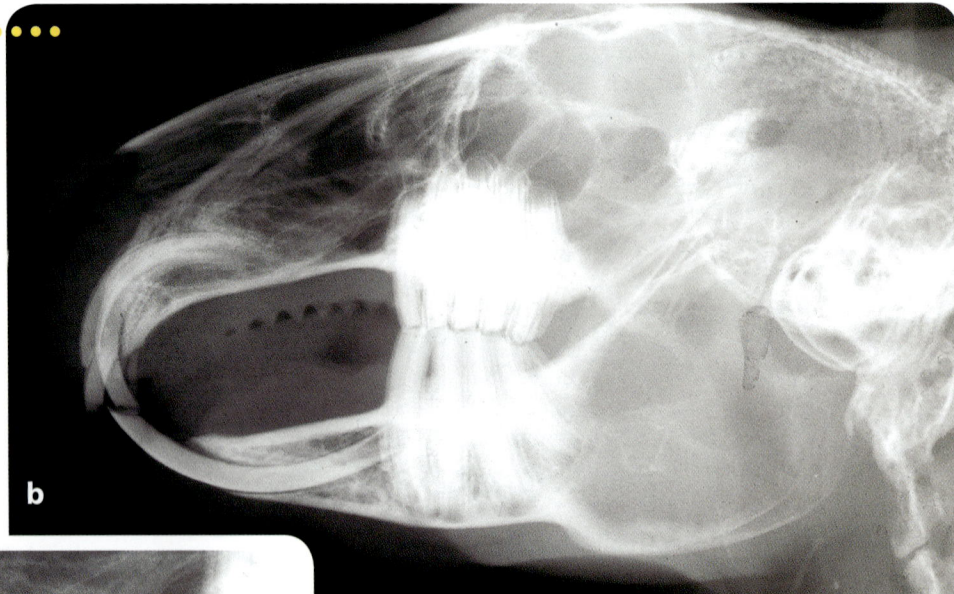

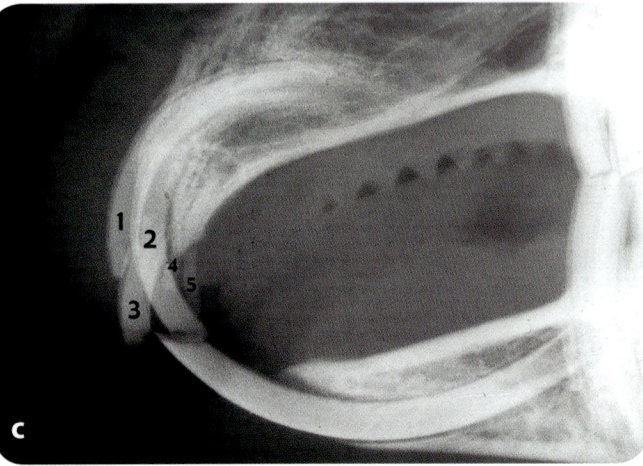

Figure 2.17a,b,c. View of the incisors (a) and lateral radiographs (b,c) of a rabbit with unusual overgrowth of a supernumerary first maxillary incisor tooth presumably formed due to traumatic disruption of the germinal tissue.
In (c) the 3 first maxillary incisor teeth are clearly visible (1, 2, 3), as well as the two secondary maxillary incisors (4, 5).

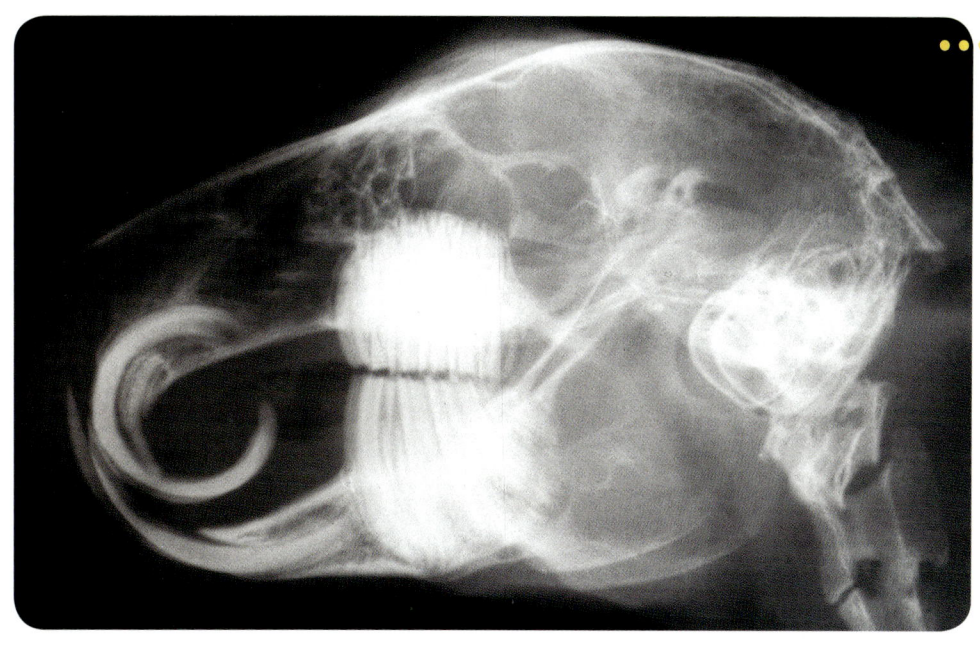

Figure 2.18. Lateral radiograph of a rabbit with severe malocclusion of maxillary incisor teeth and marked overgrowth of mandibular incisors due to congenital prognathism. Dental lesions of the cheek teeth include overall elongation (demonstrated by divergence of the lines of the diastema (see Figure 2.24) and curving of mandibular CT1.

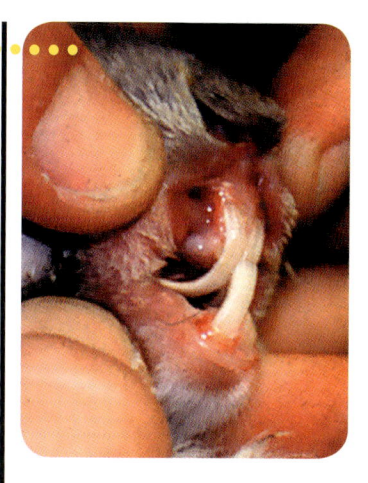

Figure 2.19. In rabbits, overgrown mandibular incisor teeth seldom produce soft tissue lesions, while overgrown maxillary incisors may damage the lips.
Soft tissue injuries may be subtle, therefore careful inspection is important.

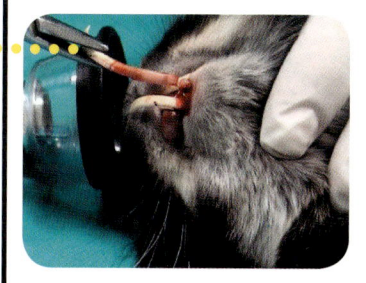

Figure 2.20. Lateral radiograph of the skull of a rabbit, post incisor extraction. The lateral view should always be obtained to confirm complete extraction of all six incisor teeth. In this case, elongation of cheek teeth is also present and must be addressed.

Figure 2.21. Indications for extraction of incisor teeth include congenital or severe acquired malocclusion, dental fractures, and endodontic and periapical disease. Incisor extraction does not normally impair feeding behavior, as lagomorphs and rodents without incisors use their lips and tongue for prehension of food.

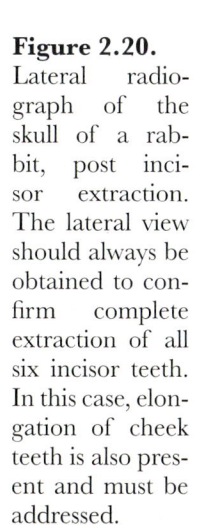

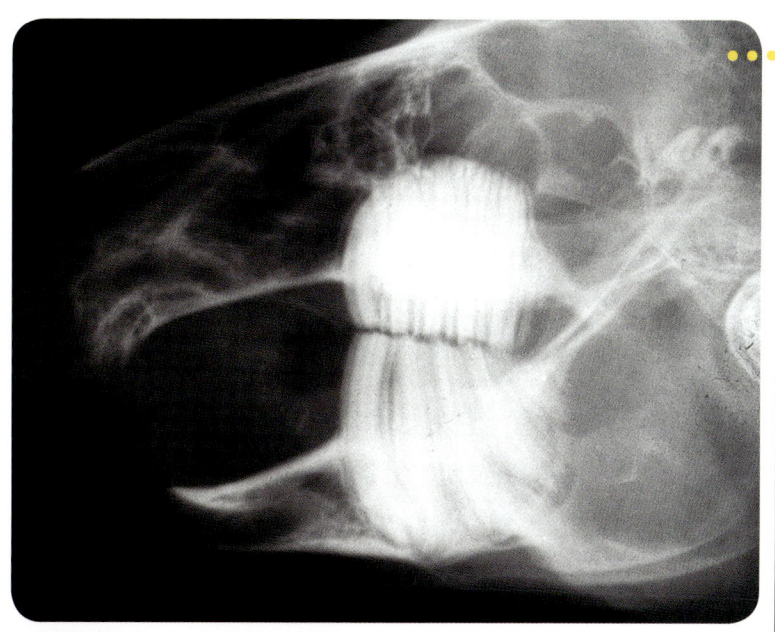

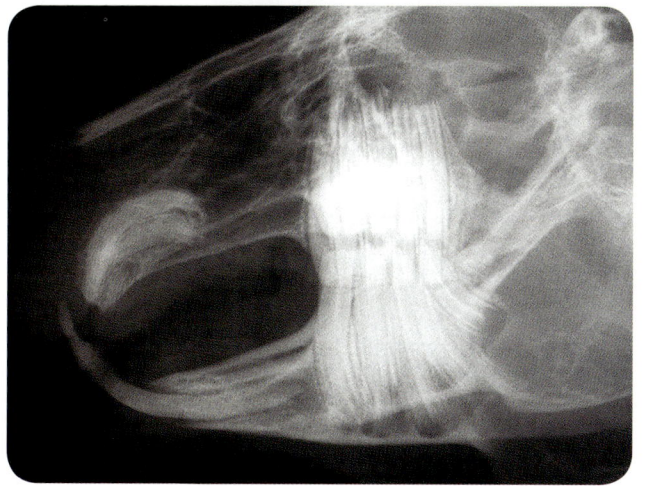

Figure 2.22. Left 30° ventral-right dorsal oblique radiograph of the skull of a rabbit made months post extraction. This projection shows both maxillary first incisors and a single mandibular incisor tooth, following incomplete removal of germinal tissue. Incomplete removal can result in complete to any degree of partial regrowth. There is no need to attempt to remove partially regrown incisors that do not erupt.

Diseases of Cheek Teeth

Acquired dental disease (ADD) produces a wide spectrum of pathologic changes, from mild to severe. Severity can be staged radiographically. Thorough evaluation and diagnosis are particularly important, as the patient's clinical signs may be mild or absent.

Early diagnosis is important for prevention of the progression of dental disease.

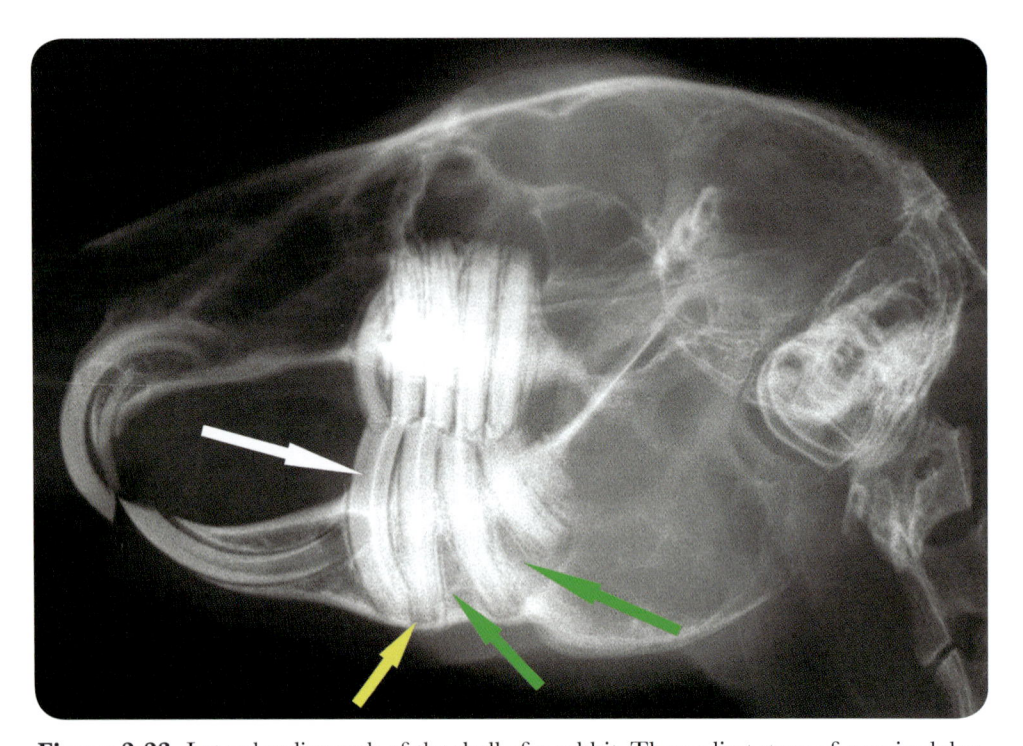

Figure 2.23. Lateral radiograph of the skull of a rabbit. The earliest stage of acquired dental disease of cheek teeth in rabbits is crown elongation. Because both the reserve crown and clinical crown begin to take up more space, abnormalities related to increased pressure begin to occur. The zigzag pattern of the cheek teeth occlusal plane is still normal, but the radiotransparent line is less visible when the mandible is at rest, even in a true lateral projection as demonstrated. Pressure on the reserve crowns begins to increase when the animal chews. Because both mandibular CT1 do not have another tooth rostral to them, they begin to curve, with increasing rostral convexity (white arrow). In some early cases, slight deformation of the ventral mandibular cortical bone due to the increased pressure may be visible (yellow arrow). Due to the abnormal convexity, the interproximal spaces of mandibular cheek teeth begin to widen (green arrows). Malocclusion of incisor teeth is usually not present at this stage.

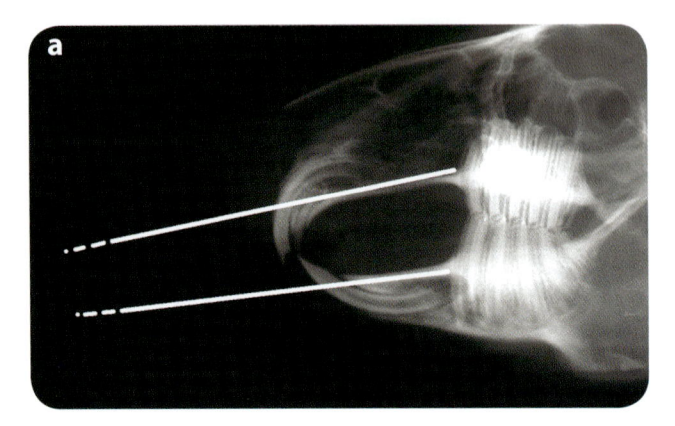

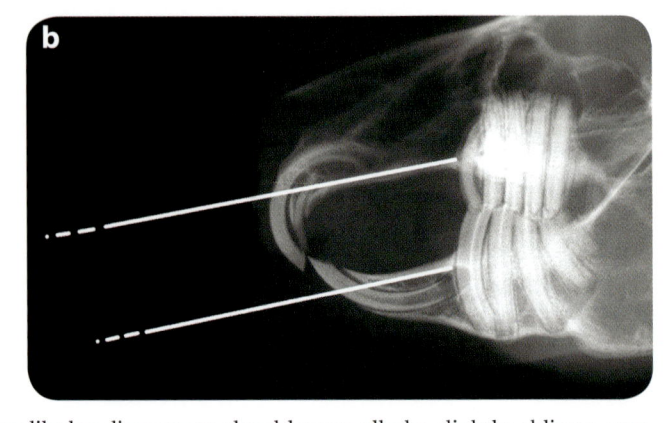

Figure 2.24a,b. The radiopaque lines of the maxillary and mandibular diastemata should normally be slightly oblique, converging in front of the incisor teeth (white lines, a). When cheek teeth are elongated, the radiopaque lines progress from normal oblique to parallel, or even begin to diverge (white lines, b). This useful radiographic evaluation is only of benefit with a true lateral projection with the cheek teeth in occlusion. This technique of comparing diastemal lines may not accurately demonstrate elongation in very small brachycephalic rabbit breeds in which diastemal lines are normally parallel.

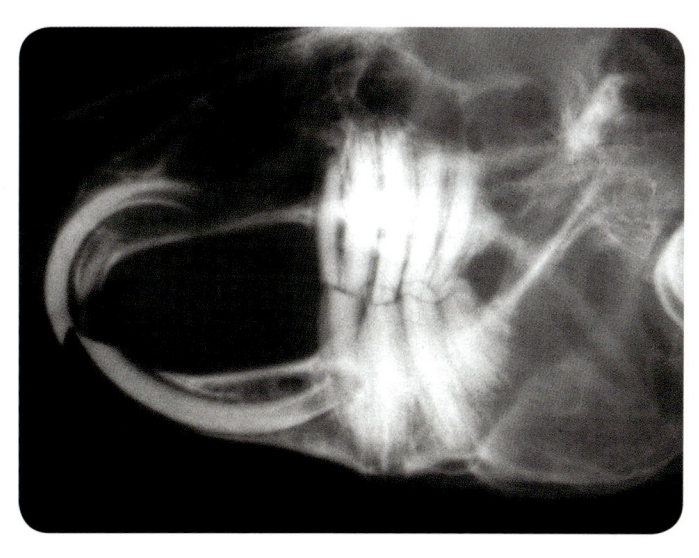

Figure 2.25. Lateral radiograph of the skull of a rabbit with ADD of cheek teeth, later stage. Abnormalities of the occlusal plane are due to excessive and irregular crown elongation, and are clearly visible, with height differences between adjacent cheek teeth of up to a few millimeters. The normal zigzag line is irregular or appears as a superimposition of several lines. Mandibular cheek teeth root deformities are also visible. Occlusion of incisor teeth is still normal at this stage. This occlusalplane deformity is described as "wave mouth." More severe alterations in crown length are often termed "step mouth."

Figure 2.26. Lateral radiograph of the skull of a rabbit with ADD of cheek teeth, later stage. Alteration of the occlusal plane is evident, with marked differences of crown length between two adjacent cheek teeth. Deformity of the mandibular ventral cortical bone is evident (arrow), with thinning of the ventral ramus adjacent to the tooth apexes.
Wave mouth and step mouth are descriptive terms for occlusal plane alterations, and both can be seen in this radiograph.

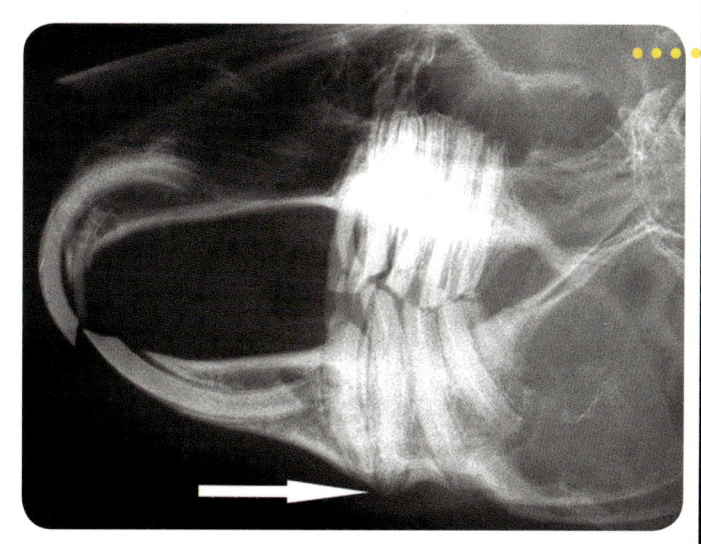

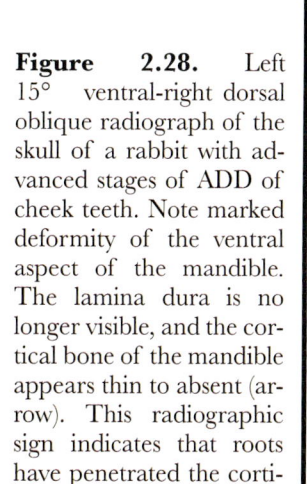

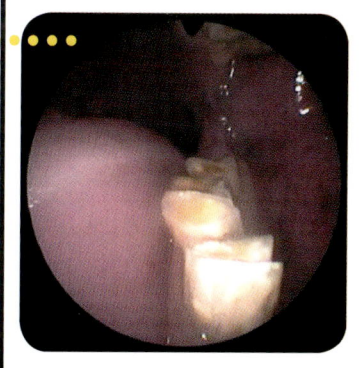

Figure 2.27. Endoscopic view of the left mandibular arcade with "step mouth." In most cases of "wave mouth" and/or "step mouth," sharp spurs do not occur, therefore clinical signs and symptoms may be mild or absent.

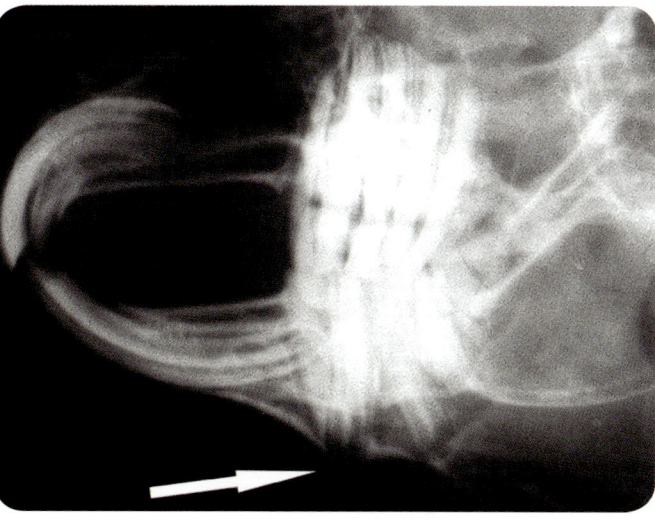

Figure 2.28. Left 15° ventral-right dorsal oblique radiograph of the skull of a rabbit with advanced stages of ADD of cheek teeth. Note marked deformity of the ventral aspect of the mandible. The lamina dura is no longer visible, and the cortical bone of the mandible appears thin to absent (arrow). This radiographic sign indicates that roots have penetrated the cortical bone. Root perforation most commonly involves CT1 and/or CT2 with perforations of the roots of CT3-5 occurring much less frequently. Detection of this abnormality requires an oblique projection, ideally less than 30°. Oblique projections with a 30° obliquity or higher may produce false impression of bone deformity.

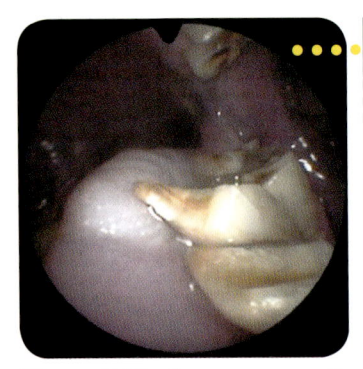

Figure 2.29. Severe overgrowth and malocclusion of the left mandibular arcade.

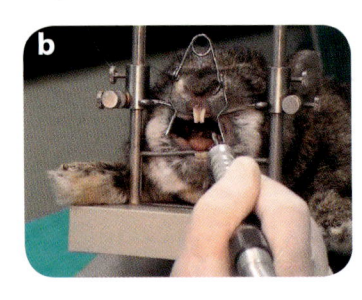

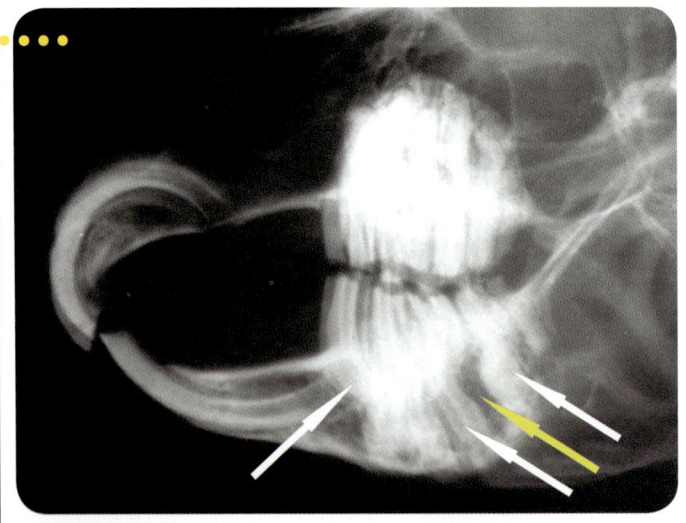

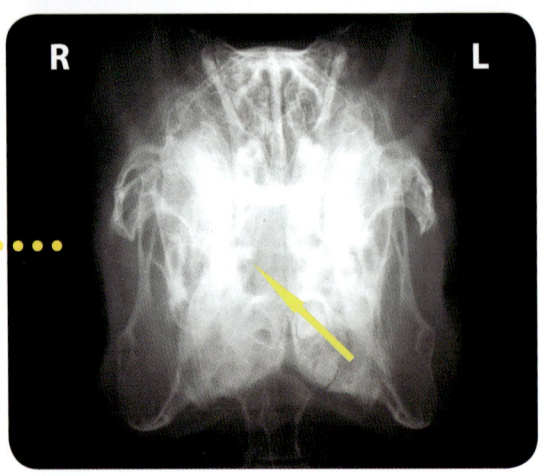

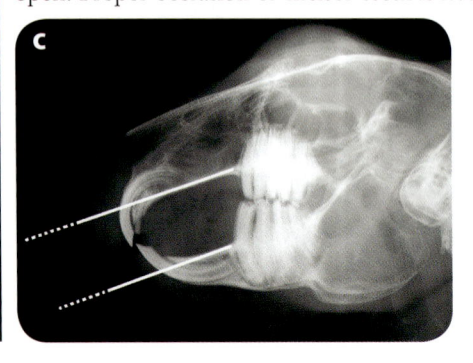

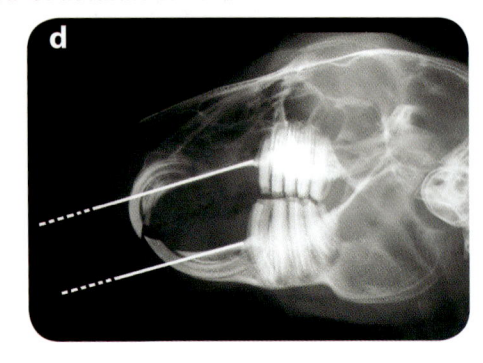

Figure 2.30. Lateral radiograph of the skull of a rabbit with ADD of cheek teeth, later stage. In addition to abnormal elongation of crowns, wave and step mouth, deformities of both clinical and reserve crown become apparent.
The crowns can curve in any plane, and the tooth may rotate within the alveolus. This leads to insufficient wearing of clinical crowns and formation of spurs. In this radiograph, deformities of mandibular CT1, CT3, and CT4 (white arrows) are clearly visible, as well as widened interproximal spaces (yellow arrow).

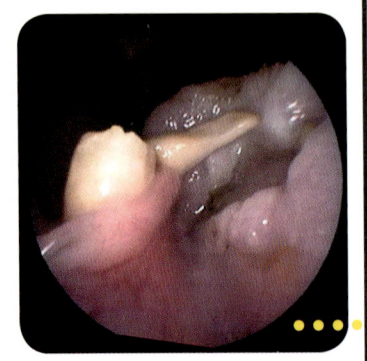

Figure 2.31. A long, sharp spur of right mandibular CT2 and subsequent ulceration of the tongue. The margins of the lesion are partially healed.

Figure 2.32. Rostrocaudal projection of the skull of a rabbit demonstrating overgrowth, malocclusion and presence of a sharp spur at the right mandibular arcade (arrow).

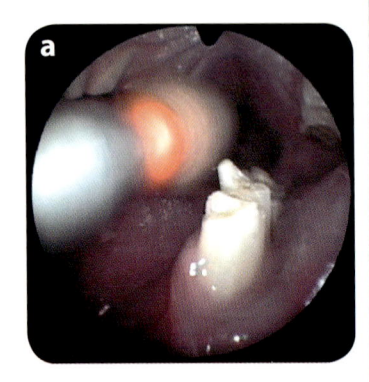

Figure 2.33a,b. ADD of cheek teeth requires reduction of elongated crowns and restoration of the occlusal plane. Spikes and elongated crowns are reduced using abrasive burs on a straight handpiece. Special tabletop mouth gags are available for proper positioning of the patient.

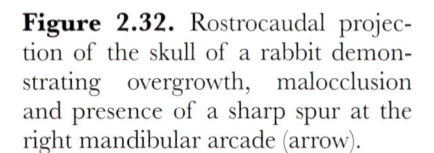

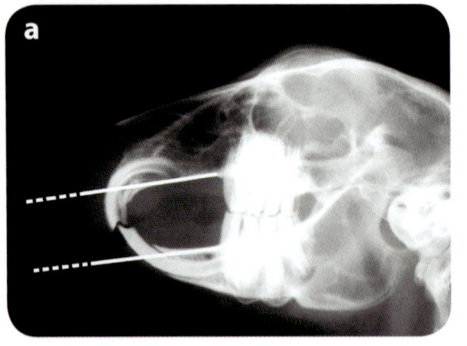

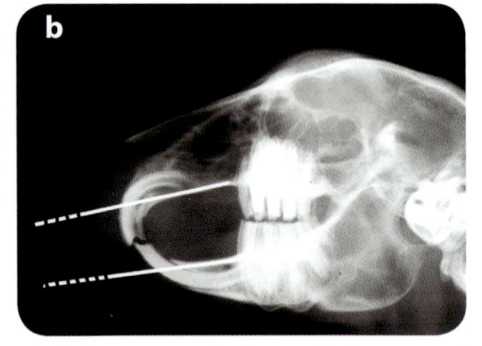

Figure 2.34a,b,c,d. Lateral projections of the skull in two cases before and after coronal reduction. In both cases prior to treatment (a) and (c) the radiopaque lines of the maxillary and mandibular diastemata are parallel or slightly diverging due to elongation of the clinical crowns of cheek teeth. After proper reduction and occlusal adjustment, the lines appear to converge normally in (b). Note that proper coronal reduction is also likely in (d), but is more difficult to evaluate because the radiograph was inadvertently taken with the mouth slightly open. Proper occlusion of incisor teeth is needed for accurate evaluation.

HEAD

Figure 2.35. Lateral radiograph of the skull of a rabbit with dental disease. A common sequela to excessive elongation of cheek teeth is fracture, especially of mandibular cheek teeth. Longitudinal fracture of mandibular CT1 is particularly common. A common result of this fracture is periapical abscess. In this case, excessive overgrowth of the mesial fragment of fractured mandibular CT1 is present. Step mouth and root deformity of CT2 are also clearly visible.

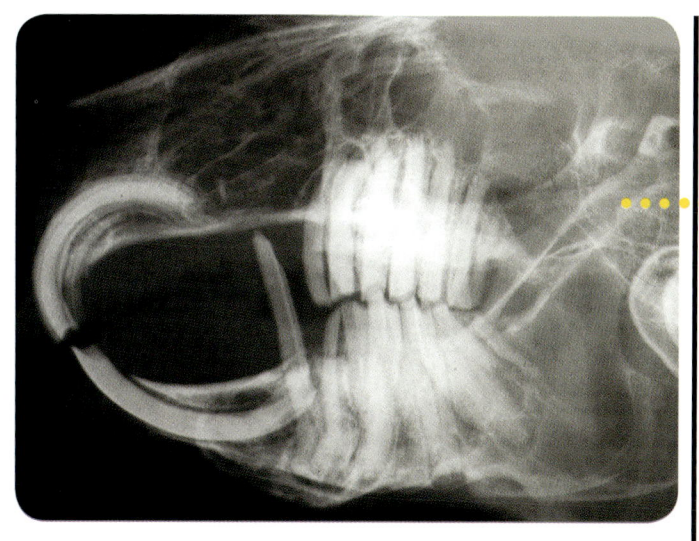

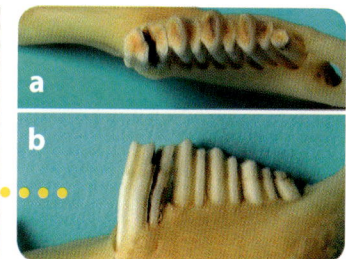

Figure 2.36a,b. Dorsal (a) and lateral (b) views of a bone specimen showing longitudinal fracture of right mandibular CT1. The clinical crown is fractured longitudinally into two visible fragments, but the fracture usually also affects the reserve portion of the crown.

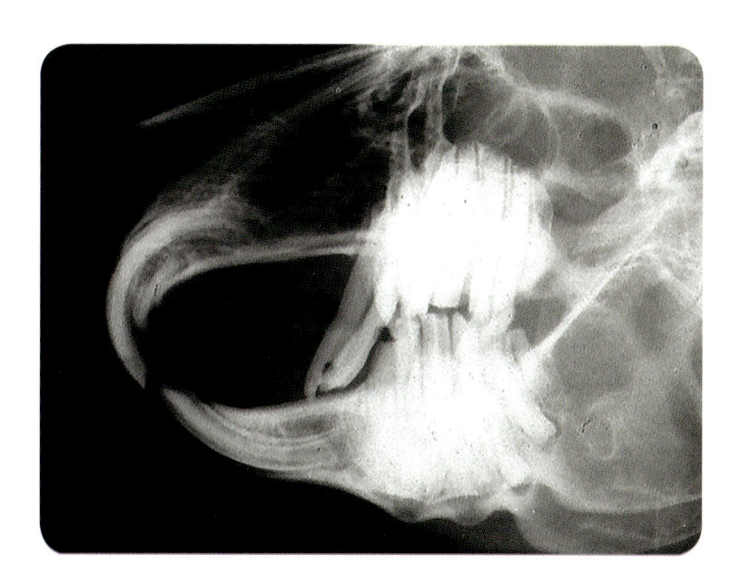

Figure 2.37. Left 15° ventral-right dorsal oblique projection of the skull of a rabbit demonstrating extensive elongation of right maxillary CT1 and CT2. Severe cheek teeth malocclusion is also visible.

Figure 2.38. Lateral radiograph of the skull of a rabbit with end-stage ADD of cheek teeth. Normal radiographic structure of the teeth is no longer visible due to advanced dentine resorption. Crowns of the mandibular cheek teeth are no longer growing and are retained as remnants. Lack of normal occlusion and wear results in overgrowth of the maxillary cheek teeth.

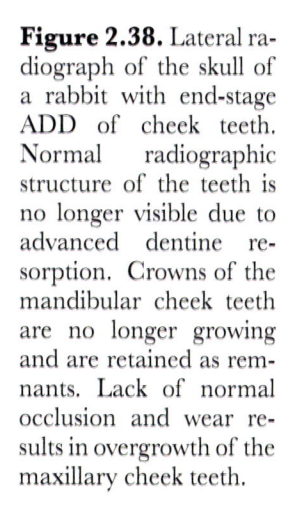

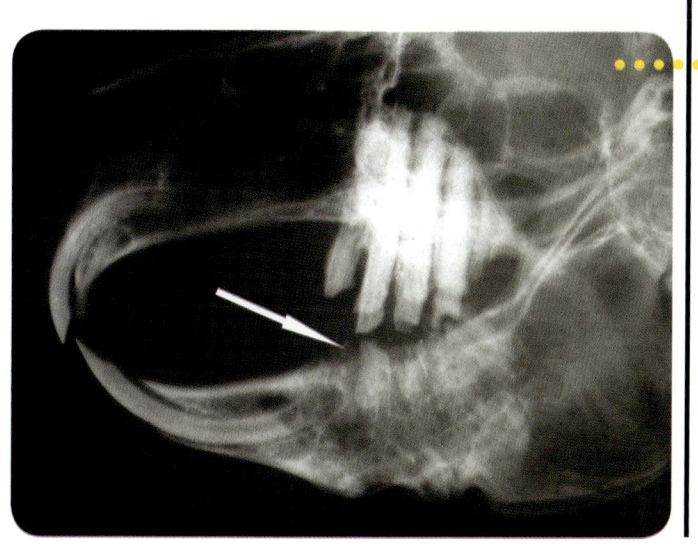

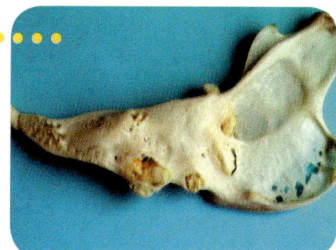

Figure 2.39. Bone specimen of hemimandible, medial aspect. Multiple cortical deformities and perforations are visible, and are a result of end-stage ADD of cheek teeth.

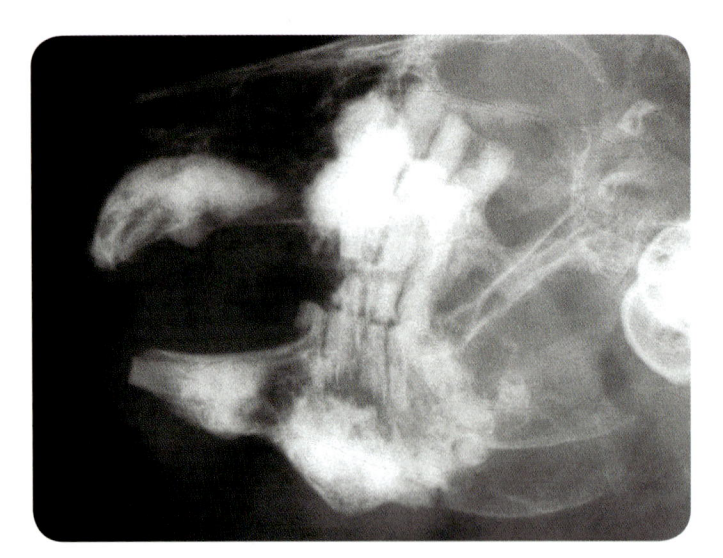

Figure 2.40. Radiograph of the skull of a rabbit with advanced ADD. In later stages of ADD, diffuse irregular mineralization occurs adjacent to areas of bone loss. The exact nature of mineralization has not been described, but may be a reparative response. End stage ADD of the incisor teeth is also visible.

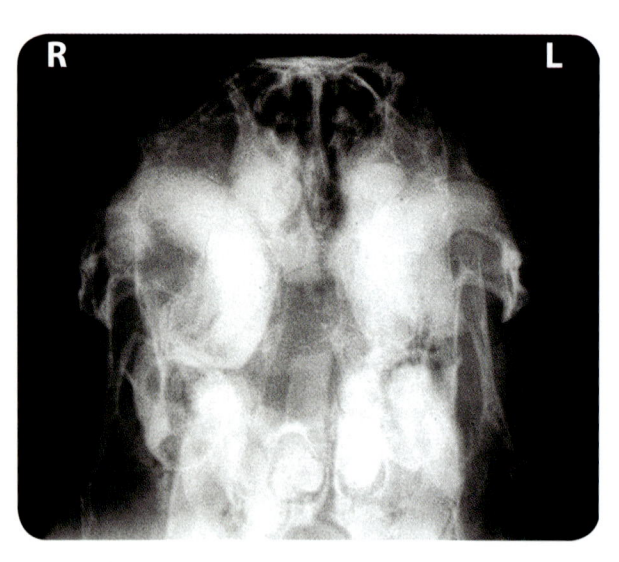

Figure 2.41. Rostrocaudal view of the skull of a rabbit with similar abnormalities to those seen in Figure 2.40.

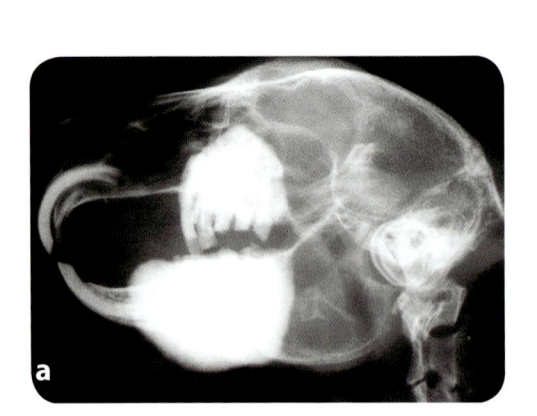

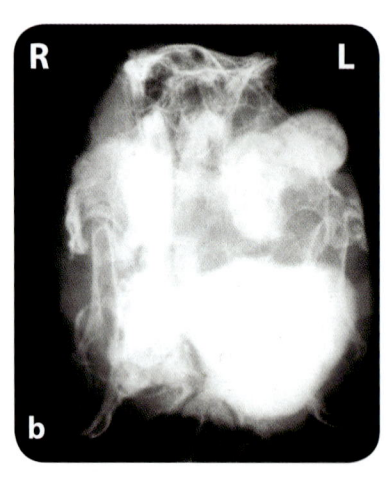

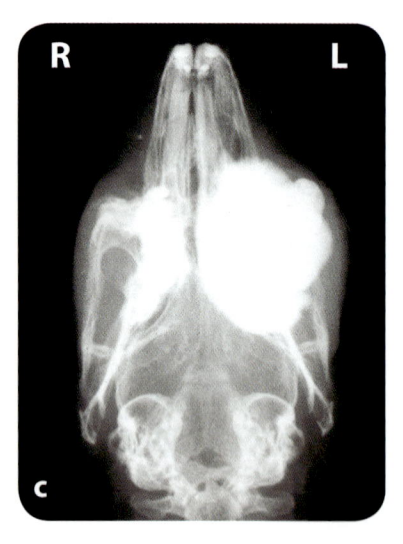

Figure 2.42a,b,c. Lateral (a), rostrocaudal (b), and ventrodorsal (c) projections of the skull of a rabbit demonstrating an unusual case of ADD with complete bilateral mineralization of the mandibular cheek teeth. There is extensive well-marginated mineralization of mandibular cheek teeth crowns and severe deformation of maxillary cheek teeth crowns and roots. Mineralization and bone deformities appear more severe on the left side.

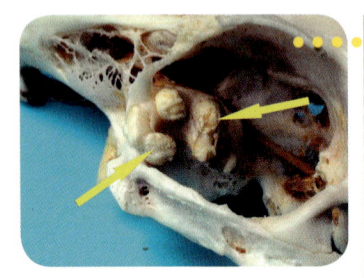

Figure 2.43. Bone specimen demonstrating severe mineralization and deformities of the left maxillary cheek teeth roots (arrows).

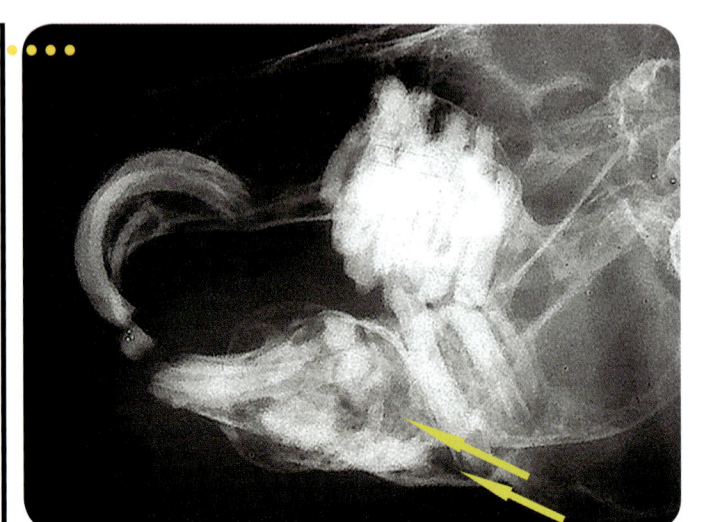

Figure 2.44. Left 30° dorsal-right ventral oblique radiograph of the skull of a rabbit with end stage ADD of incisor and cheek teeth. Note retained fractured mandibular cheek teeth and bone resorption, as well as radiolucency consistent with osteomyelitis (arrows).

Periapical infections and Osteomyelitis

Figure 2.45. Right 15° ventral-left dorsal oblique projection demonstrating periapical infection of the left mandibular incisor tooth. Radiolucency surrounding the fractured tooth, and deformation and thinning of the ventral cortical bone of the mandible are clearly visible (arrow).

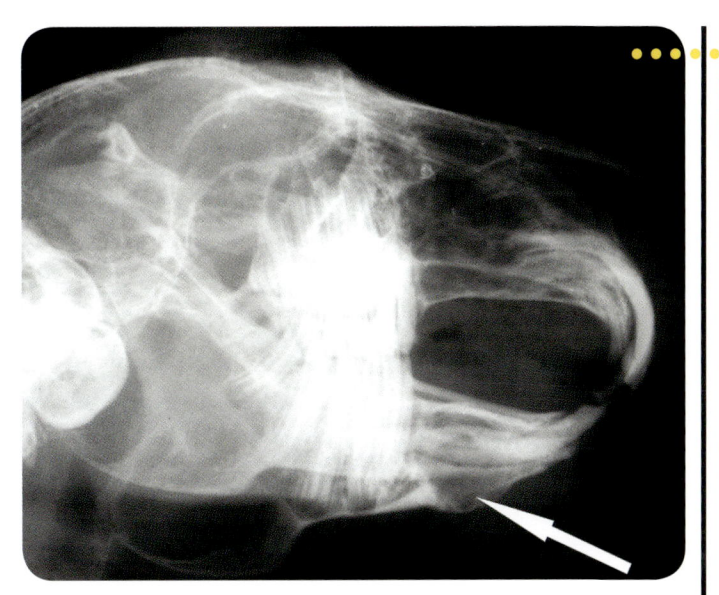

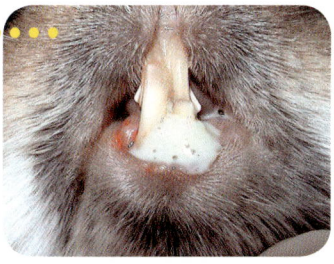

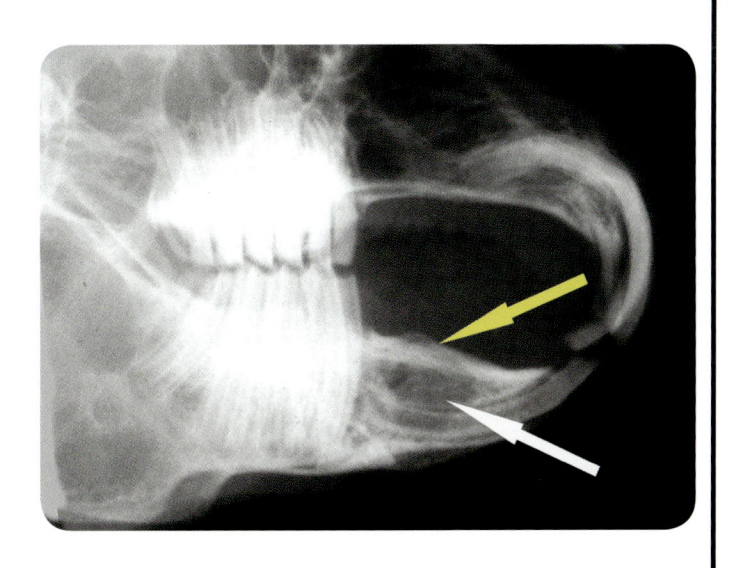

Figure 2.46. Periapical infection and fracture of the left mandibular incisor tooth following incisor malocclusion. Slight pressure on the mass under the lower lip caused purulent material to flow out the alveolar cavity.

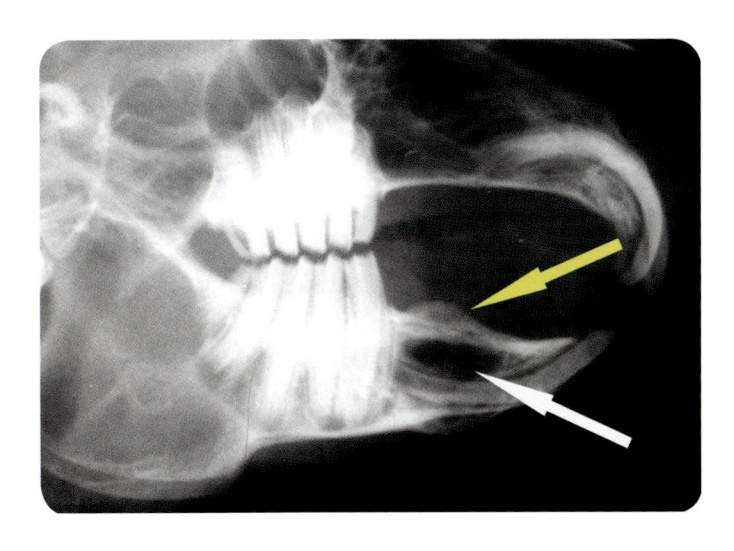

Figure 2.47. Lateral view of the same case after surgical debridement of the abscess and extraction of the fractured mandibular incisor tooth. The radiolucent area is the osteomyelitic bone cavity (white arrow). A periosteal response (yellow arrow) visible in the lateral projection was hidden in the previous oblique projection.

Figure 2.48. Lateral view of the same patient 7 weeks post surgery demonstrating remodeling and deposition of new bone in areas previously identified as osteomyelitic. The radiolucent area of osteomyelitic bone has been replaced by deposition of new bone (white arrow). The cortical bone deformation (yellow arrow) has been remodeled. Follow up radiographs 6-8 weeks post surgery help confirm appropriate healing.

Figure 2.49. Facial abscesses due to periapical infection of cheek teeth may frequently appear as large masses typically located on the ventrolateral aspect of the rostral portion of the mandible.

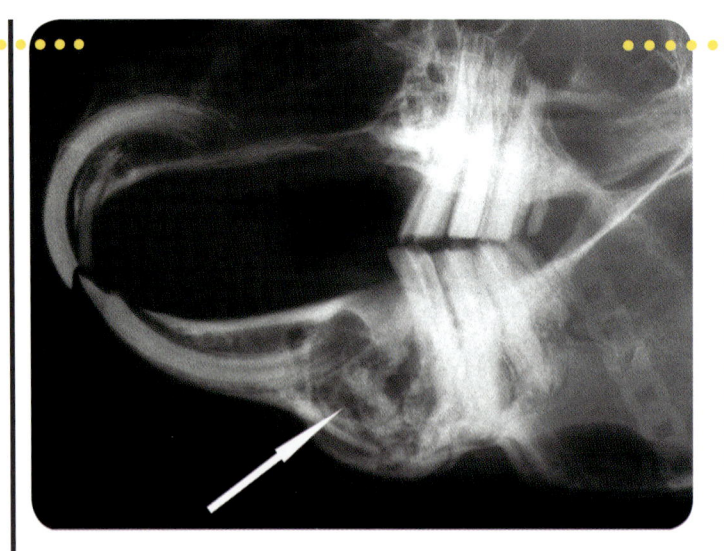

Figure 2.50. Lateral radiograph of the skull of a rabbit consistent with mandibular osteomyelitis. A fragment of mandibular CT1 (arrow) is visible in the circular osteolytic area. Note the expansile appearance and sharp sclerotic margins. Both maxillary CT1 and CT2 had been previously extracted.

Figure 2.51. Lateral radiograph of the same patient after surgical debridement and extraction of the tooth fragment and necrotic bone tissue. The osteolytic area contained purulent material.

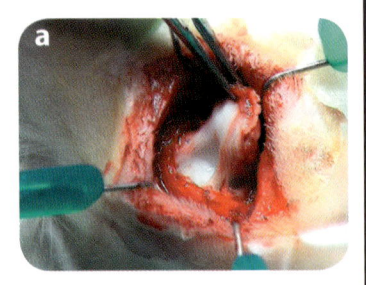

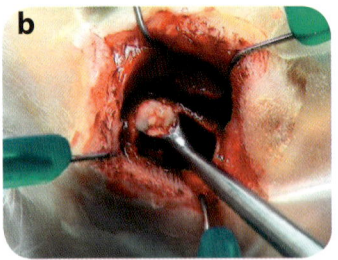

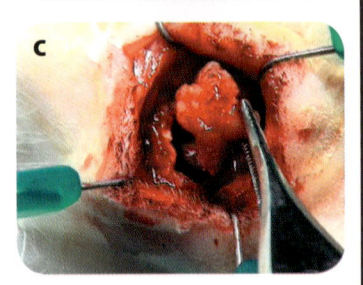

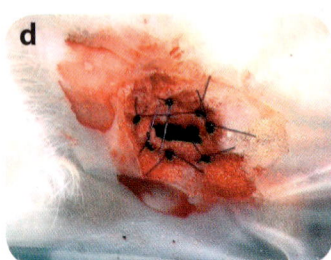

Figure 2.52a,b,c,d. Surgical treatment of facial abscesses includes removal of the capsule and the pus (a); thorough debridement of the bone cavity (b); extraction of tooth fragments or necrotic alveolar bone (c); marsupialization of soft tissues to allow postoperative flushing and treatment (d).

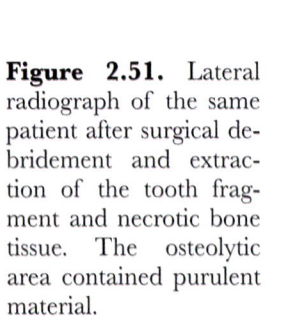

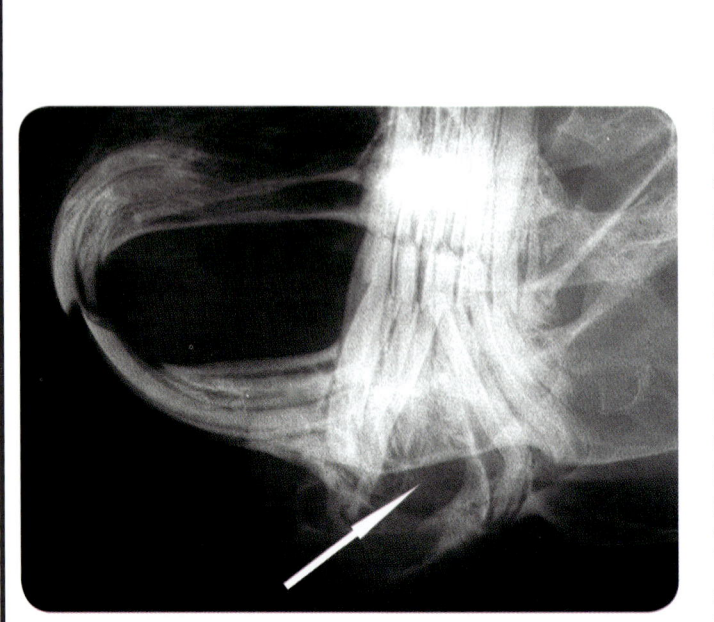

Figure 2.53. Left 30° ventral-right dorsal radiograph of the skull of a rabbit showing severe bone deformity and mandibular osteomyelitis. The radiolucent area is very large, and typical widening of the space between the two affected cheek teeth (in this case CT2 and CT3) is visible (arrow). Prognosis is guarded to poor in these cases. The owner must be advised that even aggressive treatment may fail to resolve osteomyelitis.

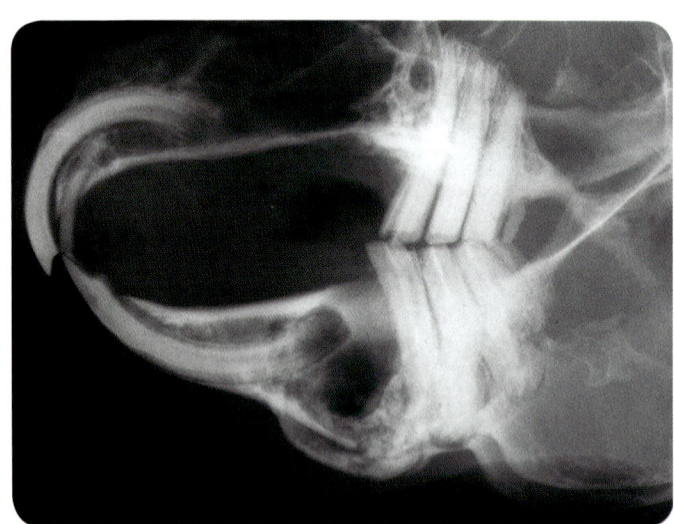

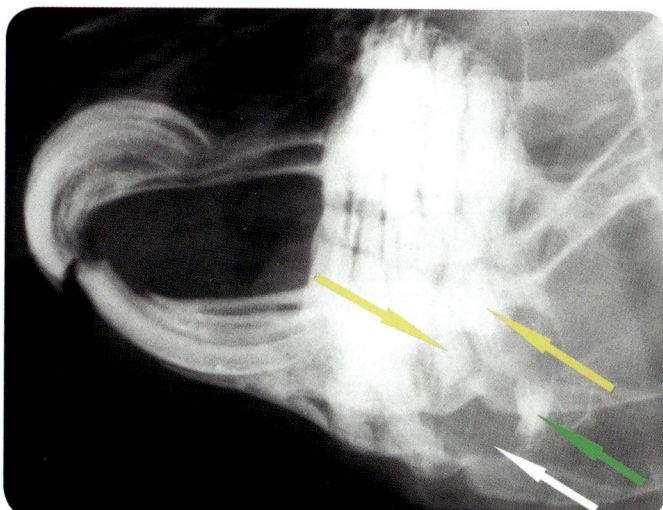

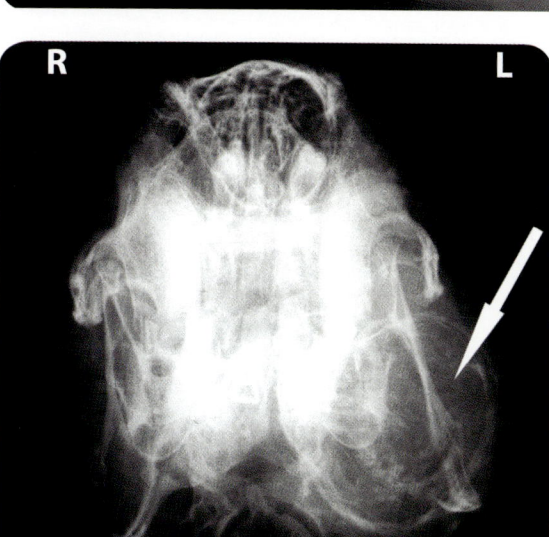

Figure 2.54. Left 15° ventral-right dorsal oblique radiograph of a rabbit with dental disease and osteomyelitis. Note the circular area of osteolysis (white arrow) in the mandibular ramus, and mandibular CT3 and CT4 (yellow arrows). A fragment of the root of CT4 is visible (green arrow). The radiolucent area represents osteomyelitis.

As the bone of the mandible is thin in this area (between the body and the masseteric fossa), it is prone to fracture during surgical debridement.

Figure 2.55. Gross specimen of a left mandible (lateral view) demonstrating osteomyelitic cavitation following periapical abscess of CT1. The thick abscess capsule is typically firmly attached to the margins of periosteal reaction. The abscess and capsule have been removed from this specimen.

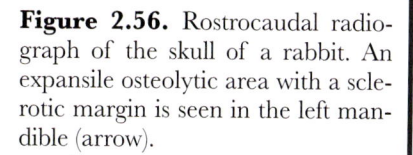

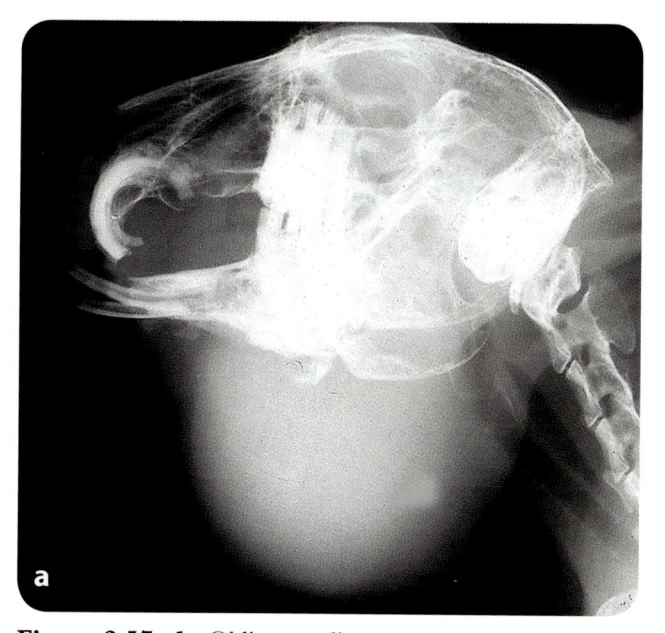

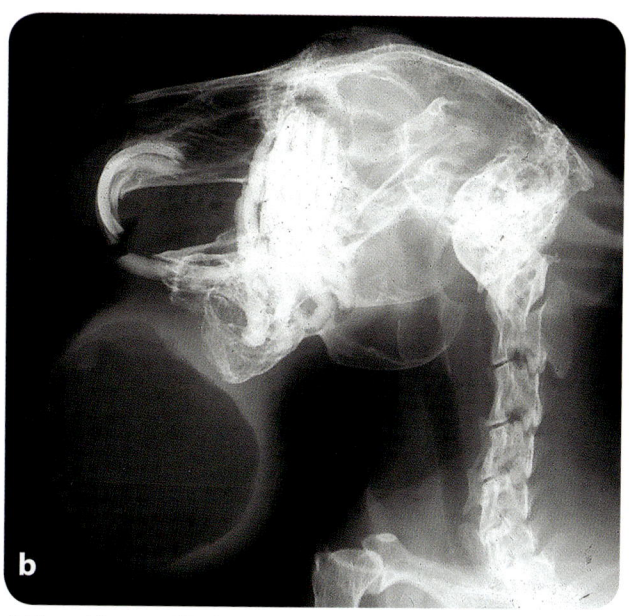

Figure 2.56. Rostrocaudal radiograph of the skull of a rabbit. An expansile osteolytic area with a sclerotic margin is seen in the left mandible (arrow).

Figure 2.57a,b. Oblique radiographs (a,b) of the skulls of two rabbits with mandibular abscesses following end-stage dental disease. The abscess in (a) involves all soft tissues of the ventral part of the head and neck. The radiolucency of the core of the abscess in (b) suggests the presence of gas-forming bacteria. Osteolysis of the rostral mandible is seen in both images.

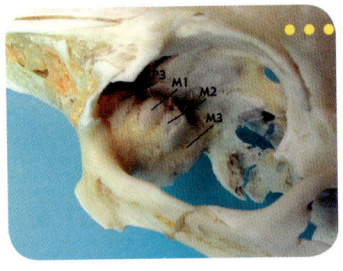

Figure 2.58. Dorsal view of the skull, left orbital fossa. The alveolar bulla contains the roots of the four most caudal maxillary cheek teeth. Their apexes can be seen protruding through the bony plate.

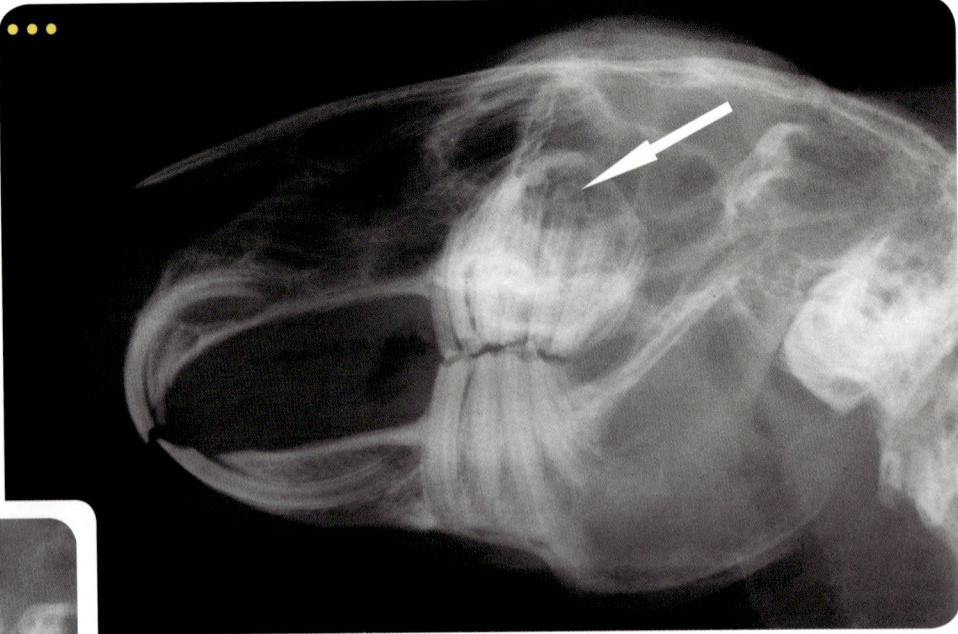

Figure 2.59. Lateral radiograph of the skull of a rabbit with periapical infection and resultant osteomyelitis following fracture of right maxillary CT4. The radiolucent area (arrow) is a bone cavity with irregular, thickened margins within the alveolar bulla. This cavity was found to contain purulent material.

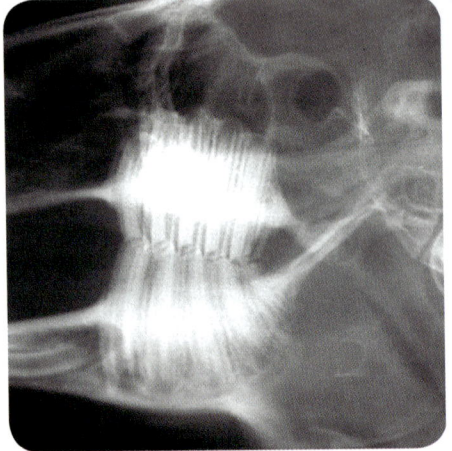

Figure 2.60. Normal radiographic appearance of the alveolar bulla.

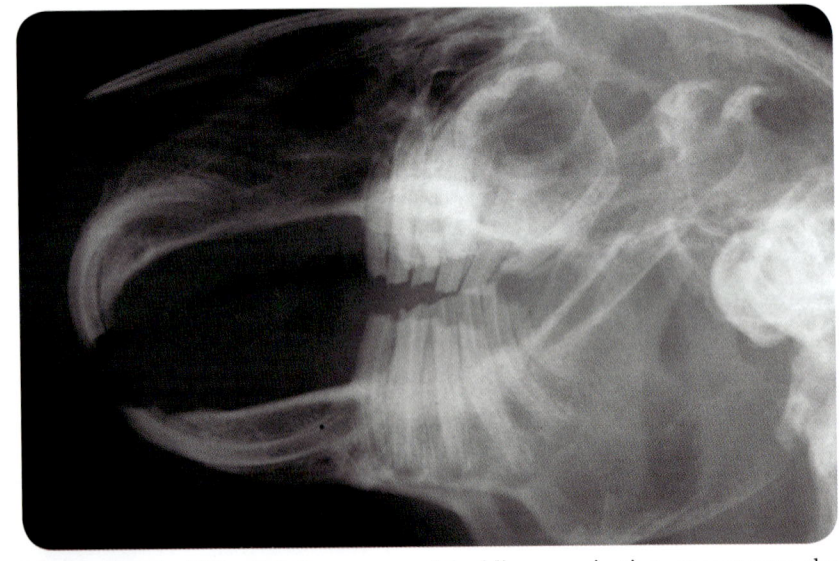

Figure 2.61. This slightly rostrocaudal oblique projection accentuates the bone cavity.

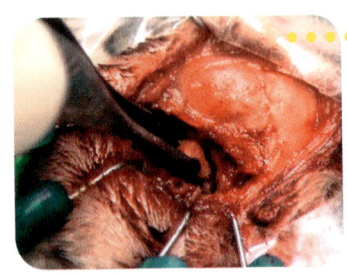

Figure 2.62. The bone cavity is filled with polymethylmethacrylate antibiotic impregnated (AIPMMA) beads). Access to this abscess site required maxillotomy.

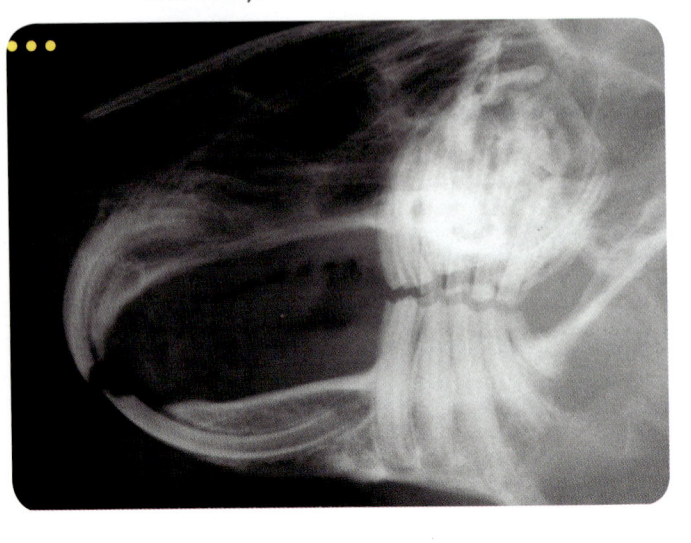

Figure 2.63. Lateral radiograph of the same patient post surgery. The bone cavity has been filled with AIPMMA beads.

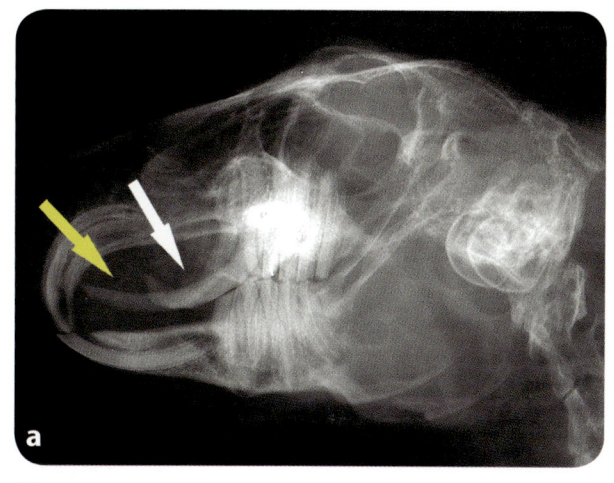

Figure 2.64a,b. Slightly oblique (a) and dorsoventral (b) radiographs of the skull of a rabbit with periapical abscess and osteomyelitis as a result of elongation and malocclusion of right maxillary CT1 (white arrows) and CT2 (yellow arrows). The green arrow depicts a radiolucent bone cavity in the rostral aspect of the zygomatic bone. Note the sclerotic rim, which is a reparative effect.

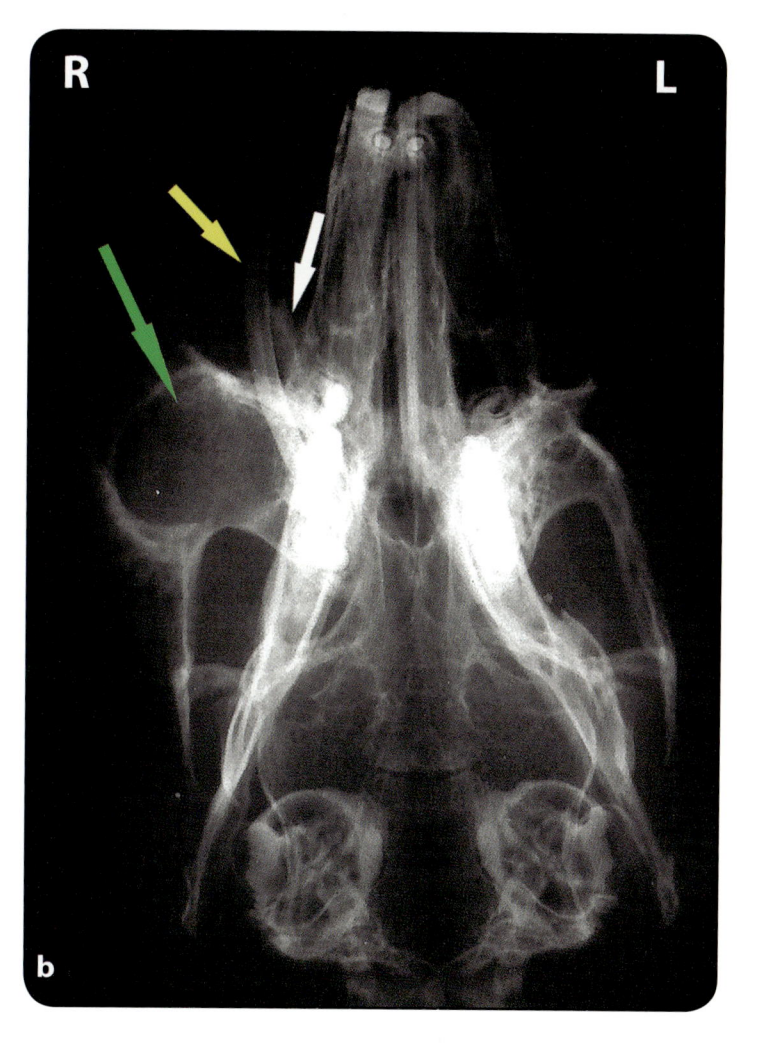

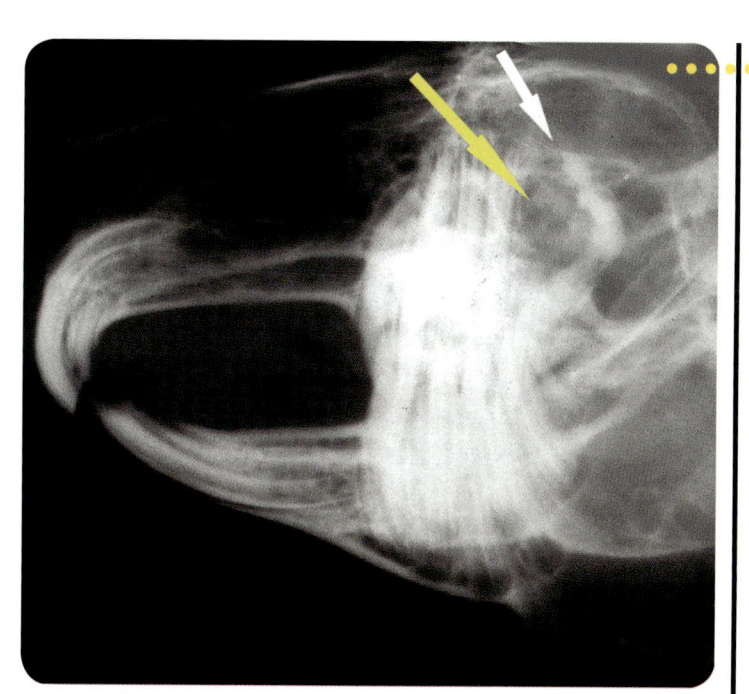

Figure 2.65. Left 30° ventral-right dorsal oblique radiograph of the skull of a rabbit with periapical infection and osteomyelitis (yellow arrow) following malocclusion and root deformity of CT5 (white arrow).

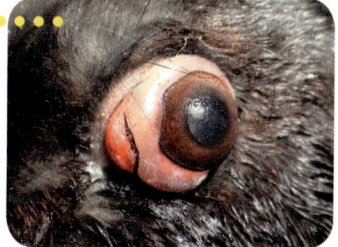

Figure 2.66. Periapical infections of maxillary cheek teeth frequently result in retrobulbar abscesses, as the caudal cheek teeth are positioned ventral to the orbit.

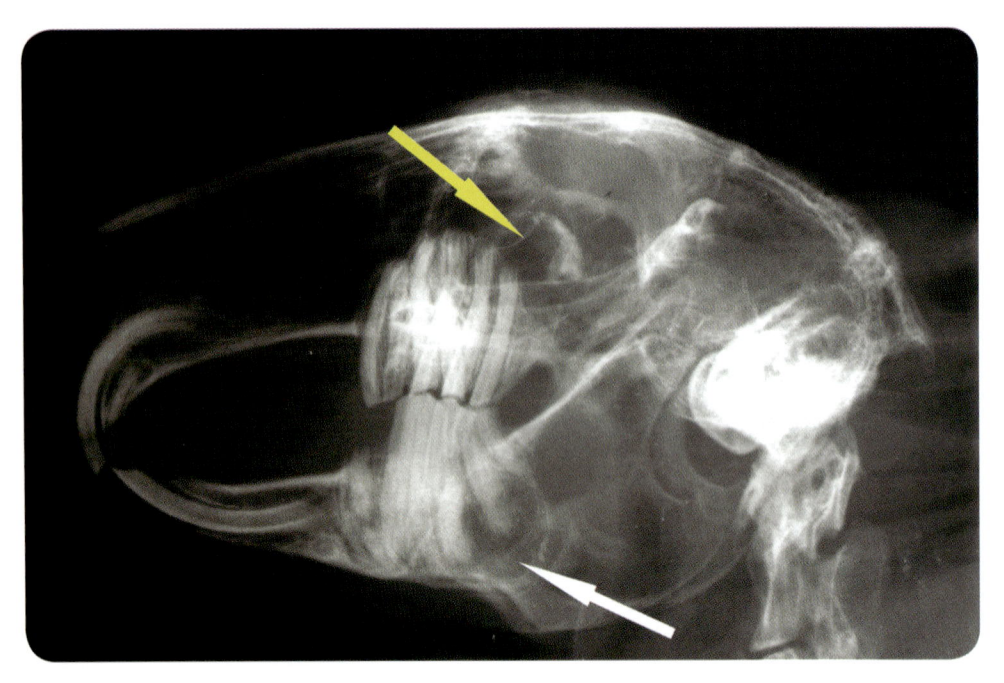

Figure 2.67. Lateral radiograph of the skull of a rabbit with ADD. Note radiographic evidence of two periapical infections: mandibular (white arrow) and maxillary (yellow arrow). Fracture of a mandibular CT1 and wave mouth are also visible. This single lateral projection does not distinguish right vs. left side involvement.

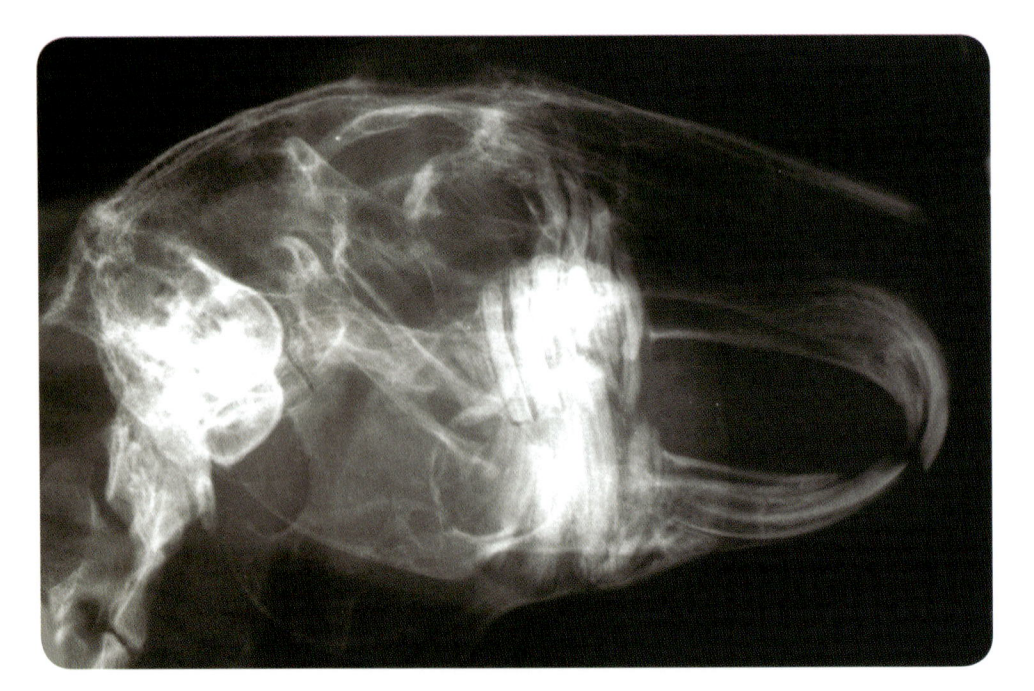

Figure 2.68. Right 15° ventral-left dorsal oblique projection of the same patient. As both osteomyelitic sites appear more dorsal than they appeared in the lateral projection, both must be located on the right side.

Miscellaneous

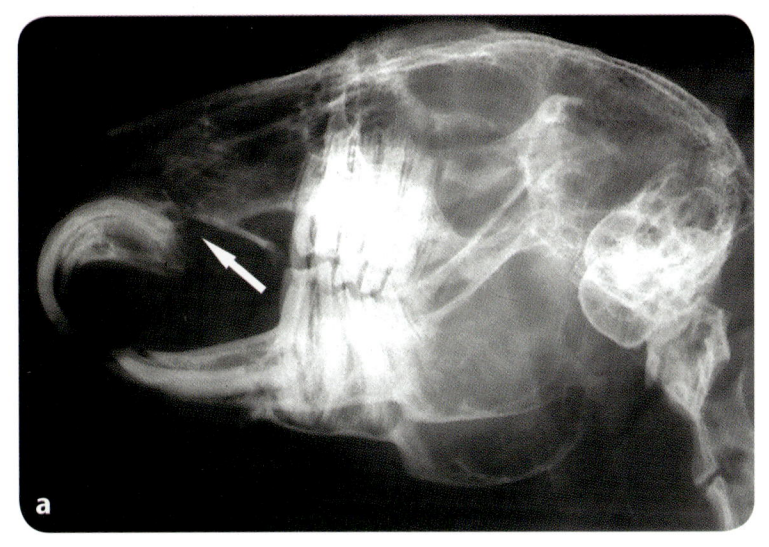

Figure 2.69a,b. Left 15° ventral-right dorsal (a) and ventrodorsal (b) projections of the skull of a rabbit with a complete fracture of incisive bones, with distraction of the bone fragments (arrows).

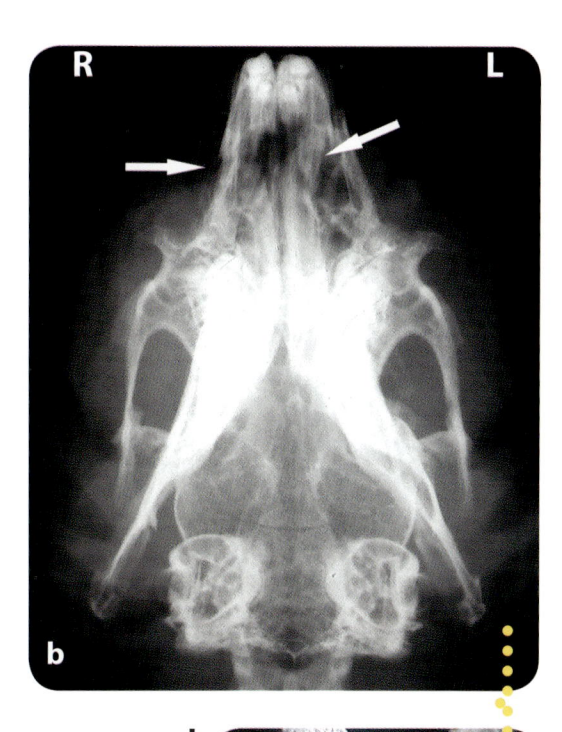

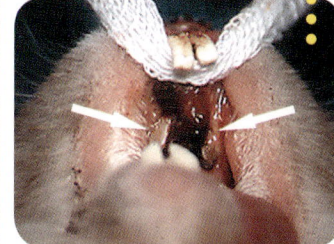

Figure 2.71. Bilateral open fracture of the incisive bones.

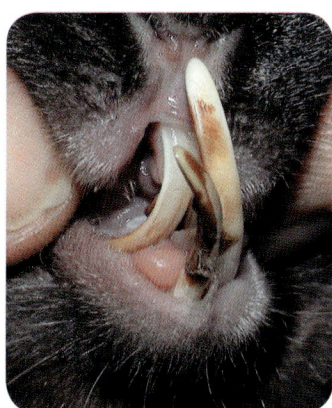

Figure 2.72. Overgrowth and malocclusion of incisor teeth can predispose to fracture of incisive bones. This typically occurs when incisors are caught on cage bars in the enclosure.

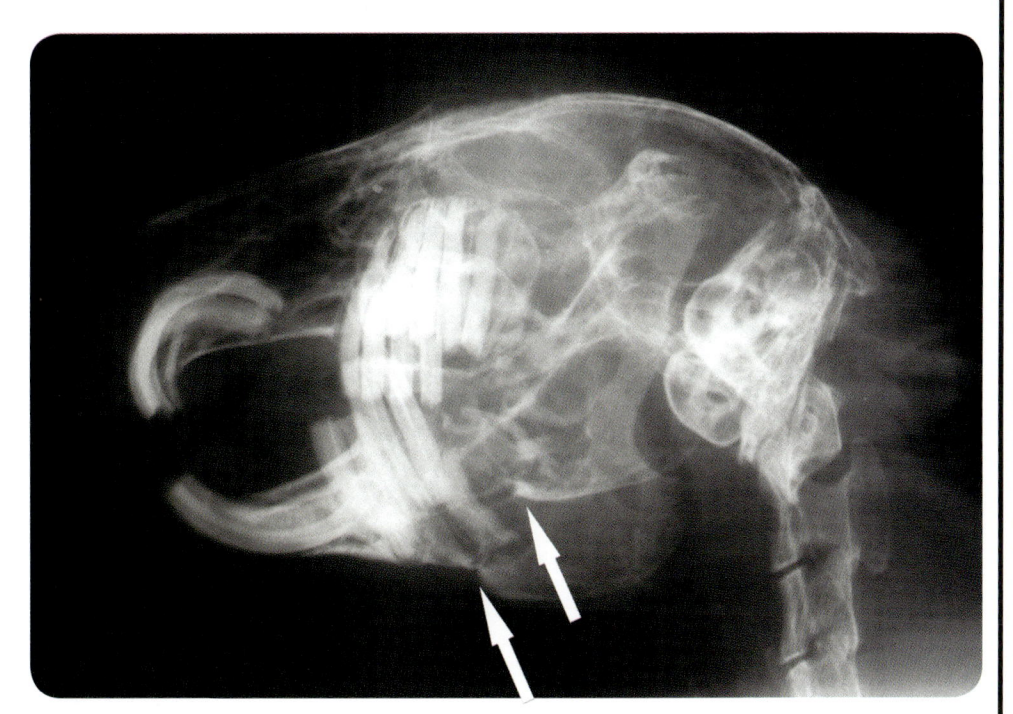

Figure 2.70. Left 20° ventral-right dorsal oblique radiograph of the skull of a rabbit with bilateral fractures of the mandible (arrows) following bilateral periapical infection and osteomyelitis of mandibular CT4 and CT5. The rabbit survived with conservative therapy and assisted feeding.

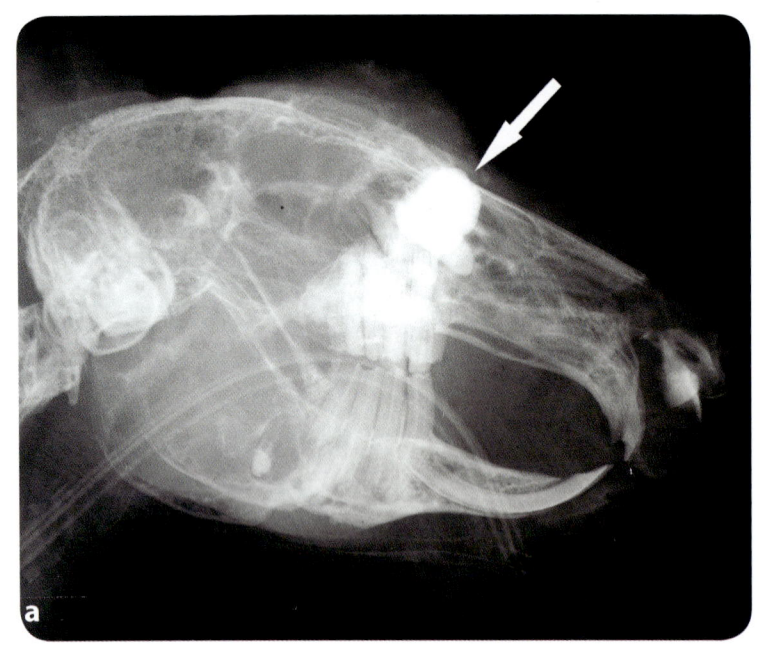

Figure 2.73a,b. Lateral (a) and ventrodorsal (b) radiographs of the skull of a rabbit with ADD demonstrating rupture of the left lacrimal sac (white arrows), and dilation of the right lacrimal sac (yellow arrow). Due to abnormalities of both lacrimal sacs, the nasolacrimal ducts are not clearly visible. However, passage is demonstrated by presence of contrast medium at both distal (nasal) openings. See also Figure 2.13-2.15.

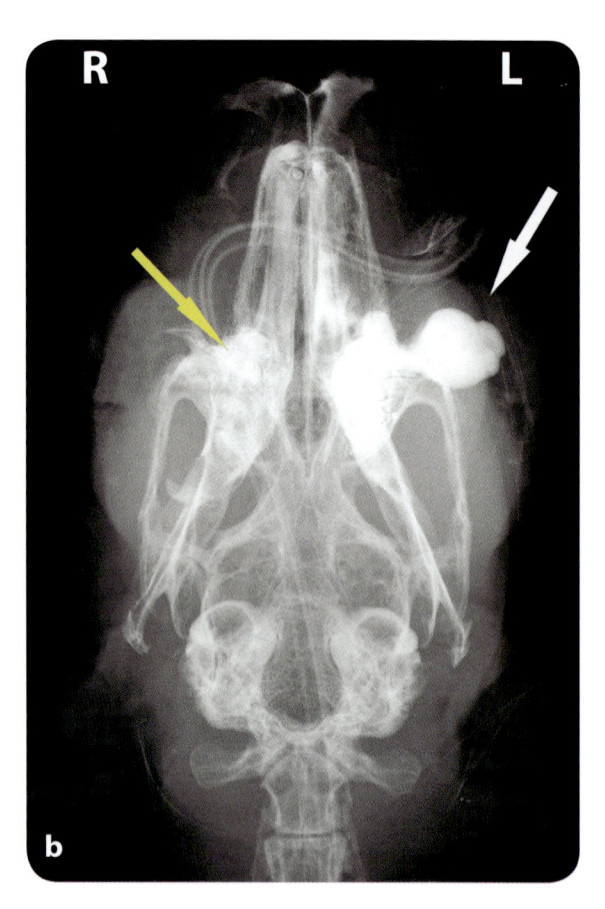

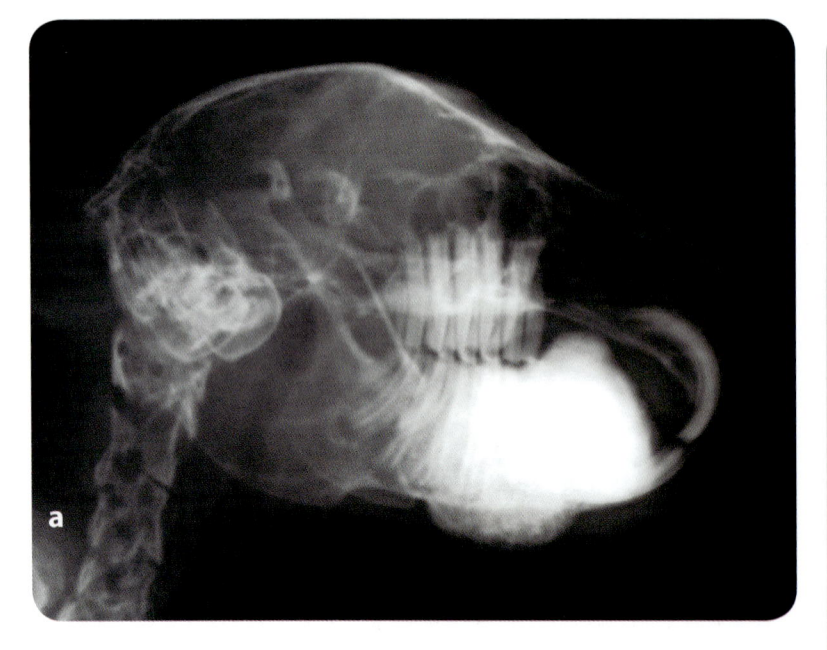

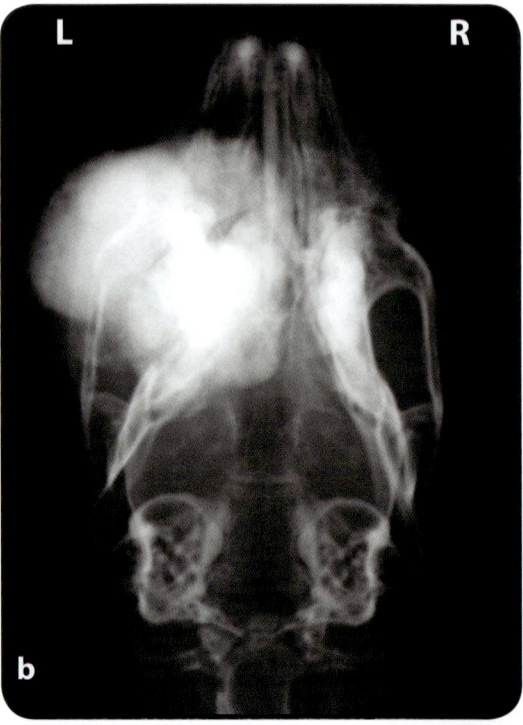

Figure 2.74a,b. Lateral (a) and ventrodorsal (b) radiographs of the skull of a rabbit showing a large smoothly marginated mineralized mass arising from the left mandible. After partial hemimandibulectomy, the mass was determined to be a cementoma. *(Courtesy of Yasutsugu Miwa, DVM)*

HEAD   75

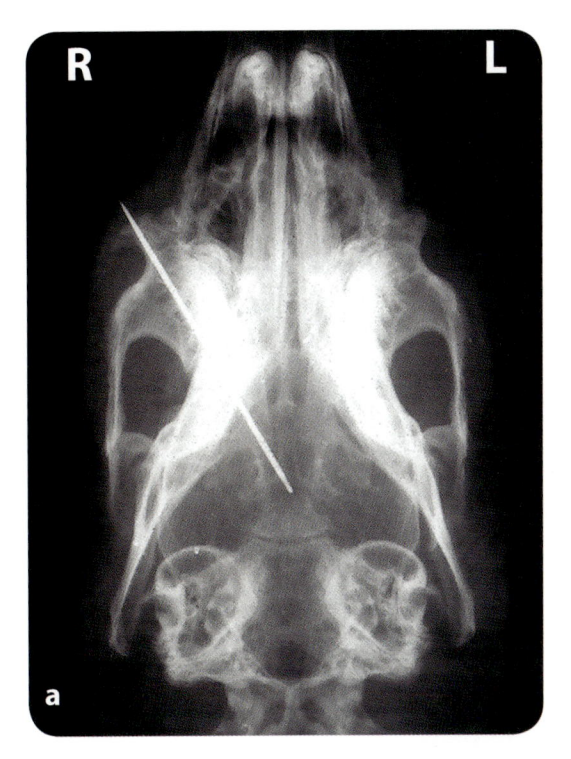

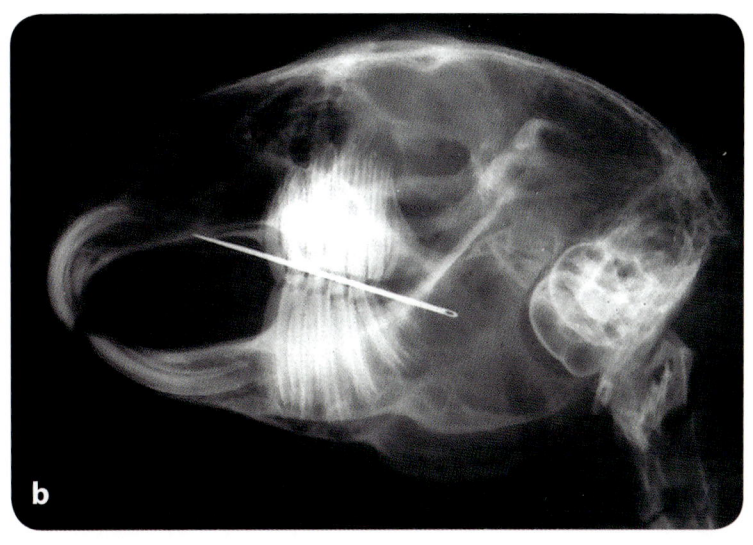

Figure 2.75a,b. Ventrodorsal (a) and left 15° ventral-right dorsal oblique (b) radiographs demonstrating an unusual foreign body in a pet rabbit, obviously identified as a sewing needle.

Figure 2.77. Otitis externa can progress to involve the middle ear. A large amount of thick purulent material is visible at the level of the ventral ear canal.

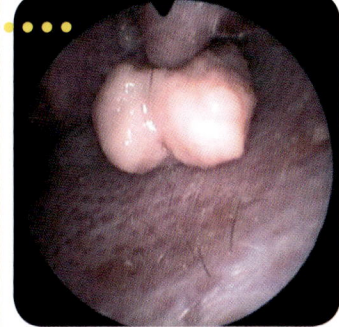

Figure 2.78. Endoscopic appearance of severe purulent otitis media.

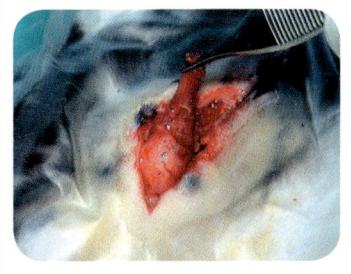

Figure 2.79. Total ear canal ablation and ostectomy of the tympanic bulla is the treatment of choice for infections not responsive to medical therapy.

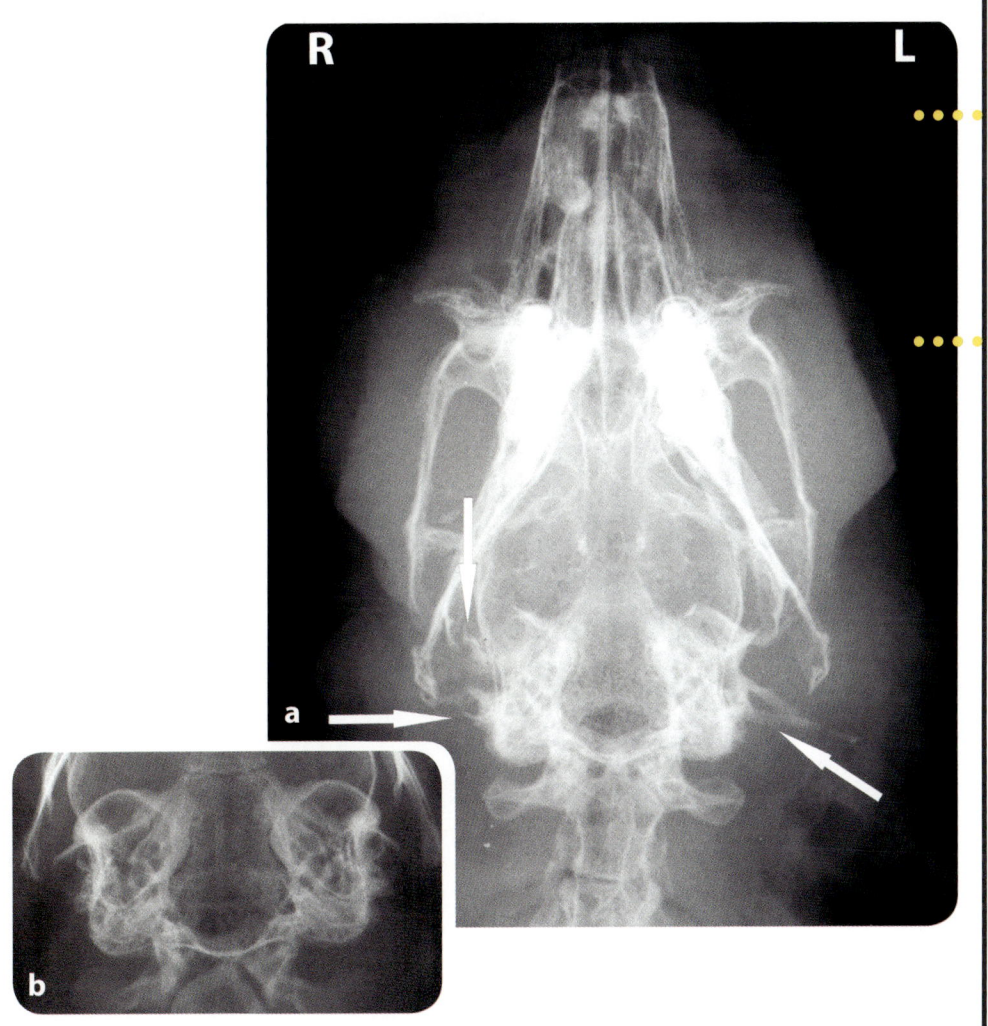

Figure 2.76a,b. Ventrodorsal projection of a rabbit with bulla osteitis (a). Notice thickening and increased opacity of the osseous bullae (arrows). An exactly symmetrical ventrodorsal projection is essential for proper evaluation.
Close up of normal tympanic bullae are shown in (b).

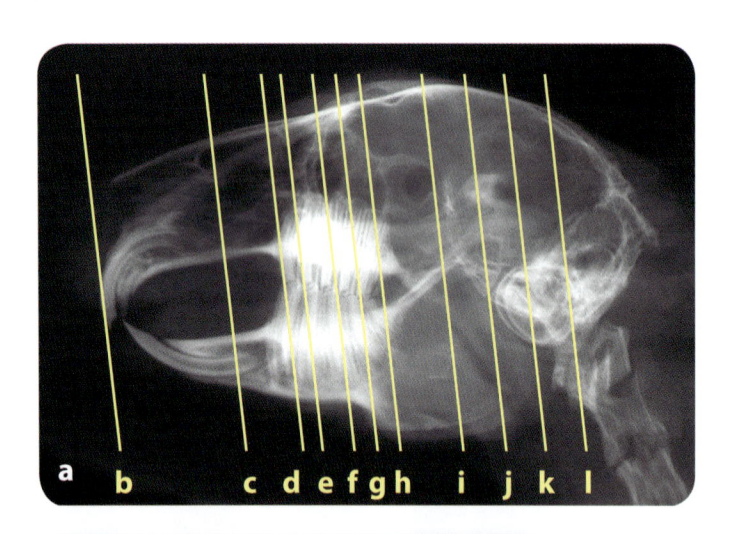

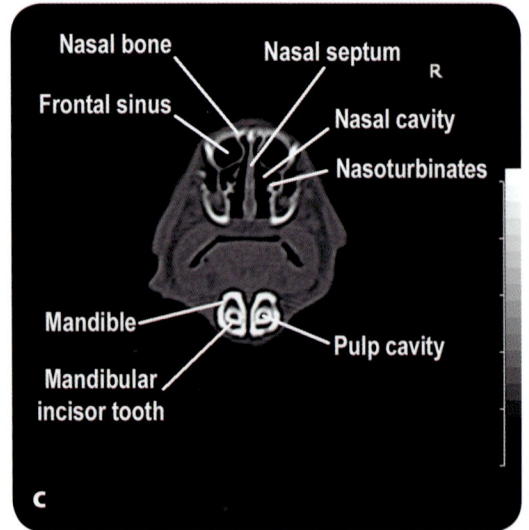

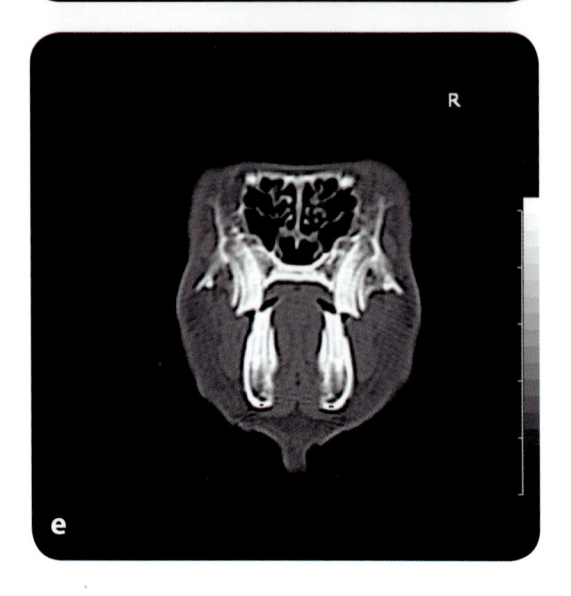

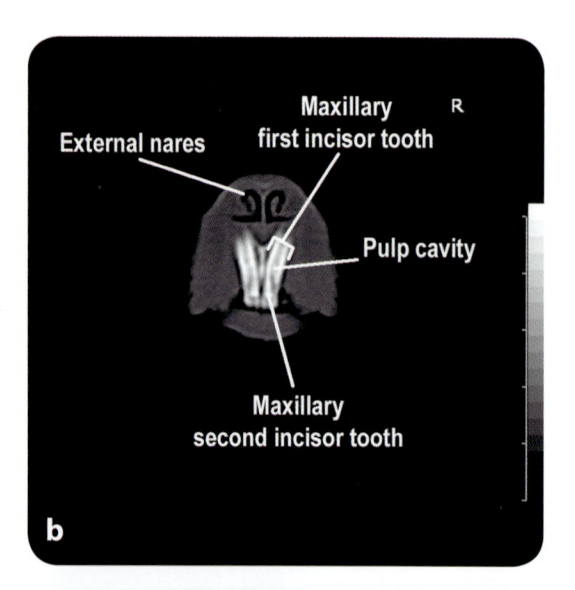

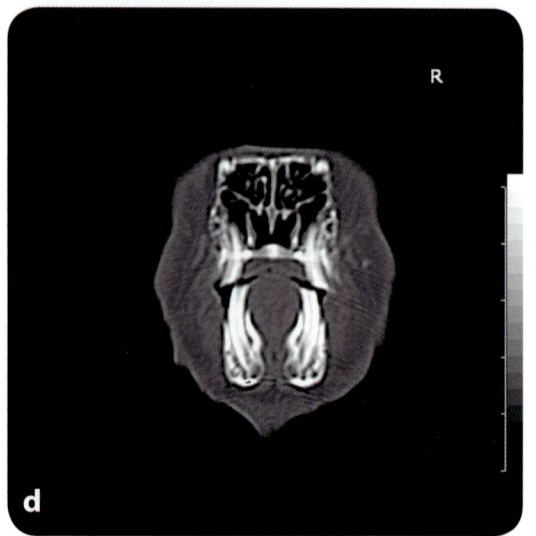

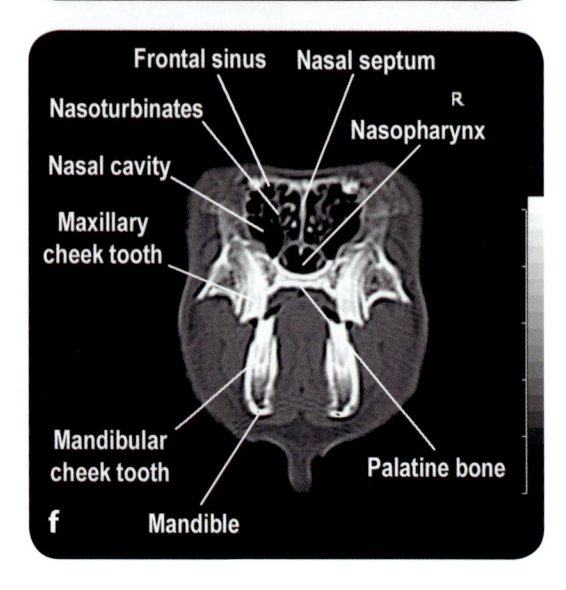

76 RABBIT

COMPUTED TOMOGRAPHY of the HEAD

V. Capello, A. Cauduro, A. Lennox

The Normal Head

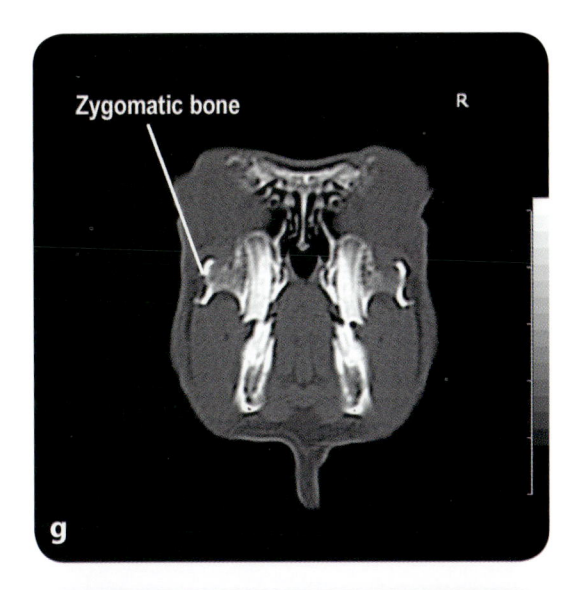

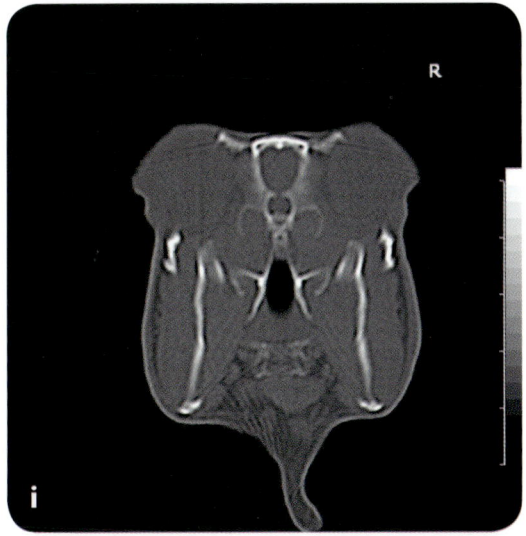

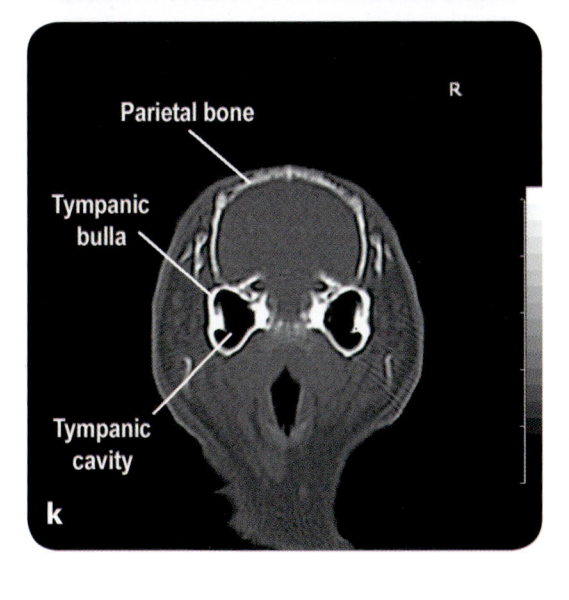

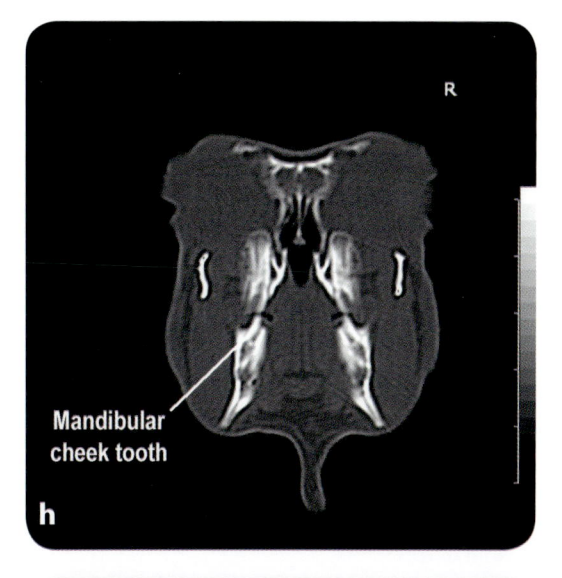

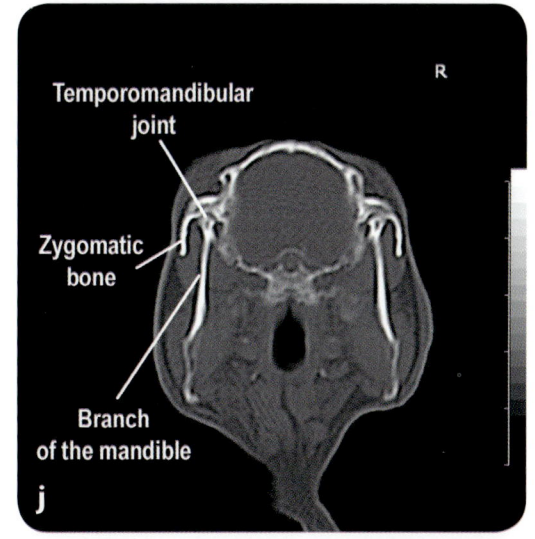

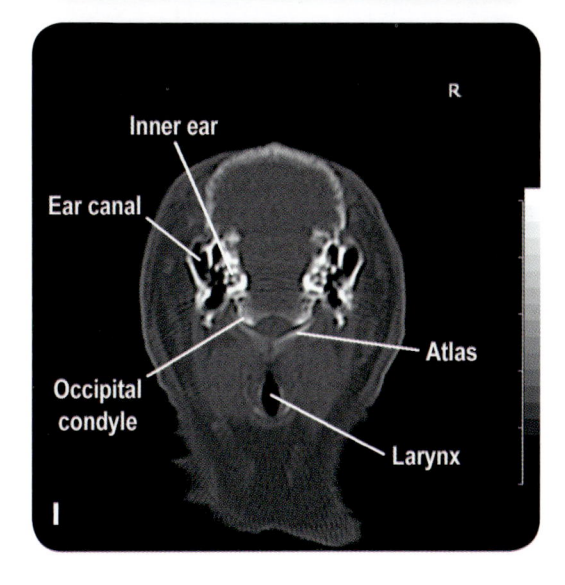

Figure 2.80a-l. Computed tomography of the normal skull of a 1.8 kg rabbit, bone window.
Scanning Parameters: scan speed: 1 sec.; mA: 125; kVp: 120; slice thickness: 3 mm.; WL: 300; WW: 1500; Image size (matrix): 512x512. Selected views from a series of 52.
The scout view demonstrating the scanning angles (a) has been adapted from a radiograph of the normal skull for demonstration purposes.

78 RABBIT

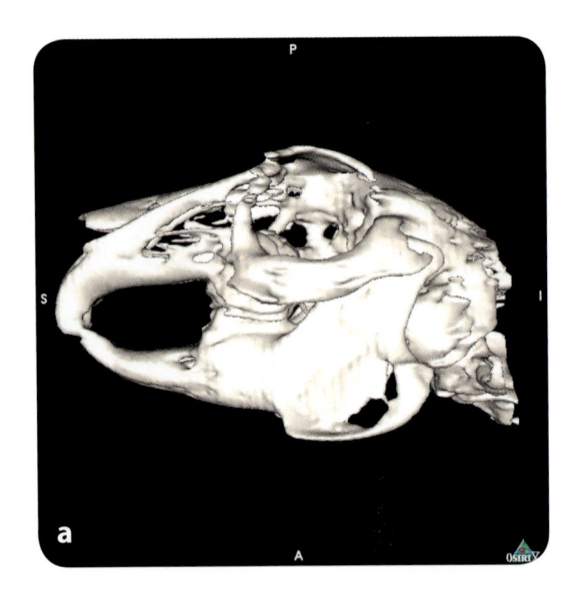

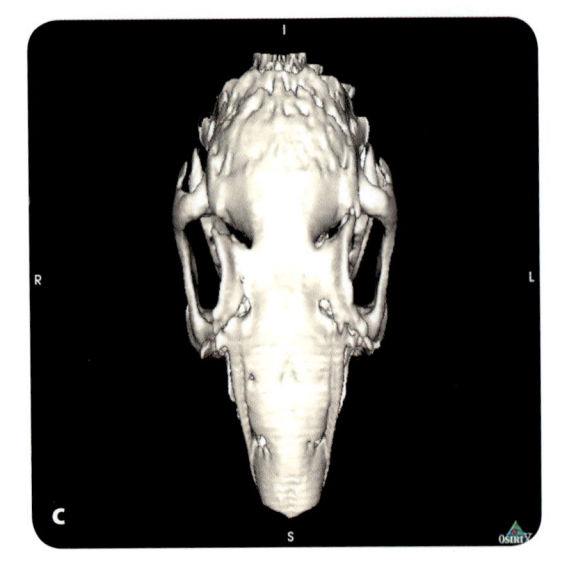

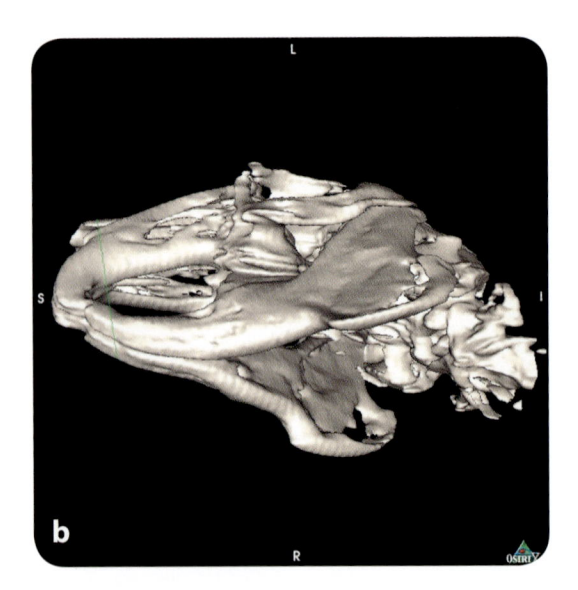

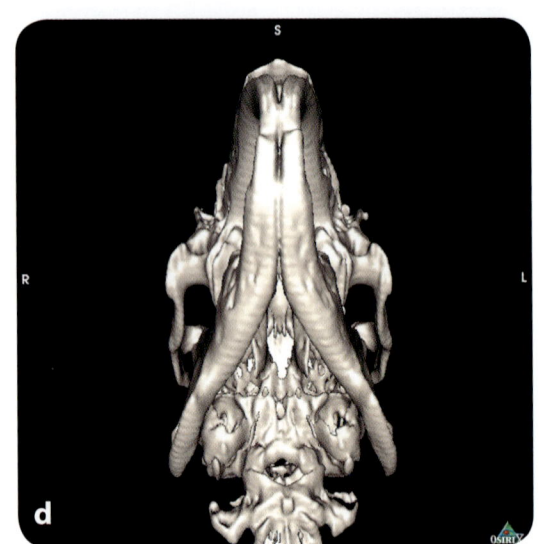

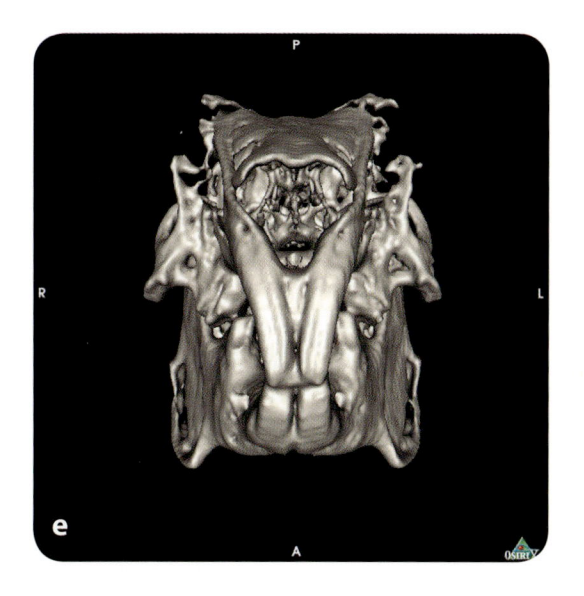

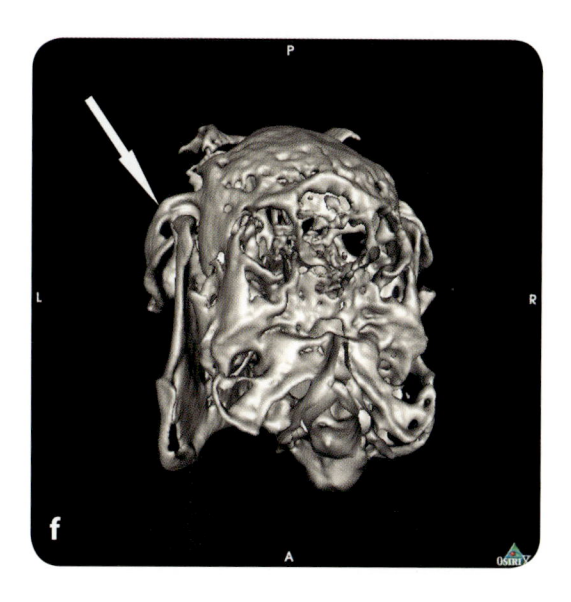

Figure 2.81a-f. 3D surface reconstruction (Shaded Surface Display - SSD) of the normal skull of a 1.8 kg rabbit. The caudal view (f) is slightly oblique to emphasize the temporomandibular joint (arrow).

Acquired Dental Disease, Periapical Infection, and Osteomyelitis

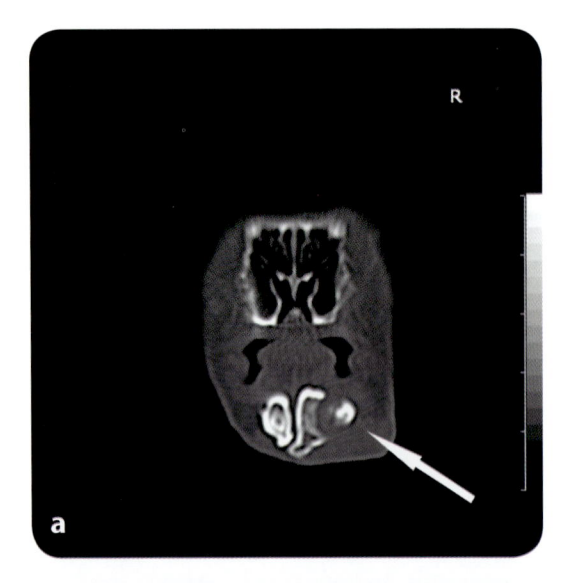

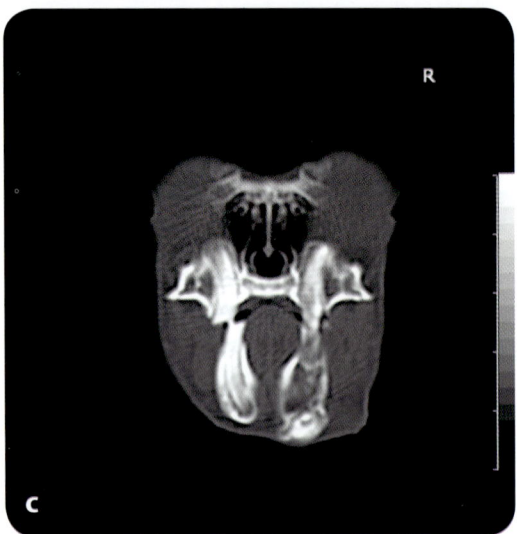

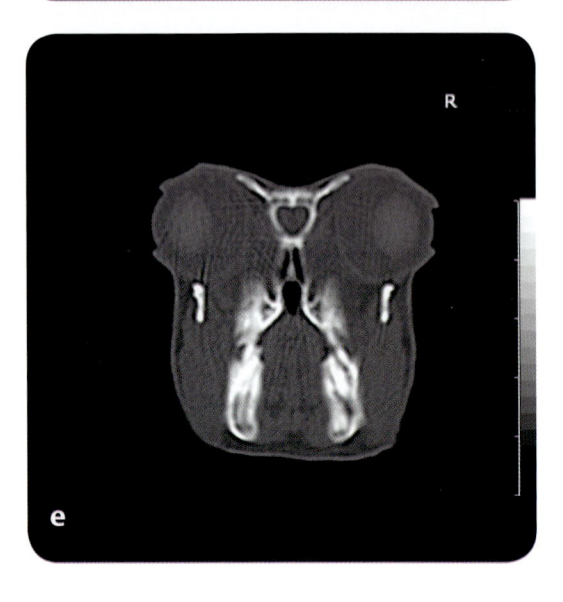

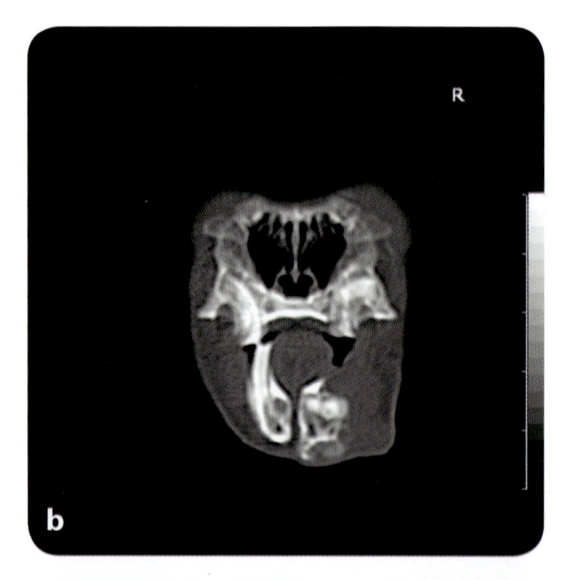

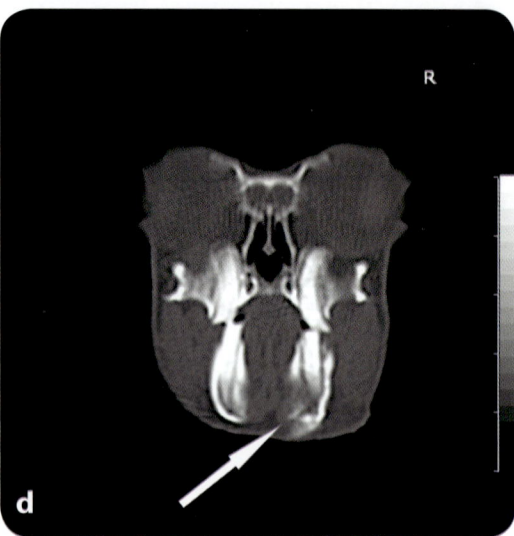

Figure 2.82a-e. Osteomyelitis of the right mandible following periapical infection of the omolateral incisor tooth, bone window.
Scanning Parameters: scan speed: 1 sec.;
mA: 125; kVp: 120; slice thickness: 3 mm.;
WL: 300; WW: 1500;
Image size (matrix): 512x512.
Fragment of the apex of the incisor tooth (a, arrow)
Comparison between the normal and the diseased body of the mandible is visible (a, b, c).
A medial fistula is present at the level of right mandibular CT3 (d, arrow).
The comparison between the normal right and left mandibular CT3 (e) helps visualize the extension of the osteomyelitic site. A slight "translated plane" effect is present in (e) (see Figure 1.101), therefore both mandibular CT3 and CT4 are partially visible.

Figure 2.83a-d. SSD of the skull in the same patient. The osteomyelitic body of the mandible is visible (a) as well as the apex and fragmented reserve crown of the right mandibular incisor tooth (arrow). The normal left mandible is shown in (b). The comparison between the normal and the diseased mandible is also visible from the rostral (c) and ventral (d) view.

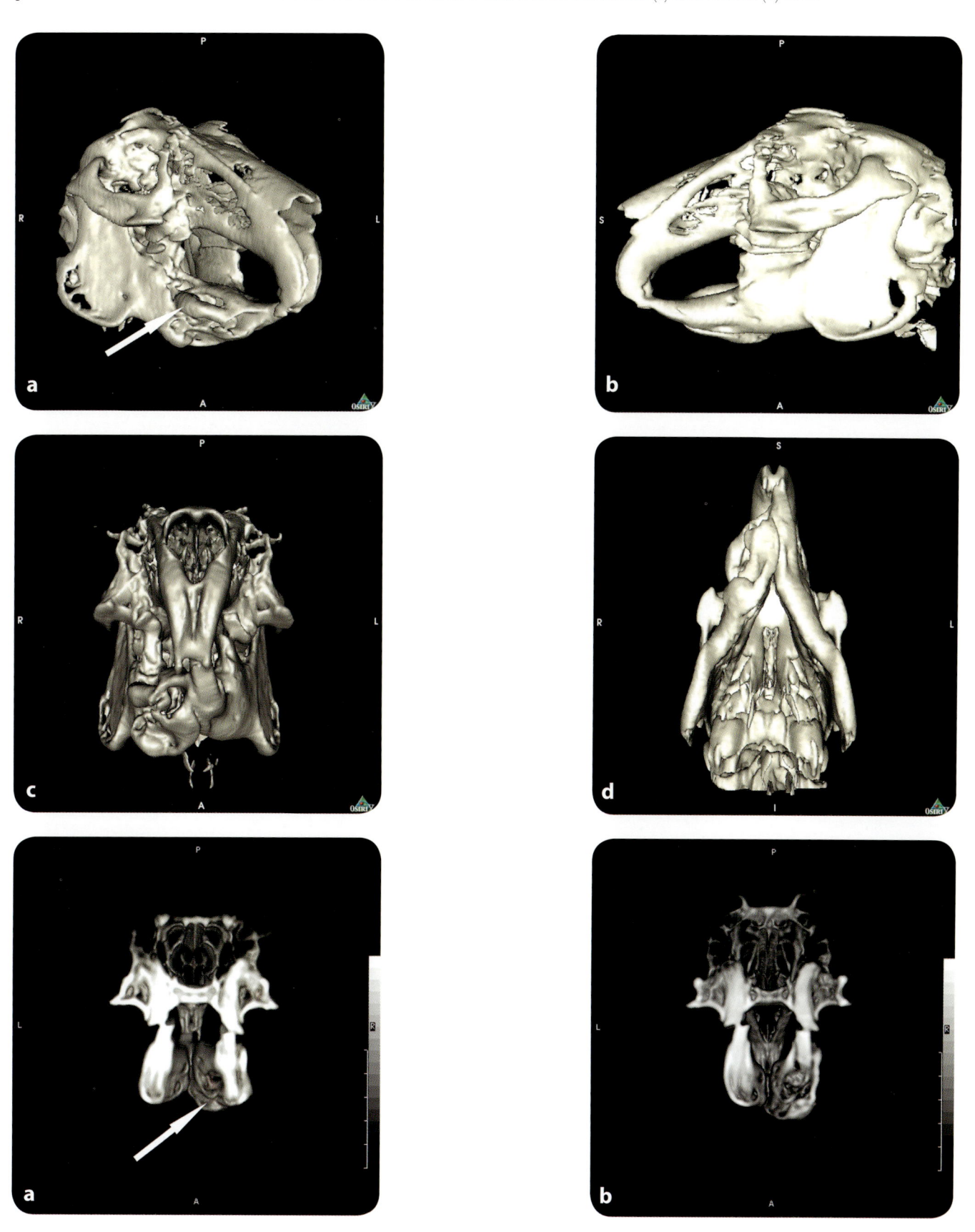

Figure 2.84a,b. 3D volume reconstructions, cropped caudal view (see Figure 1.106.). Cross sections help to understand the extension of osteomyelitis. Both SSDs and volume reconstructions help detect the presence of fistulas opening medially (a) (arrow). Prognosis for periapical infection, abscess, and osteomyelitis is guarded when medial intermandibular tissues are affected.

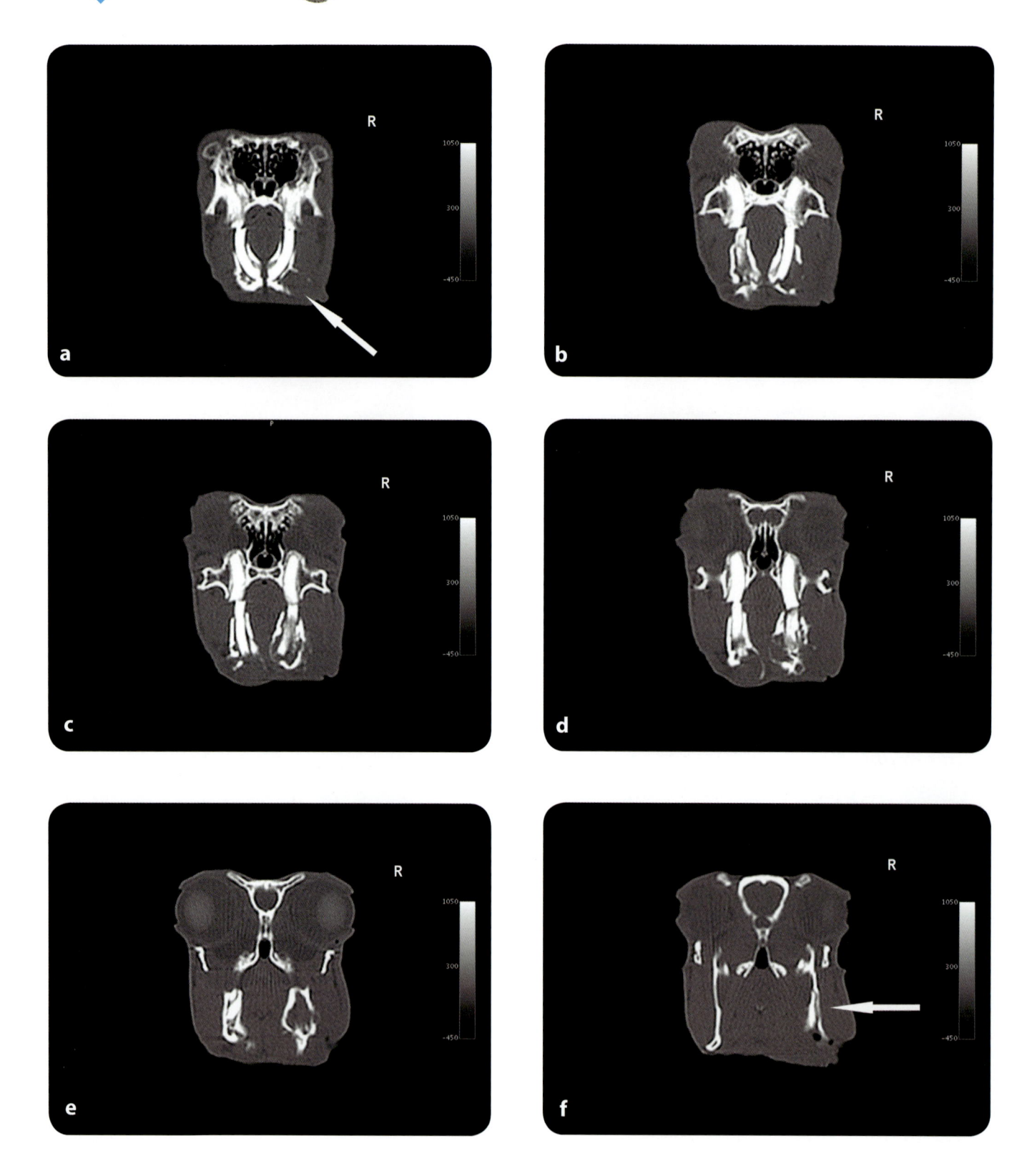

Figure 2.85a-f. CT of the skull of a 1.5 kg 2-year-old rabbit with bilateral mandibular periapical infection and osteomyelitis, bone window. Scanning Parameters: mAs: 130; kVp: 120; slice thickness: 1 mm.; WL: 300; WW: 1500; Image size (matrix): 512x512.

Cross section at mandibular CT1 (a) shows excessive elongation and abnormal curvature of both clinical and reserve crowns, and abnormal medial deviation of the apexes. Cortical bone of the ventral right mandible is missing (arrow). Bilateral mandibular periapical infection and osteomyelitis is visible in (b), (c), and (d). Left mandibular CT4 is normal, while osteomyelitis is still present on the right side at this level (e). A radiolucent line suggestive of a fracture of the branch of the mandible (arrow) is visible in (f).

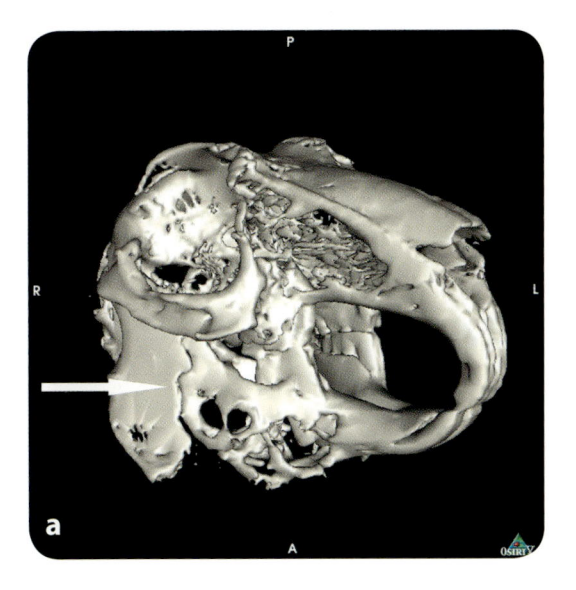

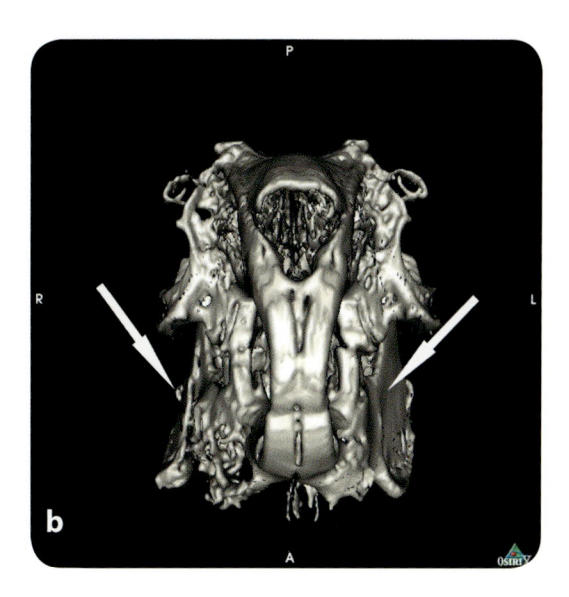

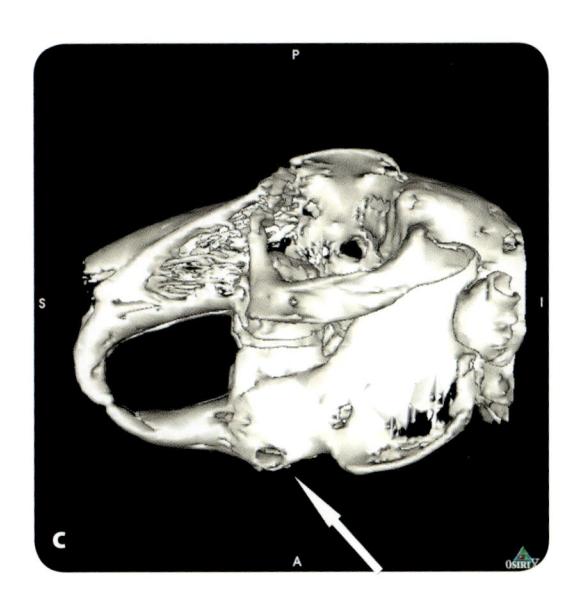

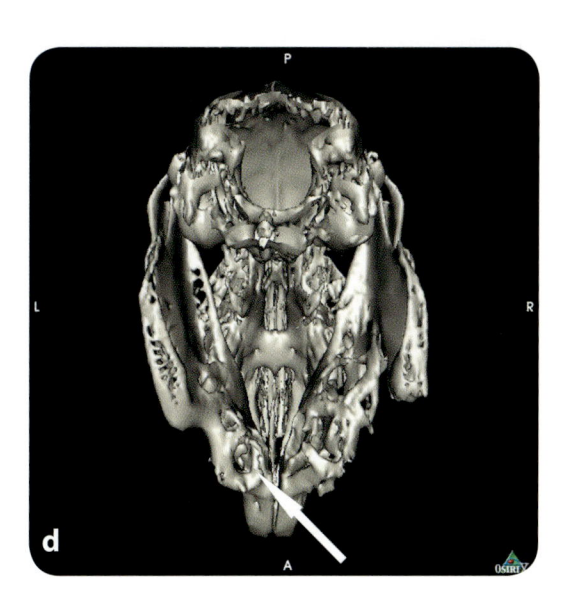

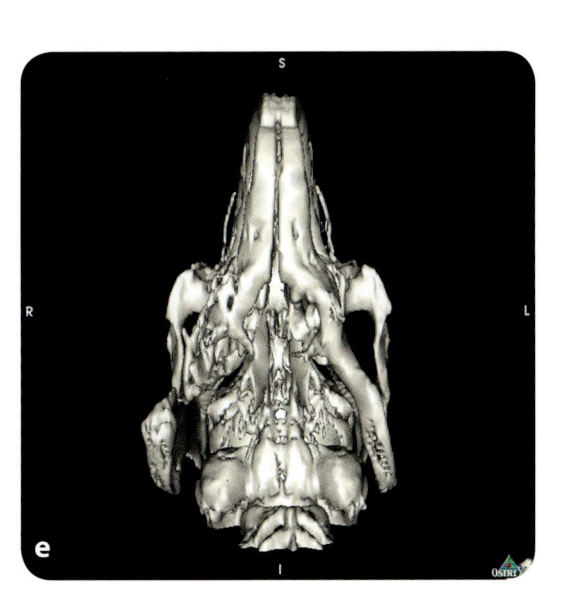

Figure 2.86a-e. SSD of the same rabbit in Figure 2.85 with osteomyelitis of both mandibles.
(a) shows bone loss extending along the right mandibular body to the ramus. A vertical fracture of the ramus is also visible (arrow). A comparison between the right and left hemimandibles is visible from a rostral perspective (b, arrows). Bony deformity and bone loss (arrow) are present in the left mandible as well (c). From oblique caudal (d) and ventral perspectives (e), the right mandible appears most severely affected. An area of bone loss consistent with a fistula is seen along the medial aspect of the left mandible (d, arrow). 3D surface reconstruction can provide additional information regarding topography which can aid in surgical planning. However, SSDs only depict the surface of structures and have a limited scale of contrast. Computed tomography using various windows and image reconstructions are superior to conventional radiographic images of the head, which are subject to summation and have limited contrast.

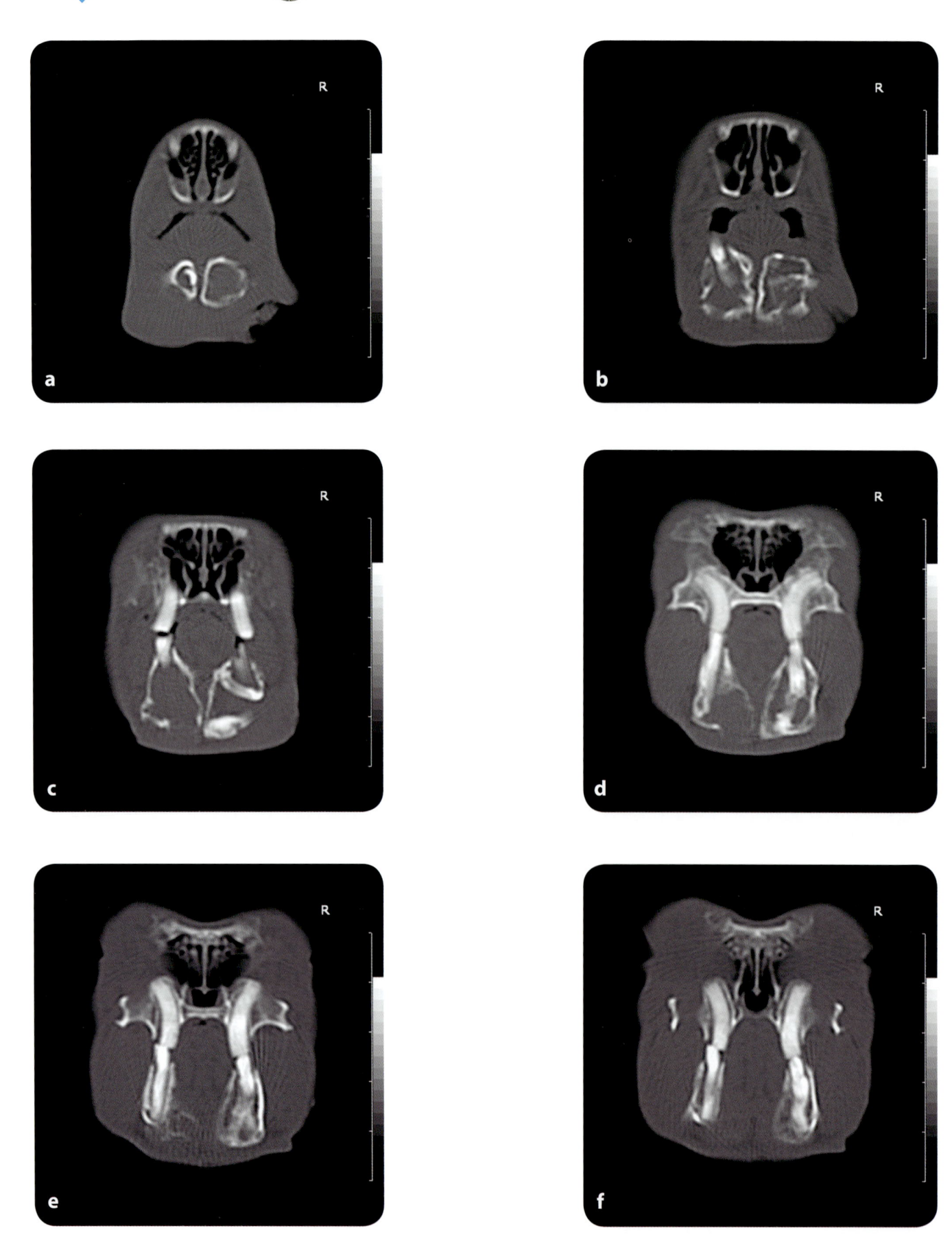

Figure 2.87a-f. CT of the skull of a 1.3 kg 2-year-old rabbit with bilateral mandibular periapical infection and osteomyelitis, bone window.
Scanning Parameters: scan speed: 1 sec.; mA: 125; kVp: 120; slice thickness: 3 mm.; WL: 300; WW: 1500; Image size (matrix): 512x512. Osteomyelitis of the right mandible is more extensive than on the left side.

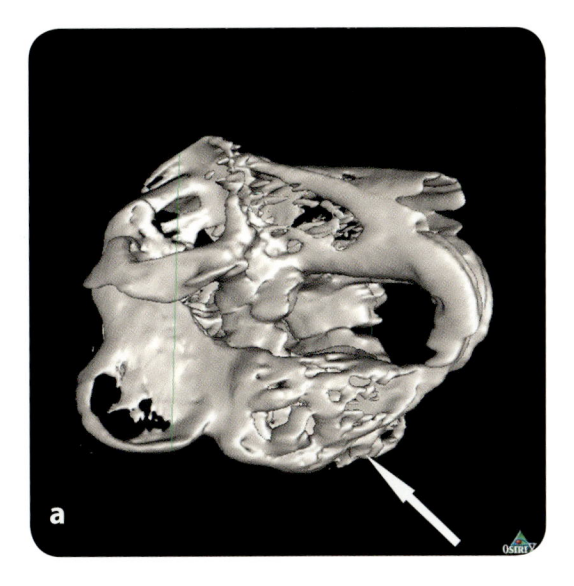

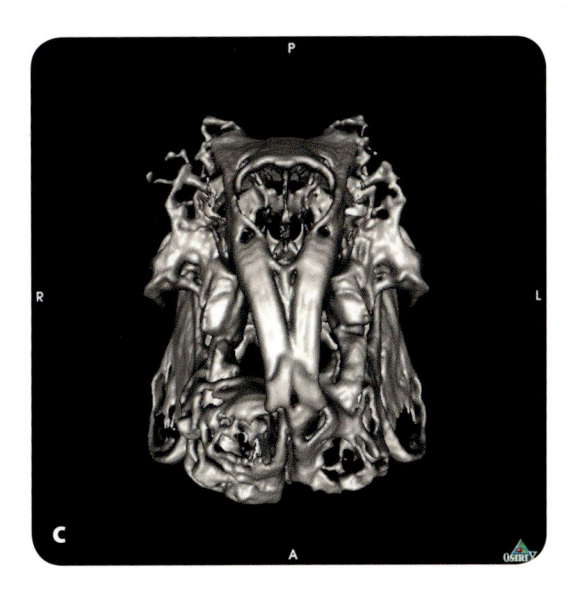

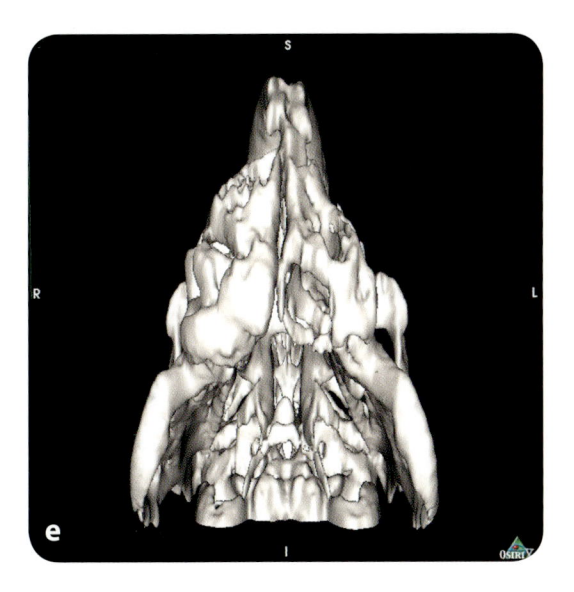

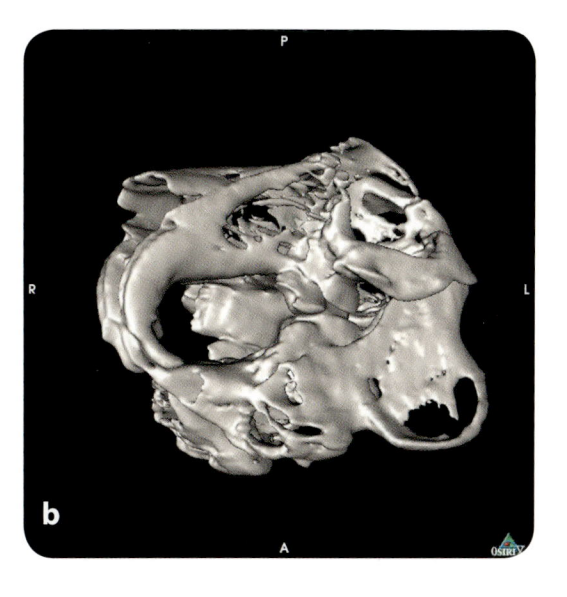

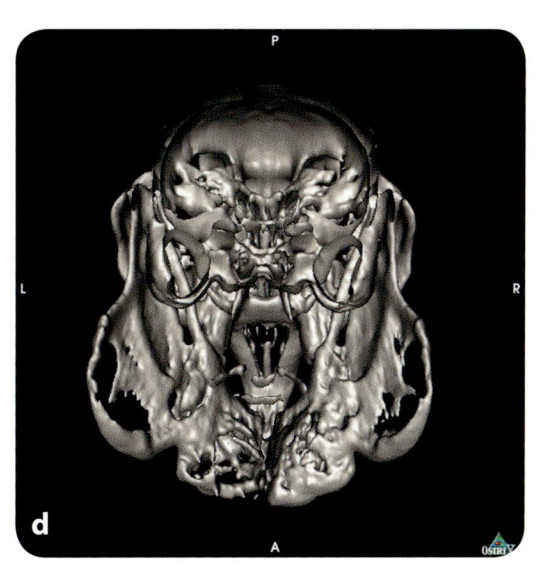

Figure 2.88a-e. SSD of the same rabbit in Figure 2.87 with osteomyelitis of the mandibles.
a) There are areas of bony lysis of the right mandible, including the incisive portion (arrow). Normal bone tissue can still be identified between the osteomyelitic site and the masseteric fossa. Bone deformity and an osteomyelitic site are also present, but to a lesser degree, in the left mandible (b). Comparison between right and left osteomyelitic sites is shown in the rostral (c), caudal (d) and ventral (e) views.

This rabbit underwent successful right hemimandibulectomy, and deep surgical debridement of the left osteomyelitic site.

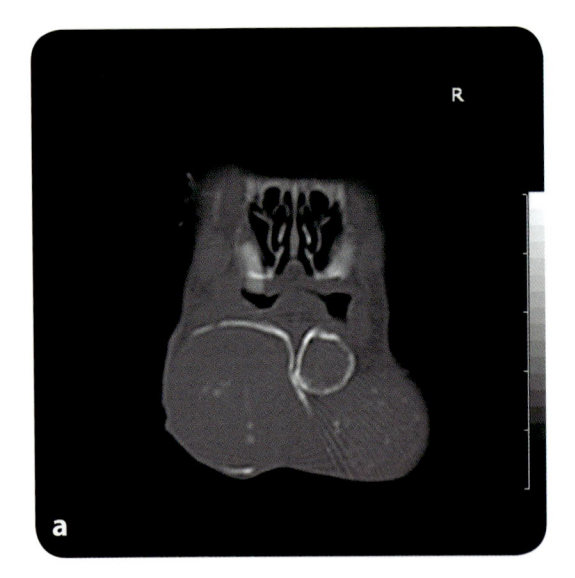

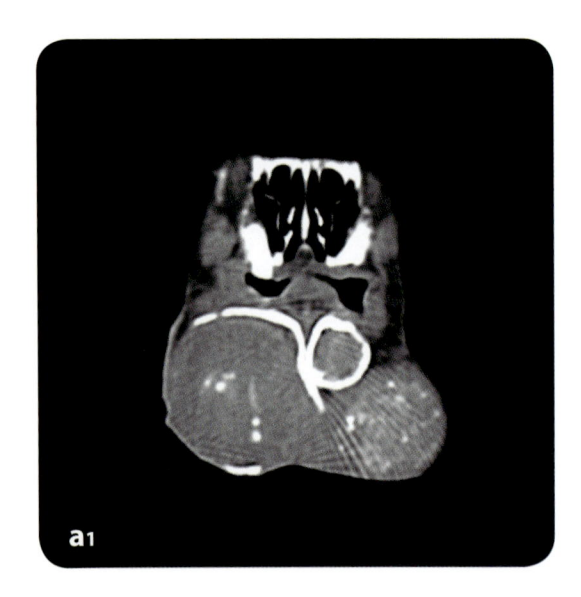

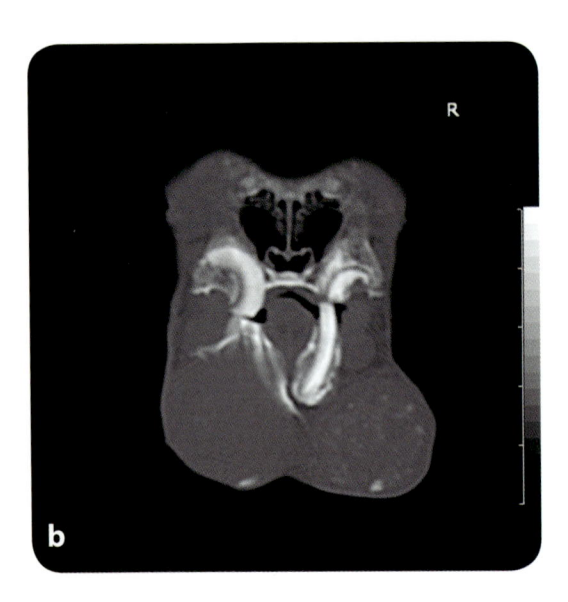

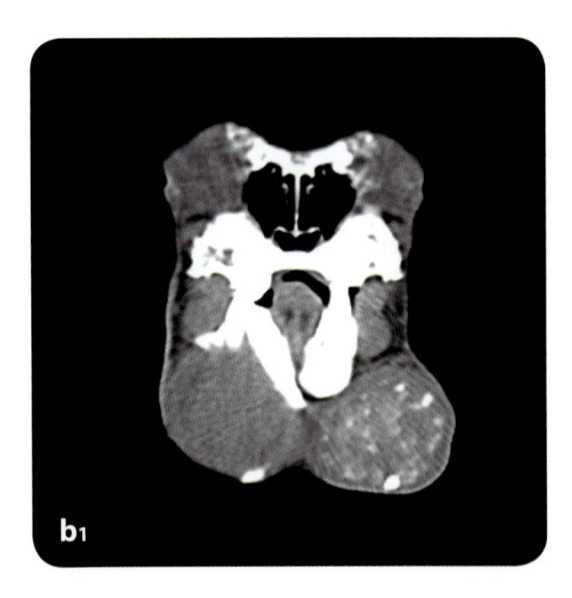

Figure 2.89a-f. CT of the skull of a rabbit with end-stage bilateral osteomyelitis of the branches of the mandibles following fracture of the mandibular incisor teeth; bone and soft tissue windows compared.
Scanning parameters: scan speed: 1 sec.; mA: 125; kVp: 120; slice thickness: 3 mm.; WL: 300; WW: 1500 (bone window); WL: 40; WW: 350 (soft tissue window); Image size (matrix): 512x512.
Complete alteration of the normal anatomical structures, bilateral abscessation, overgrowth and excessive curvature of mandibular and maxillary cheek teeth are visible in (a), (b), and (c). Normal dental and bone anatomy is present only at the level of CT4 (d). Severe osteomyelitis and periosteal reaction are visible in the SSDs, rostral (e) and ventral (f) view. The soft tissue windows (a1-d1) and double SSD (f) show the anatomical relationship between the larger abscesses and the osteomyelitic mandibles.

HEAD 87

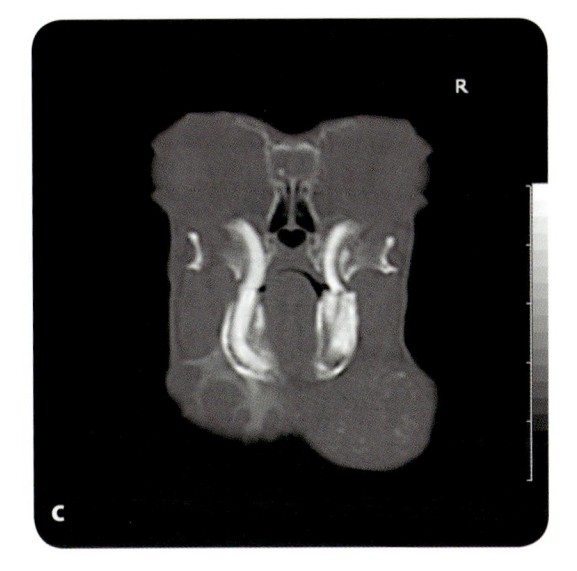

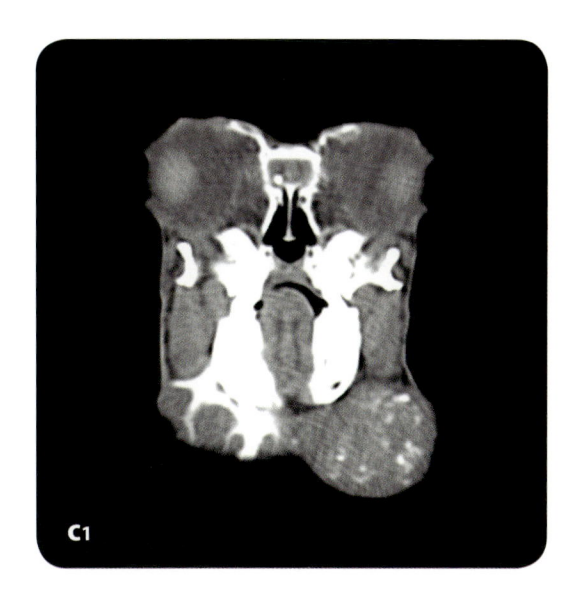

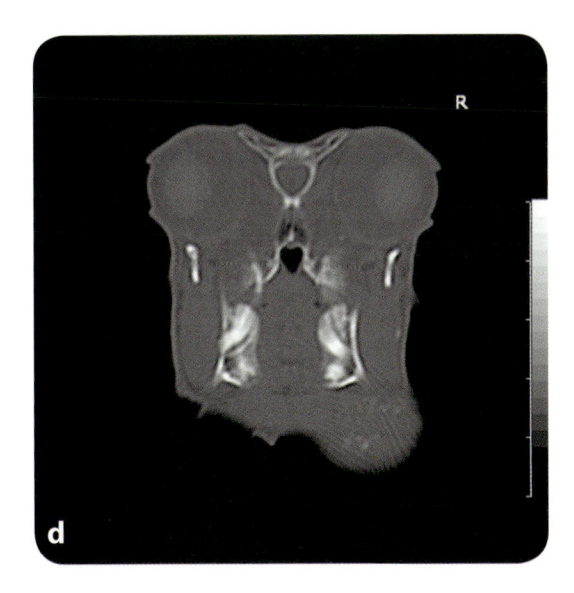

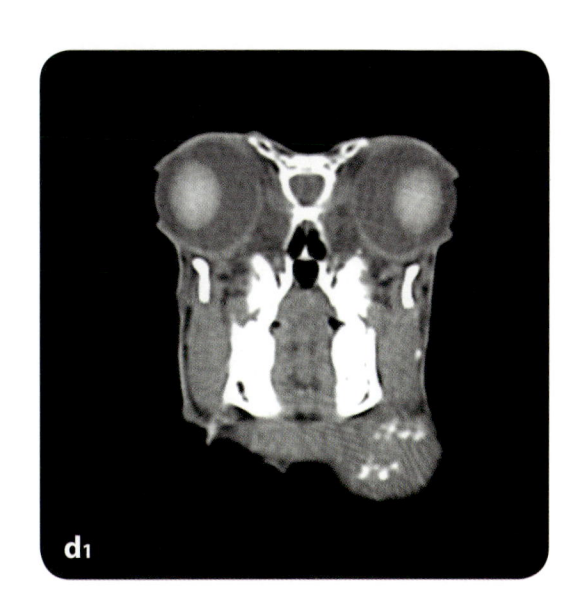

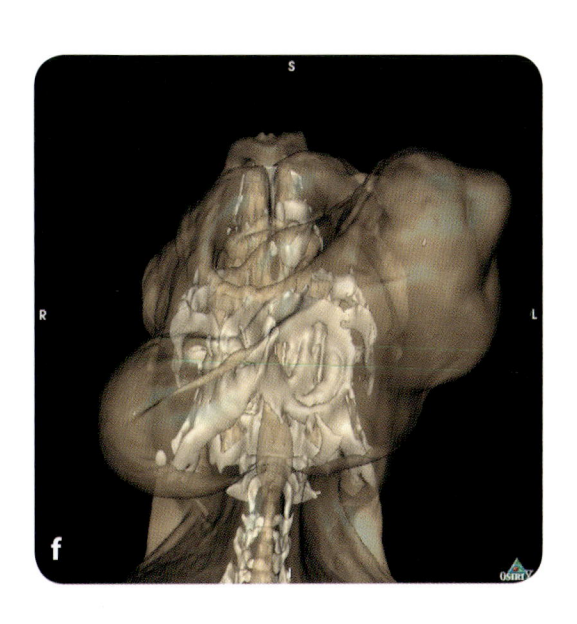

c

c1

d

d1

e

f

The NORMAL TOTAL BODY
Lateral Projection

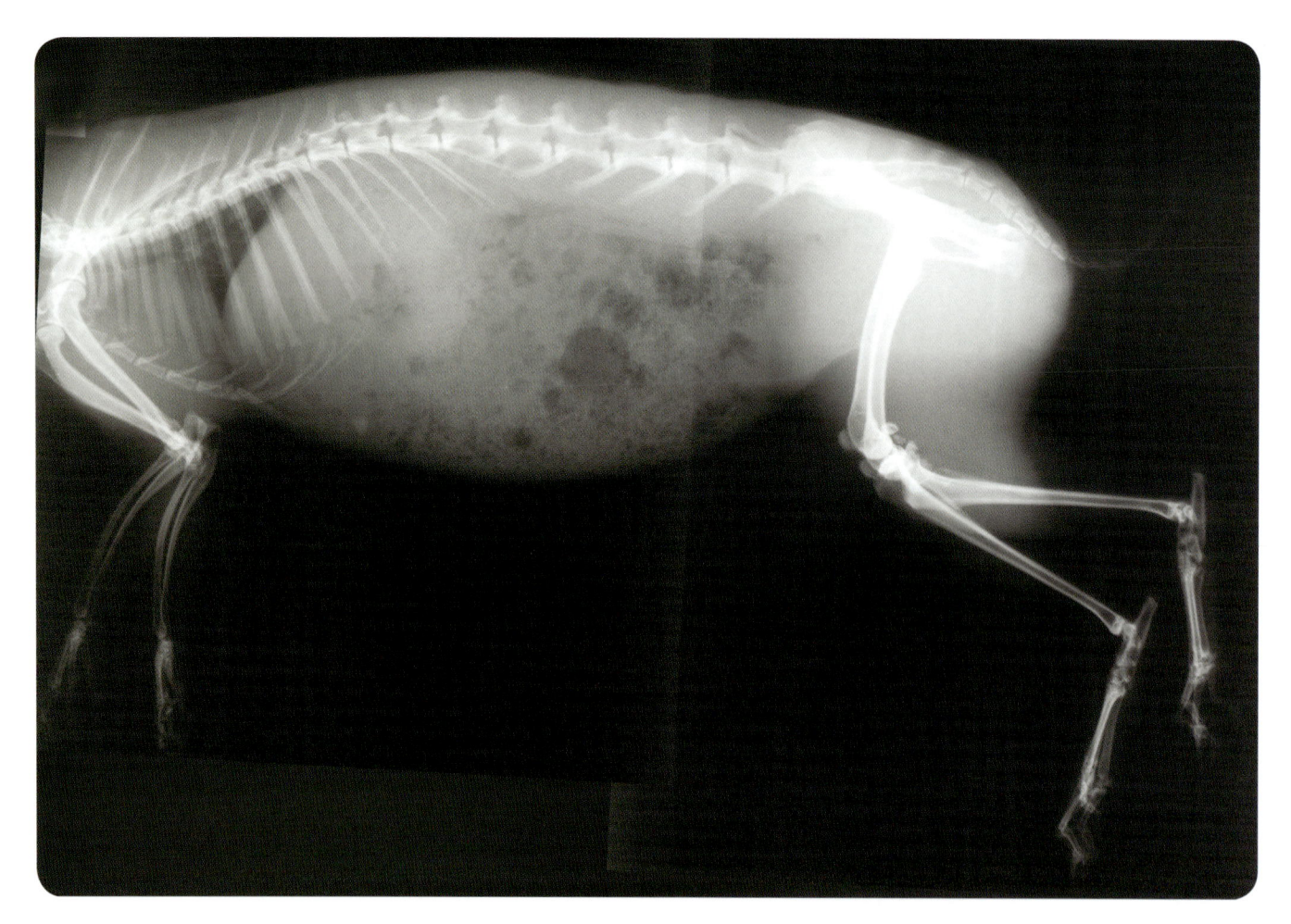

Figure 2.90. Total body radiograph of a 1.9 kg female dwarf rabbit, 50% of actual size. It is generally recommended to focus on the area of interest to optimize radiographic settings and quality. Note the small size of the thorax in comparison to the abdomen in this species.

Ventrodorsal Projection

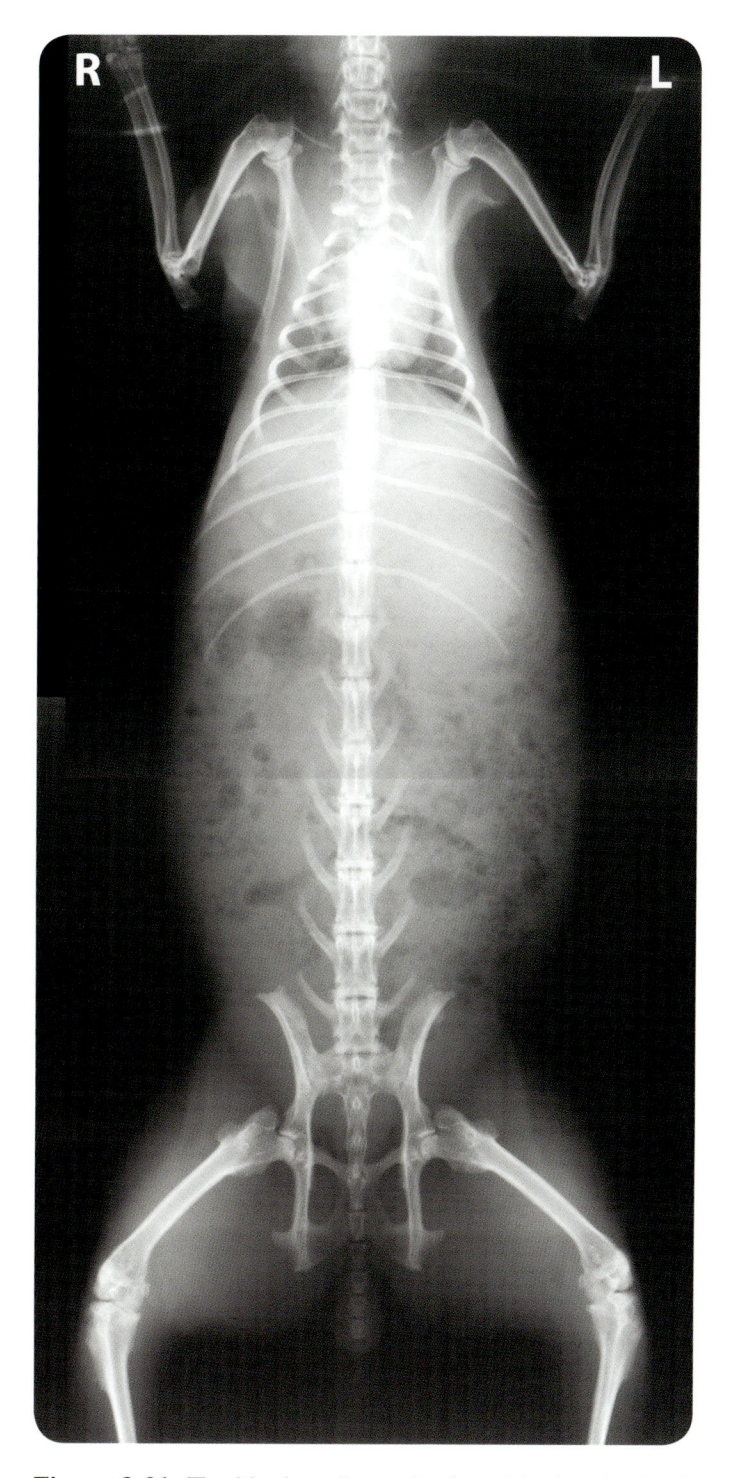

Figure 2.91. Total body radiograph of a 1.9 kg female dwarf rabbit, 50% of actual size. It is generally recommended to focus on the area of interest to optimize radiographic settings and quality. Note the small size of the thorax in comparison to the abdomen in this species.

The NORMAL THORAX
The Cervical and Thoracic Vertebral Column
Lateral Projection

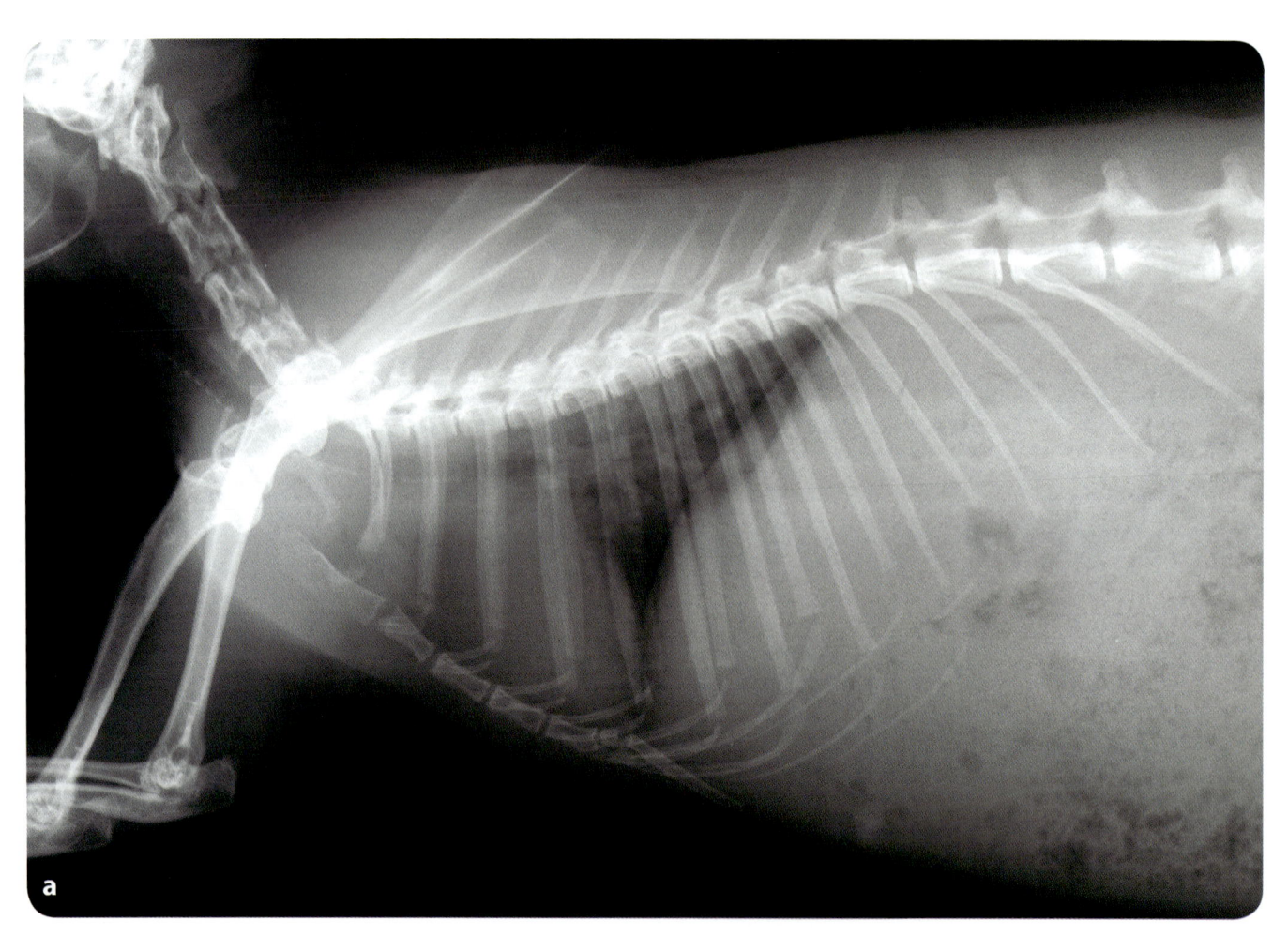

Figure 2.92a,b. Radiograph of the thorax of a 1.9 kg female dwarf rabbit. Actual size.

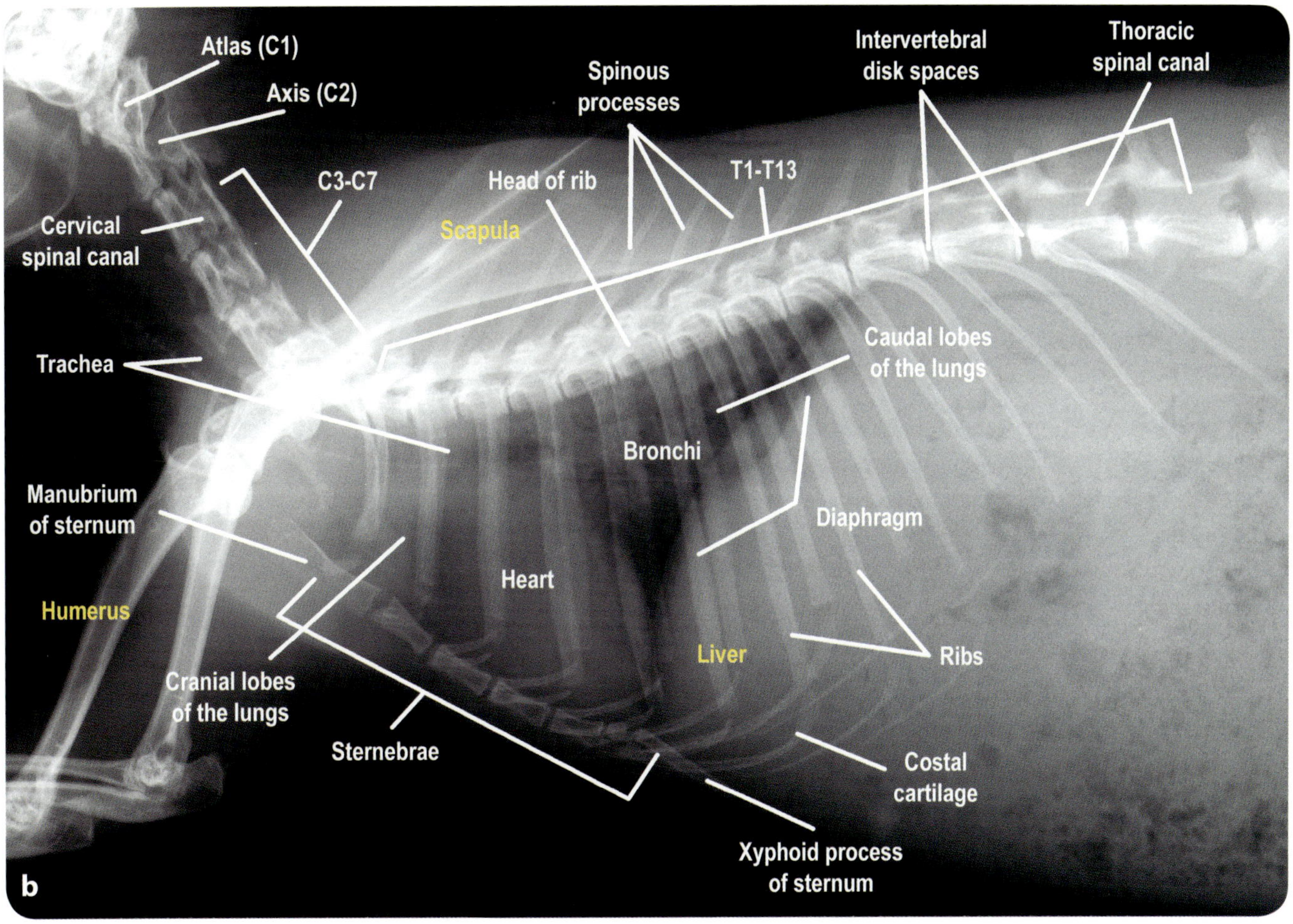
Atlas (C1)
Axis (C2)
C3-C7
Cervical spinal canal
Trachea
Manubrium of sternum
Humerus
Cranial lobes of the lungs
Sternebrae
Spinous processes
Head of rib
Scapula
T1-T13
Intervertebral disk spaces
Thoracic spinal canal
Caudal lobes of the lungs
Bronchi
Diaphragm
Heart
Liver
Ribs
Costal cartilage
Xyphoid process of sternum
b

Ventrodorsal Projection

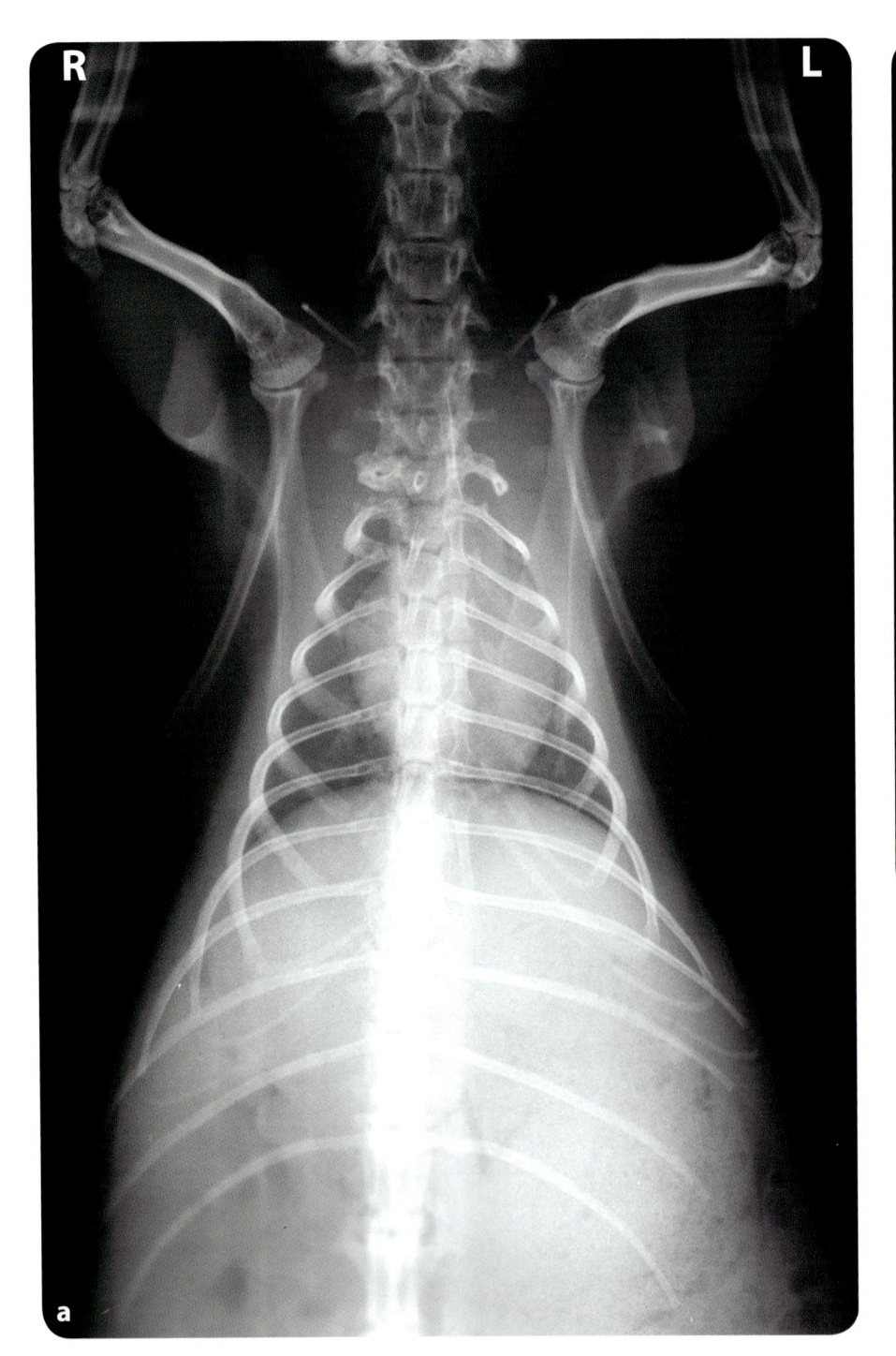

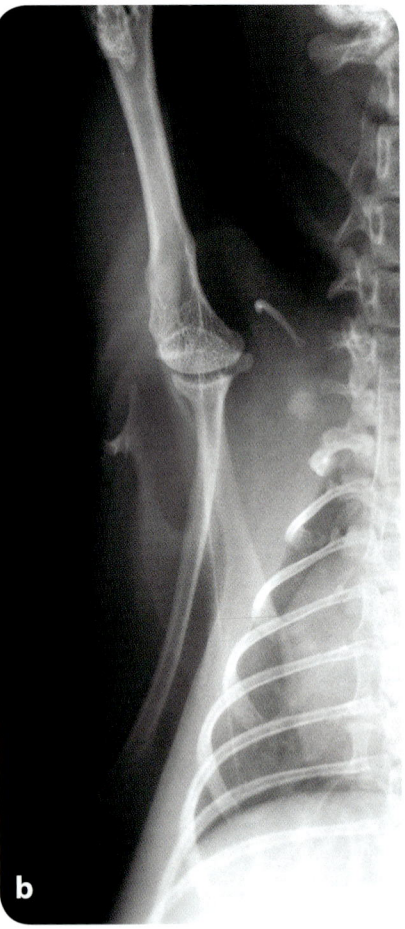

Figure 2.93a,b,c. Radiograph of the thorax of a 1.9 kg female dwarf rabbit (a,c) Actual size.
Hyperextension of the thoracic limb reduces slight superimposition of the scapula over the ipsilateral lung (b).

THORAX 93

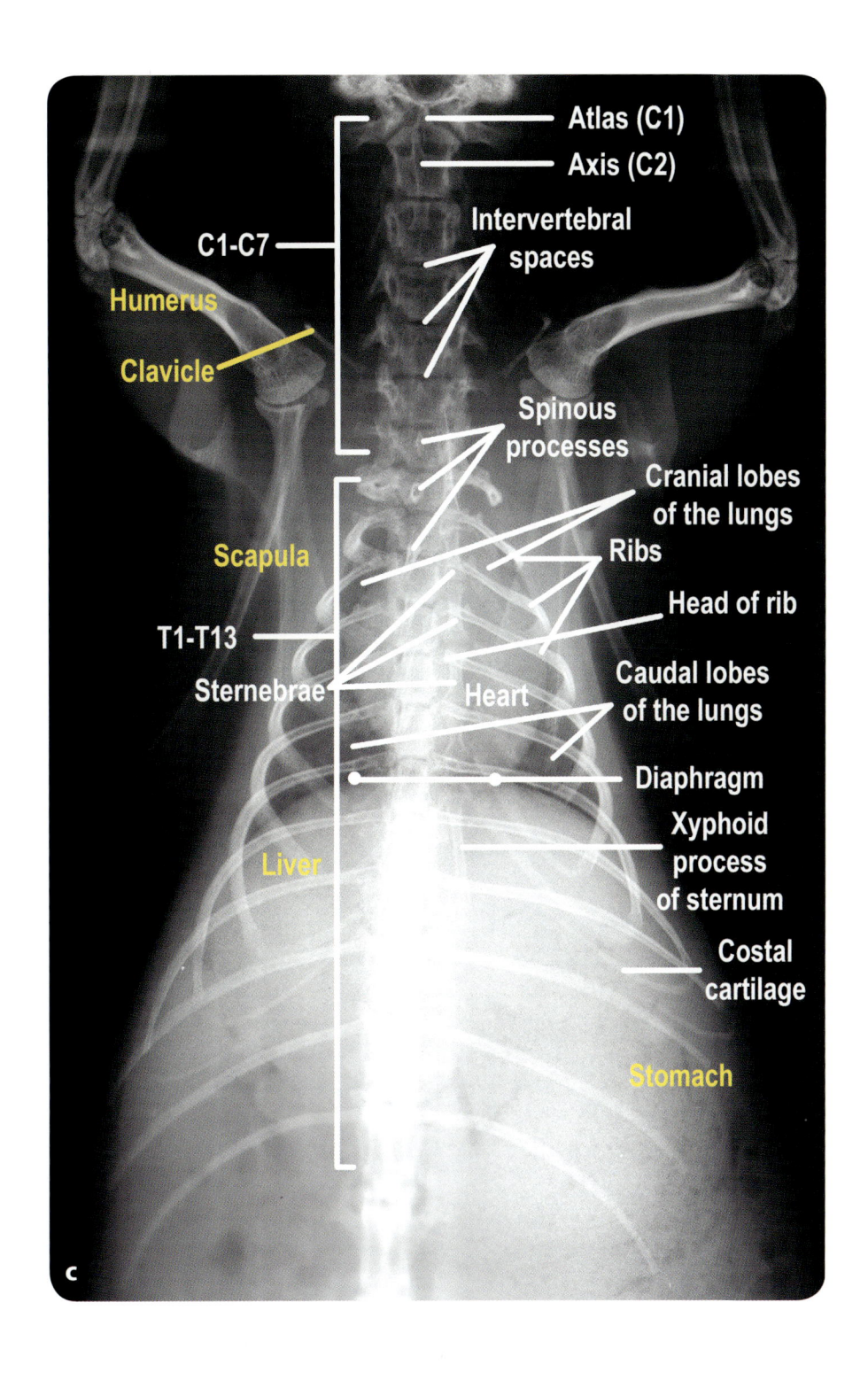

ABNORMALITIES of the THORAX
Diseases of the Lungs

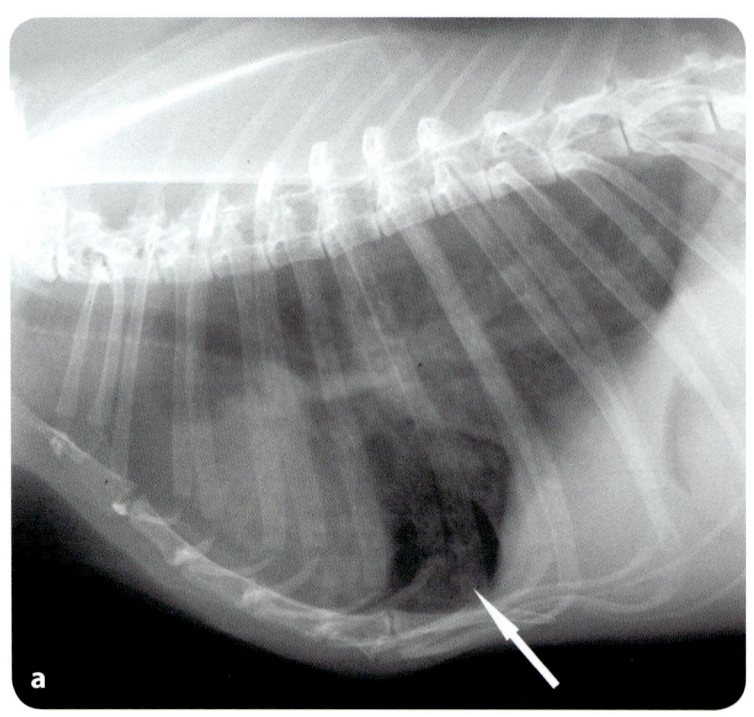

Figure 2.94a,b. Lateral (a) and ventrodorsal (b) radiographs of a rabbit with respiratory distress. A diffuse pulmonary interstitial pattern is present and is characterized by haziness of the lung field and smudging of the vasculature. Note increased pulmonary volume and caudal excursion of the diaphragm is a result of regional air trapping (lobar emphysema) (arrow). This is also seen on the ventrodorsal projection as an area of decreased opacity (arrows), which surrounds a collapsed lung lobe. The lobar vessels of the right lung are larger than those of the left lung as a result of ventilation/perfusion mismatch.

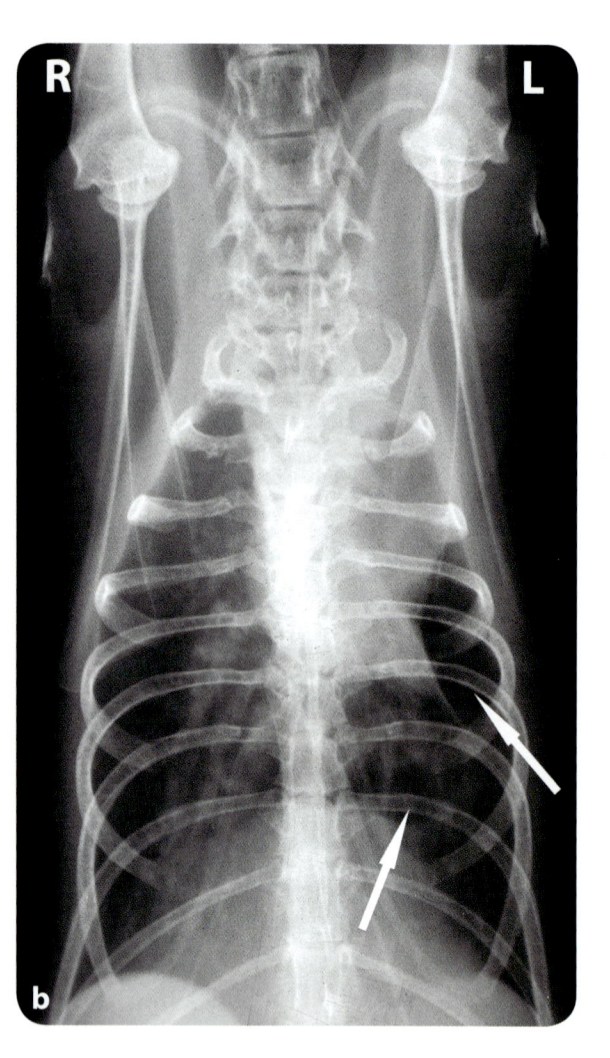

Figure 2.95. Respiratory disease is common in rabbits, and can involve both upper and lower respiratory tracts. Bacterial pneumonia can involve a wide range of pathogens.

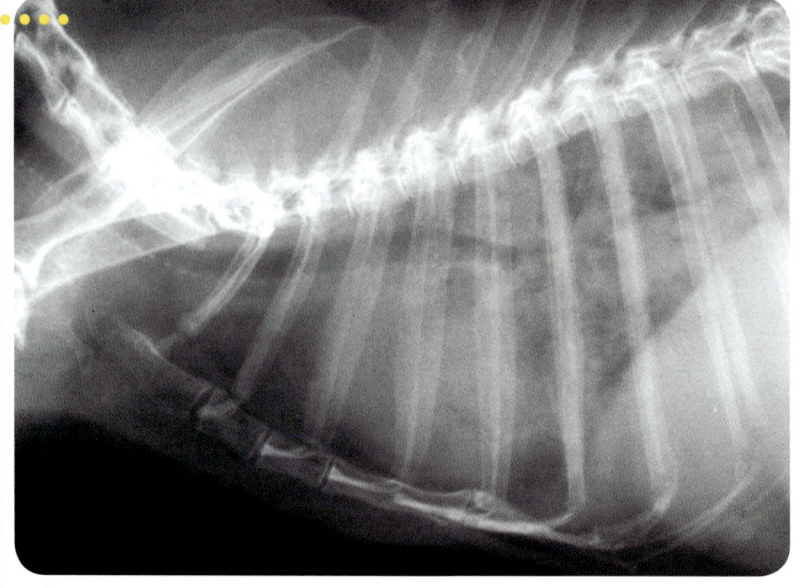

Figure 2.96. Lateral radiograph of a rabbit with respiratory distress. Radiographic changes differ from those in Figure 2.94, and consist of patchy alveolar opacities that cause effacement of the cardiac and diaphragmatic margins. These changes are typical of bronchopneumonia.

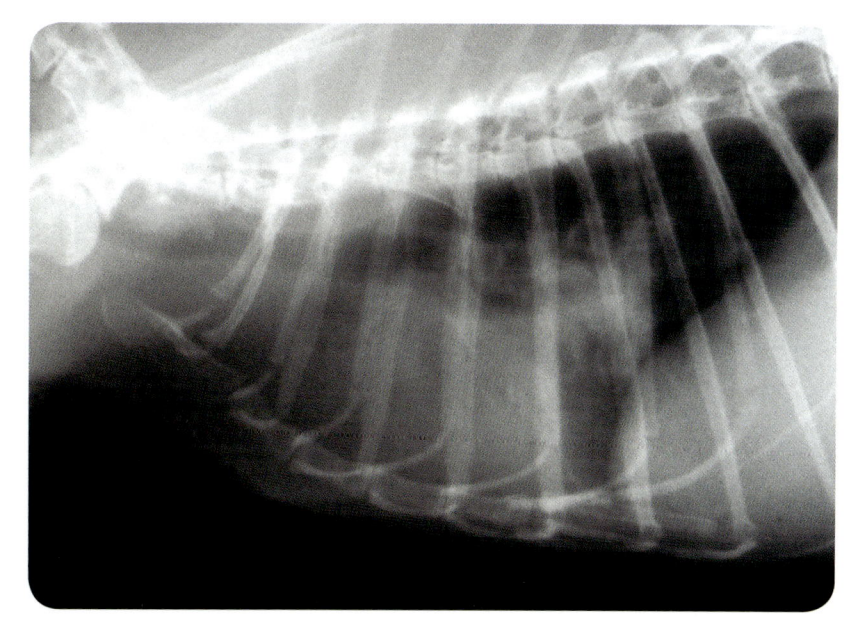

Figure 2.97. Lateral radiograph of the thorax of a rabbit with dyspnea. There is increased opacity cranially that partially obscures the cardiac silhouette. The caudal lung field is relatively normal. At necropsy, this patient was found to have a pulmonary abscess.

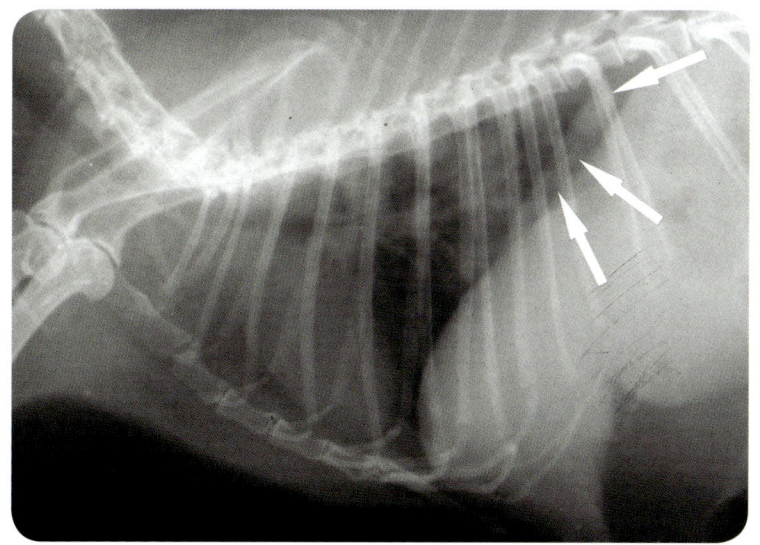

Figure 2.98. Lateral radiograph of the thorax of a rabbit with focal radiopacities in the dorsal caudal lung field (arrows). These were determined to be metastases from uterine adenocarcinoma.

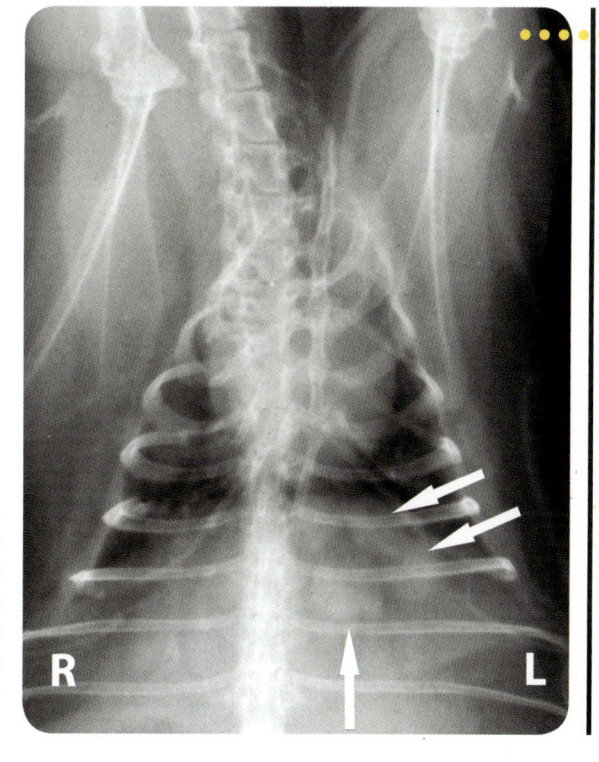

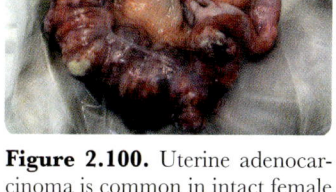

Figure 2.100. Uterine adenocarcinoma is common in intact female rabbits. End-stage metastasis to the lungs is a possible sequelae.

Figure 2.99. Ventrodorsal radiograph of the same patient as Figure 2.98. The focal radiopacities are present (arrows) but are less obvious than on the lateral radiograph. The curvature of the diaphragm causes superimposition of these lesions on the liver.

Figure 2.101a,b. Lateral (a) and ventrodorsal (b) radiographs of the thorax of a rabbit with a large solitary mass in the left hemithorax (arrows). This mass was determined to be a metastatic renal nephroblastoma.
A large, single mass usually produces less respiratory compromise than multiple masses.

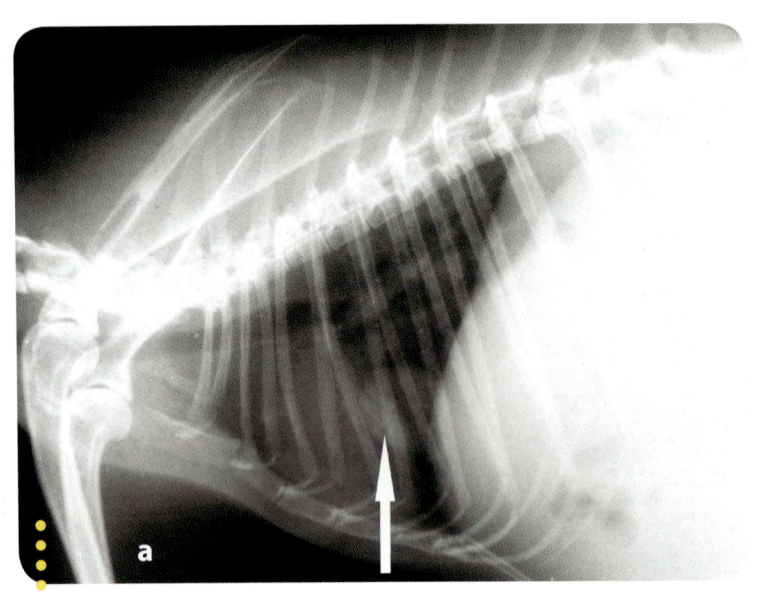

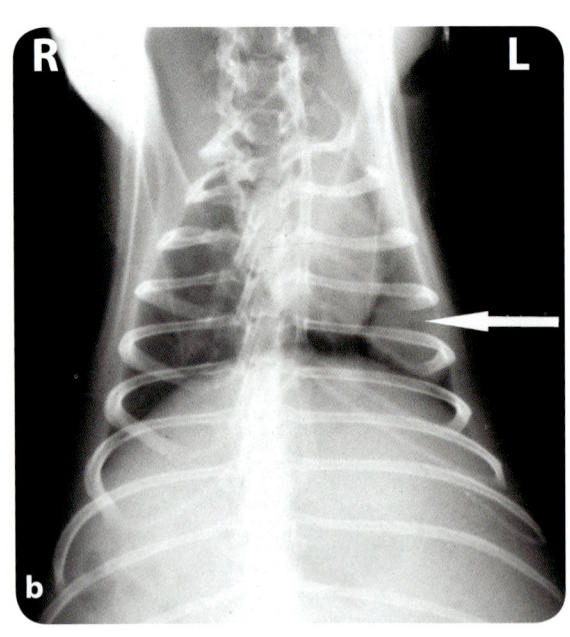

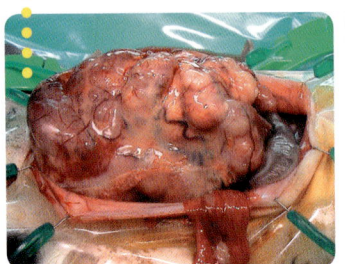

Figure 2.102. Laparotomy of the same patient demonstrating a large mass associated with the right kidney.

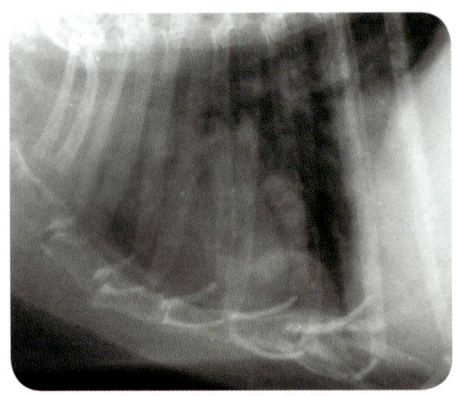

Figure 2.103. Same patient, 2-week follow-up: The thoracic mass is larger, and additional radiopacities appear caudally. Respiratory compromise had worsened.

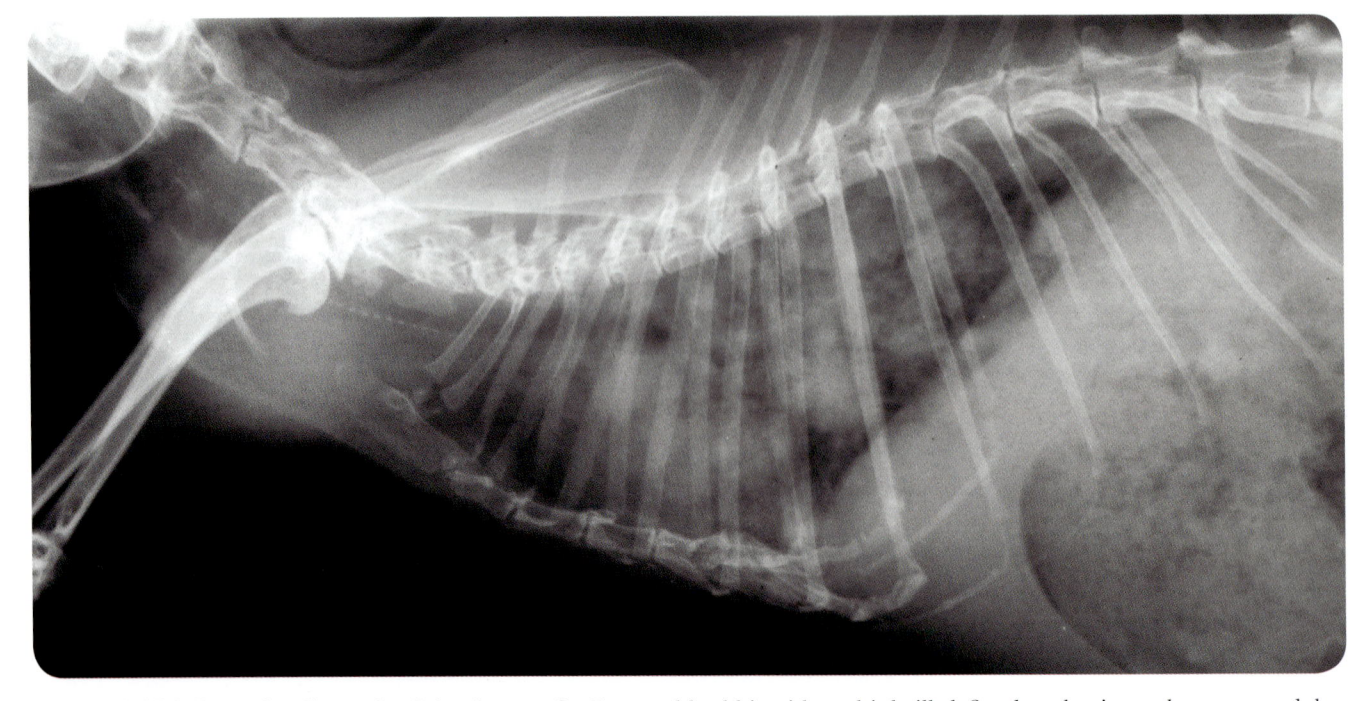

Figure 2.104. Lateral radiograph of the thorax of a 7-year-old rabbit with multiple ill-defined coalescing pulmonary nodules. These were determined to be pulmonary metastases from uterine adenocarcinoma.

Diseases of the Mediastinum

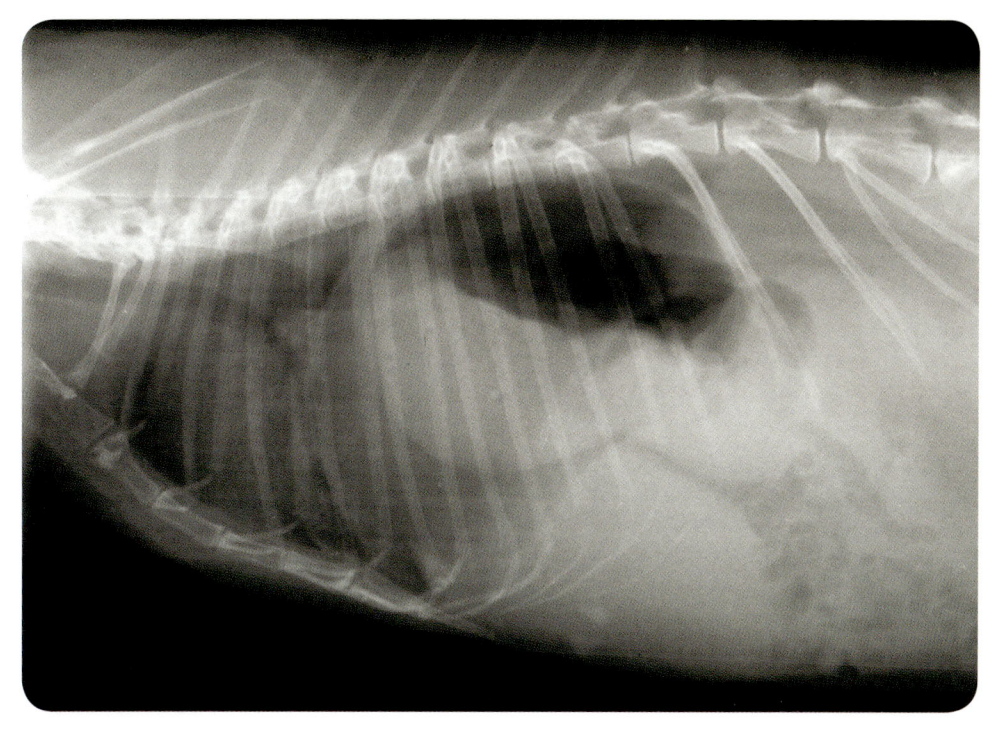

Figure 2.105. Lateral radiograph of the thorax of a rabbit with a diaphragmatic hernia. The ingesta-filled stomach and portions of the intestinal tract are visible within the thorax. There is loss of visualization of the diaphragm.

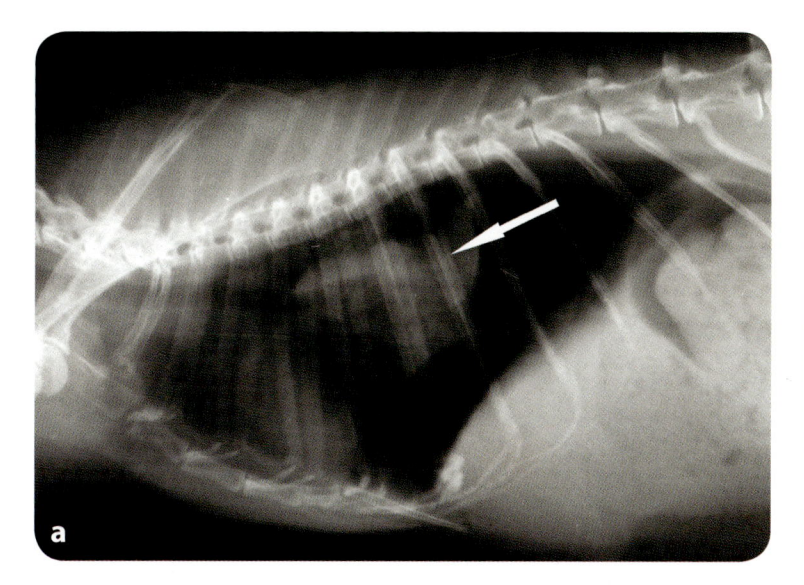

Figure 2.106a,b. Lateral (a) and ventrodorsal (b) radiographs of the thorax of a rabbit with dyspnea. Tension penumothorax is present. There is lobar collapse (arrow) surrounded by a large amount of free gas in the pleural space (a). Mediastinal shift is present in (b) with the right hemithorax most affected.

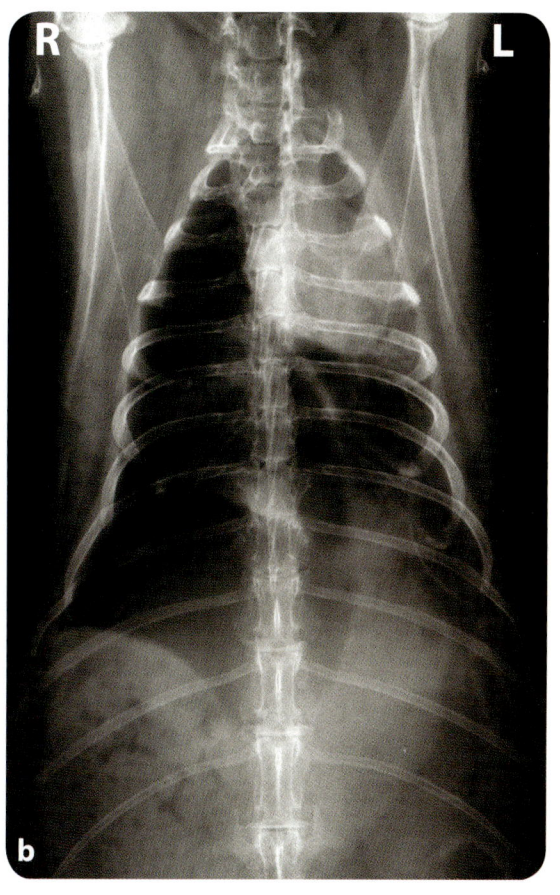

Figure 2.107a,b. Lateral (a) and ventrodorsal (b) radiographs of the thorax of a rabbit with a mediastinal mass that was later determined to be lymphoma. The cardiac silhouette is displaced caudally and partially effaced by a large cranial mediastinal mass.

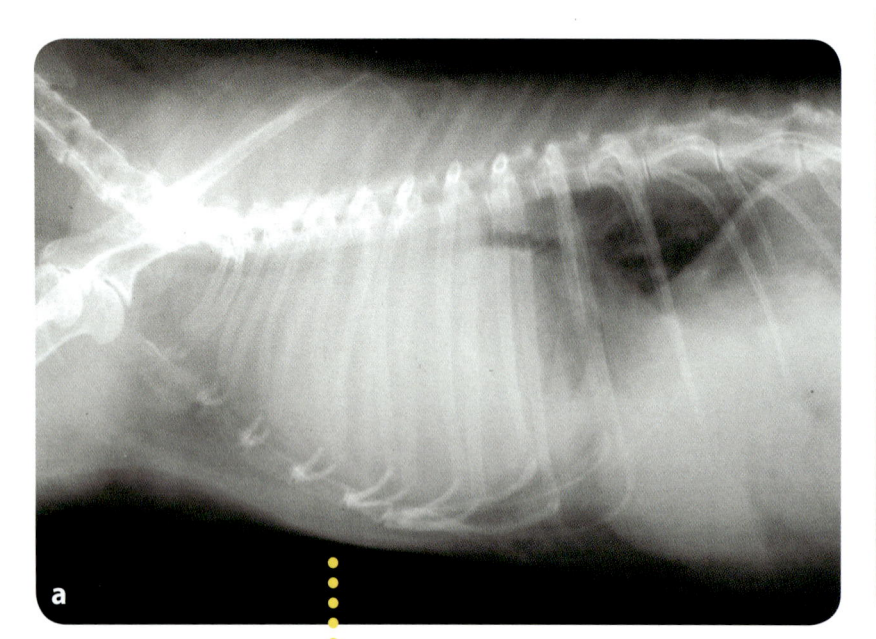

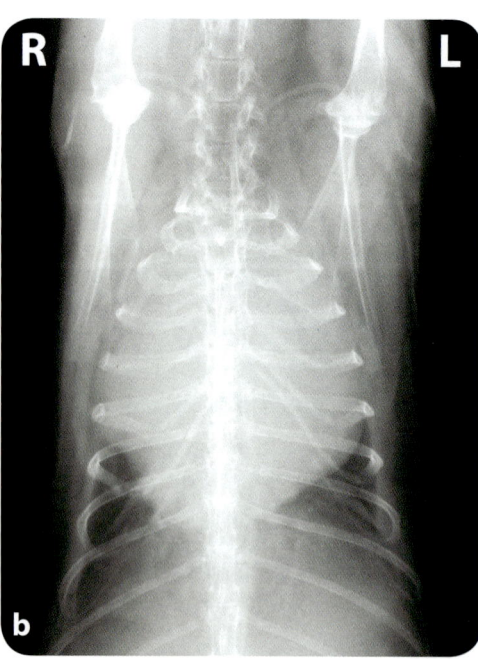

Figure 2.108. Gross specimen of the same patient demonstrating a large mediastinal mass (white arrow) cranial to the heart (yellow arrow) with small white focal metastases to the lung.

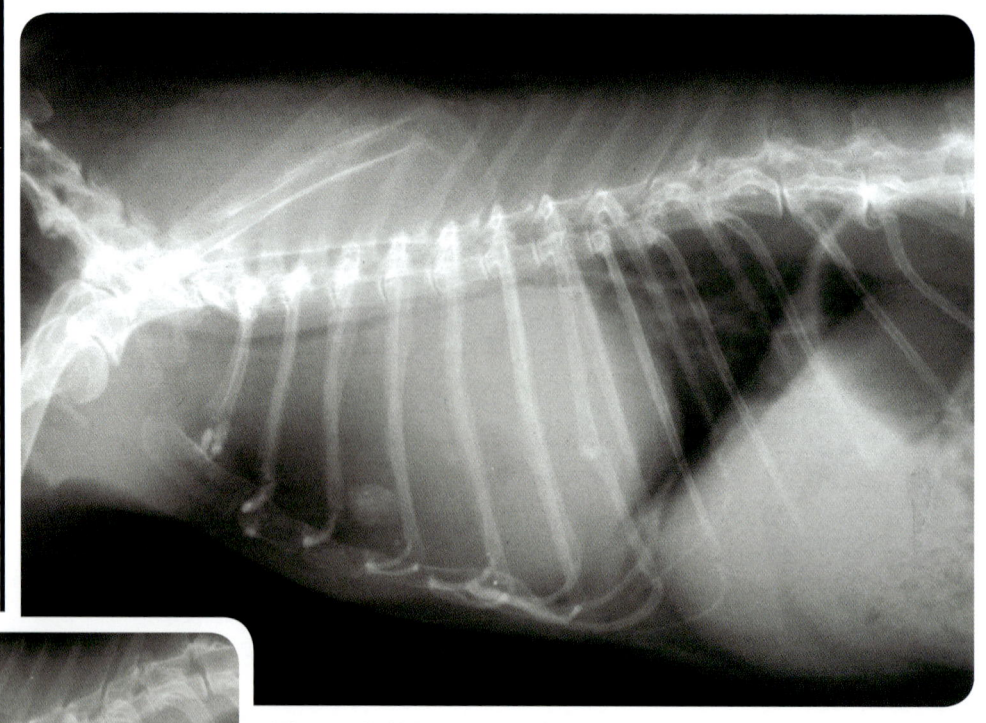

Figure 2.109. Lateral radiograph of the thorax of a rabbit with dyspnea. A large cranial mediastinal soft tissue mass is causing compression of the trachea. A mineralized nodule is present ventrally at the third intercostal space.

Figure 2.110. Same patient as above, three week follow-up. The mineralized nodule has increased in size.

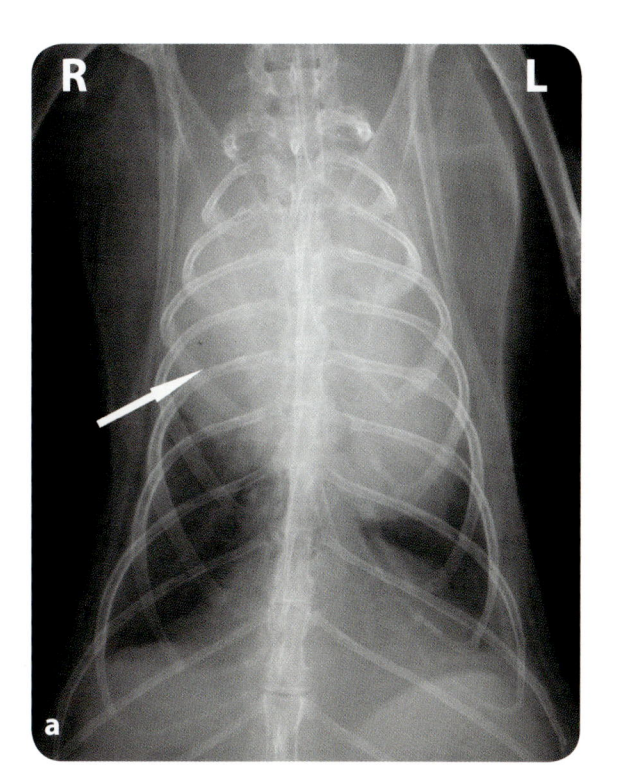

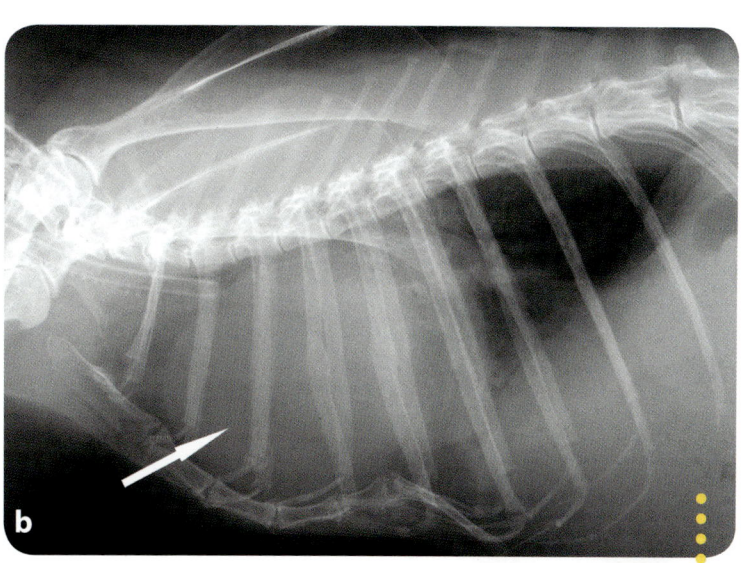

Figure 2.111a,b. Ventrodorsal (a) and lateral (b) radiographs of the thorax of a rabbit with thymoma. In (a) a cranial thoracic mass effect is present and obscures the heart. In (b) the heart is displaced caudally by a soft tissue mass (arrow). Deviation of the sternum visible on the lateral projection is the result of an earlier traumatic episode and may affect the shape of the cardiac silhouette.

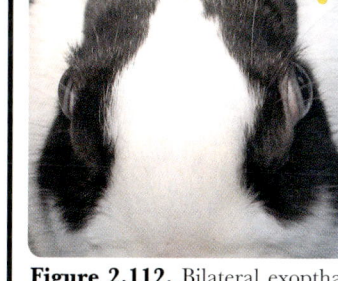

Figure 2.112. Bilateral exopthalmos is a common clinical presentation in rabbits with thymoma or other mediastinal mass.

Figure 2.113a,b. Ventrodorsal (a) and lateral (b) radiographs taken immediately post surgery. Free gas is present in the pleural space (arrows) and lobar collapse is present (yellow arrow).

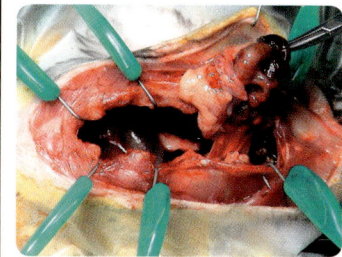

Figure 2.114. Thoracotomy and surgical removal of the thymoma.

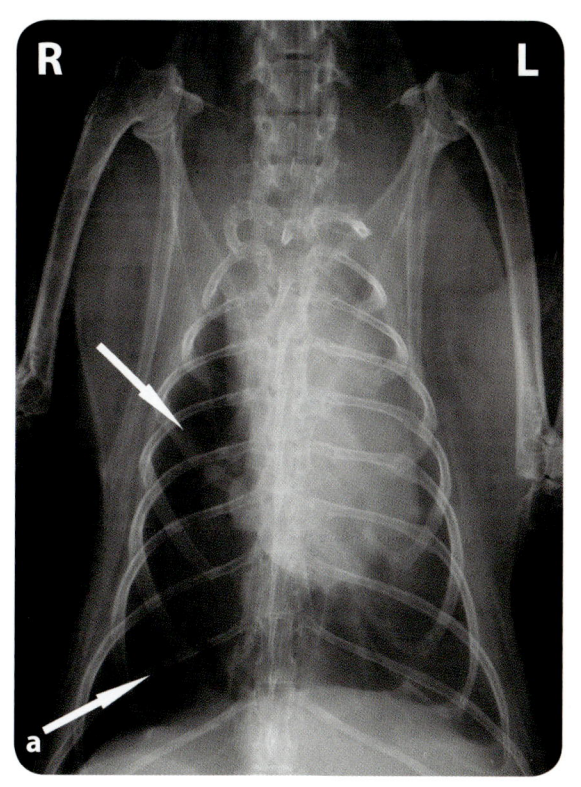

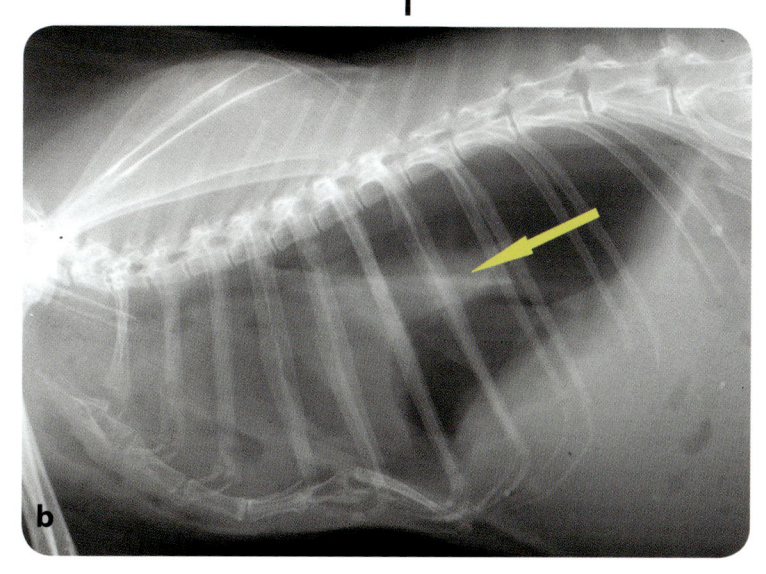

The NORMAL ABDOMEN
The Lumbar Vertebral Column
Lateral Projection

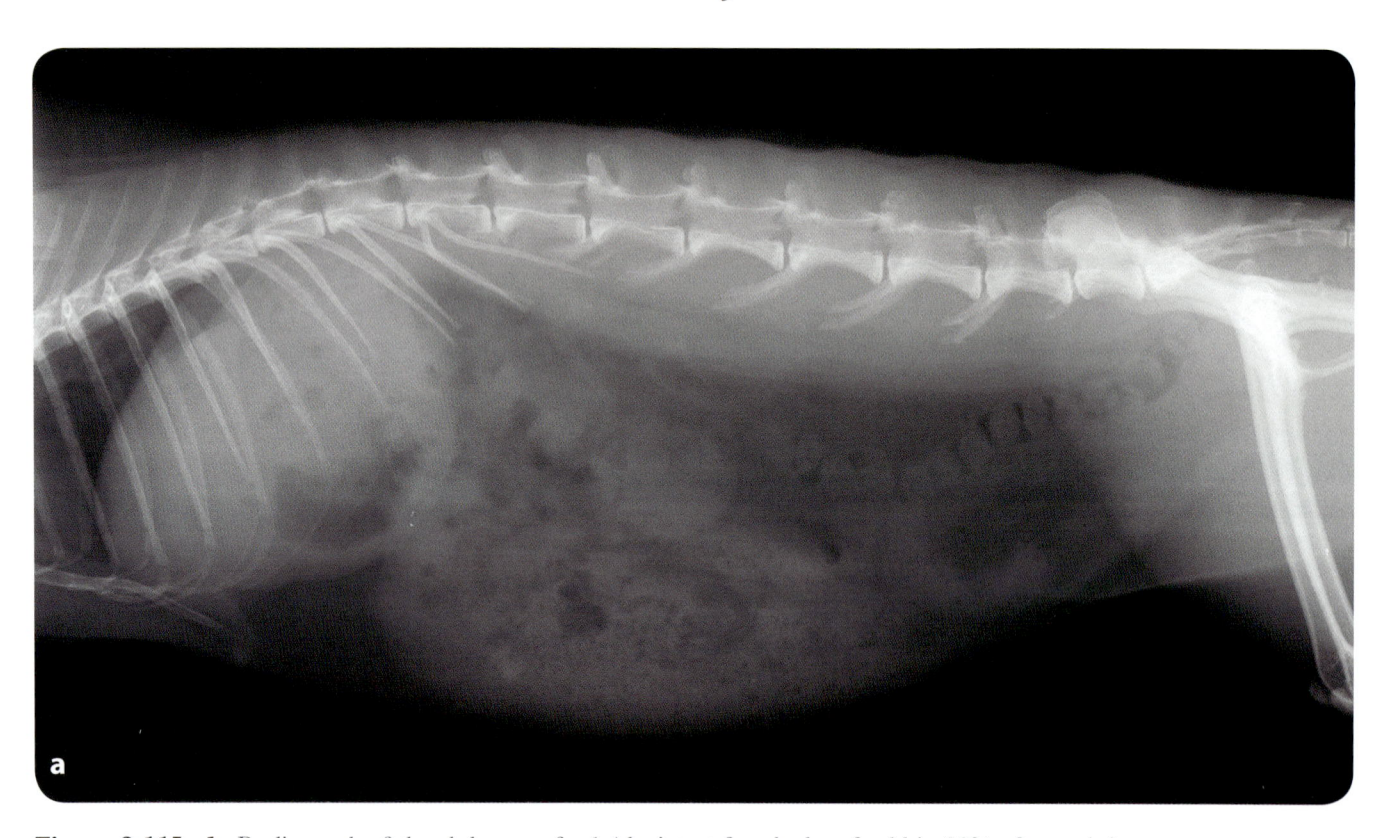

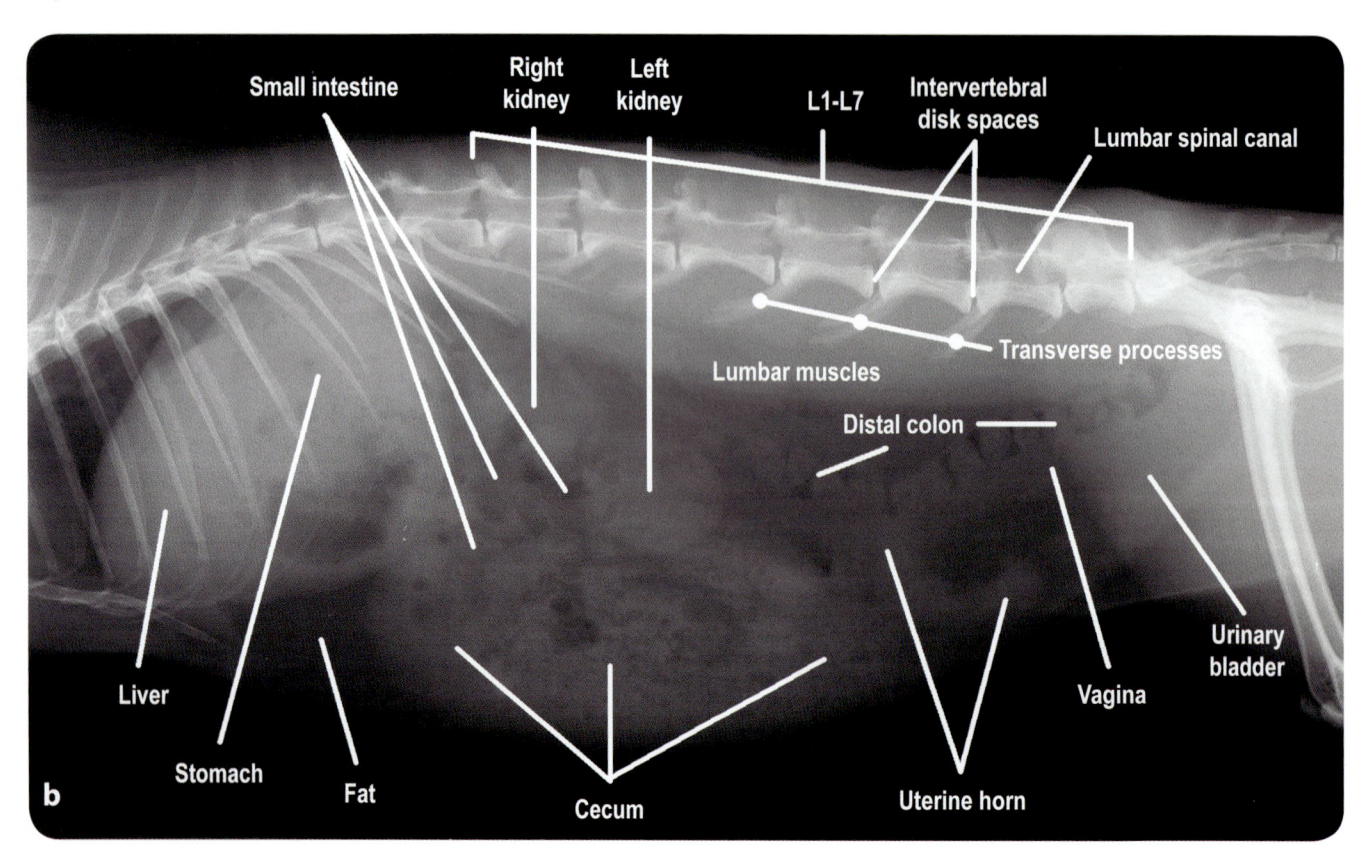

Figure 2.115a,b. Radiograph of the abdomen of a 1.4 kg intact female dwarf rabbit. 90% of actual size.

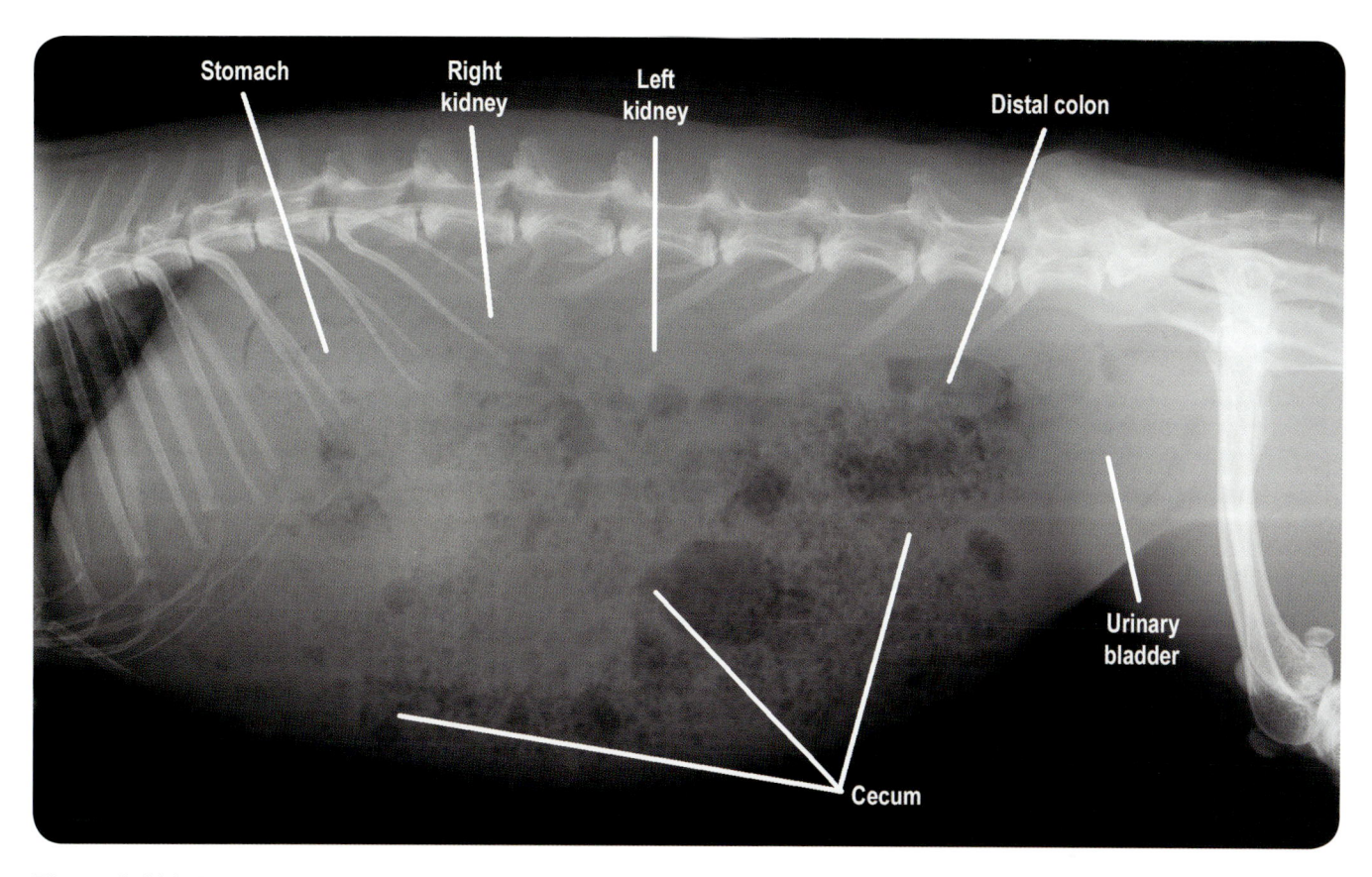

Figure 2.116. Lateral radiograph of the abdomen of a 1.9 kg female dwarf rabbit. 80% of actual size. Normal ingesta can significantly interfere with identification of other abdominal structures.

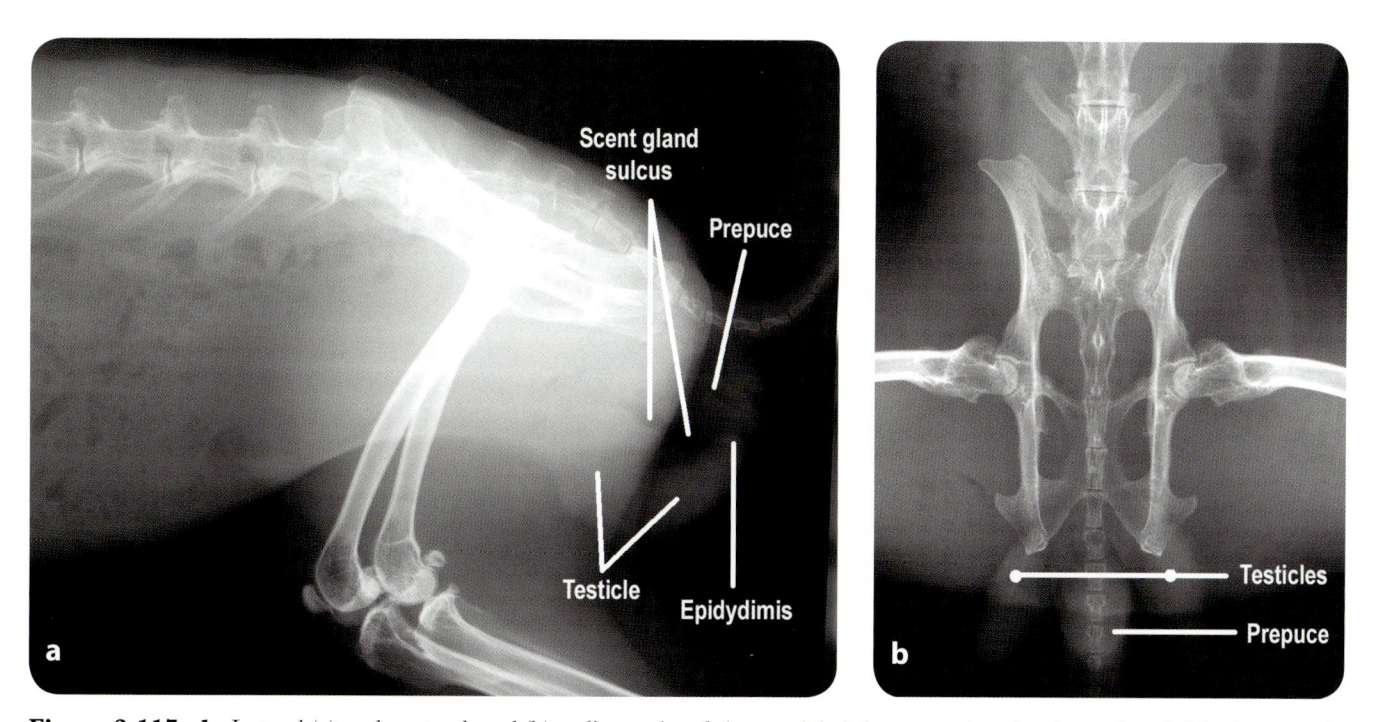

Figure 2.117a,b. Lateral (a) and ventrodorsal (b) radiographs of the caudal abdomen and perineal area in a 1.5 kg intact male dwarf rabbit, 70% of actual size.

Ventrodorsal Projection

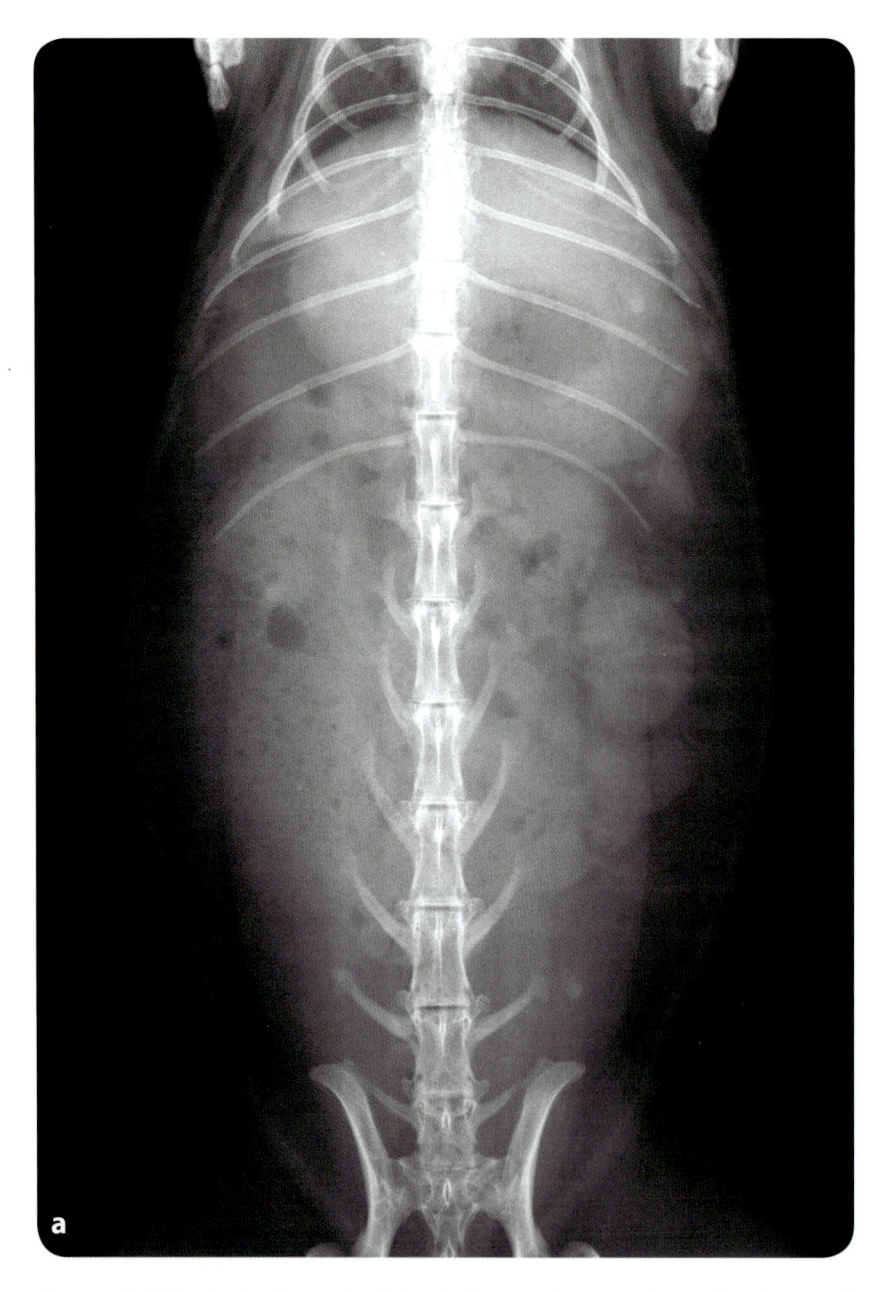

Figure 2.118a,b. Radiograph of the abdomen of a 1.4 kg intact female dwarf rabbit. 90% of actual size.

ABDOMEN 103

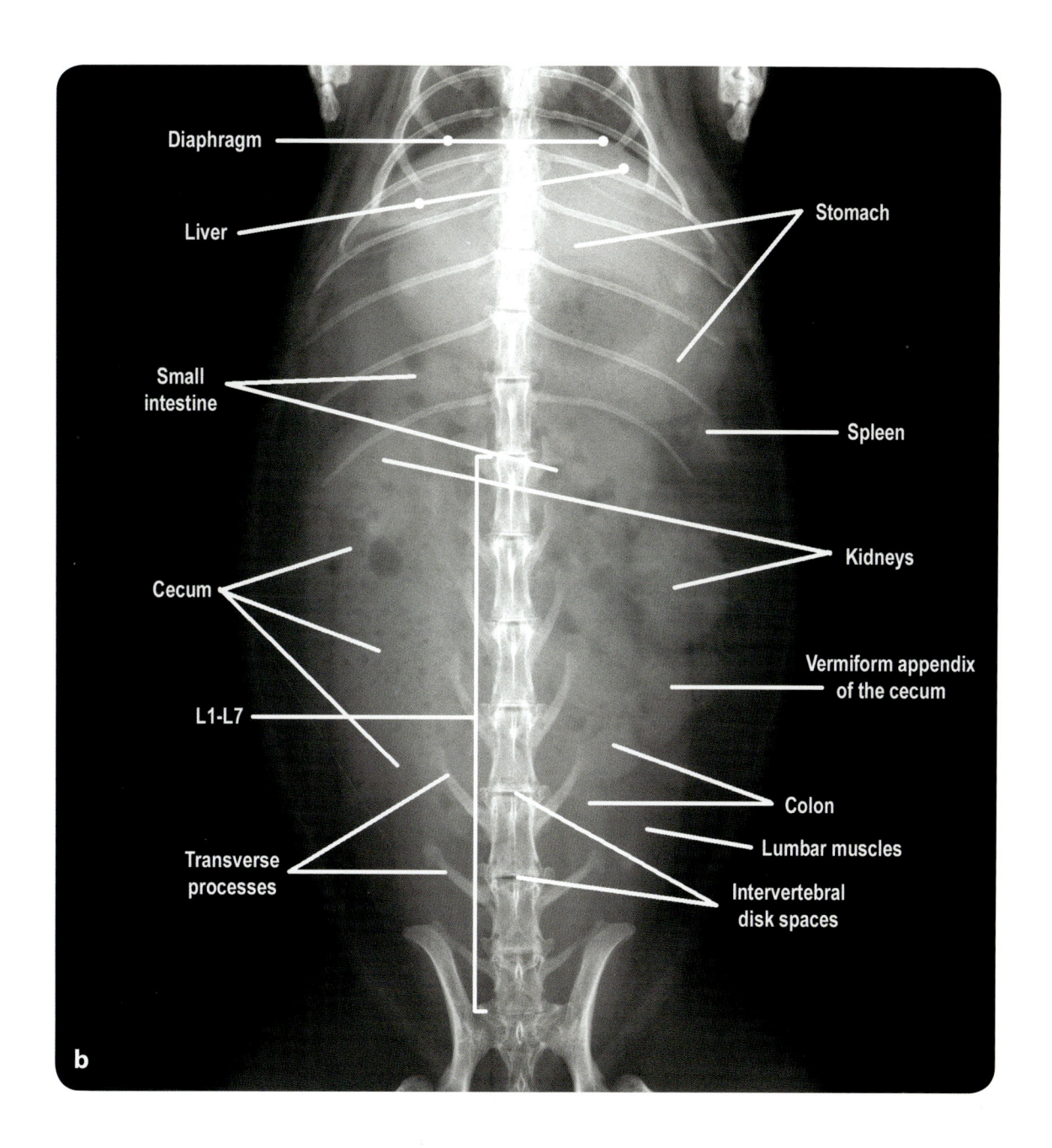

Miscellaneous

Figure 2.119. Obesity in pet rabbits is common, mostly due to inappropriate diet (especially lack of fiber) and lack of exercise.

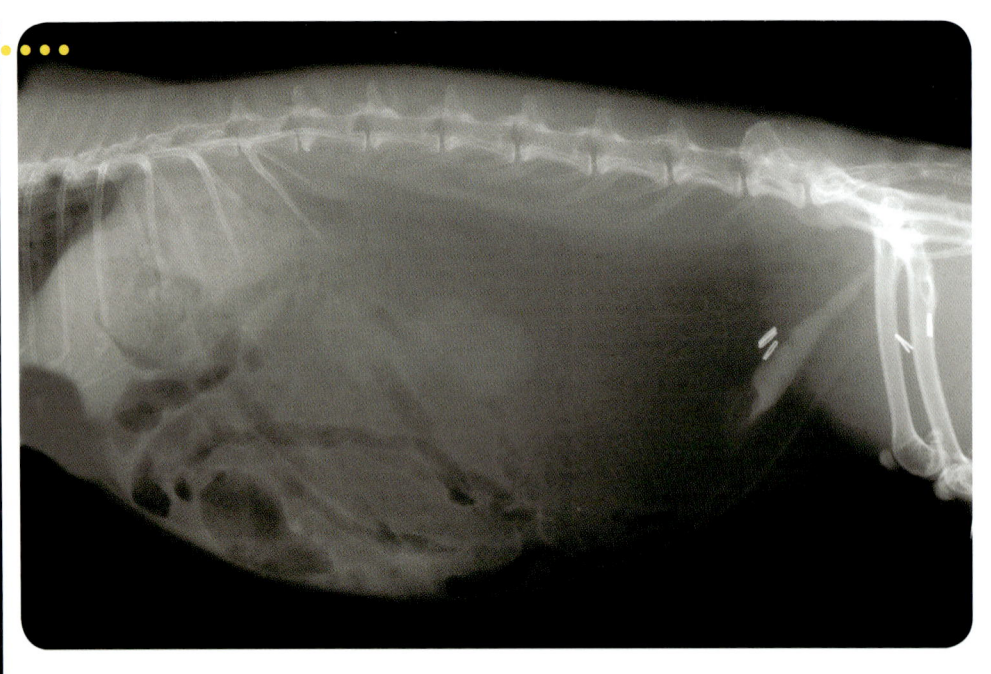

Figure 2.120. Lateral radiograph of the abdomen of an obese neutered male rabbit. Abundant retroperitoneal and intrabdominal fat displaces the viscera ventrally. Note ventral displacement of the kidneys and intestines. The urinary bladder contains mineralized sludge. Superimposed hemoclips are visible.

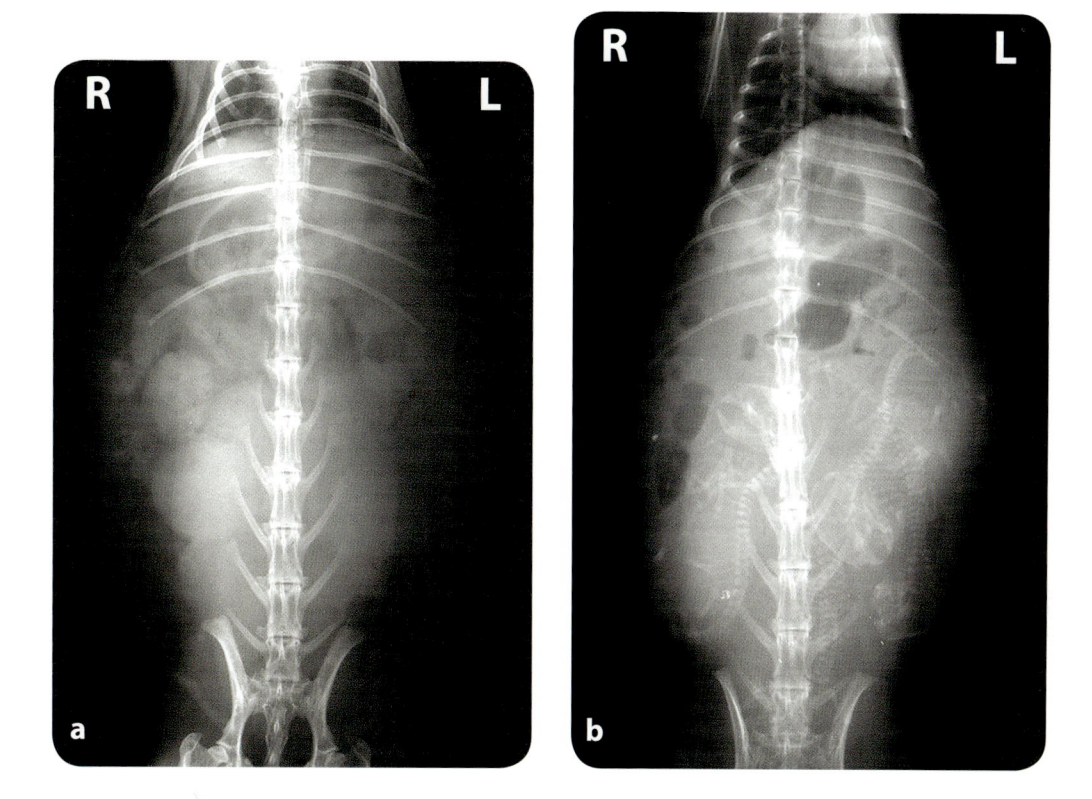

Figure 2.121a,b. Ventrodorsal radiographs of the abdomen of a rabbit at the 18th day of pregnancy (a). At this stage there is uterine enlargement, but no ossification of the skeleton of the fetuses, therefore they are not readily identifiable.
Ventrodorsal projection of the abdomen of a different rabbit at day 23 of pregnancy (b). Three fetuses are visible. Diastasis of the ischiopubic symphysis does not occur in the rabbit as it does in the guinea pig.

ABNORMALITIES of the ABDOMEN
Diseases of the Stomach

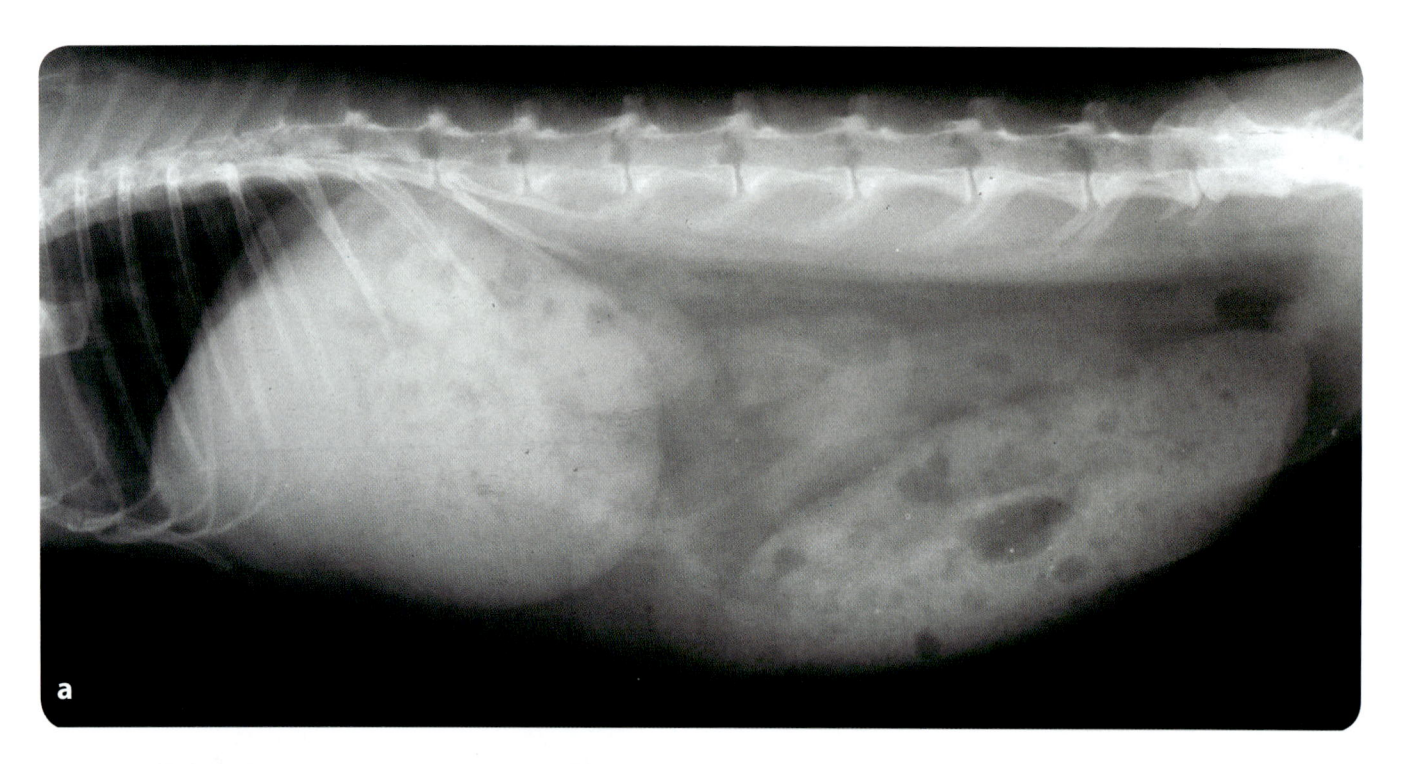

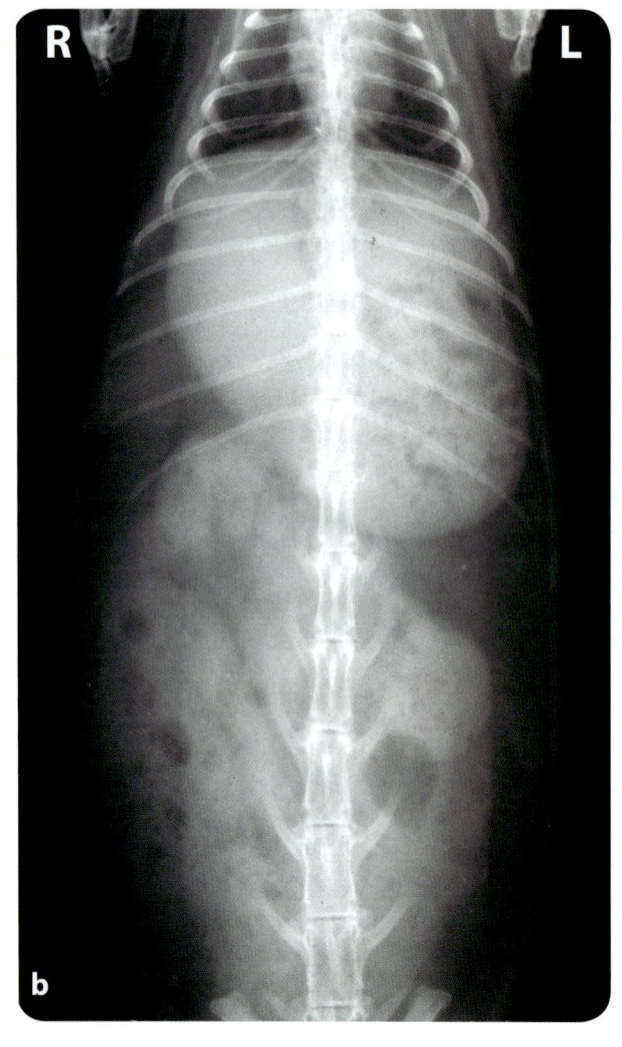

Figure 2.122a,b. Lateral (a) and ventrodorsal (b) radiographs of a rabbit demonstrating a filled gastrointestinal tract, which may represent recent heavy food intake, or possibly delayed emptying of the gastrointestinal tract. However, no specific abnormalities suggesting gastrointestinal disease, such as abnormal gas patterns, are present. A patient with this radiographic appearance should be monitored both clinically and radiographically.

ABDOMEN

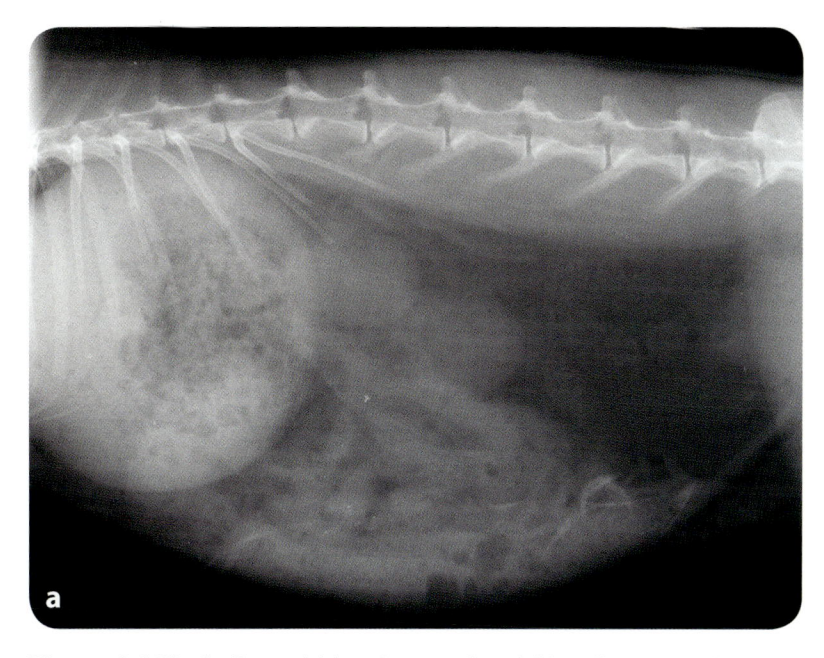

Figure 2.123a,b. Lateral (a) and ventrodorsal (b) radiographs of the abdomen of a rabbit with gastric dilation which gives the stomach a rounded appearance, as in the lateral radiograph. The overfilled stomach takes on a more typical shape and appearance in the ventrodorsal projection.

Figure 2.125. Motility of the gastrointestinal tract of the rabbit is dependant upon high fiber foods such as hay. However, many owners feed inappropriate diets such as fruit, grains, and seeds.

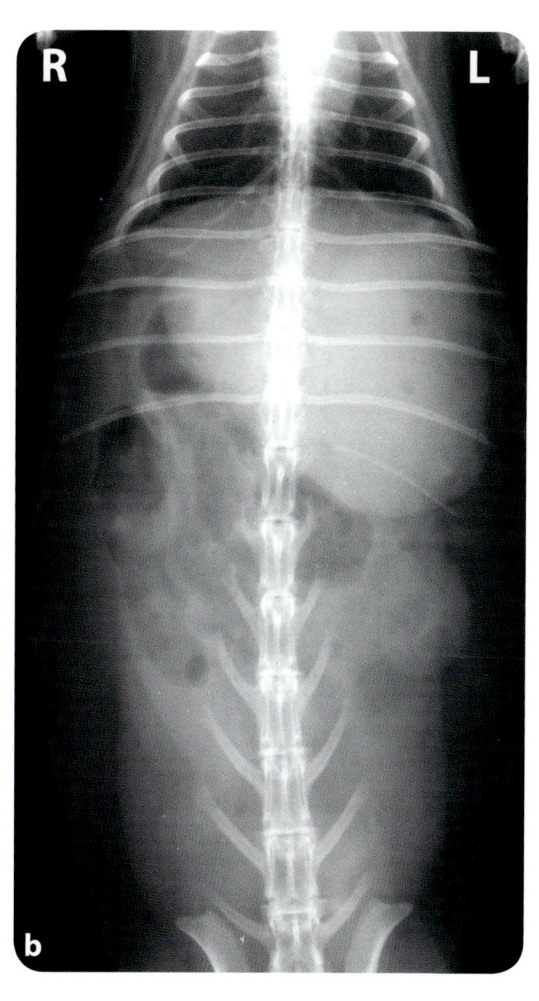

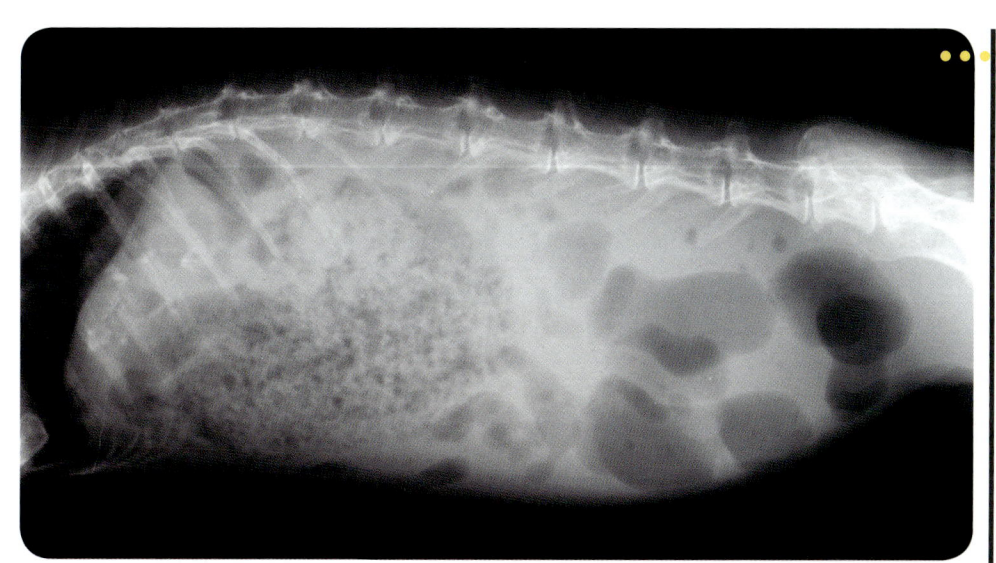

Figure 2.124. Lateral radiograph of the abdomen of a rabbit with gastric impaction as a result of inappropriate diet, in this case a seed and grain-based mixture. Overfilling of the stomach and the presence of excessive intestinal gas suggest gastrointestinal disturbance in this species.

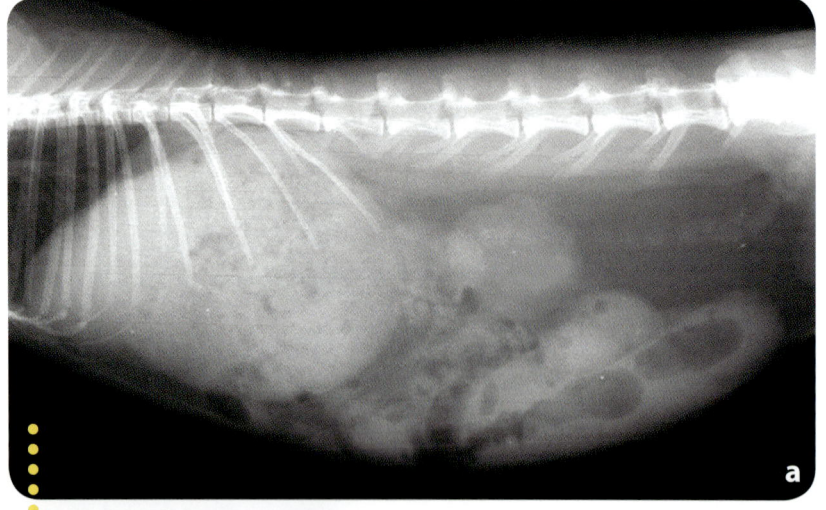

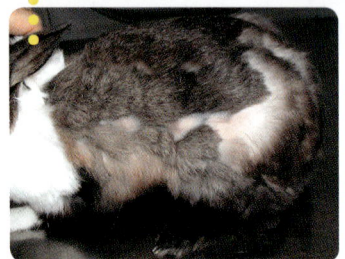

Figure 2.126a,b. Lateral (a) and ventrodorsal (b) radiographs of the abdomen of a rabbit with suspected trichobezoars. Self-grooming and ingestion of hair is normal in the rabbit. Trichobezoars are abnormal accumulations of hair, and are thought to be a result of reduced gastrointestinal motility. The provision of a high fiber diet (hay) is linked to reduction of formation of trichobezoars. Most trichobezoars can be managed with medical therapy, including correction of underlying malnutrition and disease processes, motility enhancing drugs, fluids, and hand feeding.

Figure 2.127. Molt in a rabbit. Many practitioners have noticed an increase in the incidence of trichobezoars around the time of heavy loss of haircoat.

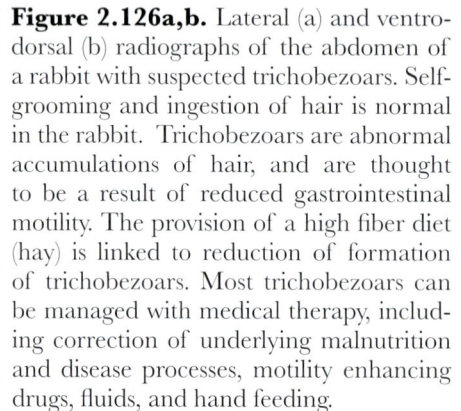

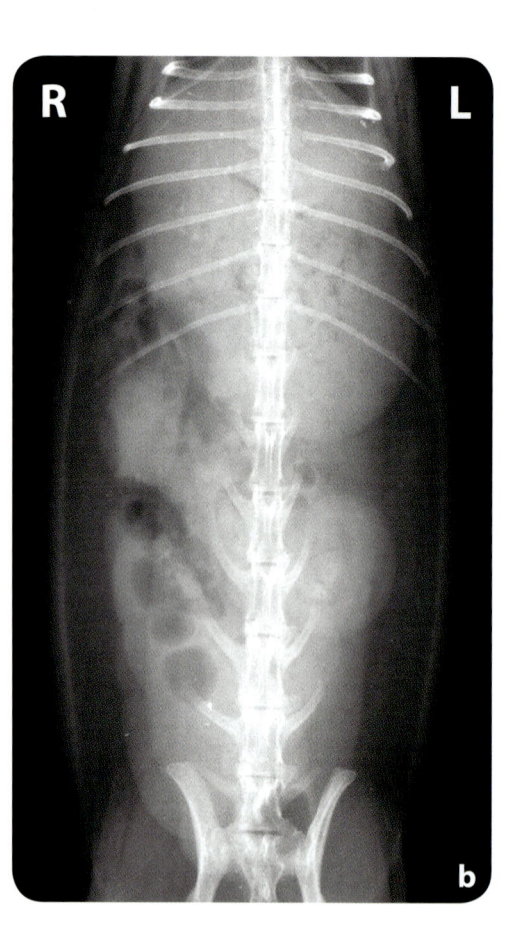

Figure 2.128. The presence of feces linked together with hair is strong evidence of excessive fur ingestion.

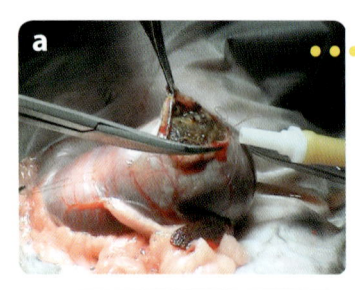

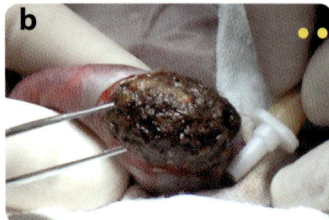

Figure 2.130a,b. Gastrotomy is indicated in those cases not responding to medical therapy.

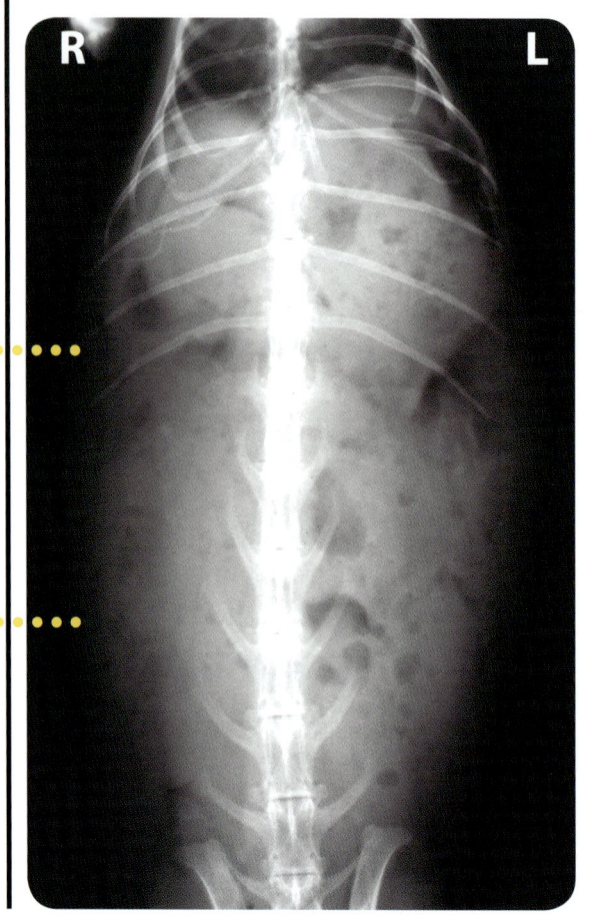

Figure 2.129. Ventrodorsal radiograph of the abdomen of a rabbit demonstrating radiopaque, irregularly shaped densities in the stomach, found at surgery to be trichobezoars.

Accumulations of foreign materials other than hair and food in the stomach are relatively uncommon in the rabbit. Contrast radiography of the gastrointestinal tract is potentially useful to confirm suspected gastrointestinal obstruction or foreign body.

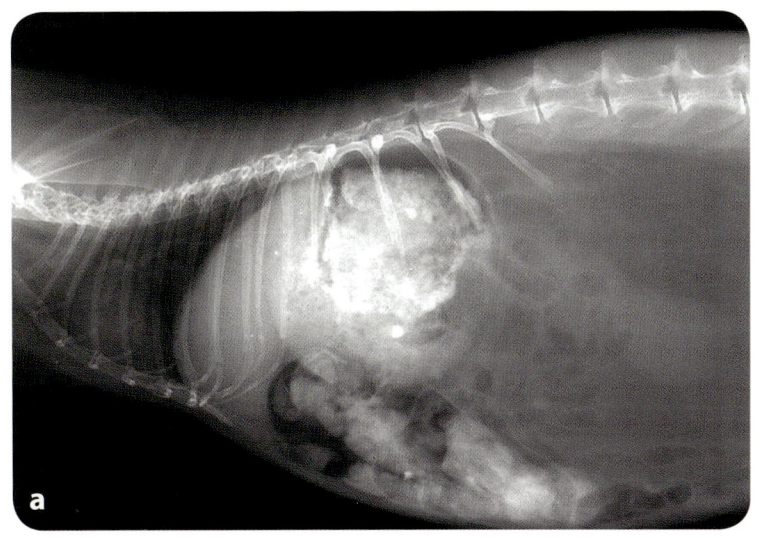

Figure 2.131a,b. Lateral (a) and ventrodorsal (b) radiographs of the abdomen of a rabbit with impaction of the stomach and the cecum following ingestion of foreign material, which was later identified as paper. Irregular radiopacities suggest the presence of very thick material. Medical therapy is often unrewarding in severe cases such as this.

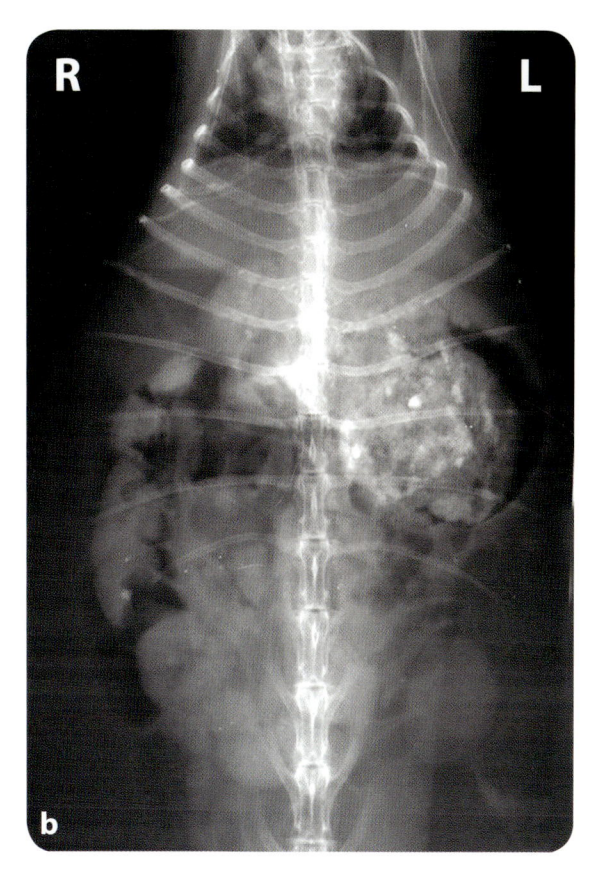

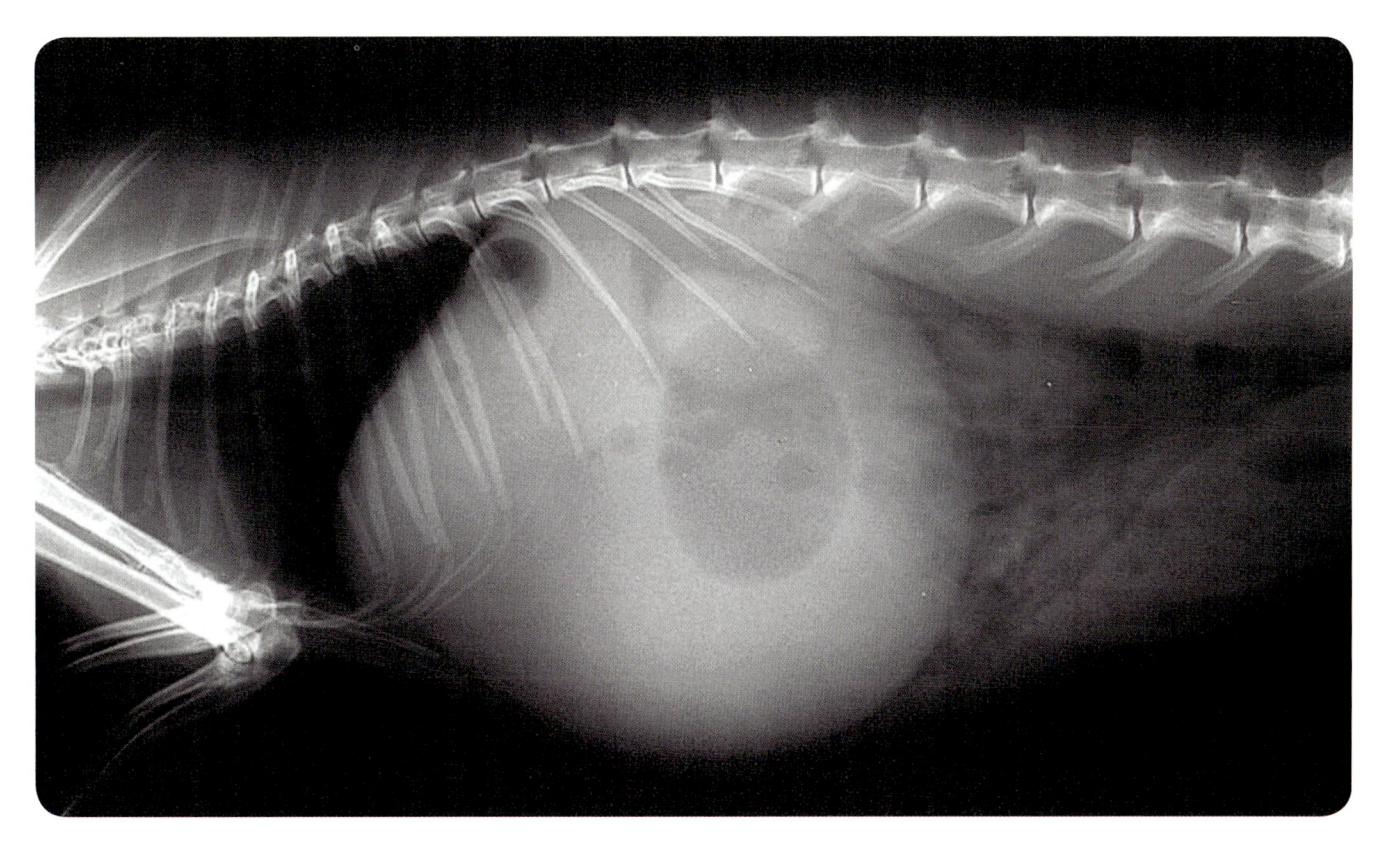

Figure 2.132. Lateral radiograph of the abdomen of a rabbit with marked gastric dilation and gas accumulation. Note compression of the diaphragm by the enlarged stomach.

Diseases of the Intestine

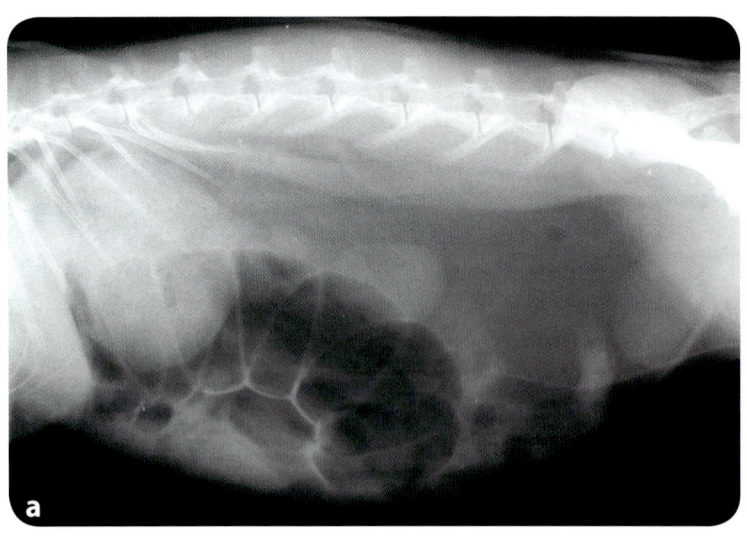

Figure 2.133a,b. Lateral (a) and ventrodorsal (b) radiographs of the abdomen of a rabbit demonstrating cecal gas. The haustra of the cecum are clearly visible (see also Figure 2.137) The stomach contains only a small amount of food, suggesting anorexia. In the ventrodorsal projection the cecum is deviated cranially and obscures the stomach (arrow) The amount of cecal gas in this case is consistent with primary enteritis.

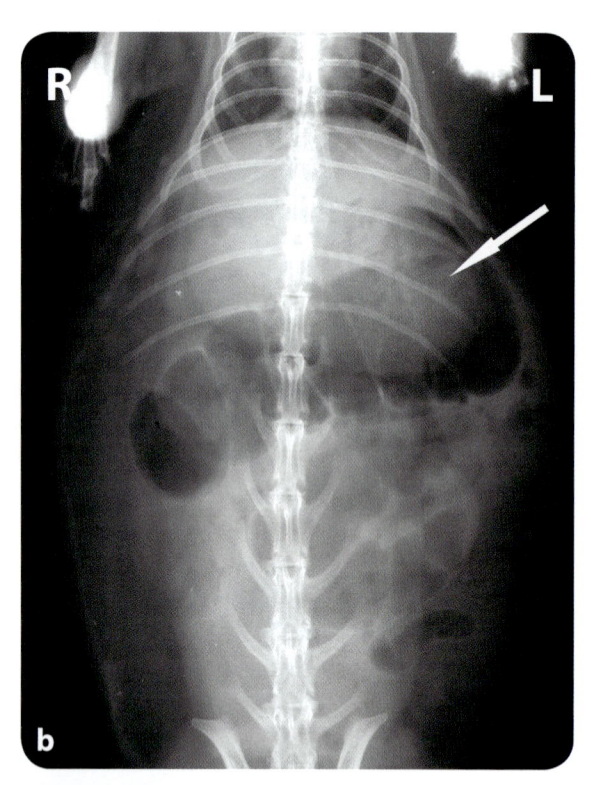

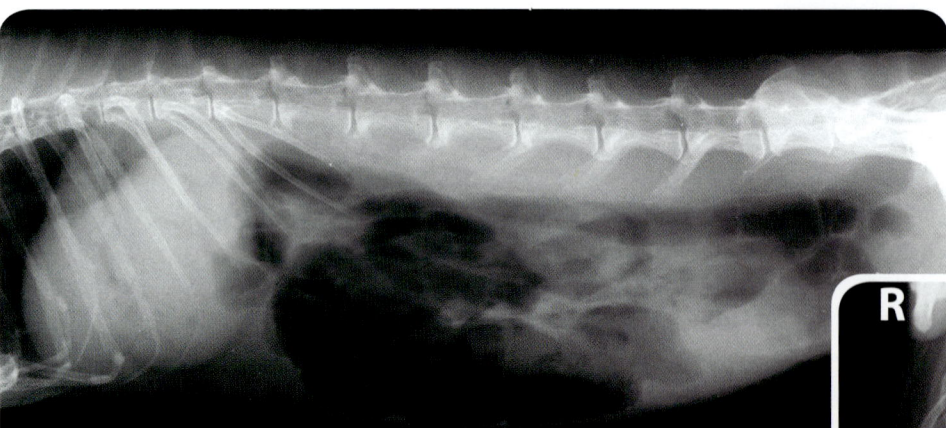

Figure 2.134a,b. Lateral (a) and ventrodorsal (b) radiographs of the abdomen of a rabbit showing marked accumulations of cecal gas. Simultaneous distention and impaction of the stomach with excessive cecal gas is another common presentation in rabbits with gastrointestinal disease.

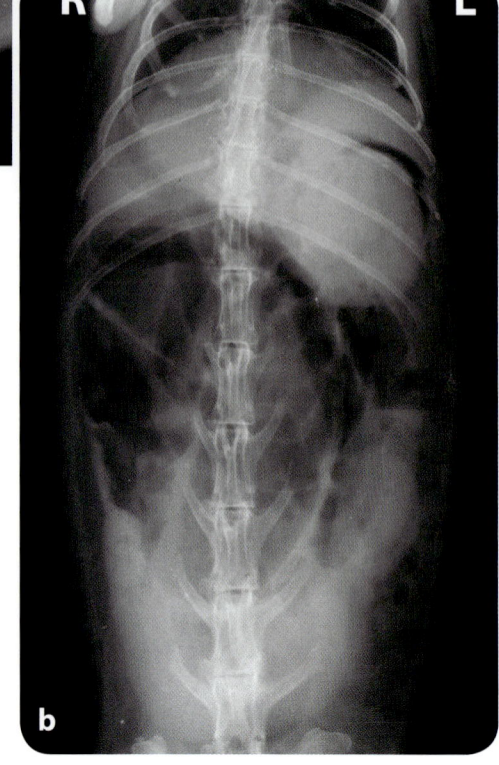

ABDOMEN

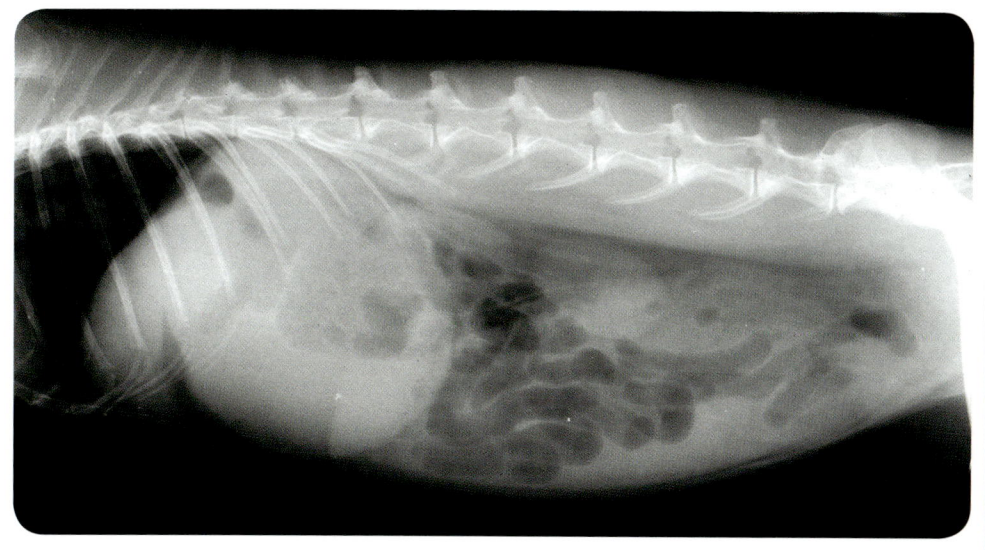

Figure 2.135. Lateral radiograph of the abdomen of a rabbit with increased gas accumulation in the small intestine. The stomach is filled, but there is no gas in the cecum. This is indicative of severe enteritis, often mucoid. Prognosis is guarded.

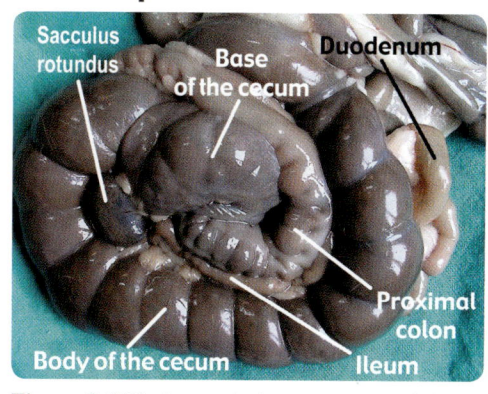

Figure 2.136. Anatomical arrangement of the intestines of the rabbit.

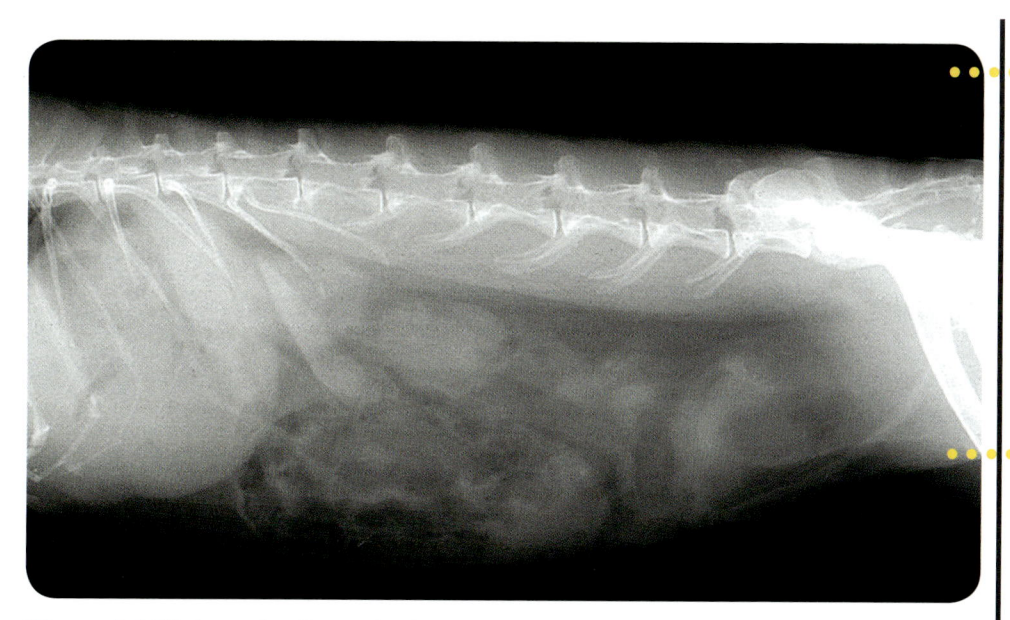

Figure 2.137. Lateral radiograph of the abdomen of a rabbit demonstrating increased gas accumulation in the colon. The small haustra of the descending colon are visible. Rabbits with intestinal gas accumulation often present with diarrhea.

Figure 2.138. True diarrhea in the rabbit. Liquid feces stain the fur of the perineum.

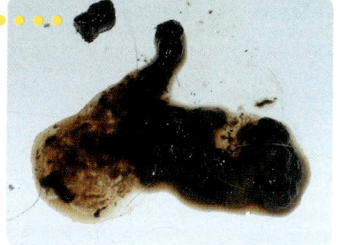

Figure 2.139. Feces typically seen in rabbits with mucoid enteritis.

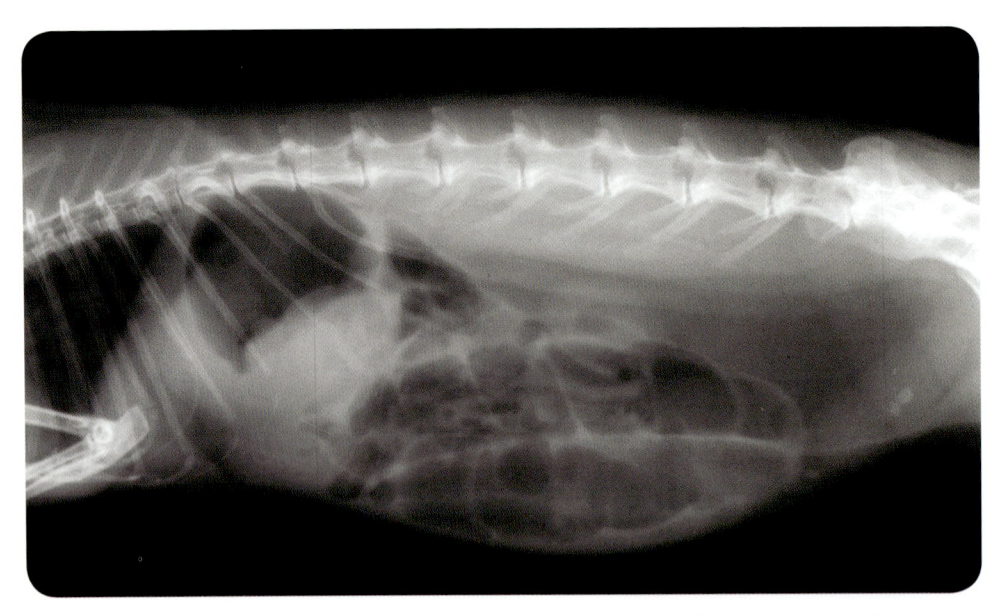

Figure 2.140. Lateral radiograph of the abdomen of a rabbit showing increased gas accumulation in the stomach and cecum. The haustra of the cecum are easily visualized. Note opacities in the caudoventral abdomen consistent with uroliths.

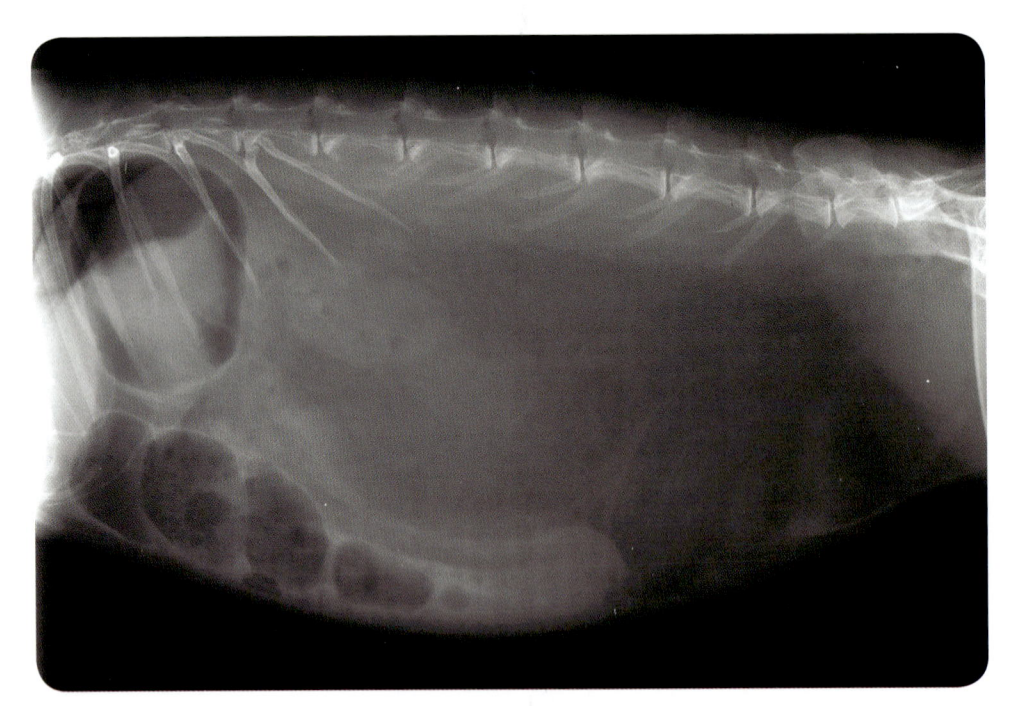

Figure 2.141. Lateral radiograph of the abdomen of an obese rabbit demontrating gastrointestinal gas. Intrabdominal fat displaces the cecum cranially. The haustra are easily visualized. Note that fat has a unique radiopacity, less than soft tissue but greater than air.

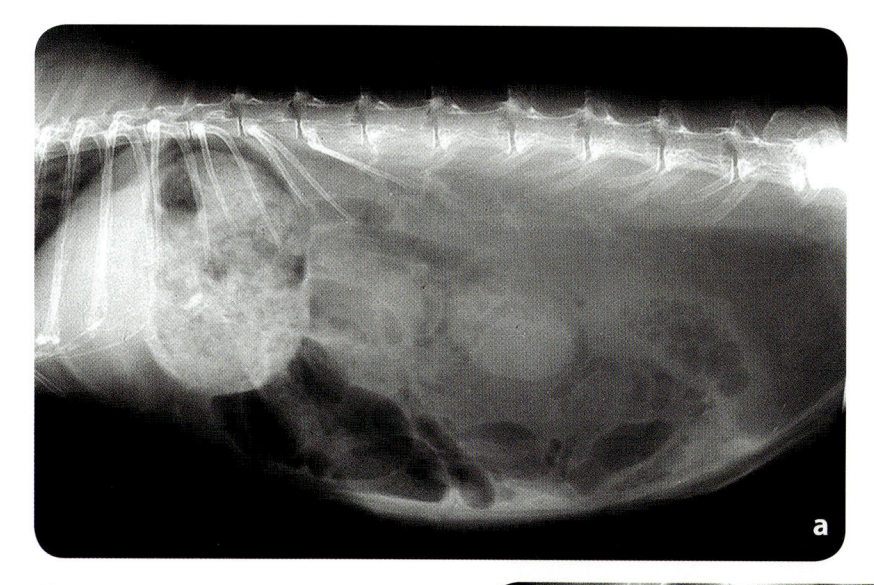

Figure 2.142a,b. Lateral (a) and ventrodorsal (b) projections of a rabbit with gastric impaction and cecal gas. Note the radiodensity of gastric contents is suggestive of an inappropriate diet. This condition often responds to medical therapy, including motility regulators, fluids, and hand-feeding of a high fiber support diet.

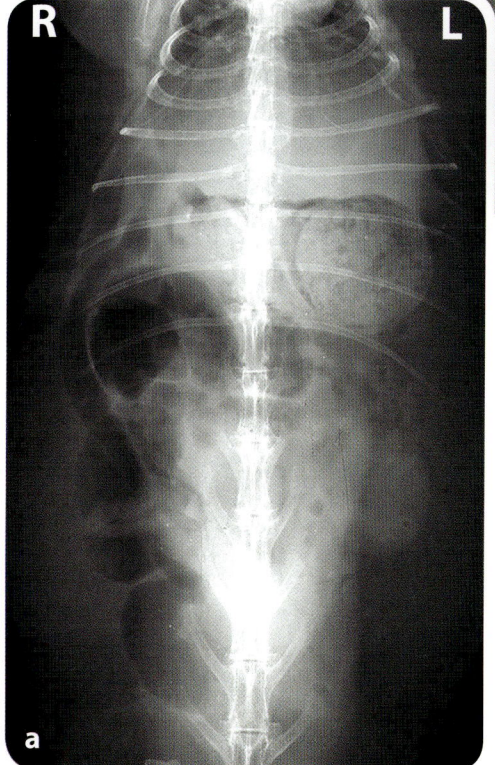

Figure 2.143a,b. Follow up, same patient, 2 days (a) and one week (b) post treatment. In (b), gastric gas is no longer present, and intestinal gas is reduced. Note the appearance of fat in this obese rabbit, which increases intra-abdominal contrast.

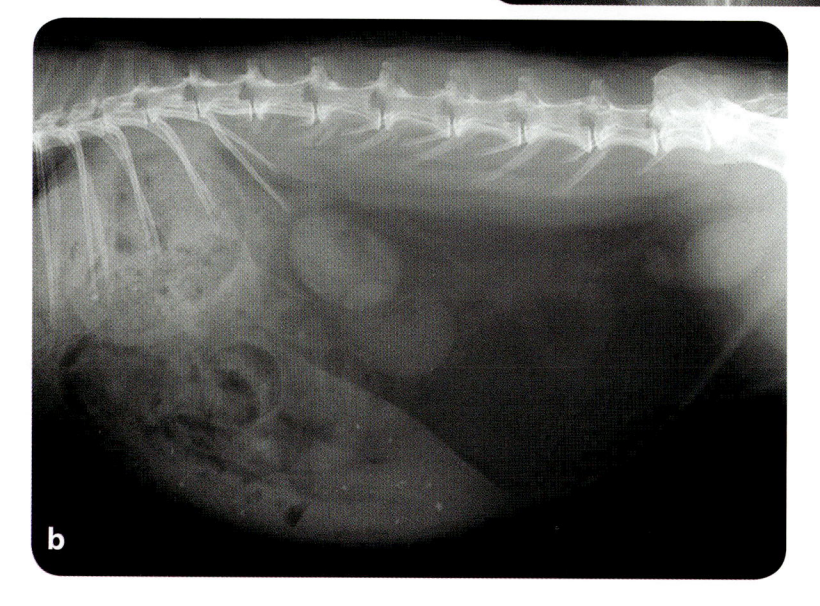

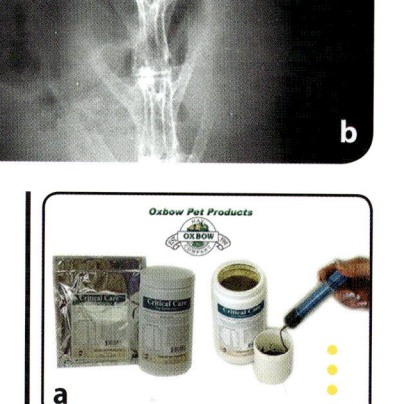

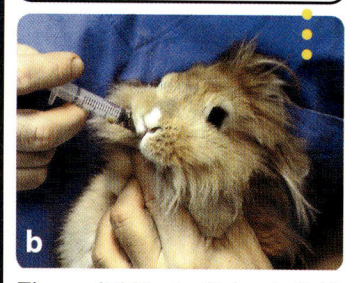

Figure 2.144a,b. Oxbow's Critical Care for herbivores is an excellent high fiber food for support feeeding.

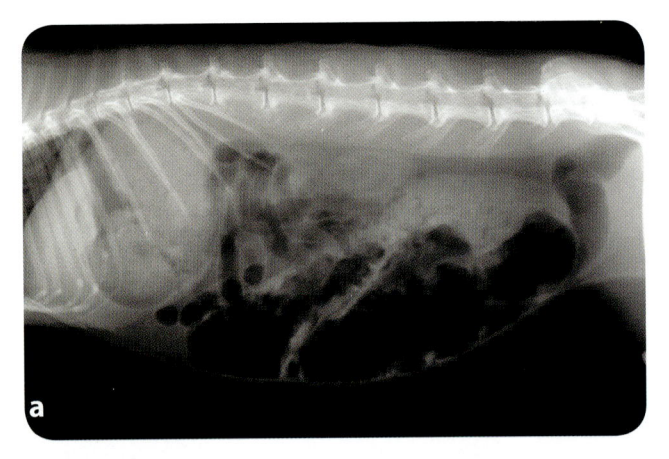

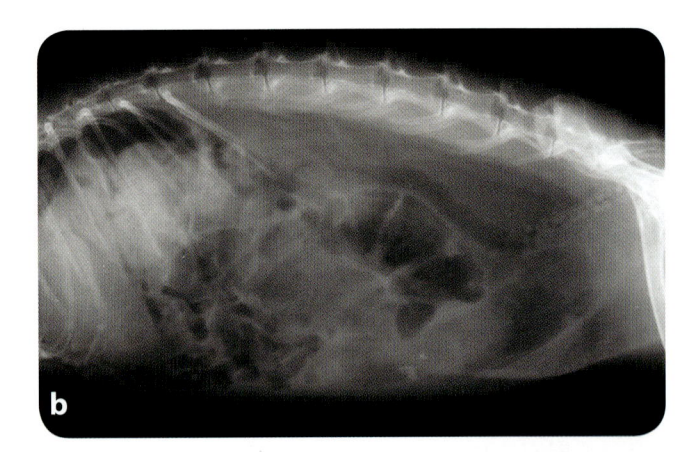

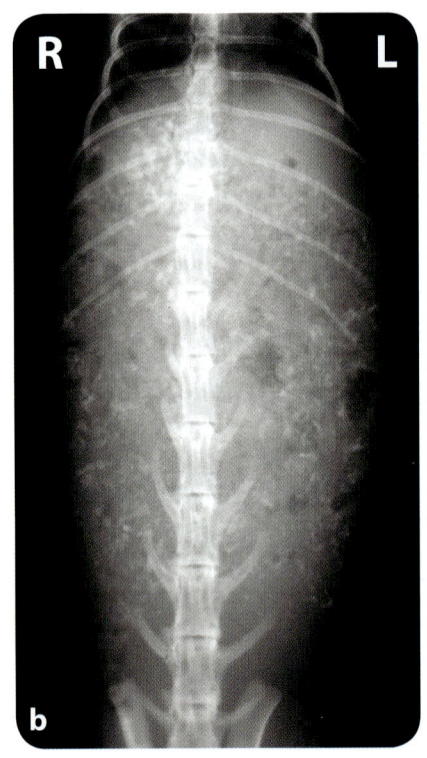

Figure 2.145a,b. Lateral radiograph of the abdomen of a rabbit with a severe case of gastric impaction and cecal gas (a). Follow-up three days after initiating medical therapy (b).

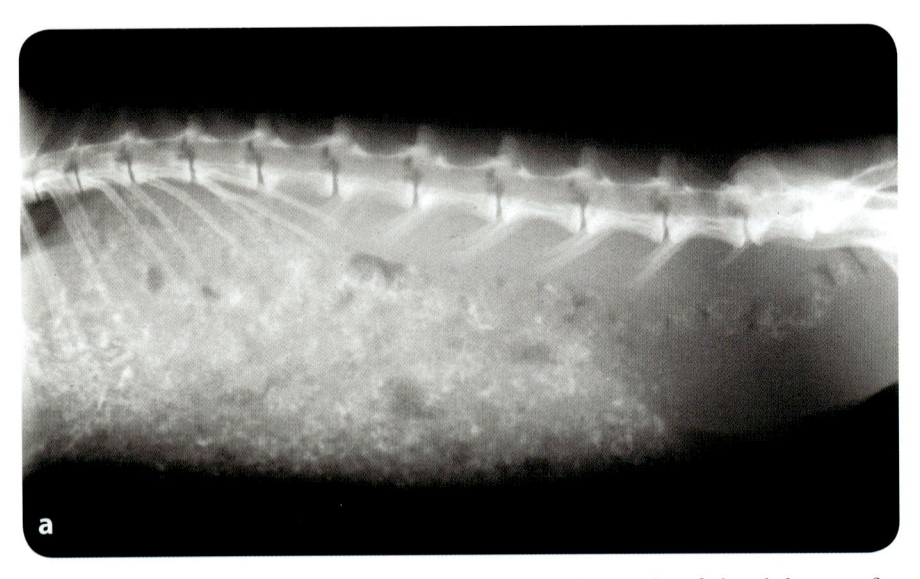

Figure 2.146a,b. Lateral (a) and ventrodorsal (b) radiographs of the abdomen of a rabbit with gastrointestinal impaction as a result of ingestion of cat litter. This granular radiopaque pattern is typical for this type of material.

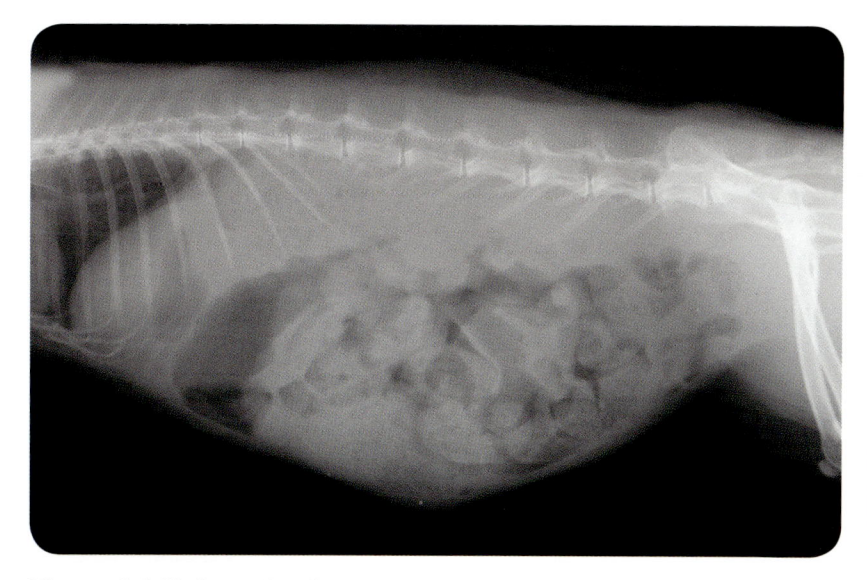

Figure 2.147. Lateral radiograph of the abdomen of a rabbit with anorexia and palpable large, firm intestinal contents. There is severe intestinal impaction due to the presence of many 1-3 cm diameter opaque intraluminal masses. Note the presence of gas, which is acting as negative contrast. At exploratory surgery, the masses were determined to be very firm ingesta impacting the cecum.

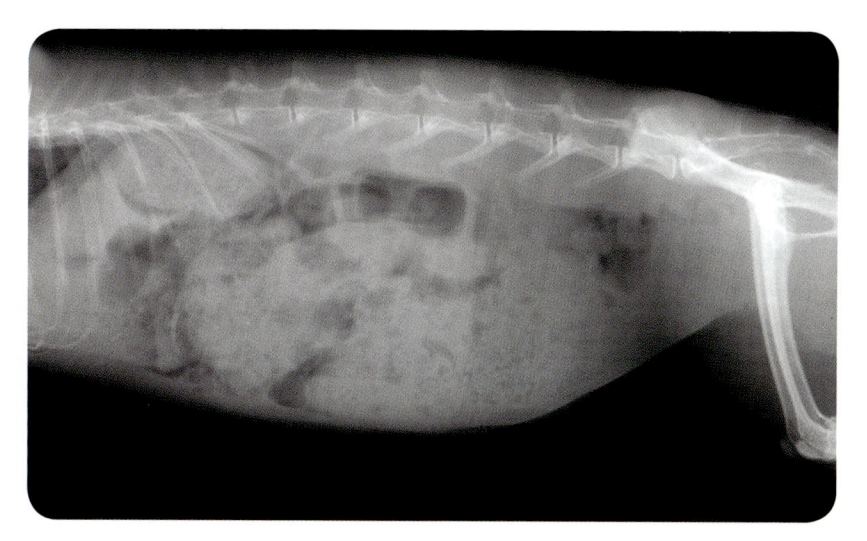

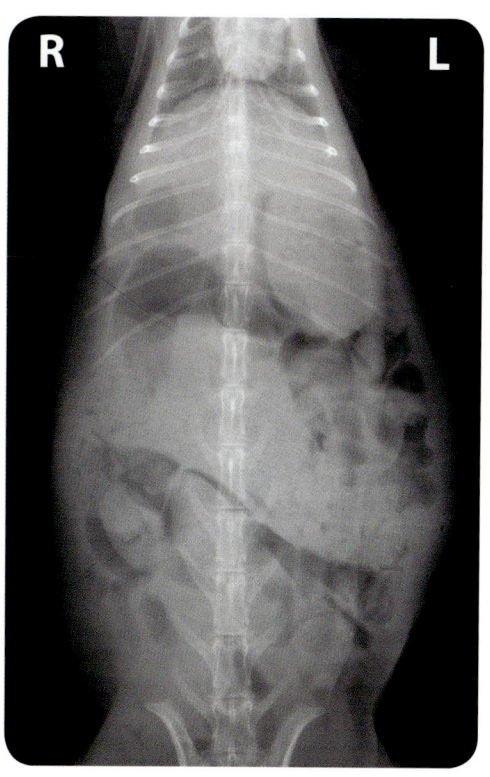

Figure 2.148. Ventrodorsal view, same patient. Gastric ingesta are more radiopaque than normal. Gas distention of the small intestine is also present. Opaque masses of various shapes and size are present in the cecum.

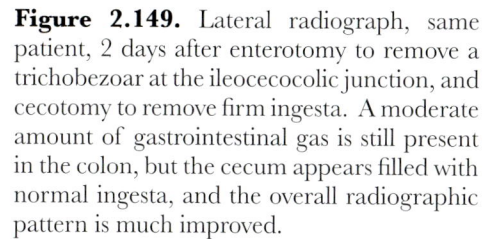

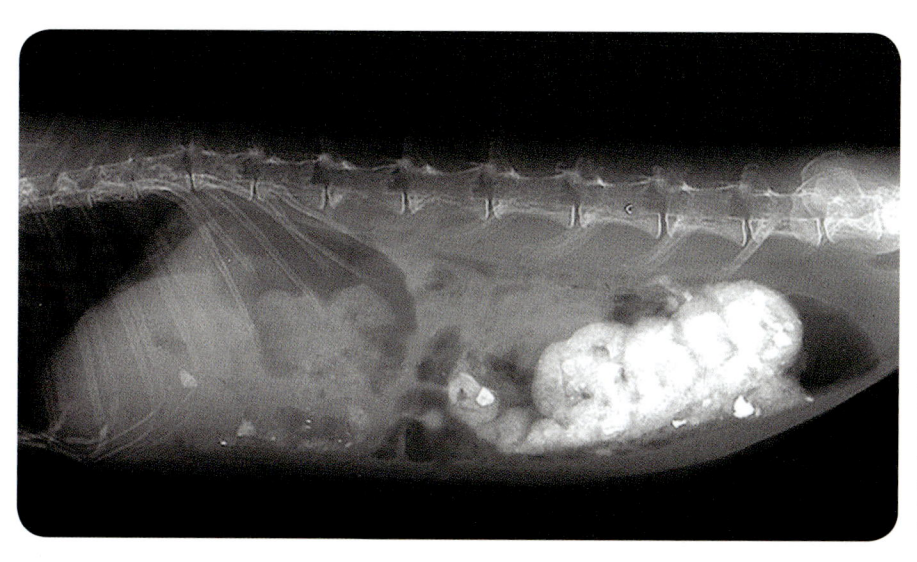

Figure 2.149. Lateral radiograph, same patient, 2 days after enterotomy to remove a trichobezoar at the ileocecocolic junction, and cecotomy to remove firm ingesta. A moderate amount of gastrointestinal gas is still present in the colon, but the cecum appears filled with normal ingesta, and the overall radiographic pattern is much improved.

Figure 2.150. Lateral radiograph of the abdomen of another rabbit with severe impaction of the cecum. This rabbit had ingested paper, dirt, and sand. Note marked radiopacity of cecal material due to high mineral content.

Diseases of the Liver

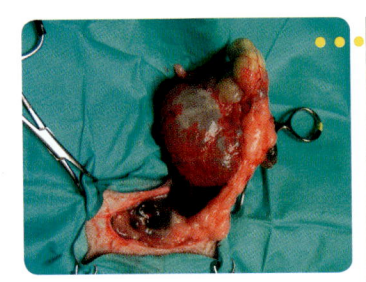

Figure 2.151. Interoperative appearance of one of the two hepatic cysts. *Cysticercus pisiformis* is the larval stage of the dog and fox tapeworm *Taenia pisiformis*, and rabbits act as the intermediate host. Cysts are typically found in the mesentary.

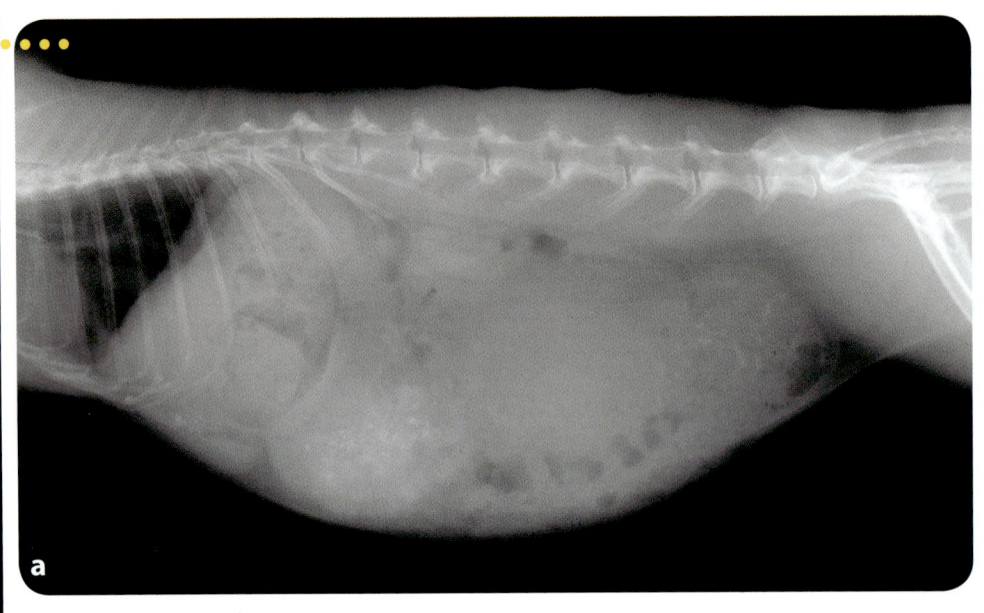

Figure 2.152a,b. Lateral (a) and ventrodorsal (b) radiographs of a 4-year-old rabbit with two large palpable abdominal masses. The ventrodorsal projection (b) shows two cyst-like masses containing focal mineralizations. Surgery and histopathology revealed these were *Cysticercus pisiformis* cysts associated with the liver.

ABDOMEN 117

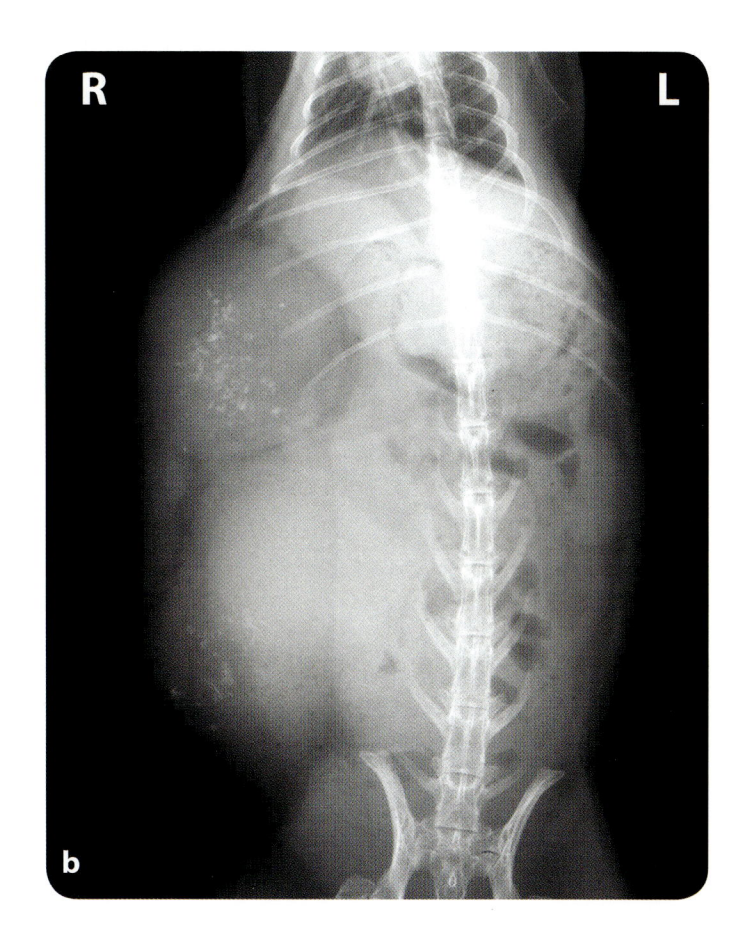

Diseases of the Kidneys and Ureters

Uroliths in rabbits are common, are usually composed of calcium carbonate, and can appear anywhere in the urinary tract. Nephrolithiasis is frequently bilateral and can affect renal function. Apparent cases of unilateral disease must be monitored carefully for development of disease in the contralateral kidney/ureter. Prognosis is guarded to poor, and is related to renal function. Blood urea nitrogen, creatinine, and phosphorus are often extremely elevated, especially in bilateral disease.

Figure 2.153. Rabbits with bilateral nephrolithiasis often present with anorexia, depression, severe, chronic weight loss, and dehydration.

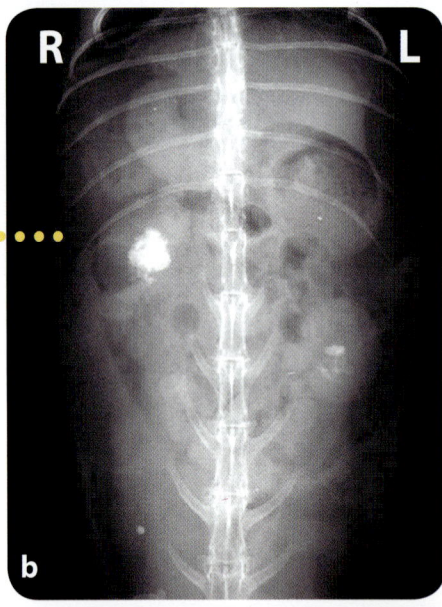

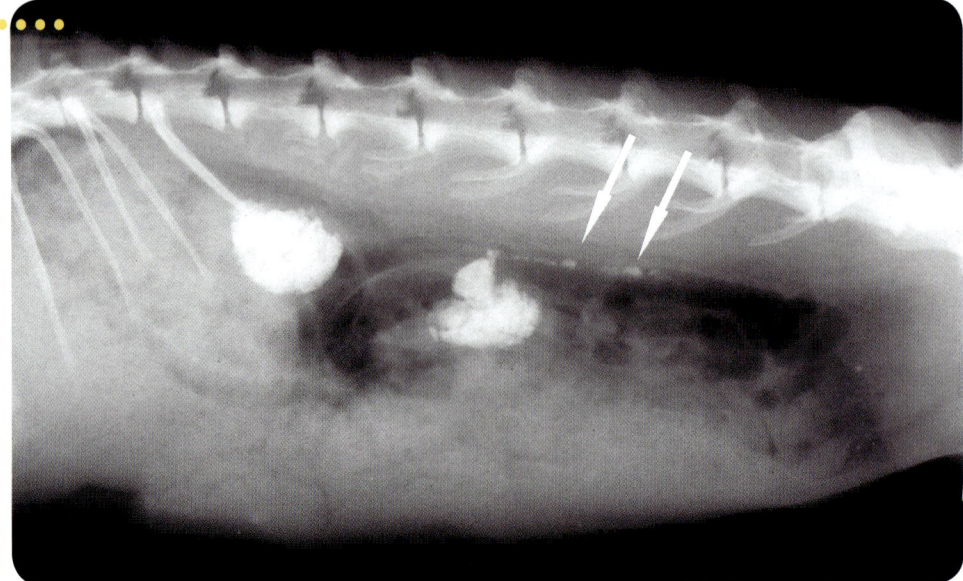

Figure 2.154a,b. Lateral (a) and ventrodorsal (b) projections of the abdomen of a rabbit with bilateral nephrolithiasis. The right kidney contains the larger nephrolith.

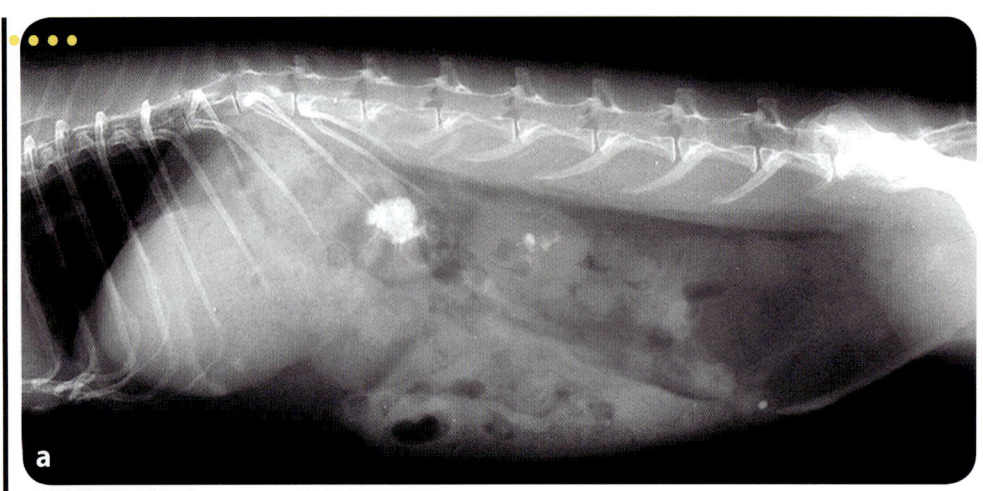

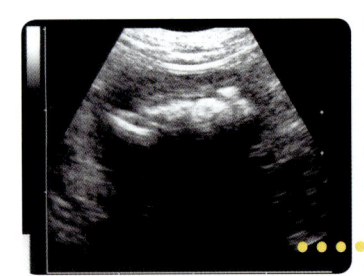

Figure 2.155. In this sonogram, intense hyperechogenicity and adjacent acoustic shadowing indicate nephrolithiasis.
(Courtesy of Claudio Bussadori, DVM)

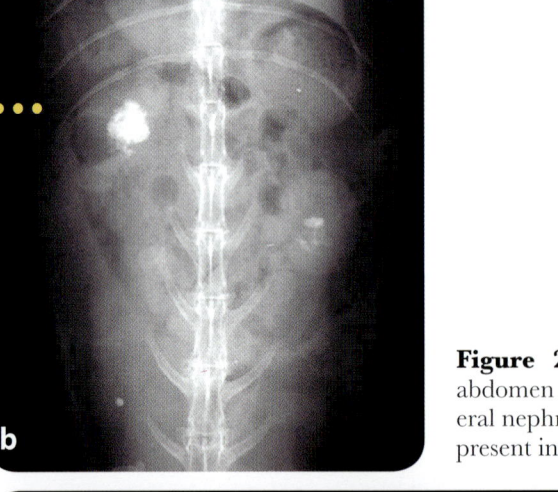

Figure 2.156. Lateral radiograph of the abdomen of a rabbit with more severe bilateral nephrolithiasis. Two small uroliths are also present in a ureter (arrows).

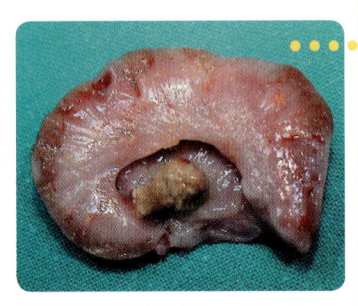

Figure 2.157. Gross appearance of nephrolithiasis. The kidney is enlarged and irregular in texture and appearance. The renal pelvis is dilated and contains the nephrolith. Nephroliths are usually softer and more friable than those in the urinary bladder.

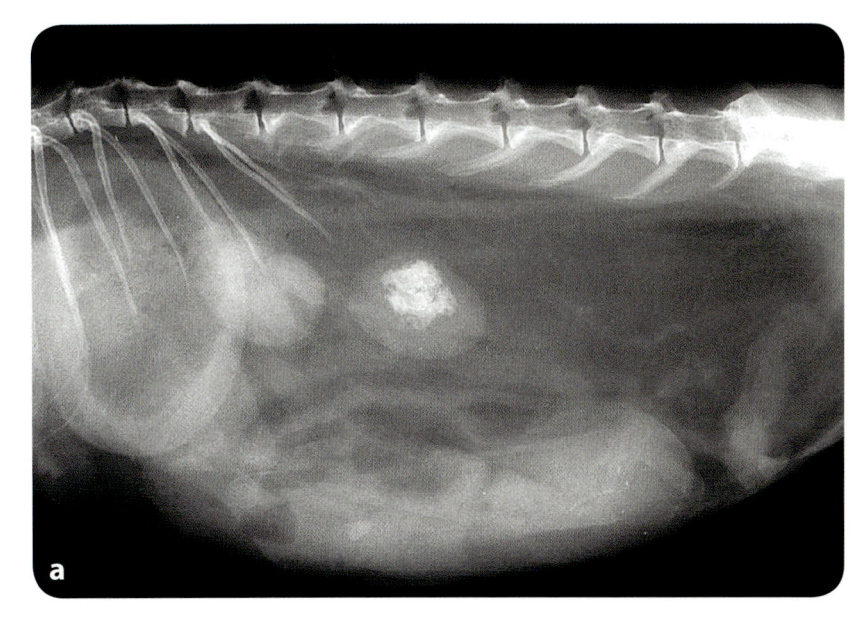

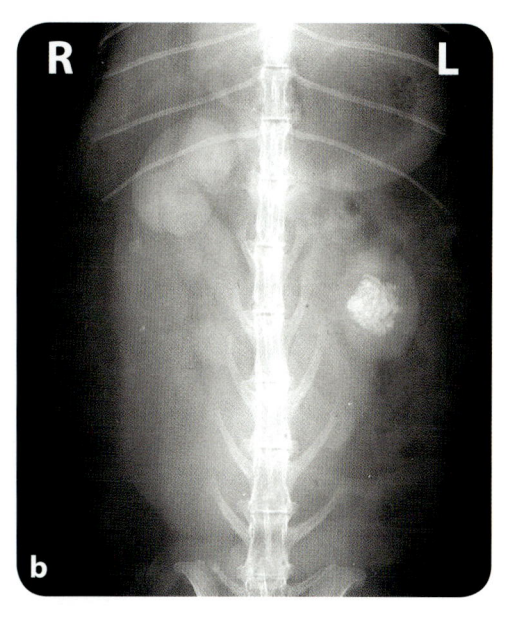

Figure 2.158a,b. Lateral (a) and ventrodorsal (b) radiographs of the abdomen of a rabbit demonstrating unilateral nephrolithiasis of the left kidney. The right kidney appears normal. Prior to decisions regarding surgical treatment of unilateral nephrolithiasis (nephrotomy, nephrectomy), function of the apparently normal kidney must be carefully evaluated.

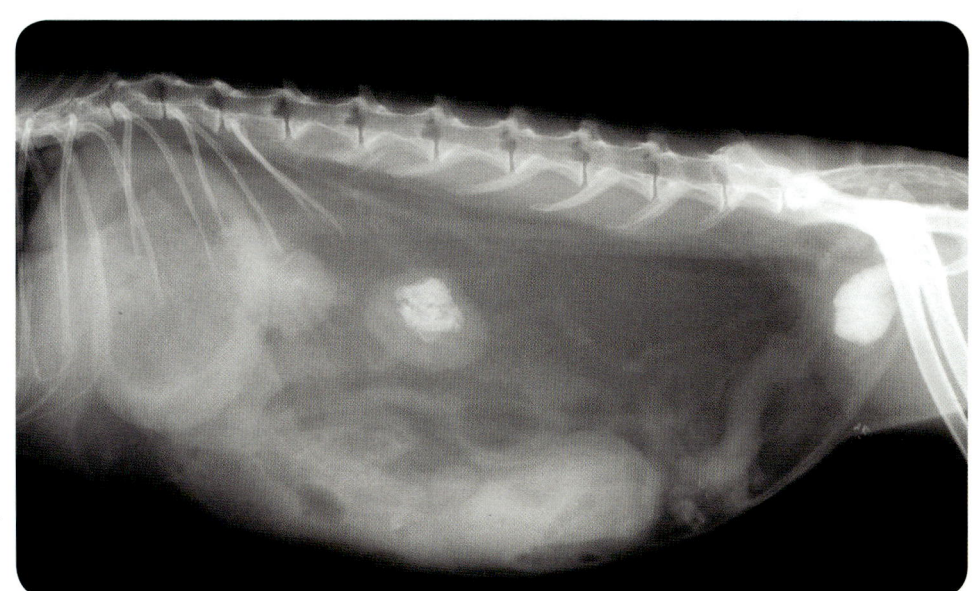

Figure 2.159. Excretory urogram, same patient. This study may give a relative indication of renal function. This lateral projection was made 15 minutes after injection of contrast medium. Contrast medium is visualized in the urinary bladder, suggesting normal function of at least one kidney.

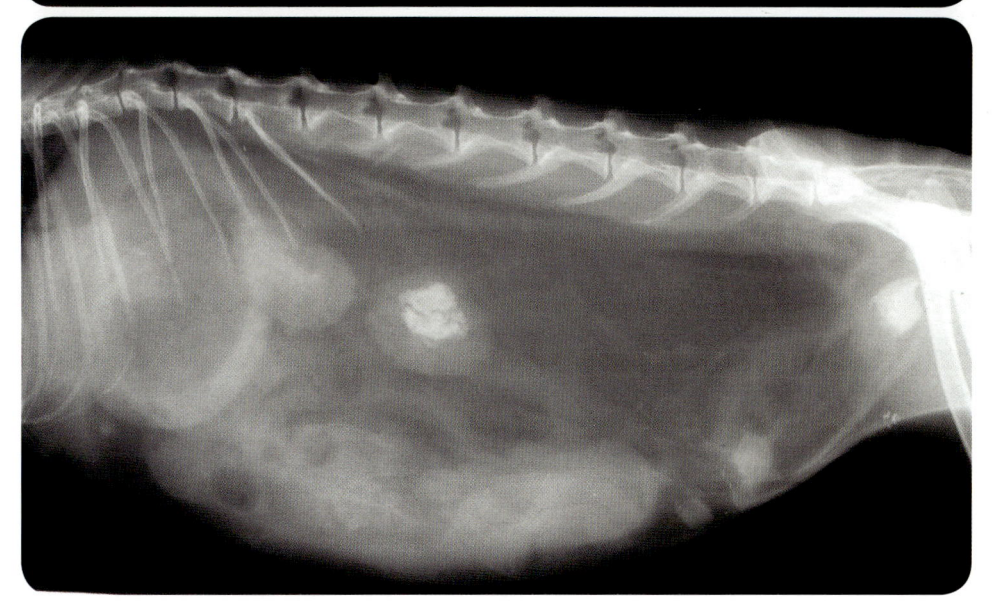

Figure 2.160. 30 minutes after the injection of contrast medium, same patient. In this case of unilateral nephrolithiasis, as one kidney appears to be functioning adequately, surgical options of nephrectomy or nephrotomy to remove the urolith(s) from the affected renal pelvis may be considered.

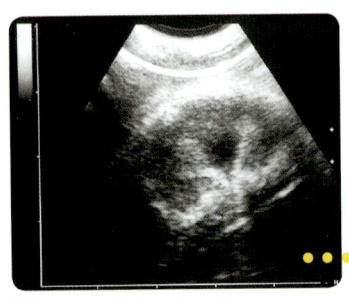

Figure 2.161. In this sonogram of diffuse renal calcinosis, intense hyperechogenicity is apparent in both the renal cortex and collecting system.
(Courtesy of Claudio Bussadori, DVM)

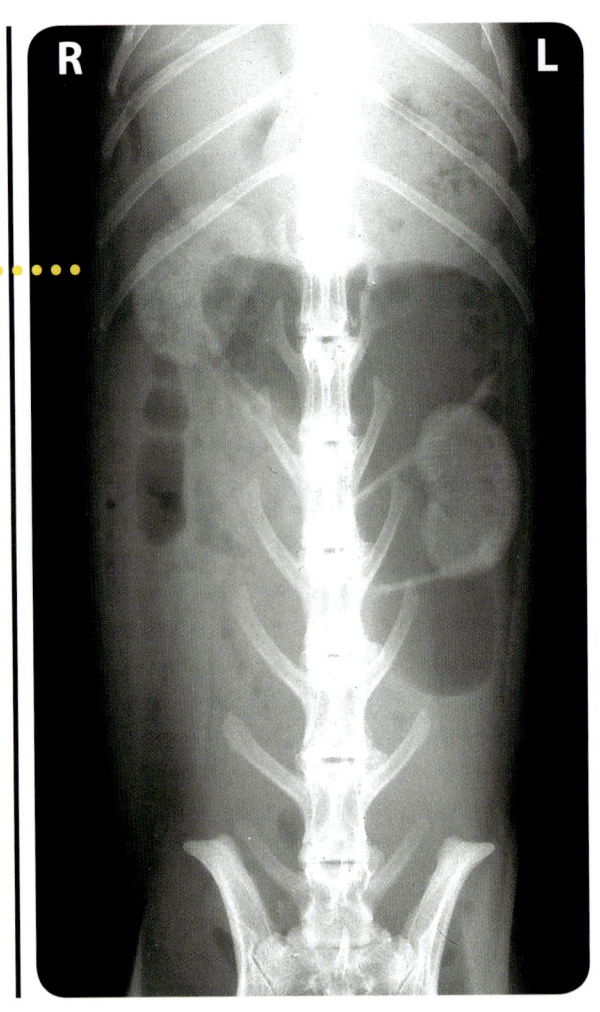

Figure 2.162. Ventrodorsal radiograph of the abdomen of a rabbit demonstrating diffuse bilateral renal mineralization. There is increased renal opacity, which is more pronounced peripherally.

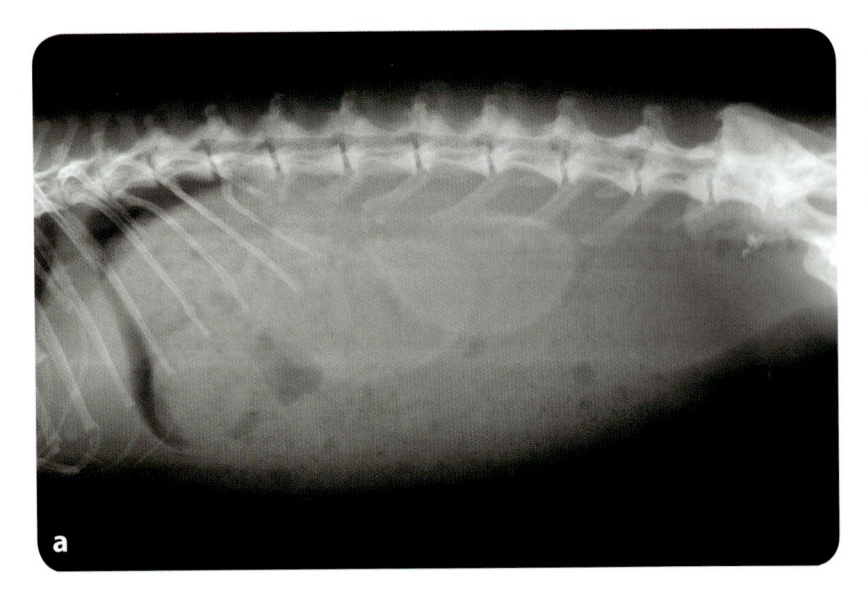

Figure 2.163a,b. Lateral (a) and ventrodorsal (b) radiographs demonstrating bilateral renomegaly with unusual cortical mineralization. There is loss of abdominal contrast and detail, consistent with peritoneal effusion. On physical examination, the rabbit was thin, and kidneys were readily palpable. BUN and creatinine were within normal limits in this patient, which survived 8 months after diagnosis. Necropsy was declined.

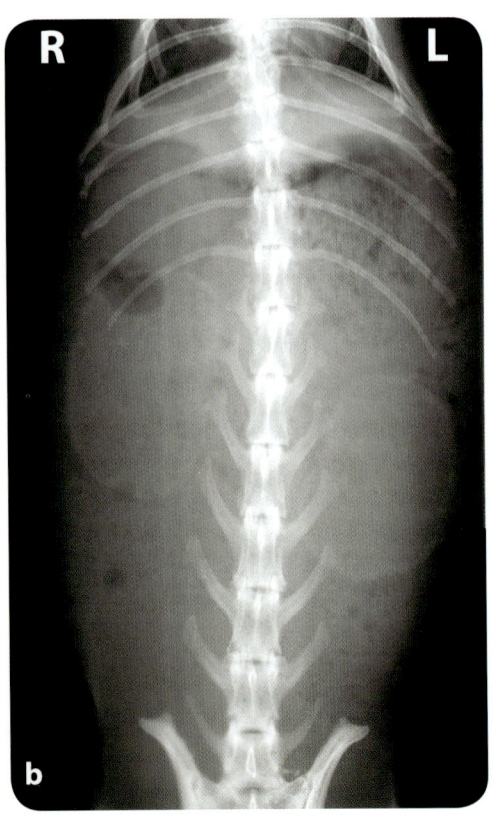

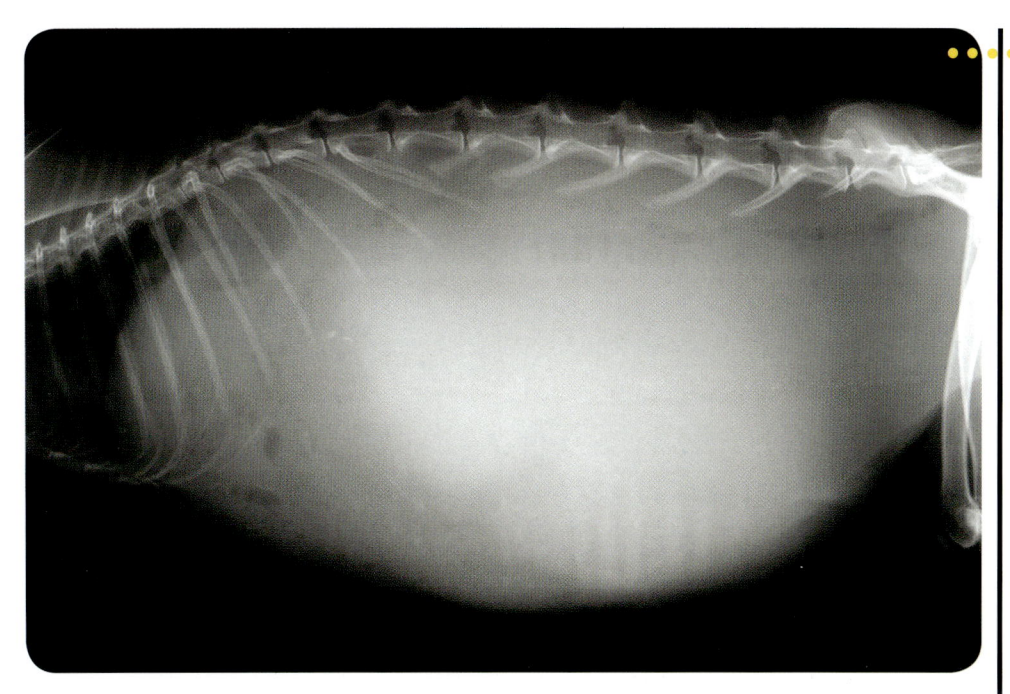

Figure 2.164. Lateral abdominal radiograph of a rabbit with renal neoplasia. There is abdominal distention with loss of visceral detail and contrast. Exploratory surgery revealed a large mass associated with the right kidney.

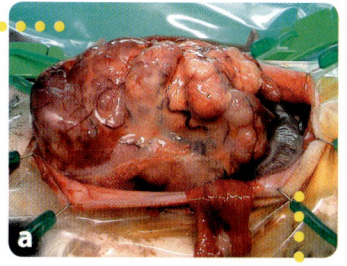

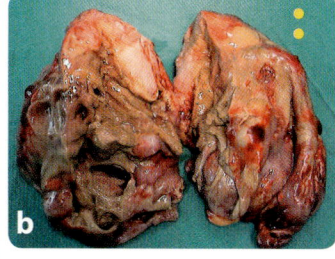

Figure 2.165a,b. Surgical excision of a 20 cm neoplastic right kidney (a) and incised gross specimen (b).

Diseases of the Urinary Bladder and Urethra

Figure 2.166a,b. Lateral (a) and ventrodorsal (b) projections showing mineralized opacity throughout the bladder and urethra. Rabbits normally excrete calcium carbonate, which forms precipitates in the urine. These precipitates are usually of no consequence, but excessive accumulation may occur giving the appearance of a contrast cystogram ("bladder sludge"). This condition is thought to be caused by urine retention, which has a variety of underlying causes, including obesity, illness, pododermatitis, and arthritis.

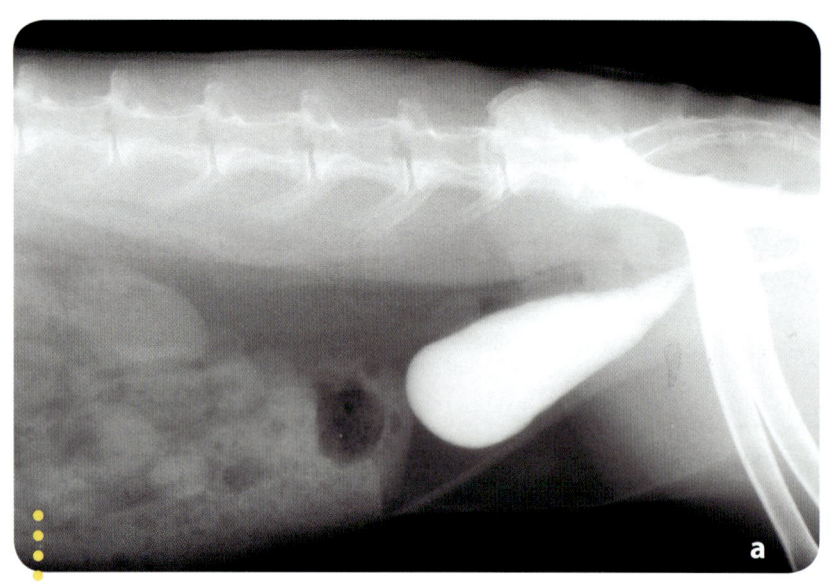

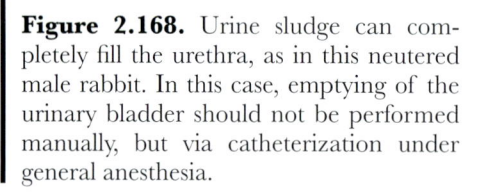

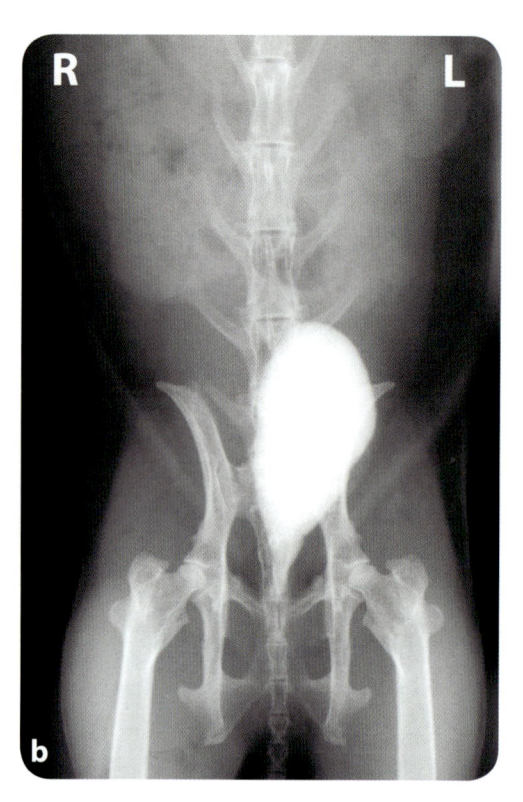

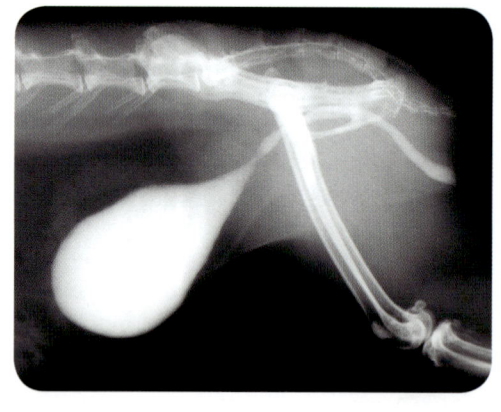

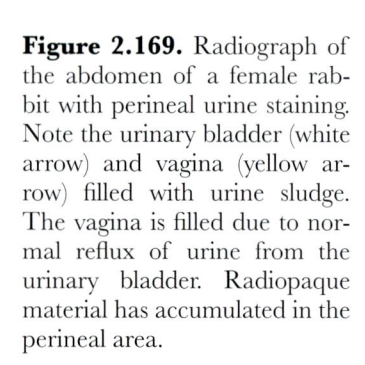

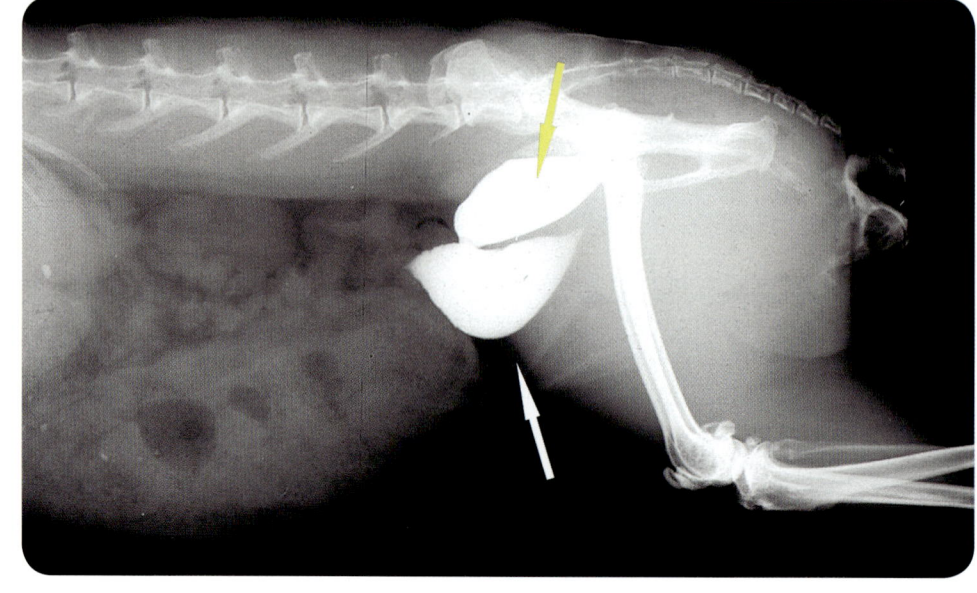

Figure 2.167. The dense, mineral-containing urine can have the appearance of liquid concrete. Sludge can often be expressed out of the bladder manually. The rabbit is usually able to eliminate the more fluid portions of the urine, but the sludge is retained in the bladder lumen.

Figure 2.168. Urine sludge can completely fill the urethra, as in this neutered male rabbit. In this case, emptying of the urinary bladder should not be performed manually, but via catheterization under general anesthesia.

Figure 2.169. Radiograph of the abdomen of a female rabbit with perineal urine staining. Note the urinary bladder (white arrow) and vagina (yellow arrow) filled with urine sludge. The vagina is filled due to normal reflux of urine from the urinary bladder. Radiopaque material has accumulated in the perineal area.

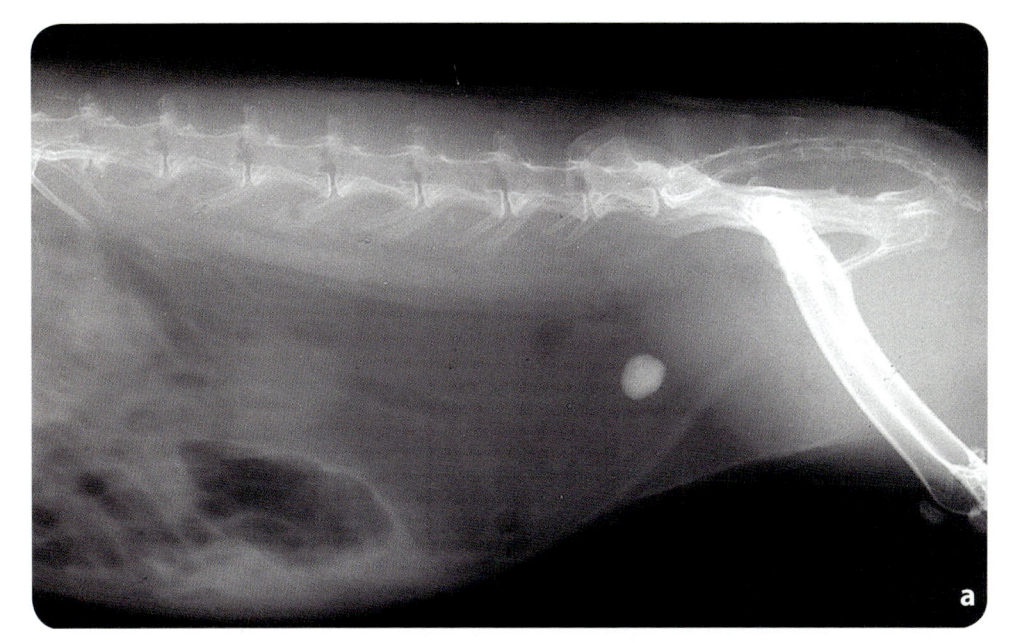

Figure 2.170a-e. Radiographs of the abdomen of several rabbits with cystic calculi. These are common in rabbits, are composed of calcium salts (carbonate, oxalate, phosphate), and are therefore radiopaque. Uroliths are typically single, and can be very large in relation to patient size.

A small urolith is shown in a neutered male (a). Small uroliths may be passed spontaneously, especially through the relatively large urethra of the female rabbit.

More than one urolith may be present (b,c,d,e). Larger uroliths (d,e) are unlikely to be passed spontaneously in eiher sex.

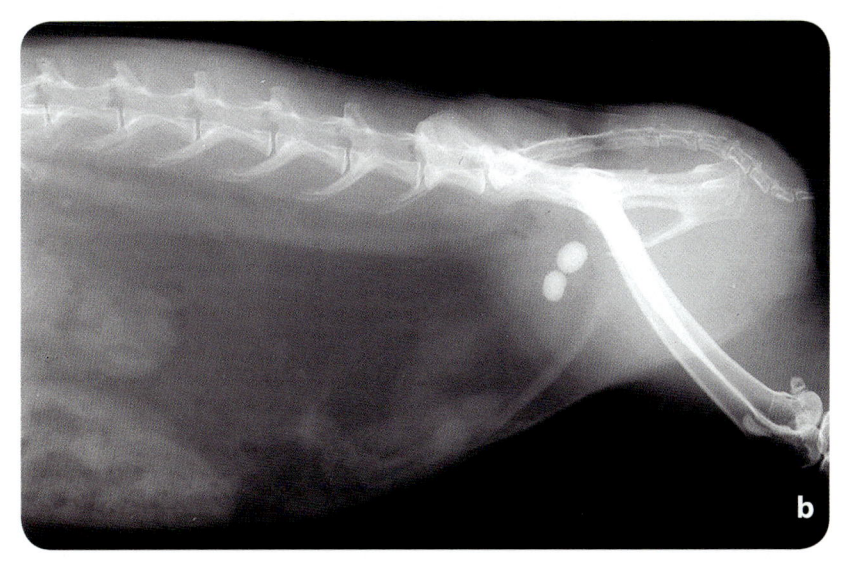

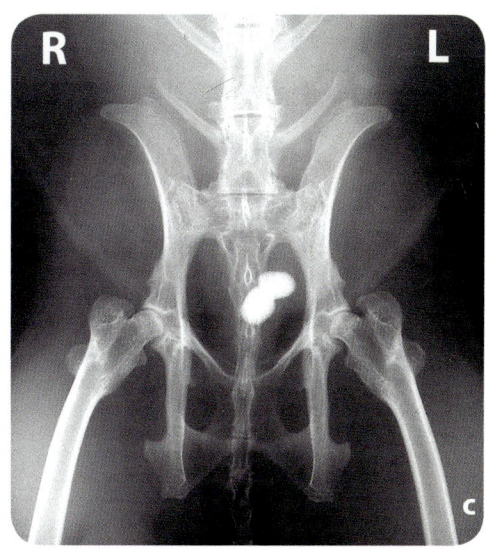

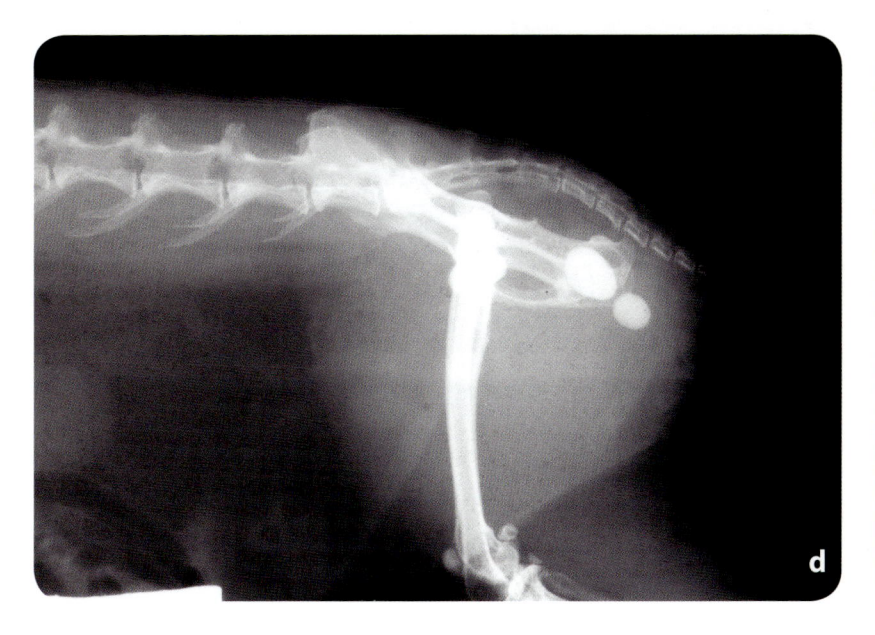

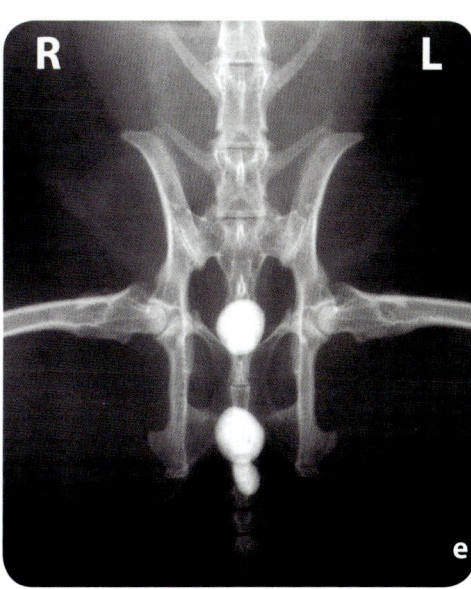

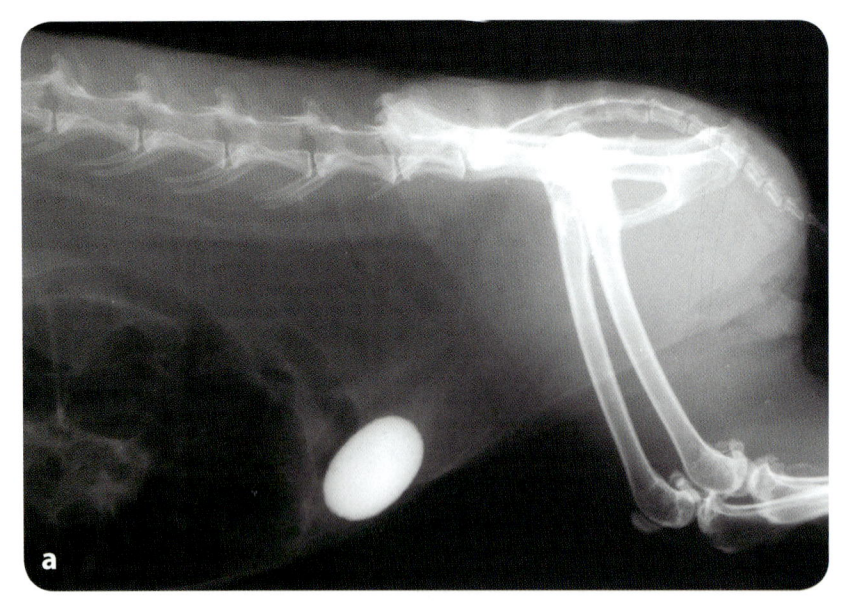

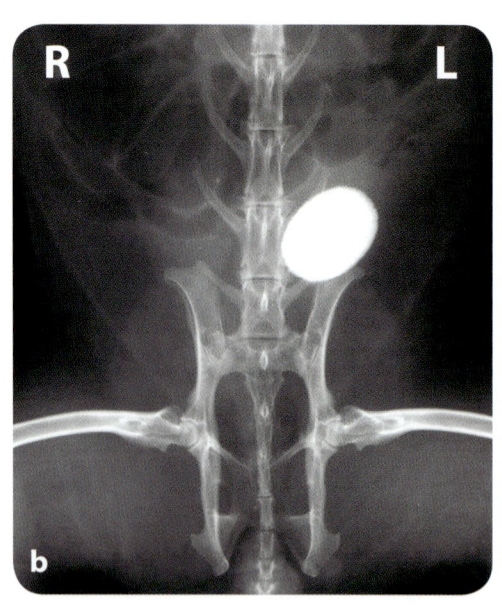

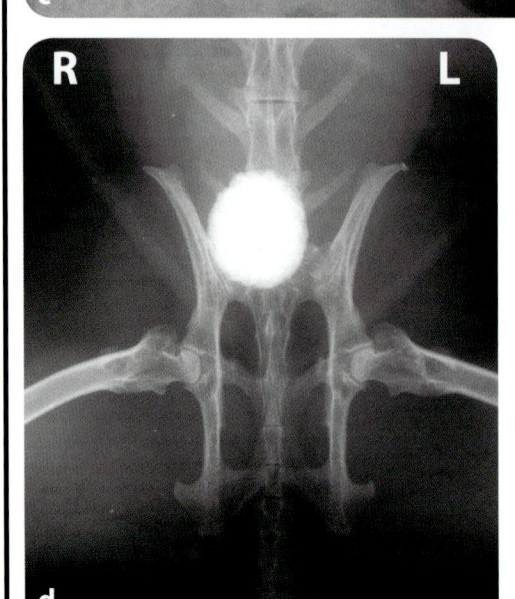

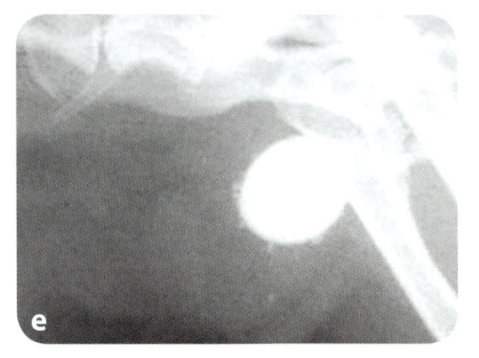

Figure 2.171a-e. Lateral (a) and ventrodorsal (b) radiographs of rabbits with cystic calculi. Some can become very large (a, b, c, d) due to continuous formation. The large urocystolith in (c) is irregularly marginated. Note numerous microuroliths surrounding a large single urolith (d, e).

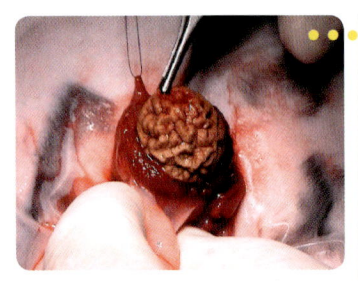

Figure 2.172. Cystotomy is treatment of choice for large stones unlikely to pass spontaneously.

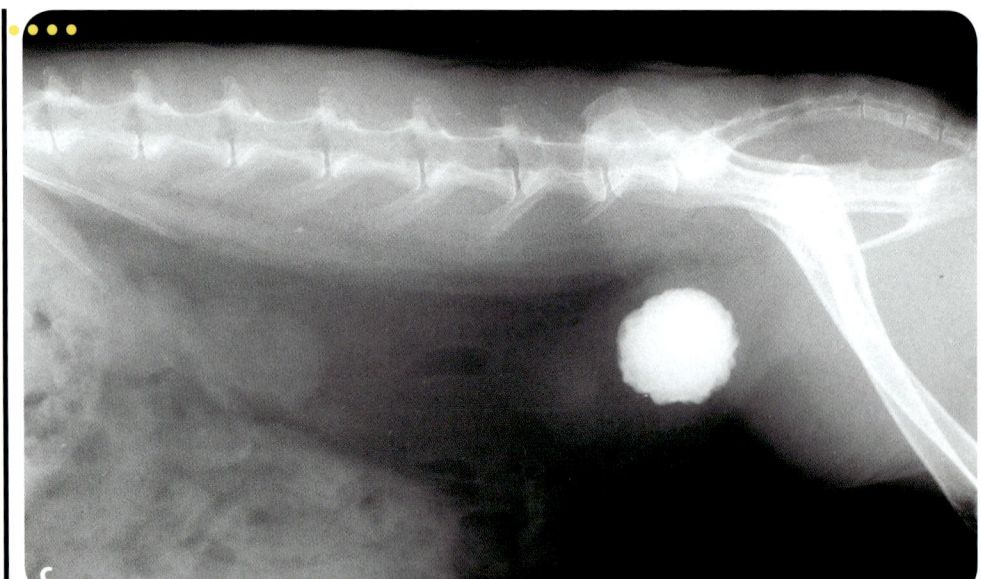

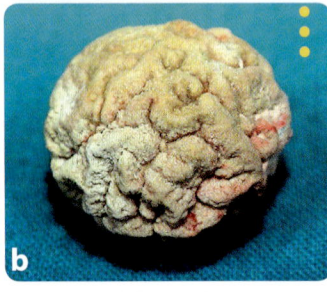

Figure 2.173a,b. Gross appearance of a large urocystolith demonstrating aggregation of thousands of small uroliths (a). In some cases, the smooth surface of the urolith forms irregular folds due to contact with the thickened mucosa of the bladder during stone formation (b).

ABDOMEN

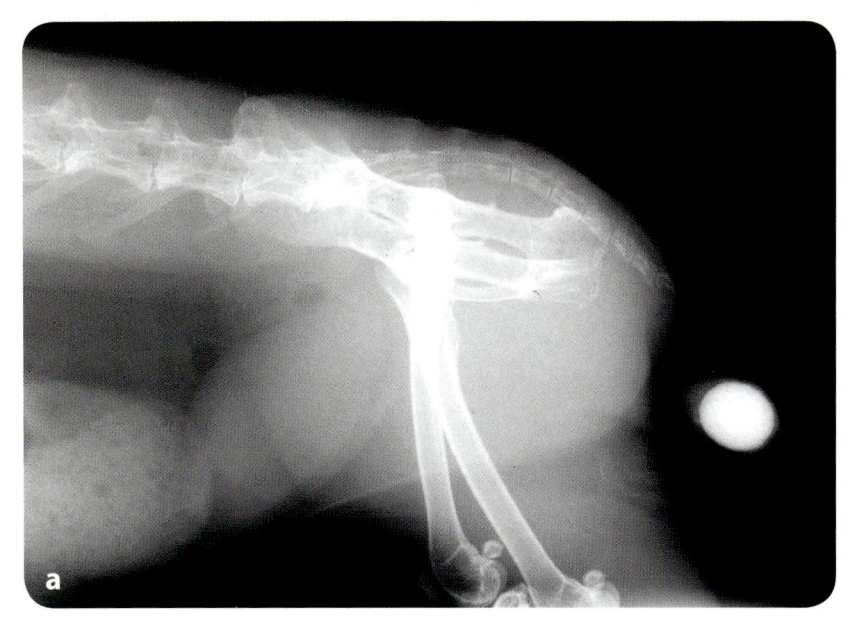

Figure 2.174a,b,c. Lateral (a,c) and ventrodorsal (b) radiographs of the abdomen and perineum of two male rabbits. Uroliths occasionally move into the urethra. In rabbits, the urethra is usually filled with a single large urolith rather than multiple small stones. Abdominal radiographs should always include the entire pelvis, or distally-located uroliths such as these will not be seen.

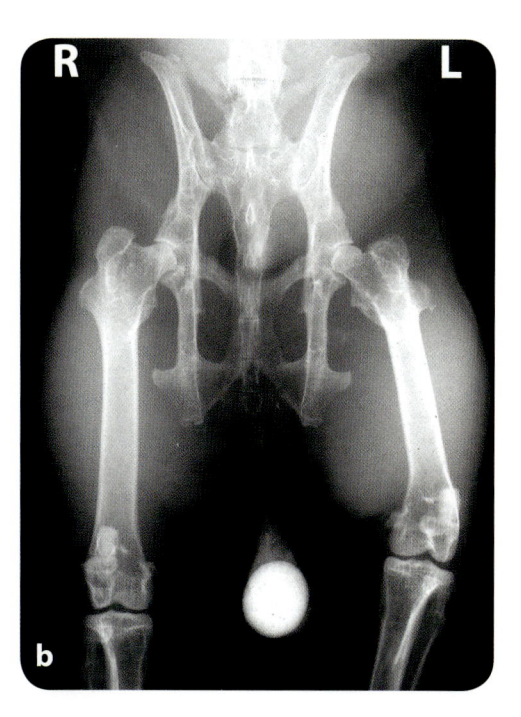

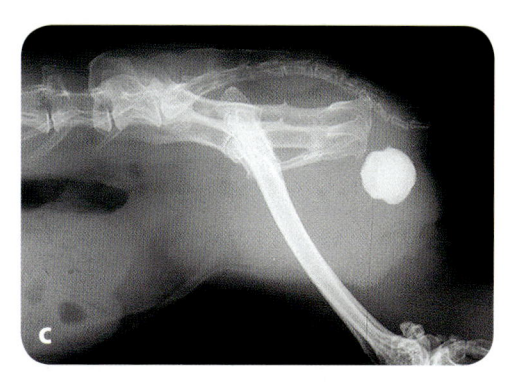

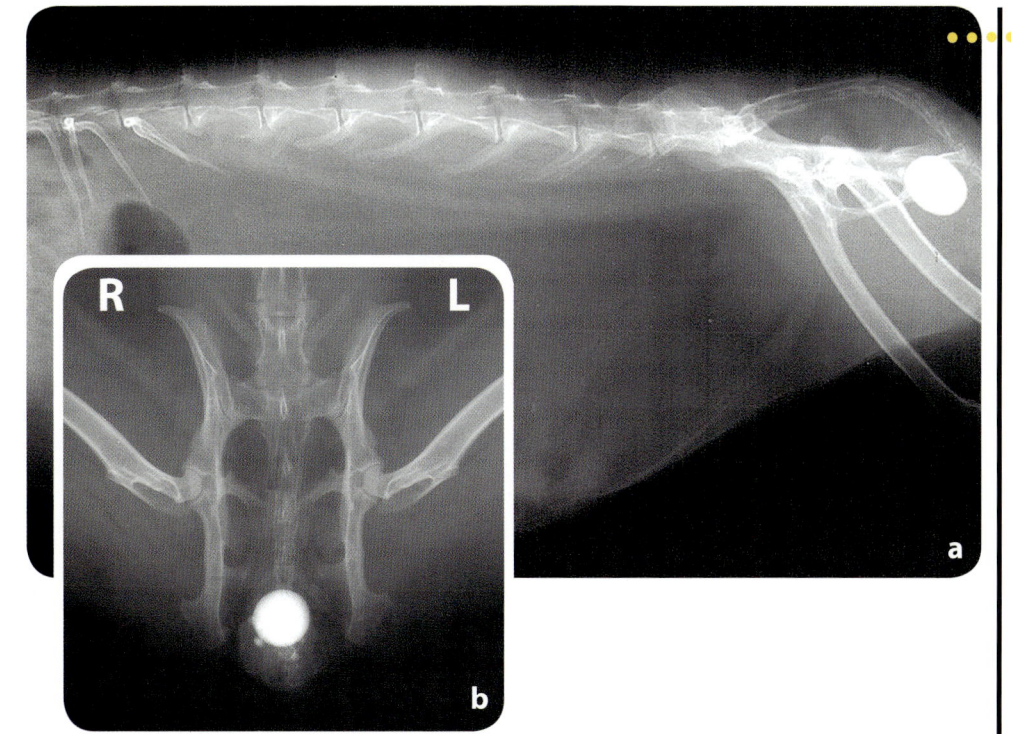

Figure 2.175a,b. Lateral (a) and ventrodorsal (b) radiographs showing a large urolith with two adjacent smaller uroliths in a female rabbit.

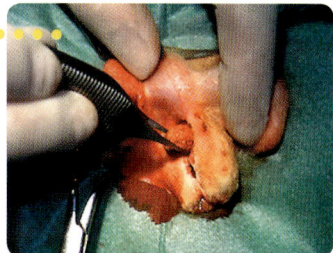

Figure 2.176. Despite large size, these uroliths often do not produce complete obstruction, and the rabbit is able to urinate in small amounts around the stone. Urethrotomy is the treatment of choice and has a good prognosis in pet rabbits.

Diseases of the Uterus and Vagina

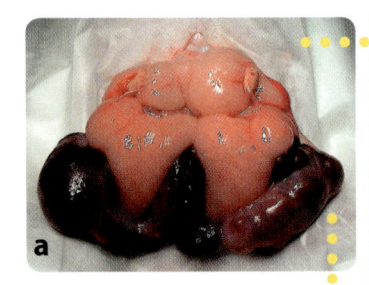

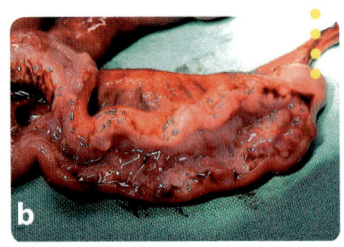

Figure 2.177a,b. Endometrial cystic hyperplasia is common in intact unbred female rabbits. Intraoperative appearance (a) and gross specimen (b).

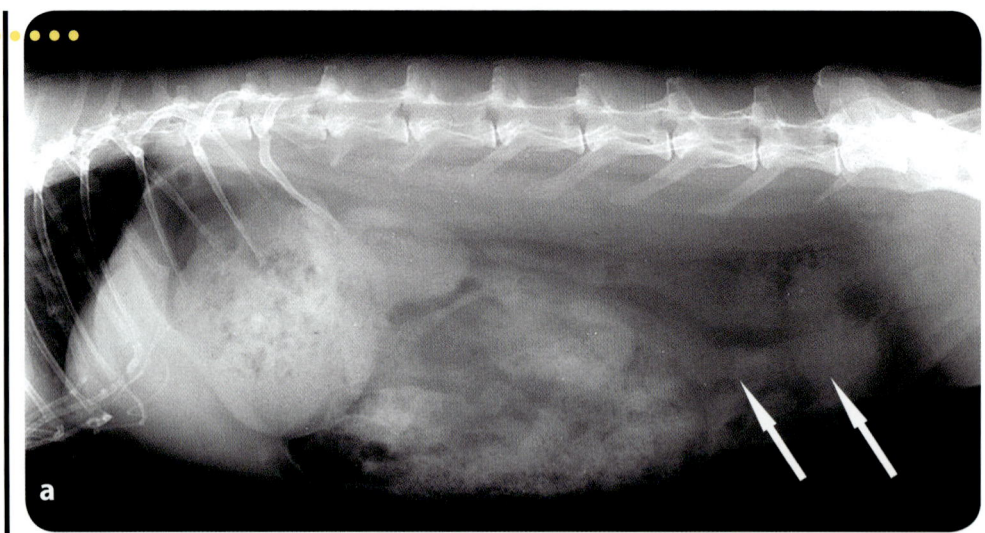

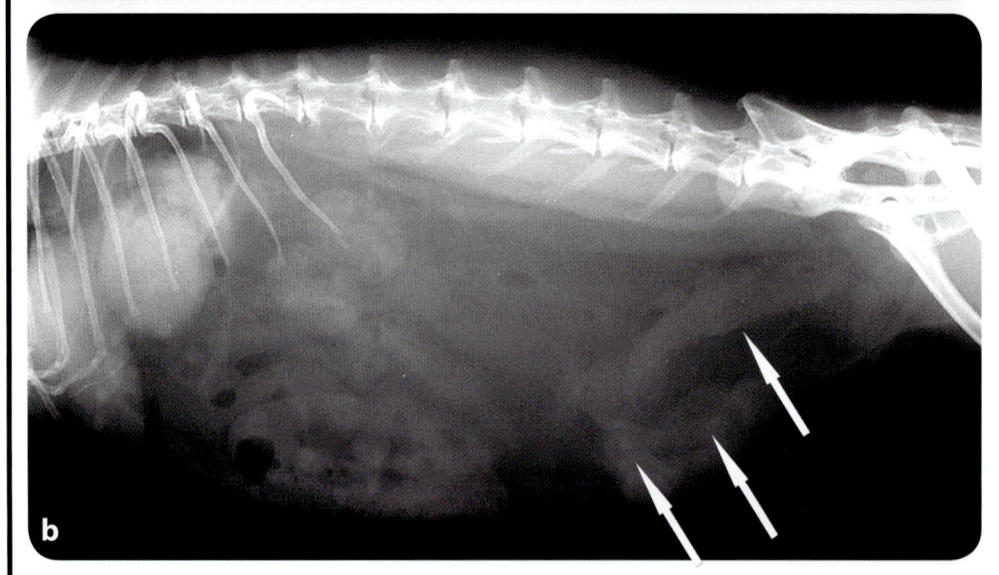

Figure 2.178a,b. Lateral radiographs of two female rabbits with uterine enlargement confirmed at surgery. Radiographic signs include cranial displacement of the intestines by a tubular soft tissue structure in the caudal abdominal cavity. The tubular opacity is inconsistent with small intestine (arrows). Secondary gastrointestinal impaction is visible in (a).

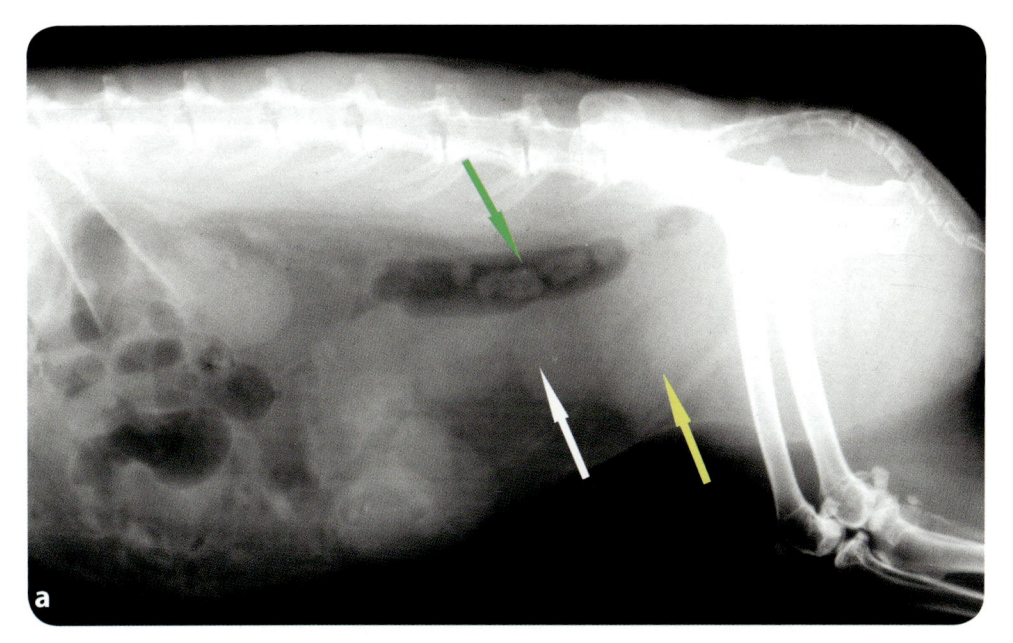

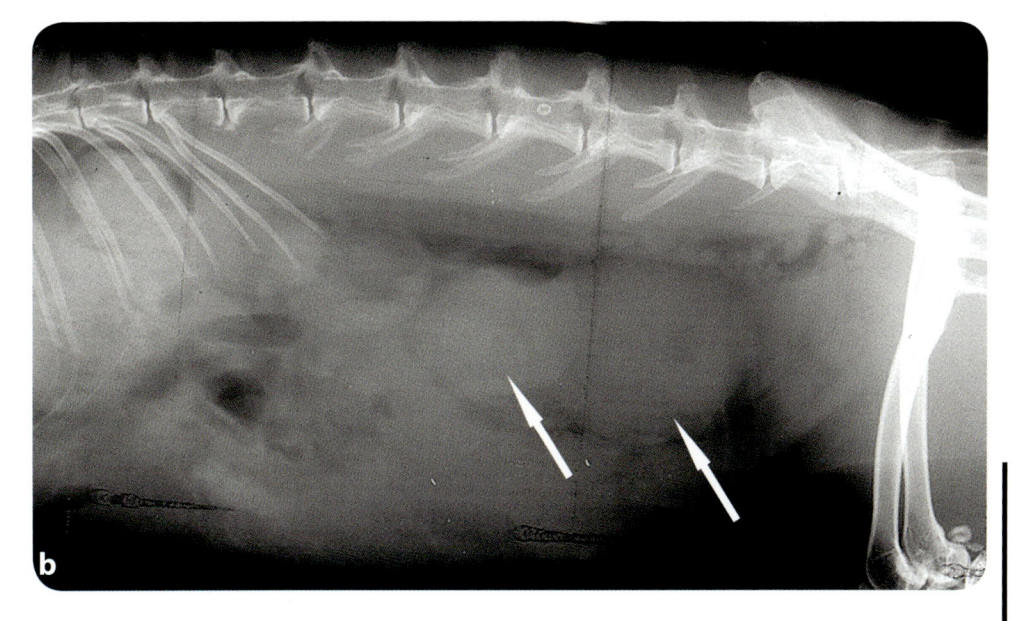

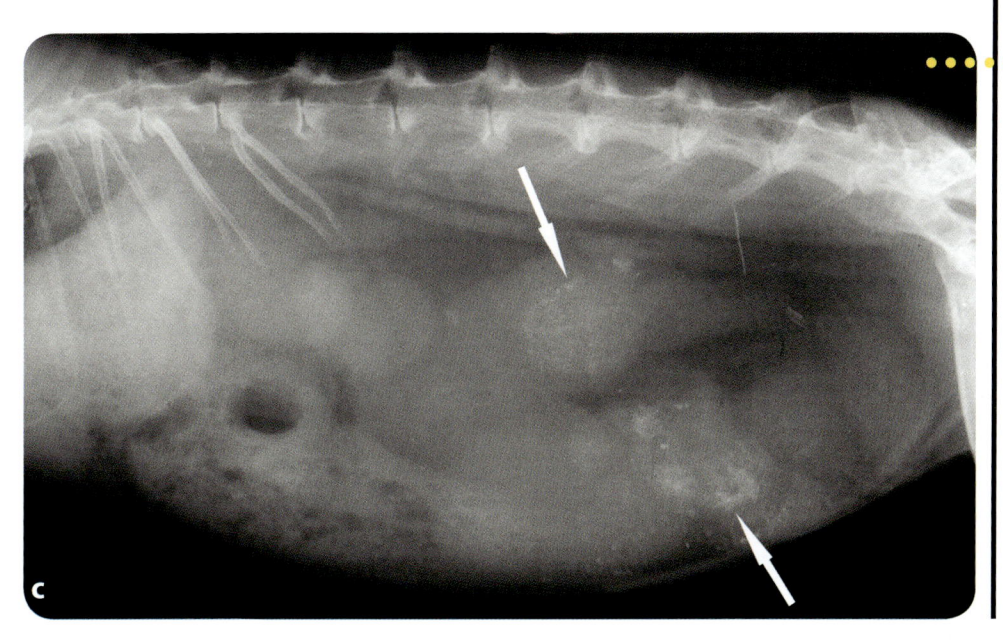

Figure 2.179a,b,c.
Lateral radiographs (a,b,c) of rabbits determined at surgery to have uterine neoplasia.

In (a) a mineralized tubular soft tissue mass (white arrow) extends cranially and is displacing the urinary bladder (yellow arrow) ventrally and the colon dorsally (green arrow). Uterine masses typically displace the urinary bladder and colon.

In (b) a bilobed soft tissue mass is seen in the caudal part of the abdomen (arrows).

In (c) a complex mass with discrete mineralization is found in the caudal part of the abdomen (arrows).

Additional imaging procedures such as ultrasound examination are often needed to confirm uterine origin of large masses.

Clinical signs in rabbits with uterine neoplasia often include weight loss and hematuria. Physical examination often reveals a palpable caudal abdominal mass.

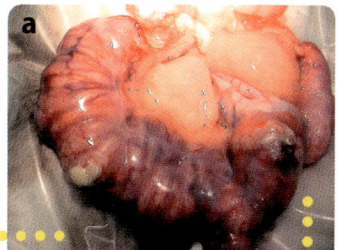

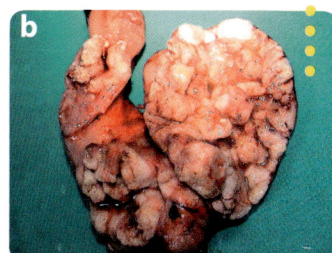

Figure 2.180a,b. Uterine adenocarcinoma shown intraoperatively (a) and the sectioned gross specimen (b).

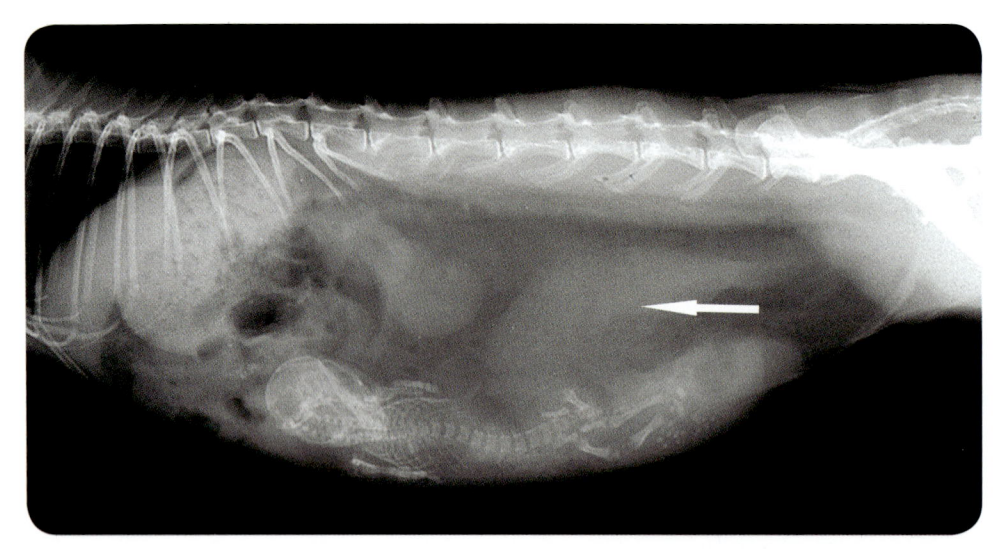

Figure 2.181. Lateral radiograph of the abdomen of a rabbit with an extrauterine pregnancy and abortion. The location of the fetus in the cranioventral abdomen is abnormal, and the fetus is malpositioned, suggesting nonviability. The uterine body is enlarged (arrow).

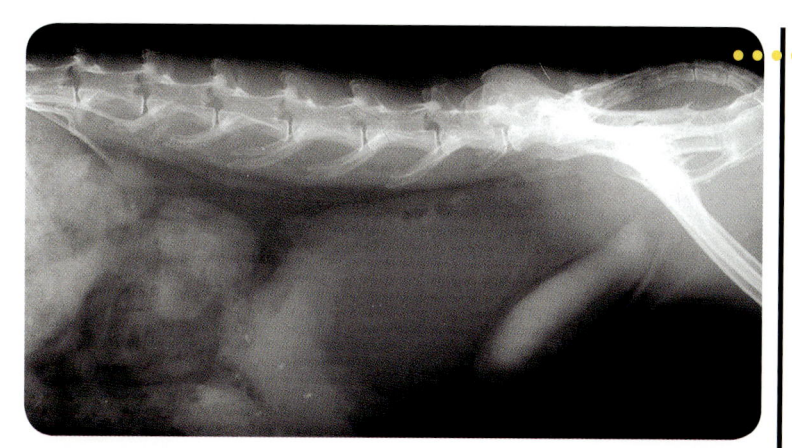

Figure 2.182. Lateral radiograph of the abdomen of a female rabbit with marked enlargement of the caudal genital tract. Urinary bladder is identified by the presence of radiopaque urine. The enlarged soft tissue structure dorsocranial to the bladder is the fluid-vagina. If the vagina contained urine, it would be expected to have the same radiodense appearance as urine in the bladder.

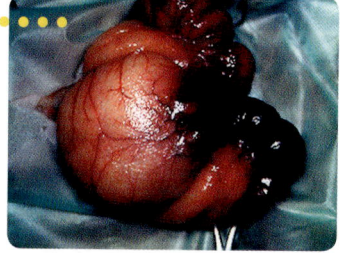

Figure 2.183. Intraoperative appearance of vagina distended with purulent material

ABNORMALITIES of the VERTEBRAL COLUMN
Diseases of the Thoracic Vertebral Column

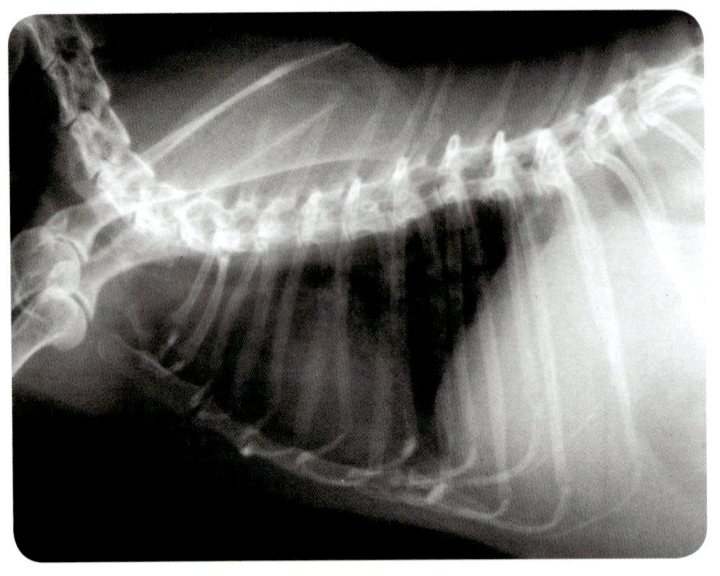

Figure 2.184. Lateral radiograph of the thorax of a rabbit demonstrating vertebral deformity at T5-T8. Deformities of the vertebral column are common in the rabbit and may be caused by metabolic bone disease. These lesions may be diagnosed as incidental findings on radiographs of the thorax or abdomen.

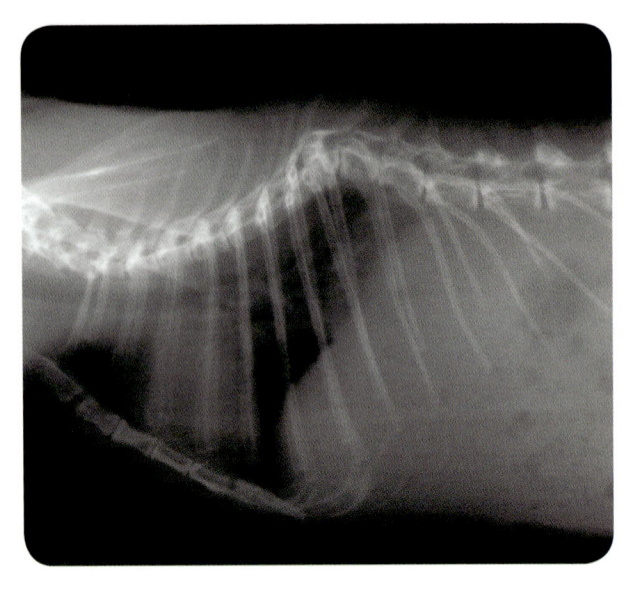

Figure 2.185. Lateral radiograph of a rabbit with kyphosis of the thoracic vertebral column. Severe deformities can affect the gait.

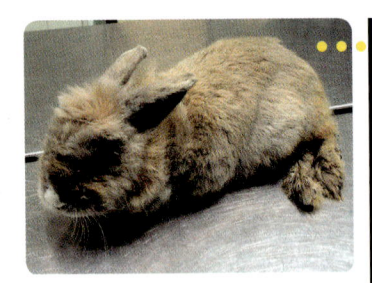

Figure 2.186. Spinal cord lesions are often a consequence of fracture of one or more vertebrae. Rabbits present with varying degrees of neurologic deficits depending on the severity and location of the fracture.

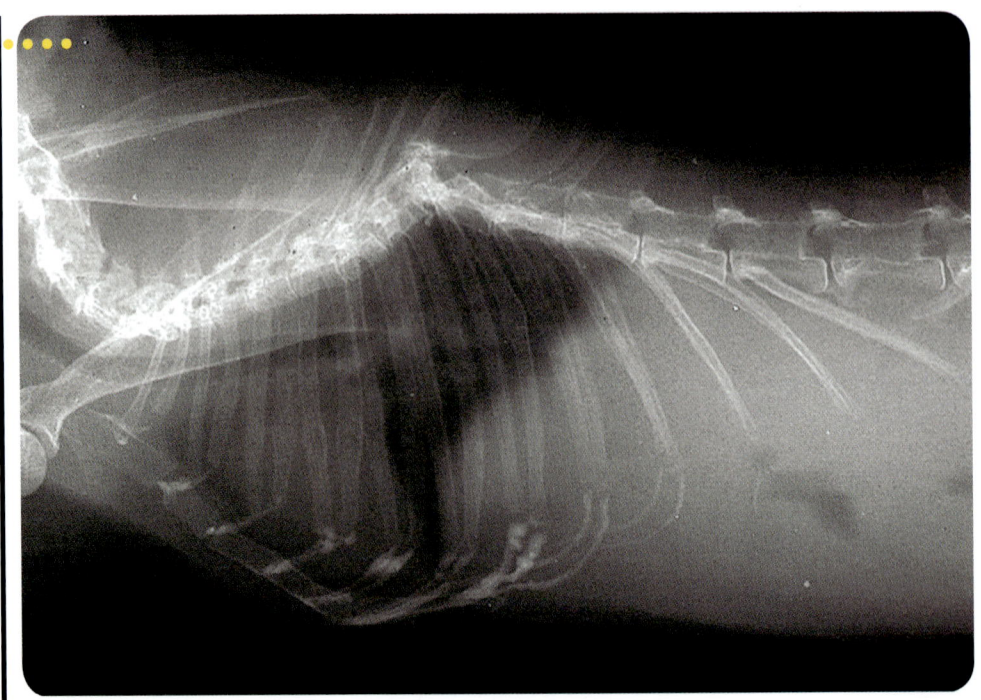

Figure 2.187. Lateral radiograph of a rabbit demonstrating a compression fracture of T8 causing malalignment of the vertebral canal. Spinal cord compression is common with these lesions.

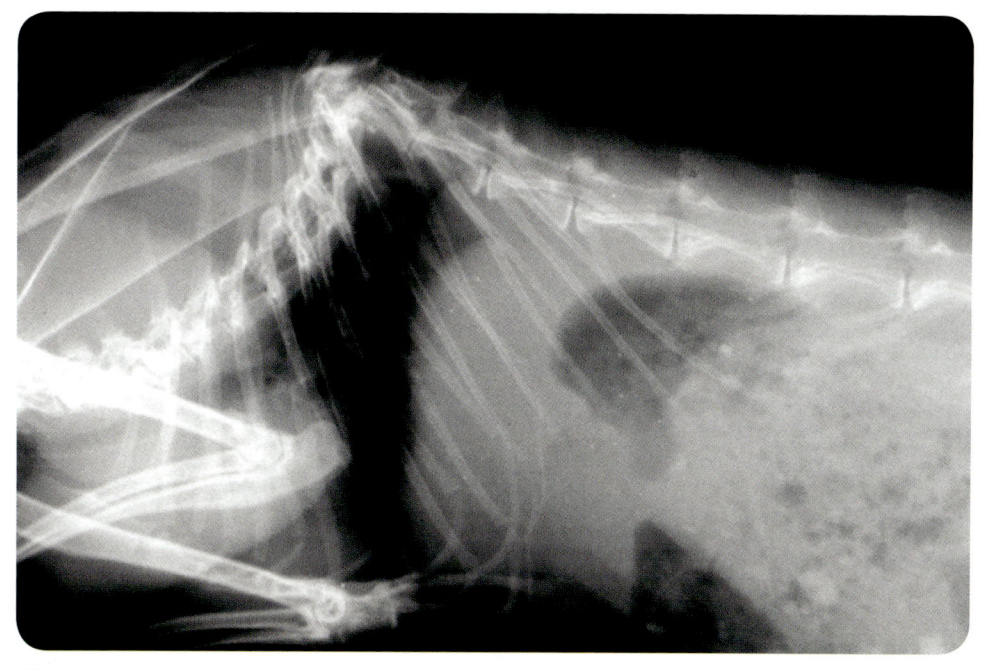

Figure 2.188. Lateral radiograph of the vertebral column of a rabbit with severe compression fracture of T9 and luxation of the vertebral column.

Diseases of the Lumbosacral and Caudal Vertebral Column

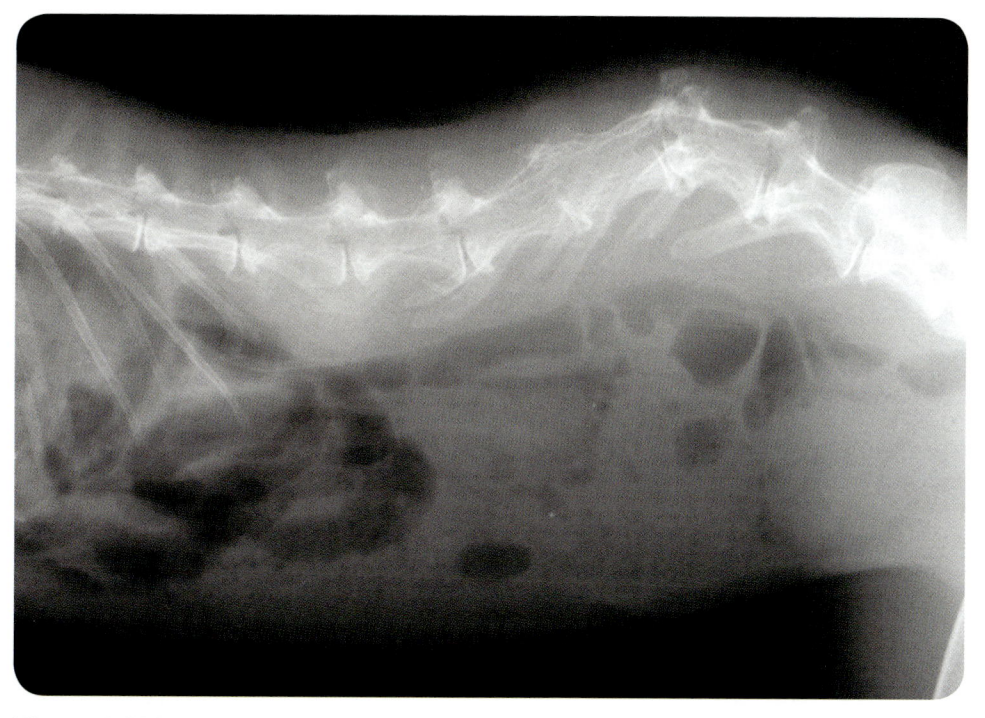

Figure 2.189. Lateral radiograph of the vertebral column of a rabbit with kyphosis at L4 and L5. L3-4 are partially fused.

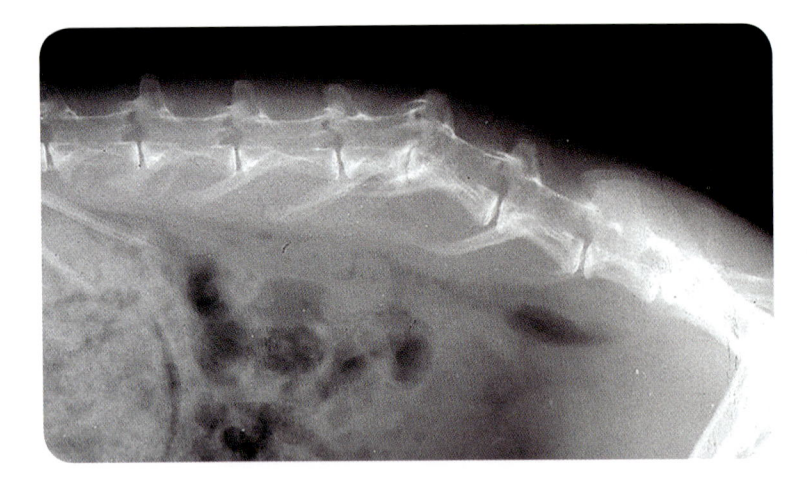

Figure 2.190. Lateral radiograph of the spine of a rabbit demonstrating an incomplete compression fracture of L4 and subluxation of L4-L5. This is the type of lesion typically produced as a consequence of improper restraint, when the patient struggles and kicks violently with the rear legs. The long-term clinical consequence of this type of lesion is variable.

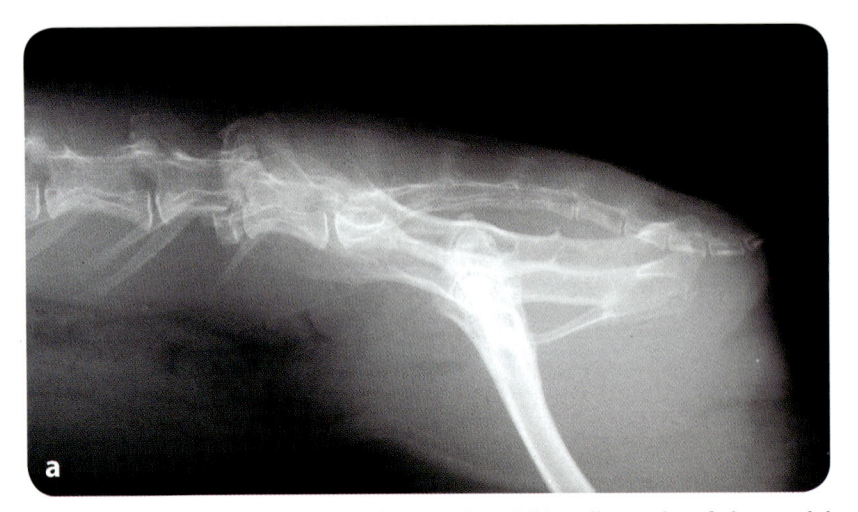

Figure 2.191a,b. Lateral (a) and ventrodorsal (b) radiographs of the caudal spine of a rabbit with a vertebral fracture of L6 with displacement of fragments. These lesions are more difficult to detect in the ventrodorsal projection. However, L6 appears much shorter than the two adjacent vertebrae (L5 and L7).

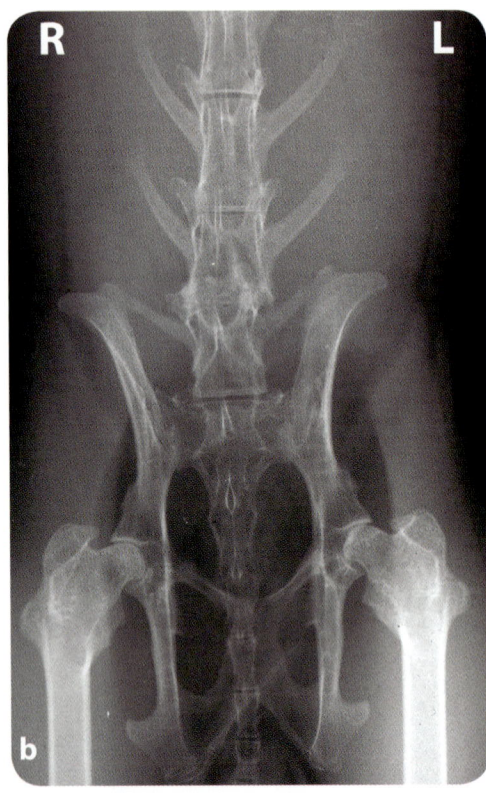

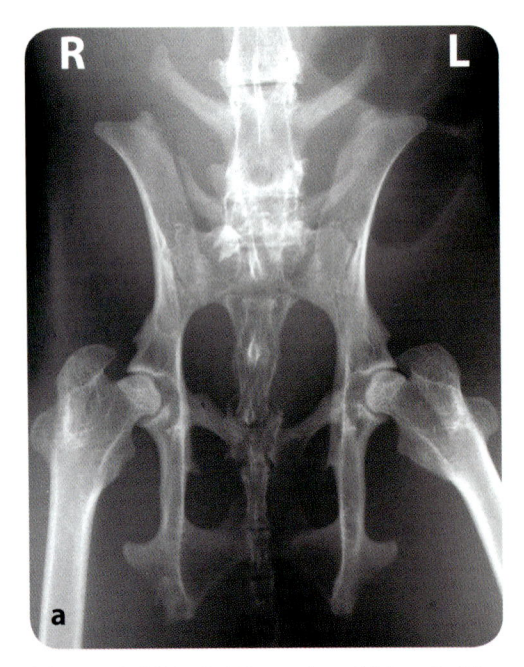

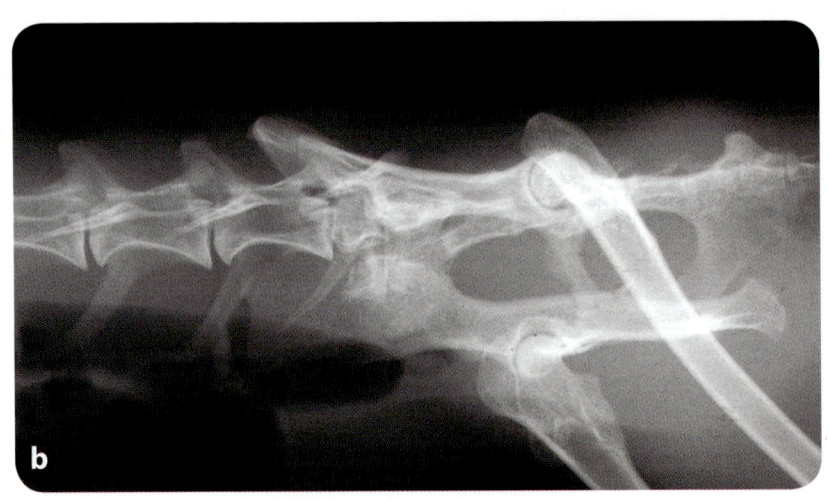

Figure 2.192a,b. Ventrodorsal (a) and oblique (b) projections showing a vertebral fracture of L7. Displacement and overriding of the fragments are visible in the oblique projection. The fracture itself is also visible in the ventrodorsal projection.

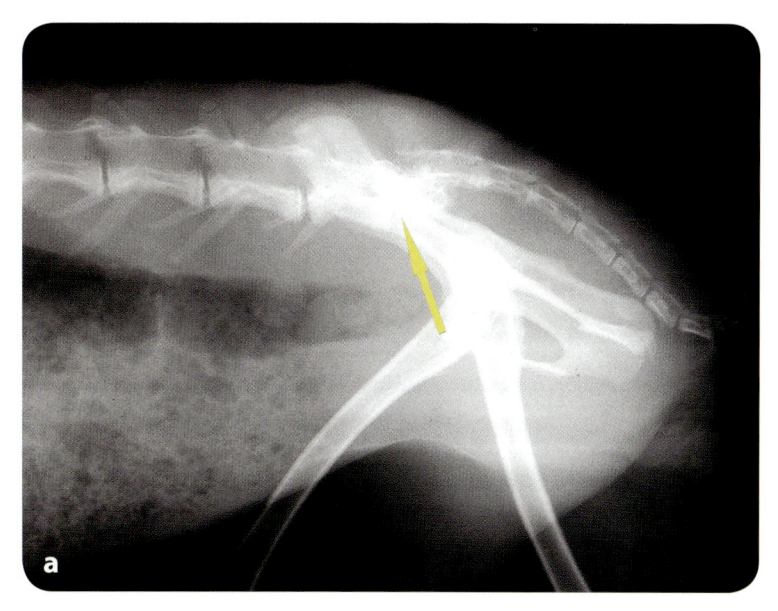

Figure 2.193a,b. Lateral (a) projection of the pelvis of a rabbit with a transverse fracture of the sacrum with dorsal displacement of the distal segment (a). The displacement is also causing a luxation of the vertebral column, (note "step" in the normal contour of the lumbosacrum; arrow). The fracture line is not seen on the ventrodorsal projection (b) because the fracture is overriding. However, there is slight axial malalignment (arrows) of the dorsal spinous processes and the sacrum is shortened when compared to normal (see Figures 2.222, 2.225).

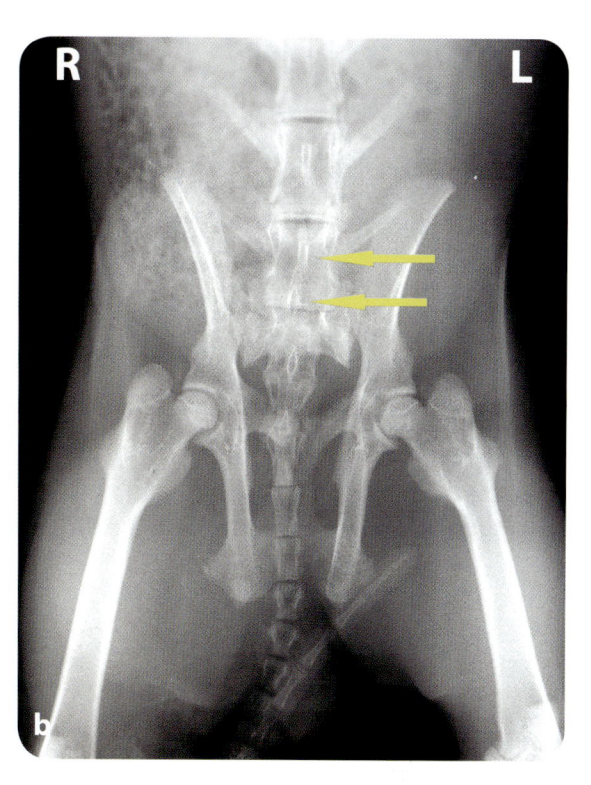

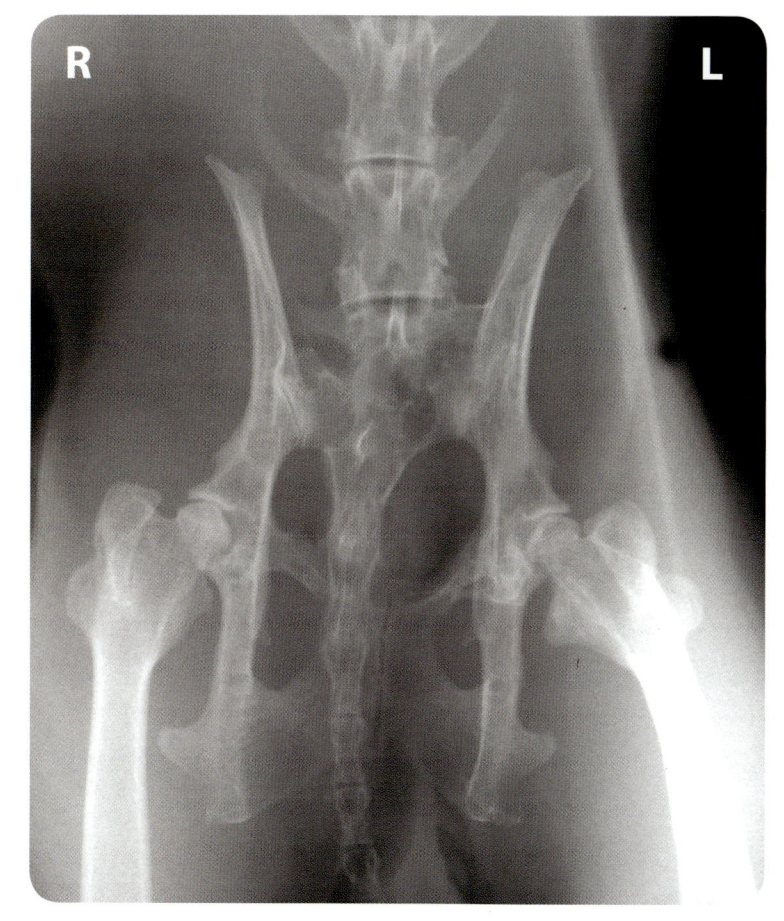

Figure 2.194. Ventrodorsal projection of the pelvis showing osseous remodeling and malalignment of the sacrum and L7 as a result of previous fracture.

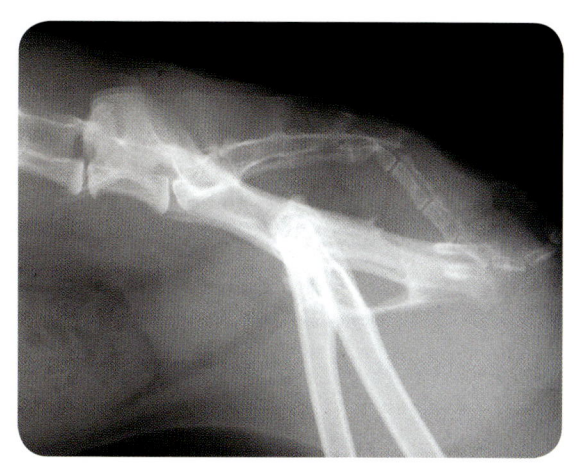

Figure 2.195. Lateral radiograph of the pelvis showing a fracture of Cd1, likely a result of trauma.

Myelography of the Normal Vertebral Column
Lateral Projection

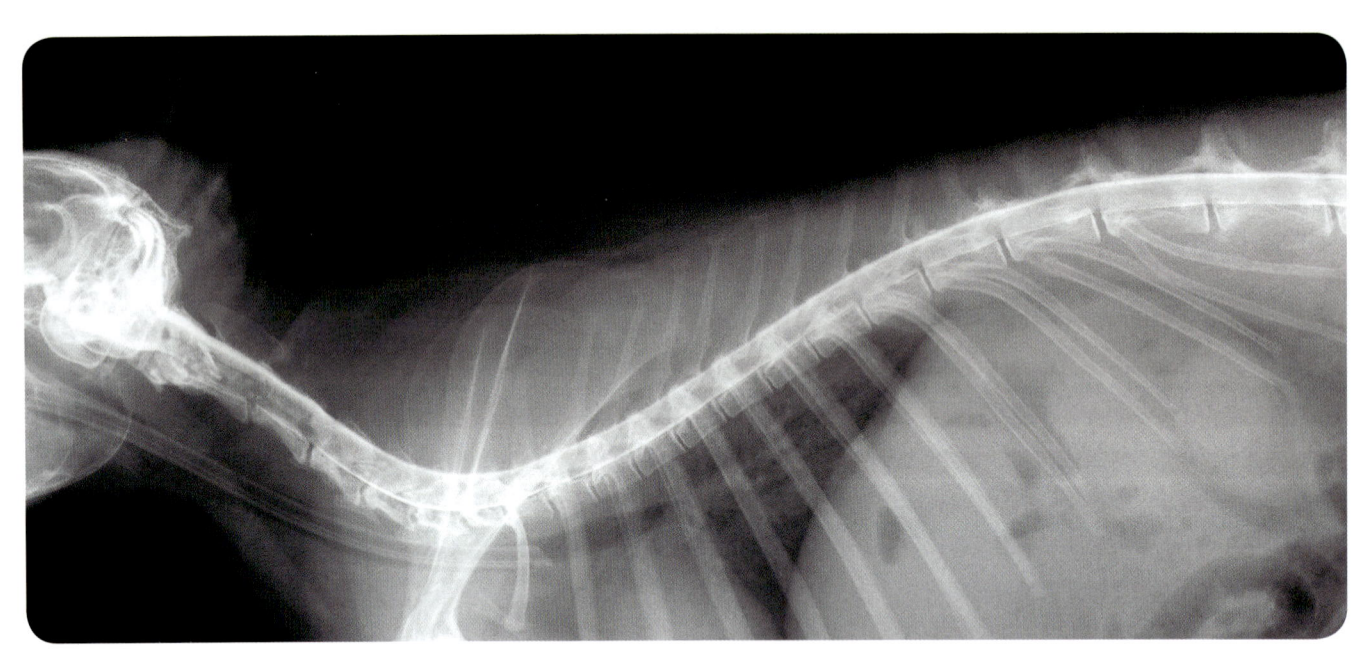

Figure 2.196. Normal myelogram of the cervical and thoracic vertebral column of a 1.5 kg dwarf rabbit, actual size.
(Courtesy of Stefania Gianni, DVM)

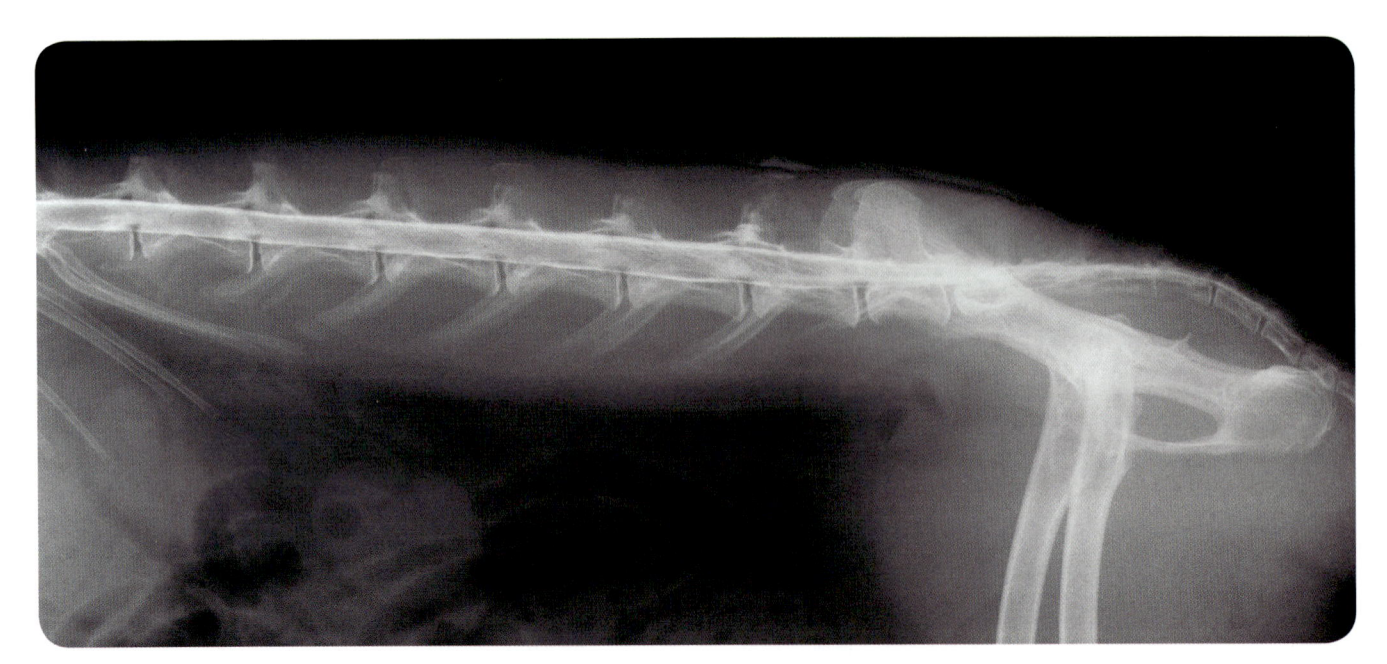

Figure 2.197. Normal myelogram of the lumbar tract of a 1.5 kg dwarf rabbit, actual size.
(Courtesy of Stefania Gianni, DVM)

Ventrodorsal Projection

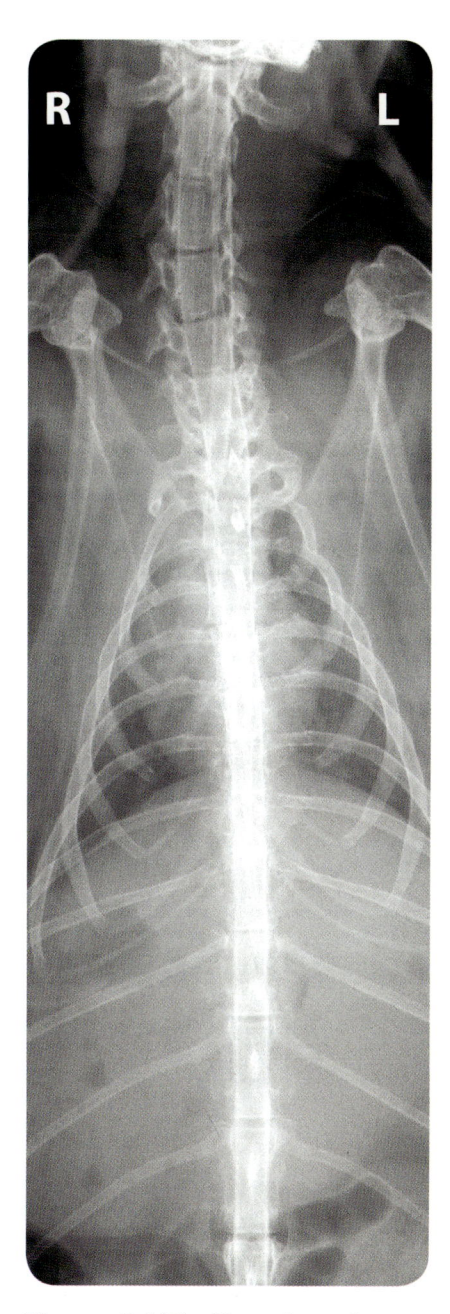

Figure 2.198. Normal myelogram of the thoracic tract of a 1.5 kg dwarf rabbit, actual size.
(Courtesy of Stefania Gianni, DVM)

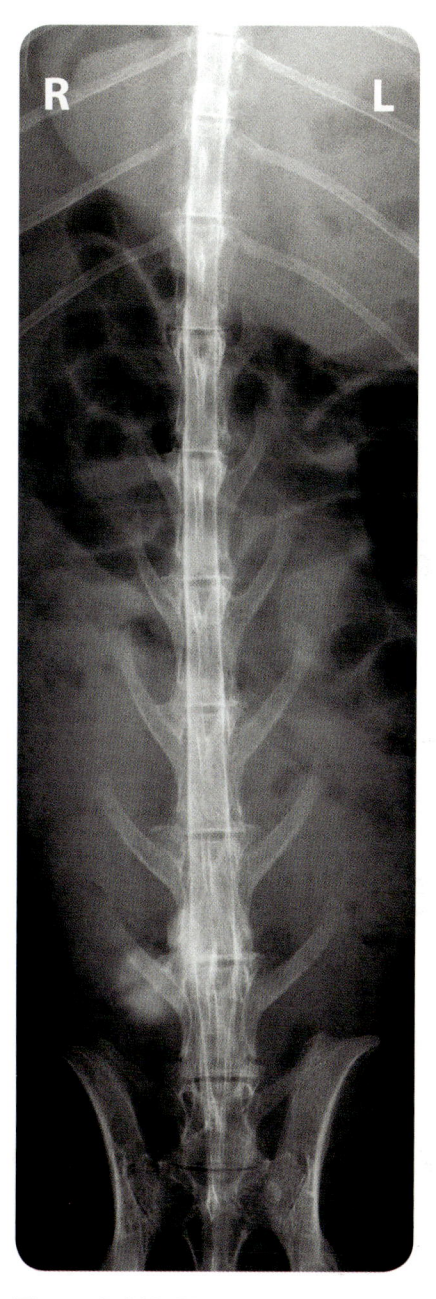

Figure 2.199. Normal myelogram of the lumbar tract of a 1.5 kg dwarf rabbit, actual size.
(Courtesy of Stefania Gianni, DVM)

Myelography of the Abnormal Vertebral Column

Lateral Projection

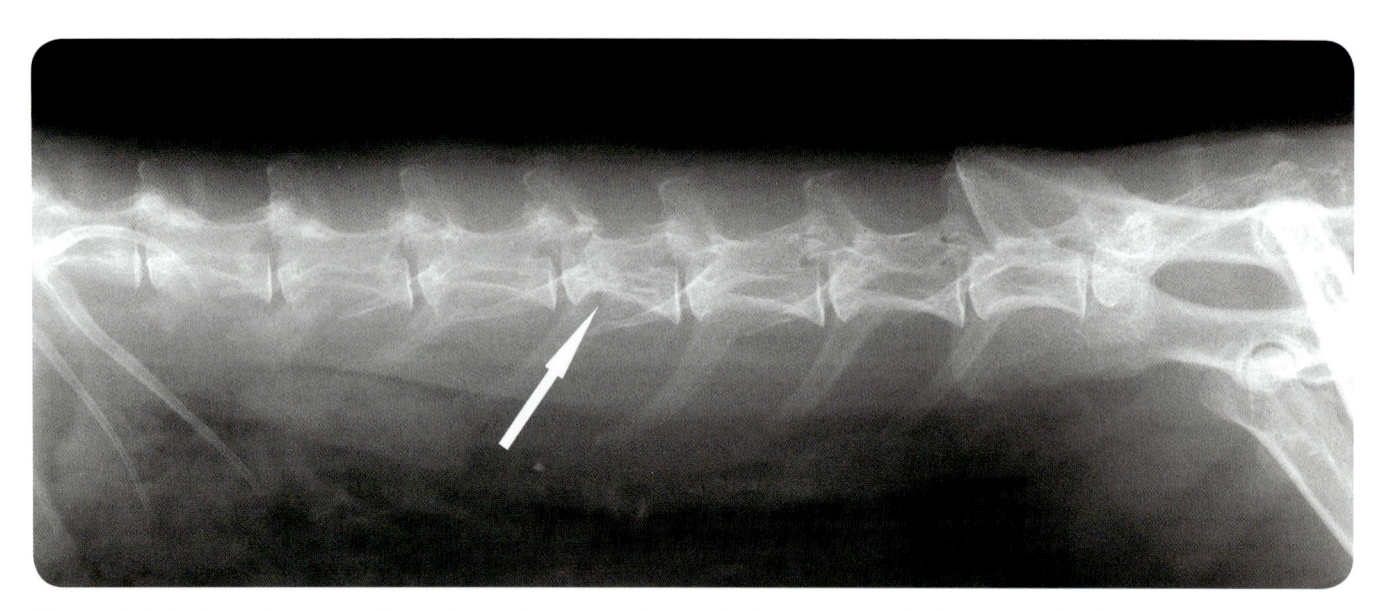

Figure 2.200. Lateral survey radiograph of a fracture of L4 in a 2.2 kg dwarf rabbit, following a fall from a terrace.
(Courtesy of Stefania Gianni, DVM)

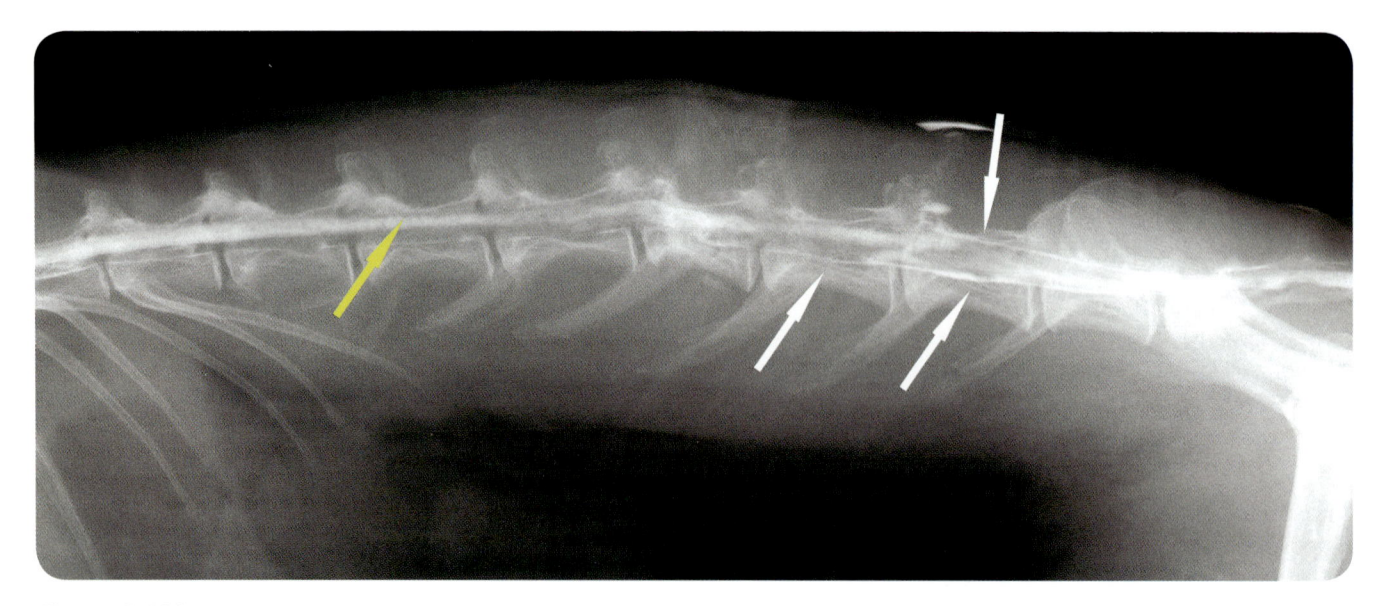

Figure 2.201. Lateral myelogram of the same patient as Figure 2.200. Contrast medium injected at the L5-L6 intervertebral space shows a normal pattern caudal to the fracture (thin radiopaque lines, white arrows), which then enters the central foramen of the spinal cord at the point of fracture (thick radiopaque line, yellow arrow) implying severe damage to the spinal cord parenchyma.
(Courtesy of Stefania Gianni, DVM)

Ventrodorsal Projection

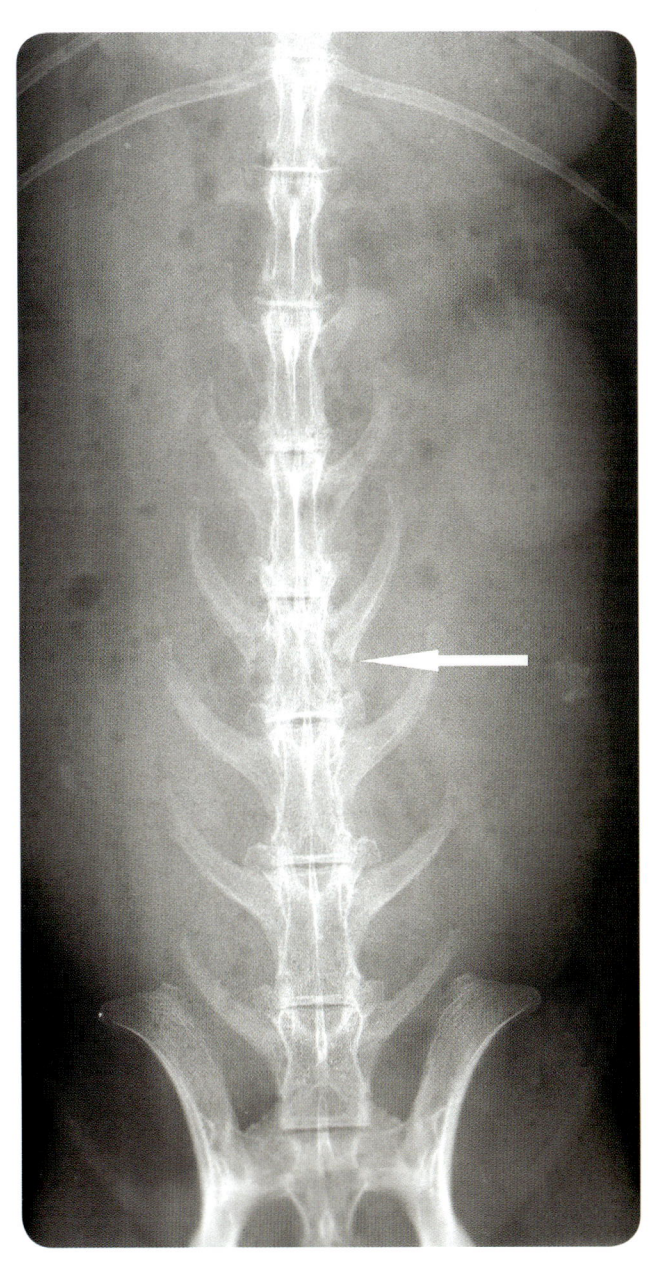

Figure 2.202. Ventrodorsal survey radiograph of the same patient as in Figures 2.200-2.201. There is a fracture of L4.
(Courtesy of Stefania Gianni, DVM)

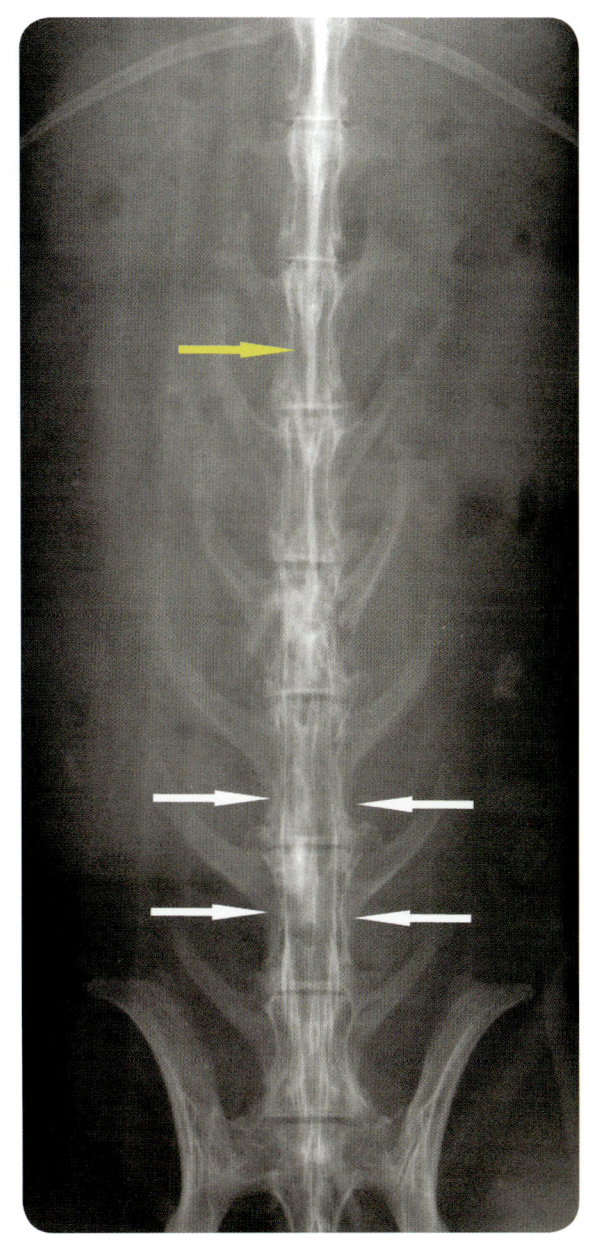

Figure 2.203. Ventrodorsal myelogram of the same patient. Contrast columns (thin radiodense lines, white arrows) are normal caudal to the fracture. At L4 the contrast medium opacifies the central canal of the spinal cord (thick radiopaque line, yellow arrow), as in Figure 2.201.
(Courtesy of Stefania Gianni, DVM)

The NORMAL THORACIC LIMB
Lateral Projection

Due to varying soft tissue thickness, the superimposition of the thorax, and the relatively small size of the manus, it is impossible to obtain an optimal single radiograph of the entire thoracic limb. Different kVp settings are needed for the scapula and the humerus, and for the radioulnar segment and the manus.

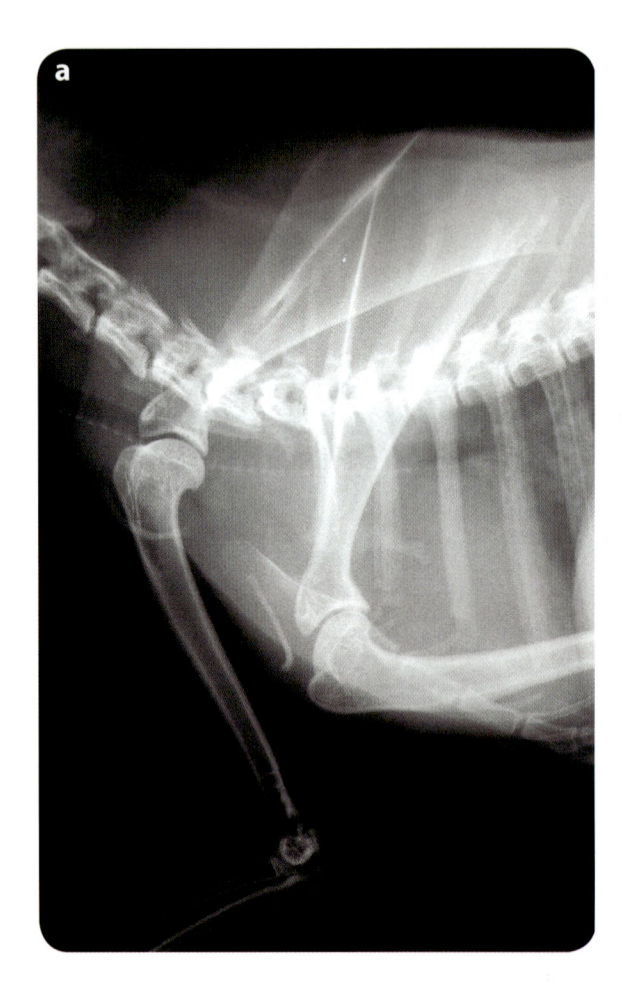

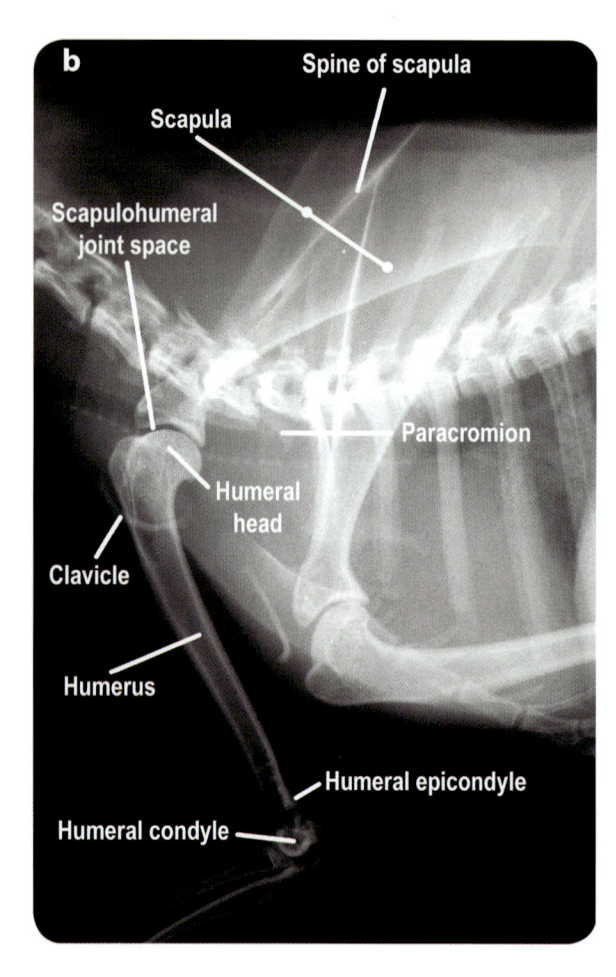

Figure 2.204a,b. Lateral radiograph of the proximal thoracic limb of a 1.9 kg dwarf rabbit, actual size.

THORACIC LIMB 139

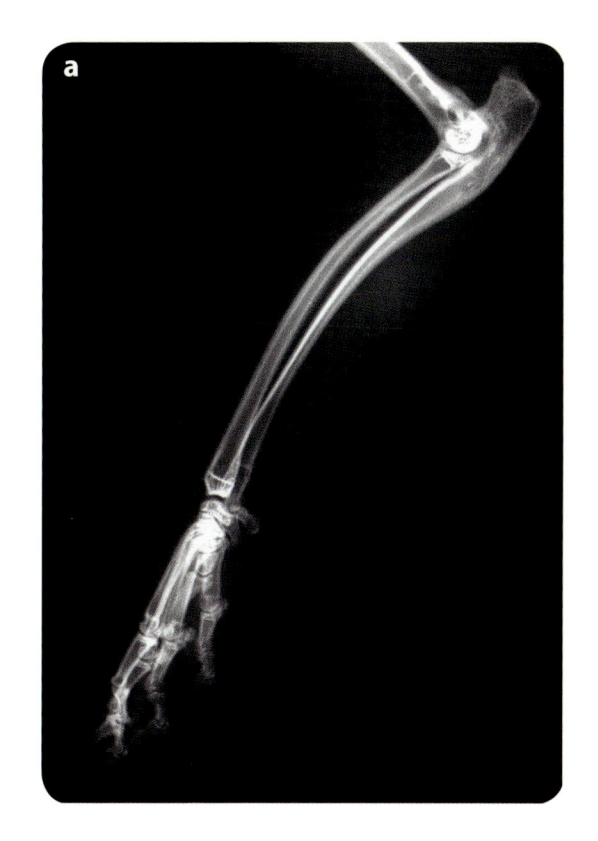

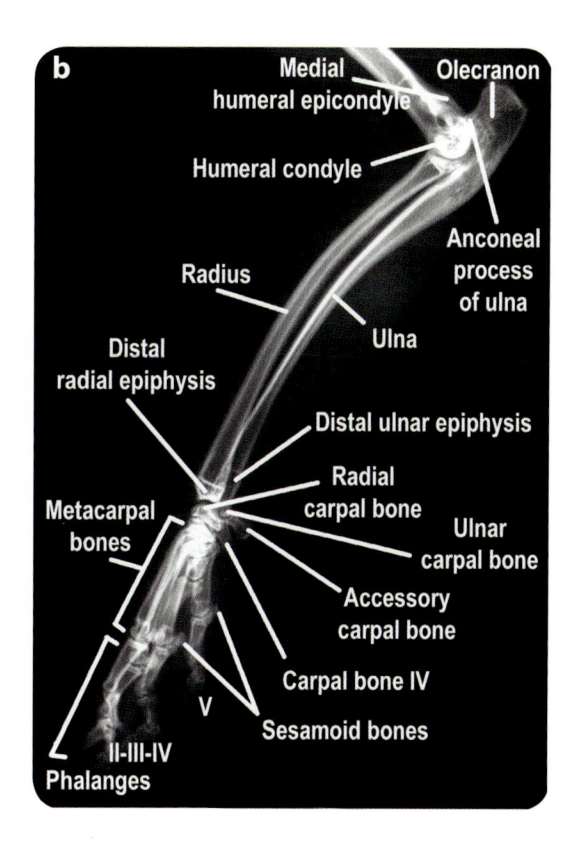

Figure 2.205a,b. Radiograph of the distal thoracic limb of a 1.9 kg dwarf rabbit, actual size.

Caudocranial Projection of the Proximal Thoracic Limb

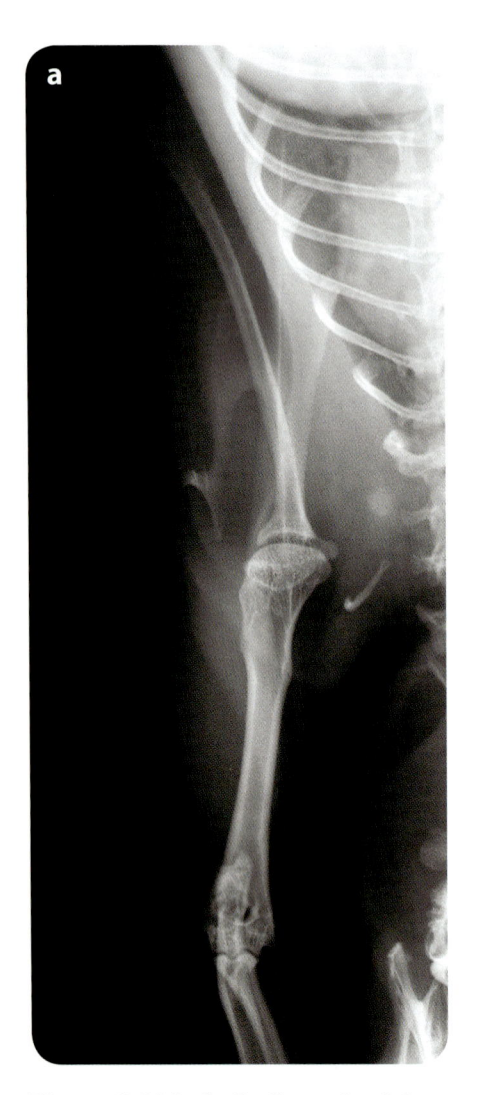

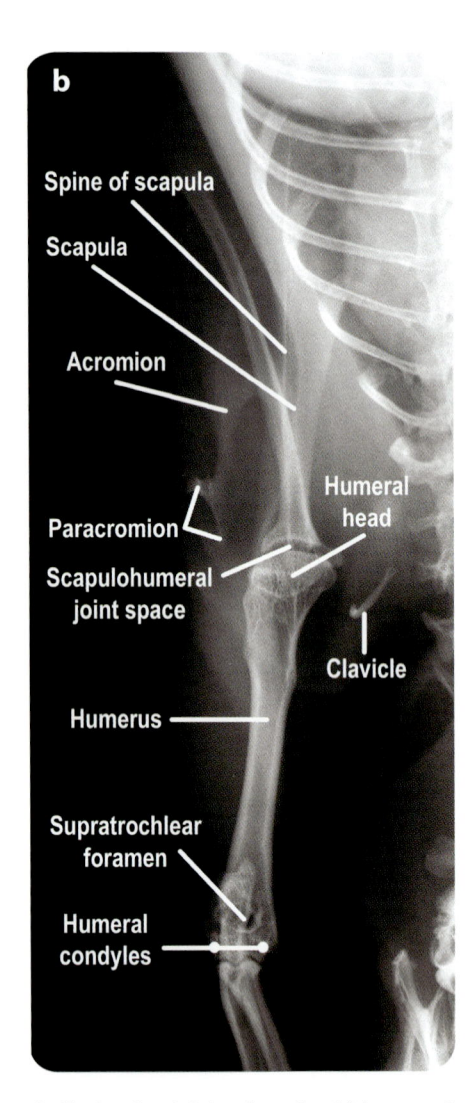

Figure 2.206a,b. Radiograph of the proximal thoracic limb of a 1.9 kg dwarf rabbit, actual size.

Craniocaudal Projection of the Distal Thoracic Limb

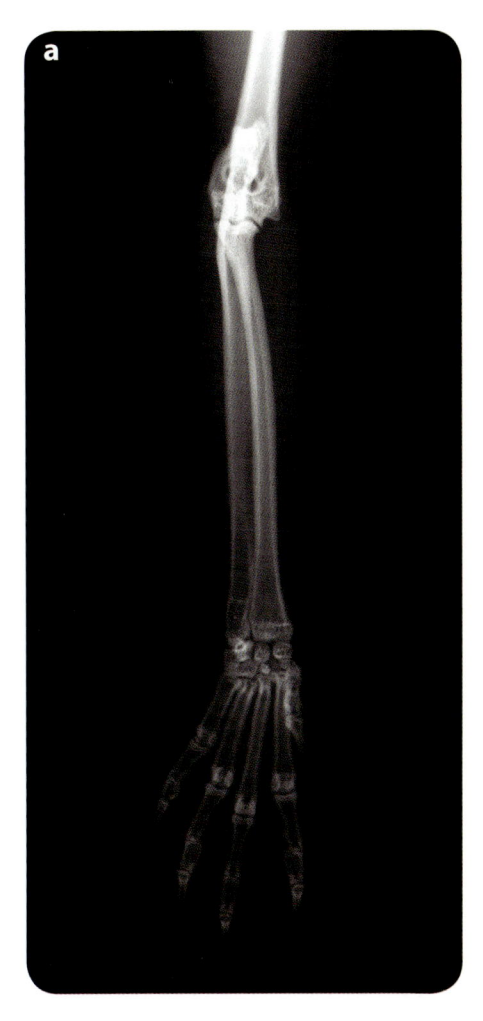

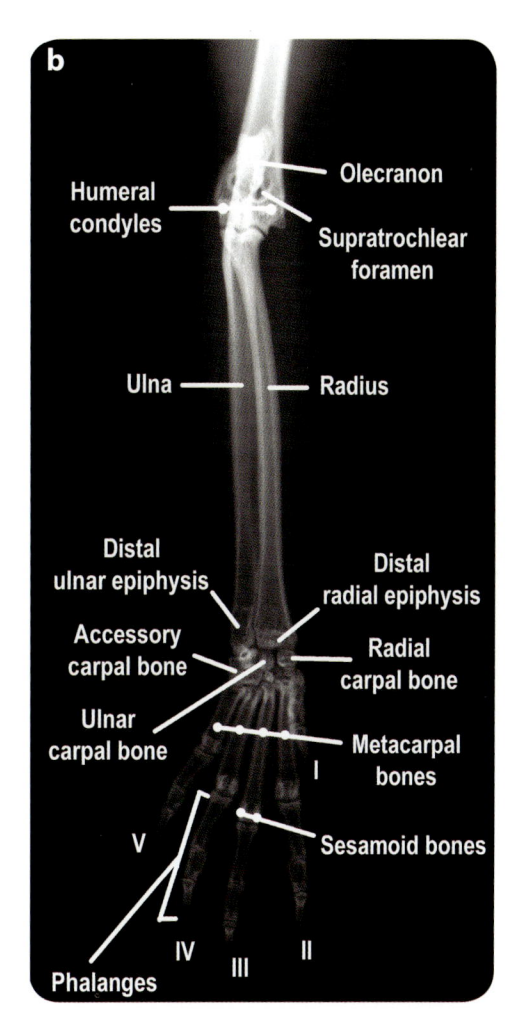

Figure 2.207a,b. Radiograph of the distal thoracic limb of a 1.9 kg dwarf rabbit, actual size.

ABNORMALITIES of the THORACIC LIMB
Diseases of the Humerus

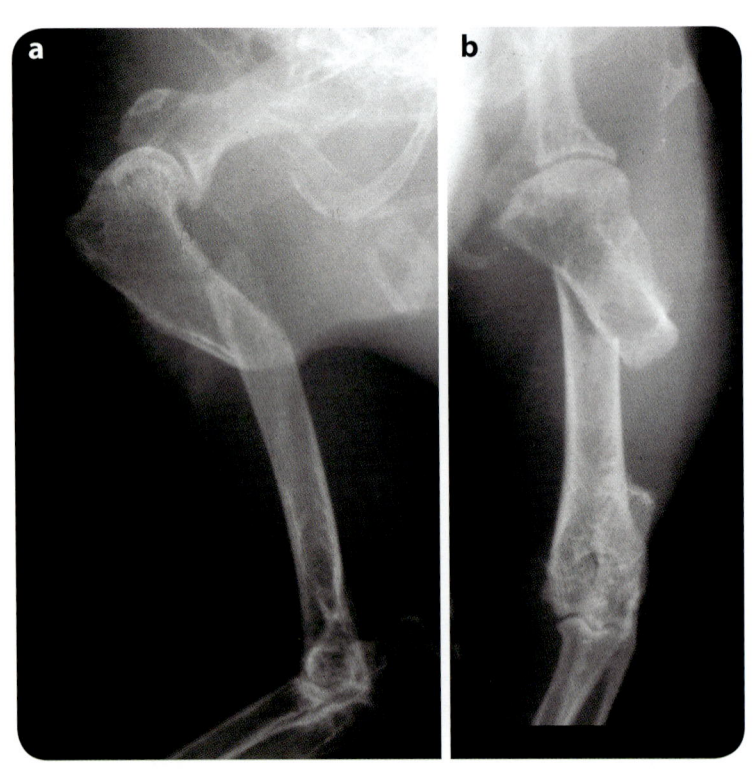

Figure 2.208a,b. Lateral (a) and craniocaudal (b) projections of the thoracic limb of a rabbit with a proximal diaphyseal oblique fracture. This pathologic fracture is a consequence of underlying metabolic bone disease, suggested by abnormally thin bone cortices. It is important to carefully check patients with humeral fractures for the presence of thoracic trauma.

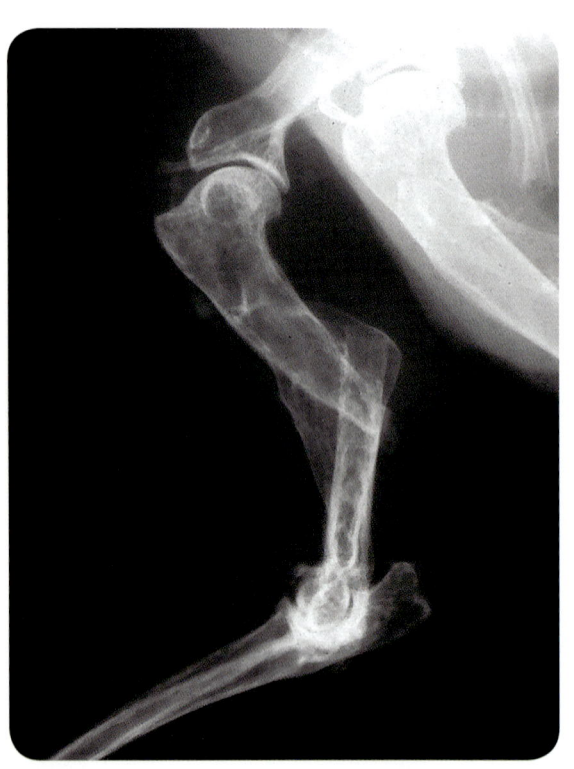

Figure 2.209. Lateral radiograph of the thoracic limb of a rabbit showing spontaneous bone healing in an untreated diaphyseal fracture. Malaligment of the humeral bone fragments is a consequence of muscle contraction, and has resulted in a malunion

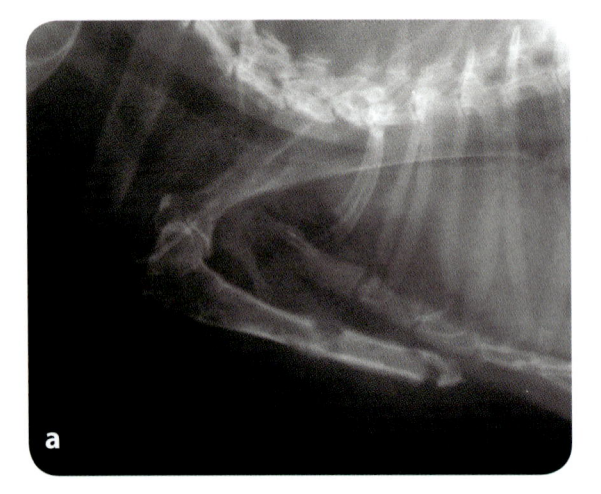

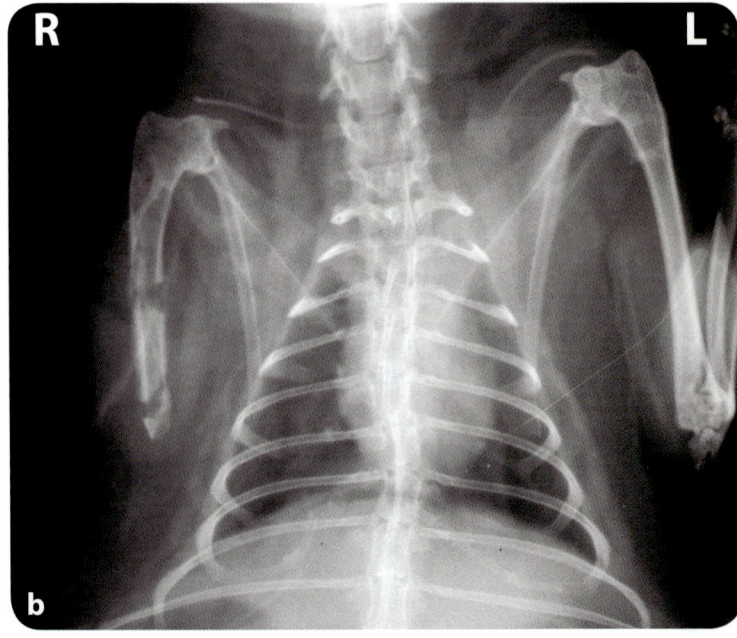

Figure 2.210a,b. Lateral (a) and craniocaudal (b) radiographs of the thoracic limb of a rabbit demonstrating osteomyelitis of the humerus as a complication following amputation of the limb distal to the elbow. Note global radiolucency in the humeral diaphysis. Amputation of the limb should be performed more proximally to prevent the patient from using the stump to ambulate (see Figure 2.211).

THORACIC LIMB

Figure 2.211. Lateral postoperative radiograph following amputation of the humerus by disarticulation at the scapulohumeral joint.

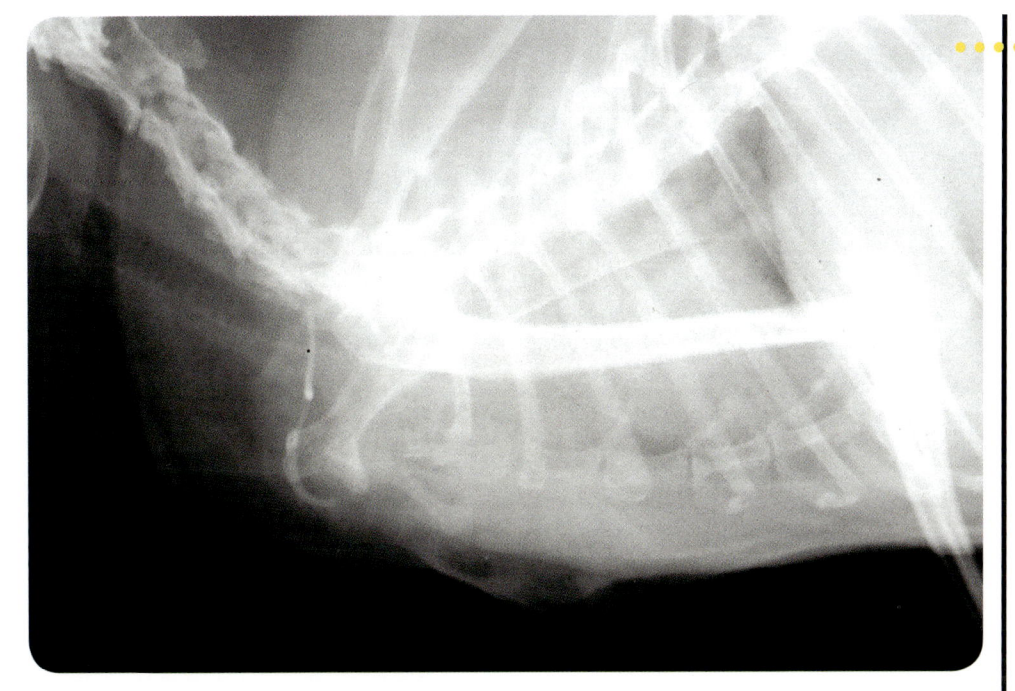

Figure 2.212. Amputation of the thoracic limb is often associated with a good prognosis. Success rate is optimized when the patient is provided with proper footing and obesity is avoided.

Diseases of the Radius, Ulna, and Elbow Joint

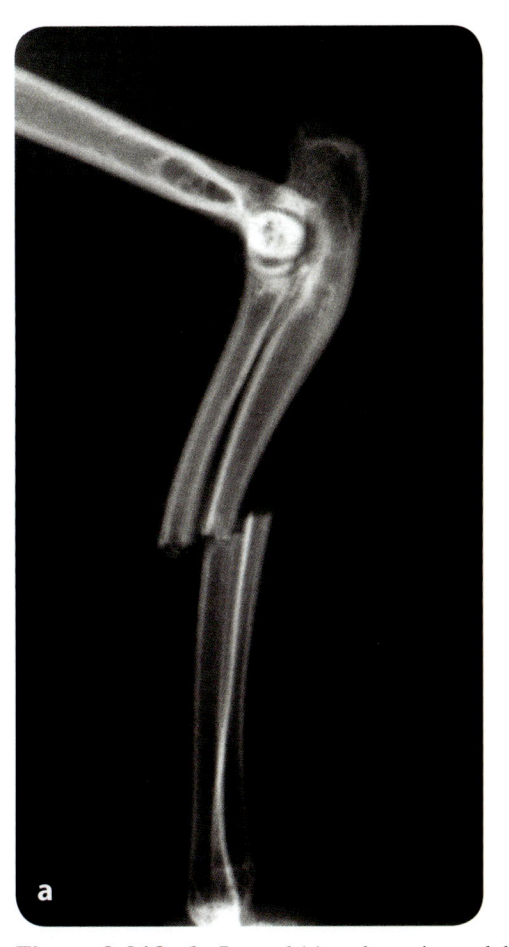

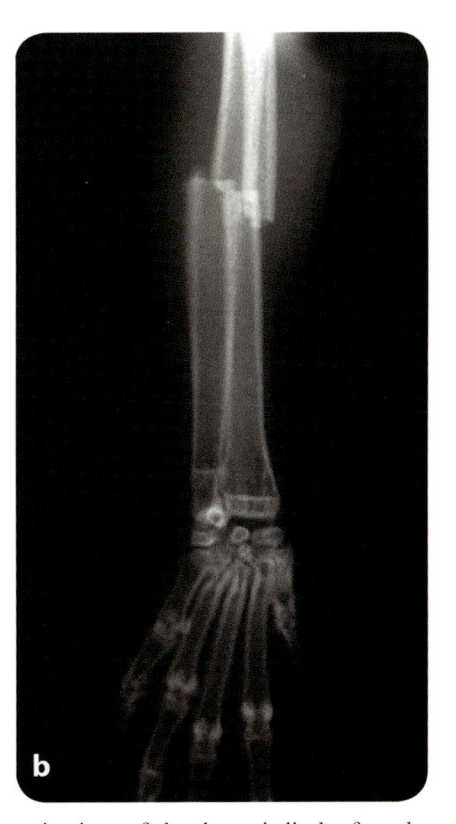

Figure 2.213a,b. Lateral (a) and craniocaudal (b) projections of the thoracic limb of a rabbit with moderately displaced transverse fractures of the diaphyses of the radius and ulna.

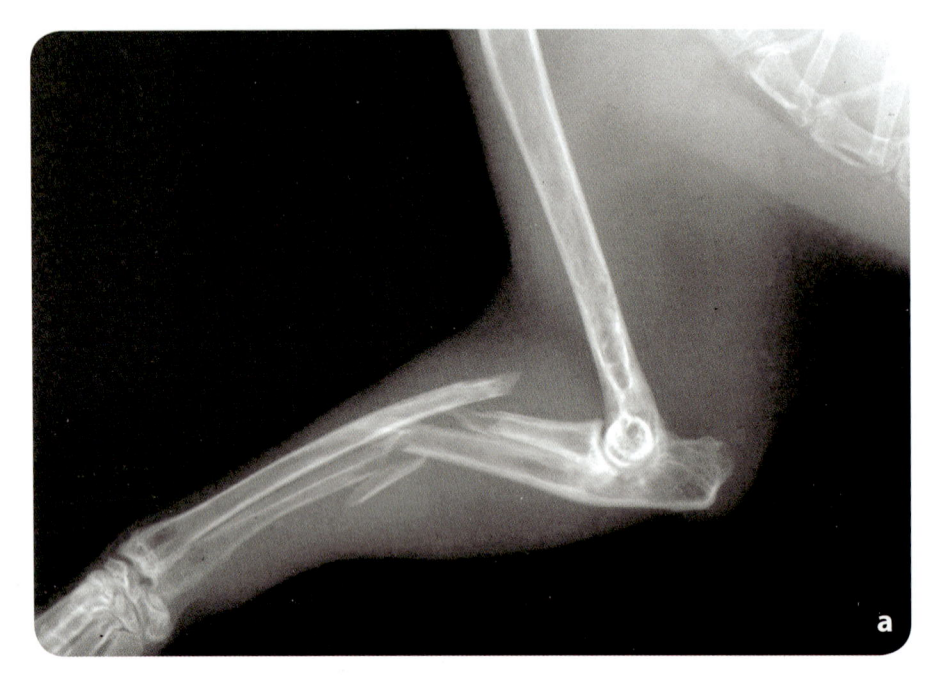

Figure 2.214a,b. Lateral (a) and craniocaudal (b) radiographs of the thoracic limb of a rabbit with an oblique, proximal metaphyseal fracture of the radius and comminuted, metaphyseal fracture of the ulna. Soft tissue swelling surround the fracture site.

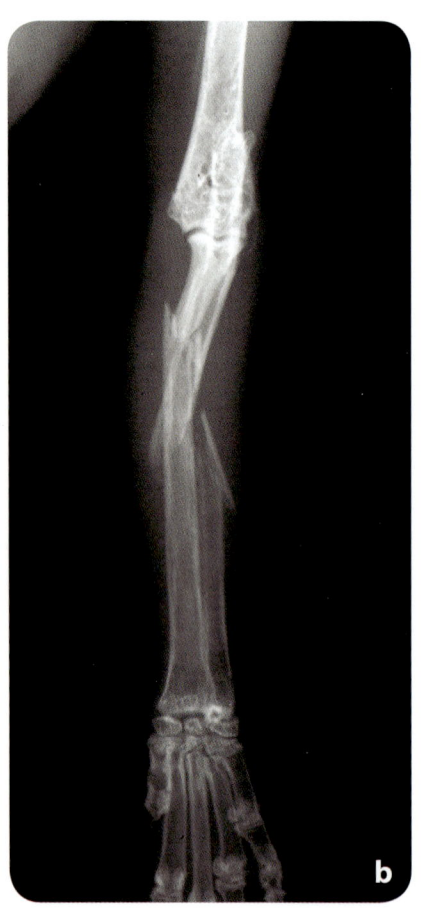

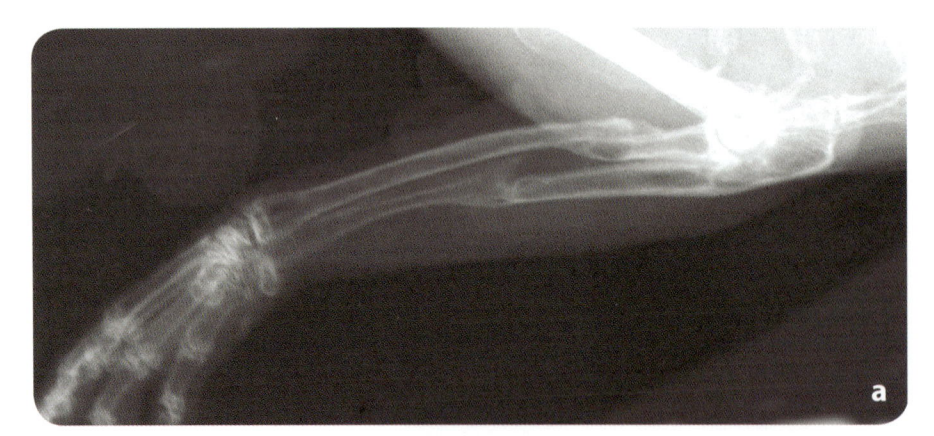

Figure 2.215a,b. Lateral (a) and craniocaudal (b) radiographs of the same patient 6 weeks post application of a splint showing evidence of healing by osseous callus. There is mild malalignment, which will resolve over time with bony remodeling.

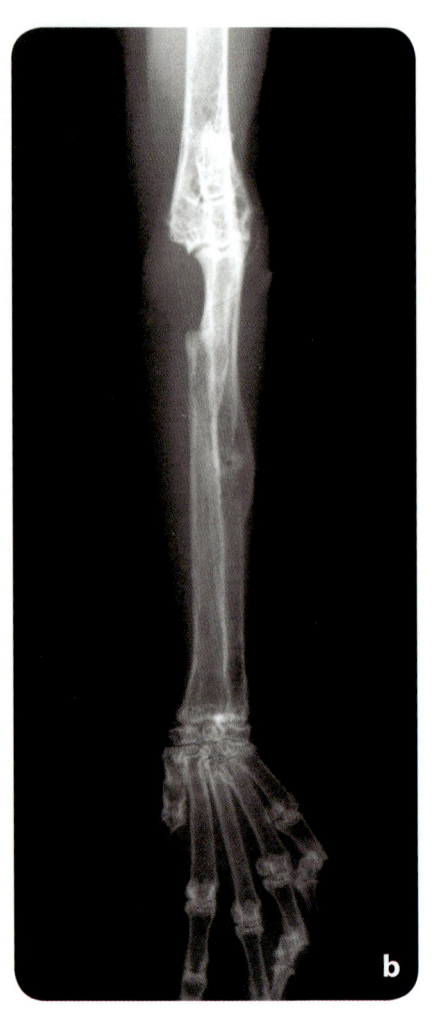

THORACIC LIMB 145

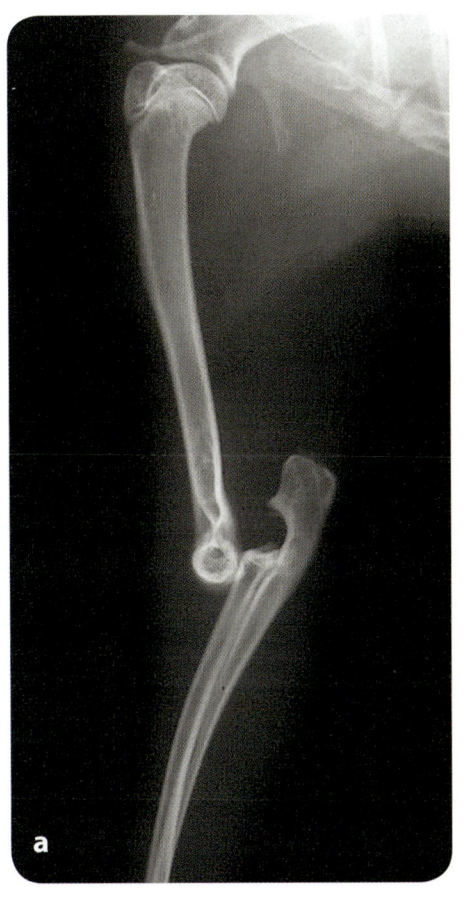

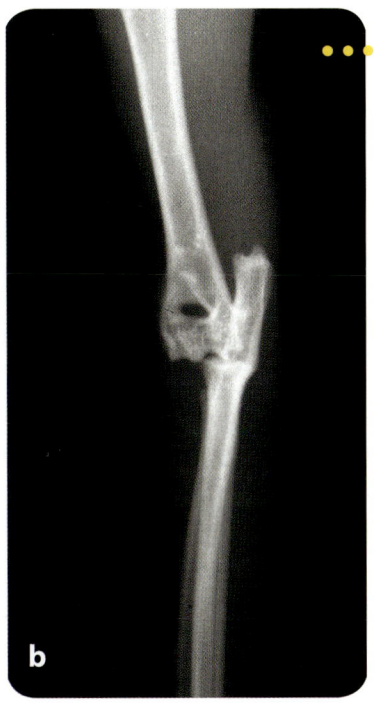

Figure 2.217. Typical clinical appearance of elbow luxation. The rabbit is unable to bear weight. In contrast, rabbits with radioulnar fractures often return to some degree of weight-bearing within a few days after injury.

Figure 2.216a,b. Lateral (a) and oblique (b) projections of the thoracic limb of a rabbit with luxation of the elbow. Abnormal relative position of the bones makes it difficult to obtain a true craniocaudal projection. This traumatic lesion is common in pet rabbits, especially in younger rabbits.

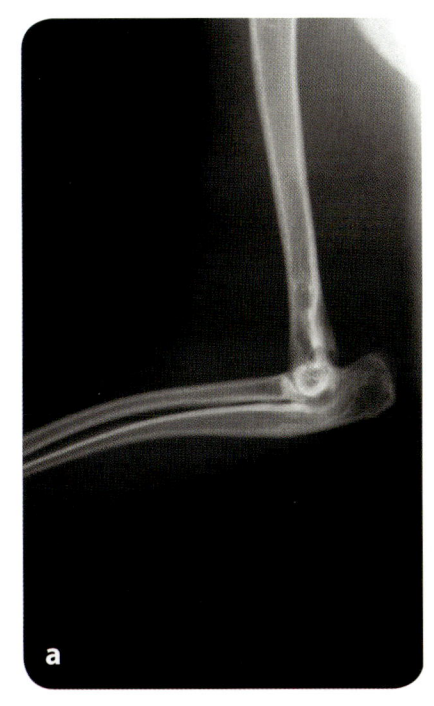

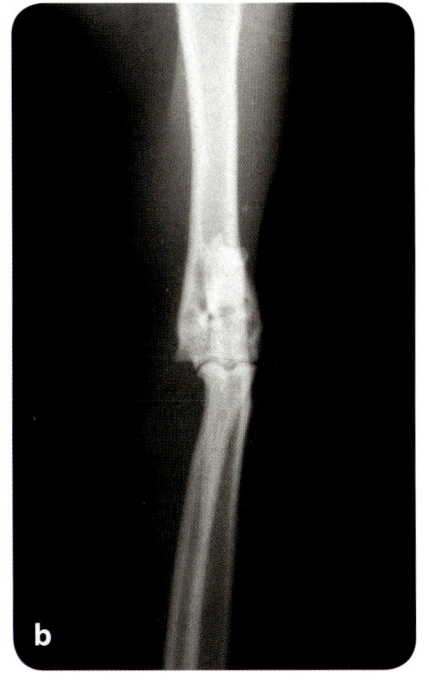

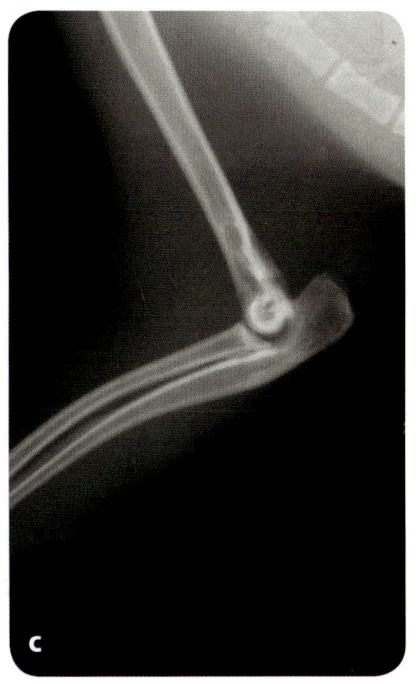

Figure 2.218a,b,c. Lateral (a) and craniocaudal (b) radiographs of the thoracic limb of a rabbit after successful manual reduction. Post procedure radiographs in two projections (a, b) must be obtained in order to confirm proper reduction. The craniocaudal or caudocranial view with the joint in extension is the most difficult to obtain but will most readily demonstrate inadequate reduction. Failed reduction can also be demonstrated on the lateral projection (c). Note the joint space appears wider than normal (see Figure 2.205). Prognosis after manual reduction of the luxation under anesthesia is very good, and cage rest alone without bandaging is usually adequate to prevent re-luxation.

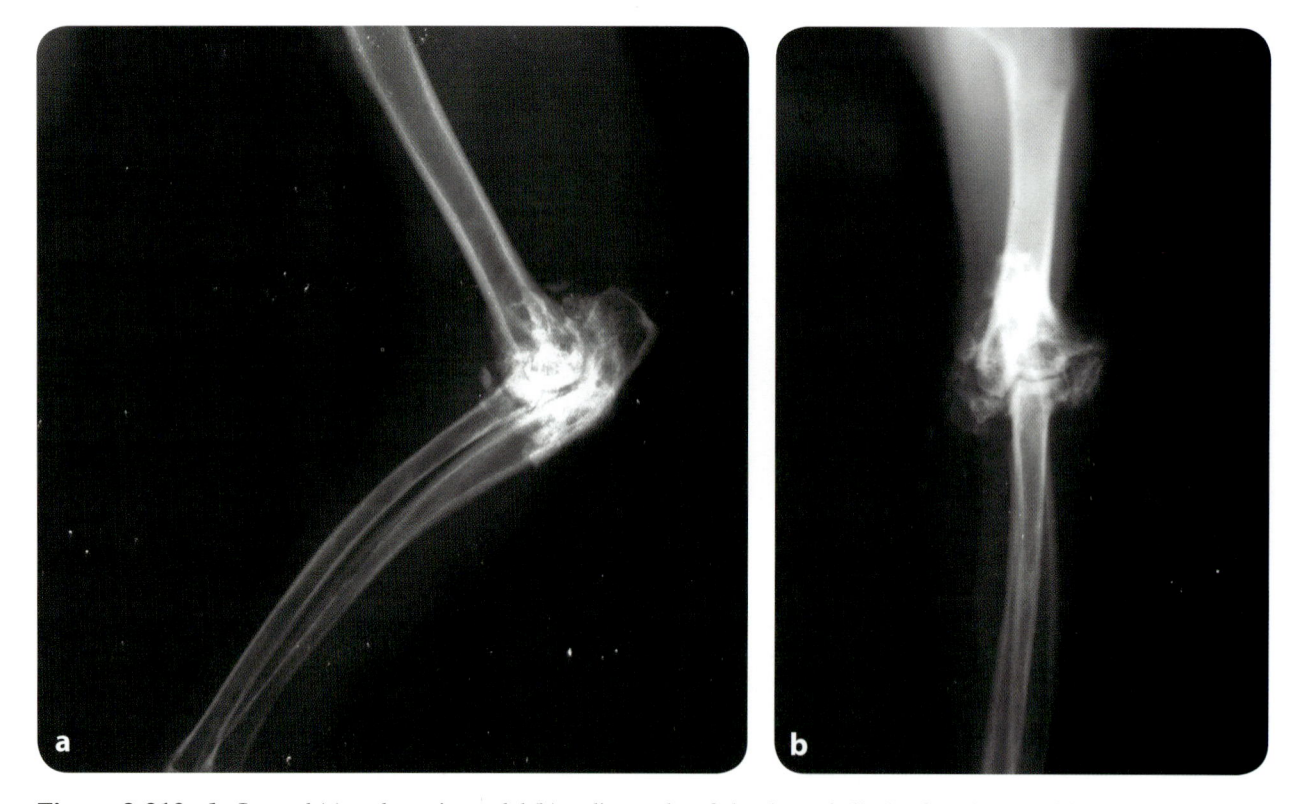

Figure 2.219a,b. Lateral (a) and craniocaudal (b) radiographs of the thoracic limb of an 8-year-old rabbit with severe degenerative arthropathy of the elbow joint. Ankylosis of the joint is developing. This was an incidental finding as the rabbit showed no symptoms of lameness.

Diseases of the Carpus, Metacarpus, and Phalanges

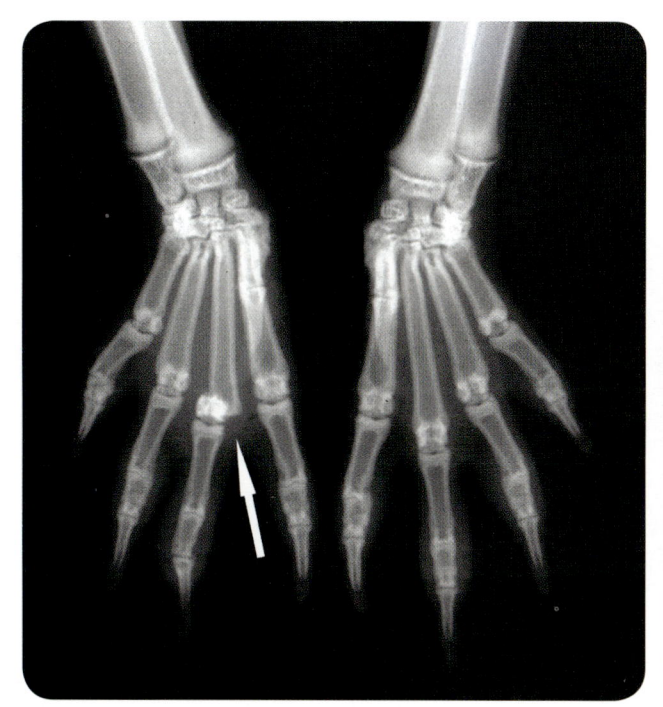

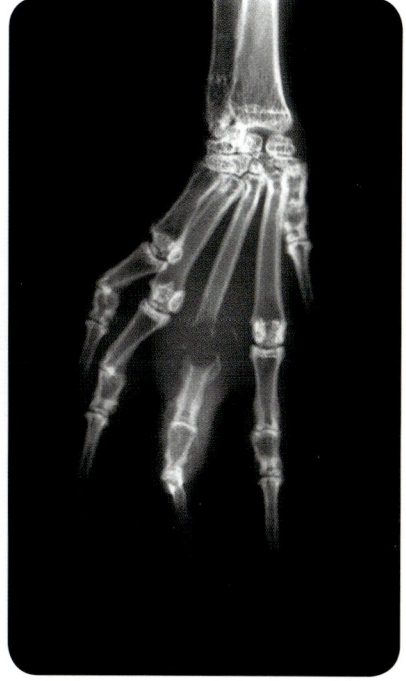

Figure 2.220 Dorsopalmer radiograph of the carpus of a rabbit with a Salter-Harris type I fracture of the third metacarpal bone of the right manus (arrow). Note the normal superimposition of digits 1 and 2.

Figure 2.221 Dorsopalmar radiograph showing fracture of the distal epiphysis of metacarpal 3 and proximal epiphysis of P1 of digit 3. This rabbit was accidentally stepped on by the owner.

The NORMAL PELVIC LIMB
Lateral Projection

Due to varying soft tissue density, superimposition of the abdomen, and the relatively small size of the distal pelvic limb, it is impossible to obtain an optimal single radiograph of the entire pelvic limb. Different kVp settings are required for the pelvis and the femur, than for the tibia and the distal limb. Actual size in a 1.9 kg dwarf rabbit.

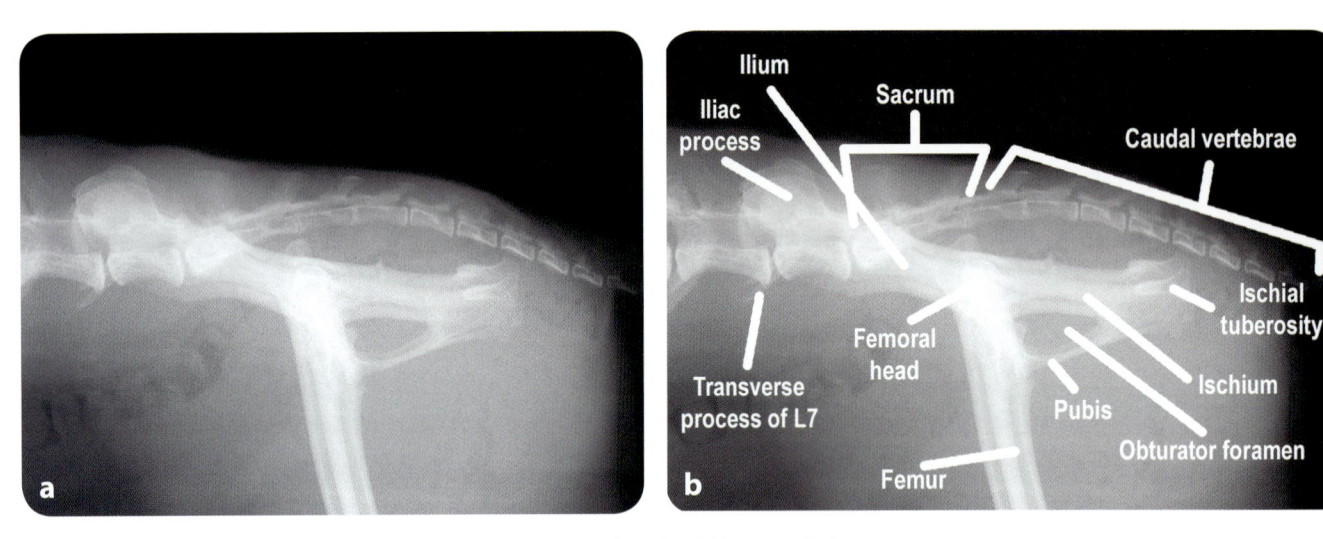

Figure 2.222a,b. Radiograph of the pelvis of a 1.9 kg dwarf rabbit, actual size.

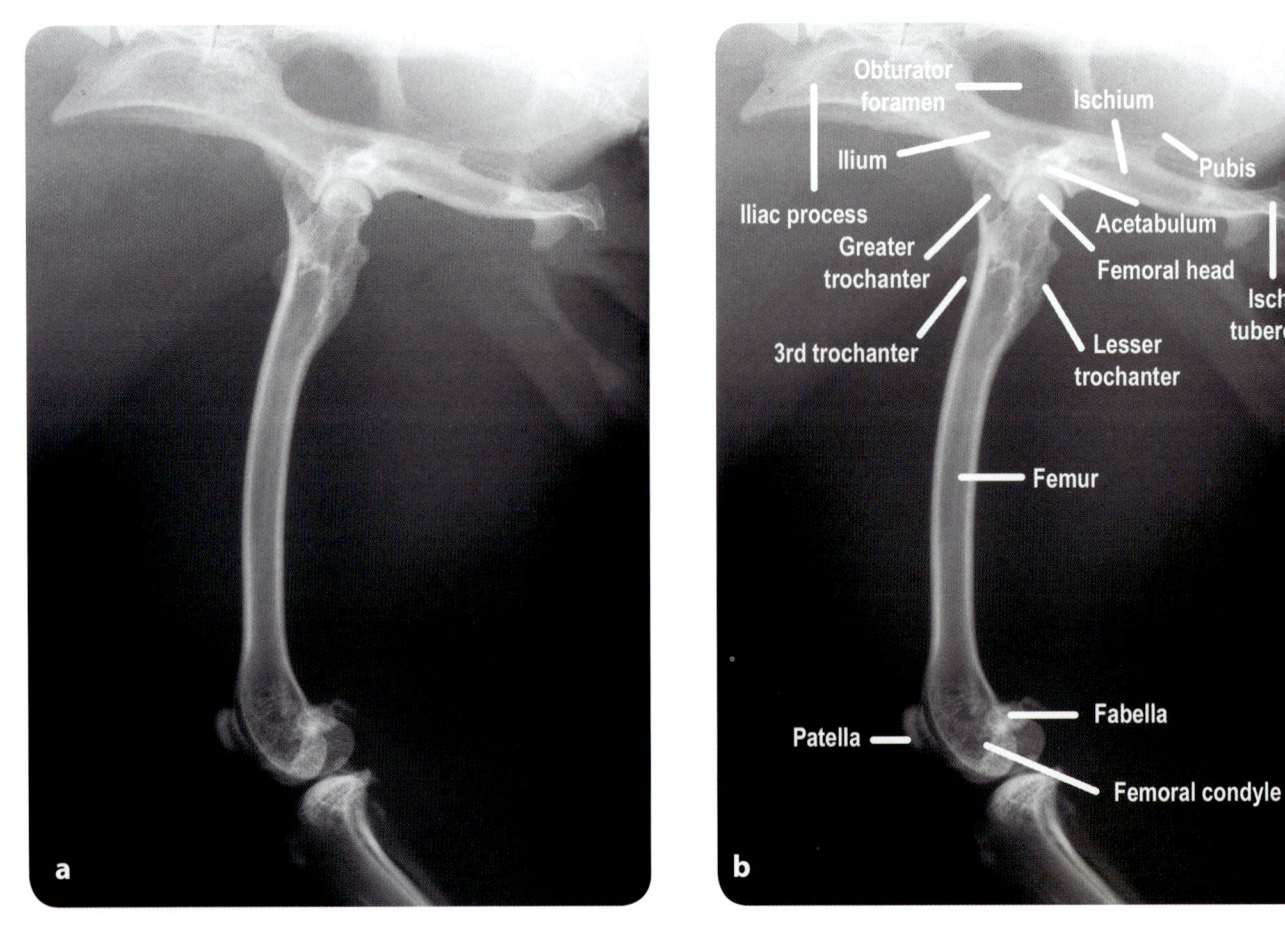

Figure 2.223a,b. Radiograph of the proximal pelvic limb of a 1.9 kg dwarf rabbit, actual size.

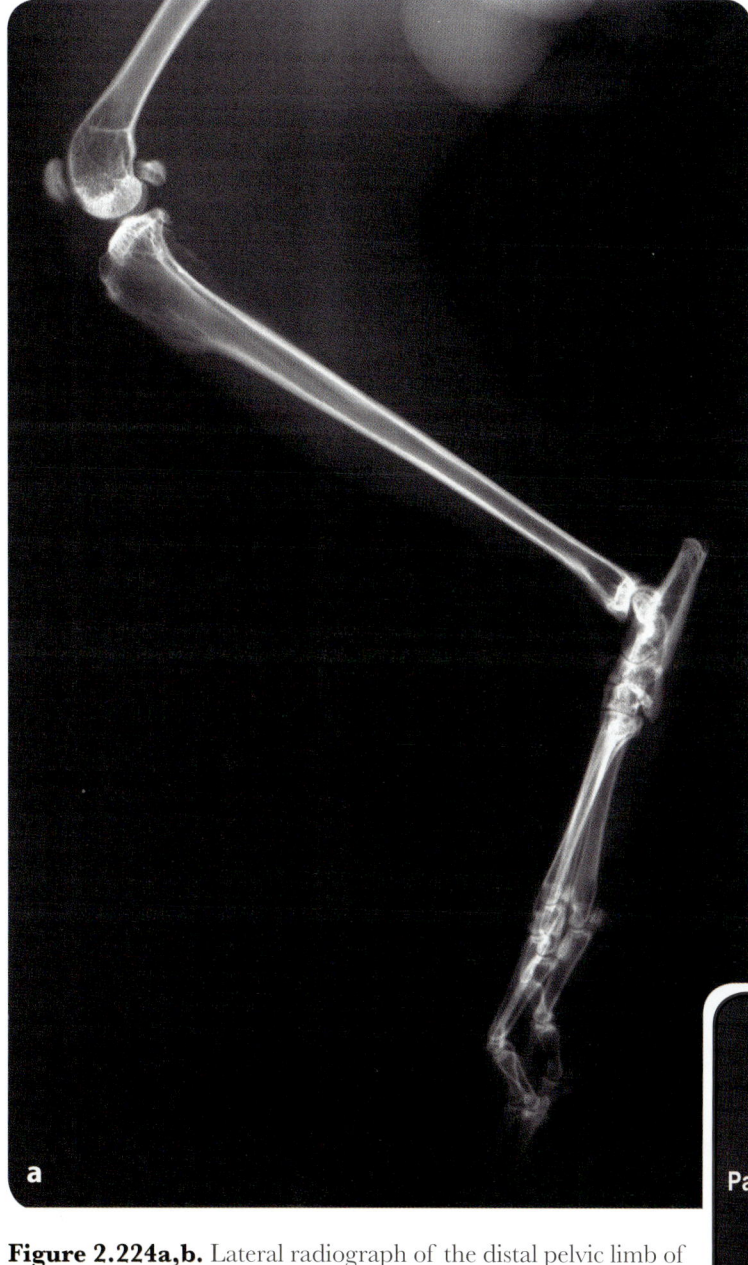

Figure 2.224a,b. Lateral radiograph of the distal pelvic limb of a 1.9 kg dwarf rabbit, actual size (a) and 75% of actual size (b).

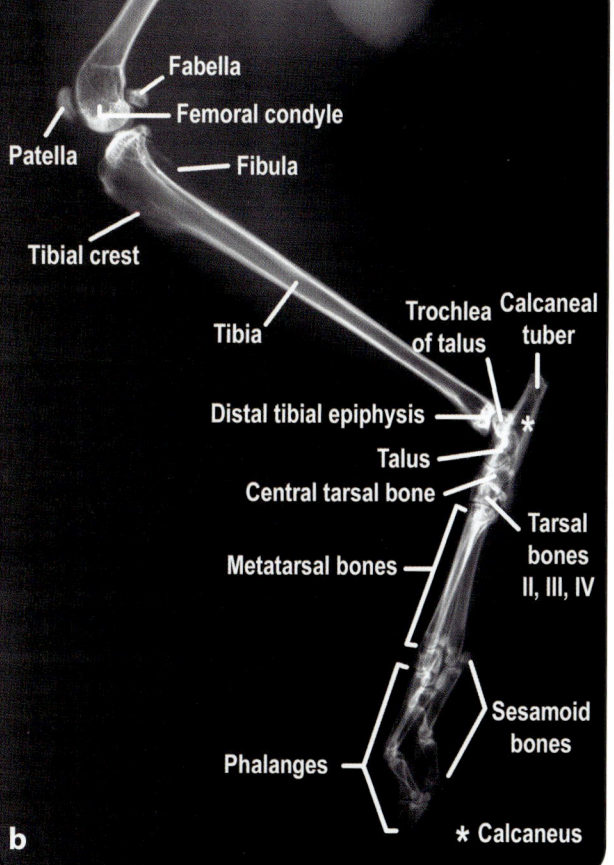

Ventrodorsal Projection of the Pelvis
Craniocaudal Projection of the Proximal Pelvic Limb

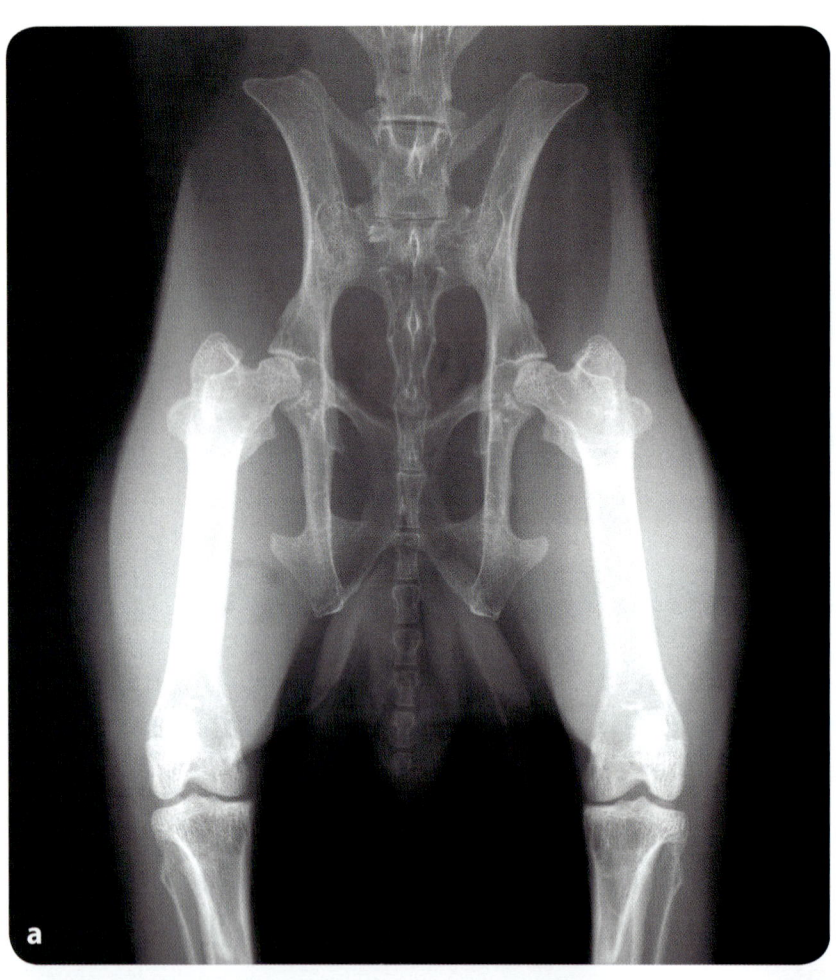

Figure 2.225a,b. Ventrodorsal projection of the pelvis and craniocaudal projection of the proximal pelvic limb of a 1.9 kg dwarf rabbit, actual size (a) and 75% of actual size (b).

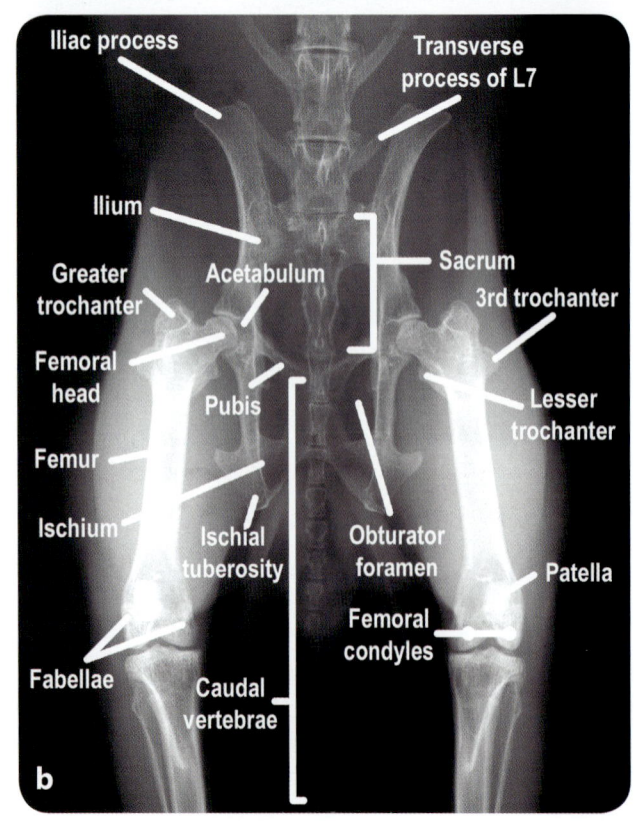

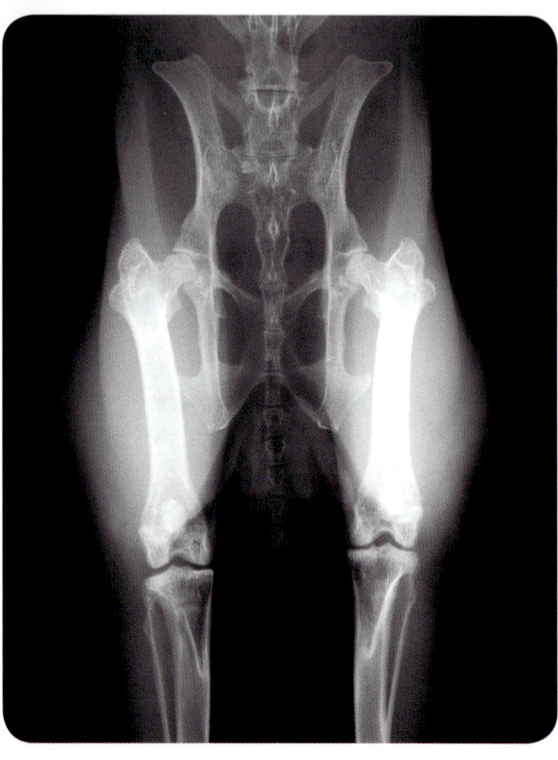

Figure 2.226. Ventrodorsal projection of the pelvis with medially-rotated femurs.

Craniocaudal Projection of the Distal Pelvic Limb

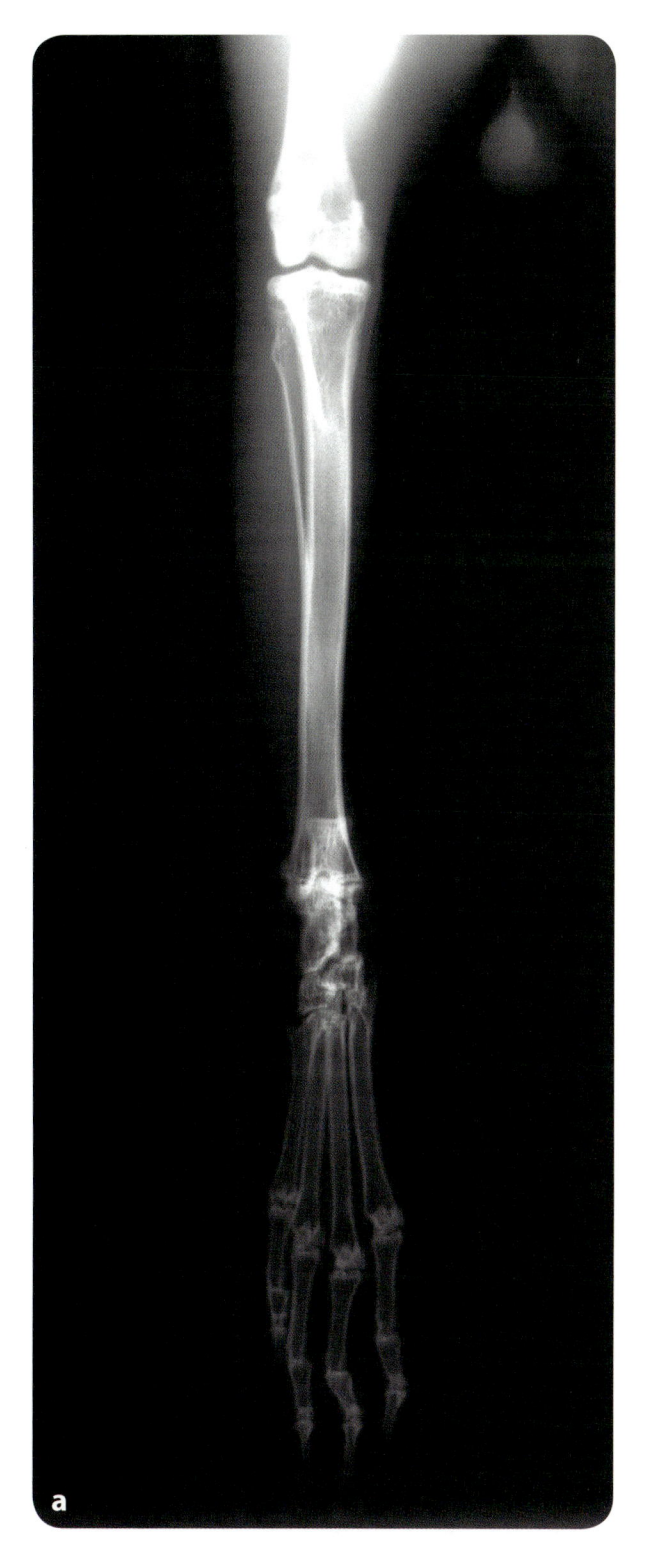

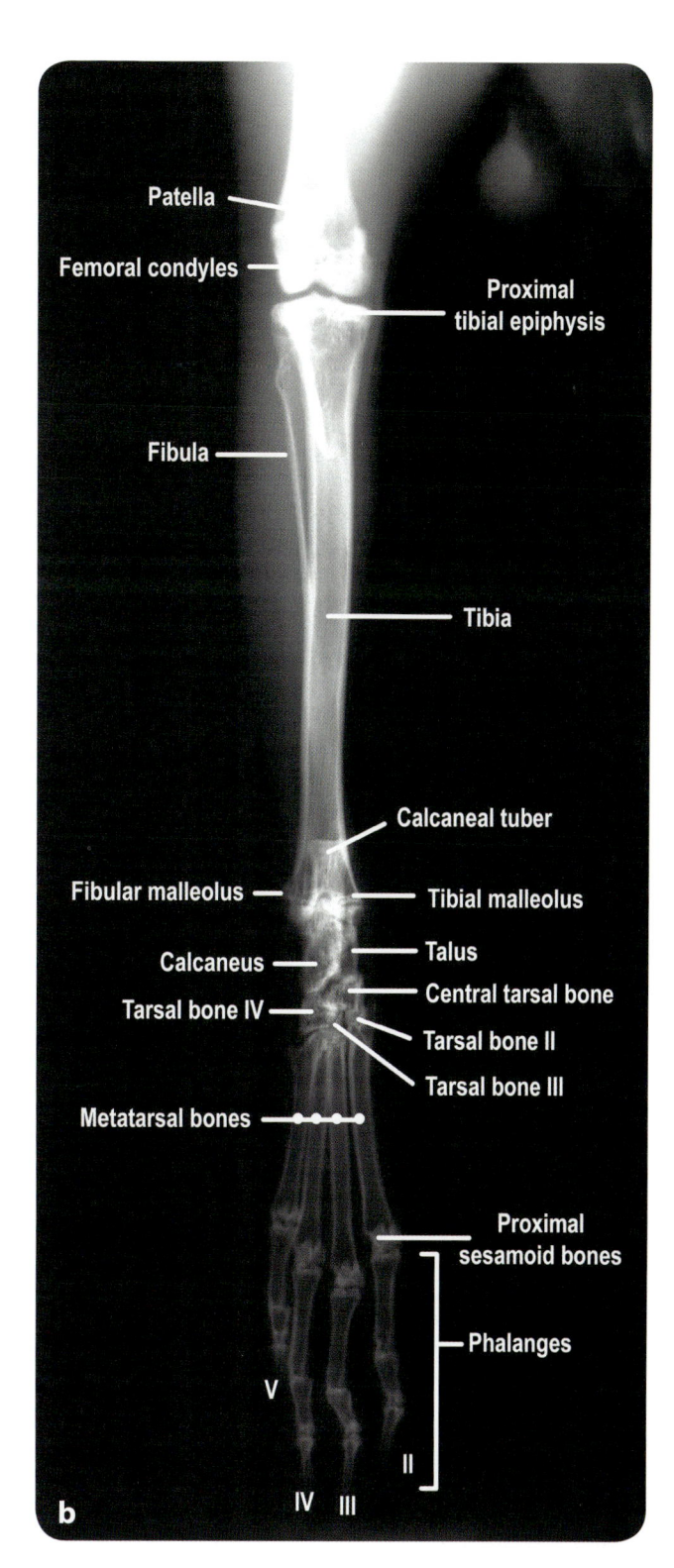

Figure 2.227a,b. Radiograph of the distal pelvic limb of a 1.9 kg dwarf rabbit, actual size.

ABNORMALITIES of the PELVIC LIMB
Diseases of the Pelvis

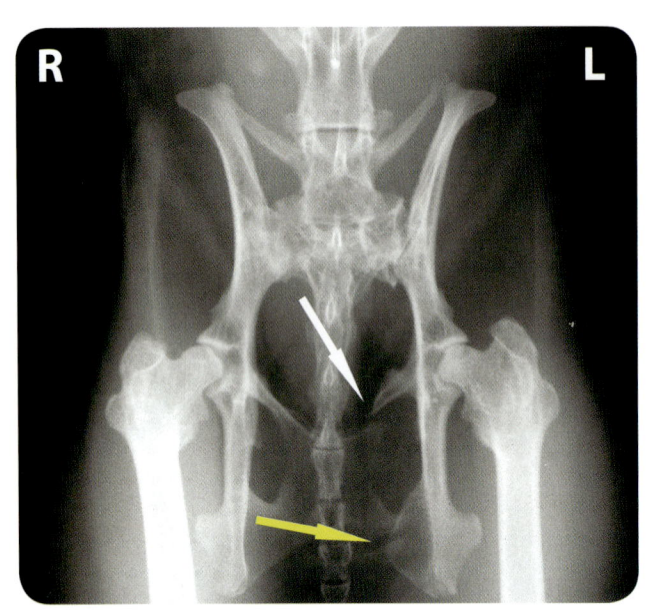

Figure 2.228. Ventrodorsal radiograph of the pelvis of a rabbit with a fracture of the ramus of the left pubic bone (white arrow) and the ischiatic arch (yellow arrow).

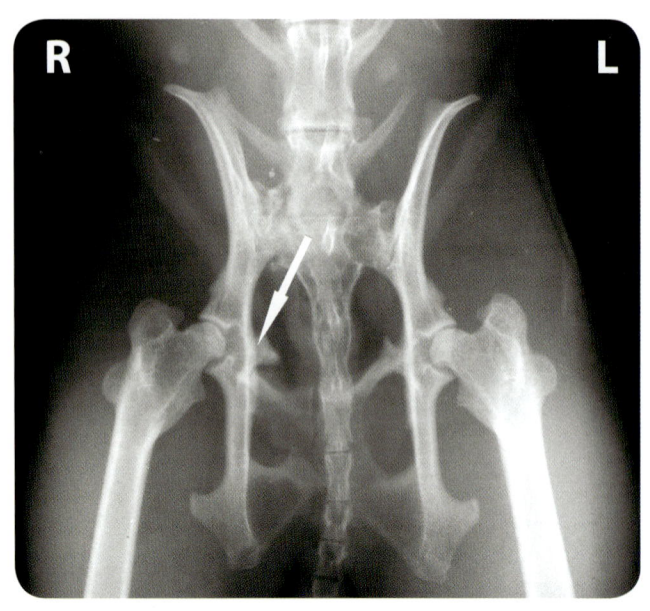

Figure 2.229. Ventrodorsal radiograph of the pelvis of a rabbit with a displaced fracture of the right pubic bone (arrow).

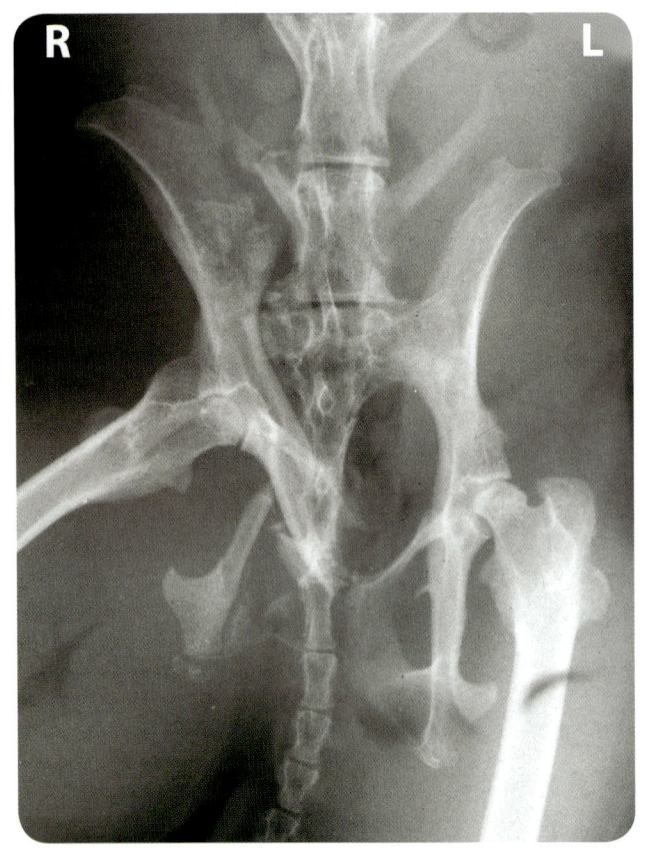

Figure 2.230. Ventrodorsal radiograph of the pelvis of a rabbit with complex fractures of the right hemipelvis. There are comminuted fractures of the right ischium and pubis, and displacement of the right ischium. This suggests sacroiliac fracture/luxation is also present.

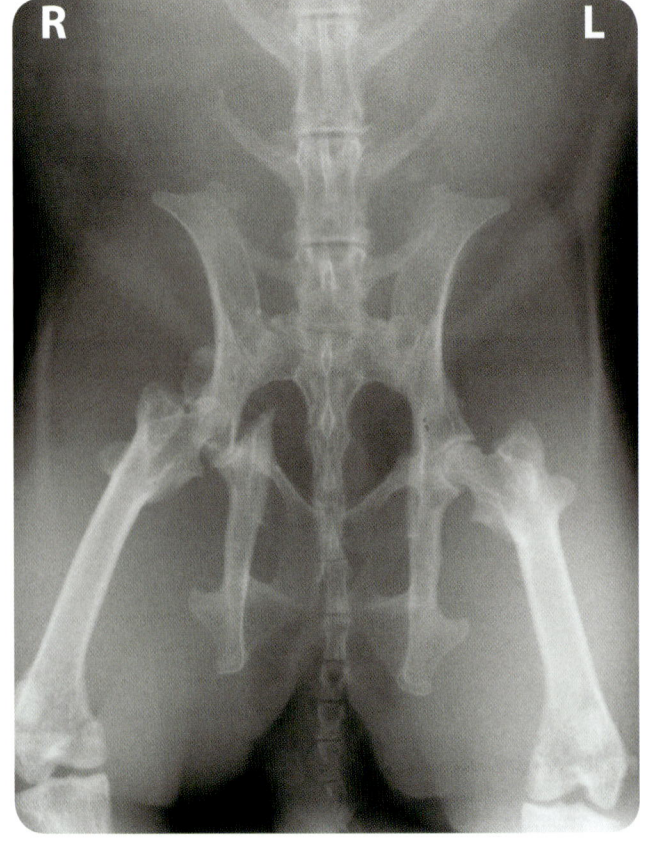

Figure 2.231. Ventrodorsal radiograph of the pelvis of a rabbit with an acetabular fracture and subluxation of the right hip joint. From the clinical standpoint, this fracture is more severe than that shown in Figure 2.230, because the coxofemoral joint is affected.

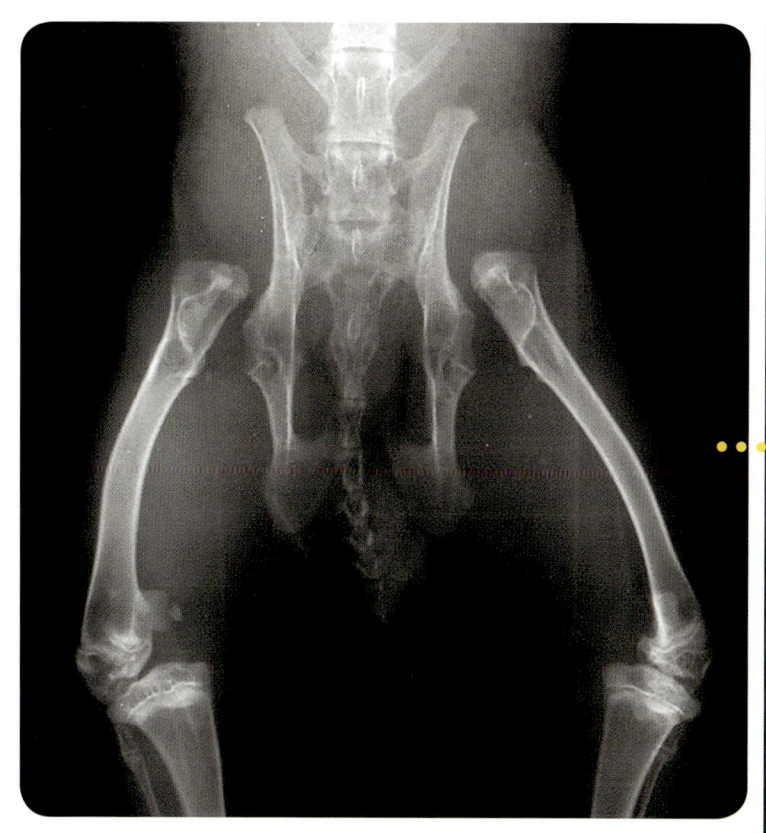

Figure 2.232. Ventrodorsal radiograph of the pelvis of a young rabbit with bilateral hip luxation as a result of hip dysplasia. There is malformation of the acetabula and the femoral heads. Secondary bilateral dysplasia of the femorotibial joints is also visible.

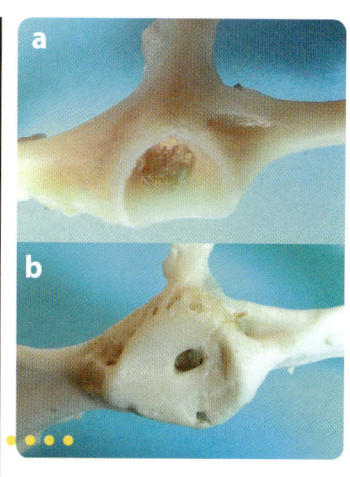

Figure 2.233a,b. Comparison between a normal acetabulum (a) and the dysplastic acetabulum shown in the radiograph (b). The abnormal specimen lacks a normal concavity, and the margins are affected by severe arthrosis.

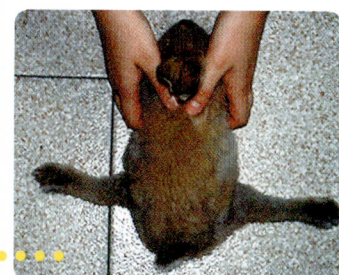

Figure 2.234. Severe leg abduction in a 3-month-old rabbit (same rabbit as in Figure 2.235) due to hip dysplasia. The age of this rabbit suggests the etiology is congenital. This clinical condition is commonly referred to as: "splay leg."

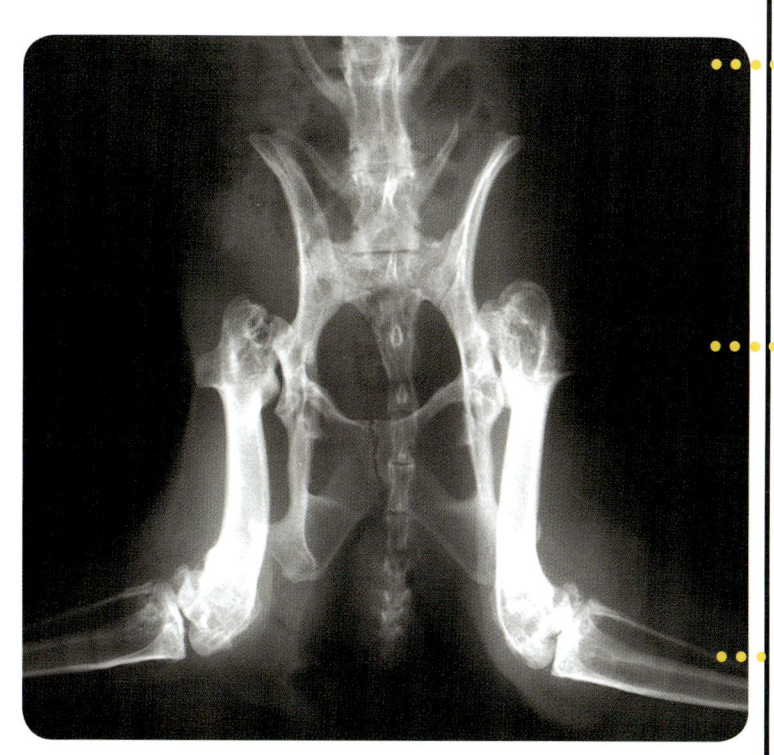

Figure 2.235. Ventrodorsal radiograph of the pelvis of a 5-year-old rabbit with bilateral hip luxation following hip dysplasia. The acetabular cavities are flatter than normal, the femoral heads have formed a pseudoarthrosis. Severe deformity of the distal femurs and bilateral dysplasia of the femorotibial joints are also visible.

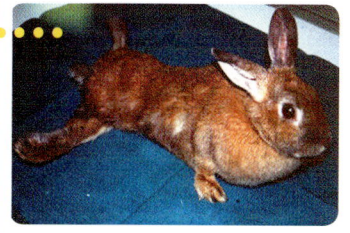

Figure 2.236. This rabbit lived 5 years with severe hip dysplasia.

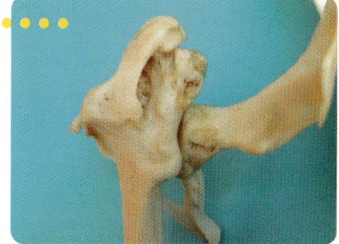

Figure 2.237. Bone specimen demonstrating pseudoarthrosis due to chronic luxation of the hip.

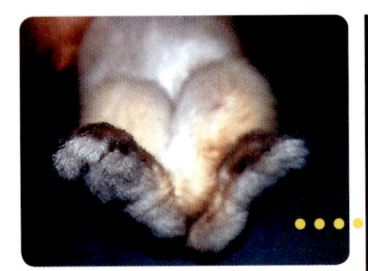

Figure 2.238. Clinical appearance of hip dysplasia with bilateral subluxation. Abnormal rotation causes this rabbit to cross the legs while in dorsal recumbency. The left foot of this rabbit is crossed to the right side of the body.

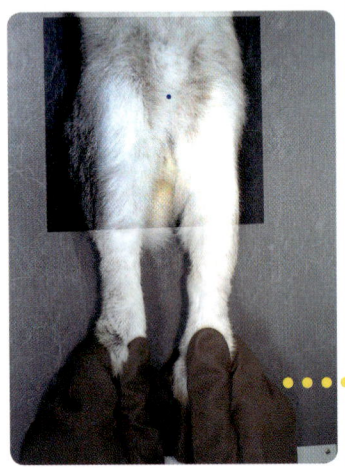

Figure 2.239. Incongruity of the hip joint is better emphasized with medial rotation of the distal femurs and the femorotibial joint, similar to the technique used for the study of hip dysplasia in dogs. The operator medially rotates the femorotibial joints manually.

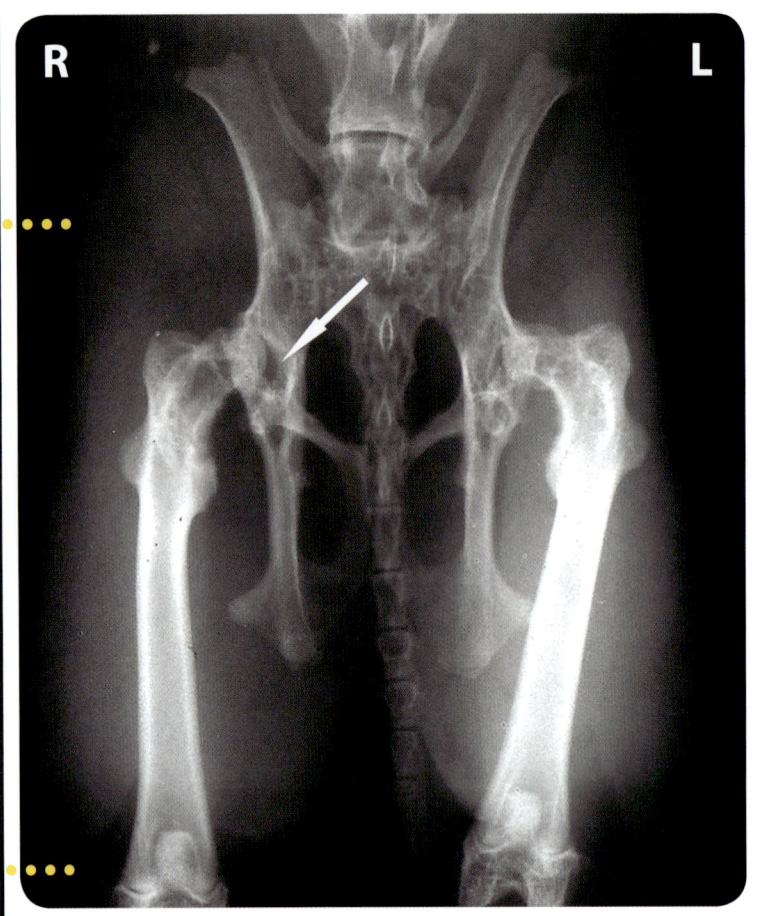

Figure 2.240. Ventrodorsal radiograph of a rabbit with bilateral hip subluxation following hip dysplasia. Note incongruity between the acetabula and the femoral heads typified by a crescent-shaped joint space (arrow).

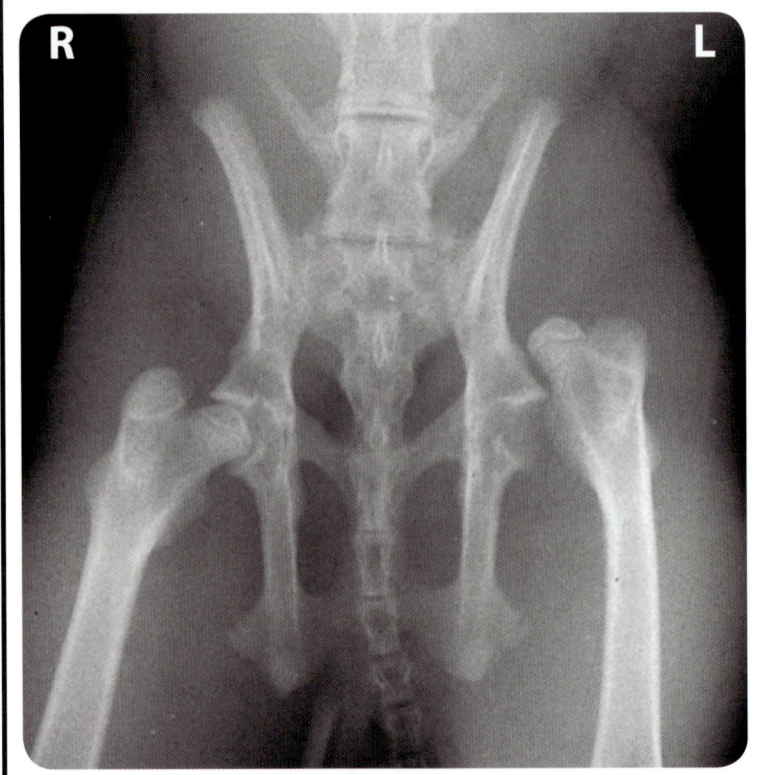

Figure 2.241. Ventrodorsal radiograph of the pelvis of a young rabbit with traumatic cranial luxation of the left hip joint. Hip dysplasia is not present, and the right hip appears normal.

PELVIC LIMB

Diseases of the Femur

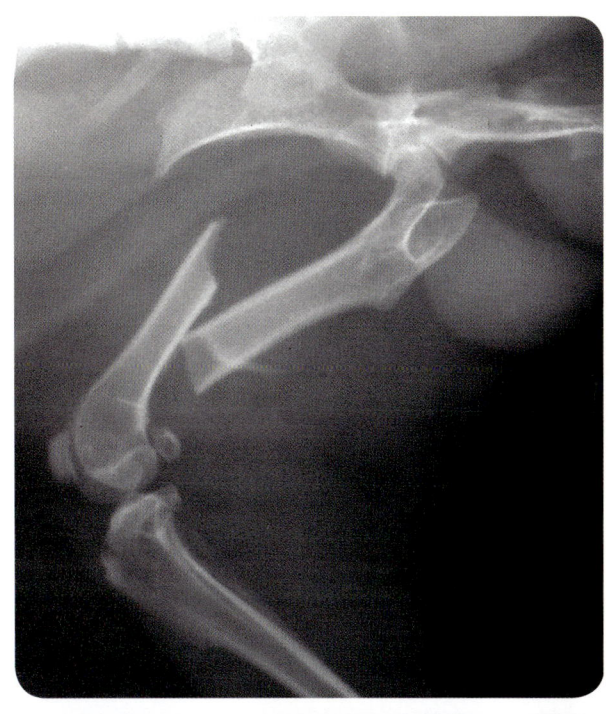

Figure 2.242. Lateral radiograph of the proximal pelvic limb of a rabbit with an oblique fracture of the diaphysis of the femur.

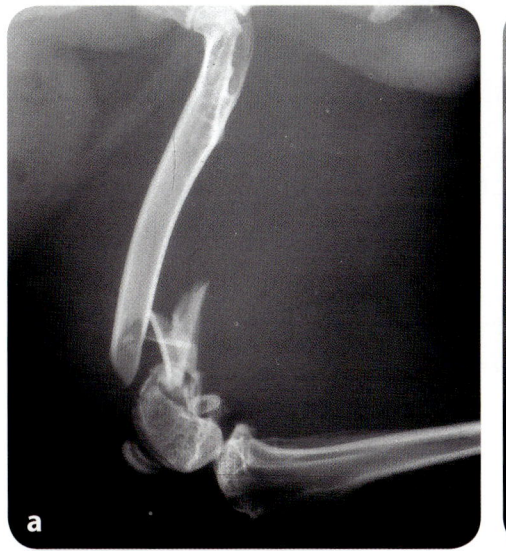

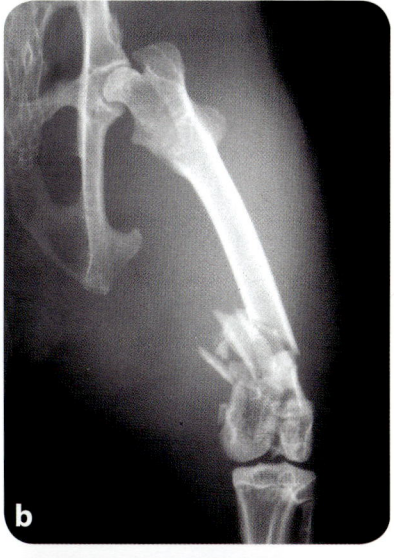

Figure 2.243a,b. Lateral (a) and craniocaudal (b) radiographs of a rabbit with a comminuted, intercondylar femoral fracture. Both radiographic views are necessary to properly assess the degree of damage, especially with intercondylar fractures.
These types of fractures require early and accurate fixation for proper healing.

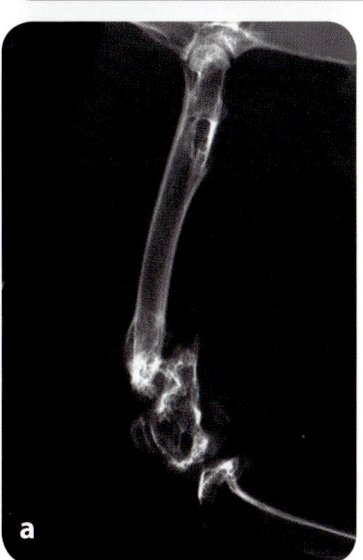

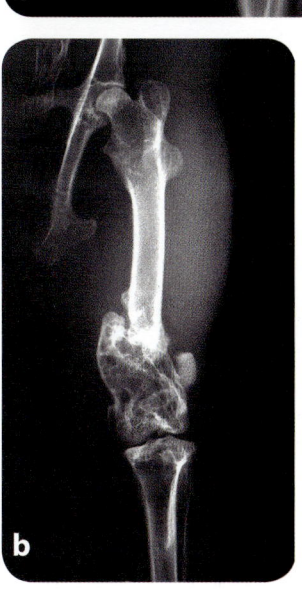

Figure 2.244a,b. Same fracture shown in Figure 2.243 after spontaneous bone healing. Lateral (a) and craniocaudal (b) views demonstrate that adequate alignment of the fragments has occurred. This resulted in minimal dysfunction of the limb in this patient.

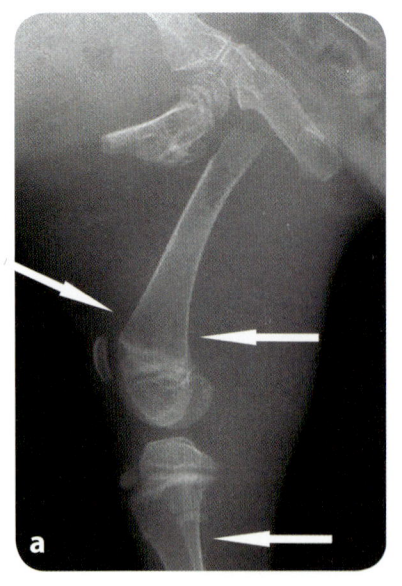

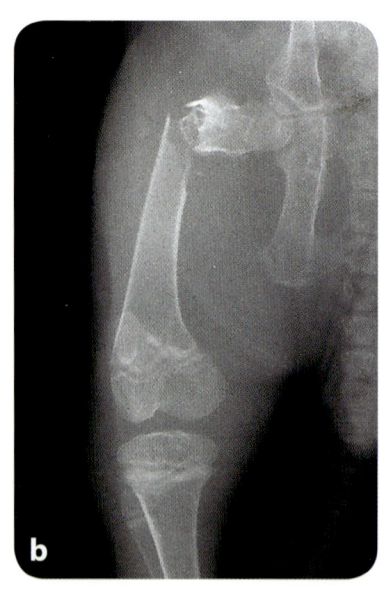

Figure 2.245a,b. Lateral (a) and craniocaudal (b) radiographs of a 5-month-old rabbit demonstrating a spiral oblique fracture of the proximal metaphysis of the femur as a consequence of metabolic bone disease. Note very thin cortices of the distal femur and the proximal tibia (arrows). Fracture repair is particularly difficult in these cases, as the bone cortices are often too fragile to support fixation devices such as pins or plates. In most cases, splinting or simple cage rest are the only options.

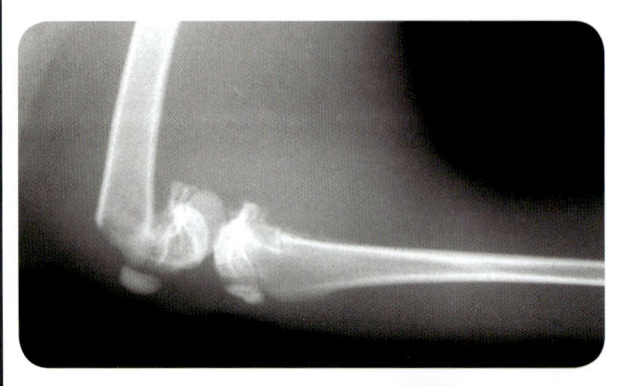

Figure 2.246. Lateral radiograph of the pelvic limb of a rabbit with a Salter-Harris type II fracture.

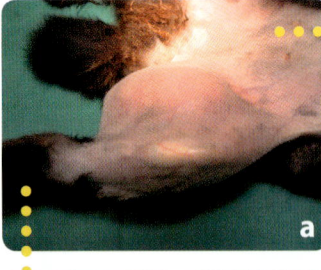

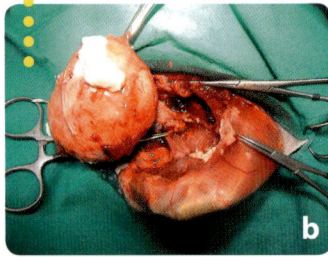

Figure 2.247a,b. Clinical appearance and surgical removal of the encapsulated abscess from the thigh muscles.

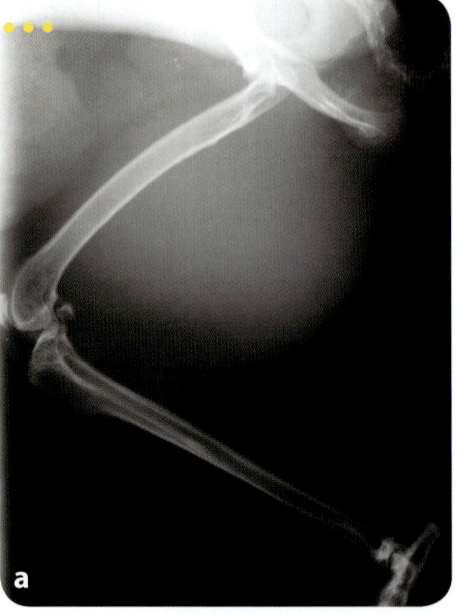

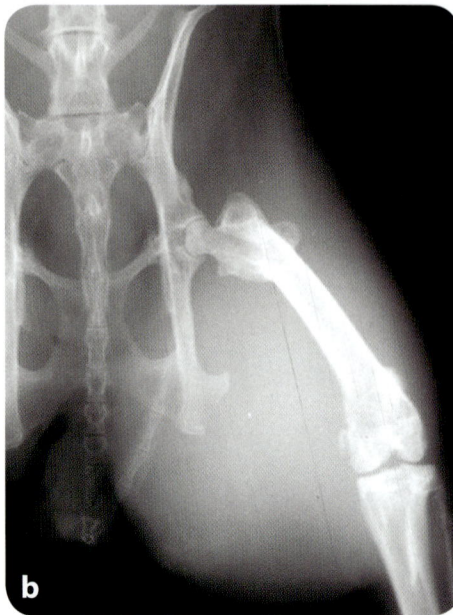

Figure 2.248a,b. Lateral (a) and craniocaudal (b) radiographs of a rabbit with a soft tissue mass associated with the thigh muscles. The mass was determined to be an abscess following intramusclular injection of a commonly-used antibiotic.

Diseases of the Tibia and Fibula

Fractures of the pelvic limbs are common in pet rabbits, and most commonly affect the tibia and fibula.
These are often a result of entrapment of the foot in the cage, and therefore commonly involve the distal diaphysis or metaphysis. For this reason, they are often comminuted and/or open.

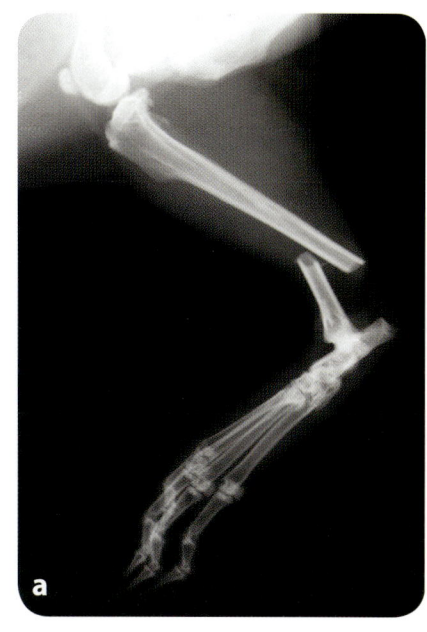

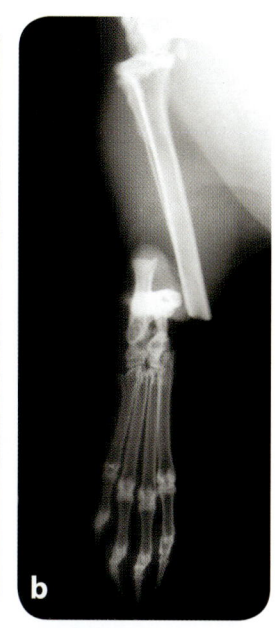

Figure 2.249a,b. Lateral (a) and craniocaudal (b) radiographs of a rabbit with a severely displaced metaphyseal fracture of the tibia and fibula.

Figure 2.250a,b. Lateral (a) and craniocaudal (b) radiographs of the pelvic limb of a rabbit with a comminuted, distal metaphyseal fracture of the tibia and fibula.
Highly comminuted fractures usually occur when the rabbit has been stepped on rather than dropped or trapped.
The short distal fragment makes fixation of these fracture challenging (see Figure 2.267).

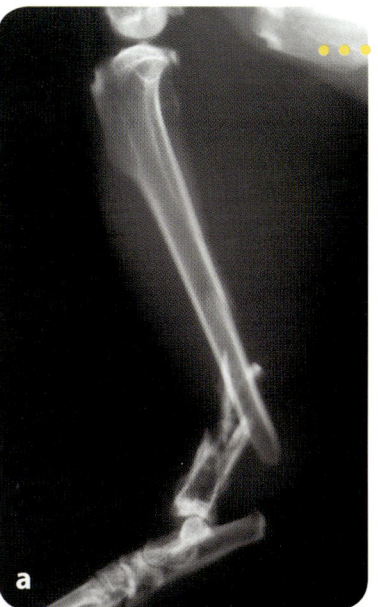

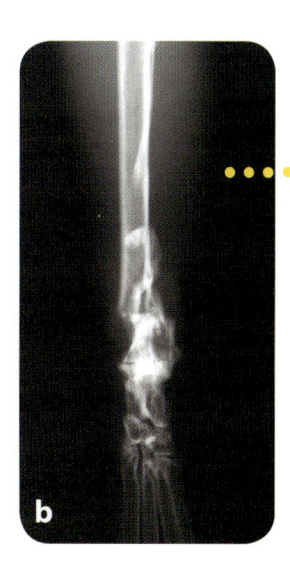

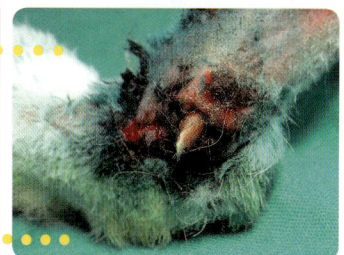

Figure 2.251. Grade III open fracture of the distal tibia. Open fractures are prone to osteomyelitis, and the risk for nonunion and other complications is high.

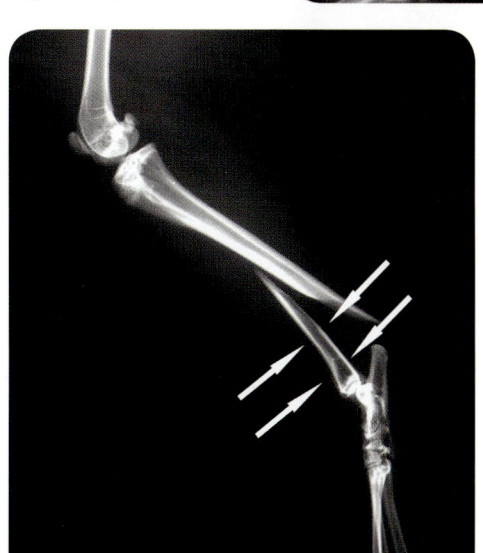

Figure 2.252. Lateral radiograph of the pelvic limb of a rabbit with a complete oblique fracture of the shaft of the tibia and fibula. Despite the presence of a long distal fragment, both cortices are present in only the most distal portion (arrows).

Figure 2.253a,b. Lateral (a) and craniocaudal (b) projections of the pelvic limb of a rabbit with a fracture of the shaft of the tibia and fibula. Several small cortical fragments are present.

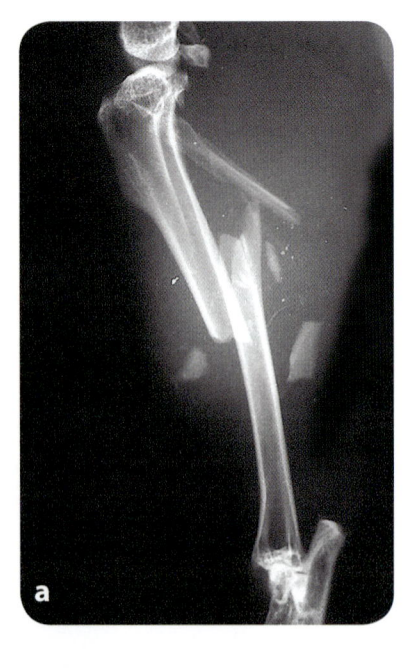

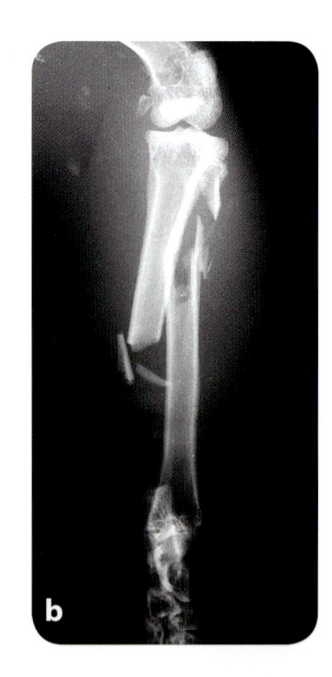

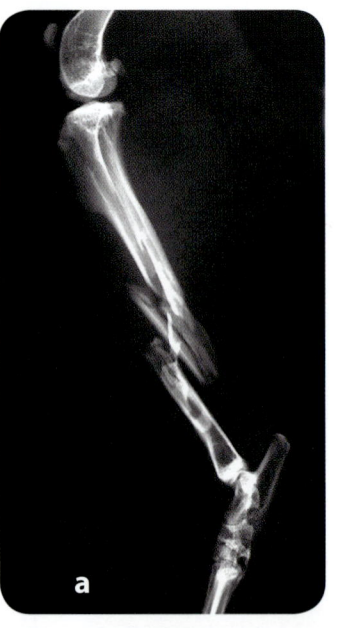

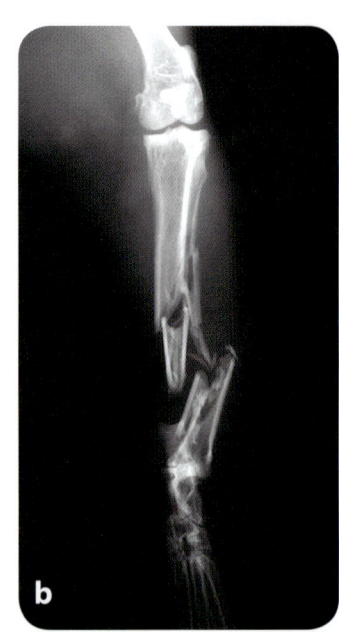

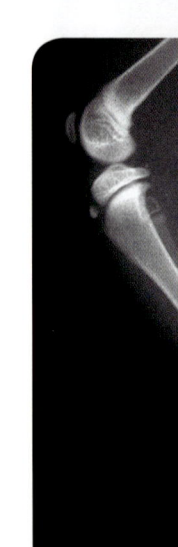

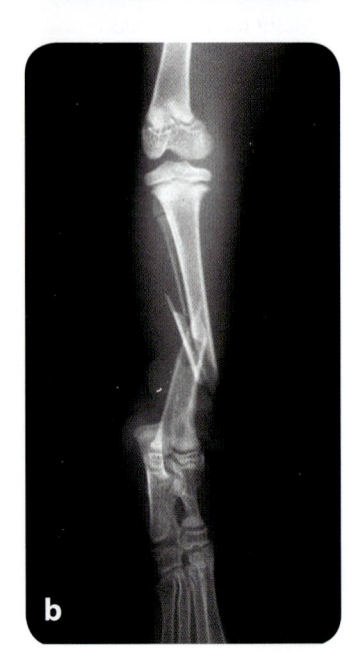

Figure 2.254a,b. Lateral (a) and craniocaudal (b) projections of the pelvic limb of a rabbit with a comminuted fracture of the diaphysis of the tibia and fibula. Highly comminuted fractures usually occur when the rabbit has been stepped on, rather than dropped or trapped.

Figure 2.255a,b. Lateral (a) and craniocaudal (b) projections of the pelvic limb of a 4-month-old rabbit with an oblique, comminuted, spiral fracture of the tibia and fibula. The open physes are clearly visible.

Figure 2.256a,b. Lateral (a) and craniocadual (b) projections of the pelvic limb of a rabbit with a proximal metaphyseal fracture of the tibia and of the distal fibula.

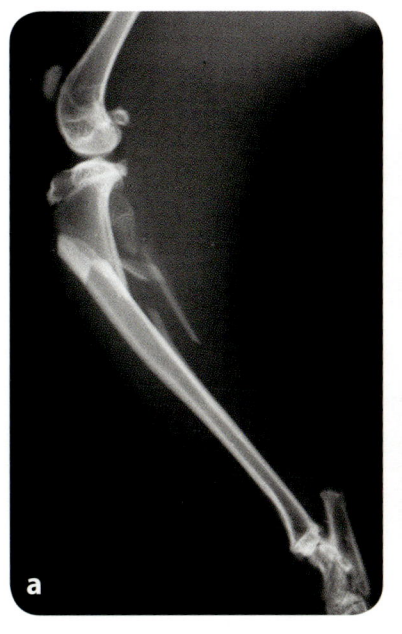

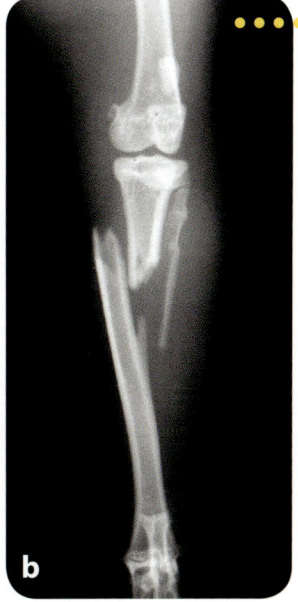

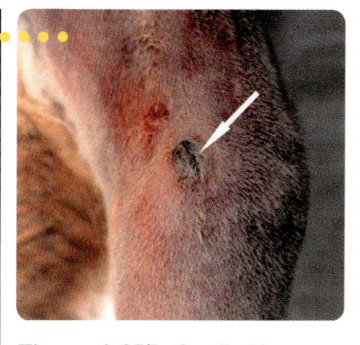

Figure 2.257. Small skin punctures produced by grade I open fractures are often hidden under fur and can easily be missed without careful shaving and inspection (arrow).

Figure 2.258a,b. Lateral (a) and craniocaudal (b) radiographs of the pelvic limb of a 7-year-old rabbit with metabolic bone disease and a pathologic fracture of the shaft of the tibia and fibula. Note thin cortices, loss of trabecular bone and reparative periosteal response (arrow). The rabbit previously underwent osteosynthesis for fracture of the tibia. Despite good bone healing, diffuse osteoporosis of the tibial segment led to a second fracture.

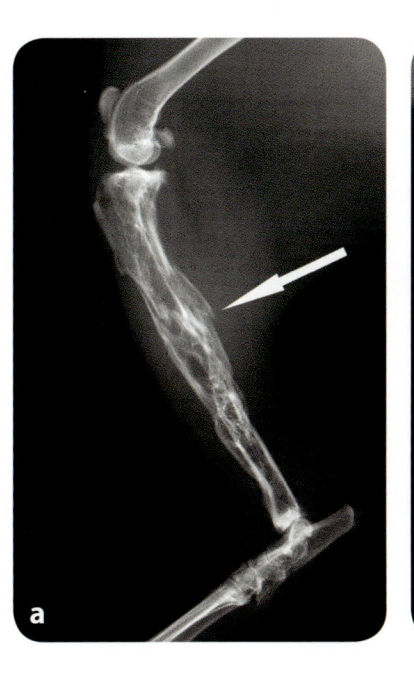

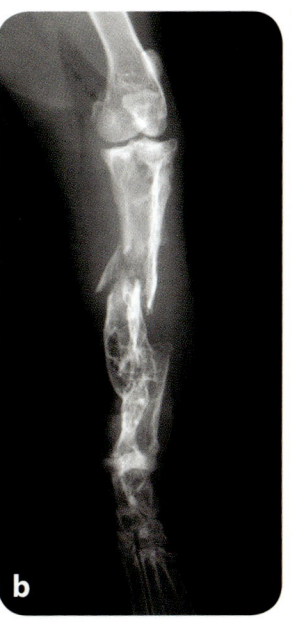

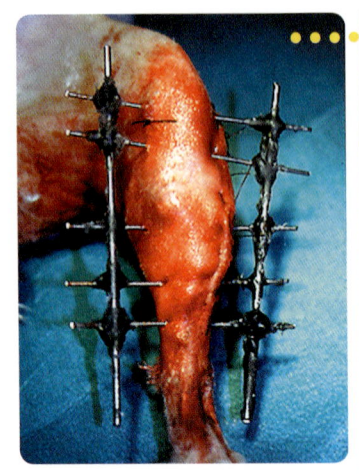

Figure 2.259. External skeletal fixation (ESF) is the treatment of choice for tibial fractures. Different configurations can be placed: a monoplanar bilateral 5 pin is demonstrated. To reduce the weight of the implant, polymethylmethacrylate is used to connect the pins with the external bars, instead of clamps.

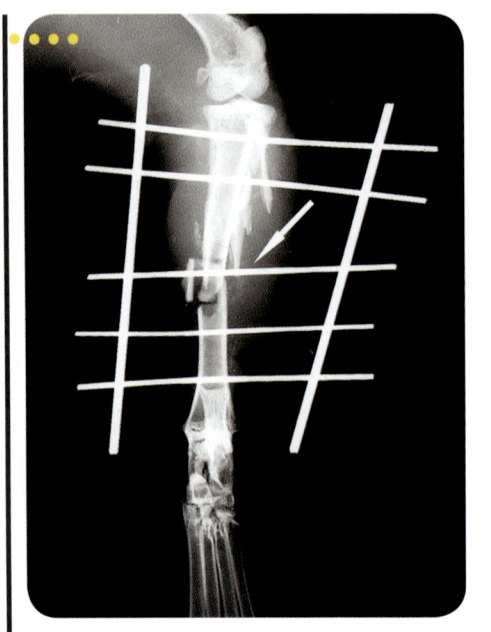

Figure 2.260. Radiographic appearance of the same case as shown in 2.253. The most distal pin in the proximal fragment (arrow) has been placed too close to the fracture site. Obtaining a true lateral projection can sometimes be difficult due to interference of the external fixator.

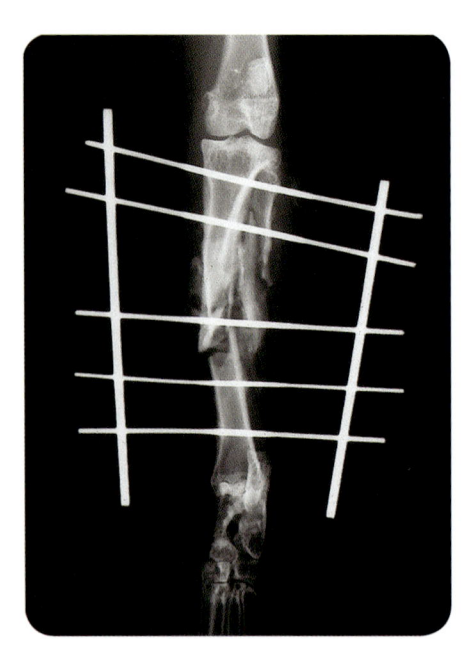

Figure 2.261. Radiographic appearance 17 days post surgery. There is exuberant periosteal reaction of the comminuted fragments and the caudolateral tibia adjacent to the fracture line.

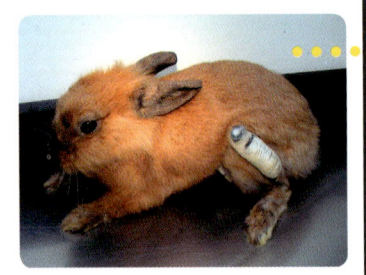

Figure 2.262. The ESF device is padded with cotton and a conforming bandage. Rabbits generally tolerate fixation devices well.

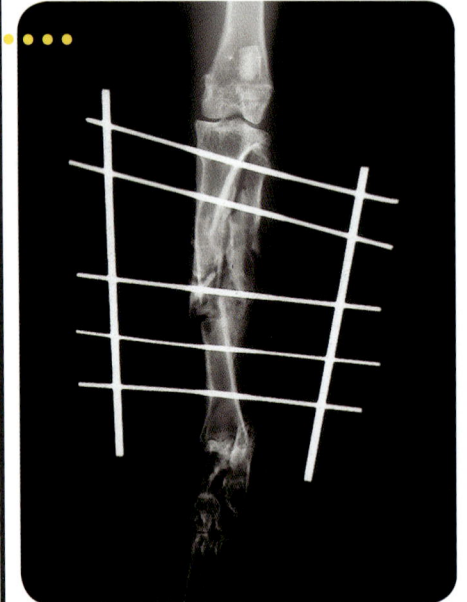

Figure 2.263. Radiographic appearance 40 days post surgery. Progressive bone healing is visible.

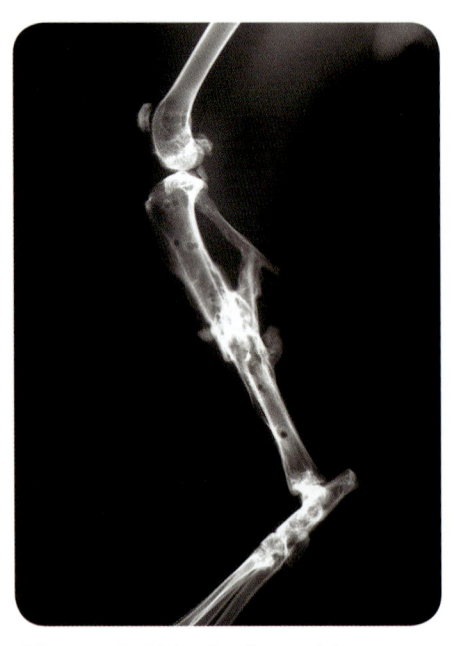

Figure 2.264. Radiographic appearance 80 days after surgery, and after removal of the ESF device at 55 days, lateral projection. Bone healing is typified by osseous remodeling at the fracture site of both the tibia and the fibula.

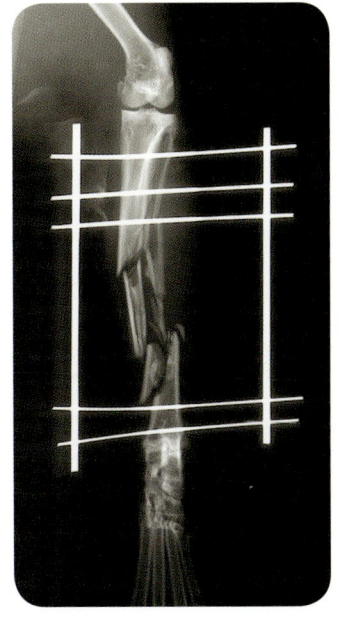

Figure 2.265. Postoperative radiograph of the same patient with a highly comminuted fracture shown in Figure 2.254.

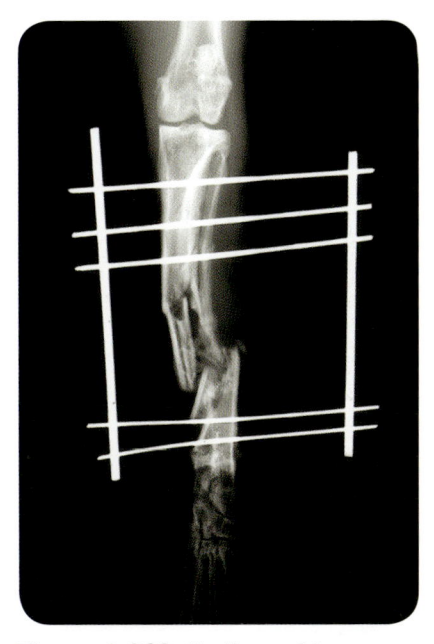

Figure 2.266. Radiographic appearance 33 days post surgery. New periosteal bone formation is present.

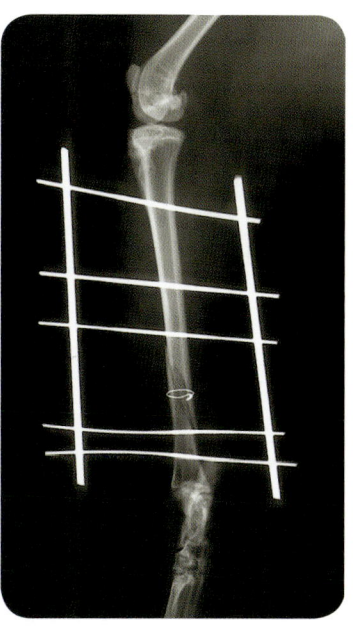

Figure 2.267. Postoperative radiograph of the oblique fracture shown in Figure 2.252, oblique projection. Other techniques can be used with external fixation. To ensure stabilization of the oblique fragments, a cerclage wire has been placed at the fracture site.

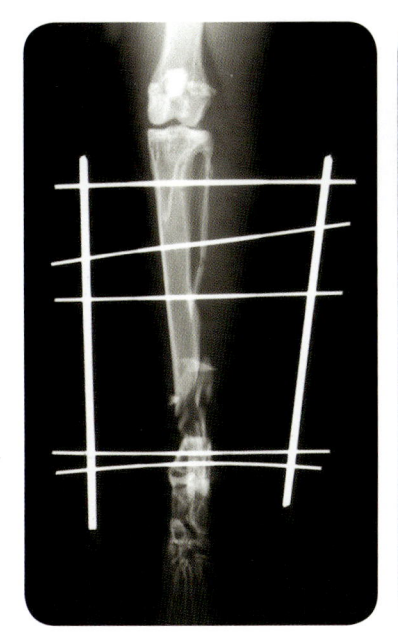

Figure 2.268. Postoperative radiograph of the same rabbit with a distal metaphyseal fracture shown in Figure 2.250. A minimum of two pins must be inserted in each fragment, penetrating both bone cortices.
Since the distal fragment is very short, the two distal pins have been inserted very close to each other.

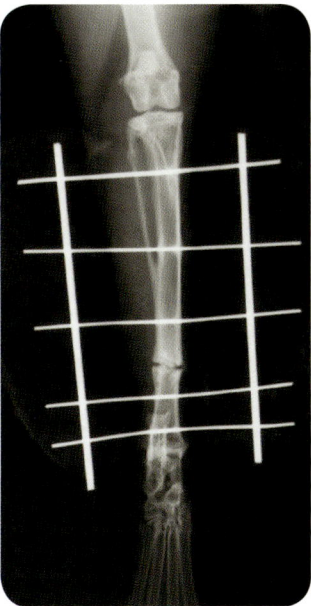

Figure 2.269. Radiographic appearance of first intention bone healing of the same rabbit as in Figure 2.249. At 35 days post surgery no periosteal reaction is visible. First intention bone healing can occur when the ESF device is very rigid and prevents micromovements at the fracture site, which stimulates osseous union with minimal callus formation. The ESF device was removed two weeks later and bone healing had occurred.

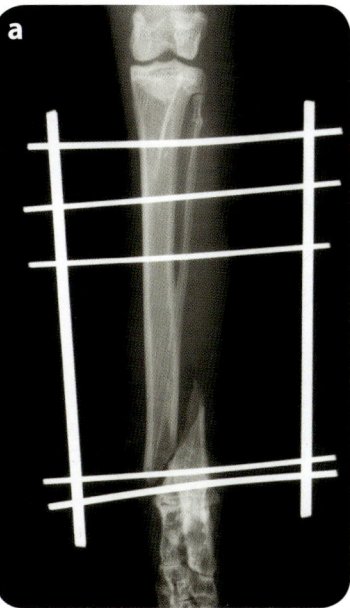

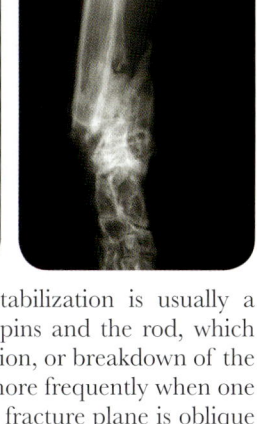

Figure 2.270a,b. Inadequate stabilization is usually a consequence of loosening of the pins and the rod, which can result in delayed union, nonunion, or breakdown of the healing fracture site. This occurs more frequently when one fragment is very short, and/or the fracture plane is oblique (a). In this case, initial stabilization was lost, as is seen in the distal segment of this epiphyseal fracture.
Despite inadequate stabilization, bone healing eventually occurred anyway (b).

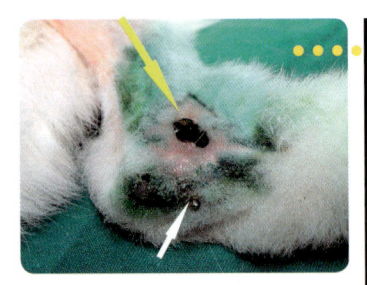

Figure 2.271. The intramedullary pin can be see protruding out of the plantar surface of the hock (white arrow). Necrotic and ulcerated skin is present at the level of the proximal cerclage wire (yellow arrow).

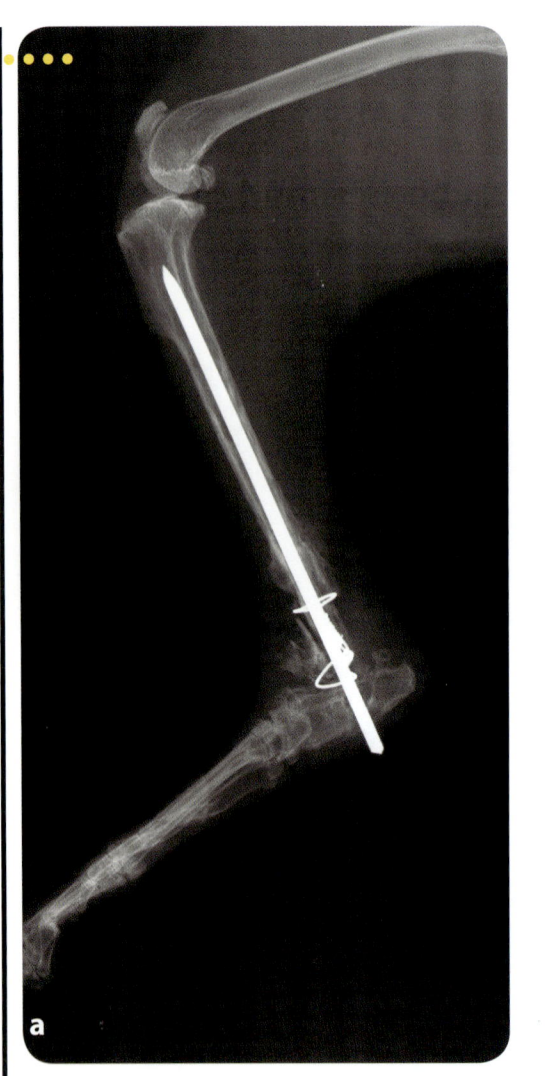

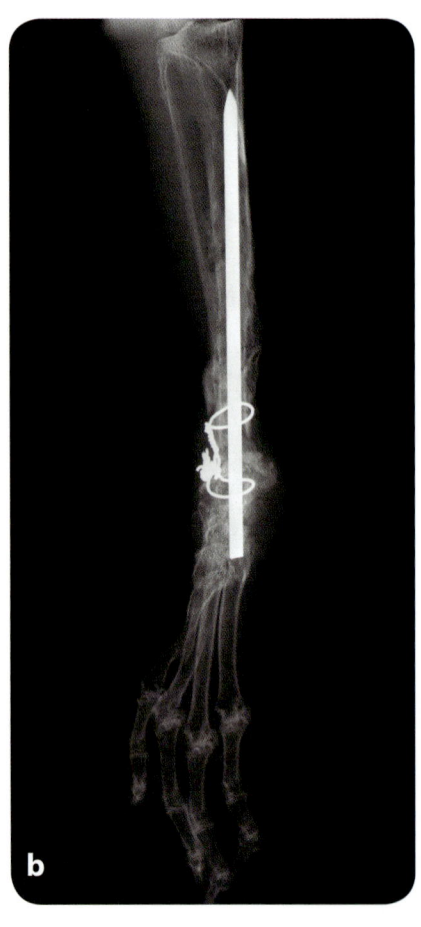

Figure 2.272a,b. Lateral (a) and craniocaudal/dorsoplantar (b) radiographs of a rabbit with osteomyelitis of the tibia and tarsal bones following improper fixation of a comminuted epiphyseal fracture of the tibia, and athrodesis of the tarsocrural joint.

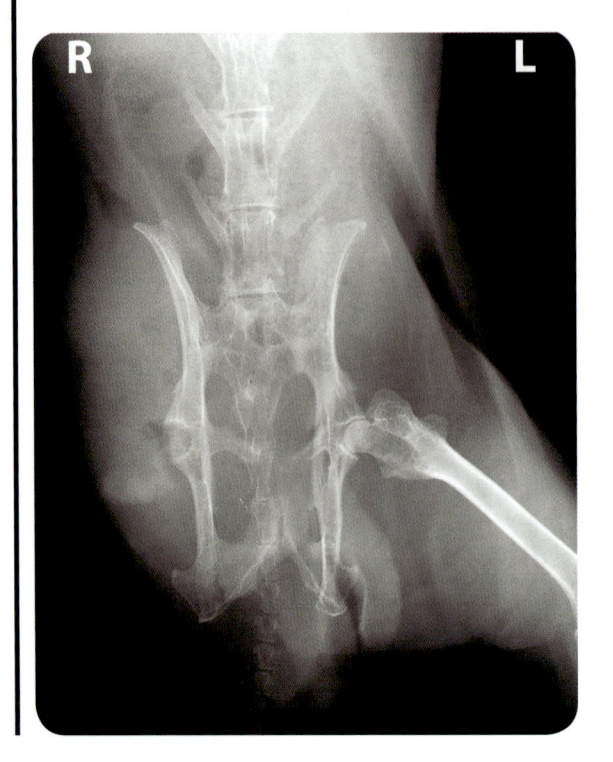

Figure 2.273. Ventrodorsal radiograph of the pelvis of the same patient as in Figure 2.271 after amputation of the pelvic limb. Proximal amputation at the coxofemoral joint prevents dissection of the femur and in the authors' experience is associated with excellent outcomes.

Diseases of the Tarsus, Metatarsus, and Phalanges

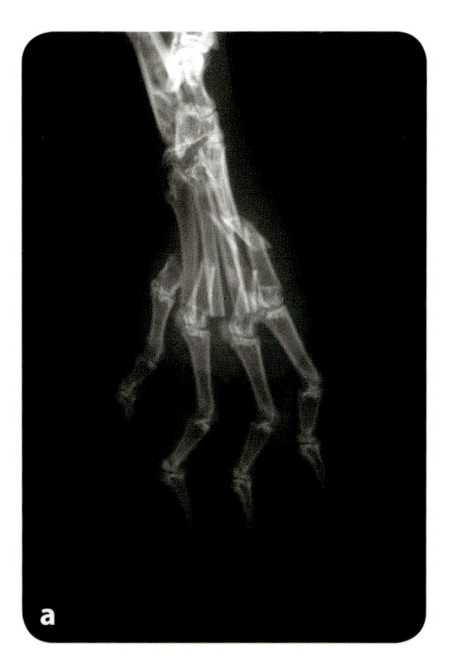

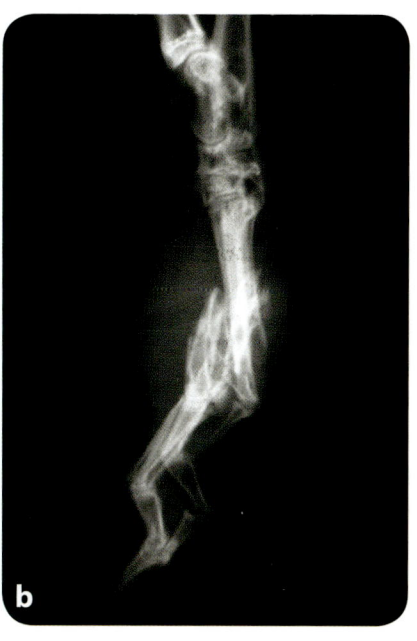

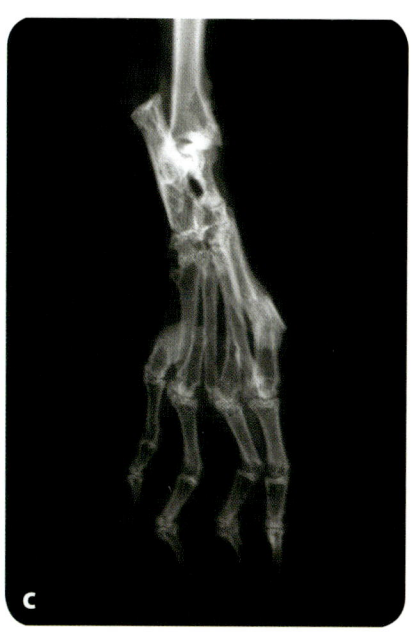

Figure 2.274a,b,c. Oblique (a) and lateral (b) projections of a rabbit with multiple fractures of the metatarsal bones. Metatarsal bones two, three, four, and five are fractured. Follow-up after two months (c) shows osseous union of the fragments. Due to size and location, these fractures are treated with simple cage rest or splinting.

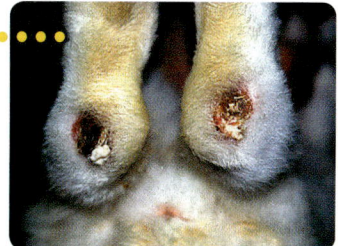

Figure 2.276. Severe bilateral ulcerative pododermatitis and purulent arthrosynovitis of the tarsal joints.

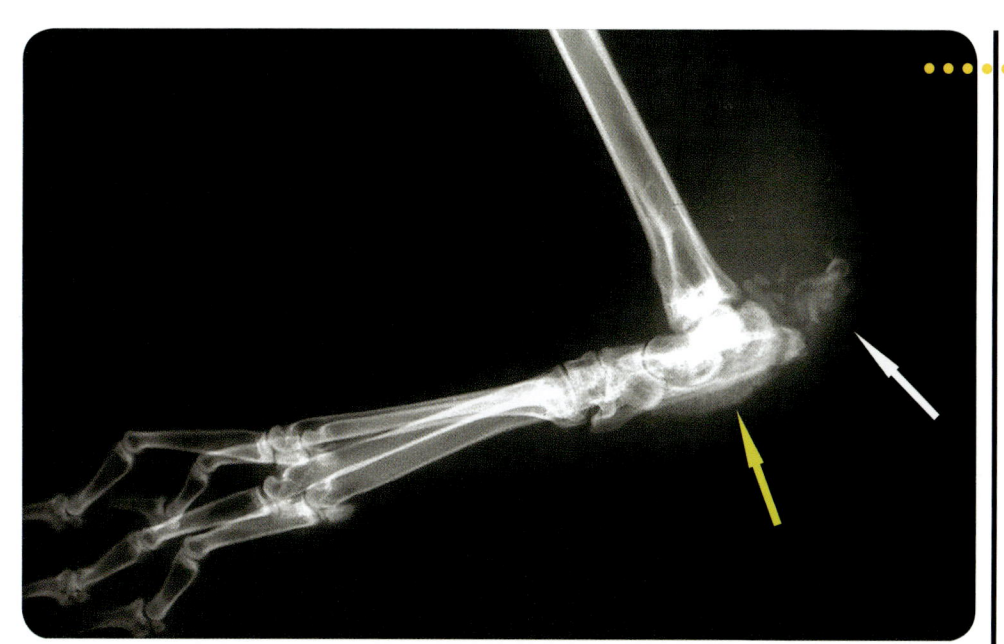

Figure 2.275. Lateral radiograph of the pelvic limb of a rabbit showing periarticular soft tissue swelling and mineralization (white arrow). The calcaneus is remodeled and there is osseous reaction along the plantar aspect of the tarsus (yellow arrow).

References:

Aiken S: Small mammal dentistry, part I: Surgical treatment of dental abscesses in rabbits. In: Quesenberry KE, Carpenter JW eds. Ferrets, Rabbits and Rodents. Clinical Medicine and Surgery 2nd ed. Philadelphia, PA: Elsevier; 2004:379-382.

Bevilacqua L, Benato L: Uterine rupture with ectopic fetuses in a holland lop rabbit. Exotic DVM. 2006;8(2):3.

Brown SA, Rosenthal LR: Question #45. In: Self Assessment Color review of Small Mammals. London: Manson Publishing;1997:41-42.

Brown SA, Rosenthal LR: Question #66. In: Self Assessment Color Review of Small Mammals. London: Manson Publishing;1997:59-60.

Capello V, Gracis M: Radiology of the skull and teeth. In: Lennox A, ed. Rabbit and Rodent Dentistry Handbook. Ames, IA: Blackwell Publishing, (Formerly Zoological Education Network, Lake Worth, FL); 2005:65-99.

Capello V, Gracis M: Dental diseases. In: Lennox A, ed. Rabbit and Rodent Dentistry Handbook. Ames, IA: Blackwell Publishing, (Formerly Zoological Education Network, Lake Worth, FL); 2005:113-164.

Capello V, Gracis M: Secondary diseases. In: Lennox A, ed. Rabbit and Rodent Dentistry Handbook. Ames, IA: Blackwell Publishing, (Formerly Zoological Education Network, Lake Worth, FL); 2005:165-186.

Capello V, Gracis M: Dental procedures. In: Lennox A, ed. Rabbit and Rodent Dentistry Handbook. Ames, IA: Blackwell Publishing, (Formerly Zoological Education Network, Lake Worth, FL); 2005:213-248.

Capello V, Gracis M: Surgical treatment of periapical abscessations. In: Lennox A, ed. Rabbit and Rodent Dentistry Handbook. Ames, IA: Blackwell Publishing, (Formerly Zoological Education Network, Lake Worth, FL); 2005:249-272.

Capello V: Diagnosis and treatment of urolithiasis in a pet rabbit. Exotic DVM. 2004; 6(2):15-22.

Capello V: Surgical treatment of otitis externa and media in pet rabbits. Exotic DVM. 2004; 6(3):15-21.

Capello V: Extraction of incisor teeth in pet rabbits. Exotic DVM. 2004; 6(4):23-30.

Capello V: Extraction of cheek teeth and surgical treatment of periodontal abscessation in pet rabbits with acquired dental disease. Exotic DVM 2004; 6(4) 31-38

Capello V: External fixation for fracture repair in small exotic mammals. Exotic DVM. 2005; 7(6):21-37.

Chambers JN, McBride MP, Hernandez-Divers SJ: Dynamic crossed-pin fixation of a distal femoral growth plate fracture in a domestic rabbit *(Oryctolagus cuniculus)*. J Exotic Mam Med Surg. 2005; 3(2): 4-5.

Crossley DA: Oral biology and disorders of lagomorphs. Vet Clin N Am Exotic Anim Prac. 2003; 6:629-659.

Deeb B: Section 2. Rabbits. Respiratory disease and pasteurellosis. In: Quesenberry KE, Carpenter JW, eds. Ferrets, Rabbits and Rodents. Clinical Medicine and Surgery 2nd ed. Philadelphia, PA:Elsevier; 2004:172-182.

Deeb B: Section 2. Rabbits. Neurologic and muscoloskeletal diseases. In: Quesenberry KE, Carpenter JW, eds. Ferrets, Rabbits and Rodents. Clinical Medicine and Surgery 2nd ed. Philadelphia, PA:Elsevier; 2004:203-210.

Deeb B: The dyspneic rabbit. Exotic DVM. 2005; 7(1):39-42.

Divers SJ: Mandibular abscess treatment using antibiotic-impregnated beads. Exotic DVM. 2000; 2(5):15-18.

Fisher PG: Exotic mammal renal disease: causes and clinical presentation. Vet Clin N Am Exot Anim Prac. 2006; 9:33-67 (2006).

Girling SJ: Mammalian imaging and anatomy. In: Meredith A, Redrobe S, eds. BSAVA Manual of Exotic Pets 4th ed. Quedgeley, Glouchester: British Small Animal VeterinaryAssociation; 2005:1-12.

Gobel T: Transurethral uroendoscopy in the female rabbit. Exotic DVM. 2002; 4(5) 23-27.

Guzman Sanchez-Migallon D, Mayer J, Gould J, Azuma C: Radiation therapy for the treatment of thymoma in rabbits *(Oryctolagus cuniculus)*. J Exotic Pet Med. 2006; 15(2):138-144.

Harcourt-Brown FM: Dental disease. In: Textbook of Rabbit Medicine. Philadelphia, PA: Butterworth-Heinemann, imprint of Elsevier Science; 2002:165-205.

Harcourt-Brown FM: Abscesses. In: Textbook of Rabbit Medicine. Philadelphia, PA: Butterworth-Heinemann, imprint of Elsevier Science; 2002:206-223.

Harcourt-Brown FM: Digestive disorders. In: Textbook of Rabbit Medicine. Philadelphia, PA: Butterworth-Heinemann, imprint of Elsevier Science; 2002:249-221.

Harcourt-Brown FM: Ophtalmic diseases. In: Textbook of Rabbit Medicine. Philadelphia, PA: Butterworth-Heinemann, imprint of Elsevier Science; 2002:229-306.

Harcourt-Brown FM: Neurological and locomotor disorders. In: Textbook of Rabbit Medicine. Philadelphia, PA: Butterworth-Heinemann, imprint of Elsevier Science; 2002:307-323.

Harcourt-Brown FM: Cardiorespiratory diseases. In: Textbook of Rabbit Medicine. Philadelphia, PA: Butterworth-Heinemann, imprint of Elsevier Science; 2002:324-333.

Harcourt-Brown FM: Urogenital diseases. In: Textbook of Rabbit Medicine. Philadelphia, PA: Butterworth-Heinemann, imprint of Elsevier Science; 2002:335-351.

Harcourt-Brown FM: Treatment of facial abscesses in rabbits. Exotic DVM. 1999; 1(3): 83-88.

Harcourt-Brown FM: Update on metabolic bone disease in rabbits. Exotic DVM. 2002; 4(3):43-46.

Harcourt-Brown FM: Dacryocystitis in rabbits. Exotic DVM. 2002; 4(3):47-49.

Harcourt-Brown FM: Intestinal obstruction in rabbits. Exotic DVM. 2002; 4(3):51-53.

Harcourt-Brown FM, Harcourt-Brown N: Surgical removal of a mediastinal mass in a rabbit. Exotic DVM. 2002; 4(3):59-60.

Harcourt-Brown FM: Radiology of rabbits: part 1. Soft tissue. Exotic DVM. 2004; 6(2):27-29.

Harcourt-Brown FM: Radiology of rabbits: part 2. Hard tissue. Exotic DVM. 2004; 6(2):30-32.

Harcourt-Brown N: Approach to selected orthopedic disorders in rabbits. Exotic DVM. 2004; 6(2) 33-36.

Hernandez-Divers SJ: Molar disease and abscesses in rabbits. Exotic DVM. 2001; 3(3):65-69.

Huston SM, Quesenberry KE: Section 2. Rabbits. Cardiovascular and lymphoproliferative diseases. In: Quesenberry KE, Carpenter JW, eds. Ferrets, Rabbits and Rodents. Clinical Medicine and Surgery 2nd ed. Philadelphia, PA:Elsevier; 2004:211-220.

Jenkins JR: Section 2. Rabbits. Gastrointestinal diseases. In: Quesenberry KE, Carpenter JW, eds. Ferrets, Rabbits and Rodents. Clinical Medicine and Surgery 2nd ed. Philadelphia, PA: Elsevier; 2004:161-171.

Kapatkin A: Orthopedics in small mammals. In: Quesenberry KE, Carpenter JW, eds. Ferrets, Rabbits and Rodents. Clinical Medicine and Surgery 2nd ed. Philadelphia, PA:Elsevier; 2004:383-391.

Mayer J, Azuma C: The use of radiation therapy in exotic cancer patients. Exotic DVM. 2006; 8(3):38-43.

Mehler SJ, Bennet RA: Surgical oncology of exotic animals. Vet Clin N Am Exot Anim Prac. 2004;7 :783-805.

Meredith A, Crossley DA: Rabbits. In: Meredith A, Redrobe S, eds. BSAVA Manual of Exotic Pets 4th ed. Quedgeley, Glouchester: British Small Animal Veterinary Association; 2005:79-92.

Morrisey JK, McEntee M: Therapeutic options for thymoma in the rabbit. Sem Avian Exotic Pet Med. 2005; 14(3):175-181.

Paré JA, Paul-Murphy J: Section 2. Rabbits. Disorders of the reproductive and urinary systems. In: Quesenberry KE, Carpenter JW, eds. Ferrets, Rabbits and Rodents. Clinical Medicine and Surgery 2nd ed. Philadelphia, PA:Elsevier; 2004:183-193.

Popesko P, Rijtovà V, Horàk J: A Colour Atlas of Anatomy of Small Laboratory Animals. Vol. I: Rabbit, Guinea Pig. Bratislava: Príroda Publishing House; 1990.

Reusch B: Rabbit gastroenterology. Vet Clin N Am Exotic Pet Prac. 2005; 8:351-375.

Silverman S, Tell LA: Domestic Rabbit *(Oryctolagus cuniculus)*. In: Radiology of Rodents, Rabbits and Ferrets. An Atlas of Normal Anatomy and Positioning. Philadelphia, PA: Elsevier Saunders; 2005:159-230.

Sjoberg JG: Hematuria in the rabbit. Exotic DVM. 2004; 6(4):23-30.

Stauber E, Finch N, Caplazi P: Diaphragmatic kidney herniation in a rabbit. Exotic DVM. 2005; 8(1):11-12.

Stefanacci JD, Hoefer HL: Radiology and ultrasound. In: Quesenberry KE, Carpenter JW, eds. Ferrets, Rabbits and Rodents. Clinical Medicine and Surgery 2nd ed. Philadelphia, PA:Elsevier; 2004:395-413.

Verstraete FJM: Advances in diagnosis and treatment of small exotic mammal dental disease. Seminars Avian Exotic Pet Med. 2003;12(1):37-48.

Yasutsugu M: Mandibulectomy for treatment of oral tumors (cementoma and chondrosarcoma) in two rabbits. Exotic DVM. 2006; 8(3):18-22.

GUINEA PIG

The NORMAL HEAD
Lateral Projection

Figure 3.1a,b. 180% of actual size. The most common indication for skull radiographs in this species is for evaluation of dental disease. Both the occlusal plane of the cheek teeth and the apex of the mandibular incisors are difficult to evaluate in this view. The apex of the mandibular incisors extends lingual to the second cheek tooth, which obscures it on a true lateral projection (red circle). The apex of the maxillary incisor teeth is visible just mesial to the root of the first cheek teeth. The mandibular cheek teeth apexes are very close to the ventral cortex of the mandible. Unlike in rabbits and chinchillas, the occlusal plane is not visible in the lateral projection of the skull.

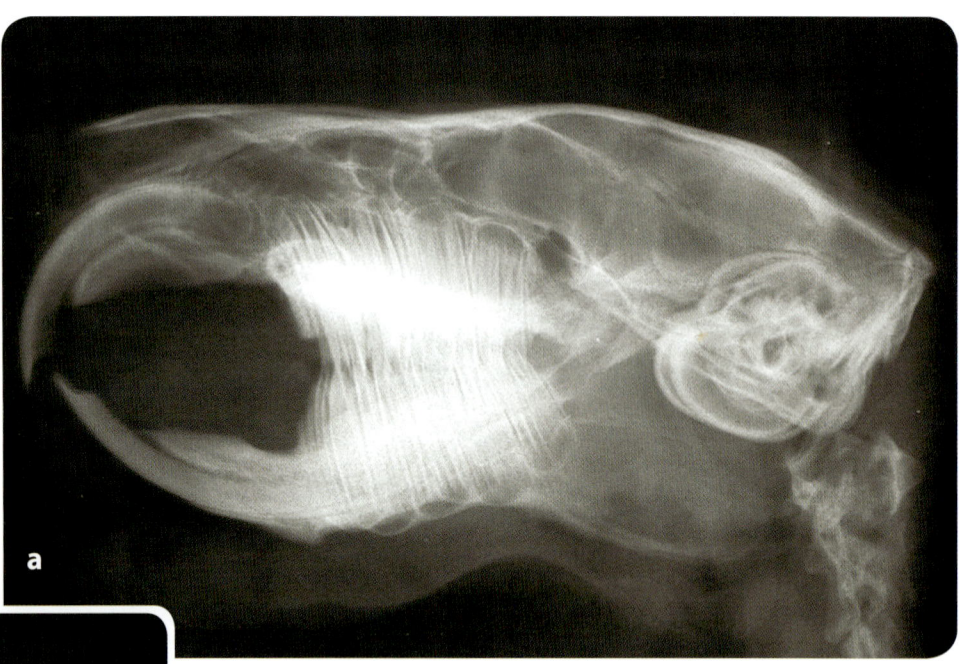

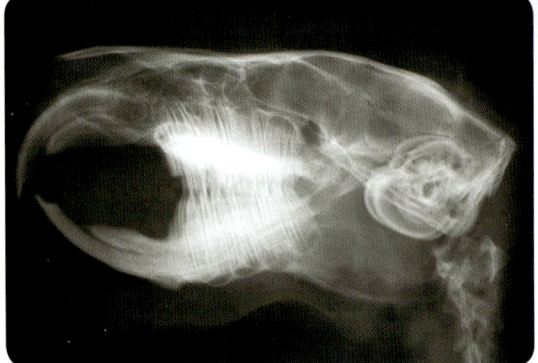

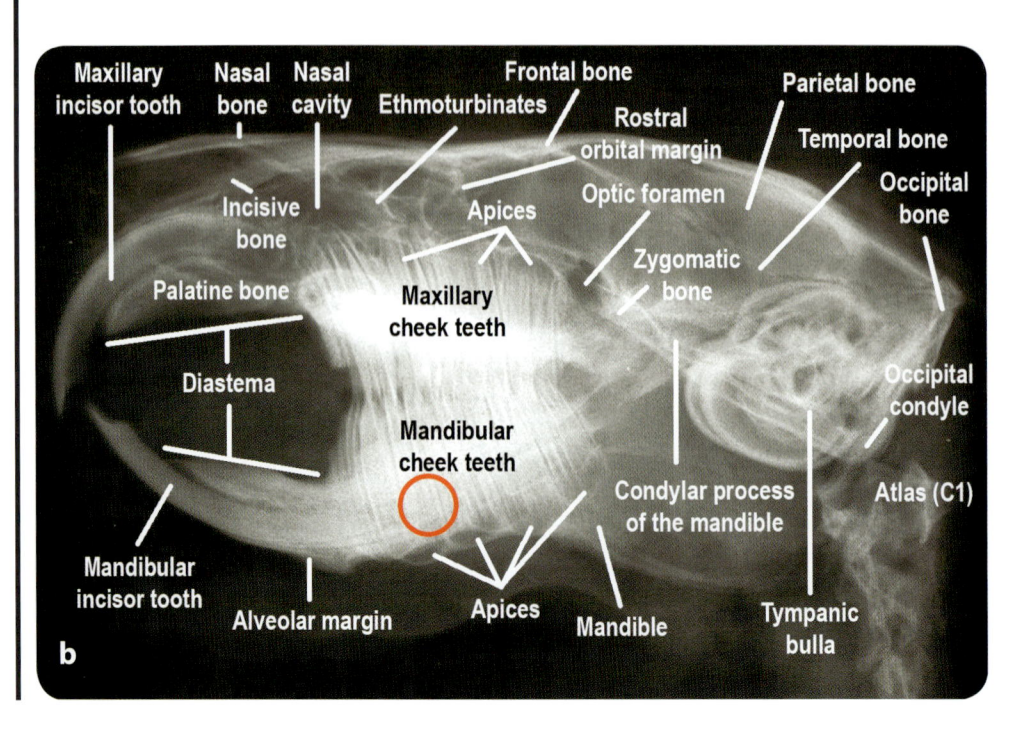

Figure 3.2. Actual size of the skull of a 1.1 kg male guinea pig.

Oblique Projection

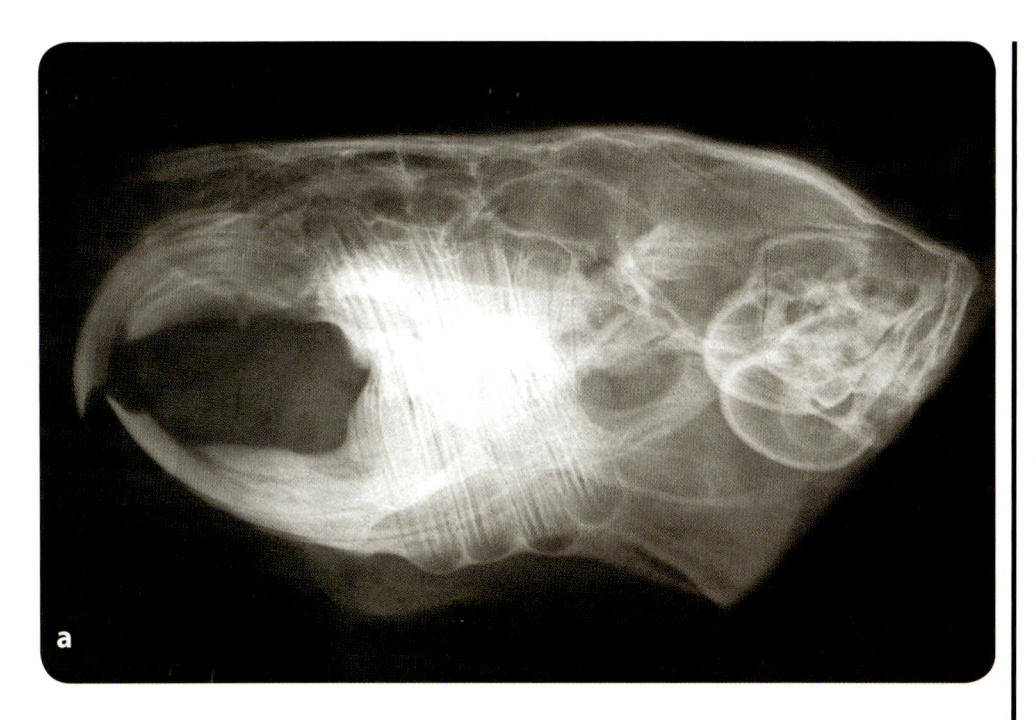

Figure 3.3a,b. Oblique projections with slight obliquity from the true lateral position allow better visualization of the mandibular cheek teeth roots on one side and the maxillary cheek teeth roots on the opposite side. These views are also helpful to evaluate the apex of the incisor teeth, which is obscured on a lateral view (red circle). The contralateral oblique view should always be obtained to compare findings. Oblique views are made from a true lateral position with the head rotated axially.

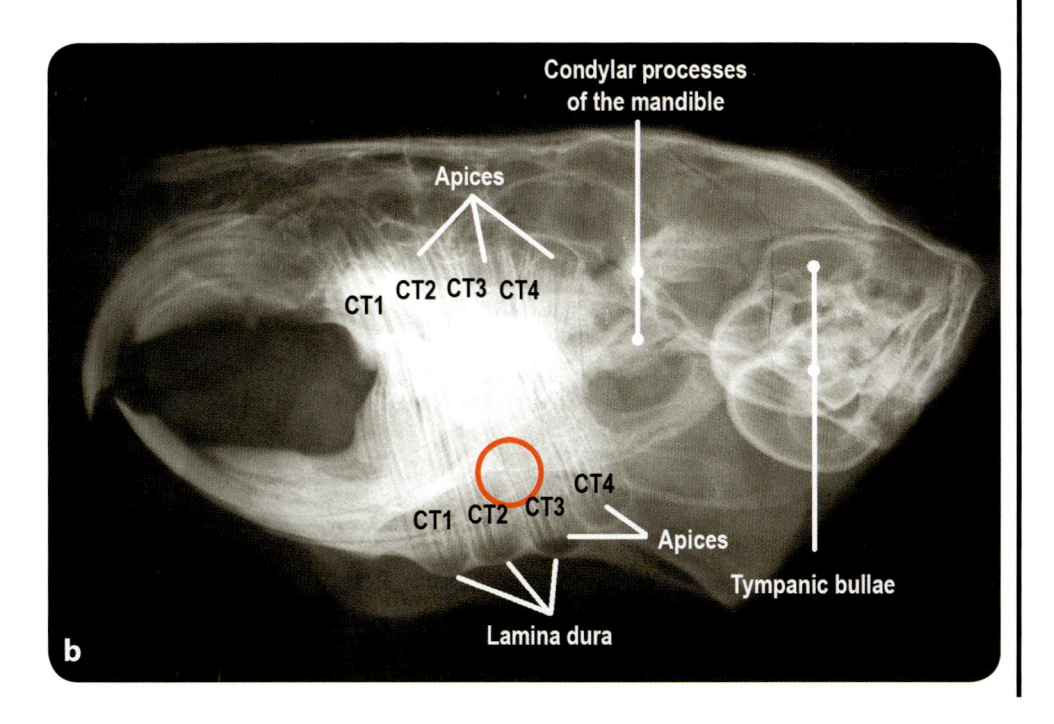

Ventrodorsal Projection

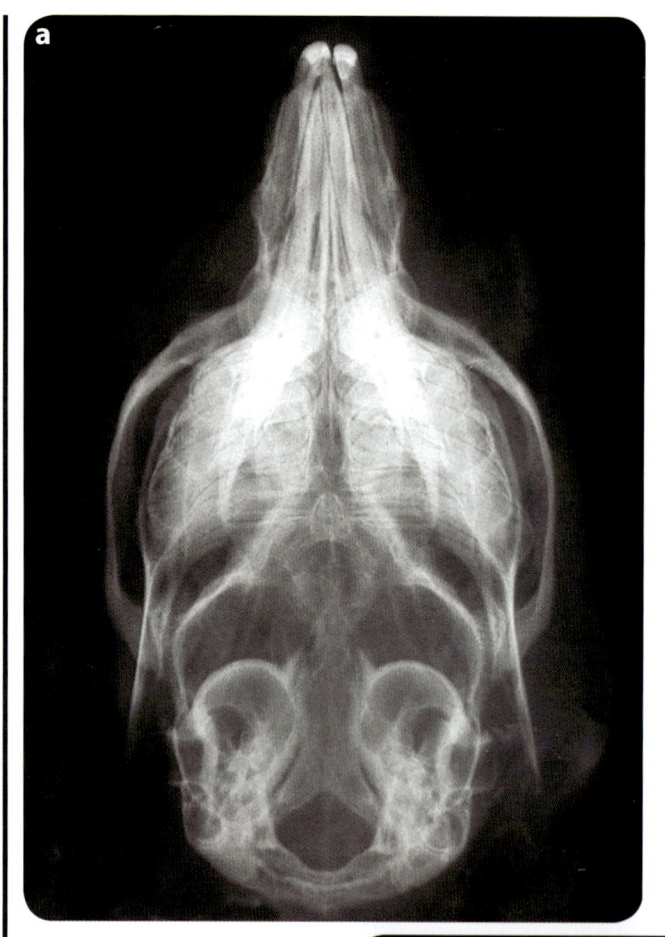

Figure 3.4a,b. This view is rarely useful for diagnosis of dental disease due to superimposition of dental and cranial structures. However, severe dental overgrowth may alter the normal bony profile. This is the most useful projection for evaluation of the tympanic bullae, and may help identify other bony lesions of the skull.

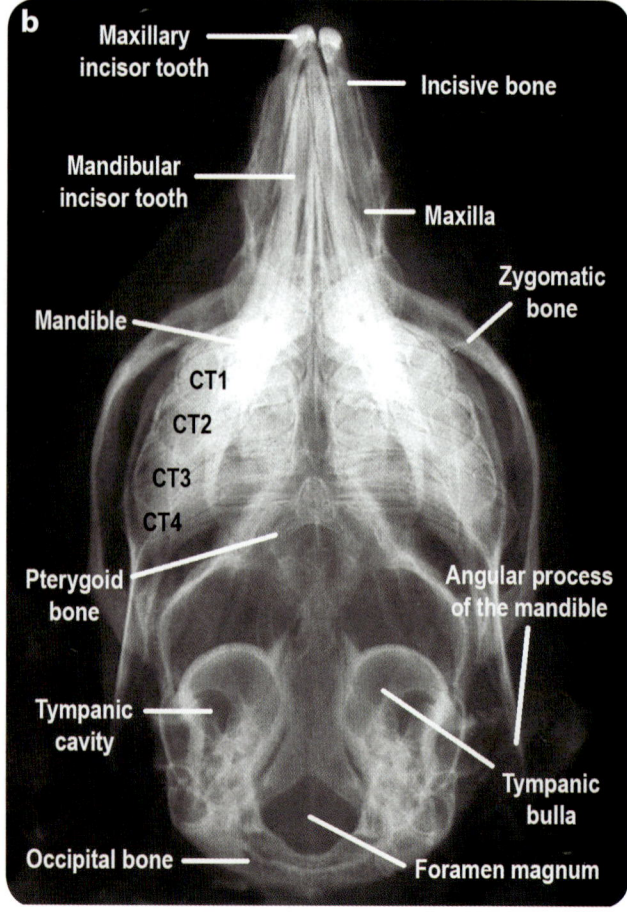

Rostrocaudal Projection

Figure 3.5a,b. Note the marked vestibular curvature of the maxillary cheek teeth, and the lingual curvature of the mandibular cheek teeth. The occlusal plane is therefore oblique in a buccodorsal-linguoventral direction. The temporomandibular joint and the mandibular symphysis are also clearly visible.
This is the only projection useful to evaluate the occlusal plane in the guinea pig.

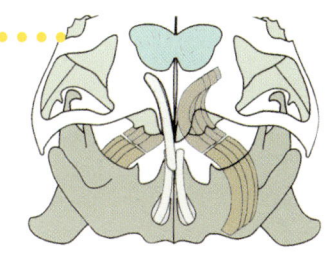

Figure 3.6. Diagram of cheek teeth of guinea pigs, rostrocaudal view. Illustration modified from Crossley DA: Clinical aspects of rodent dental anatomy. J Vet Dent 1995, 12: 131-135.

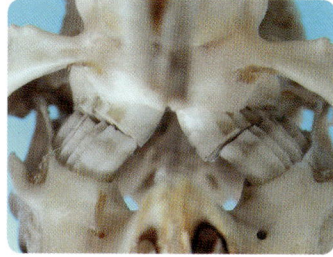

Figure 3.7. Close-up of a rostral view of the cheek teeth, incisors removed. The oblique occlusal plane is clearly visible. The mandibular teeth are curved with a pronounced buccal convexity, and the maxillary teeth with a prominent palatal convexity. This results in a 30 degrees oblique occlusal plane that slopes from buccal to lingual, dorsal to ventral.

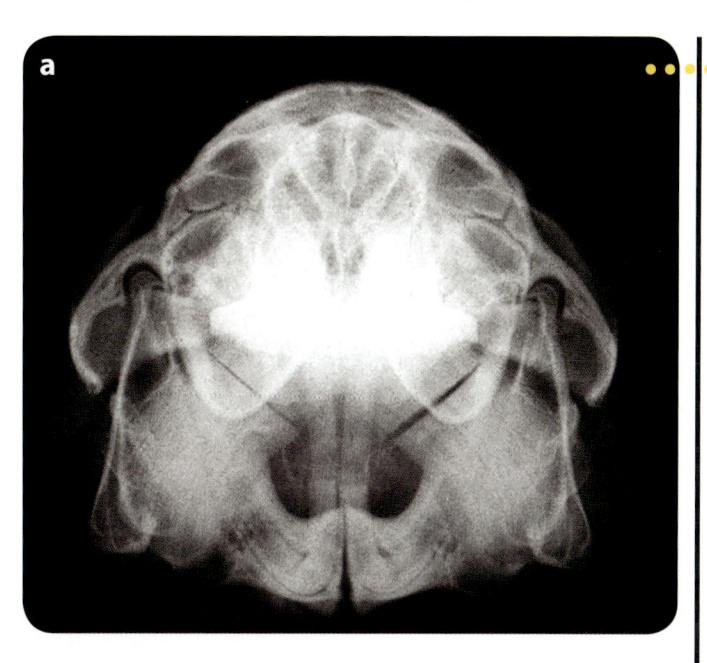

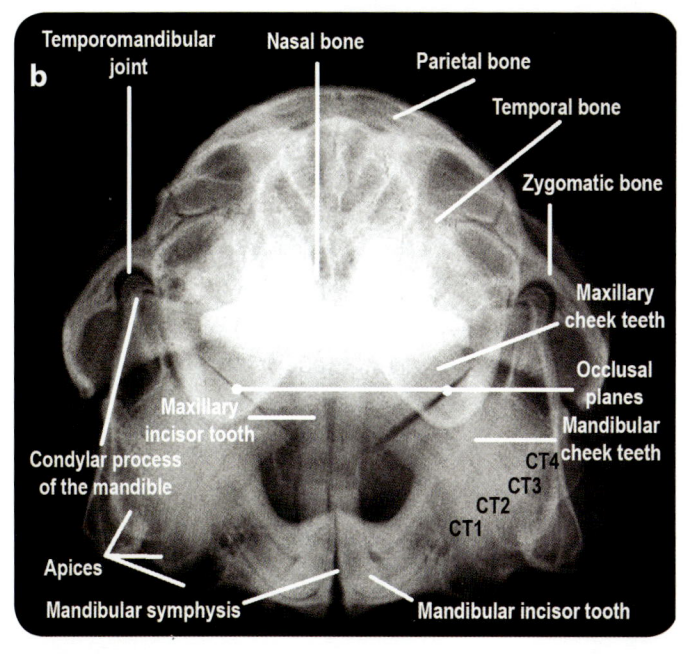

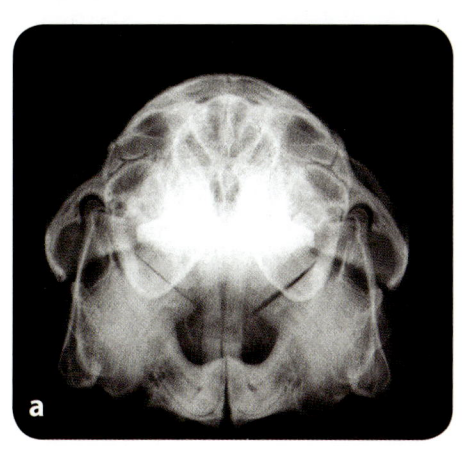

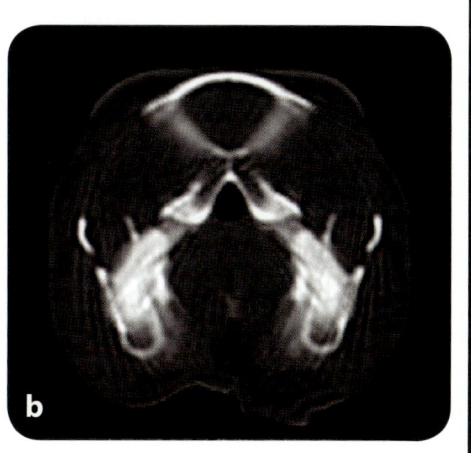

Figure 3.8a,b. Comparison between the radiographic rostrocaudal projection and the CT scan. The CT image is superior for visualization of the shape and orientation of the teeth, and the tooth roots.

ABNORMALITIES of the HEAD
Diseases of Incisor Teeth

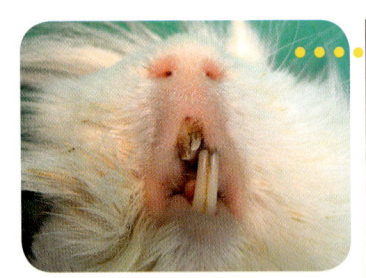

Figure 3.9. Malocclusion of incisor teeth is often secondary to severe malocclusion of cheek teeth. The mandible is deviated to the left, and the normal occlusal plane and function of incisor teeth is lost.

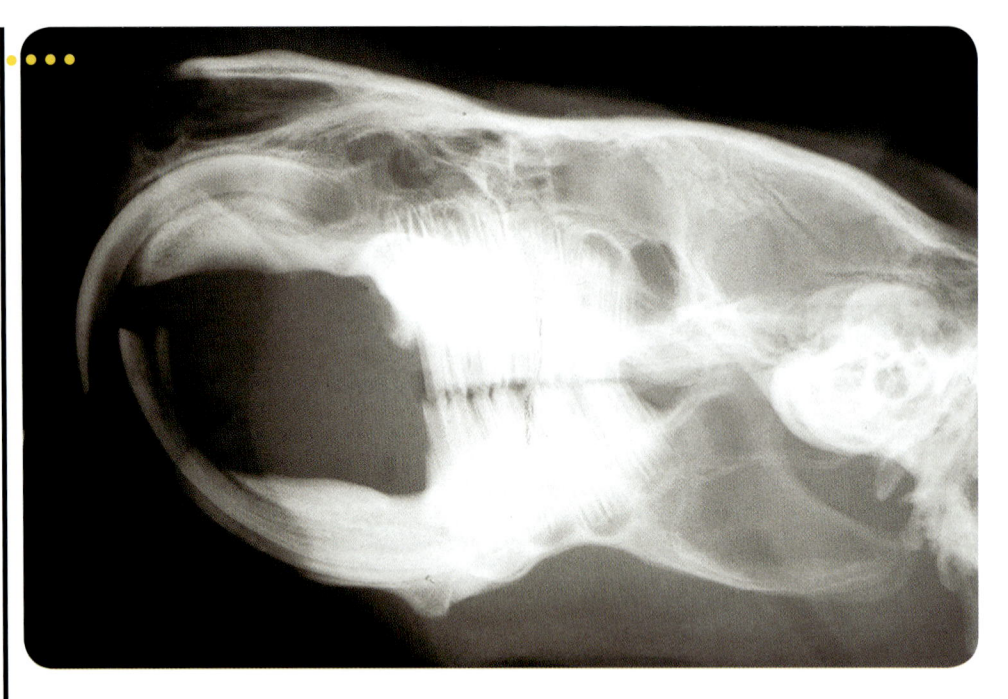

Figure 3.10. Lateral radiograph of the skull of a guinea pig with severe malocclusion of the mandibular incisor teeth. Unlike in rabbits, maloccluded mandibular incisors do not usually deviate rostrally to grow over the maxillary incisors. Malocclusion and abnormal occlusal plane of the cheek teeth are also visible in this patient (see Figure 3.15).

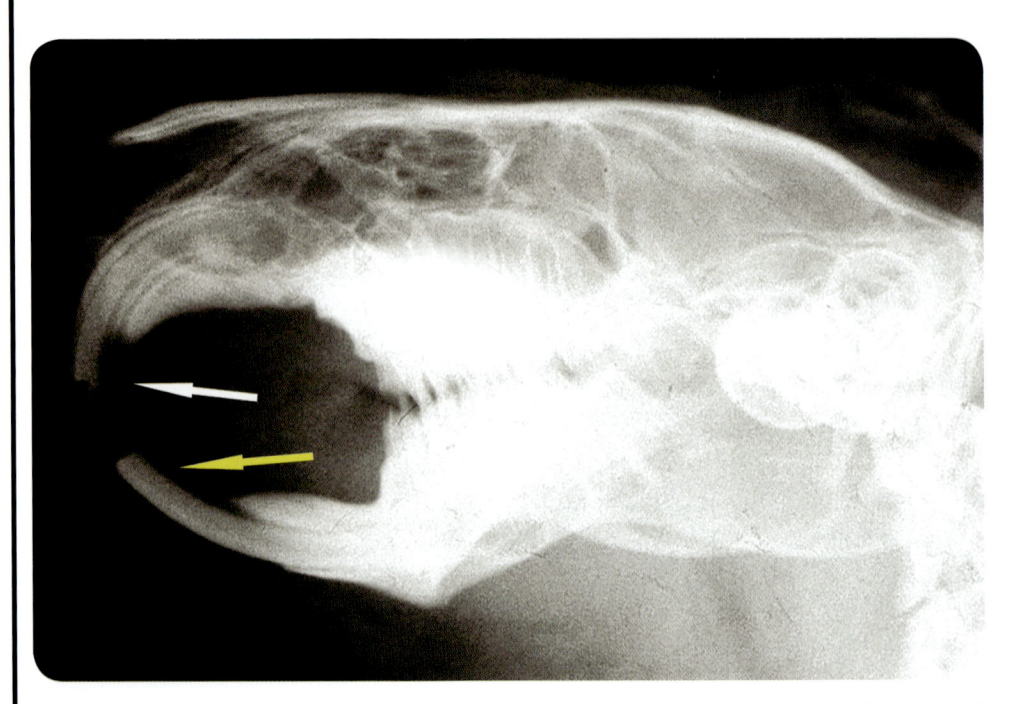

Figure 3.11. Radiograph of the skull of a guinea pig demonstrating coronal fractures of both maxillary incisor teeth (white arrow). There is slight mandibular malocclusion demonstrated by the loss of the normal chisel-shaped occlusal plane (yellow arrow).

Figure 3.12. Lateral radiograph of the skull of a guinea pig with malocclusion of mandibular incisor teeth and fracture of the alveolus.

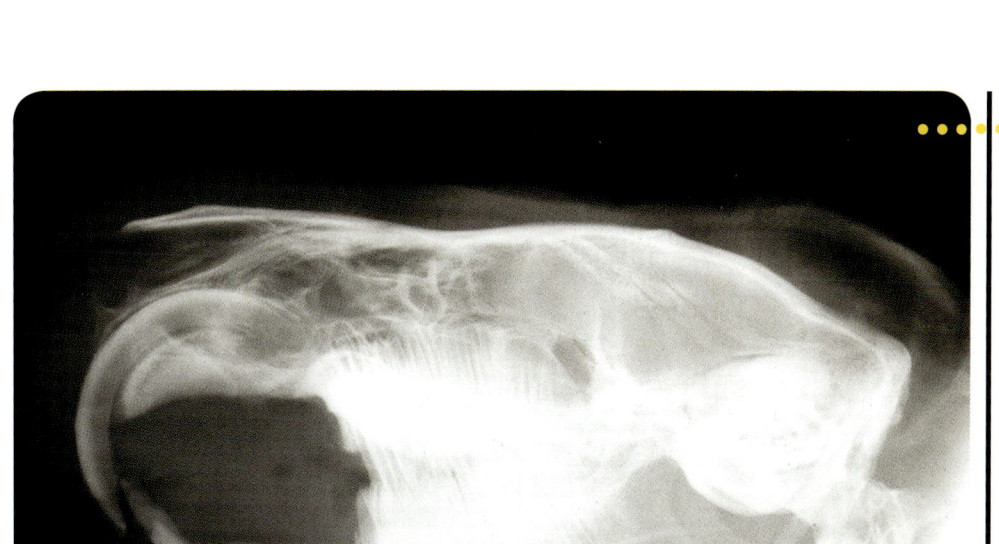

Figure 3.14. Clinical appearance of the fracture of the right mandibular incisor alveolus and exposure of the bone.

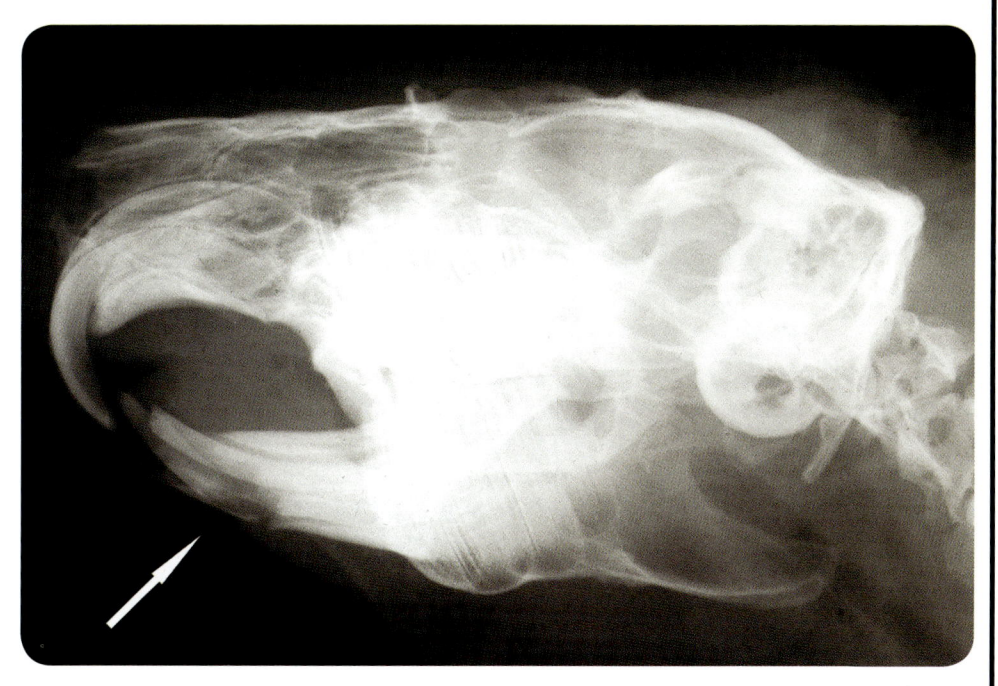

Figure 3.13. Left 30° ventral-right dorsal oblique radiograph of the same guinea pig as Figure 3.12. The oblique projection is superior for visualization of the missing cortical bone of the alveolus (arrow).

Diseases of Cheek Teeth

Excessive crown elongation and malocclusion of cheek teeth are common in guinea pigs. Presenting signs and symptoms are usually more severe in guinea pigs than in rabbits. Guinea pigs often present with reduced food intake or complete anorexia, which is sometimes reported by owners as sudden onset. For this reason, concurrent radiographic abnormalities of the gastrointestinal tract are a common finding. The underlying cause of tooth elongation in guinea pigs is likely inadequate wear due to lack of dietary fiber. Congenital malocclusion and metabolic bone disease have not been reported as a cause of dental disease in guinea pigs.

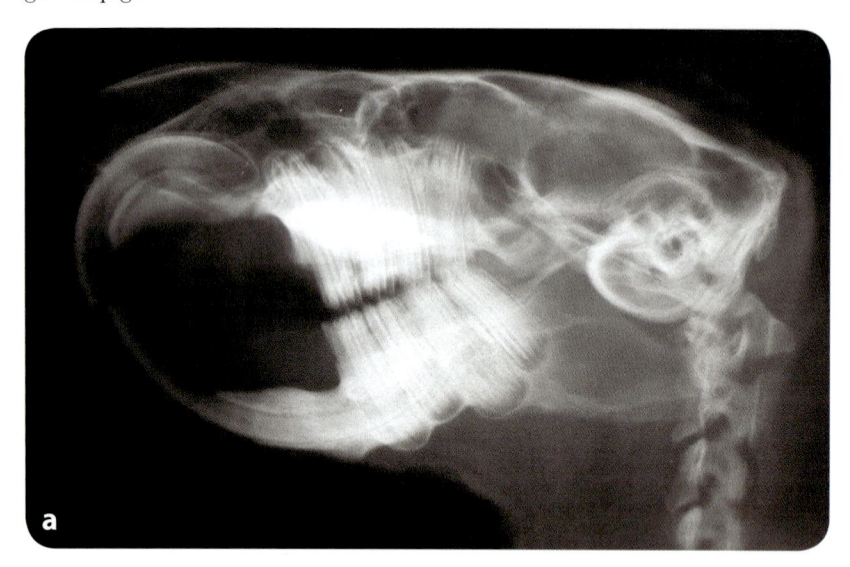

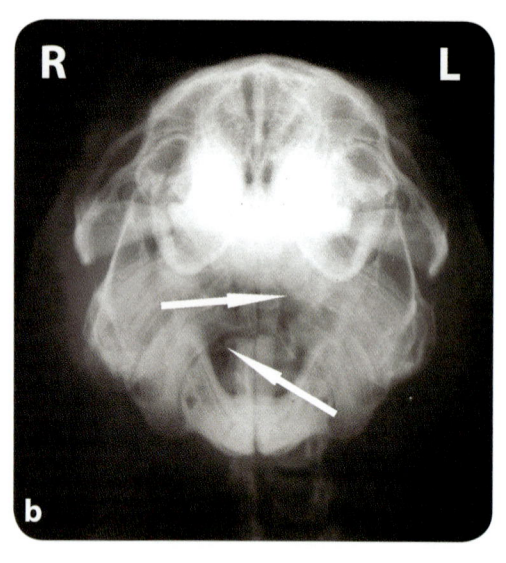

Figure 3.15a,b. Lateral (a) and rostrocaudal (b) radiographs of a guinea pig with cheek teeth elongation. It is difficult to appreciate radiographic dental abnormalities of cheek teeth such as crown elongation, wave mouth or step mouth in this species. The most obvious evidence of cheek teeth disease is the appearance of a flat occlusal plane on a true lateral projection (a). In normal guinea pigs, the occlusal plane is sloped and is not visible on a true lateral projection. Overgrowth of cheek teeth forces the mouth open wider than normal, eventually producing elongation of mandibular incisor teeth. Overgrown incisors often still maintain a normal chisel-shape. The rostrocaudal projection (b) also demonstrates excessive elongation of cheek teeth (arrows).

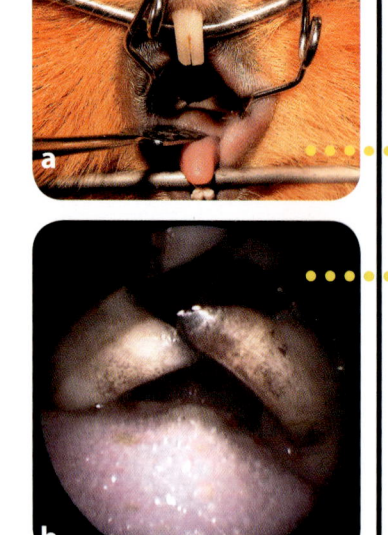

Figure 3.16a,b. Gross (a) and endoscopic (b) views of guinea pigs with bilateral overgrowth of the lingual margins of mandibular CT1. The condition of the guinea pig in (b) is more severe. CT meet and entrap the tongue, which severely impairs both chewing and swallowing.

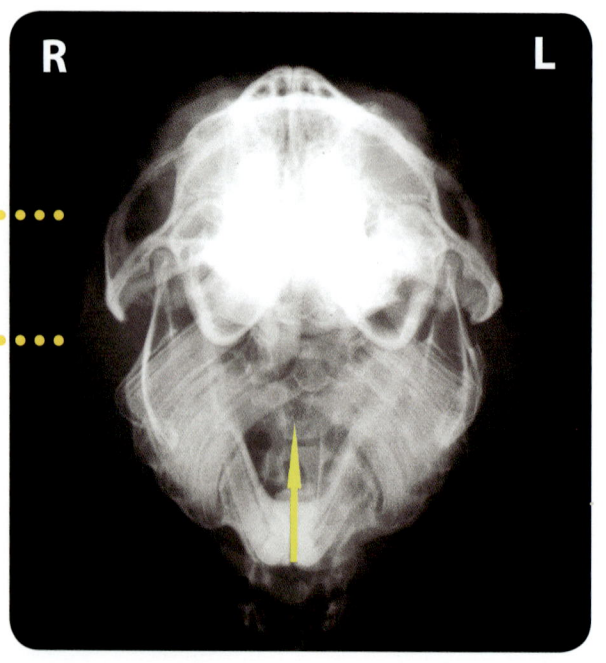

Figure 3.17. Rostrocaudal projection of the skull of a guinea pig with another common pattern of malocclusion: crown elongation of one or more rostral mandibular cheek teeth. This condition is often difficult to detect radiographically. However, the rostrocaudal view above demonstrates an advanced stage leading to a complete bridge-like malocclusion.

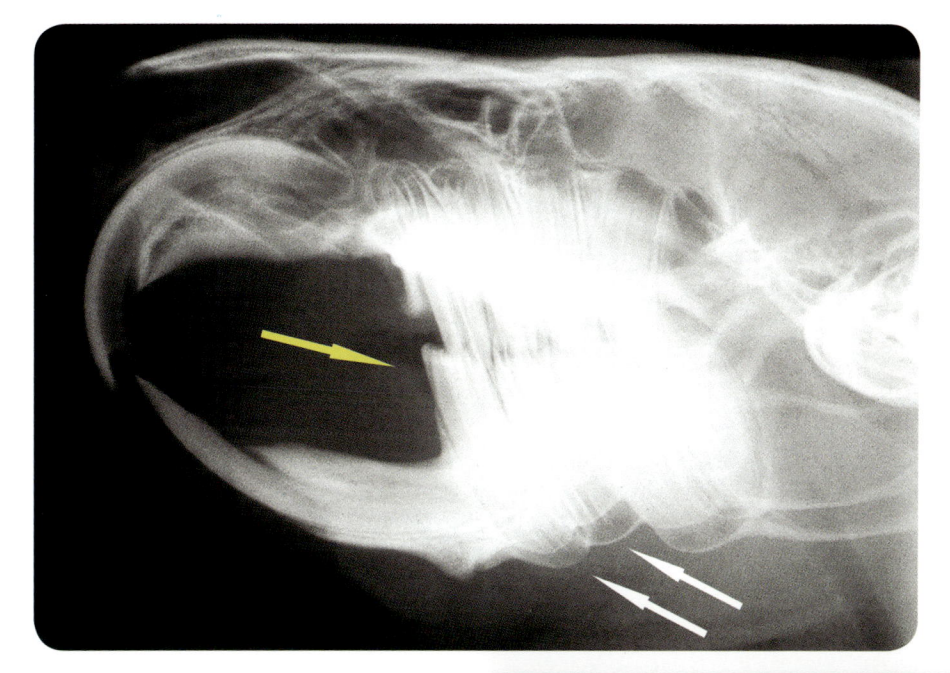

Figure 3.18. Lateral radiograph of the skull of a guinea pig with dental disease.
Crown elongation and malocclusion of mandibular cheek teeth also cause elongation of the apexes and deformity of the ventral cortical bone of the mandible, similar to the disease process in rabbits. Deformities of the cortical bone (white arrows) and coronal malocclusion (yellow arrow) are visible. Deformities of this type are more subtle in guinea pigs, and often more difficult to diagnose.

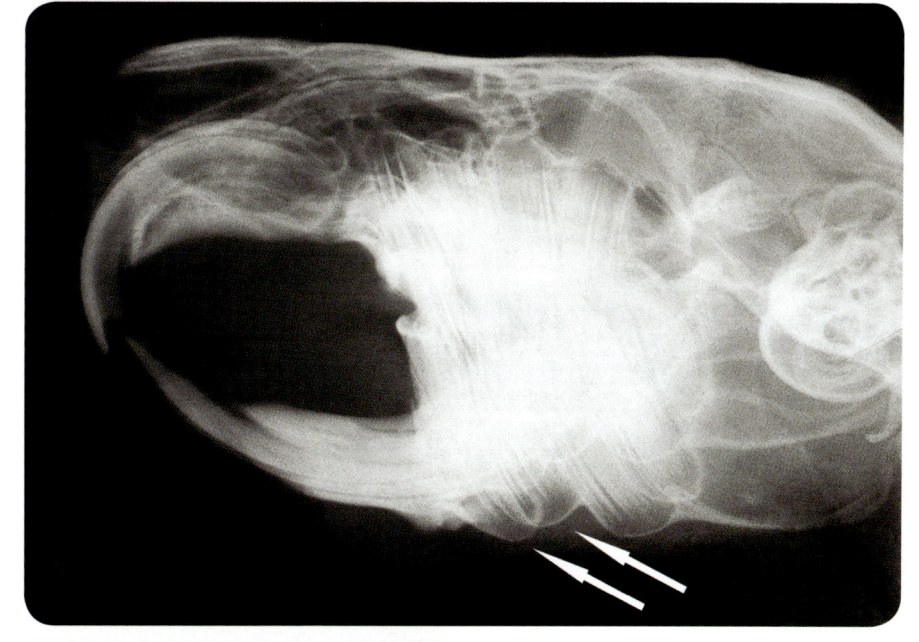

Figure 3.19. Left 15° ventral-right dorsal oblique projection of the skull of a guinea pig, highlighting deformity of the mandibular ventral cortical bone (arrows). Good quality oblique projections are critical for evaluation. When axial rotation is excessive, both normal and abnormal apexes can appear distorted.

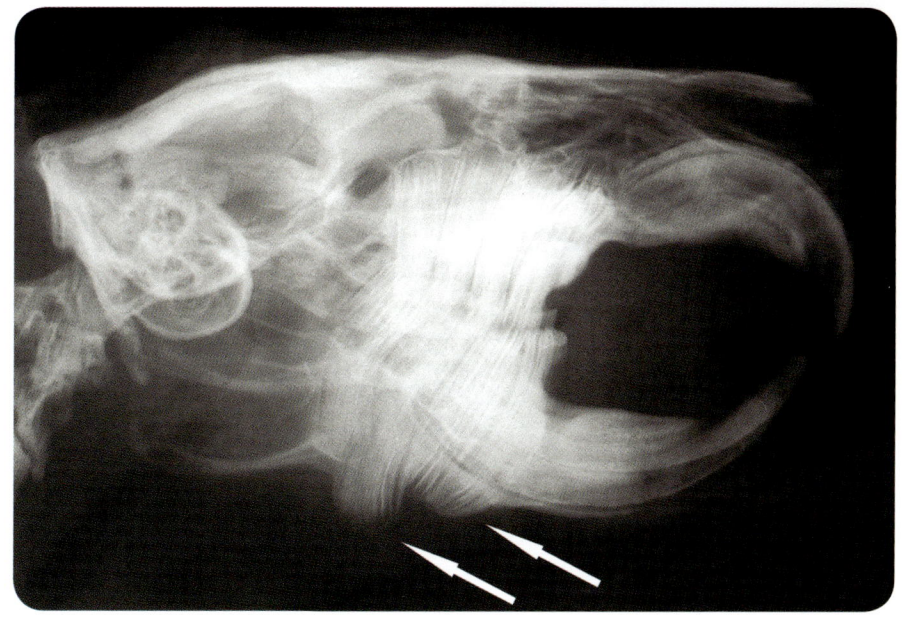

Figure 3.20. Right 15° ventral-left dorsal oblique view of the skull of a guinea pig demonstrating more severe apex deformation.

Periapical Infections and Osteomyelitis

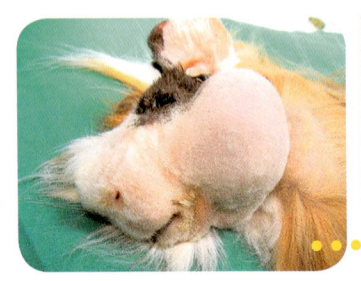

Figure 3.21. Abscesses secondary to acquired dental disease occur in guinea pigs, but with less frequency than in the rabbit. They may be very large.

Due to the structure and position of the masticatory muscles of the guinea pig, abscesses usually develop under the masseter muscle and are located more caudally than most mandibular abscesses of rabbits. Deviation of the masticatory muscles causes subluxation of the mandible, with subsequent malocclusion of incisor teeth and disruption of the normal cheek teeth occlusal plane.

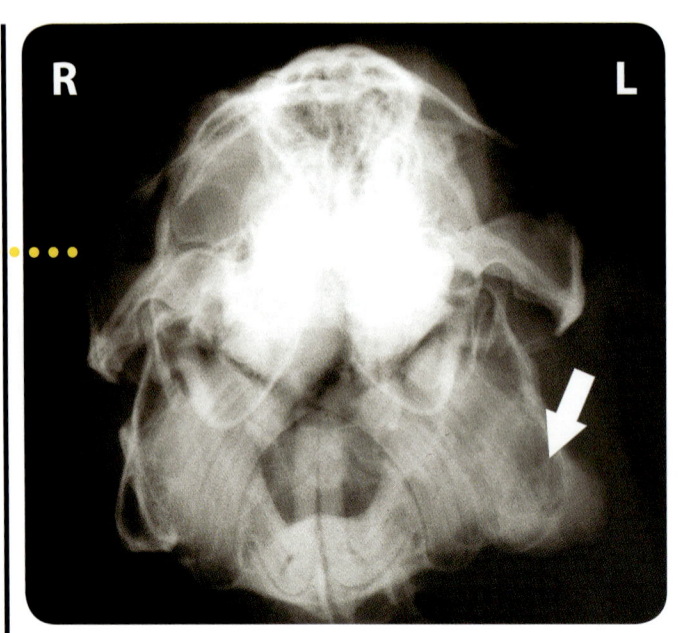

Figure 3.22. Rostrocaudal radiograph of a guinea pig with anorexia.
Radiolucency secondary to periapical infection of left mandibular CT4 and osteomyelitis is clearly visible (arrow). Impaired chewing motions and right lateral subluxation of the mandible have resulted in a severely abnormal cheek teeth occlusal plane.

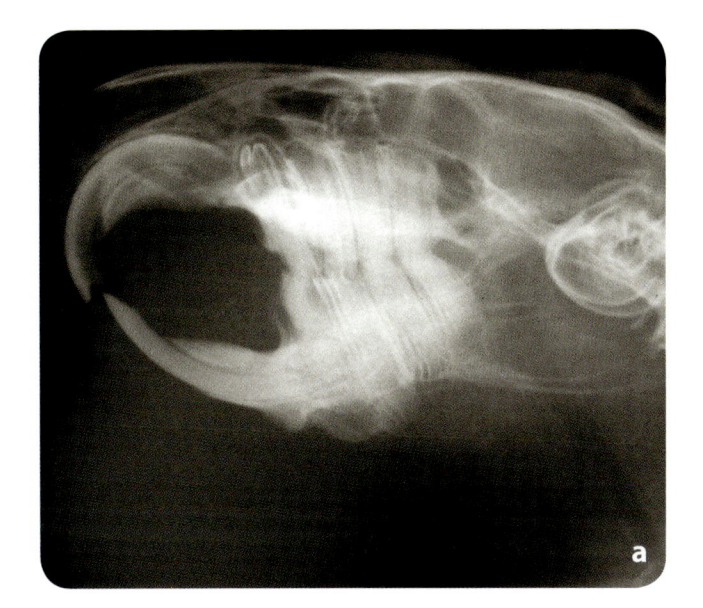

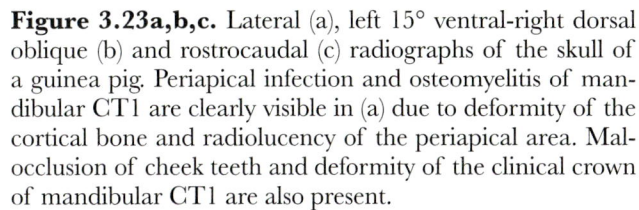

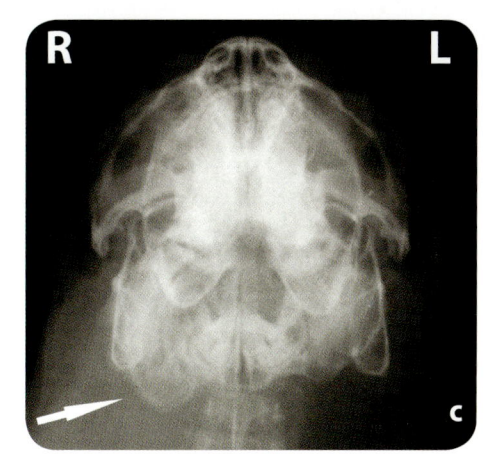

Figure 3.23a,b,c. Lateral (a), left 15° ventral-right dorsal oblique (b) and rostrocaudal (c) radiographs of the skull of a guinea pig. Periapical infection and osteomyelitis of mandibular CT1 are clearly visible in (a) due to deformity of the cortical bone and radiolucency of the periapical area. Malocclusion of cheek teeth and deformity of the clinical crown of mandibular CT1 are also present.
Oblique (b) and rostrocaudal (c) projections help demonstrate on which side of the jaw the abscess is located.

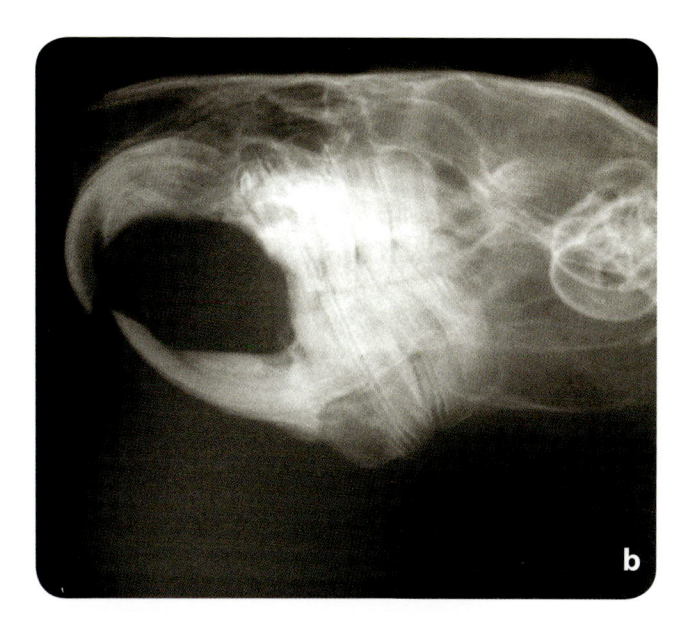

COMPUTED TOMOGRAPHY of the HEAD

V. Capello, A. Cauduro, A. Lennox

The Normal Head

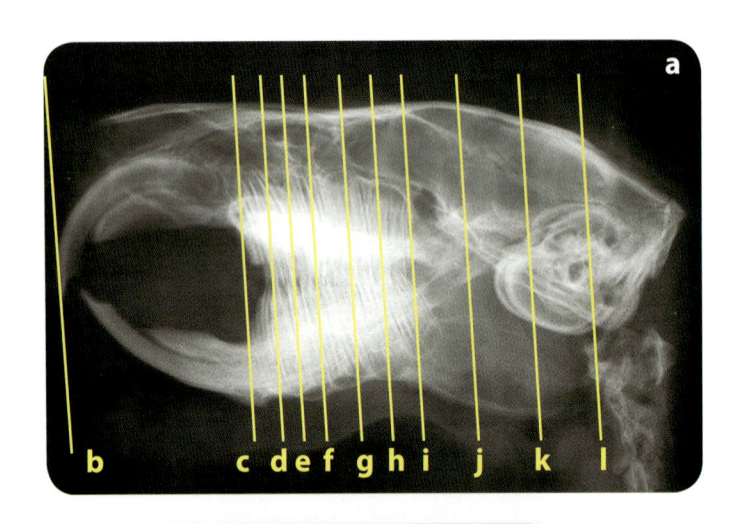

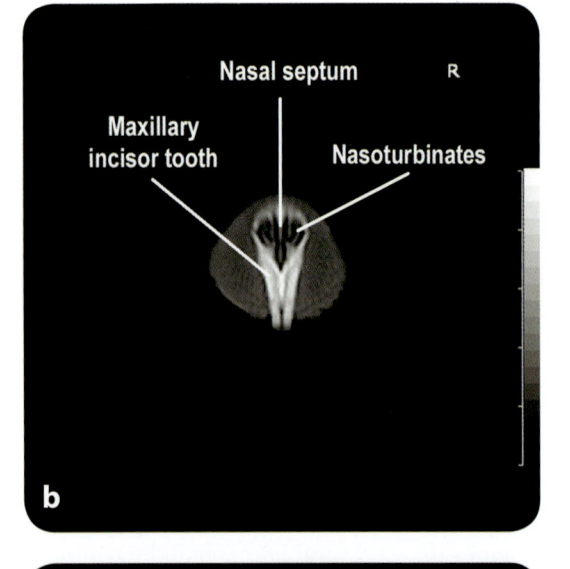

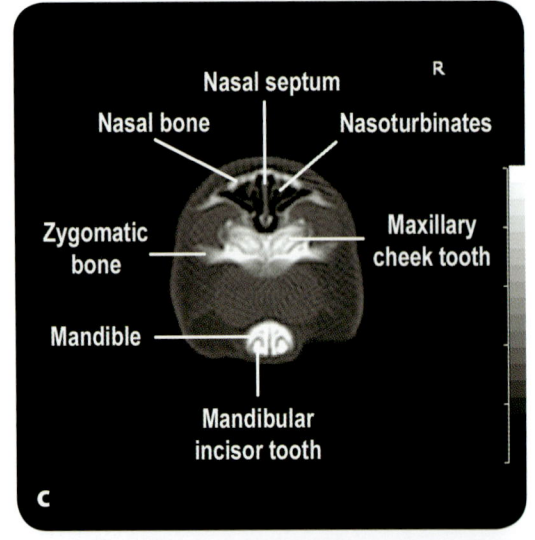

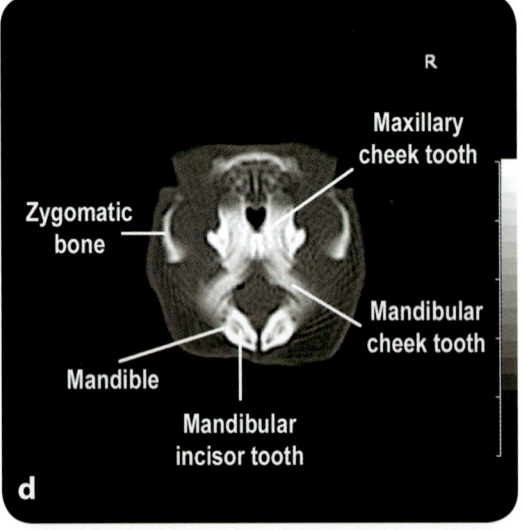

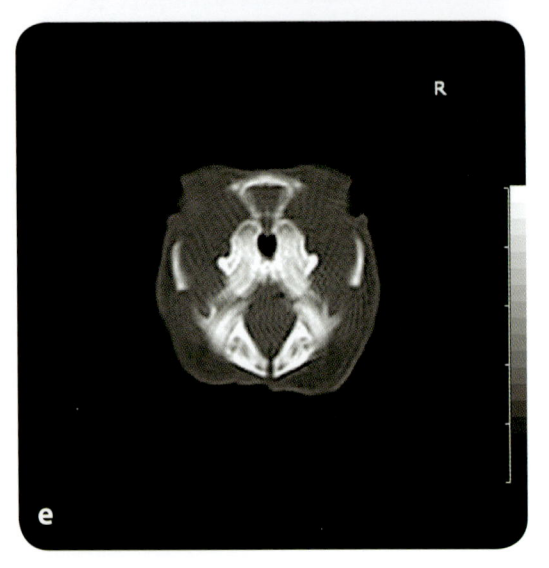

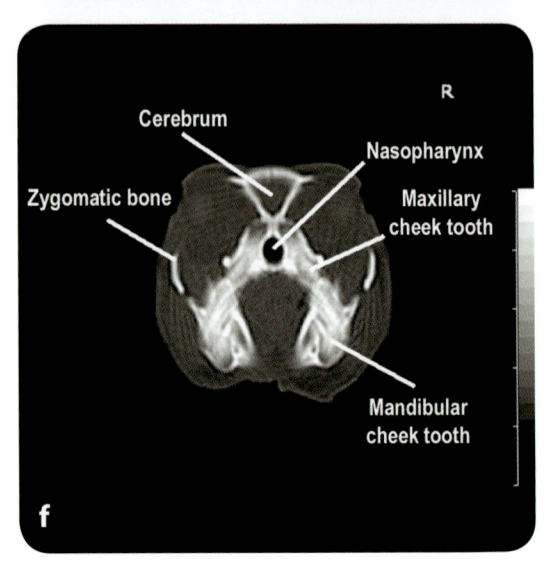

HEAD 179

Figure 3.24a-l. Computed tomography of the normal skull of a 1.5 kg guinea pig, bone window.. Scanning Parameters: scan speed: 1 sec.; mA: 125; kVp: 120; slice thickness: 3 mm; WL: 300; WW: 1500; Image size (matrix): 512x512. Selected views from a series of 70.
The scout view demonstrating the scanning angles (a) has been adapted from a radiograph of the normal skull for demonstration purposes.

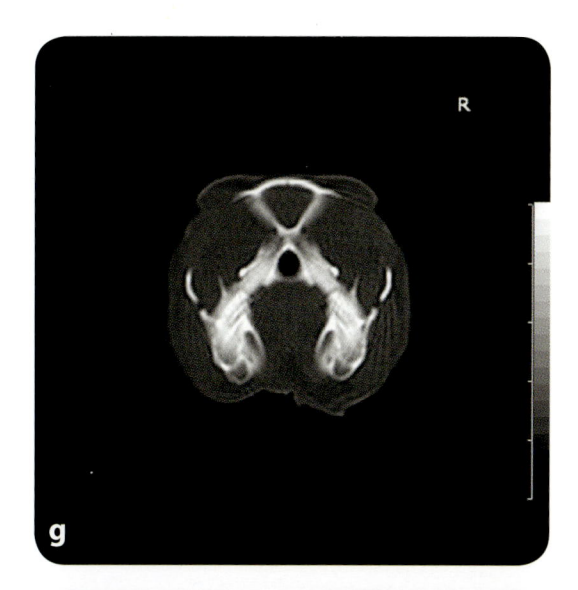

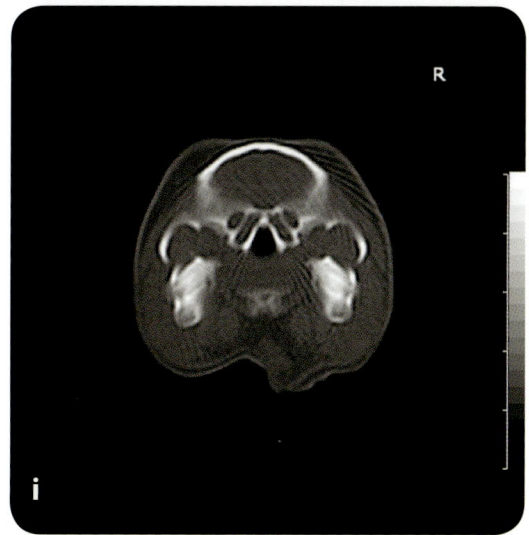

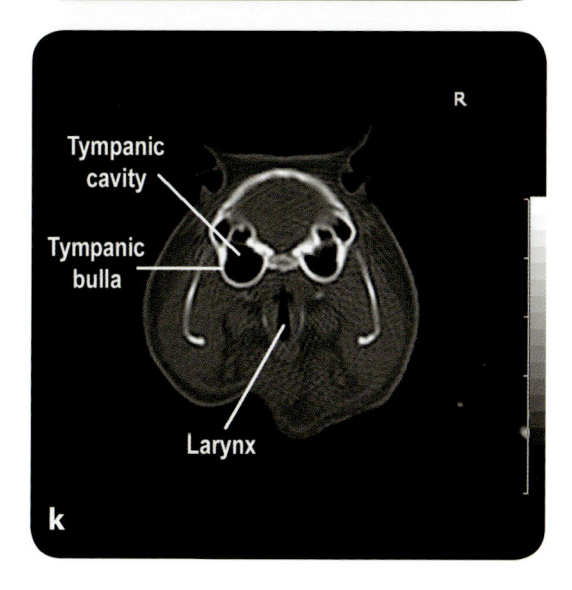

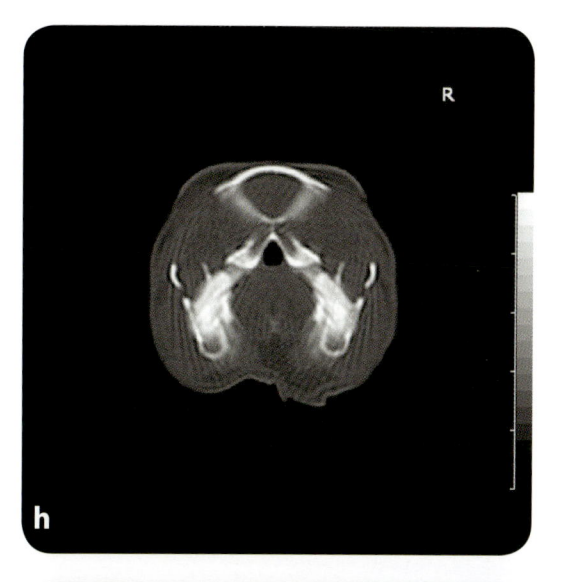

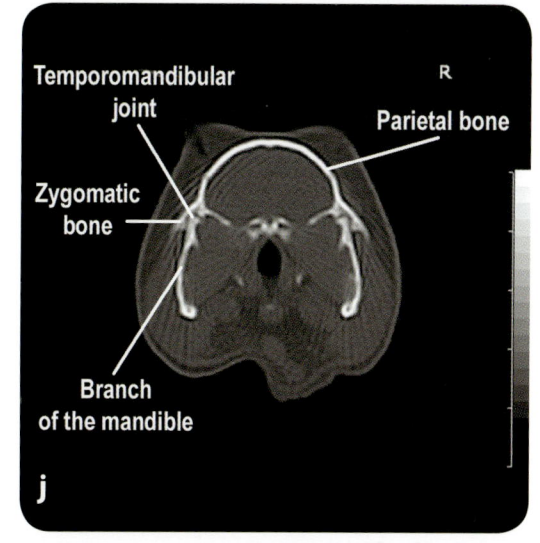

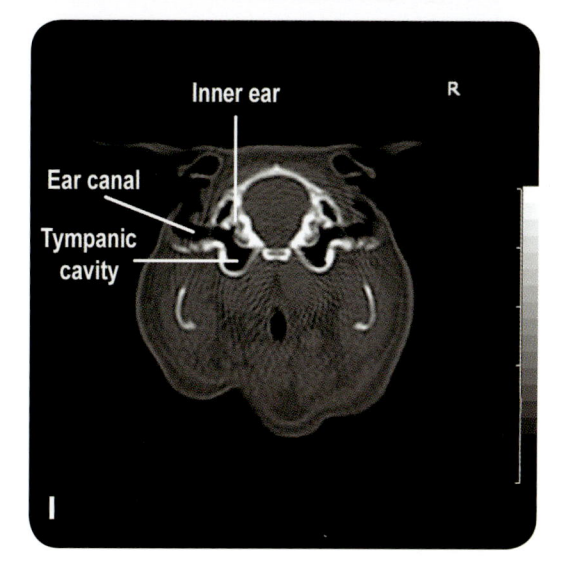

180 GUINEA PIG

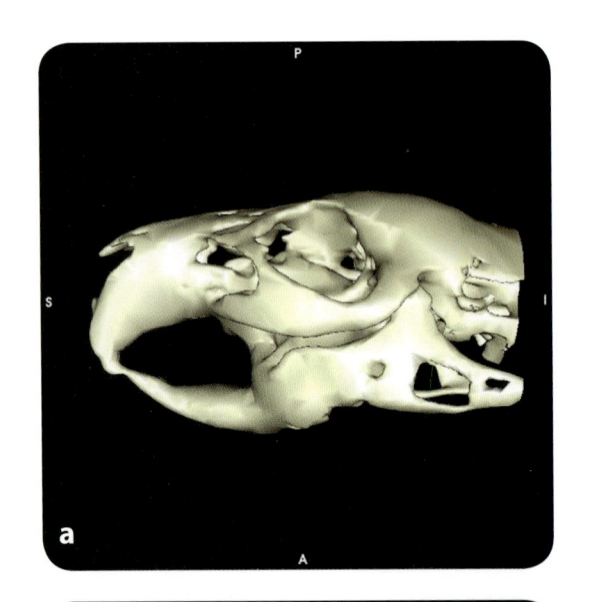

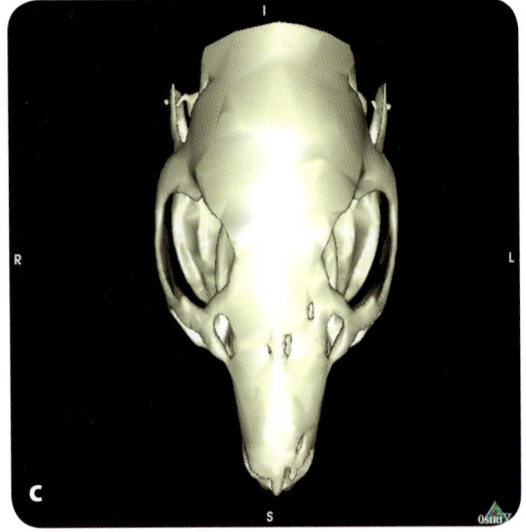

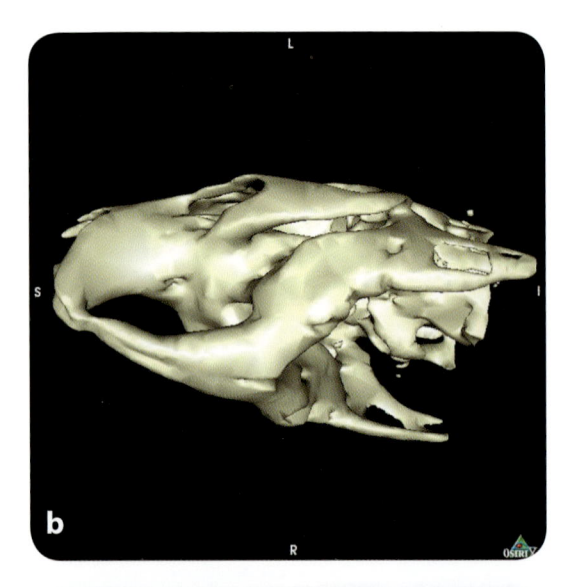

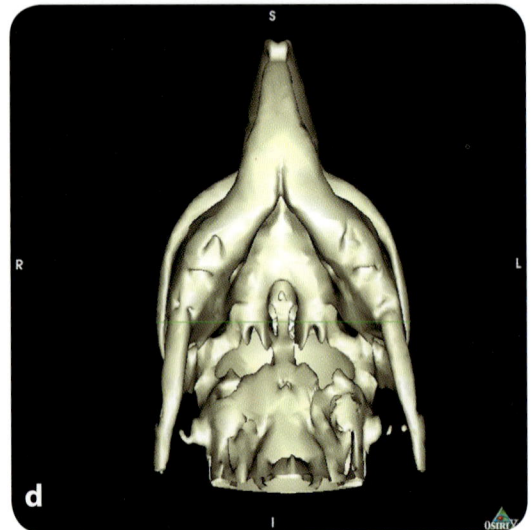

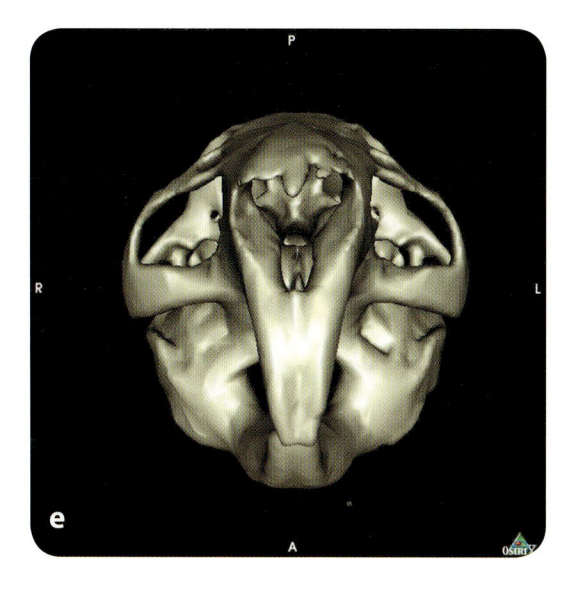

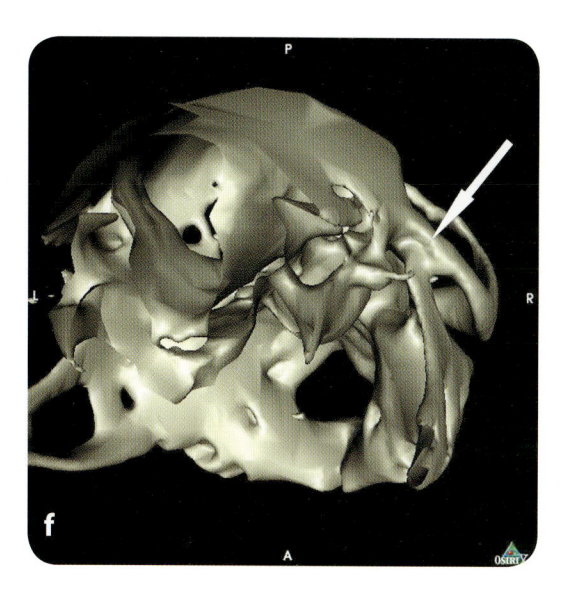

Figure 3.25a-f. Shaded surface display (SSD) of the normal skull of a 1.1 kg guinea pig. The caudal view (f) is slightly oblique to emphasize the temporomandibular joint (arrow).

Acquired Dental Disease, Periapical Abscessation, and Osteomyelitis

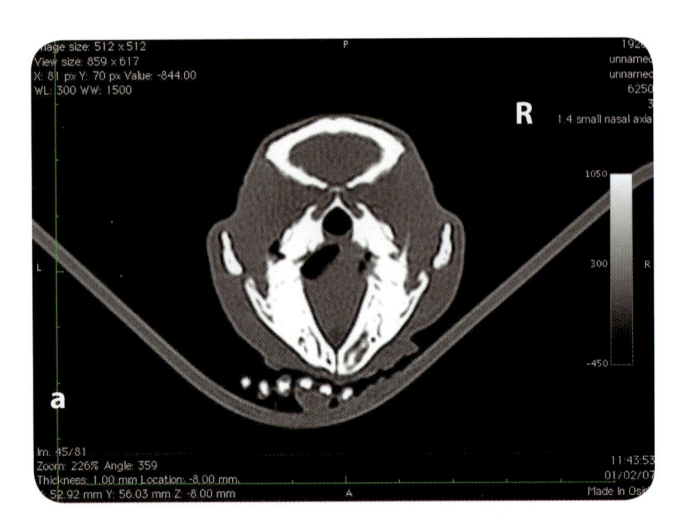

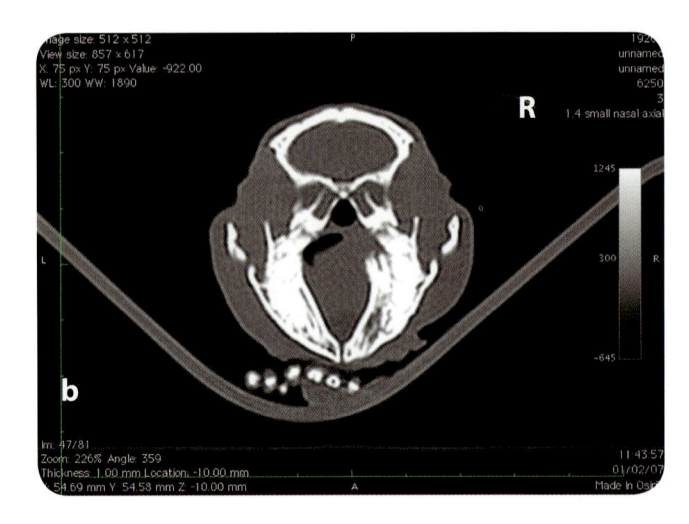

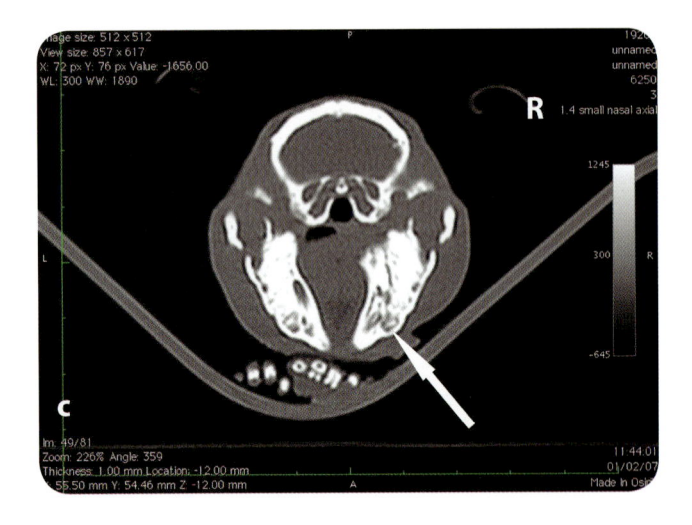

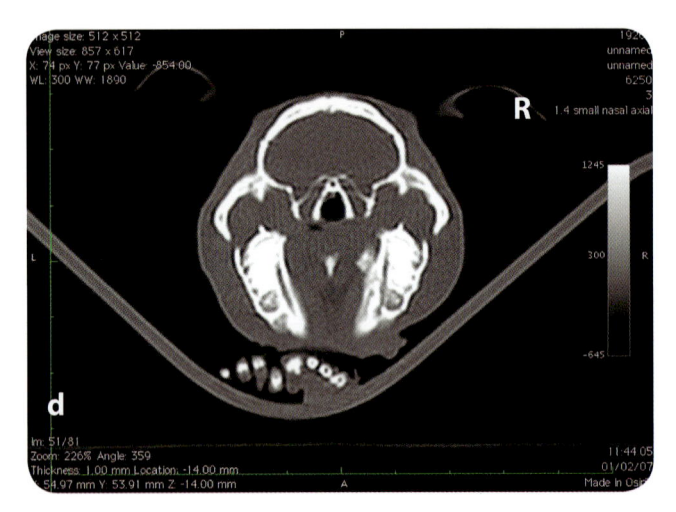

Figure 3.26a-d. CT of a 3-year-old guinea pig with marked deformity of the right mandibular arcade due to an expansile bony lesion associated with the tooth roots. At necropsy the lesion appeared to be proliferative bone, and was not neoplastic. Scanning Parameters: scan speed: 80 mAs; kVp: 120; slice thickness: 1 mm.; WL: 300; WW: 1890; Image size (matrix): 512x512. A lesion is visible in a-c, but not d.

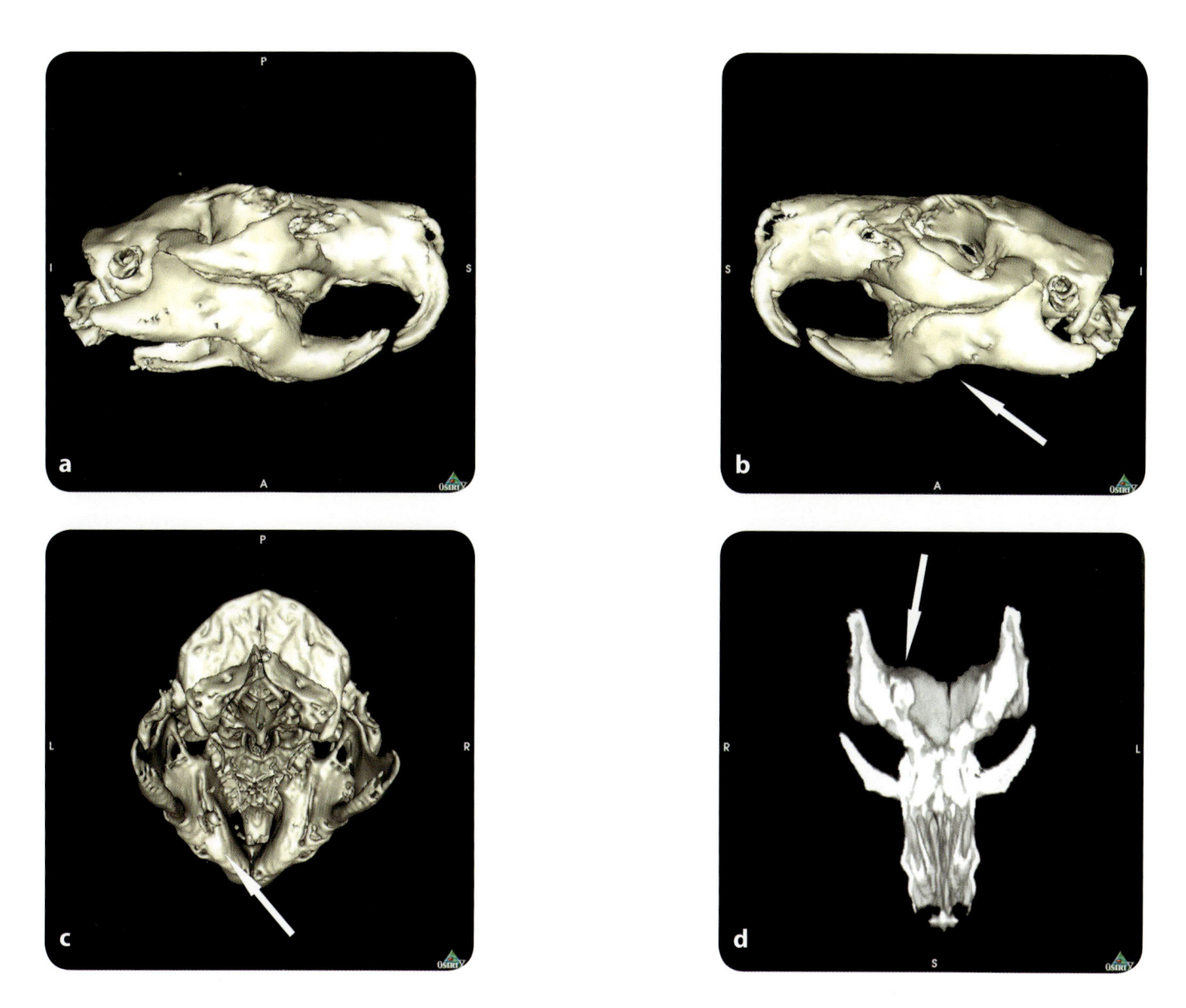

Figure 3.27a-d. SSD of the skull of the same guinea pig in Figure 3.26. Lateral view of the right mandible (a) does not demonstrate deformities or other abnormalities. A slight deformity of the apexes of mandibular cheek teeth is visible in the left lateral view (b). The caudal view (c) shows the bone lesion at the lesion site (arrow). In the cropped 3D Volume Reconstruction (VR) (d) the lesion is visible (arrow).

The NORMAL TOTAL BODY
Lateral Projection

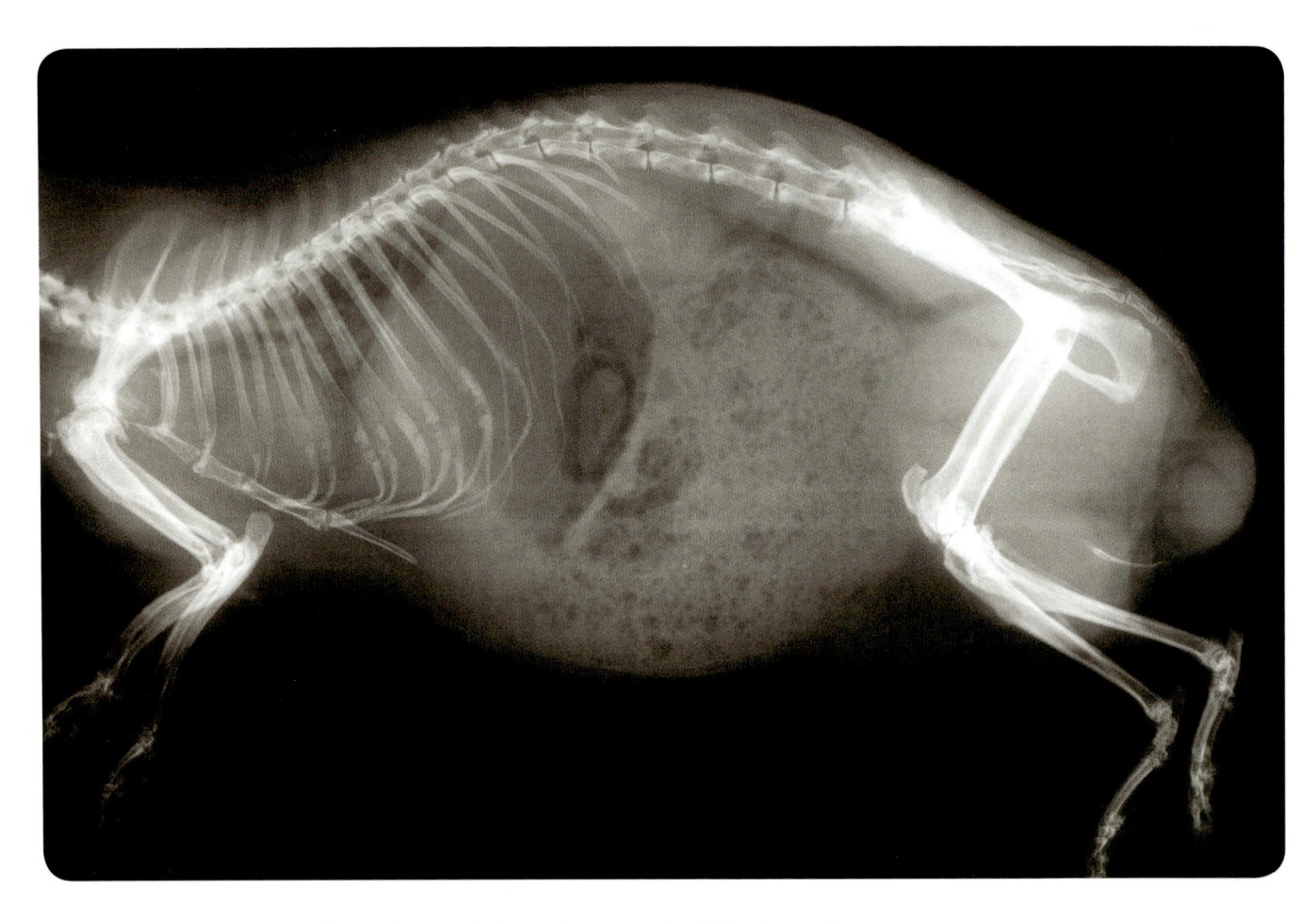

Figure 3.28. Total body radiograph of an 850 g male guinea pig, 80% of actual size.

Ventrodorsal Projection

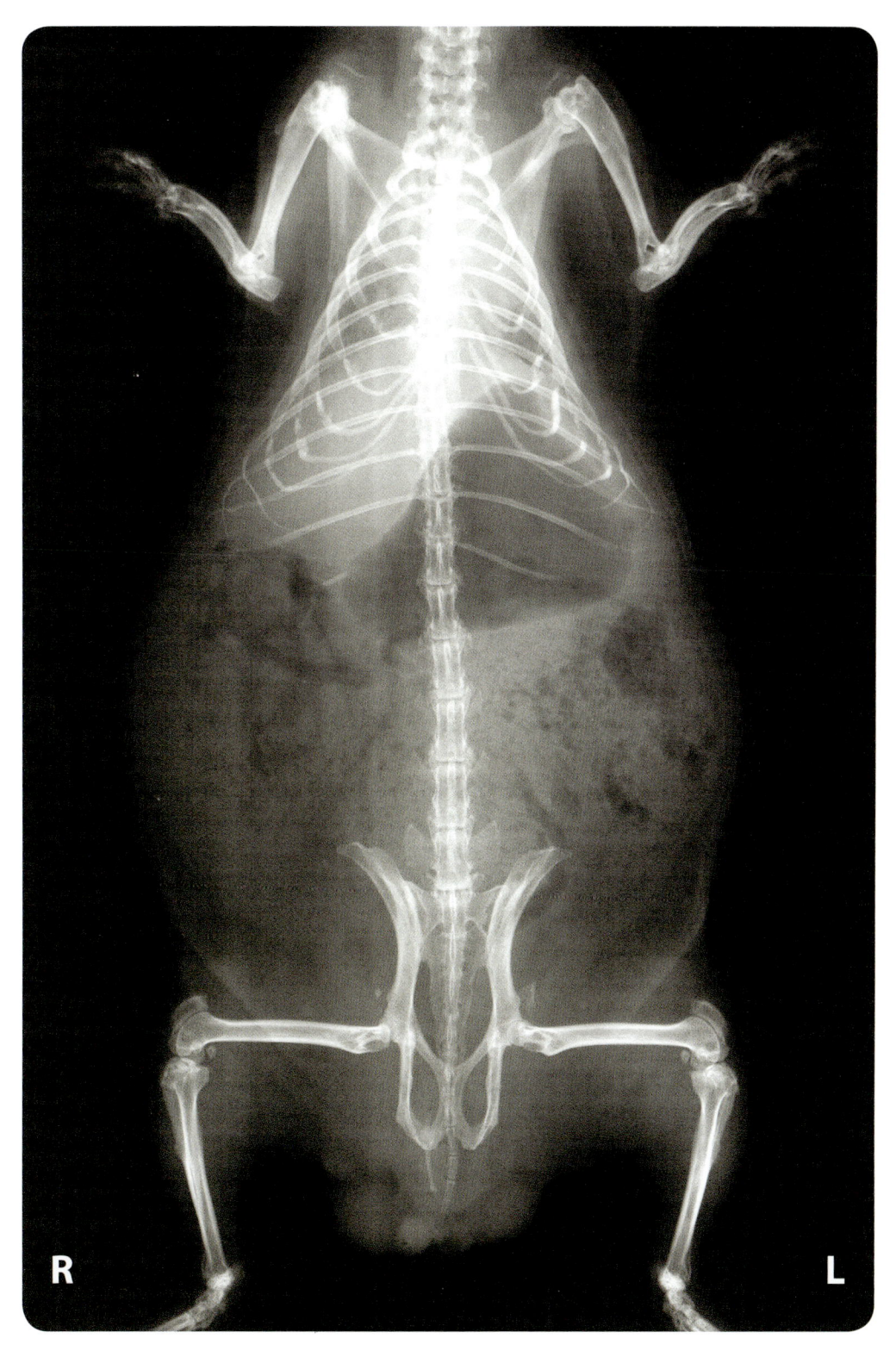

Figure 3.29. Total body radiograph of an 850 g male guinea pig, 80% of actual size.

The NORMAL THORAX
The Cervical and Thoracic Vertebral Column
Lateral Projection

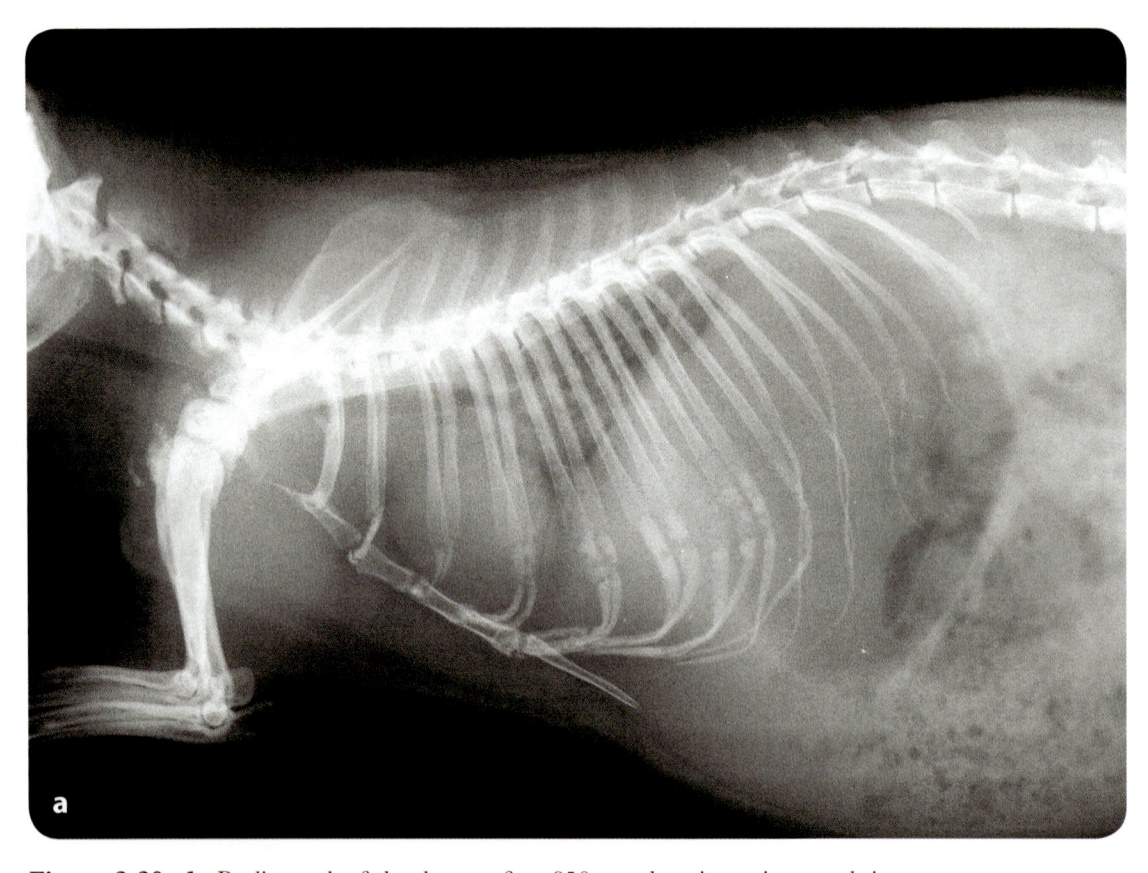

Figure 3.30a,b. Radiograph of the thorax of an 850 g male guinea pig, actual size.

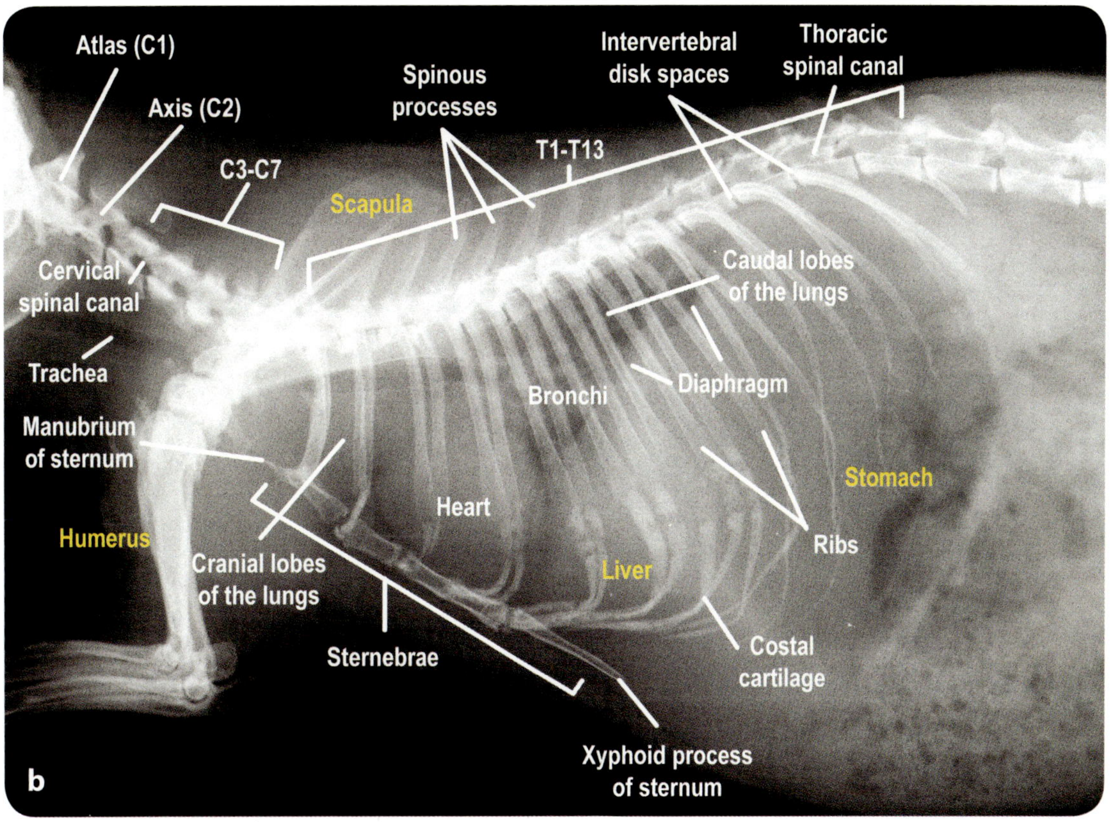
Atlas (C1)
Axis (C2)
C3-C7
Spinous processes
T1-T13
Intervertebral disk spaces
Thoracic spinal canal
Scapula
Cervical spinal canal
Trachea
Caudal lobes of the lungs
Diaphragm
Bronchi
Manubrium of sternum
Stomach
Heart
Humerus
Ribs
Cranial lobes of the lungs
Liver
Sternebrae
Costal cartilage
Xyphoid process of sternum
b

Ventrodorsal Projection

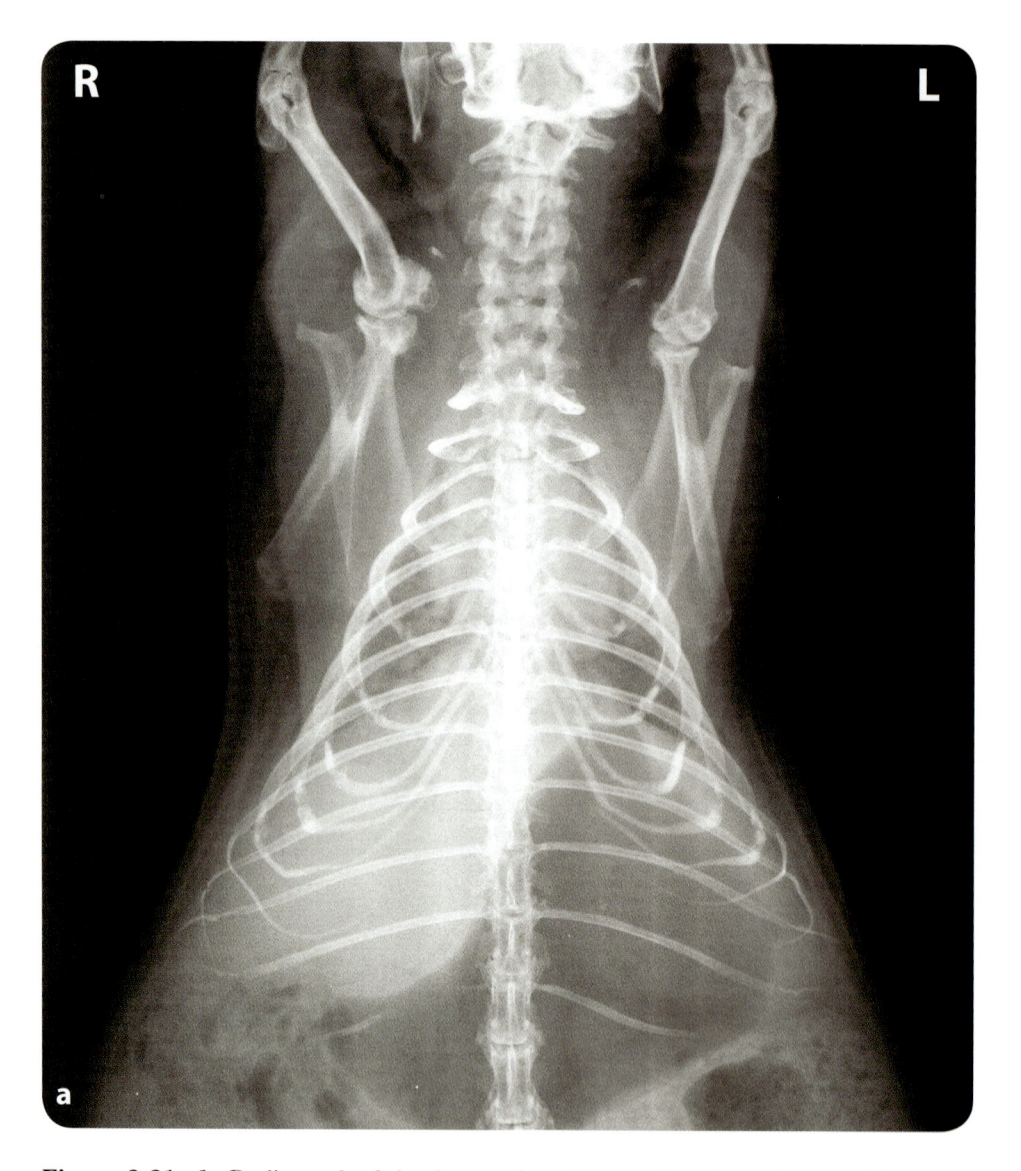

Figure 3.31a,b. Radiograph of the thorax of an 850 g male guinea pig, actual size.

THORAX 189

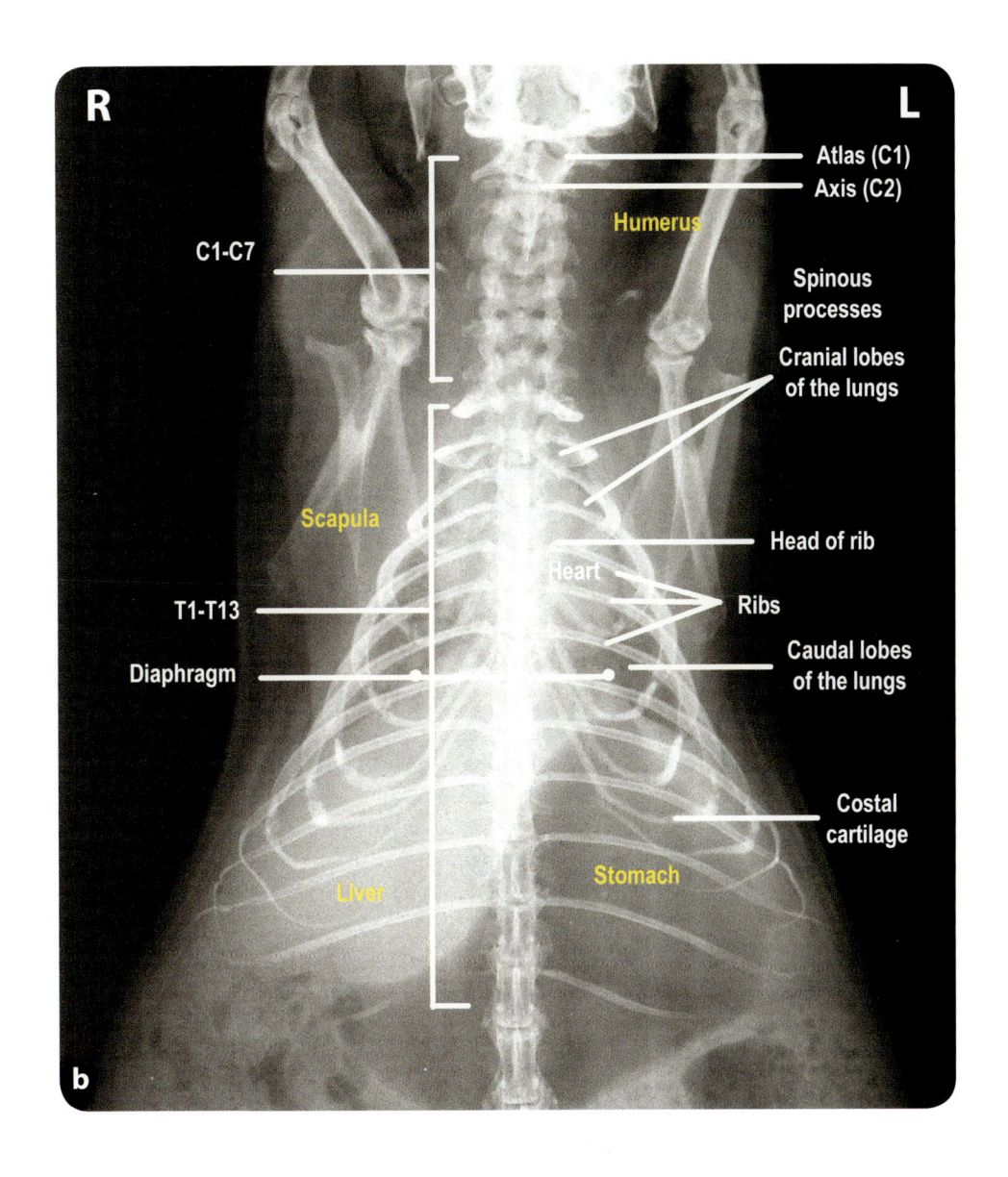

R | L
- Atlas (C1)
- Axis (C2)
- Humerus
- C1–C7
- Spinous processes
- Cranial lobes of the lungs
- Scapula
- Head of rib
- Heart
- T1–T13
- Ribs
- Diaphragm
- Caudal lobes of the lungs
- Costal cartilage
- Liver
- Stomach

b

ABNORMALITIES of the THORAX
Diseases of the Lungs

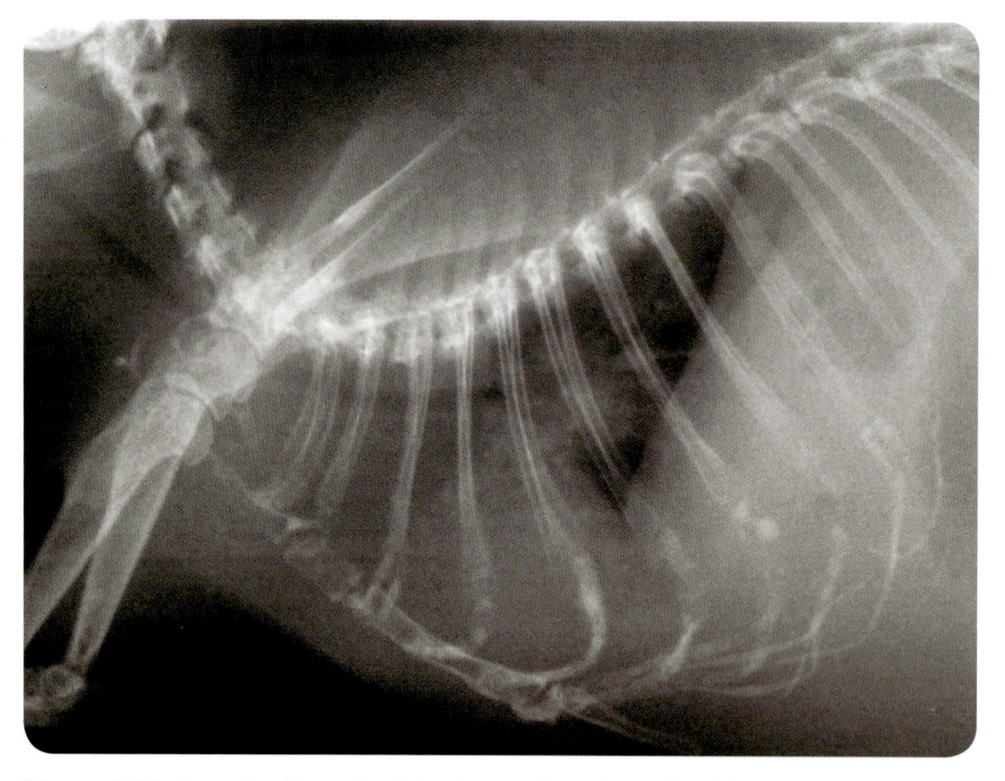

Figure 3.32. Lateral radiograph of the thorax of a guinea pig with severe respiratory distress. Note the area of increased density associated with the hilus of the lung.

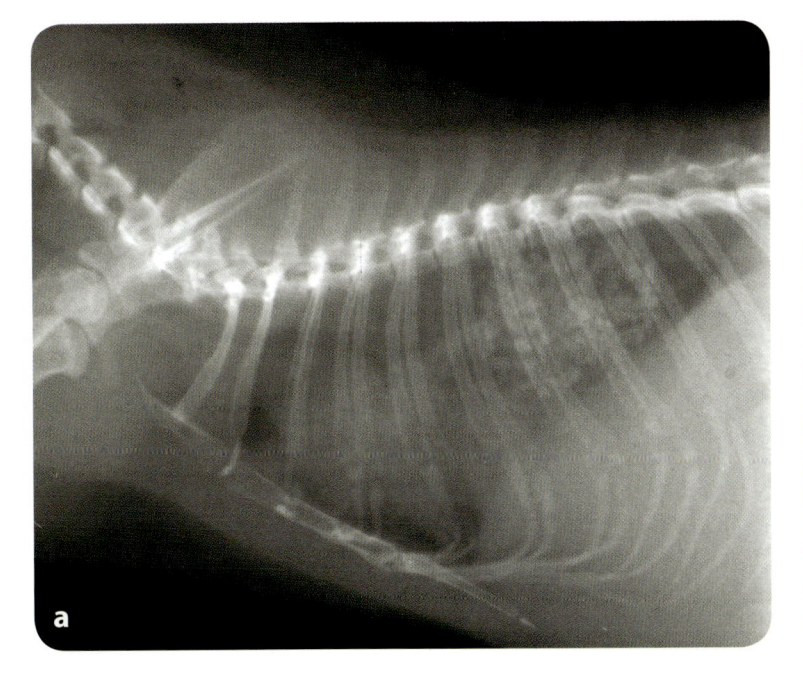

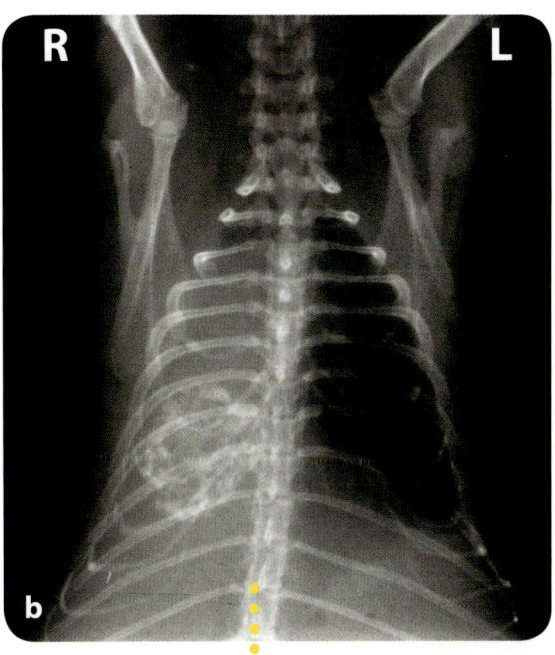

Figure 3.33a,b. Lateral (a) and ventrodorsal (b) radiographs of the thorax of a guinea pig with moderate respiratory distress that did not respond to antibiotic therapy. Note the large, discrete mineralized mass in the right caudodorsal hemithorax.

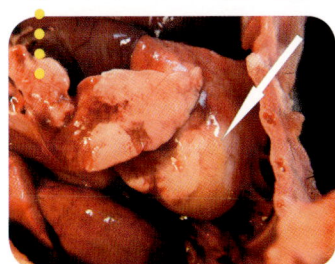

Figure 3.34. Necropsy specimen of the same patient demonstrating the large, firm nodule consistent with neoplasia (arrow).

Diseases of the Chest

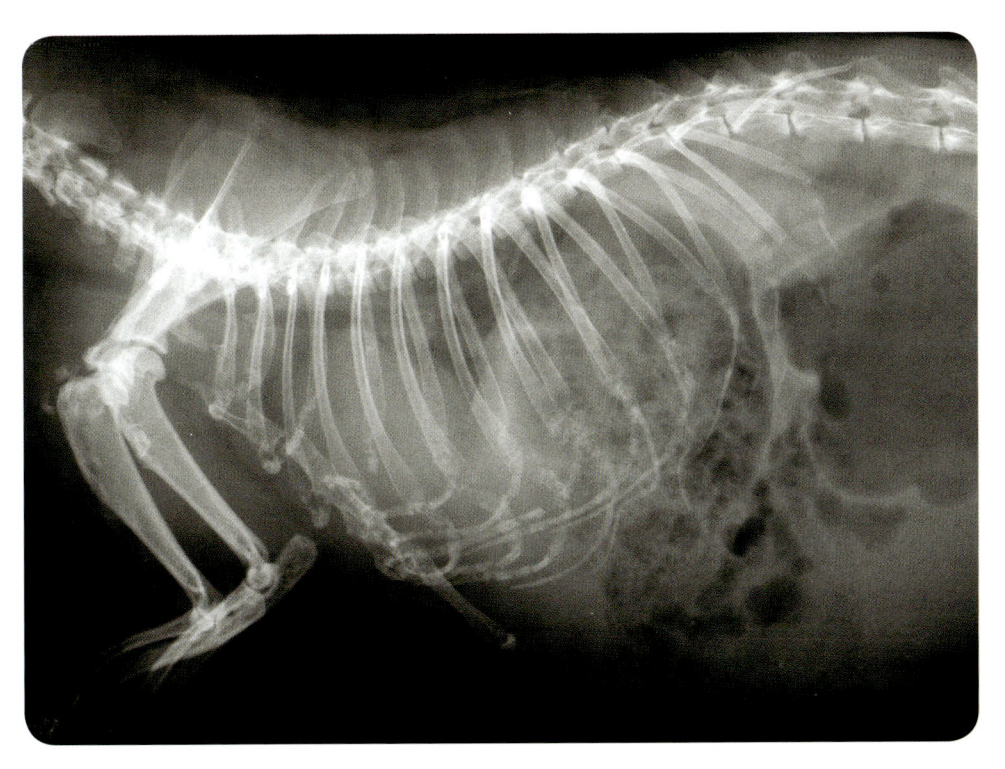

Figure 3.35. Lateral radiograph of the thorax of a 6-year-old guinea pig with mineralization of the ribs and sternebrae, and fractures of the caudal ribs. Also note luxation of the first two sternebrae.
These radiologic findings are occasionally encountered in older patients, and usually have no clinical significance.

The NORMAL ABDOMEN
The Lumbar Vertebral Column
Lateral Projection

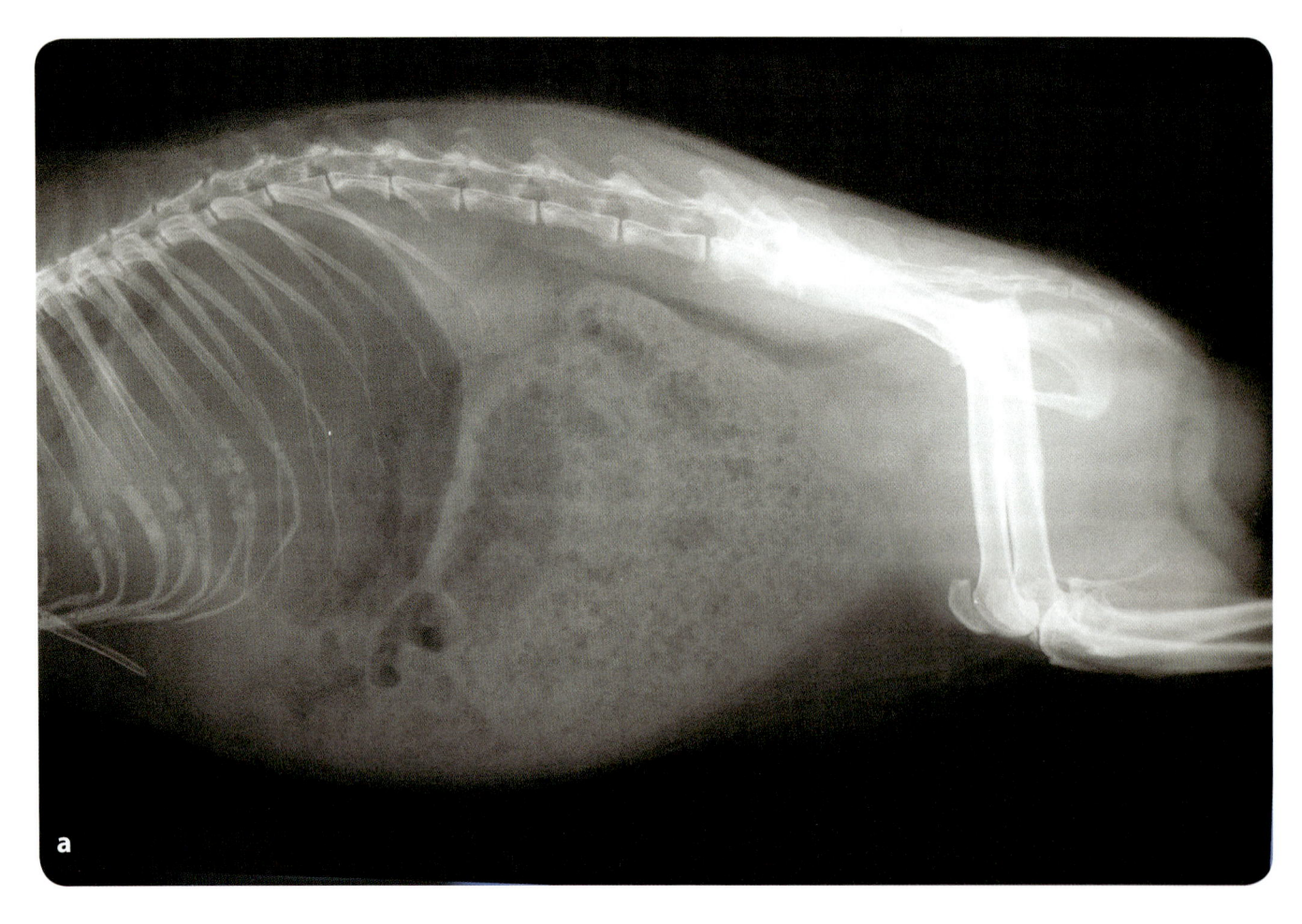

Figure 3.36a,b. Radiograph of the abdomen of an 850 g male guinea pig, actual size. Normal ingesta can significantly interfere with identification of abdominal structures.

ABDOMEN

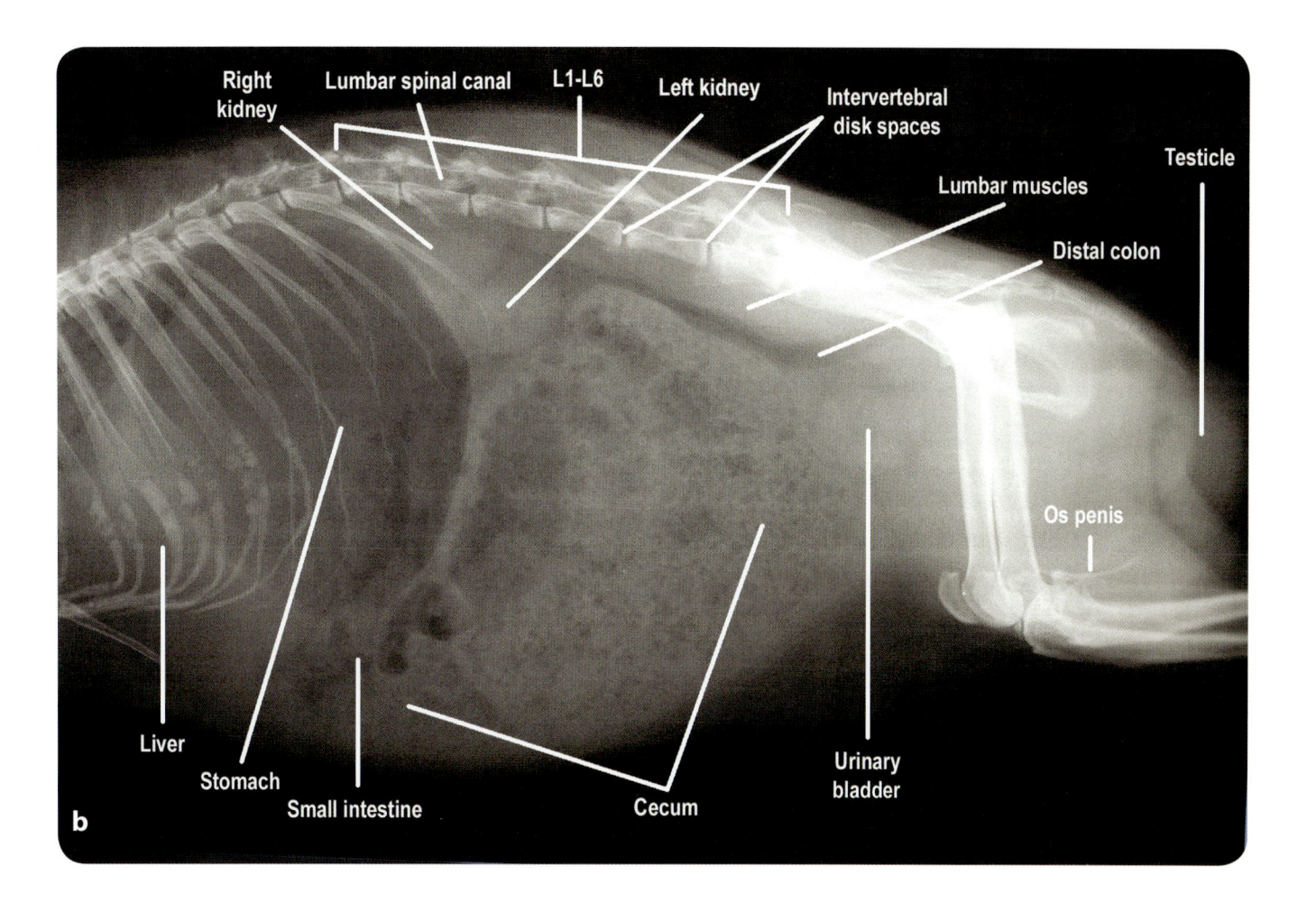

Right kidney — Lumbar spinal canal — L1-L6 — Left kidney — Intervertebral disk spaces — Lumbar muscles — Distal colon — Testicle — Os penis — Urinary bladder — Cecum — Small intestine — Stomach — Liver

b

Ventrodorsal Projection

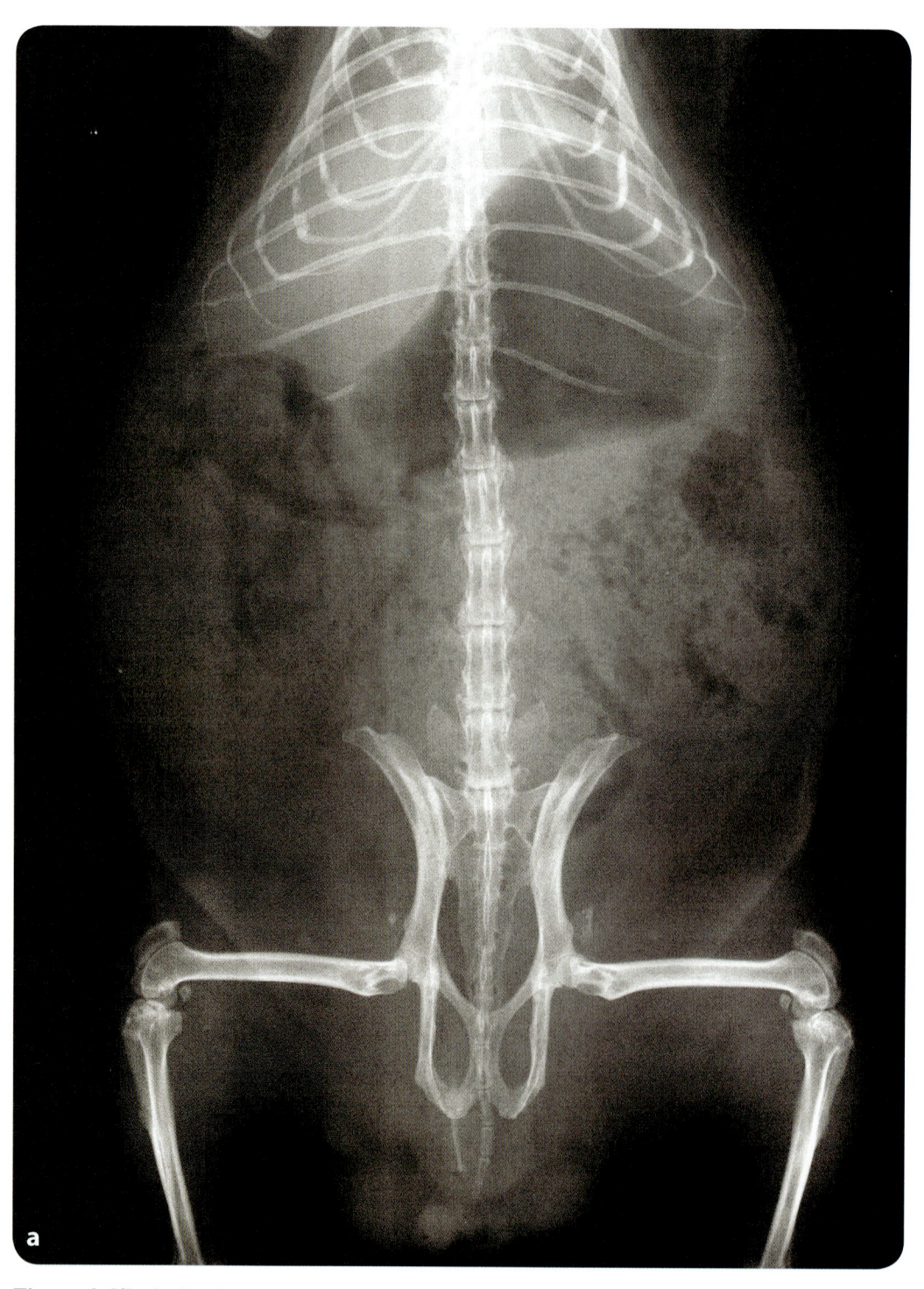

Figure 3.37a,b. Radiograph of the abdomen of an 850 g male guinea pig, actual size.

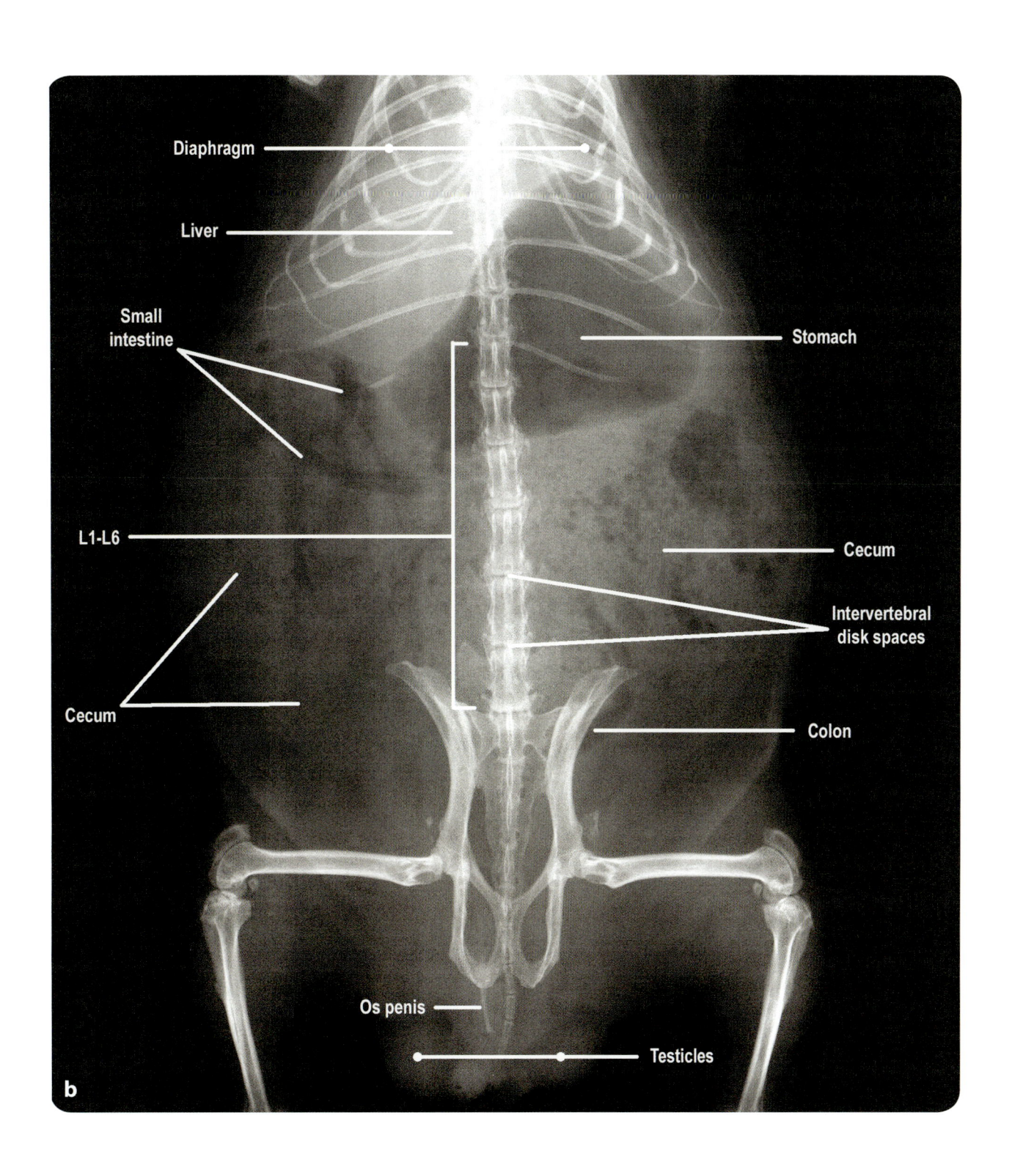

- Diaphragm
- Liver
- Small intestine
- L1-L6
- Cecum
- Stomach
- Cecum
- Intervertebral disk spaces
- Colon
- Os penis
- Testicles

b

Miscellaneous

Pregnancy in the guinea pig. Fetuses of porcupine-like rodents are very large in comparison to species of similar size. Calcification of the skeleton occurs early in fetal development, often giving the impression the fetuses are more developed than they actually are.

In female guinea pigs, the ischiopubic symphysis dilates to allow passage of the large head of the fetus through the birth canal. A physiologic peculiarity of guinea pigs is that the ischiopubic symphysis fuses at 7-8 months if the female has not yet been bred, which is reported to contribute to dystocia. However, clinical experience has demonstrated that female guinea pigs bred after this time-period often experience normal parturition. Radiographs are useful during pregnancy to determine the number of fetuses, check for signs of fetal death, and to check the status of ischiopubic symphysis. The most reliable indicator of impending parturition is dilation of the symphysis.

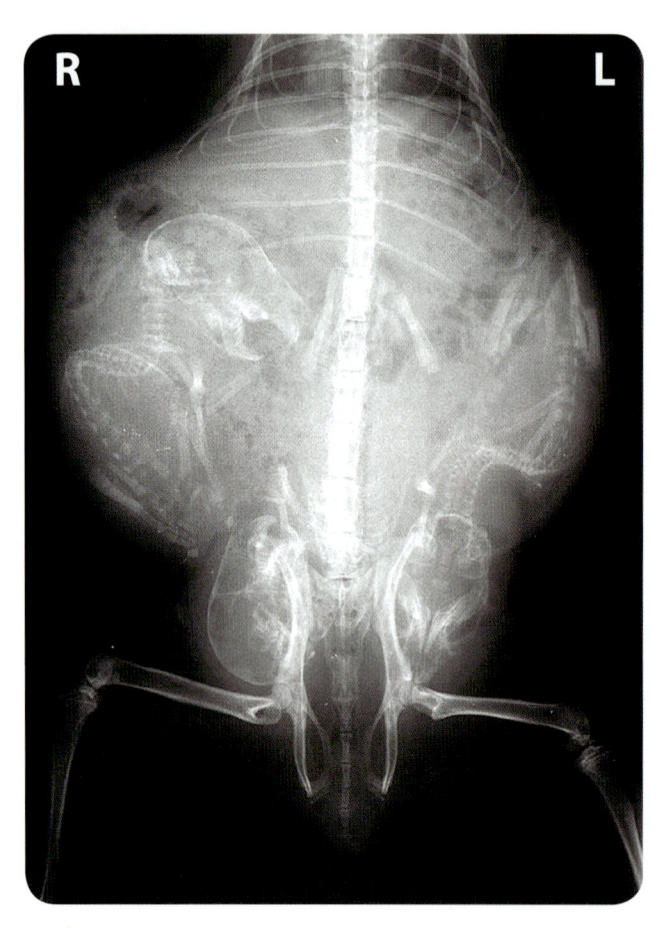

Figure 3.38. Ventrodorsal radiograph of the abdomen of a pregnant guinea pig. Note the presence of three fetuses. The ischiopubic diastasis is visible, but not dilated.

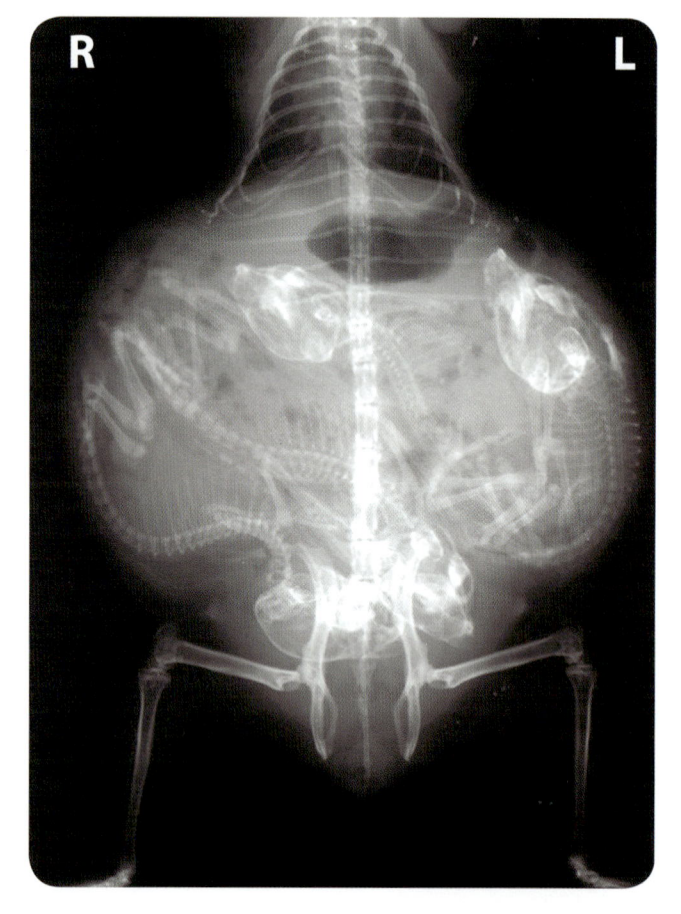

Figure 3.39. Ventrodorsal radiograph of the abdomen of a pregnant guinea pig. Note the presence of four fetuses. Fetal death is more common when there are more than 3 fetuses, especially in the sow's first pregnancy. There is no evidence of fetal death in this image. The ischiopubic diastasis is more dilated than in Figure 3.63, but delivery is likely not imminent.

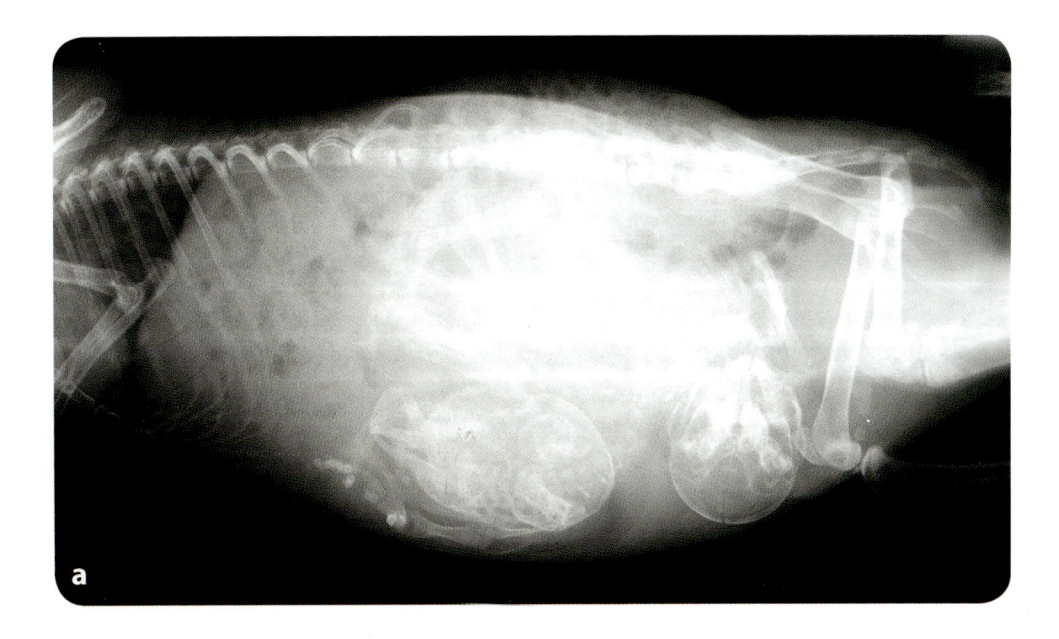

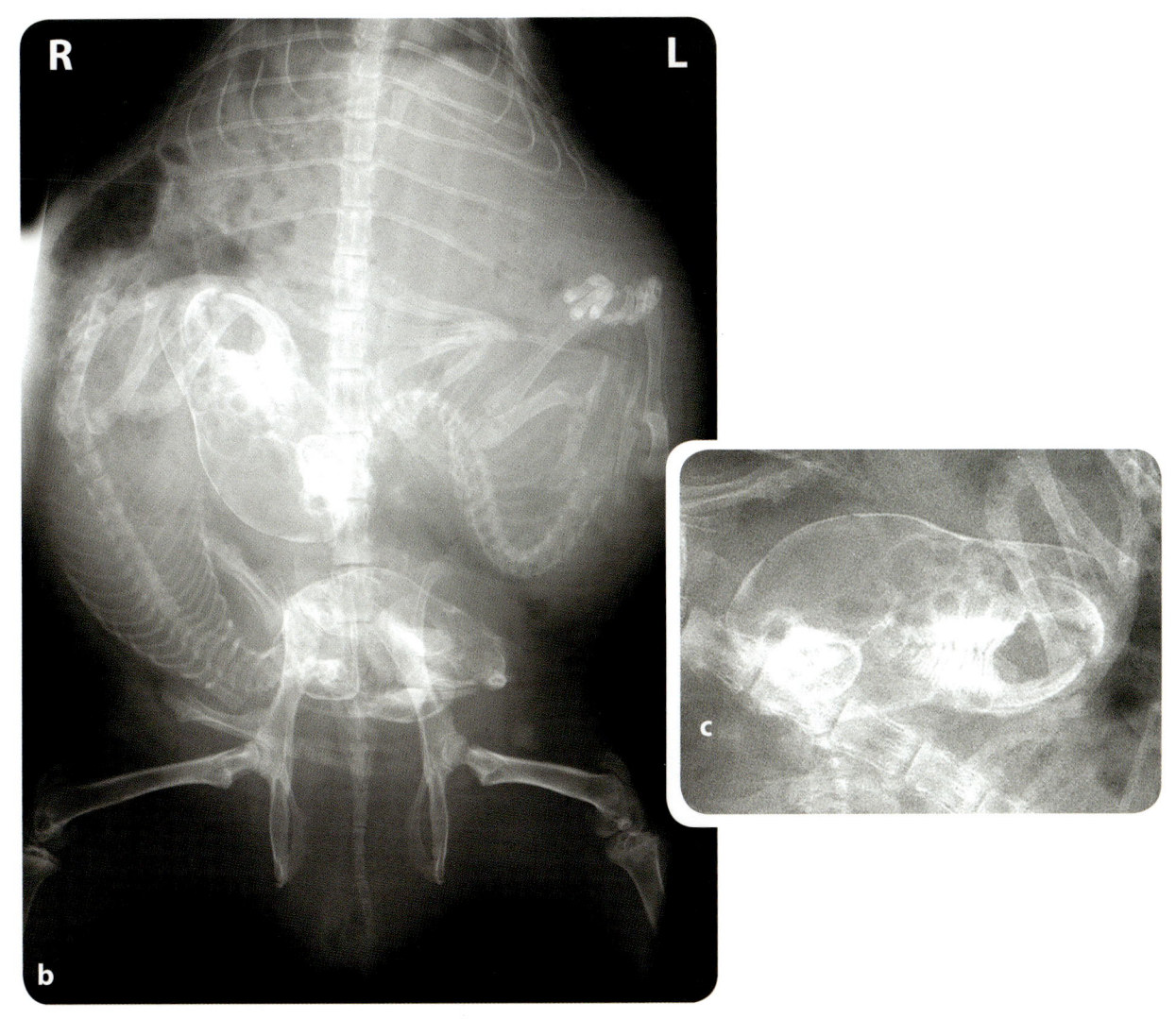

Figure 3.40a,b,c. Lateral (a) and ventrodorsal (b) radiographs of the abdomen of a pregnant guinea pig less than 24 hours prior to parturition. The ischiopubic symphysis is completely dilated (b) and the guinea pig is ready for delivery.
The skull of one fetus was captured in a perfect lateral projection, and the incisors and cheek teeth, which are present at birth, are clearly visible in the rotated enlarged image (c).

ABNORMALITIES of the ABDOMEN
Diseases of the Stomach

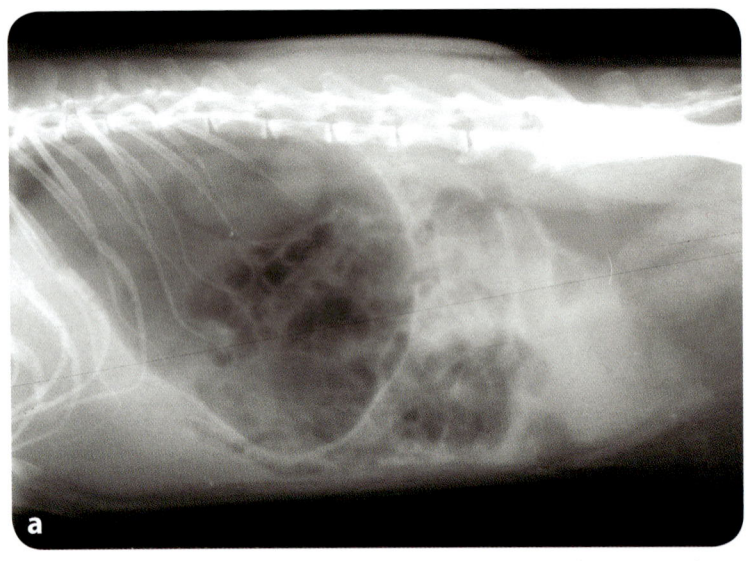

Figure 3.41a,b. Lateral (a) and ventrodorsal (b) radiographs of the abdomen of a guinea pig showing an excessive accumulation of gastric gas. The ventrodorsal projection (b) also demonstrates the presence of gas in the duodenum.

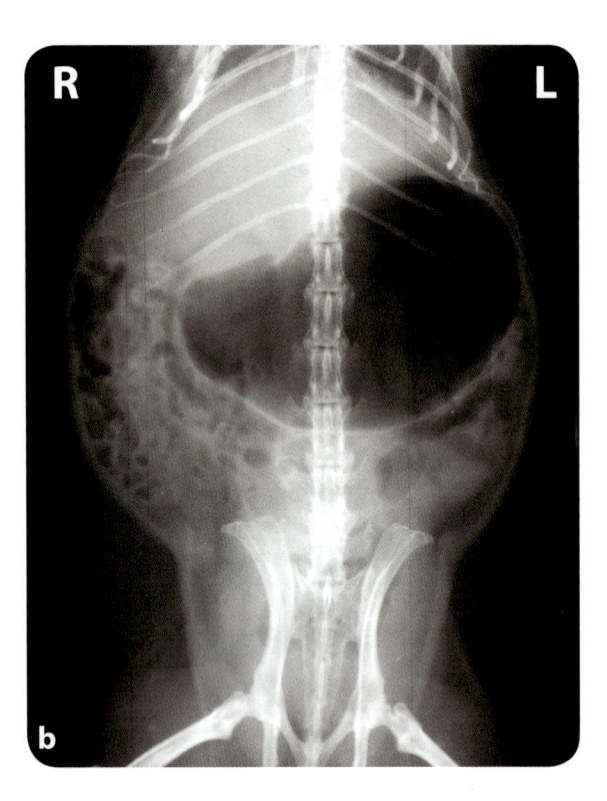

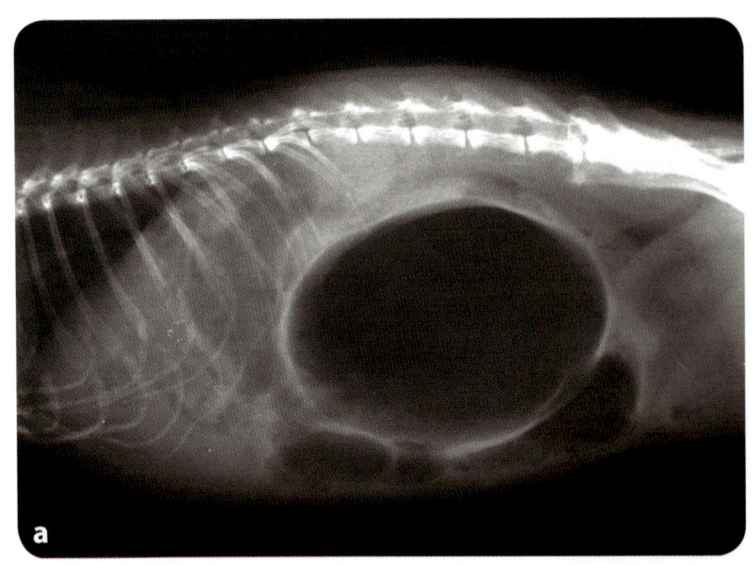

Figure 3.42a,b. Lateral (a) and ventrodorsal (b) radiographs of the abdomen of a guinea pig with more severe gastric gas. The stomach is distended and round in appearance. Gas is also present in the cecum.

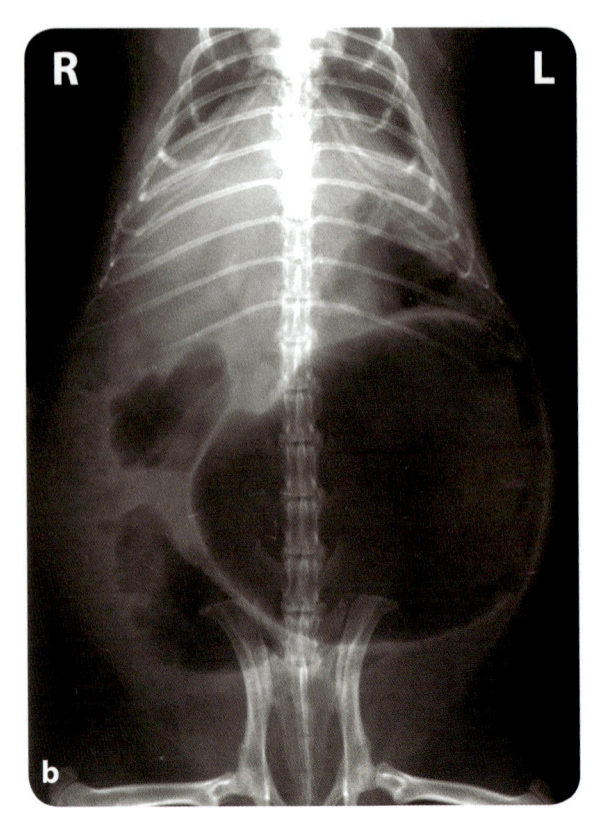

ABDOMEN 199

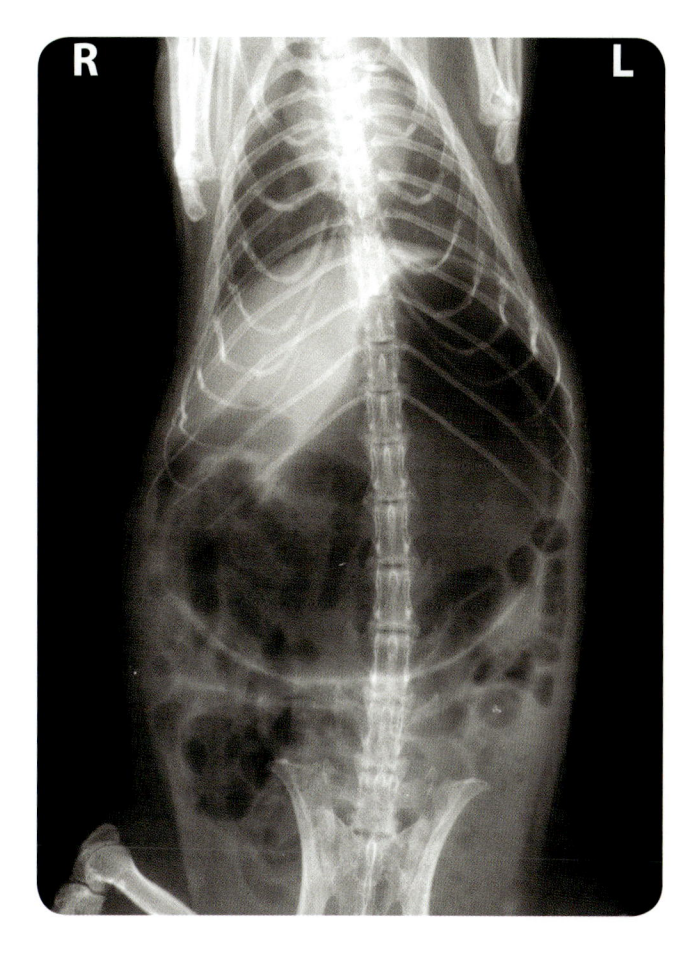

Figure 3.43. Ventrodorsal radiograph of the abdomen of a guinea pig with gastric gas distention. Note deviation of the liver to the right and compression of the diaphragm. Intestinal gas is also present.

Figure 3.44. Gastric trichobezoars are common in guinea pigs, and are likely a result of low fiber diets, skin disease and possibly behavioral problems. Long-haired Peruvian cavies are more prone to this condition.

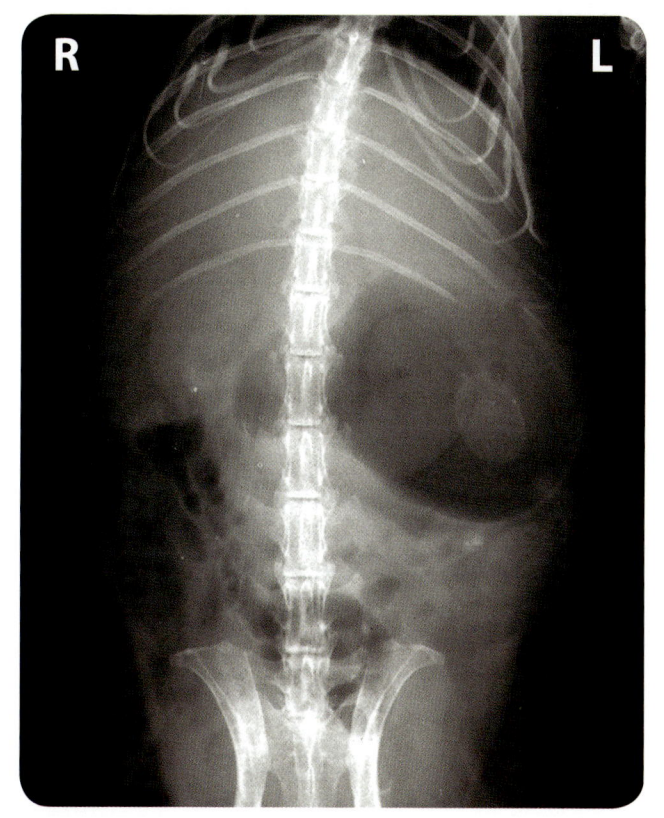

Figure 3.45. Ventrodorsal radiograph of the abdomen of a guinea pig with gas distention of the stomach. Note the presence of a small oval density which likely represents a small bezoar.

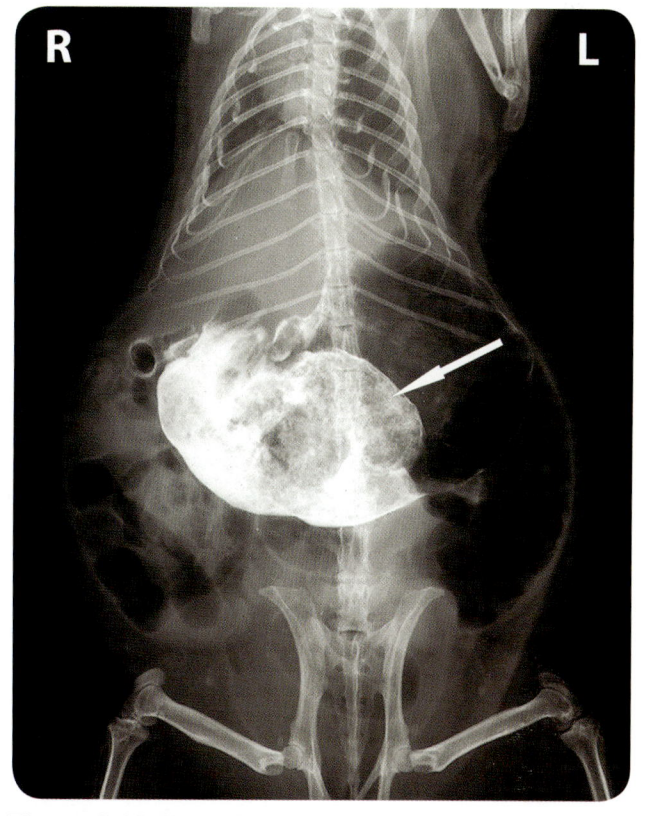

Figure 3.46. Ventrodorsal radiograph of the abdomen of a guinea pig after administration of barium sulfate. Gastric gas often provides negative contrast to aid in identification of bezoars, but occasionally positive contrast studies are needed.

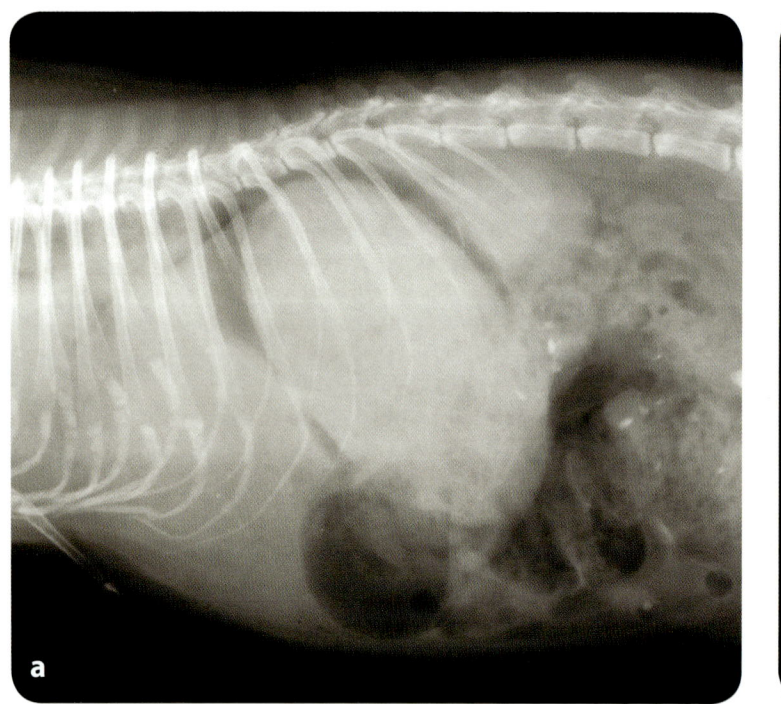

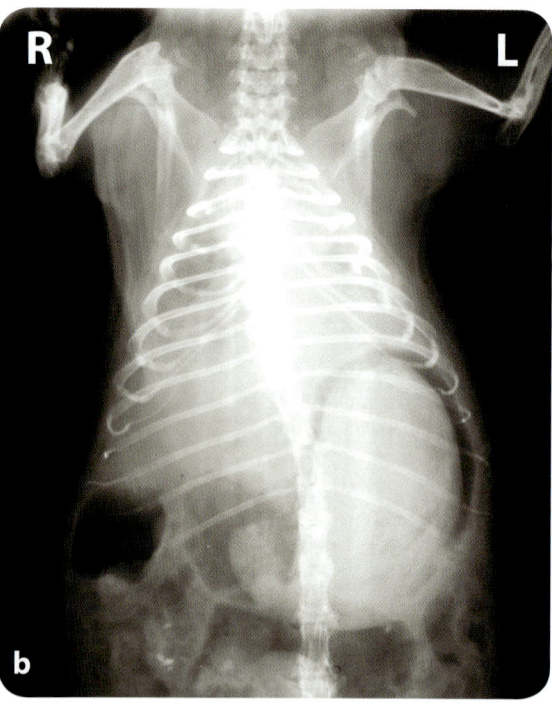

Figure 3.47a,b. Lateral (a) and ventrodorsal (b) radiographs of the abdomen of a guinea pig with a large bezoar filling the entire stomach. Note the presence of gas in the duodenum and cecum.

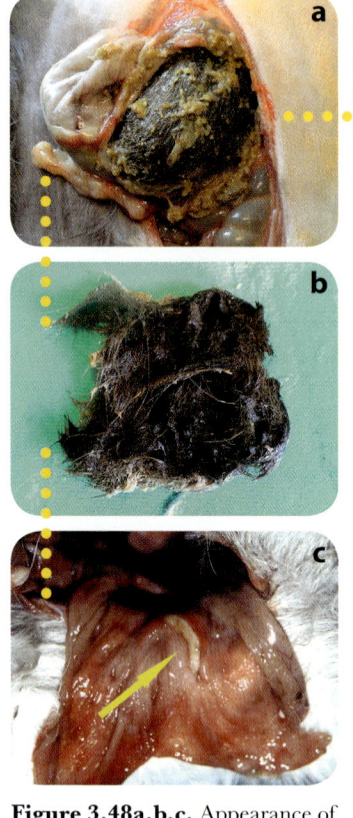

Figure 3.48a,b,c. Appearance of the previous case at necropsy (a); appearance of the dissected trichobezoar (b) and the ulceration of the gastric mucosa (c, arrow).

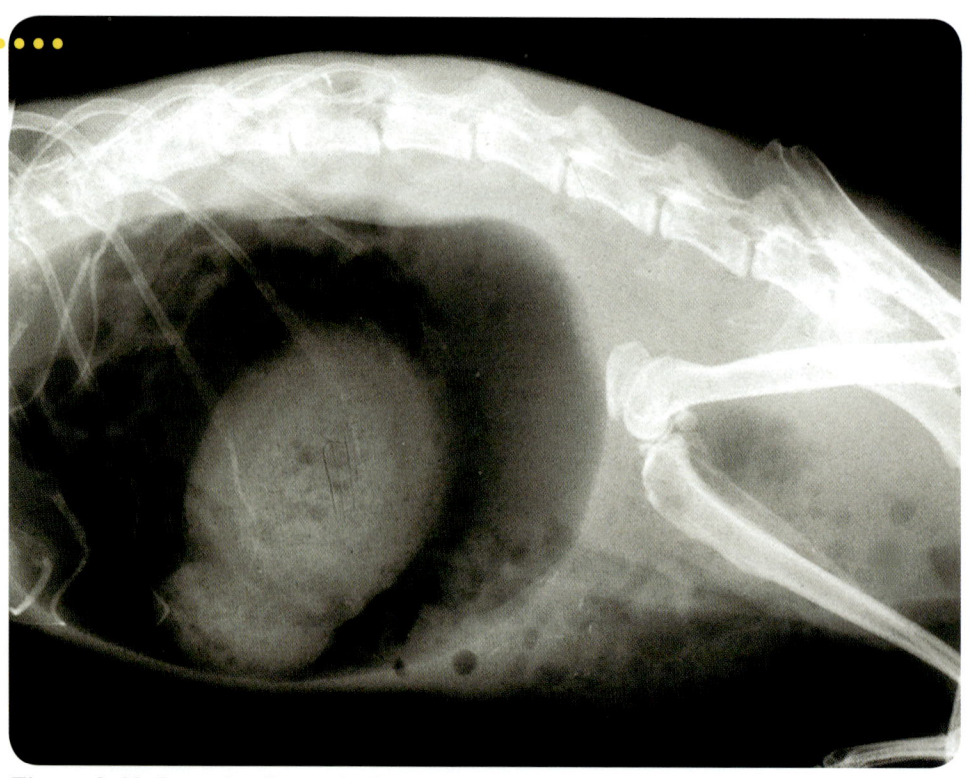

Figure 3.49. Lateral radiograph of the abdomen of a guinea pig demonstrating a very large bezoar clearly visible due to marked gas distention of the stomach. Gas frequently acts as a negative contrast medium.

Diseases of the Intestine

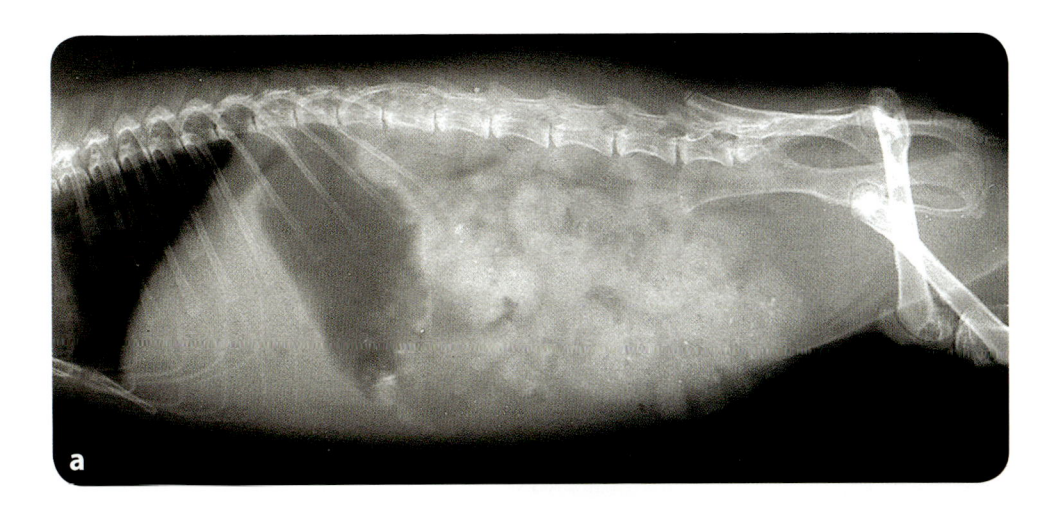

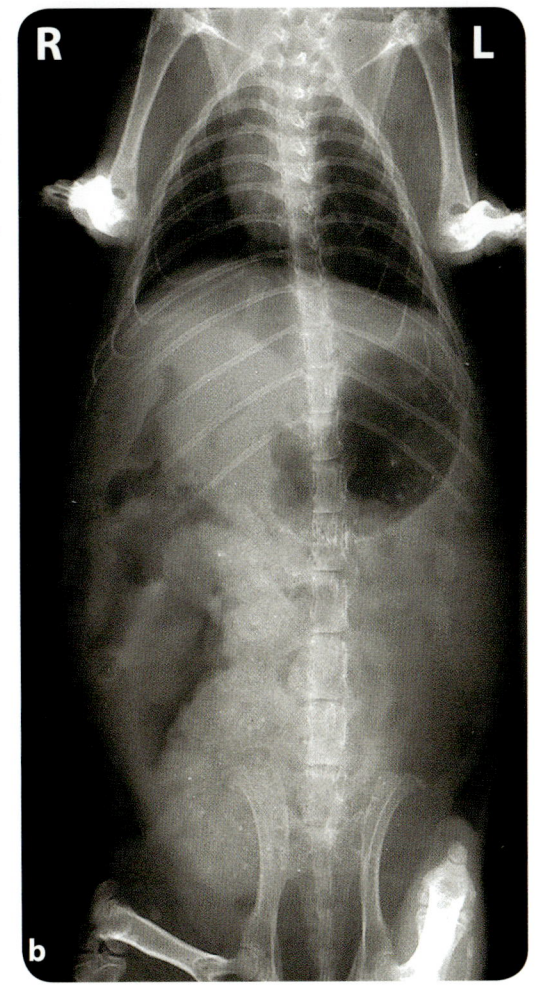

Figure 3.50a,b. Lateral (a) and ventrodorsal (b) radiogaphs of the abdomen of a guinea pig with impaction of the cecum, which is usually a consequence of feeding inappropriate low fiber diets. The haustra of the cecum are highlighted by radiodense dry ingesta.
Moderate gas distention of the stomach is also visible.

Diseases of the Kidneys and Ureters

Urolithiasis is common in guinea pigs, and can occur throughout the urinary tract. The most common stone type is calcium oxalate, which is radiopaque and easily identified. The underlying causes of urolithiasis in guinea pigs are currently under investigation.

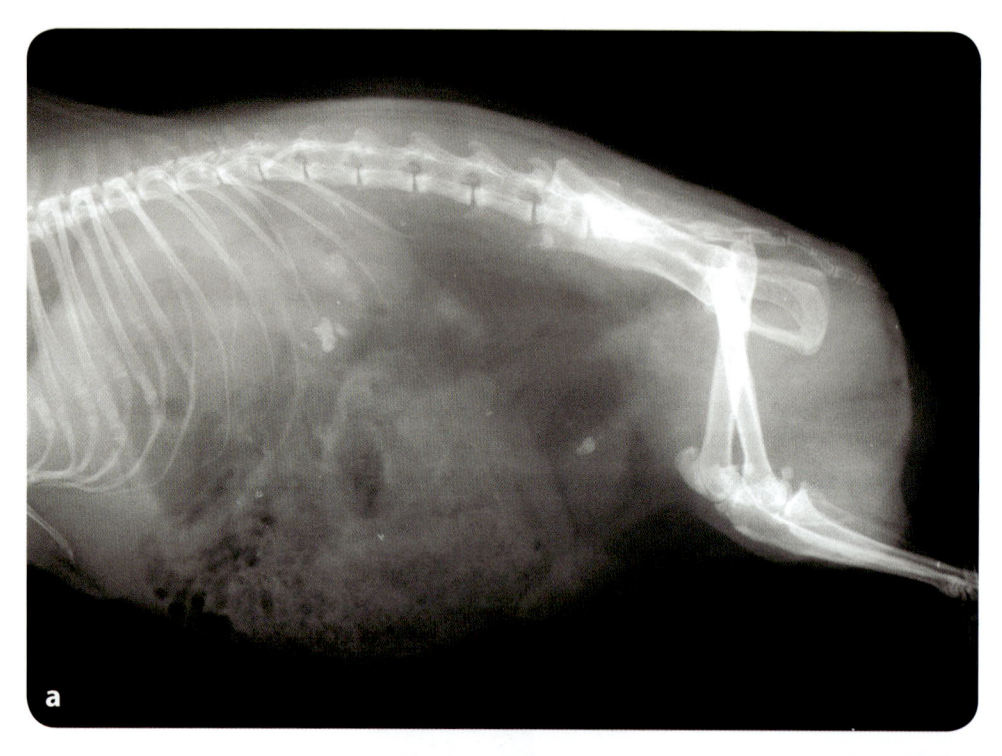

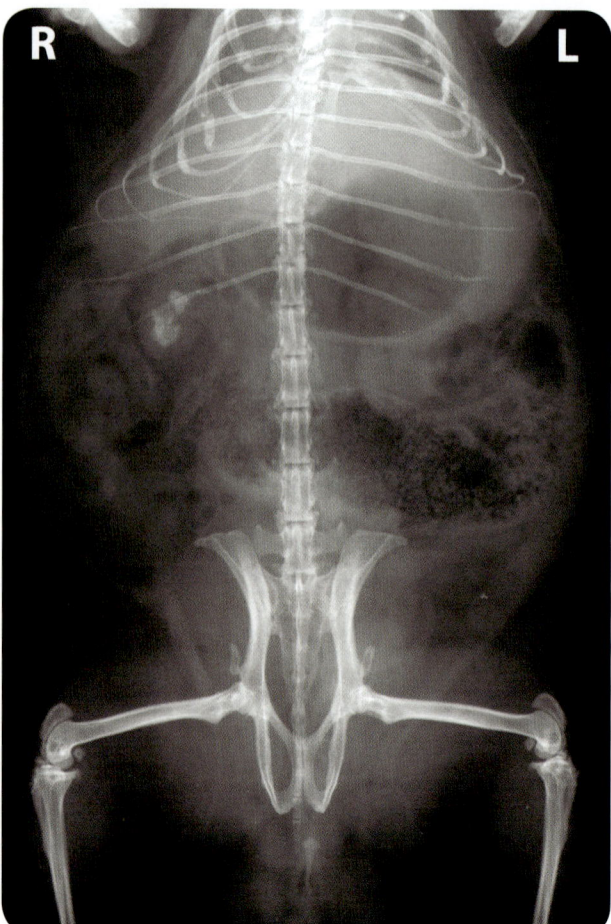

Figure 3.51a,b.
Lateral (a) and ventrodorsal (b) radiographs of the abdomen of a male guinea pig. Note unilateral nephrolithiasis of the right kidney. A small cystolith is also visible on the lateral view (a). It is obscured by superimposition of the wing of the ilium in the ventrodorsal projection.

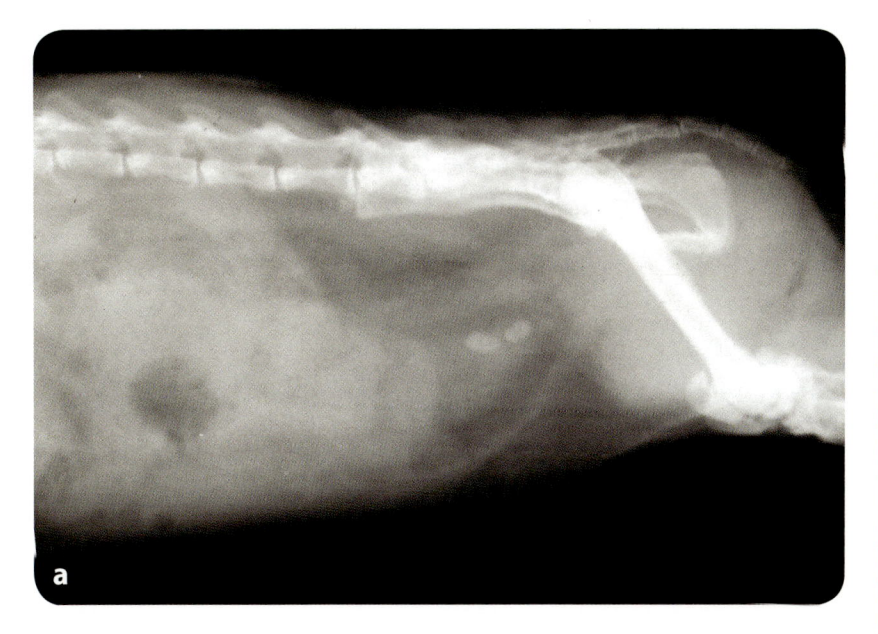

Figure 3.52a,b. Lateral (a) and ventrodorsal (b) radiographs of the abdomen of a guinea pig. It is often difficult to differentiate cystoliths from ureteroliths in the lateral projection due to the superimposition of the urinary bladder over the ureters. Ureteroliths are typically located in the distal ureters, in close proximity to the urinary bladder.

In the ventrodorsal projection (b), the ureterolith appears lateral to the urinary bladder, which aids in correct identification (white arrows). Note bilateral ossification of the distal tendons as they insert at the greater trochanters (yellow arrows). This is a common finding in older guinea pigs.

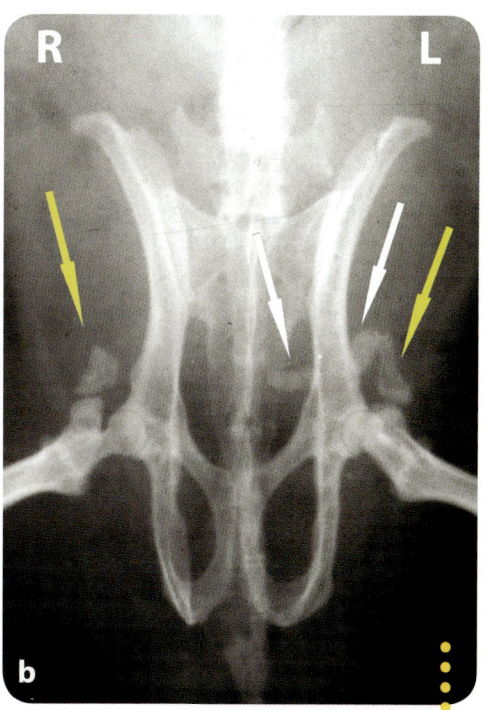

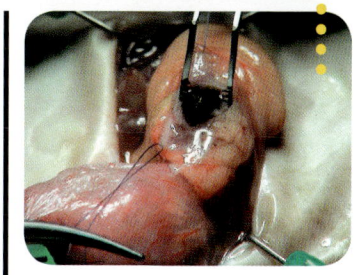

Figure 3.54. Removal of a urolith from the distal ureter through ureterotomy. The pink structure is the urinary bladder.

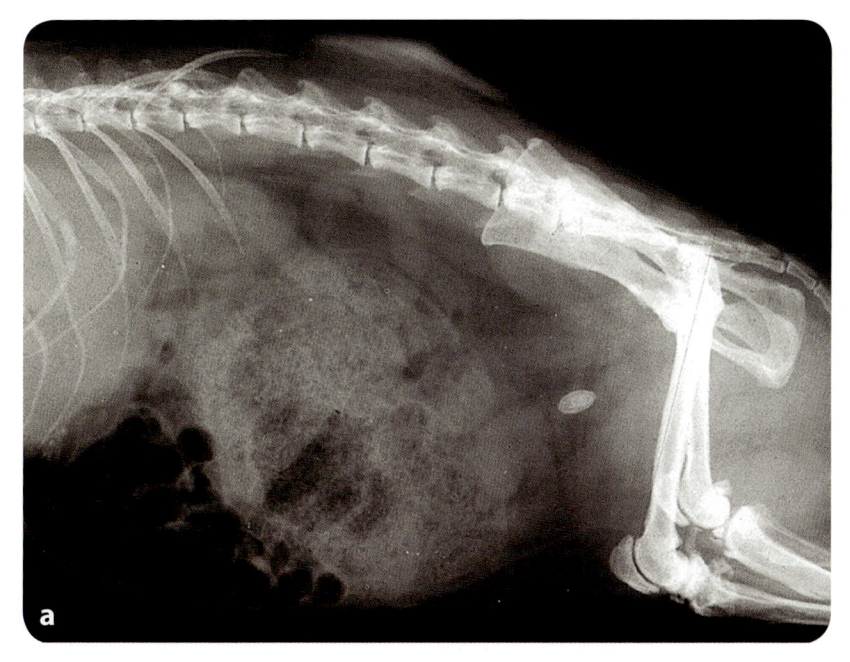

Figure 3.53a,b. Lateral (a) and ventrodorsal (b) projections of a guinea pig with a ureterolith. While this stone appears to be within or near the urinary bladder in the ventrodorsal projection, it was actually located in the distal left ureter (see Figure 3.49). This is an artifact, due to the slight obliquity of the ventrodorsal projection. Note the presence of excessive intestinal gas.

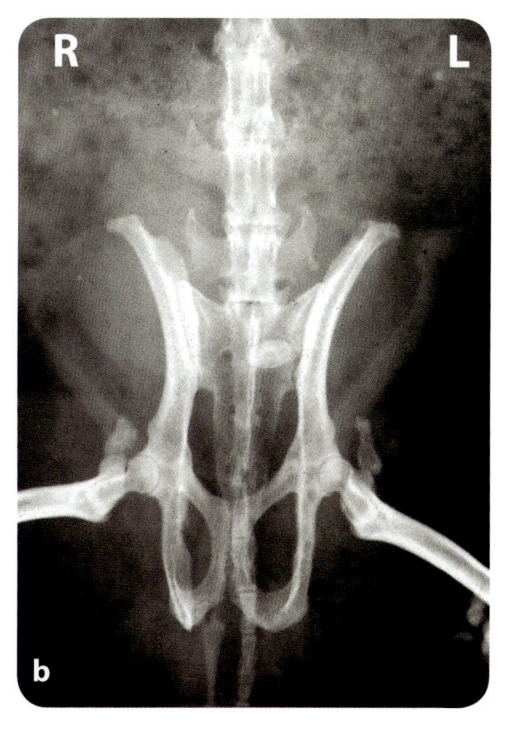

Diseases of the Urinary Bladder and Urethra

Figure 3.55a,b. Lateral (a) and ventrodorsal (b) projections of the abdomen of a guinea pig with hypermineralization (bladder sludge) partially outlining the ureter. The ventrodorsal projection shows sludge, and uroliths in the right ureter and urinary bladder. Ureteroliths are often cylindrical to oval in shape, while cystoliths are usually spherical.

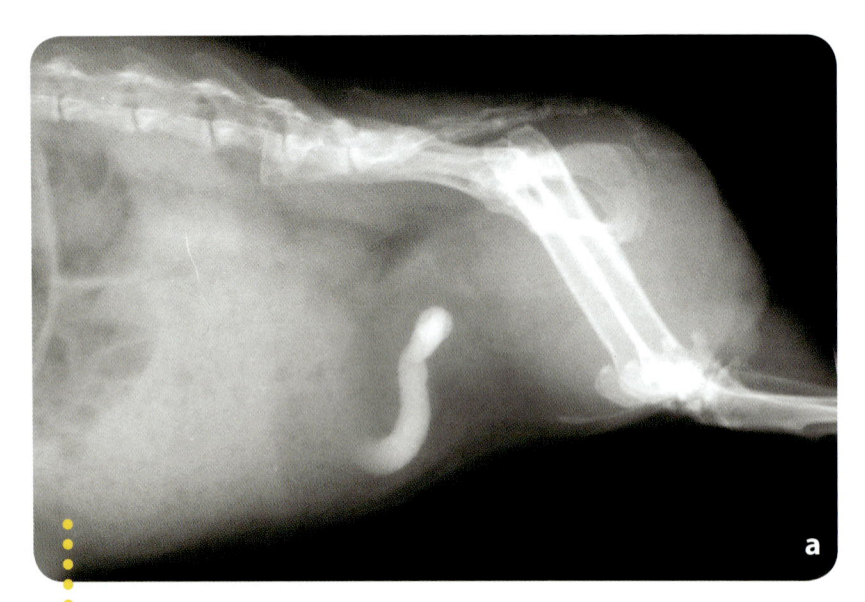

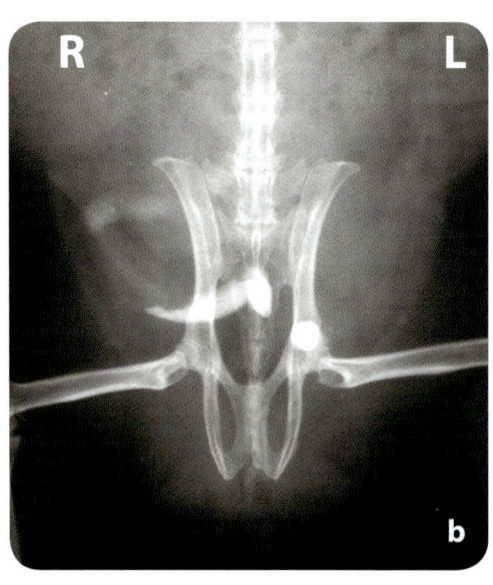

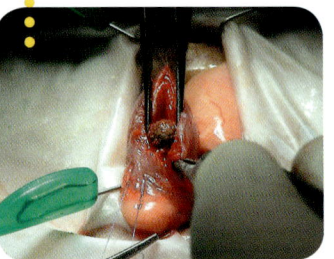

Figure 3.56. Cystotomy for removal of cystoliths in a guinea pig.

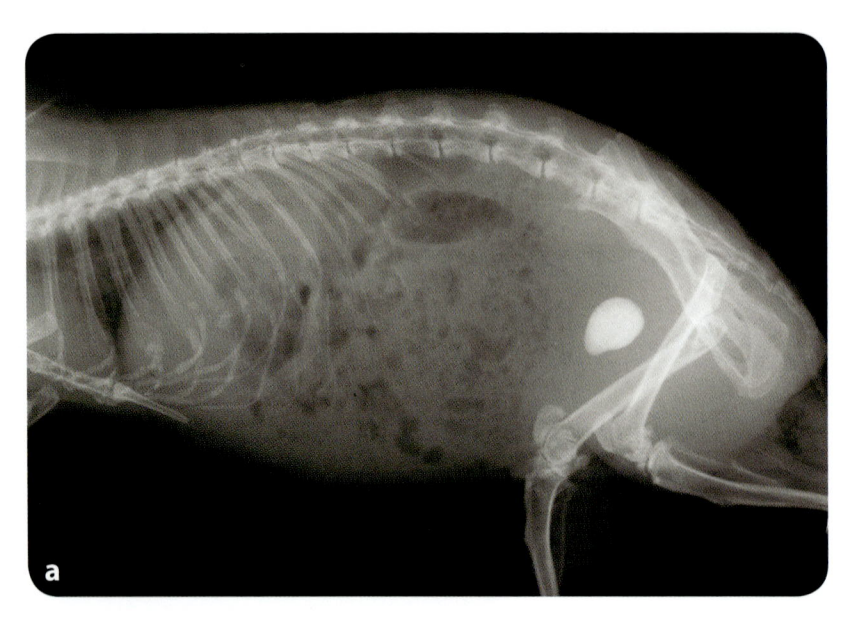

Figure 3.57a,b. Lateral (a) and ventrodorsal (b) radiographs of the abdomen of a male guinea pig demonstrating a single large cystolith.

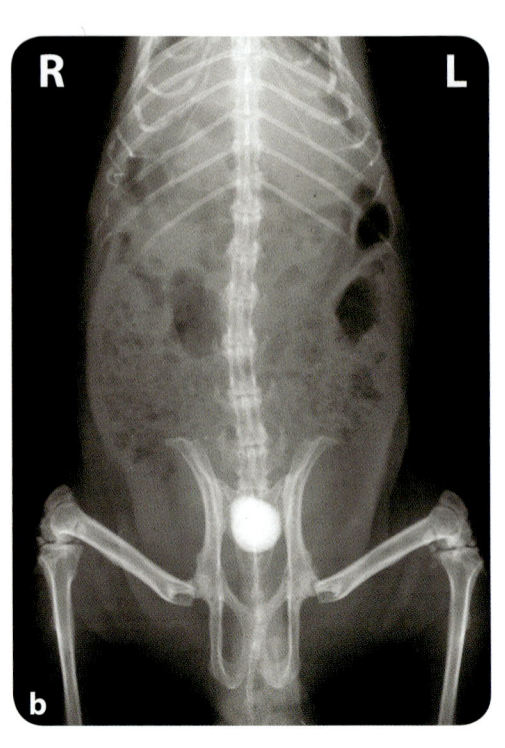

Figure 3.58. Ventodorsal radiograph of the abdomen of a female guinea pig demonstrating urethral calculi.

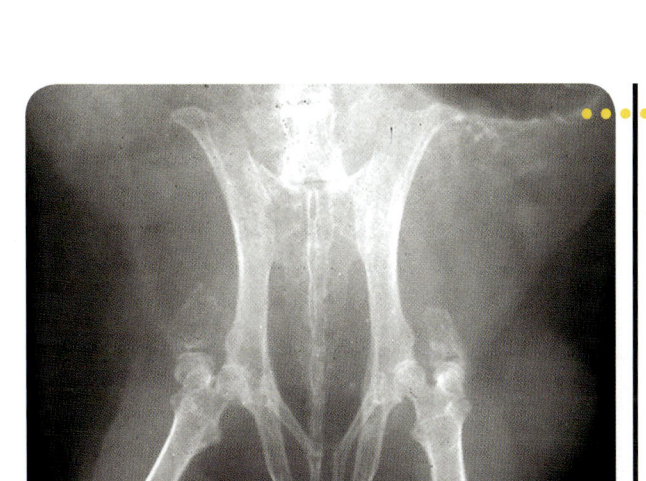

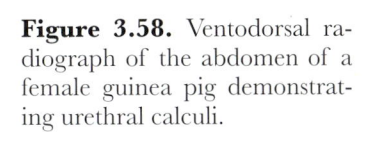

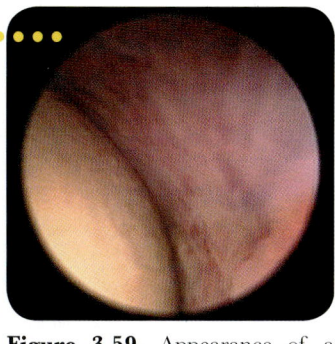

Figure 3.59. Appearance of a urethral calculus in a female guinea pig viewed via urethroscopy.

Diseases of the Female Genital Tract

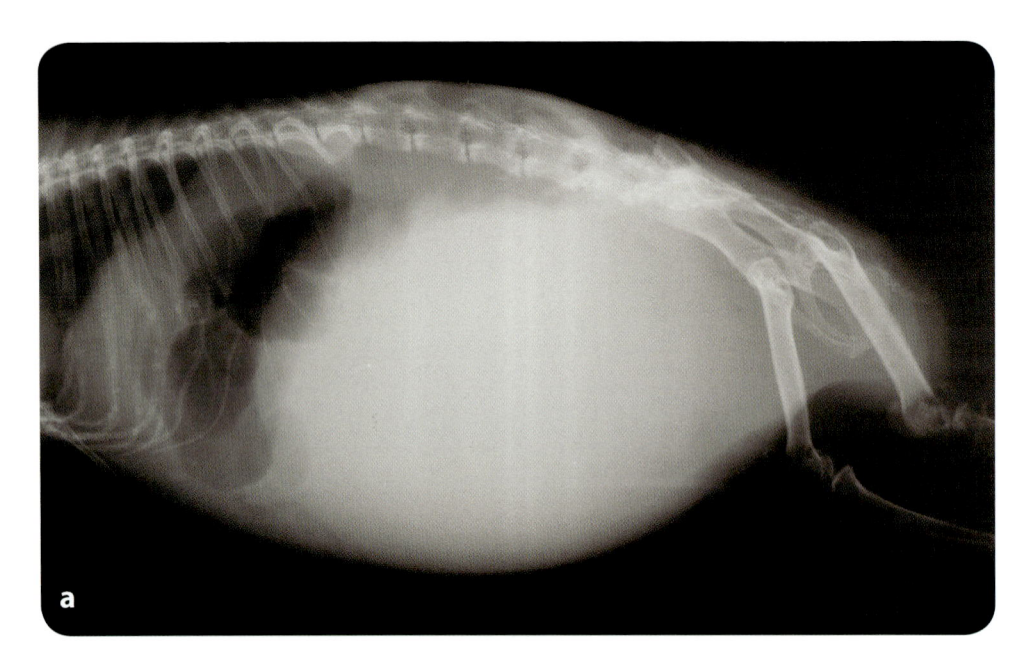

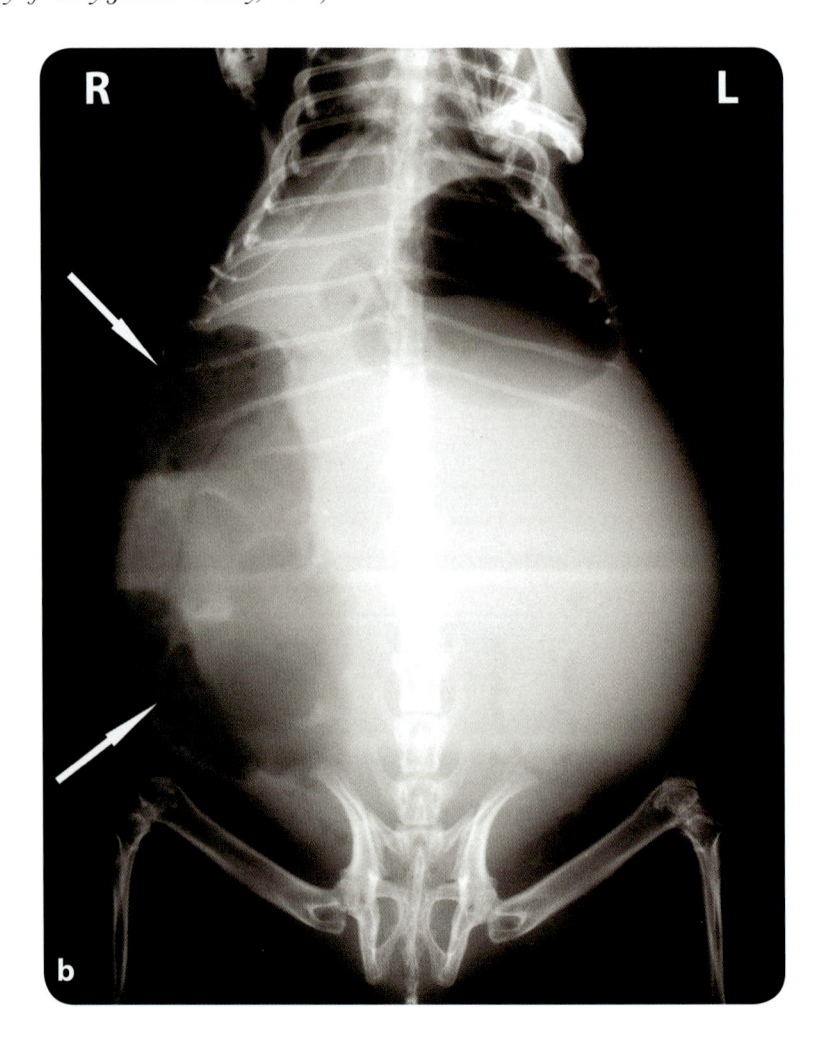

Figure 3.60a,b. Lateral (a) and ventrodorsal (b) radiographs of a guinea pig with a large ovarian cyst identified at surgery. The fluid-filled structure fills most of the abdominal cavity obscuring visualization of other organs. Note lateral and cranial displacement of the cecum (arrows).
(Courtesy of Cathy Johnson-Delaney, DVM)

ABDOMEN 207

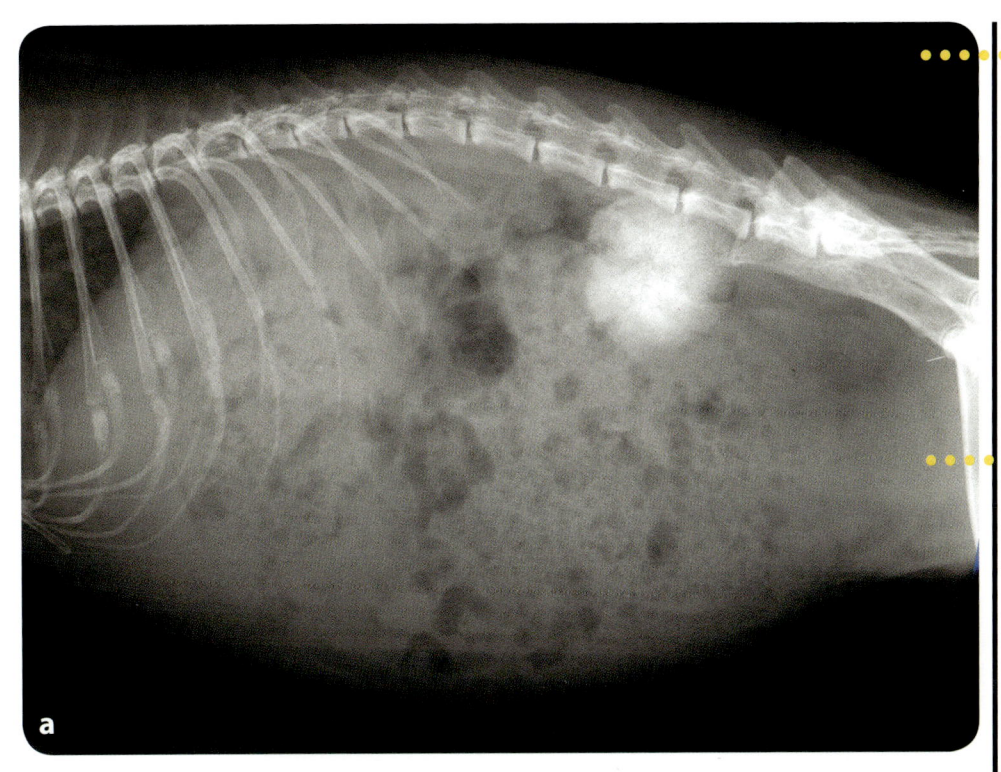

Figure 3.61a,b. Lateral (a) and ventrodorsal (b) radiographs of a 4-year-old intact female guinea pig determined to have uterine neoplasia. The mineralized abdominal mass is suggestive of uterine disease in older intact female guinea pigs. Distention of the cecum with food obscures the uterus, making radiographic diagnosis of uterine disease difficult. Ultrasonography is extremely useful in these cases.

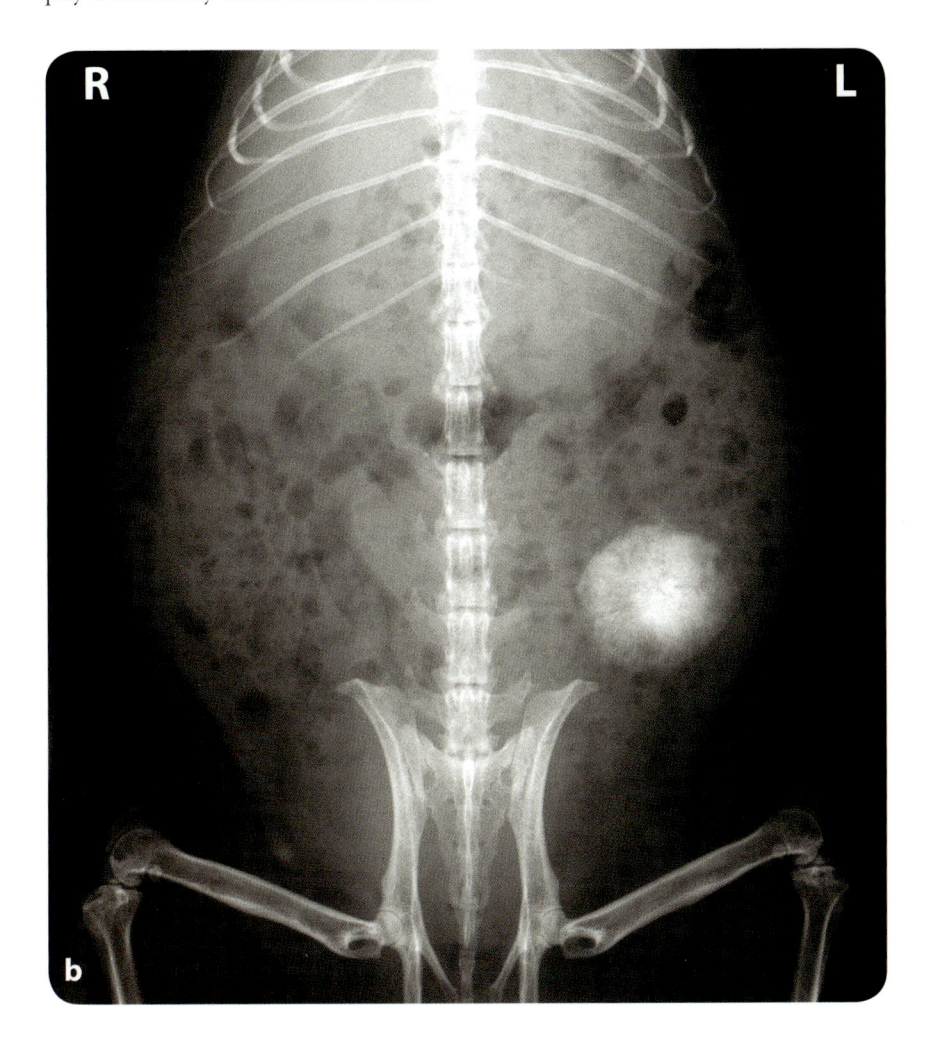

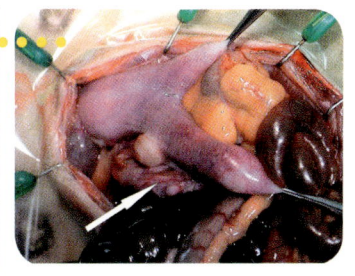

Figure 3.62. Intraoperative appearance of the uterus and associated mass (arrow)

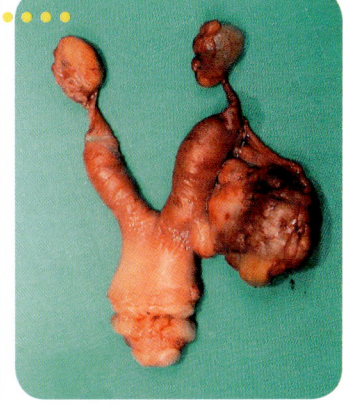

Figure 3.63. Appearance of the reproductive tract removed at surgery. Note ovarian cysts, enlarged uterine horns and a large mass associated with the left uterine horn. Histopathology revealed endometrial hyperplasia, and uterine and cervical adenocarcinoma.

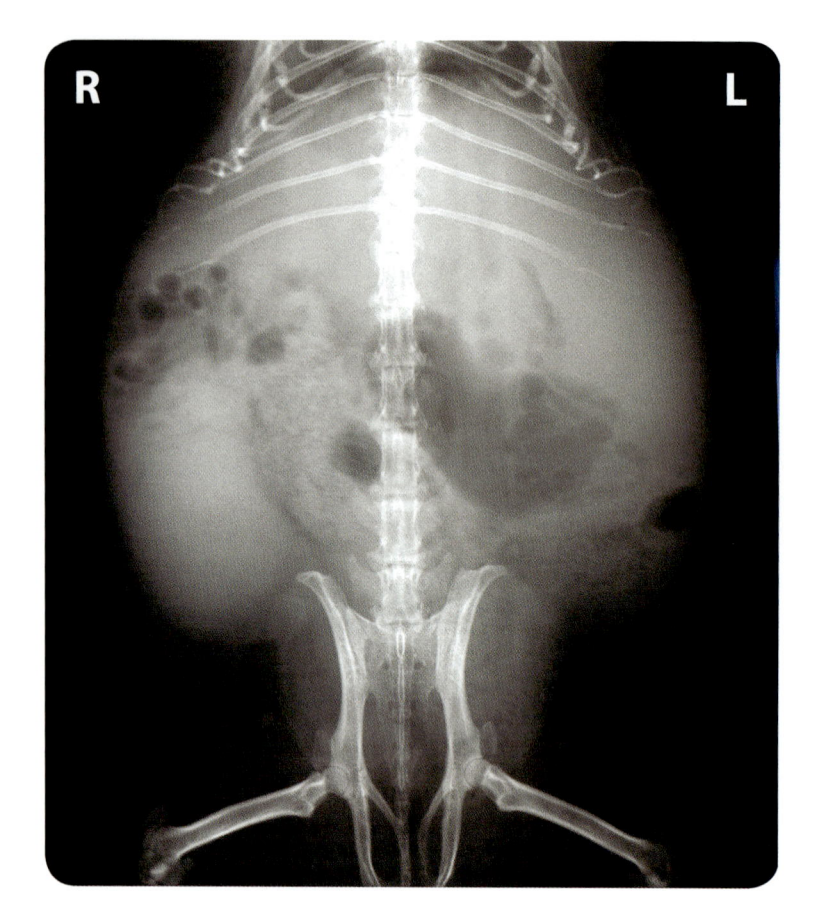

Figure 3.64. Ventrodorsal radiograph of a 3-year-old intact female guinea pig with enlargement of the right uterine horn. The uniform radiopacity visible in the cranial abdomen is indicative of peritoneal effusion.

Miscellaneous

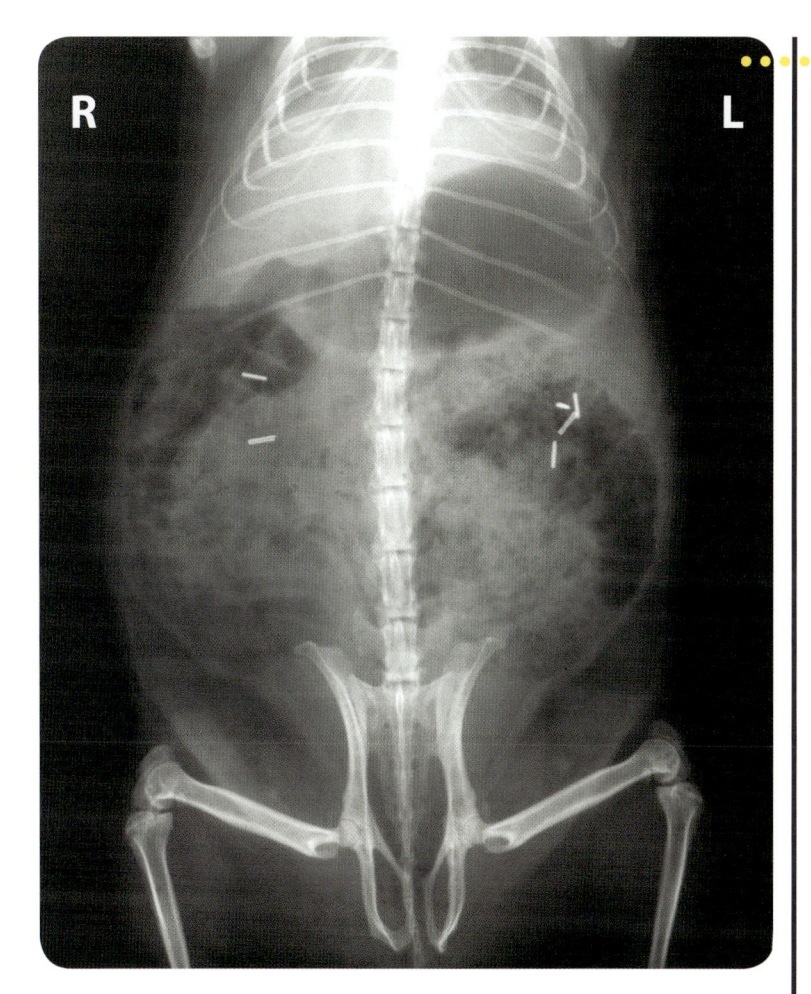

Figure 3.65. Ventrodorsal radiograph of the abdomen of a female guinea pig demonstrating the presence of metal hemostatic clips post ovariectomy.

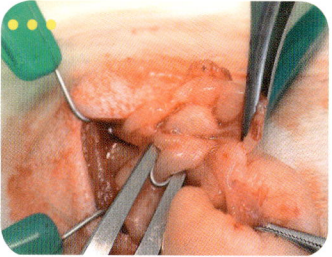

Figure 3.66. Positioning of a hemostatic clip during ovariectomy via a flank approach.

The NORMAL THORACIC LIMB
Lateral Projection

Due to varying soft tissue density, the superimposition of the thorax, and the relatively small size of the manus, it is impossible to obtain an optimal single radiograph of the entire thoracic limb. Different kVp settings are needed for the scapula and the humerus, and for the radioulnar segment and the manus.

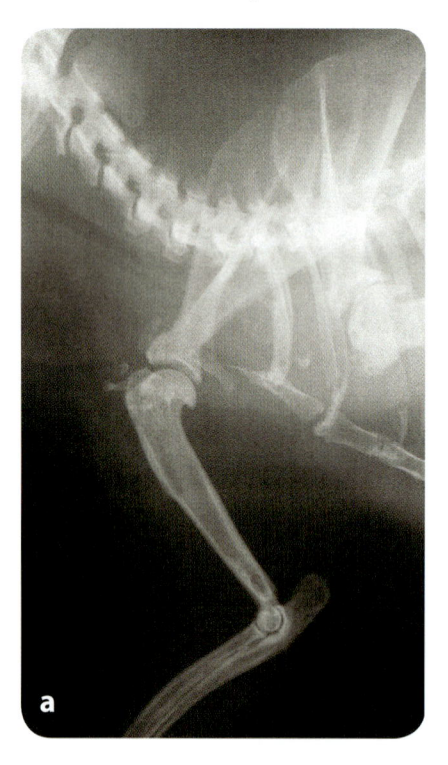

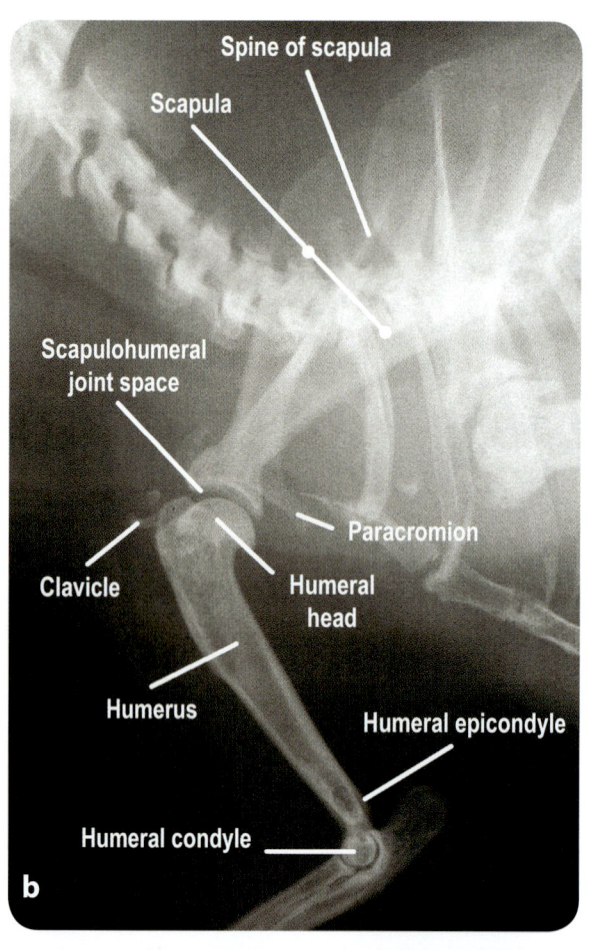

Figure 3.67a,b. Radiograph of the proximal thoracic limb of an 850 g male guinea pig, actual size (a) and 140% of actual size (b).

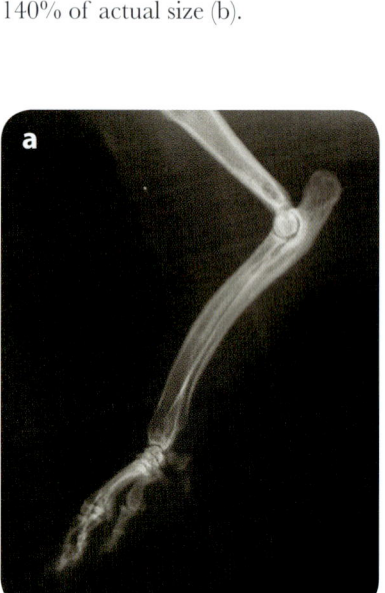

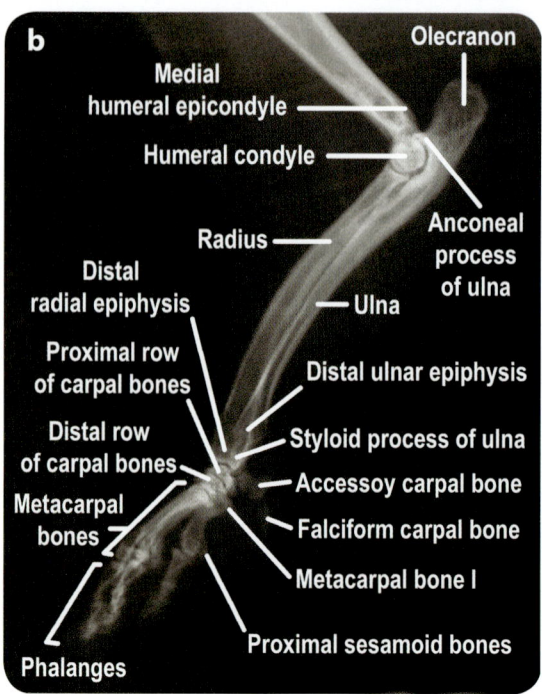

Figure 3.68a,b. Radiograph of the distal thoracic limb of an 850 g male guinea pig, actual size (a), and 140% of actual size (b).

Caudocranial Projection of the Proximal Thoracic Limb

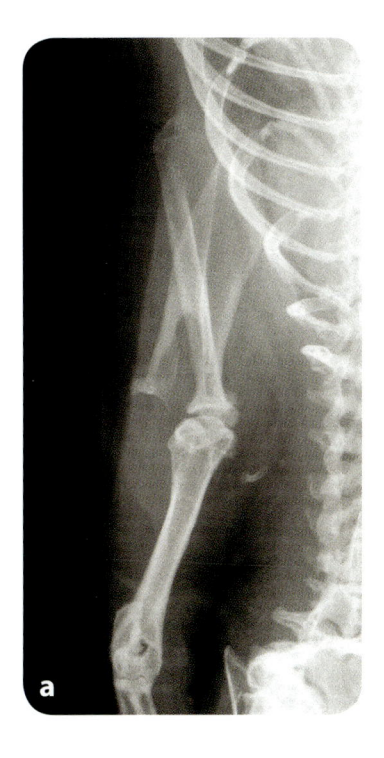

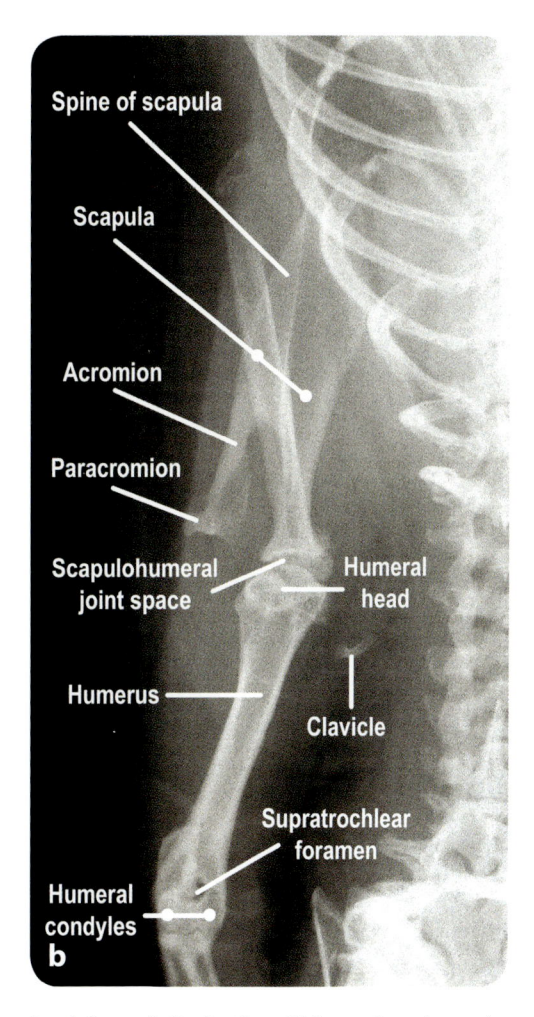

Figure 3.69a,b. Radiograph of the proximal thoracic limb of an 850 g male guinea pig, actual size (a) and 140% of actual size (b).

Craniocaudal Projection of the Distal Thoracic Limb

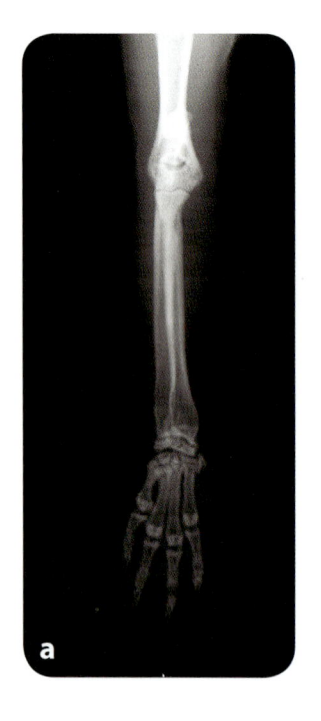

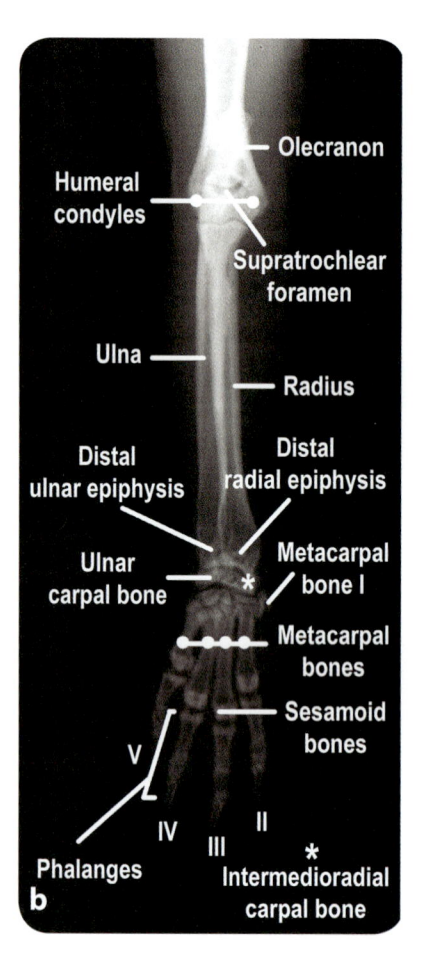

Figure 3.70a,b. Radiograph of the distal thoracic limb of an 850 g male guinea pig, actual size (a), and 140% of actual size (b).

ABNORMALITIES of the THORACIC LIMB

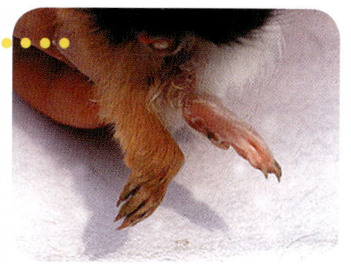

Figure 3.72. Clinical appearance of lateral deviation of the right carpus.

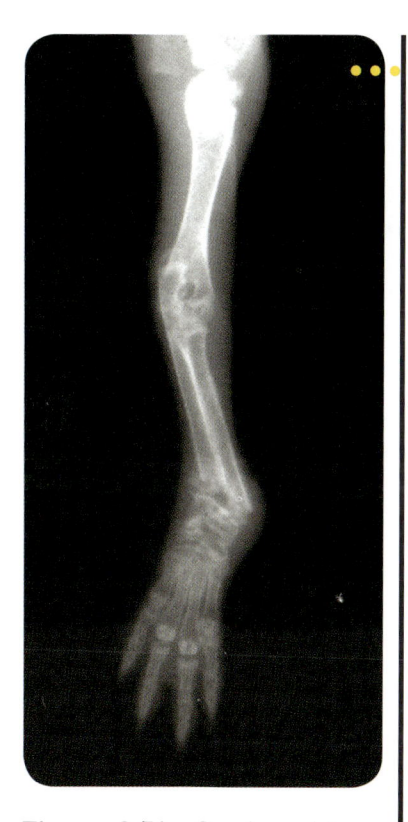

Figure 3.71. Craniocaudal radiograph of the thoracic limb of a guinea pig with osteodystrophy of the right carpus. This is likely secondary to hypovitaminosis C.

THE NORMAL PELVIC LIMB
Lateral Projection

Due to varying soft tissue density, superimposition of the abdomen, and the relatively small size of the distal pelvic limb, it is impossible to obtain an optimal single radiograph of the entire pelvic limb. Different kVp settings are required for the pelvis and the femur, and for the tibia and the distal limb.

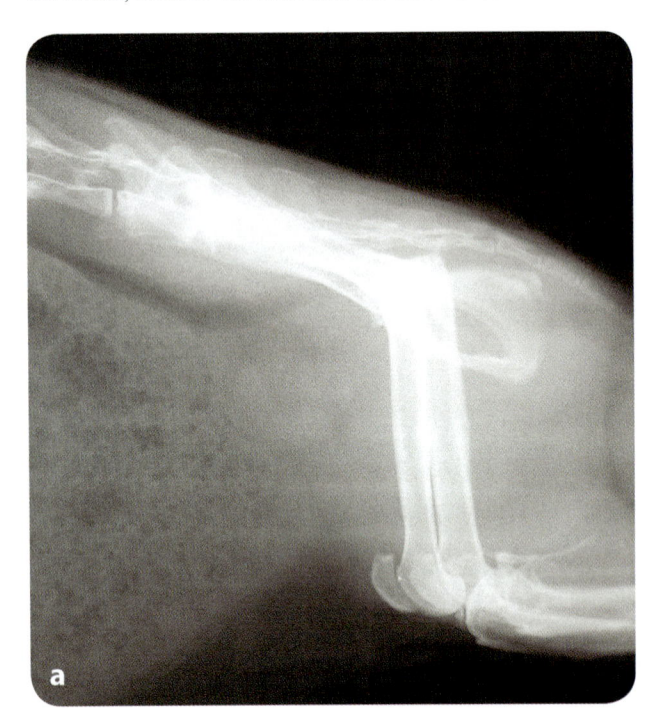

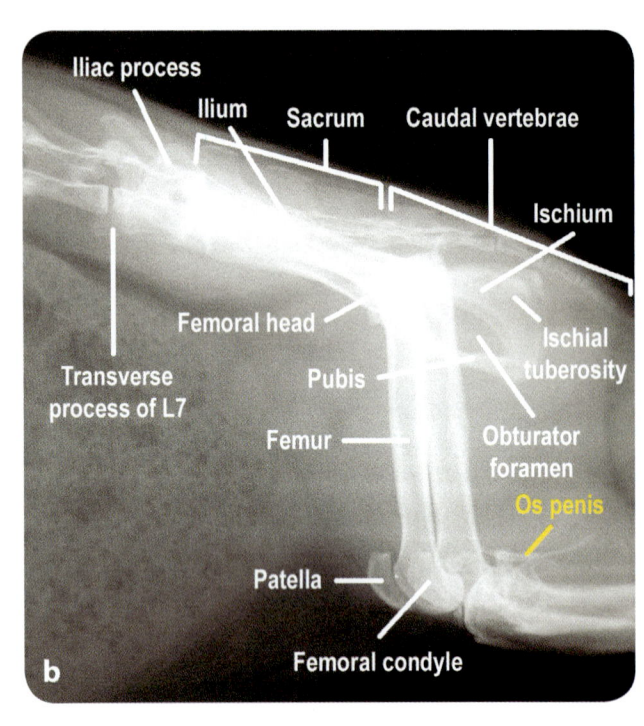

Figure 3.73a,b. Lateral radiograph of the pelvis of an 850 g male guinea pig, actual size.

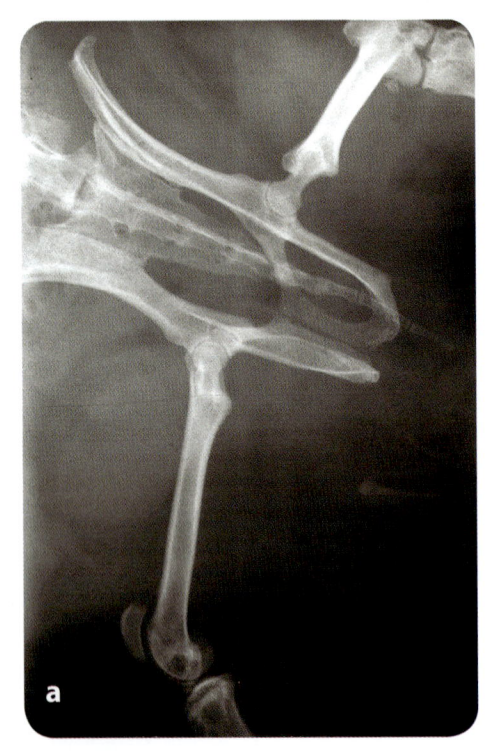

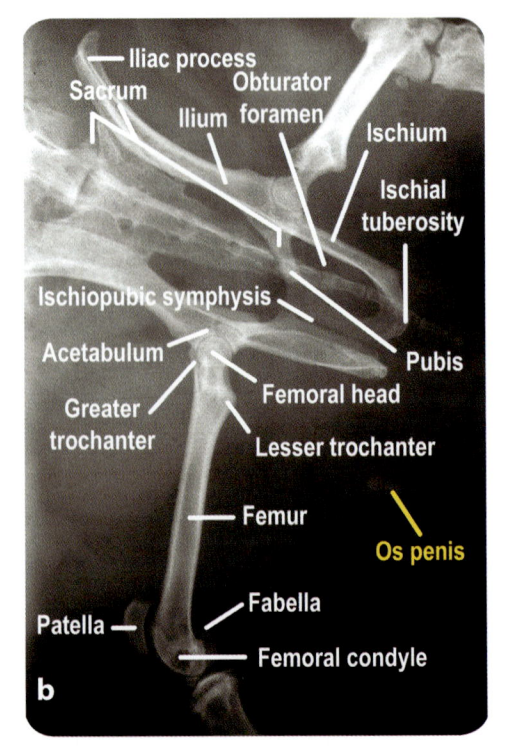

Figure 3.74a,b. Lateral radiograph of the femur and oblique projection of the pelvis of an 850 g male guinea pig, actual size.

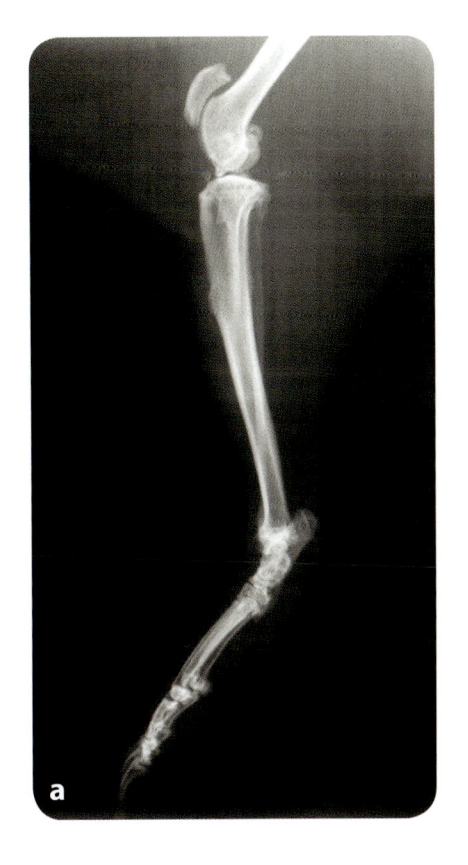

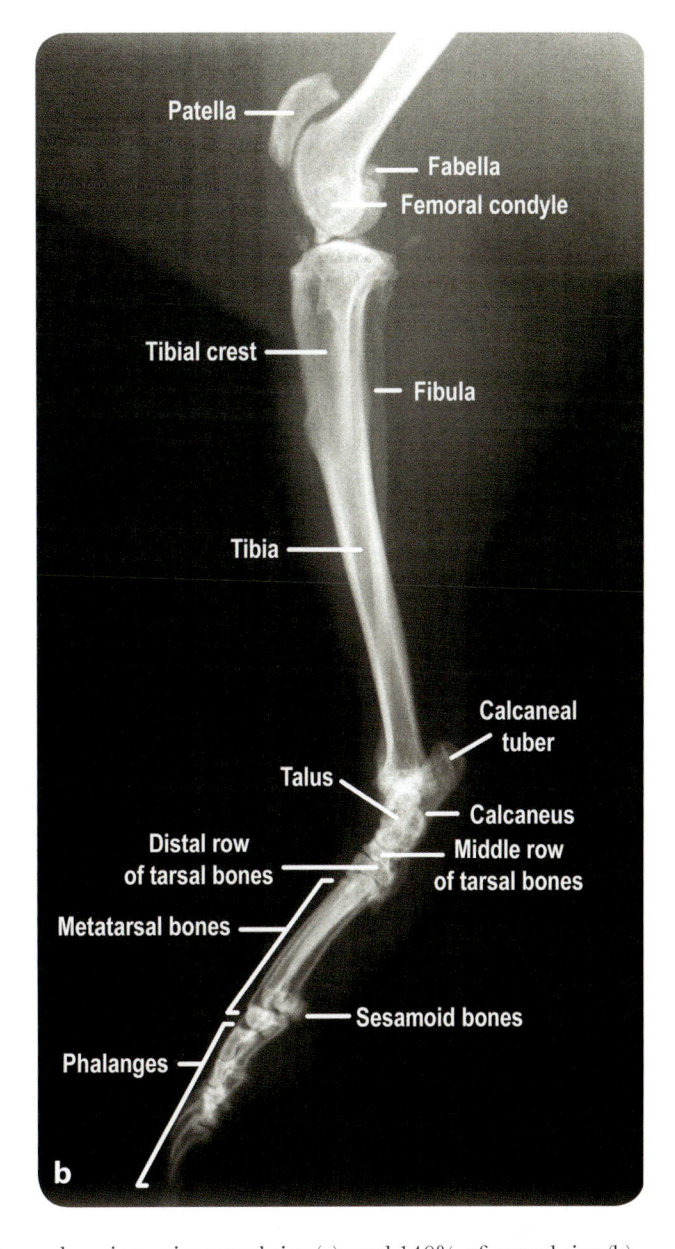

Figure 3.75a,b. Radiograph of the distal pelvic limb of an 850 g male guinea pig, actual size (a), and 140% of actual size (b).

Ventrodorsal Projection of the Pelvis
Craniocaudal Projection of the Proximal Pelvic Limb

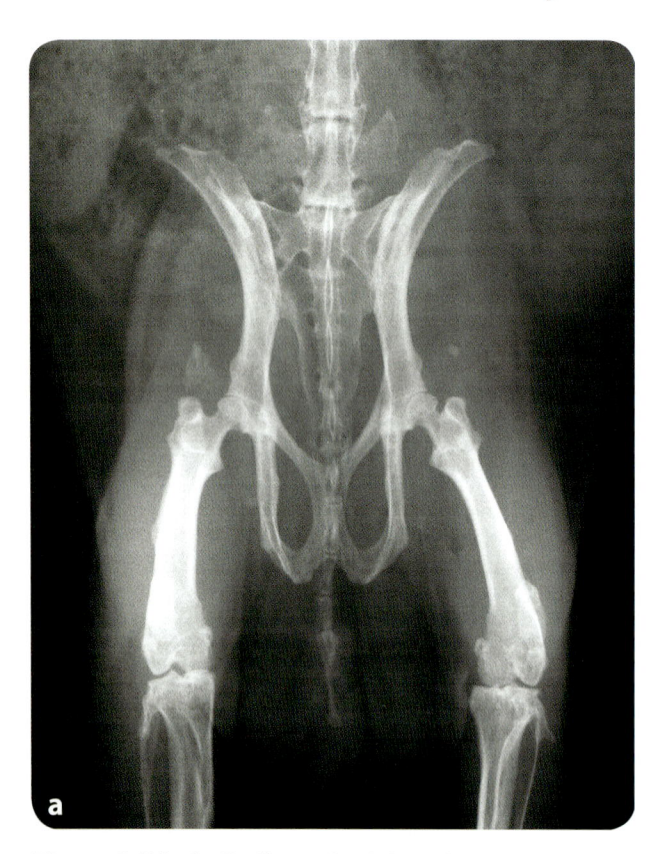

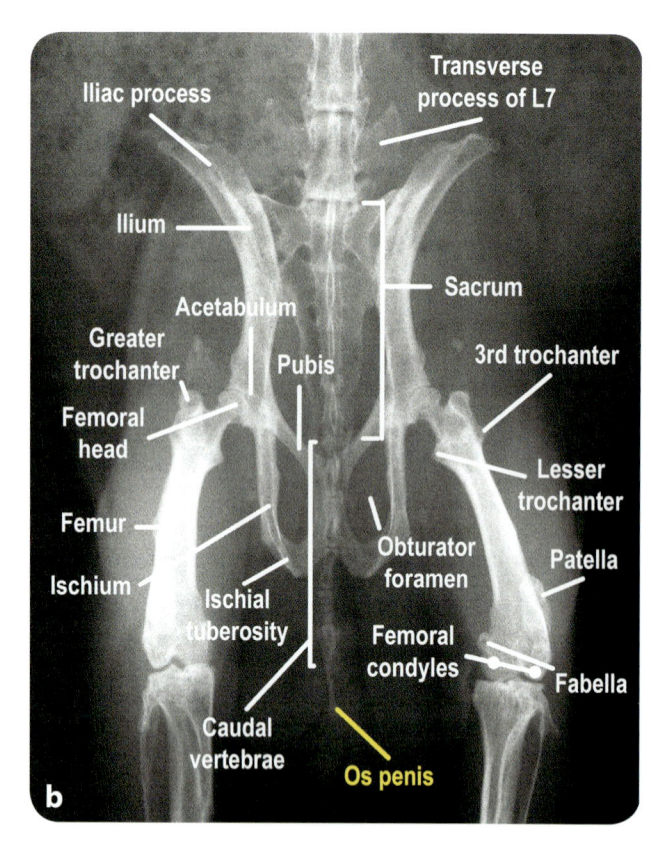

Figure 3.76a,b. Radiograph of the pelvis and the proximal pelvic limbs of an 850 g male guinea pig, actual size.

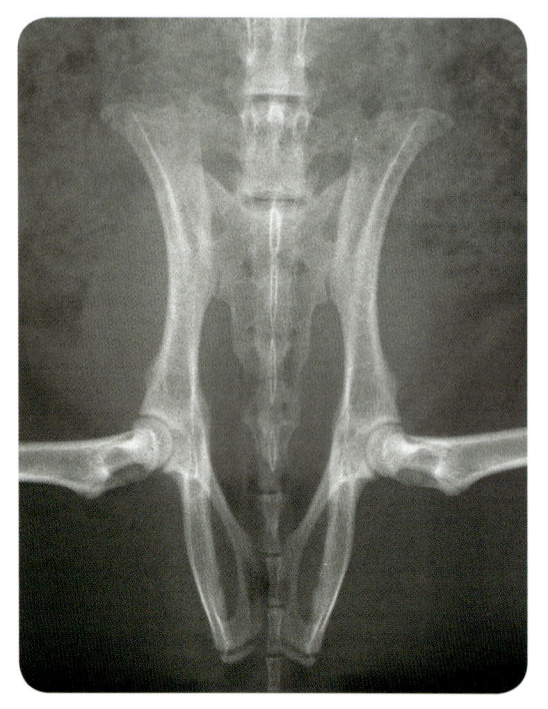

Figure 3.77. Pelvis of a 3-month-old female guinea pig, 150% of actual size. Note the open physis of the ischiatic tuberosities. In the female, the ischiopubic symphysis remains open until about 8 months of age. This is difficult to identify in a symmetrical ventrodorsal projection due to superimposition of the caudal vertebrae.

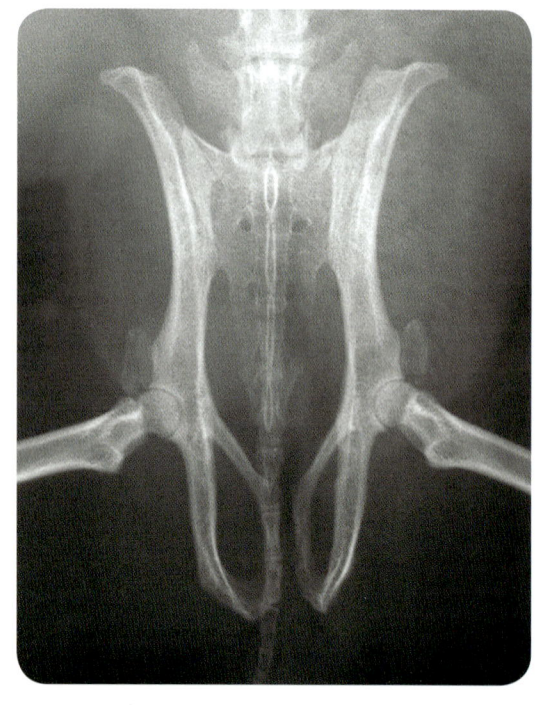

Figure 3.78. This slightly oblique projection of the same patient emphasizes the open ischiopubic symphysis. It remains open throughout life if the guinea pig gives birth before approximately 8 months of age.

Craniocaudal Projection of the Distal Pelvic Limb

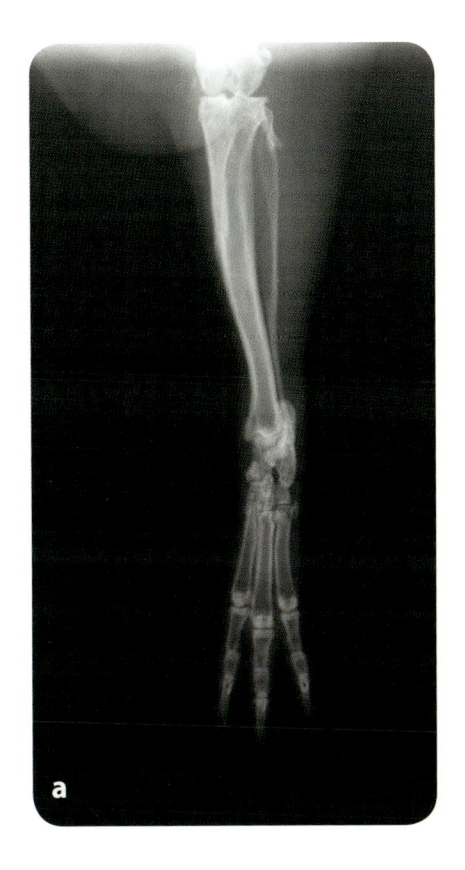

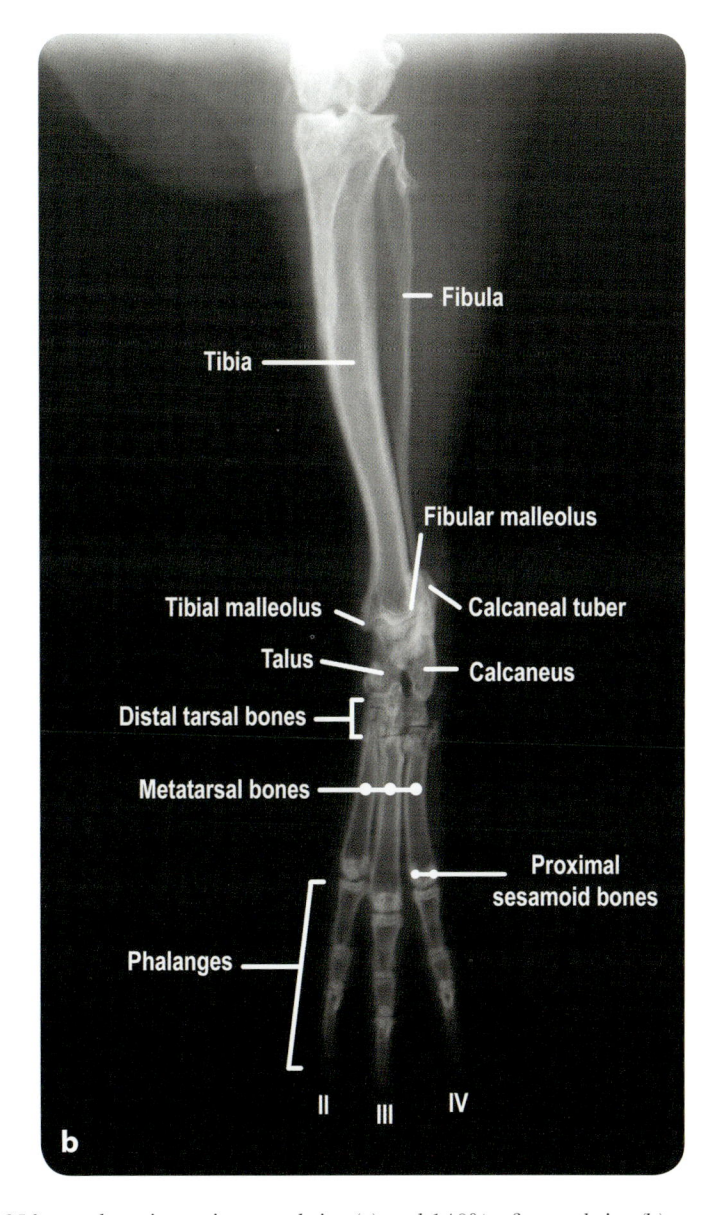

Figure 3.79a,b. Radiograph of the distal pelvic limb of an 850 g male guinea pig, actual size (a) and 140% of actual size (b).

ABNORMALITIES of the PELVIC LIMB
Diseases of the Femur

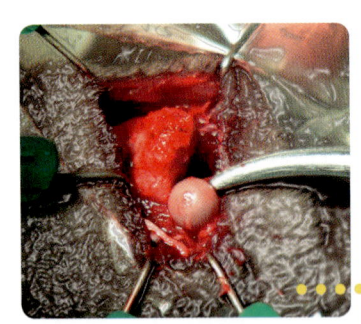

Figure 3.80. Femoral head ostectomy in the same patient.

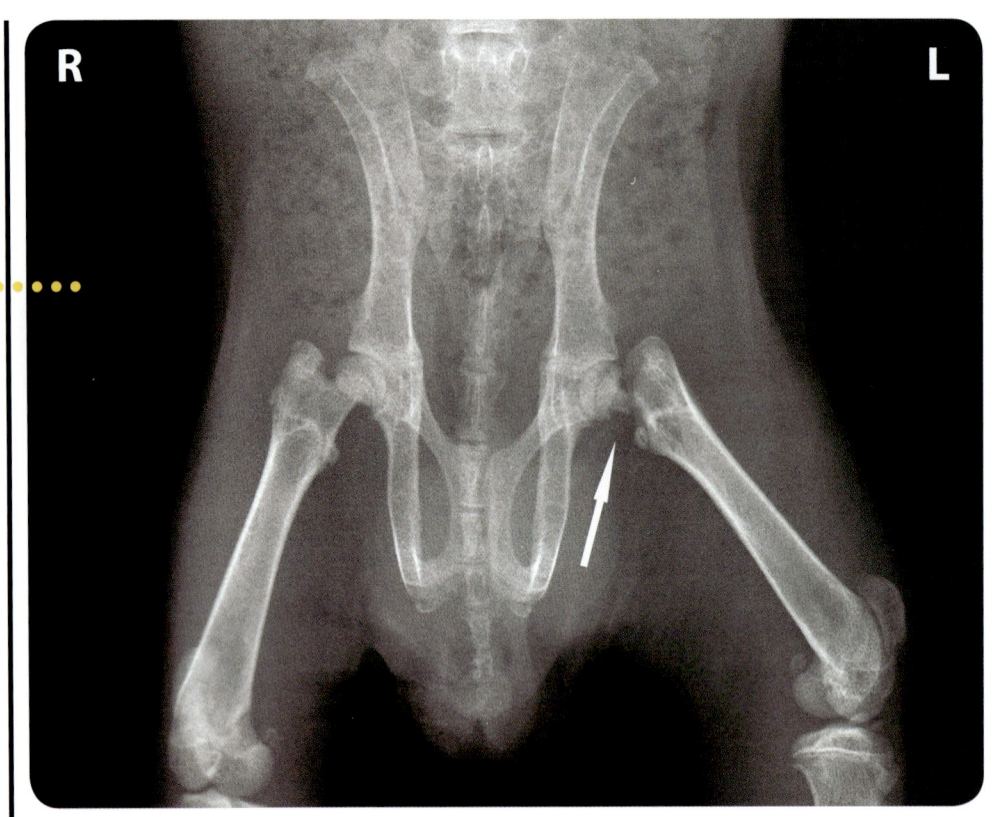

Figure 3.81. Ventrodorsal radiograph of the pelvis of a young guinea pig with a fracture of the left femoral head and neck as a result of an unknown traumatic episode (arrow). The guinea pig could bear weight but was significantly lame on the affected limb.

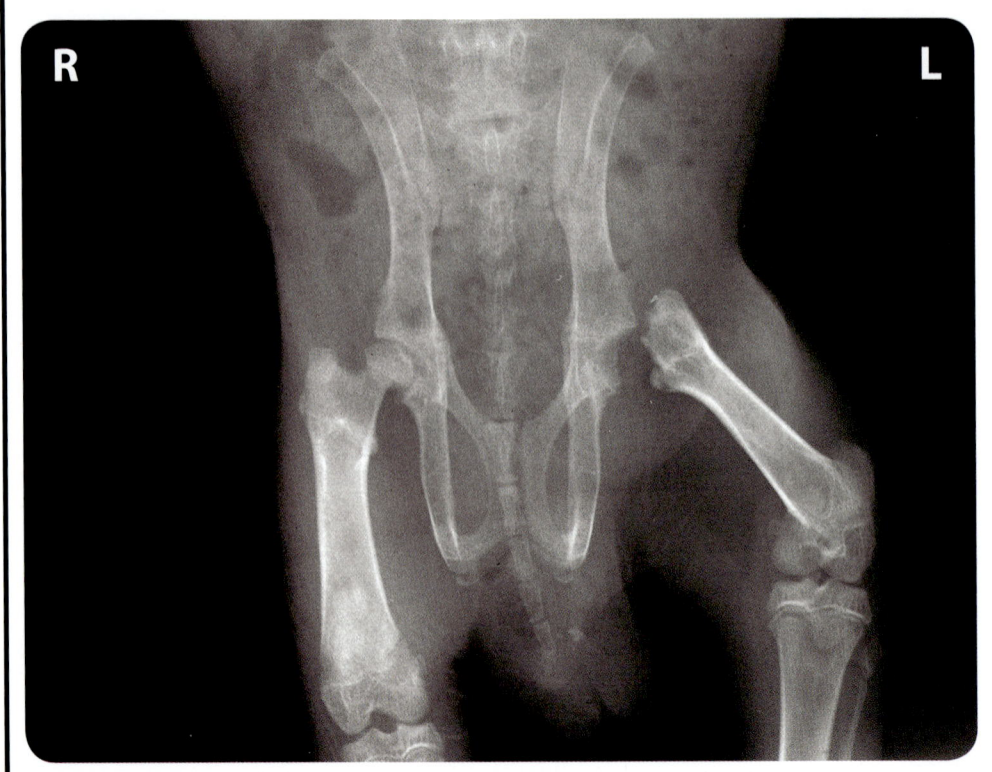

Figure 3.82. Same patient after femoral head ostectomy. The guinea pig was much improved at one week post surgery, and ambulating normally at three weeks.

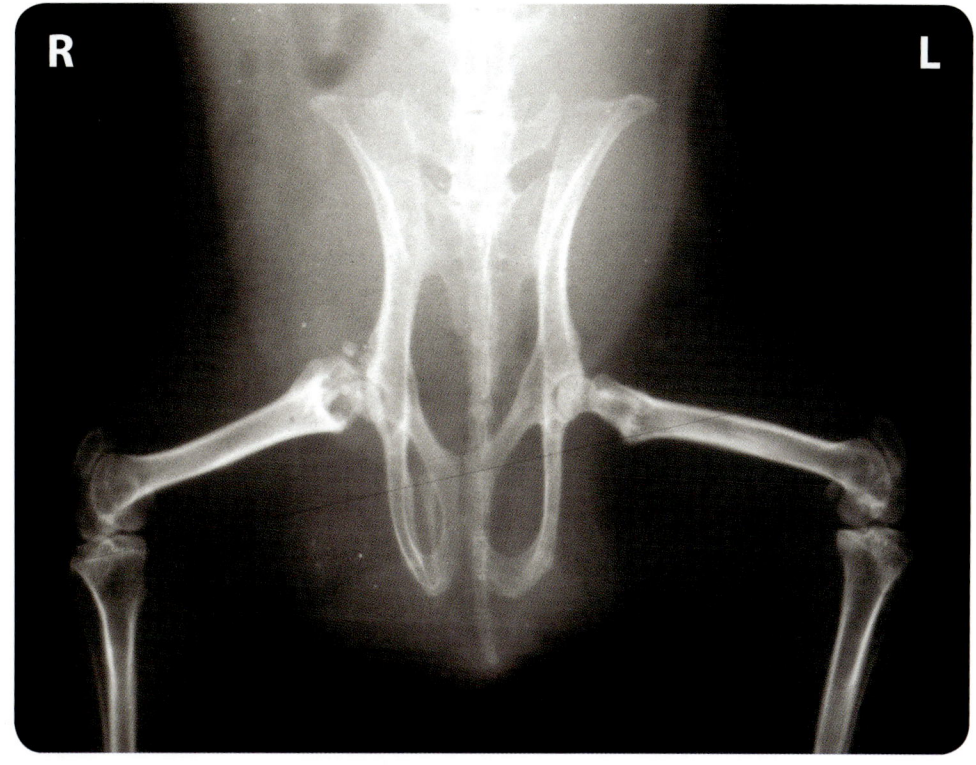

Figure 3.83. Ventrodorsal radiograph of the pelvis of a guinea pig demonstrating previous fracture of the right femoral head and neck as a result of trauma. Note evidence of arthropathy and osteolysis.

Diseases of the Stifle Joint

Abnormalities of the stifle joint are common in guinea pigs, and may be related to vitamin C deficiency.

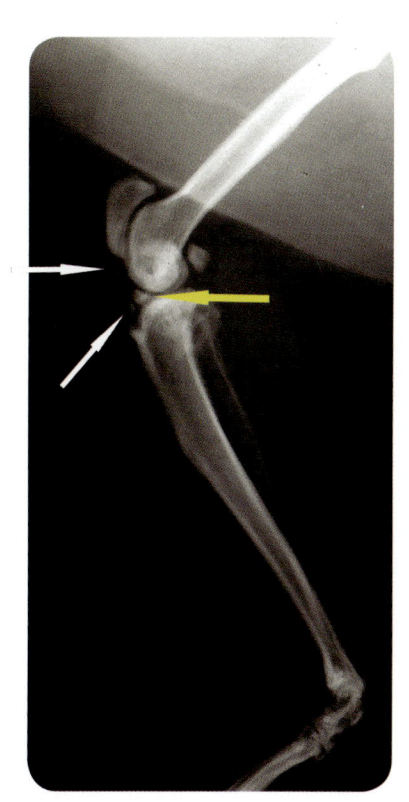

Figure 3.84. Ossification of the meniscus (yellow arrow) and of the ligament of patella (white arrows).

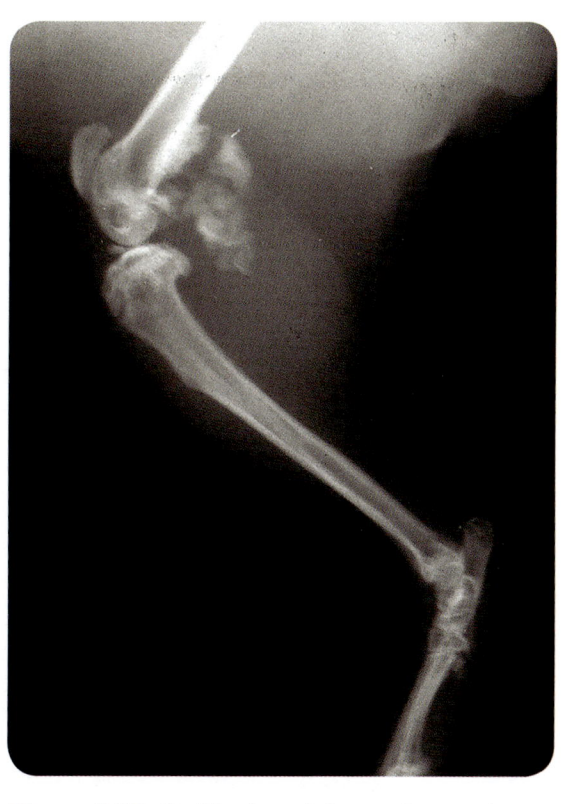

Figure 3.85. Ossification of the meniscus and proximal tendons of the gastrocnemius muscle.

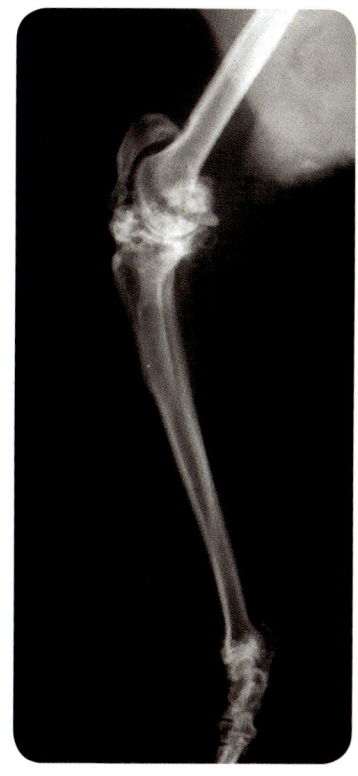

Figure 3.86. Severe degenerative arthropathy and ankylosis of the stifle joint.

Diseases of the Tibia and Fibula

Figure 3.87. External fixation is the treatment of choice for tibial fractures. Guinea pigs tolerate this method of fixation very well.

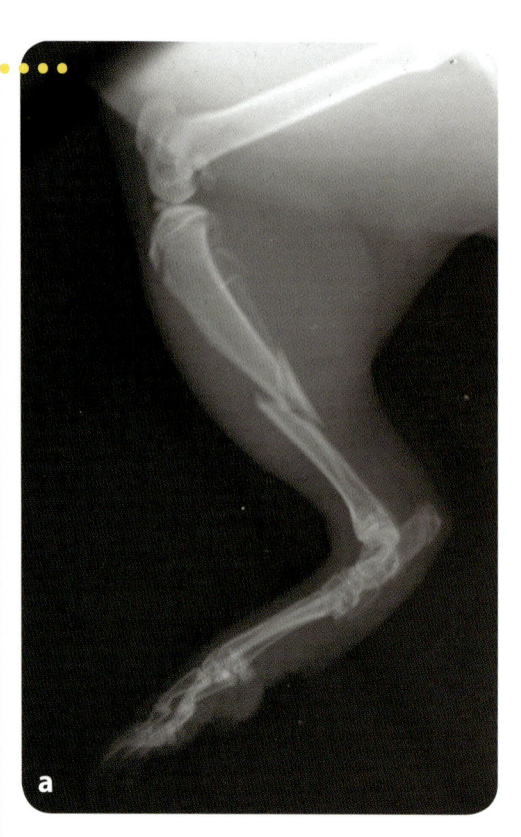

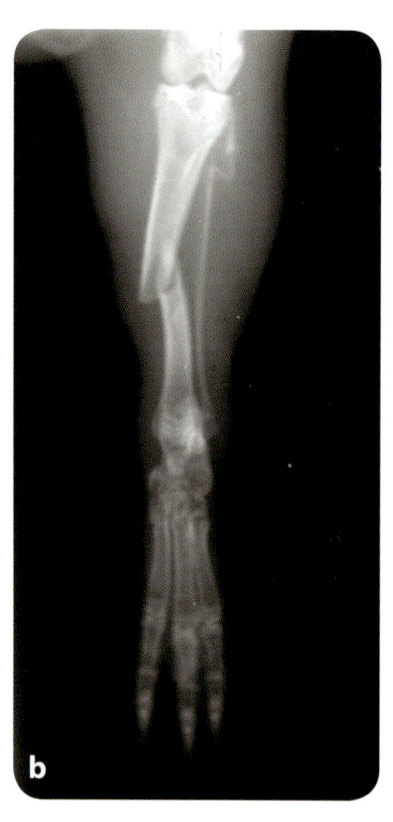

Figure 3.88a,b. Lateral (a) and craniocaudal (b) radiographs of a guinea pig with an oblique, comminuted fracture of the tibia and fracture of the proximal fibula.

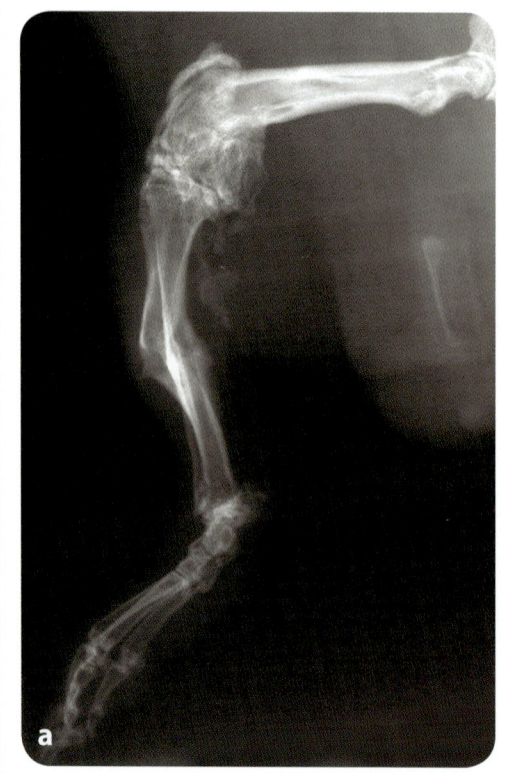

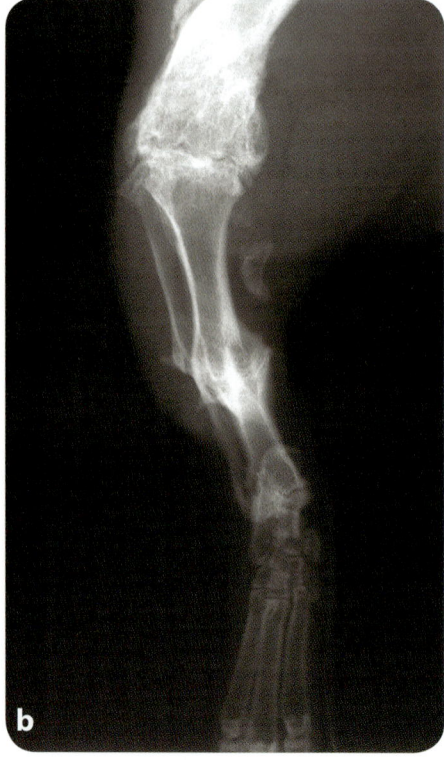

Figure 3.89. Lateral (a) and craniocaudal (b) radiographs of a guinea pig showing spontaneous bone healing of midshaft fractures of the tibia and the fibula. Prolonged lameness and disuse led to degenerative arthropathy and ankylosis of the stifle joint.

Diseases of the Tarsus, Metatarsus, and Phalanges

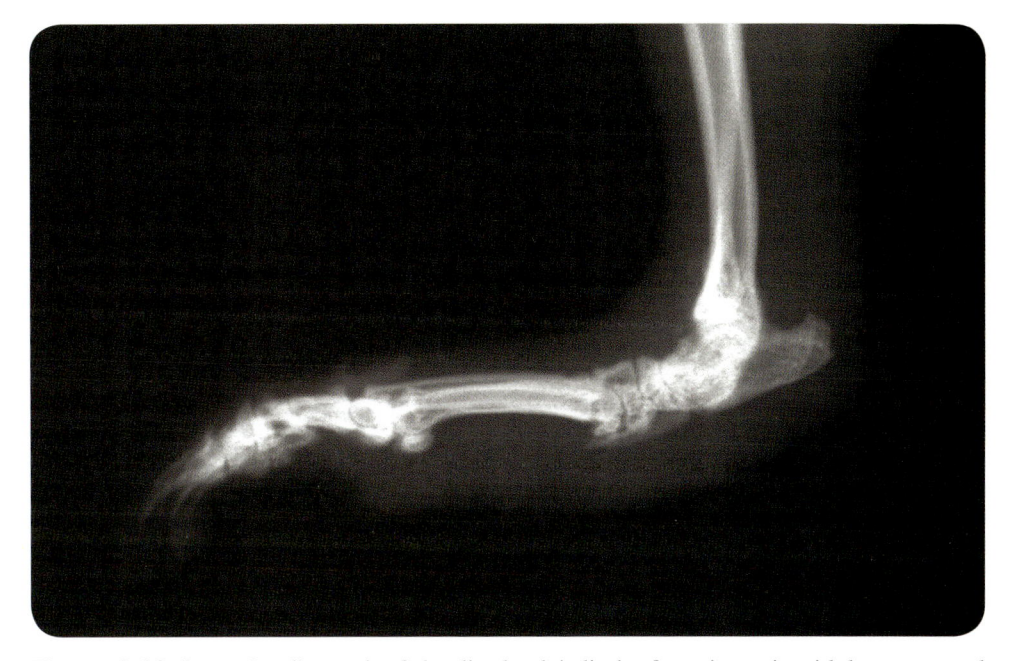

Figure 3.90. Lateral radiograph of the distal pelvic limb of a guinea pig with lameness and pain upon palpation of the distal limb. Note mineralization of the flexor tendons of the phalanges. The cause of this type of lesion is unknown.

References:
  Brown SA, Rosenthal LR: Question #169. In: Self Assessment Color Review of Small Mammals. London: Manson Publishing; 1997: 145-6.
  Brown SA, Rosenthal LR: Question #209. In: Self Assessment Color Review of Small Mammals. London: Manson Publishing; 1997:181-2.
  Capello V, Gracis M: Radiology of the skull and teeth. In: Lennox A, ed. Rabbit and Rodent Dentistry Handbook. Ames, IA:Blackwell Publishing, (Formerly Zoological Education Network, Lake Worth, FL); 2005:65-99.
  Capello V: Dental diseases. In: Lennox A, ed. Rabbit and Rodent Dentistry Handbook. Ames, IA:Blackwell Publishing, (Formerly Zoological Education Network, Lake Worth, FL); 2005:113-163
  Capello V: Secondary diseases. In: Lennox A, ed. Rabbit and Rodent Dentistry Handbook. Ames, IA:Blackwell Publishing, (Formerly Zoological Education Network, Lake Worth, FL); 2005:165-185.
  Capello V, Gracis M: Dental procedures. In: Lennox A, ed. Rabbit and Rodent Dentistry Handbook. Ames, IA:Blackwell Publishing, (Formerly Zoological Education Network, Lake Worth, FL); 2005:213-247.
  Capello V, Gracis M: Surgical treatment of periapical abscessations. In: Lennox A, ed. Rabbit and Rodent Dentistry Handbook. Ames, IA:Blackwell Publishing, (Formerly Zoological Education Network, Lake Worth, FL); 2005:249-273.
  Capello V: Dental diseases and surgical treatment in pet rodents. Exotic DVM. 2003;5(3):21-27.
  Crossley DA: Clinical aspects of rodent dental anatomy. J Vet Dent. 1995;12(4):131-135.
  Eatwell K: Ovarian and uterine disease in guinea pigs: a review of 5 cases. Exotic DVM. 2003;5(5):37-39.
  Girling SJ: Mammalian imaging and anatomy. In: Meredith A, Redrobe S, eds. BSAVA Manual of Exotic Pets 4th ed. Quedgeley, Glouchester: British Small Animal VeterinaryAssociation; 2005:1-12.
  Hawkins MG: Diagnostic evaluation of urinary tract calculi in guinea pigs. Exotic DVM. 2006; 8(3):43-47.
  Hoefer HL: Guinea pig urolithiasis. Exotic DVM. 2004;6(2):23-25.
  Legendre LFJ: Oral disorders of exotic rodents. Vet Clin N Am Exotic Anim Prac. 2003;6:601-628.
  Popesko P, Rijtovà V, Horàk J: A Colour Atlas of Anatomy of Small Laboratory Animals. Vol. I: Rabbit, Guinea Pig. Bratislava: Príroda Publishing House; 1990.
  Ruelokke ML, Arnbjerg J: Retrobulbar abscess secondary to molar overgrowth in a guinea pig. Exotic DVM. 2003;5(2):10-16.
  Ruelokke ML, Arnbjerg J, Martensen MR: Assessing gastrointestinal motility in guinea pigs using contrast radiography. Exotic DVM. 2003;5(2):10-16.
  Silverman S, Tell LA: Domestic guinea pig *(Cavia porcellus)*. In: Radiology of Rodents, Rabbits and Ferrets. An Atlas of Normal Anatomy and Positioning. Philadelphia, PA: Elsevier Saunders; 2005:105-157.
  Stefanacci JD, Hoefer HL: Radiology and ultrasound. In: Quesenberry KE, Carpenter JW, eds. Ferrets, Rabbits and Rodents. Clinical Medicine and Surgery 2nd ed. Philadelphia, PA:Elsevier;2004:395-413.

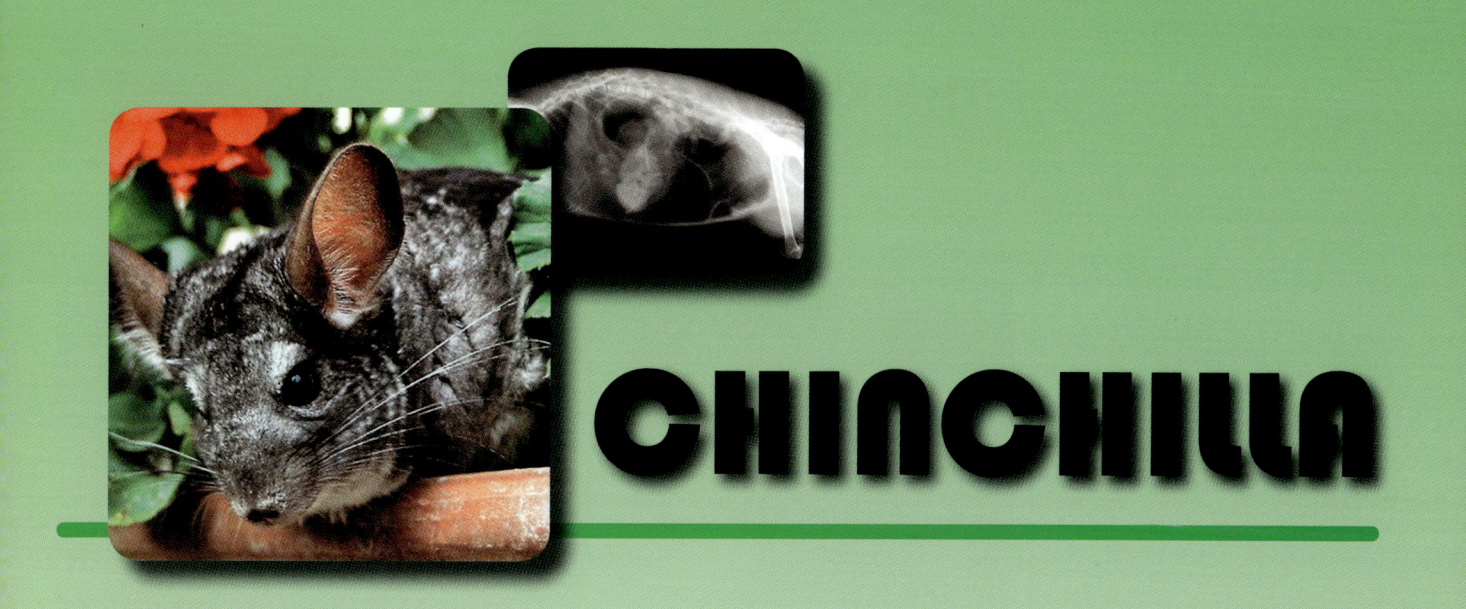

CHINCHILLA

The NORMAL HEAD
Lateral Projection

Figure 4.1a,b. 180% of actual size. The prominent tympanic bullae occupy most of the caudal aspect of the skull. At rest, cheek teeth are in occlusion, while maxillary and mandibular incisors are slightly separated. On a lateral radiograph, normal cheek teeth meet in a flat, horizontal occlusal plane, and form a regular, smooth palisade. Cheek teeth are relatively short, especially the most caudal mandibular and maxillary cheek teeth. The apexes of the first three mandibular cheek teeth are close to the ventral cortex of the mandible. Mandibular incisor teeth are very long and slightly curved. Their roots are positioned lingually to the cheek teeth, and their apex reaches the second or third cheek tooth (red circle).

The maxillary incisor teeth are even more curved, nearly forming a 1/2 circle. Their apex is located near the palate, about 2/3 the length of the maxillary diastema (yellow circle).

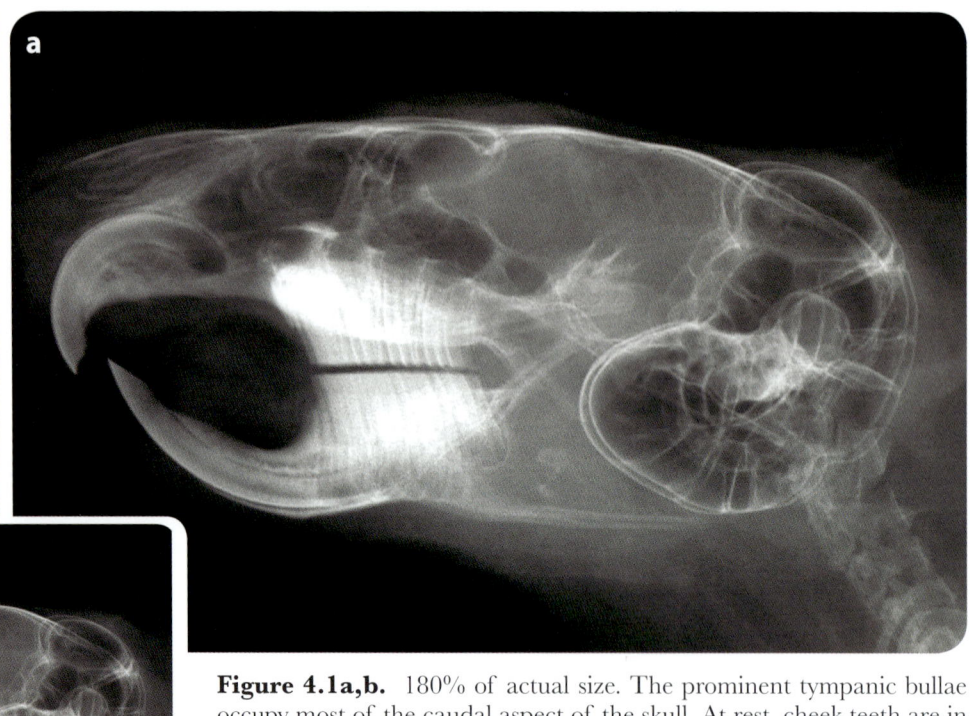

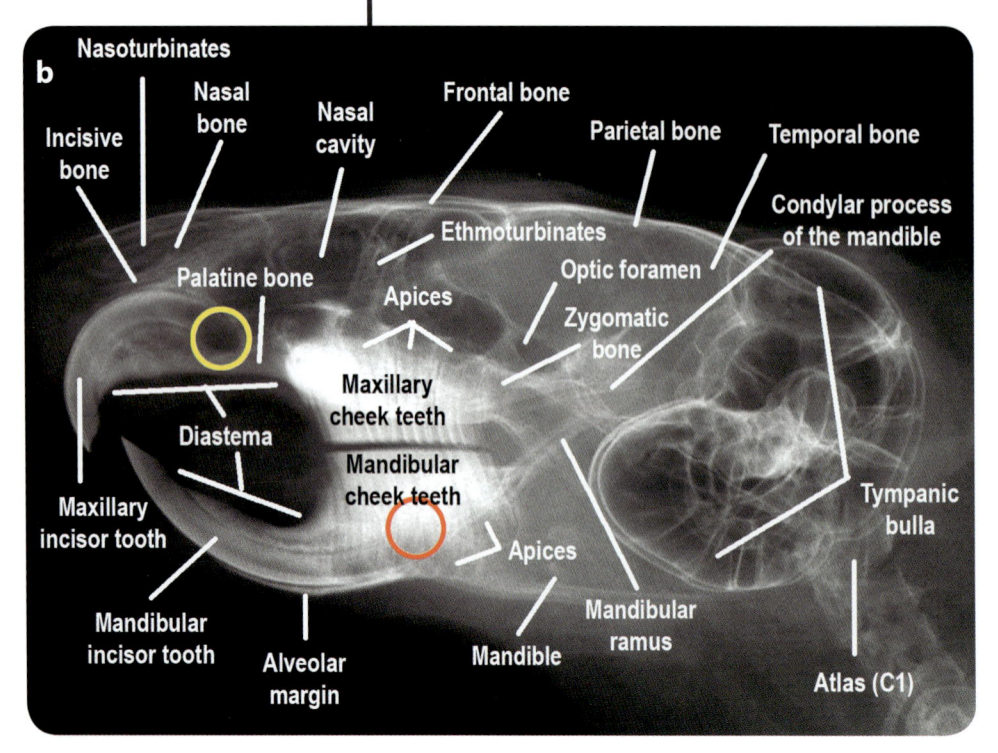

Figure 4.2. Actual size of the skull of a 500 g male chinchilla.

Oblique Projection

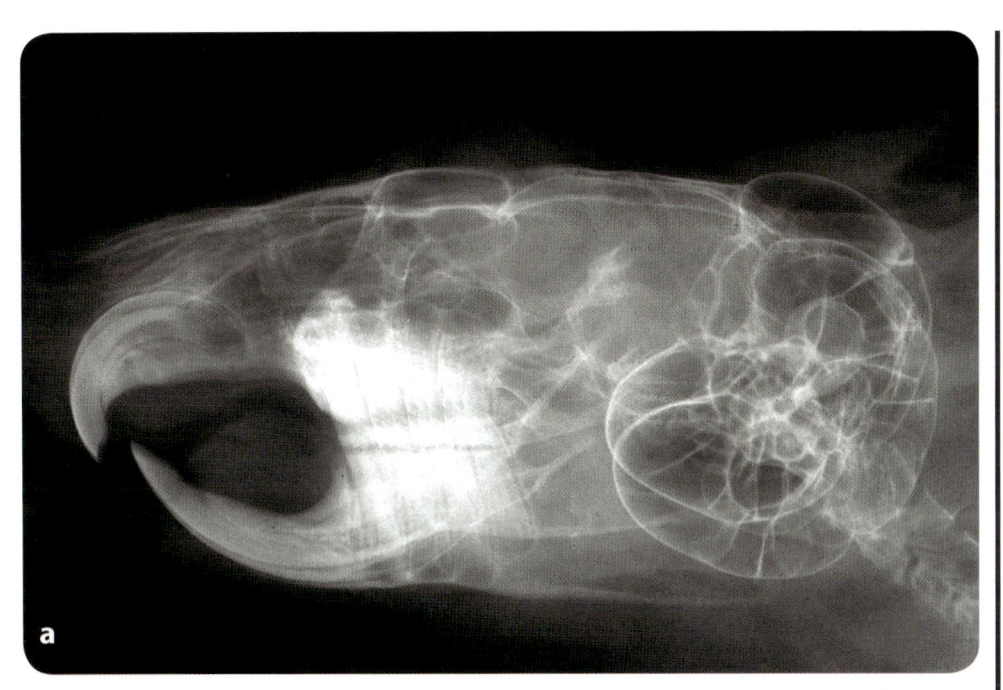

Figure 4.3a,b. Oblique views allow better visualization of the mandibular cheek teeth roots of one side and the maxillary cheek teeth roots of the opposite side. This view is also useful for evaluation of the apex of the incisor teeth (red circle). The contralateral view should always be obtained to compare findings.

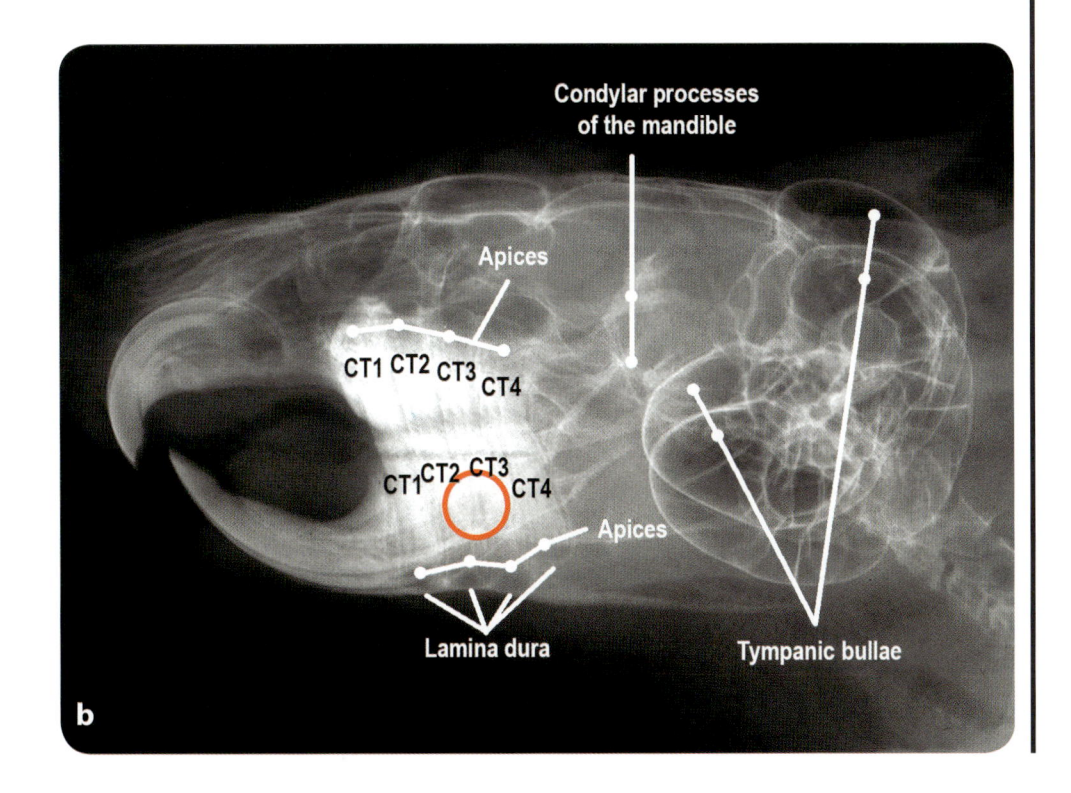

Ventrodorsal Projection

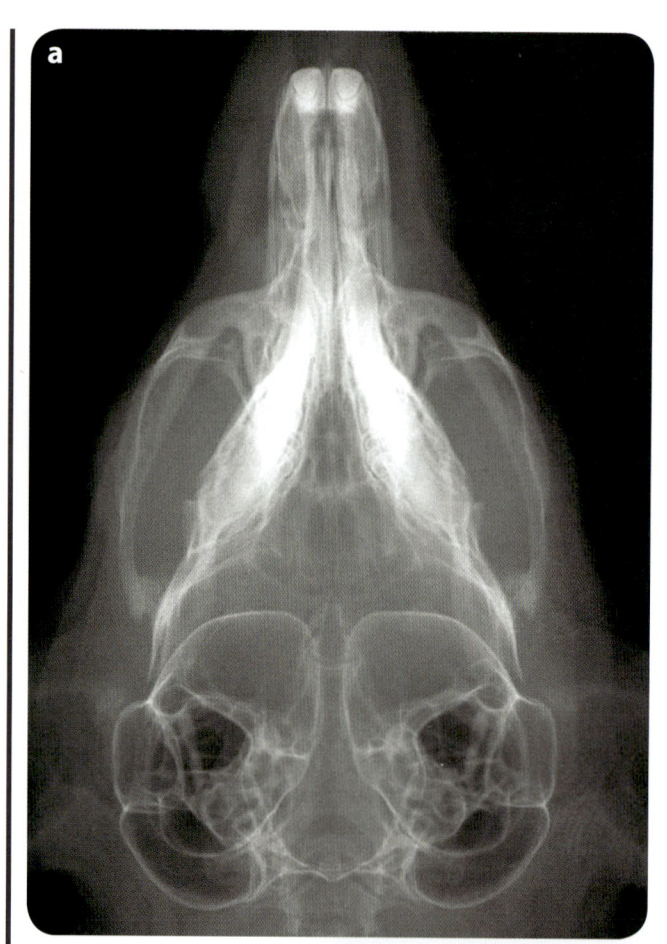

Figure 4.4a,b. Ventrodorsal radiographs of the skull of a chinchilla.

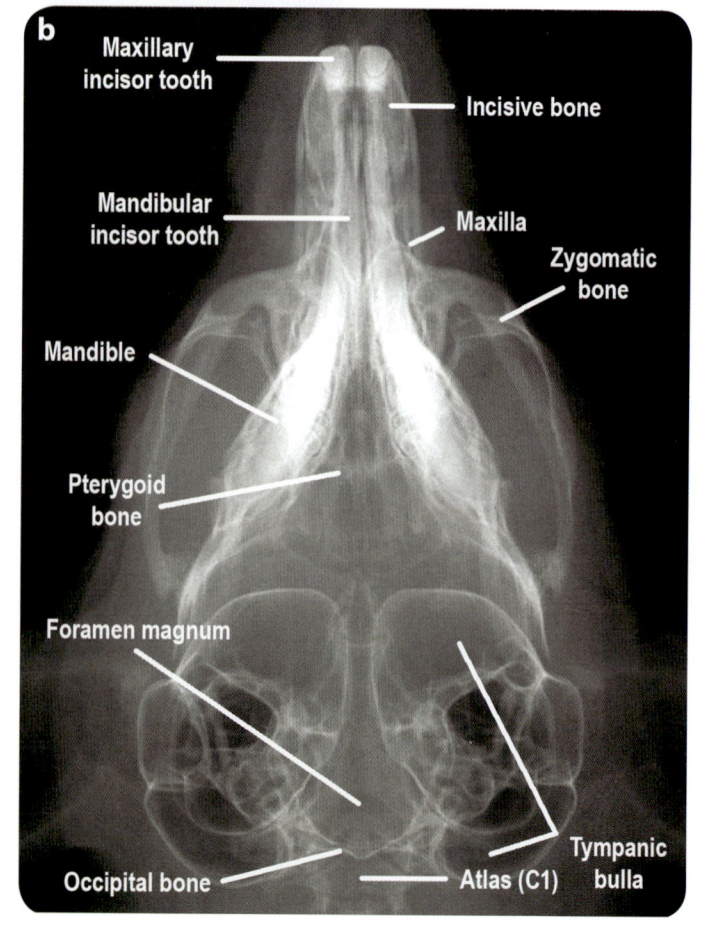

Rostrocaudal Projection

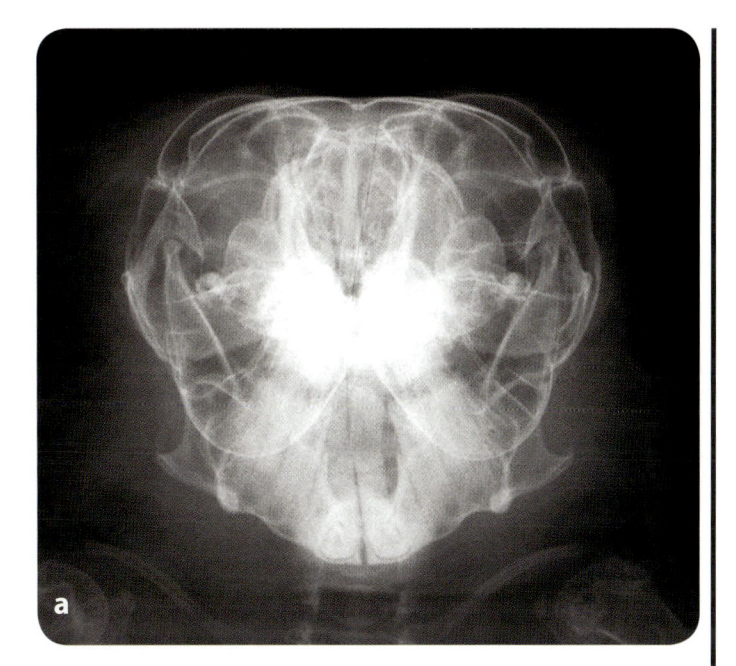

Figure 4.5a,b. Note the proximity of the mandibular cheek teeth to the ventral cortex of the mandible. The occlusal plane of the cheek teeth is nearly horizontal, with a slight ventro-lingual slope. The temporomandibular joint and the mandibular symphysis are clearly visible.

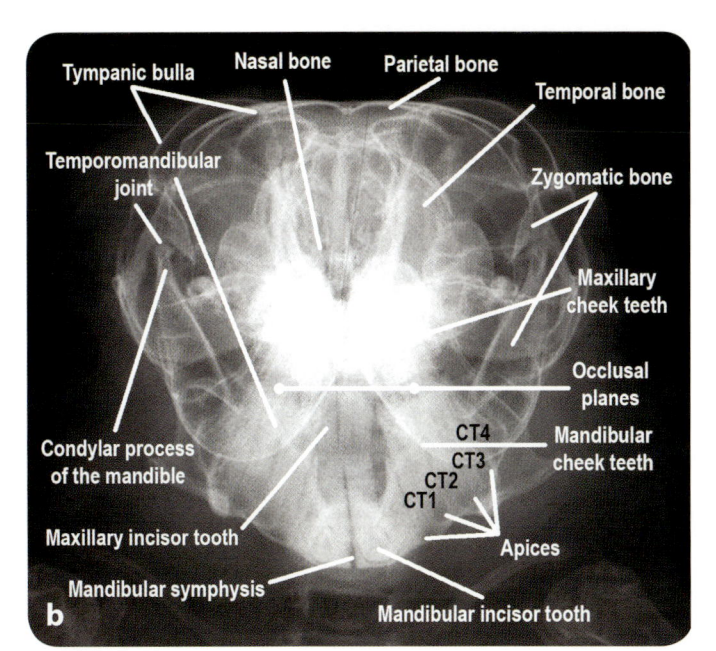

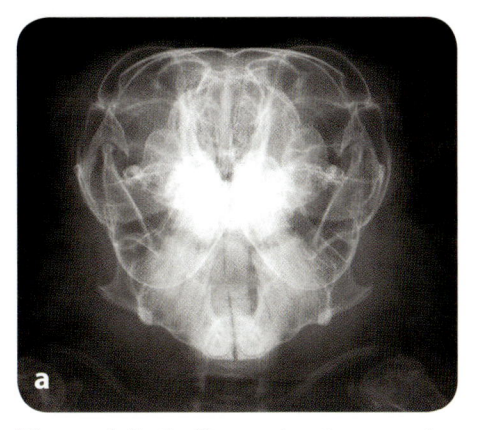

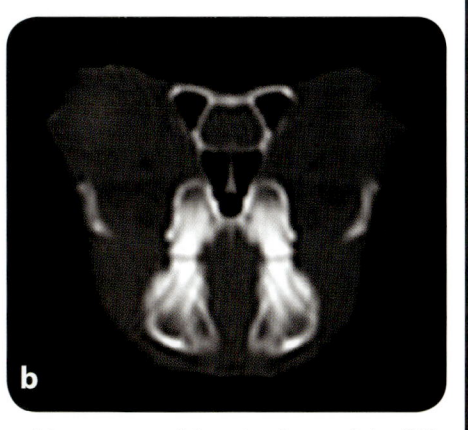

Figure 4.6a,b. Comparison between the radiographic rostrocaudal projection and the CT scan. The CT image is superior for visualization of the shape and orientation of the teeth, and the tooth roots.

ABNORMALITIES of the HEAD
Diseases of Incisor Teeth

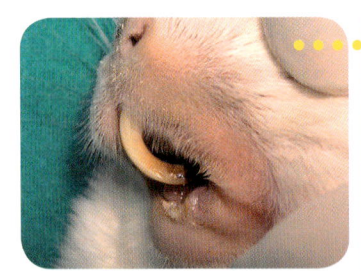

Figure 4.7. Same patient demonstrating severe elongation of the left maxillary incisor tooth. This deformity interferes with movement of the tongue and lips.

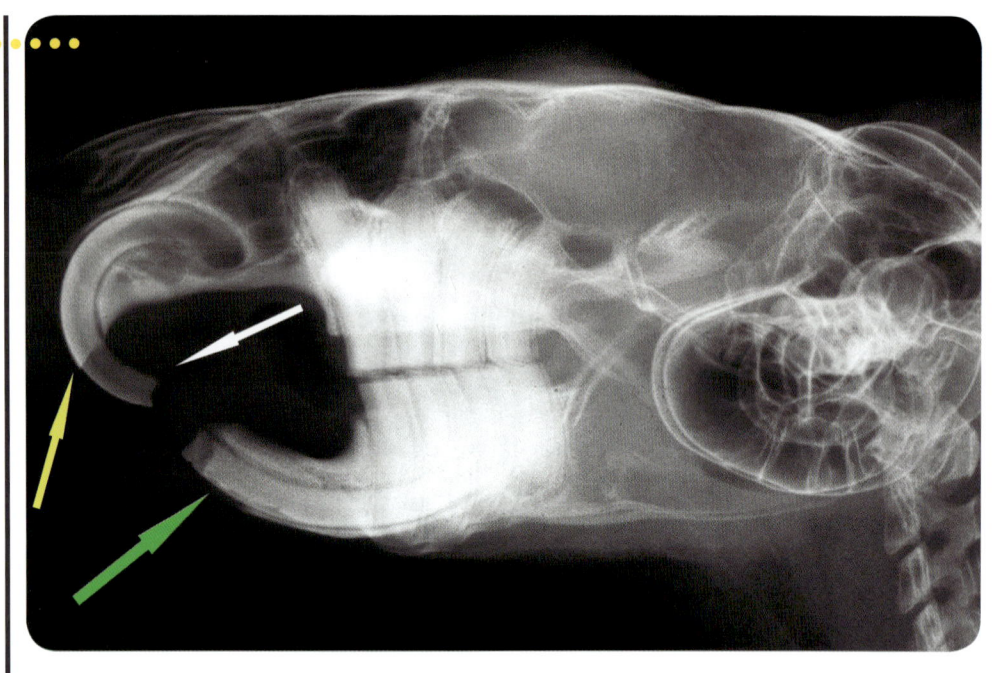

Figure 4.8. Lateral radiographs of the skull of a chinchilla with overgrowth of the left maxillary incisor (white arrow), fracture of the right maxillary incisor (yellow arrow) and fracture of both mandibular incisor teeth (green arrow).

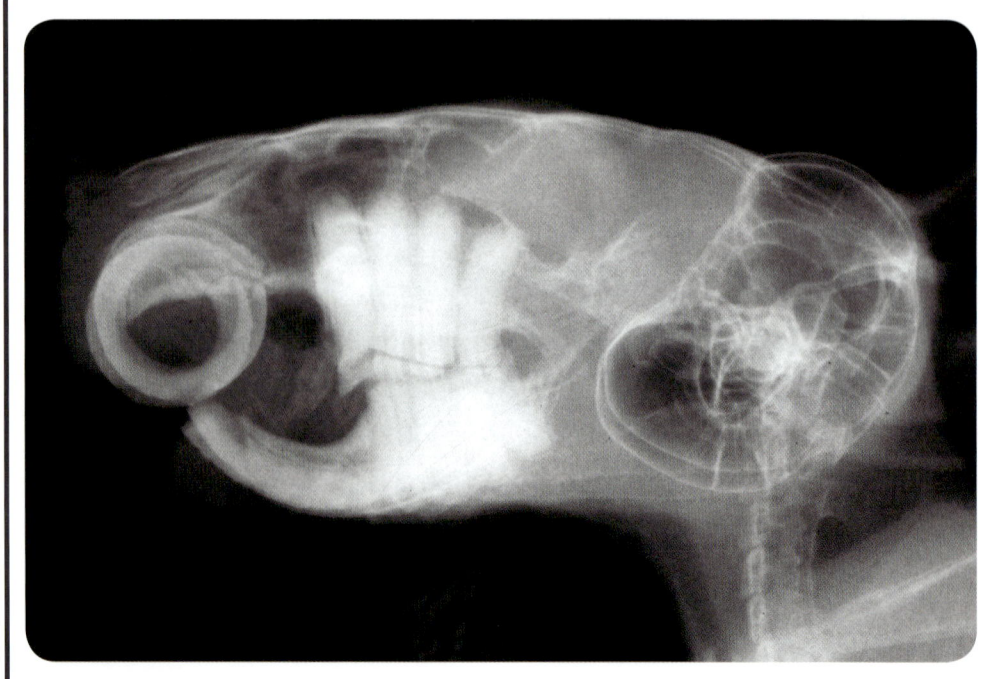

Figure 4.9. Severe malocclusion of maxillary incisor teeth. The 360° overgrowth was beginning to damage the palate.
Crown elongation and wave mouth of maxillary cheek teeth are also visible.

Diseases of Cheek Teeth

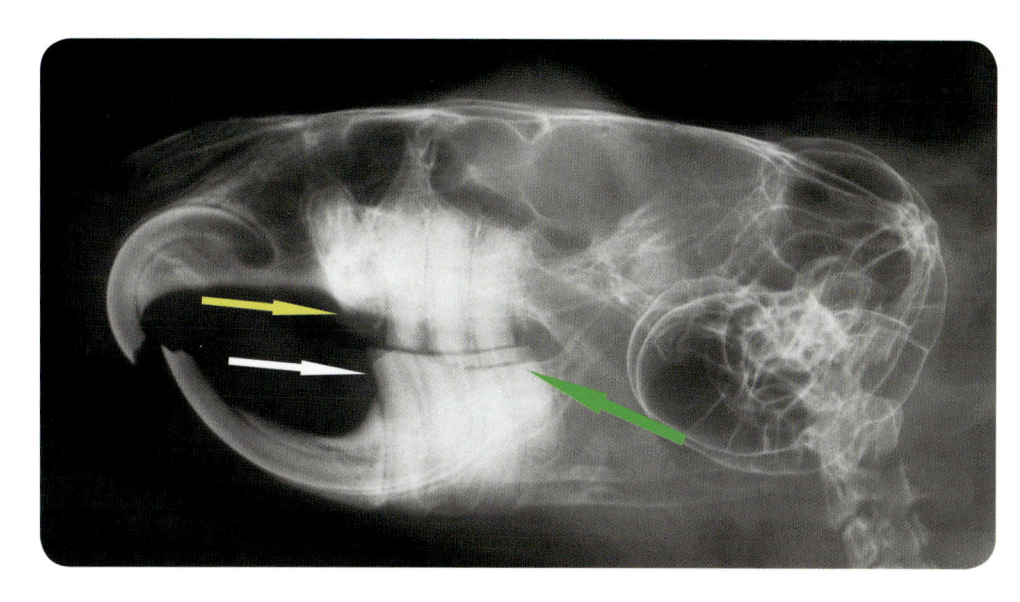

Figure 4.10. Slightly oblique radiograph of the skull of a chinchilla. As in the rabbit, the earliest indication of dental disease of cheek teeth is often wave mouth. Initial overgrowth of mandibular CT1 (white arrow) is present, and the occlusal plane of the right mandibular cheek teeth arcade is not completely flat, but concave (green arrow). The slightly oblique projection accentuates this lesion. The crown of right maxillary CT1 is also shorter than normal (yellow arrow). At this stage, the patient is often asymptomatic, but represents the ideal opportunity for dental correction, before disease worsens.

Figure 4.11. Lateral radiograph of the same patient (Figure 4.10), after occlusal adjustment. The right mandibular arcade has been burred, and flat occlusal plane restored. This near-perfect lateral view demonstrates that the right and left mandibular arcades are now even. Previous cheek tooth elongation and stretching of the masticatory muscles produces a slight gap between cheek teeth arcades in this anesthetized patient. The difference can be best appreciated by comparing the occlusal position of incisor teeth in this and in the previous radiograph.

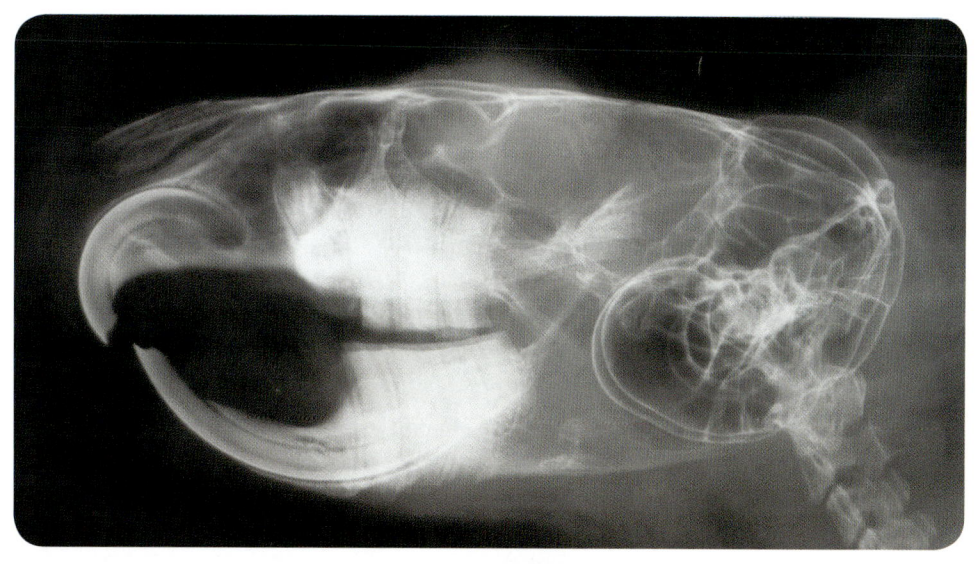

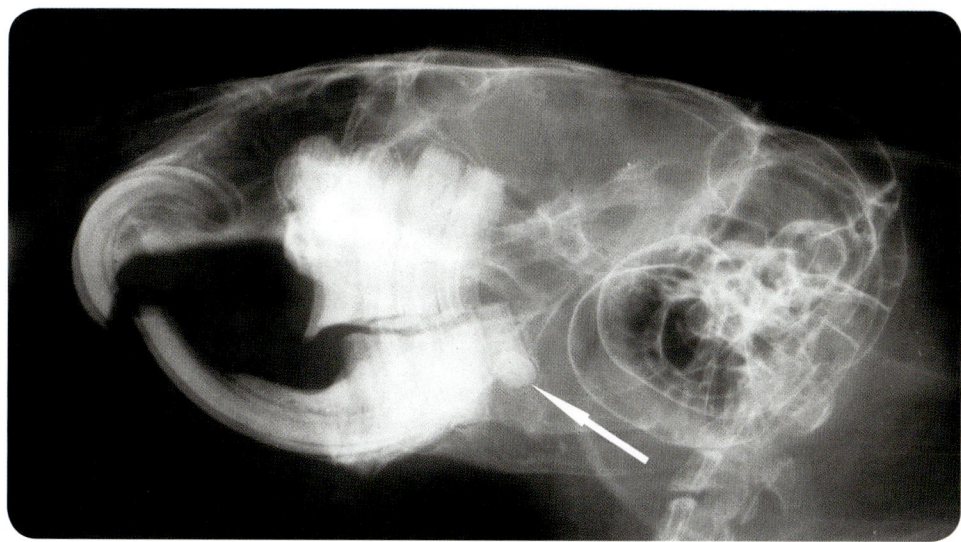

Figure 4.12. Excessive elongation and wave mouth of maxillary cheek teeth. Deformities of the apexes are also visible, including marked curvature of the apex of left mandibular CT4 to near horizontal position (arrow).
See Figures 4.14 and 4.15.

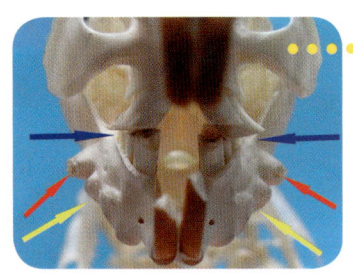

Figure 4.14. Severe malocclusion in chinchillas leads to apical deformities of the cheek teeth. The natural degree of curvature in this species results in maxillary and mandibular cortical bone deformities developing more laterally than in rabbits. Typical firm swellings can be palpated on the ventrolateral aspect of the mandible, rather than on the ventral aspect as in rabbits. Apical deformities can result in perforation of the mandibular cortical bone, with resulting exposure of the apexes. Despite this tendency, chinchillas appear much less prone to development of periapical abscesses, soft tissue involvement and osteomyelitis than rabbits.
Gross specimen demonstrating excessive elongation of the crowns and malocclusion of cheek teeth (blue arrows), apical bone deformity (yellow arrows) and perforated cortical bone (red arrows).

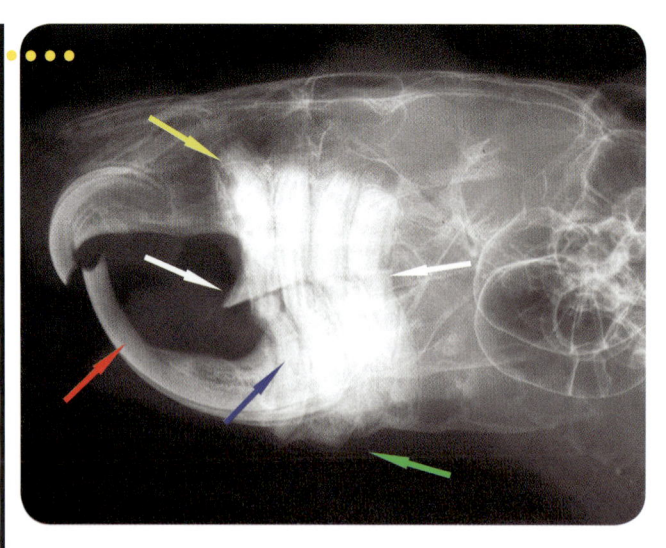

Figure 4.13. Left 15° ventral-right dorsal oblique radiograph of the skull of a chinchilla with severe cheek teeth malocclusion. Abnormalities include uneven cheek teeth occlusal plane; coronal elongation of maxillary CT1 and mandibular cheek teeth arcade (white arrows); elongation of the apex of maxillary CT1 (yellow arrow); abnormal appearance and partial resorption of mandibular CT1 (blue arrow); deformity of the mandibular ventral cortical bone (green arrow). Malocclusion of mandibular incisor teeth, with excessive elongation of crowns and abnormal occlusal plane are also visible (red arrow).

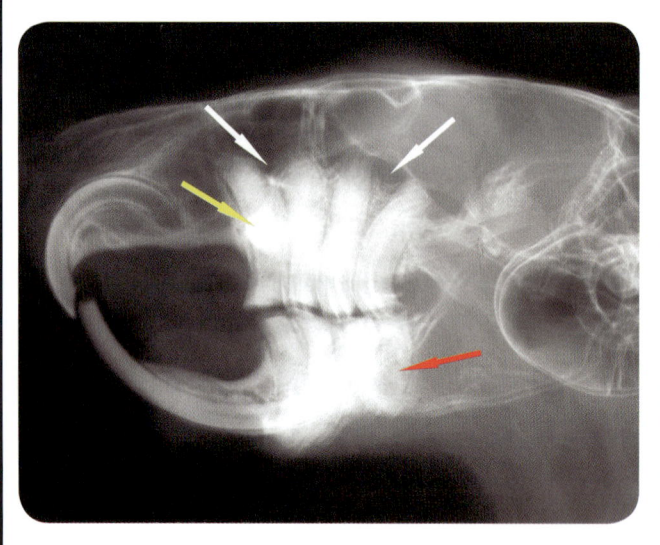

Figure 4.15. Lateral radiograph of the skull of a chinchilla with maloccluded maxillary cheek teeth. Note marked curvature, with widened interproximal spaces (white arrows). The yellow arrow indicates the apex of CT1, which appears positioned horizontally as opposed to a normal more vertical orientation.
Reabsorption of mandibular CT3 and CT4 is also visible (red arrow).

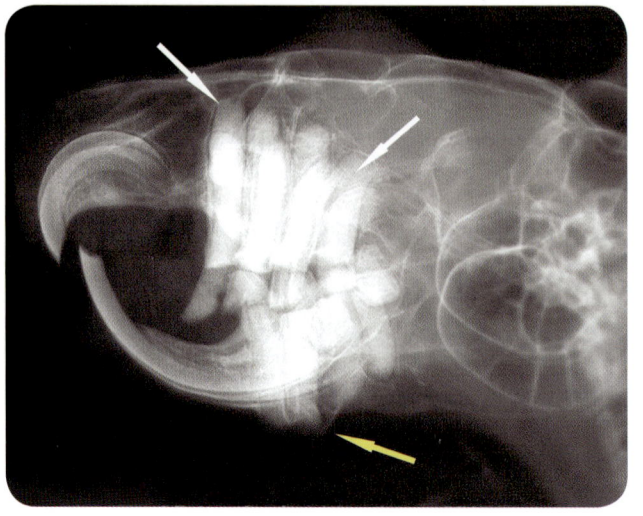

Figure 4.16. Left 15° ventral-right dorsal oblique radiograph of the skull of a chinchilla with advanced malocclusion of cheek teeth. Note severe elongation of the crowns and apexes of maxillary cheek teeth, abnormal occlusal planes, widened interproximal spaces, and excessive elongation of mandibular incisor teeth.
This oblique projection highlights deformation of cheek teeth apexes (white arrows) and mandibular deformities (yellow arrow).

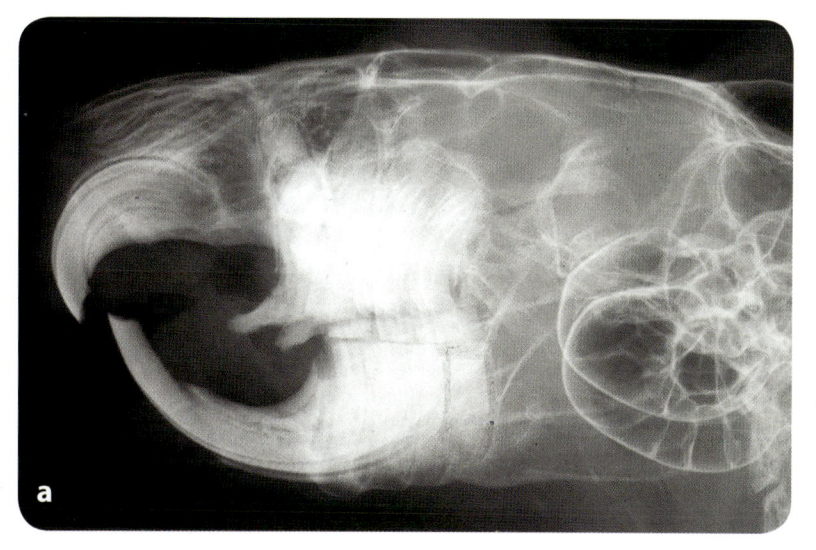

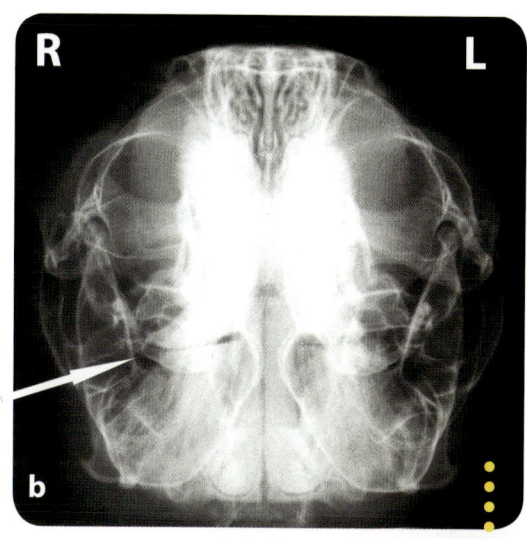

Figure 4.17a,b. Left 15° ventral-right dorsal oblique (a) and rostrocaudal (b) radiographs of the skull of a chinchilla with excessive elongation of crowns, buccal deviation and abnormal sharp edge of both maxillary CT1, especially on the right (b, arrow).

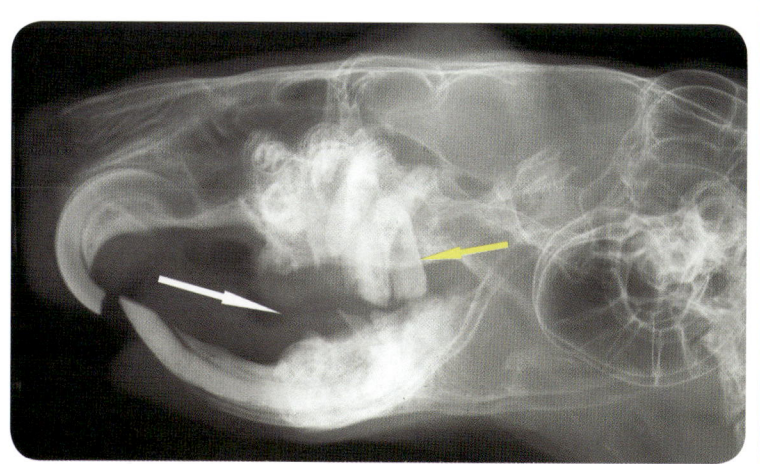

Figure 4.18. Lateral radiograph of the skull of a chinchilla with end-stage acquired dental disease. Note complete wearing of mandibular cheek teeth crowns (white arrow). The only remaining clinical crowns are two maloccluded maxillary CT4 (yellow arrow). Incisor teeth are normal. Despite end-stage disease, this animal was able to eat a normal diet for over a year.

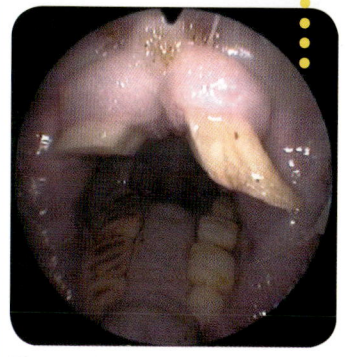

Figure 4.19. Endoscopic view of severe unilateral elongation of maxillary CT crowns. Tooth elongation in chinchillas is frequently accompanied by an increase in height of both the alveolar crest and the gingival margin.

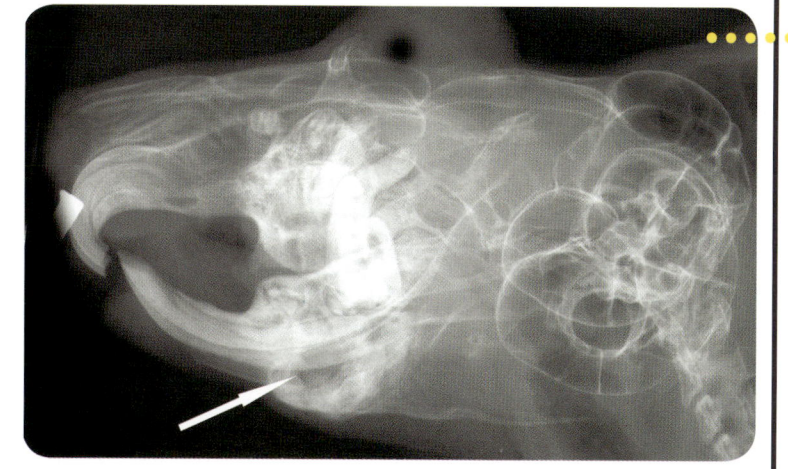

Figure 4.20. Left 15° ventral-right dorsal radiograph of a chinchilla with a mandibular abscess (arrow) as a result of end-stage dental disease. The core of the osteomyelitic site is radiolucent and is surrounded by sclerotic bone.

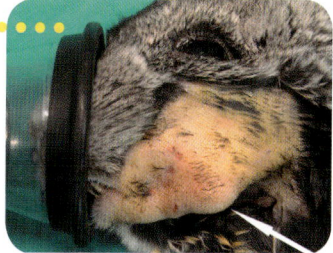

Figure 4.21. Mandibular abscesses typically appear as a firm swelling (arrow).

COMPUTED TOMOGRAPHY of the HEAD

V. Capello, A. Cauduro

The Normal Head

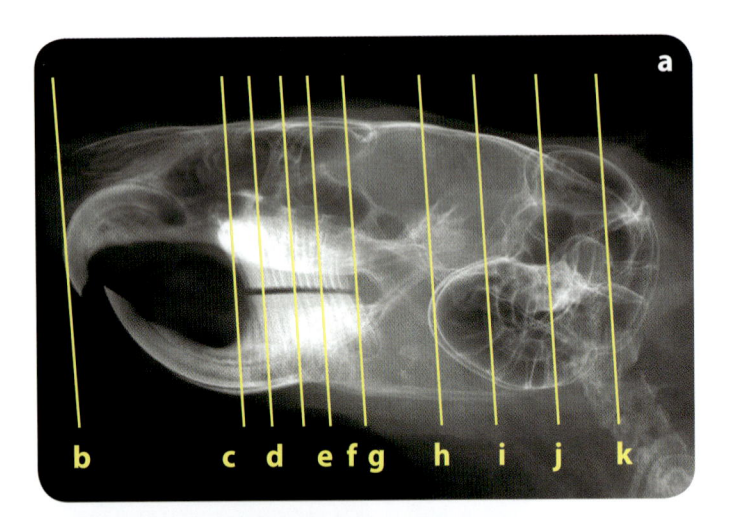

Figure 4.22a-k. Computed tomography of the normal skull of a 500 g chinchilla.
Scanning Parameters: scan speed: 1 sec.;
mA: 125; kVp: 120; slice thickness: 1 mm;
WL: 300; WW: 1500;
Image size (matrix): 512x512.
Selected views from a series of 70.
The scout view demonstrating the scanning angles (a) has been adapted from a radiograph of the normal skull for demonstration purposes.

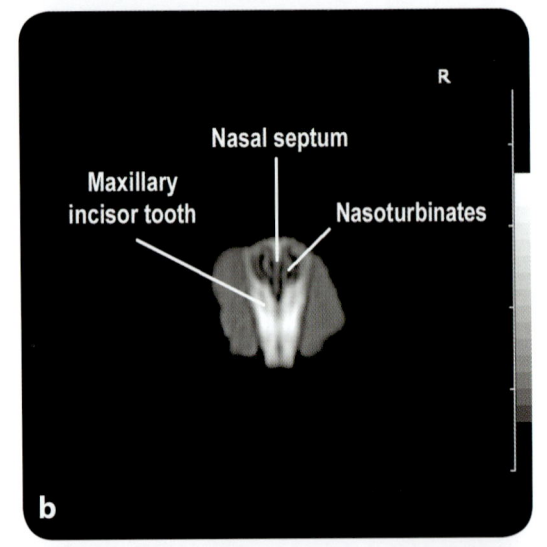

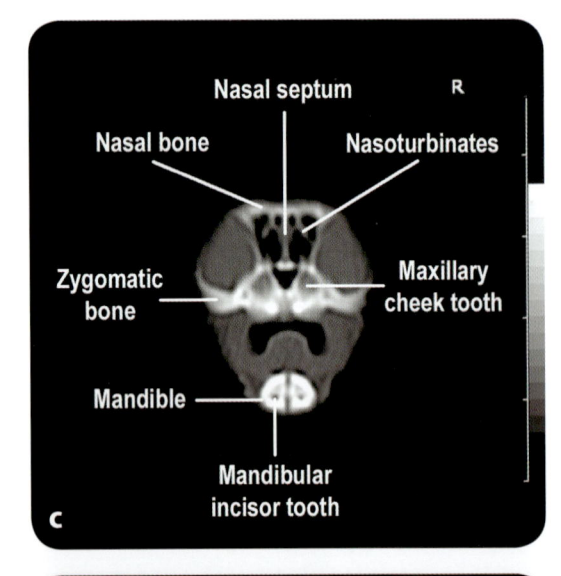

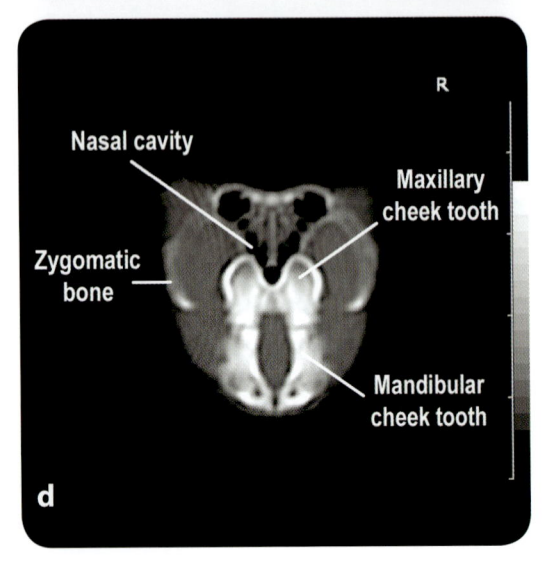

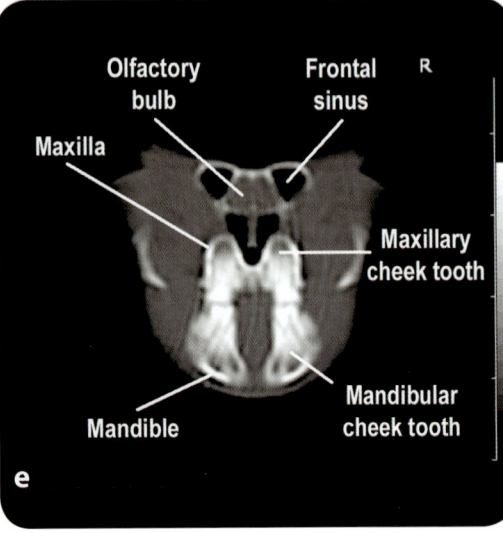

HEAD 233

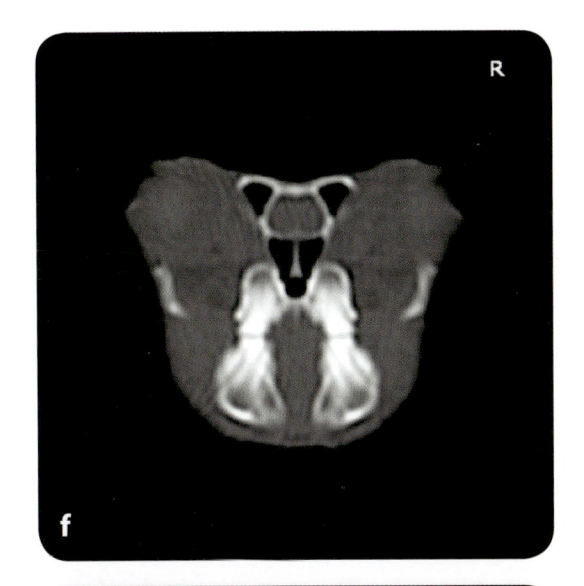

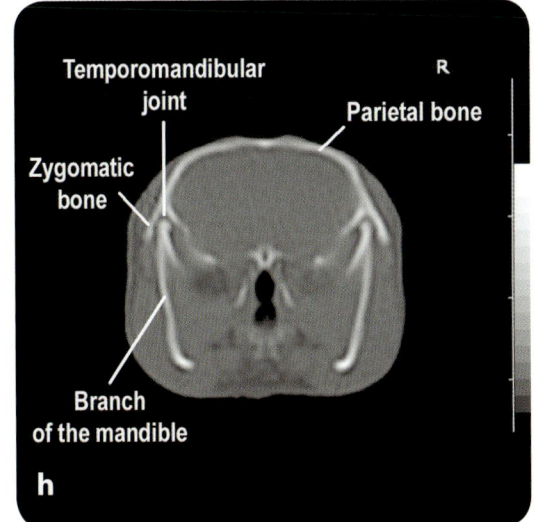

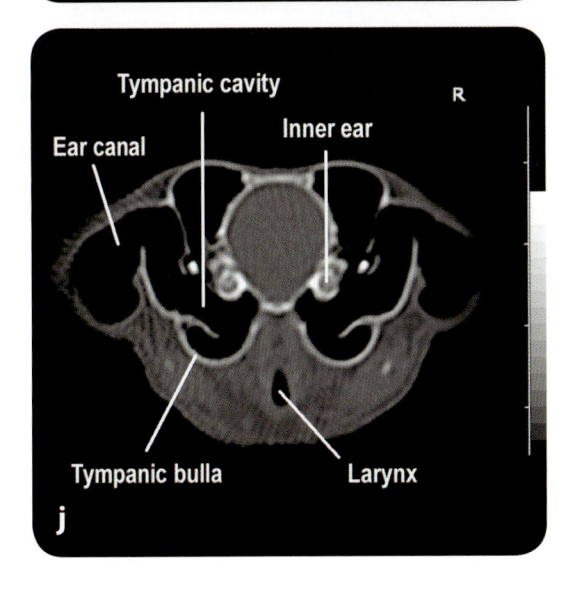

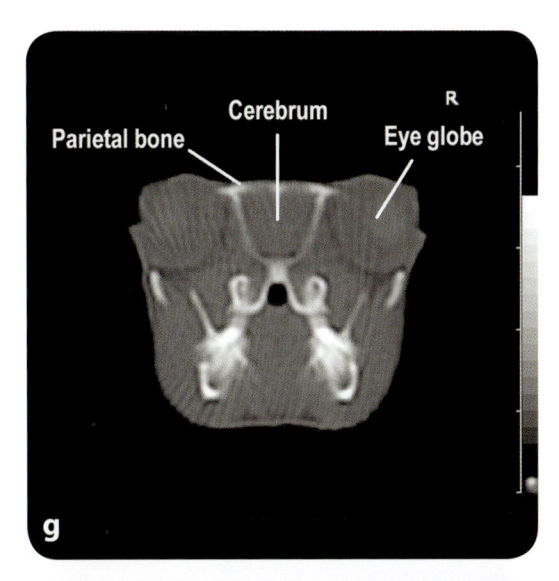

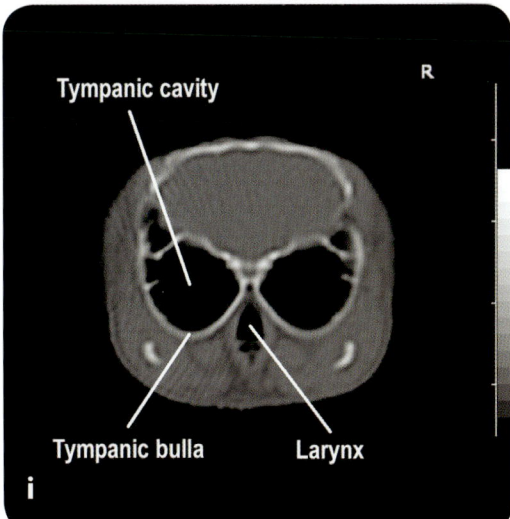

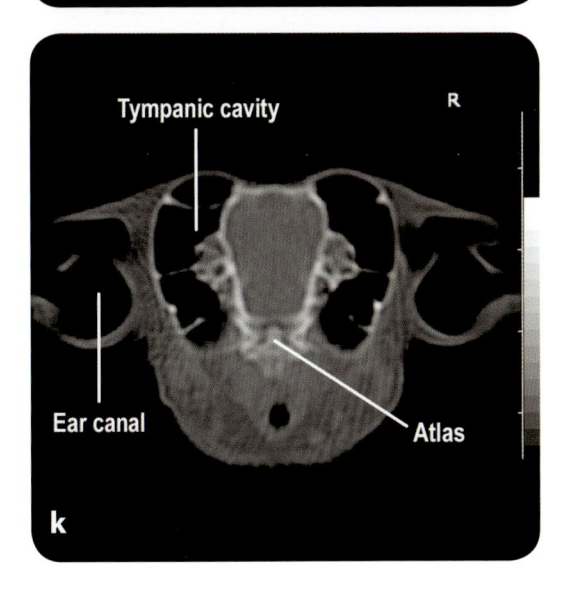

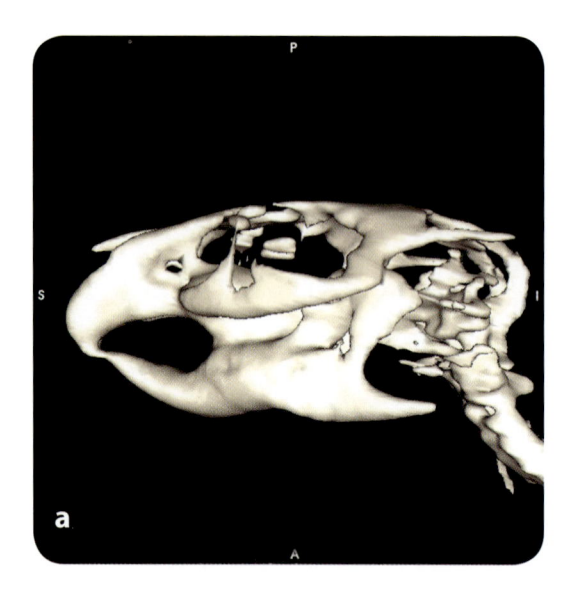

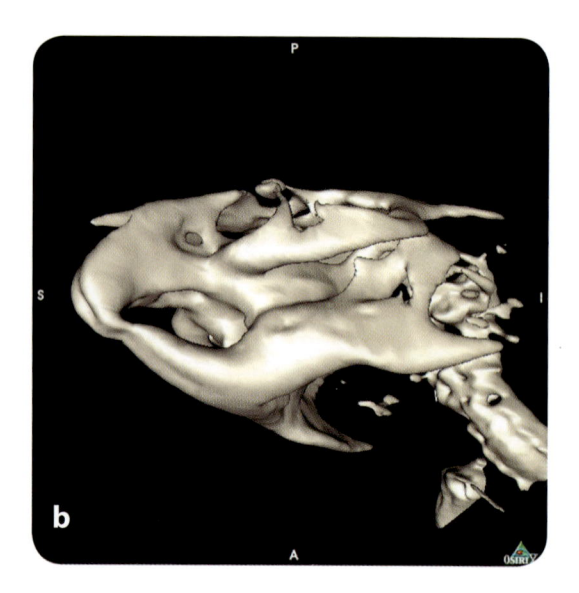

Figure 4.23a-d. Shaded surface display (SDD) of the normal skull of a 500 g chinchilla.

HEAD 235

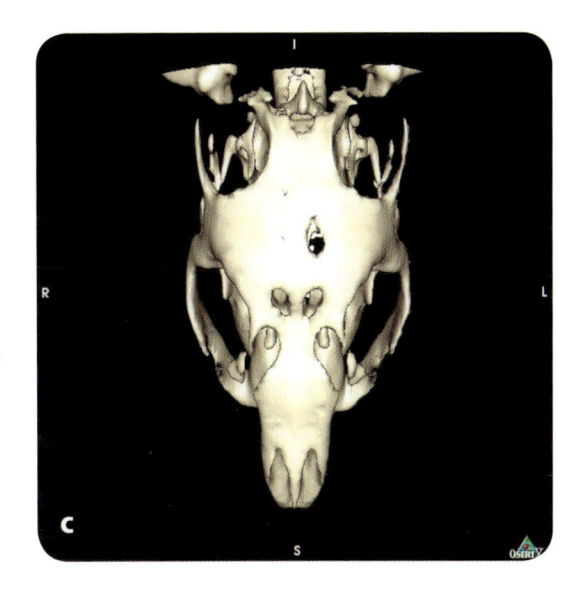

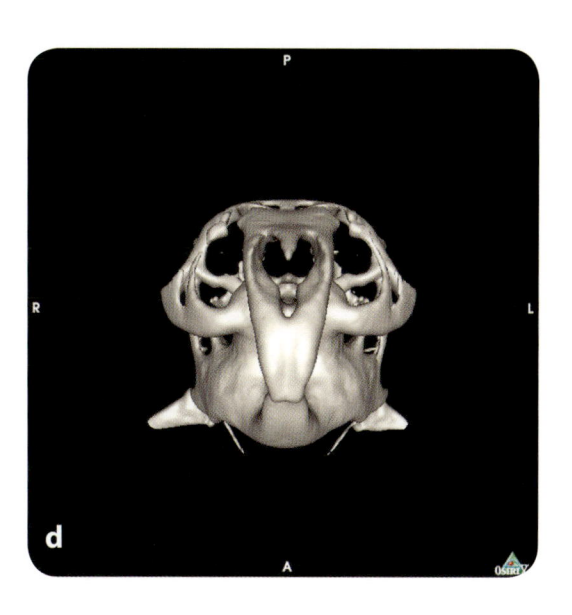

c

d

Acquired Dental Disease

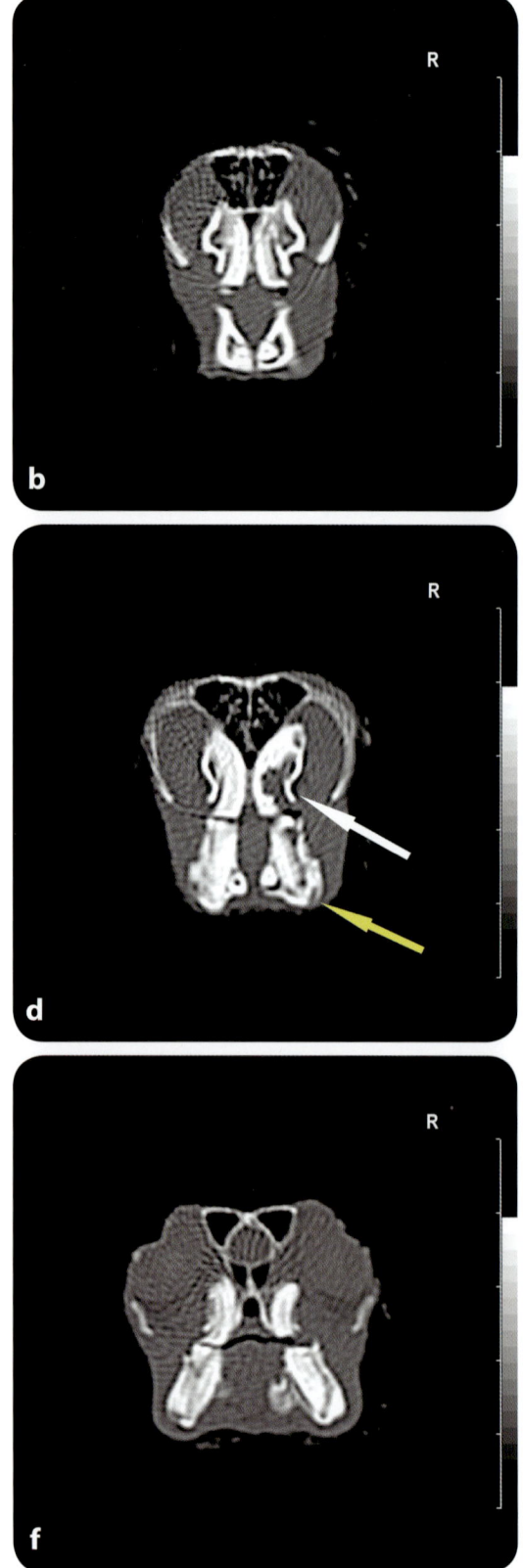

Figure 4.24a-f. Severe acquired dental disease (ADD) in a 4-year-old chinchilla.
Scanning Parameters: scan speed: 1 sec.; mA: 100; kVp: 120; slice thickness: 1 mm; WL: 300; WW: 1500; Image size (matrix): 512x512.
Elongation and curving of maxillary cheek teeth are visible in (b, c, d, e). Curving of the reserve crowns and the apexes typical of ADD is visible, especially on the right side in (c, d, e). Caries of right maxillary CT2 are visible in (d) and of right mandibular CT3 (e) (arrows). Typical deformities and bone perforation of mandibular cheek teeth are visible, especially in (d) (yellow arrow).

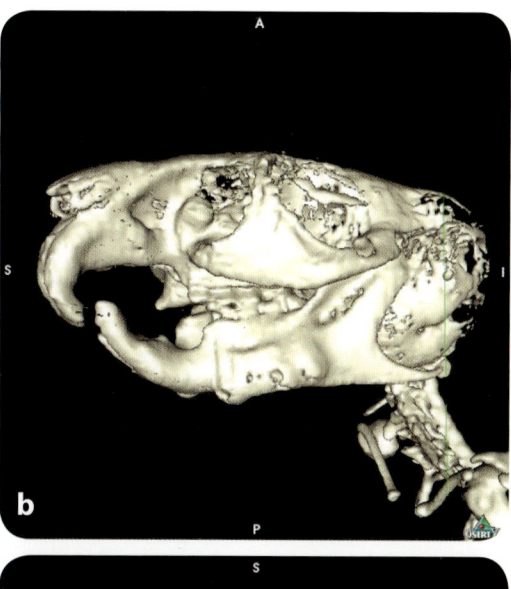

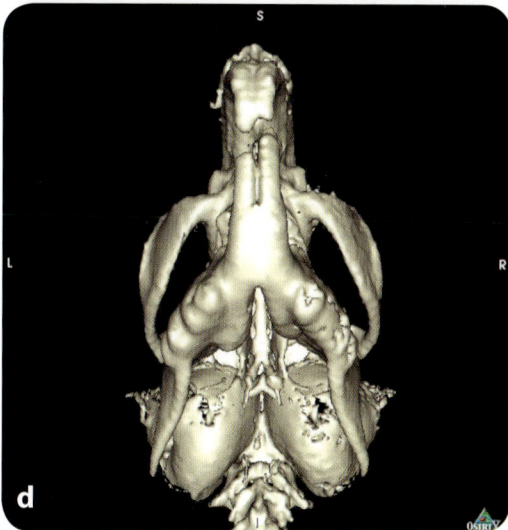

Figure 4.25a-d. SSD of the skull of the same chinchilla in Figure 4.24.

Apical deformities of cheek teeth, typical of ADD in chinchillas (see Figures 4.13-4.16) are visible in both lateral views (a, b). Perforation is also visible on the left lateral view (b).

Severe apical elongation and deformity of right maxillary CT1 and CT2 are highlighted with this imaging technique. This can permit earlier diagnosis of ADD which may be contributing to vague clinical signs such as epiphora.

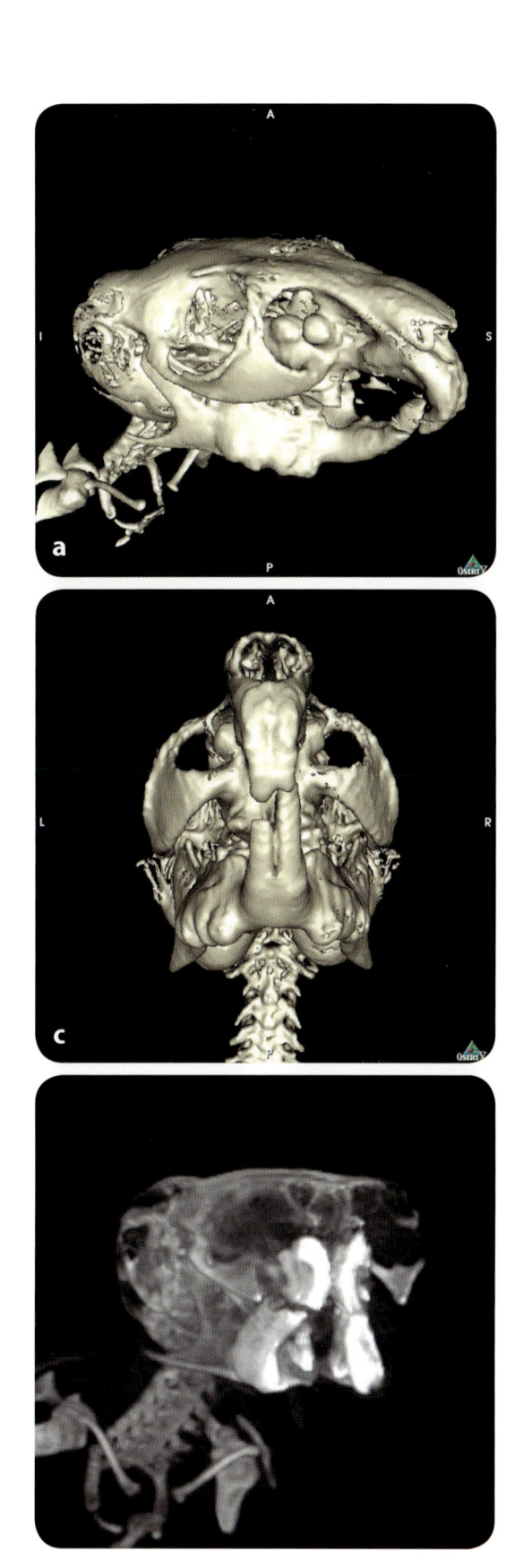

Figure 4.26. 3D volume rendering (VR) image cropped at the interproximal space between mandibular CT2 and CT3. Note the deformity of the apexes of all cheek teeth.

The NORMAL TOTAL BODY
Lateral Projection

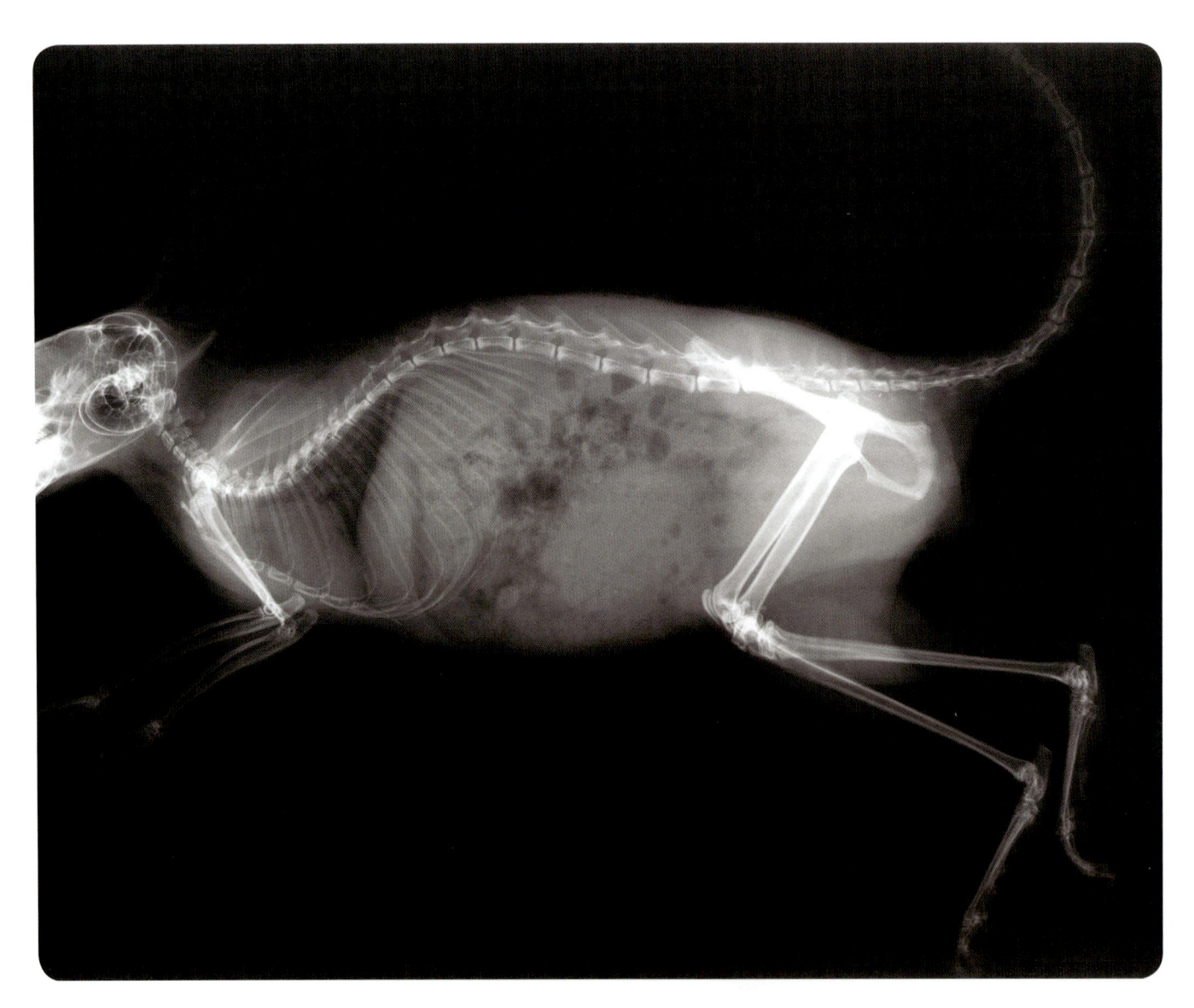

Figure 4.27. Total body radiograph of a 550 g female chinchilla, 70% of actual size.

Ventrodorsal Projection

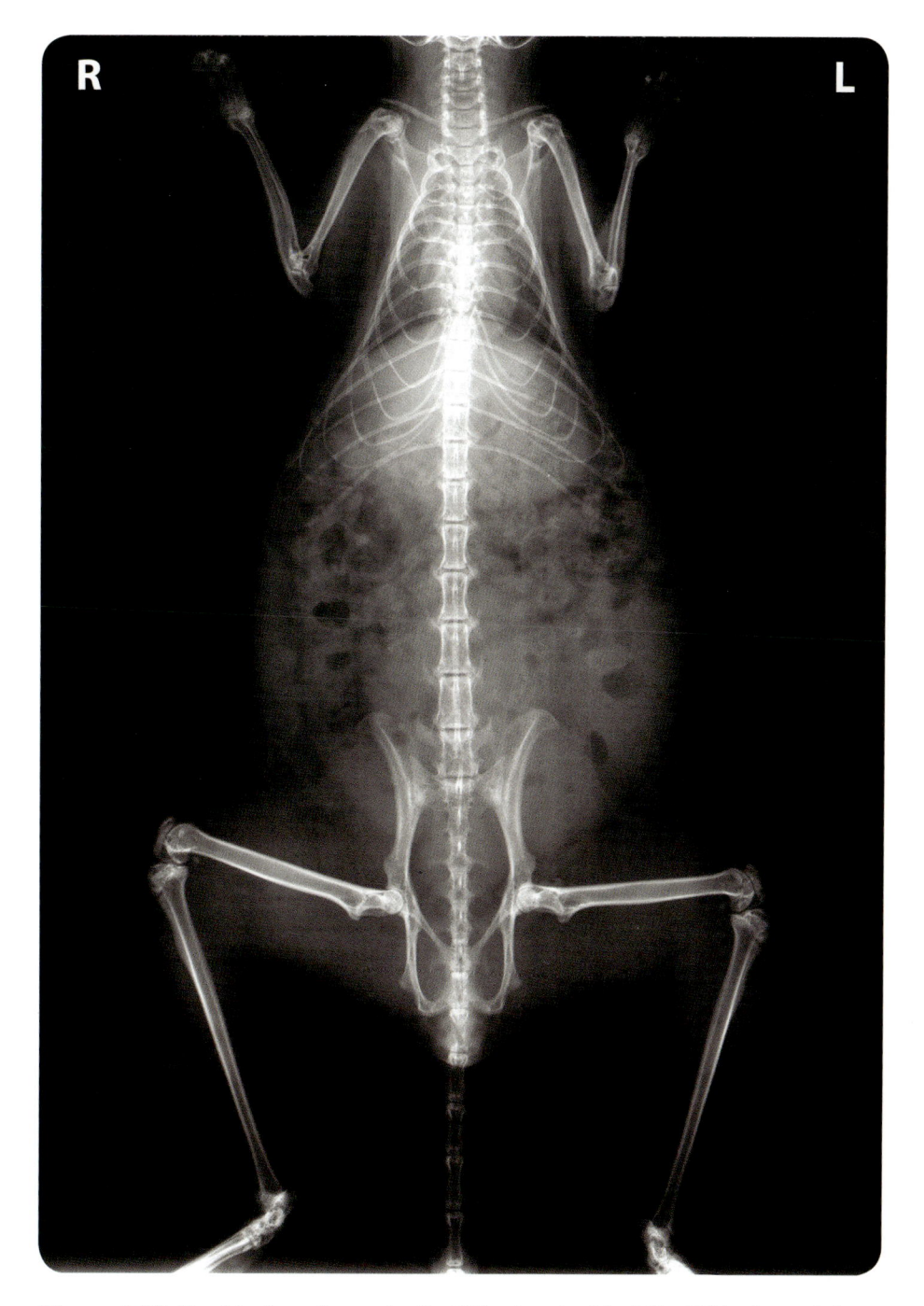

Figure 4.28. Total body radiograph of a 550 g female chinchilla, 70% of actual size.

The NORMAL THORAX
The Cervical and Thoracic Vertebral Column
Lateral Projection

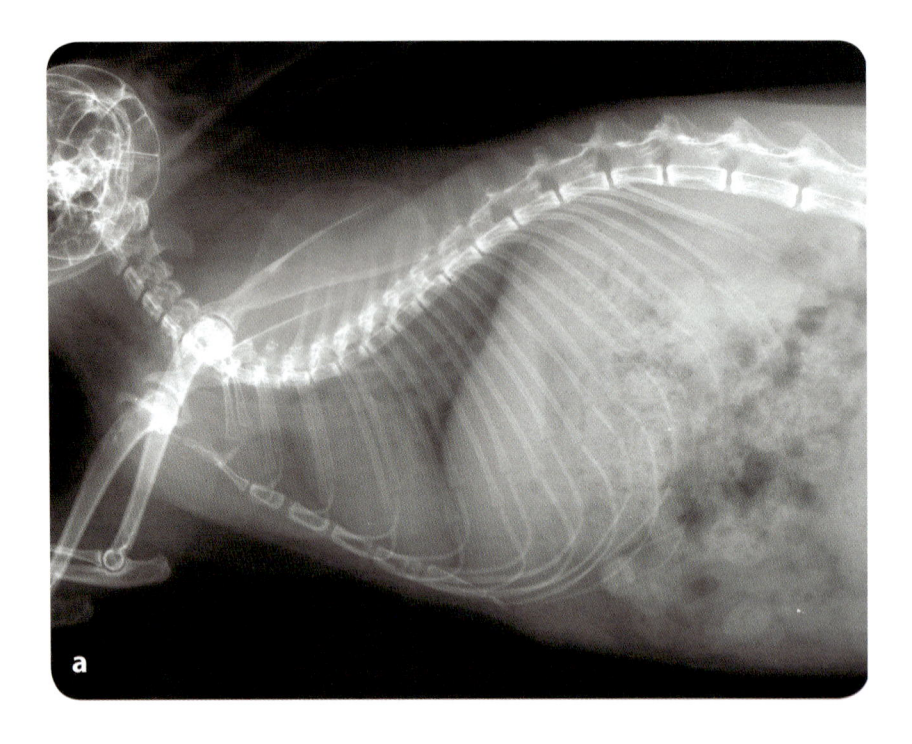

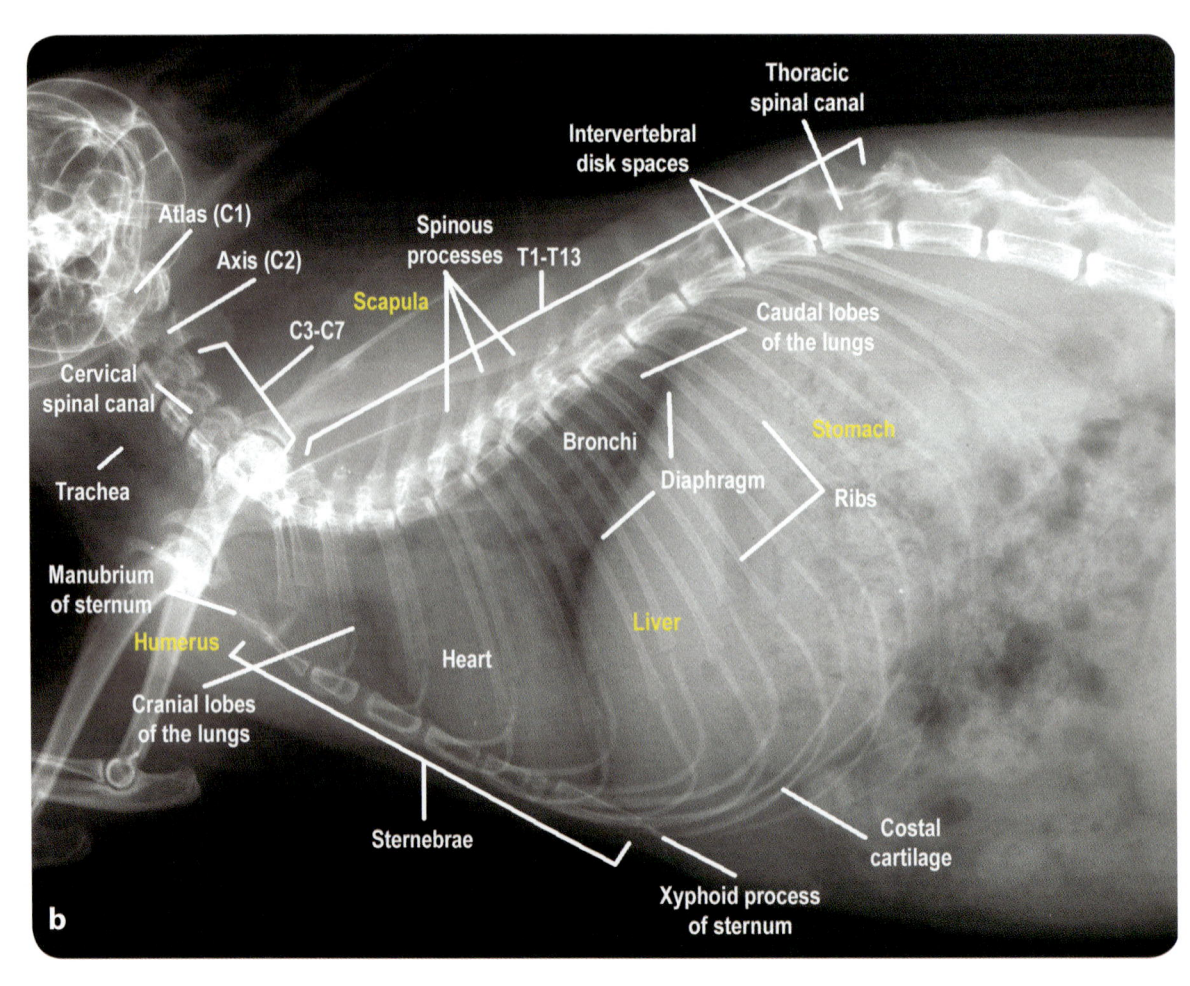

Figure 4.29a,b. Radiograph of the thorax of a 550 g female chinchilla, actual size (a), and 140% of actual size (b).

Ventrodorsal Projection

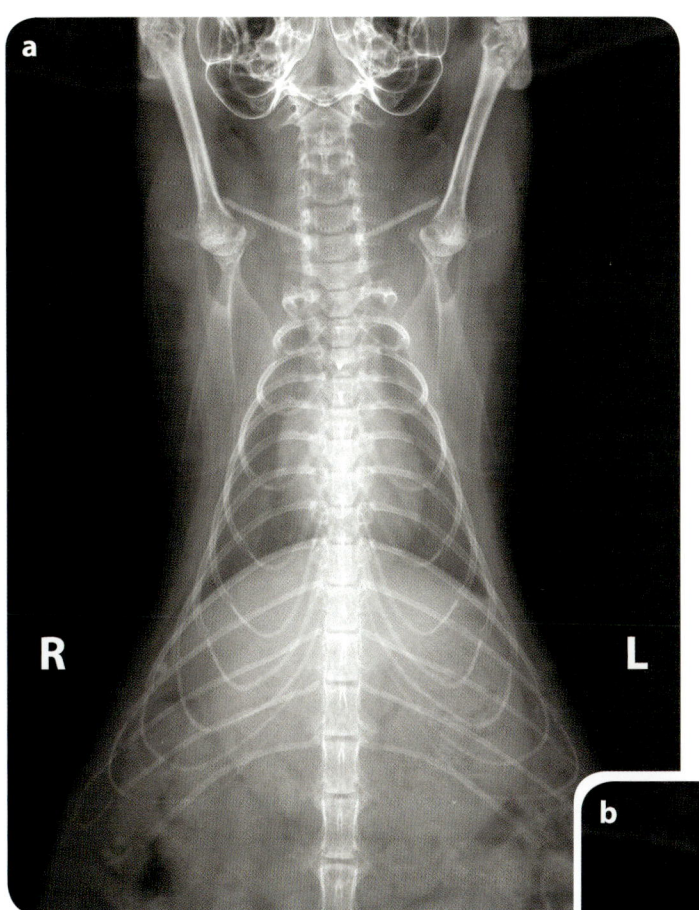

Figure 4.30a,b. Radiograph of the thorax of a 550 g female chinchilla, actual size.

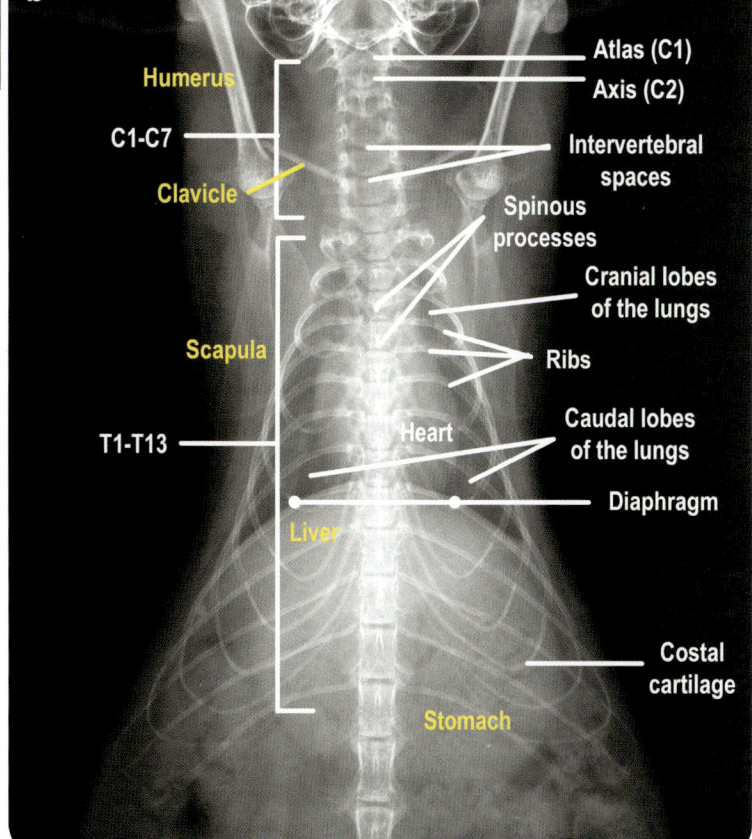

The NORMAL ABDOMEN
The Lumbar Vertebral Column
Lateral Projection

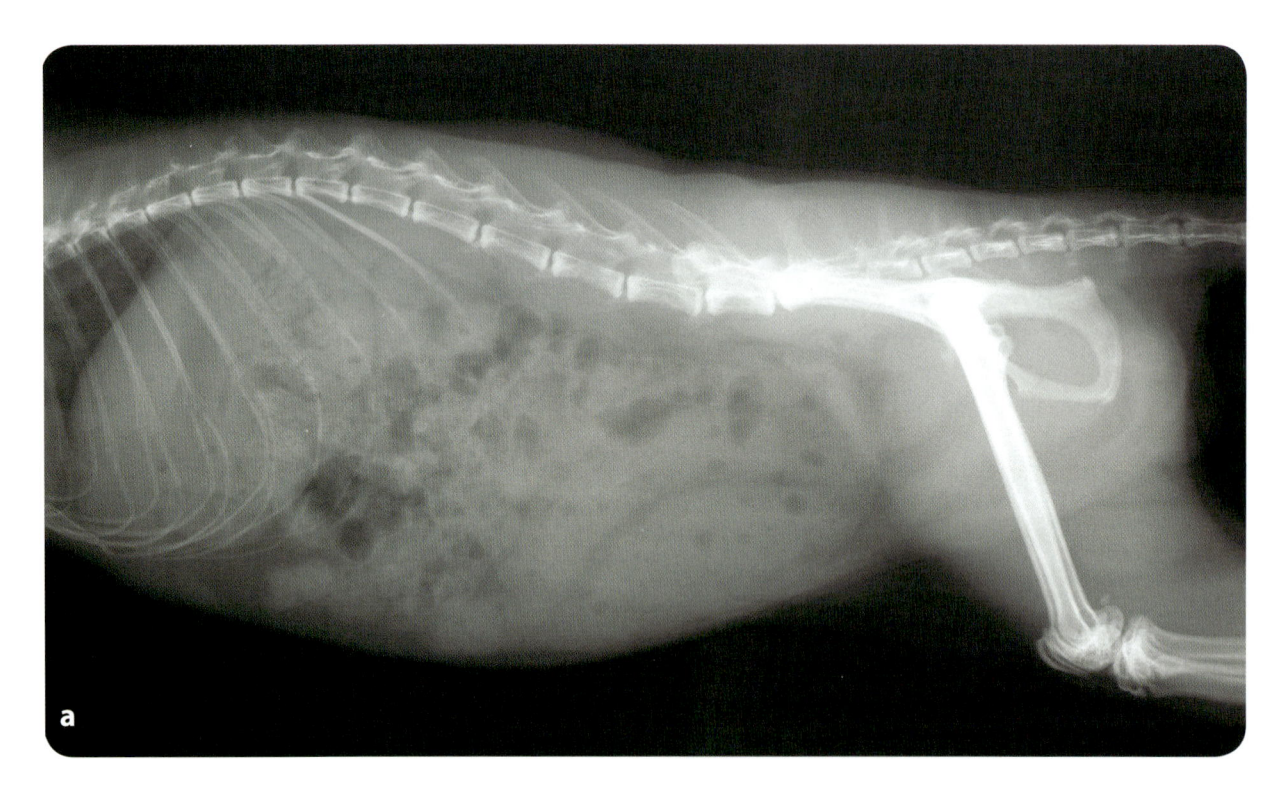

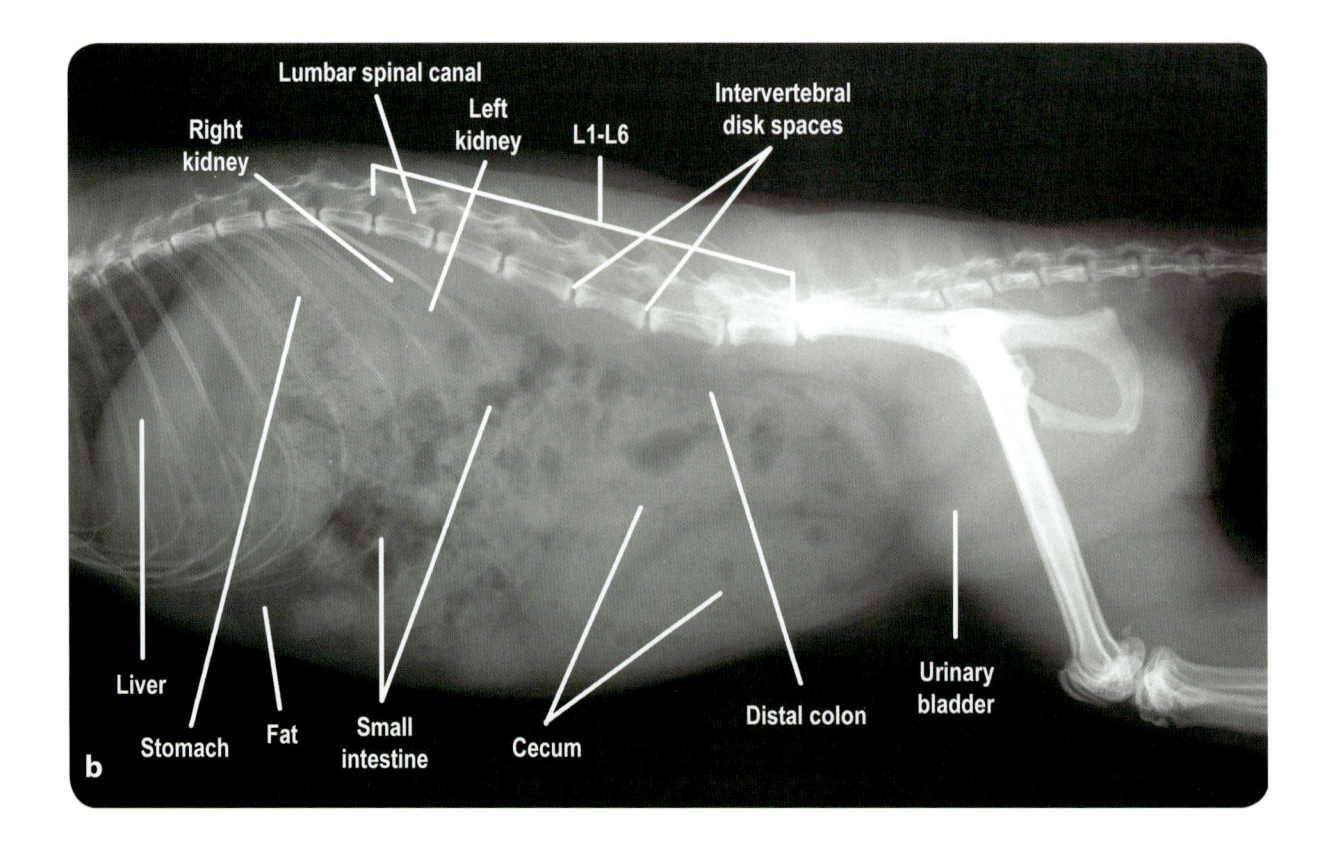

Figure 4.31a,b. Radiograph of the abdomen of a 550 g female chinchilla, actual size.

ABDOMEN 243

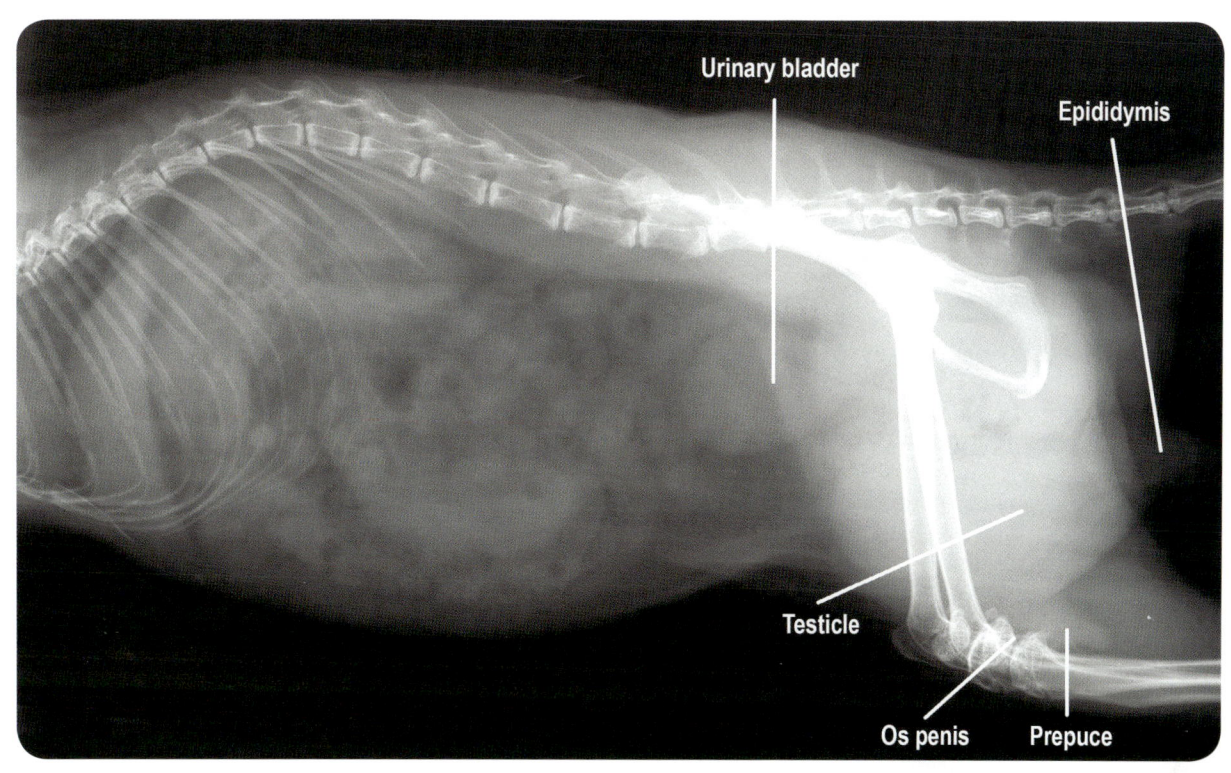

Figure 4.32a,b. Radiograph of the abdomen of a 500 g male chinchilla, actual size. The os penis is covered by the proximal tibiae (b).

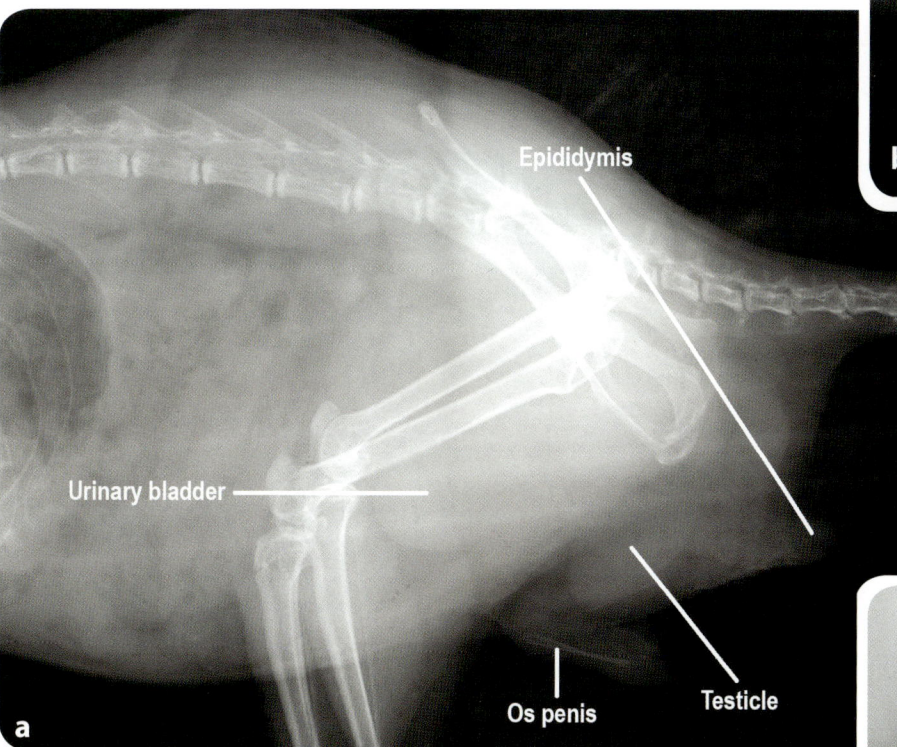

Figure 4.33a,b. Radiograph of a male chinchilla with pelvic limbs deflected cranially to highlight the os penis (a), also shown in (b).

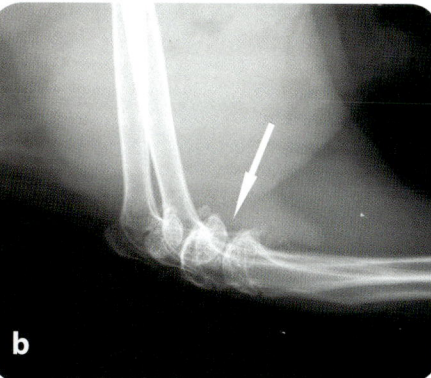

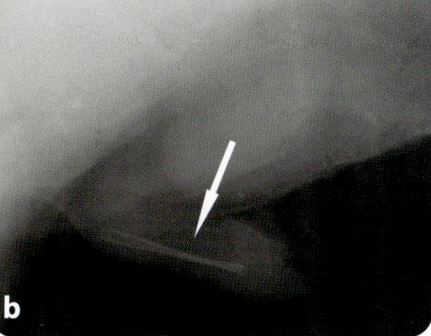

Ventrodorsal Projection

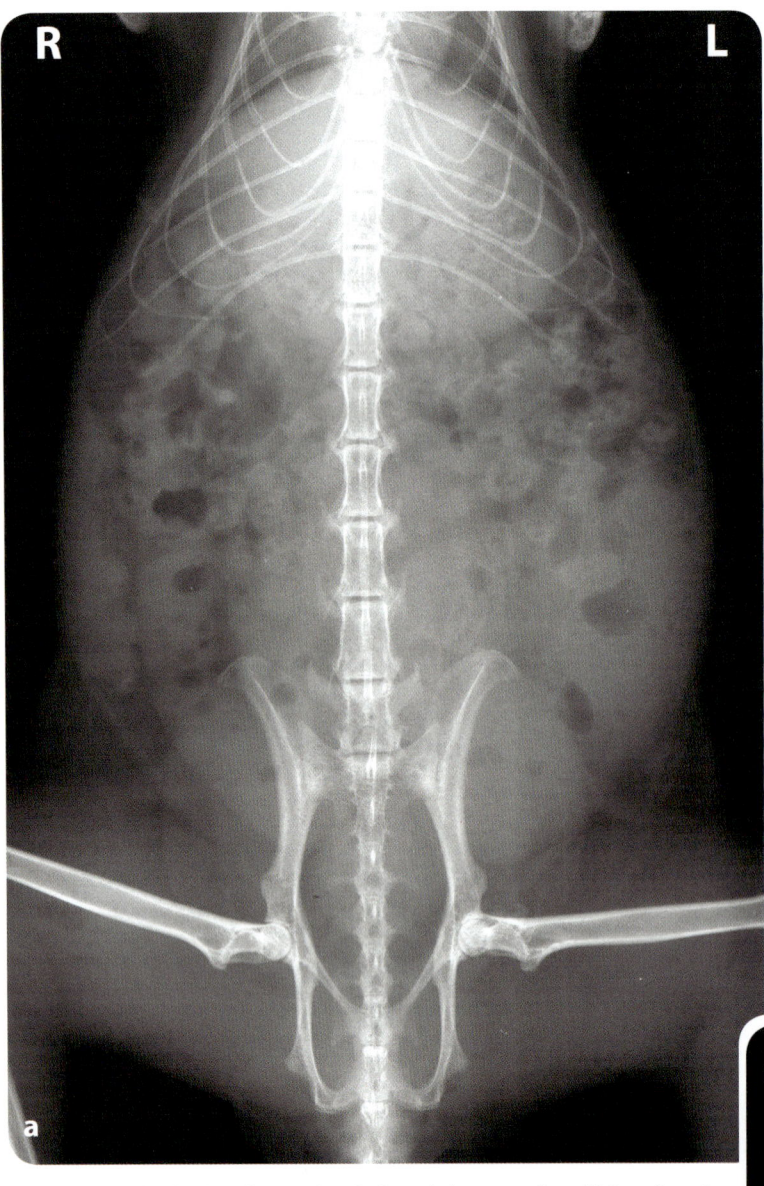

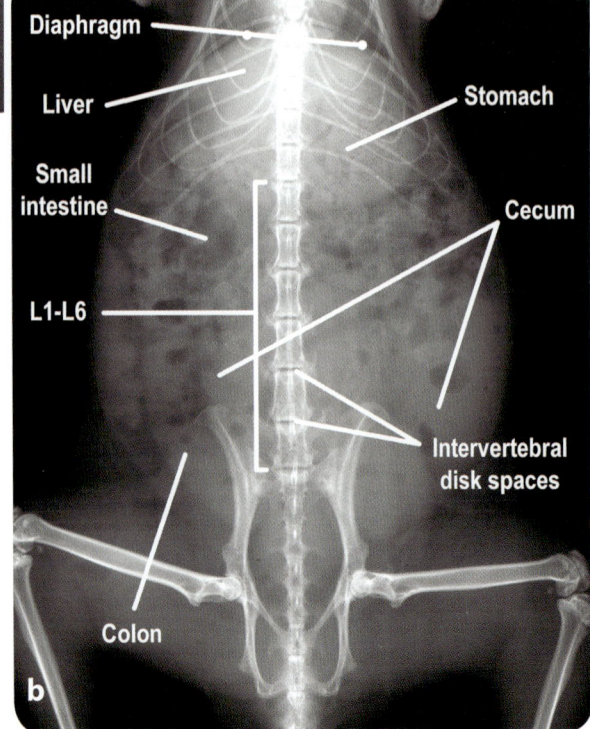

Figure 4.34a,b. Radiograph of the abdomen of a 550 g female chinchilla, actual size (a) and 80% of actual size (b).

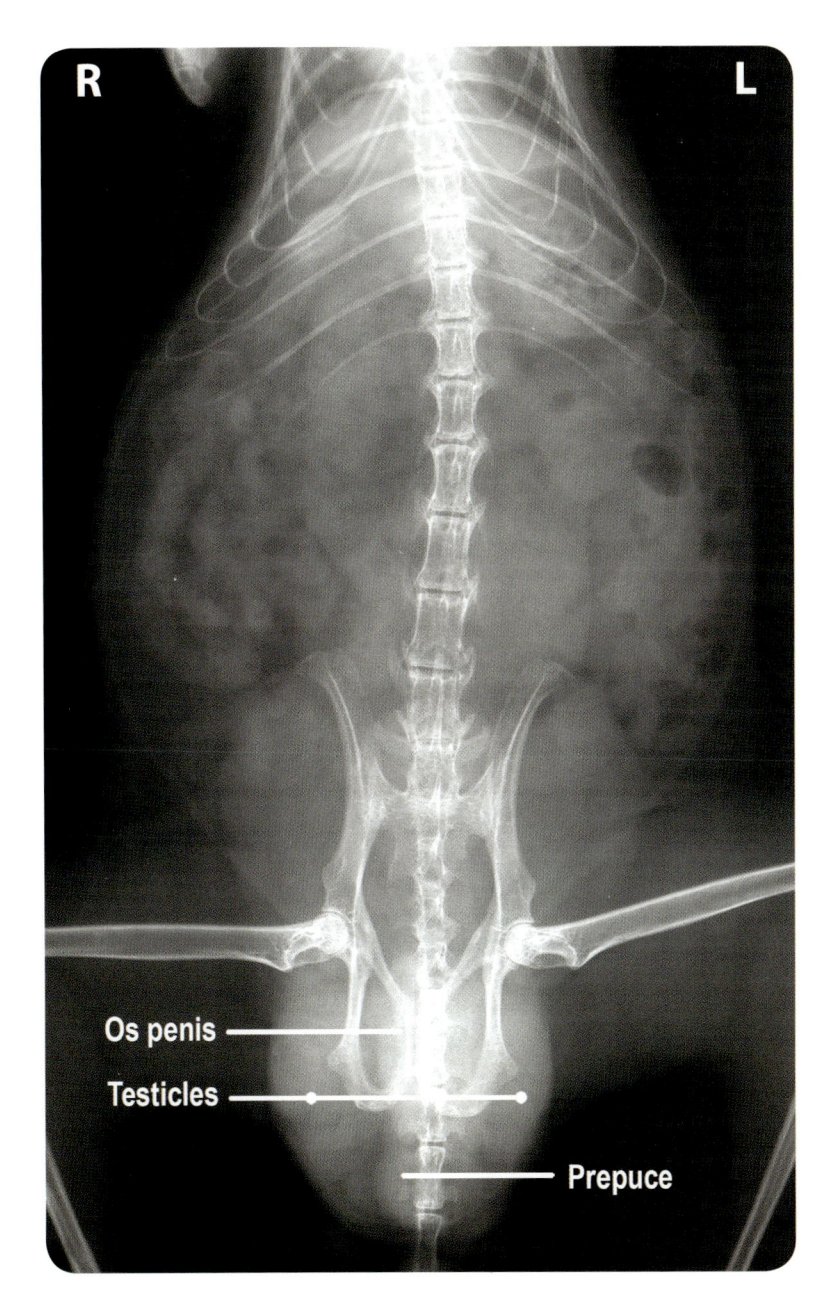

Figure 4.35. Radiograph of the abdomen of a 500 g male chinchilla, actual size.

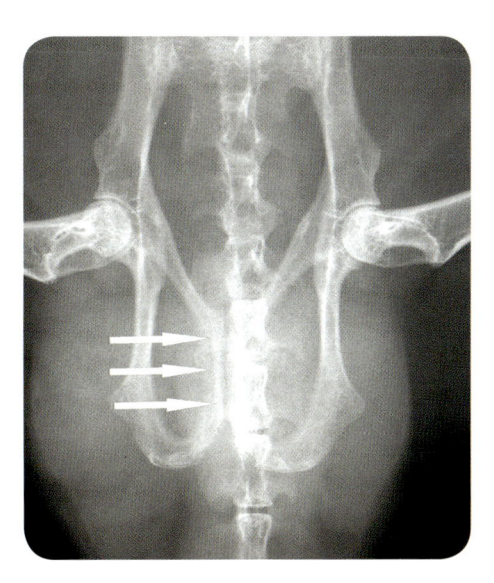

Figure 4.36. Highlight of the pelvis demonstrating the os penis.

Miscellaneous

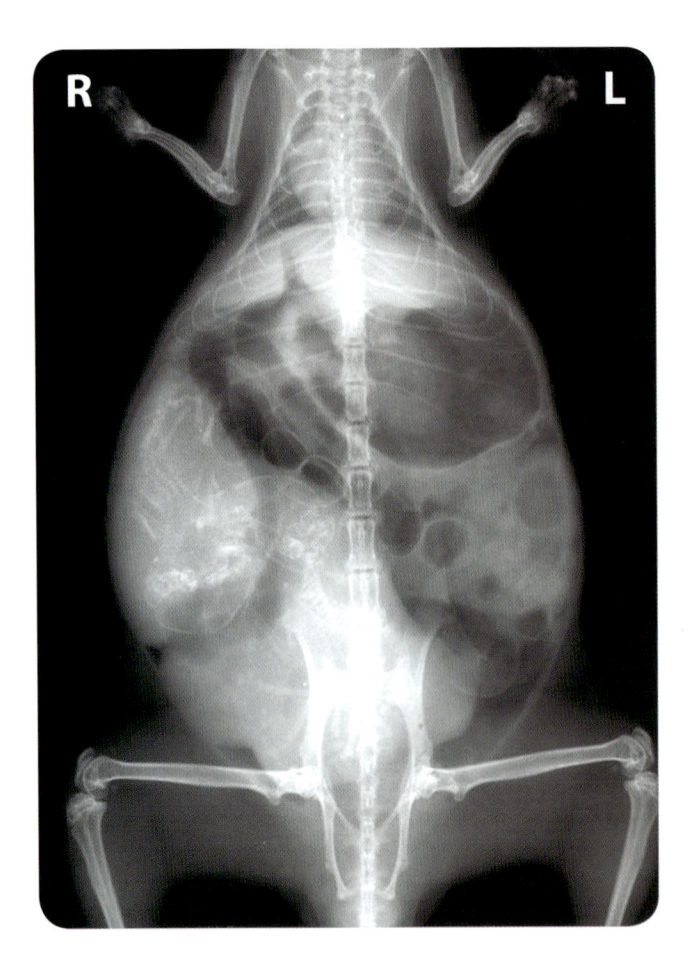

Figure 4.37. Ventrodorsal radiograph of a pregnant chinchilla, approximately 45 to 50 days gestation, with two visible fetuses. Similarly to guinea pigs, ossification of the skeleton of the fetuses occurs early. Note gastrointestinal gas accumulation which was apparently causing no clinical symptoms.

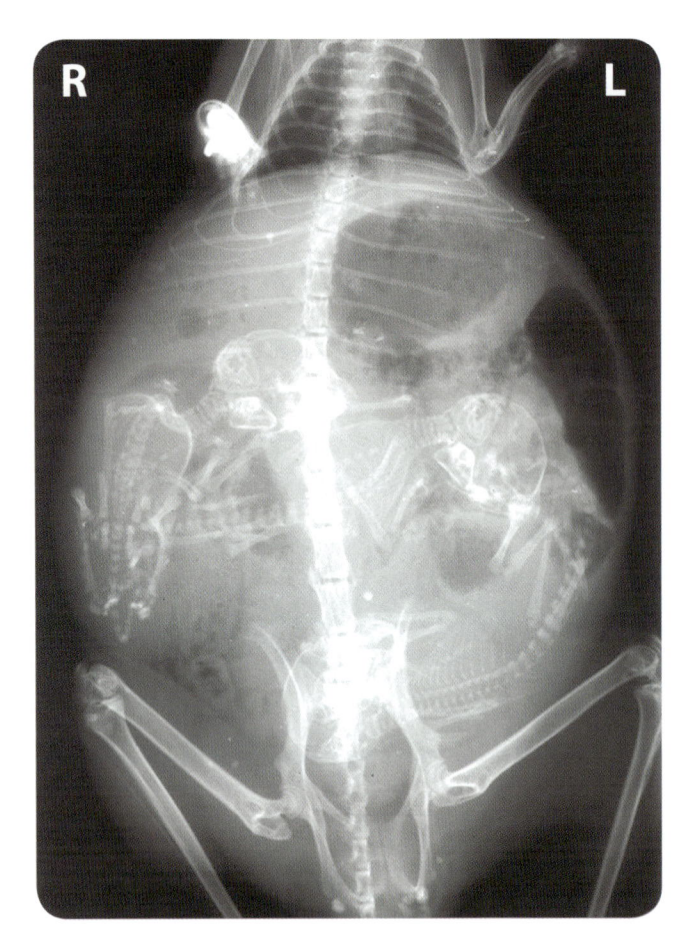

Figure 4.38. Ventrodorsal radiograph of a pregnant chinchilla, later gestation. Three fetuses are present. Note that in the chinchilla, diastasis of the ischiopubic symphysis does not occur as it does in guinea pigs, despite the relatively large size of the young.

Figure 4.39. Ventrodorsal radiograph of the abdomen of a pregnant chinchilla. Two fetuses are present, but radiographic evidence of fetal death is present in the more cranial of the two. This fetus is much smaller, and the skull bones have collapsed.

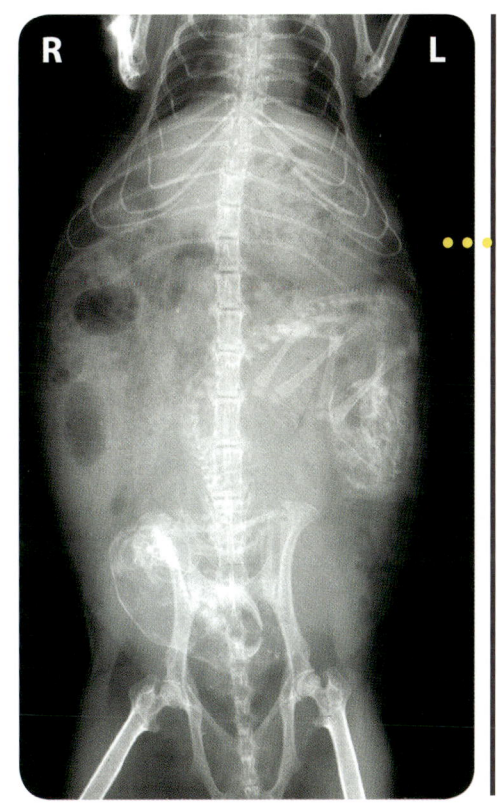

Figure 4.40. Chinchillas give birth to precocious young, completely furred, with open eyes and teeth already present. This young nursing chinchilla is 24 hours old.

ABNORMALITIES of the ABDOMEN
Diseases of the Gastrointestinal Tract

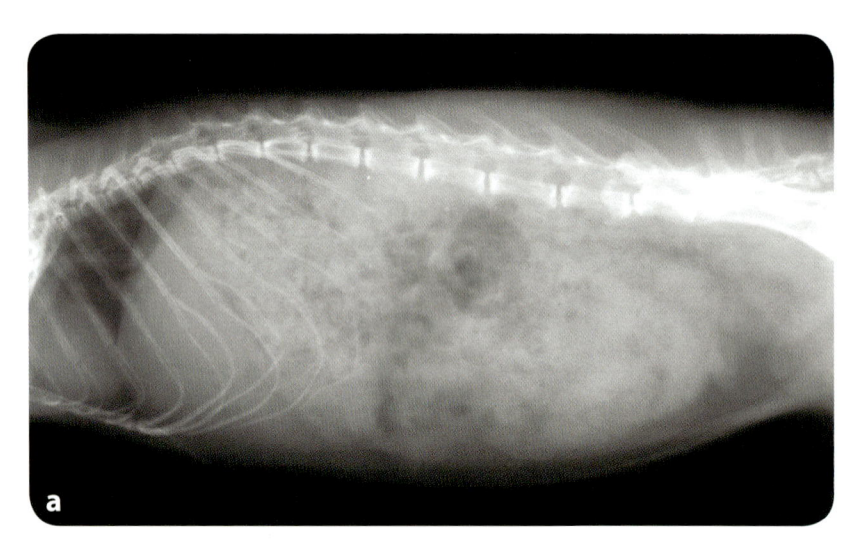

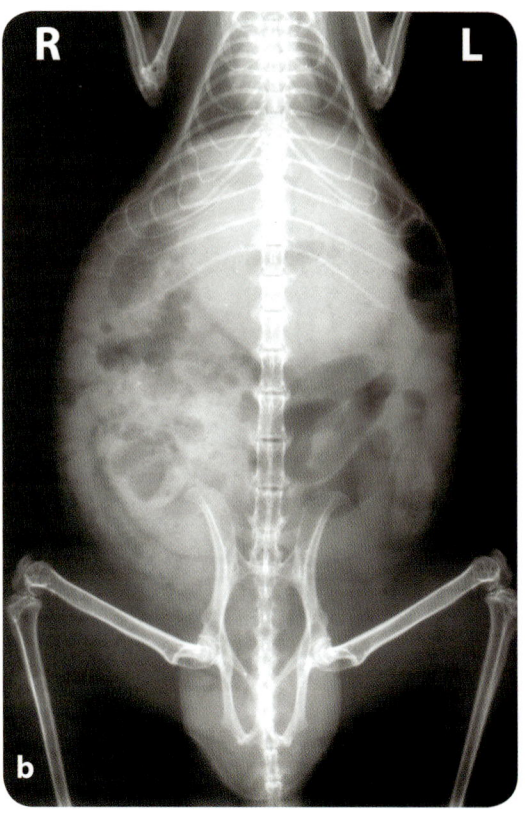

Figure 4.41a,b. Lateral (a) and ventrodorsal (b) radiographs of the abdomen of a chinchilla with a gastrointestinal tract completely filled with food. There is no abnormal gas present. Therefore, it is unclear if this radiograph demonstrates recent heavy food intake or a pathologic condition. This patient should be monitored radiographically and clinically.

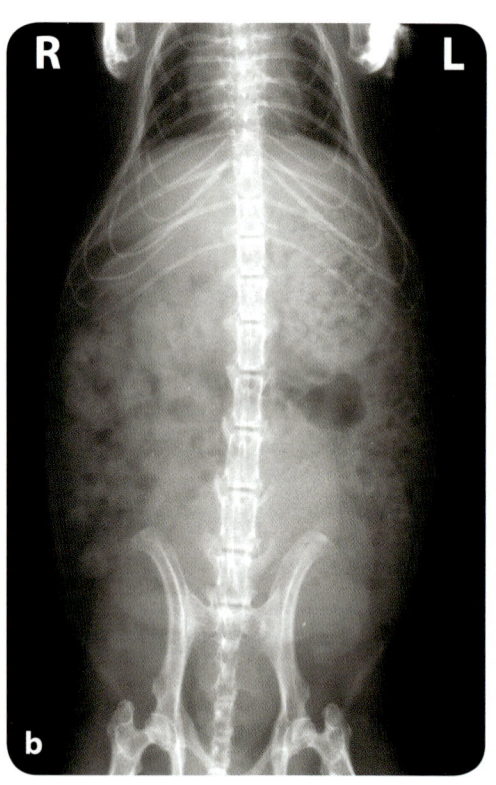

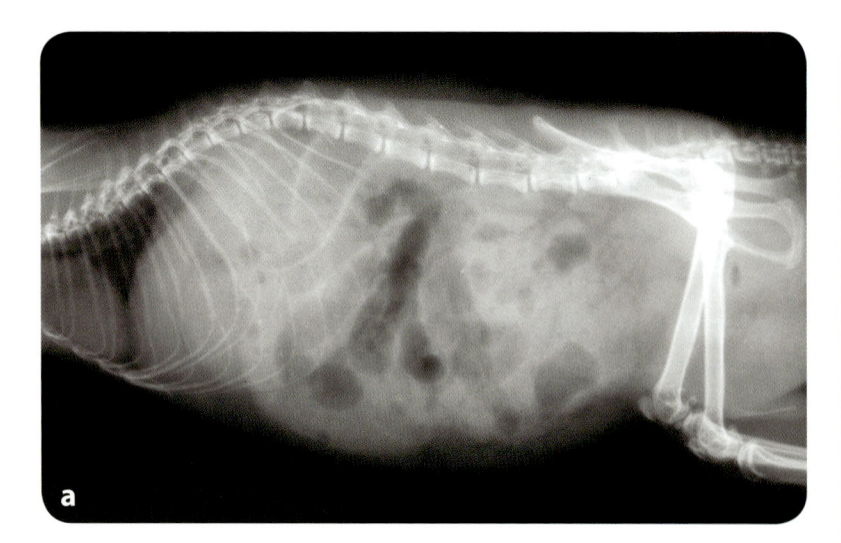

Figure 4.42a,b. Lateral (a) and ventrodorsal (b) radiographs of a chinchilla with gastrointestinal impaction and moderate intestinal gas. The stomach is completely filled with ingesta, and appears slightly radiopaque, which is emphasized in (b). Intestinal gas may be a result of decreased gastrointestinal motility.

ABDOMEN 249

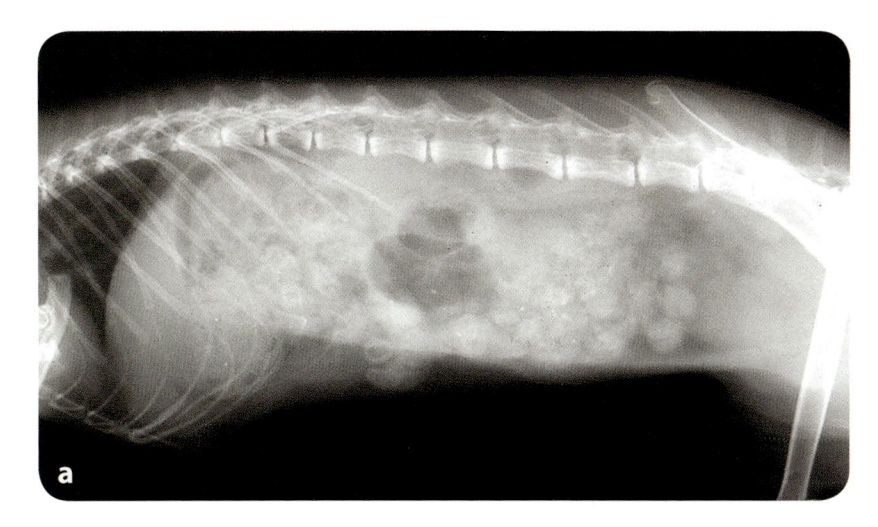

Figure 4.43a,b. Lateral (a) and ventrodorsal (b) radiographs of the abdomen of a chinchilla with gastrointestinal impaction as a result of inappropriate diet and acquired dental disease. The abdominal silhouette is more slender than normal in this chronically underweight animal. Fecal pellets are more dense than normal, and fill most of the intestine. Note radiopaque stomach contents. Abnormal gas accumulation is also present in the cecum.

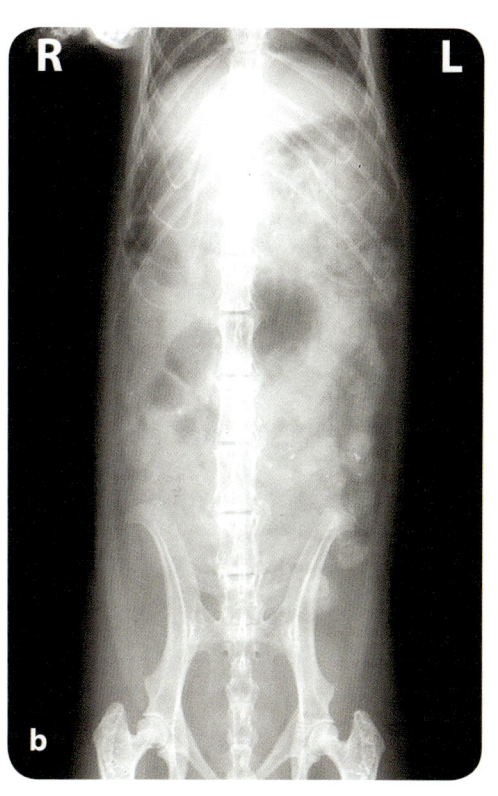

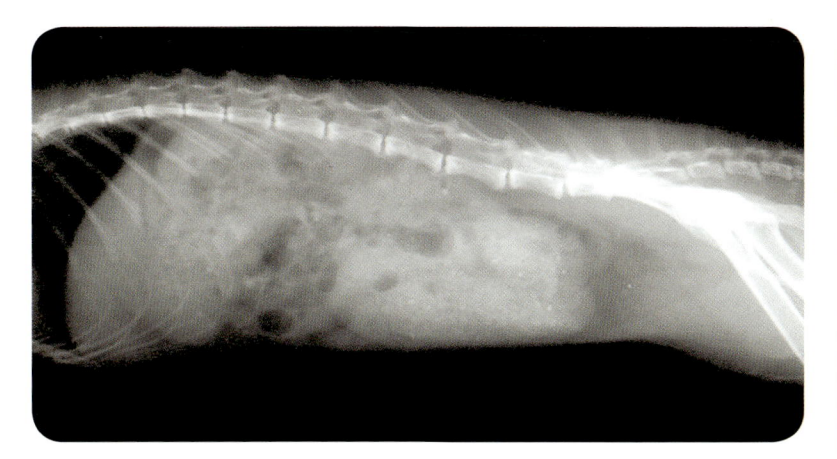

Figure 4.44. Lateral radiograph of the abdomen of a chinchilla with gastrointestinal impaction. Ingesta in the cecum appears abnormally radiodense.

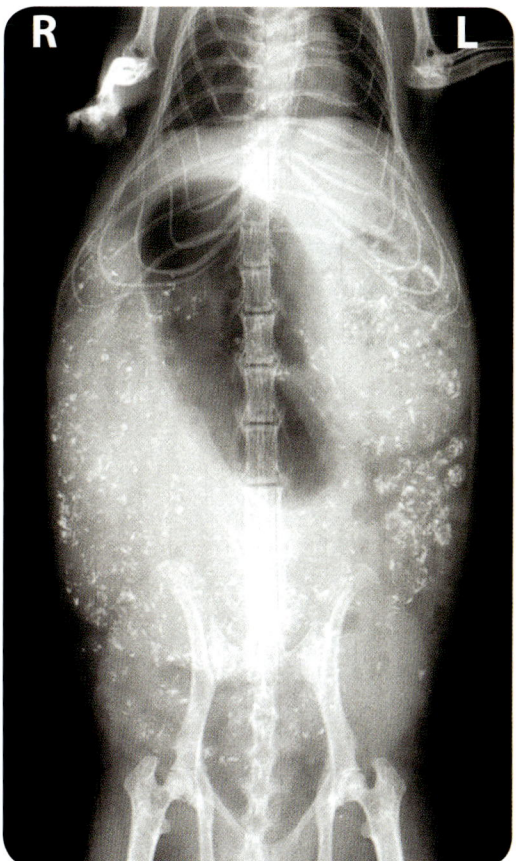

Figure 4.45. Ventrodorsal radiograph of the abdomen of a chinchilla with gastrointestinal impaction due to ingestion of inappropriate bedding (cat litter), which appears very radiopaque.

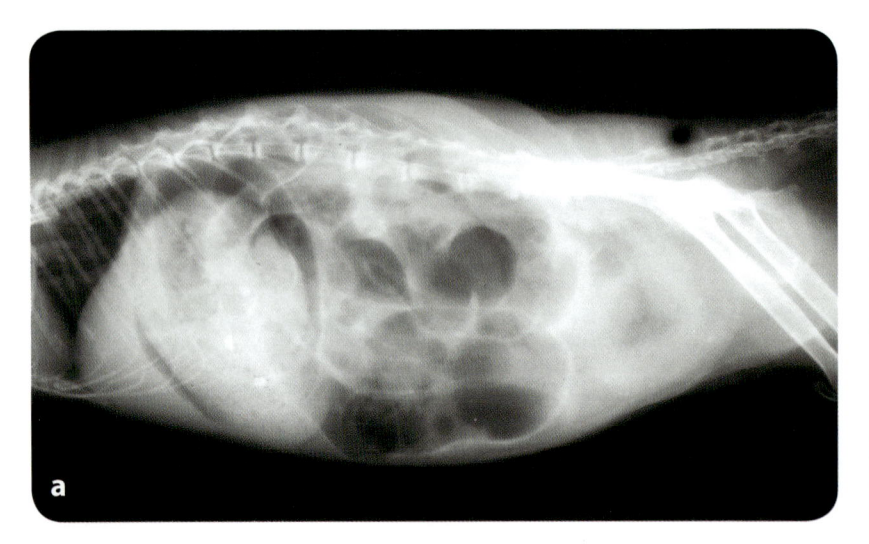

Figure 4.46a,b. Lateral (a) and ventrodorsal (b) radiographs of a chinchilla with marked gastric impaction and marked gaseous distension of the cecum.

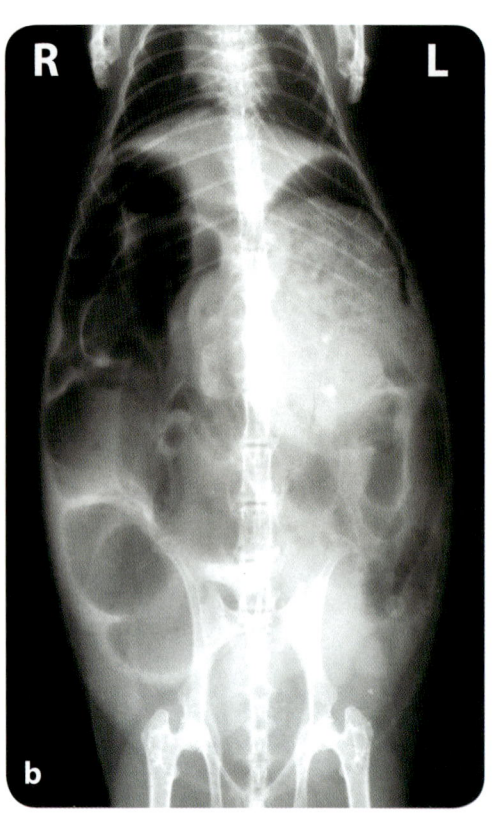

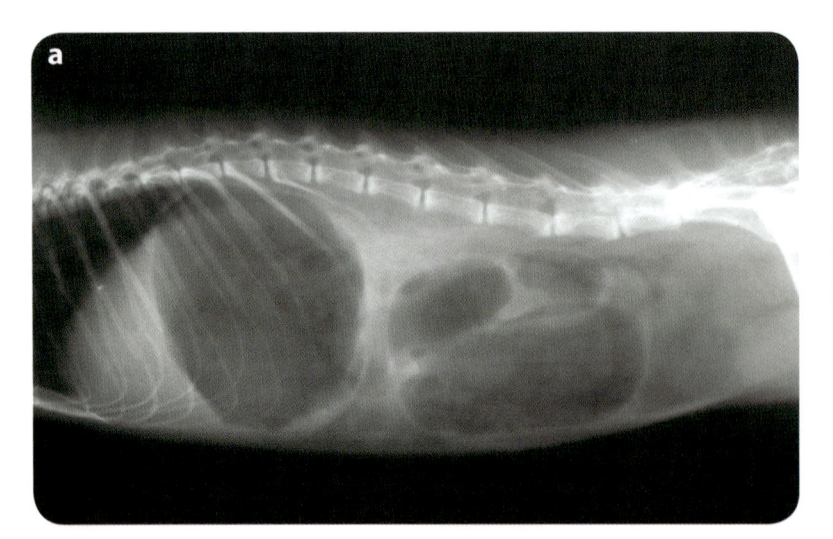

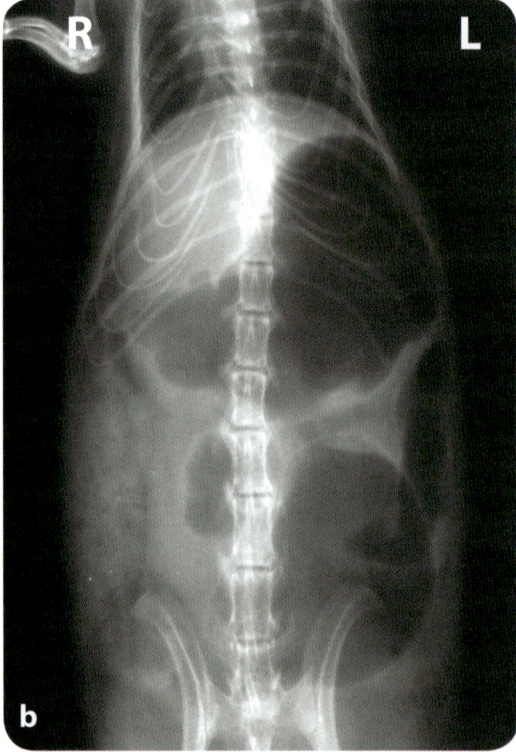

Figure 4.47a,b. Lateral (a) and ventrodorsal (b) radiographs of the abdomen of a chinchilla with a large accumulation of gas in the stomach and lower gastrointestinal tract. This patient was anorexic secondary to severe acquired dental disease.

ABDOMEN 251

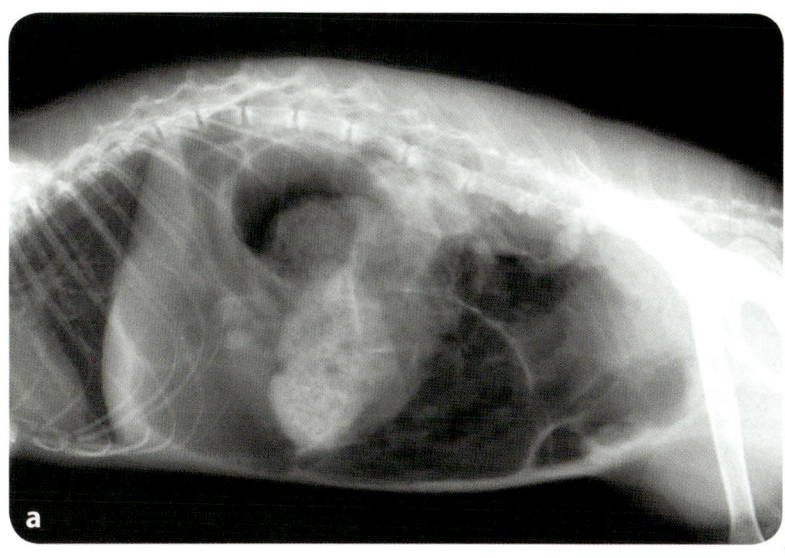

Figure 4.48a,b. Lateral (a) and ventrodorsal (b) radiographs of the abdomen of a chinchilla with severe distention of the gastrointestinal tract with gas. The cecum is severely distended, with air acting as a negative contrast medium, outlining abnormally radiodense ingesta.

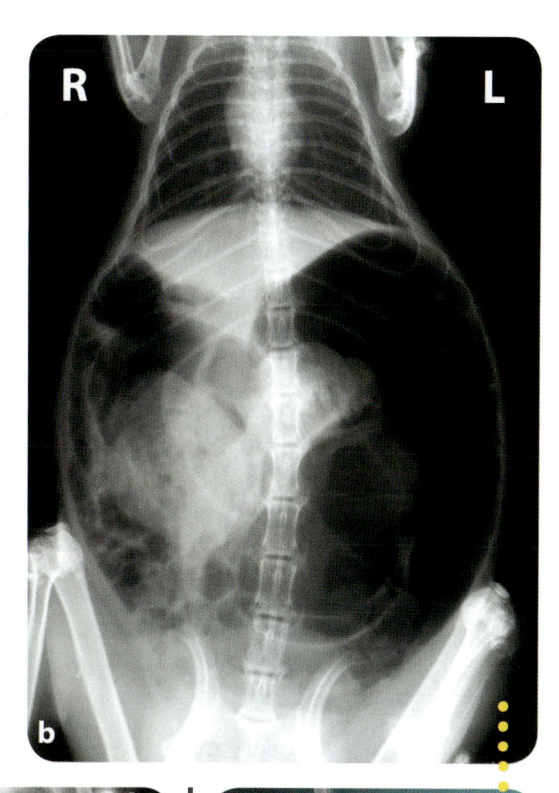

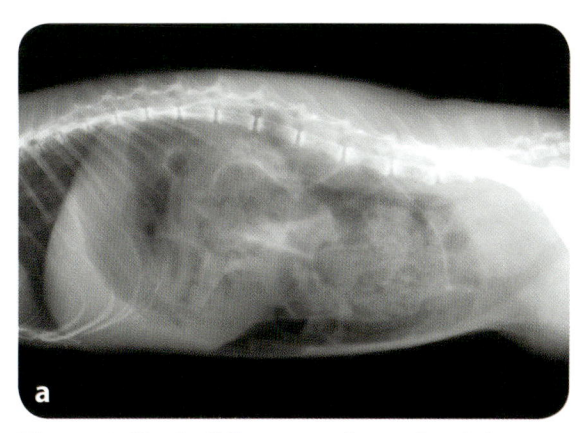

Figure 4.49a,b. Follow-up radiographs of the same patient, lateral (a) and ventrodorsal (b), after 3 days of medical therapy, including fluids, hand feeding and motility enhancing drugs. Gas accumulation is significantly reduced.

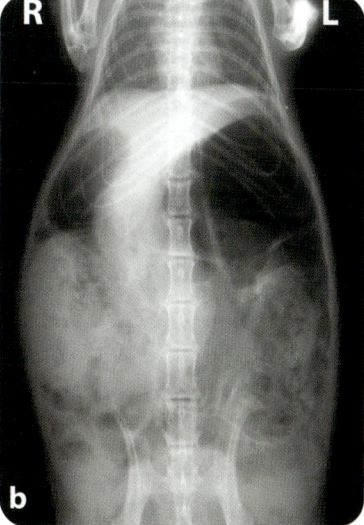

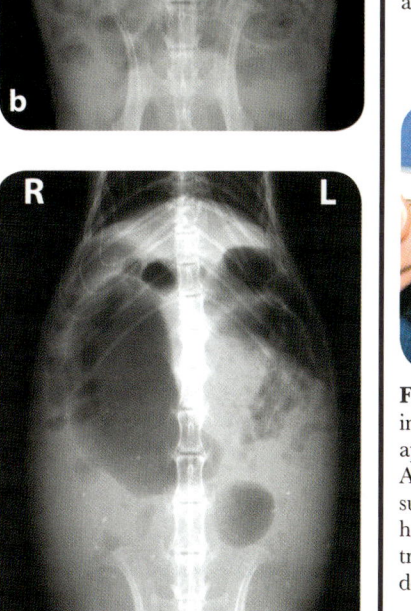

Figure 4.50. Clinical appearance of a severely depressed and dehydrated chinchilla with gastrointestinal disease. Reduced activity and hunched posture are indications of abdominal discomfort and pain.

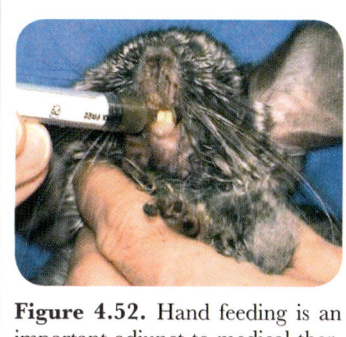

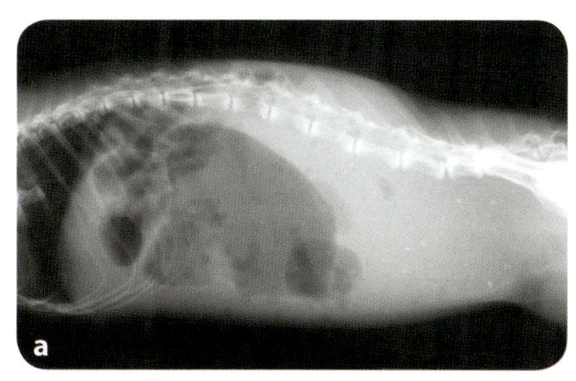

Figure 4.51a,b. Follow-up radiographs of the same patient, lateral (a) and ventrodorsal (b), after 7 days of medical therapy. Gas distention of the cecum is present but significantly reduced. Uniform radiodensity of the ingesta is due to oral rehydration and force feeding.

Figure 4.52. Hand feeding is an important adjunct to medical therapy, especially in anorexic patients. A high fiber hand feeding product such as Herbivore Critical Care helps promote peristalsis of the GI tract (Oxbow Pet Products, Murdoch, NE, www.oxbowhay.com).

Diseases of the Female Genital Tract

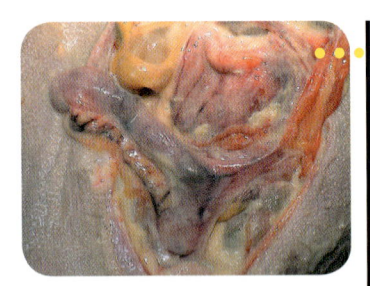

Figure 4.53. Necropsy of the same patient. The distended left uterine horn has ruptured, and purulent material fills the abdominal cavity. The right uterine horn is enlarged as well.

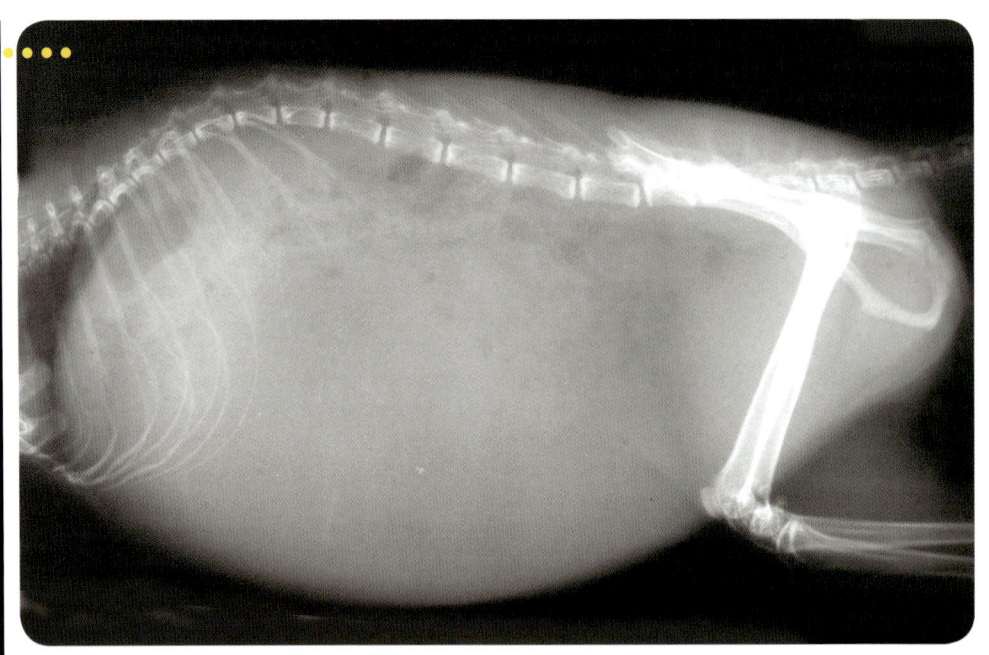

Figure 4.54. Lateral radiograph of the abdomen of a female chinchilla with abdominal distention. There is complete loss of abdominal contrast due to the presence of peritoneal fluid.

ABNORMALITIES of the VERTEBRAL COLUMN

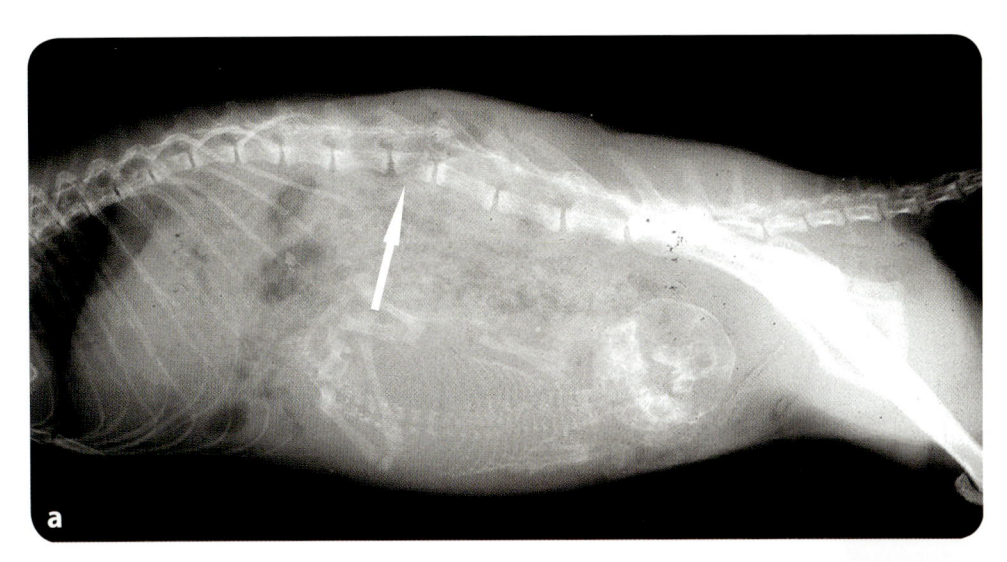

Figure 4.55a,b. Lateral (a) and ventrodorsal (b) radiographs of the abdomen of a pregnant chinchilla with a fracture of L3 (white arrow), following a fall from a height. The ventrodorsal projection demonstrates fracture associated with subluxation of L3-L4 (yellow arrow). The single fetus was later aborted. The vertebral lesion produced mild paresis, which eventually resolved.

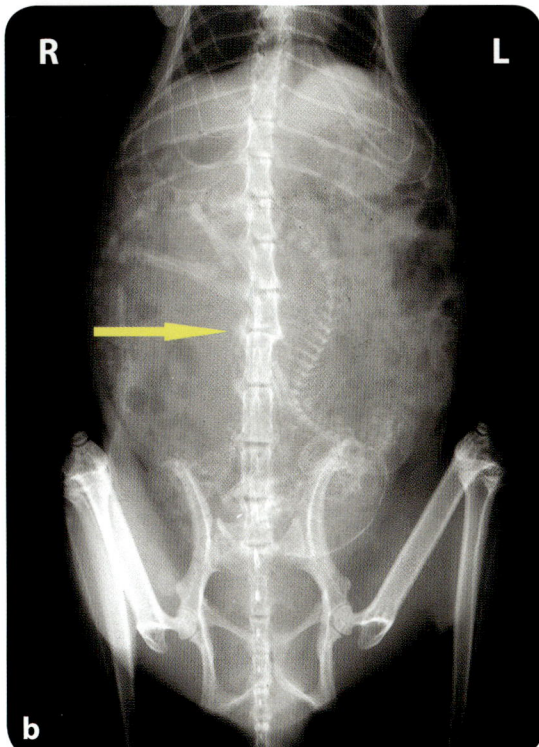

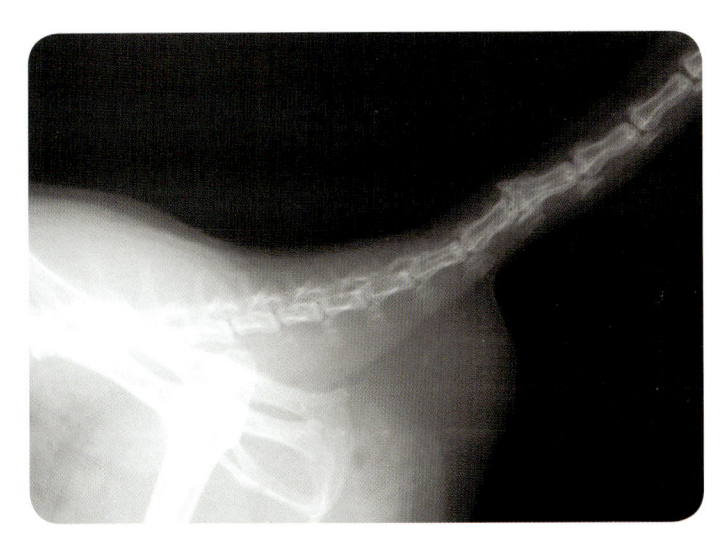

Figure 4.56. Lateral radiograph of the pelvis and tail of a chinchilla with fractures of several caudal vertebrae as a result of improper restraint.

The NORMAL THORACIC LIMB
Lateral Projection

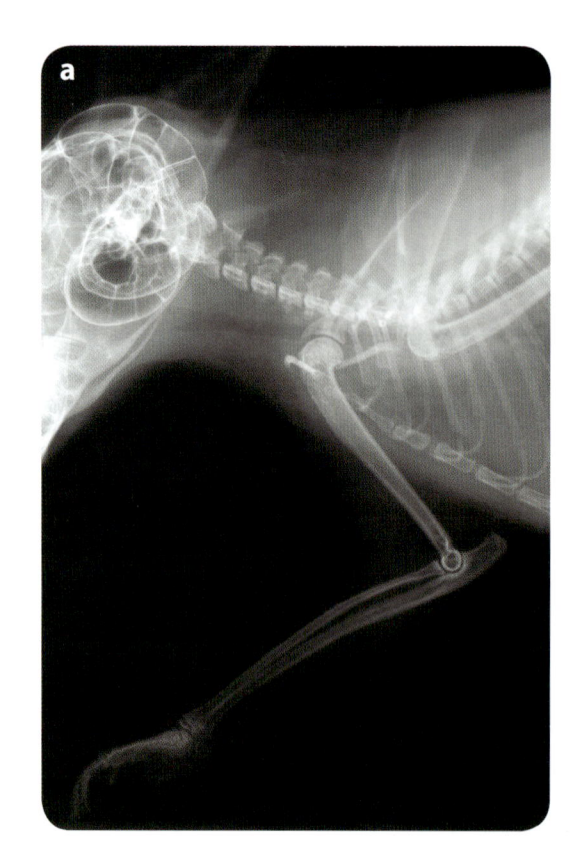

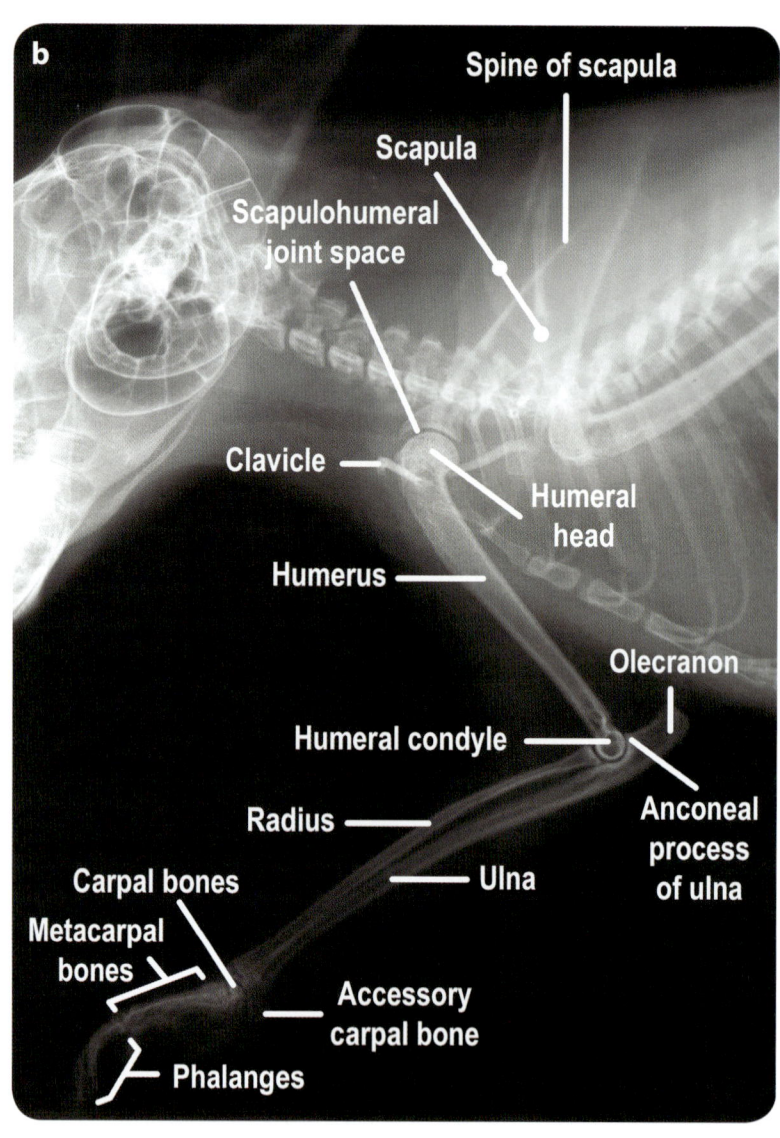

Figure 4.57a,b. Radiograph of the thoracic limb of a 550 g female chinchilla, actual size (a) and 140% of actual size (b).

Craniocaudal Projection

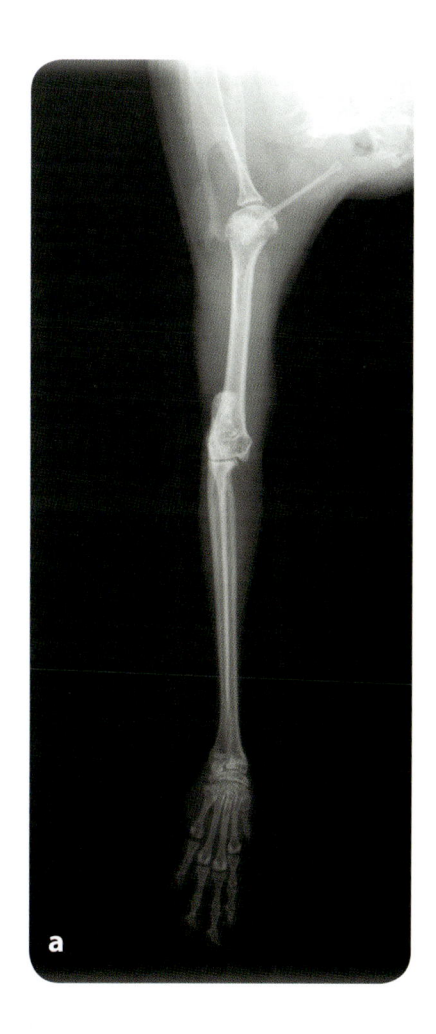

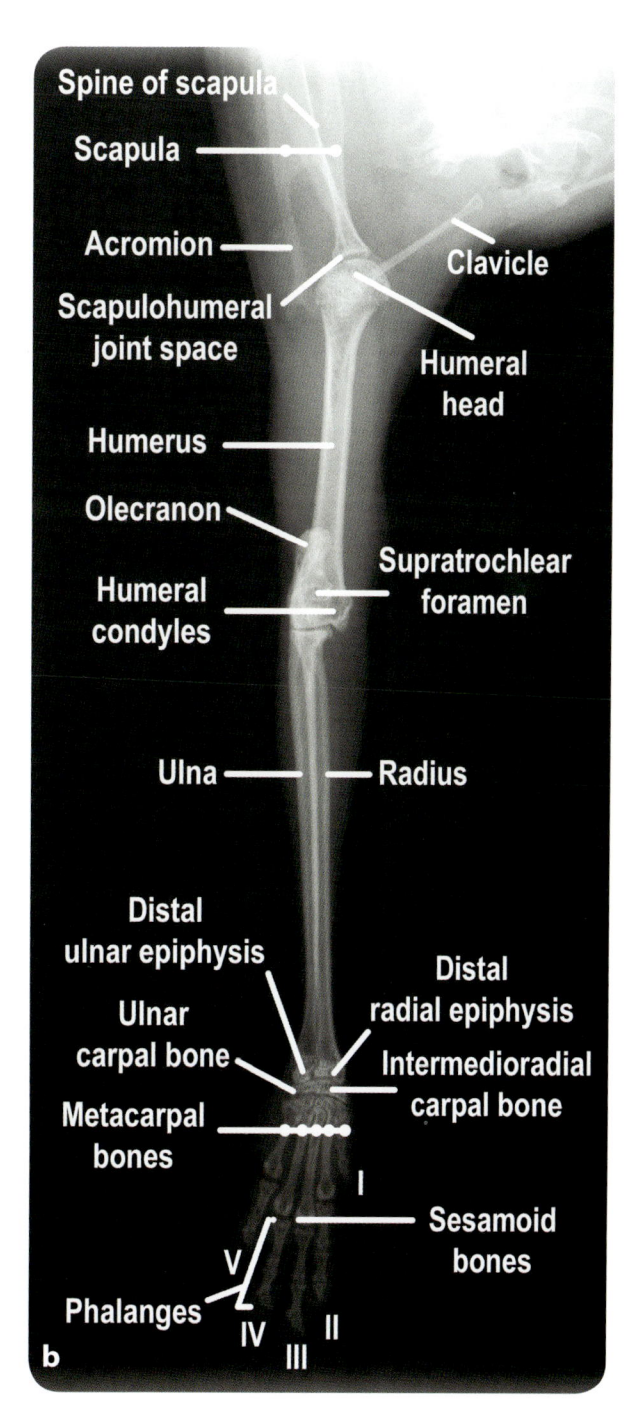

Figure 4.58a,b. Radiograph of the thoracic limb of a 550 g female chinchilla, actual size (a), and 140% of actual size (b).

The NORMAL PELVIC LIMB
Lateral Projection

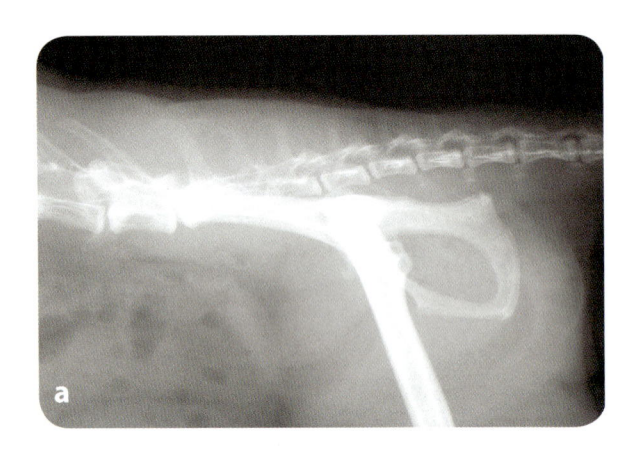

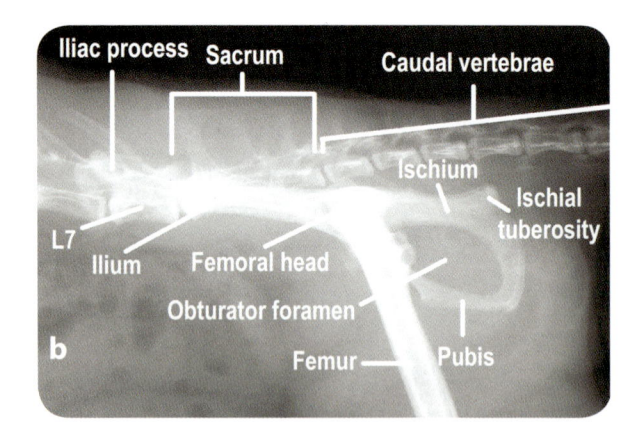

Figure 4.59a,b. Radiograph of the pelvis of a 550 g female chinchilla, actual size.

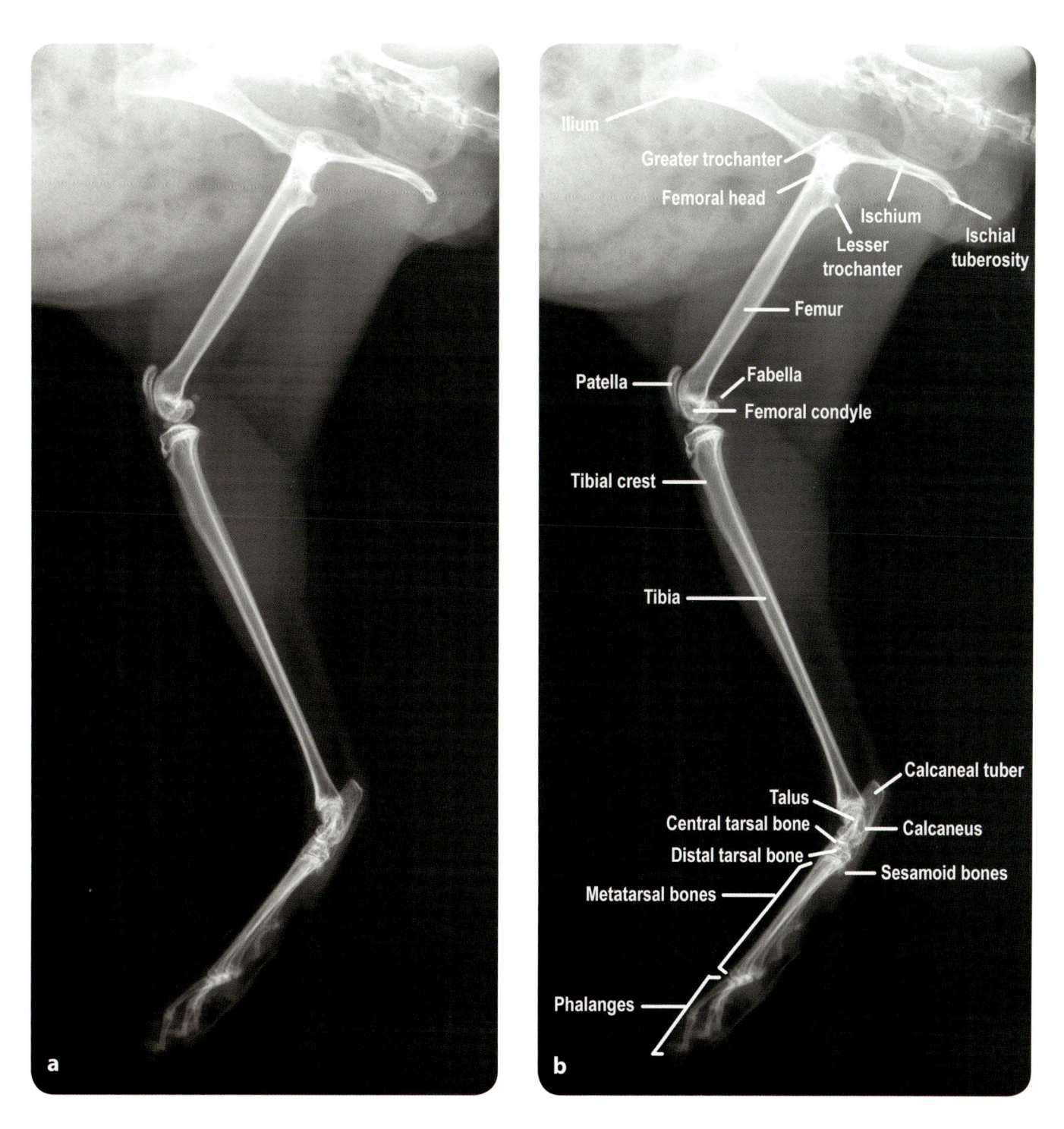

Figure 4.60a,b. Radiograph of the pelvic limb of a 550 g female chinchilla, actual size.

Ventrodorsal Projection of the Pelvis
Craniocaudal Projection of the Proximal Pelvic Limb

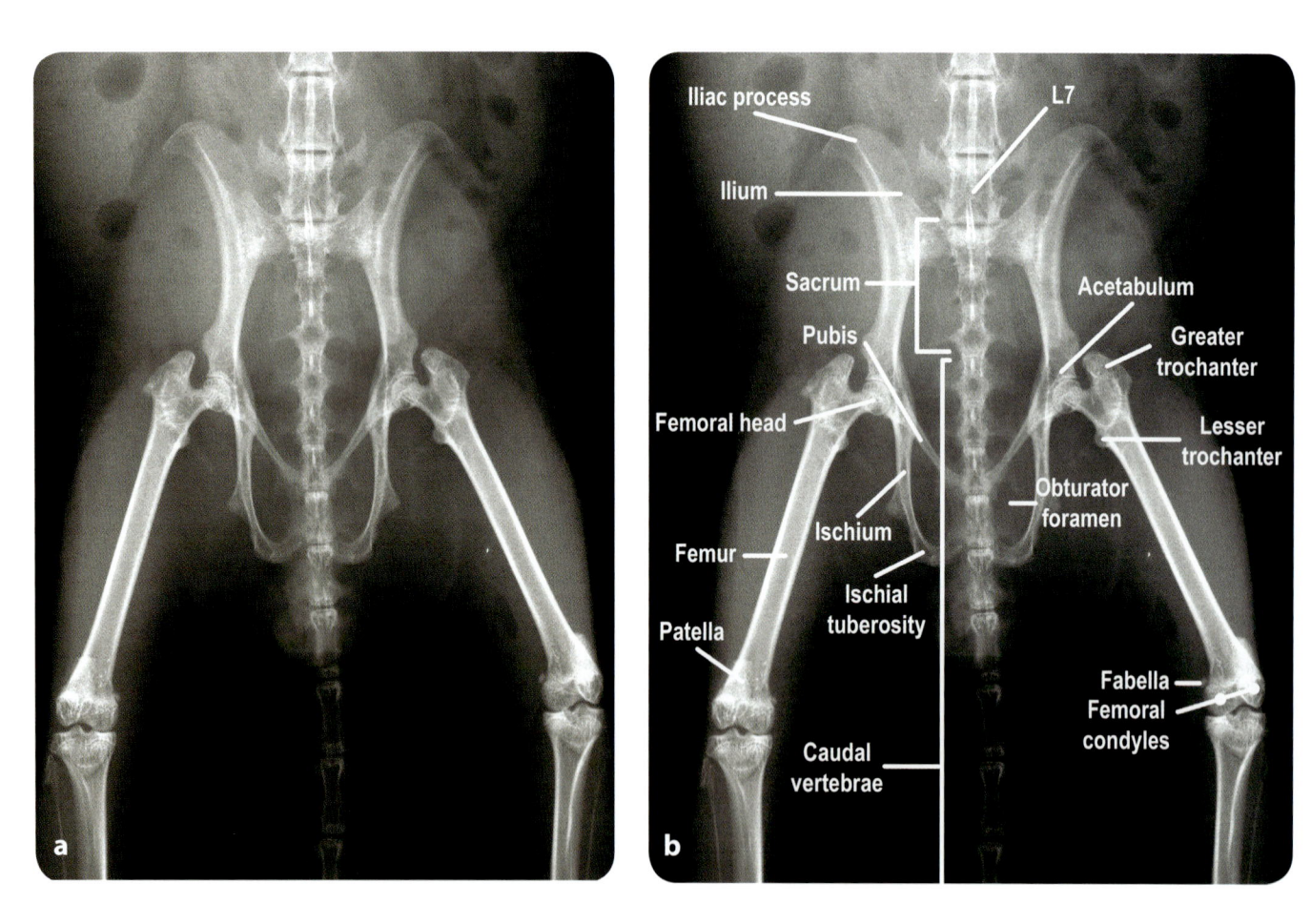

Figure 4.61a,b. Radiograph of the pelvis and proximal pelvic limb of a 550 g female chinchilla, actual size.

Craniocaudal Projection of the Distal Pelvic Limb

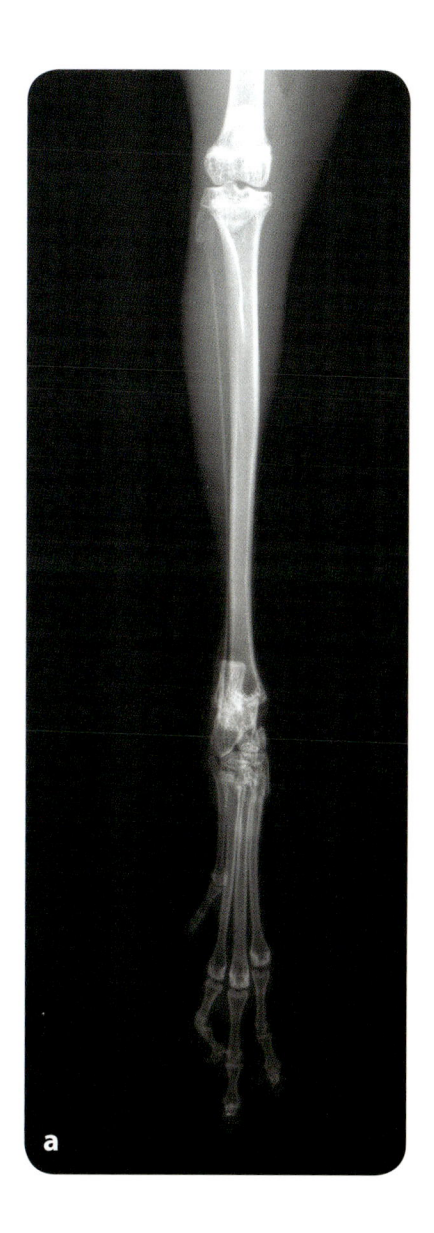

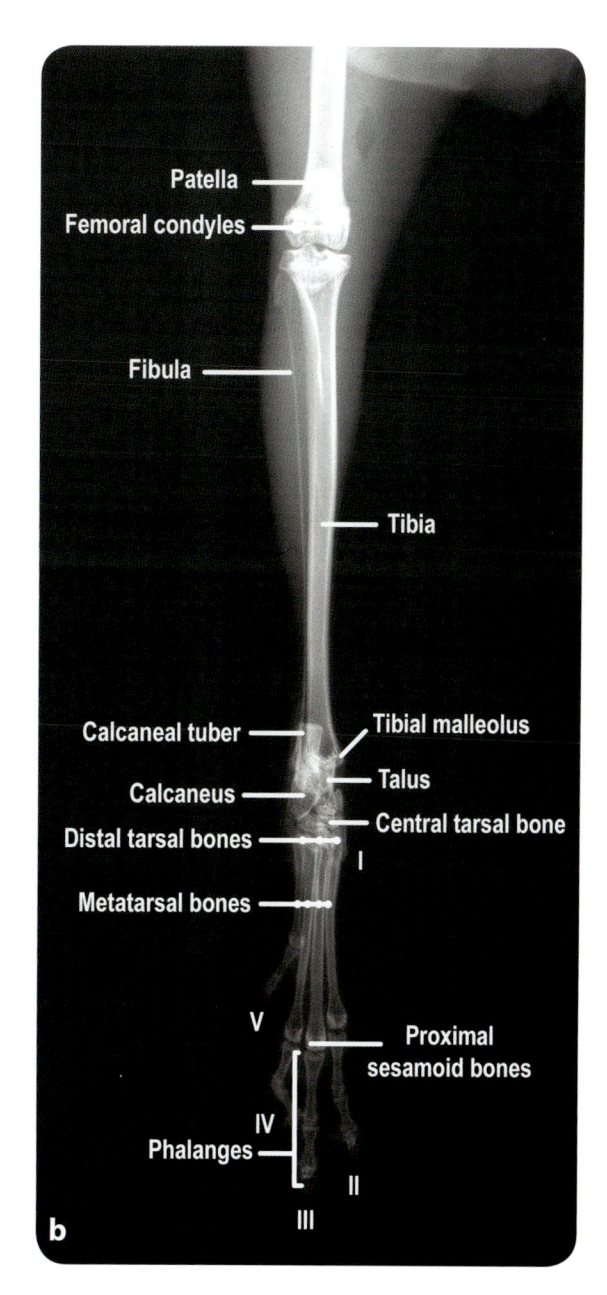

Figure 4.62a,b. Radiograph of the distal pelvic limb of a 550 g female chinchilla, actual size.

ABNORMALITIES of the PELVIC LIMB
Diseases of the Pelvis

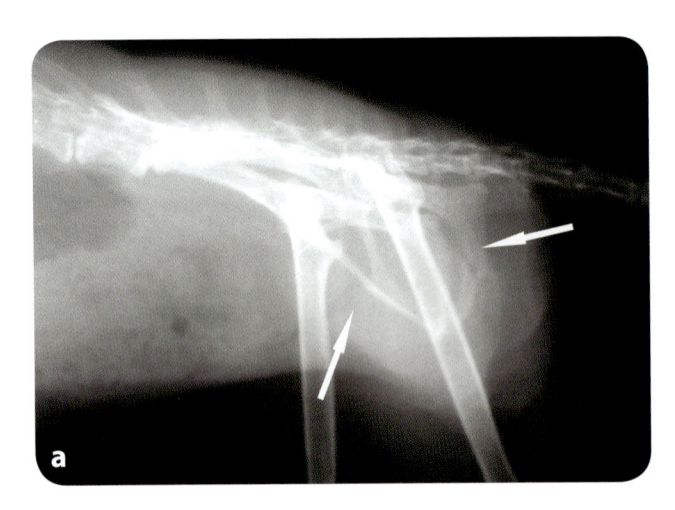

Figure 4.63a,b. Lateral (a) and ventrodorsal (b) radiographs of the pelvis of a chinchilla with a fracture of the ischiatic and pubic bones (arrows). The fracture of the ischium is more visible in the lateral projection; the pubic fracture is more visible in the ventrodorsal projection.

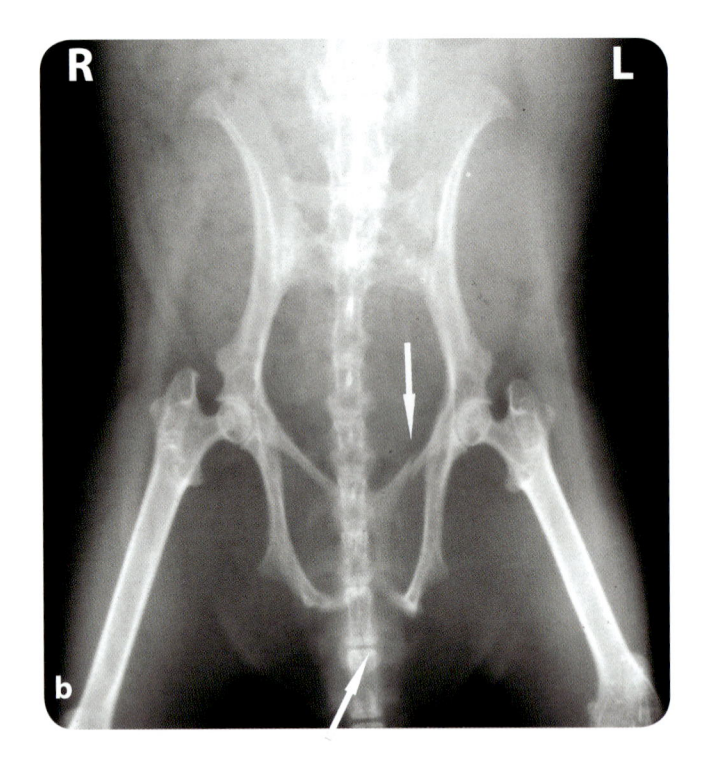

Diseases of the Tibia and Fibula

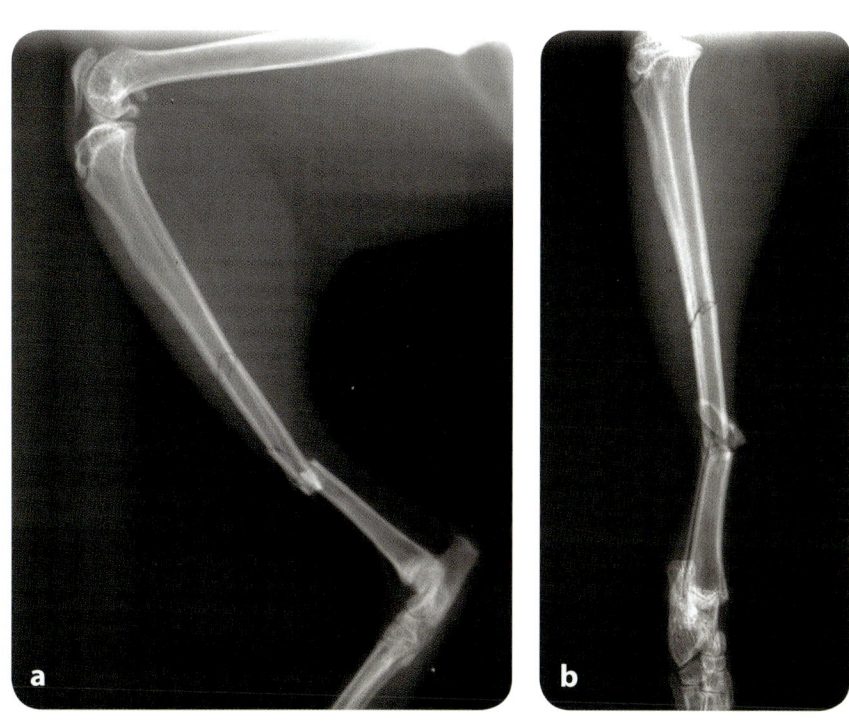

Figure 4.64a,b. Lateral (a) and craniocaudal (b) radiographs of the pelvic limb of a chinchilla with a comminuted distal metaphyseal fracture of the tibia with minimal displacement, and nondisplaced fracture of the shaft of the fibula.

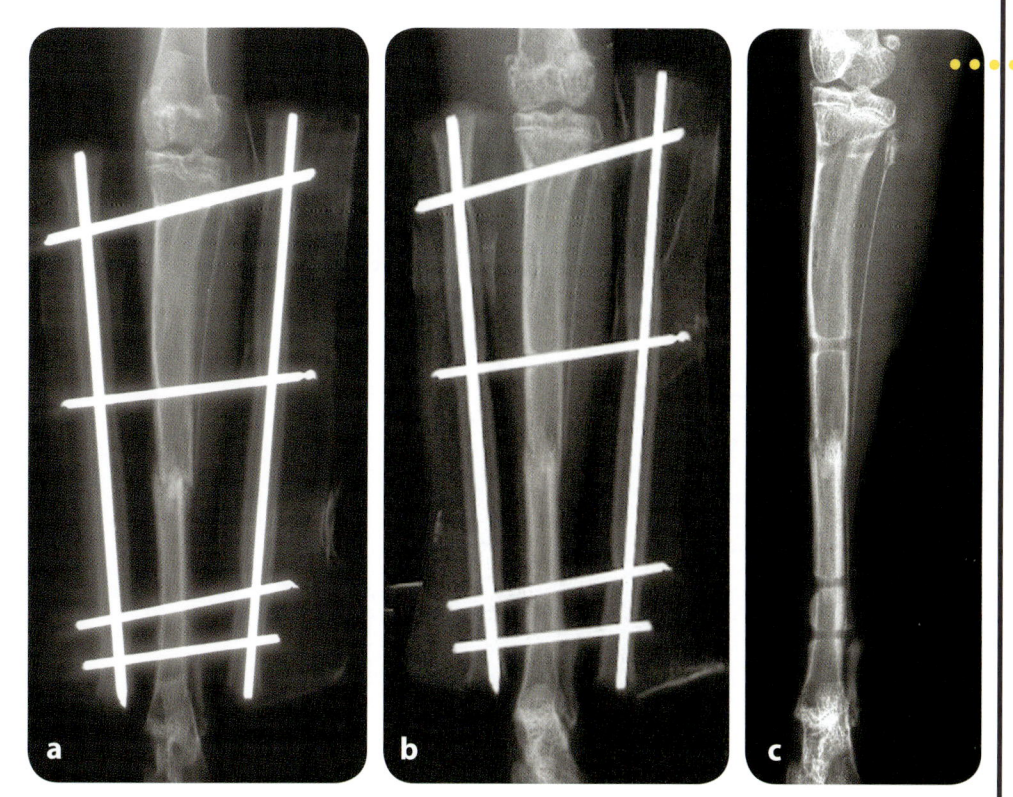

Figure 4.65a,b,c. Craniocaudal radiograph of the tibia of a chinchilla after insertion of a 2+2 1 mm pin external fixator device (a). Same patient 6 weeks post surgery showing normal bone healing at the fracture site (b). Radiographic appearance soon after removal of the external fixator (c). The pin tracts will eventually fill in with new bone.
External fixation is the treatment of choice for fractures of the tibia.

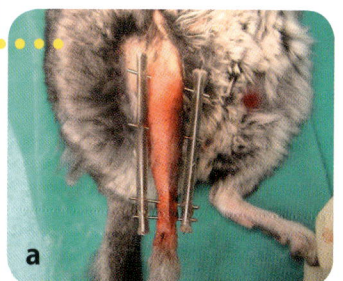

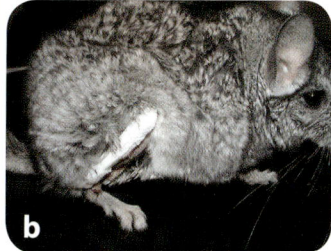

Figure 4.66a,b. External fixation devices are well tolerated by chinchillas, but external bars must be adequately protected with padding to prevent damage to soft tissues.

References:
Brenner SZG, Hawkins MG, Tell LA, Hornof WJ, Plopper CG, Verstraete FJM: Clinical anatomy, radiography, and computer tomography of the chinchilla skull. Comp Cont Ed. 2005; 27:933-944.

Brown SA, Rosenthal LR: Question #36. In: Self Assessment Color Review of Small Mammals. London: Manson Publishing; 1997:33-34.

Brown SA, Rosenthal LR: Question #77. In: Self Assessment Color Review of Small Mammals. London: Manson Publishing; 1997:39-40.

Capello V, Gracis M: Radiology of the skull and teeth. In: Lennox A, ed. Rabbit and Rodent Dentistry Handbook. Ames, IA: Blackwell Publishing, (Formerly Zoological Education Network, Lake Worth, FL); 2005:65-69.

Capello V: Dental diseases. In: Lennox A, ed. Rabbit and Rodent Dentistry Handbook. Ames, IA: Blackwell Publishing, (Formerly Zoological Education Network, Lake Worth, FL); 2005:113-163.

Capello V: Secondary diseases. In: Lennox A, ed. Rabbit and Rodent Dentistry Handbook. Ames, IA: Blackwell Publishing, (Formerly Zoological Education Network, Lake Worth, FL); 2005:165-189.

Capello V, Gracis M: Dental procedures. In: Lennox A, ed. Rabbit and Rodent Dentistry Handbook. Ames, IA: Blackwell Publishing, (Formerly Zoological Education Network, Lake Worth, FL); 2005:213-247.

Capello V, Gracis M: Surgical treatment of periapical abscessations. In: Lennox A, ed. Rabbit and Rodent Dentistry Handbook. Ames, IA: Blackwell Publishing, (Formerly Zoological Education Network, Lake Worth, FL); 2005:249-237.

Capello V: Dental diseases and surgical treatment in pet rodents. Exotic DVM. 2003; 5(3):21-27.

Capello V: External fixation for fracture repair in small exotic mammals. Exotic DVM. 2005; 7(6):21-37.

Crossley DA, Jackson A, Yates J, Boydell IP: Use of computer tomography to investigate cheek tooth abnormalities in chinchillas *(Chinchilla laniger)*. J Small An Pract 1998; 39:385-389.

Crossley DA: Dental disease in chinchillas in the UK. J Small Anim Pract 2001; 42:12-19.

Girling SJ: Mammalian imaging and anatomy In: Meredith A, Redrobe S, eds. BSAVA Manual of Exotic Pets 4th ed. Quedgeley, Glouchester: British Small Animal Veterinary Association; 2005:1-12.

Hoefer HL, Crossley DA: Chinchillas. In: Meredith A, Redrobe S, eds. BSAVA Manual of Exotic Pets 4th ed. Quedgeley, Glouchester: British Small Animal VeterinaryAssociation; 2005:65-75.

Kapatkin A: Orthopedics in small mammals. In: Quesenberry KE, Carpenter JW, eds. Ferrets, Rabbits and Rodents. Clinical Medicine and Surgery 2nd ed. Philadelphia, PA: Elsevier; 2004:383-391.

Legendre LFJ: Oral disorders of exotic rodents. Vet Clin Exot Anim. 2003; 6:601-628.

Silverman S, Tell LA: Domestic chinchilla *(Chinchilla lanigera)*. In: Radiology of Rodents, Rabbits and Ferrets. An Atlas of Normal Anatomy and Positioning. Philadelphia, PA: Elsevier Saunders; 2005:67-104.

Stefanacci JD, Hoefer HL: Radiology and ultrasound. In: Quesenberry KE, Carpenter JW, eds. Ferrets, Rabbits and Rodents. Clinical Medicine and Surgery 2nd ed. Philadelphia, PA: Elsevier; 2004:395-419.

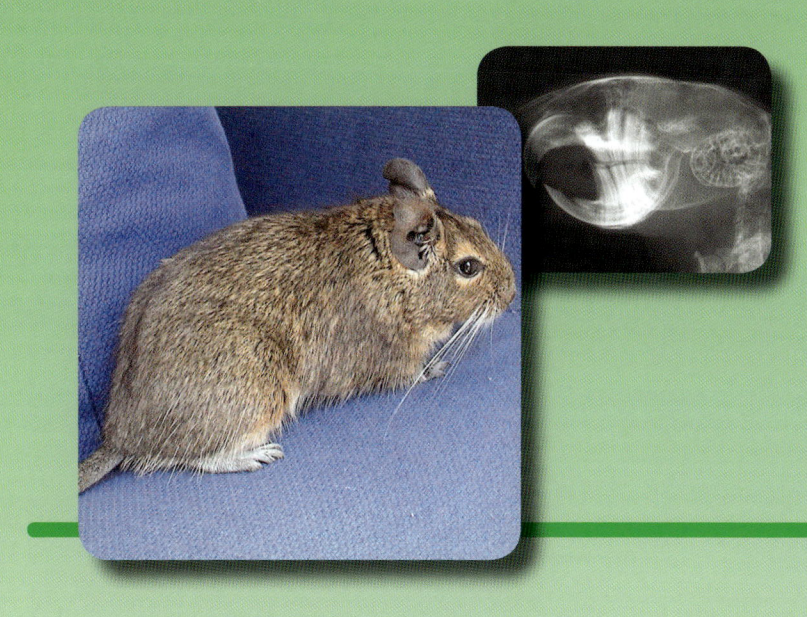

DEGU

The NORMAL HEAD
Lateral Projection

Figure 5.1a,b. 240% or actual size. The normal cheek teeth occlusal plane in the degu is horizontal. Incisor teeth are very large. The apex of the maxillary incisor teeth extends to CT1 (yellow circle). The mandibular incisor teeth extend the entire length of the mandible, and their apex is located caudal to the last cheek tooth (red circle). In this image, the mandible appears slightly rostrally subluxated. This is indicated by the imperfect occlusion between maxillary and mandibular cheek teeth.

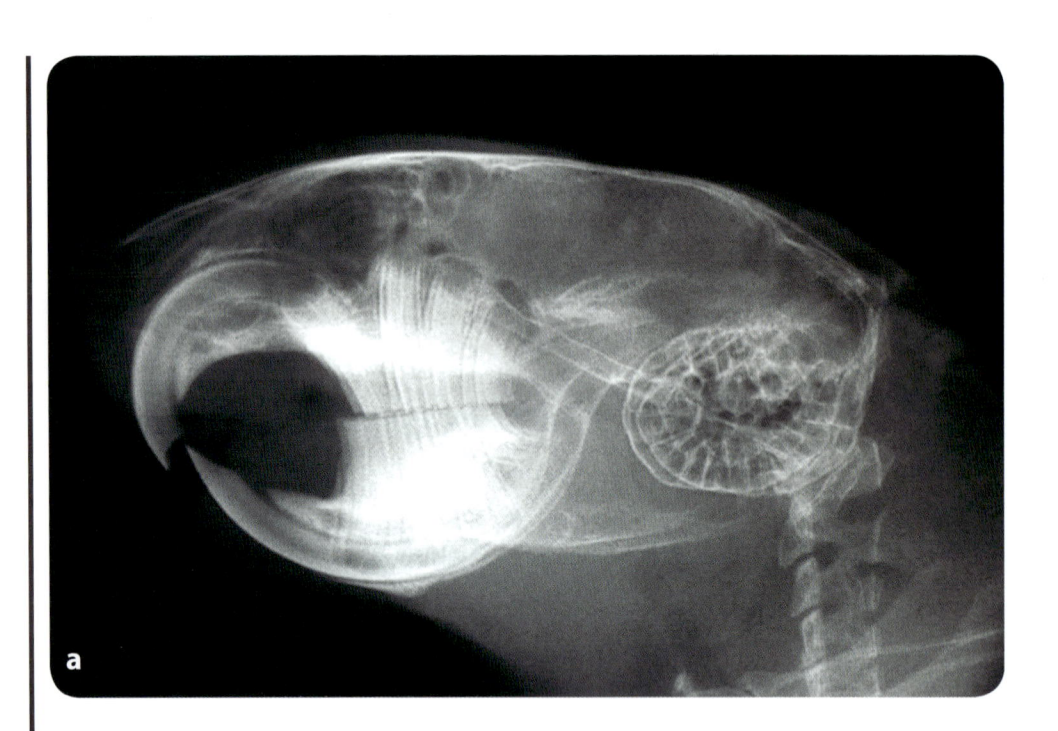

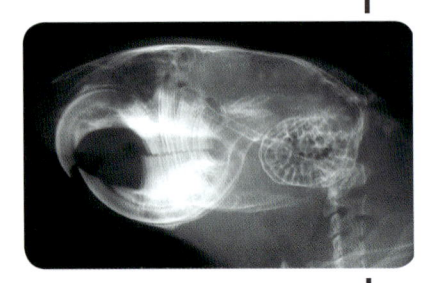

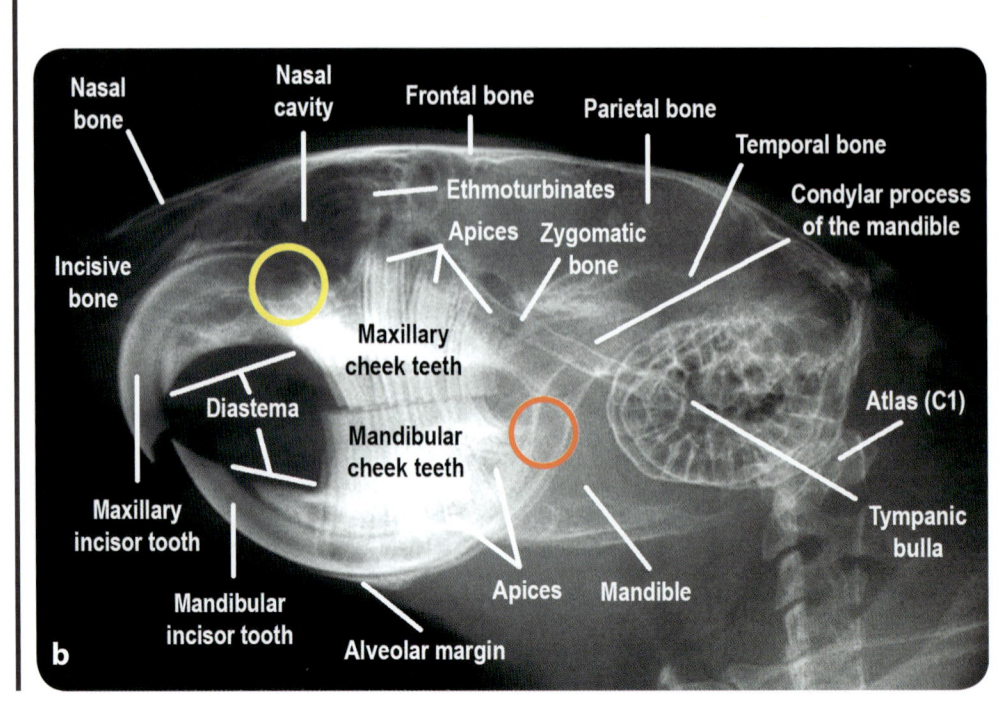

Figure 5.2. Actual size of the skull of a 200 g male degu.

Oblique Projection

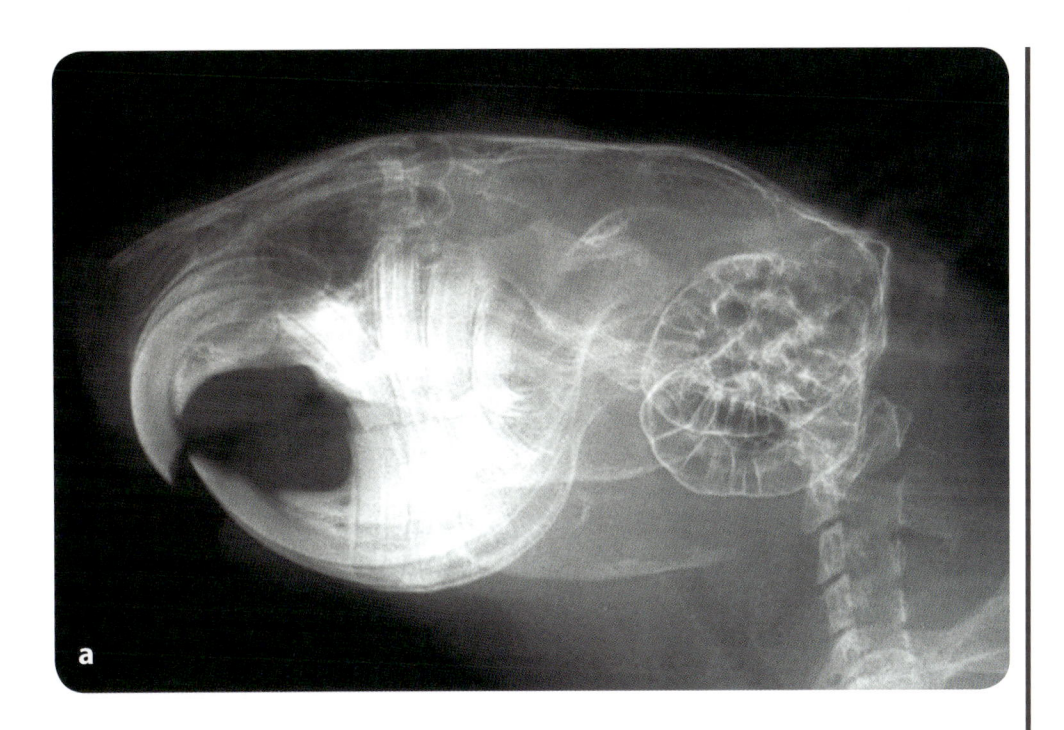

Figure 5.3a,b. Left 15° ventral-right dorsal oblique radiograph of the skull of a degu. The apex of the mandibular incisor tooth is caudal to the last mandibular cheek tooth (red circle.)

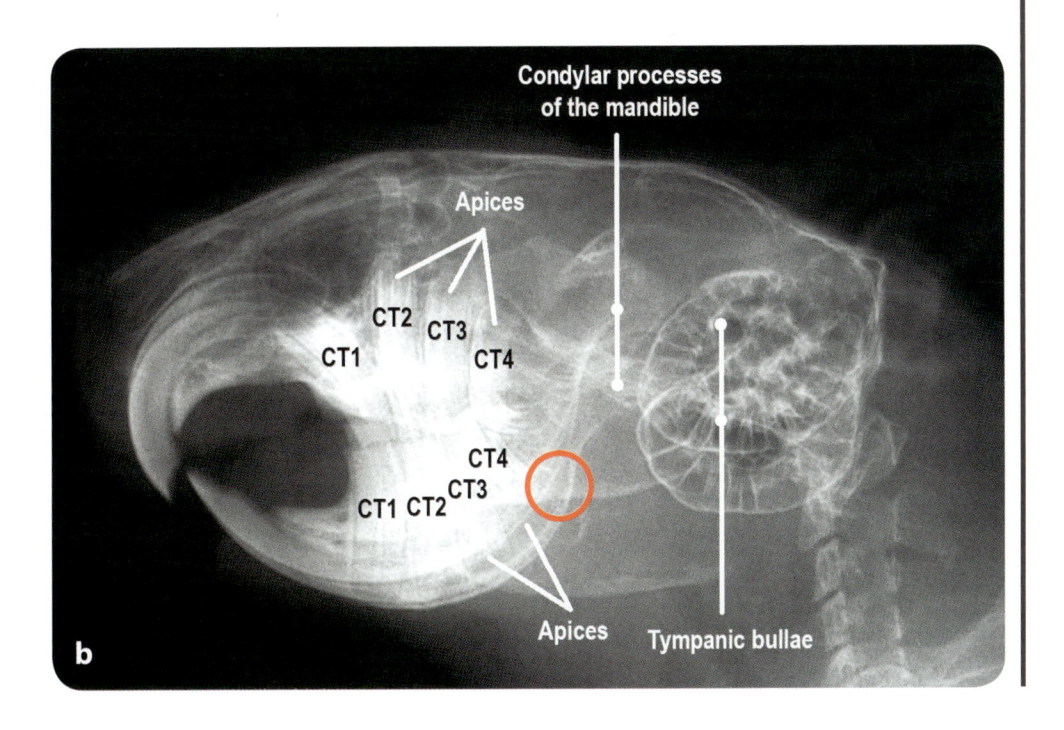

Ventrodorsal Projection

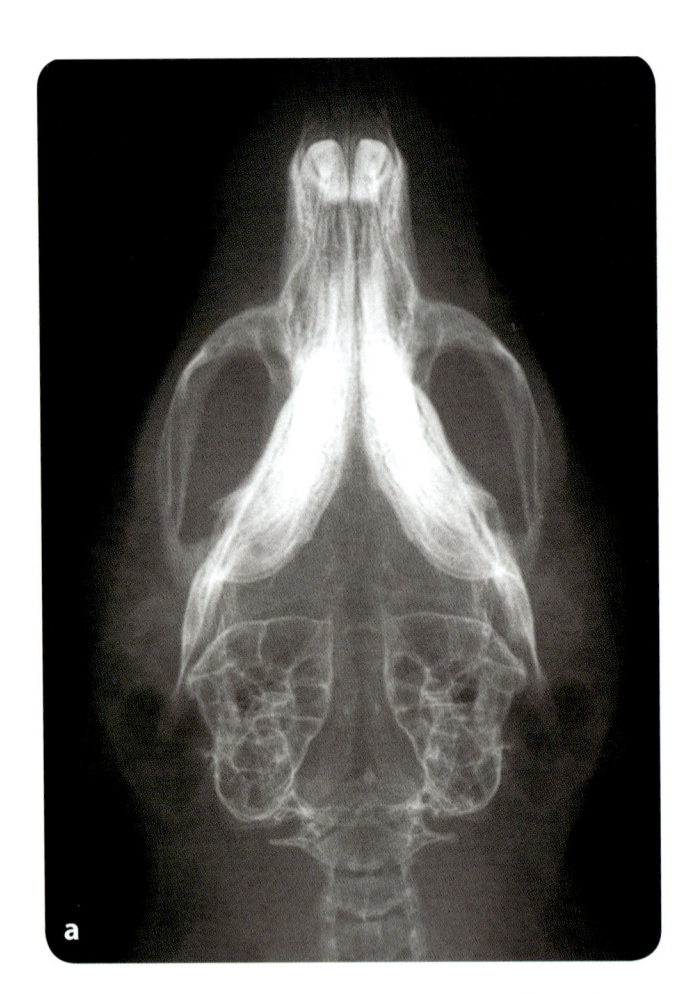

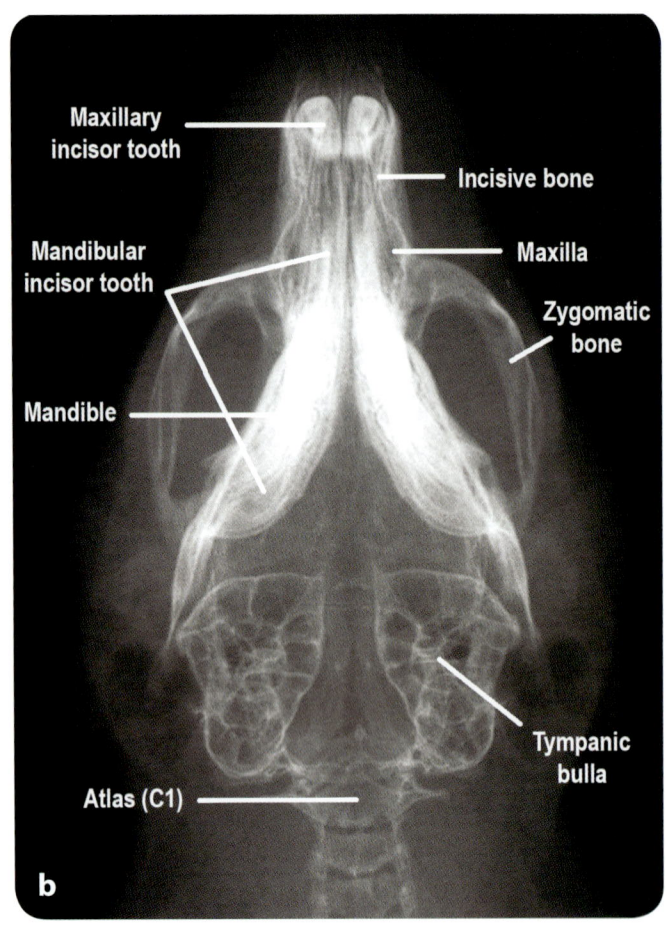

Figure 5.4a,b. Ventrodorsal view of the skull of a degu.

Rostrocaudal Projection

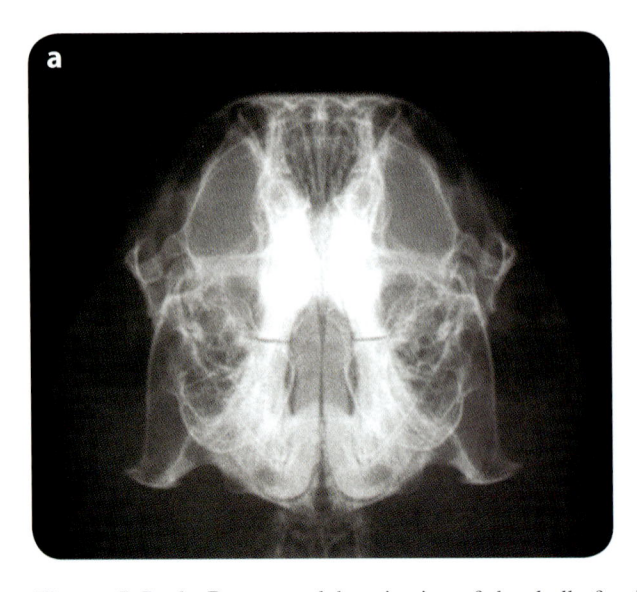

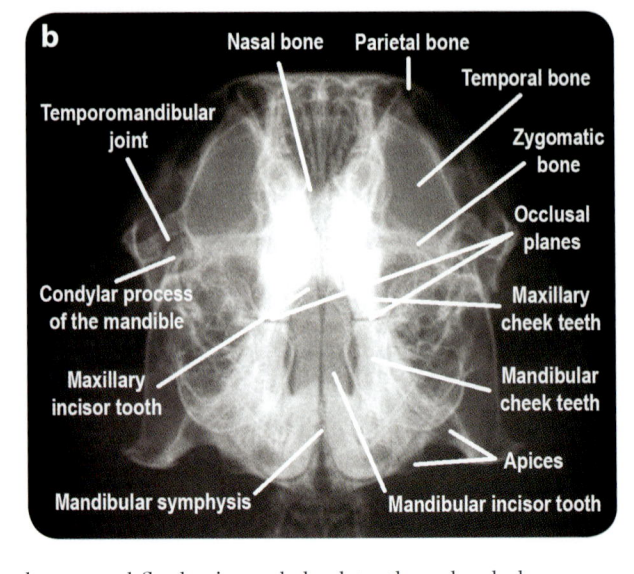

Figure 5.5a,b. Rostrocaudal projection of the skull of a degu. Note the normal flat horizontal cheek teeth occlusal plane.

ABNORMALITIES of the HEAD
Diseases of Cheek Teeth

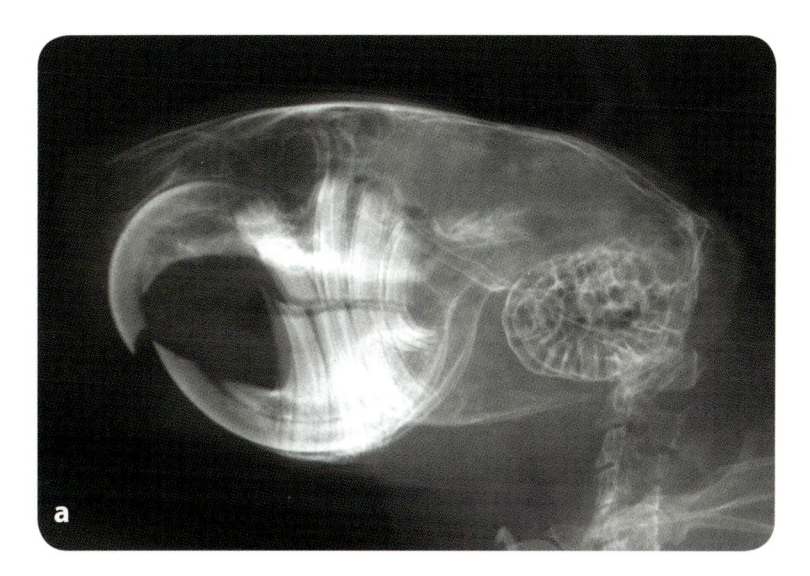

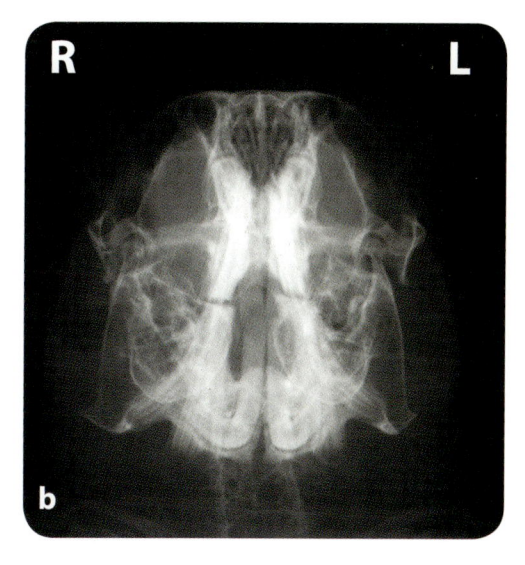

Figure 5.6a,b. Lateral (a) and rostrocaudal (b) views of a degu with acquired dental disease of the cheek teeth. Uneven mandibular cheek teeth produce alteration of the normal horizontal occlusal plane which is clearly visible on the lateral view. This condition corresponds to "wave mouth" of rabbits. The rostrocaudal projection demonstrates excessive length of both right maxillary and left mandibular arcades. A spur is also visible on left mandibular CT1 or CT2.

The NORMAL WHOLE BODY SKELETON
Lateral Projection

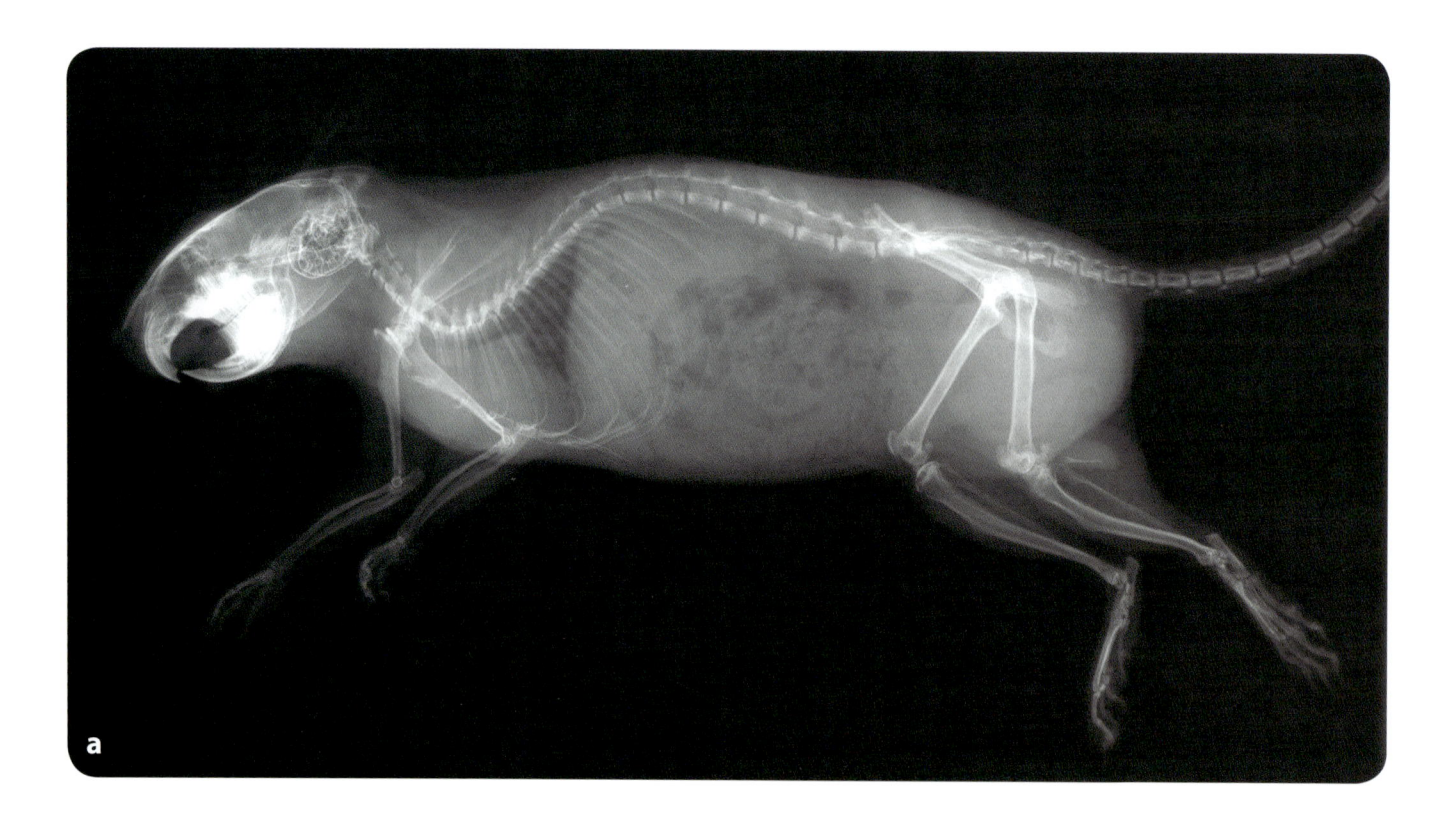

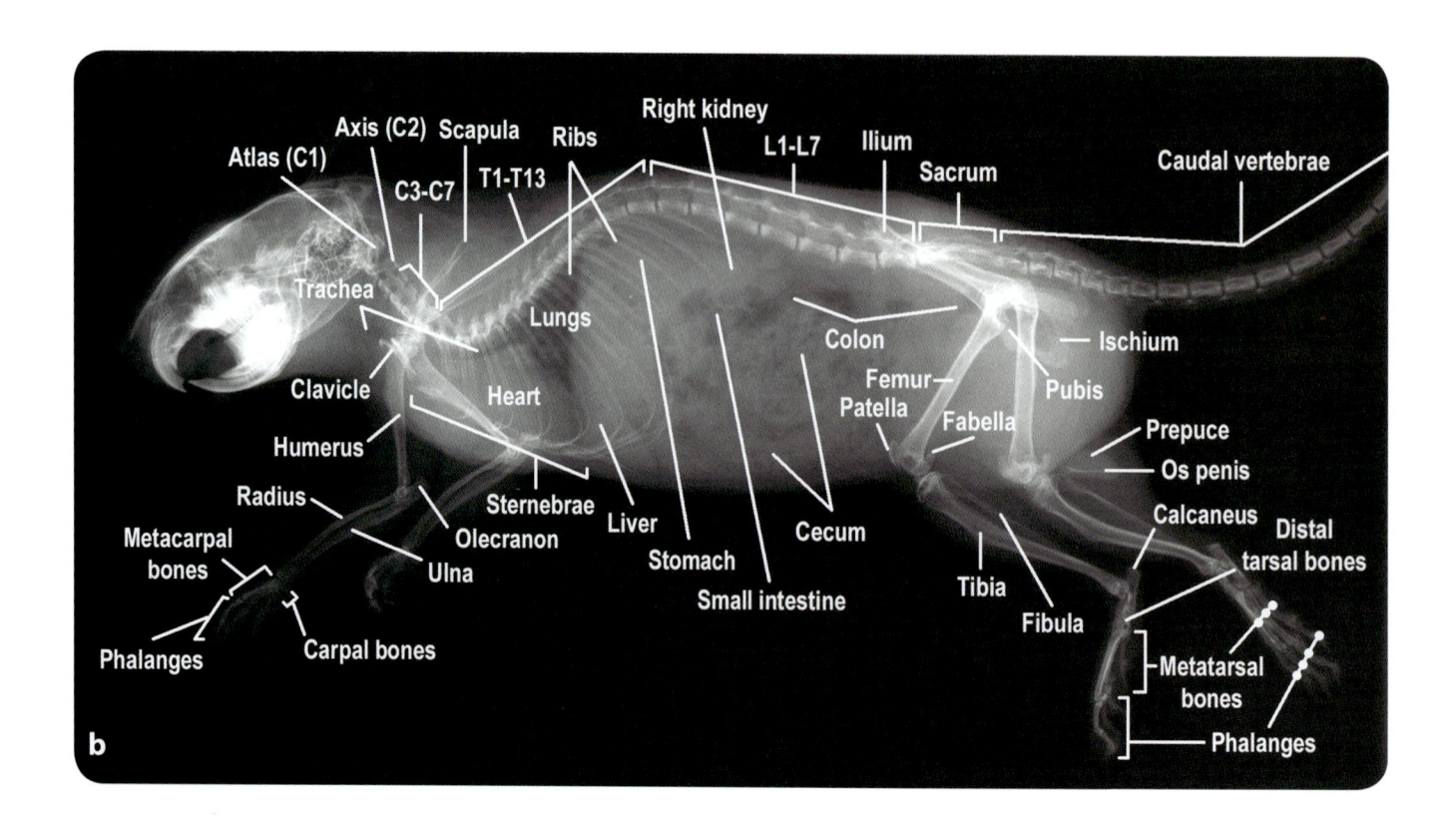

Figure 5.7a,b. Whole body skeleton of a 200 g neutered male degu, 85% of actual size.

Ventrodorsal Projection

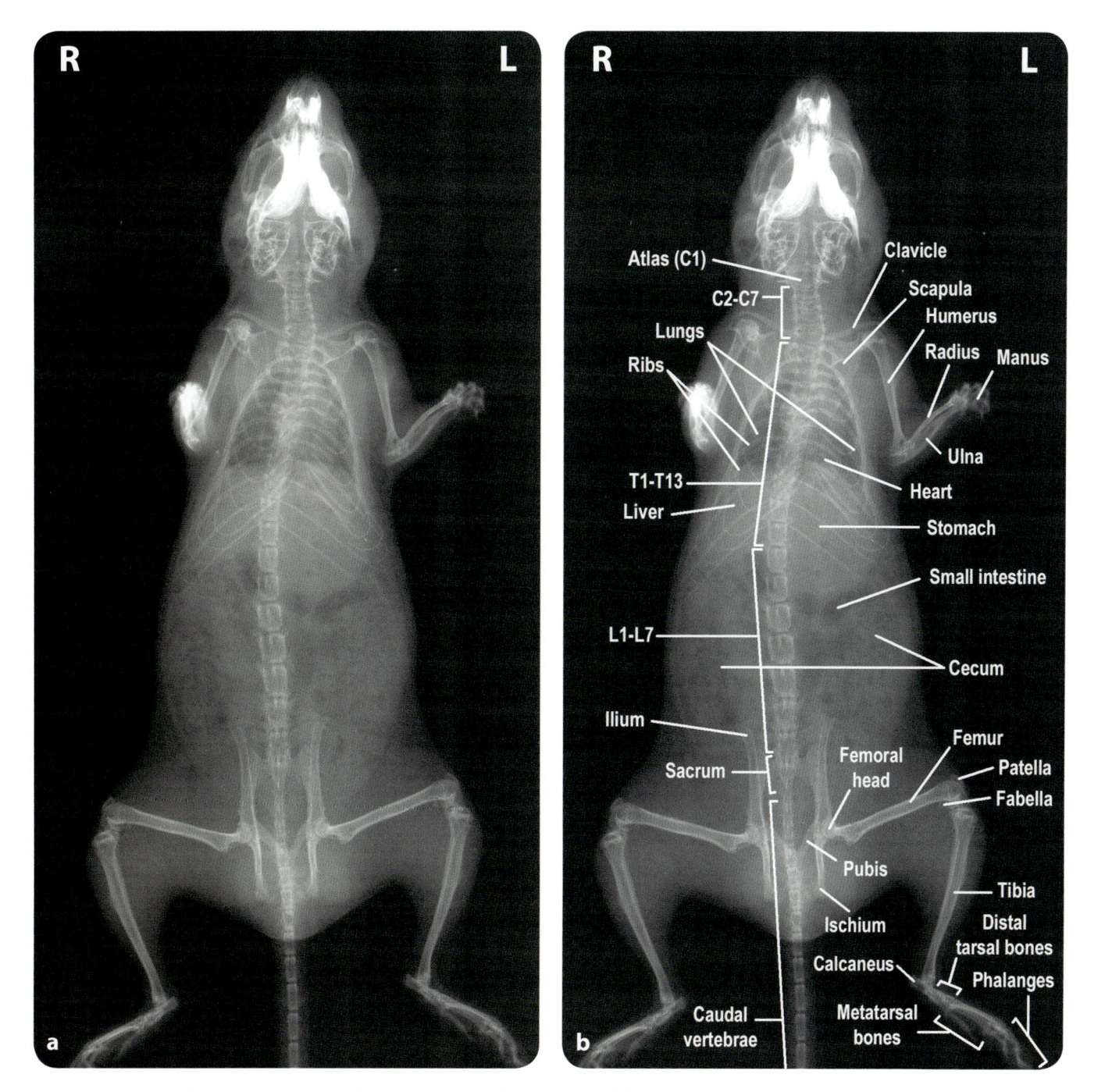

Figure 5.8a,b. Whole body skeleton of a 200 g neutered male degu, 85% of actual size.

Miscellaneous

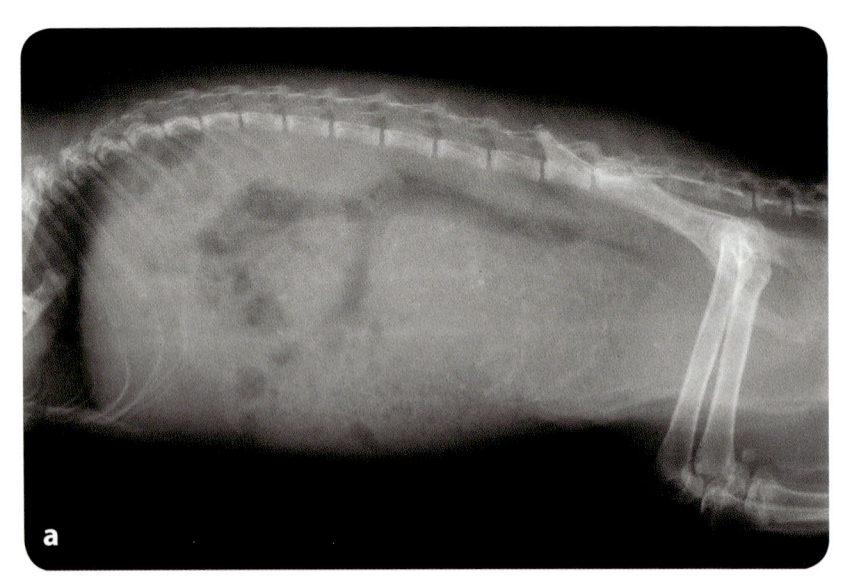

Figure 5.9a,b. Lateral (a) and ventrodorsal (b) radiographs of the abdomen of a female degu pregnant approximately 30 days. The lateral projection clearly demonstrates a single skull and several bones, but the exact number of fetuses is difficult to determine. Two skulls are visible in the ventrodorsal projection. The gestation period of the degu is about 3 months.

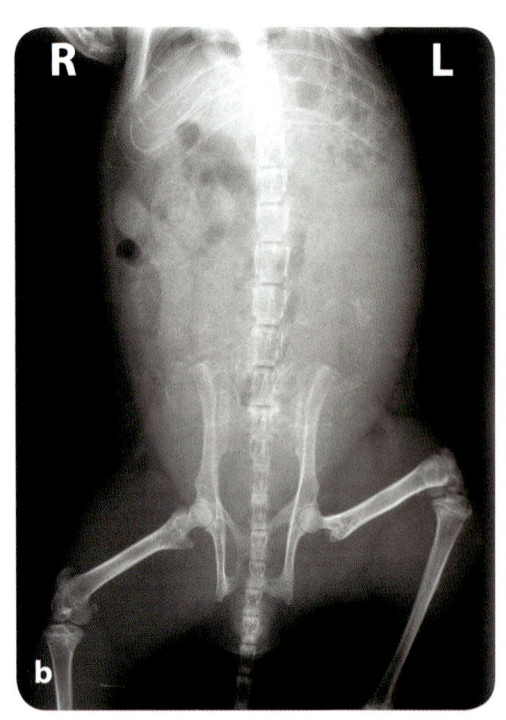

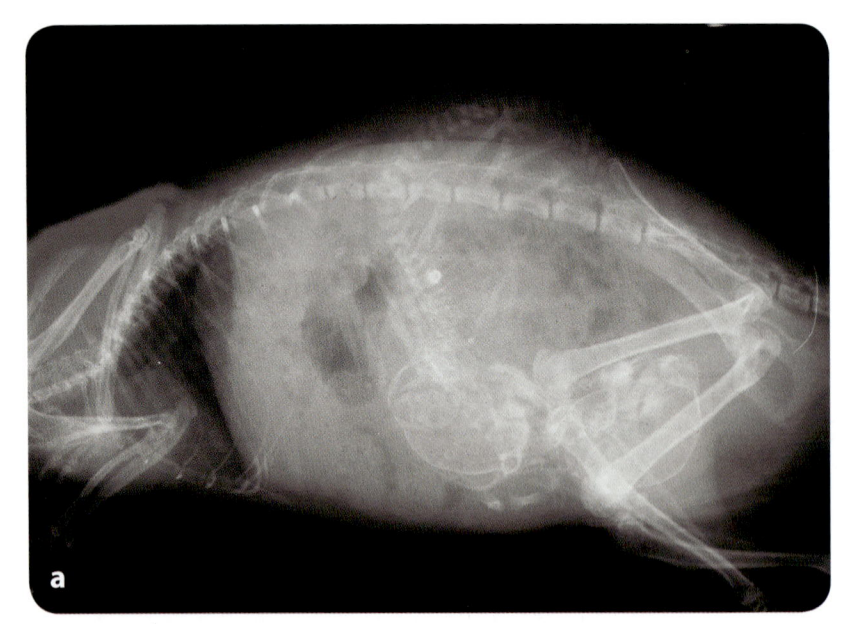

Figure 5.10a,b. Lateral (a) and ventrodorsal (b) radiographs of a pregnant degu, about two days prior to delivery, manual restraint. Two fetuses are clearly visible. The ventrodorsal projection is not perfectly symmetrical as the animal was not anesthetized. As in chinchillas, the pubic symphysis is not open.

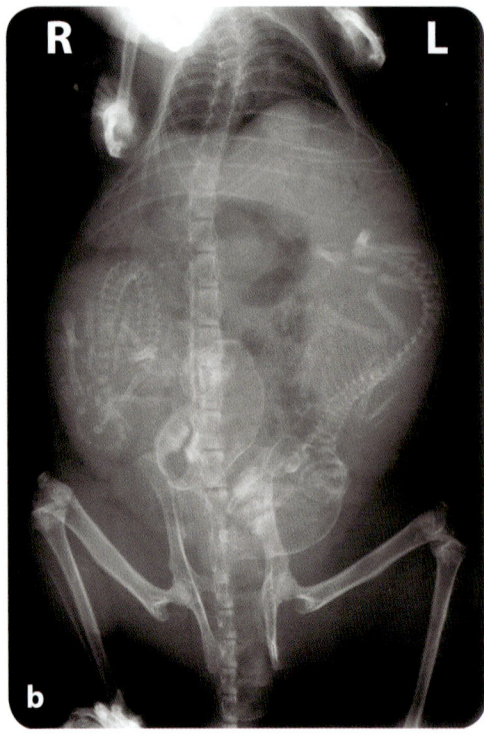

MISCELLANEOUS 273

Figure 5.11. Lateral projection of a 20-day-old degu with a fracture of T11.

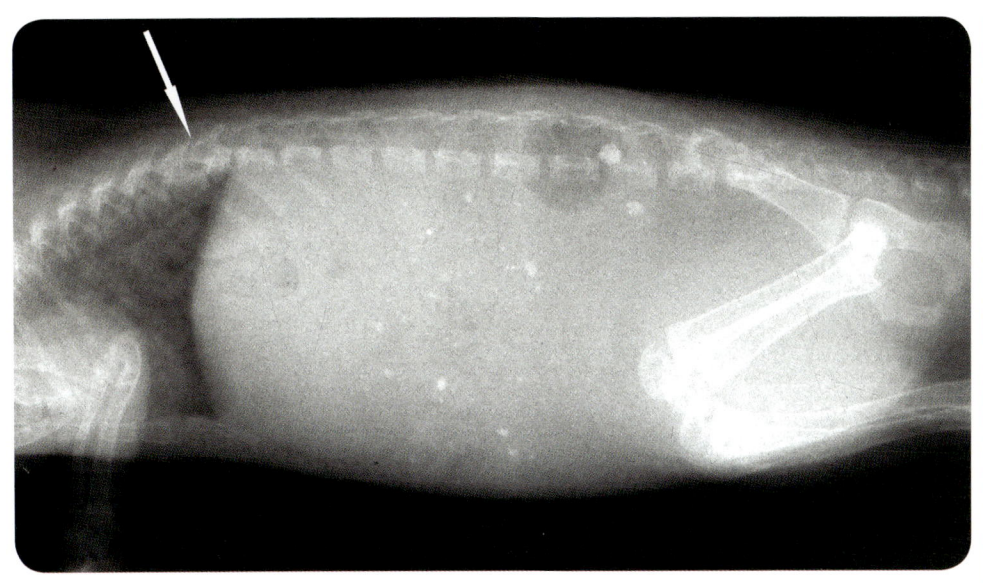

Figure 5.12. Clinical appearance of a paraparetic 20-day-old degu. Deep pain sensitivity was still present at the time of the injury. The degu recovered after 48 hours of supportive therapy.

References:
Capello V, Gracis M: Radiology of the skull and teeth. In: Lennox A, ed. Rabbit and Rodent Dentistry Handbook. Ames, IA: Blackwell Publishing, (Formerly Zoological Education Network, Lake Worth, FL); 2005:65-99.

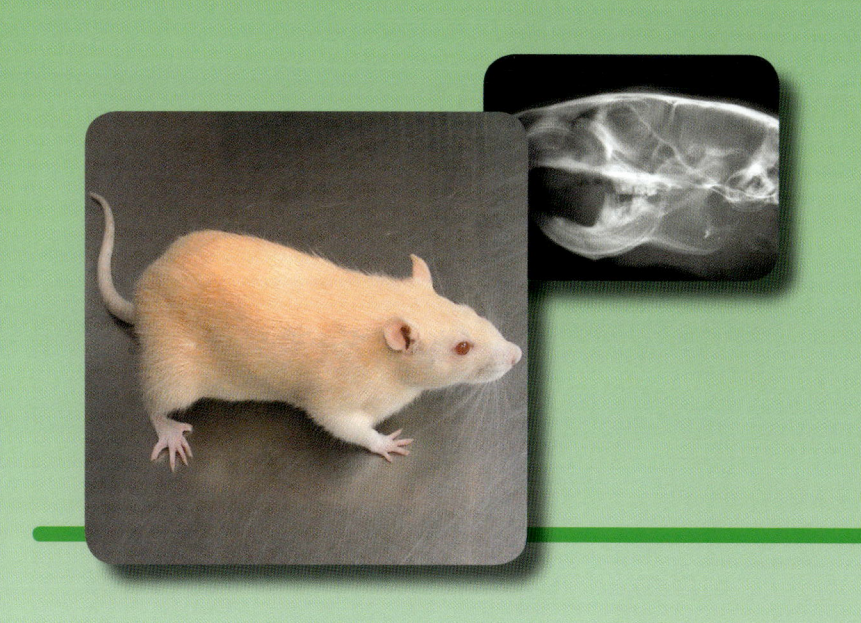

RAT

The NORMAL HEAD
Lateral Projection

Figure 6.1a,b. 230% of actual size. Rats have well developed incisor teeth. The apex of the maxillary incisor teeth is about half the length of the diastema, at some distance from the palate (yellow circle). The mandibular incisor teeth are ventral to the cheek teeth, and the apex is located distal to the last cheek tooth (red circle). In this image, the mandible appears slightly rostrally subluxated. When the jaw is at rest, cheek teeth are in occlusion while incisor teeth remain at some distance, giving the appearance of mandibular retrognathism. Rats are able to voluntarily subluxate the mandible in a rostral direction.

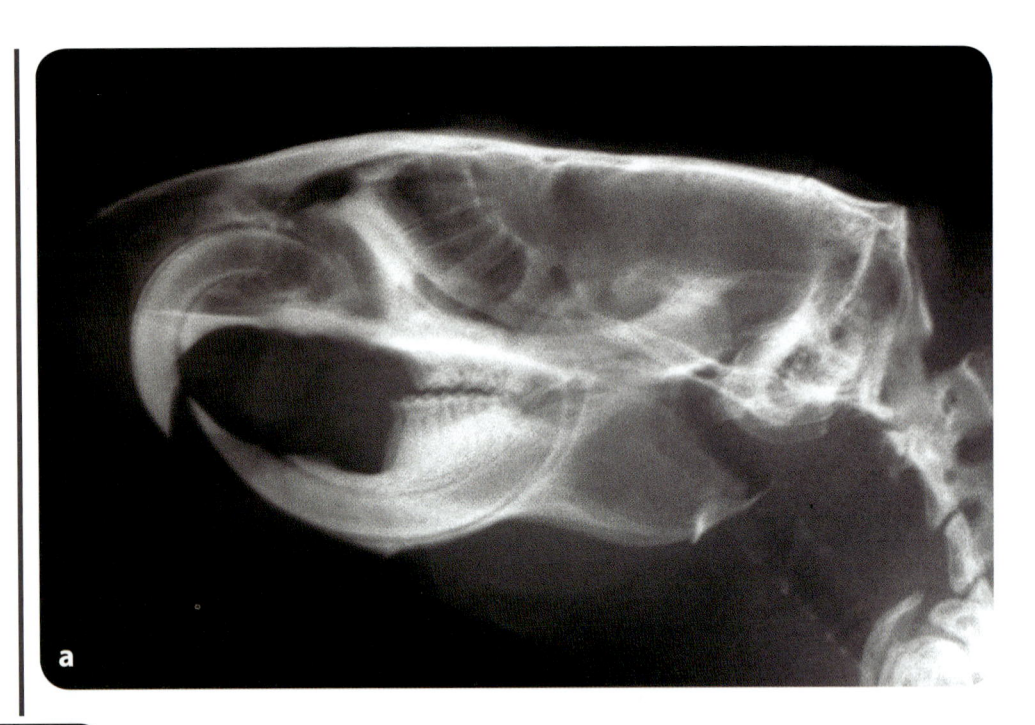

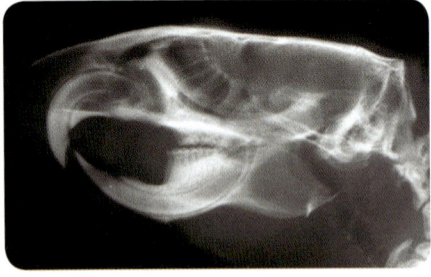

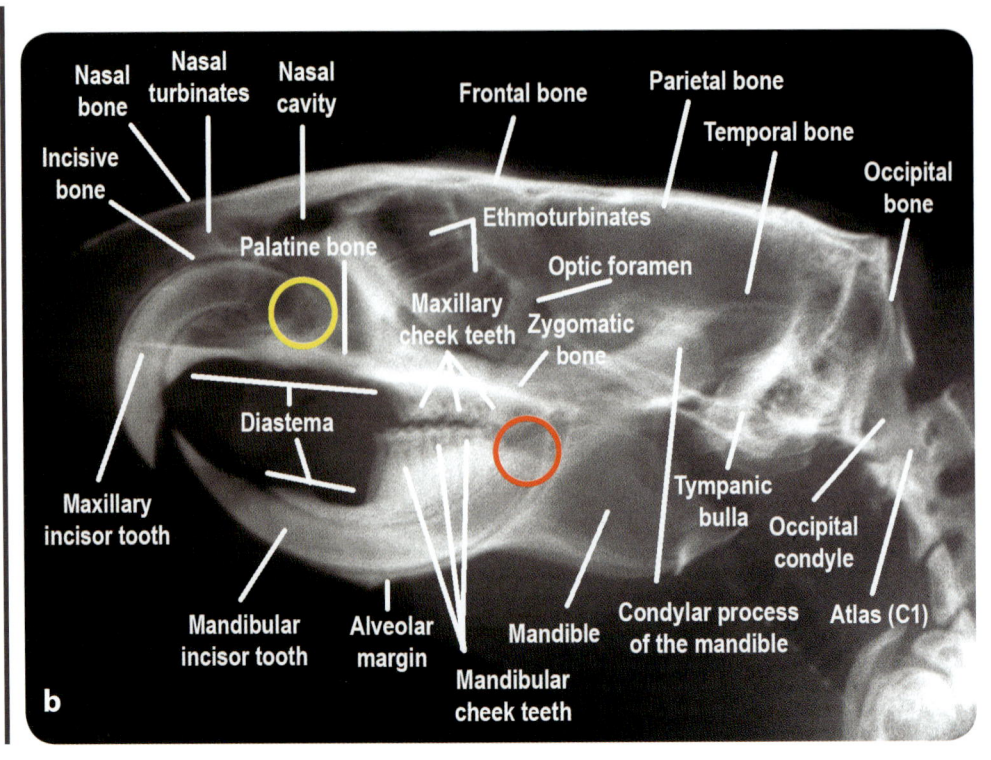

Figure 6.2. Actual size of the skull of a 400 g male rat.

HEAD 277

Oblique Projection

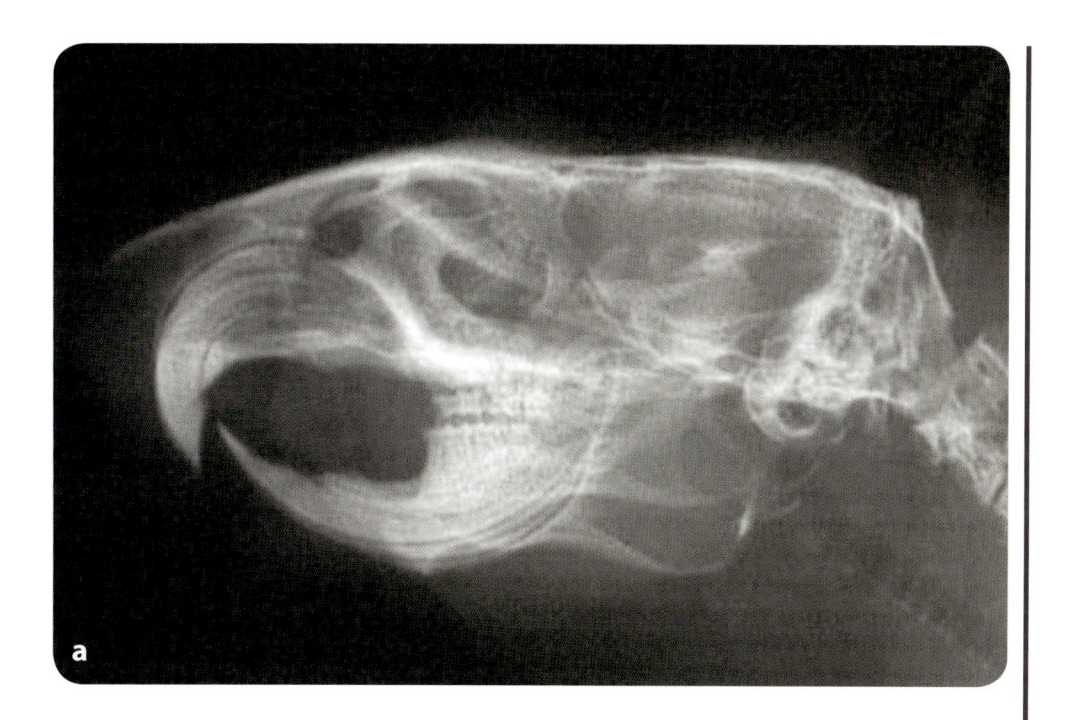

Figure 6.3a,b. Left 30° ventral-right dorsal oblique projection of the skull. The apex of the mandibular incisor tooth is indicated by the red circle.

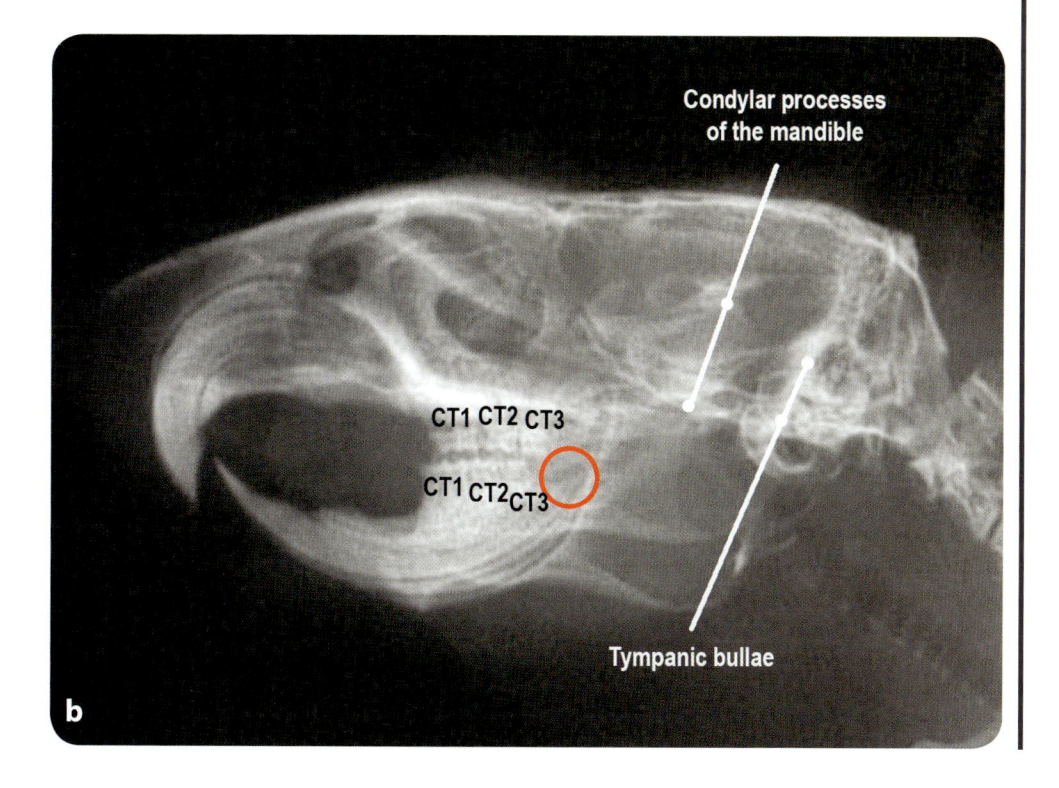

Ventrodorsal Projection

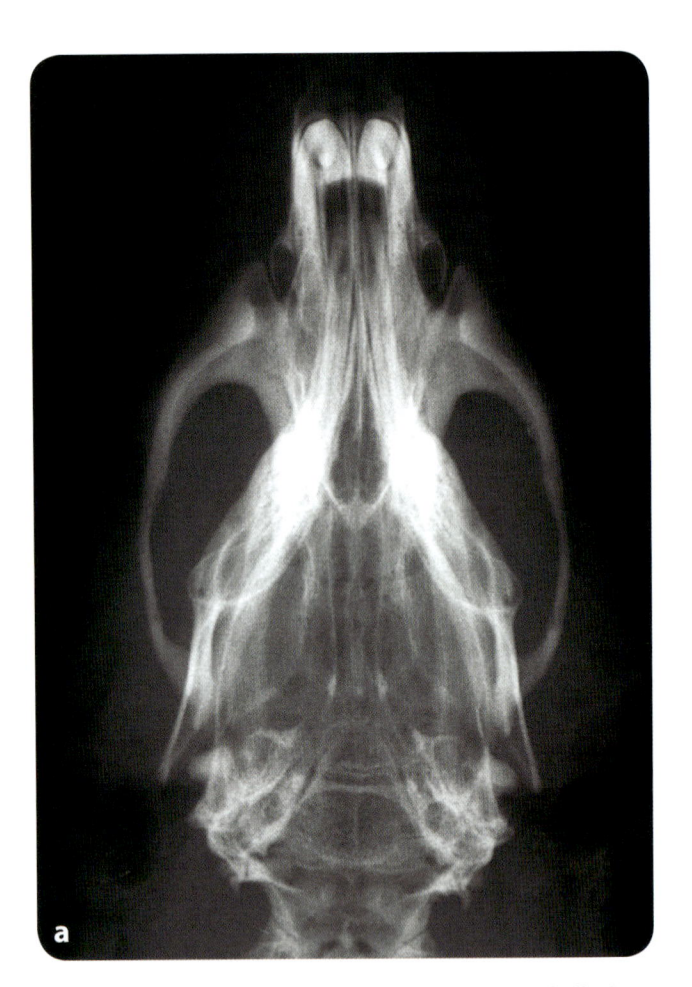

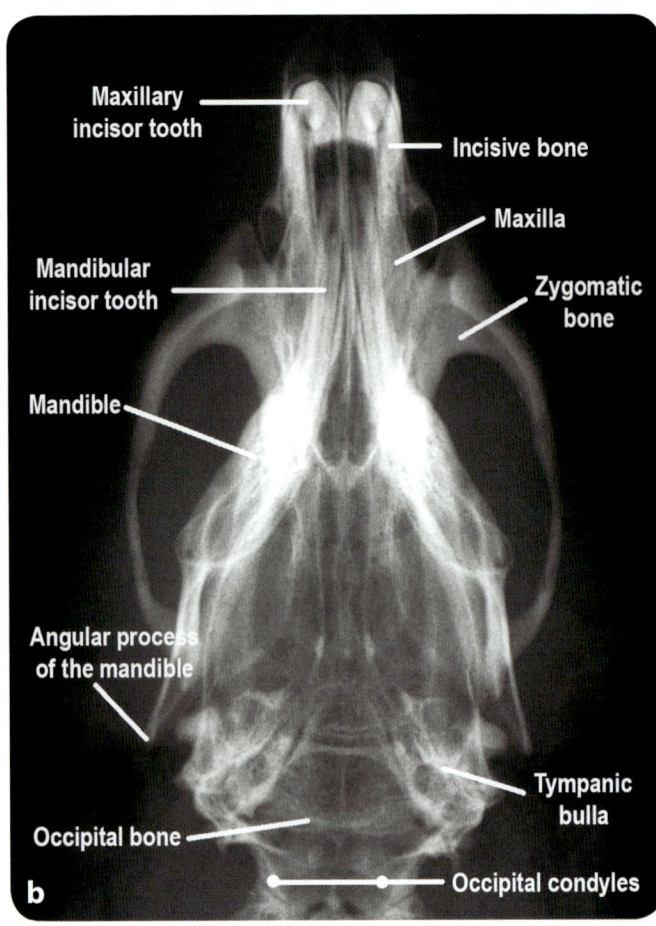

Figure 6.4a,b. Ventrodorsal radiograph of the skull of a rat.

Rostrocaudal Projection

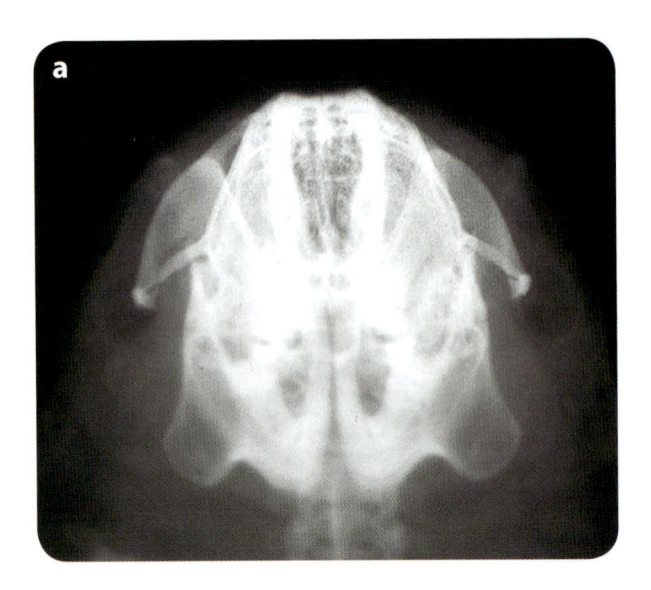

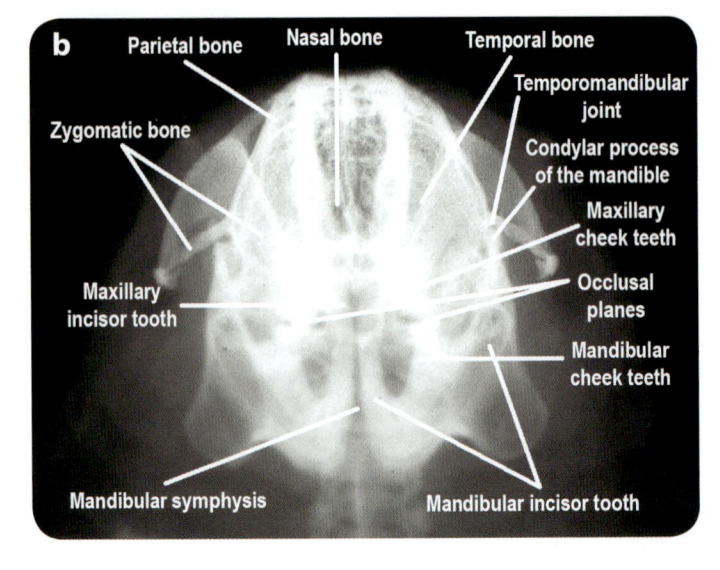

Figure 6.5a,b. Rostrocaudal radiograph of the skull of a rat.

ABNORMALITIES of the HEAD
Diseases of Cheek Teeth

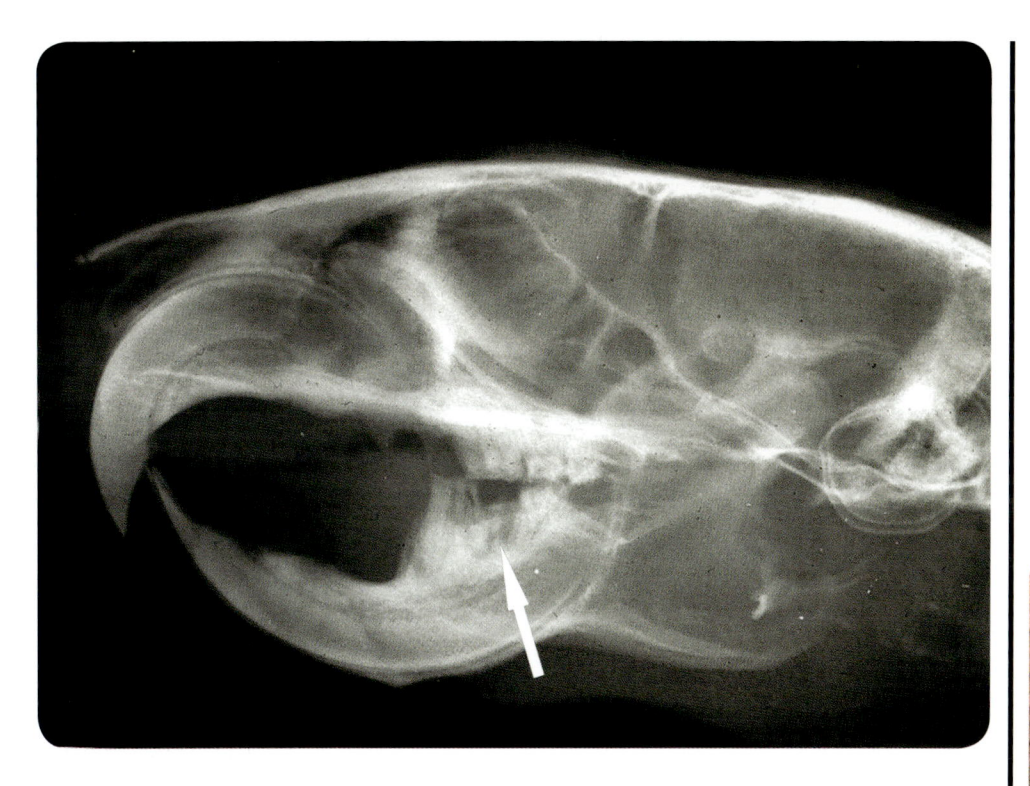

Figure 6.6. Lateral radiograph of the skull demonstrating a fracture of mandibular CT2 (arrow).

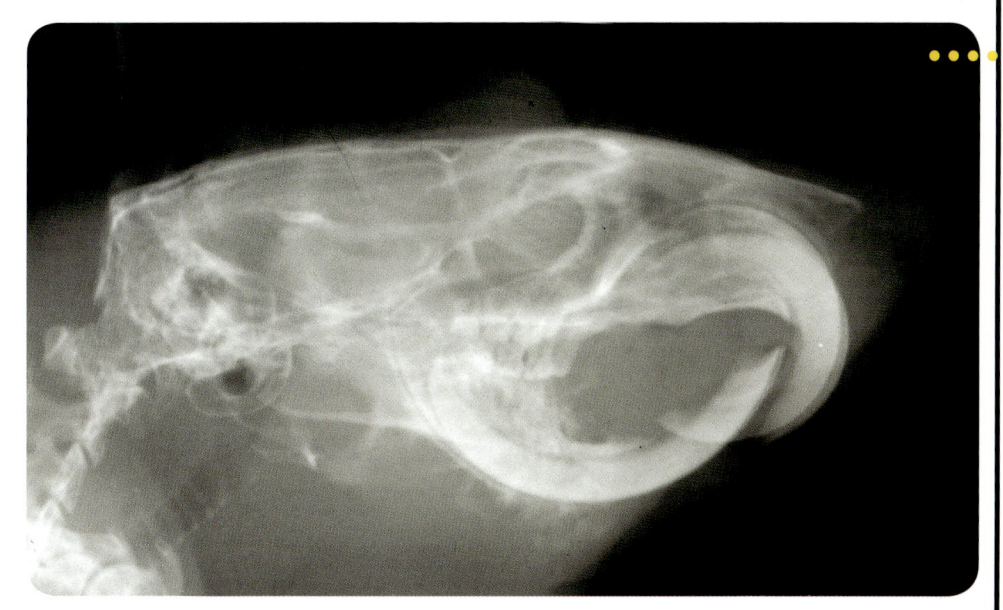

Figure 6.7. Right 30° ventral-left dorsal oblique projection of the skull of a rat with malocclusion of maxillary incisor teeth and severe acquired dental disease of both right and left mandibular cheek teeth arcades. Periosteal reaction is visible at the ventral margin of the mandible. Swelling of soft tissues suggests the presence of an abscess.

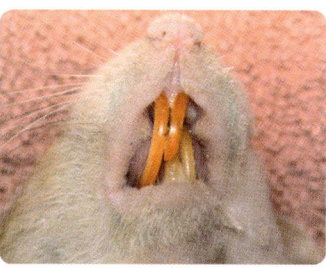

Figure 6.8. Same case as Figure 6.7 demonstrating overgrowth of maxillary incisor teeth with malocclusion.

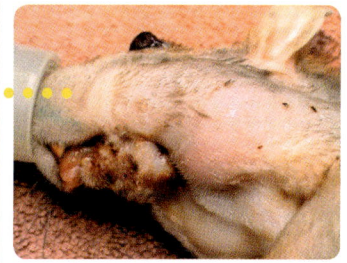

Figure 6.9. Mandibular abscess with a fistulous tract from the abscess site.

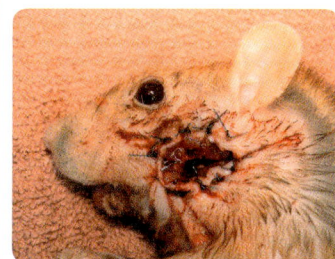

Figure 6.10. Clinical appearance after surgical debridement of the abscess and marsupialization.

The NORMAL TOTAL BODY
Lateral Projection

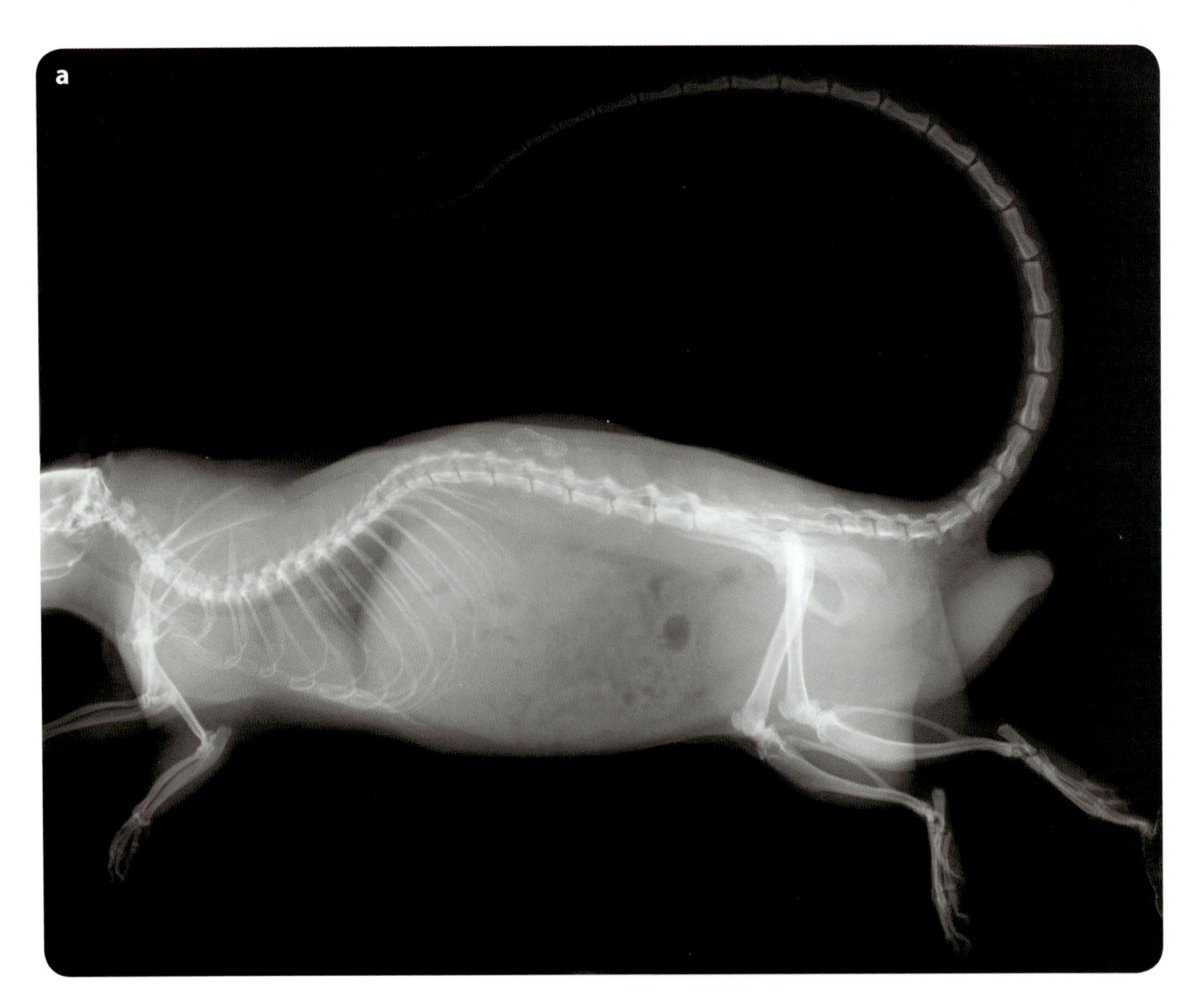

Figure 6.11a,b. Total body radiograph of a 450 g male rat, 70% of actual size.

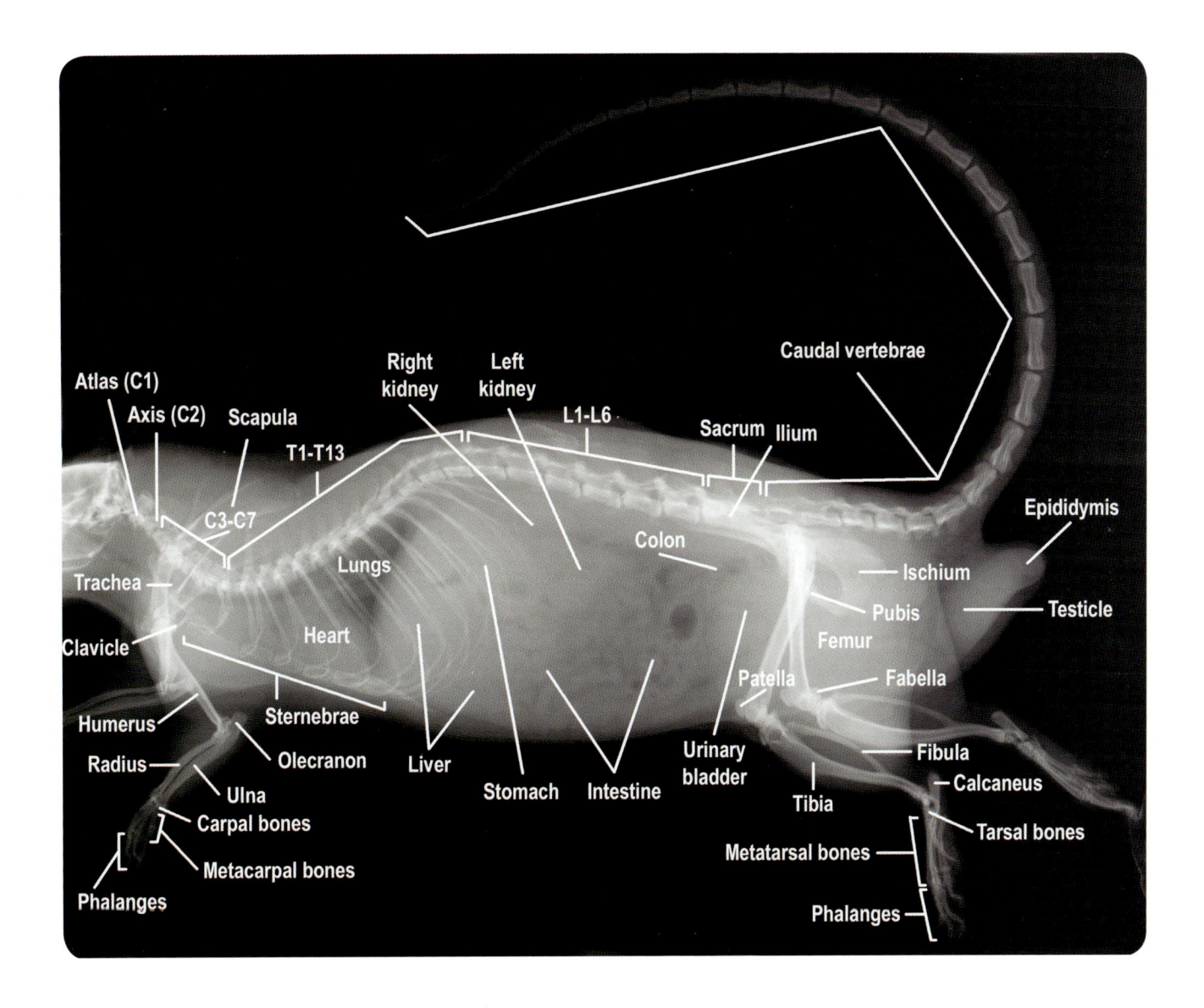

TOTAL BODY 281

Ventrodorsal Projection

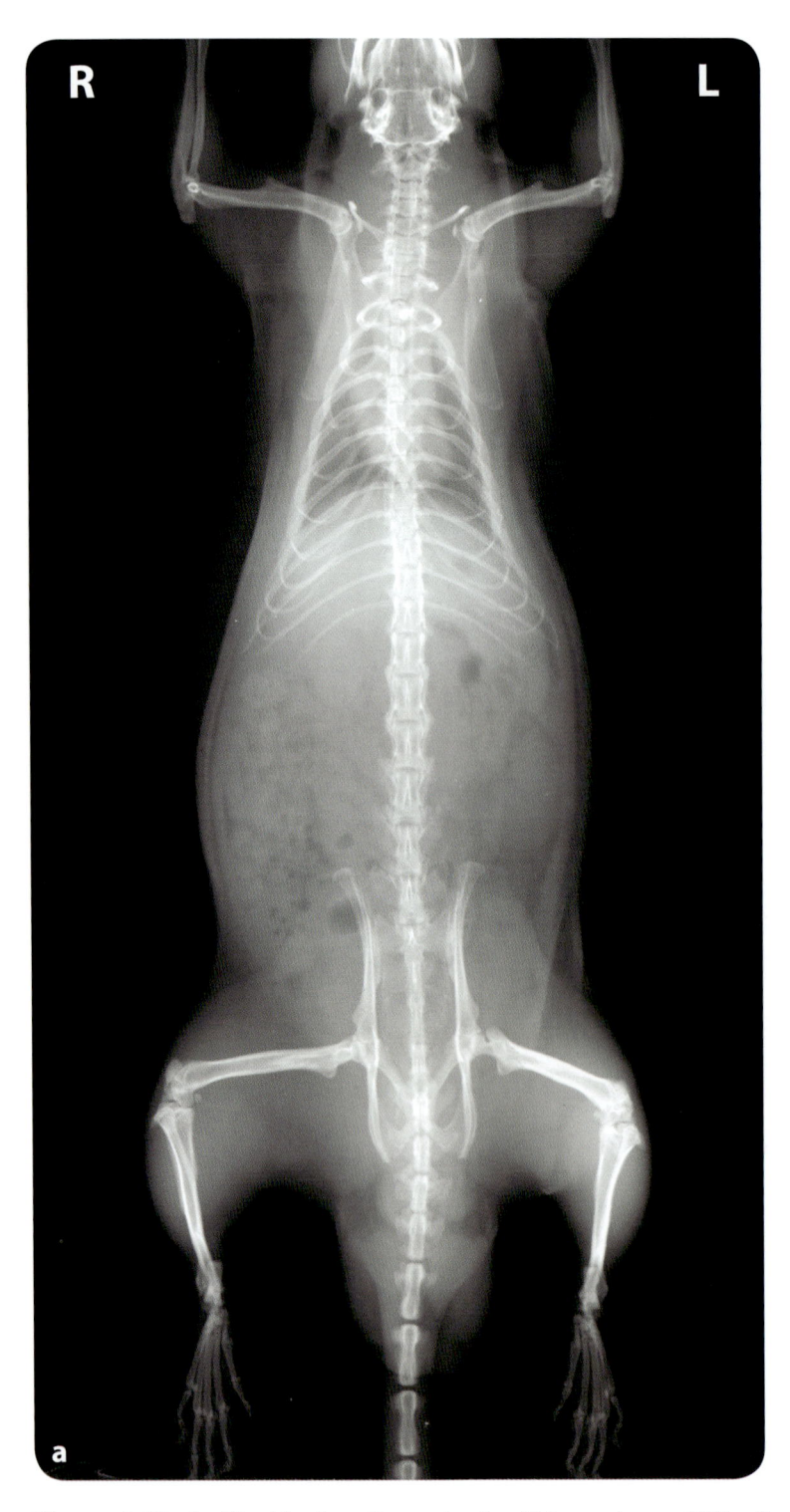

Figure 6.12a,b. Total body radiograph of a 450 g male rat, 70% of actual size.

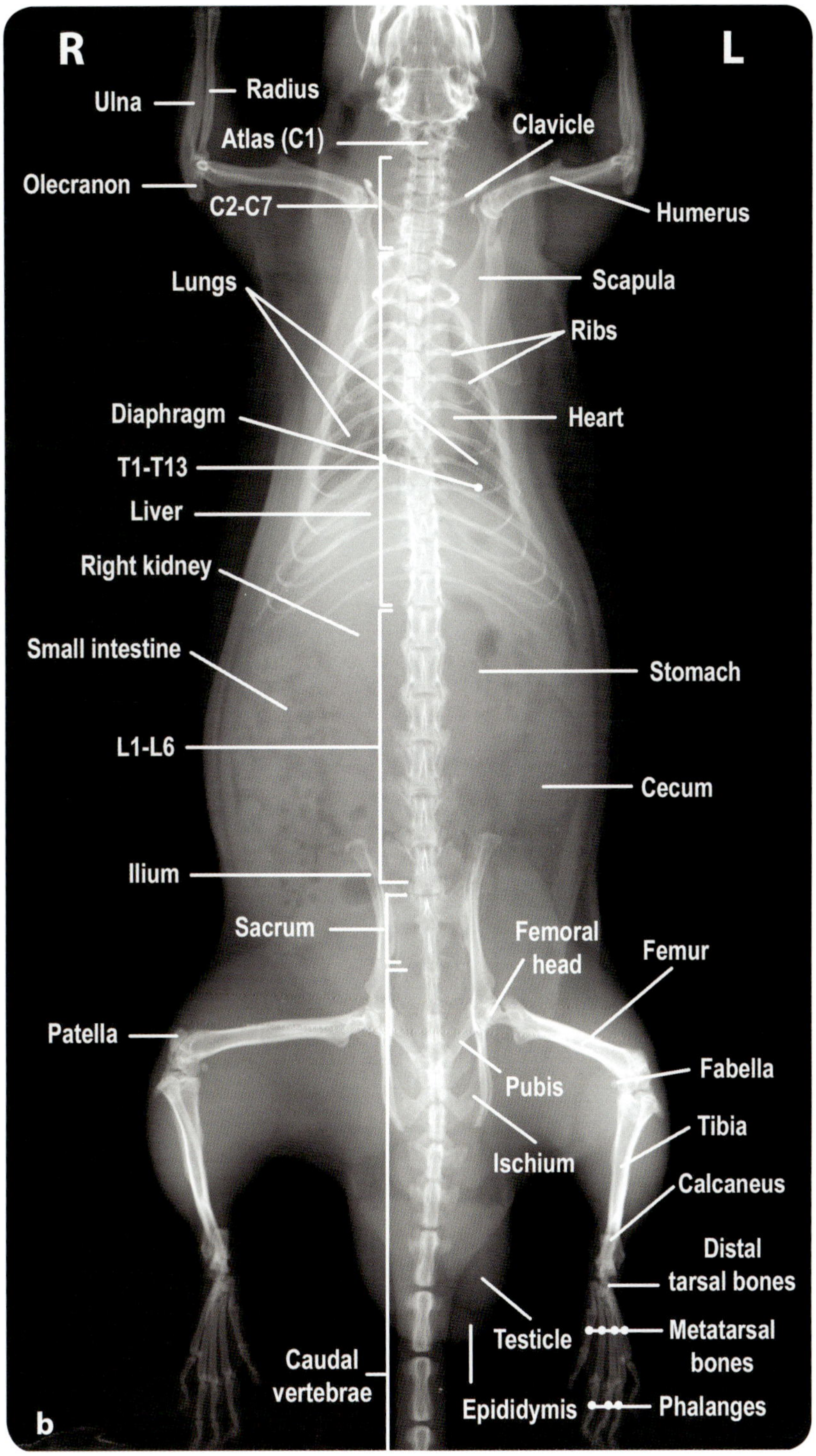
R
L
Ulna
Radius
Atlas (C1)
Clavicle
Olecranon
C2-C7
Humerus
Lungs
Scapula
Ribs
Diaphragm
Heart
T1-T13
Liver
Right kidney
Small intestine
Stomach
L1-L6
Cecum
Ilium
Sacrum
Femoral head
Femur
Patella
Fabella
Pubis
Tibia
Ischium
Calcaneus
Distal tarsal bones
Testicle
Metatarsal bones
Caudal vertebrae
Epididymis
Phalanges
b

MISCELLANEOUS ABNORMALITIES
Diseases of the Thorax

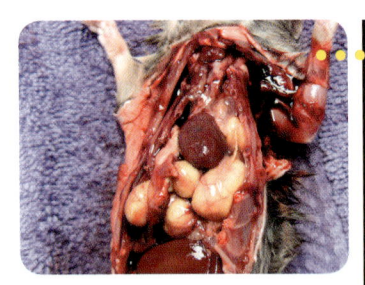

Figure 6.13. Appearance of the same patient at necropsy. Note multiple lung abscesses typical of severe bacterial pneumonia.

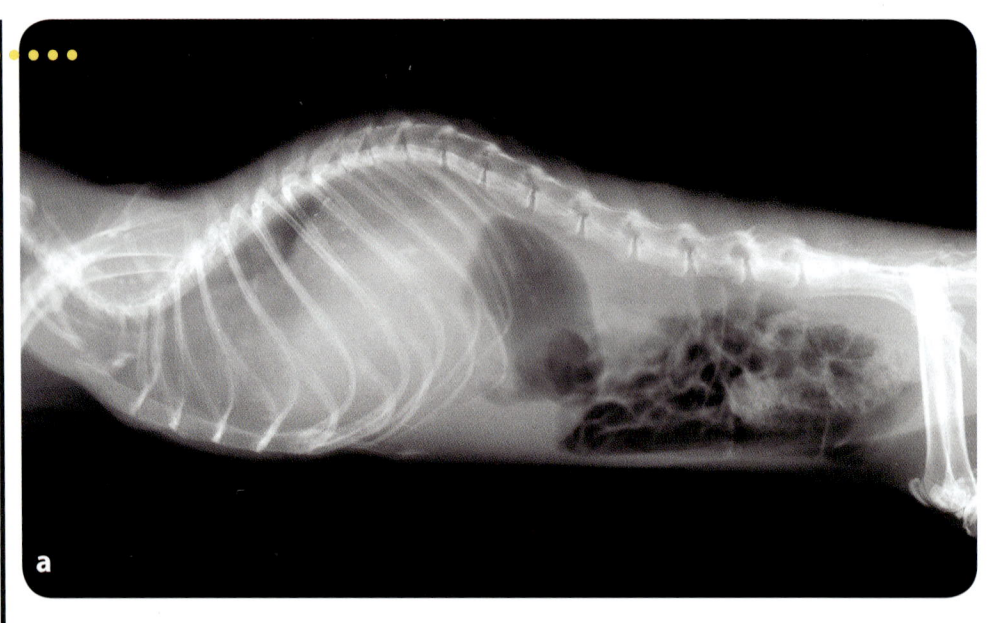

Figure 6.14a,b. Lateral (a) and ventrodorsal (b) radiographs of a rat with severe respiratory distress. There is increased opacity within the thorax, and pulmonary volume is decreased. Marked accumulation of gas in the stomach and intestines is likely due to aerophagia.

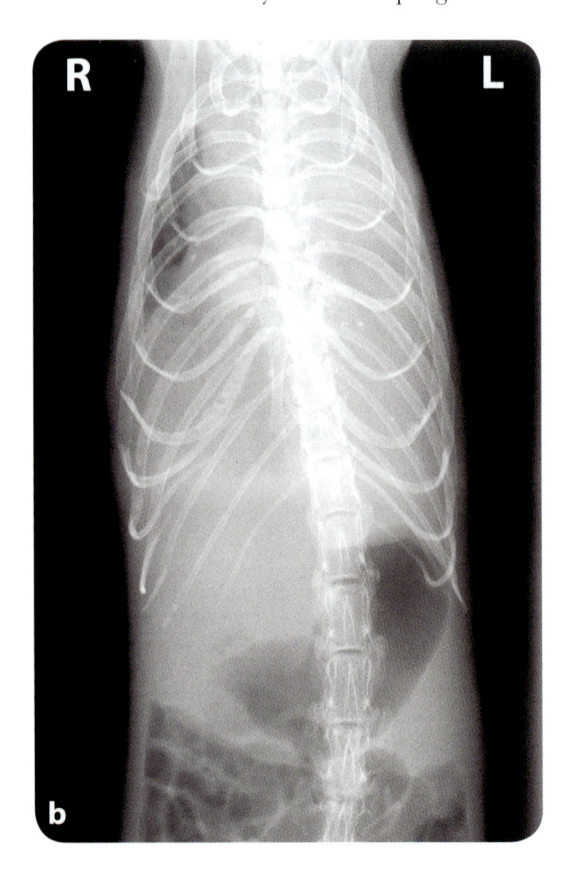

Diseases of the Abdomen

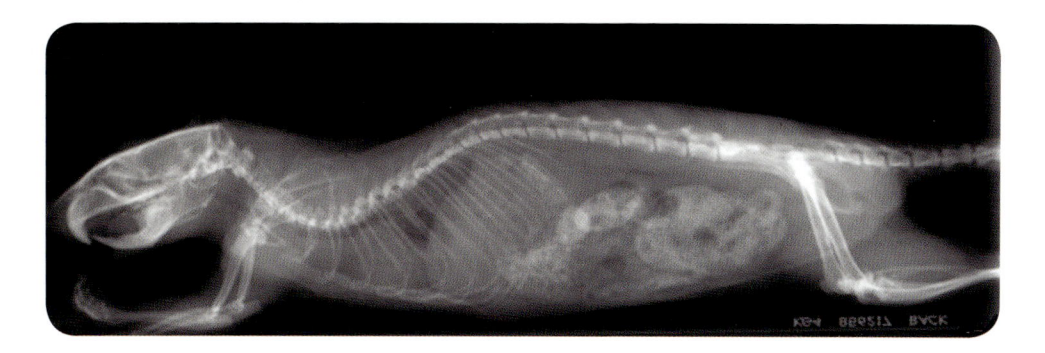

Figure 6.15. Lateral radiograph of a rat that had presented for mildly increased respiratory rate. Note marked impaction of the intestines with radiodense material. The owner suspected the rat was consuming a new bedding. Follow up radiographs taken 48 hours later demonstrated passage of the radiodense material. As respiratory rate was now normal, it was assumed increased respiratory rate was due to gastrointestinal discomfort.

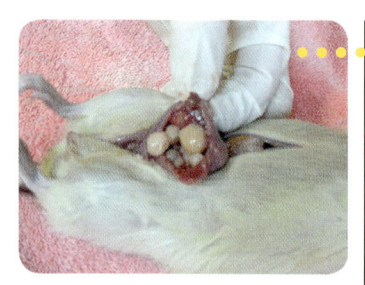

Figure 6.16. Appearance of the cystouroliths at necropsy.

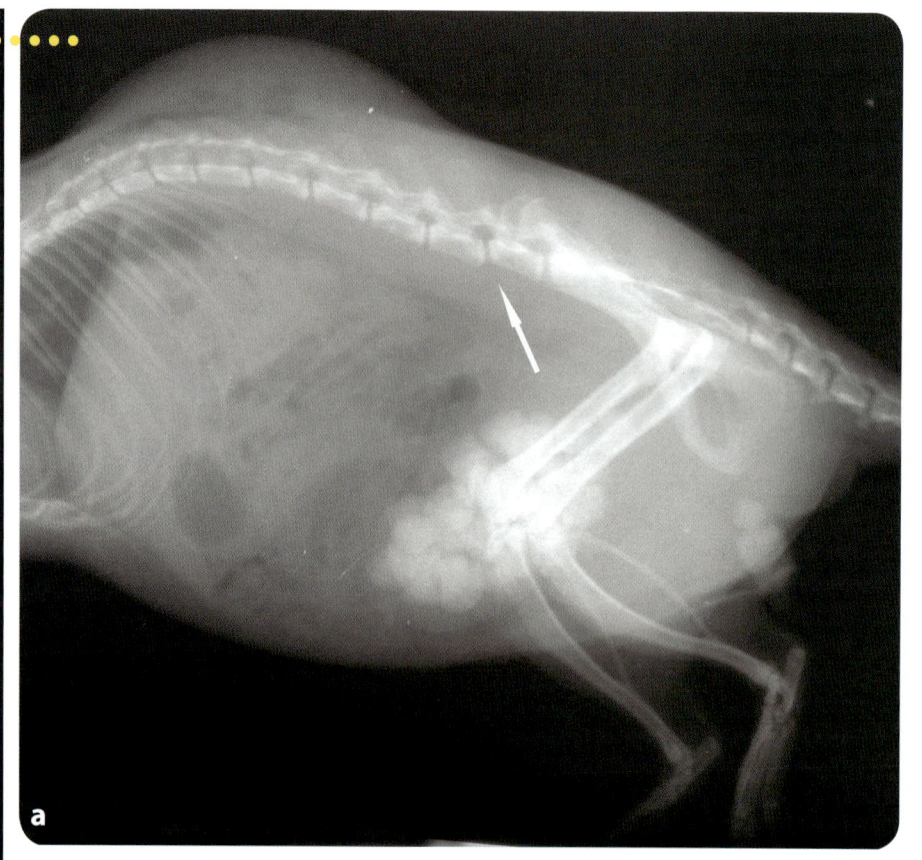

Figure 6.17a,b. Lateral (a) and ventrodorsal (b) radiographs of a male rat with dysuria and severe depression (actual size). Note severe cystolithiasis and urethrolithiasis, with multiple calculi. The pelvic limbs were not hyperextended caudally in order to highlight the urethral stones. Vertebral subluxation of L6-L7 of unknown origin (arrow) is also visible on the lateral projection.

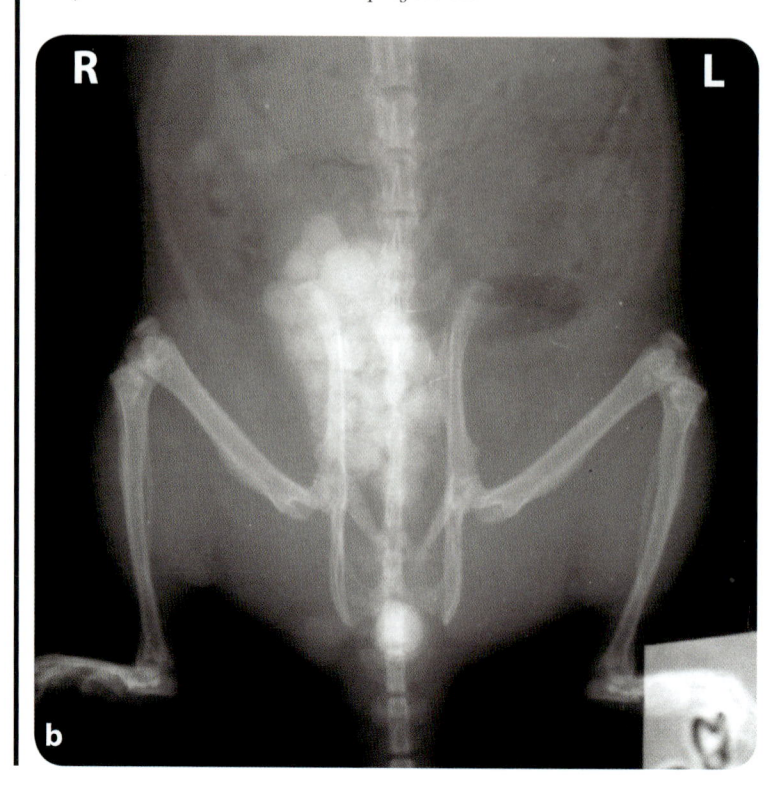

Diseases of the Limbs

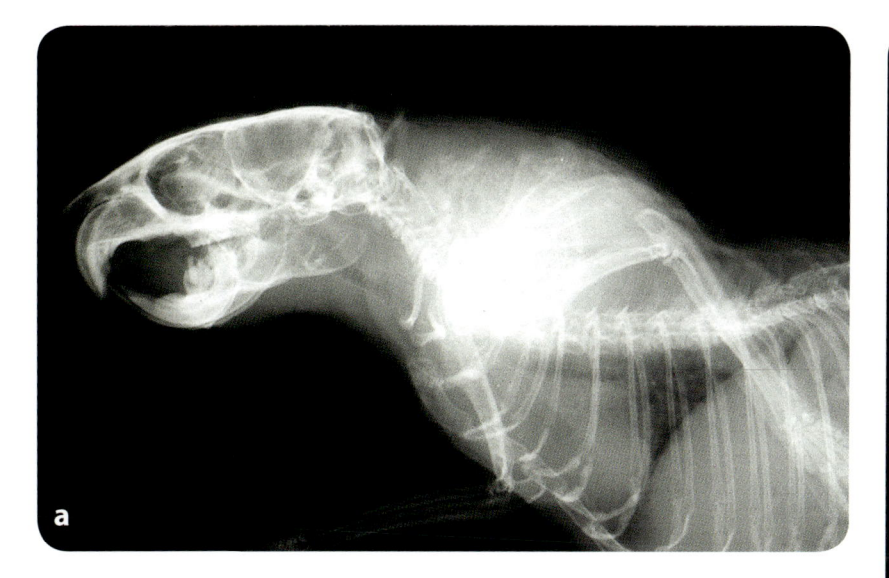

Figure 6.18a,b. Lateral (a) and ventrodorsal (b) radiographs of a 3-year-old rat with a solid soft tissue mass associated with the right scapula. There is periarticular soft tissue swelling in the right shoulder with mineralization. Severe acquired dental disease of mandibular cheek teeth is also visible. At necropsy the solid mass was consistent with neoplasia.

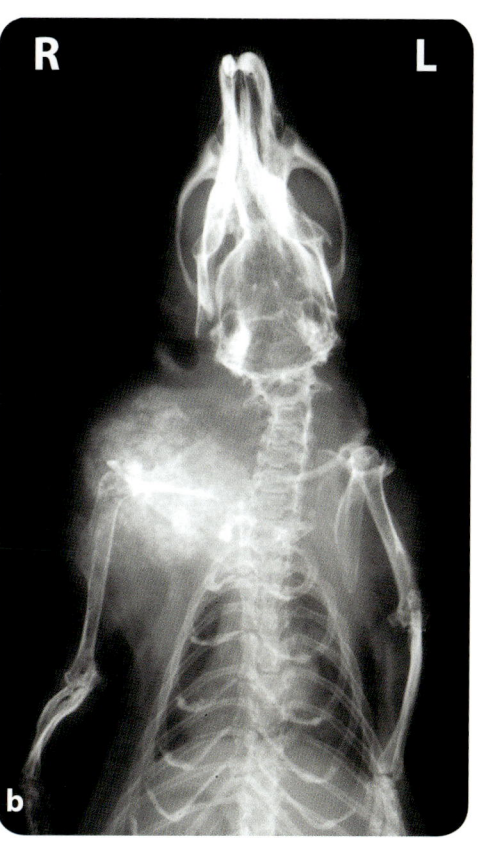

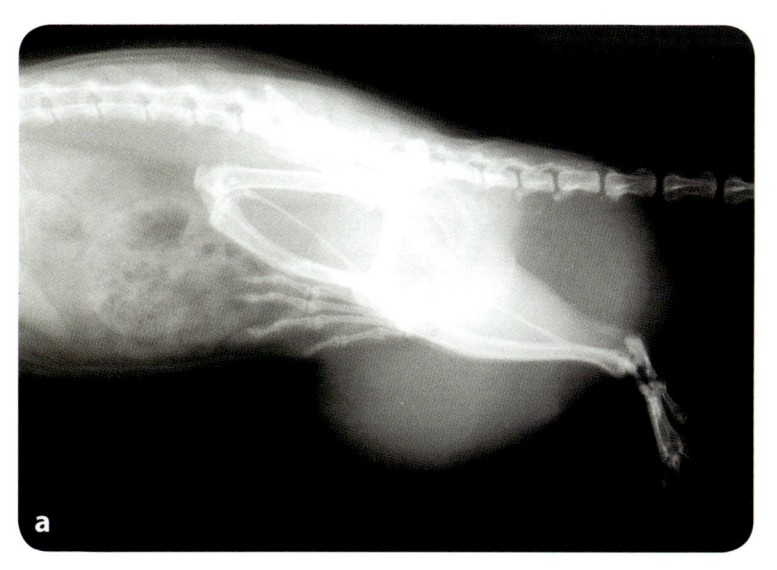

Figure 6.19a,b. Lateral (a) and ventrodorsal (b) radiographs of the abdomen and pelvic limbs of a 2-year-old female rat with a mass associated with the soft tissues of the right medial pelvic limb. At surgery this mass appeared to be neoplastic.

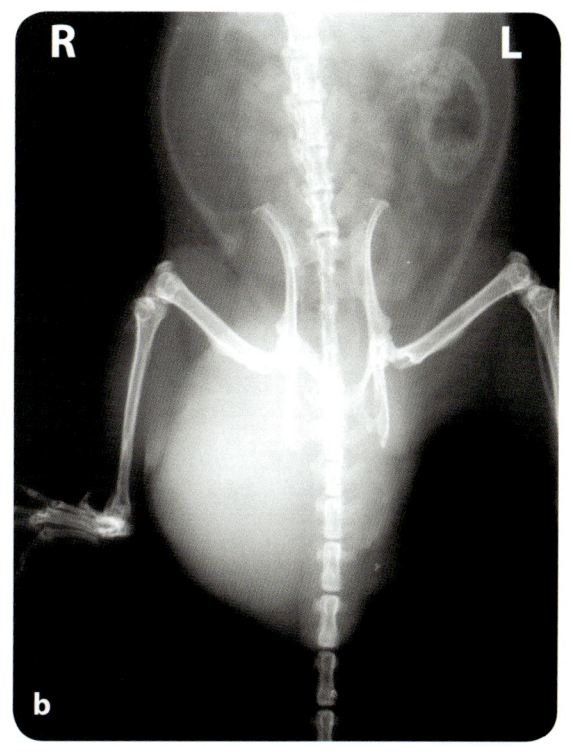

References:
Brown SA, Rosenthal LR: Question #110. In: Self Assessment Color Review of Small Mammals. London: Manson Publishing; 1997:97-98.
Capello V: Dental diseases. In: Lennox A, ed. Rabbit and Rodent Dentistry Handbook. Ames, IA: Blackwell Publishing, (Formerly Zoological Education Network, Lake Worth, FL); 2005:113-163.
Capello V, Gracis M: Radiology of the skull and teeth. In: Lennox A, ed. Rabbit and Rodent Dentistry Handbook. Ames, IA: Blackwell Publishing, (Formerly Zoological Education Network, Lake Worth, FL); 2005:65-99.
Deeb B: Respiratory diseases in pet rats. Exotic DVM. 2005; 7(1):31-33.
Donnelly TM: Disease problems of small rodents. In: Quesenberry KE, Carpenter JW, eds. Ferrets, Rabbits and Rodents. Clinical Medicine and Surgery 2nd ed. Philadelphia, PA: Elsevier; 2004:299-315.
Girling SJ: Mammalian imaging and anatomy. In: Meredith A, Redrobe S, eds. BSAVA Manual of Exotic Pets 4th ed. Quedgeley, Glouchester: British Small Animal VeterinaryAssociation; 2005:1-12.
Mehler SJ, Bennet RA: Surgical oncology of exotic animals. Vet Clin Exot Anim. 2004; 7:783-805.
Popesko P, Rijtovà V, Horàk J: A Colour Atlas of Anatomy of Small Laboratory Animals. Vol. II: Rat, Mouse, Hamster. Bratislava: Príroda Publishing House; 1990.
Silverman S, Tell LA: Norway rat *(Rattus norvegicus)*. In: Radiology of Rodents, Rabbits and Ferrets. An Atlas of Normal Anatomy and Positioning. Philadelphia, PA: Elsevier Saunders; 2005:19-43.
Stefanacci JD, Hoefer HL: Radiology and ultrasound. In: Quesenberry KE, Carpenter JW, eds. Ferrets, Rabbits and Rodents. Clinical Medicine and Surgery 2nd ed. Philadelphia, PA: Elsevier; 2004:319-413.

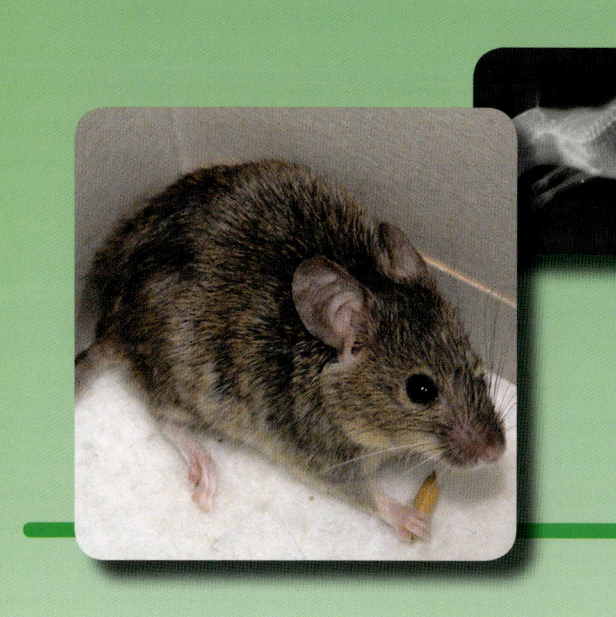

MOUSE

The NORMAL WHOLE BODY SKELETON
Lateral Projection

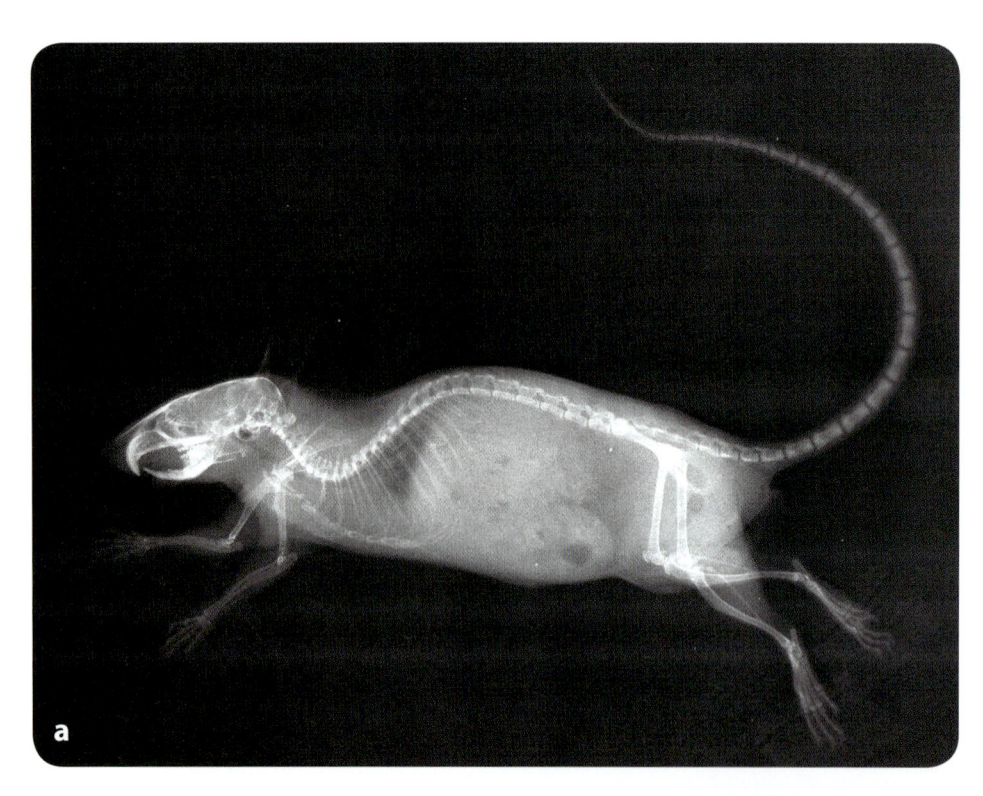

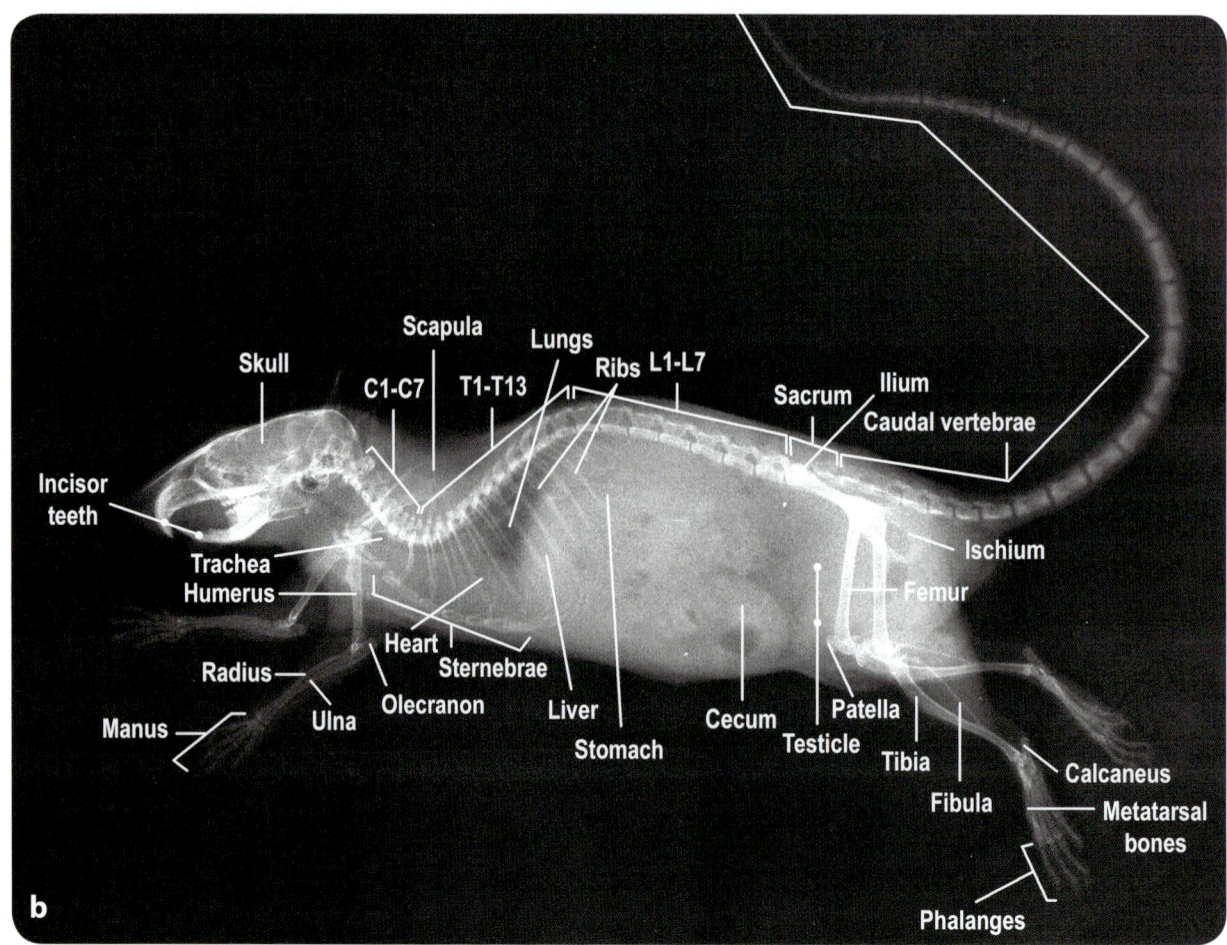

Figure 7.1a,b. Whole body skeleton of a 40 g male mouse, actual size (a) and 130% of actual size (b).

Ventrodorsal Projection

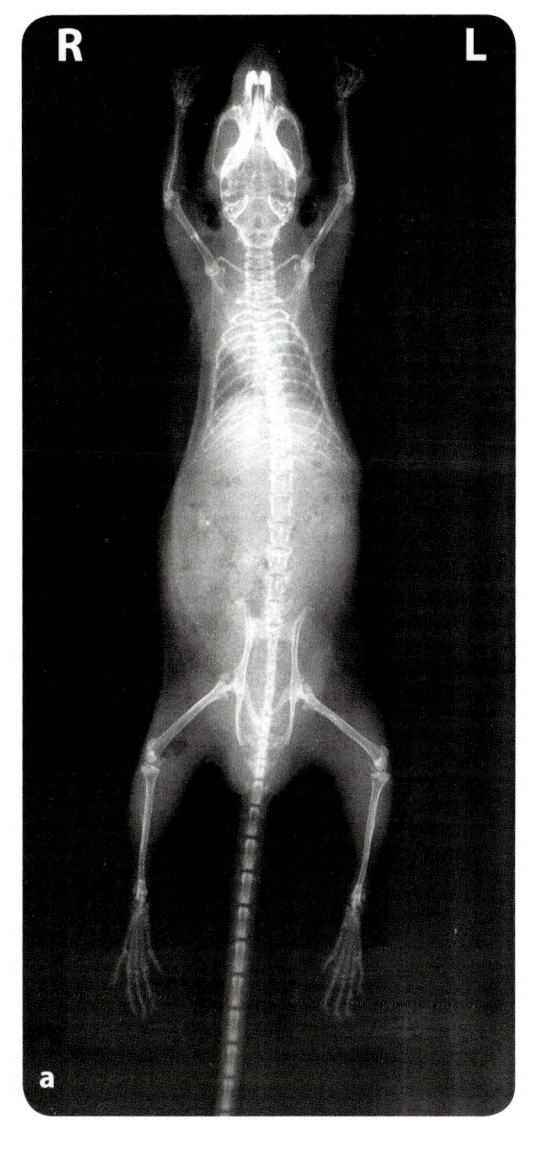

Figure 7.2a,b. Whole body skeleton of a 40 g male mouse, actual size (a) and 130% of actual size (b).

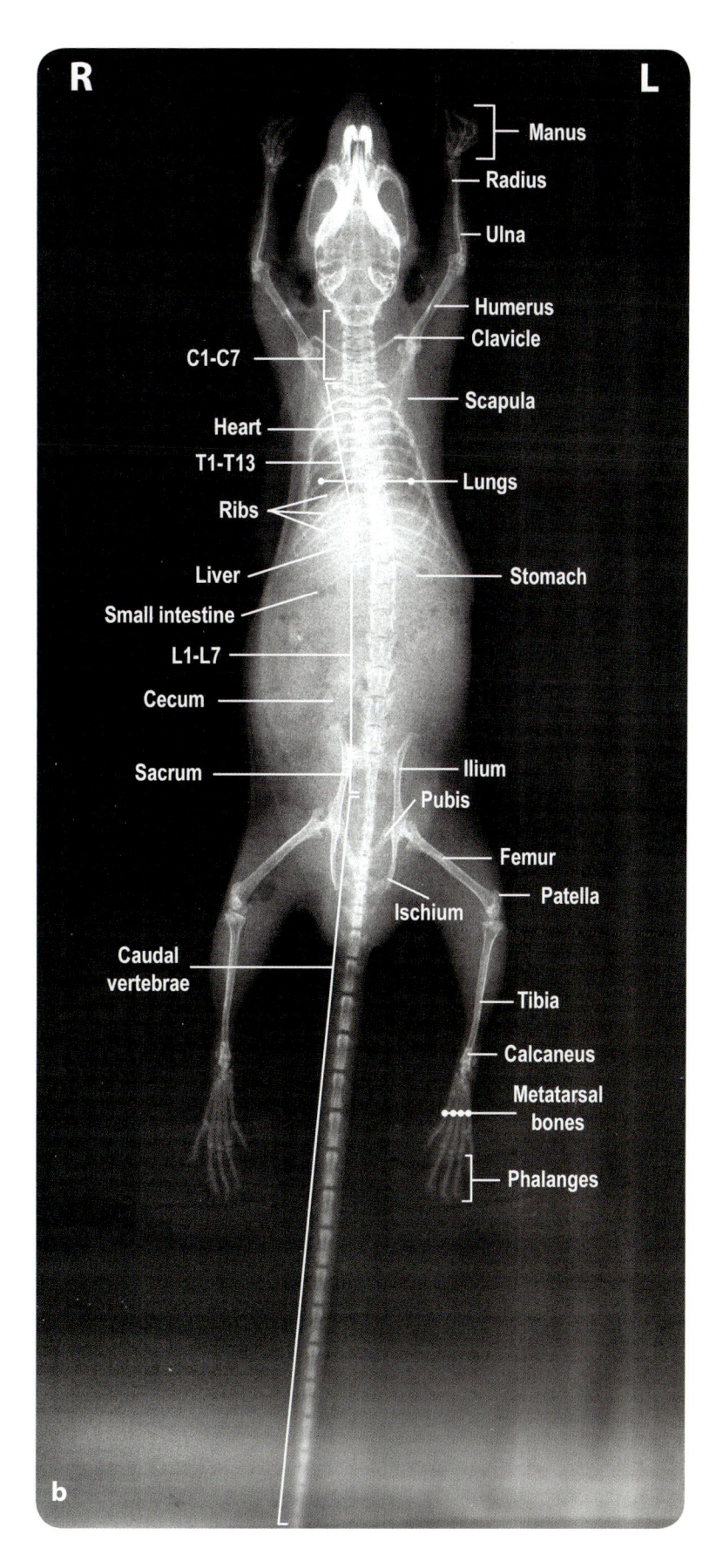

MISCELLANEOUS ABNORMALITIES
Diseases of the Abdomen

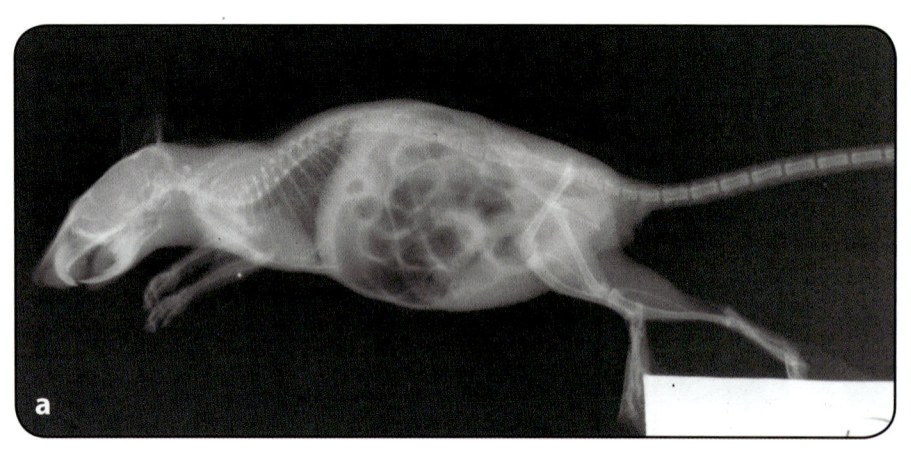

Figure 7.3a,b. Lateral (a) and ventrodorsal (b) views of a mouse with abdominal distention and depression, actual size. Note severe gas distension of the entire intestinal tract as a result of enteritis.
(Courtesy of Cathy Johnson-Delaney, DVM)

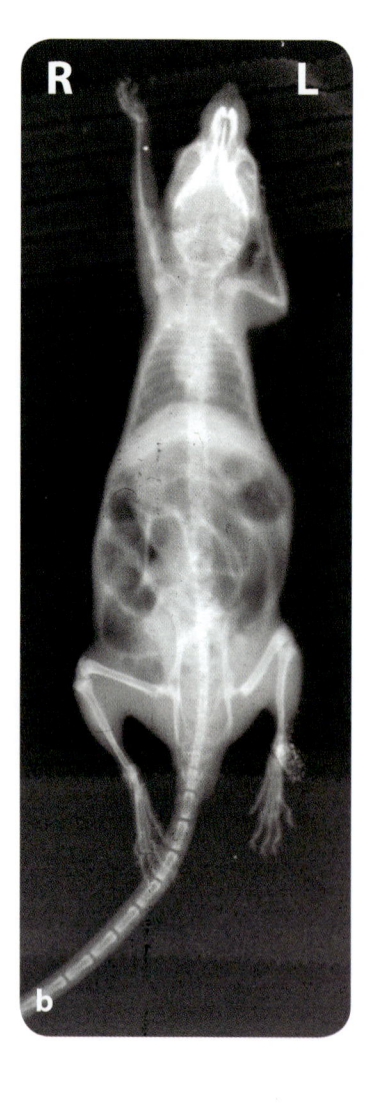

References:
　Popesko P, Rijtová V, Horák J: A Colour Atlas of Anatomy of Small Laboratory Animals. Vol. II: Rat, Mouse, Hamster. Bratislava: Príroda Publishing House; 1990.
　Silverman S, Tell LA: Laboratory mouse *(Mus musculus)*. In: Radiology of Rodents, Rabbits and Ferrets. An Atlas of Normal Anatomy and Positioning. Philadelphia, PA: Elsevier Saunders; 2005:9-17.
　Smith AN, Burke H, Heatley JJ, Beard DM, Weiss RC, Blue JT: Chemotherapy for lymphosarcoma in a pet mouse. Exotic DVM 2004; 6(5): 5-8.

HAMSTERS

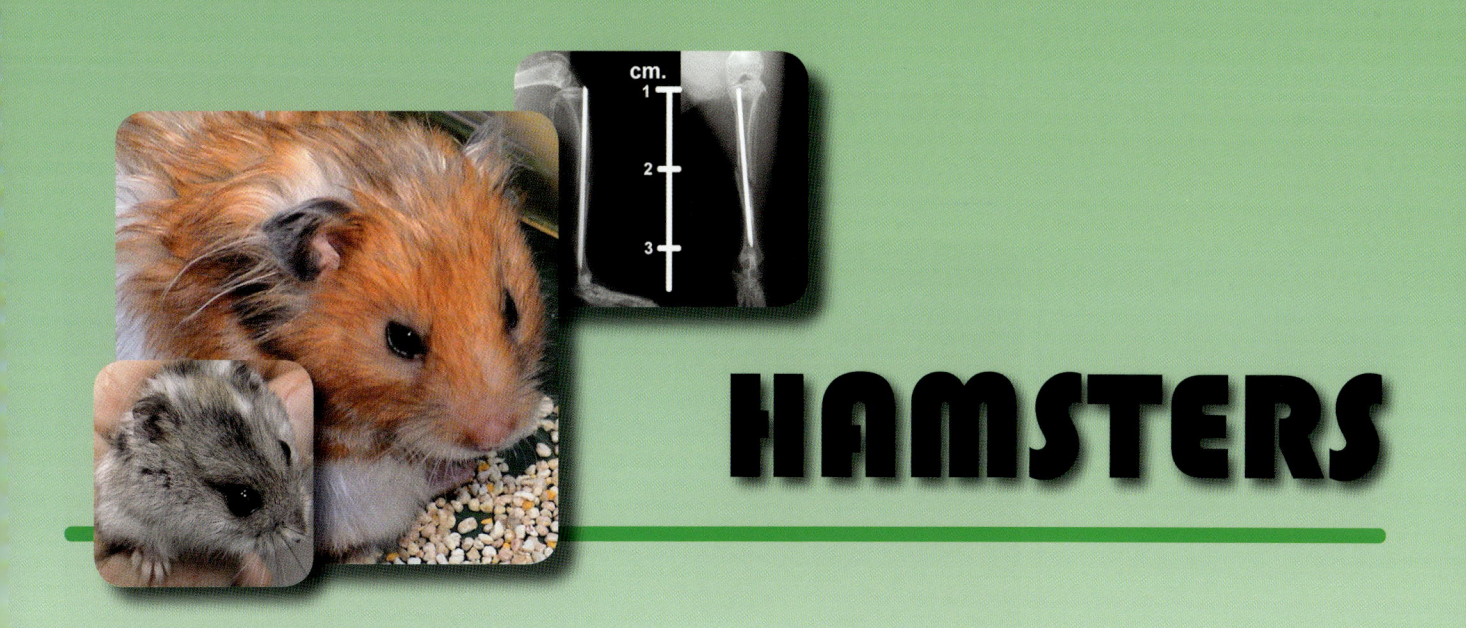

The NORMAL HEAD
Lateral Projection

Figure 8.1a,b. 300% of actual size. The apex of the maxillary incisor teeth is normally located at the rostral 1/3 of the maxillary diastema (yellow circle). Large mandibular incisor teeth extend within the mandible ventral to the molar teeth, with the apex located caudal to the last cheek tooth (red circle). In this image, the mandible appears slightly rostrally subluxated. When the jaw is at rest, cheek teeth are in occlusion, while incisor teeth remain separated, giving the appearance of mandibular retrognathism.

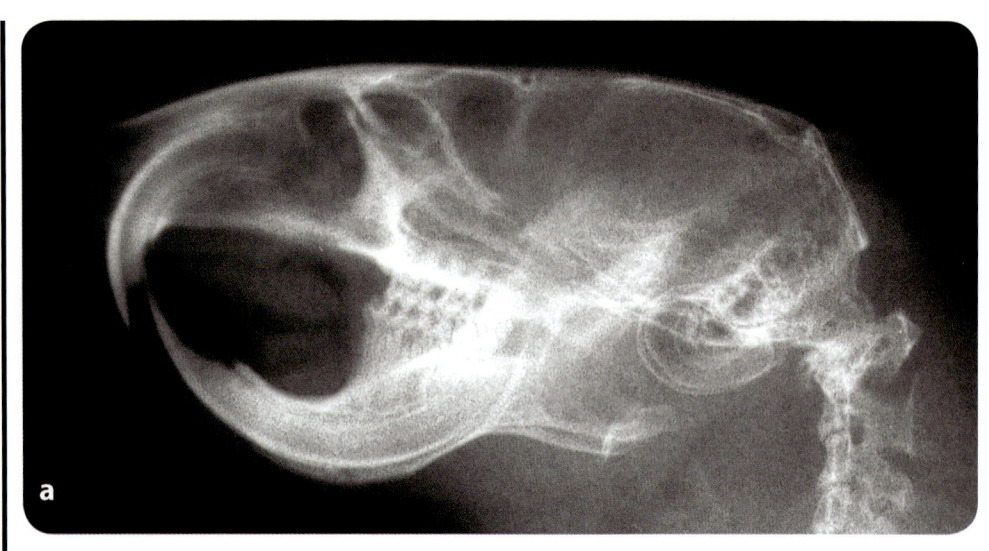

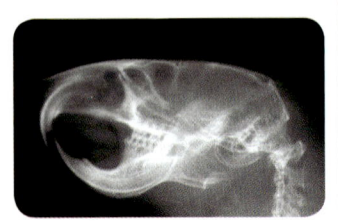

Figure 8.2. Actual size of the skull of a 100 g male golden hamster.

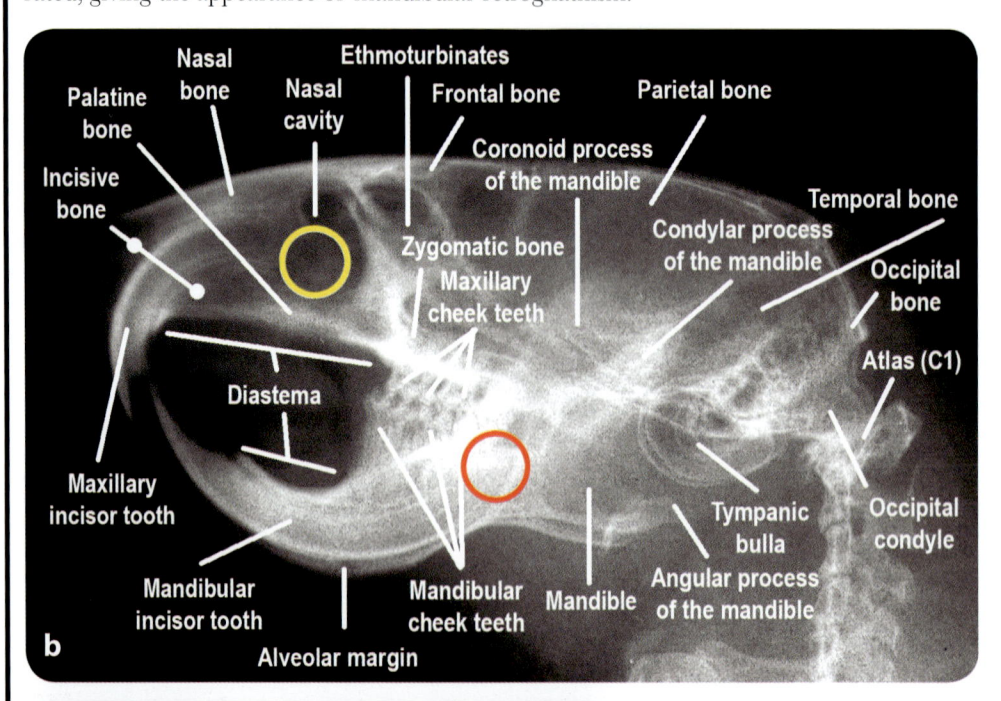

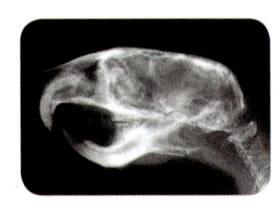

Figure 8.3. Actual size of the skull of a 40 g male Russian hamster.

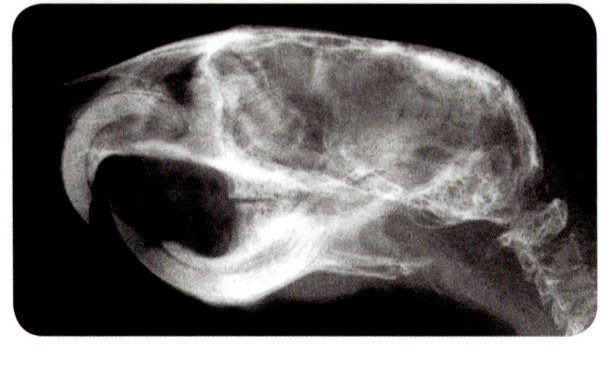

Figure 8.4. Radiograph of the skull of a 40 g male Russian hamster.

HEAD 299

Oblique Projection

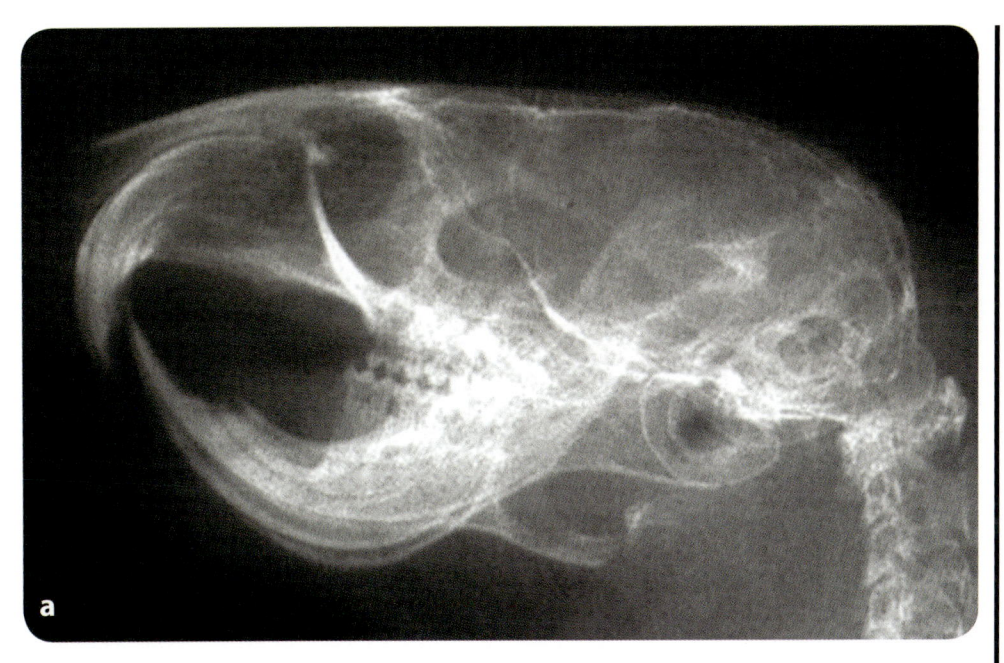

Figure 8.5a,b. Left 15° ventral-right dorsal oblique radiograph of the skull of a hamster. Note the apex of the mandibular incisor tooth indicated by the red circle.

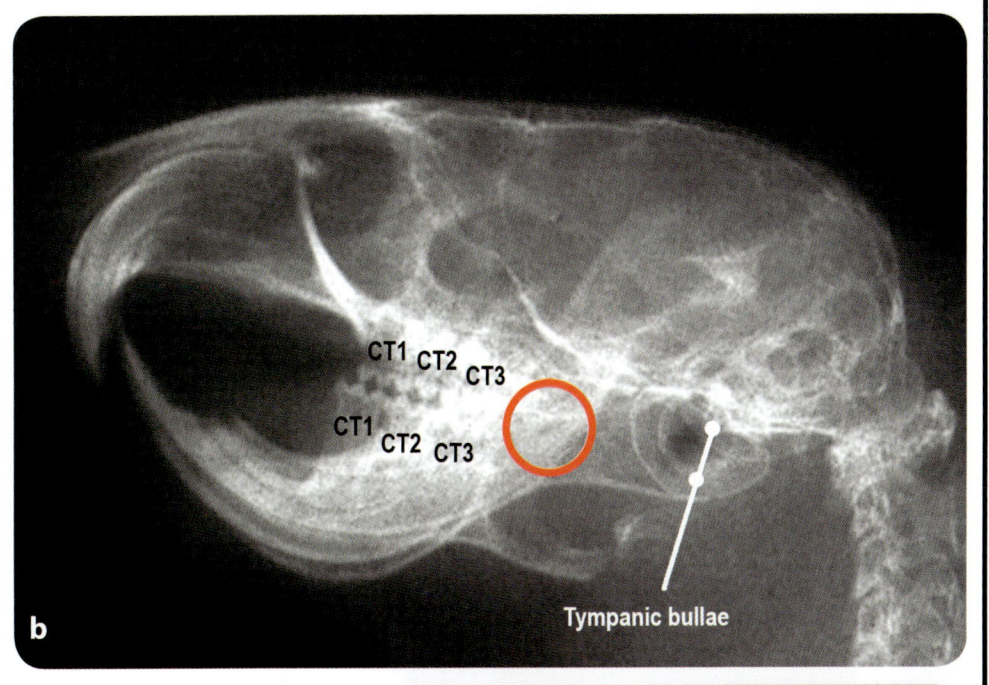

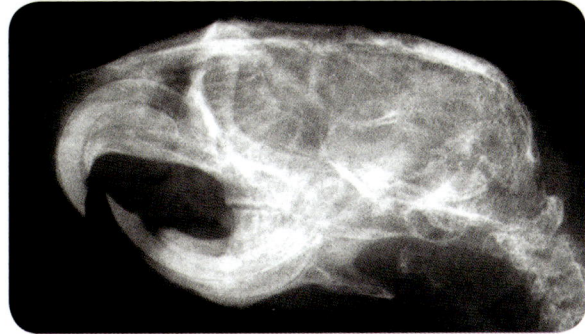

Figure 8.6. Radiograph of the skull of a 40 g male Russian hamster.

Ventrodorsal Projection

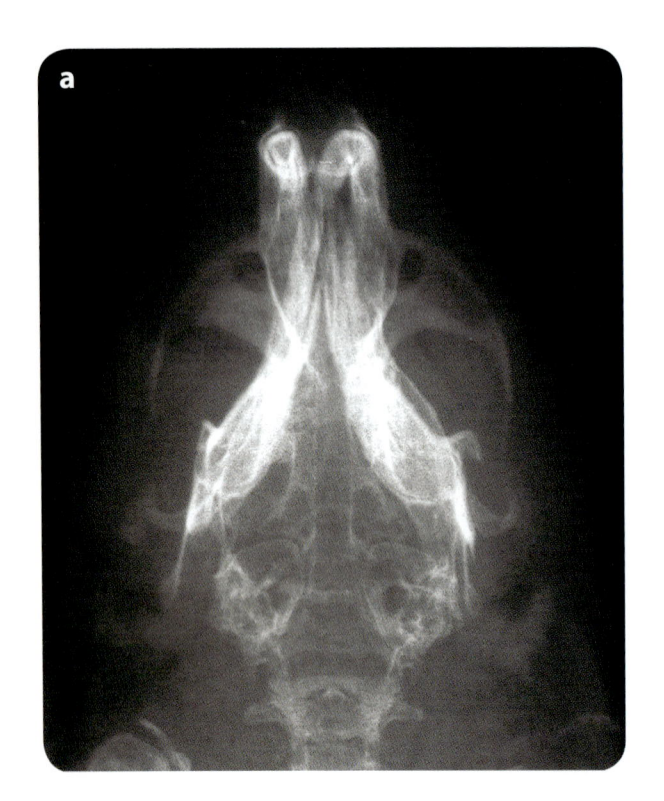

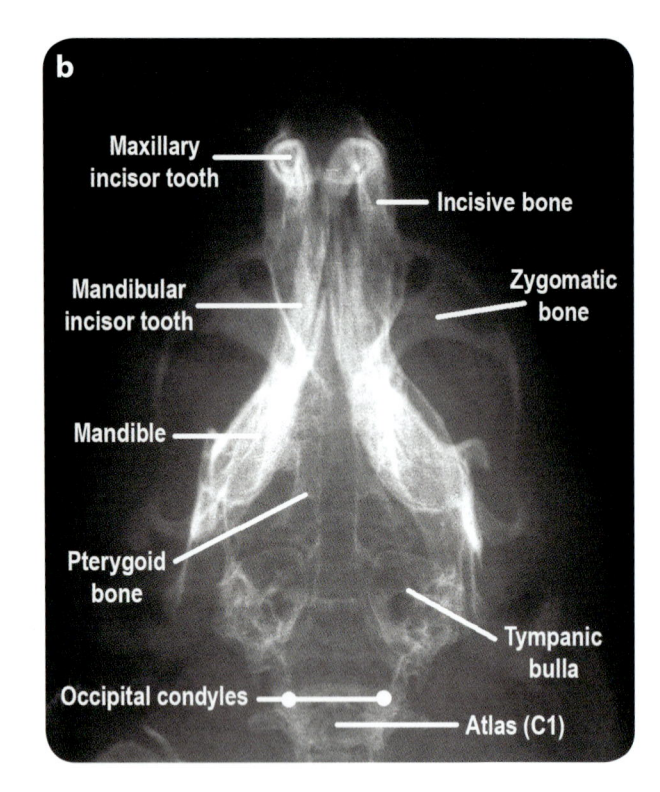

Figure 8.7a,b. Radiograph of the skull of a 100 g male golden hamster.

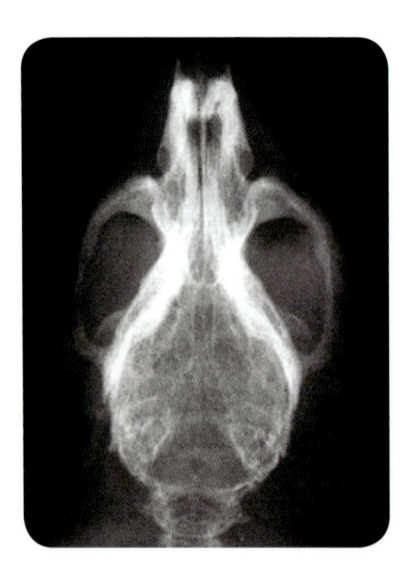

Figure 8.8. Radiograph of the skull of a 40 g male Russian hamster.

HEAD

Rostrocaudal Projection

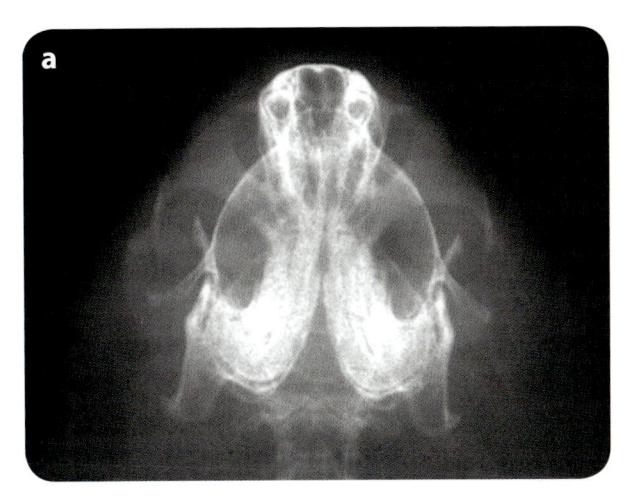

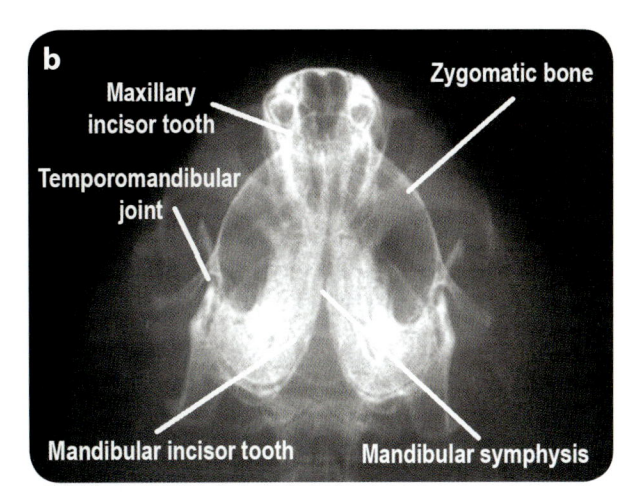

Figure 8.9a,b. Radiograph of the skull of a 100 g male golden hamster.

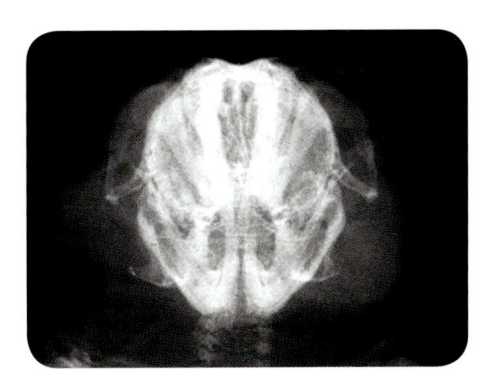

Figure 8.10. Radiograph of the skull of a 40 g male Russian hamster.

ABNORMALITIES of the HEAD
Diseases of Incisor Teeth

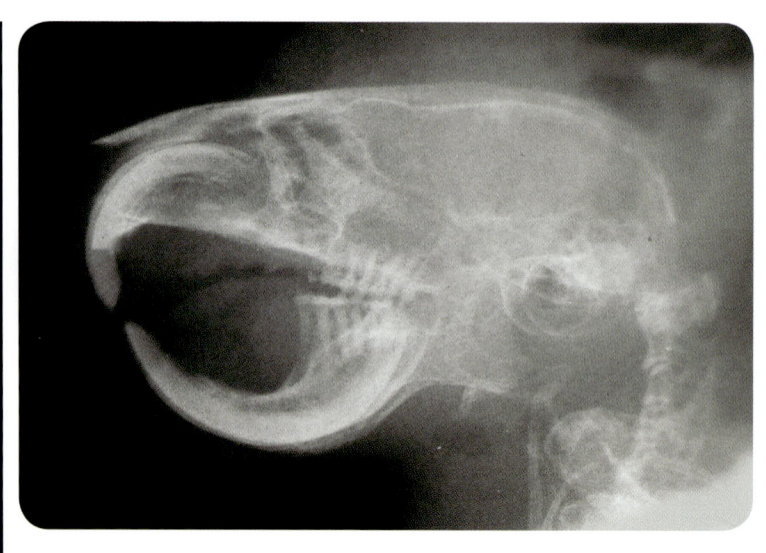

Figure 8.11. Lateral radiograph of the skull of a golden hamster with a fracture of a maxillary incisor tooth. Note the comparative lengths of the maxillary incisors.

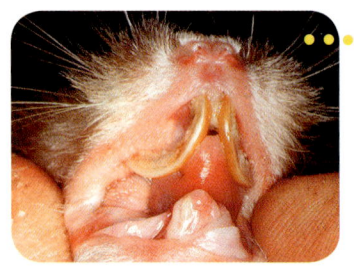

Figure 8.12. Severe malocclusion of the maxillary incisors and fracture of the mandibular incisors. Growth of the fractured mandibular incisors is impaired.
Various patterns of maxillary incisor malocclusion may occur. The most common presentation is curved elongation of the incisors, often with secondary lesions of the lips, tongue and hard palate.

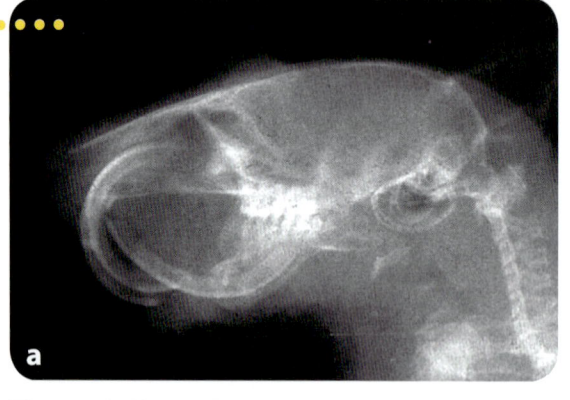

Figure 8.13a,b. Lateral (a) and ventrodorsal (b) radiographs of a young golden hamster with severe malocclusion of both maxillary and mandibular incisor teeth.

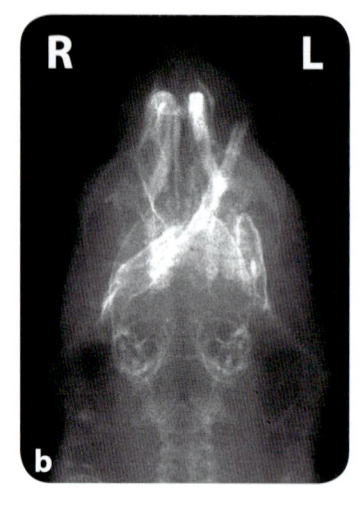

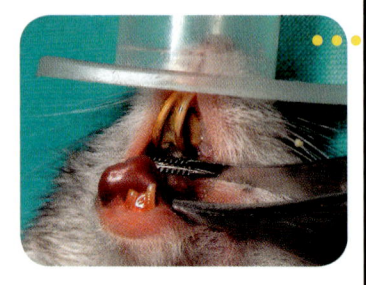

Figure 8.14. Same patient. The elongated left maxillary incisor has created a mucosal lesion of the soft palate.

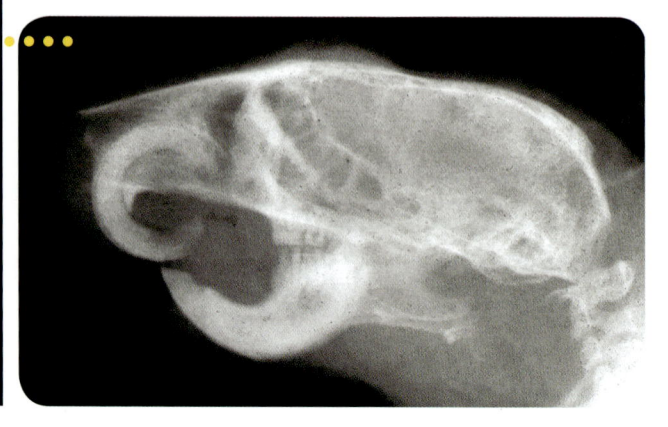

Figure 8.15. Lateral radiograph of the skull of a Russian hamster with severe elongation and malocclusion of maxillary incisor teeth, and fractured mandibular incisors.

Diseases of Cheek Teeth

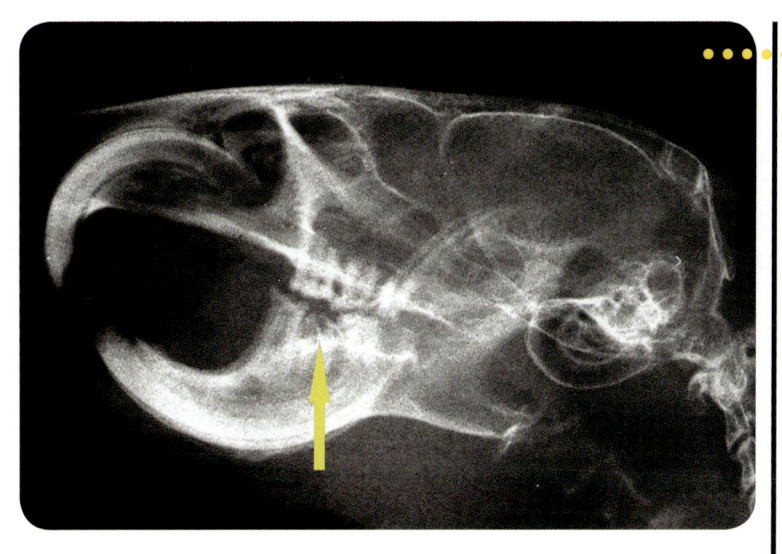

Figure 8.16. Lateral radiograph of the skull of a golden hamster with a fracture of mandibular CT2 (arrow).

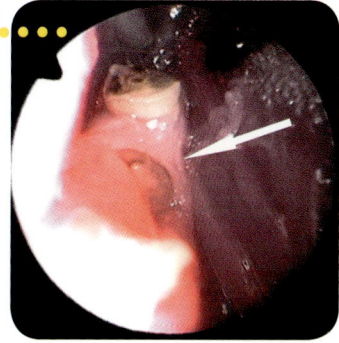

Figure 8.17. Endoscopic view of the same patient with fracture of mandibular CT2 (arrow).

The NORMAL WHOLE BODY SKELETON
The GOLDEN HAMSTER
Lateral Projection

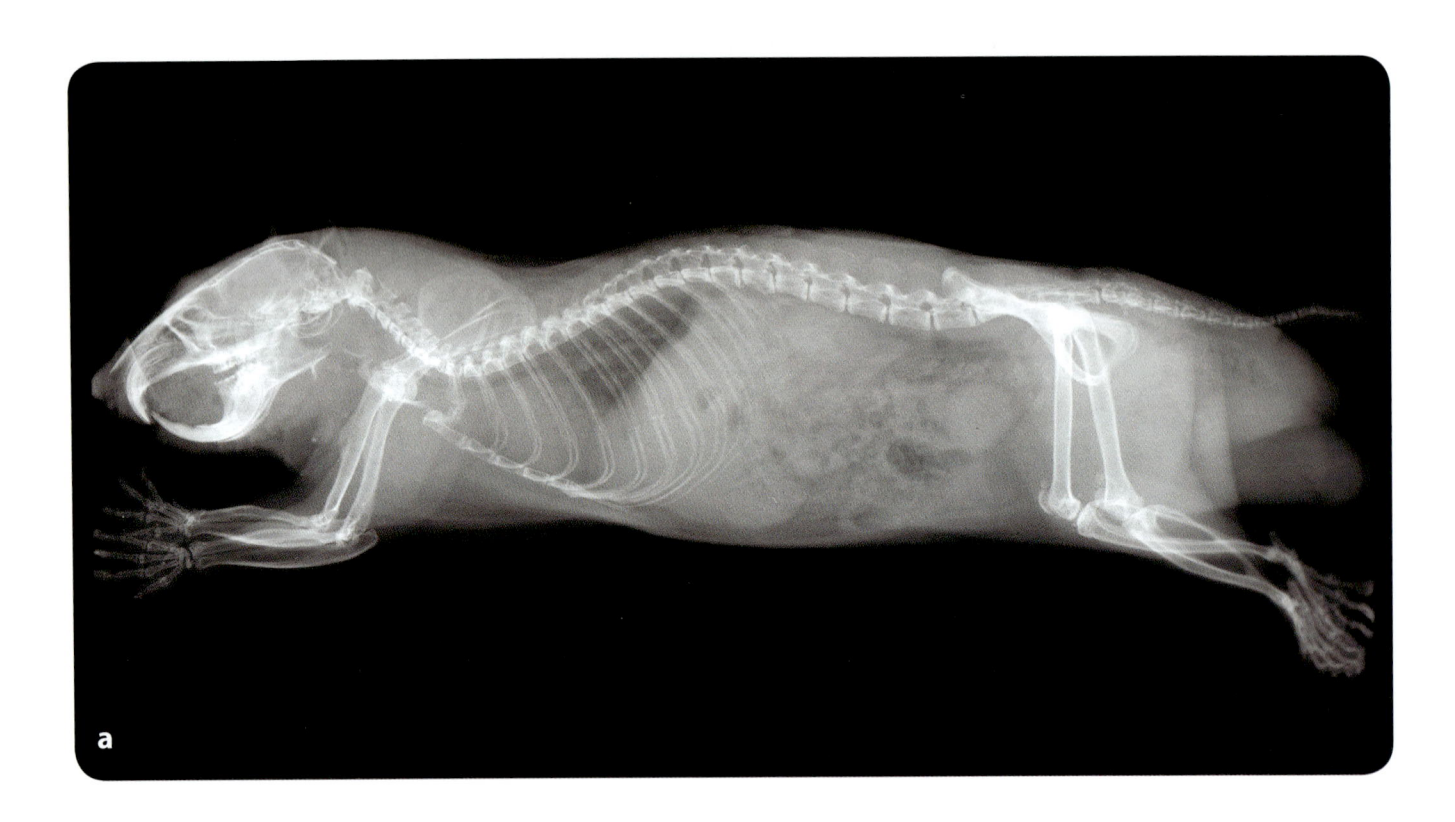

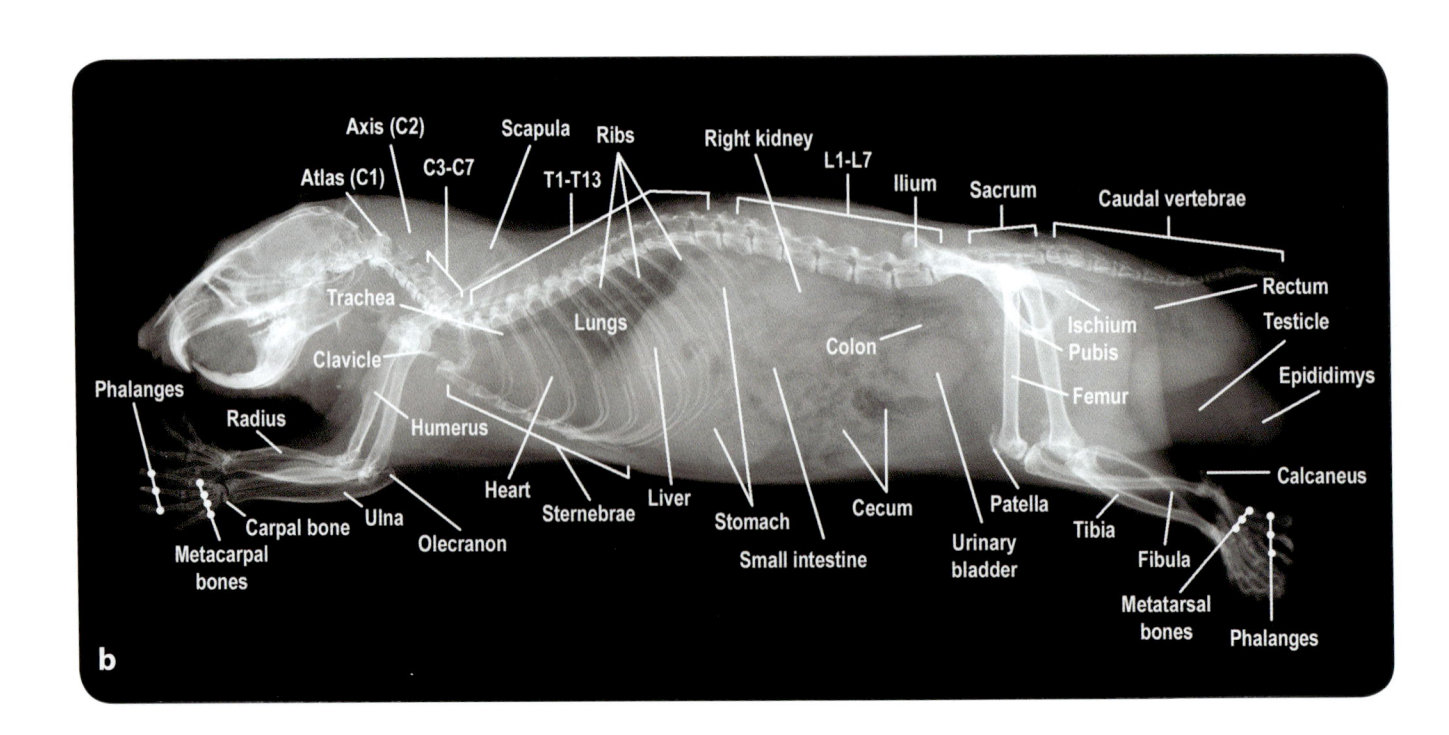

Figure 8.18a,b. Whole body skeleton of a 110 g male Golden hamster, actual size.

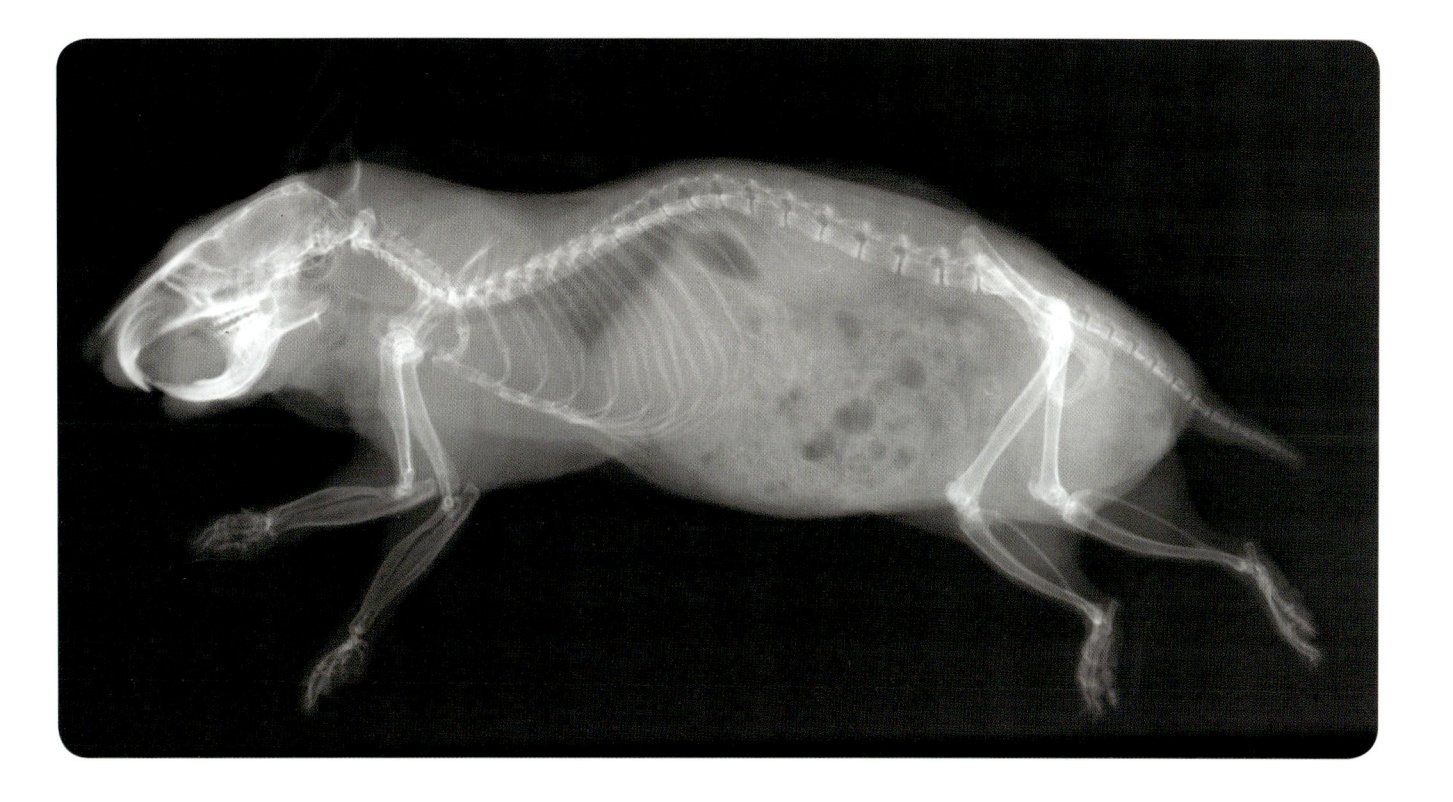

Figure 8.19. Whole body skeleton of a 150 g female golden hamster, actual size.

Ventrodorsal Projection

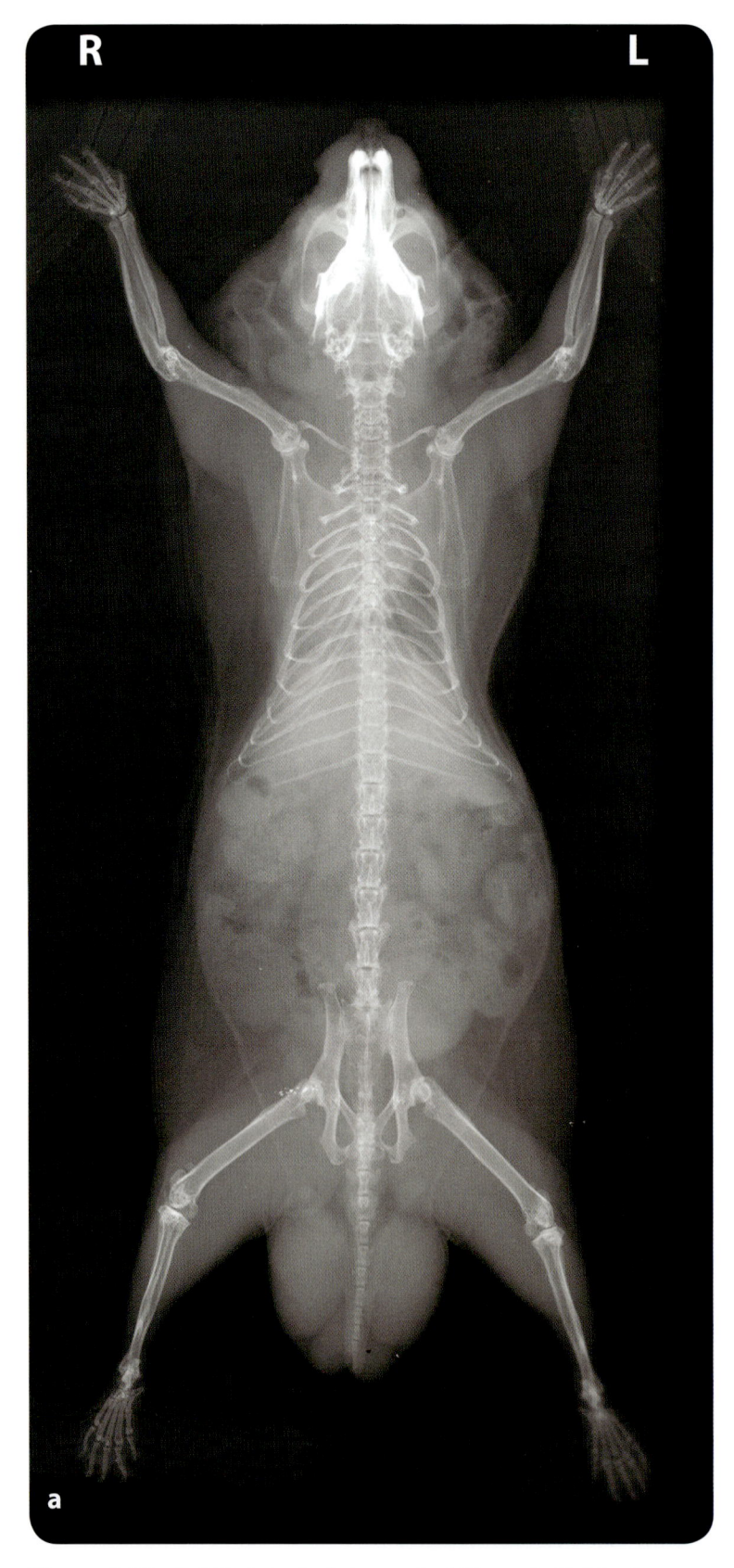

Figure 8.20a,b. Whole body skeleton of a 110 g male Golden hamster, actual size.

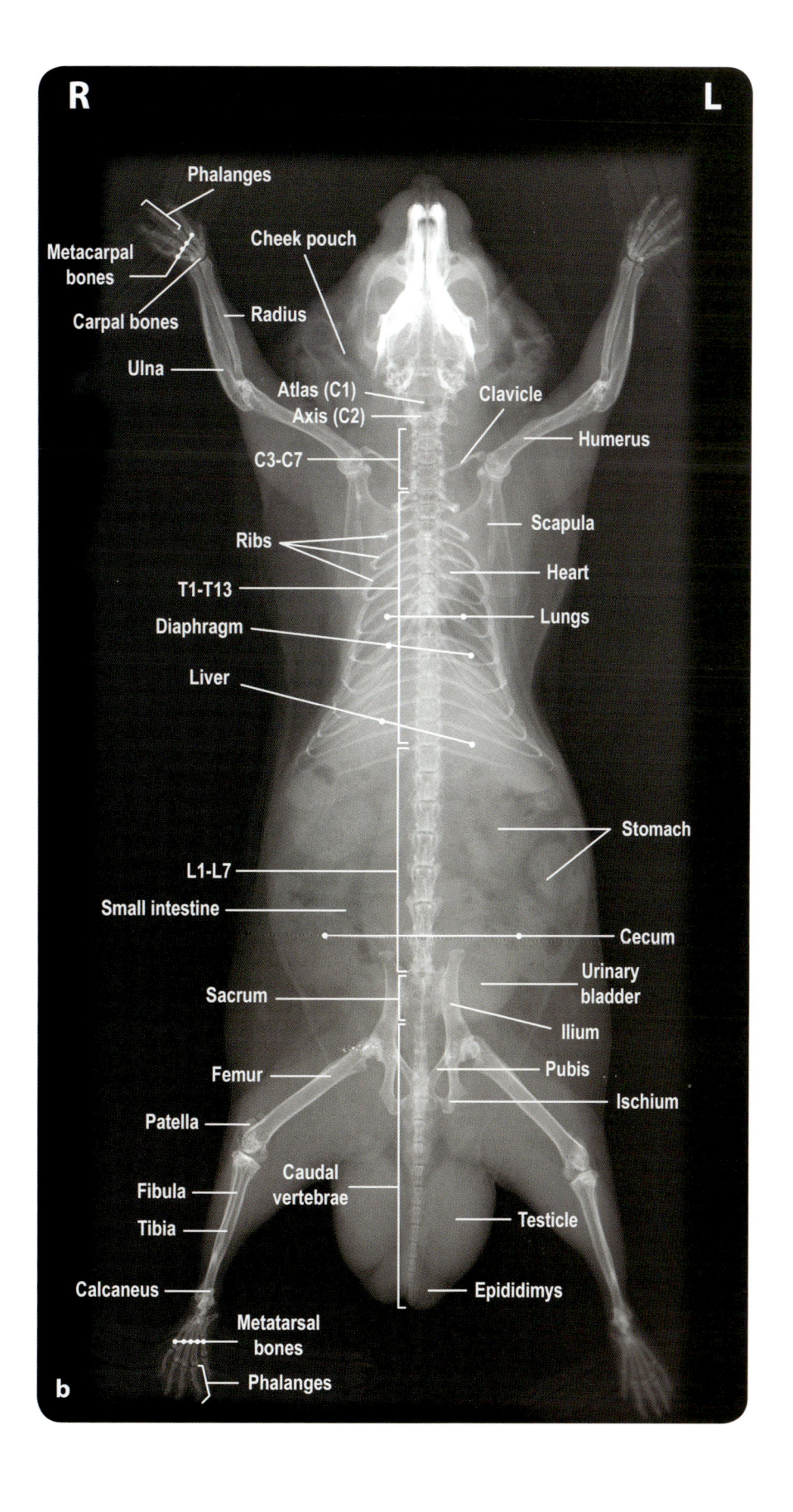

WHOLE BODY SKELETON 307

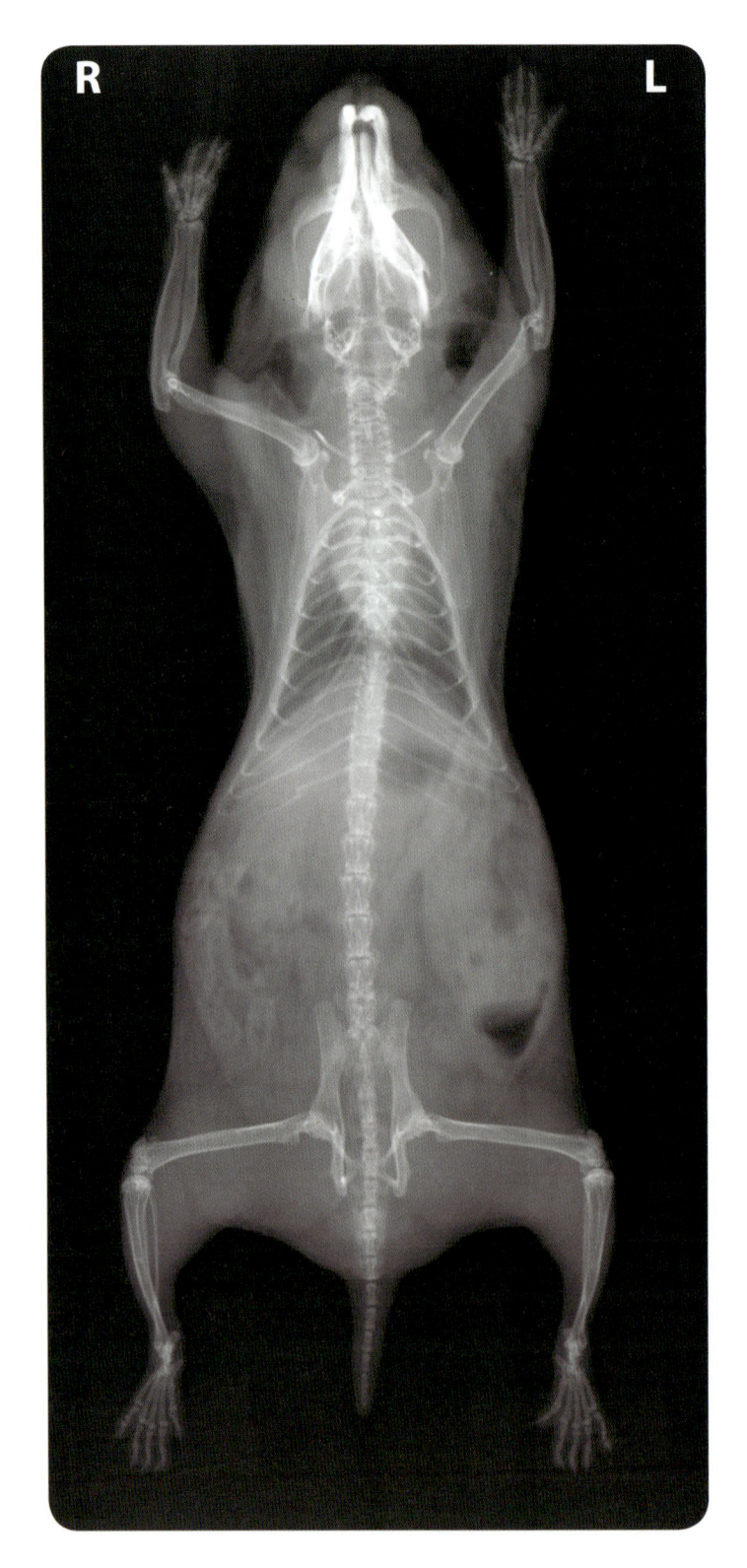

Figure 8.21a,b. Whole body skeleton of a 150 g female golden hamster, actual size.

Miscellaneous

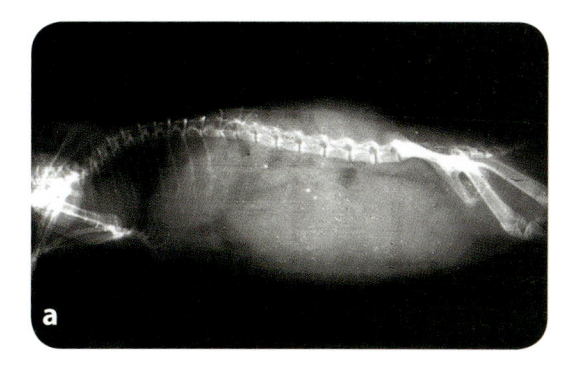

Figure 8.22a,b. Lateral (a) and ventrodorsal (b) radiographs of a pregnant golden hamster near partuition. Enlargement of the uterine body is visible in both lateral and ventrodorsal projections.

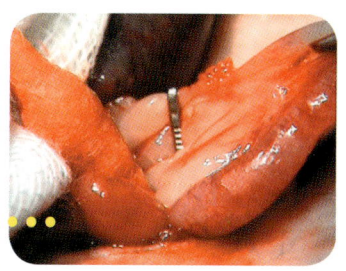

Figure 8.23. Elective ovariectomy or ovariohysterectomy is occasionally indicated in female hamsters. Positioning of a hemostatic clip during ovariectomy.

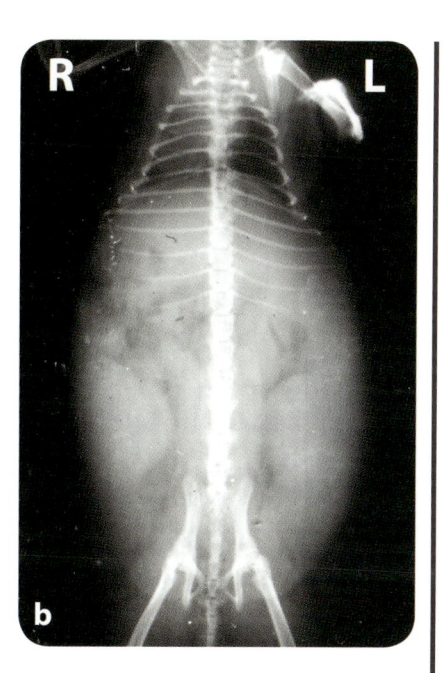

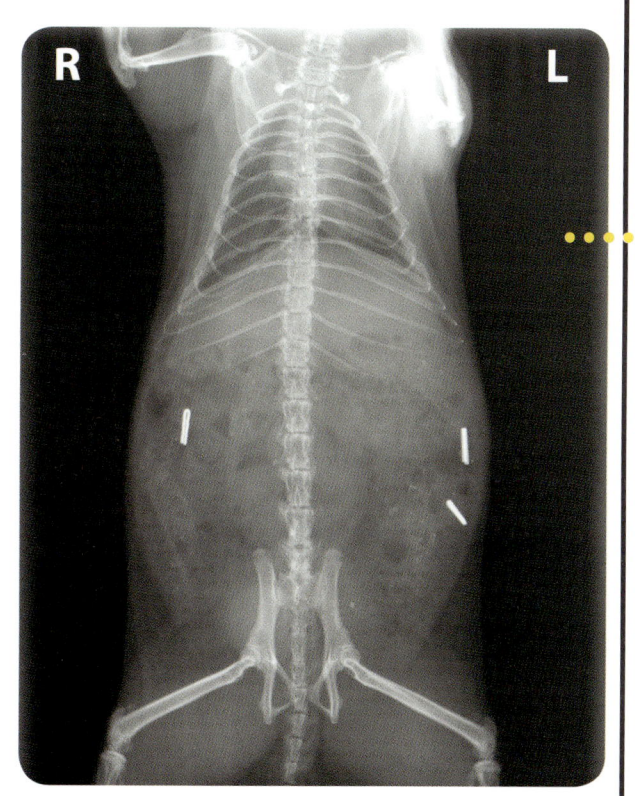

Figure 8.24. Ventrodorsal radiograph of a golden hamster demonstrating hemostatic clips used during ovariectomy.

The NORMAL WHOLE BODY SKELETON
The RUSSIAN HAMSTER
Lateral Projection

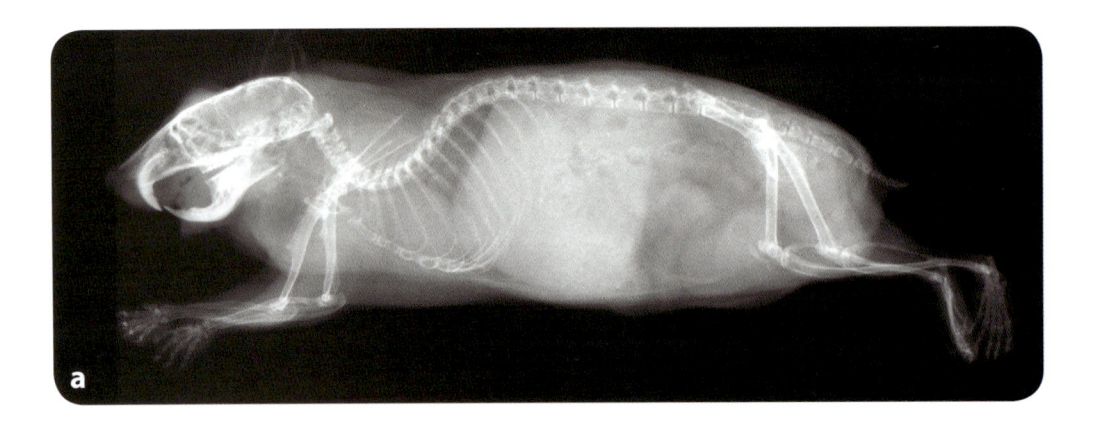

Figure 8.25a,b. Whole body skeleton of a 45 g male Russian hamster, actual size (a) and 130% of actual size (b).

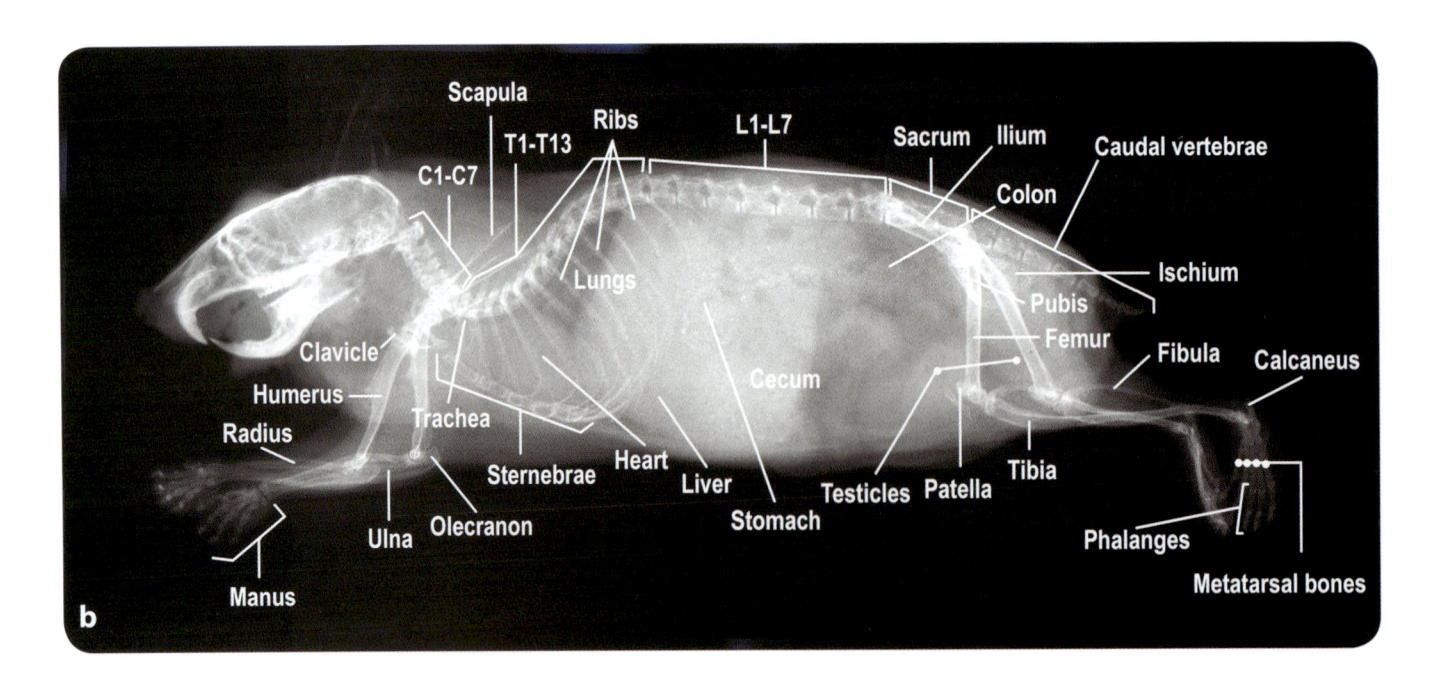

Ventrodorsal Projection

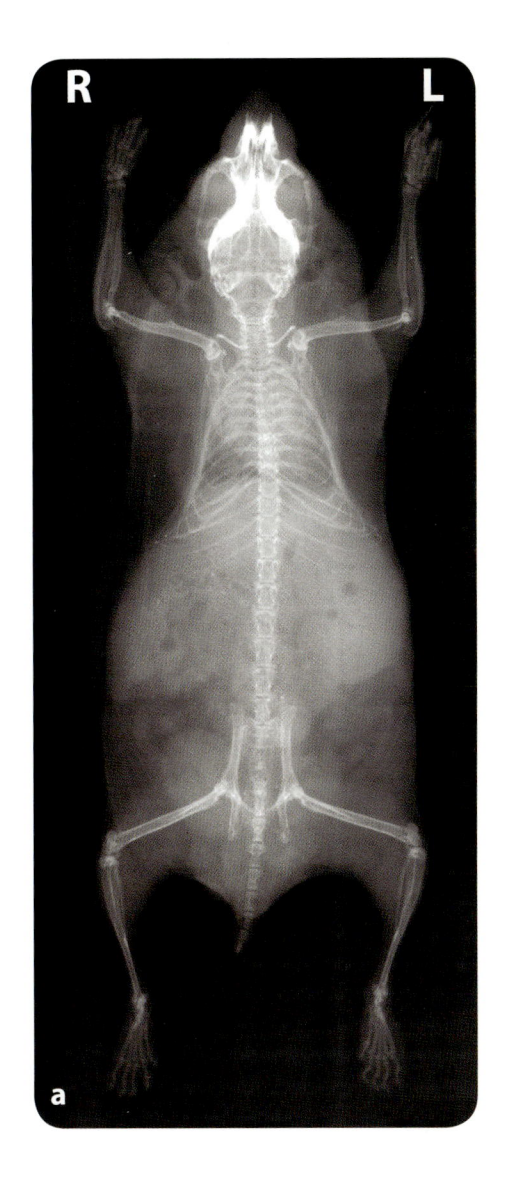

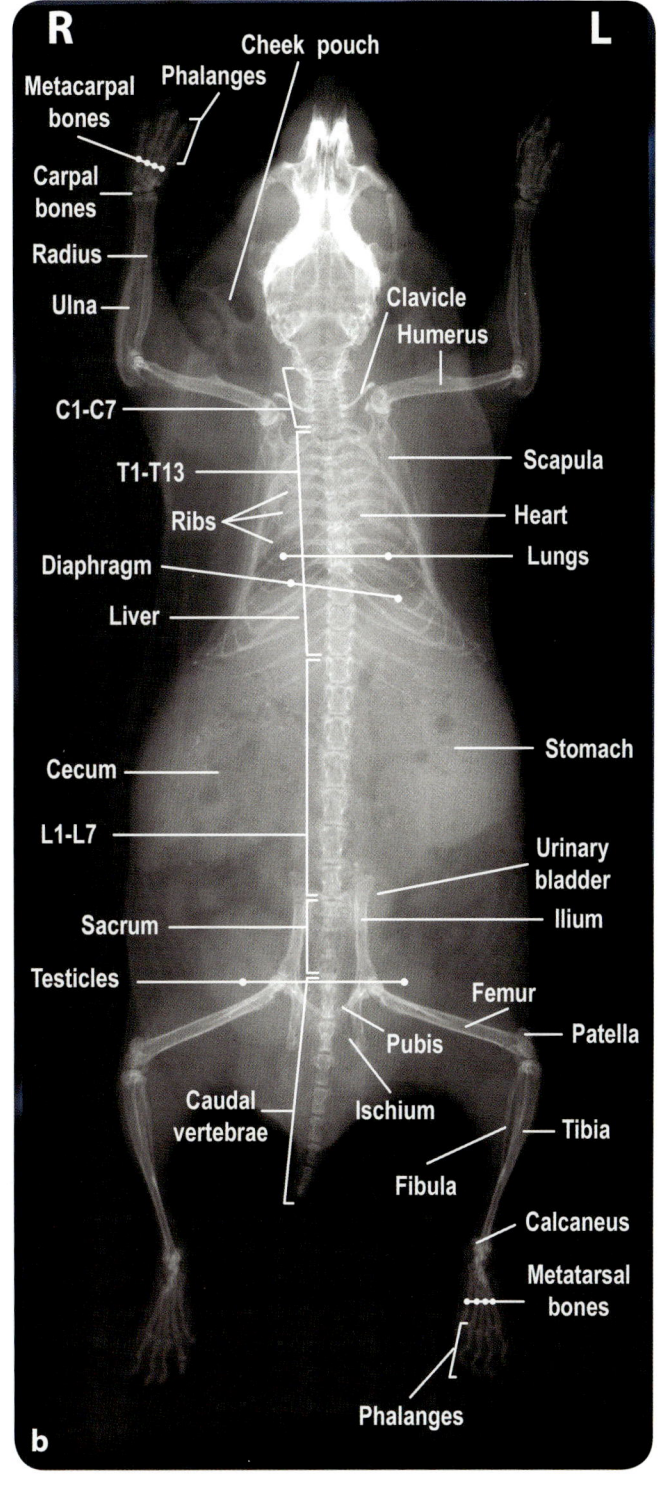

Figure 8.26a,b. Whole body skeleton of a 45 g male Russian hamster, actual size (a) and 130% of actual size (b).

ABNORMALITIES of the ABDOMEN

Diseases of the Abdominal Cavity

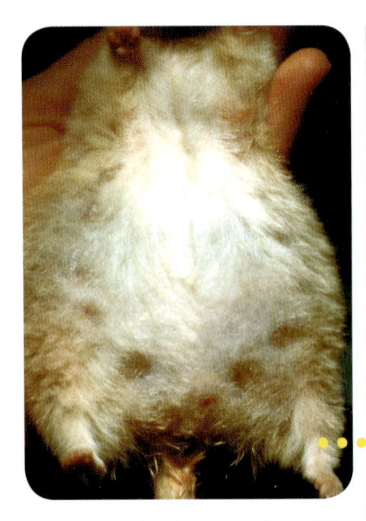

Figure 8.27. Same patient. The abdomen was uniformly distended. In this case, no masses were palpable. Further diagnostic testing was declined.

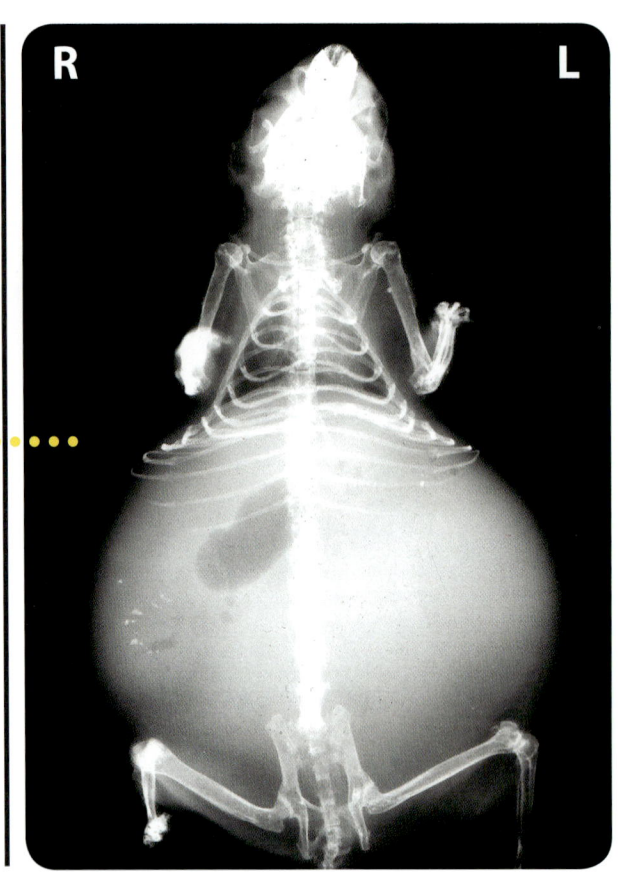

Figure 8.28. Ventrodorsal radiograph of the whole body of a 2-year-old female golden hamster with abdominal distention. Marked ascites obscures abdominal details with the exception of the gas filled stomach.
Multiple diseases can produce ascites in small rodents, including organ failure, cystic disease, neoplasia, and cardiac disease.

Diseases of the Gastrointestinal Tract

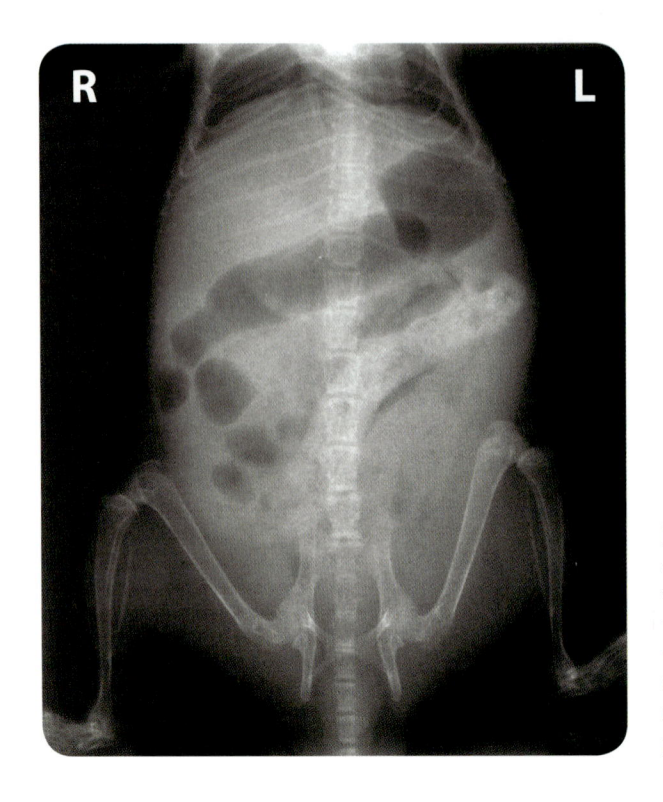

Figure 8.29. Ventrodorsal radiograph of the abdomen of a young golden hamster with intestinal impaction and gas distension. The radiopaque intestinal material is clearly visible, as is gas distension of the intestine proximal to the impaction.

ABDOMEN 313

Figure 8.30. Ventrodorsal radiograph of the abdomen of a hamster with severe enteritis. Note marked gas distension of the entire intestinal tract.

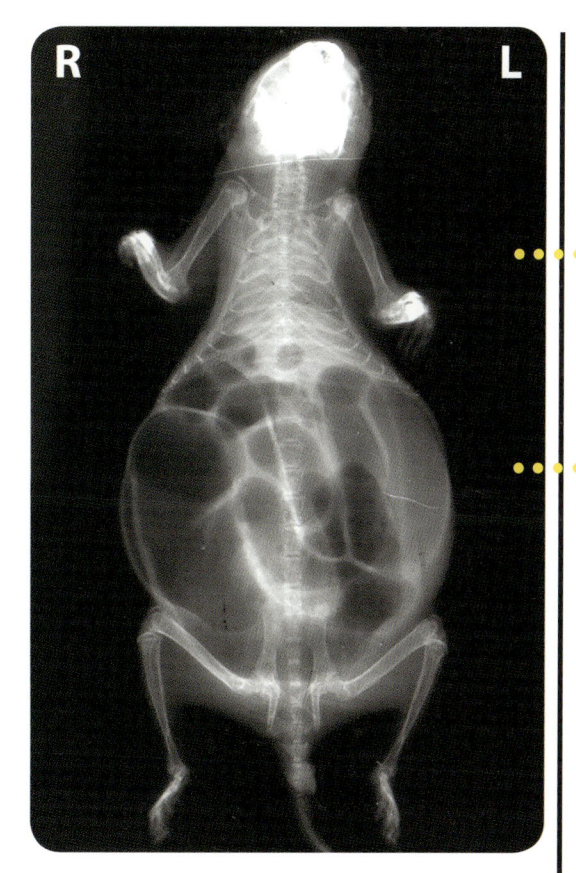

Figure 8.31. Hamsters with enteritis typically present with a characteristic hunched posture as a result of abdominal pain.

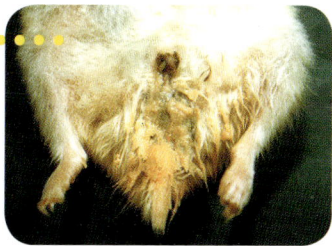

Figure 8.32. Enteritis complex is often referred to as "wet tail," and has numerous underlying etiologies. Typical clinical signs include liquid, yellowish diarrhea pasted to the perineum.

Figure 8.33. Ventrodorsal radiograph of a male golden hamster with hepatic cysts identified at sugery. Two large cysts occupy most of the right and left abdominal cavity. Intestines (represented by gas) are deflected to the left and caudally.
Polycystic disease occurs in many organs, with the liver most commonly affected. Size of the cysts can range from a few millimeters to several cm in diameter. Clinical signs are related to space occupying abdominal mass, and the hamster is often presented for enlarged abdomen. Dyspnea is also a common clinical finding.

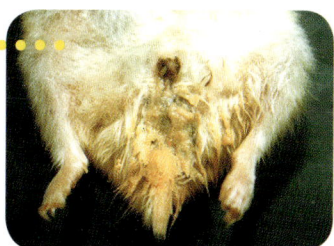

Figure 8.34. Round, fluid-filled, hepatic cysts are easily diagnosed via ultrasonography.

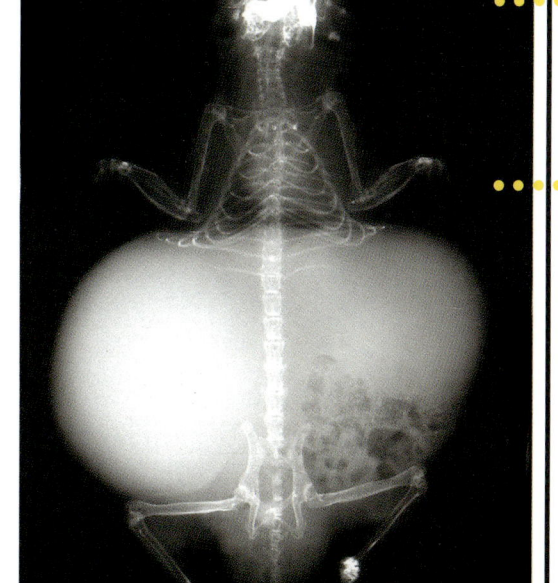

Figure 8.35. The fur has been partially shaved to better visualize the enlarged abdomen. Note abdominal distention is non-uniform, and appears as separate, distinct enlargements.

Diseases of the Urogenital Tract

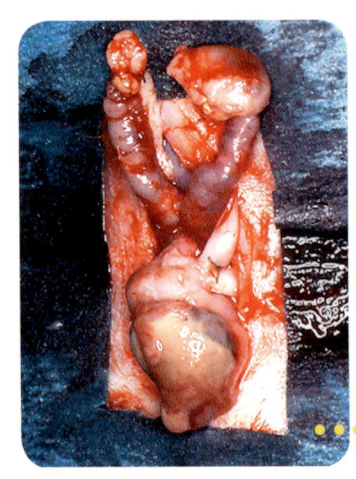

Figure 8.36. Surgical appearance of a large uterine mass confirmed as uterine fibroma by histopathology. The mass was palpable on physical examination.

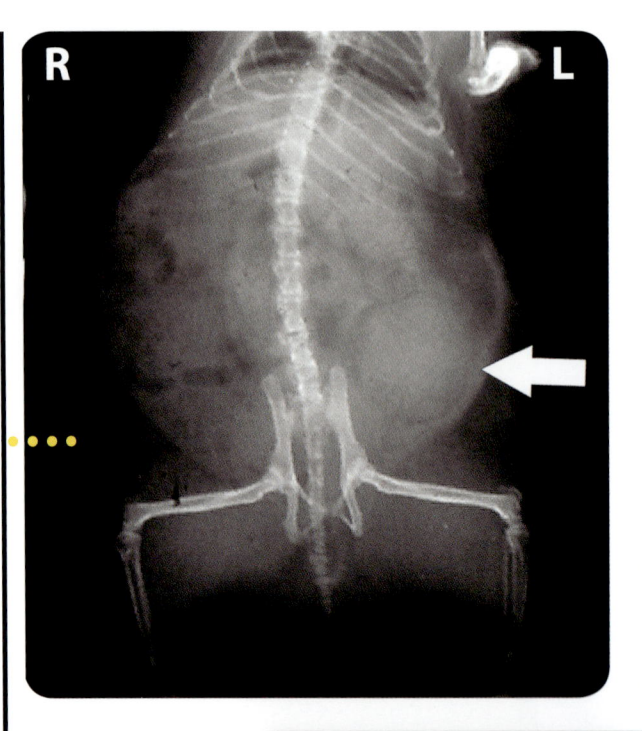

Figure 8.37. Ventrodorsal radiograph of a female golden hamster with a rounded density. The location in this female hamster suggests a uterine mass (arrow).

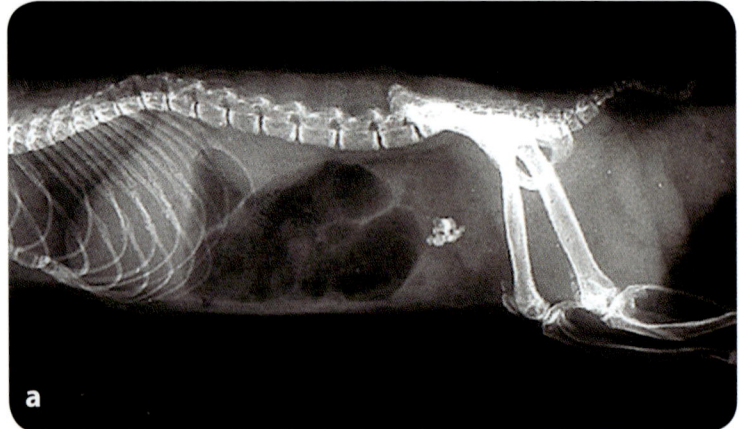

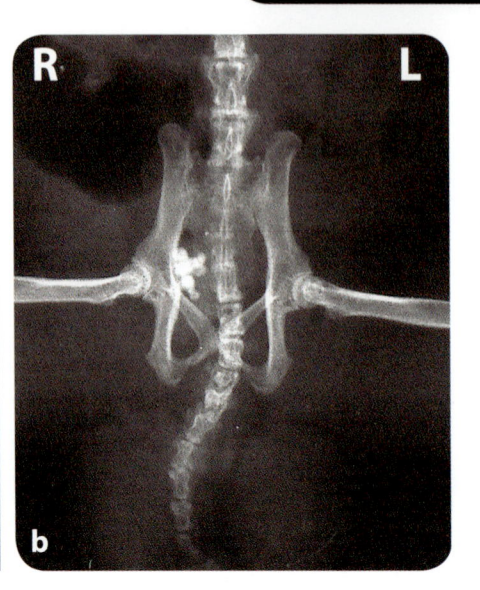

Figure 8.38a,b. Lateral (a) and ventrodorsal (b) radiographs of a 3-year old male golden hamster with cystolithiasis. 48 hours of anorexia resulted in excessive intestinal gas. Three cases of urolithiasis have been reported in the literature in golden hamsters, and stone composition was reported as magnesium and phosphate.

ABNORMALITIES of the VERTEBRAL COLUMN

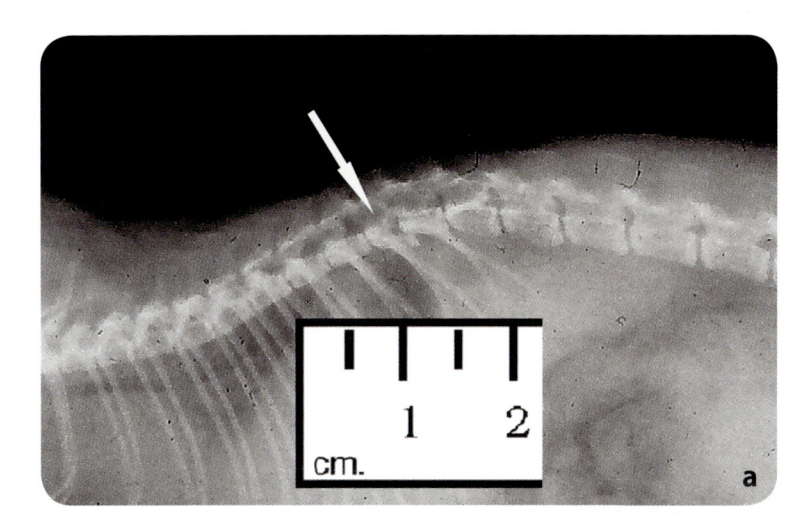

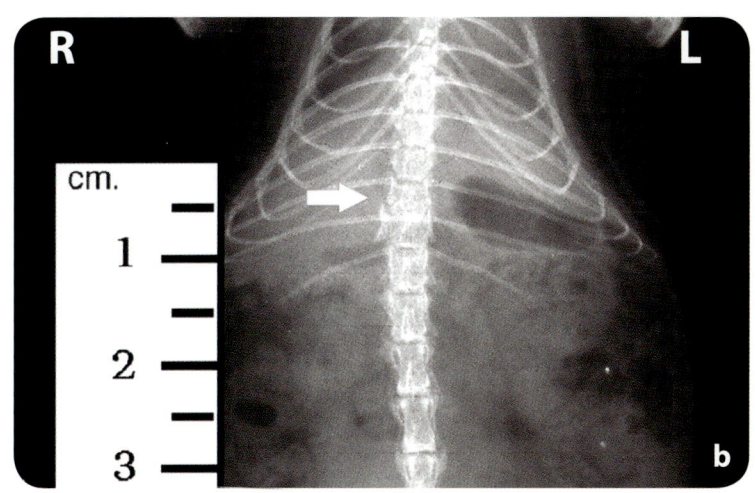

Figure 8.39a,b. Lateral (a) and ventrodorsal (b) radiographs of a golden hamster demonstrating a fracture of T12 (arrows).

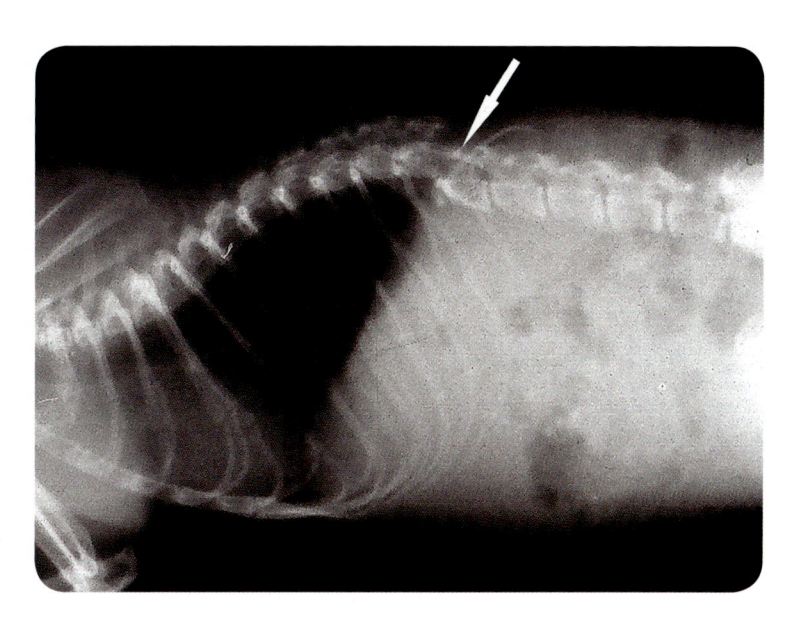

Figure 8.40. Lateral radiograph of a golden hamster with a luxation of T13-L1 (arrow).

ABNORMALITIES of the THORACIC LIMB

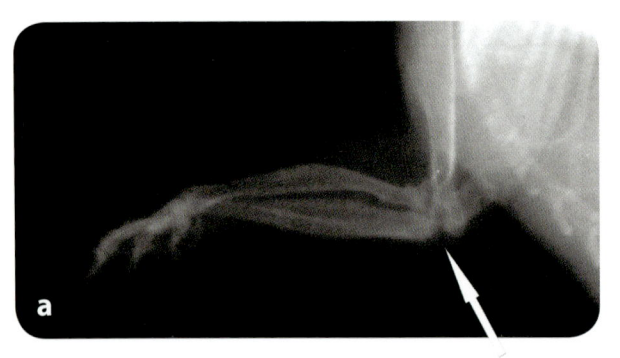

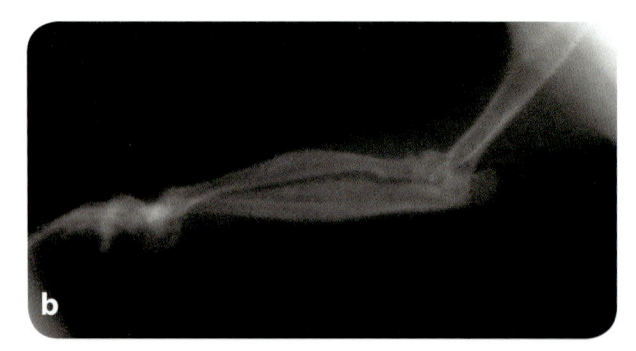

Figure 8.41a,b. Lateral radiographs (a,b) of the thoracic limb of a golden hamster with a fracture of the olecranon, seen in (a). Compare with the normal contralateral limb (b).

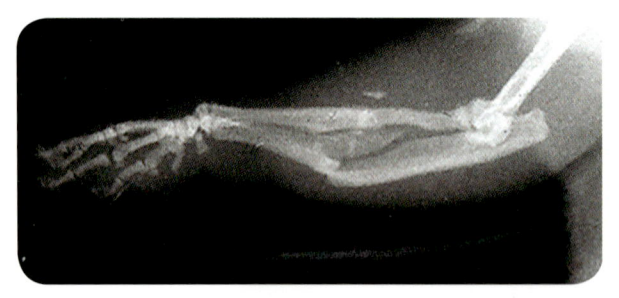

Figure 8.42. Lateral radiograph of the thoracic limb of a hamster with a comminuted fracture of the ulna.

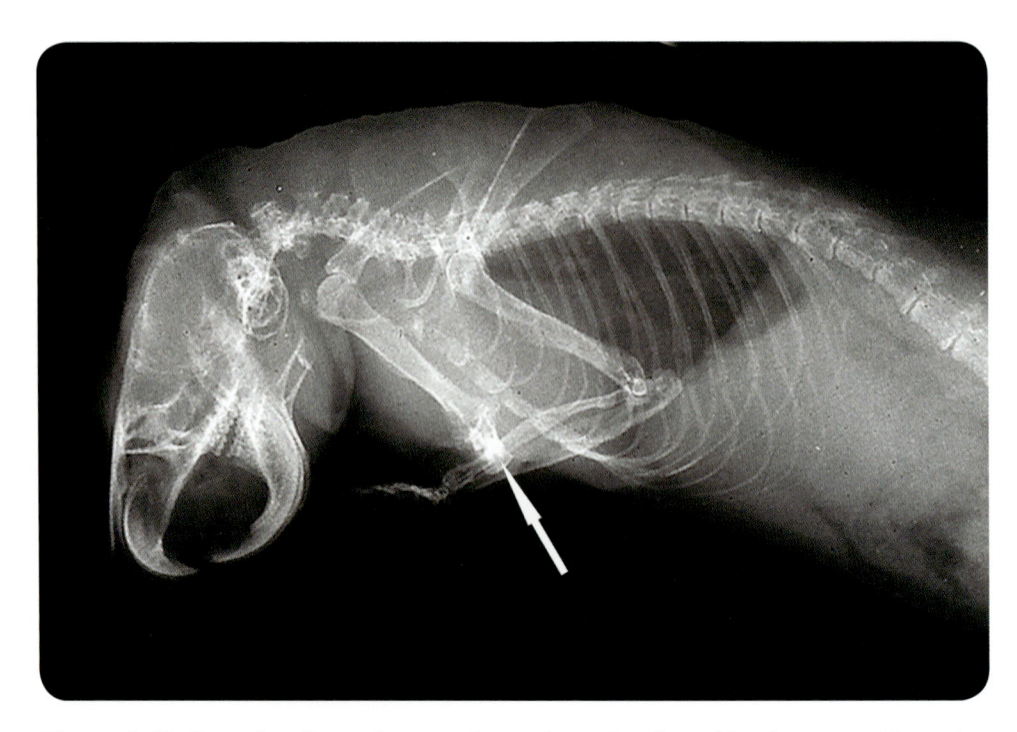

Figure 8.43. Lateral radiograph, manual restraint only, of a golden hamster with a missing distal forelimb, assumed to be congenital agenesis. Traumatic loss of limb early in life is a more likely diagnosis considering the presence of a near normal appearing proximal ulnar fragment (arrow).

THORACIC LIMB 317

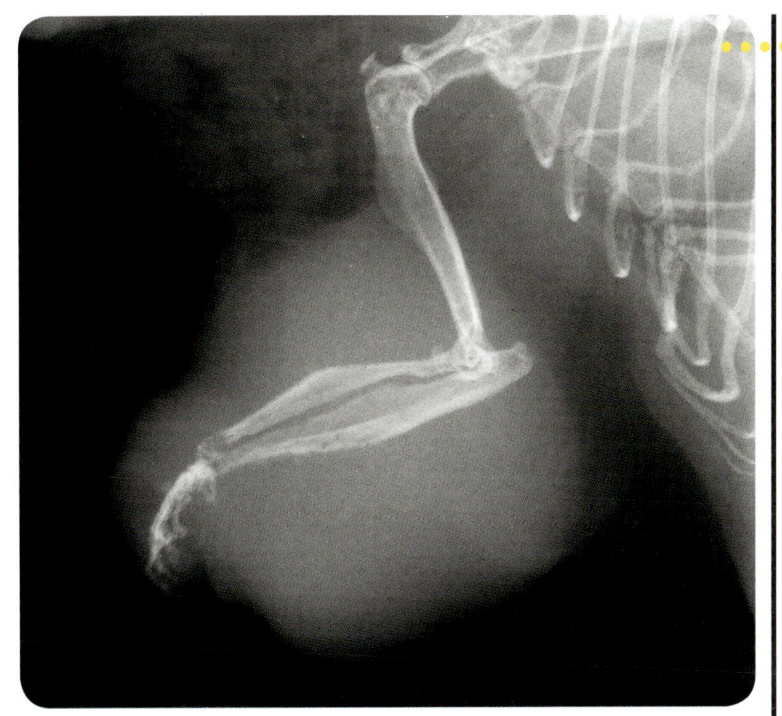

Figure 8.44. Lateral radiograph of the thoracic limb of a 2-year-old female hamster with neoplasia of the forelimb. Osteolysis of both the radius and ulna are clearly visible.

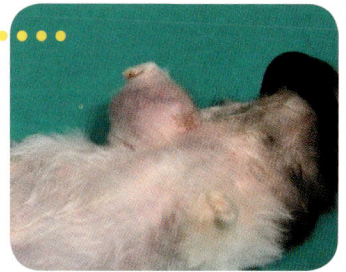

Figure 8.45. Same patient in dorsal recumbency prepared for limb amputation.

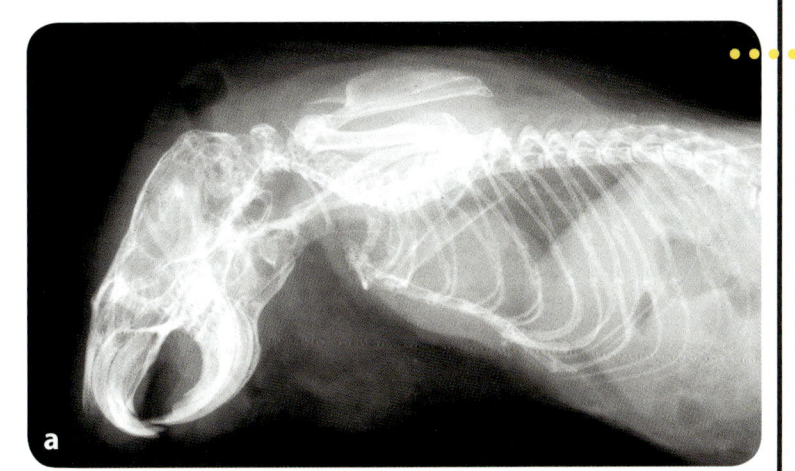

Figure 8.46a,b. Lateral (a) and ventrodorsal (b) radiographs of the same patient after amputation of the thoracic limb through disarticulation of the scapular-humeral-clavicular joint. Note the clavicula has been left intact (b).

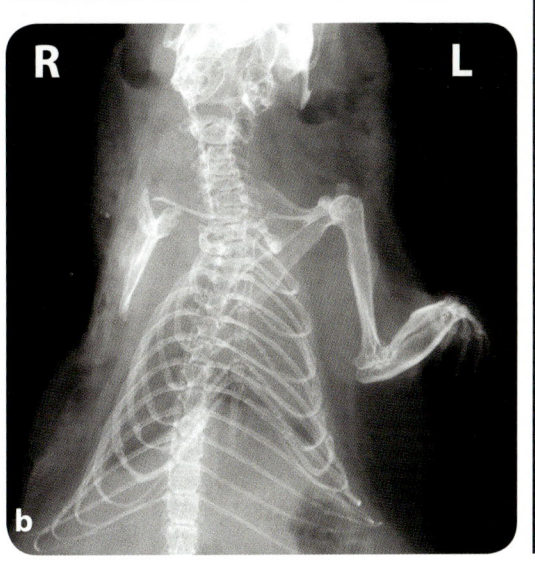

Figure 8.47. Follow up 10 days post surgery. Limb ampuation was well tolerated by this patient.

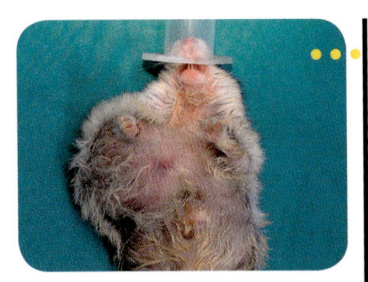

Figure 8.48. Anesthetized Russian hamster in dorsal recumbency prepared for radiology. Note the relative size of the mass when compared to the size of the patient.

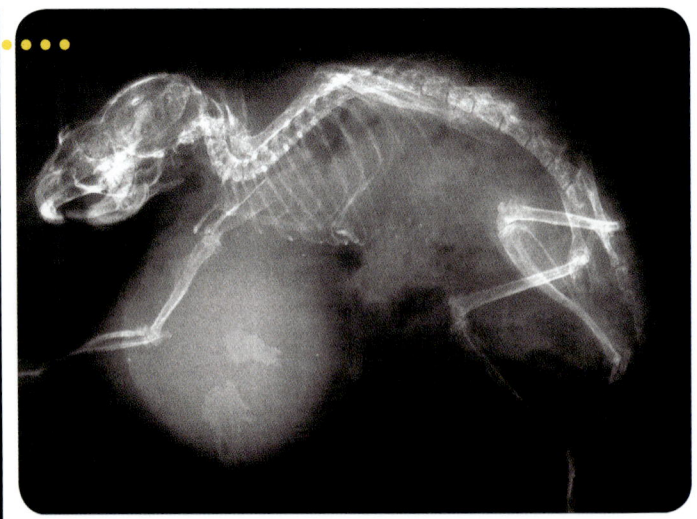

Figure 8.49. Lateral radiograph of the whole body of a male Russian hamster with a soft tissue mass of the thoracic limb determined at surgery to be most consistent with neoplasia.

ABNORMALITIES of the PELVIC LIMB

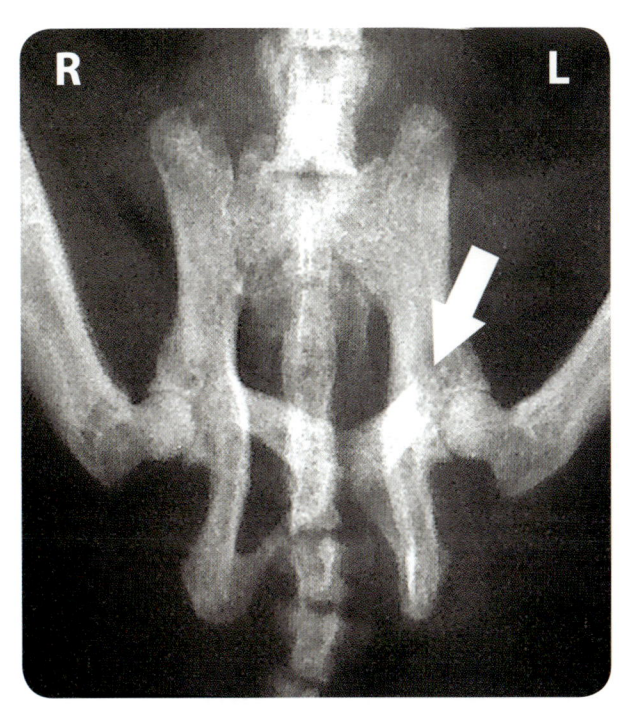

Figure 8.50. Ventrodorsal radiograph of the pelvis of a golden hamster with a fracture of the pubis (arrow). Pelvic fractures are often a consequence of a fall.

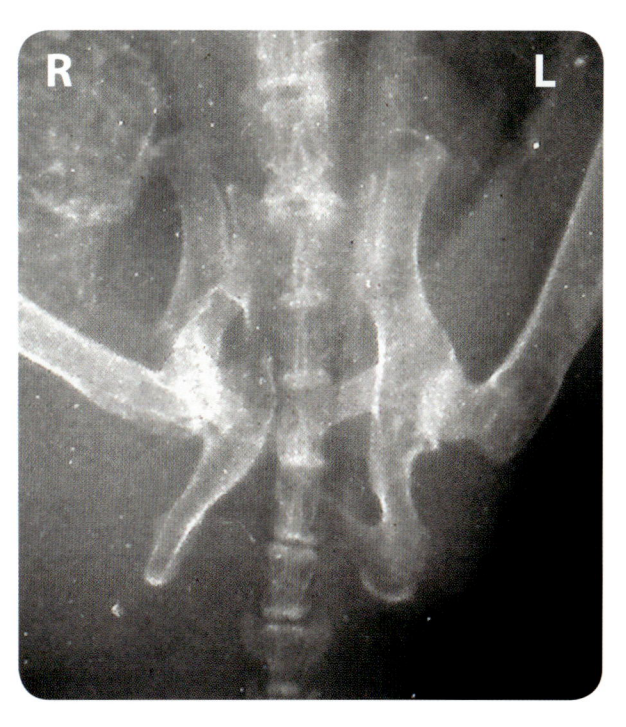

Figure 8.51. Ventrodorsal radiograph of the pelvis of a golden hamster with a severe acetabular fracture.

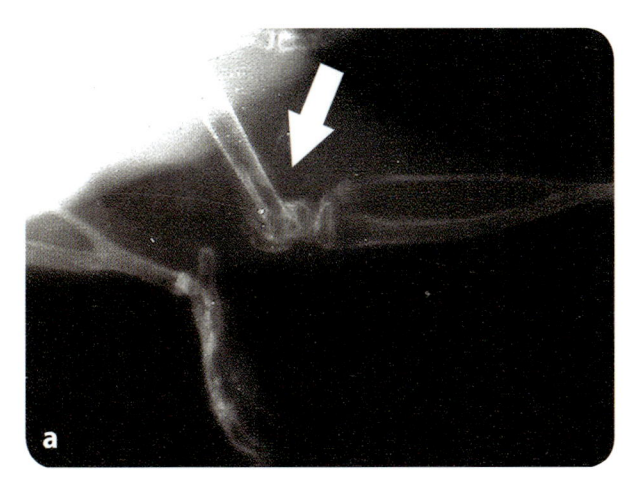

Figure 8.52a,b. Lateral (a) and craniocaudal (b) radiographs of the pelvic limb of a golden hamster with a distal metaphyseal fracture of the femur (arrow).

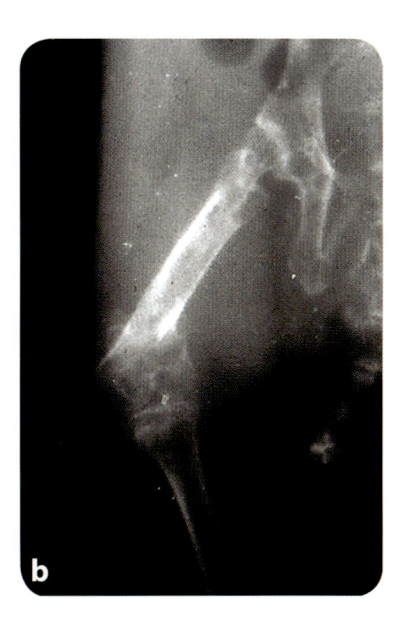

Fractures of the tibia and fibula, especially of the distal metaphysis, are the most common limb fractures in hamsters, often due to entrapment of the foot in the cage or wheel bars. These fractures are often open.

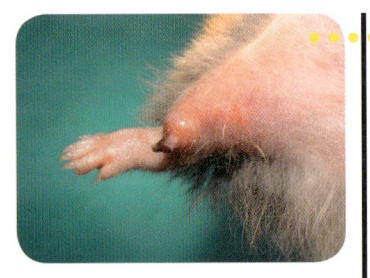

Figure 8.53. Same patient as in Figure 8.54. The hamster was presented several days after initial trauma, therefore soft tissue healing had already begun. The exposed distal aspect of the proximal fragment of the tibia appears necrotic.

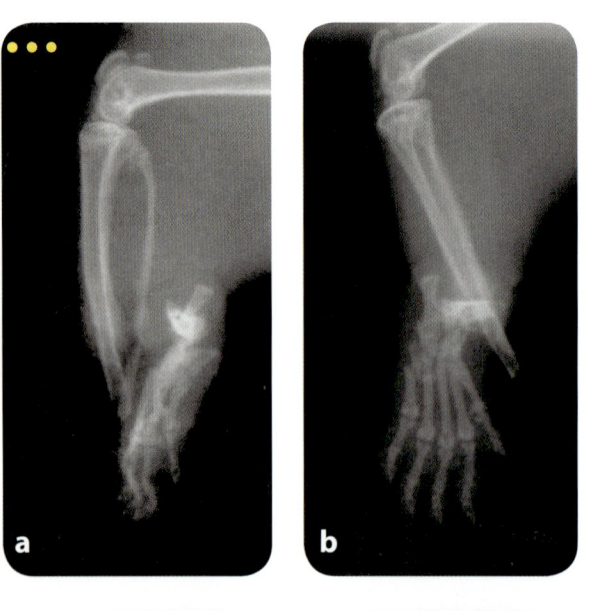

Figure 8.54a,b. Lateral (a) and dorsoventral (b) radiographs of the pelvic limb of a golden hamster with a grade 3 open distal metaphyseal fracture of the tibia and fibula. This fracture is most apparent radiographically in the dorsoventral projection, as the distal aspect of the proximal fragment is relatively unprotected by soft tissues.

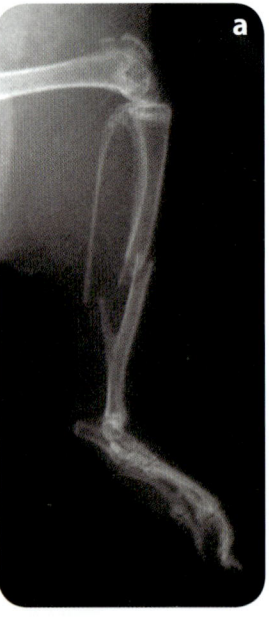

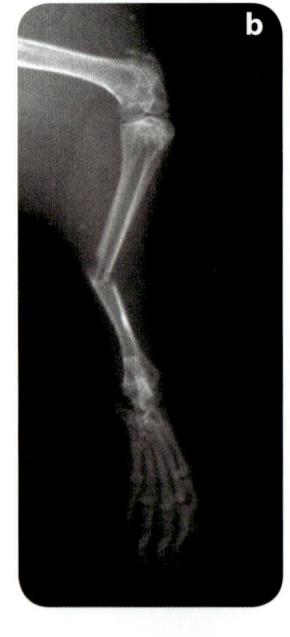

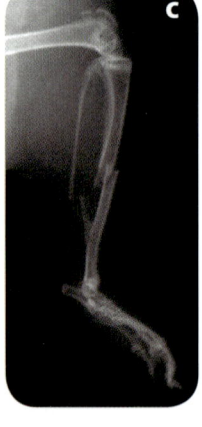

Figure 8.55a,b,c. Lateral (a), craniocaudal (b), and actual size lateral projections (c) of a golden hamster with a midshaft fracture of the tibia and distal fracture of the fibula, non-comminuted.

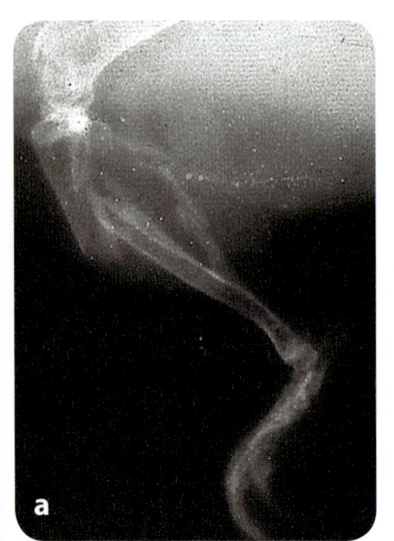

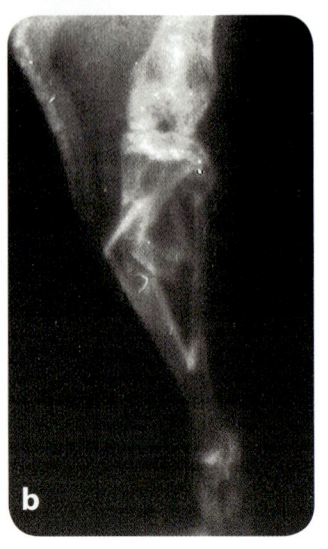

Figure 8.56a,b. Lateral (a) and dorsoventral (b) radiographs of the pelvic limb of a golden hamster with a proximal metaphyseal fracture of the tibia and fibula.

PELVIC LIMB   321

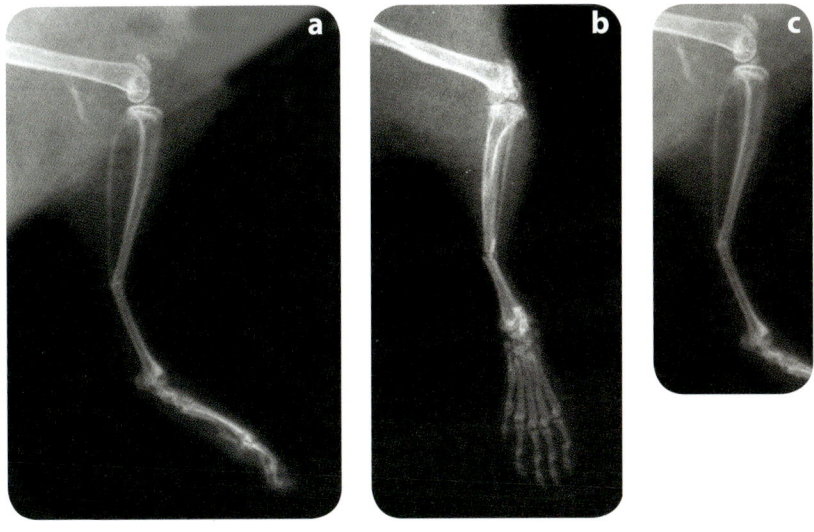

Figure 8.57a,b,c. Lateral (a), craniocaudal (b) and actual size lateral (c) radiographs of a Russian hamster with a non-comminuted midshaft fracture of the tibia and distal fracture of the fibula.

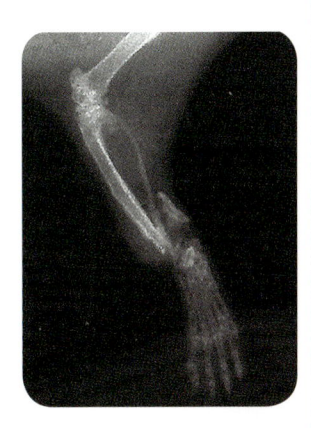

Figure 8.58. Radiograph of the pelvic limb of a Russian hamster with a fracture of the tibia and fracture of the tarsal bone.

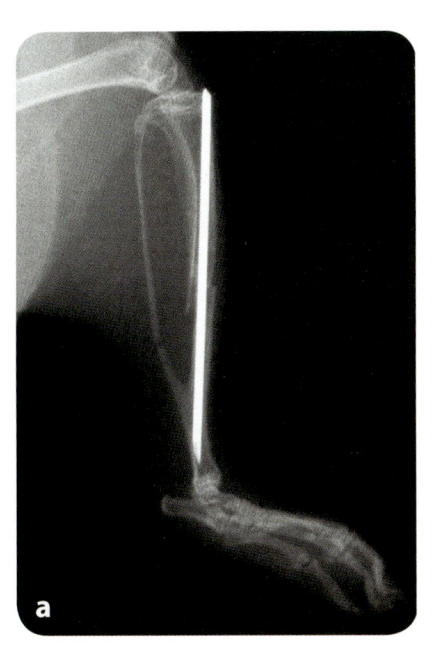

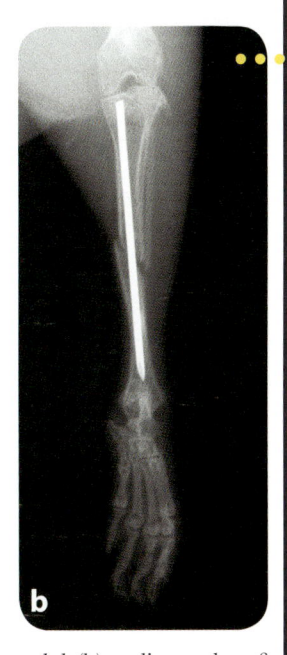

Figure 8.59a,b. Lateral (a) and craniocaudal (b) radiographs of the same patient after intramedullary pinning. Note reduction of the tibial fracture as a result of adequate reduction of the fibula in (a).

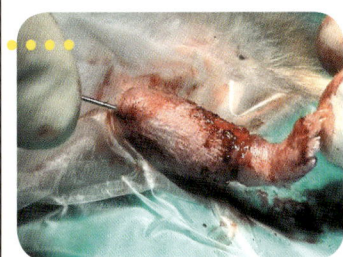

Figure 8.60. Pinning of a distal metaphyseal open fracture after debridement of the fracture site.

Intramedullary pinning of midshaft to distal metaphyseal, non-comminuted fractures is possible in very small patients, and often the most practical method of repair. 24 to 22 g hypodermic needles are commonly used as IM pins. Resulting slight rotation of the foot during healing is usually of no significant consequence.

While often not the method of choice for repair of the tibia in most larger mammal species, adequate bone healing is often achieved.

Placement of splints or external fixator devices is often not practical in patients of this size.

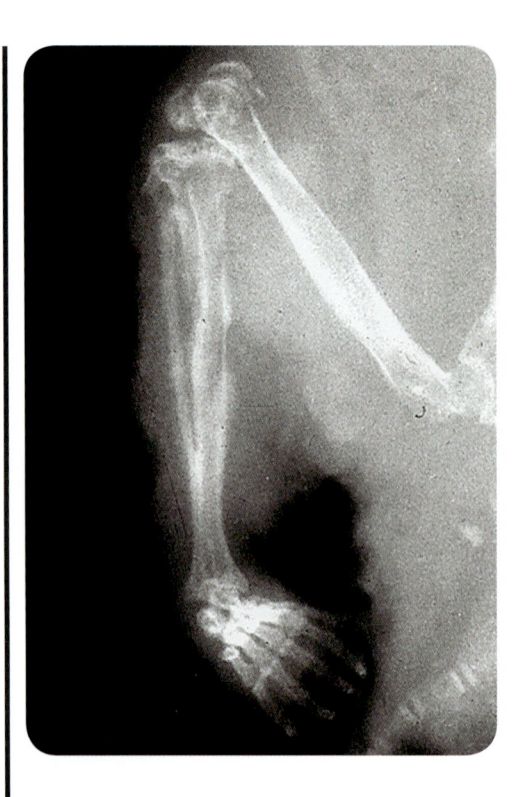

Figure 8.61. Ventrodorsal radiograph of the pelvic limb of a golden hamster with osteomyelitis of the tibia and fibula as a result of severe bite wounds from a cagemate. Oseteolysis of the tibia is clearly visible.

Figure 8.62. Clinical appearance of a large mass associated with the left pelvic limb in the same hamster.

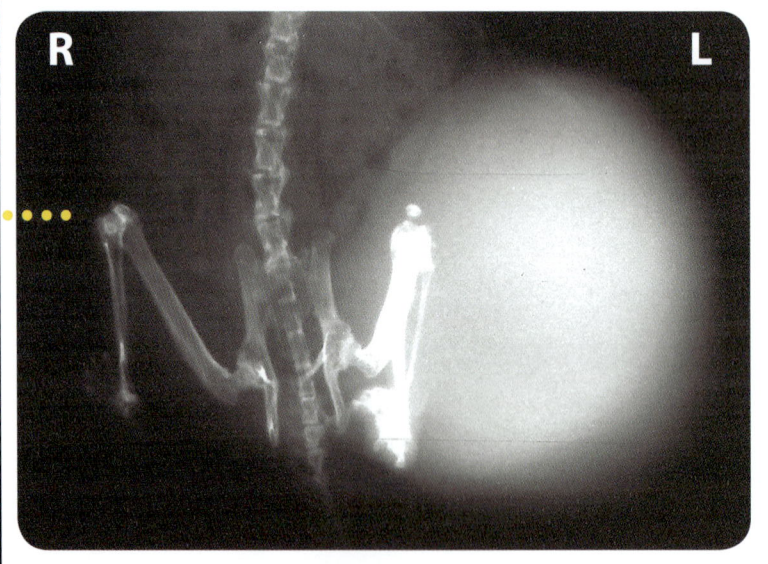

Figure 8.63. Dorsoventral radiograph of the pelvis and pelvic limbs of a hamster with a large mass associated with the left pelvic limb. The radiodense appearance of the bones of the affected limb is due to the superimposition with abnormally radiodense soft tissues.

Figure 8.64. Follow-up 10 days post surgery. The hamster was able to ambulate well, and occasionally used the base of the tail for support.

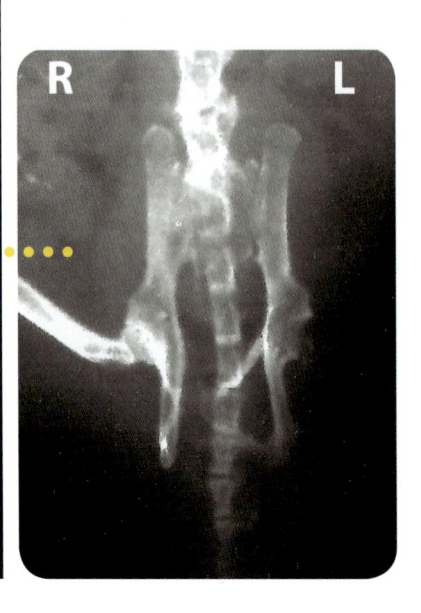

Figure 8.65. Ventrodorsal radiogaph of the same patient as in Figure 8.63, taken after amputation of the limb at the level of the coxofemoral joint.

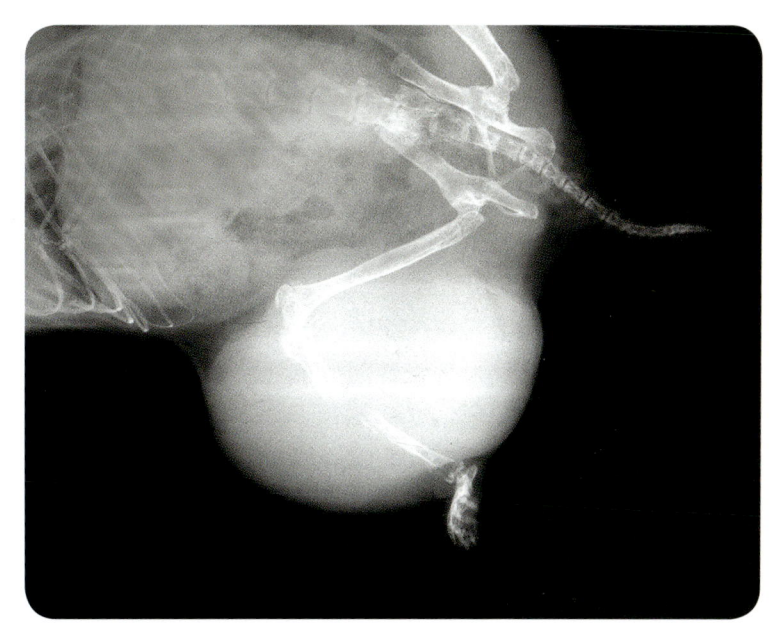

Figure 8.66. Lateral radiograph of the pelvic limb of a hamster with a large associated mass. Note the large soft tissue mass and osteolysis of the midshaft of the tibia.

References:

Bennett RA: Section 4. Small rodents. Soft tissue surgery. In: Quesenberry KE, Carpenter JW, eds. Ferrets, Rabbits and Rodents. Clinical Medicine and Surgery 2nd ed. Philadelphia, PA: Elsevier; 2004:316-328.

Capello V: Pet hamster medicine and surgery part II: clinical evaluation and therapeutics. ExoticDVM. 2001; 3(4):33-39.

Capello V: Pet hamster medicine and surgery part III: Infectious, parasitic and metabolic diseases. ExoticDVM 2002; 3(6):27-32.

Capello V: Dental diseases. In: Lennox A, ed. Rabbit and Rodent Dentistry Handbook. Ames, IA: Blackwell Publishing, (Formerly Zoological Education Network, Lake Worth, FL); 2005:113-163.

Capello V: Secondary diseases. In: Lennox A, ed. Rabbit and Rodent Dentistry Handbook. Ames, IA: Blackwell Publishing, (Formerly Zoological Education Network, Lake Worth, FL); 2005:165-185.

Capello V: Dental diseases and surgical treatment in pet rodents. ExoticDVM. 2003; 5(3):21-27.

Capello V, Gracis M: Radiology of the skull and teeth. In: Lennox A, ed. Rabbit and Rodent Dentistry Handbook. Ames, IA: Blackwell Publishing, (Formerly Zoological Education Network, Lake Worth, FL); 2005:65-69.

Fisher PG: Exotic mammal renal disease: causes and clinical presentation. Vet Clin N Am Exotic Anim Prac. 2006; 9:33-67.

Girling SJ: Mammalian imaging and anatomy. In: Meredith A, Redrobe S, eds. BSAVA Manual of Exotic Pets 4th ed. Quedgeley, Glouchester: British Small Animal VeterinaryAssociation; 2005:1-12.

Popesko P, Rijtovà V, Horàk J: A Color Atlas of Anatomy of Small Laboratory Animals. Vol. II: Rat, Mouse, Hamster. Bratislava: Príroda Publishing House; 1990.

Silverman S, Tell LA: Syrian (Golden) Hamster *(Mesocricetus auratus)*. In: Radiology of Rodents, Rabbits and Ferrets. An Atlas of Normal Anatomy and Positioning. Philadelphia, PA: Elsevier Saunders; 2005: 45-65.

PRAIRIE DOG
and other Squirrel-like Rodents

The NORMAL HEAD
Lateral Projection

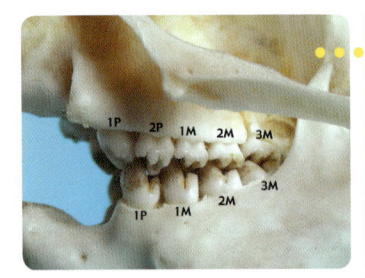

Figure 9.1. Cheek teeth. Prairie dogs are herbivorous ground squirrels. However, the structure of the cheek teeth is similar to that of primates, with bunodont cusps and ridges that create a rough occlusal surface.

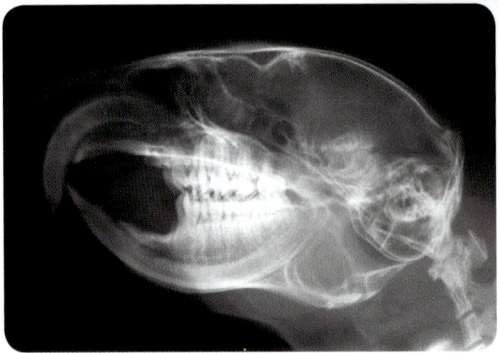

Figure 9.3. Actual size of the skull of a 1 kg male prairie dog.

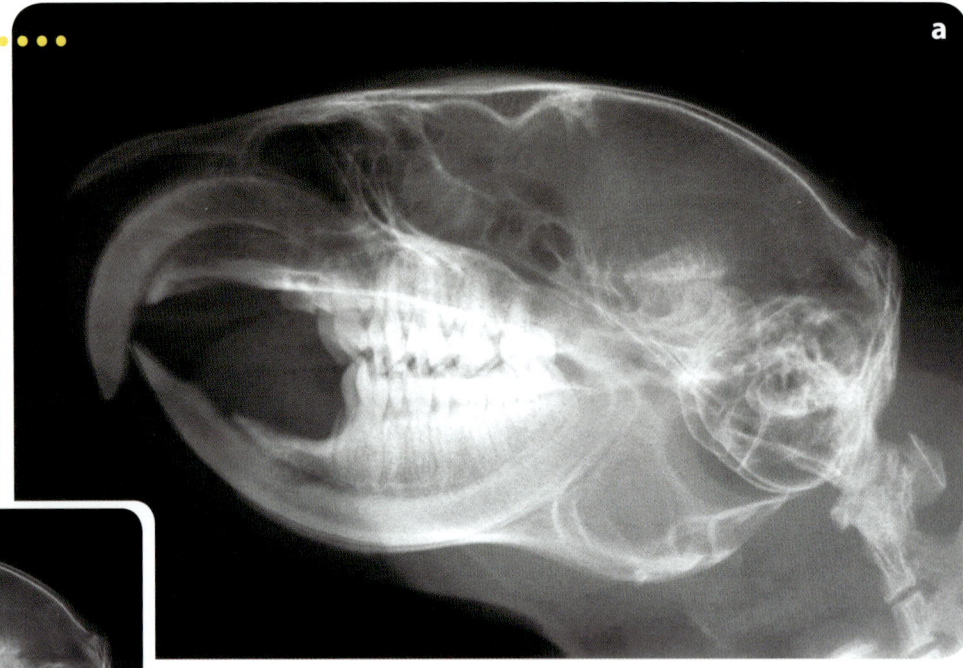

Figure 9.2a,b. 200% of actual size. When the jaw is at rest, incisors and cheek teeth are in occlusion. Incisor teeth are very large. The apex of the maxillary incisor teeth is apical to the first cheek tooth (yellow circle). The mandibular incisor teeth run ventral to the cheek teeth, with the apex located caudal to the last cheek tooth (red circle).

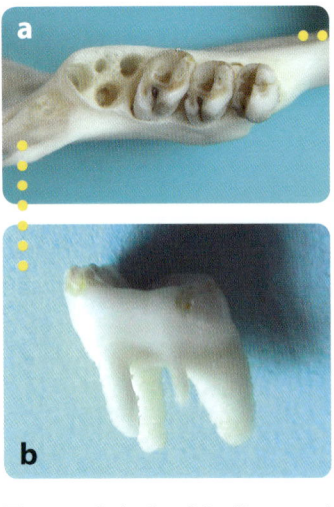

Figure 9.4a,b. Maxillary and mandibular cheek teeth have multiple roots. Shown is the mandible, dorsal view (a), and extracted mandibular CT4 (b).

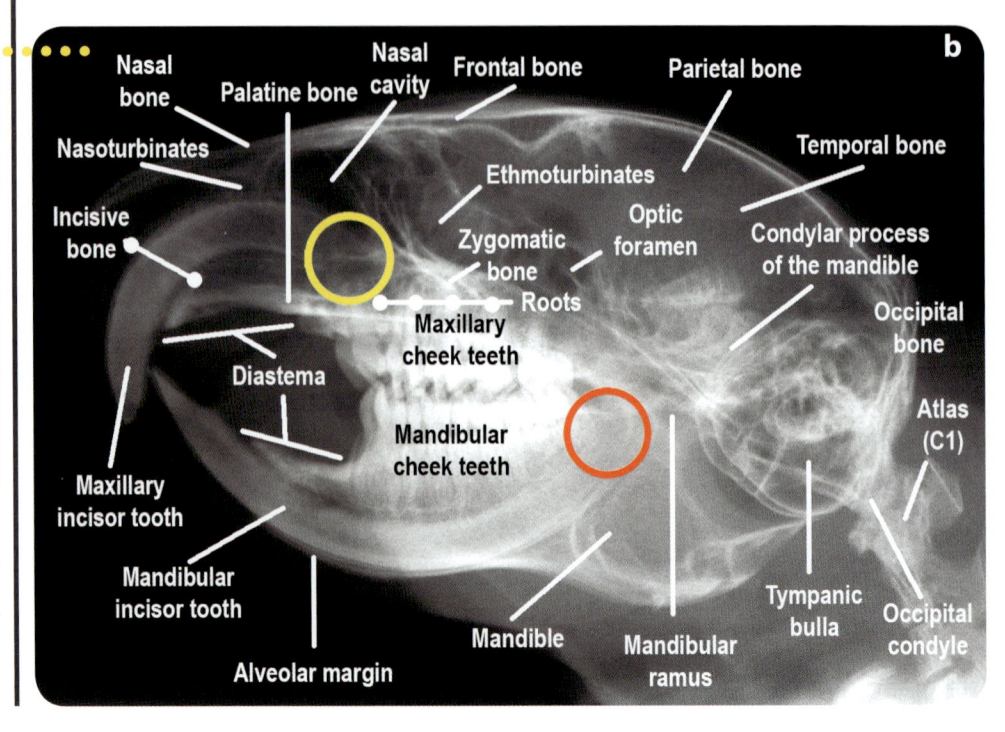

Oblique Projection

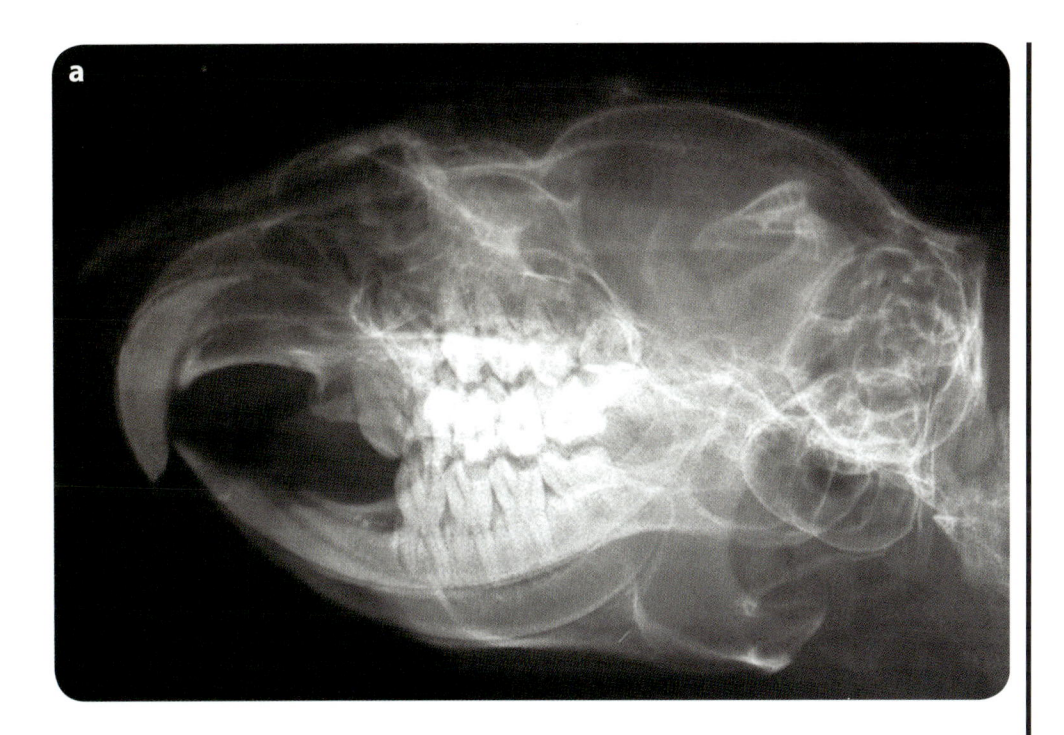

Figure 9.5a,b. Left 15° ventral-right dorsal oblique view of the skull. The apex of the mandibular incisor tooth is indicated by the red circle.

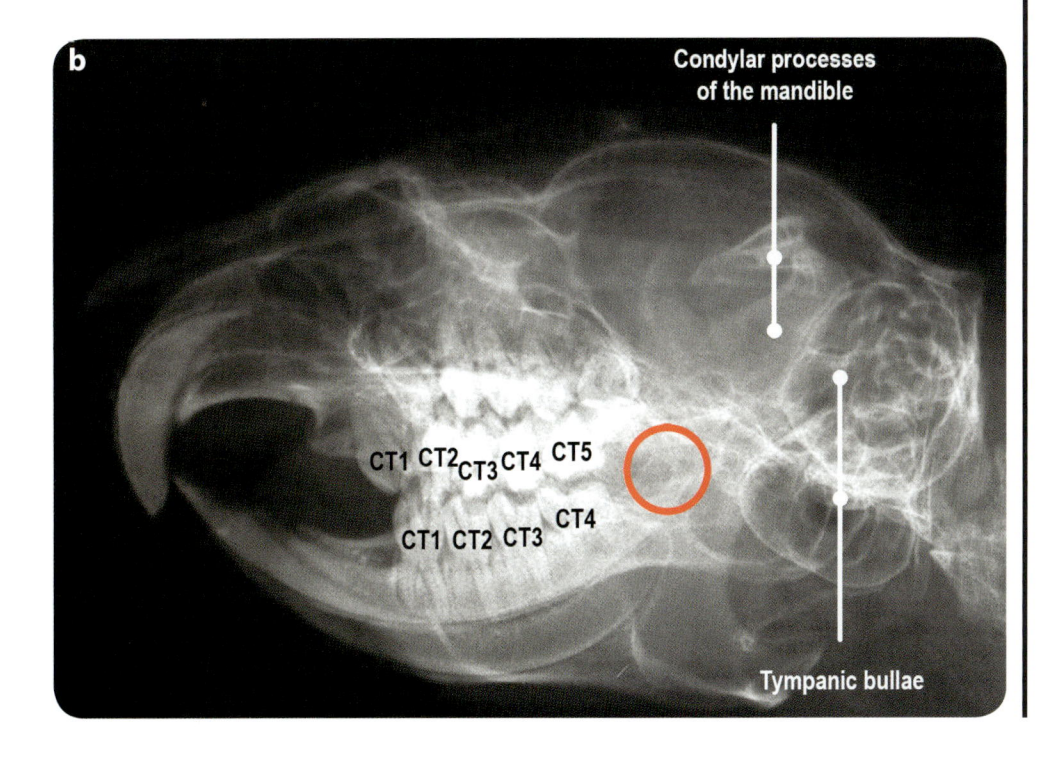

Ventrodorsal Projection

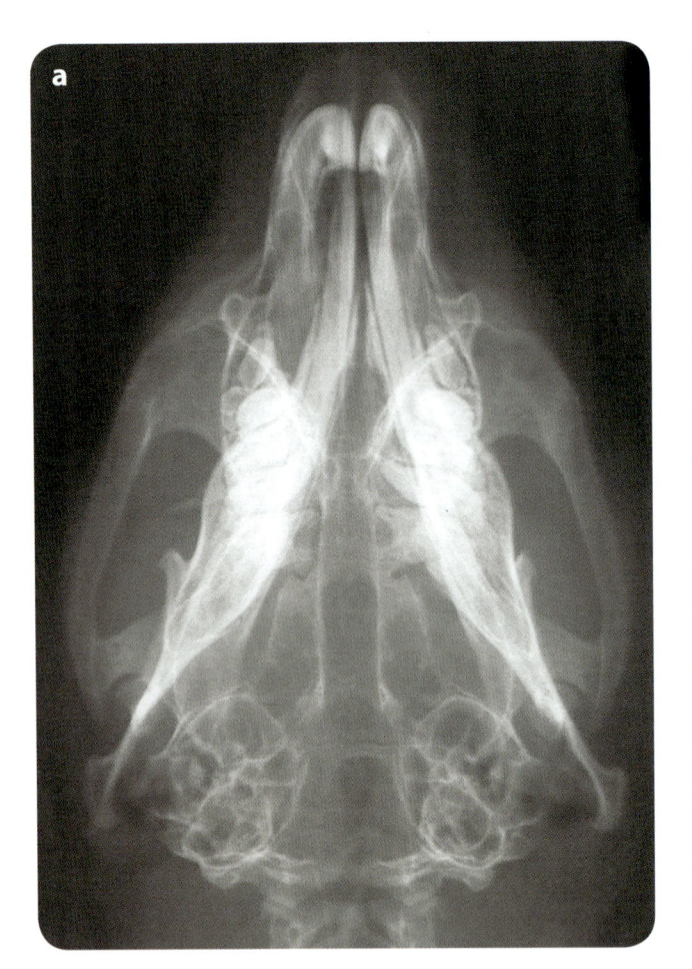

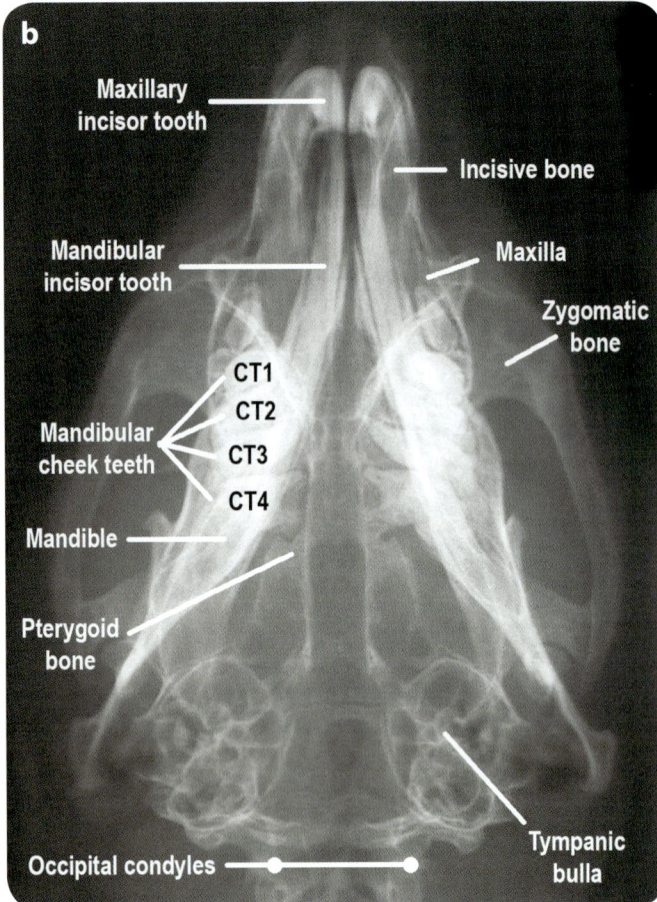

Figure 9.6a,b. Ventrodorsal view of the skull of a prairie dog.

Rostrocaudal Projection

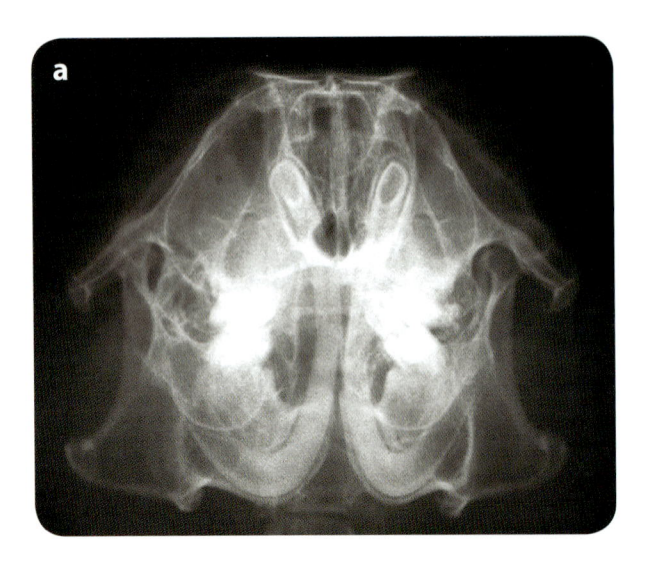

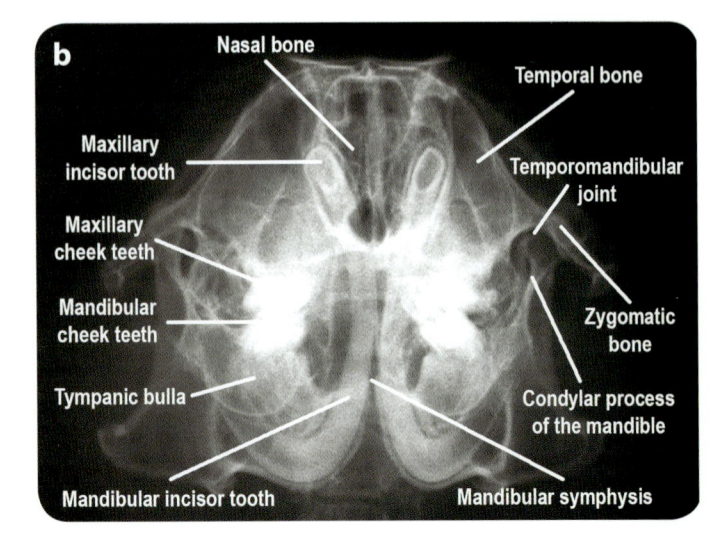

Figure 9.7a,b. Rostrocaudal projection of the skull of a prairie dog.

Intraoral Projections

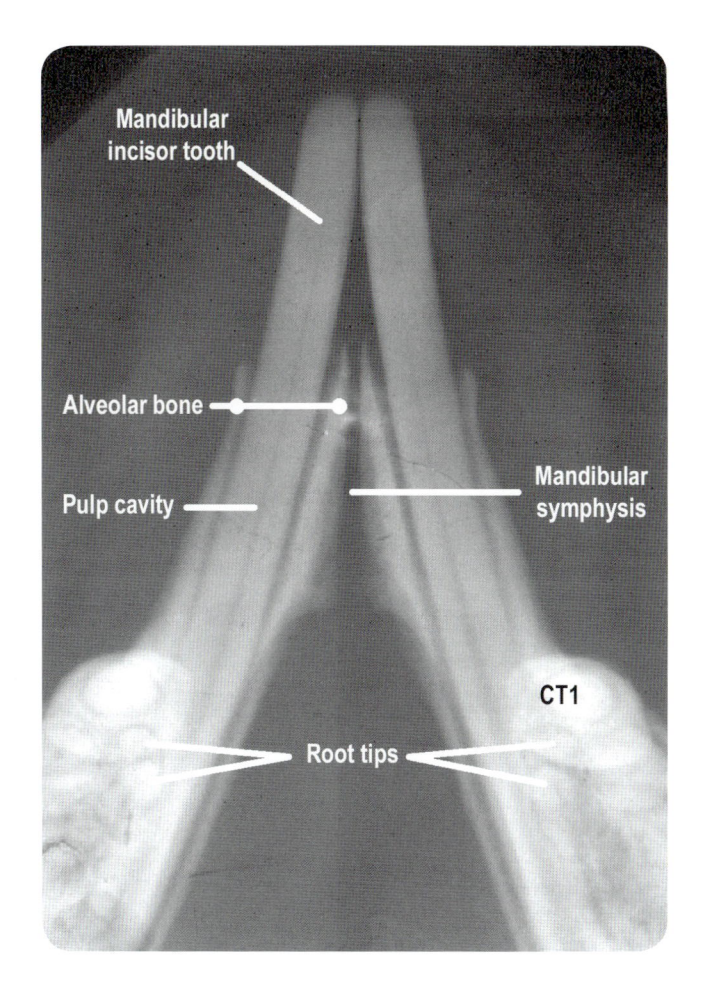

Figure 9.8. Intraoral occlusal view of the mandibular incisor teeth. In larger animals, radiographic film may be inserted more caudally into the mouth to include the apical region of the incisor teeth. This projection may also be used to visualize and locate cheek teeth root tips.
(Courtesy of Margherita Gracis, DVM)

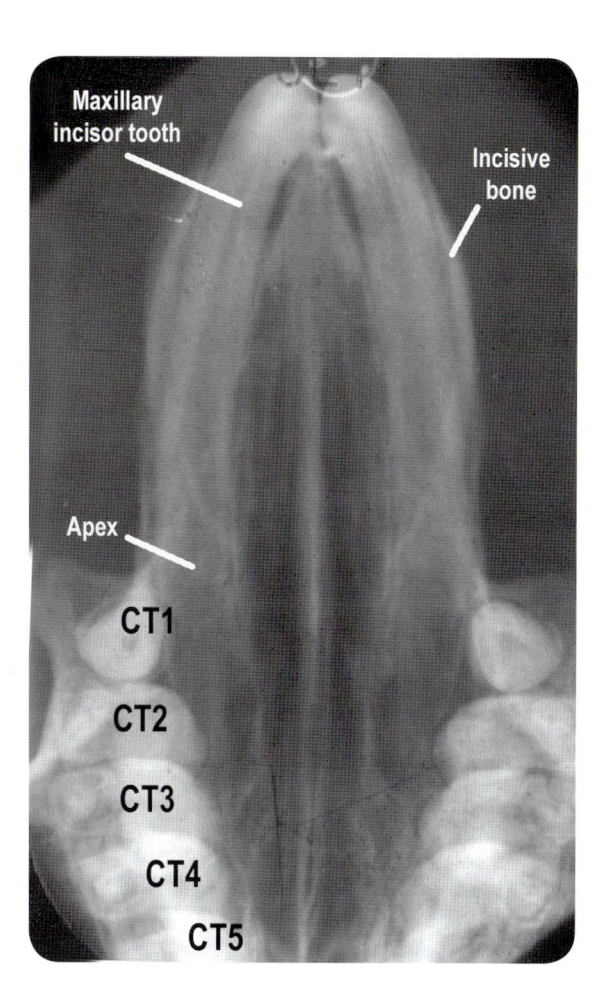

Figure 9.9. Intraoral occlusal view of the maxillary incisor teeth. CT5 is partially included in this view.
(Courtesy of Margherita Gracis, DVM)

ABNORMALITIES of the HEAD
Diseases of Incisor Teeth

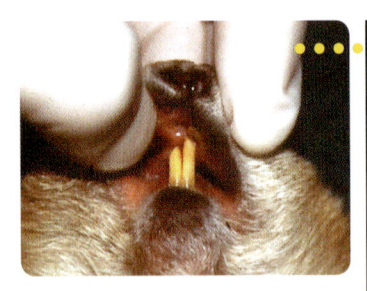

Figure 9.10. Fracture of both maxillary incisor teeth at the gingival margin. Overgrowth of the left mandibular incisor is also visible. Fractures of the incisor teeth occur commonly in captive prairie dogs. Fractures are secondary to trauma from falls, or continuous chewing of cage bars, especially when cage size is inadequate. Maxillary incisor teeth are most commonly affected. In some cases this lesion occurs so commonly owners incorrectly believe this to be normal shedding of deciduous teeth.

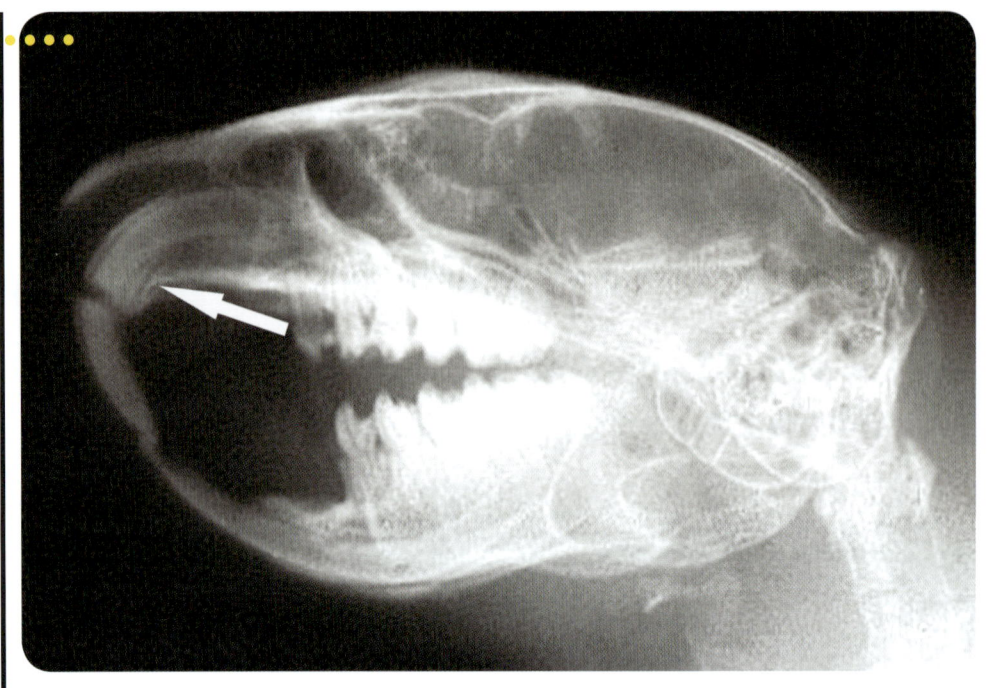

Figure 9.11. Lateral radiograph of the skull of a prairie dog with a complete fracture of both maxillary incisor teeth (arrow). The fracture line often does not involve the erupted, visible portions of the teeth (clinical crowns). Fractured teeth are often retained in place, attached to the gingiva.

Pseudo-odontoma is a dysplastic disease affecting the roots of incisor teeth, particularly the maxillary incisors of squirrel-like rodents such as the prairie dog. Repeated trauma, fractures, or improper trimming ultimately intereferre with eruption of the teeth and may result in malocclusion. Apical growth continues, causing deformation of the germinal tissue and surrounding structures such as the incisive bone.

Early changes include abnormalities and irregularities of the new dentine. Ultimately tooth eruption ceases, and resultant root deformation acts as a space-occupying mass leading to progressive nasal obstruction.

This is not a true neoplastic disease, therefore the former definition "odontoma" is incorrect.

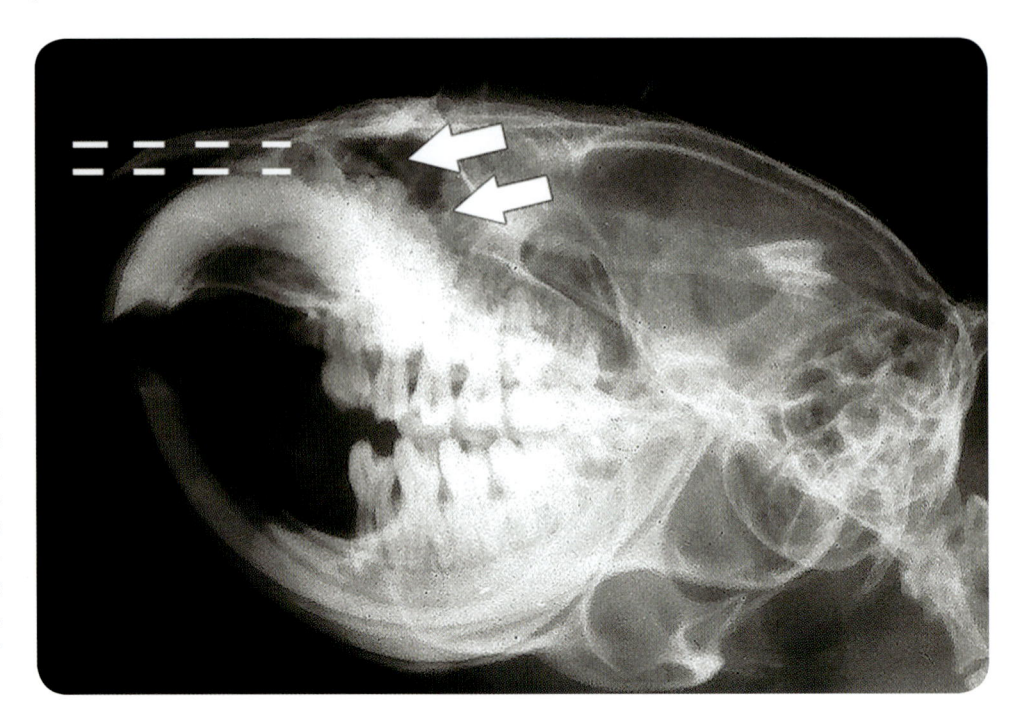

Figure 9.12. Lateral radiograph of the skull of a prairie dog with pseudo-odontoma. The apex of the maxillary incisors is irregular and hyperplastic (arrows). Size of the nasal cavity is reduced (white lines, see Fig.9.2a) and air flow is limited, resulting in sneezing and varying degrees of dyspnea.

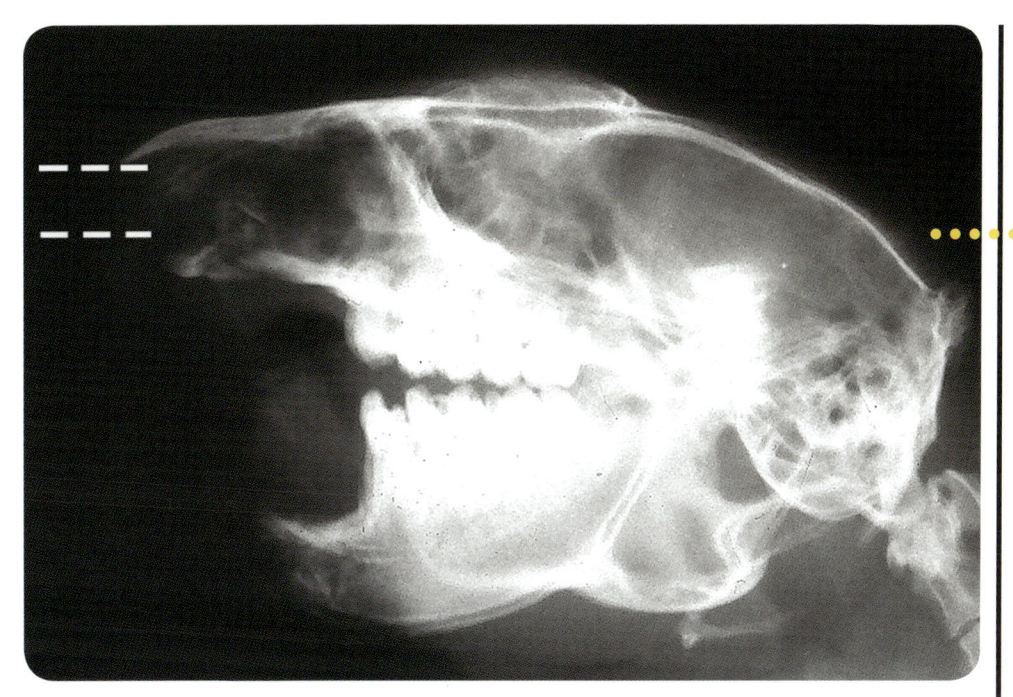

Figure 9.13. Lateral projection of the skull of the same patient, postoperative skull radiograph. Obstruction of the nasal cavity is greatly reduced after extraction of the affected incisors (white lines).

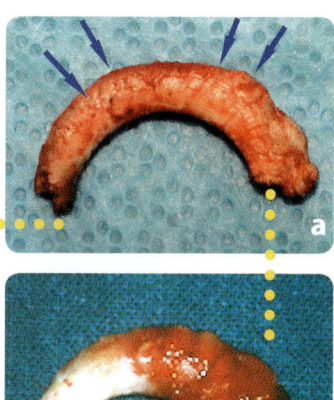

Figure 9.14a,b. Maxillary incisor teeth extracted from two separate prairie dogs with pseudo-odontoma. Folding of new dentin (arrows) and apex deformity are visible in (a). This occurs as a result of incisor tooth growth arrest or impairment.
The apex of the incisor in (b) appears thicker than normal due to hyperplasia of surrounding hard and soft tissues.

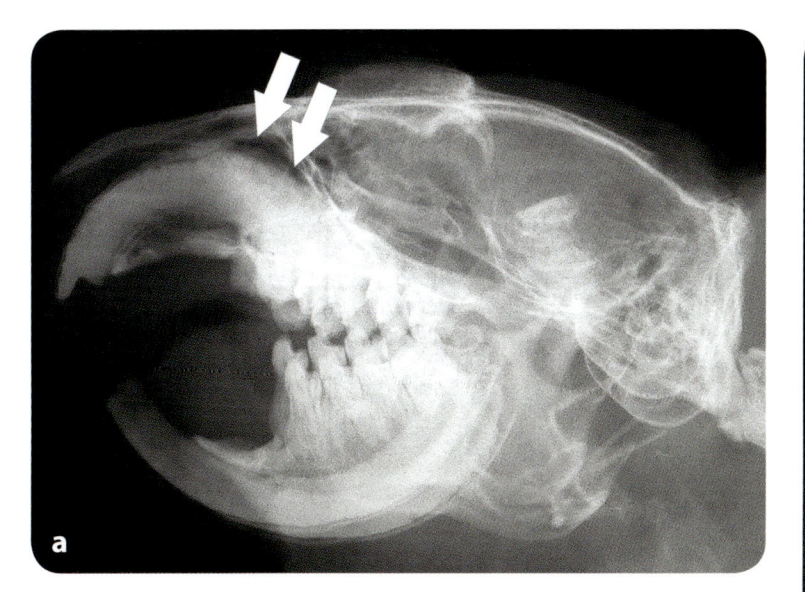

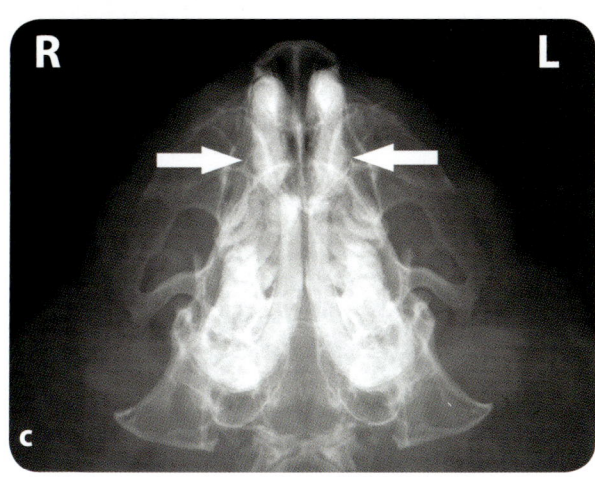

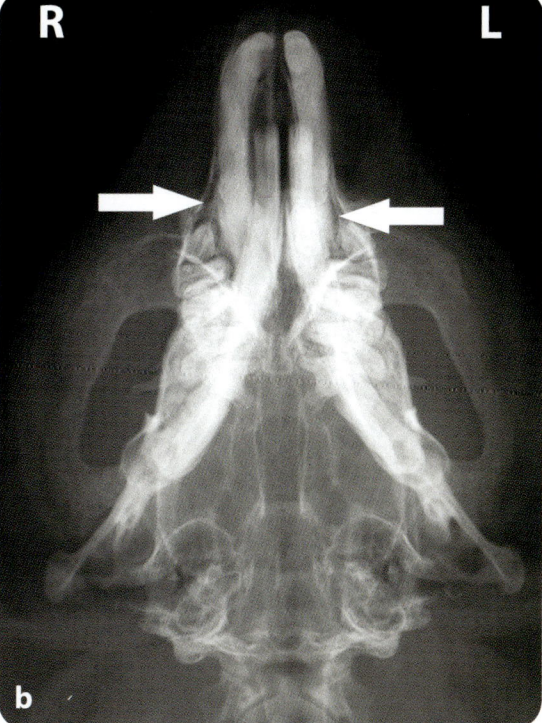

Figure 9.15a,b,c. Lateral (a), ventrodorsal (b) and rostrocaudal (c) projections of the skull of another prairie dog with pseudo-odontoma demonstrating less severe radiographic signs than the clinical case described above. Deformites of the maxillary incisor teeth are indicated in (a) (white arrows).
Deformities of the roots of maxillary incisors are also clearly visible on both ventrodorsal (b) and rostrocaudal (c) projections (arrows).

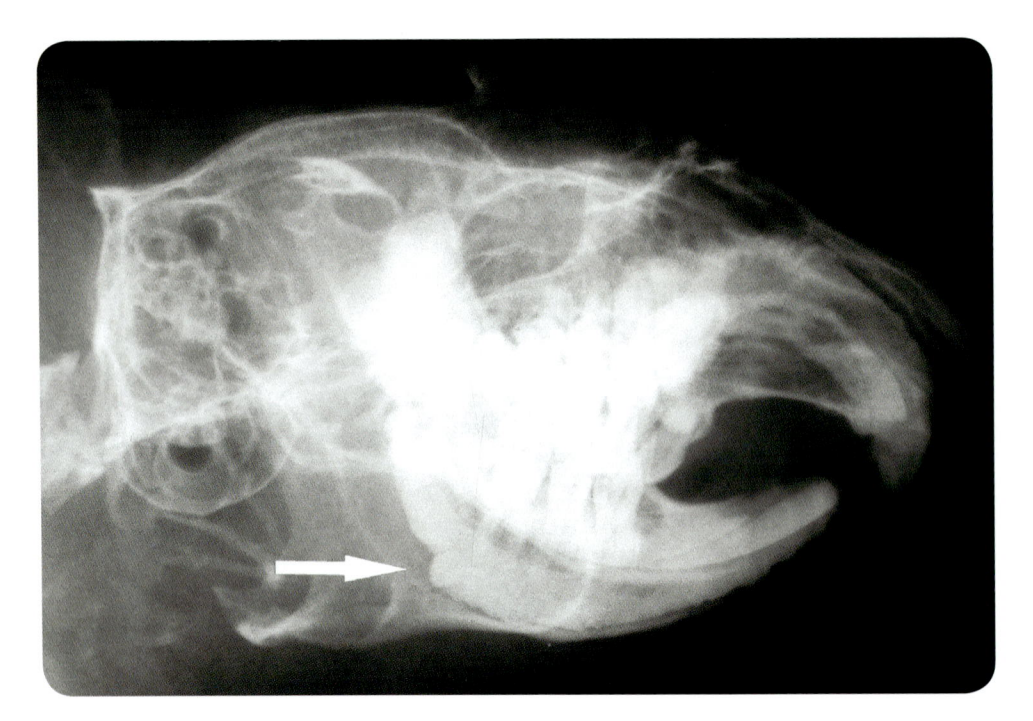

Figure 9.16. Right 15° ventral-left dorsal oblique view of the skull of a prairie dog with pseudo-odontoma. Note severe deformity of the apex of the left mandibular incisor tooth (arrow).

Diseases of Cheek Teeth

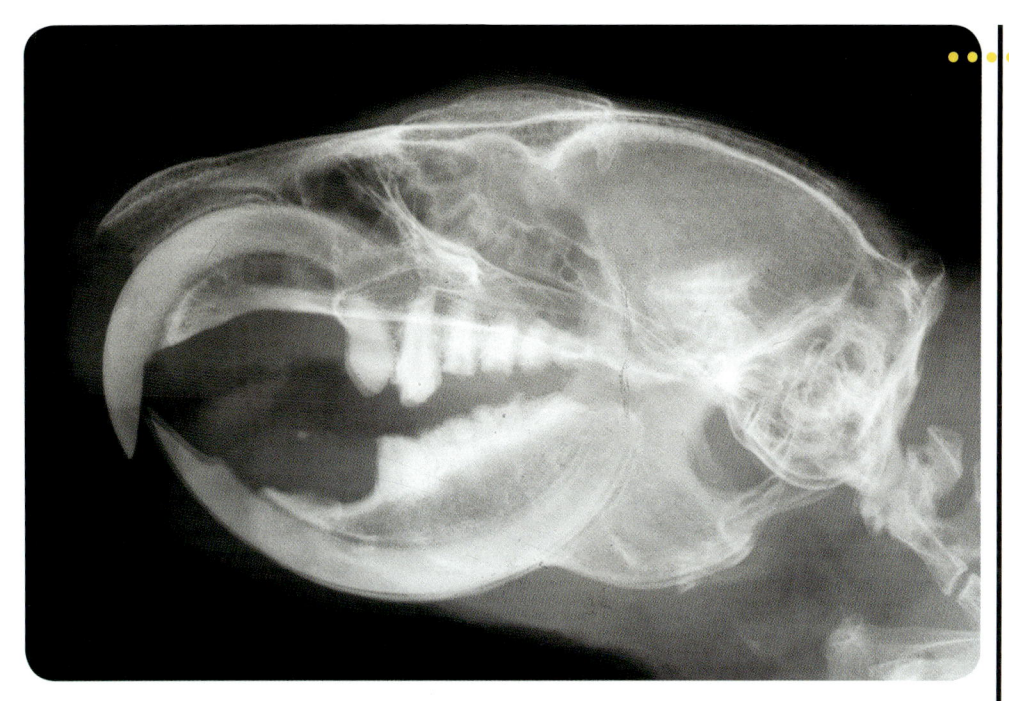

Figure 9.17. Lateral view of the skull of a prairie dog demonstrating excessive wearing of cheek teeth. With the exception of two maxillary premolar teeth, the crowns are worn to the level of the gingiva. Excessive wearing and cavities of cheek teeth are common in older prairie dogs. Note that the incisor teeth are normal.

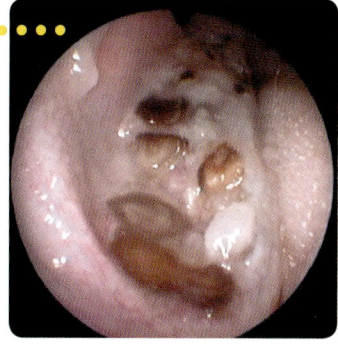

Figure 9.18. Endoscopic appearance of the right mandibular arcade in the same prairie dog. Excessive wearing of the crowns is clearly visible.

The NORMAL TOTAL BODY
Lateral Projection

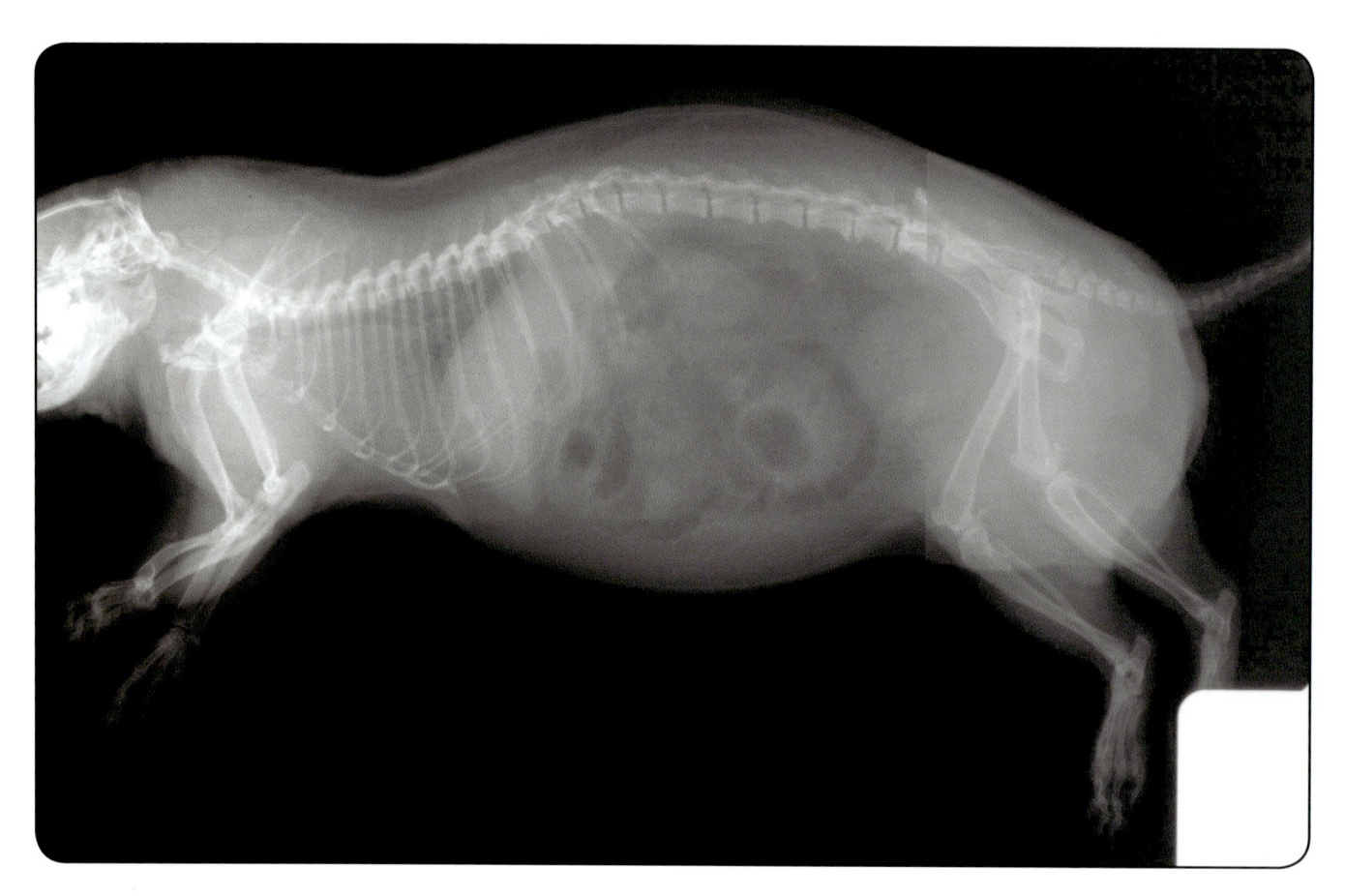

Figure 9.19. Total body radiograph of a 1.1 kg male prairie dog, 60% of actual size.

Ventrodorsal Projection

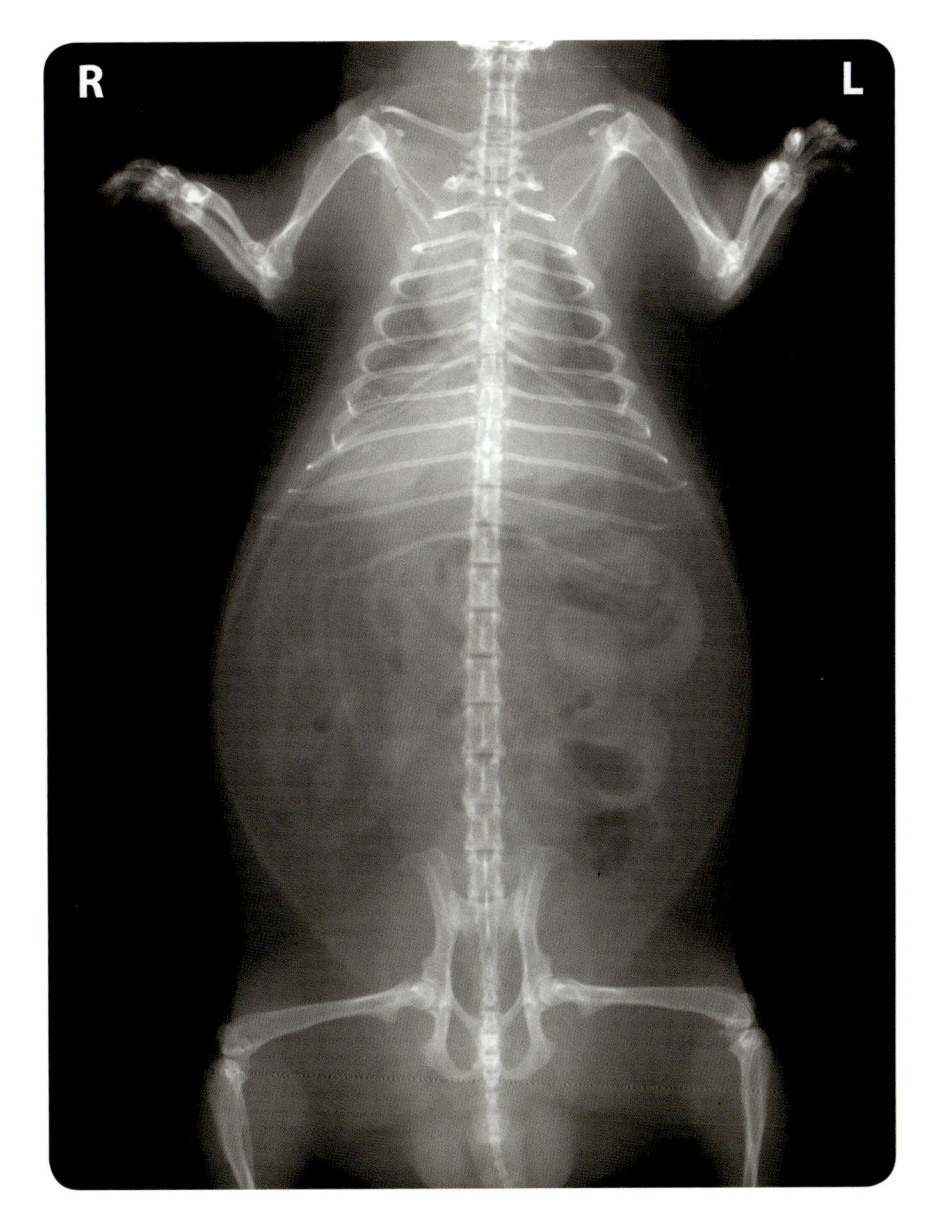

Figure 9.20. Total body radiograph of a 1.1 kg male prairie dog, 60% of actual size.

The NORMAL THORAX
The Cervical and Thoracic Vertebral Column
Lateral Projection

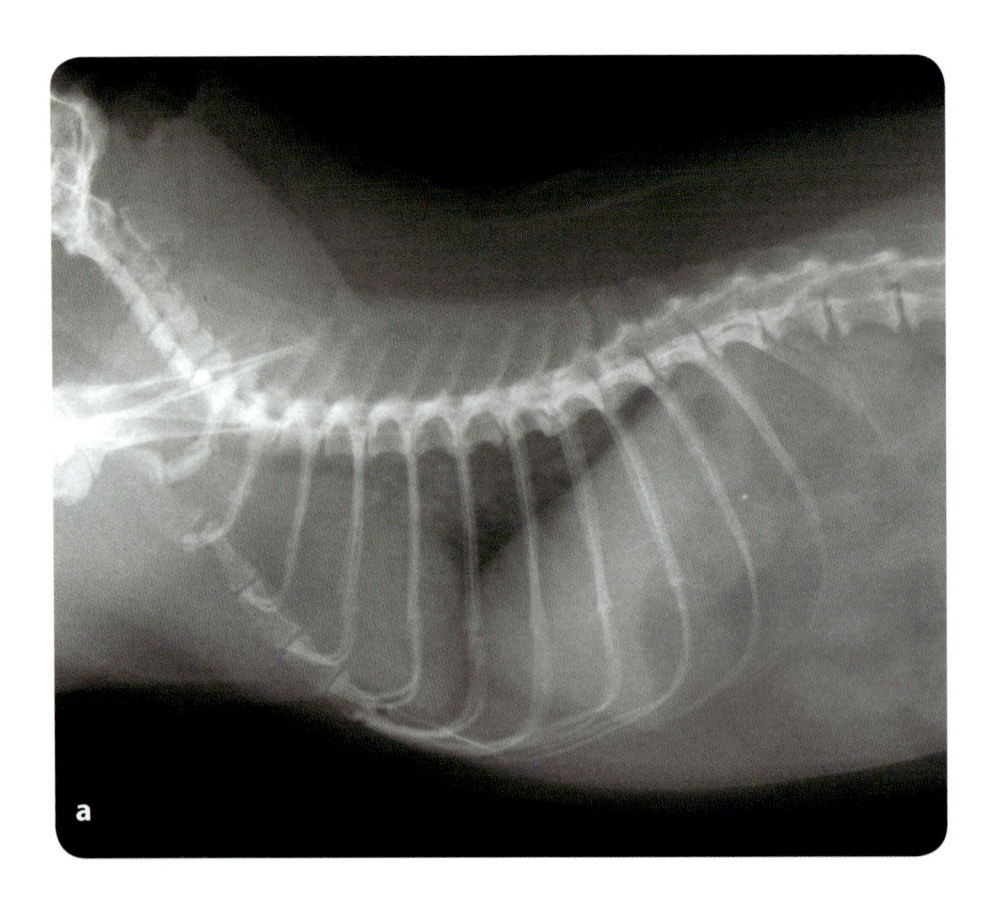

Figure 9.21a,b. Radiograph of the thorax of a 1.1 kg male prairie dog, actual size.

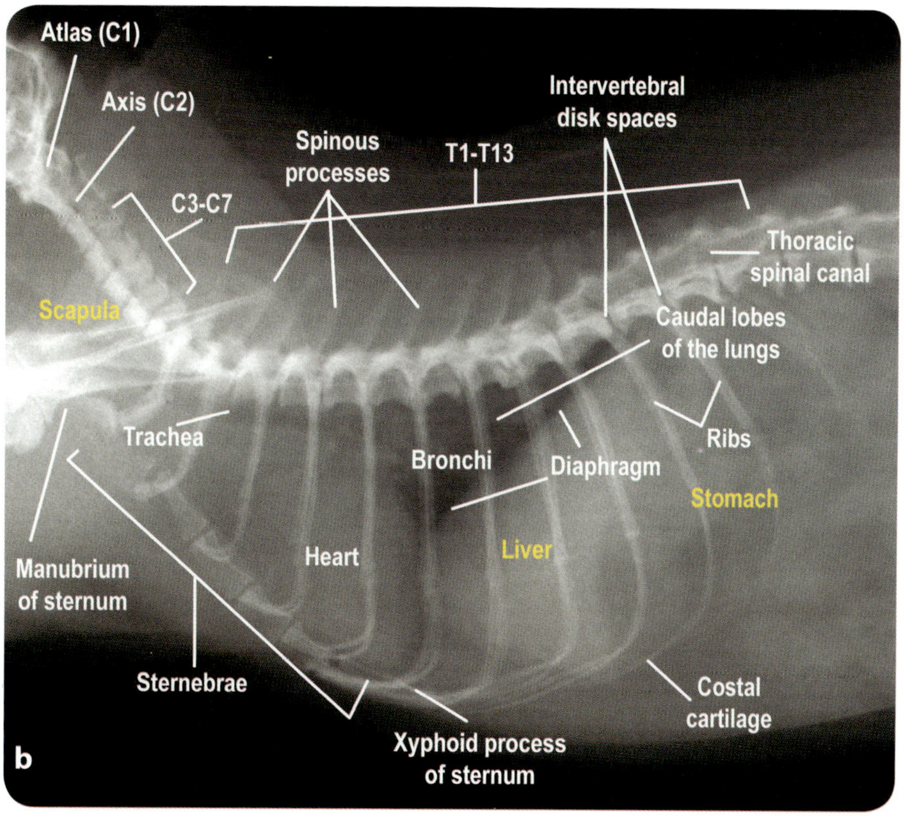
Atlas (C1)
Axis (C2)
C3-C7
Spinous processes
T1-T13
Intervertebral disk spaces
Thoracic spinal canal
Scapula
Caudal lobes of the lungs
Trachea
Ribs
Bronchi
Diaphragm
Stomach
Heart
Liver
Manubrium of sternum
Sternebrae
Costal cartilage
Xyphoid process of sternum
b

Ventrodorsal Projection

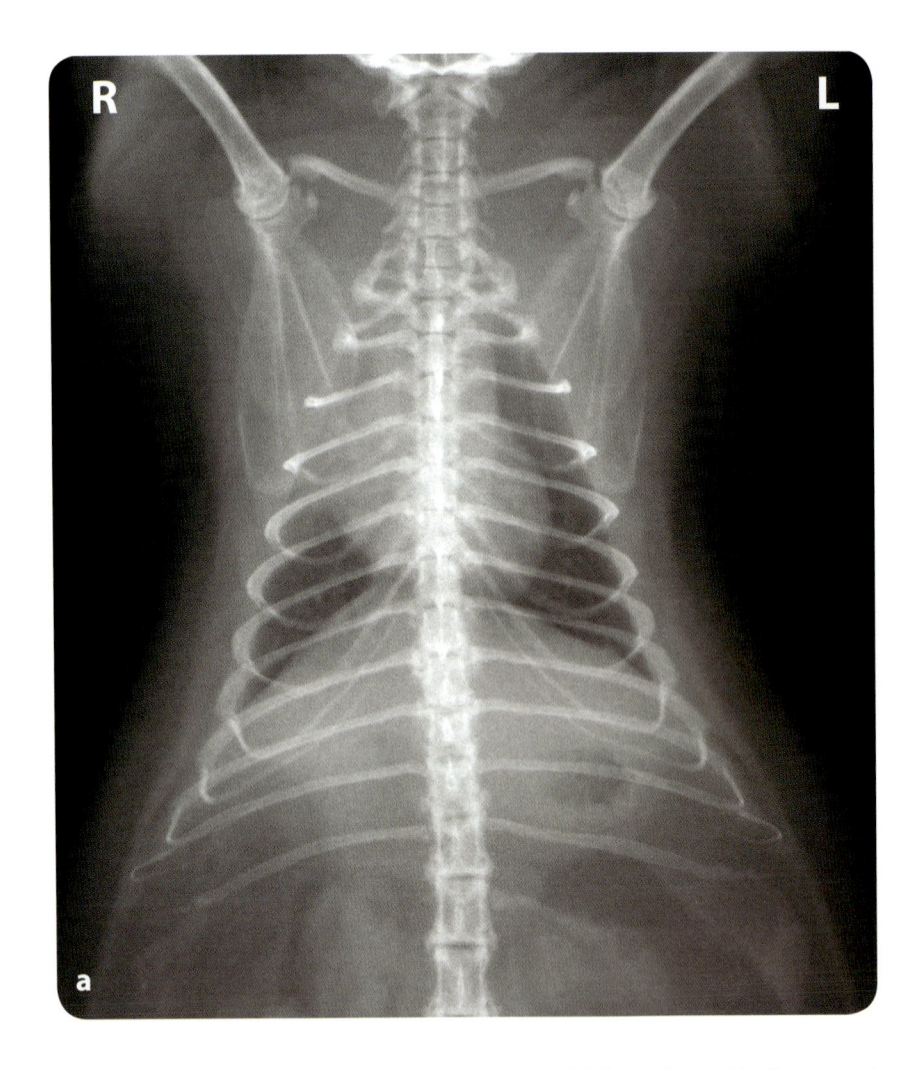

Figure 9.22a,b. Radiograph of the thorax of a 1.1 kg male prairie dog, actual size.

THORAX

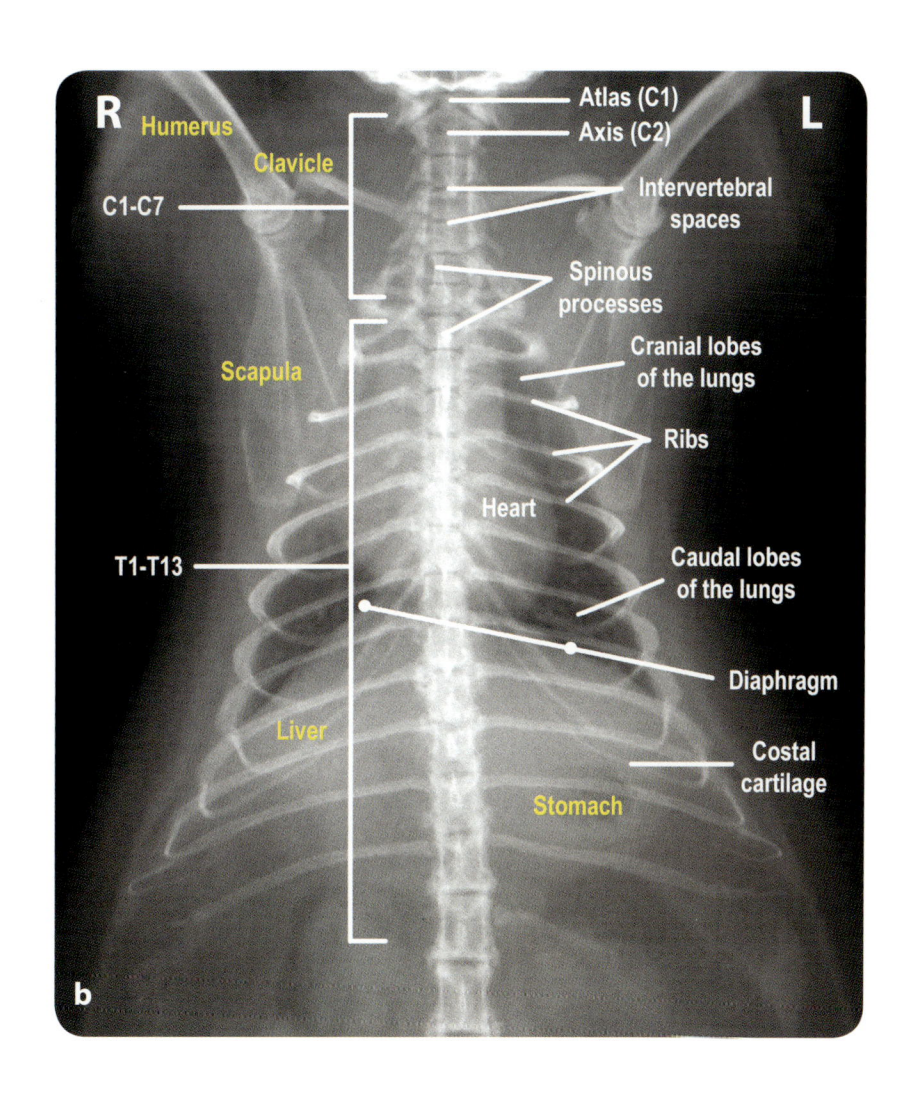

The NORMAL ABDOMEN
The Lumbar Vertebral Column
Lateral Projection

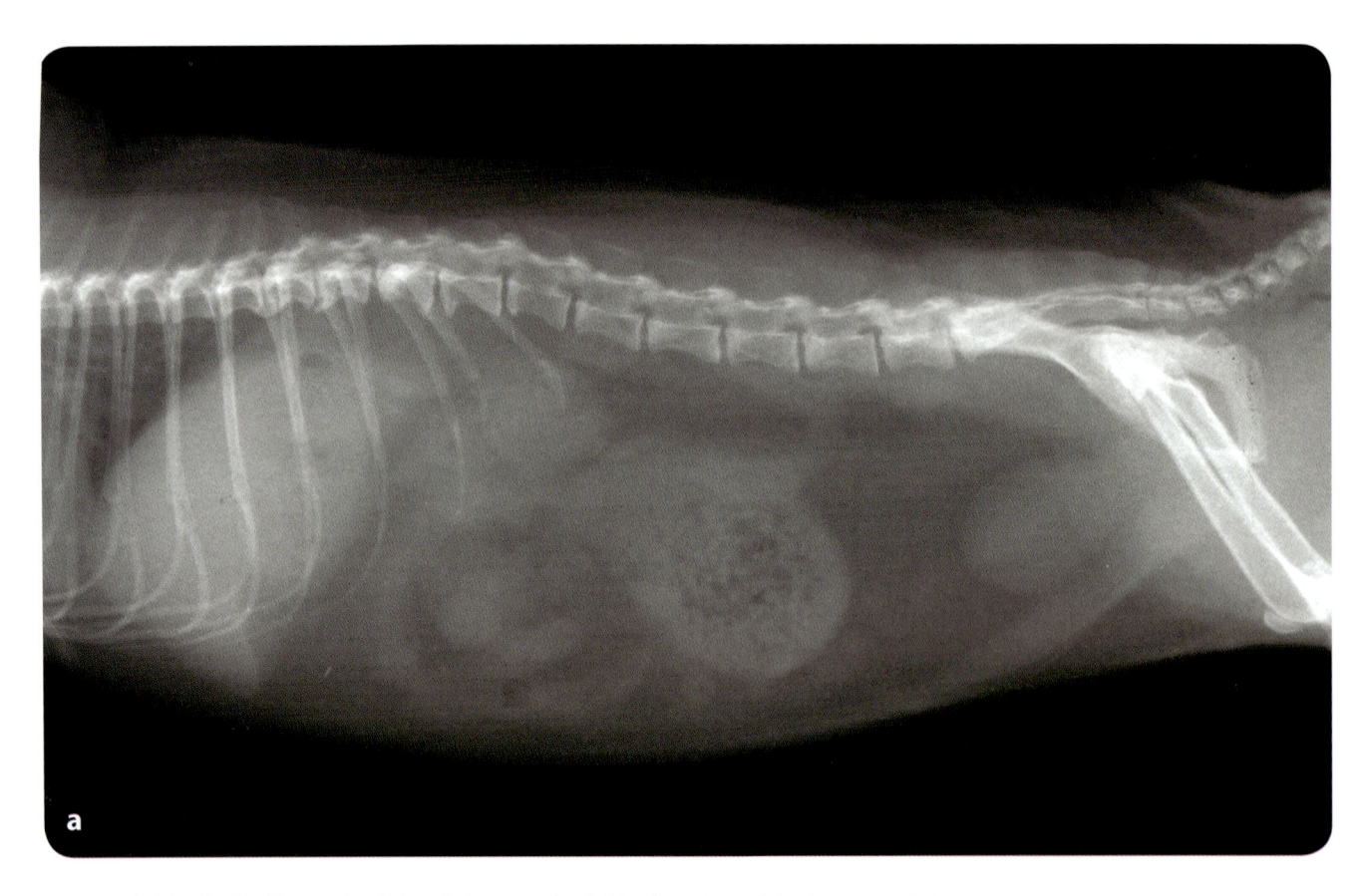

Figure 9.23a,b. Radiograph of the abdomen of a 1.0 kg female prairie dog, actual size.

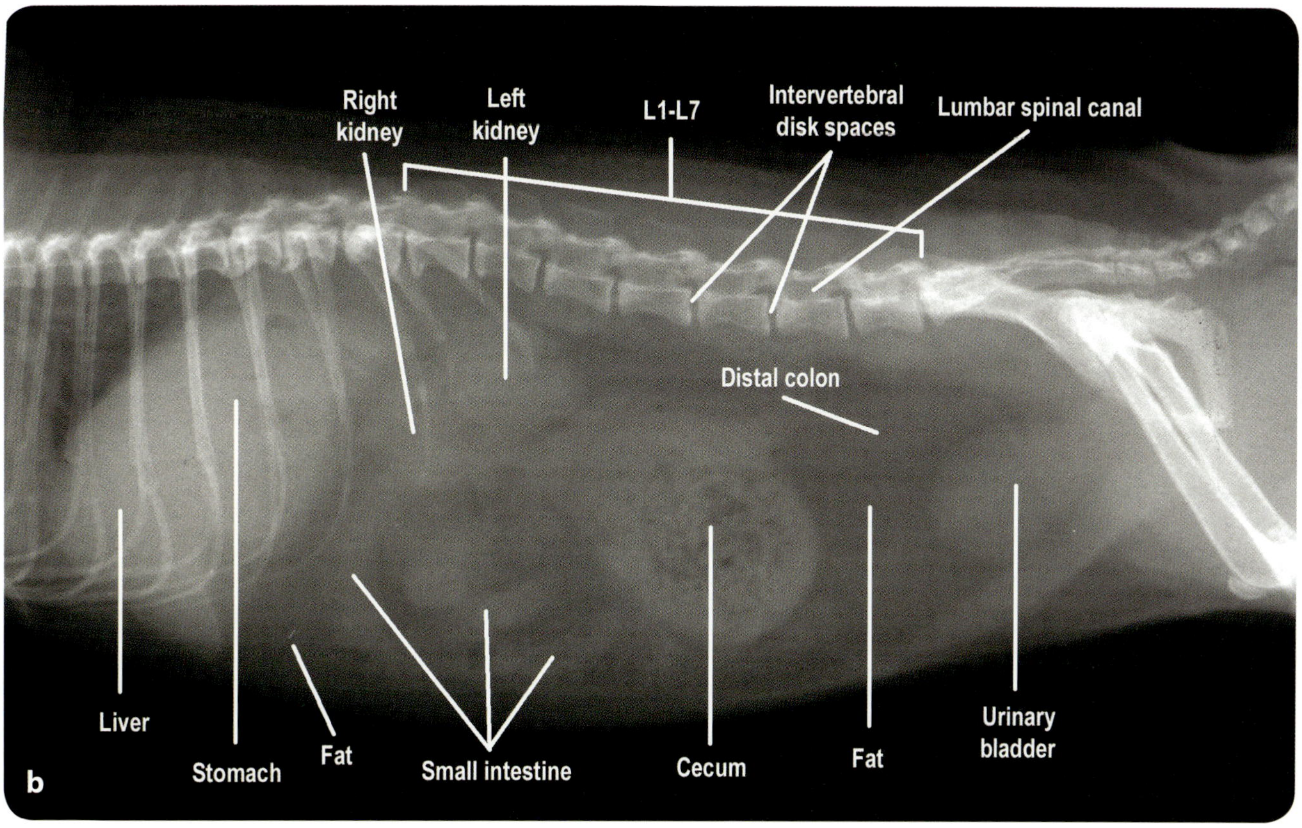
Right kidney
Left kidney
L1-L7
Intervertebral disk spaces
Lumbar spinal canal
Distal colon
Liver
Stomach
Fat
Small intestine
Cecum
Fat
Urinary bladder
b

Ventrodorsal Projection

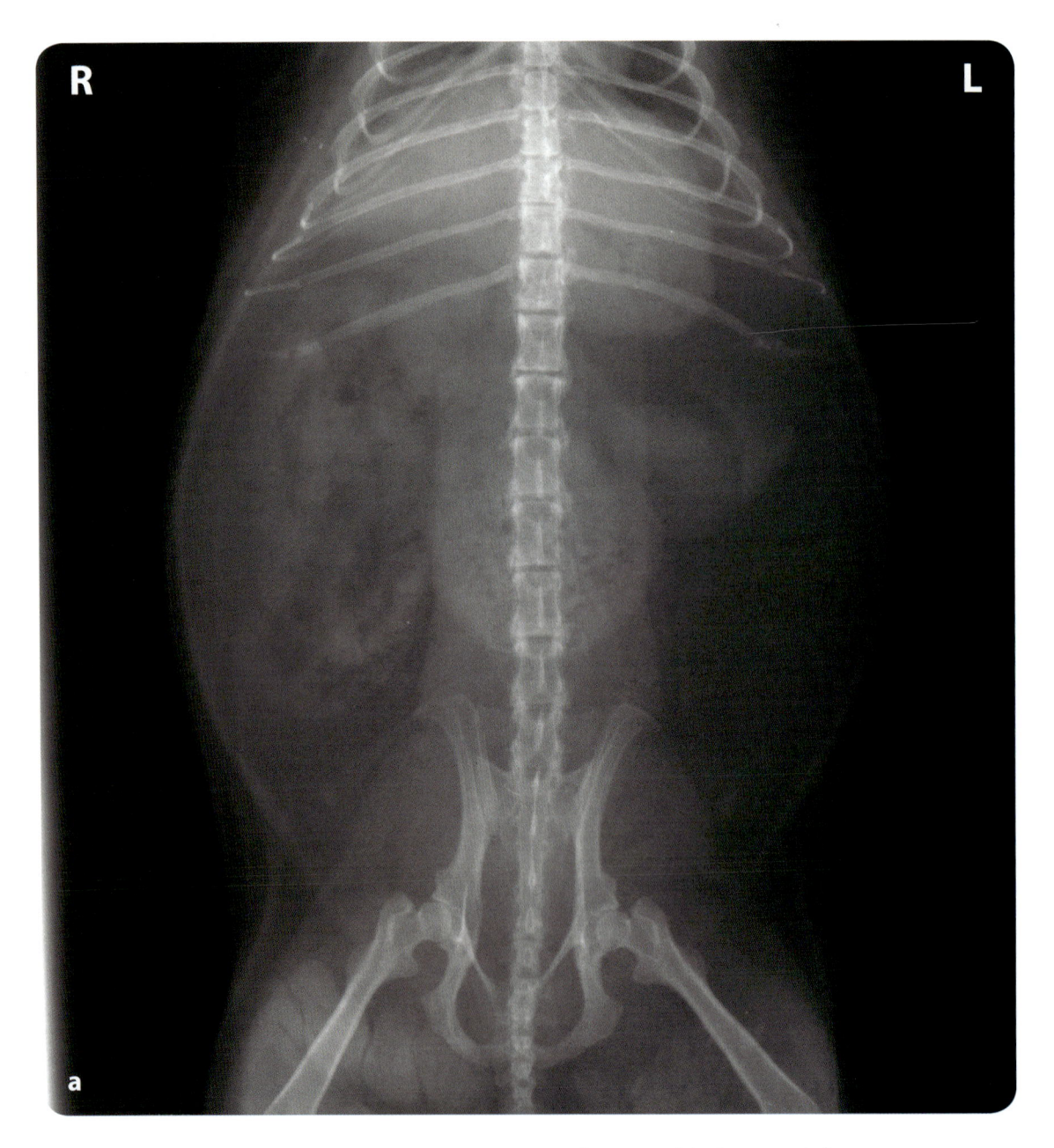

Figure 9.24a,b. Radiograph of the abdomen of a 1.0 kg female prairie dog, actual size.

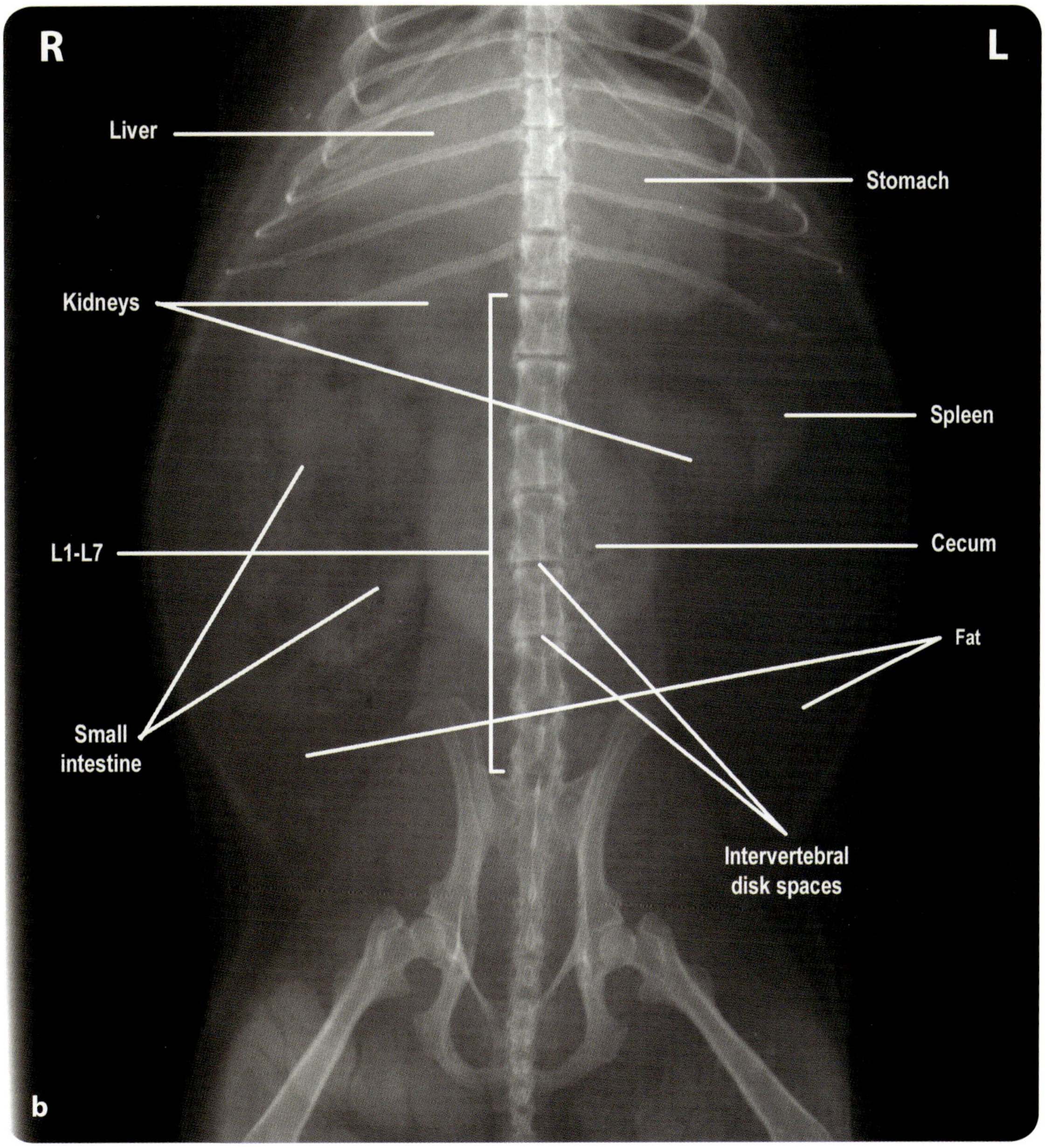
R
L
Liver
Stomach
Kidneys
Spleen
L1-L7
Cecum
Fat
Small intestine
Intervertebral disk spaces
b

ABNORMALITIES of the ABDOMEN
Diseases of the Intestine

Figure 9.25. Severe abdominal distension in a prairie dog. Gastrointestinal disease is common in prairie dogs, as owners frequently feed diets inappropriate for a strict herbivore.

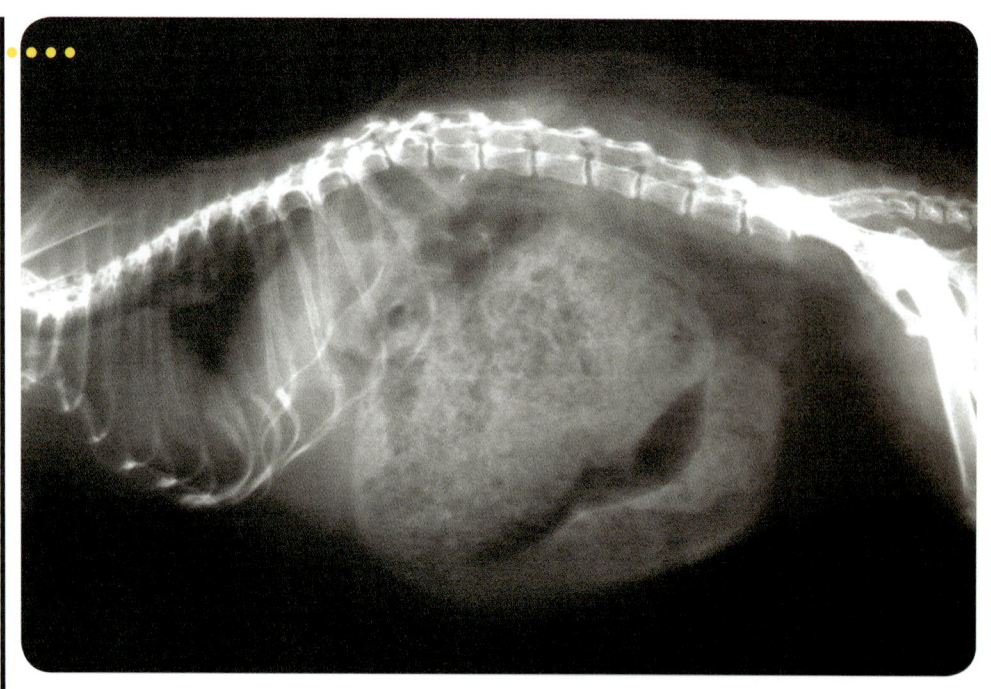

Figure 9.26. Lateral radiograph of the abdomen of a prairie dog with severe impaction of the cecum as a result of inappropriate diet and obesity. The cecum is filled with radiodense ingesta.

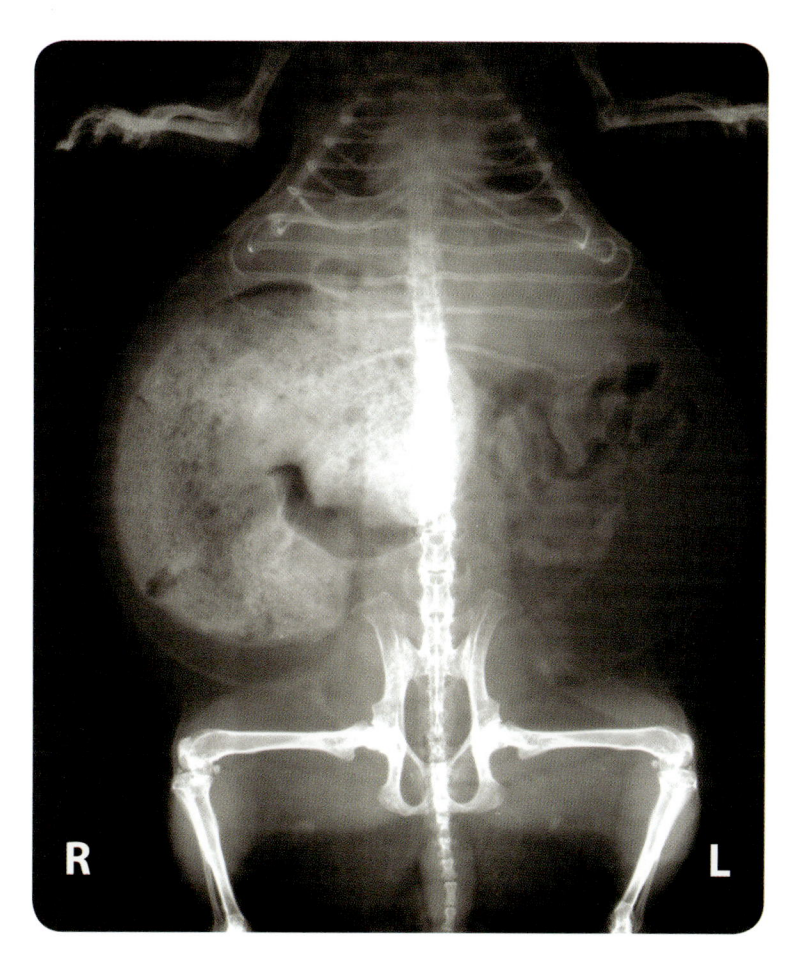

Figure 9.27. Ventrodorsal projection of the same prairie dog showing impaction of the cecum. Densities representing ingesta are emphasized by the presence of gas, which acts as a natural contrast medium.

The NORMAL THORACIC LIMB
Lateral Projection

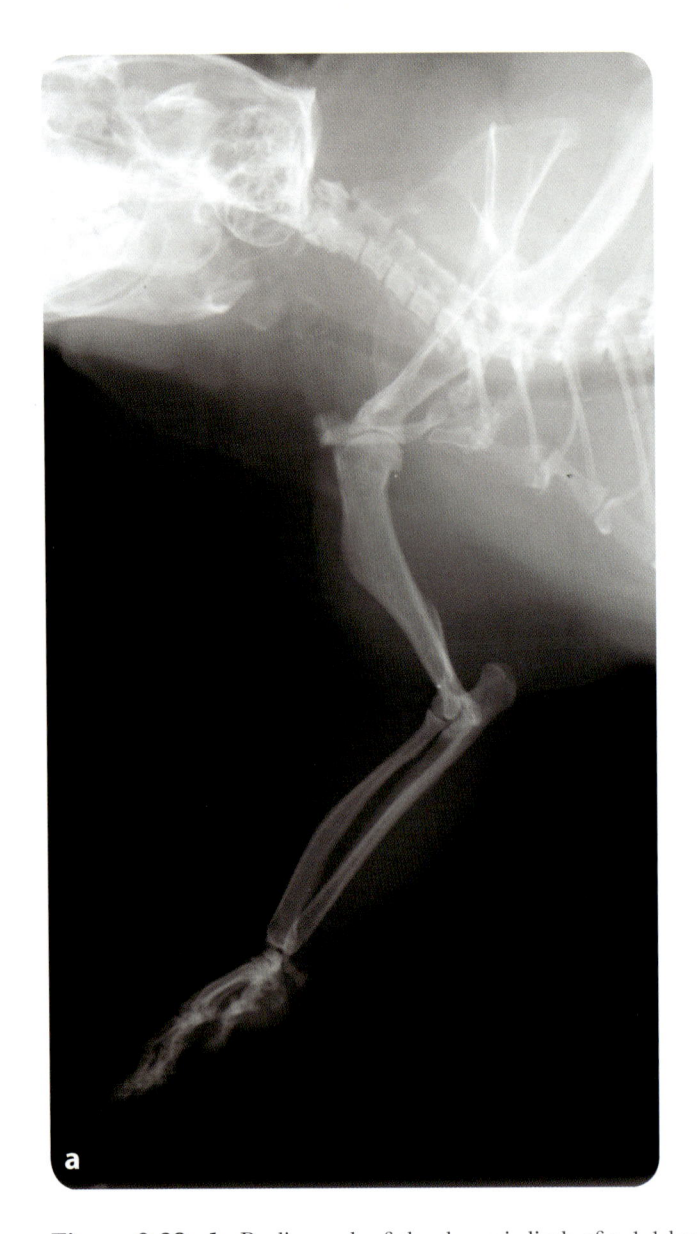

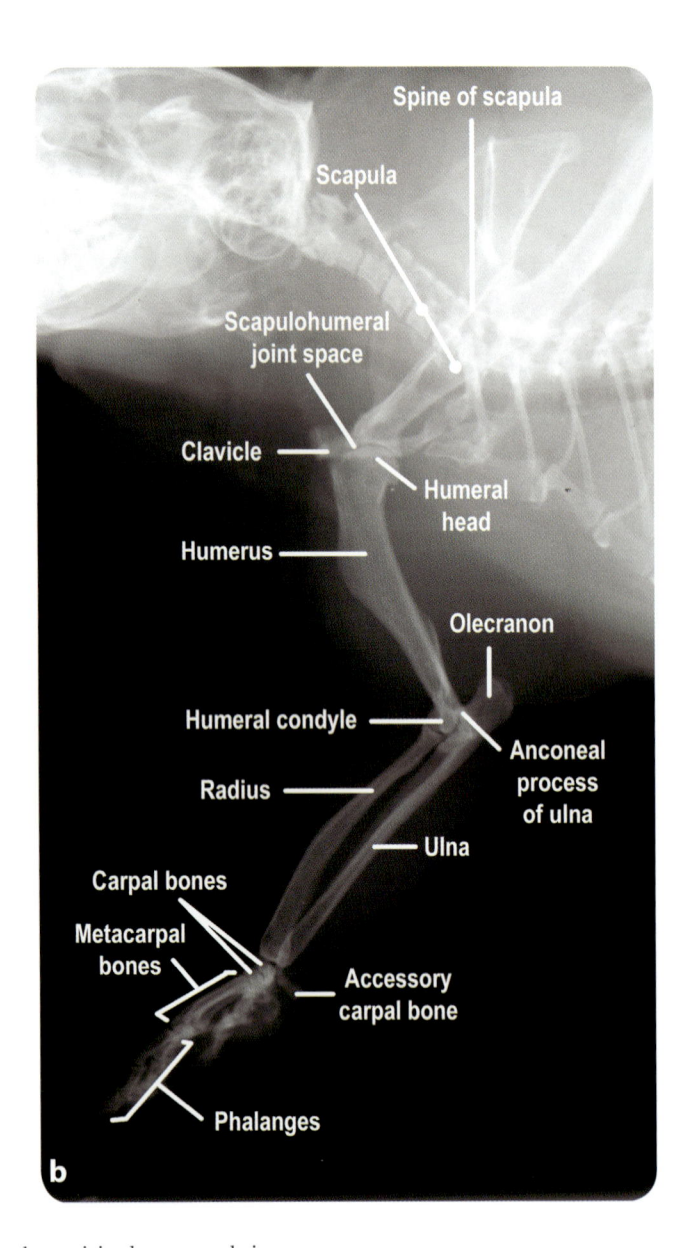

Figure 9.28a,b. Radiograph of the thoracic limb of a 1.1 kg male prairie dog, actual size.

Caudocranial Projection

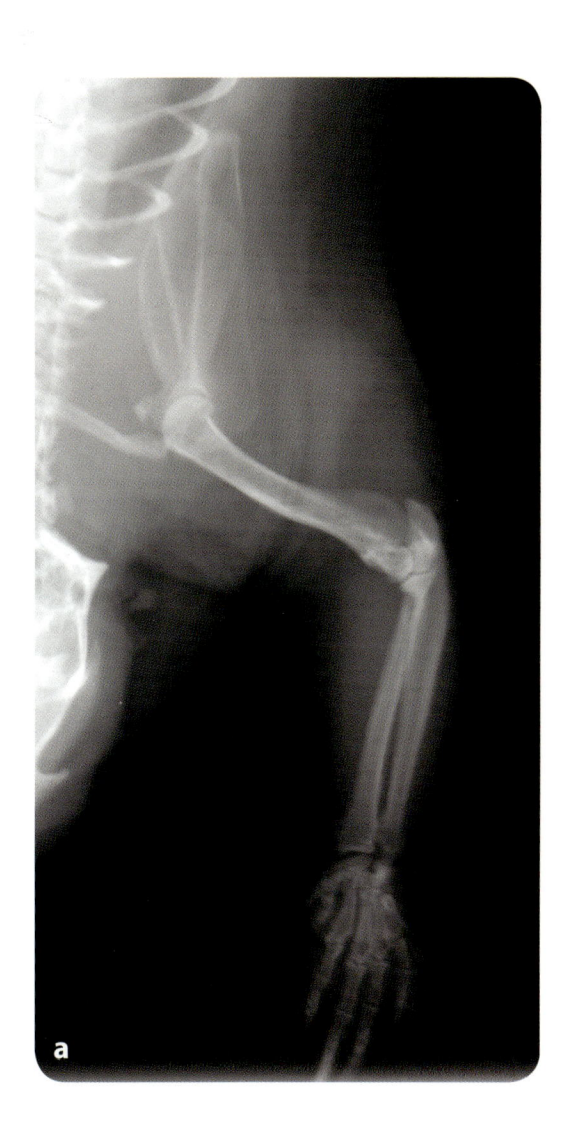

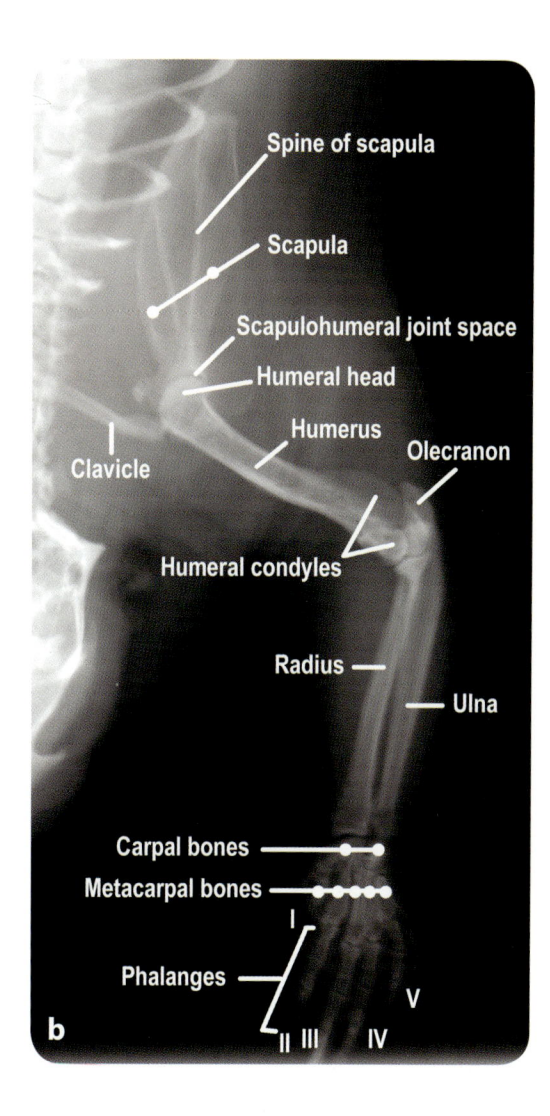

Figure 9.29a,b. Radiograph of the thoracic limb of a 1.1 kg male prairie dog, actual size. Note that in some brachymorphic species is impossible to obtain a perfect caudocranial (or craniocaudal) view of every portion of the limb. In this view, all portions of the limb with the exception of the carpus are actually oblique.

The NORMAL PELVIC LIMB
Lateral Projection

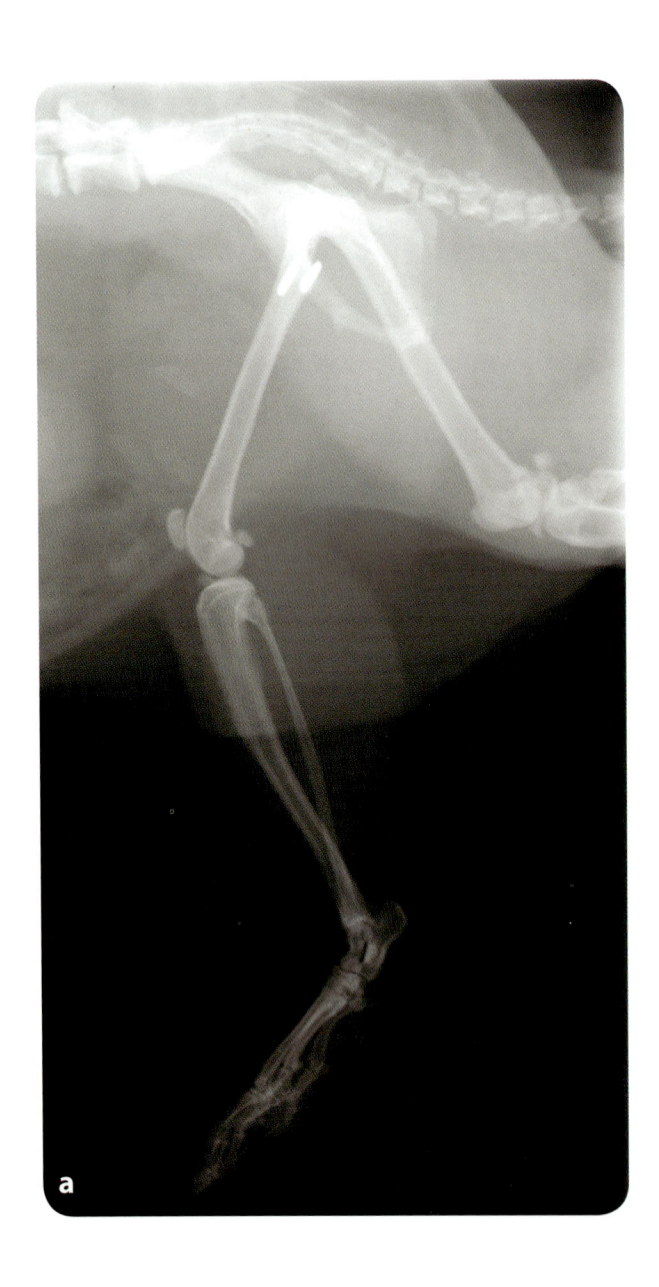

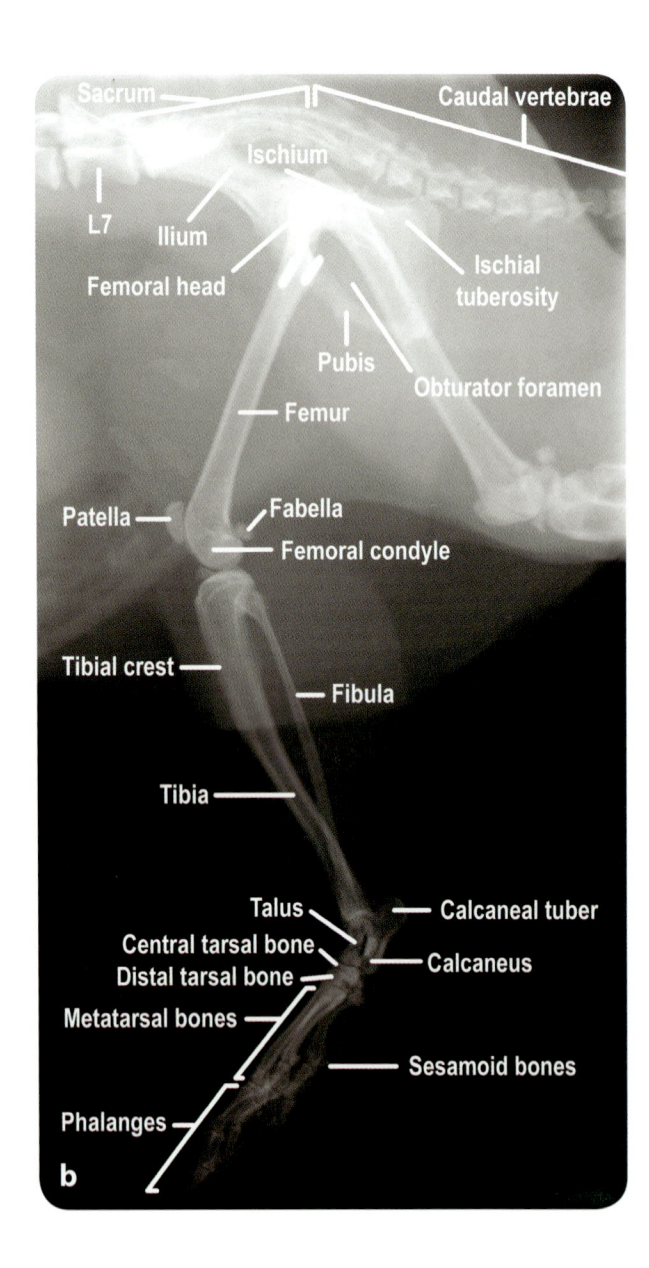

Figure 9.30a,b. Radiograph of the pelvic limb of a 1.1 kg male prairie dog, actual size. Note hemoclips used for castration superimposed upon the femur.

Craniocaudal Projection

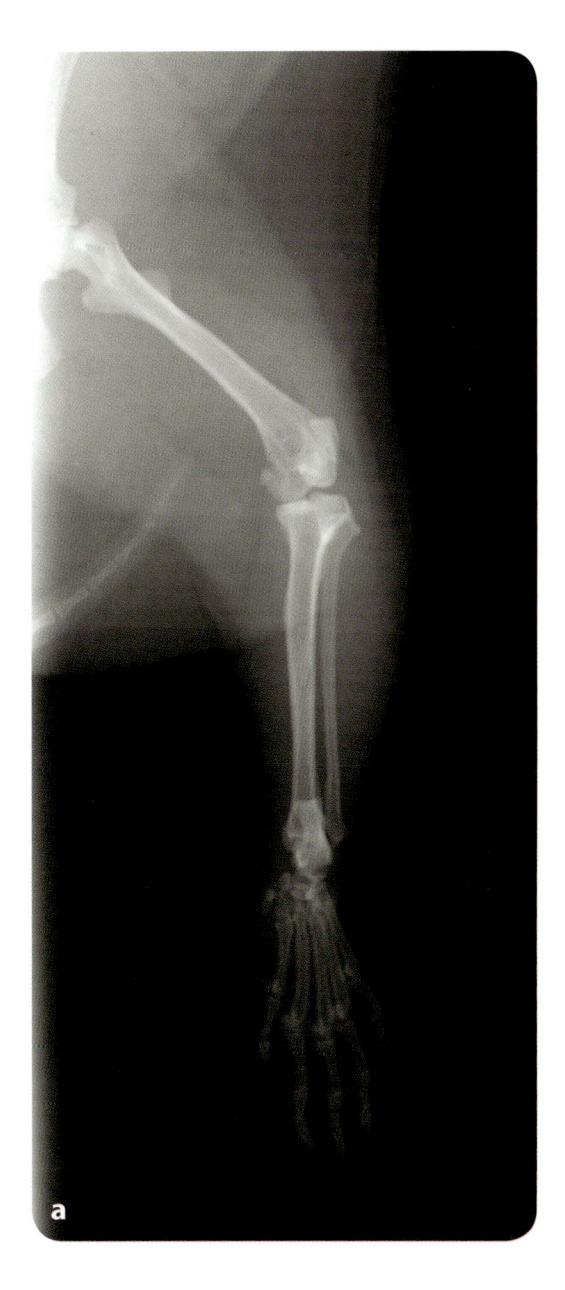

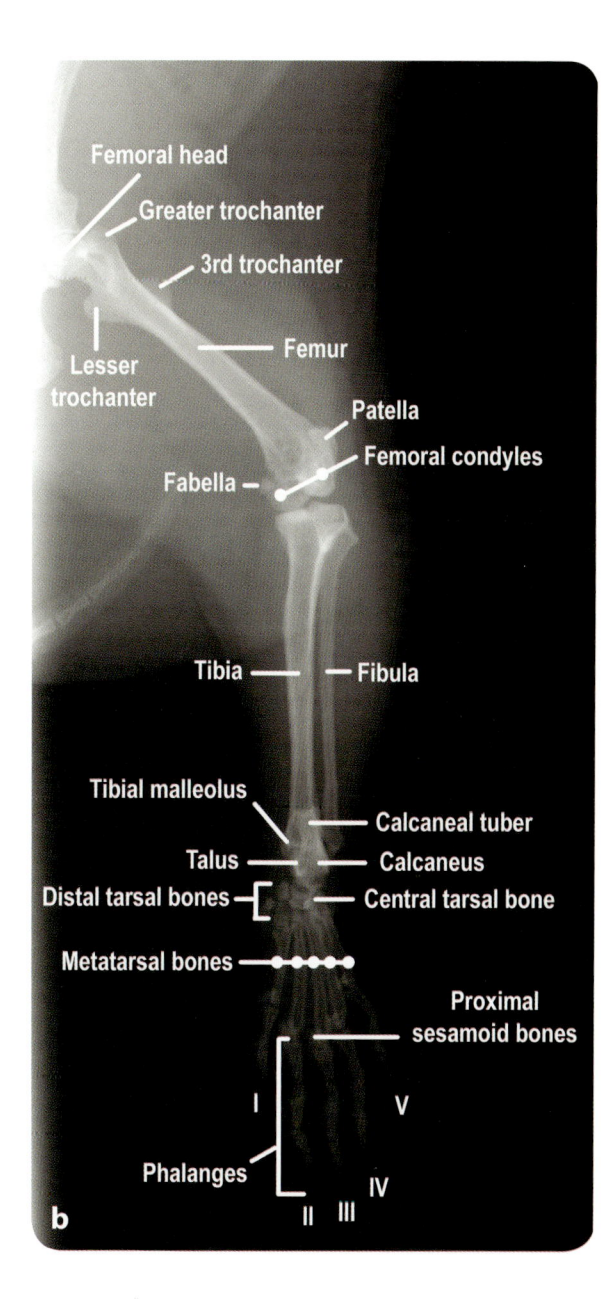

Figure 9.31a,b. Radiograph of the pelvic limb of a 1.1 kg male prairie dog, actual size. As described above for the thoracic limb, the proximal portion is actually oblique.

The CHIPMUNK
The NORMAL WHOLE BODY SKELETON
Lateral Projection

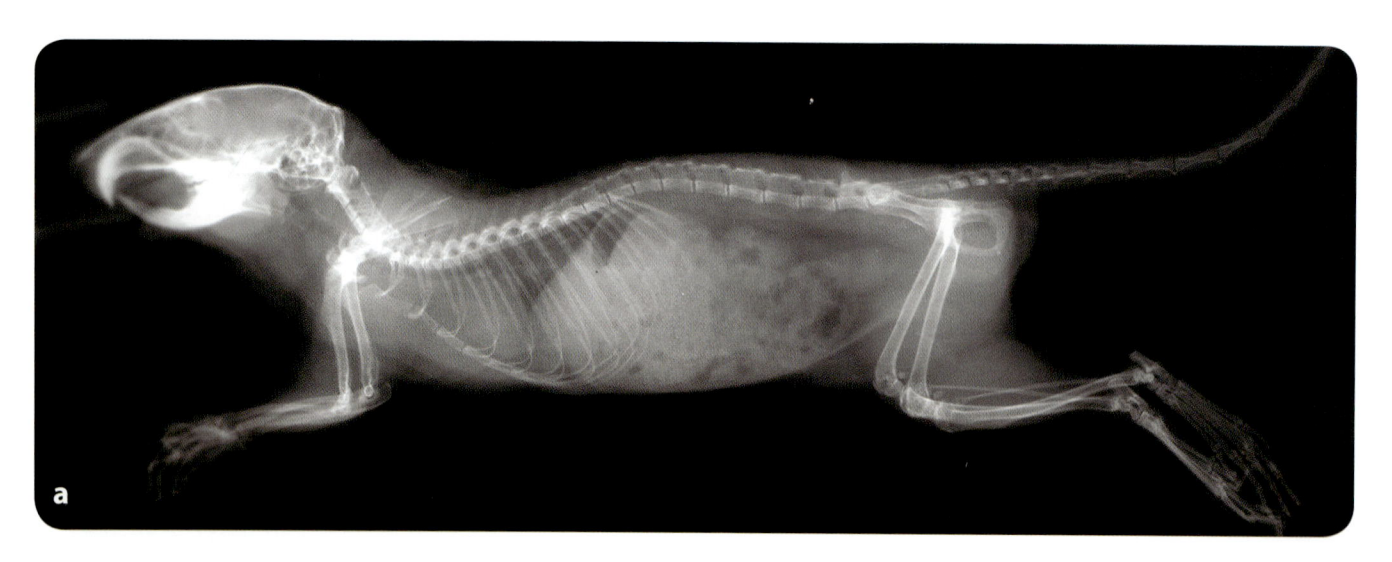

Figure 9.32a,b. Whole body skeleton of a 75 g male chipmunk, 90% of actual size.

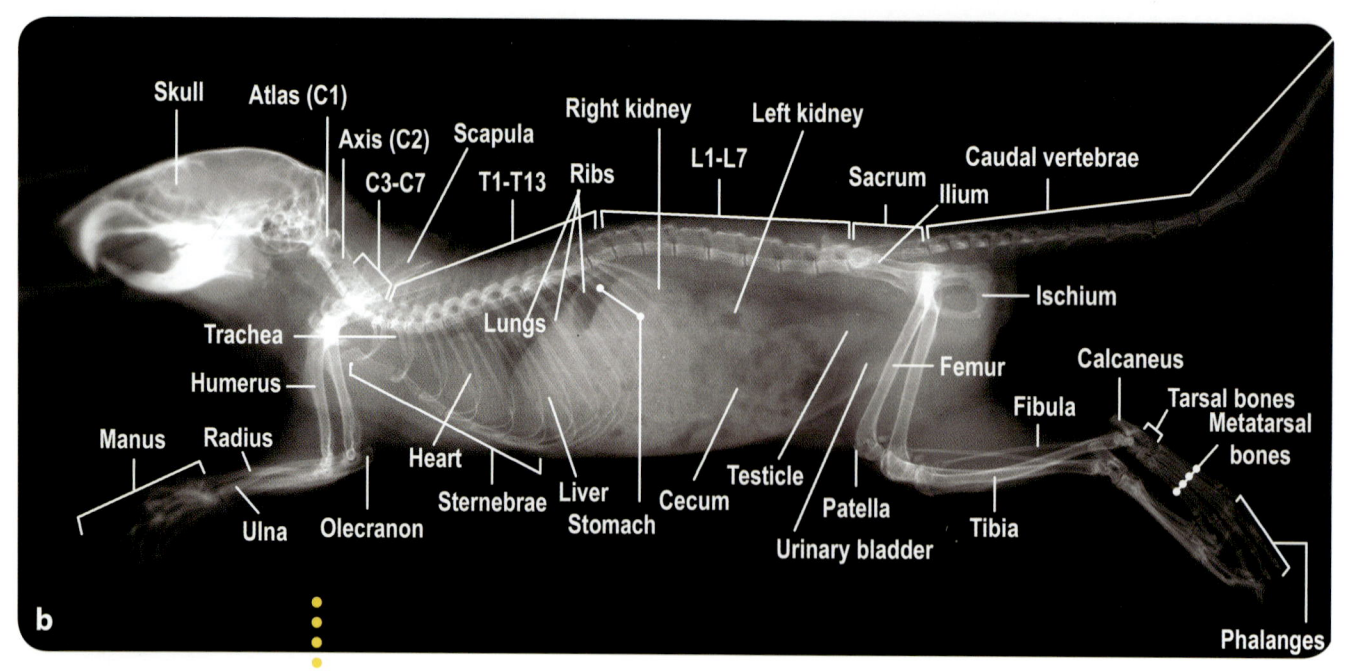

Figure 9.33. There are many species of chipmunks. *Tamias striatus*, and *Eutamias sibiricus* are occasionally encountered as a pet species in Europe.

Ventrodorsal Projection

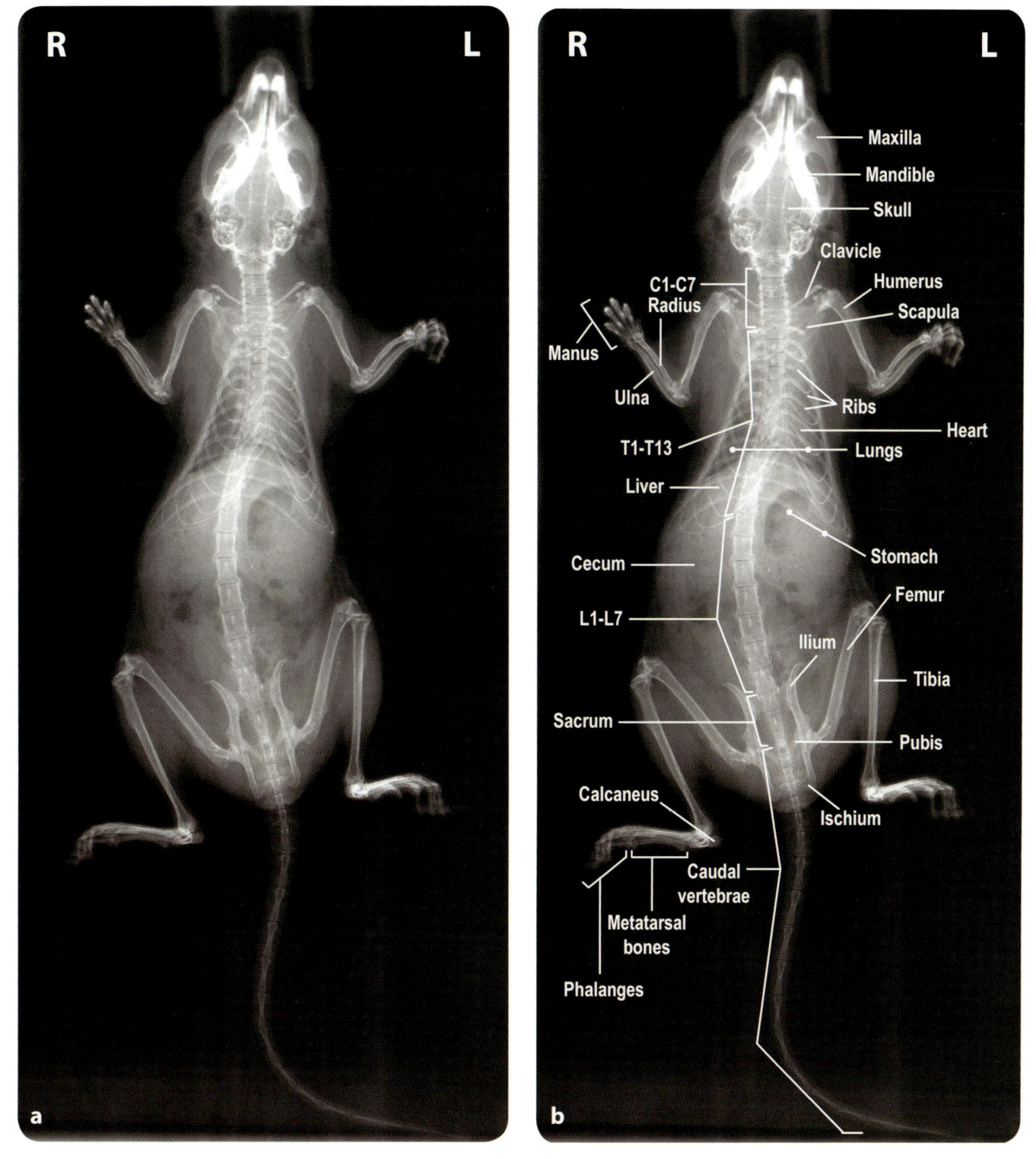

Figure 9.34a,b. Whole body skeleton of a 75 g male chipmunk, 90% of actual size.

ABNORMALITIES of the HEAD
of other Squirrel-like Rodents

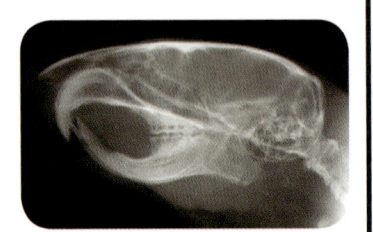

Figure 9.35. Actual size of the skull of a 80 g male chipmunk.

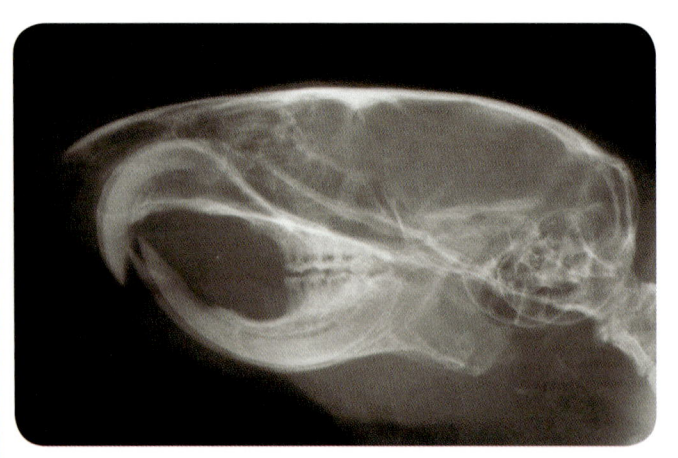

Figure 9.36. Radiograph of the normal skull of a 80 g male chipmunk. The apparent asymmetry of the mandibular incisor teeth is because this projection is slightly oblique in a cranio-caudal direction (see the slight asymmetry of the tympanic bullae).

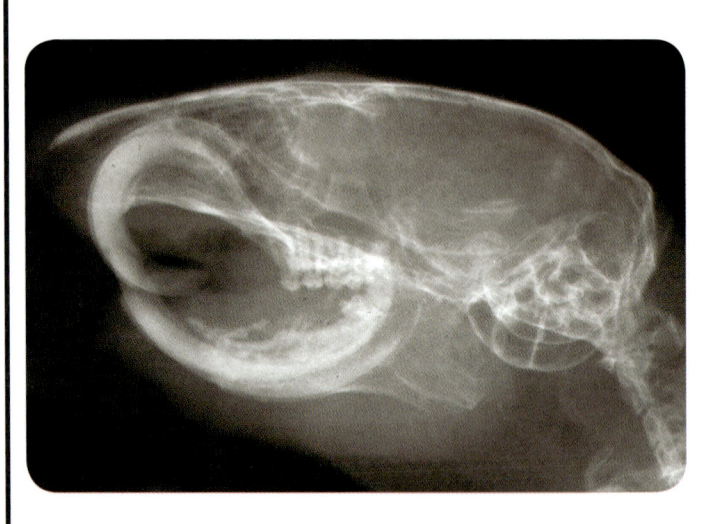

Figure 9.37. Lateral view of the skull of a chipmunk demonstrating severe malocclusion of maxillary incisor teeth and excessive wearing of cheek teeth.

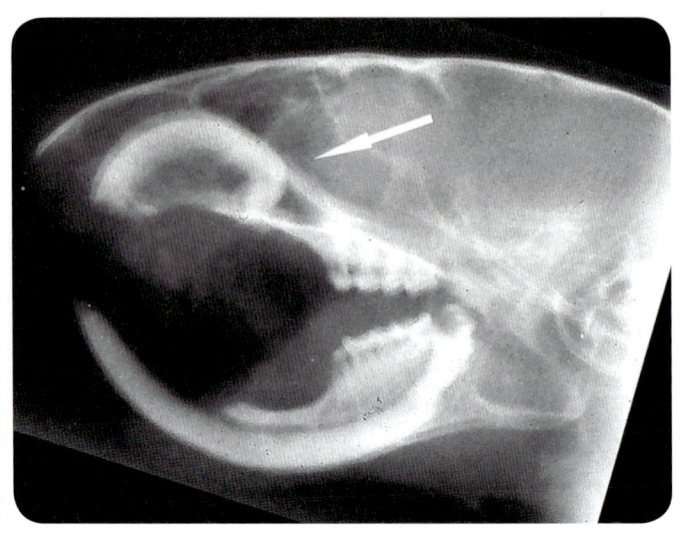

Figure 9.38. Lateral radiogaph of the skull of a chipmunk with fractures of the maxillary incisors and abnormal growth and deformity of the apexes (arrow). Note the abnormal apexes of the mandibular incisors as well. Differential diagnosis includes pseudo-odontoma. This chipmunk was presented for severe dyspnea and suspected respiratory disease.

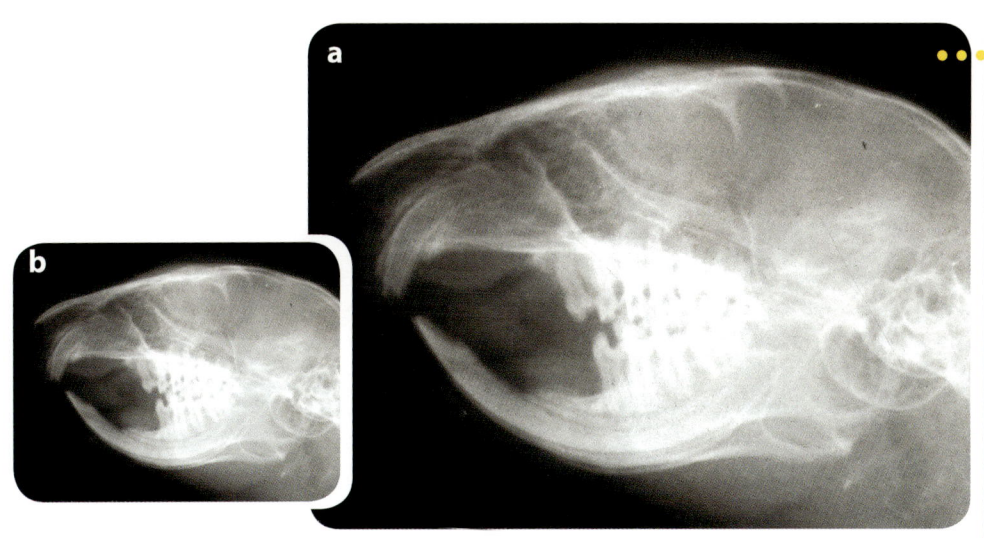

Figure 9.39a,b. Lateral view of the skull of a citellus showing a fracture of the maxillary incisor teeth, and actual size of the skull (b).

Figure 9.40. The European citellus *(Citellus citellus)*, also called dwarf prairie dog, is similar to, but smaller than prairie dogs native to the US. The citellus is occasionally encountered as a pet species in Europe.

MISCELLANEOUS ABNORMALITIES
of other Squirrel-like Rodents

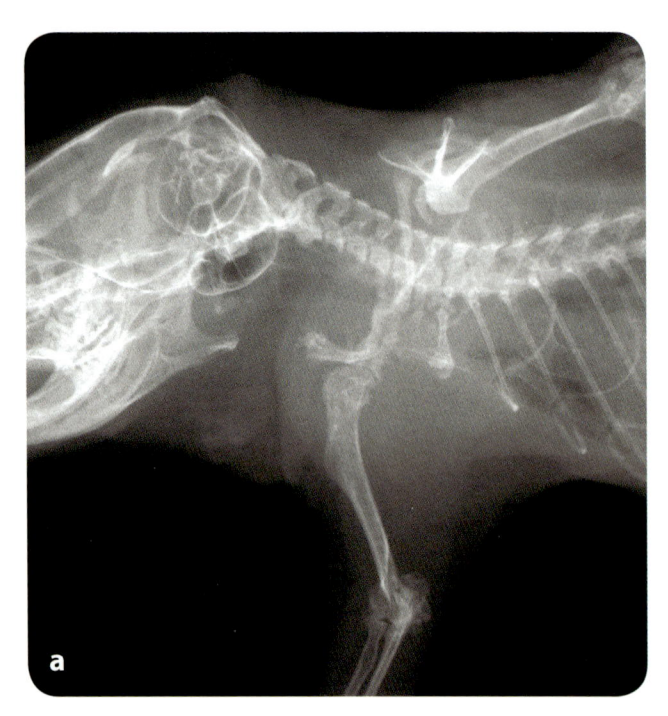

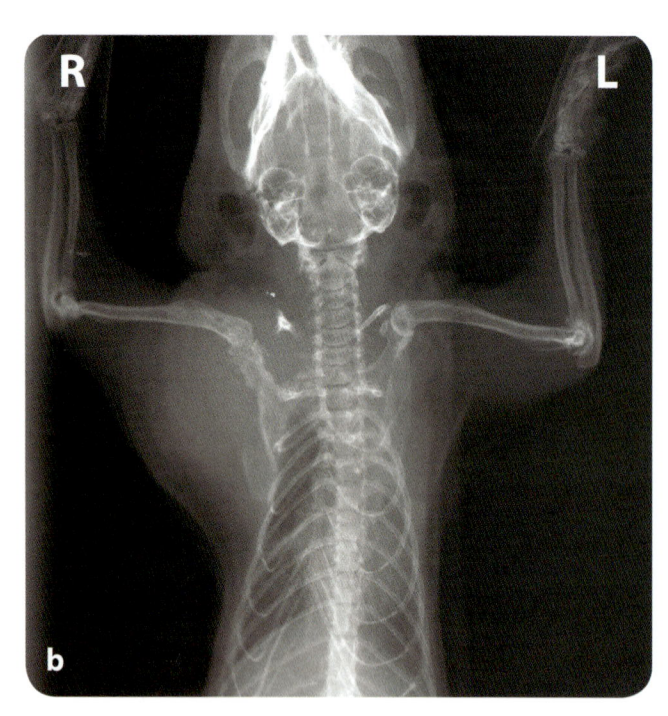

Figure 9.41a,b. Lateral with left limb retracted (a) and ventrodorsal (b) radiographs of a chipmunk with neoplasia of the right thoracic limb. Abnormalities of bone and soft tissues are more apparent on the ventrodorsal projection, and are emphasized by comparison with the normal left thoracic limb. Abnormal changes of the clavicula are also visible.

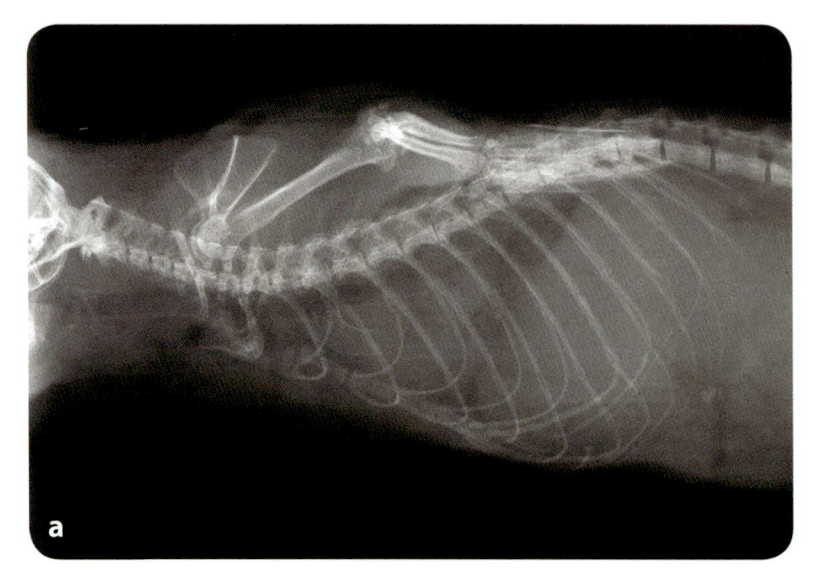

Figure 9.42a,b. Lateral (a) and ventrodorsal (b) follow up radiographs post amputation of the right thoracic limb performed via disarticulation of the scapula. This animal tolerated limb amputation well.

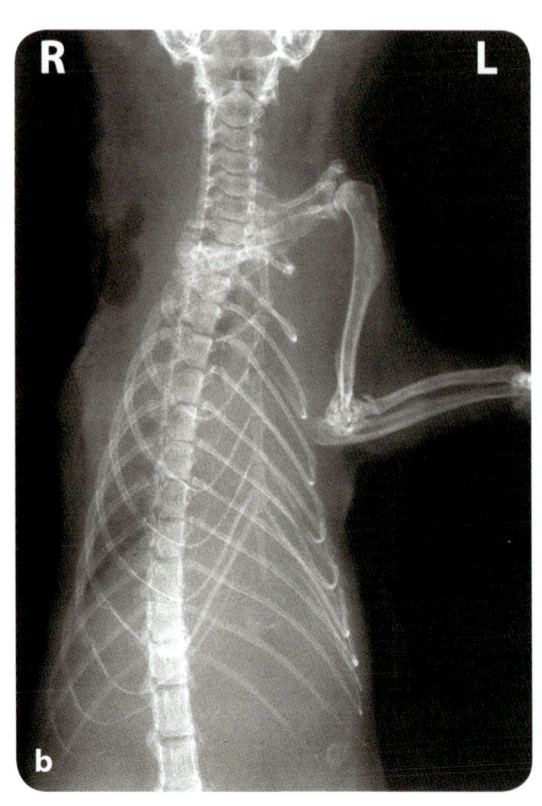

MISCELLANEOUS 355

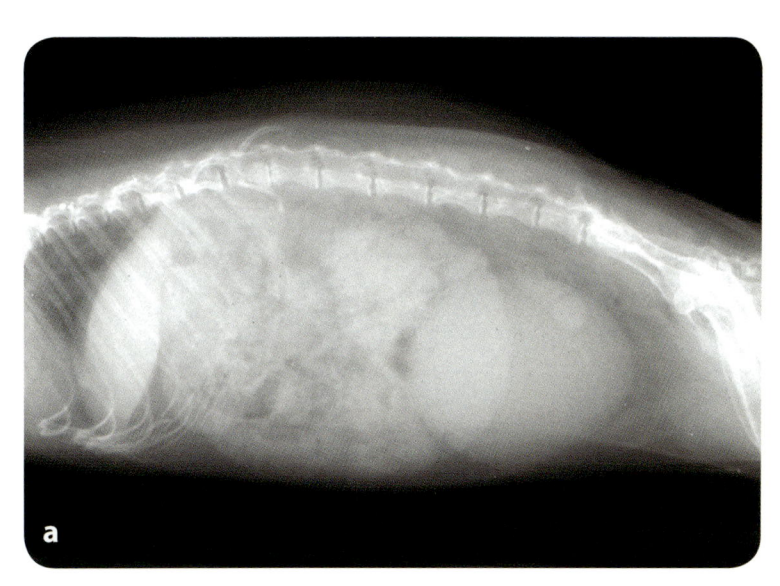

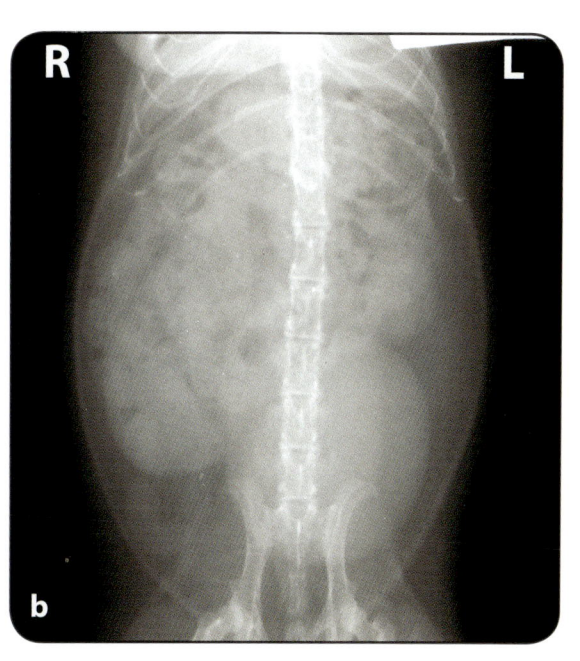

Figure 9.43a,b. Lateral (a) and ventrodorsal (b) radiographs of a male citelus with abnormal distension of the urinary bladder secondary to obstruction of the urethra with a copulatory plug. The bladder appears as a large round structure in the caudal abdomen, and deflects gastrointestinal loops cranially.

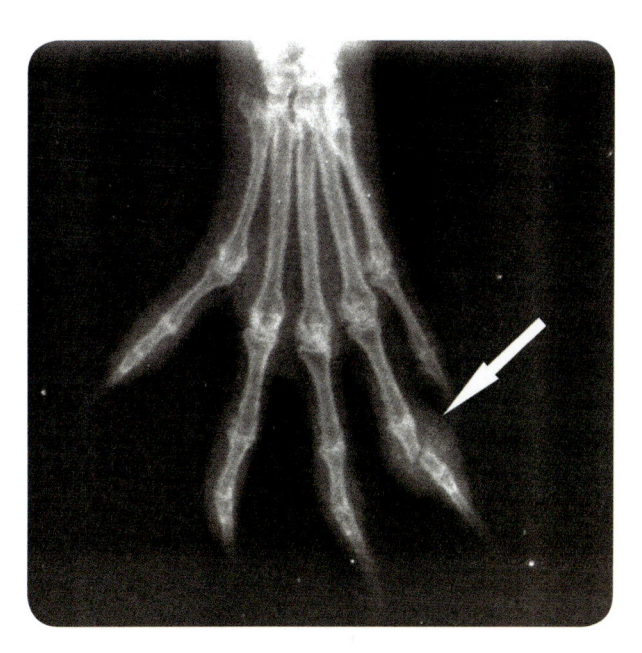

Figure 9.44. Dorsoventral radiograph of the distal thoracic limb of a citellus with a fracture of the second phalanx of digit 2 (arrow). Note swelling of the soft tissues at the fracture site.

References:
Capello V: Dental diseases. In: Lennox A, ed. Rabbit and Rodent Dentistry Handbook. Ames, IA: Blackwell Publishing, (Formerly Zoological Education Network, Lake Worth, FL); 2005:113-163.
Capello V: Incisor extraction to resolve clinical signs of odontoma in a prairie dog. ExoticDVM. 2002; 4(1):9.
Capello V: Dental diseases and surgical treatment in pet rodents. ExoticDVM. 2003; 5(3):21-27.
Capello V, Gracis M: Radiology of the skull and teeth. In: Lennox A, ed. Rabbit and Rodent Dentistry Handbook. Ames, IA: Blackwell Publishing, (Formerly Zoological Education Network, Lake Worth, FL); 2005:65-69.
Crossley DA: Small mammal dentistry, part I: Dental anatomy and dental disease. In: Quesenberry KE, Carpenter JW, eds. Ferrets, Rabbits and Rodents. Clinical Medicine and Surgery 2nd ed. Philadelphia, PA: Elsevier; 2004:370-378.
Funk RS: Medical management of prairie dogs. In: Quesenberry KE, Carpenter JW, eds. Ferrets, Rabbits and Rodents. Clinical Medicine and Surgery 2nd ed. Philadelphia, PA: Elsevier; 2004:266-273.
Girling SJ: Mammalian imaging and anatomy. In: Meredith A, Redrobe S, eds. BSAVA Manual of Exotic Pets 4th ed. Quedgeley, Glouchester: British Small Animal VeterinaryAssociation; 2005:1-12.
Greenacre CB: Spontaneous tumors of small mammals. Vet Clin Exot Anim. 2004; 7:627-651.
Legendre LFJ: Oral disorders of exotic rodents. Vet Clin Exot Anim. 2003; 6:601-628.
Morera N: Osteosarcoma in a Siberian chipmunk. ExoticDVM. 2004; 6(1):11-12.
Wagner RA, Garman RH, Collins BM: Diagnosing odontomas in prairie dogs. ExoticDVM. 1999; 1(1):7-10.
Wagner R, Johnson D: Rhinotomy for treatment of odontoma in prairie dogs. ExoticDVM. 2001; 3(5):29-34.

FERRET

The NORMAL HEAD
Lateral Projection

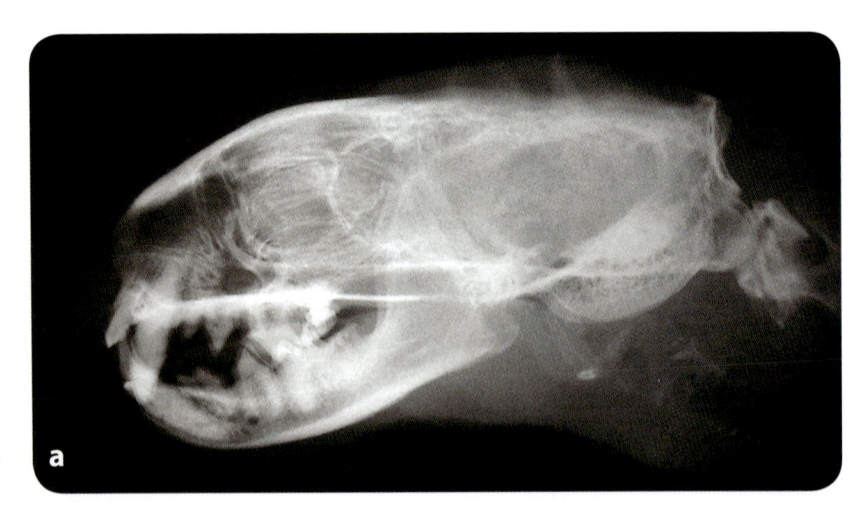

Figure 10.1a,b. Radiograph of the skull of a 1.1 kg male ferret, 140% of actual size.

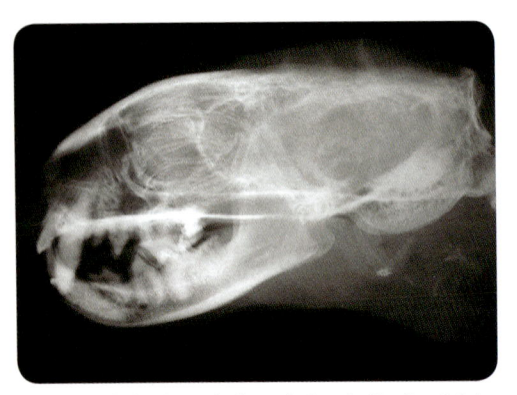

Ethmoturbinates
Nasal bone
Nasal cavity
Maxillary incisor tooth
P2 P3
P4 (Carnassial tooth)
M1
Coronoid process of the mandible
Temporomandibular joint
Parietal bone
Occipital protuberance
Occipital condyle
Atlas (C1)
Tympanic bulla
Condylar process of the mandible
Mandible
M2
M1 (Carnassial tooth)
P2 P3 P4
Mandibular canine tooth
Maxillary canine tooth

Figure 10.2. Actual size of the skull of a 1.1 kg ferret.

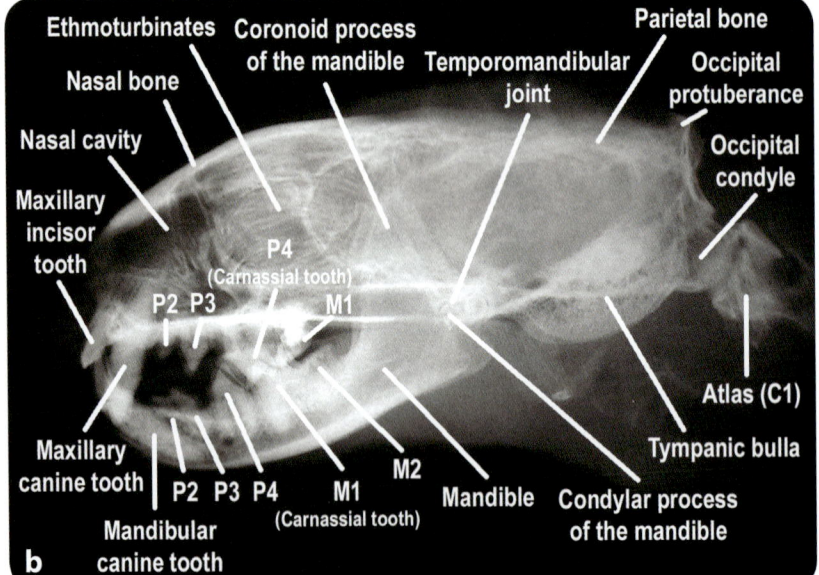

Figure 10.3. Radiograph of the skull of a 1.1 kg male ferret, open mouth view, 140% of actual size.
I=Incisor
C=Canine
P=Premolar
M=Molar
Note that only three premolars are present, as P1 has been lost in development.

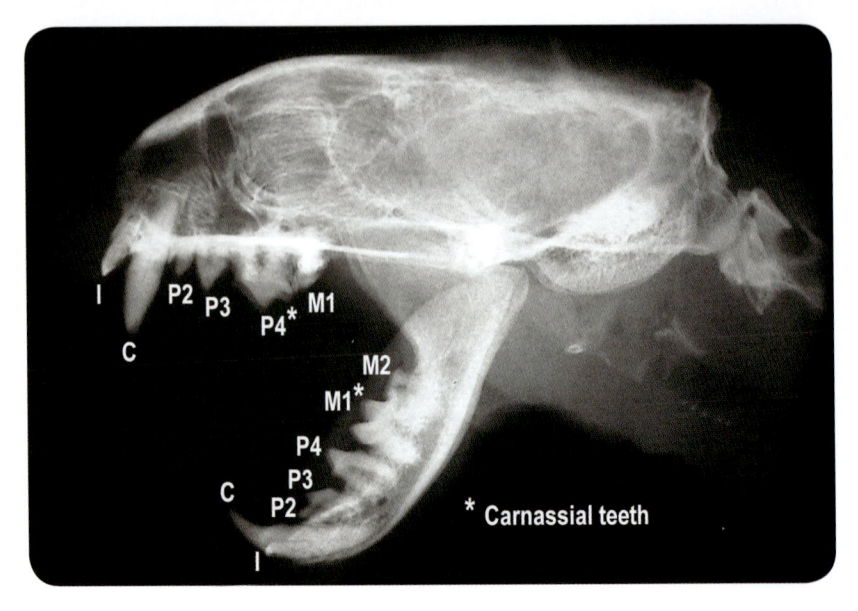

* Carnassial teeth

Oblique Projection

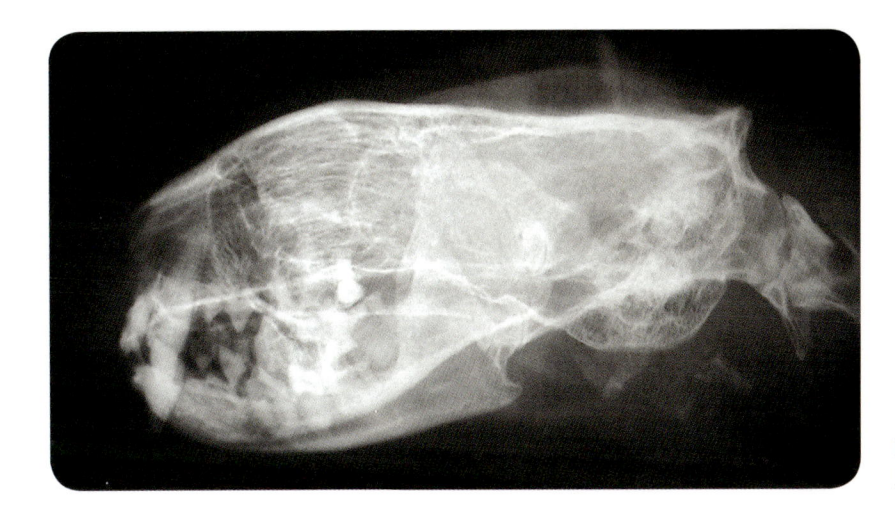

Figure 10.4. Radiograph of the skull of a 1.1 kg male ferret, 140% of actual size.

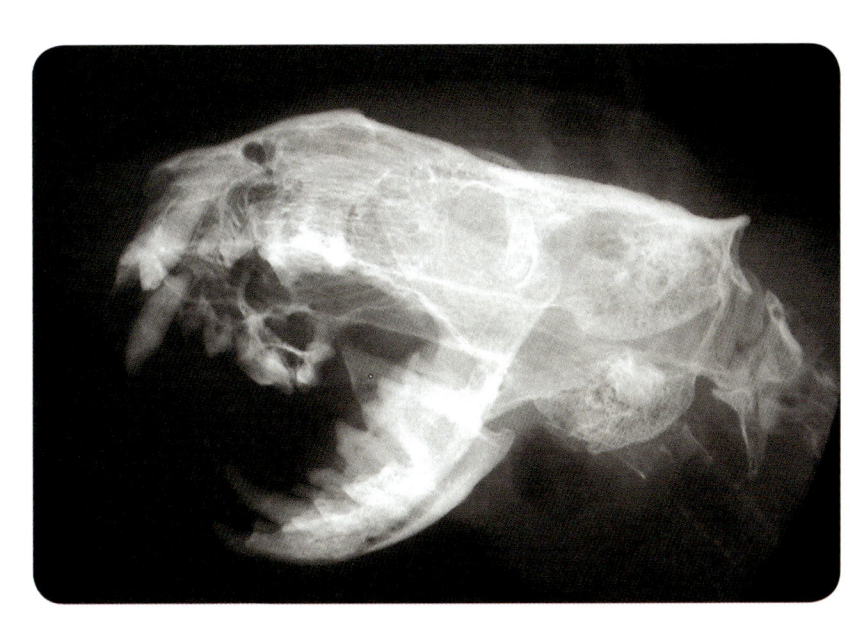

Figure 10.5. Radiograph of the skull of a 1.1 kg male ferret, open mouth view, 140% of actual size.

Ventrodorsal Projection

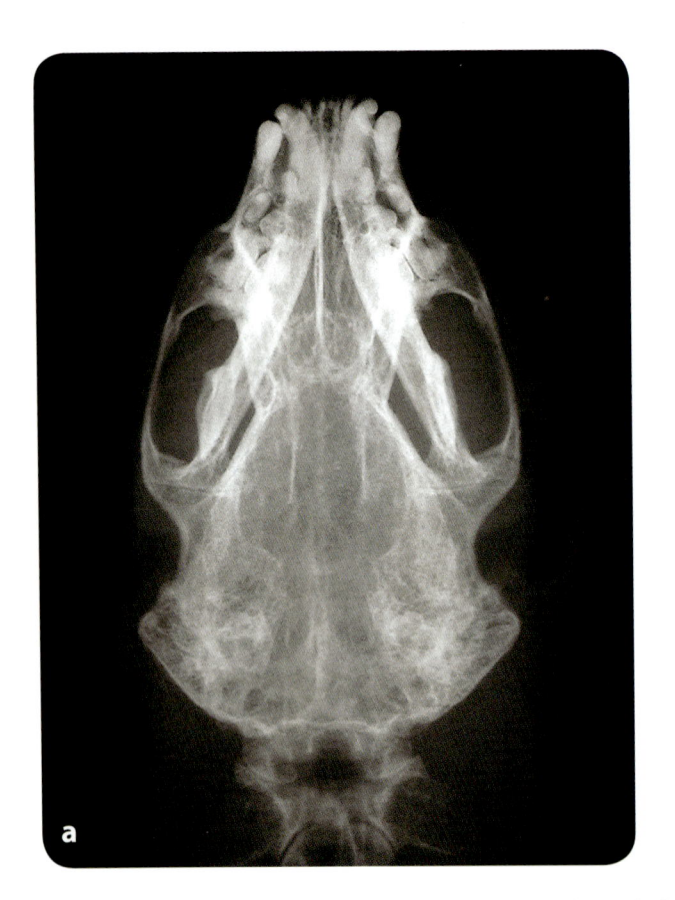

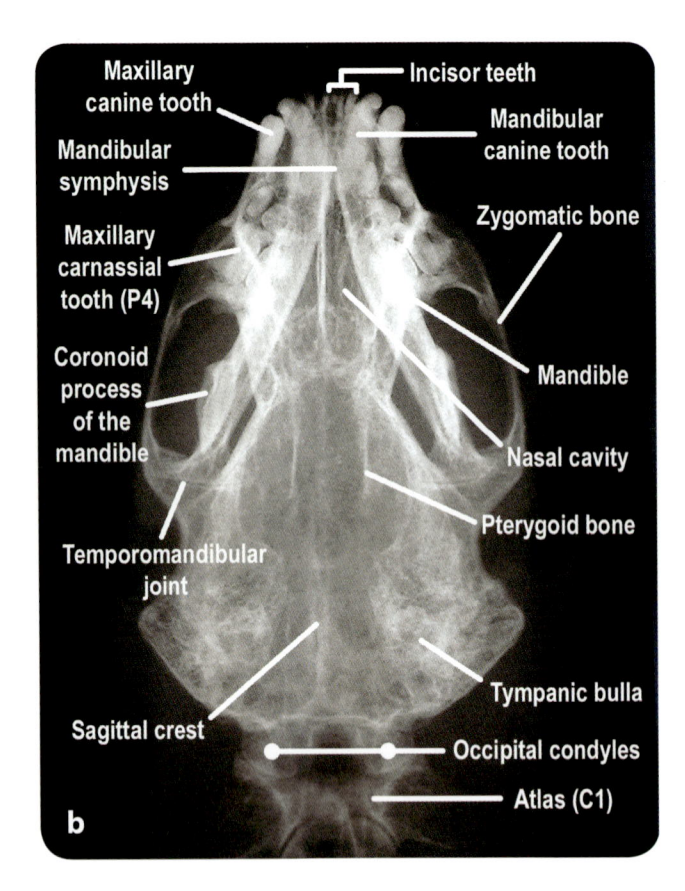

Figure 10.6a,b. Radiograph of the skull of a 1.1 kg male ferret, 115% of actual size.

Rostrocaudal Projection

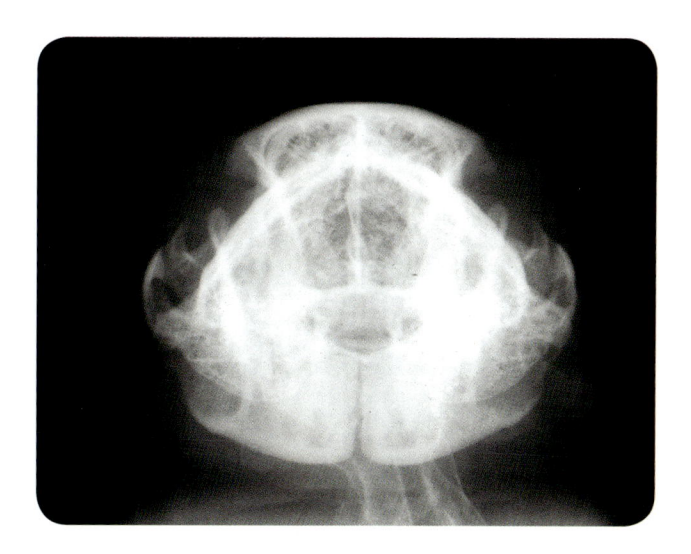

Figure 10.7. Radiograph of the skull of a 1.1 kg male ferret, 140% of actual size.

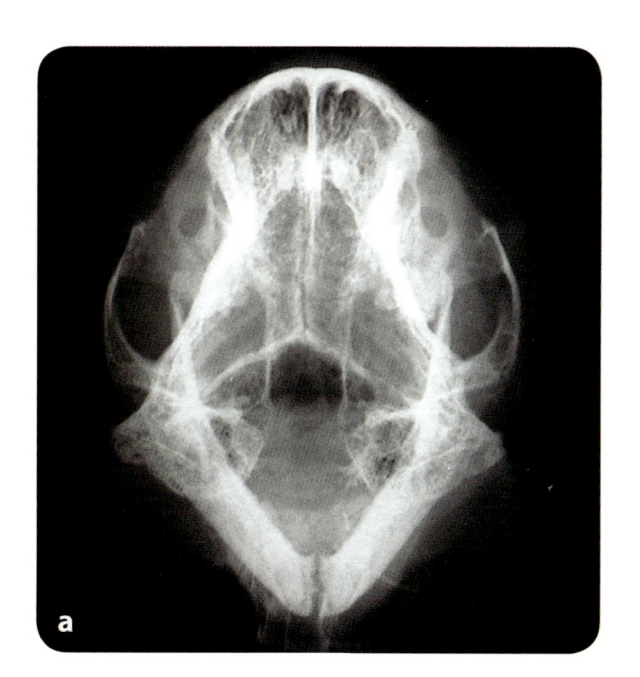

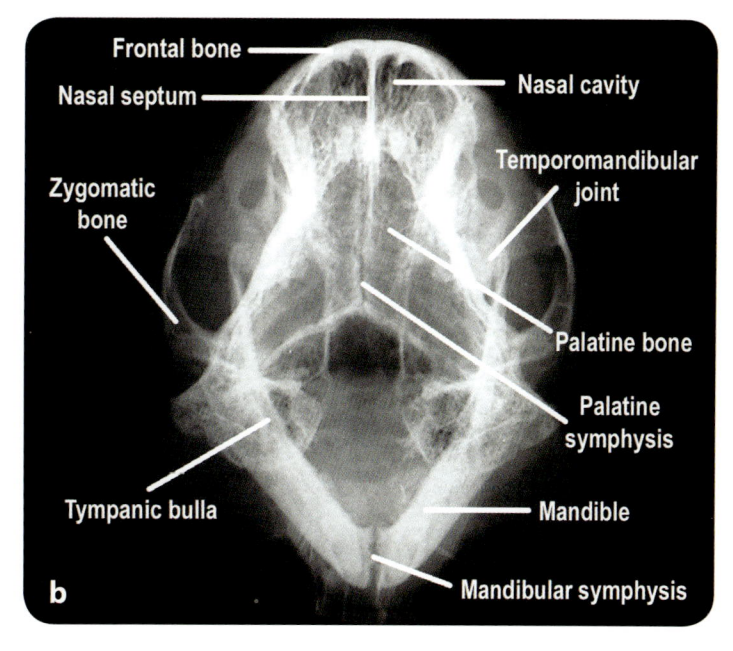

Figure 10.8a,b. Radiograph of the skull of a 1.1 kg male ferret, open mouth view 115% of actual size.

ABNORMALITIES of the HEAD

Figure 10.9a,b. Lateral (a) and ventrodorsal (b) radiographs of the skull of a 5-year-old ferret with mandibular osteosarcoma diagnosed by histopathology. Note soft tissue swelling surrounding the right mandibular ramus and aggresive bone reaction. There is new bone formation and osteolysis.
(Courtesy of Giorgio Romanelli, DVM)

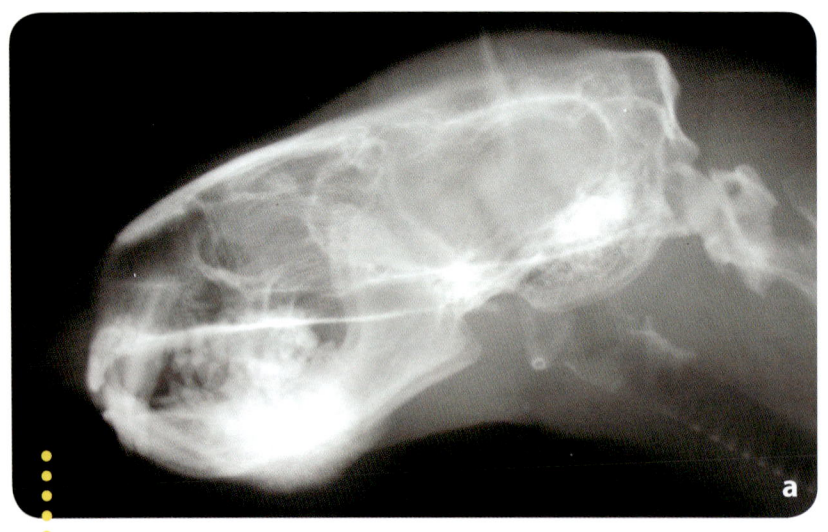

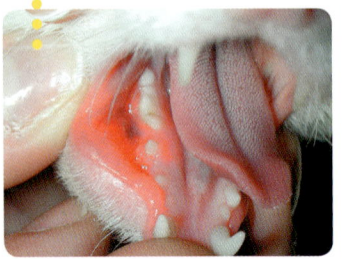

Figure 10.10. Intraoral clinical appearance of the mandibular neoplasia.
(Courtesy of Giorgio Romanelli, DVM)

Figure 10.11a,b. Lateral (a) and ventrodorsal (b) radiographs of the skull and neck of a male ferret with an abscess of the neck as a result of a bite wound from a cage mate. Note the ventral soft tissue opaque mass on the lateral projection (a). On the ventrodorsal projection, the mass has deviated the trachea to the left (b).

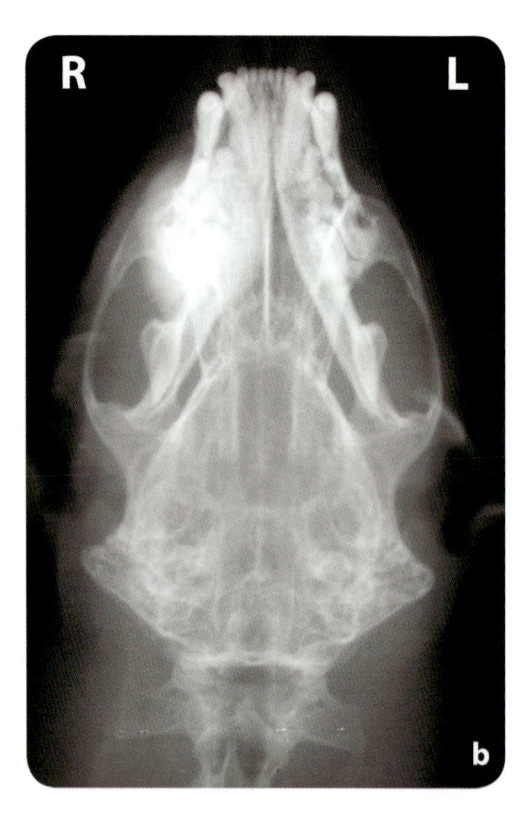

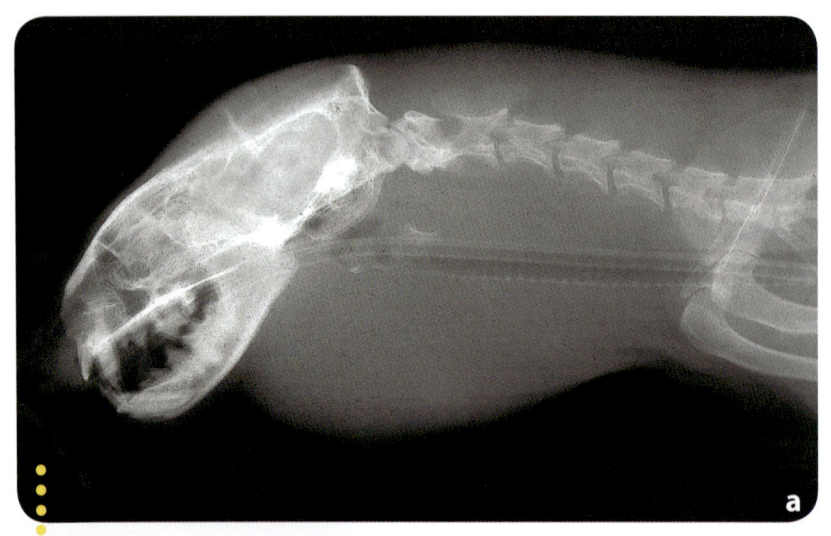

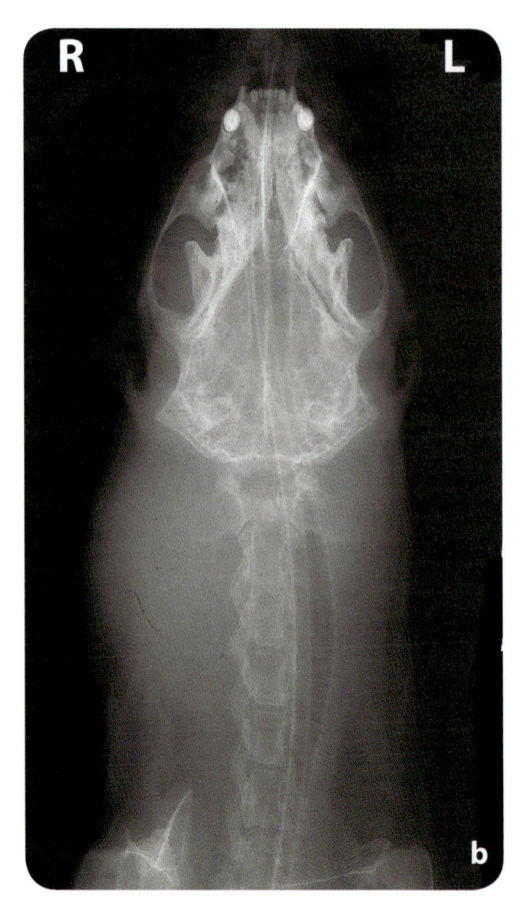

Figure 10.12. Clinical appearance of the abscess in the same patient.

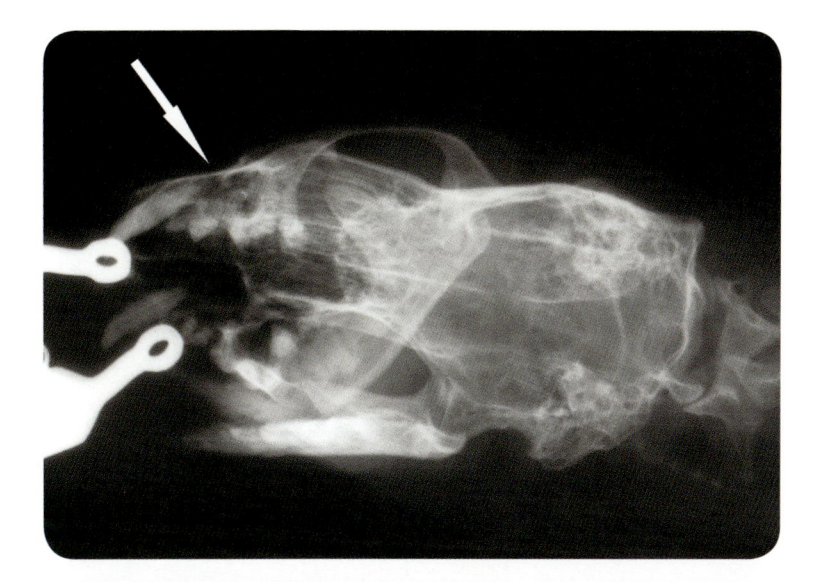

Figure 10.13. Left 45° ventral-right dorsal open mouth oblique projection showing lysis of the maxillary bone periapical to the left canine tooth (arrow).
(Courtesy of Cathy Johnson-Delaney, DVM)

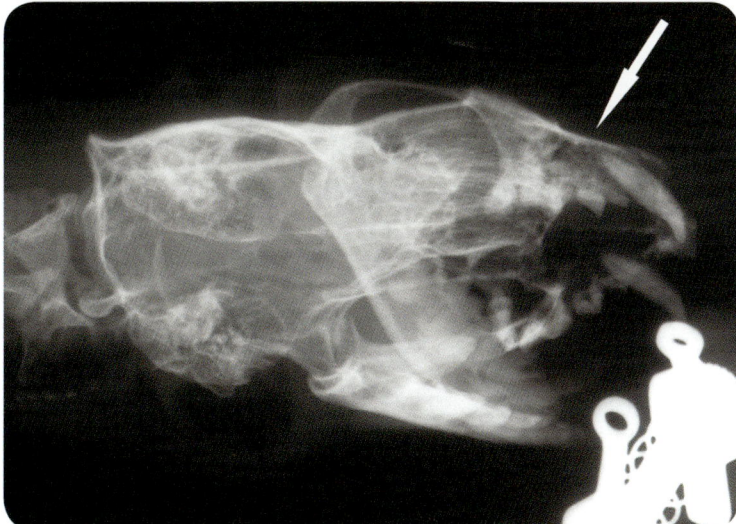

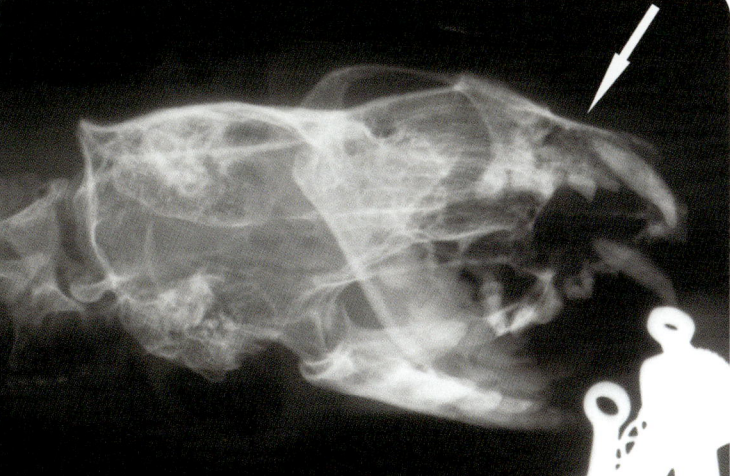

Figure 10.14. Right 45° ventral-right dorsal open mouth oblique projection in the same patient as Figure 10.13 showing a less severe lysis of the maxillary bone periapical to the right canine tooth (arrow).
(Courtesy of Cathy Johnson-Delaney, DVM)

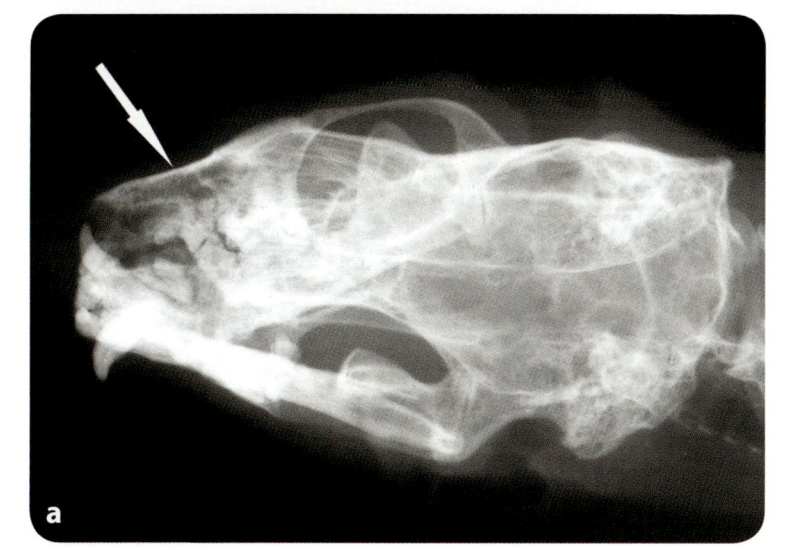

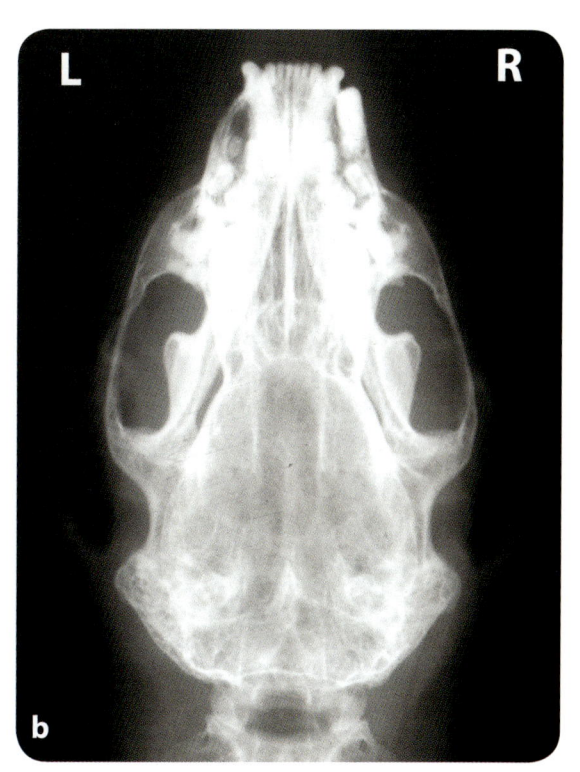

Figure 10.15a,b. Left 45° ventral-right dorsal oblique (a) and ventrodorsal (b) radiographs of the same ferret after extraction of the left maxillary canine tooth (arrow).
(Courtesy of Cathy Johnson-Delaney, DVM)

The NORMAL TOTAL BODY
Lateral Projection

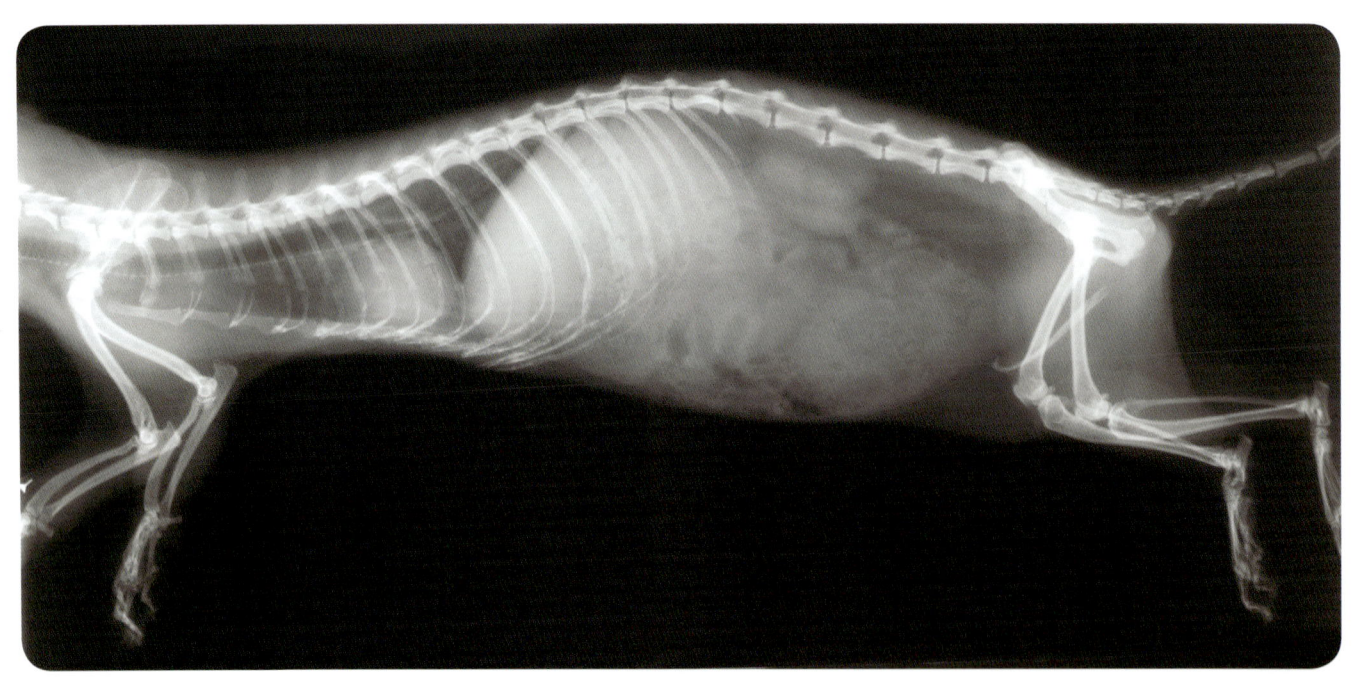

Figure 10.16. Total body radiograph of a 1.1 kg male ferret, 50% of actual size.

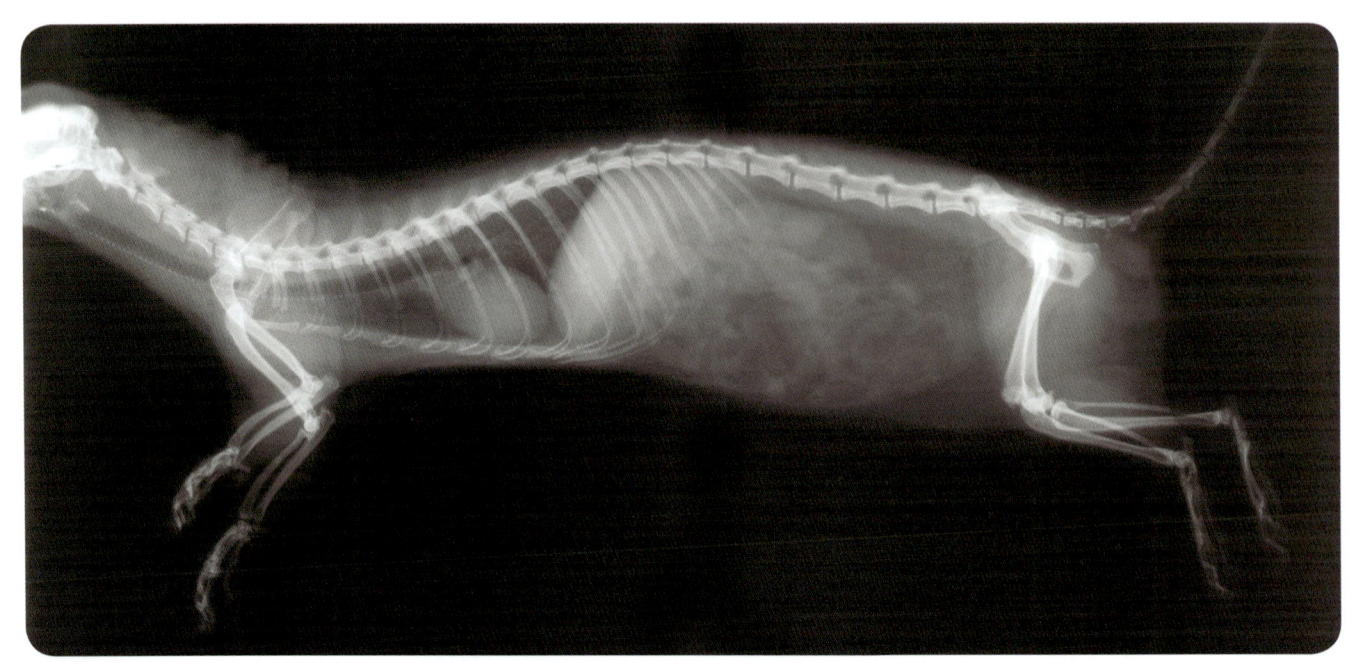

Figure 10.17. Total body radiograph of a 750 g intact female ferret, 50% of actual size.
Ferrets are sexually dimorphic, which is most obvious between intact males and females, or those neutered and spayed after puberty. The gender can be determined radiographically by the presence or absence of the os penis.

Ventrodorsal Projection

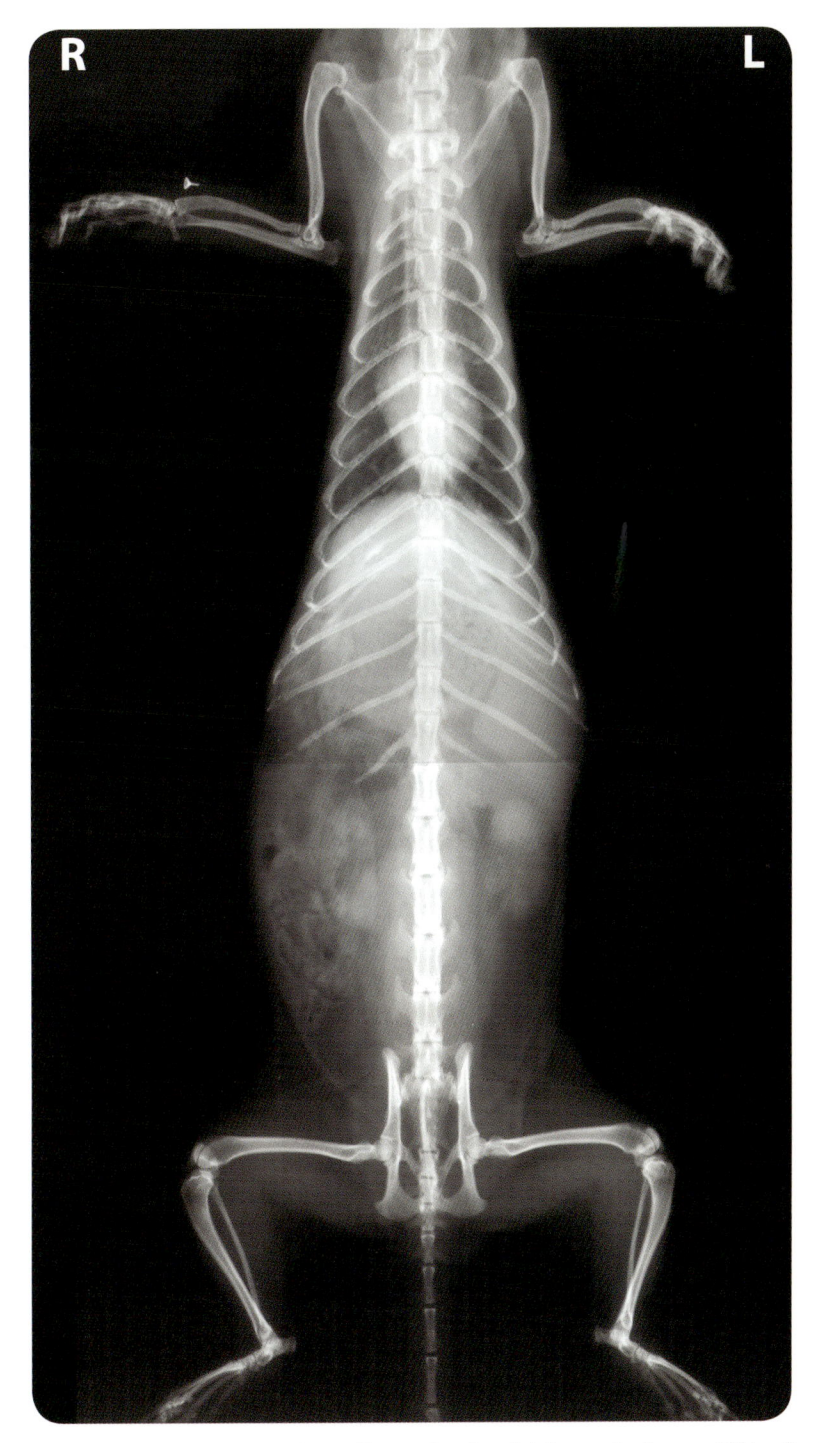

Figure 10.18. Total body radiograph of a 1.1 kg male ferret, 50% of actual size.

The NORMAL THORAX
The Cervical and Thoracic Vertebral Column
Lateral Projection

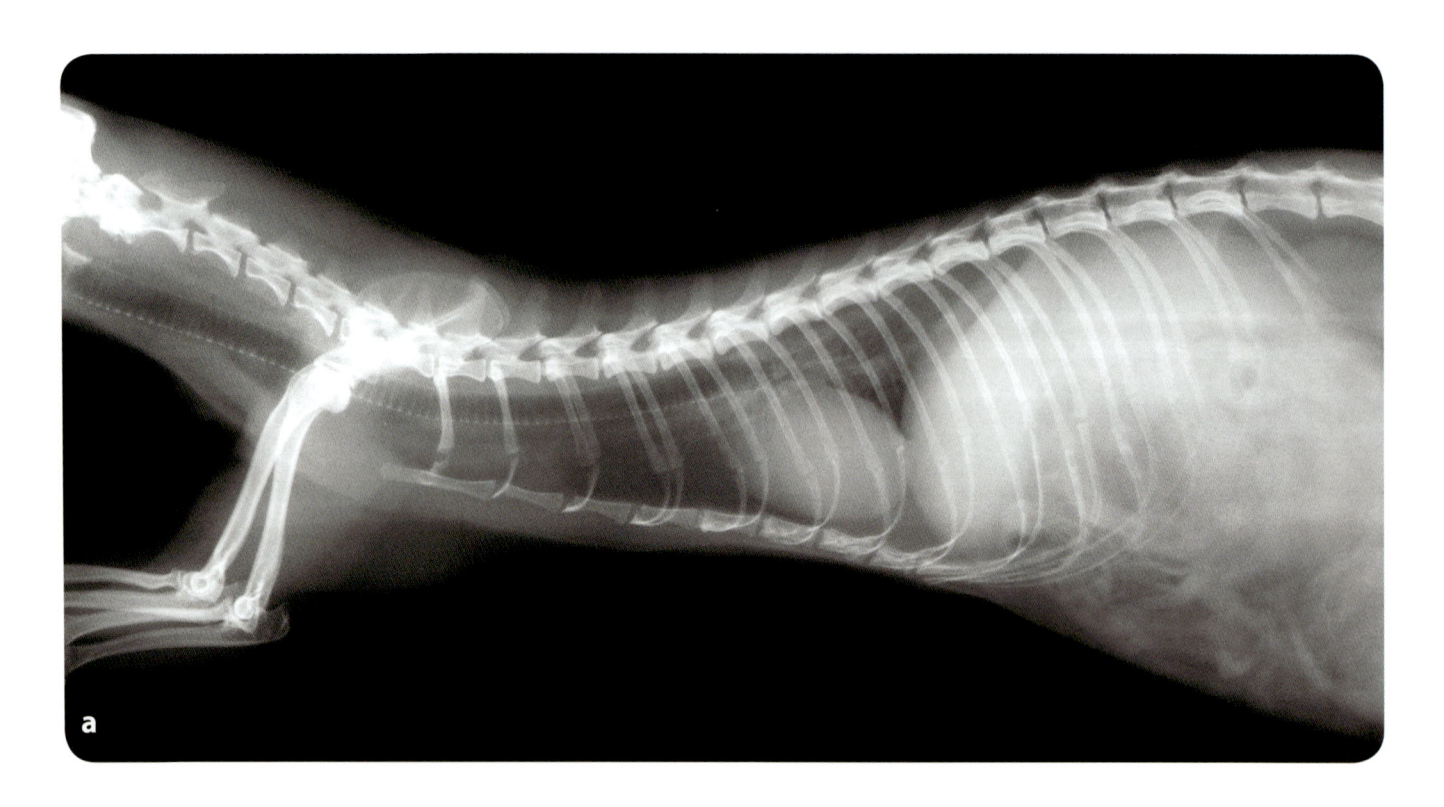

Figure 10.19a,b. Radiograph of the thorax of a 1.1 kg male ferret, 75% of actual size.

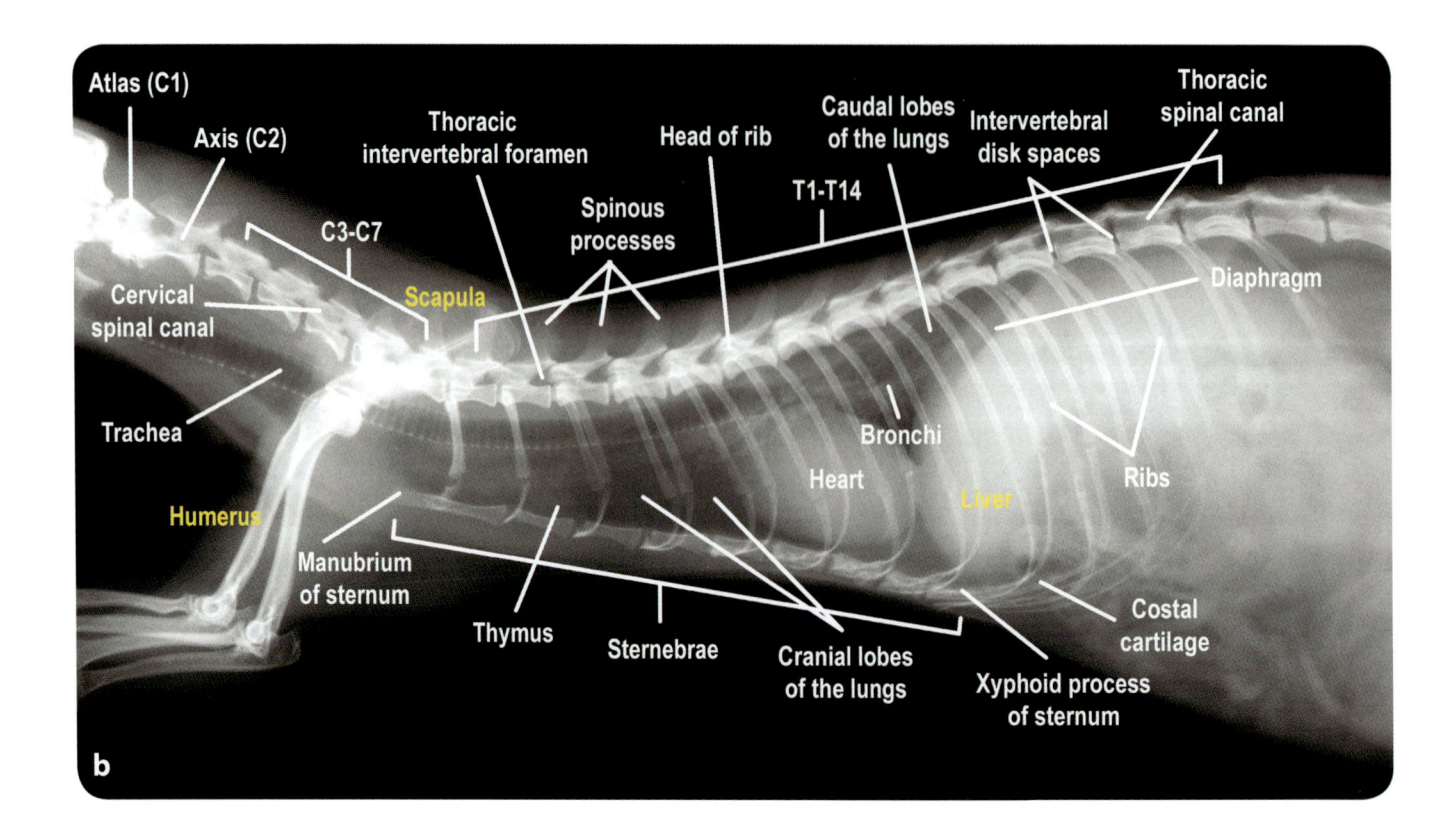

Ventrodorsal Projection

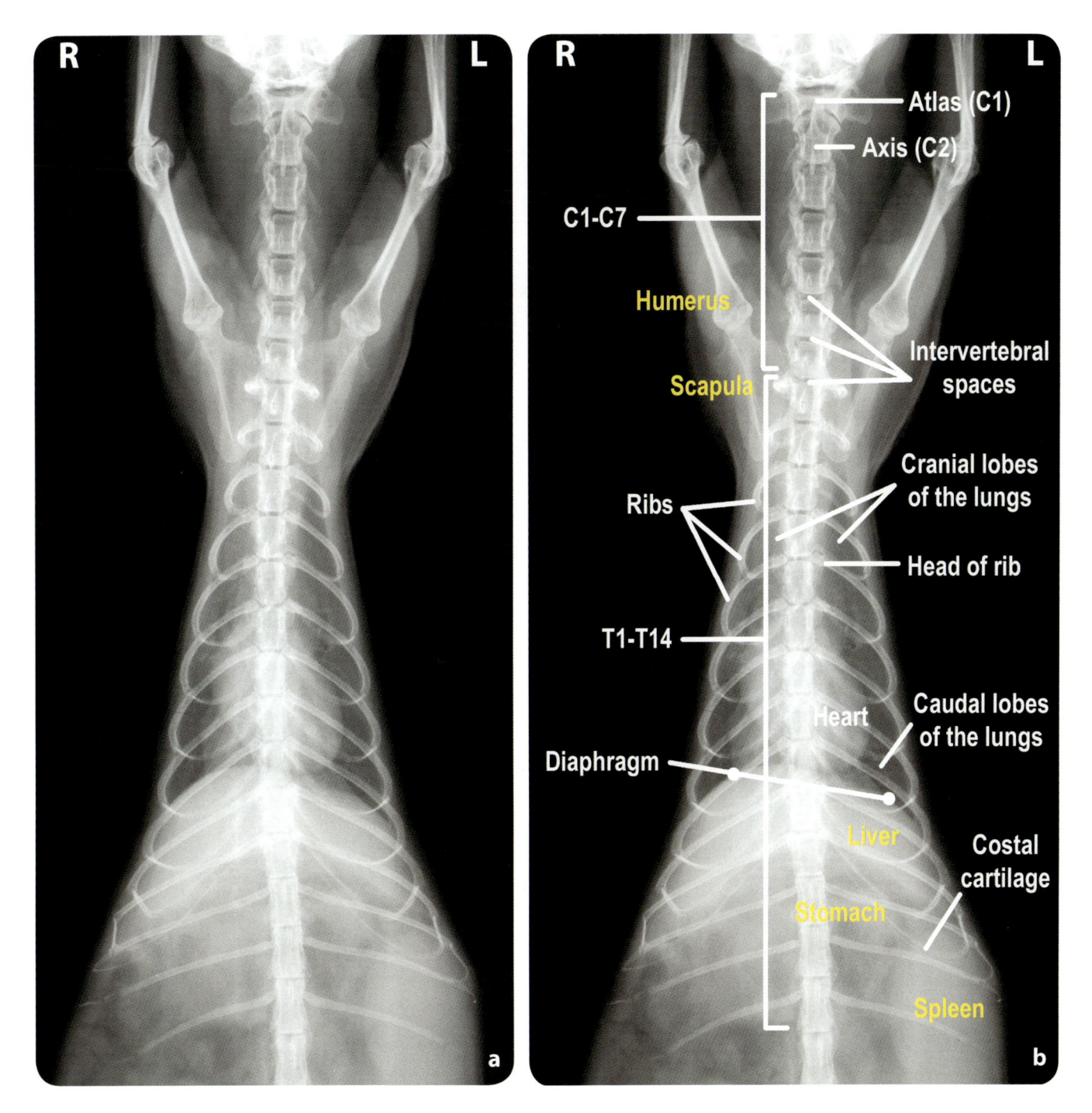

Figure 10.20a,b. Radiograph of the thorax of a 1.1 kg male ferret, 75% of actual size.

ABNORMALITIES of the THORAX
Diseases of the Lungs

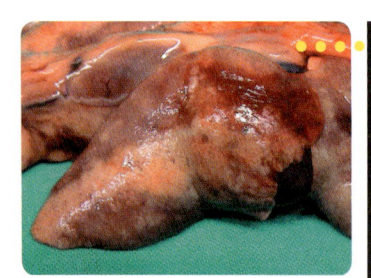

Figure 10.21. Collapsed, consolidated lung lobes at necropsy in the same patient.

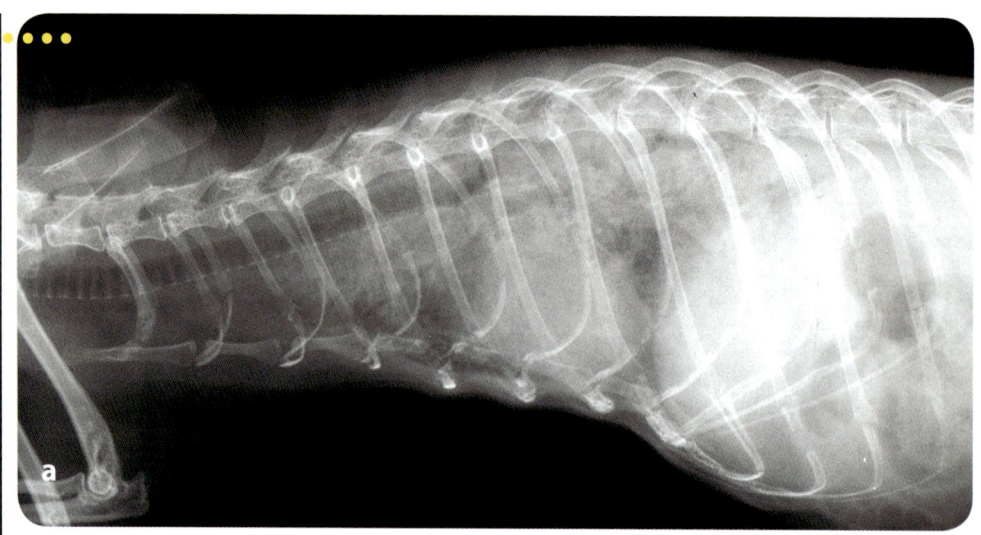

Figure 10.22a,b. Lateral (a) and ventrodorsal (b) radiographs of the thorax of a ferret with severe respiratory distress. Radiographic changes in the thorax are typified by alveolar/interstitial infiltrates. Air bronchograms can be seen on the ventrodorsal projection (arrows). Note partial effacement of the heart.

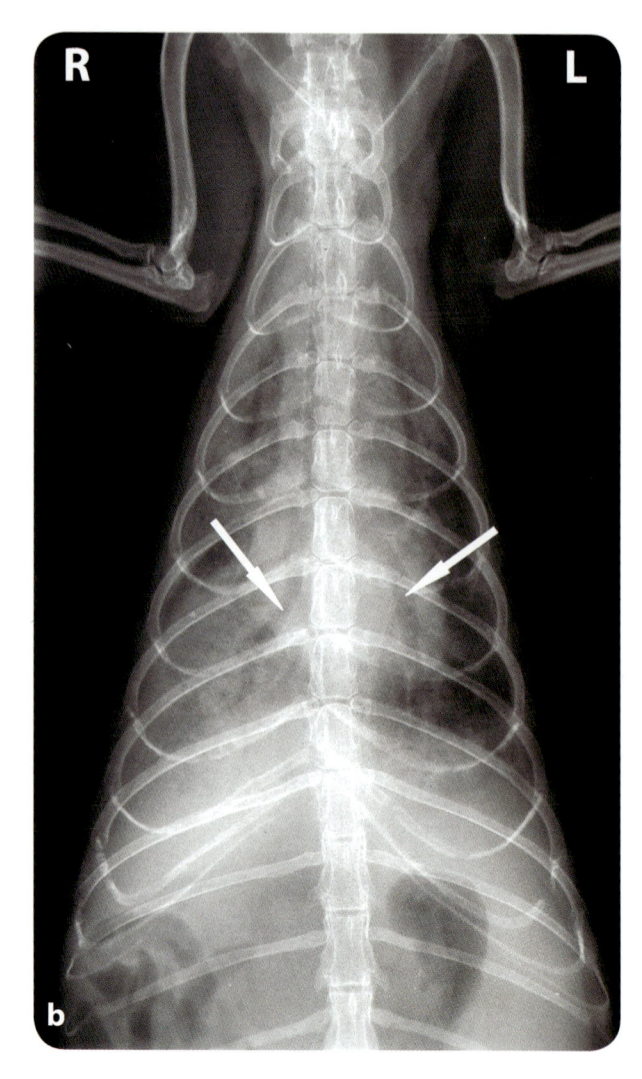

Diseases of the Heart

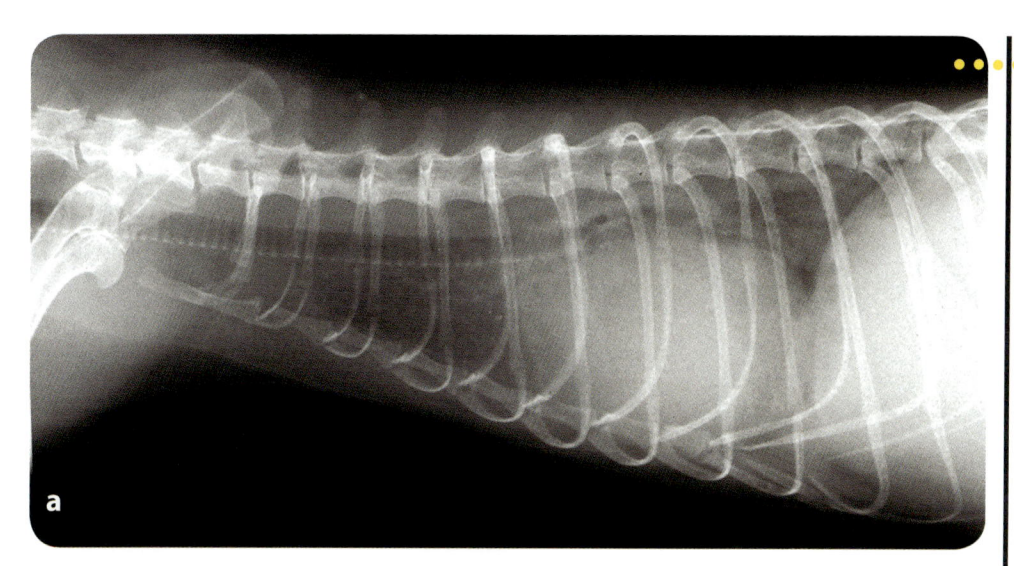

Figure 10.23a,b. Lateral (a) and ventrodorsal (b) radiographs of the thorax of a ferret with dilated cardiomyopathy following stenosis of the aorta. Radiographic changes include generalized cardiomegaly, rounding of the cardiac apex and narrowing of the carina. Mild increased interstitial/alveolar opacity is present.

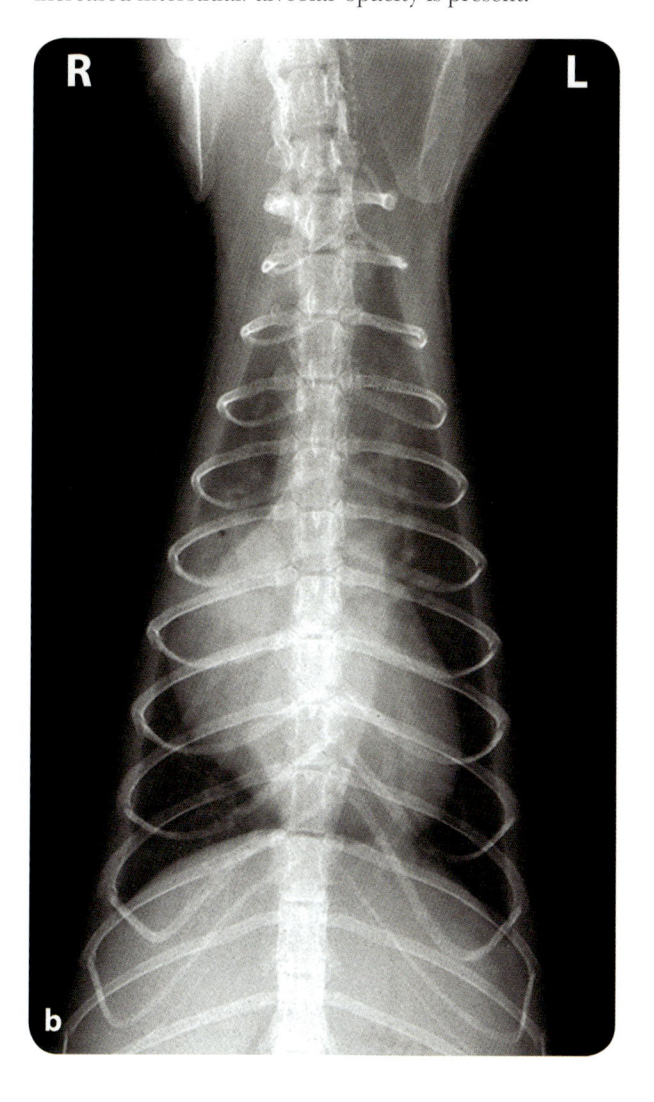

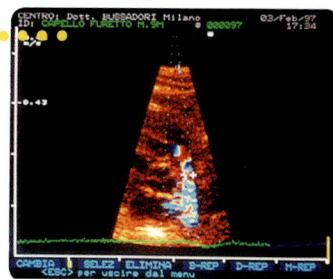

Figure 10.24. Cases of suspected cardiac disease should be followed up with ultrasonography of the thorax.
(Courtesy of Claudio Bussadori, DVM)

Diseases of the Mediastinum

Figure 10.25a,b. Lateral (a) and ventrodorsal (b) radiographs of the thorax of a 2-year-old male ferret with acute respiratory distress due to heartworm disease, diagnosed at necropsy. There is generalized increased opactiy of the thorax that obscures the heart, great vessels, and diaphram, typical of pleural effusion. Evidence of peritoneal effusion is also present.

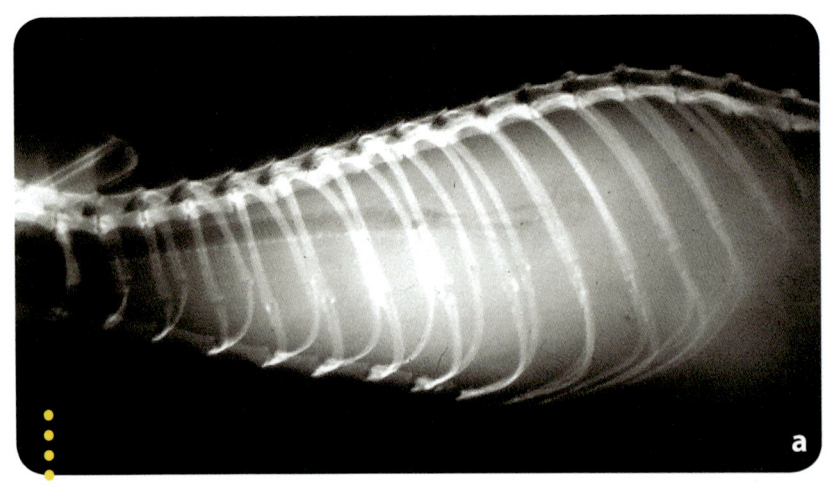

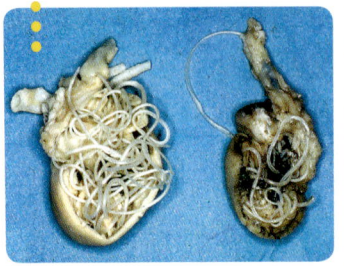

Figure 10.26. Necropsy specimens of 2 ferrets with heartworm disease, male (left) and female (right). Specimens were fixed in formalin and opened, Adult worms are easily recognized in the right atrium and ventricle .
Heartworm disease can potentially affect any carnivore. Clinical presentation in ferrets is typically acute respiratory distress.

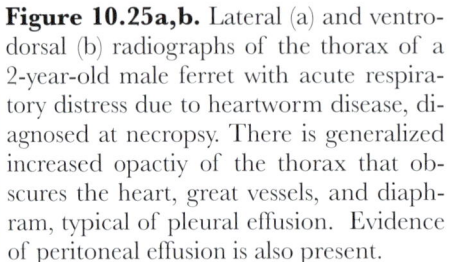

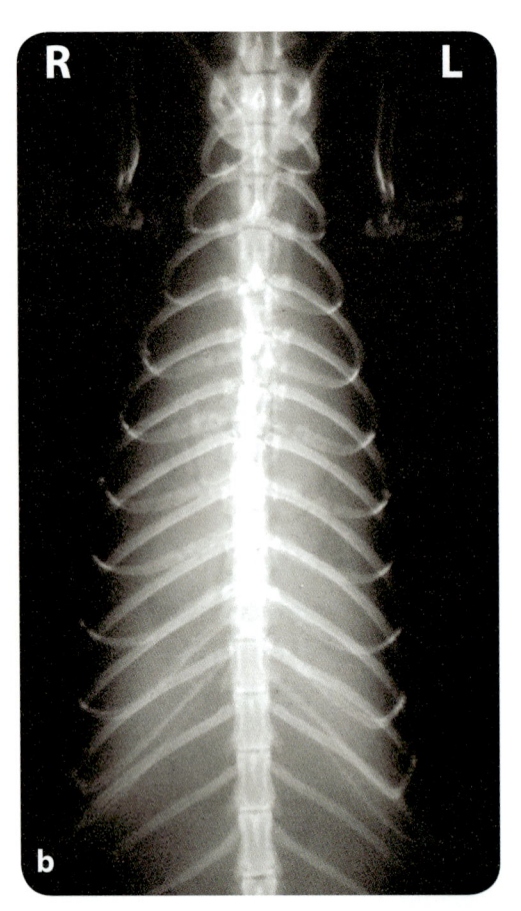

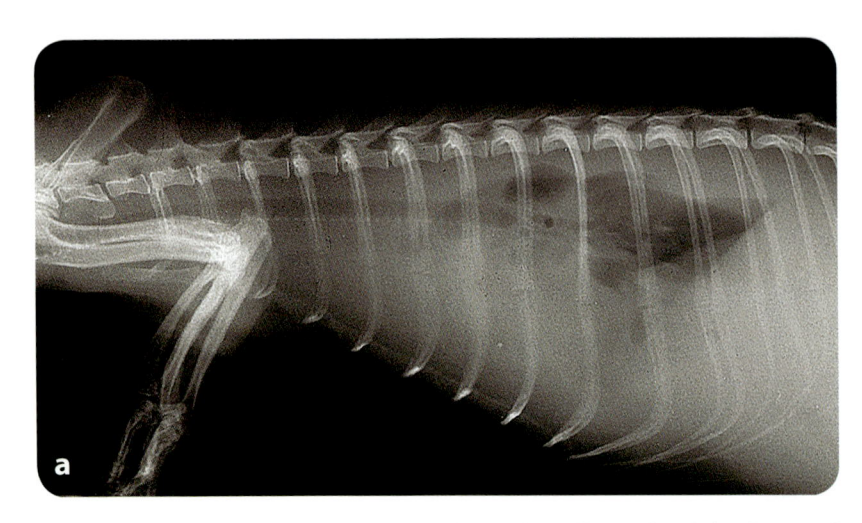

Figure 10.27a,b Lateral (a) and ventrodorsal (b) radiographs of the thorax of a 1-year-old female ferret with pleural effusion secondary to heartworm disease diagnosed at necropsy. The caudal lung lobes are surrounded by fluid opacity, and are partially collapsed. The presenting signs included collapse and respiratory distress. Thoracocentesis was performed and revealed serosanguinous fluid. Note that radiography cannot distinguish types of pleural effusion.

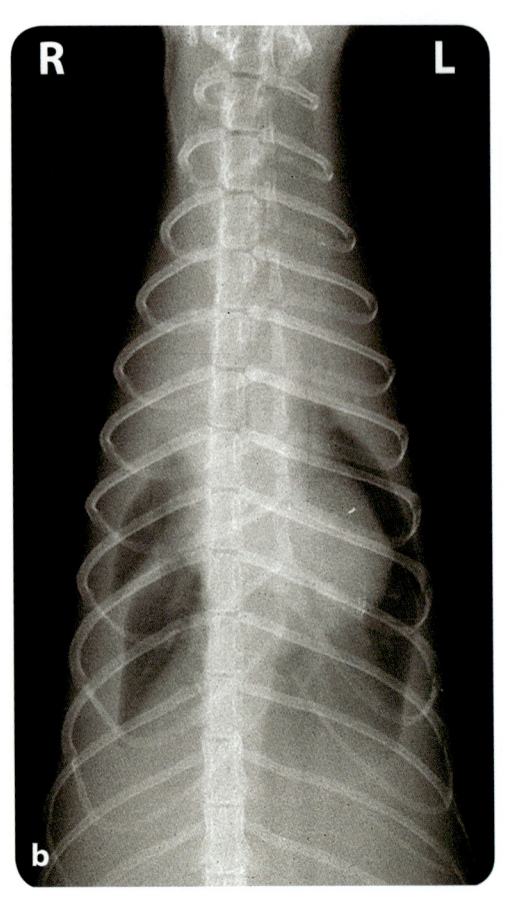

THORAX 371

Figure 10.28. Lateral radiograph of the thorax of a ferret with pleural effusion and mediastinal lymphoma diagnosed at necropsy. Generalized increased thoracic opacity has produced effacement of cardiopulmonary structures. This opactiy is due to both pleural fluid and a mediastinal mass, which have the same radiopacity. Ultrasound or CT should be considered as alternate imaging choices.

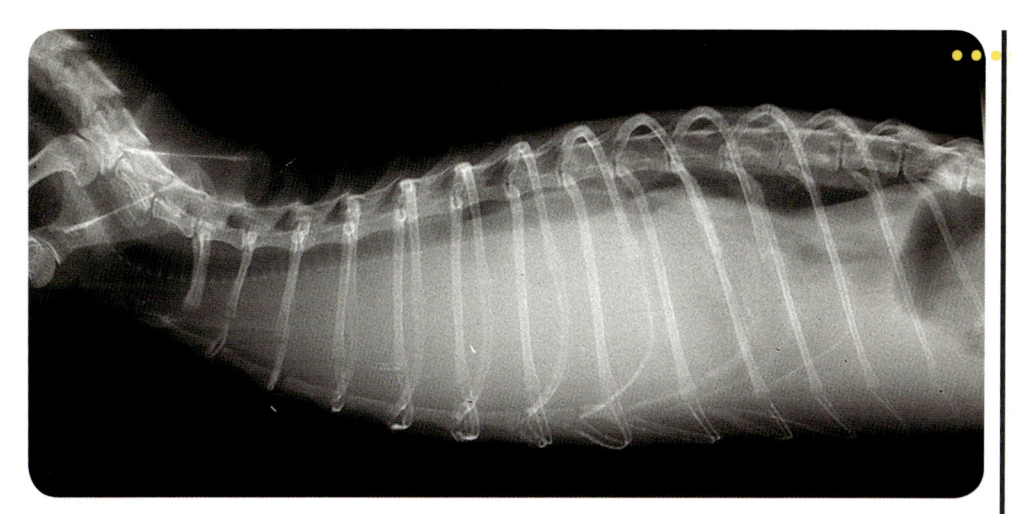

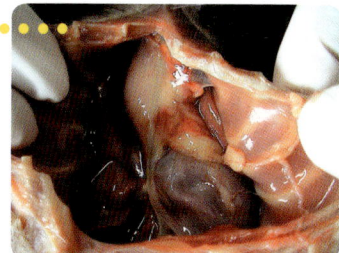

Figure 10.29. Mediastinal mass cranial to the heart in the same patient. Thoracic fluid is also visible.

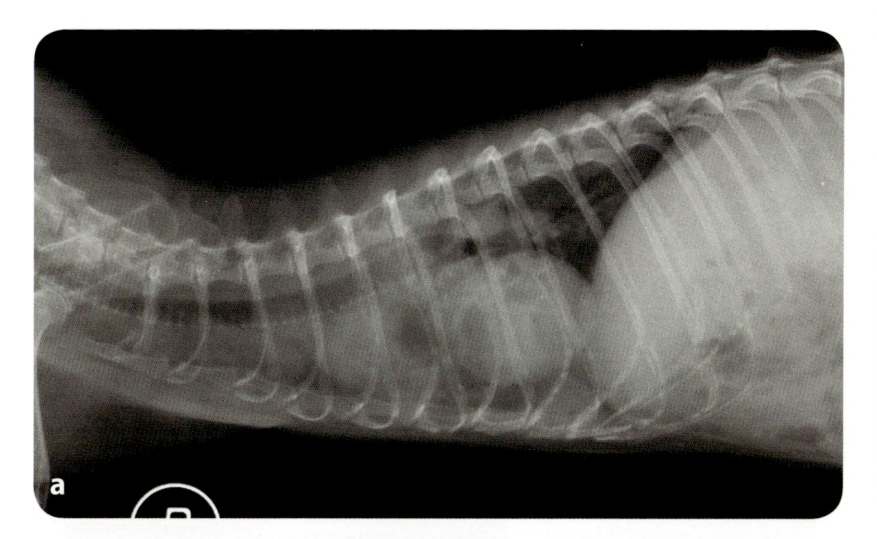

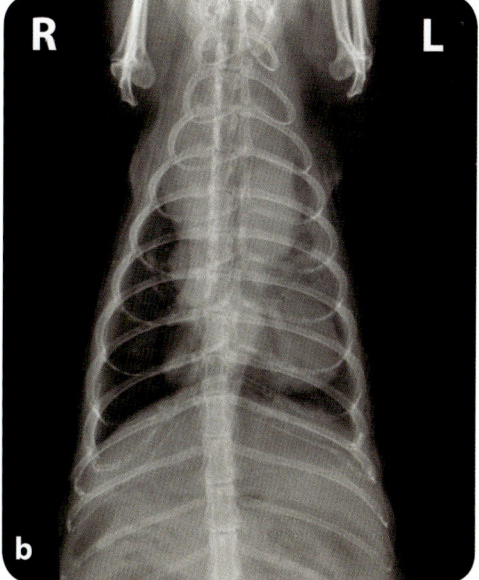

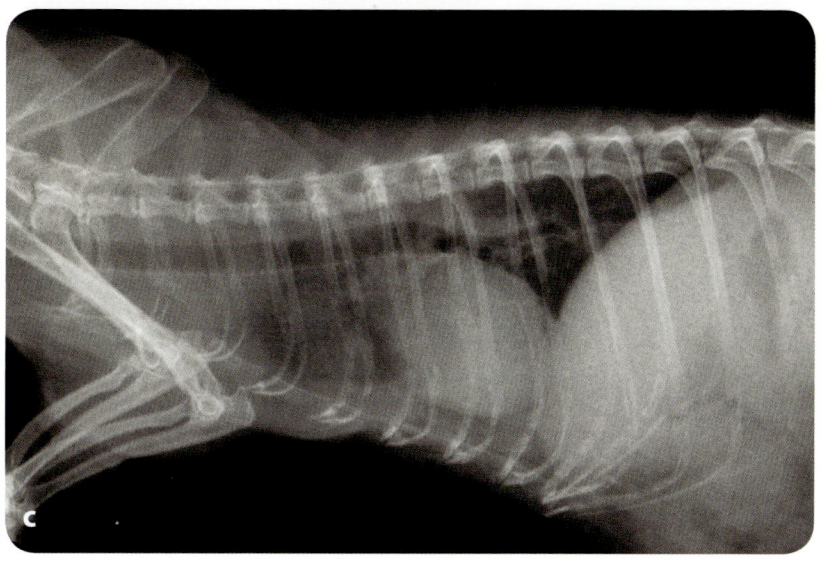

Figure 10.30a,b,c. Lateral (a) and ventrodorsal (b) radiographs of a ferret with a mediastinal mass cranial to the heart diagnosed via biopsy as lymphoma. Lateral (c) radiograph of the same patient after radiation therapy.
(Courtesy of Joerg Mayer, DVM)

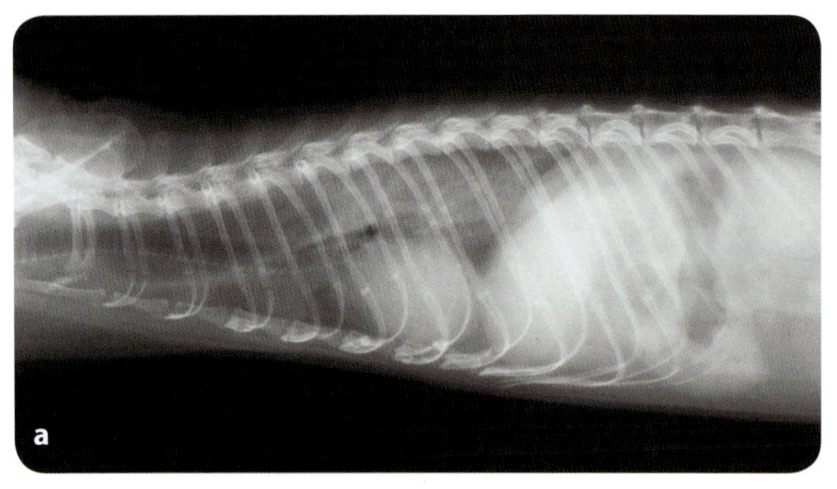

Figure 10.31a,b. Lateral (a) and ventrodorsal (b) radiographs of the thorax of a 6-year-old female ferret presenting for chronic weight loss. Note severe radiopacity of the lung field. Respiratory symptoms were not present. Note deflection of the trachea to the right on the ventrodorsal projection.

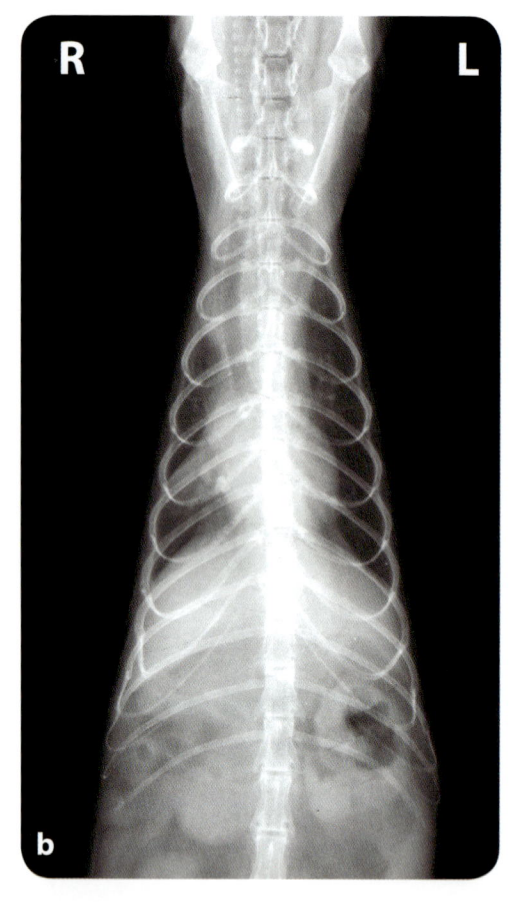

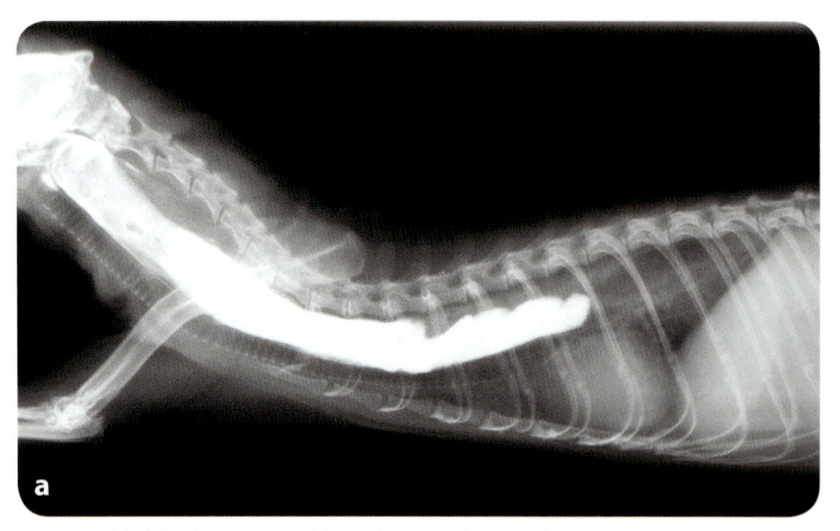

Figure 10.32a,b. Lateral (a) and ventrodorsal (b) radiographs of the same patient taken immediately after administration of positive contrast medium (barium sulfate). Megaesophagus is evident.

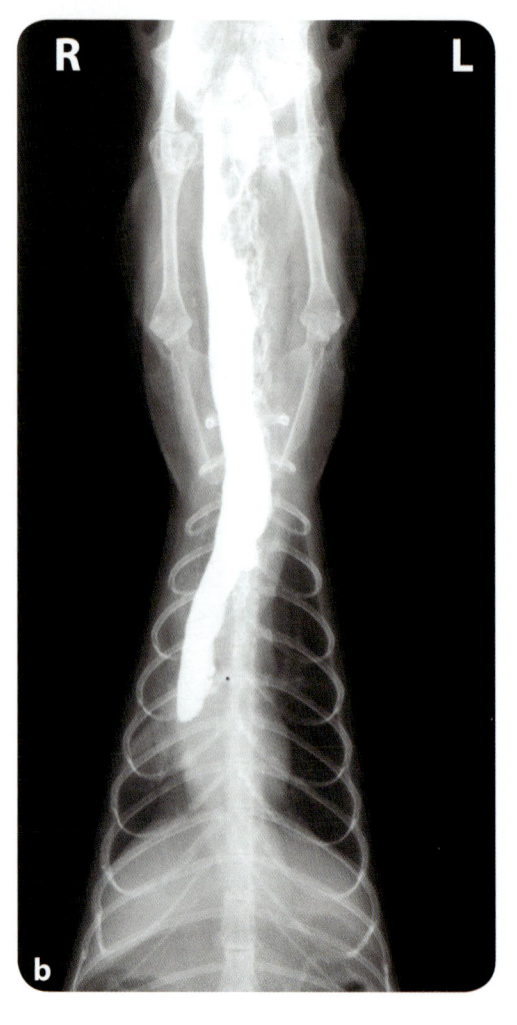

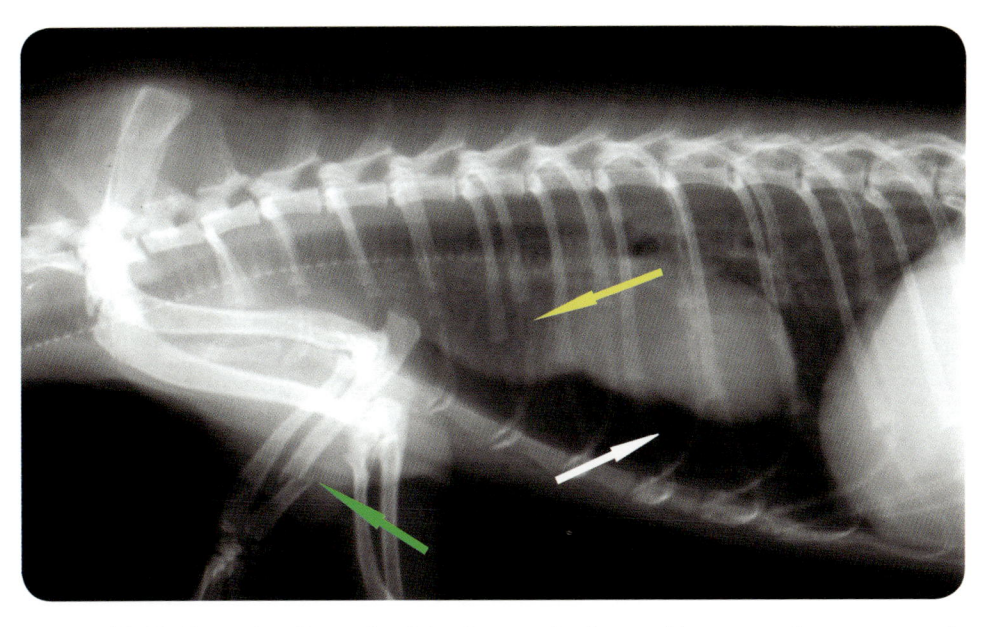

Figure 10.33. Lateral radiograph of the thorax of a ferret with pneumothorax as a result of a fall from a terrace. Air within the pleural space (white arrow) has caused pulmonary collapse (yellow arrow). The heart is deviated dorsally as a result of pulmonary collapse. A distal metaphyseal fracture of the ulna is also visible (green arrow).

Diseases of the Ribs

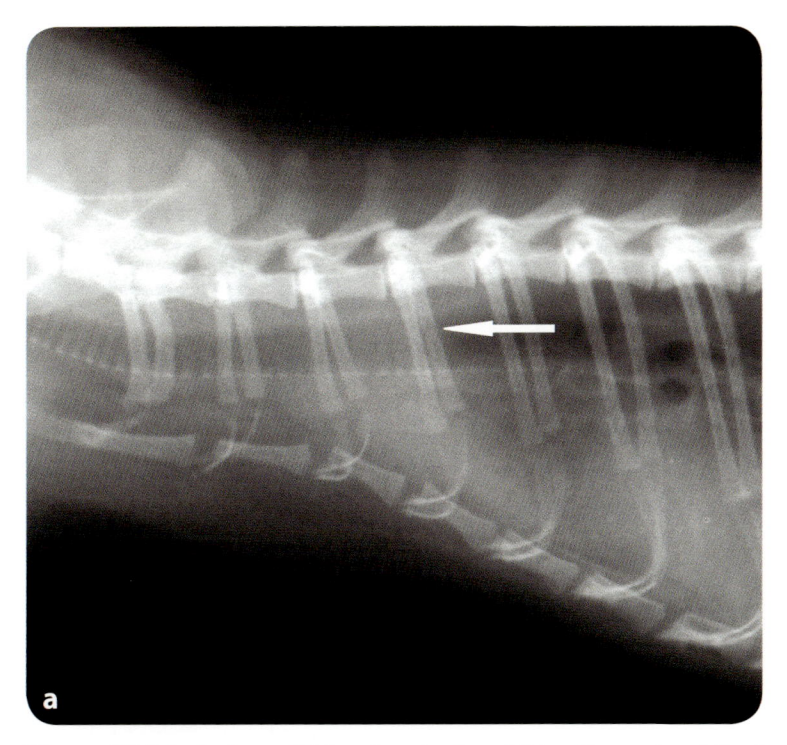

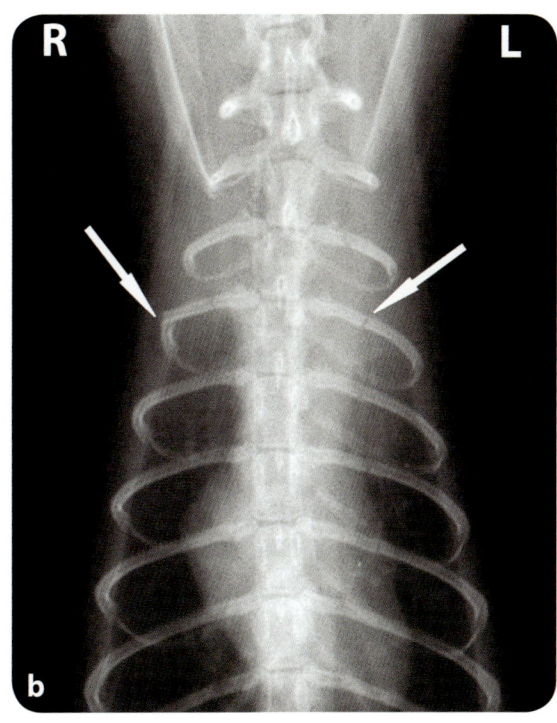

Figure 10.34a,b. Lateral (a) and ventrodorsal (b) radiographs of the thorax of a ferret with a fracture of the 4th rib pair. In the lateral projection (a) a single fracture line is visible (arrow), but the ventrodorsal projection (b) demonstrates bilateral fracture of both ribs. Note pulmonary or mediastinal lesions cranial to the fracture, which could represent a contusion or hemorrhage.

The NORMAL ABDOMEN
The Lumbar Vertebral Column
Lateral Projection

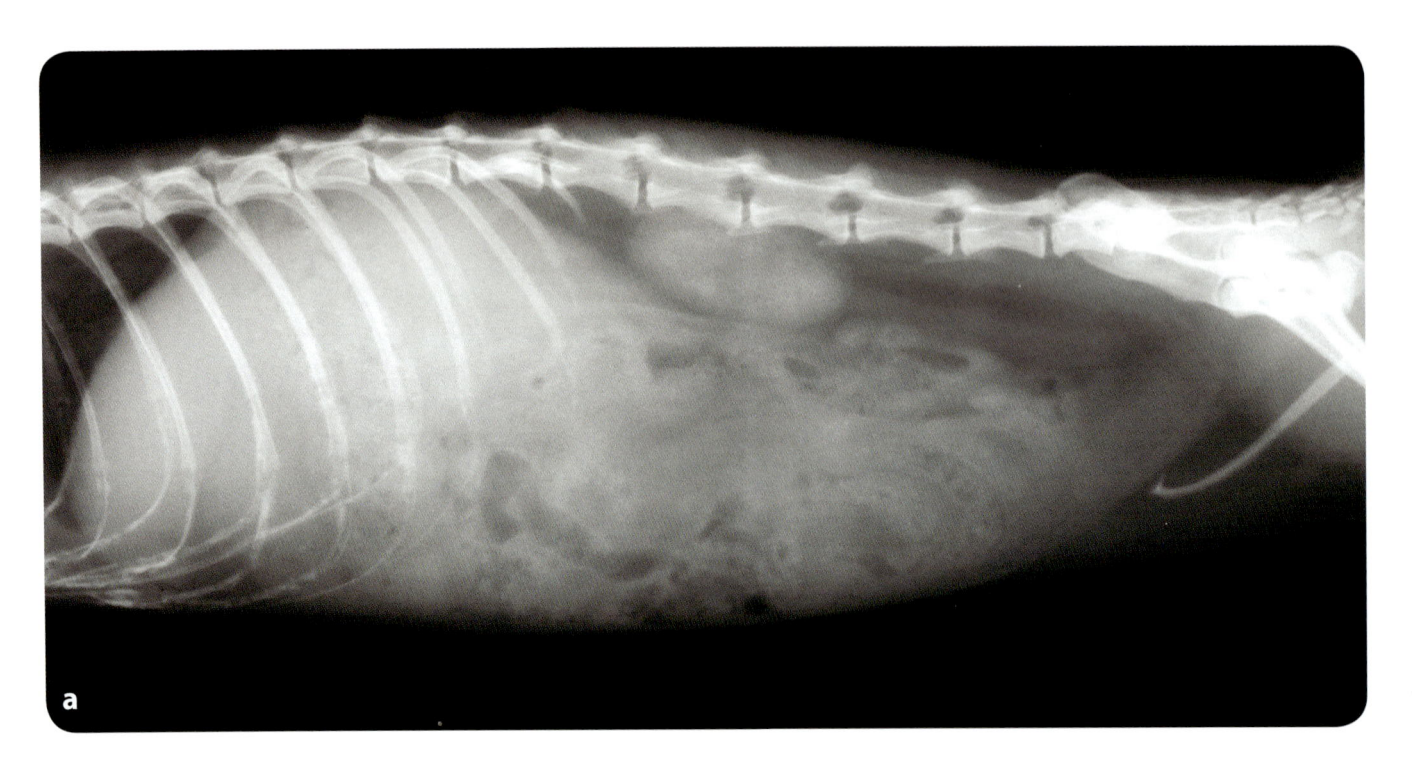

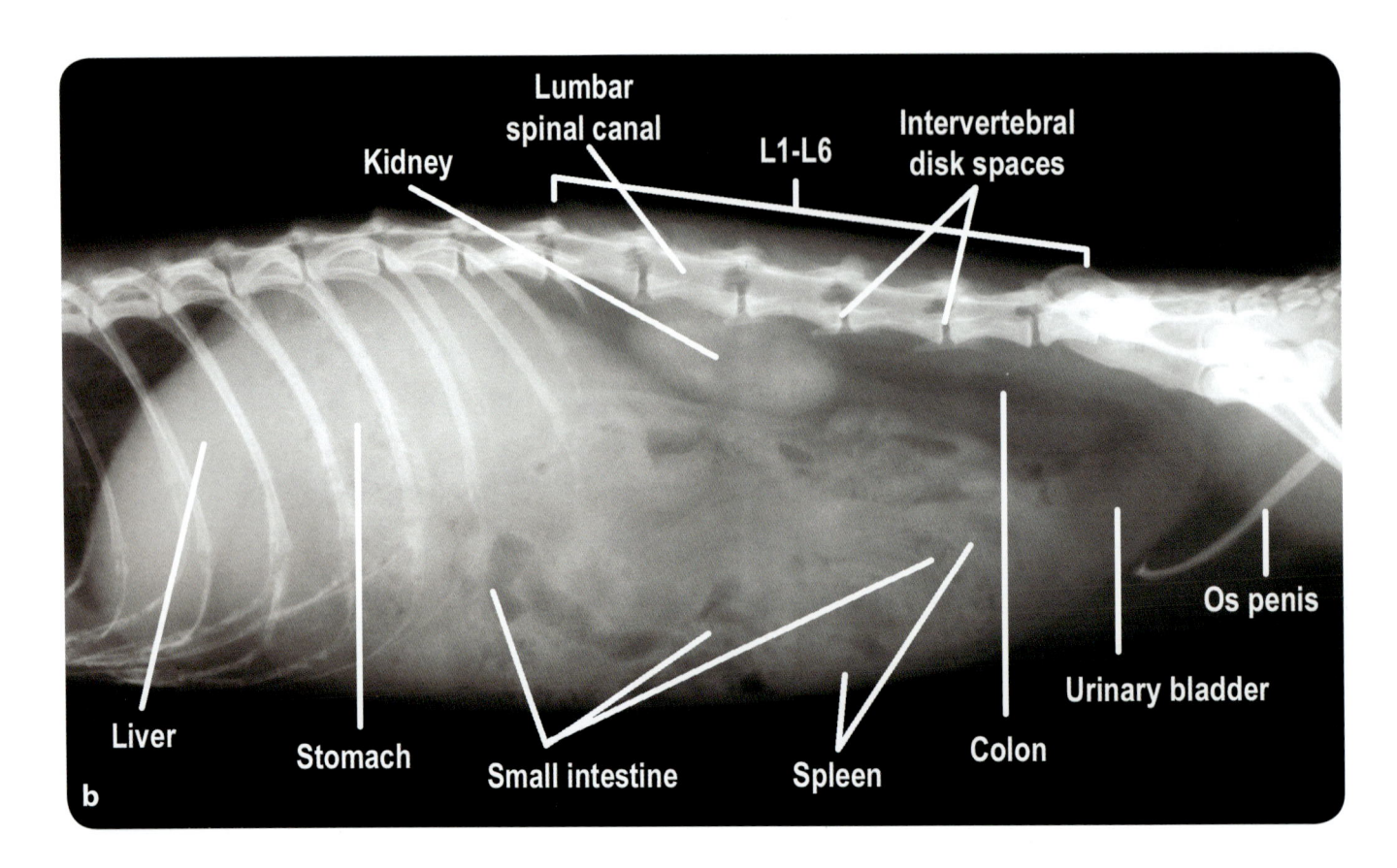

Figure 10.35a,b. Radiograph of the abdomen of a 1.1 kg male ferret, 90% of actual size.

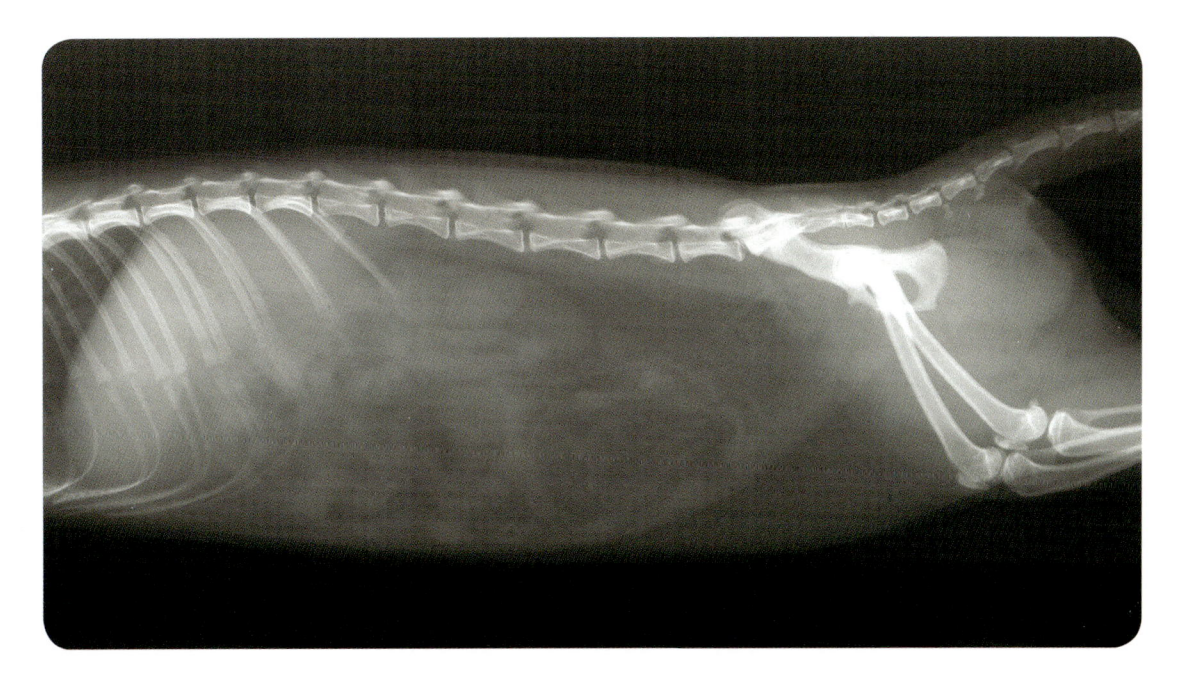

Figure 10.36. Radiograph of the abdomen of a 750 g female ferret, 90% of actual size.

Ventrodorsal Projection

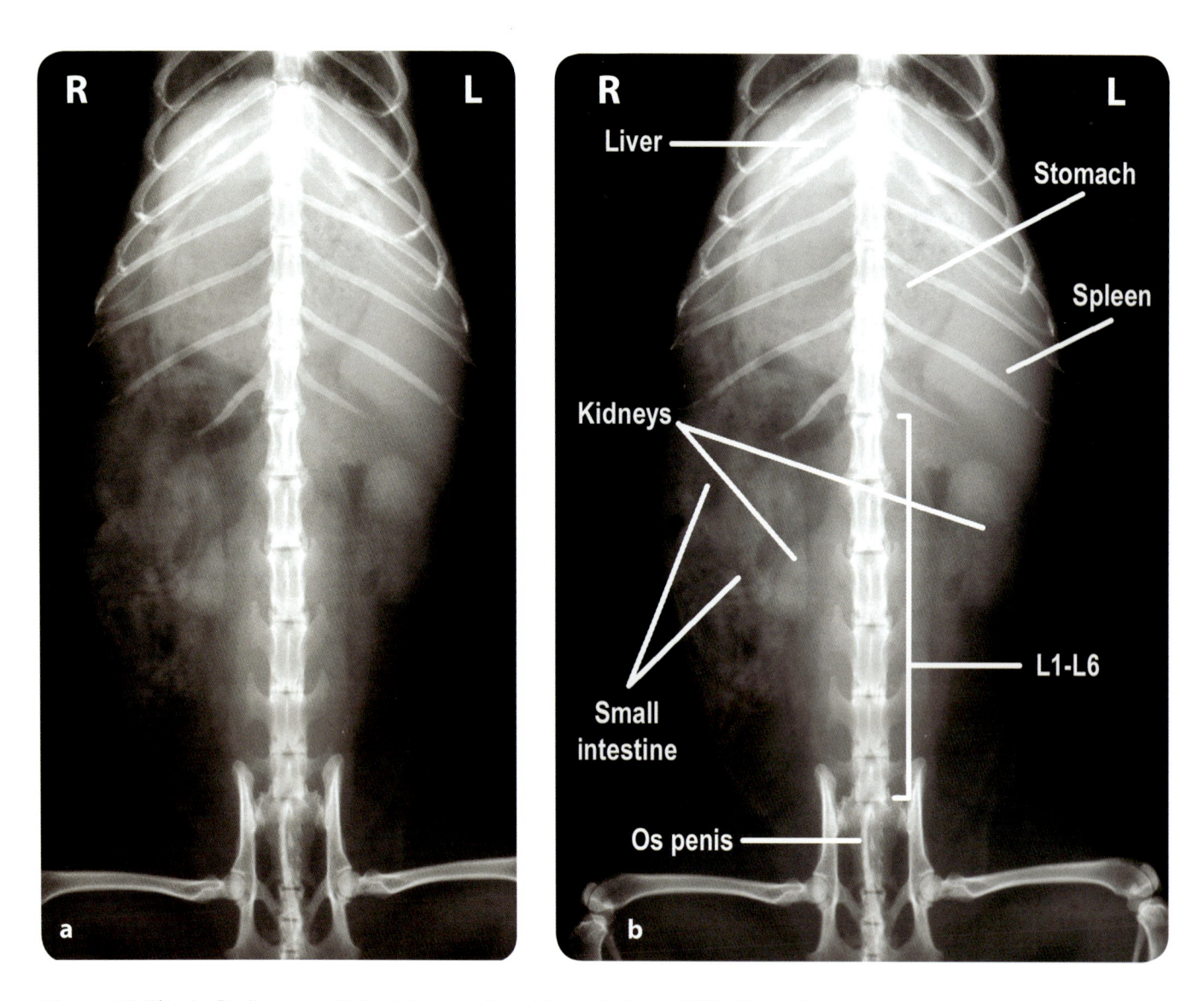

Figure 10.37a,b. Radiograph of the abdomen of a 1.1 kg male ferret, 70% of actual size.

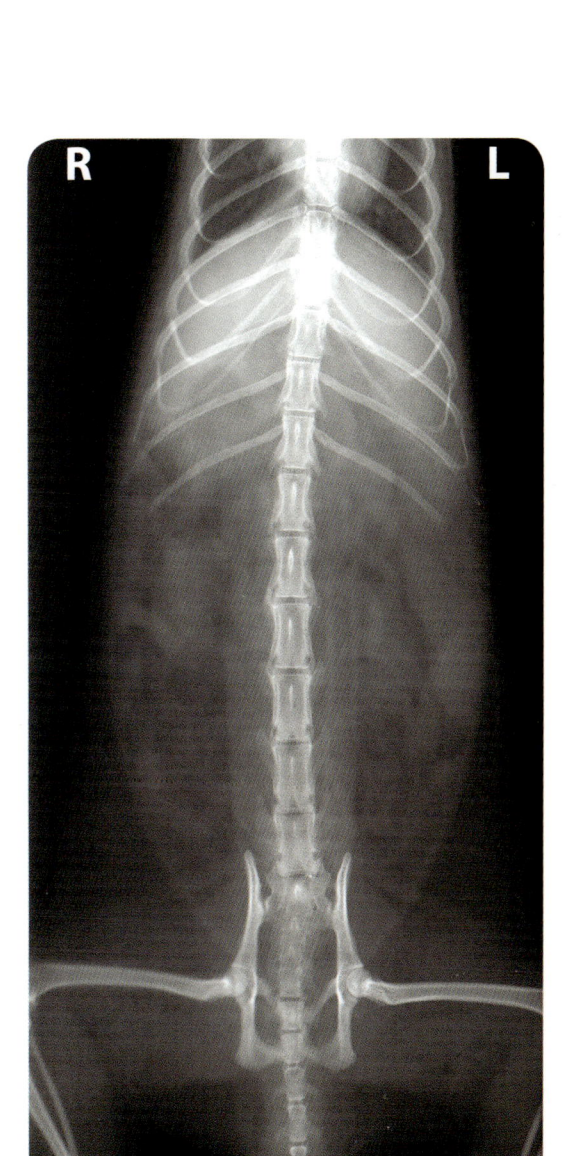

Figure 10.38. Radiograph of the abdomen of a 750 g female ferret, 90% of actual size.

ABNORMALITIES of the ABDOMEN
Diseases of the Stomach

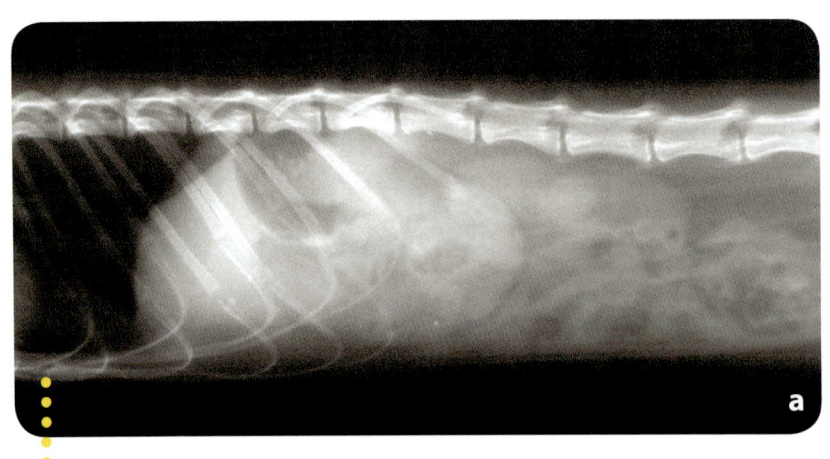

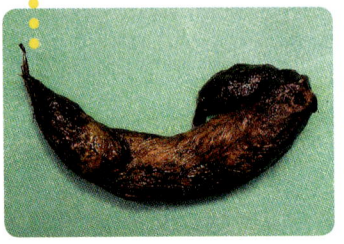

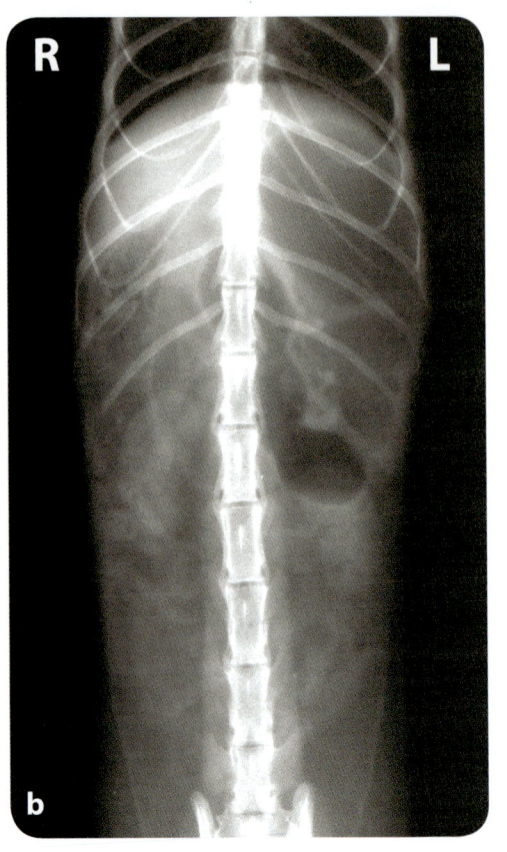

Figure 10.39a,b. Lateral (a) and ventrodorsal (b) radiographs of a ferret with reduced food intake and activity. The stomach is enlarged, contains fluid, excessive gas, and a solid soft tissue opactiy.

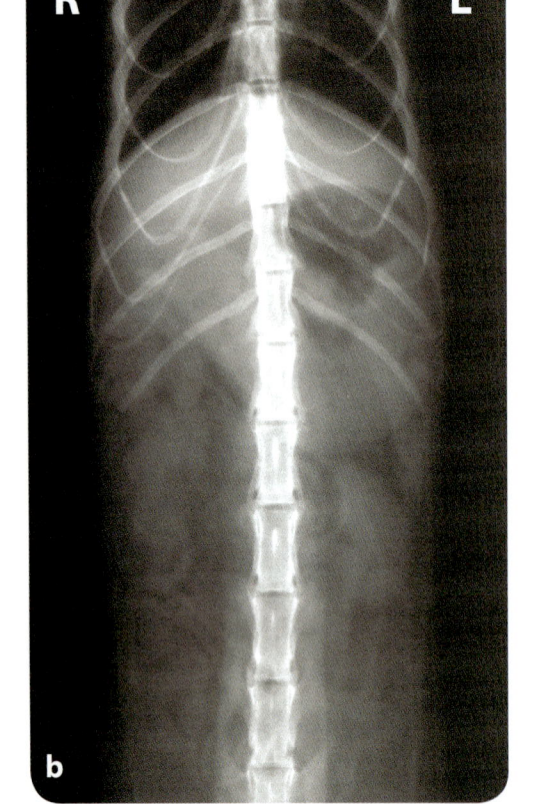

Figure 10.40. Appearance of a gastric trichobezoar of ferrets, which is often a typical "comma-shaped" mass of hair covered with mucus.
Gastric trichobezoars are relatively common in ferrets, and are likely secondary to other gastrointestinal diseases.
They can remain in the stomach for long periods of time without causing clinical signs. Usually the onset of symptoms occurs when the "tail" of the bezoar enters the pylorus and the proximal duodenum.

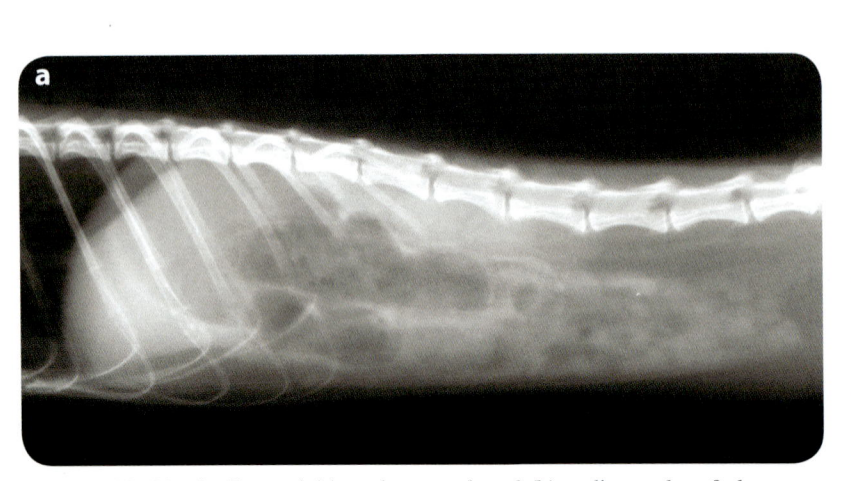

Figure 10.41a,b. Lateral (a) and ventrodorsal (b) radiographs of the same patient 24 hours later. Gastric dilation is indicative of a pyloric outflow obstruction, which in this ferret was determined at surgery to be a trichobezoar.

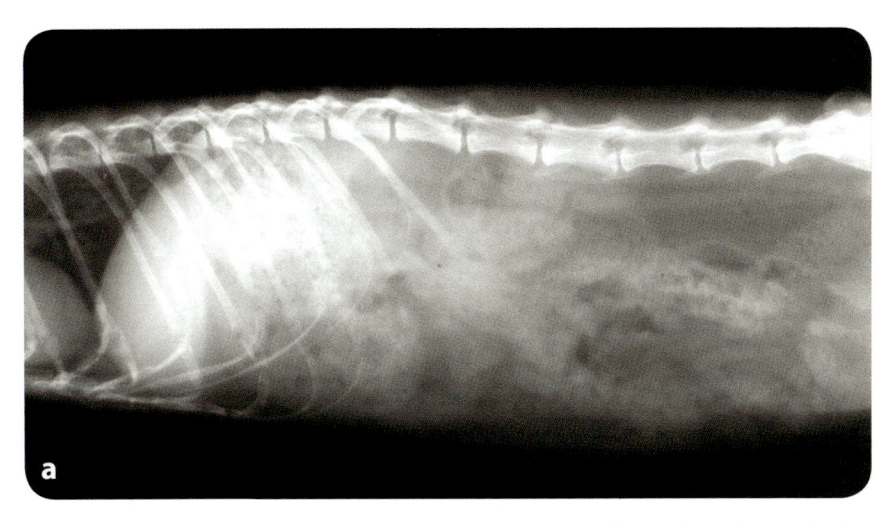

Figure 10.42a,b. Lateral (a) and ventrodorsal (b) radiographs of a male ferret with anorexia and dehydration. The stomach is distended with heterogenous radiopaque material. Distention of the colon is also present, and is an unusual radiographic finding in the ferret.

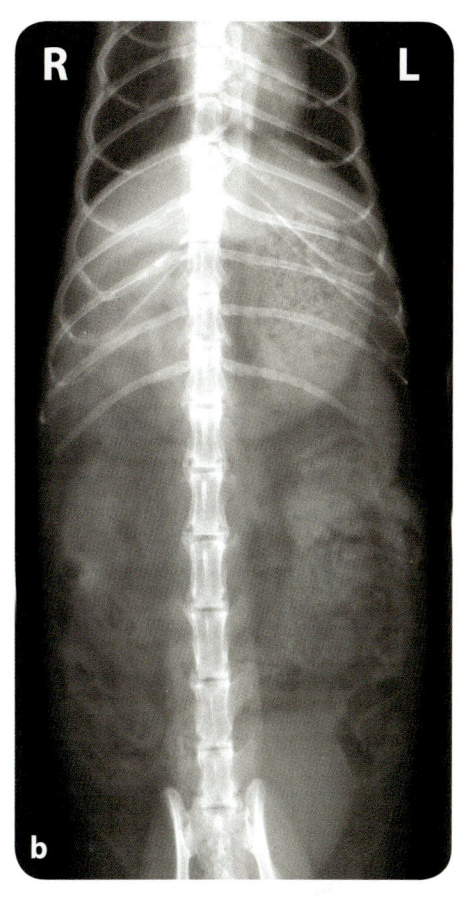

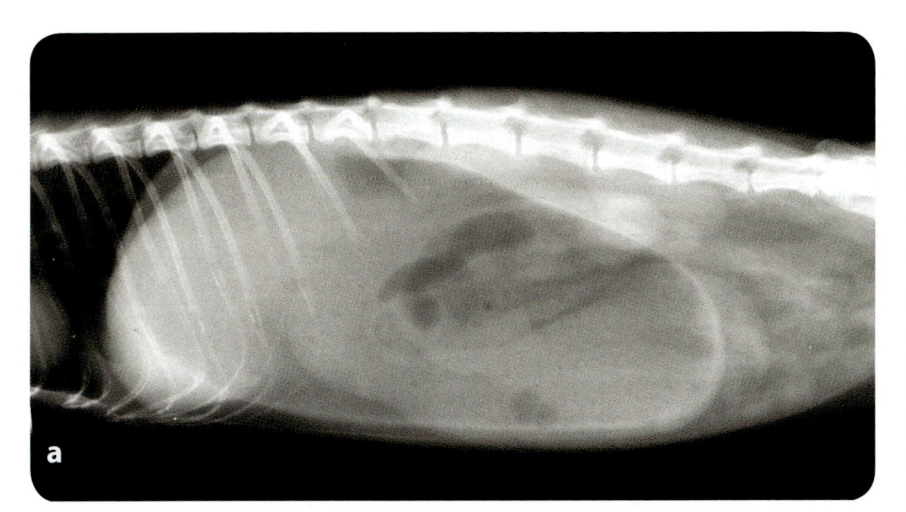

Figure 10.43a,b. Lateral (a) and ventrodorsal (b) radiographs of the abdomen of a ferret with marked gastric dilation resulting from foreign body obstruction of the small intestine. This radiographic appearance is similar to gastric dilation and volvulus (GDV) in dogs, but lacks a compartmentalization sign typical of a 180 degree rotation of the fundus and pylorus. GDV has not been reported in ferrets.

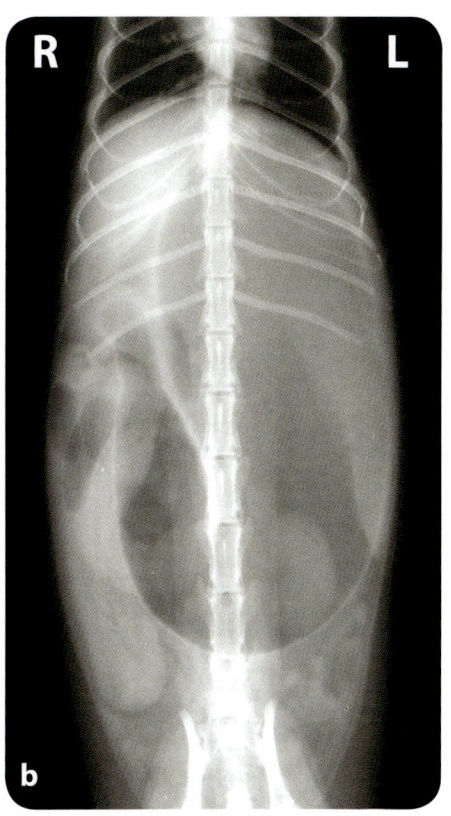

Diseases of the Intestine

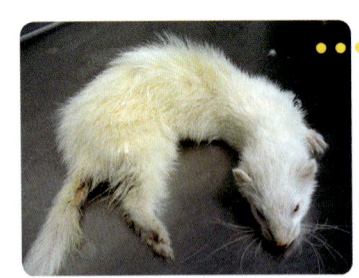

Figure 10.44. Intestinal foreign bodies are very common, especially in younger ferrets. Patients usually present weak and dehydrated, but unlike dogs and cats, vomiting is infrequently reported. However, diarrhea may be present.
Physical examination often reveals a palpable foreign body and/or associated intestinal discomfort.

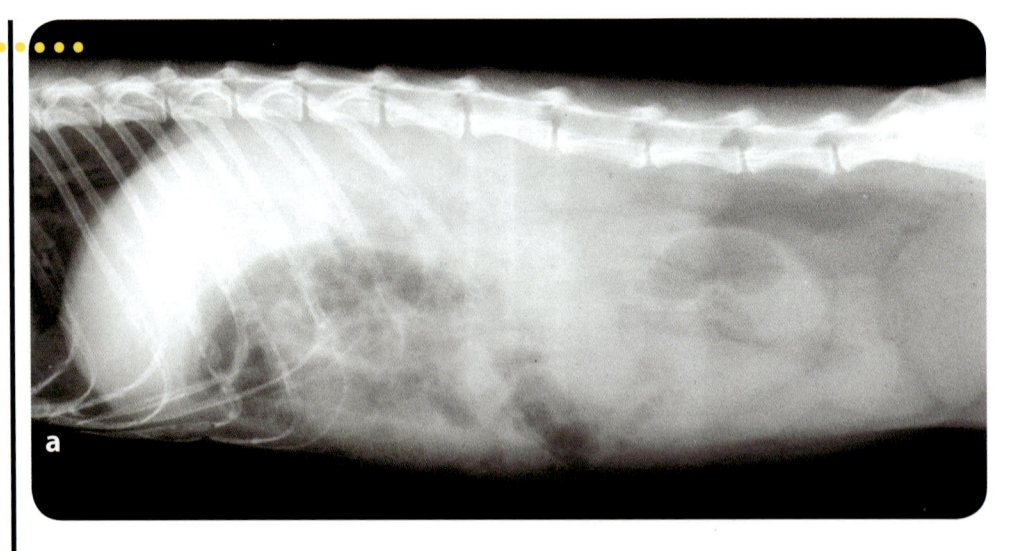

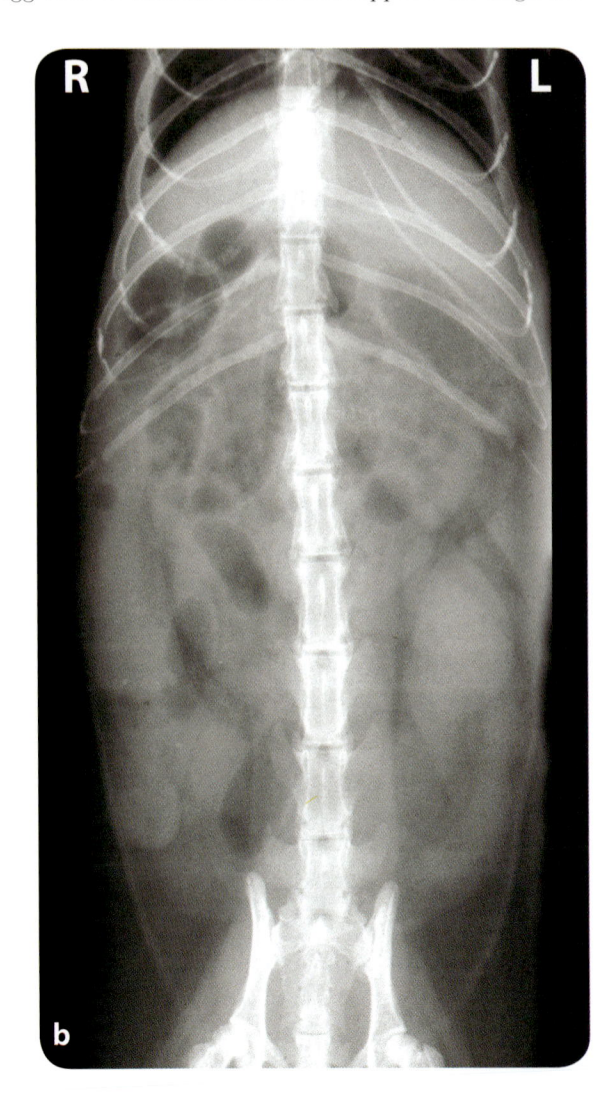

Figure 10.45a,b. Lateral (a) and ventrodorsal (b) radiogaphs of the abdomen of a young ferret with anorexia, depression, and dehydration. There are a number of abnormally gas-filled small intestinal loops. Some of the dilated loops contain chyme. Radiologic identification of a foreign body depends on size and composition but many are radiolucent. Diagnosis is usuallly made on the basis of history and physical examination findings. Presence of gas-distended small intestine is highly suggestive of obstructive ileus and supports the diagnosis.

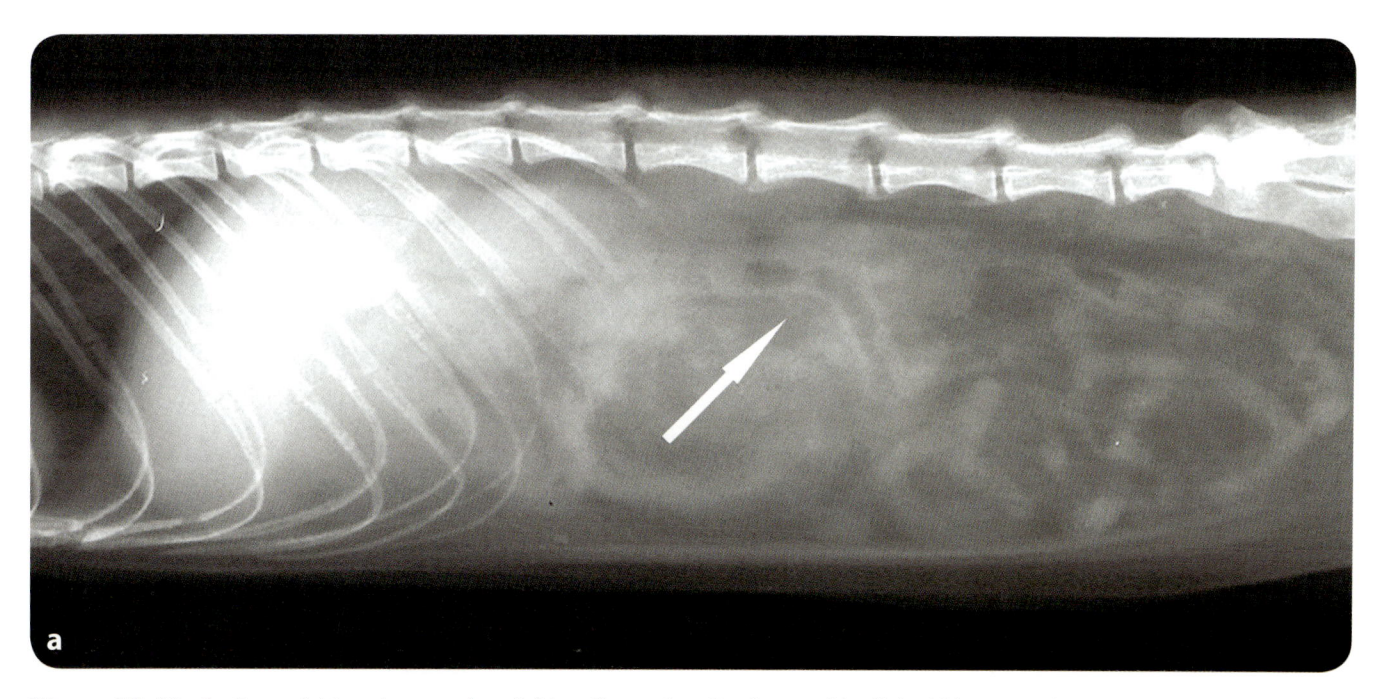

Figure 10.46a,b. Lateral (a) and ventrodorsal (b) radiographs of a ferret with clinical history and presentation suggestive of a foreign body. In most cases it is difficult to directly visualize the foreign body radiographically. In (a,b), the intestinal loop indicated by the white arrow is filled with gas, and should be viewed with suspicion.

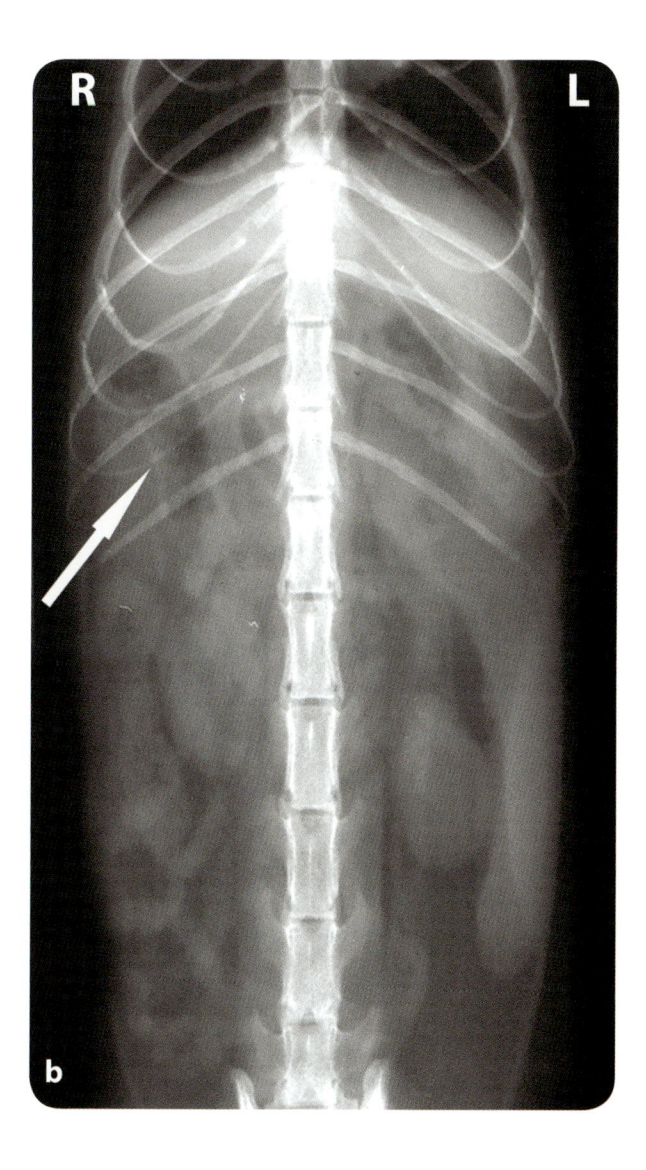

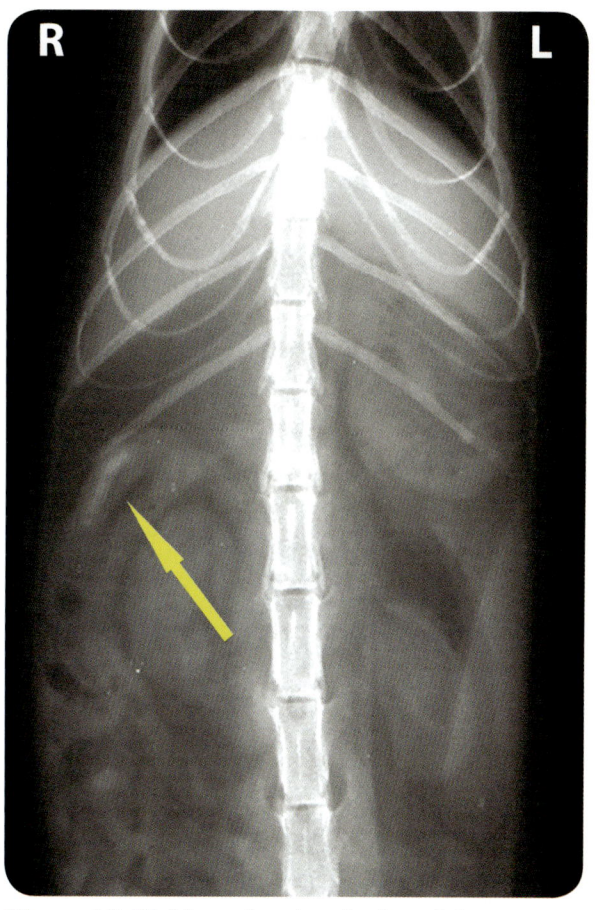

Figure 10.47. Ventrodorsal projection of the same patient 24 hours later, after supportive care and worsening of clinical signs. A small radiodense foreign body is clearly visible in the same area (yellow arrow). The material was determined at surgery to be a piece of rubber.

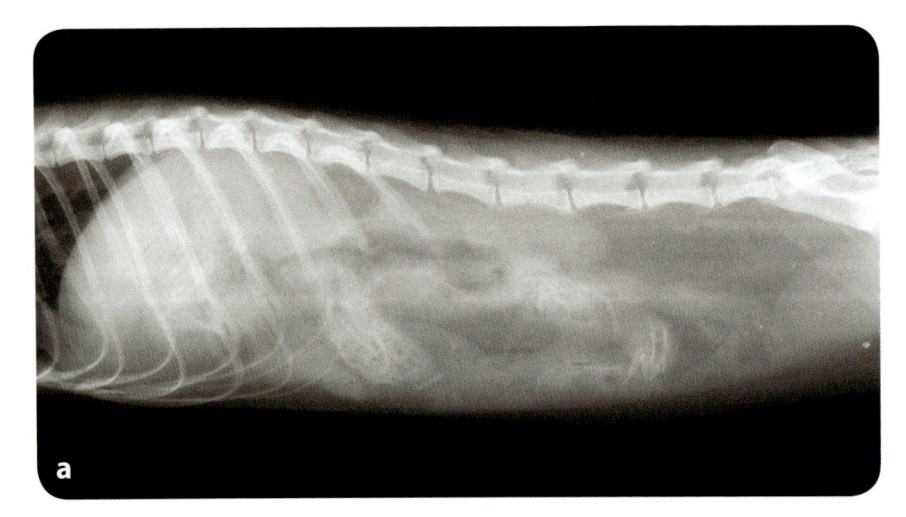

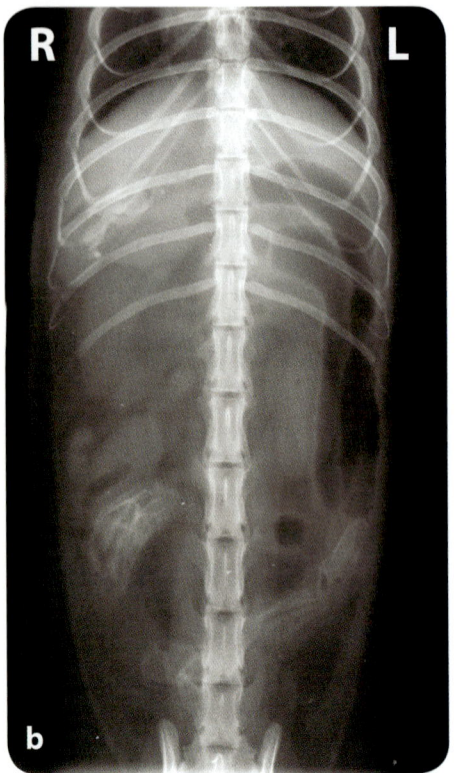

Figure 10.48a,b. Lateral (a) and ventrodorsal (b) radigoraphs of the abdomen of a young ferret that was observed to chew and swallow a rubber glove. In this case, the gastrointestinal foreign bodies are radiopaque and associated with gastric and intestinal gas.

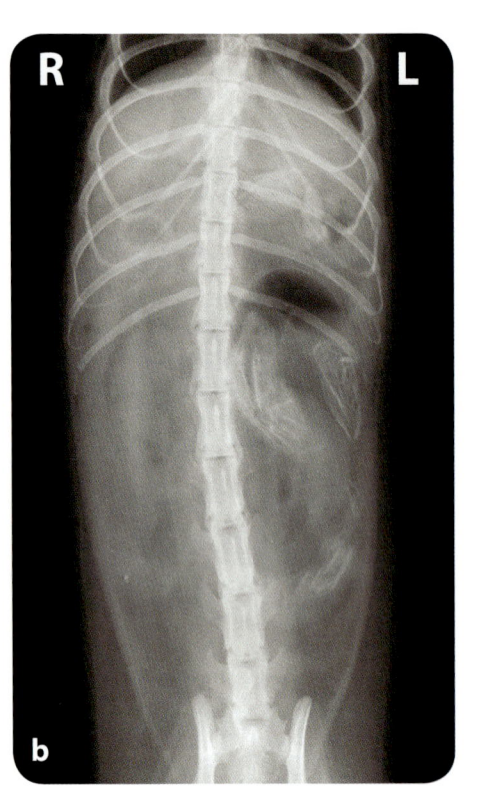

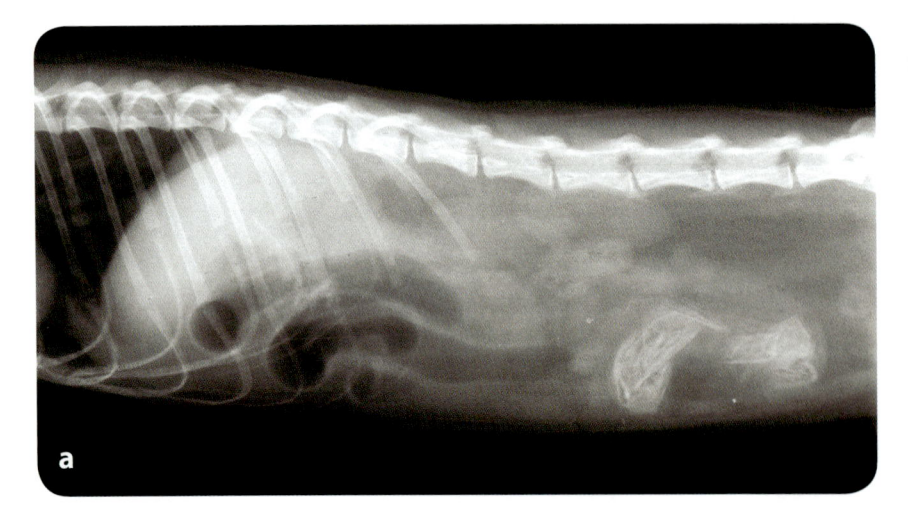

Figure 10.49a,b. Lateral (a) and ventrodorsal (b) radiographs of the same patient as in Figure 10.48, 24 hours later. The owner had elected to attempt supportive therapy in hopes the foreign material would pass naturally. Instead, clinical condition worsened. Note increased accumulation of intestinal gas secondary to obstuction. The ferret underwent surgery for removal of the rubber foreign bodies and recovered uneventfully.

Figure 10.51. Firm stool in the same patient, passed after medical therapy including fluids, laxatives, and gastrointestinal motility enhancing drugs. It is uncertain why this ferret ingested cat litter.

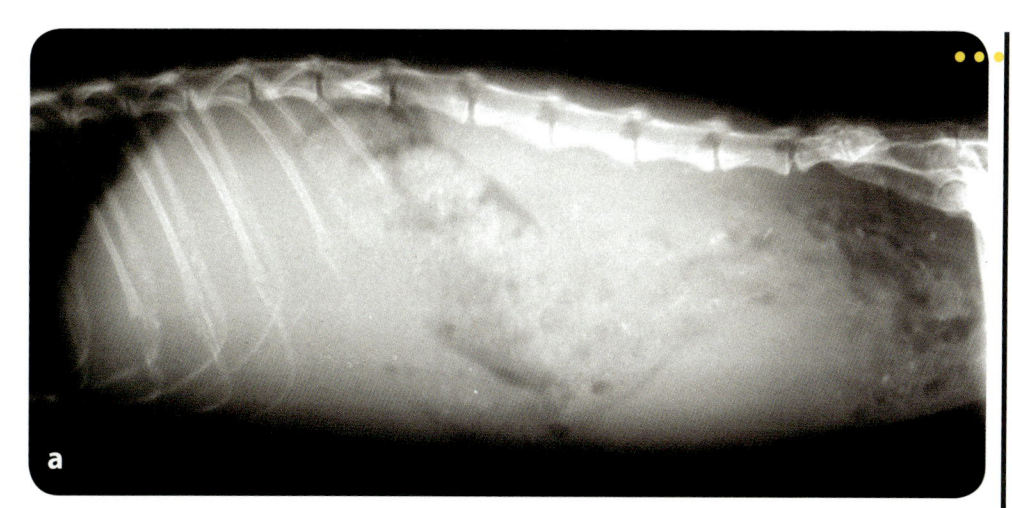

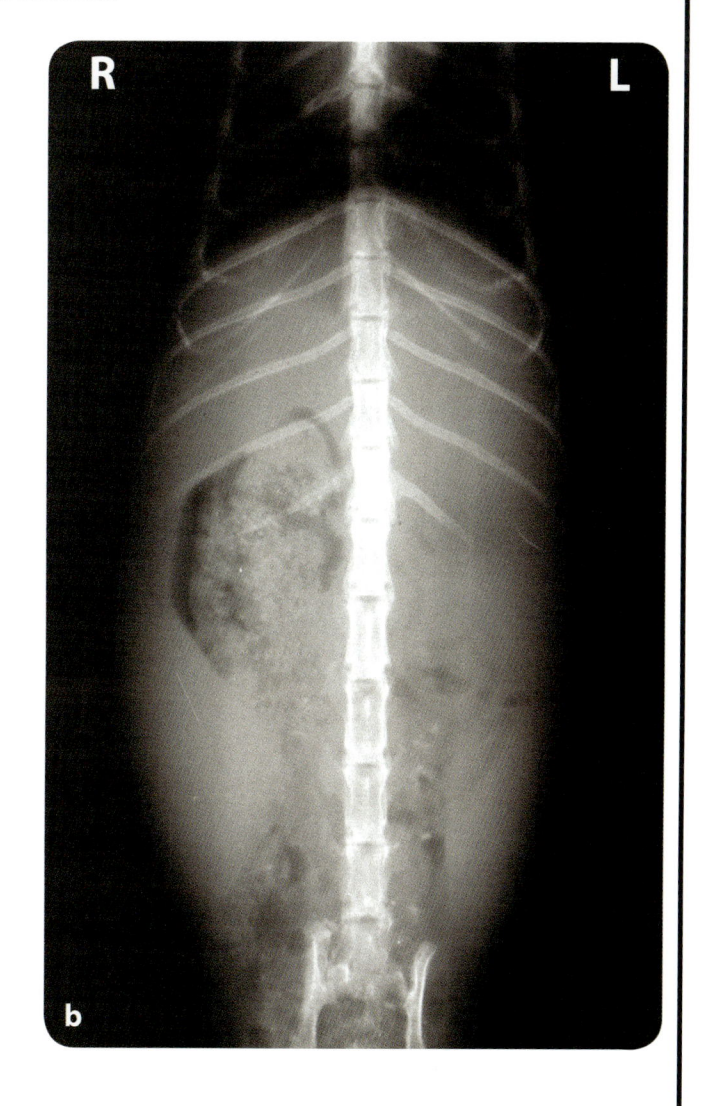

Figure 10.50a,b. Lateral (a) and ventrodorsal (b) raadiographs of the abdomen of a ferret with severe distention and impaction of the large intestine following ingestion of cat litter. The colon is filled with granular radiopaque material.
This condition is uncommon in ferrets.

Diseases of the Liver

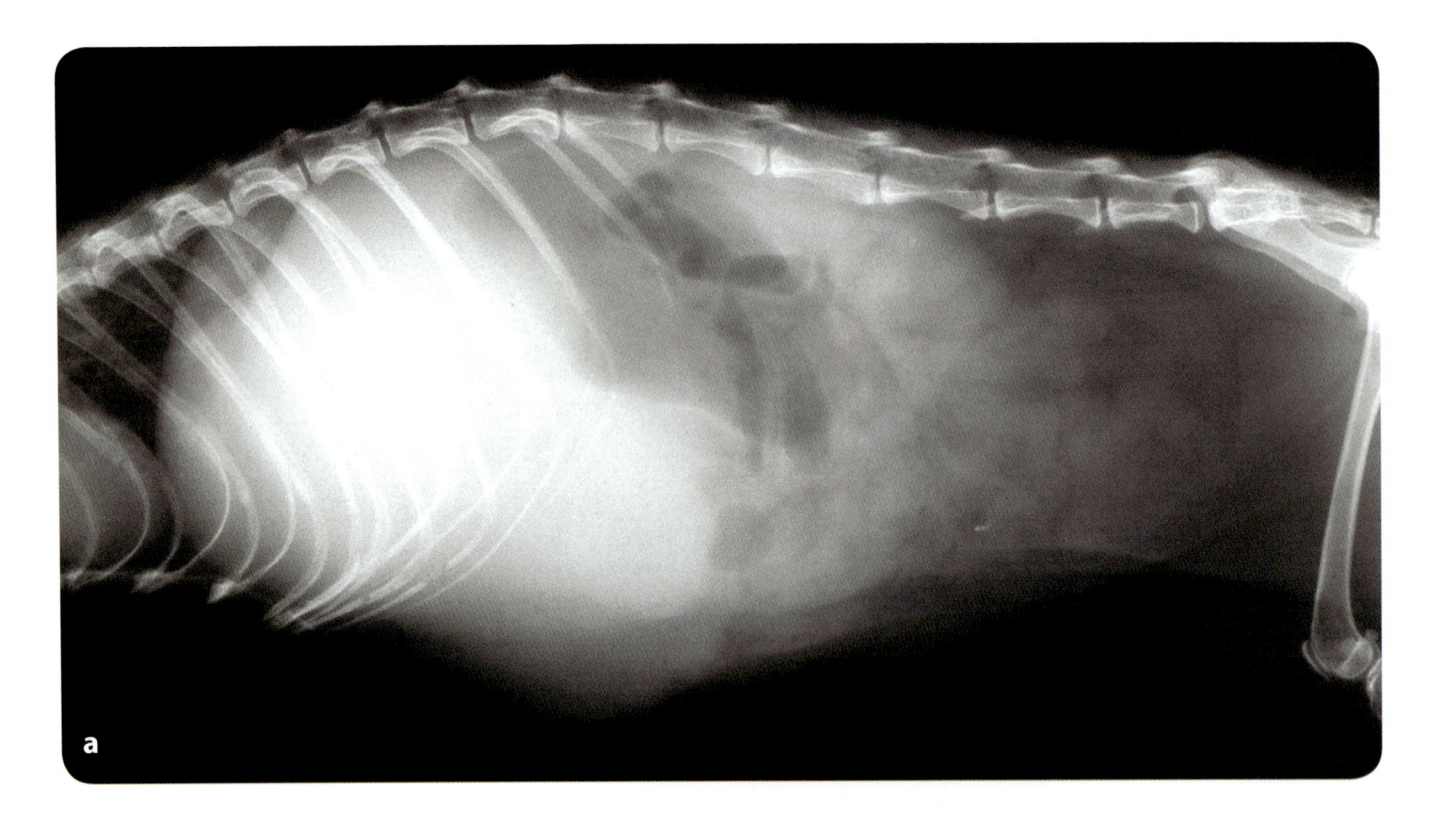

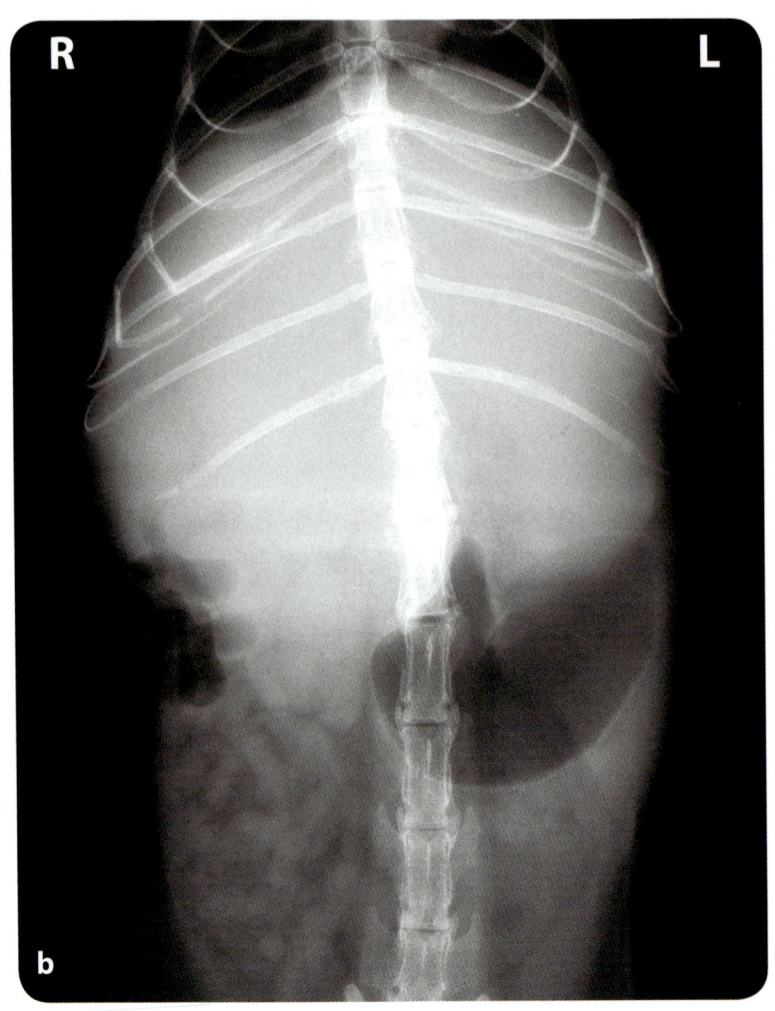

Figure 10.52a,b. Lateral (a) and ventrodorsal (b) radiographs of the abdomen of a 7-year-old ferret with hepatomegaly due to liver neoplasia diagnosed at surgery. The liver extends well beyond the caudal rib, deflecting the stomach caudally. *(Courtesy of Cathy Johnson-Delaney, DVM)*

Diseases of the Spleen

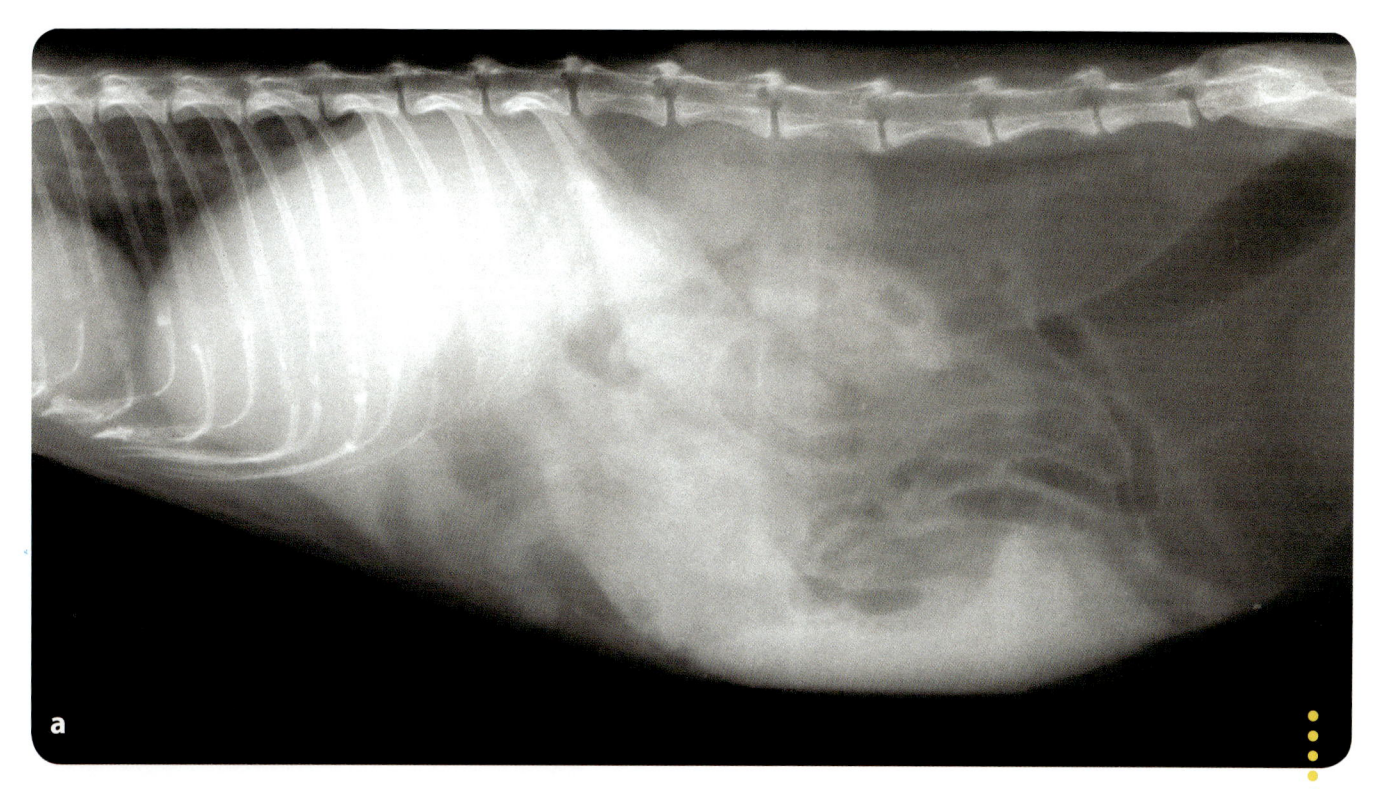

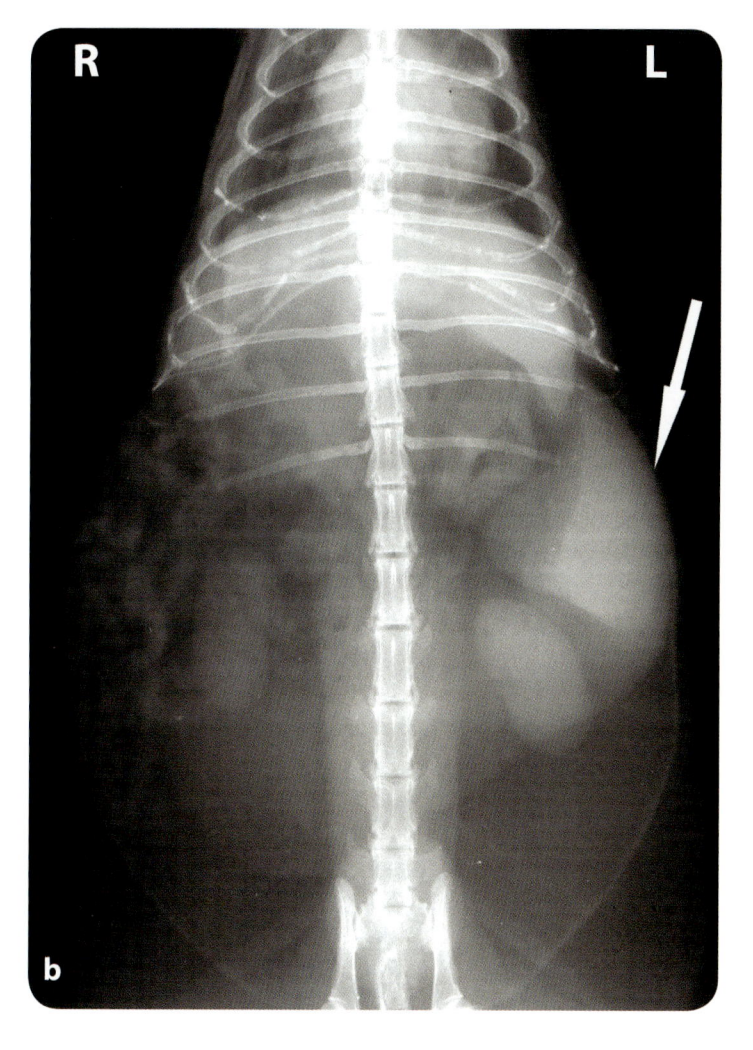

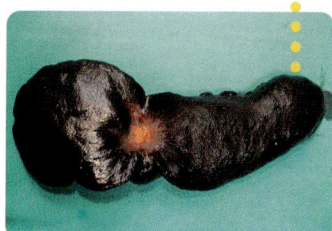

Figure 10.54. Appearance of the neoplastic spleen post splenectomy.

Figure 10.53a,b. Lateral (a) and ventrodorsal (b) radiographs of the abdomen of a ferret with splenic lymphoma diagnosed by histopathology. The spleen is large and boot-shaped on the lateral projection, and has an irregular dorsal margin. On the ventrodorsal projection the spleen is triangle-shaped with rounded margins (arrow). Mild hepatomegaly and a large amount of intraperitoneal fat are also visible.

Splenomegaly is a common finding in ferrets. The most common histopathologic diagnosis is not neoplasia but extramedullary hematopoiesis. Cause is frequently uncertain and may be related to a number of underlying disease conditions.

Diseases of the Urogenital Tract and the Adrenal Glands

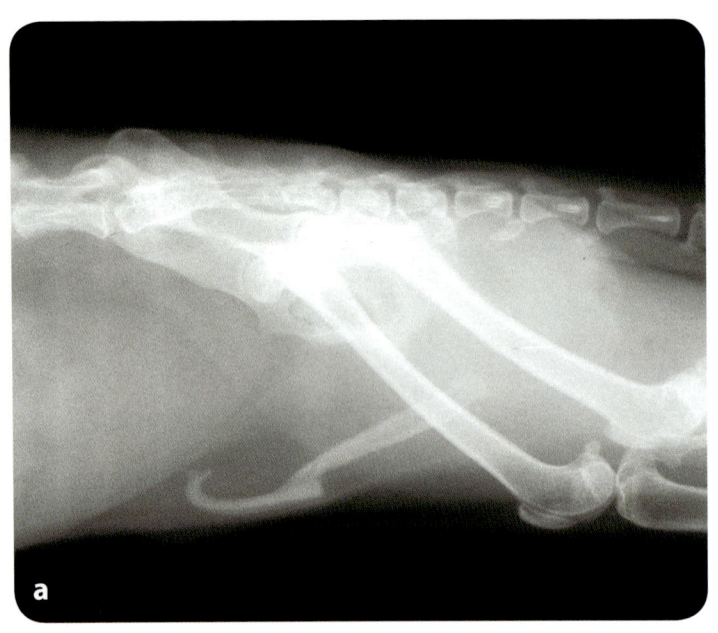

Figure 10.55a,b. Lateral (a) and ventrodorsal (b) radiographs of the pelvis of a male ferret demonstrating a rare fracture of the os penis.
(Courtesy of Cathy Johnson-Delaney, DVM)

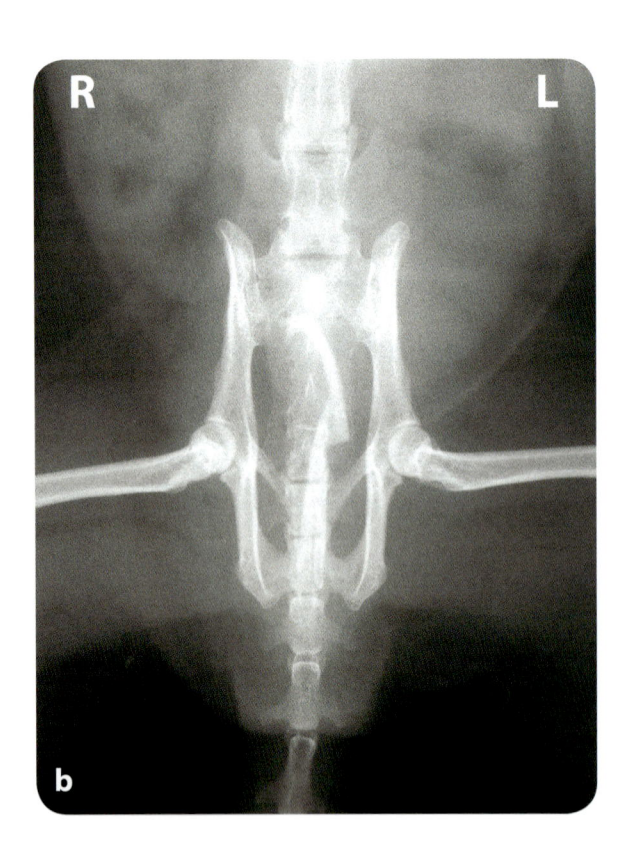

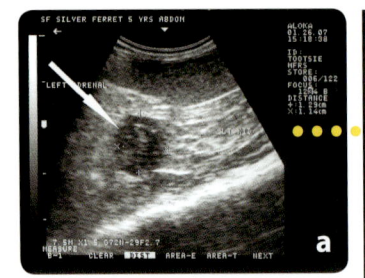

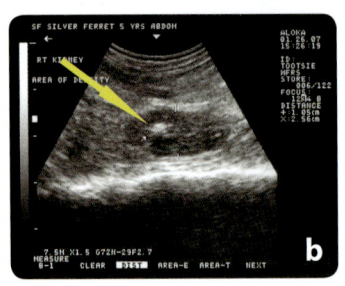

Figure 10.56.a,b. Ultrasonic images of the same case as Figure 10.57. The adrenal gland is indicated in (a) by the white arrow, and the radiodensity associated with the right kidney is indicated in (b) by the yellow arrow.
Radiology is not the diagnostic imaging of choice for diagnosis of adrenal disease. Most changes associated with adrenal disease are not visible radiographically. Ultrasonography is typically much more useful.
(Courtesy of Cathy Johnson-Delany, DVM)

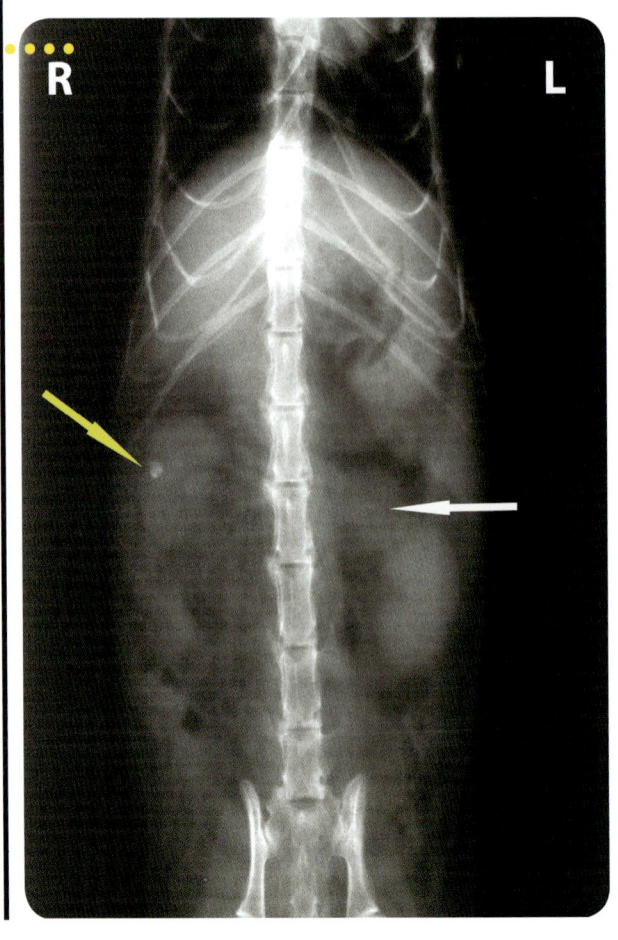

Figure 10.57. Ventrodorsal radiograph of the abdomen of a ferret demonstrating a rounded radiopaque mass confirmed at ultrasound to be the enlarged left adrenal gland (white arrow) and the radiodensity associated with the right kidney (yellow arrow).
(Courtesy of Cathy Johnson-Delaney, DVM)

Figure 10.58a,b. Lateral (a) and ventrodorsal (b) radiographs of the adomen of a ferret with a central mid-abdominal soft tissue mass. On the ventrodorsal projection, the mass is located between the kidneys (arrow). At surgery the mass was determined to be an enlarged right adrenal gland, and confirmed as adenocarcinoma by histopathology. Most neoplasms of the adrenal gland are only minimally to moderately enlarged. However marked enlargement and metastasis do occur.
Enlarged right adrenal neoplasias tend to expand medially due to the presence of the right caudal liver lobe and the kidney itself.

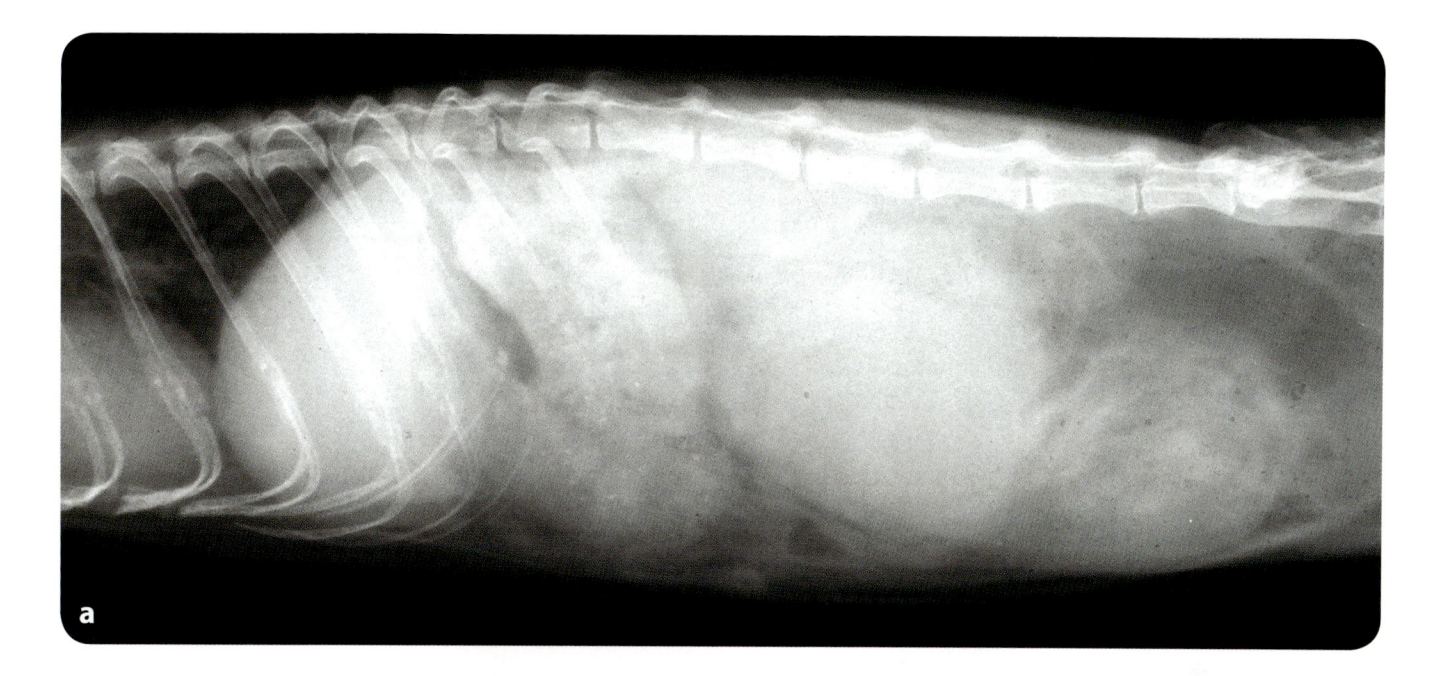

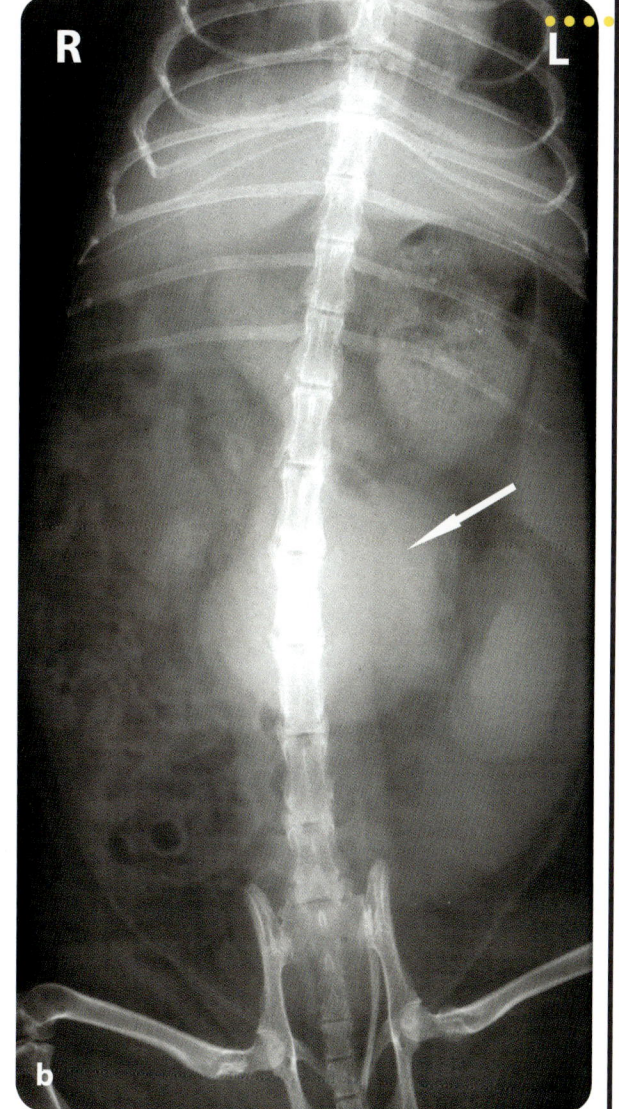

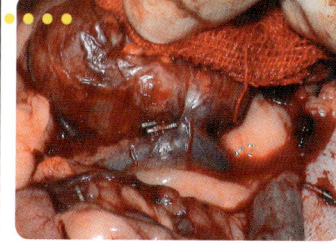

Figure 10.59. Surgical excision of the neoplastic adrenal gland in the same patient.

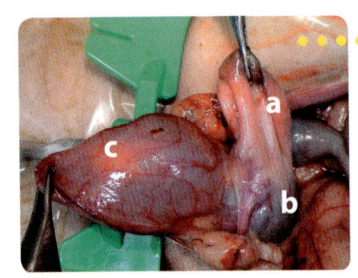

Figure 10.60. Intraoperative appearance of a prostatic cyst (a) and the relationship with the prostate gland (b) and the urinary bladder (c).

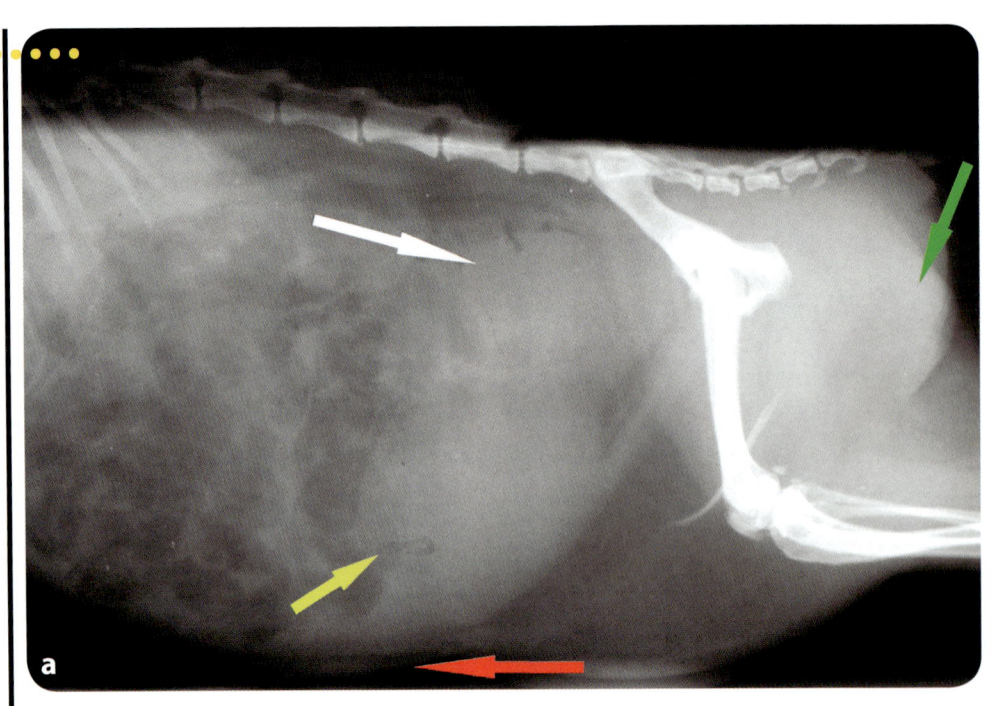

Figure 10.61a,b. Lateral (a) and ventrodorsal (b) radiographs of the abdomen of a 3.5 year old ferret with a prostatic cyst secondary to adrenal disease. There is an oval soft tissue opacity representing the prostatic cyst (white arrow) dorsal to the urinary bladder (yellow arrow). Increased soft tissue opacity caudal to the ishium (green arrow) was determined to be a paraurethral cyst. Edema of the prepuce is also present (red arrow). Note cranial displacement of the small intestine.

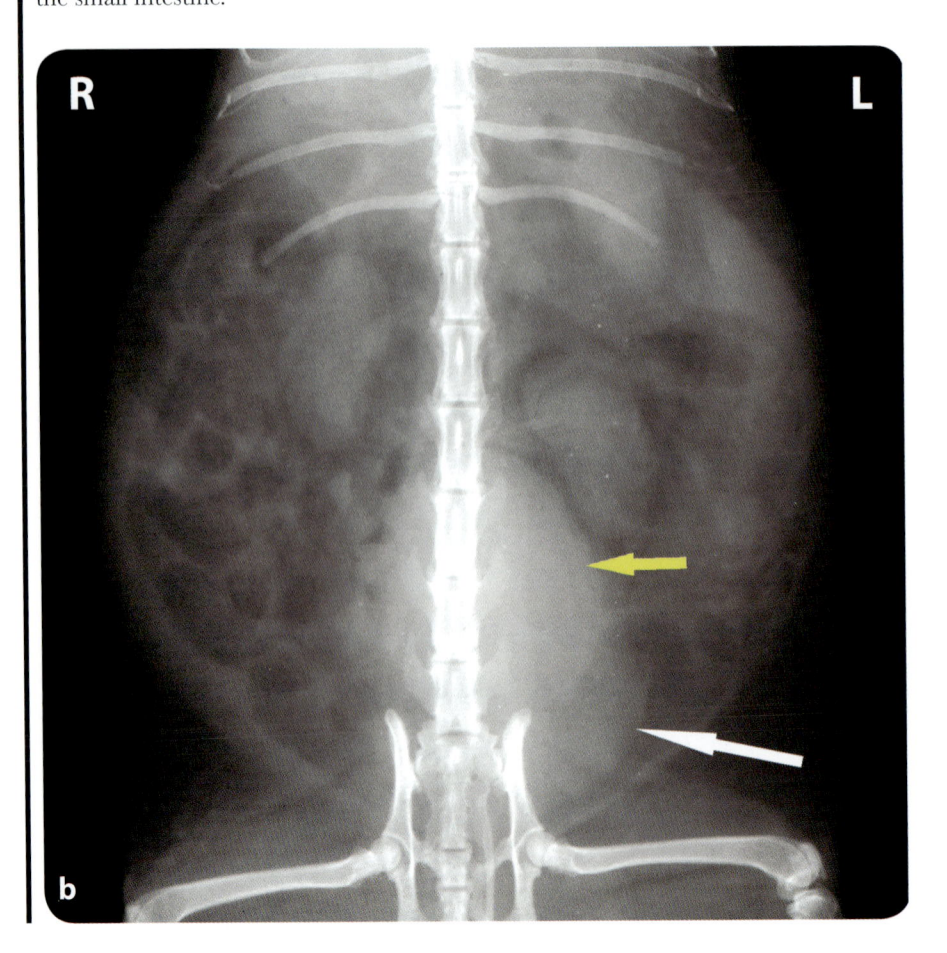

Miscellaneous

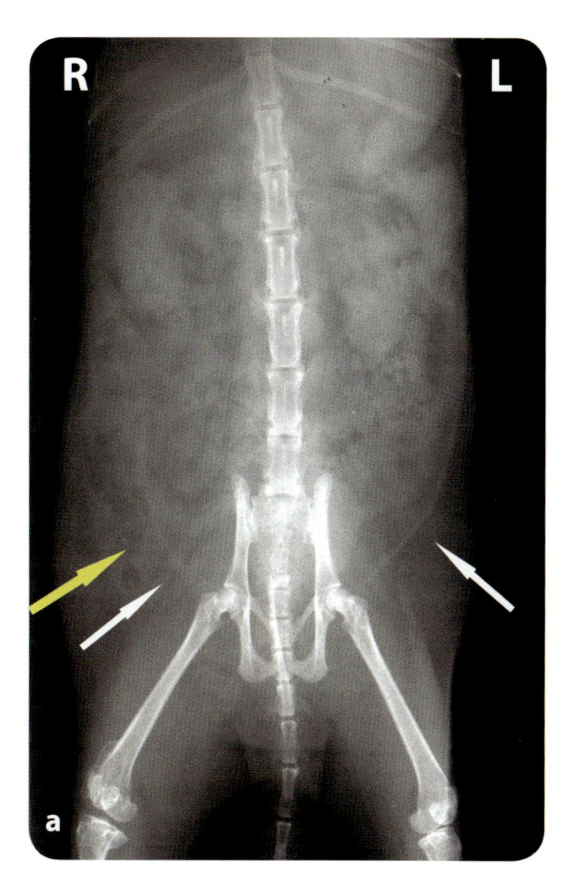

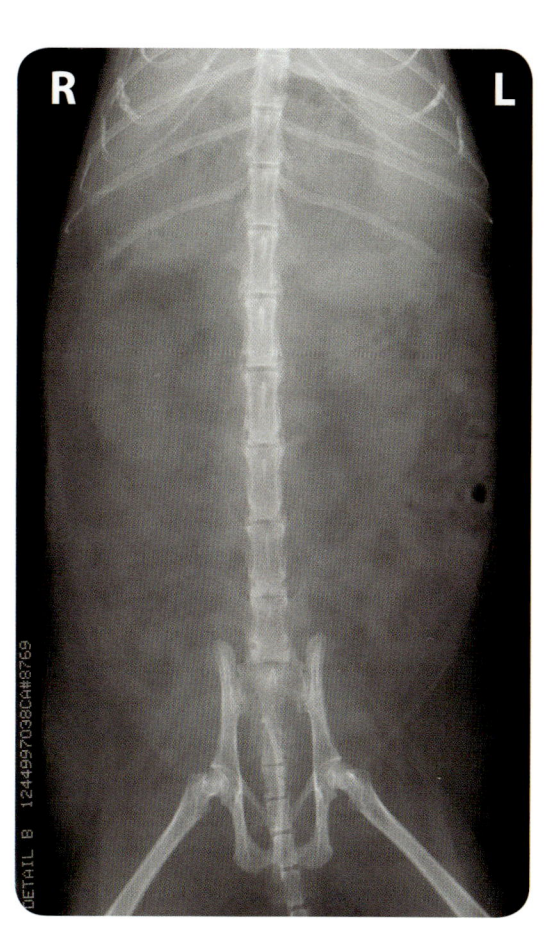

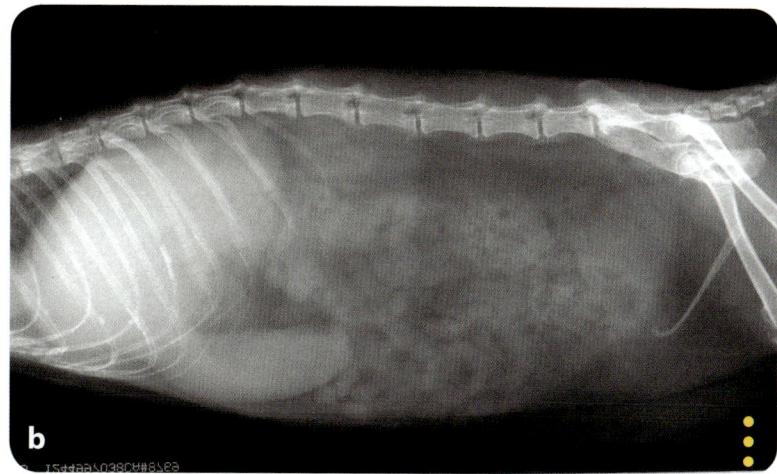

Figure 10.62a,b. Ventrodorsal (a) and lateral (b) radiographs of a young male ferret with an abdominal hernia following an attack by dogs. The herniation was present on the right side of the abdomen, despite the presence of skin wounds on the left side, presumably due to the ferret's attempt to roll during the attack. On the ventrodorsal projection, intestinal loops (yellow arrow) can be seen lateral to the abdominal musculature represented by the radiopaque line (white arrow) on the right side. Appearance is normal on the left side of the abdomen.
Note that the lateral projection was not useful for diagnosis in this patient.

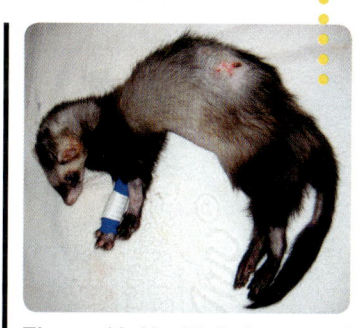

Figure 10.63. Clinical aspect of the wound on the left side of the body.

Figure 10.64. Follow-up radiograph in the same patient after surgical repair. Intestinal loops lie within the borders of the abdominal cavity.

The NORMAL THORACIC LIMB
Lateral Projection

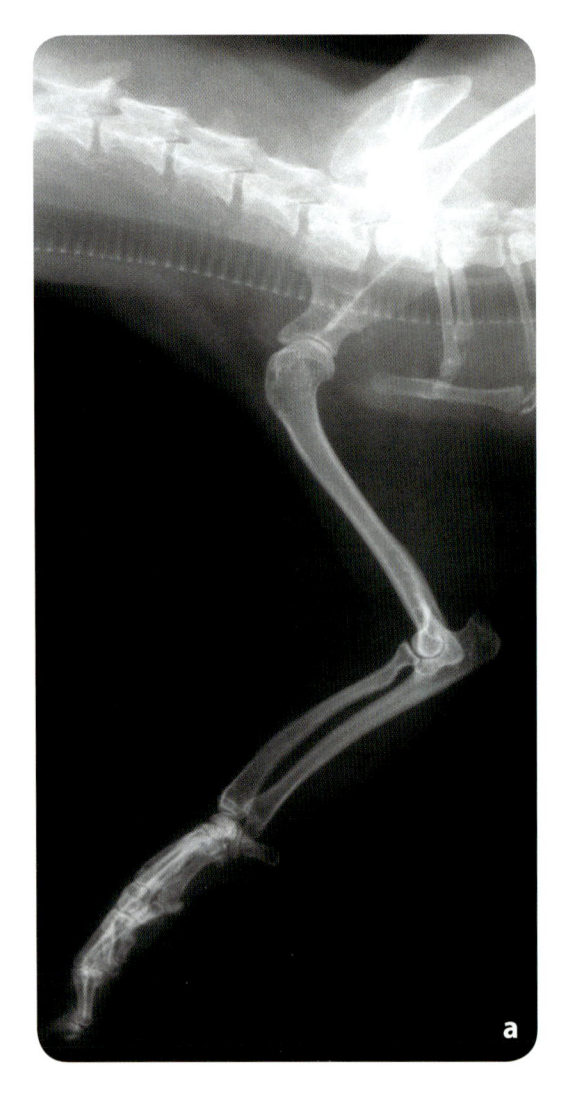

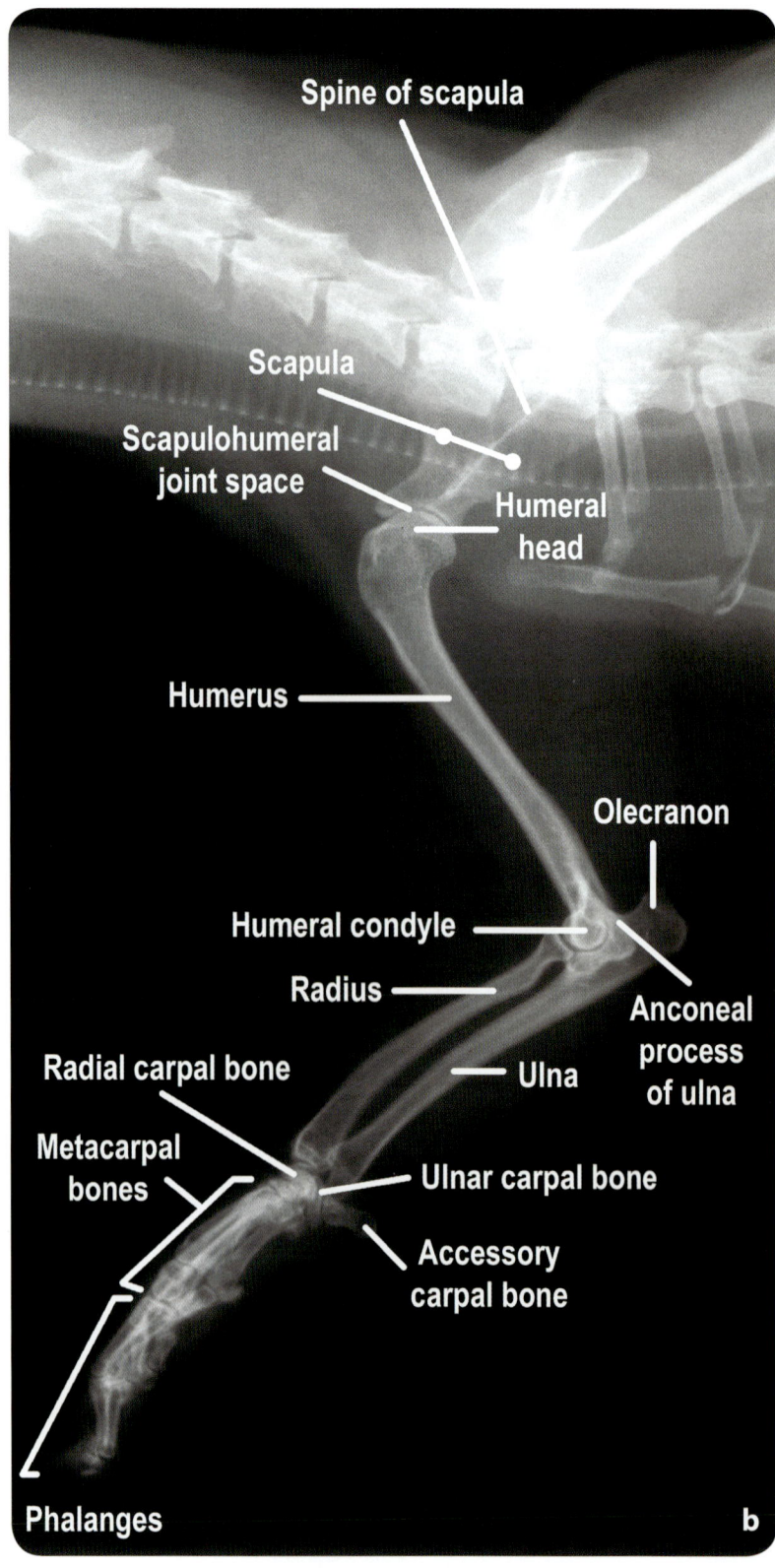

Figure 10.65a,b. Radiograph of the thoracic limb of a 1.1 kg male ferret, actual size (a) and 140% of actual size (b).

Craniocaudal Projection

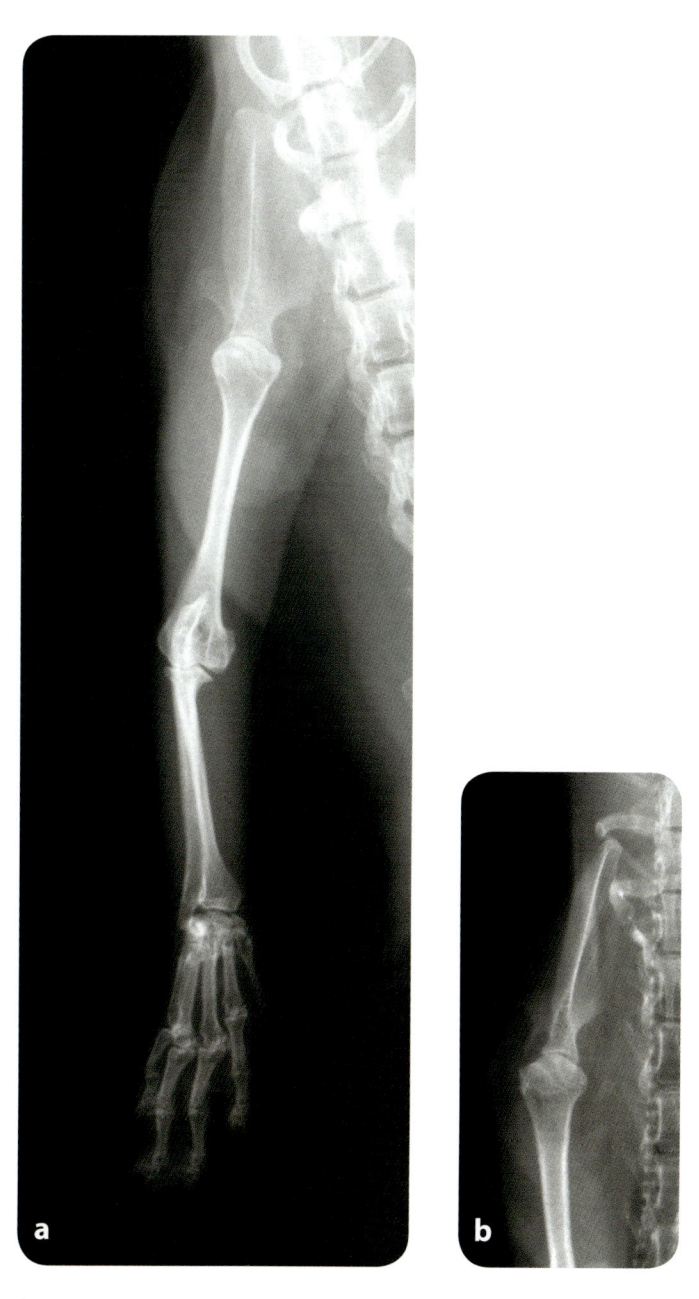

Figure 10.66a,b,c. Craniocaudal radiographs of the thoracic limb of a 1.1 kg male ferret, actual size (a) and 140% of actual size (c).
A more accurate craniocaudal view of the scapula is seen in (b).

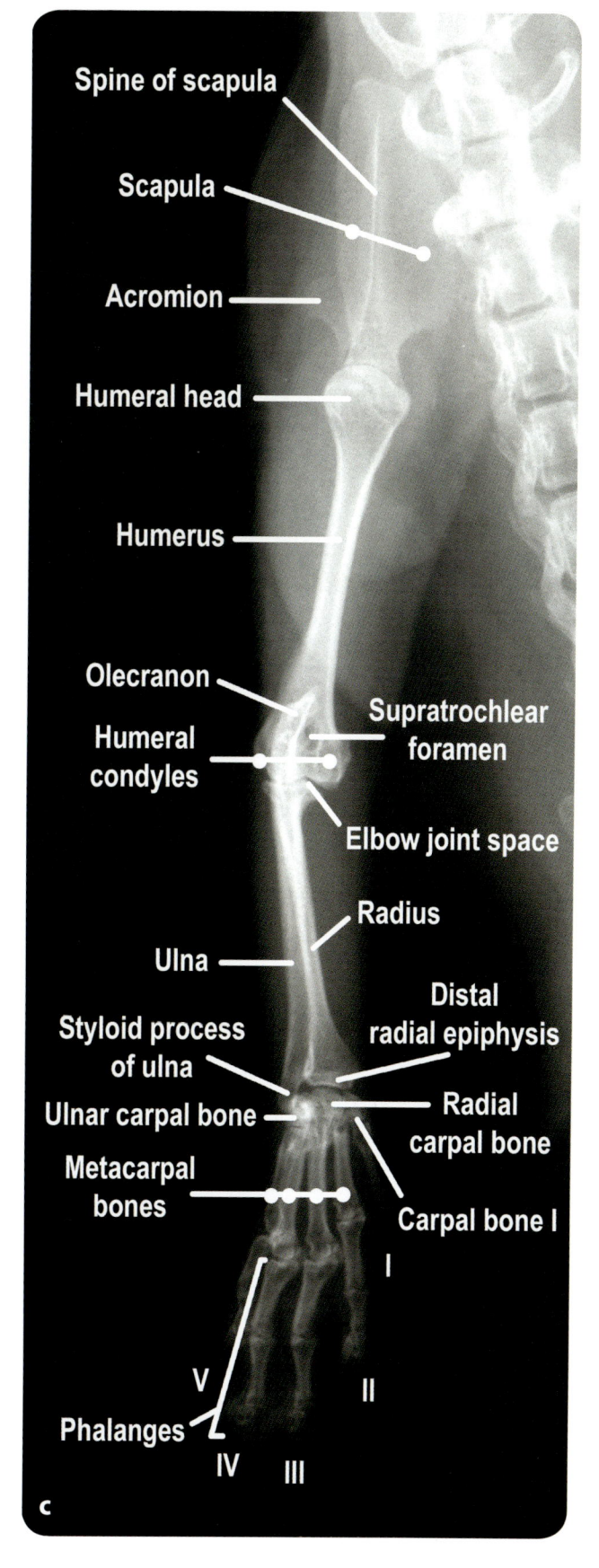

ABNORMALITIES of the THORACIC LIMB
Diseases of the Scapula

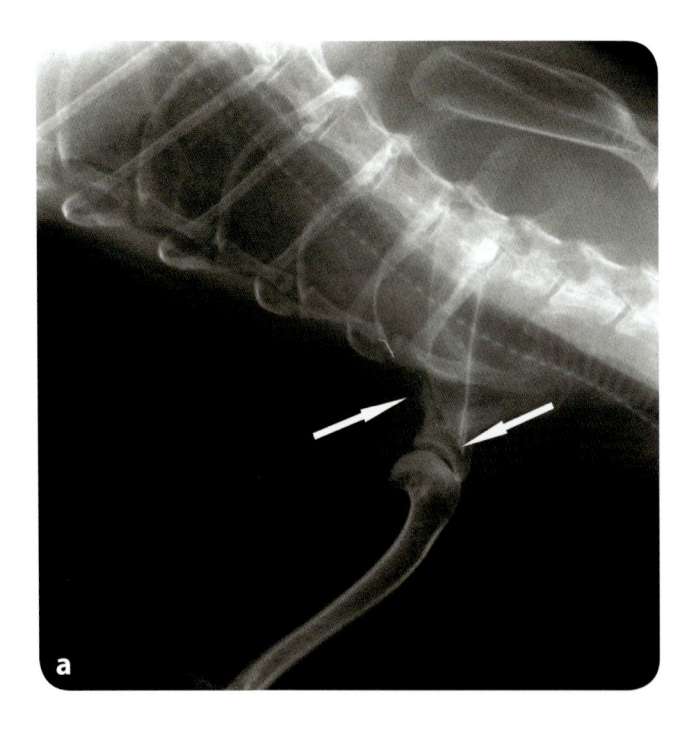

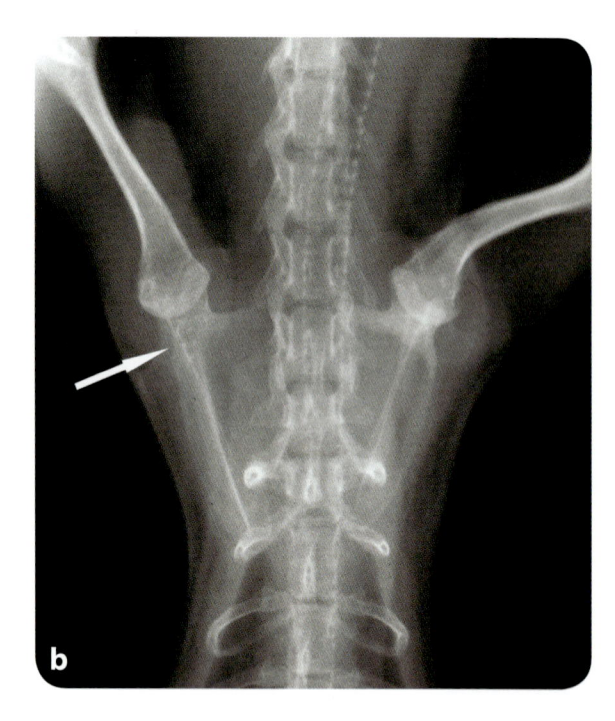

Figure 10.67a,b. Lateral (a) and ventrodorsal (b) radiographs of a ferret with an oblique fracture of the neck of the right scapula (arrows).
(Courtesy of Cathy Johnson-Delaney, DVM)

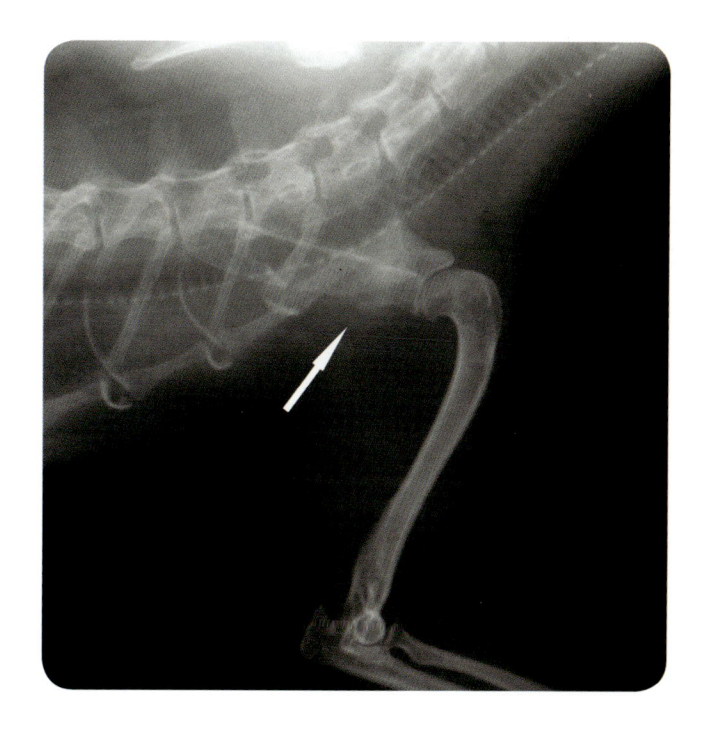

Figure 10.68. Lateral radiograph of the same patient 2 months later demonstrating partial closure of the fracture (arrow).
(Courtesy of Cathy Johnson-Delaney, DVM)

Diseases of the Radius, Ulna, and Elbow Joint

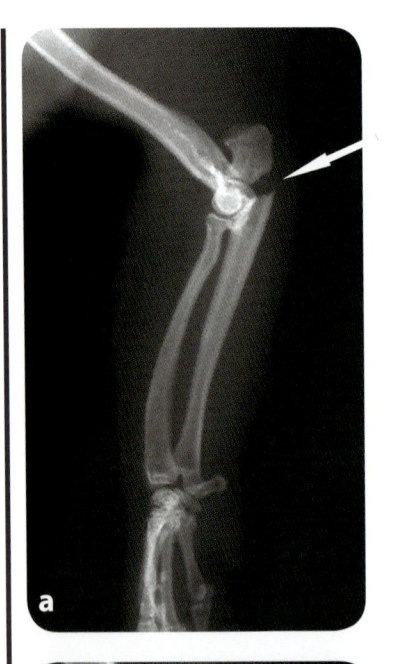

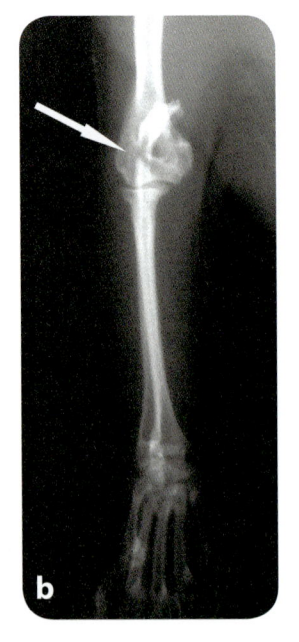

Figure 10.69a,b.
Lateral (a) and craniocaudal (b) radiographs of the thoracic limb of a ferret with a transverse fracture of the olecranon process of the ulna (arrow). The fracture is intra-articular and is subject to tension from the triceps musculature. Although internal fixation is preferred, this may be difficult due to patient size and medial curvature of the olecranon.

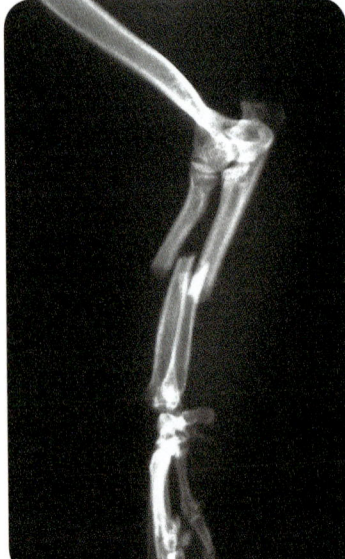

Figure 10.70.
Lateral radiograph of the thoracic limb of a ferret with a fracture of the midshaft of the radius and ulna.

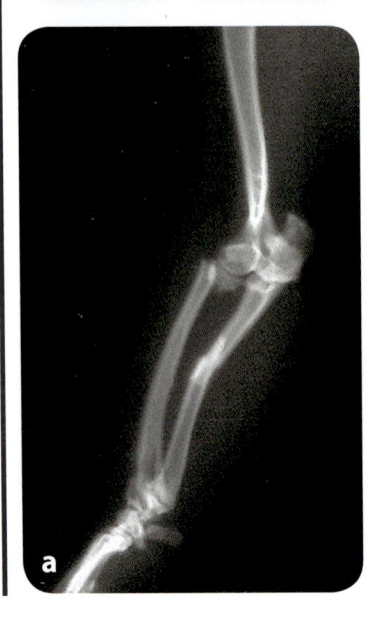

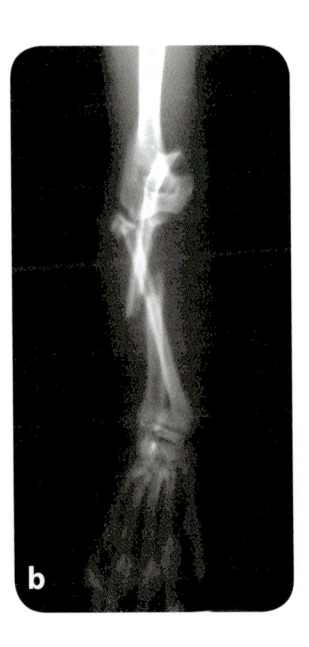

Figure 10.71a,b.
Lateral (a) and craniocaudal (b) radiographs of the thoracic limb of a ferret with a fracture of the shaft of the ulna and epiphyseal fracture of the radial head.

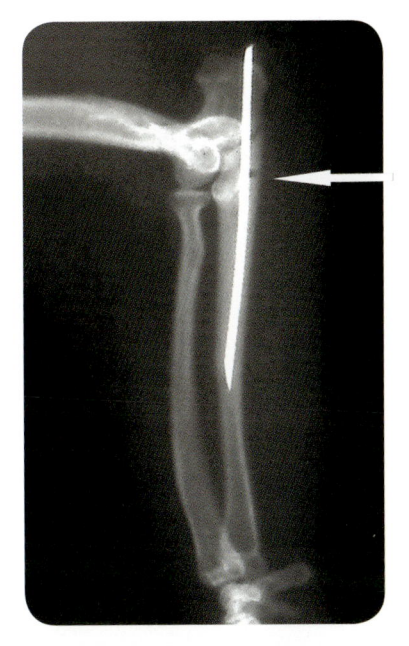

Figure 10.72. Post-operative lateral radiograph of the same patient as Figure 10.70 following osteosynthesis of the olecranon with an intramedullary pinning technique. A 21 gauge needle has been used as a pin. As the olecranon of ferrets normally curves medially the pin must be inserted slightly laterally, and not from the proximal aspect of the olecranon (see also Figure 10.75b) IM pinning alone is not adequate to counteract forces from the triceps musculature acting on the olecranon. Therefore cerclage tension suture (Vycryl 2:0) has been secured along the tip of the needle and through a small hole created in the proximal fragment of the ulna (arrow).

Figure 10.73a,b,c. Lateral post-operative radiograhs (a-c) of the same patient as Figure 10.71 showing osteosynthesis of the radius with a 2+3 pin monolateral external fixation device. Stabilization of the radius is adequate for proper stabilization of the ulna. The most proximal pin has been inserted too deep and appears to cross the ulnar shaft, but actually does not (a). Follow up after 4 weeks demonstrates adequate bone healing (b). Follow up at 6 weeks after removal of the external fixation devices.

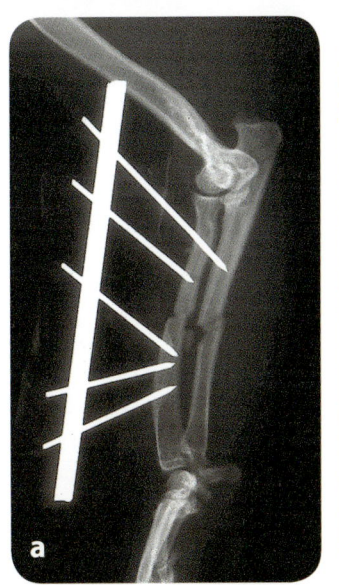

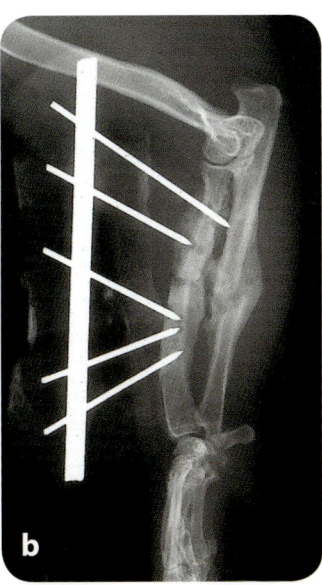

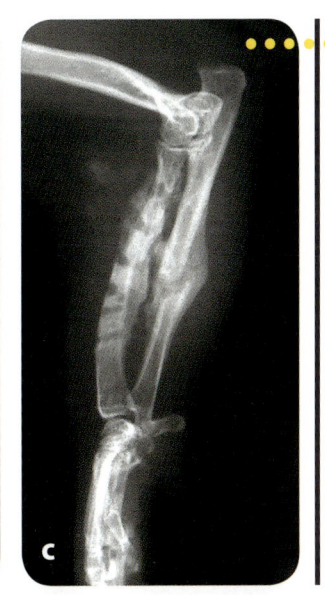

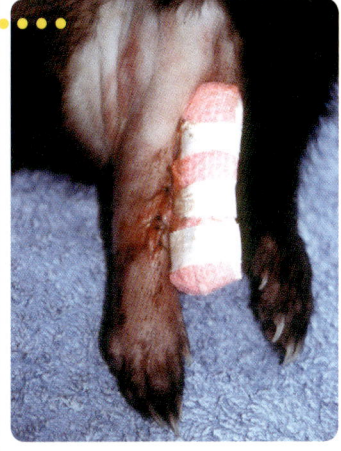

Figure 10.74. Same patient as Figures 10.73 (a-c). External fixation devices are usually well tolerated by ferrets.

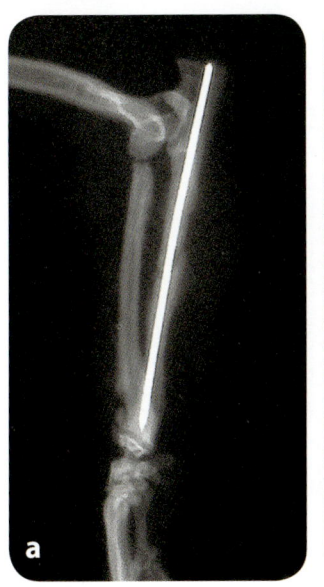

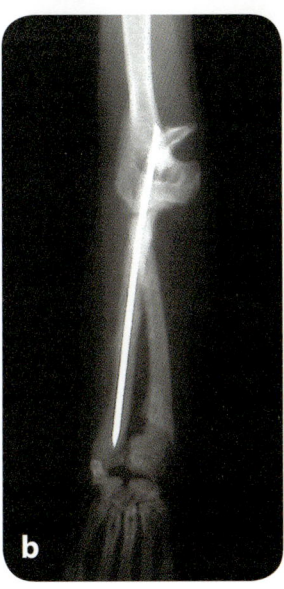

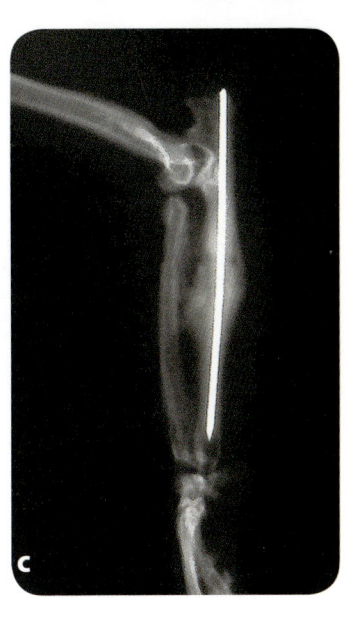

Figure 10.75a,b,c.
Lateral (a) and craniocaudal (b) radiographs of a ferret with an ulnar diaphyseal fracture repaired with an intramedullary pinning technique. The epiphyseal fragment of the radius has been lost, and for this reason slight subluxation of the radioulnar joint is visible (a).
Follow up 5 weeks after pinning (c). Adequate bone healing is visible at the fracture site. Despite the subluxation of the radioulnar joint functional recovery of the thoracic limb was complete.

Figure 10.76. Clinical appearance of luxation of the right elbow. The thoracic limb appears rotated laterally.

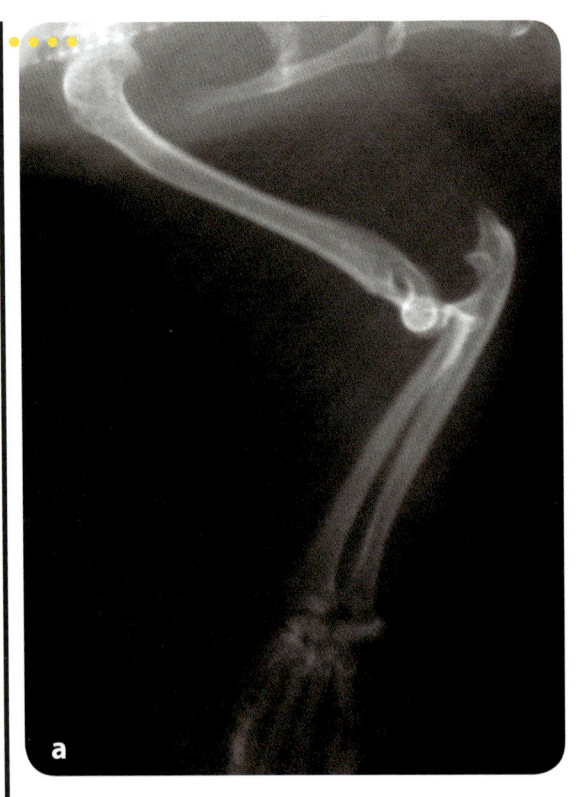

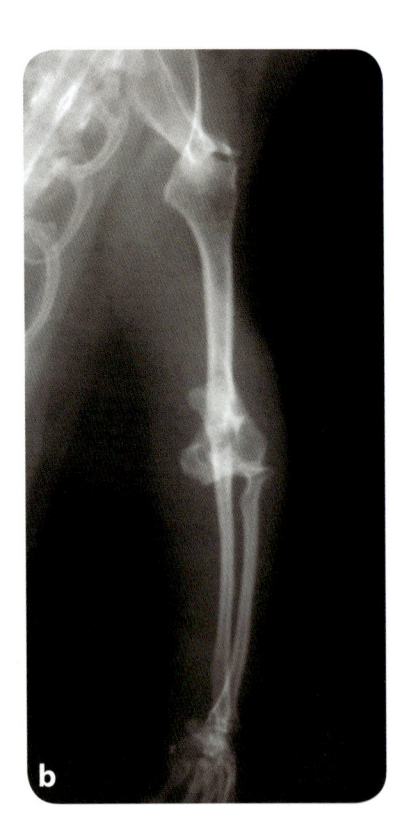

Figure 10.77a,b. Lateral (a) and craniocaudal (b) radiographs of the thoracic limb of a ferreet with luxation of the elbow. The luxation is most obvious in (a).

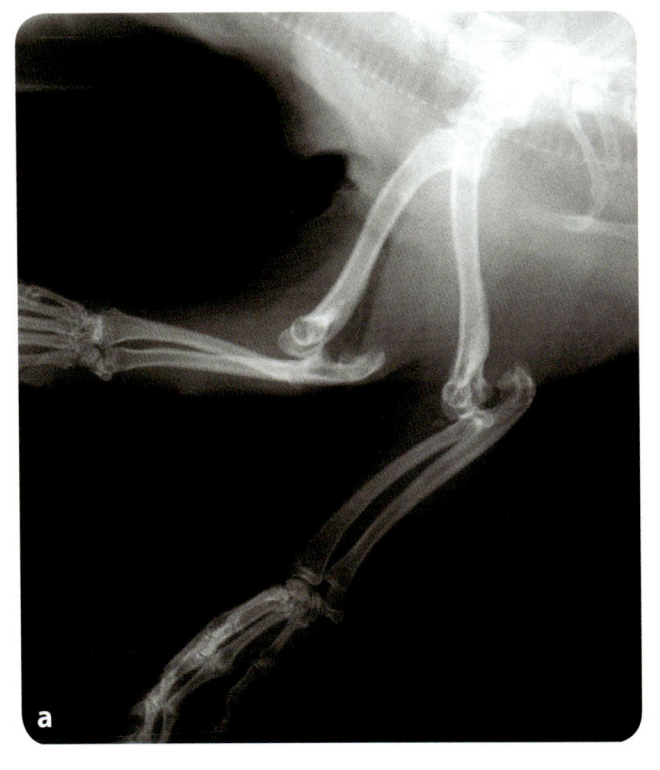

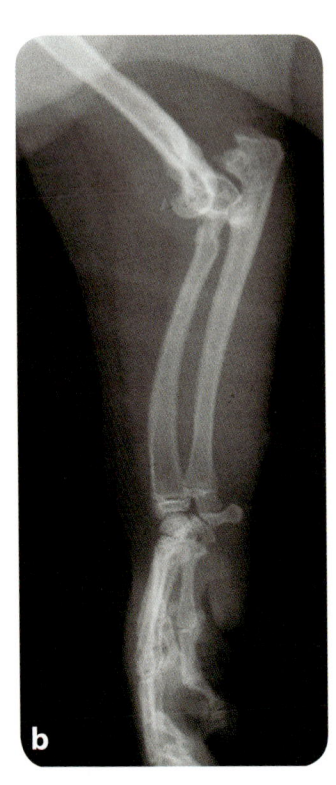

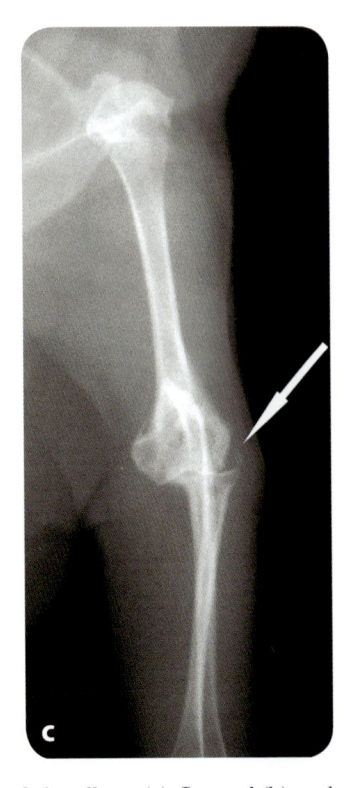

Figure 10.78a,b,c. Lateral radiograph of the thoracic limbs of a ferret with bilateral luxation of the elbow (a). Lateral (b) and craniodcaudal (c) radiographs taken post reduction demonstrate chronic subluxation following incomplete reduction. Note in (c) the articular surface of the radius is shifted laterally, and there is evidence of arthrosis (arrow).
(Courtesy of Cathy Johnson-Delaney, DVM)

THORACIC LIMB 397

Figure 10.79a,b. Lateral (a) and craniocaudal (b) radiographs of the thoracic limb of a ferret with a large plapable mass, determined at surgery to be consistent with neoplasia. Note soft tissue swelling and aggressive bone response involving the elbow.

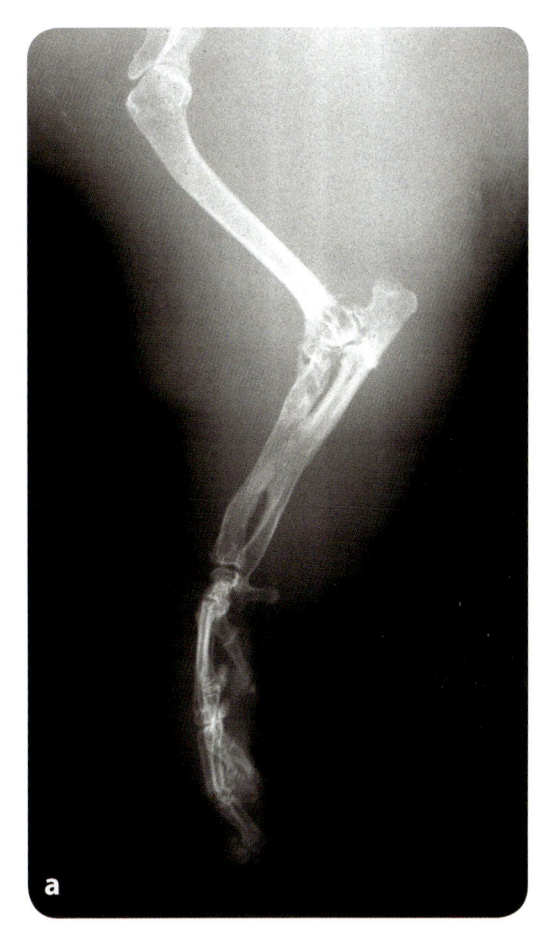

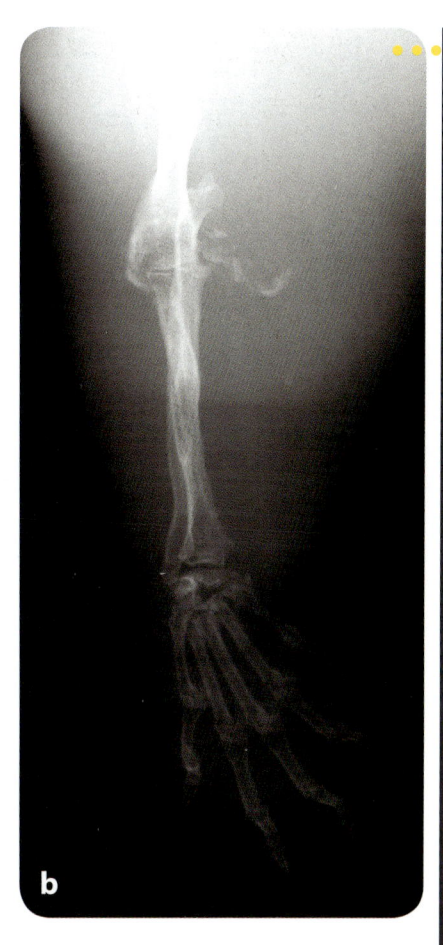

Figure 10.80. Clinical appearance of neoplasia of the right thoracic limb in a ferret.

The NORMAL PELVIC LIMB
Lateral Projection

Due to varying soft tissue density, superimposition of the abdomen, and the relatively small size of the distal pelvic limb, it is impossible to obtain an optimal single radiograph of the entire pelvic limb. Different kVp settings are required for the pelvis and the femur, and for the tibia and the distal limb.

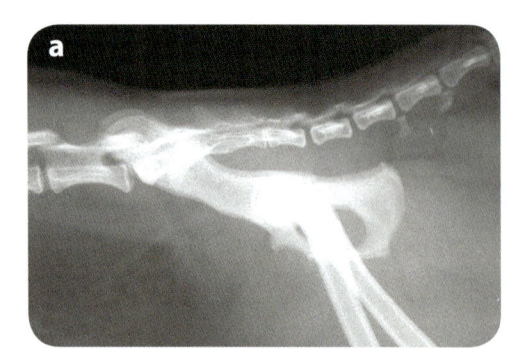

Figure 10.81a,b. Radiograph of the pelvis of a 1.1 kg male ferret, actual size (a) and 150% of actual size (b).

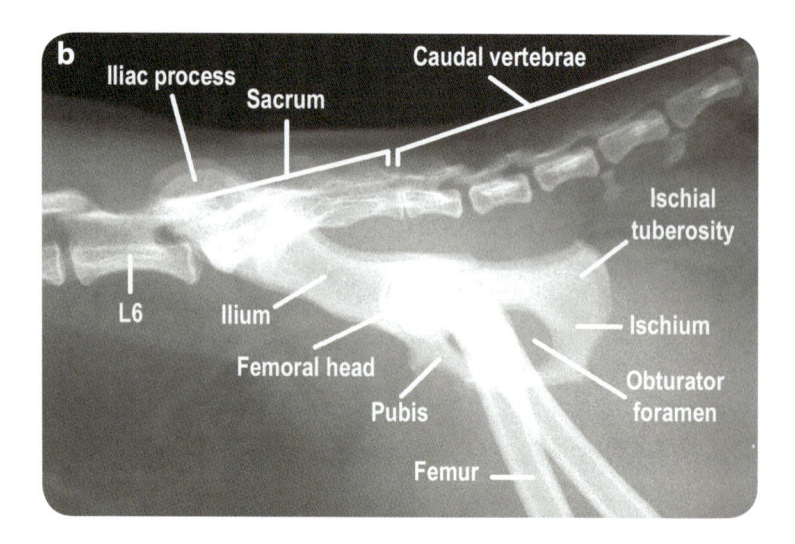

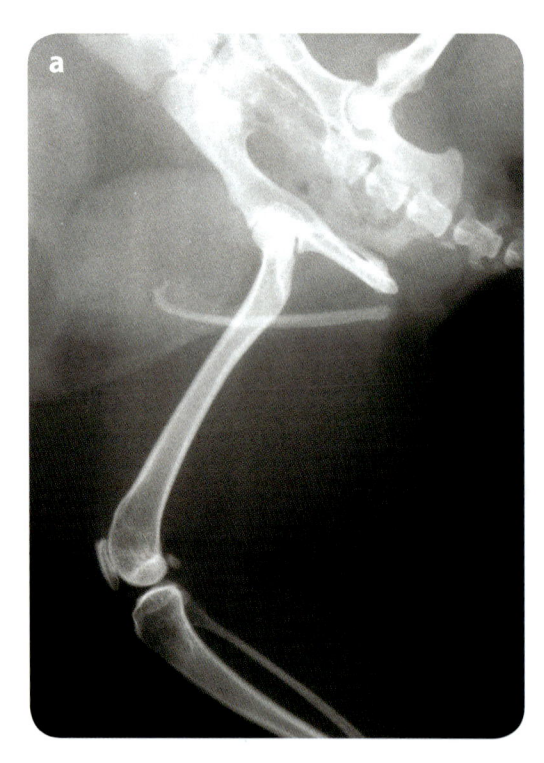

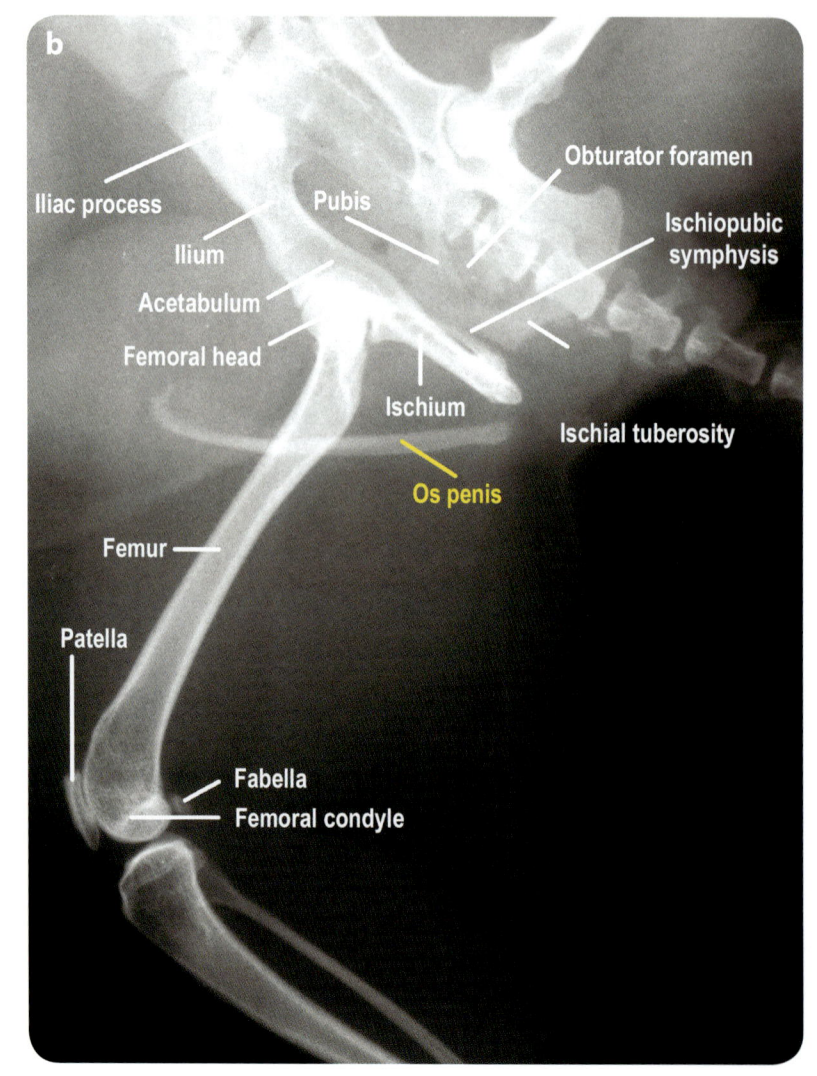

Figure 10.82a,b. Radiograph of the proximal pelvic limb of a 1.1 kg male ferret, actual size (a) and 150% of actual size (b).

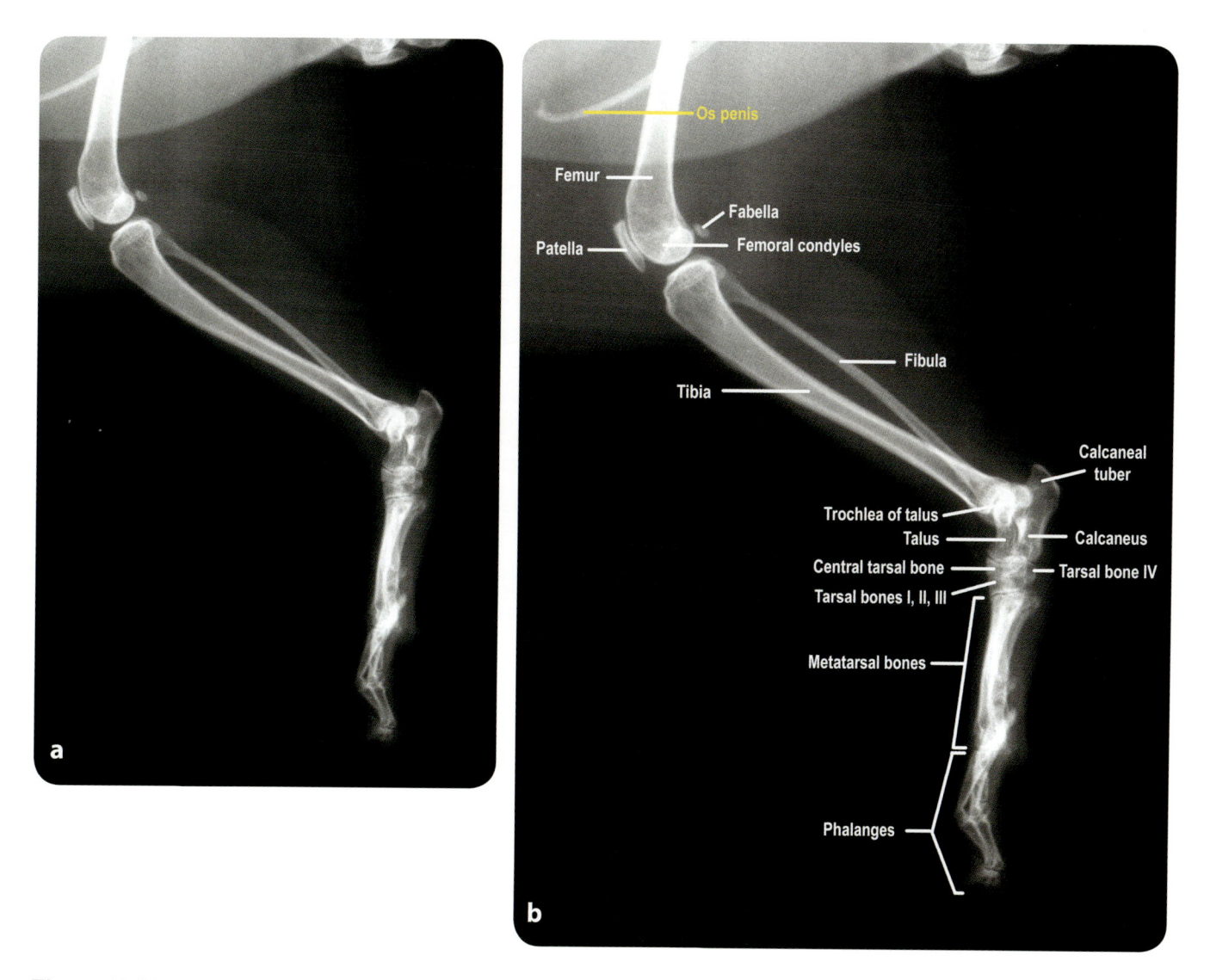

Figure 10.83a,b. Radiograph of the distal pelvic limb of a 1.1 kg male ferret, actual size (a) and 120% of actual size (b).

Ventrodorsal Projection of the Pelvis
Craniocaudal Projection of the Proximal Pelvic Limb

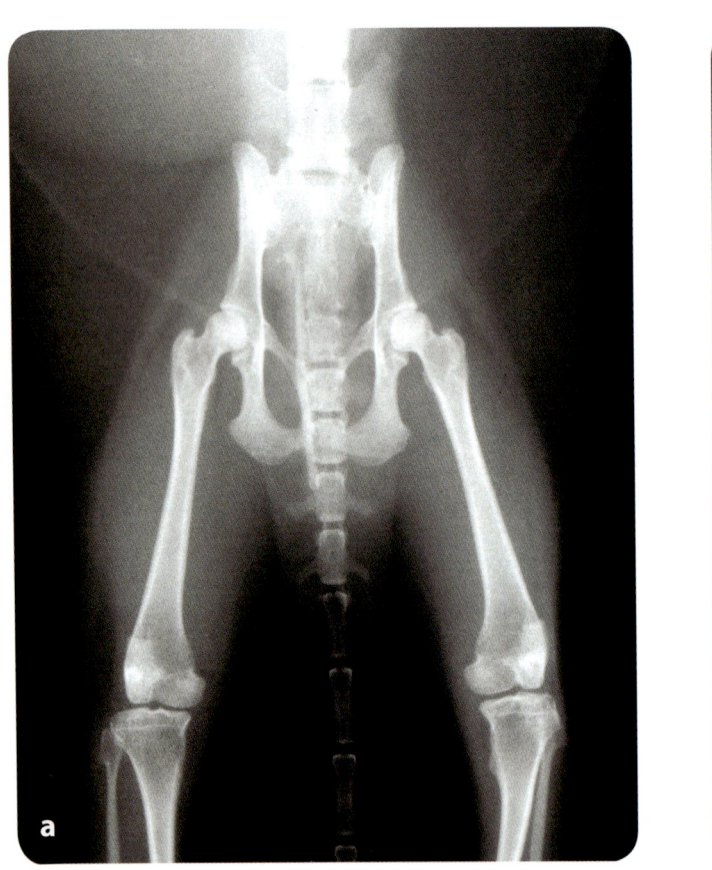

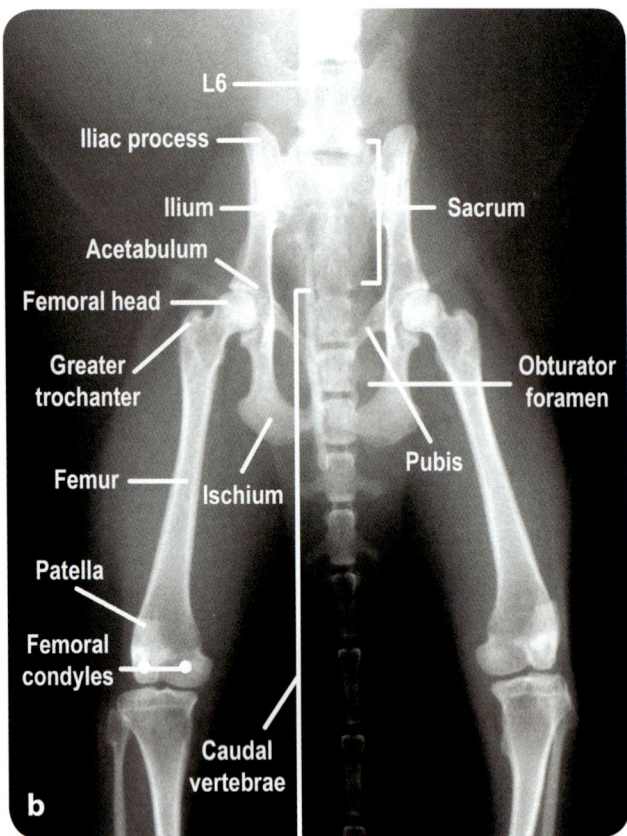

Figure 10.84a,b. Radiograph of the pelvis and the proximal pelvic limb of a 1.1 kg male ferret, actual size.

Craniocaudal Projection of the Distal Pelvic Limb

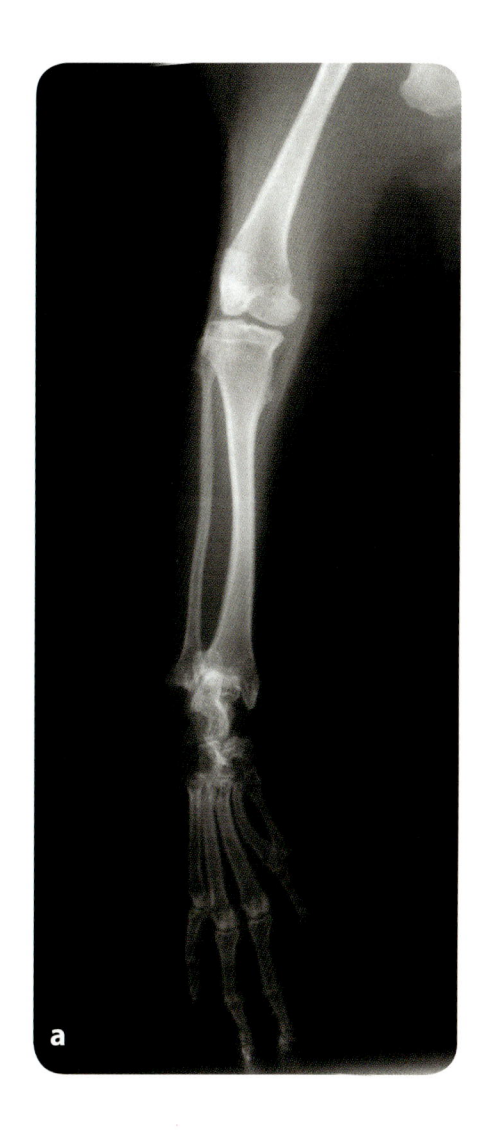

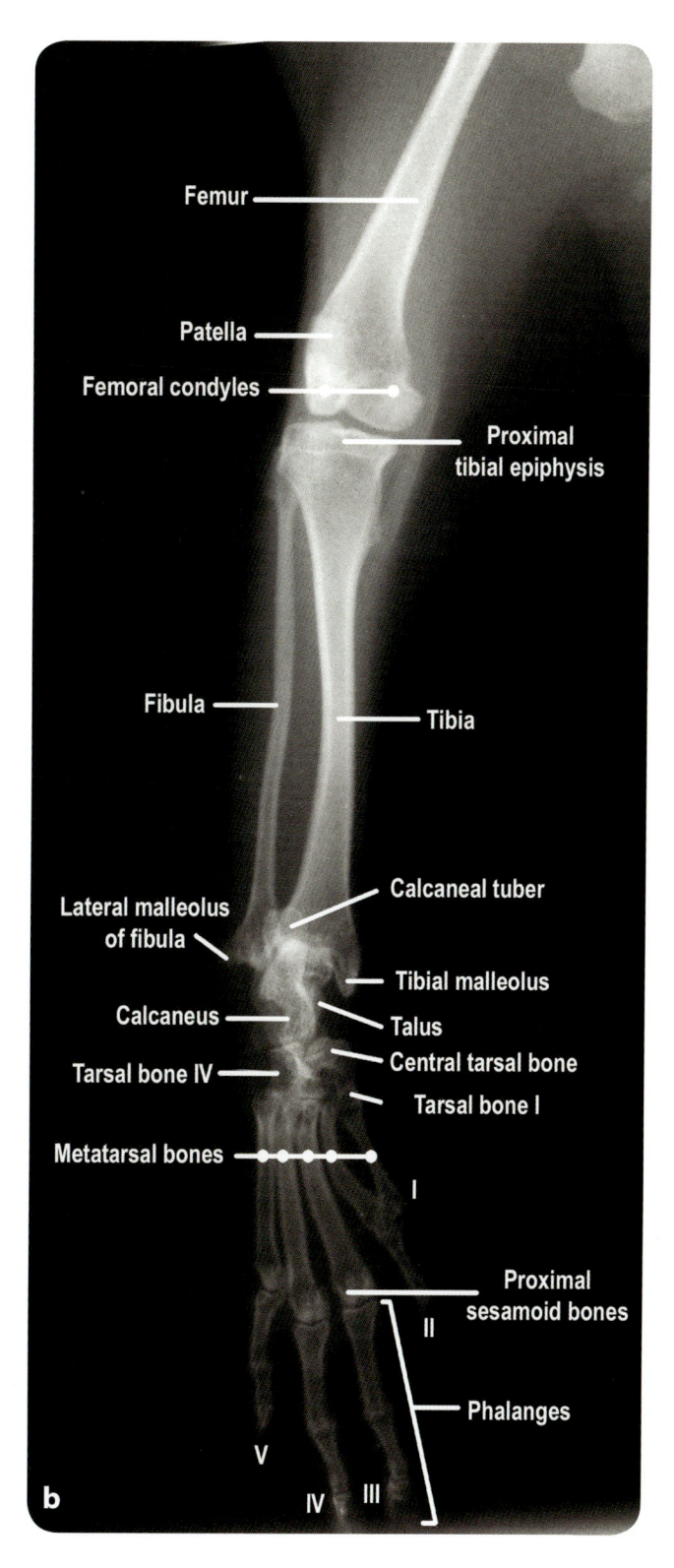

Figure 10.85a,b. Radiograph of the distal pelvic limb of a 1.1 kg male ferret, actual size (a) and 150% of actual size (b).

ABNORMALITIES of the PELVIC LIMB
Diseases of the Pelvis

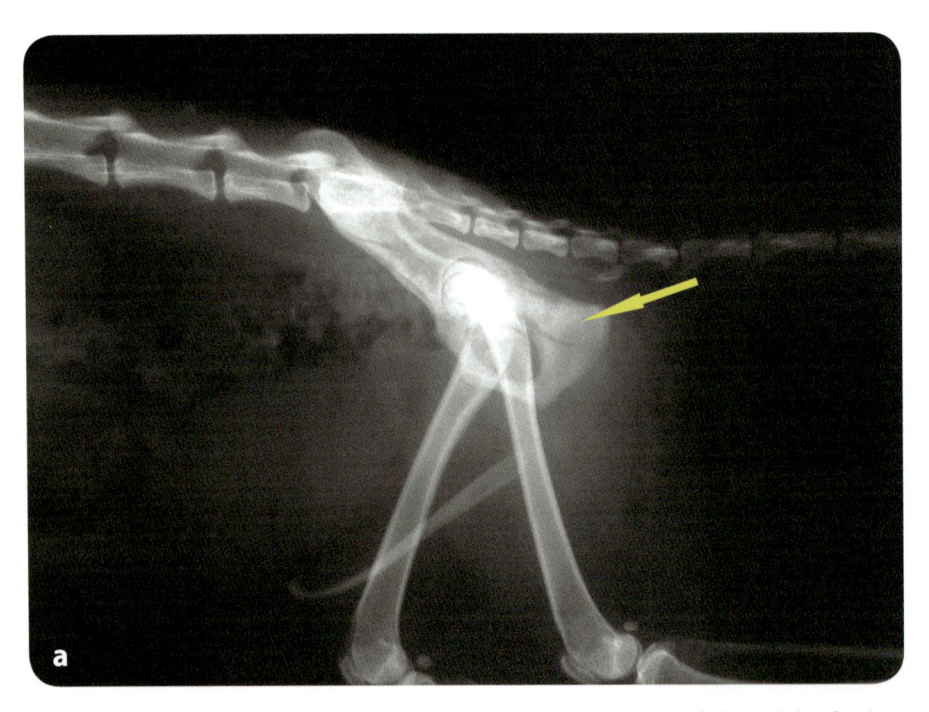

Figure 10.86a,b. Lateral (a) and ventrodorsal (b) radiographs of the pelvis of a 1.5-year-old male ferret with a comminuted fracture of the left ilum (white arrow) and the left ischiatic bone (yellow arrow) The urinary bladder is intact.
(Courtesy of Cathy Johnson-Delaney, DVM)

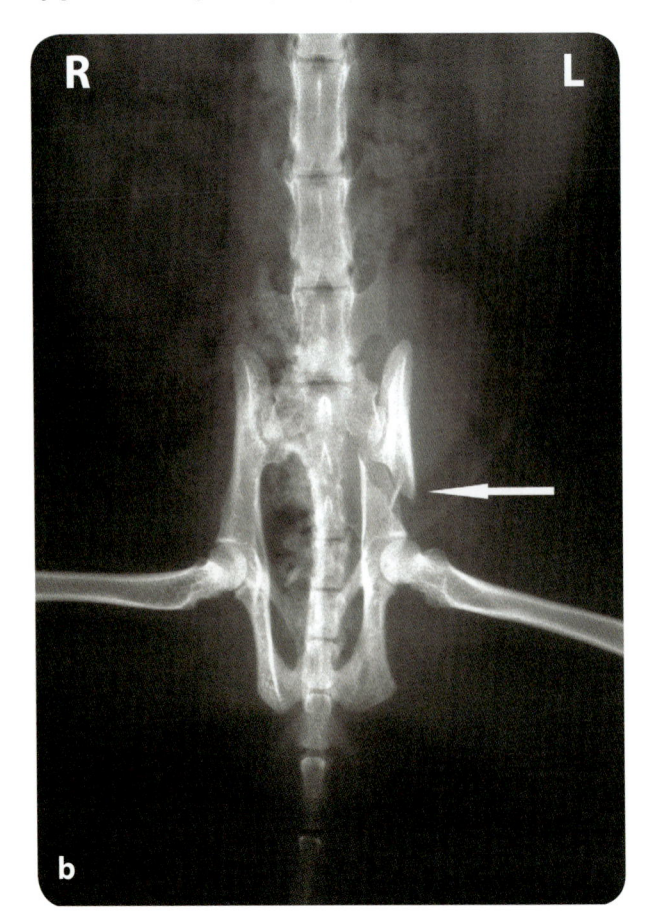

Diseases of the Femur

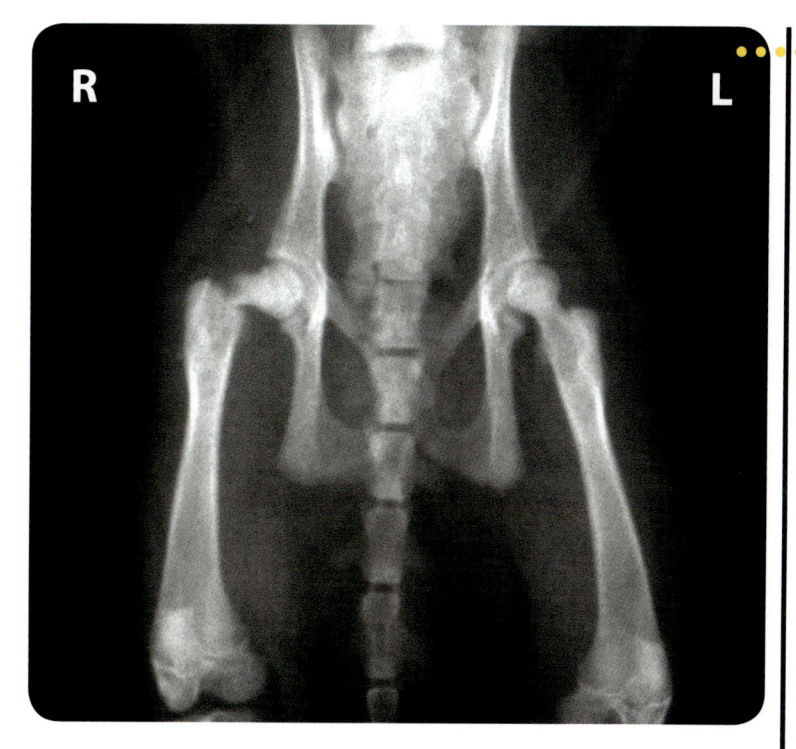

Figure 10.87. Ventrodorsal projection of the pelvis and femurs of a ferret with a fracture of the right femoral neck. Preferred treatment is femoral head ostectomy. Attempts at internal fixation generally result in degenerative joint disease.

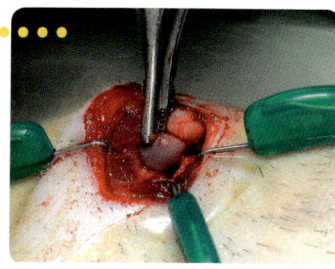

Figure 10.88. Surgical removal of the femoral head in the same patient.

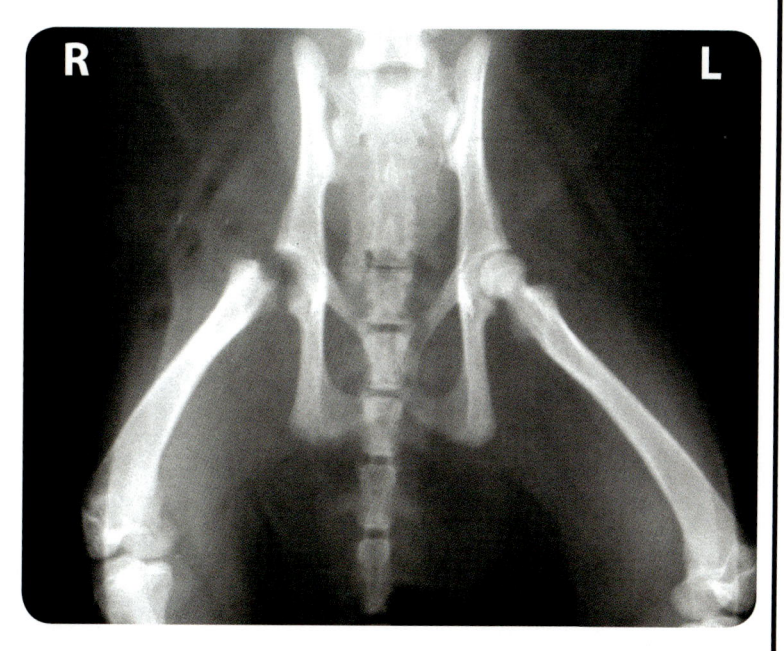

Figure 10.89. Postoperative radiograph after surgical removal of the femoral neck and head. Note absence of the femoral head and smooth margin of the proximal aspect of the femur.

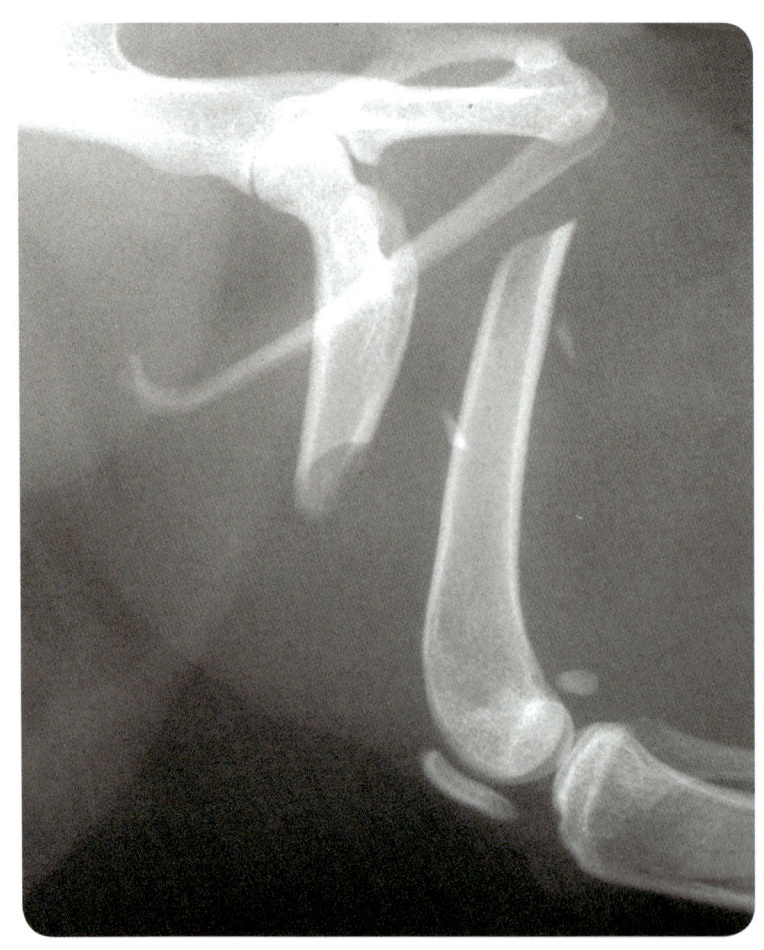

Figure 10.90. Lateral radiograph of the femur of a 1.6 kg 1-year-old ferret with an oblique comminuted fracture of the femoral shaft. In male ferrets, the well developed os penis may interfere with visualization of femoral fractures in this projection.

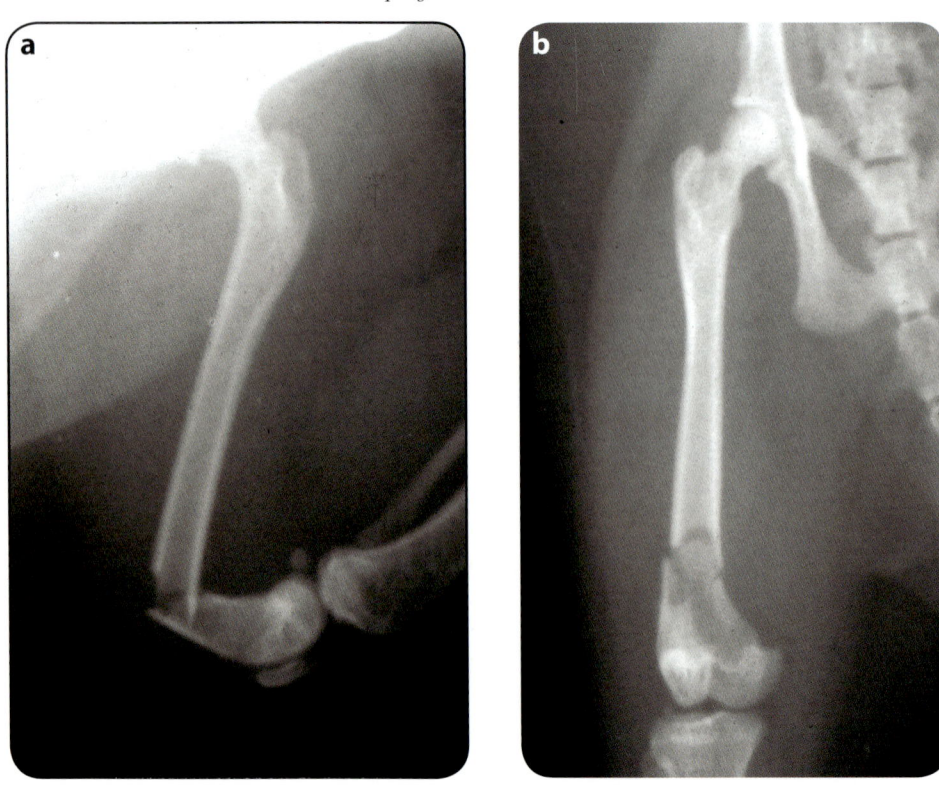

Figure 10.91a,b. Lateral (a) and ventrodorsal (b) radiographs of the femur of a ferret with a distal metaphyseal fracture. Note that displacement is most evident in the lateral view.

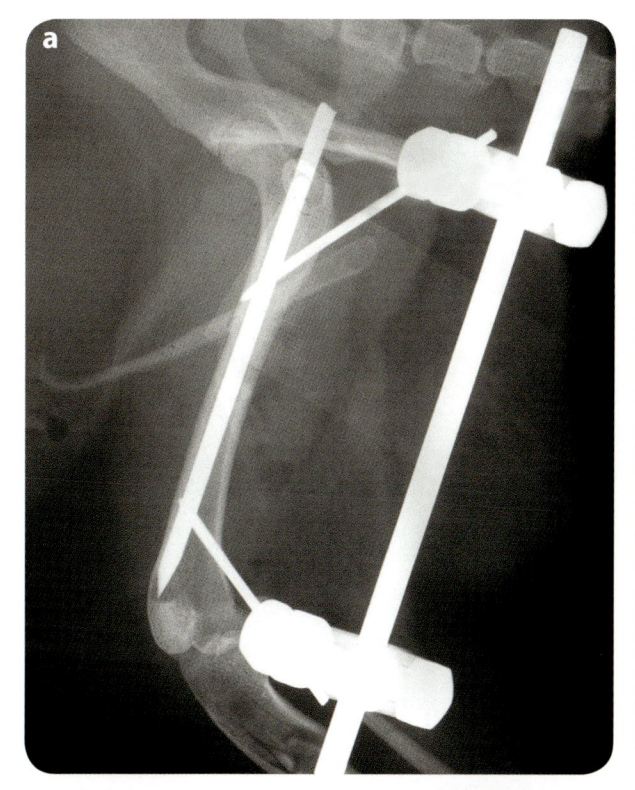

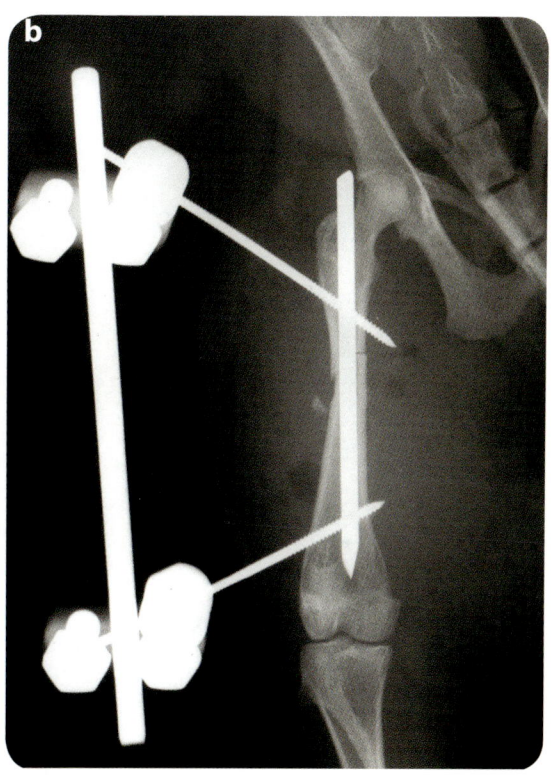

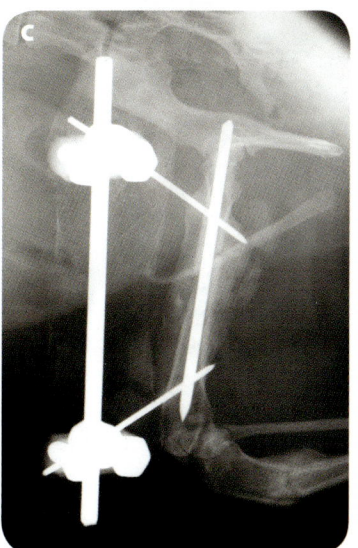

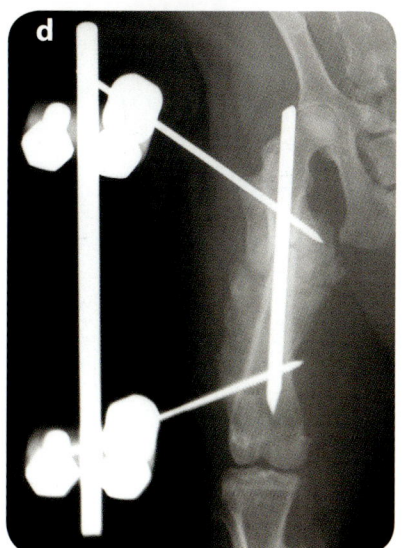

Figure 10.92a,b,c,d. Post-operative radiographs following repair of the fracture shown in Figure 10.90 using a 2-pin monoplanar configuration of external fixation in conjunction with intramedullary pinning. When the goal of the external fixator is simply to counteract rotational forces, it may not be necessary to adhere to the basic principle of 2 pins per bone segment. Two 1.2 threaded exernal fixation pins and a 2-mm intramedullary pin have been placed.
Lateral and craniocaudal projections immediately (a,b) and 5 weeks (c,d) post surgery.
In (c,d) note marked periosteal callus is present, which is not uncommon in young animals. A transverse, non-displaced proximal metaphyseal fracture of the tibia was also present. This fracture was stabilized by the intact fibula, and is partially healed (c)

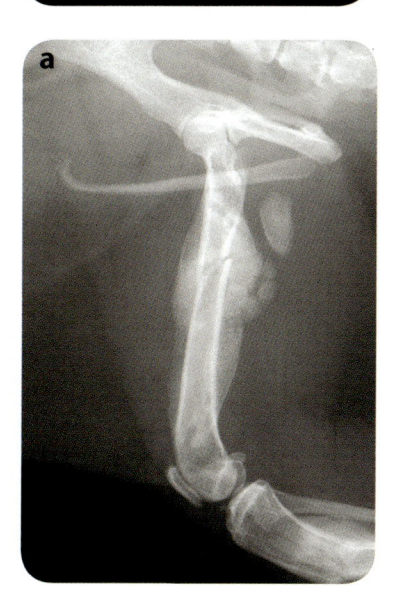

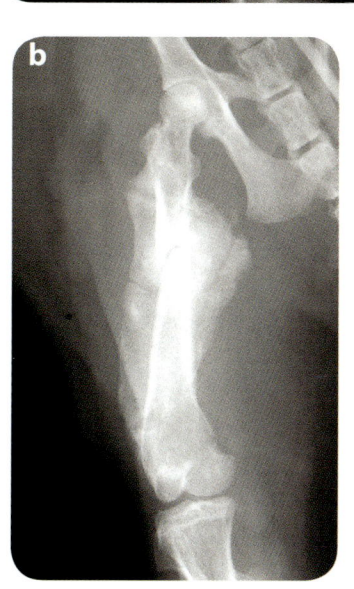

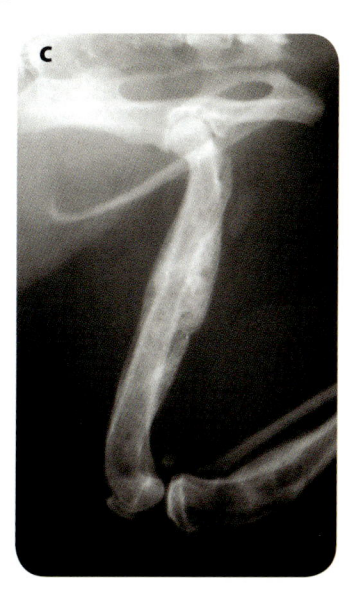

Figure 10.93a,b,c. Radiographic appearance after removal of all implants, 7 weeks post surgery, lateral (a) and craniocaudal (b) projections.
Radiographic appearance 2 months after removal of the implants (c). The callus is becoming more organized, which is a normal sequence in fracture healing.

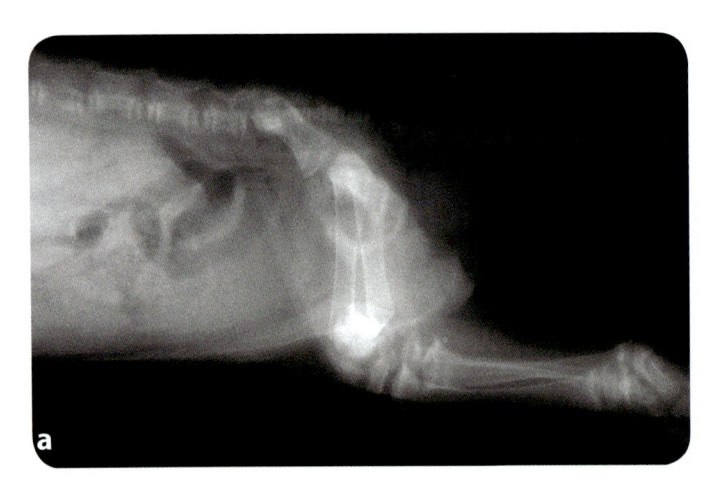

Figure 10.94. Lateral radiograph of the pelvis and pelvic limbs of a 3-month-old female ferret with metabolic bone disease. Note the very thin cortical bones of the femurs. Presenting signs included general lameness.

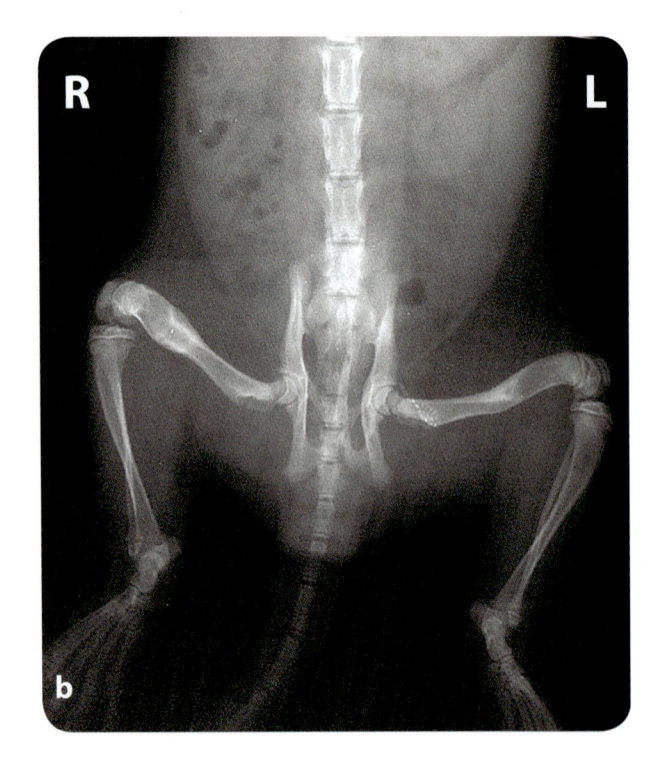

Figure 10.95. Ventrodorsal radiograph of the same patient taken after one month of dietary correction (p/d Prescription Diet, Hills Pet Nutrition Inc., Topeka KS). Note femoral shaft deformities due to healing of pathologic fractures.

Diseases of the Tibia and Fibula

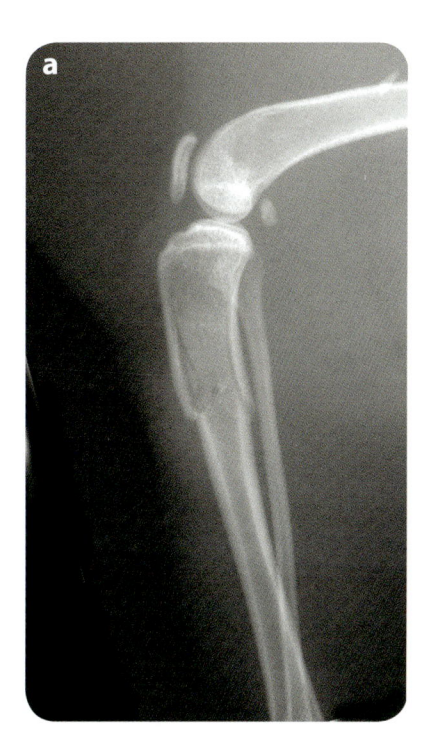

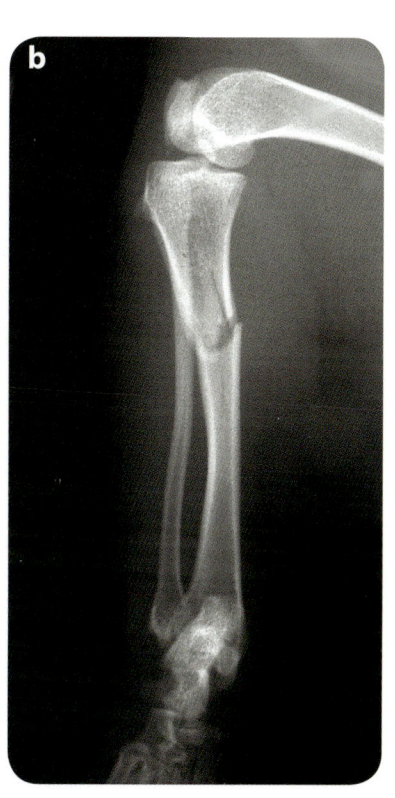

Figure 10.96a,b. Radiographs of a transverse, minimally displaced proximal diaphyseal fracture of the tibia in the same patient in Figures 10.92, 10.93.
The fibula is intact, providing adequate natural stabilization for the fracture.

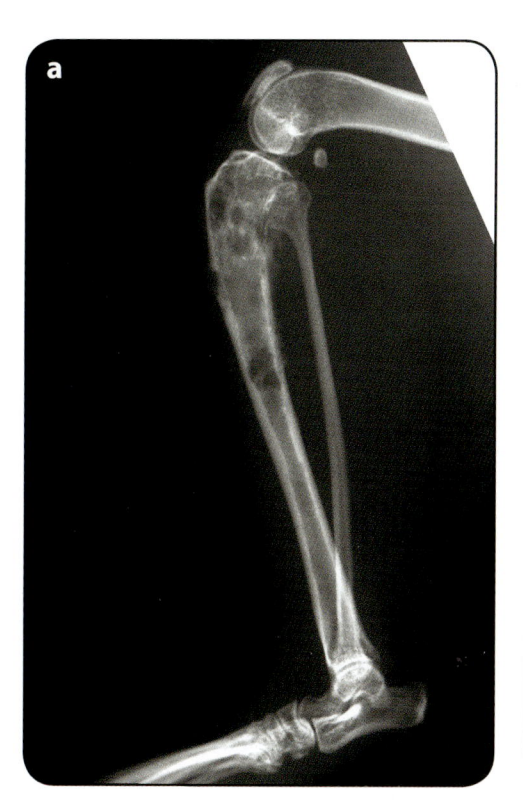

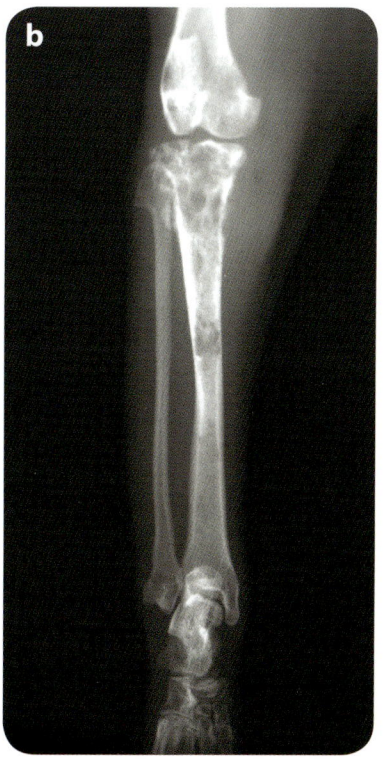

Figure 10.97a,b. Lateral (a) and craniocadual (b) radiographs of a lame ferret. Multiple lytic areas are seen including cortical thining, trabecular loss and zones of global lysis suggesting bone neoplasia.

Diseases of the Tarsus, Metatarsus, and Phalanges

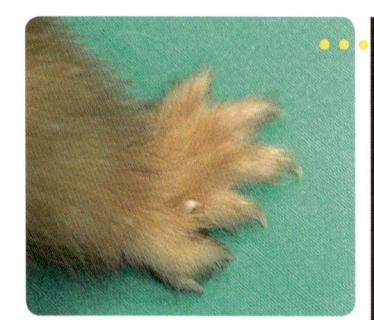

Figure 10.98. Clinical appearance of the foot.

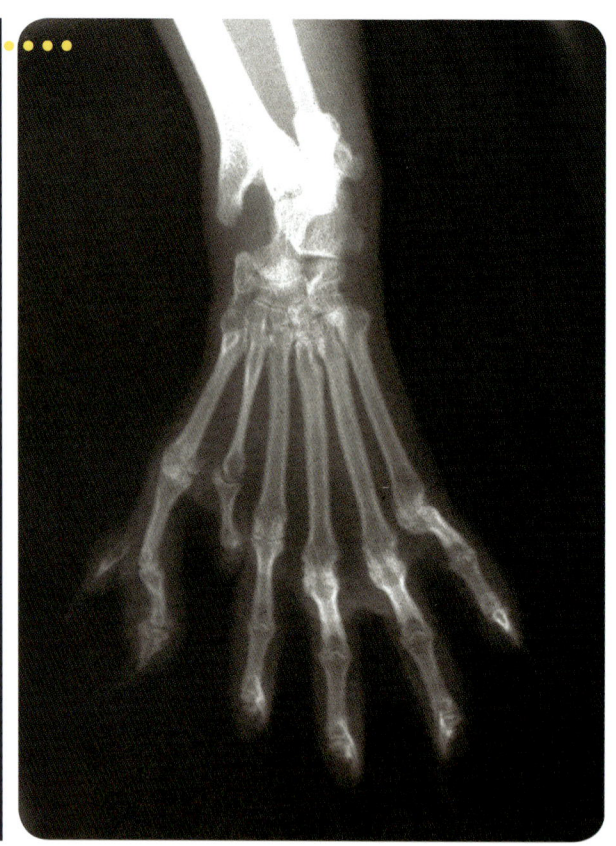

Figure 10.99. Dorsoplantar radiograph of a ferret with an unusual case of polydactylism. A complete supranumerary metatarsal bone (with just two phalanges) is present between digit 4 and digit 5. A supranumerary 3rd phalanx with toenail is present at digit 5.

References:
Antinoff N, Hahn K: Ferret oncology: diseases, diagnostics, and therapeutics. Vet Clin N Am Exotic Anim Prac. 2004;7:579-625.
Brown SA, Rosenthal LR: Question #30. In: Self Assessment Colour Review of Small Mammals. London: Manson Publishing; 1997:27-28.
Brown SA, Rosenthal LR: Question #56. In: Self Assessment Colour Review of Small Mammals. London: Manson Publishing; 1007:51-52.
Brown SA, Rosenthal LR: Question #102. In: Self Assessment Colour Review of Small Mammals. London: Manson Publishing; 1997:89-90.
Capello V: External fixation for fracture repair in small exotic mammals. Exotic DVM. 2005;7(6):21-37.
Castanheira de Matos RE, Morrisey JK: Common procedures in the pet ferret. Vet Clin N Am Exotic Anim Prac. 2006;9:347-365.
Fisher PG: Urethrostomy and penile amputation to treat urethral obstruction and preputial masses in male ferrets. ExoticDVM. 2002; 3(6):21-25.
Fisher PG: Exotic mammal renal disease: diagnosis and treatment. Vet Clin N Am Exotic Anim Prac. 2006; 9:69-96.
Girling SJ: Mammalian imaging and anatomy. In: Meredith A, Redrobe S, eds. BSAVA Manual of Exotic Pets 4th ed. Quedgeley, Glouchester: British Small Animal VeterinaryAssociation;2005:1-12.
Hess L: Ferret lymphoma: the old and the new. Sem Av Exotic Pet Med. 2005;14(3):199-204.
Hoefer HL: Section 1. Ferrets. Gastrointestinal diseases, part I. In: Quesenberry KE, Carpenter JW, eds. Ferrets, Rabbits and Rodents. Clinical Medicine and Surgery 2nd ed. Philadelphia, PA:Elsevier;2004:25-33.
Kapatkin A: Orthopedics in small mammals. In: Quesenberry KE, Carpenter JW, eds. Ferrets, Rabbits and Rodents. Clinical Medicine and Surgery 2nd ed. Philadelphia, PA:Elsevier;2004:383-391.
Kottwitz J: Horizontal beam radiography in ferrets. Exotic DVM. 2004;6(1):37-41.
Orcutt CJ: Ferret urogenital disease. Vet Clin N Am Exotic Anim Prac. 2003, 6:113-138.
Petrie JP: Section 1. Ferrets. Cardiovascular and other diseases, part I. In: Quesenberry KE, Carpenter JW, eds. Ferrets, Rabbits and Rodents. Clinical Medicine and Surgery 2nd ed. Philadelphia, PA:Elsevier;2004:58-66.
Pollock CG: Section 1. Ferrets. Urogenital diseases. In: Quesenberry KE, Carpenter JW, eds. Ferrets, Rabbits and Rodents. Clinical Medicine and Surgery 2nd ed. Philadelphia, PA:Elsevier;2004:41-49.
Powers LV: Pyothorax in a ferret: ExoticDVM, 1999;1(1):32.
Powers LV, Winkler K, Garner M, et al.: Omentalization of prostatic abscesses and large cysts in ferrets *(Mustela putorius furo)*. J Exotic Pet Med. 2007;16(3):186-194.
Rosenthal KL: Section 1. Ferrets. Respiratory diseases. In: Quesenberry KE, Carpenter JW, eds. Ferrets, Rabbits and Rodents. Clinical Medicine and Surgery 2nd ed. Philadelphia, PA:Elsevier;2004:72-78.
Silverman S, Tell LA: Domestic Ferret *(Mustela putorius)*. In: Radiology of Rodents, Rabbits and Ferrets. An Atlas of Normal Anatomy and Positioning. Philadelphia, PA: Elsevier Saunders; 2005:231-289.
Stefanacci JD, Hoefer HL: Radiology and ultrasound. In: Quesenberry KE, Carpenter JW, eds. Ferrets, Rabbits and Rodents. Clinical Medicine and Surgery 2nd ed. Philadelphia, PA:Elsevier;2004:395-413.
Wyre NR, Hess L: Clinical technique: ferret thoracocentesis. Sem Avian Exotic Pet Med. 2005;14(1):22-25.

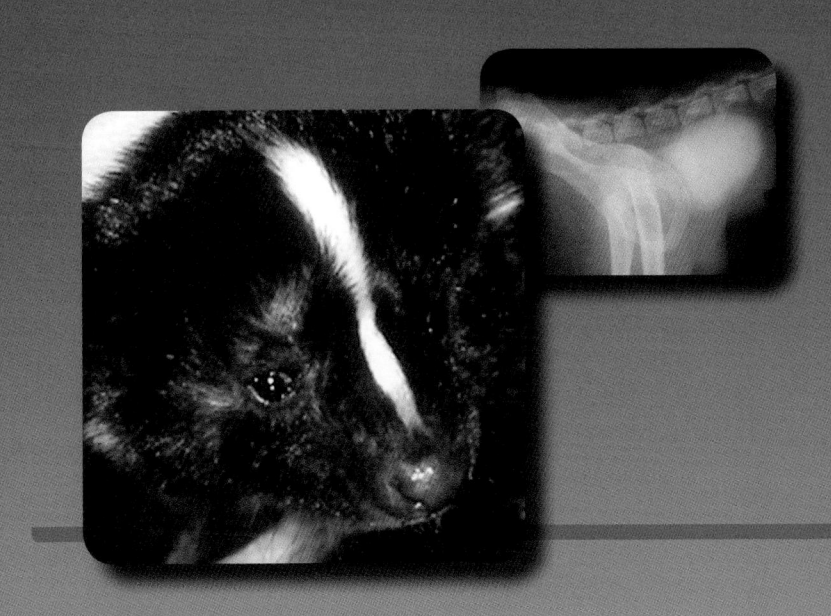

SKUNK

The NORMAL HEAD

Lateral Projection

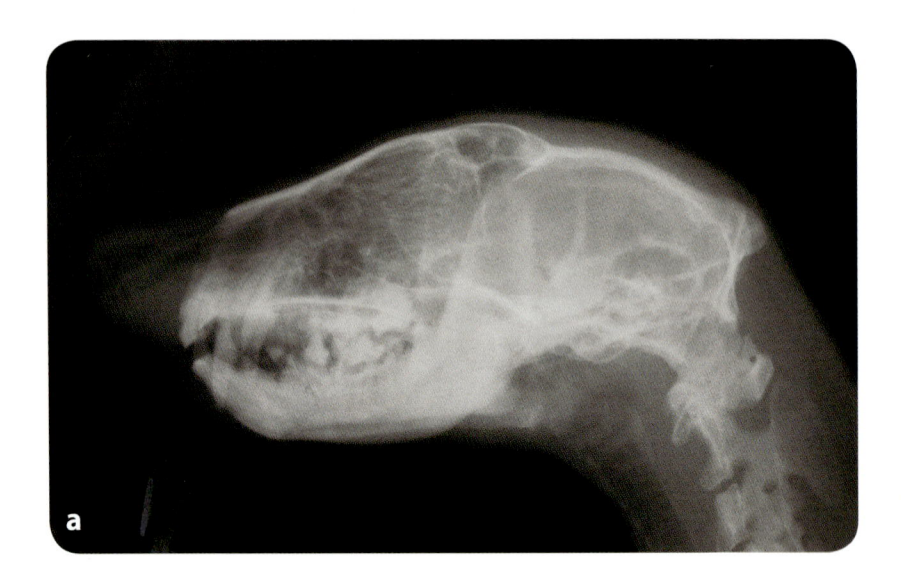

Figure 11.1a,b. Radiograph of a 3.5 kg wild intact sub-adult male skunk, actual size (a) and 150% of actual size (b).

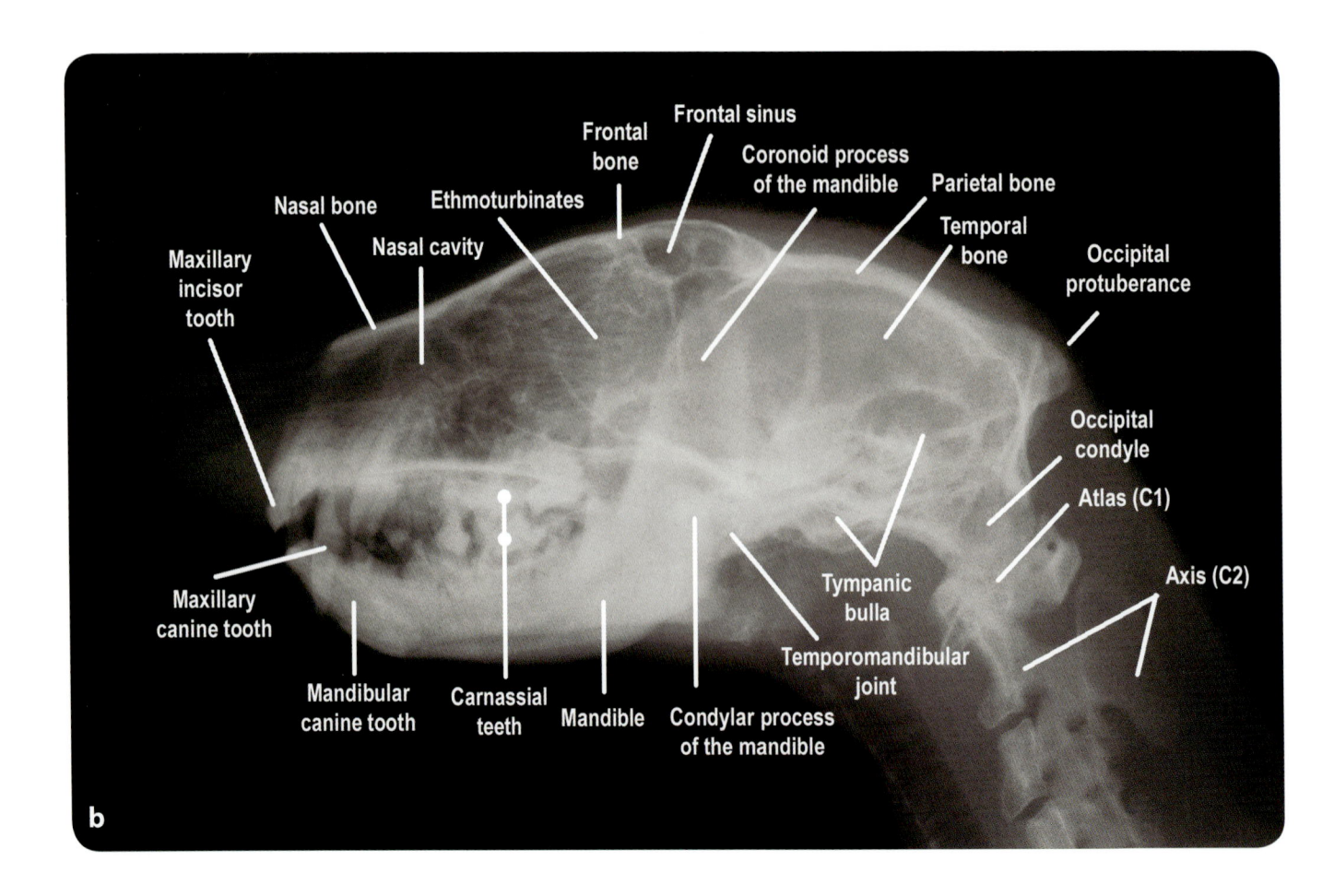

The NORMAL THORAX
The Cervical and Thoracic Vertebral Column
Lateral Projection

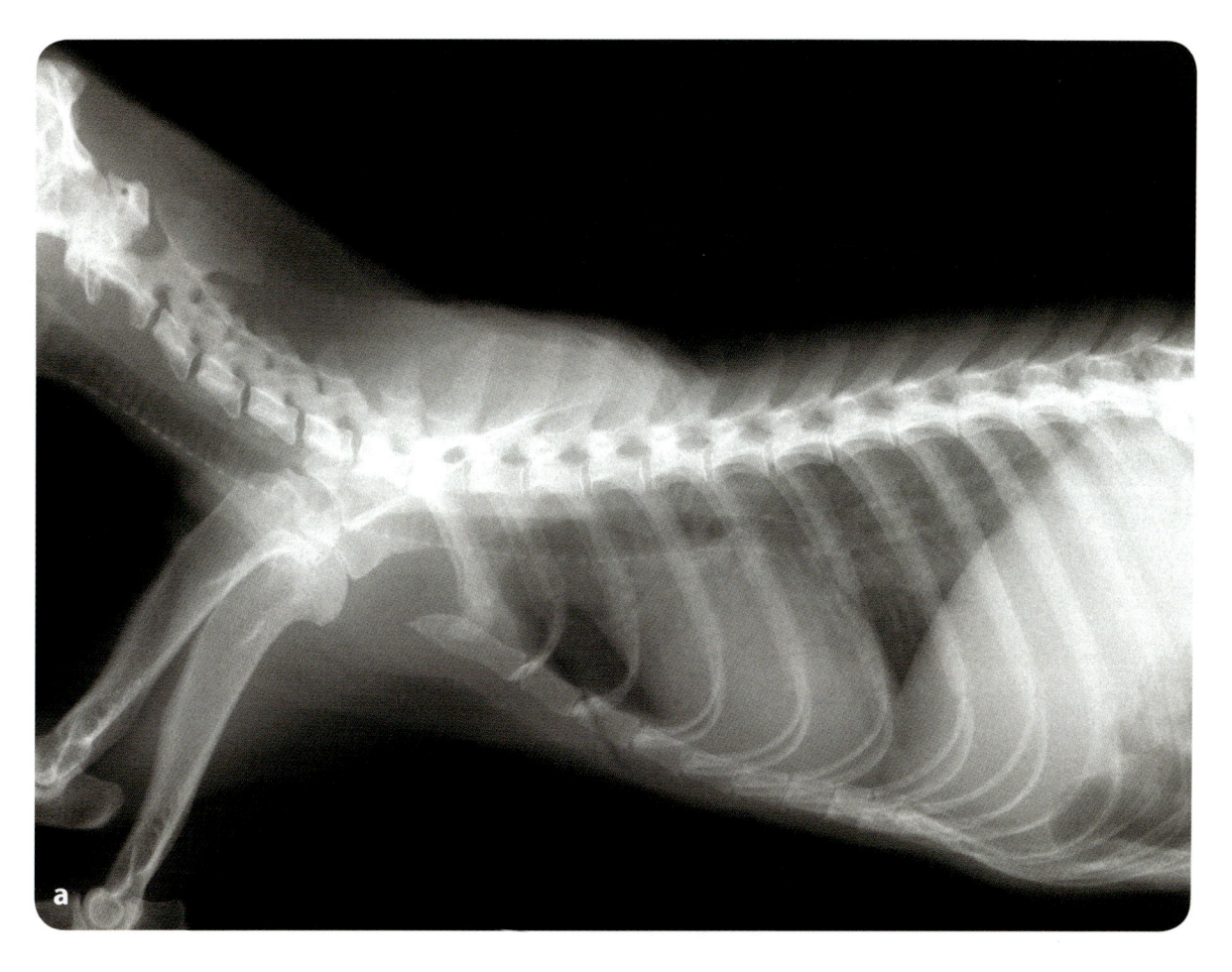

Figure 11.2a,b.
Radiograph of a 3.5 kg wild intact subadult male skunk, actual size (a) and 70% of actual size (b).

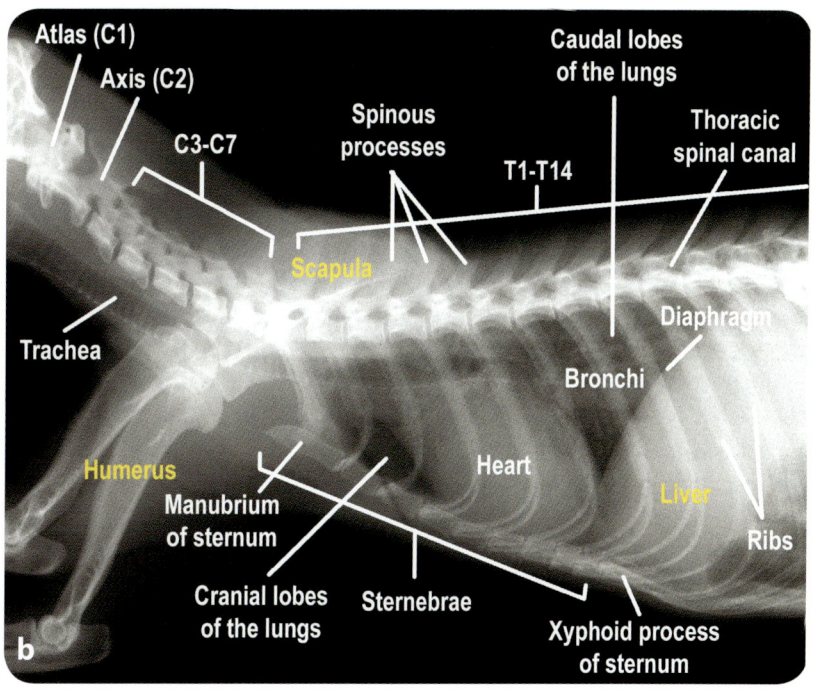

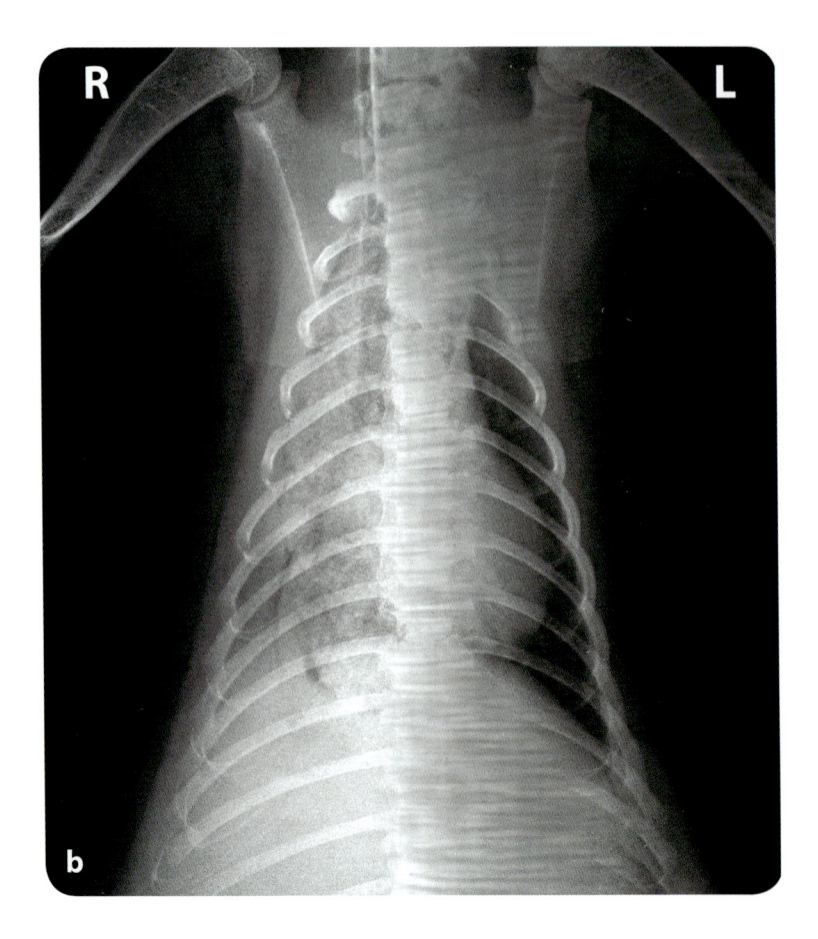

ABNORMALITIES of the THORAX
Diseases of the Mediastinum

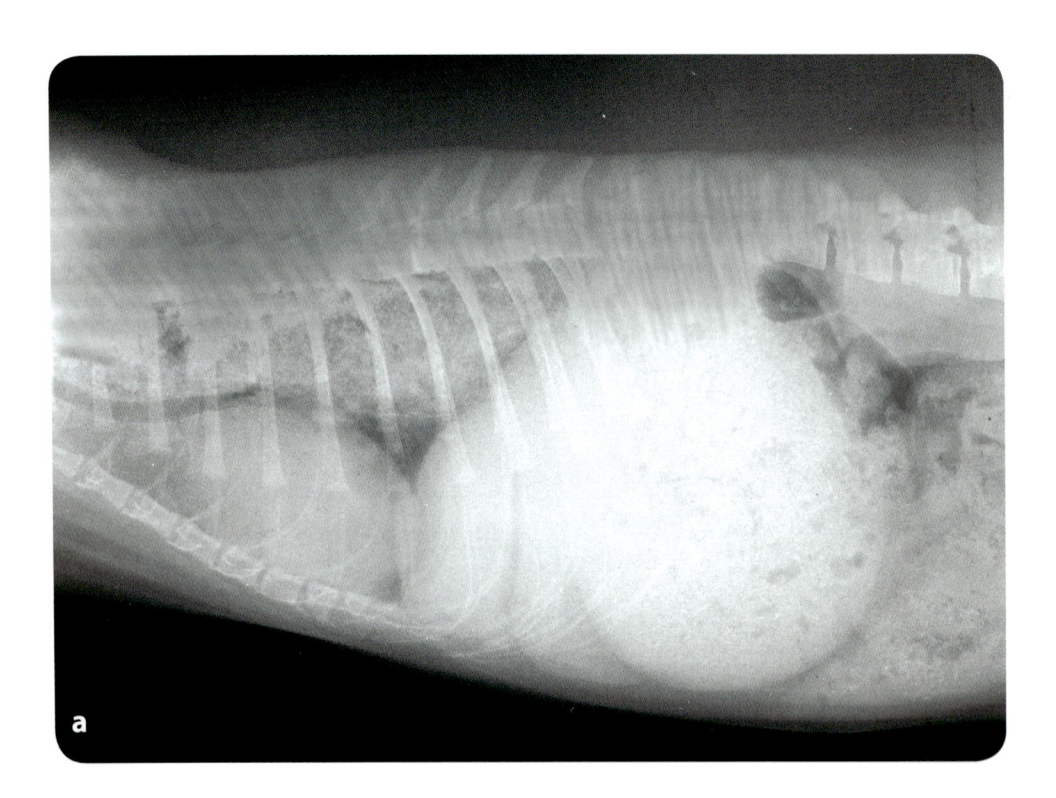

Figure 11.3a,b. Lateral (a) and ventrodorsal (b) projections of the thorax of a clinically normal 2-year-old male skunk. In (a) a tubular heterogenous opacity displaces the trachea and heart base ventrally and also obscures the dorsal aspect of the lung field. In (b) the same heterogenous opacity is confirmed to be within the mediastinum and is consistent with megaesophagus. The distended esophagus has caused partial collapse of the right lung.

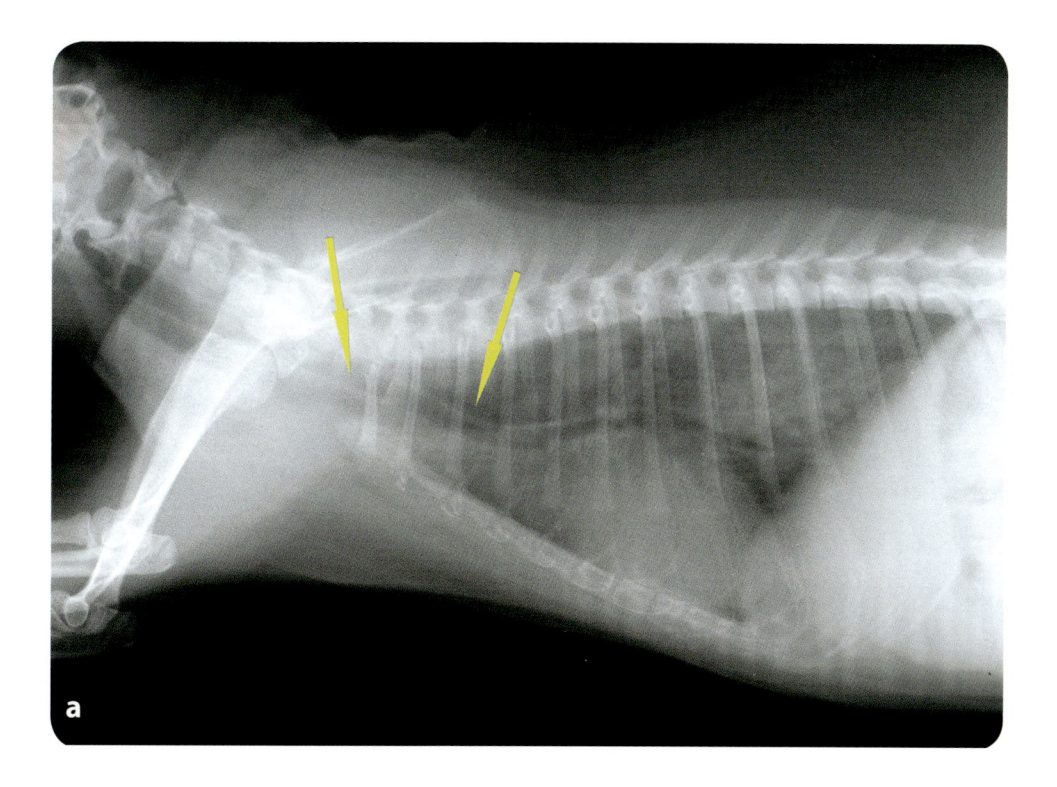

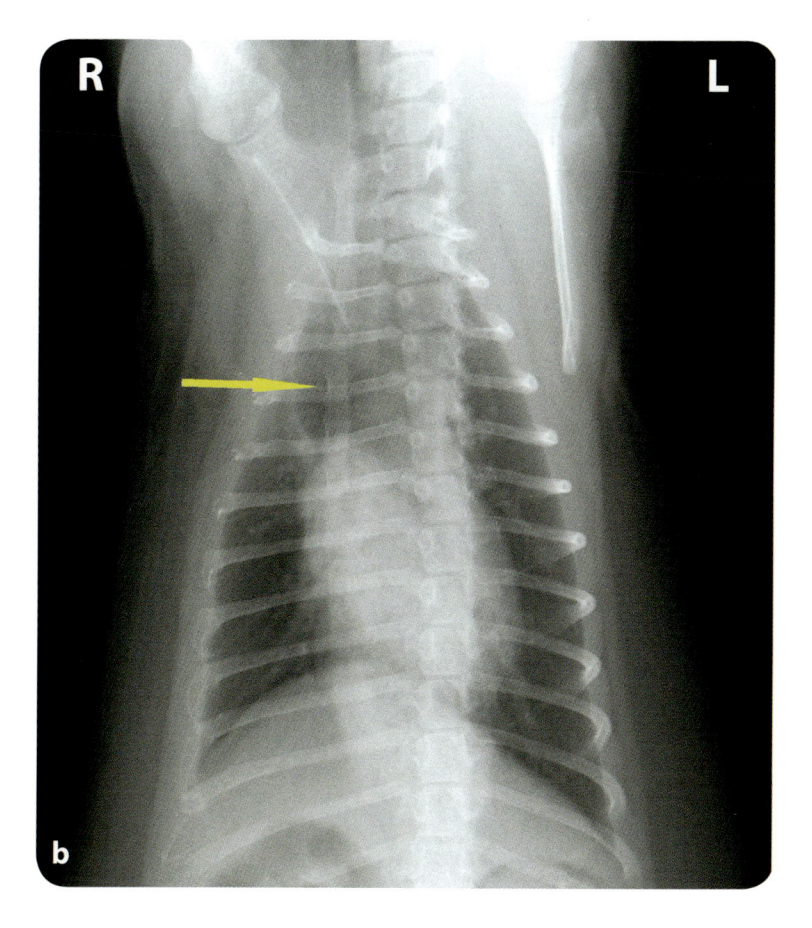

Figure 11.4a,b. Radiographs of the same patient as in Figure 11.3 after a 3-hour fast. Lateral (a) and ventrodorsal (b) projections reveal a dilated air filled esophagus (arrows). The abnormal pattern seen in Figure 11.3b has resolved because the esophagus has been cleared of food.

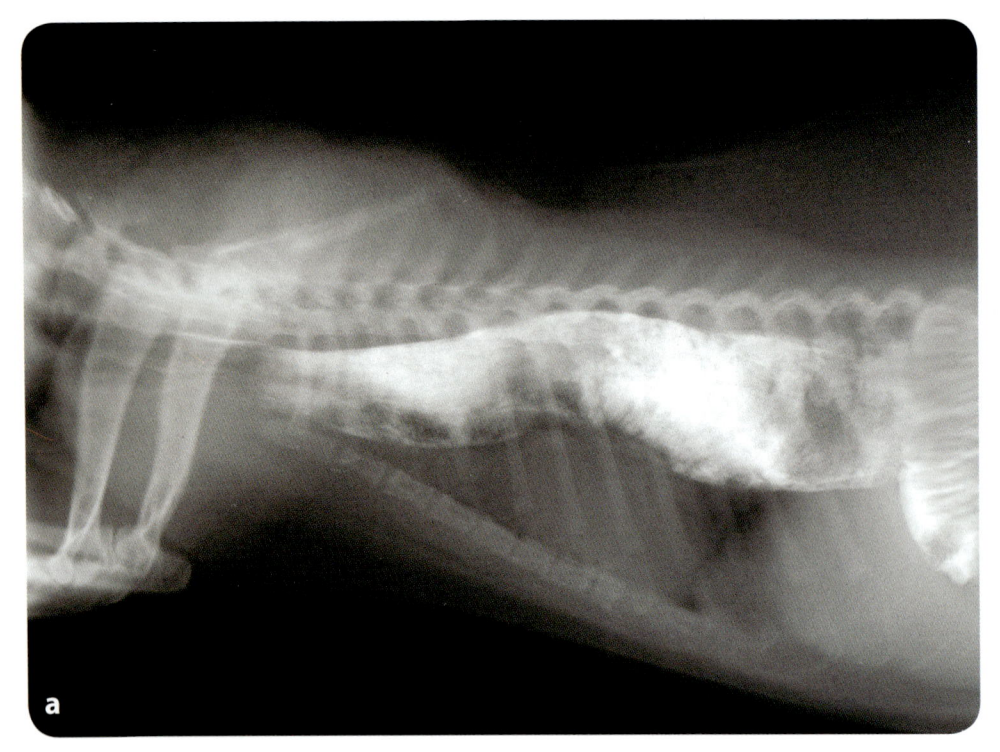

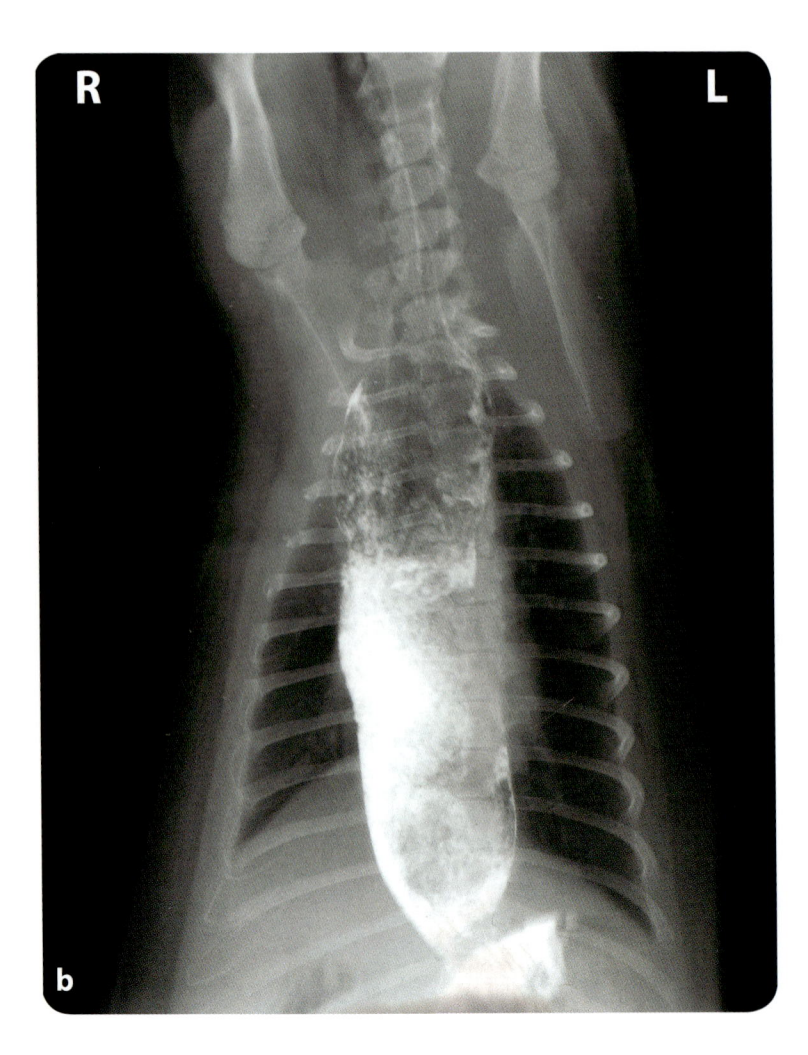

Figure 11.5a,b. Lateral (a) and ventrodorsal (b) projections of the same patient made immediatley after administration of barium sulfate. Retention of contrast material within the dilated esophagus confirms megaesophagus as suspected on the survey images. Cause of megaesophagus in skunks is unknown.

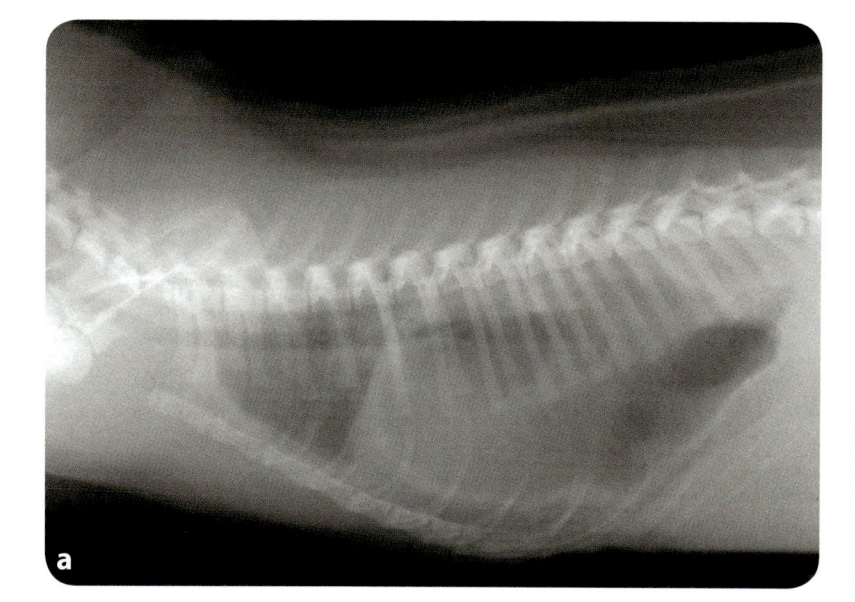

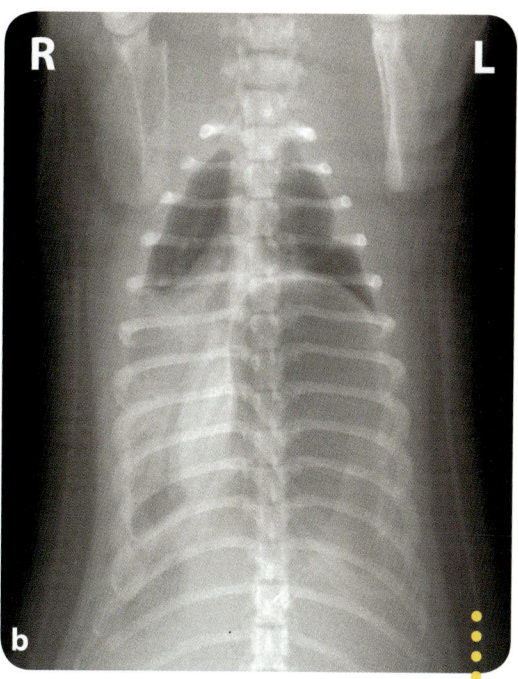

Figure 11.6a,b. Lateral (a) and ventrodorsal (b) projections of the thorax of a 4-month-old skunk presented for marked dyspnea. (a) reveals a distended hollow viscus within the thorax that is displacing the lung cranially. On (b) the distended viscus is primarily on the left side and the left crus of the diaphragm is not visible. This study is typical of a diaphragmatic hernia with the stomach displaced into the pleural cavity.
The owners suspect the skunk may have fallen from a second floor landing.

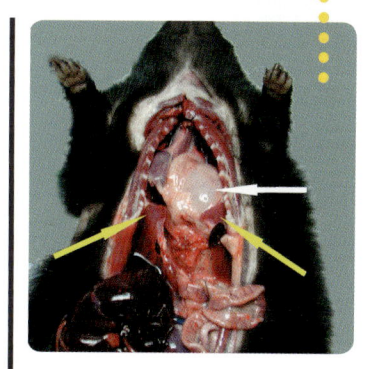

Figure 11.7. Necropsy specimen showing the stomach herniated into the thoracic cavity (white arrow), cranial to the tear in the diaphragm (yellow arrows).

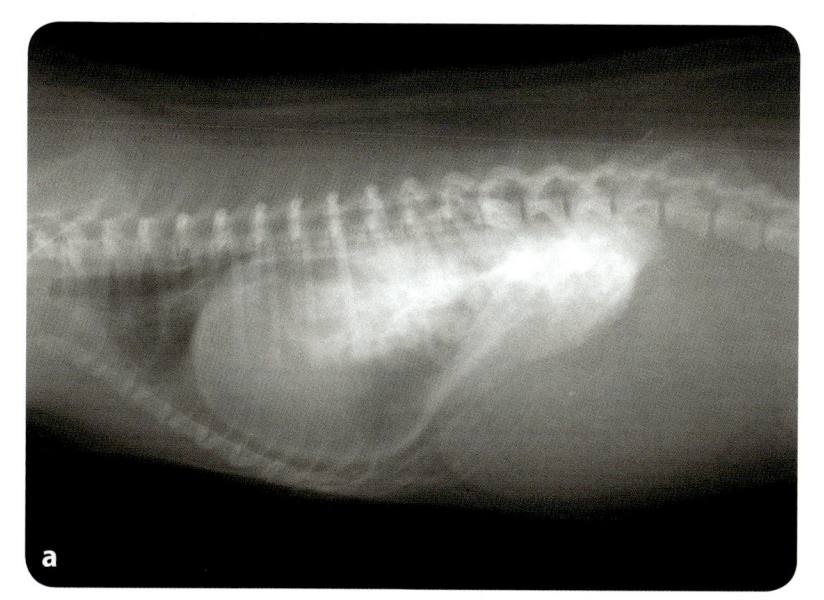

Figure 11.8a,b. Lateral (a) and ventrodorsal (b) projections of the same patient following oral administration of barium sulfate. The stomach is herniated into the thoracic cavity (arrows). Although an upper gastrointestinal contrast study was useful in this case, it should be noted that this procedure is only of benefit if stomach or intestine are herniated into the thorax. Because many hernias contain only liver and other solid organs, a celiogram or ultrasound examination may be the procedure of choice for confirmation.

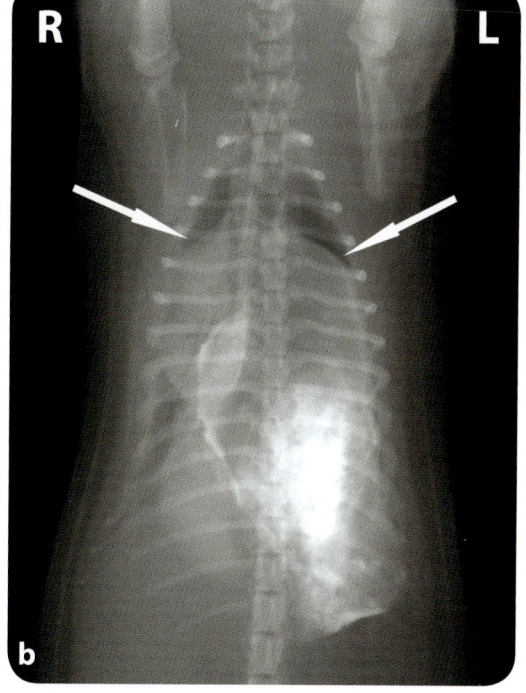

The NORMAL ABDOMEN
The Lumbar Vertebral Column
Lateral Projection

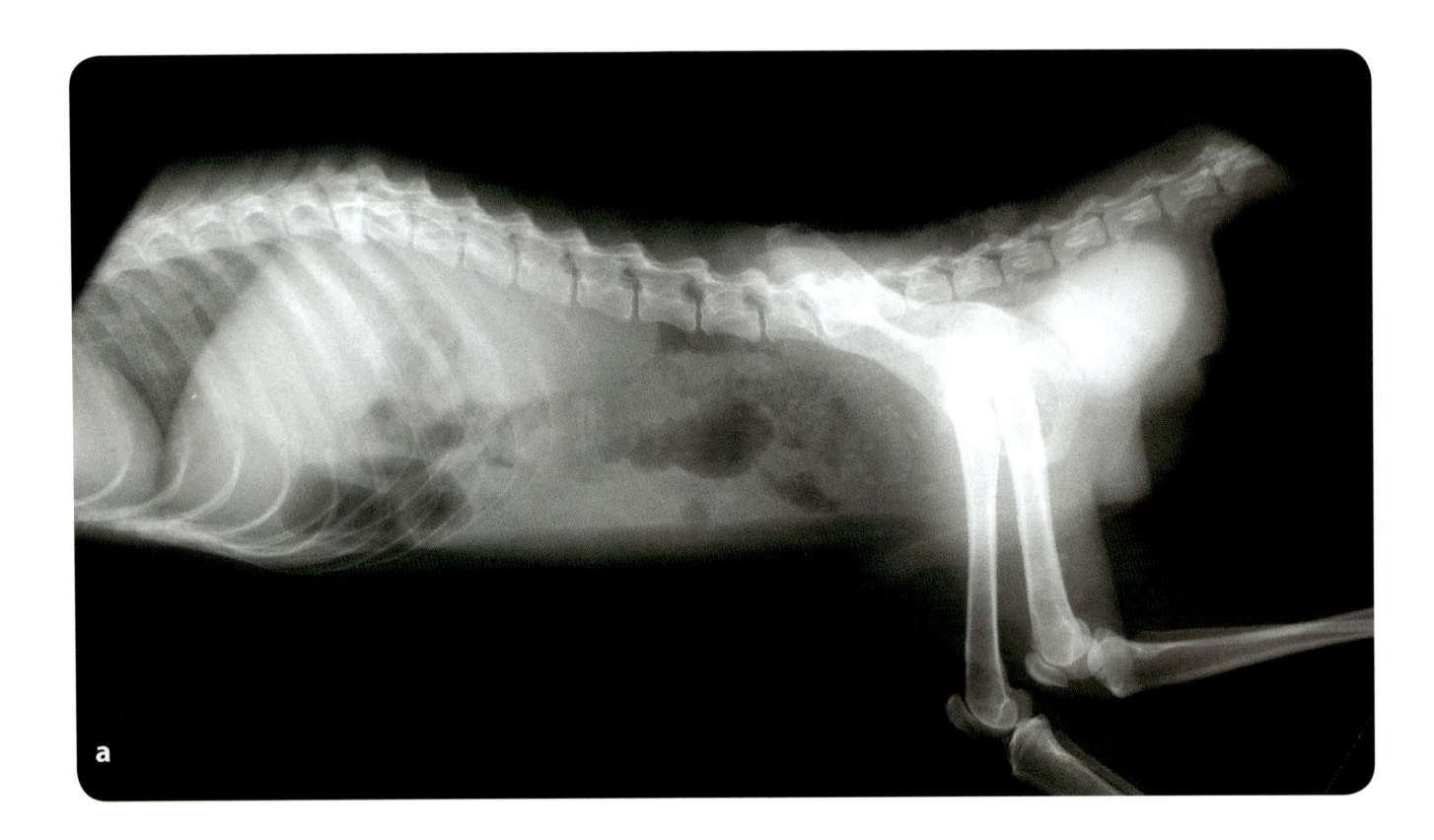

Figure 11.9a,b. Radiograph of a 3.5 kg wild intact sub-adult male skunk, 70% of actual size. Note the size of normal anal sacs in this species.

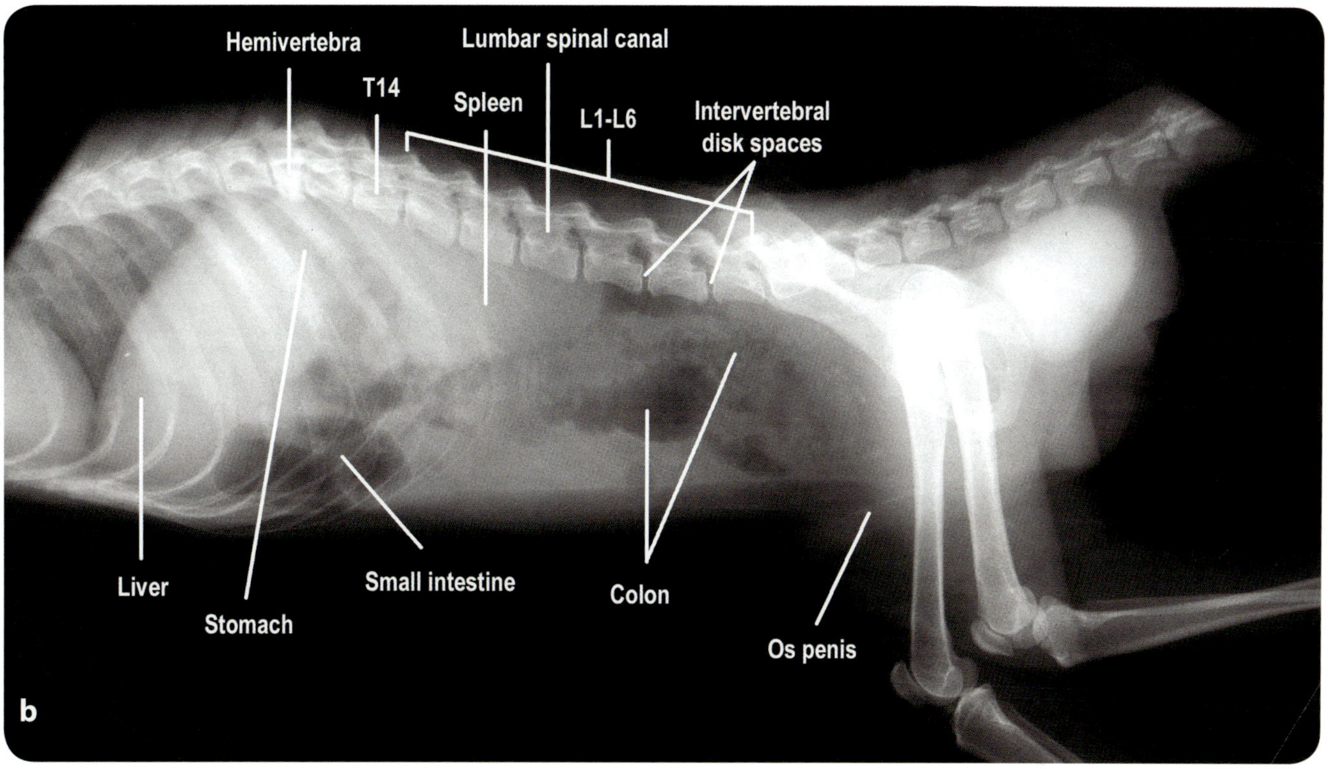
Hemivertebra
Lumbar spinal canal
T14
Spleen
L1-L6
Intervertebral
disk spaces
Liver
Stomach
Small intestine
Colon
Os penis
b

The NORMAL THORACIC LIMB
Lateral Projection

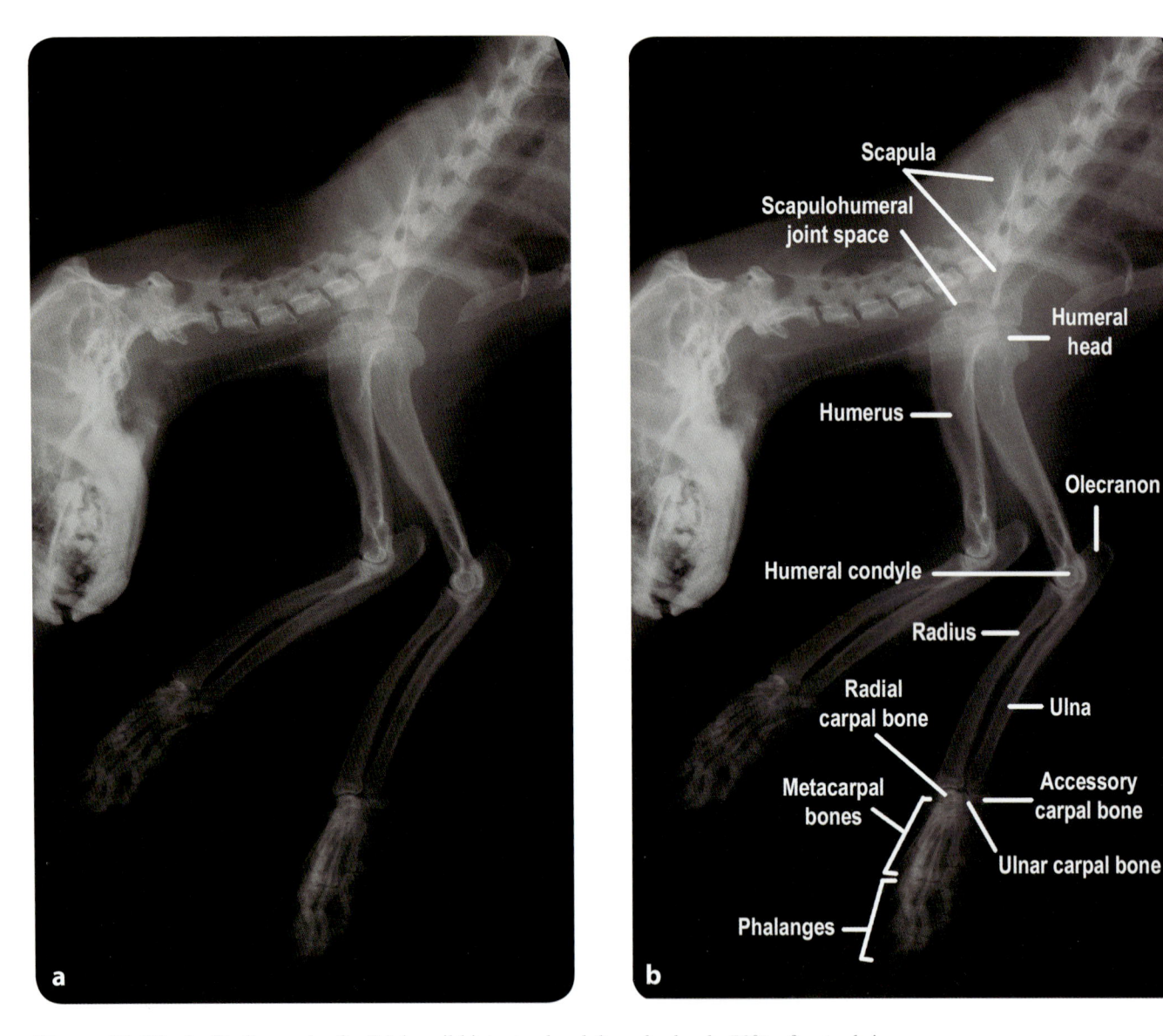

Figure 11.10a,b. Radiograph of a 3.5 kg wild intact sub-adult male skunk, 70% of actual size.

THORACIC LIMB

ABNORMALITIES of the THORACIC LIMB

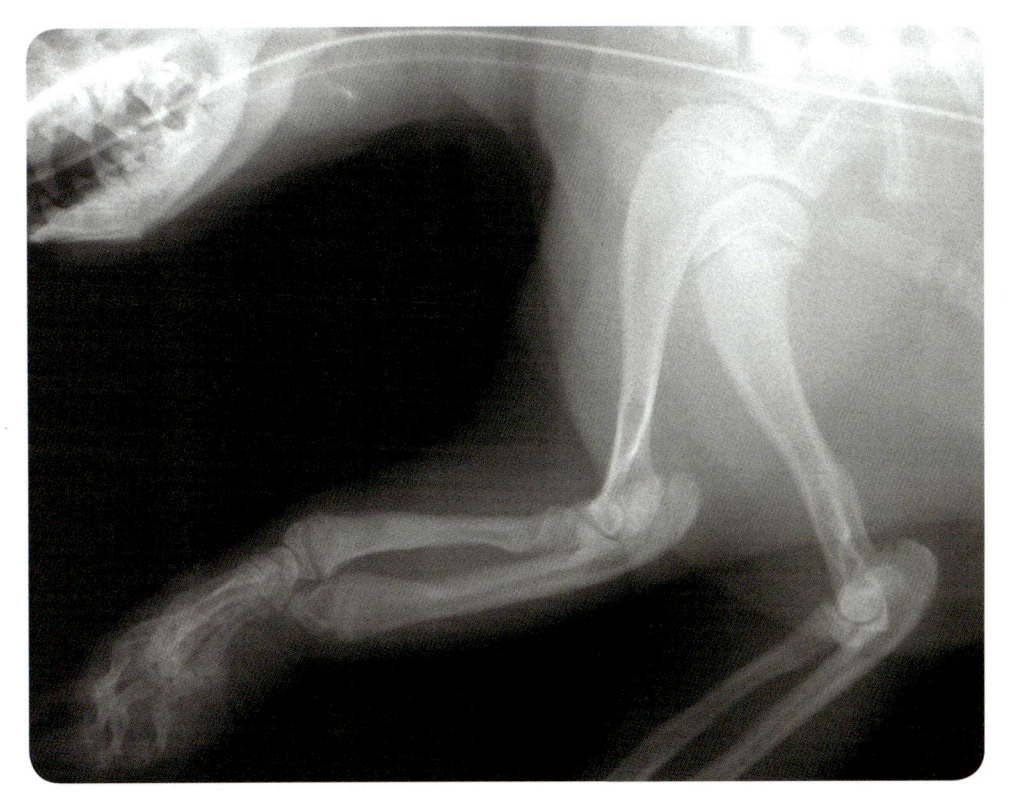

Figure 11.11. Lateral view of the thoracic limb of an 8-month-old skunk showing deformation of the radius and ulna due to metabolic bone disease.
Metabolic bone disease is common in pet skunks, and is often a result of feeding these omnivorous species vegetable-based diets.

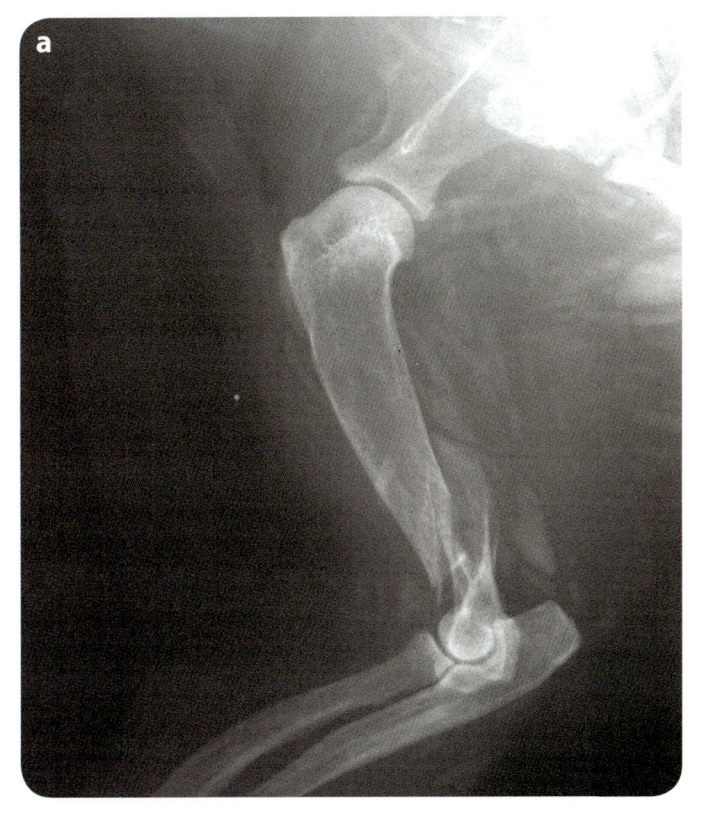

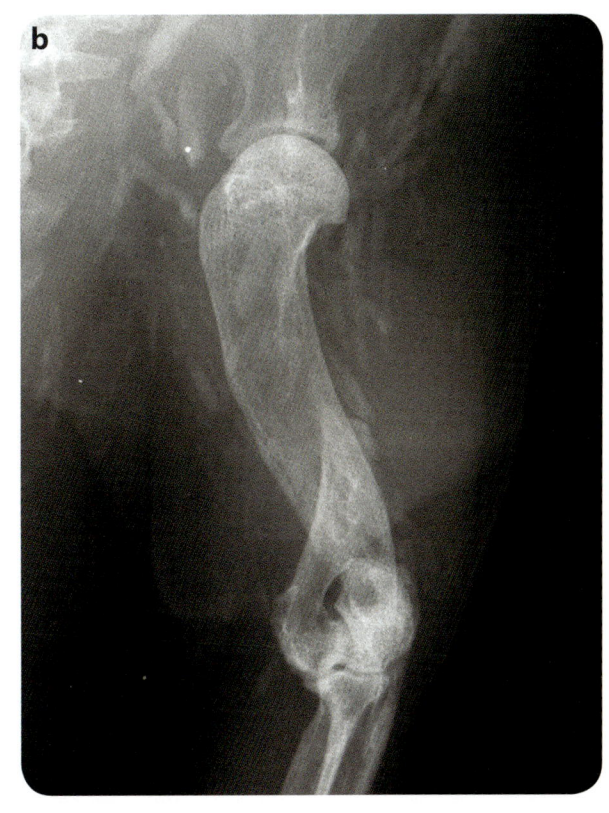

Figure 11.12a,b. Lateral (a) and craniocaudal (b) radiographs of the thoracic limb of a 3-year-old female skunk with metabolic bone disease. Note the distal metaphyseal, spiral fracture of the humerus and overall thin bone cortices. Increased soft tissue opacity around the bones is a result of thick dirty haircoat.

# The NORMAL PELVIC LIMB
Lateral Projection

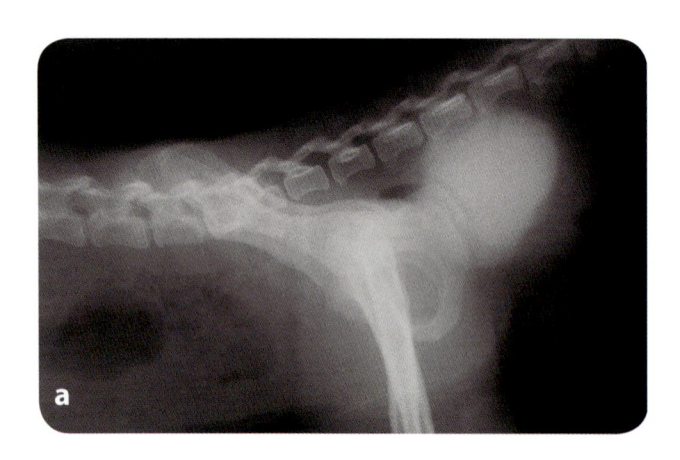

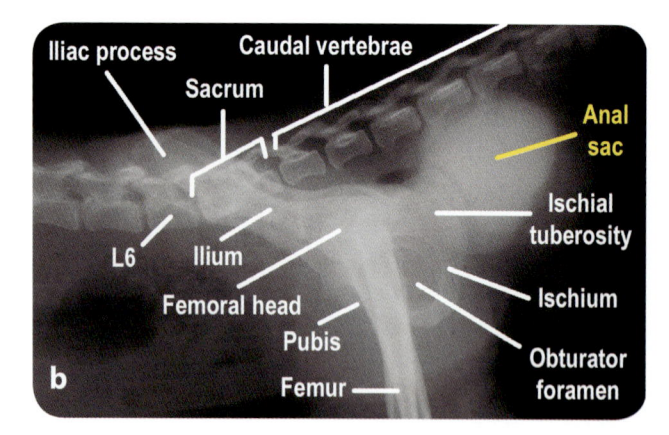

Figure 11.13a,b. Radiograph of a 3.5 kg wild intact sub-adult male skunk, 70% of actual size. Note the large anal sacs and their relationship with the pelvis.

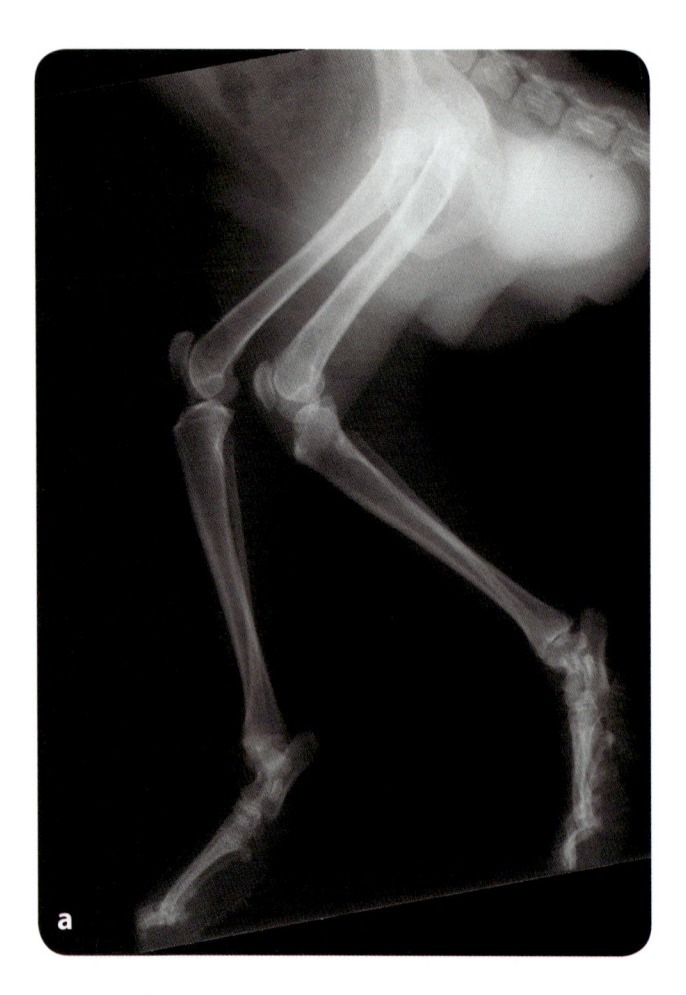

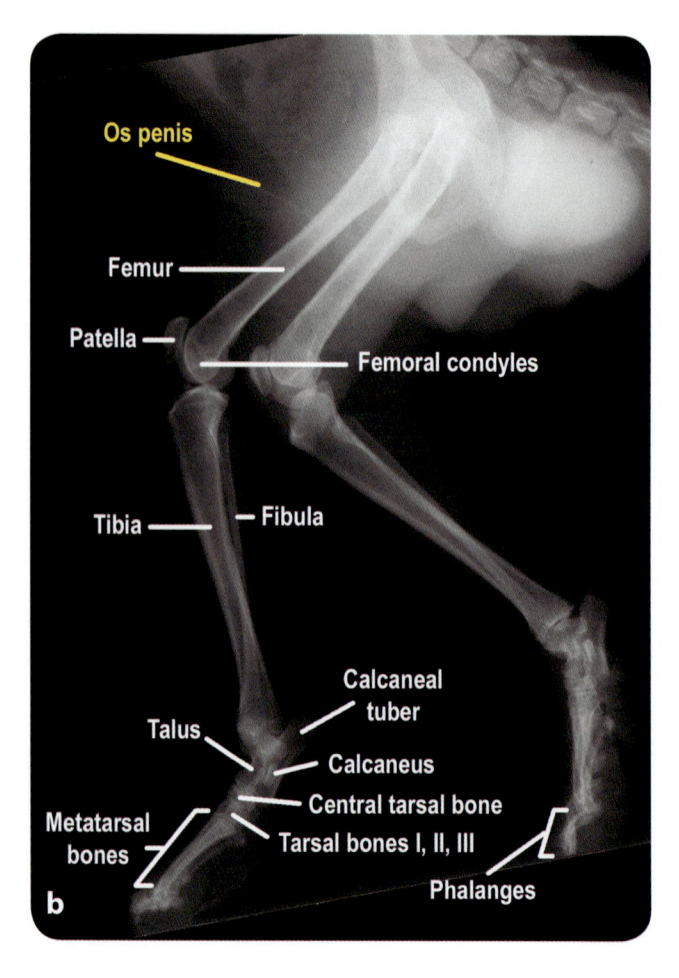

Figure 11.14a,b. Radiograph of a 3.5 kg wild intact sub-adult male skunk, 70% of actual size.

Ventrodorsal Projection of the Pelvis

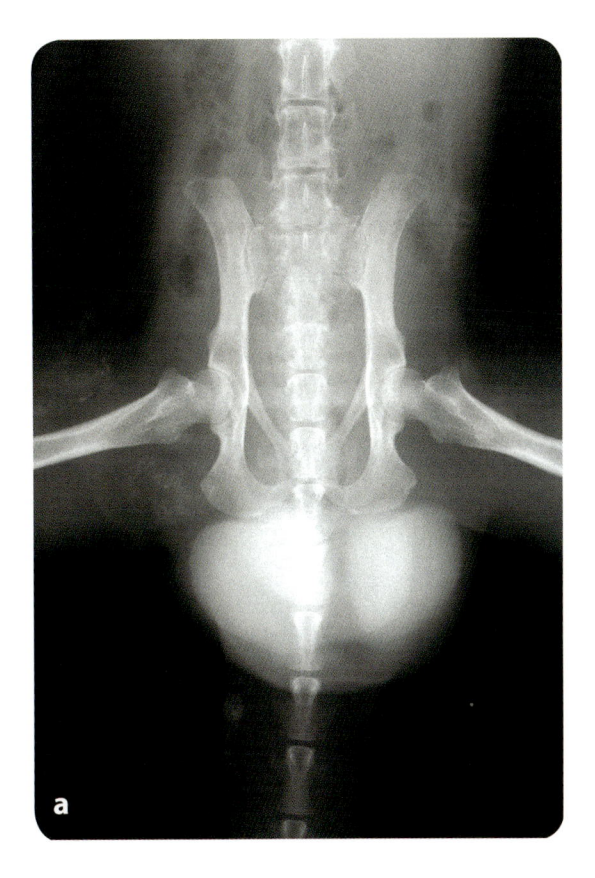

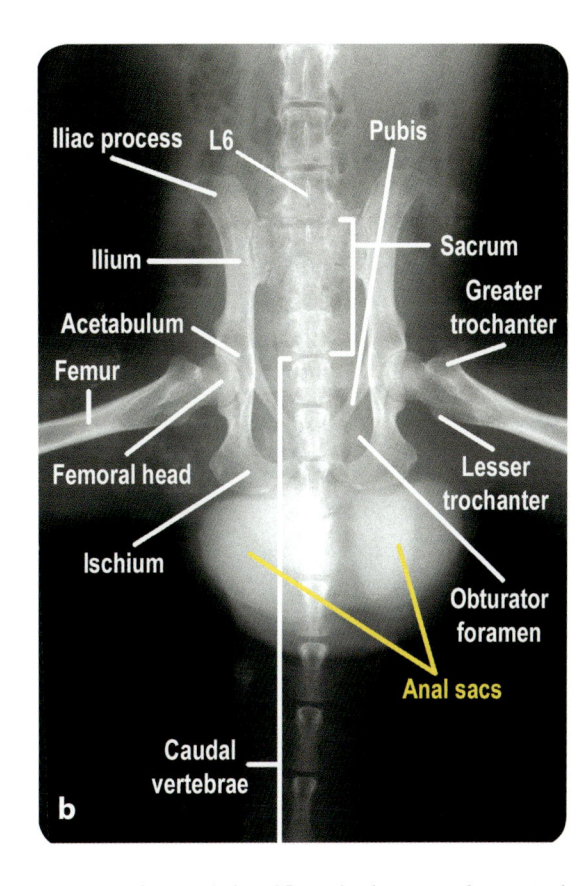

Figure 11.15a,b. Radiograph of a 3.5 kg wild intact sub-adult male skunk, 70% of actual size. Note the large anal sacs and their relationship with the pelvis.

Craniocaudal Projection of the Distal Pelvic Limb

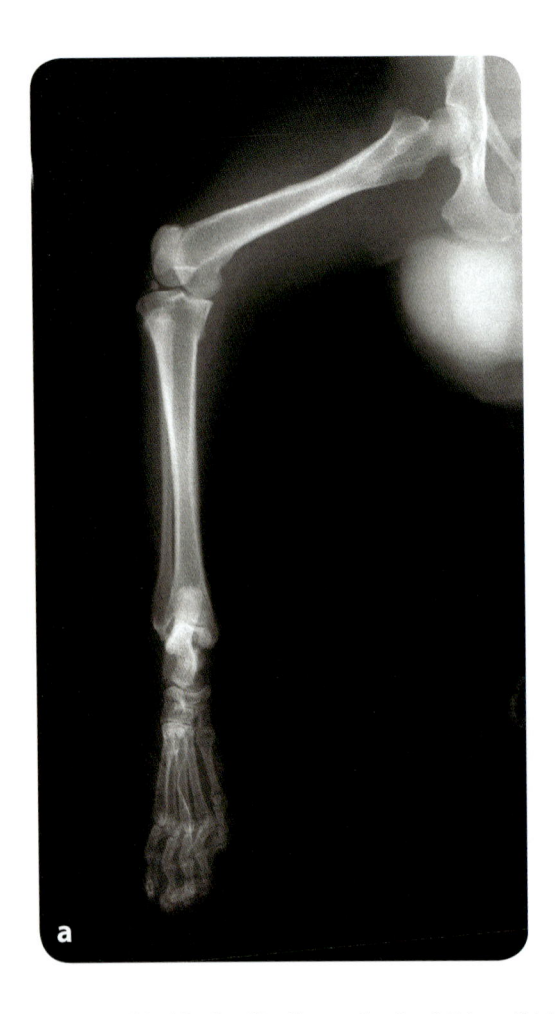

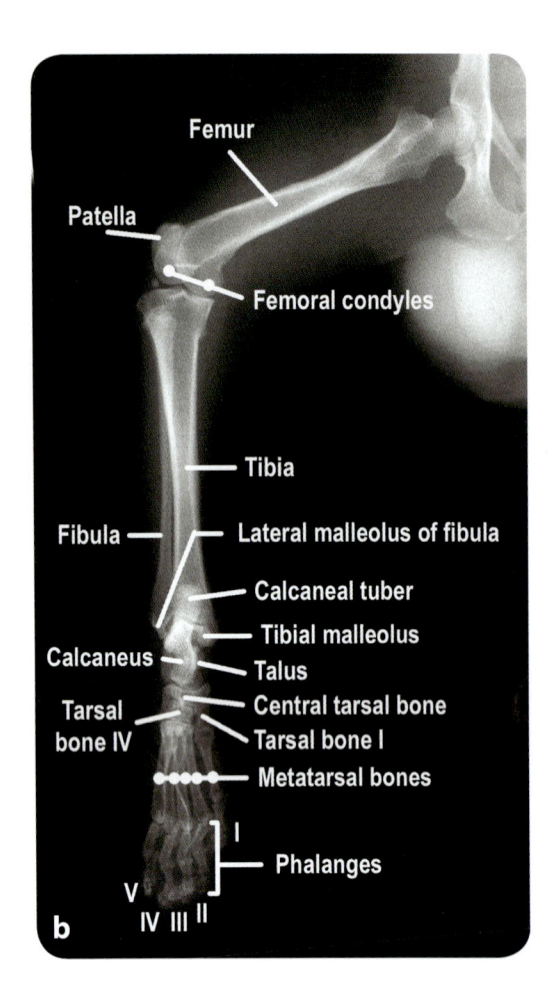

Figure 11.16a,b. Radiograph of a 3.5 kg wild intact sub-adult male skunk, actual size.

ABNORMALITIES of the PELVIC LIMB
Diseases of the Femur

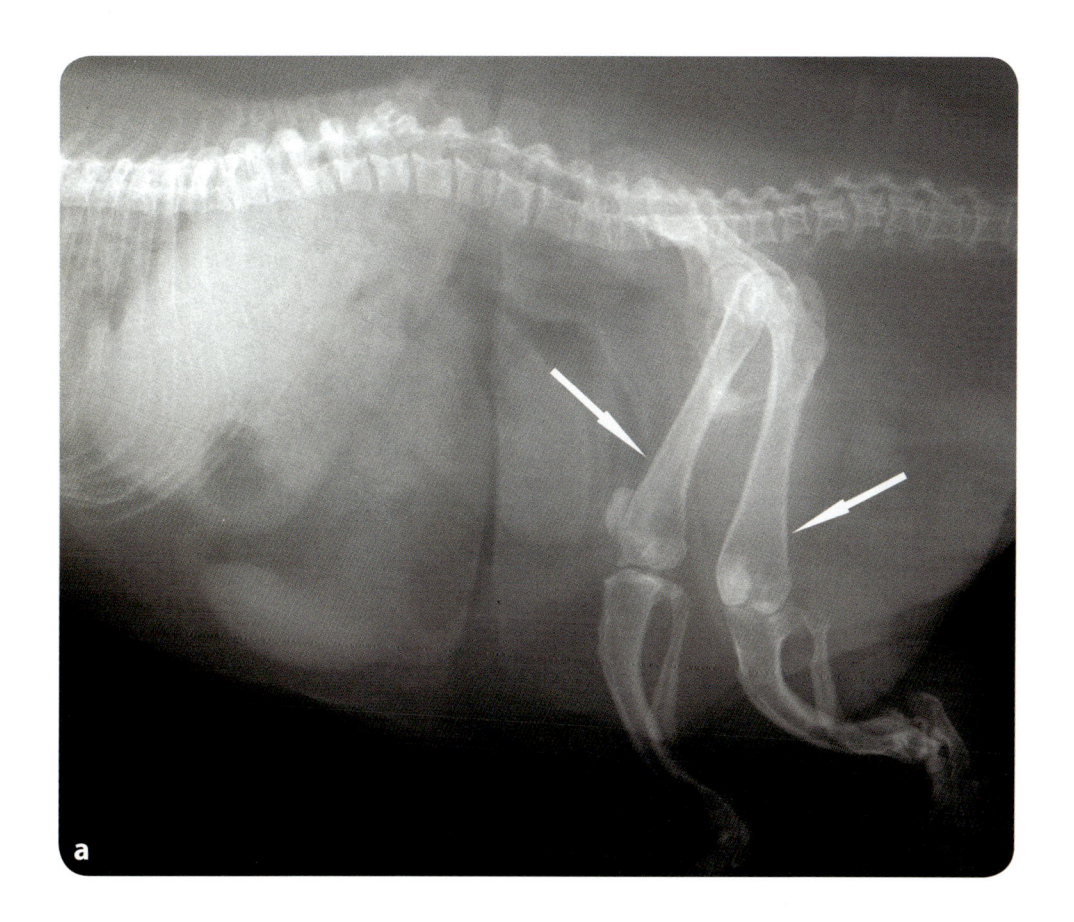

Figure 11.17a,b. Lateral (a) and ventrodorsal (b) projections of the spine and pelvic limbs of an obese 3-year-old male skunk. Note thin cortices of the distal femurs (white arrows), small pathologic fractures of the femur (yellow arrows), and angular limb deformities of the tibia and fibula.

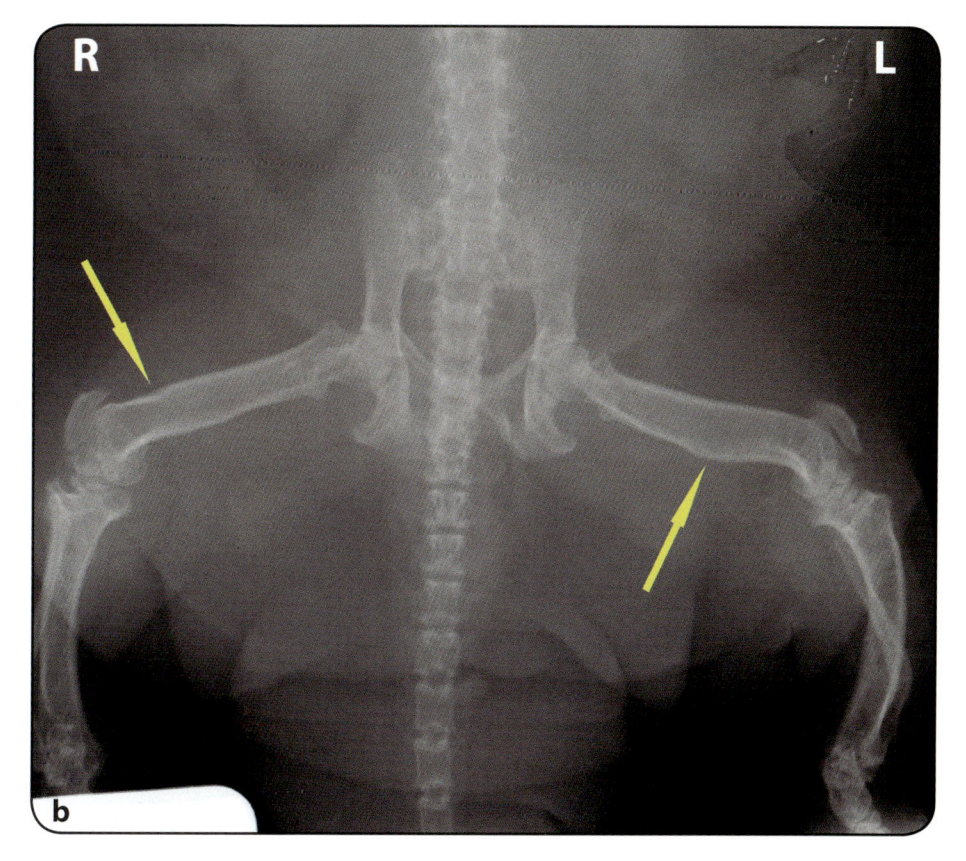

Diseases of the Tibia and Fibula

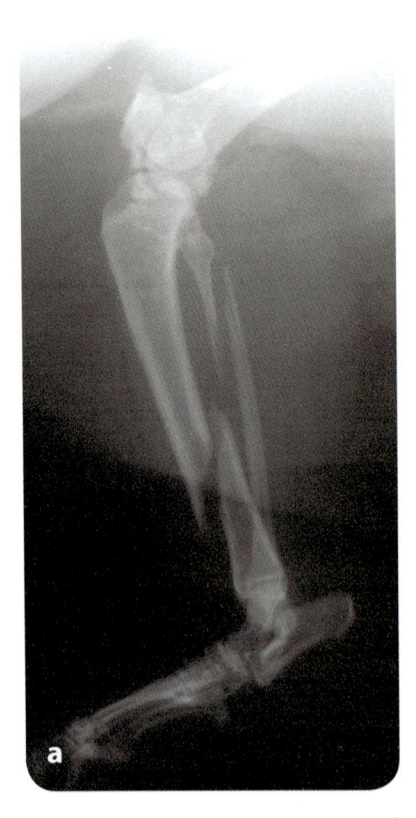

Figure 11.18. Lateral projection of the pelvic limb of a male skunk with a spiral fracture of the tibial and fibular shafts.

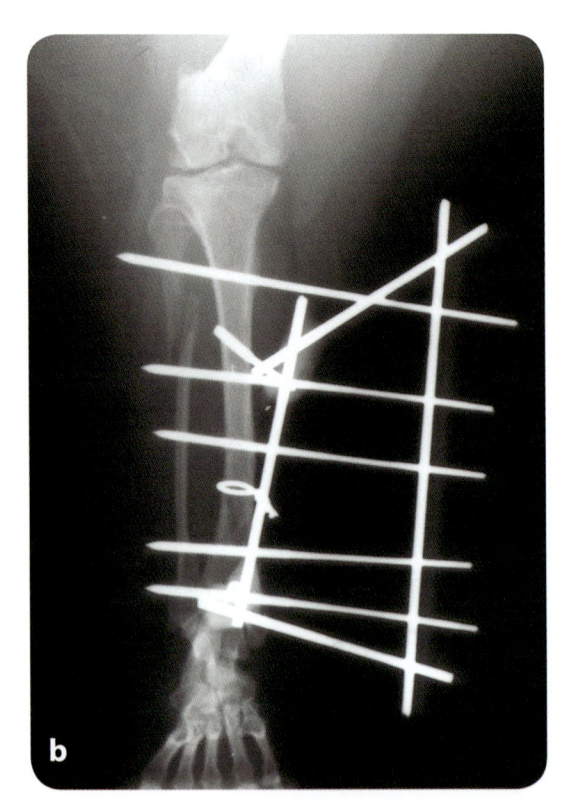

Figure 11.19. Craniocaudal projection of the same patient following osteosynthesis of the tibial fracture. A biplanar external fixator configuration and cerclage wire were used: monoplanar 1+1 pin plus a monoplanar 3+2 pin. The fibular fracture was stabilized by repair of the tibial fracture.

Miscellaneous

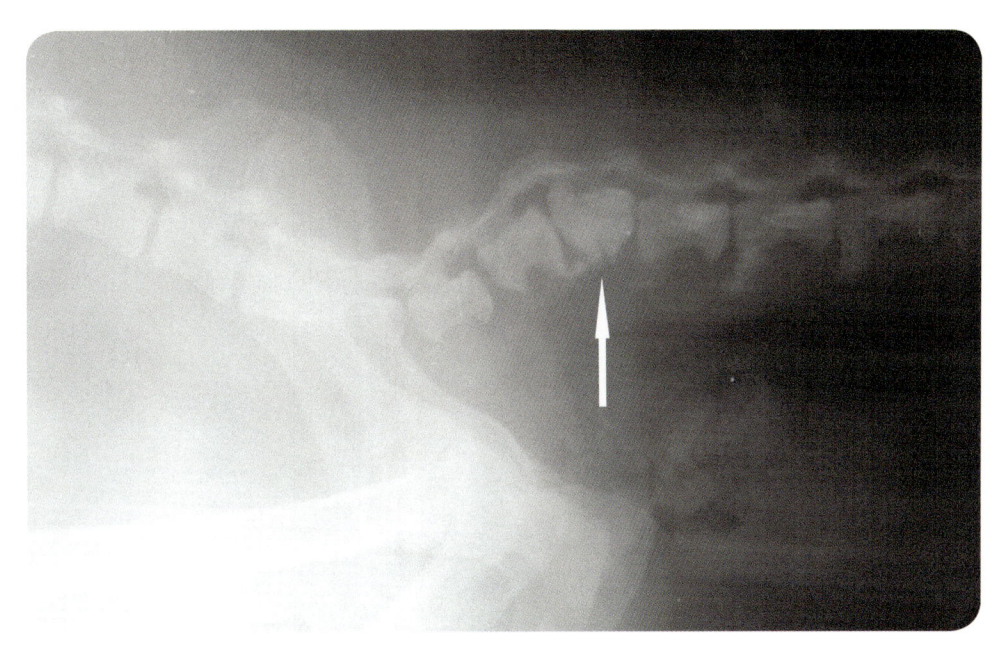

Figure 11.20. Lateral projection of the pelvis and tail of a skunk with a compression fracture of the third caudal vertebra (arrow) and subluxation of Cd 2-3 and Cd 3-4.

References:
Hanley CS, Wilson GH, Hernandez-Divers SJ: Secondary nutritional hyperparathyroidism associated with vitamin D deficiency in two domestic skunks *(Mephitis mephitis)*. Vet Rec. 2004; 115(8):233-7.
Kramer MH, Lennox AM: What veterinarians need to know about skunks. Exotic DVM. 2003; 5(1):36-39.
Schneider R. Hypocalcemia in a skunk. Exotic DVM. 2003; 5(1):5-6.

SUGAR GLIDER

The NORMAL WHOLE BODY SKELETON
Lateral Projection

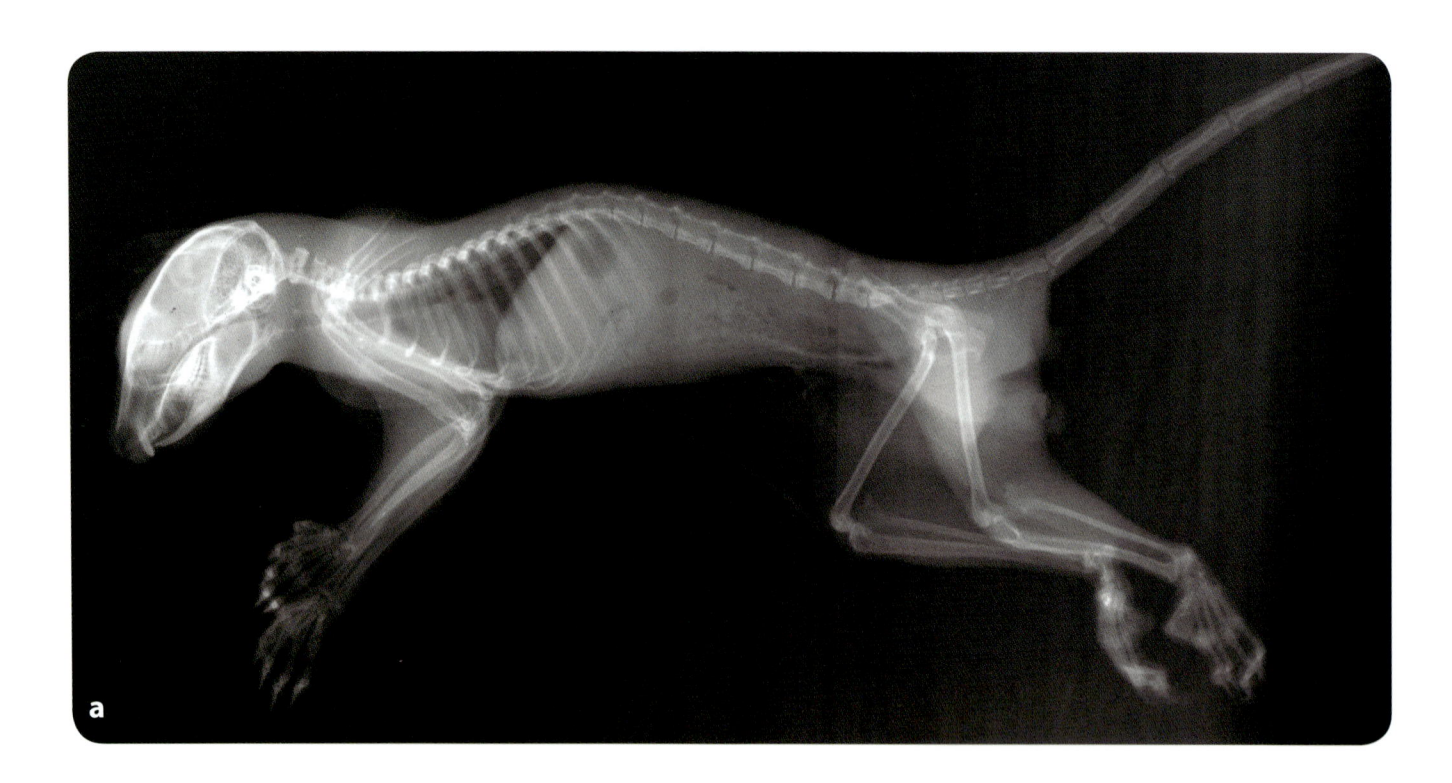

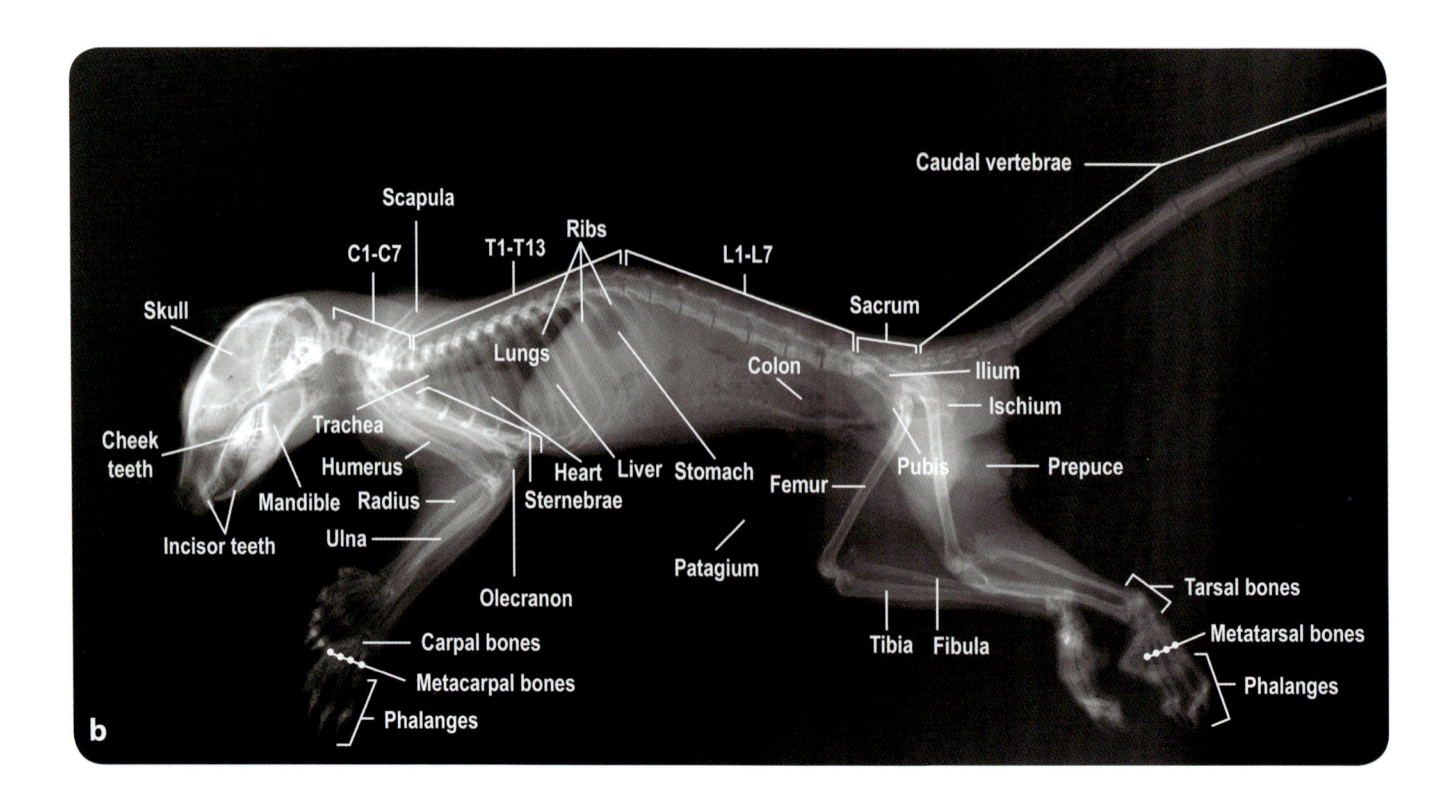

Figure 12.1a,b. Whole body skeleton of a 85 g neutered male sugar glider, actual size (a), and 90% of actual size (b). The eu-pubic bones are atrophied or absent in marsupial moles and *Petaurus sp* (the gliders).

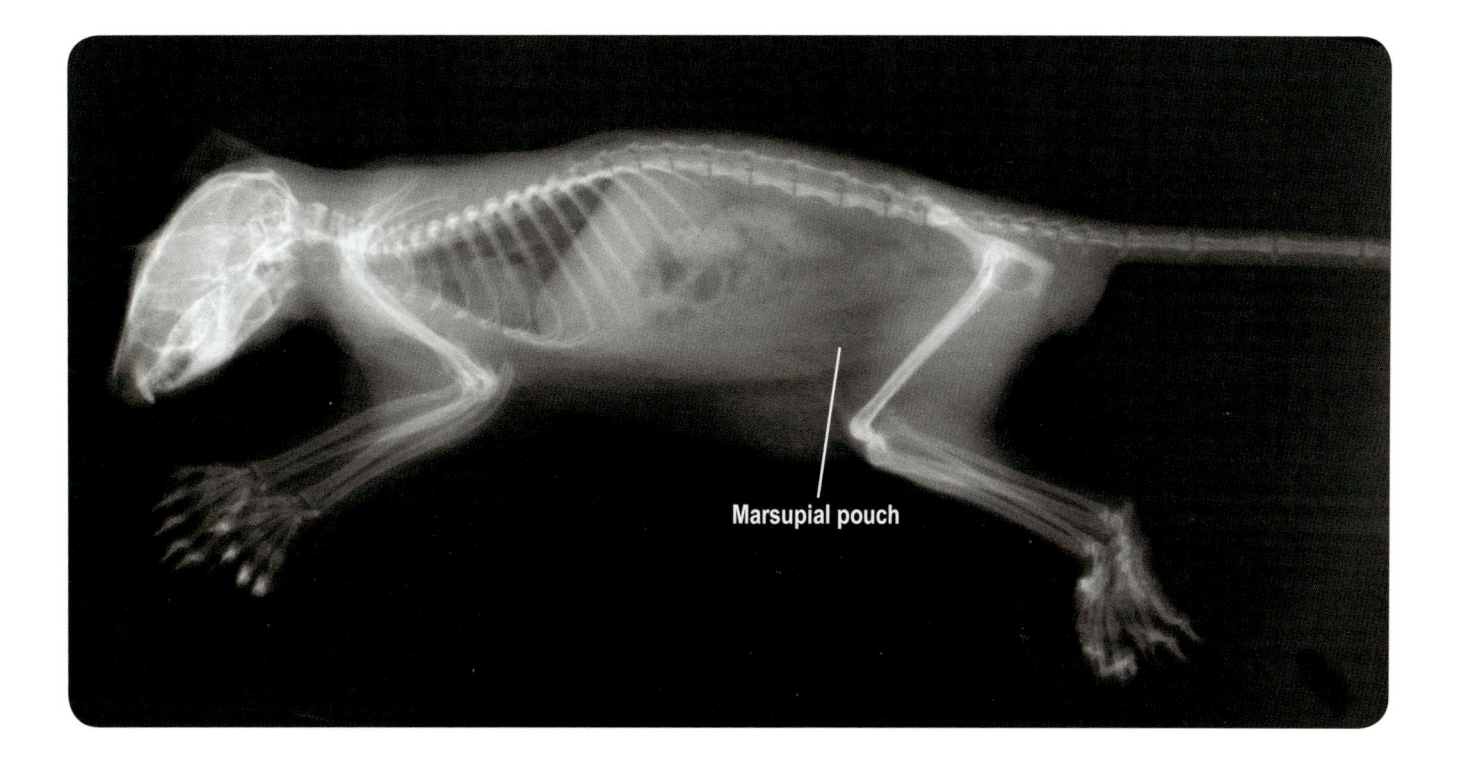

Figure 12.2. Whole body skeleton of a 75 g female sugar glider, actual size.

Ventrodorsal Projection

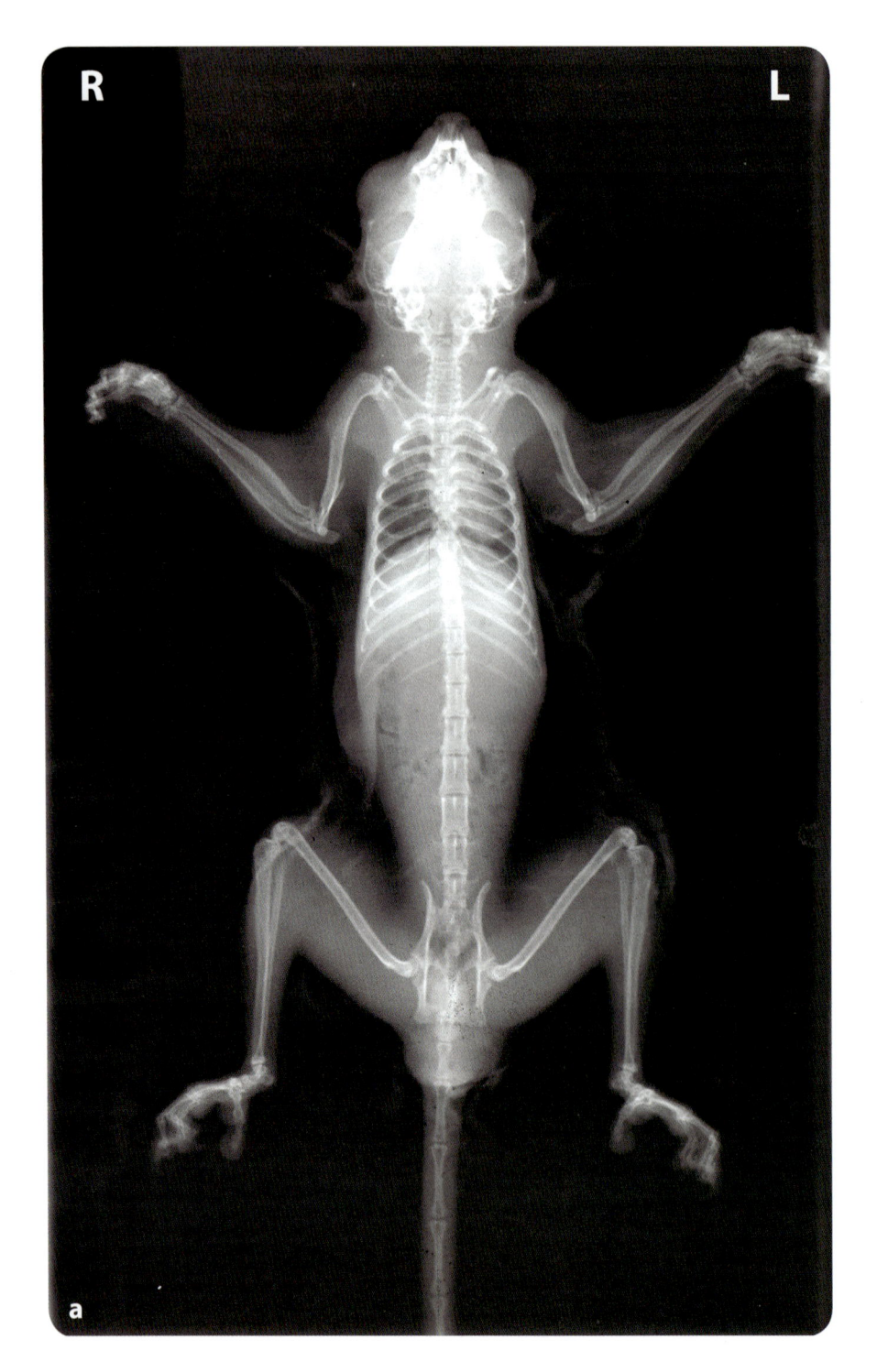

Figure 12.3a,b. Whole body skeleton of a 85 g neutered male sugar glider, actual size.

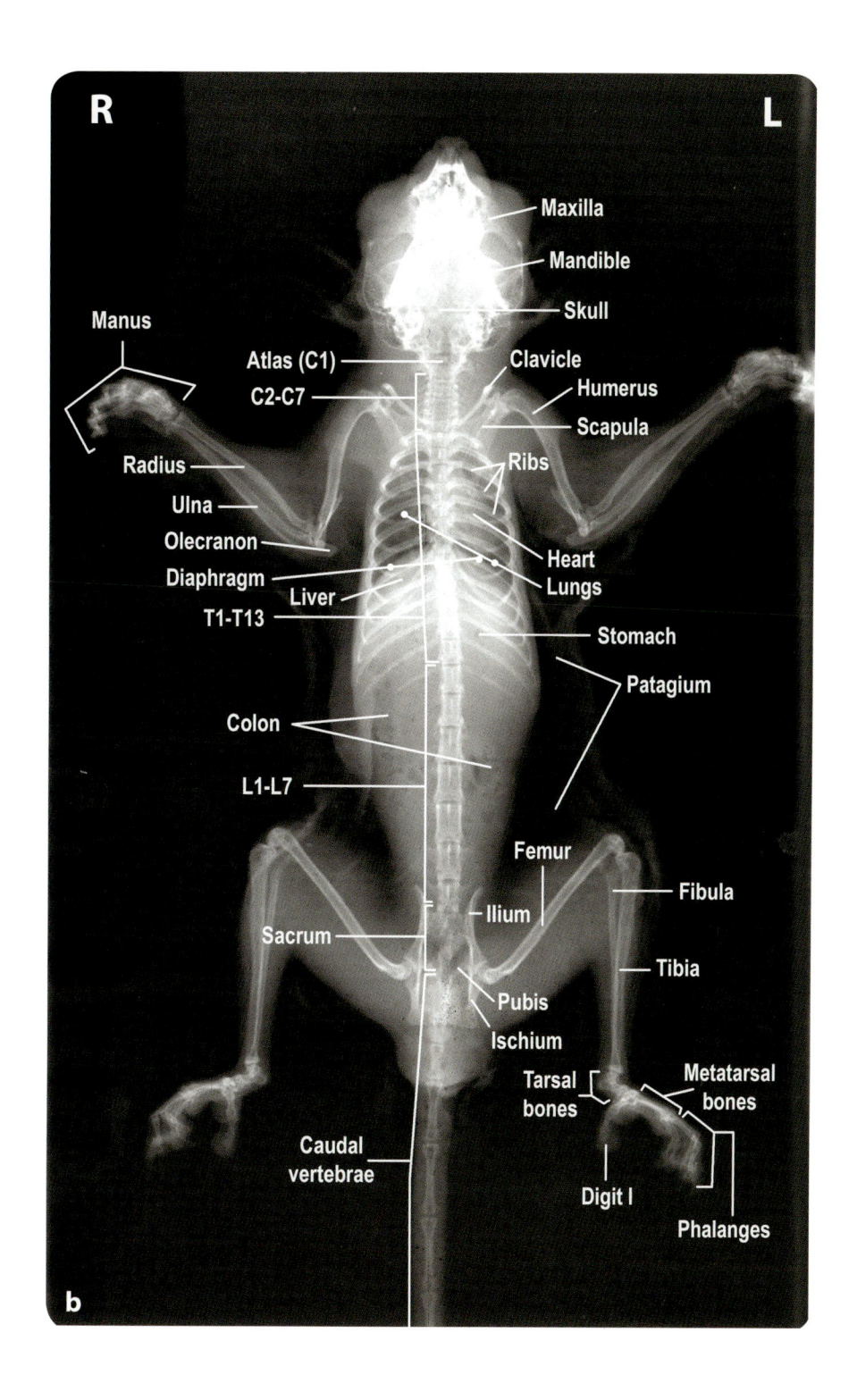

WHOLE BODY SKELETON

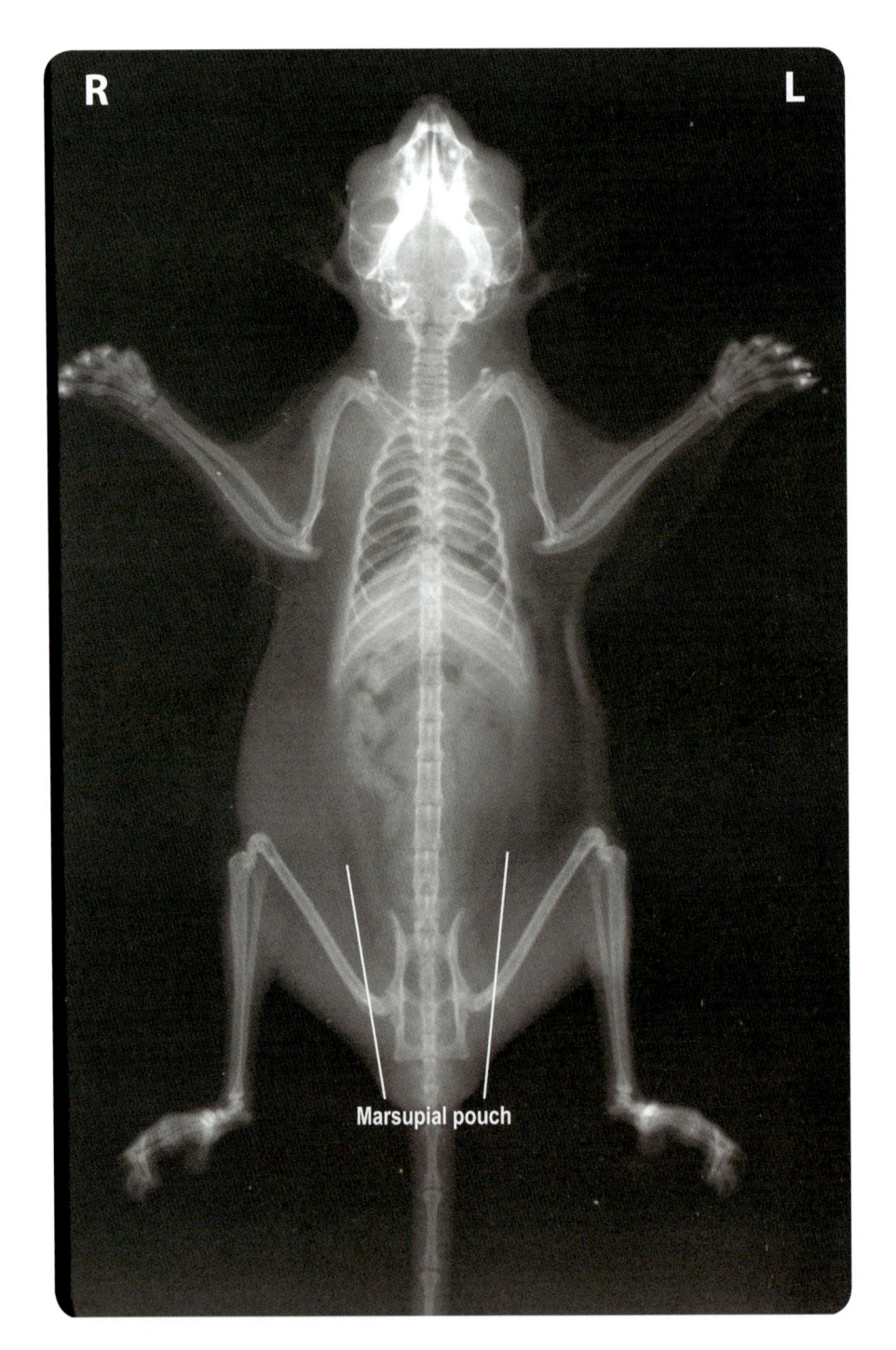

Figure 12.4. Whole body skeleton of a 75 g female sugar glider, actual size. Note the symmetrical radiolucent areas representing the pouch.

References:

Booth R: Sugar gliders. Sem Av Exotic Pet Med. 2003; 12(4):228-231.

Ness RD, Booth R: Sugar gliders. In: Quesenberry KE, Carpenter JW, eds. Ferrets, Rabbits and Rodents. Clinical Medicine and Surgery 2nd ed. Philadelphia, PA: Elsevier; 2004:330-338.

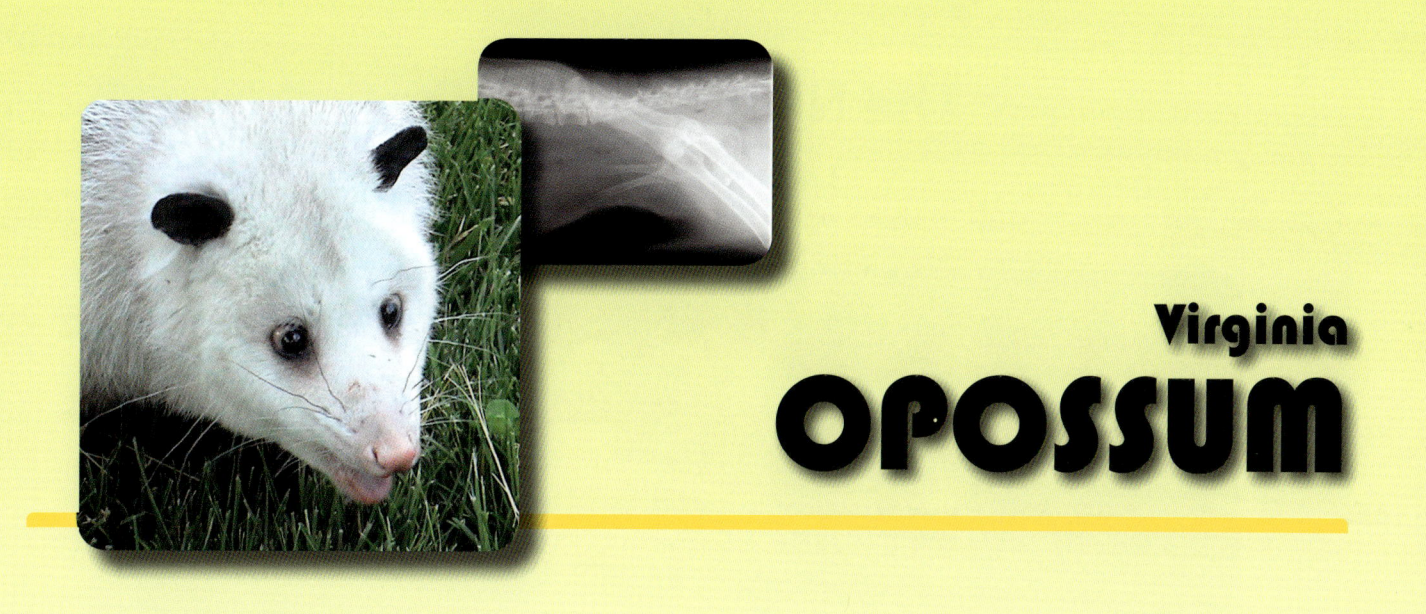

Virginia OPOSSUM

The NORMAL HEAD
Lateral Projection

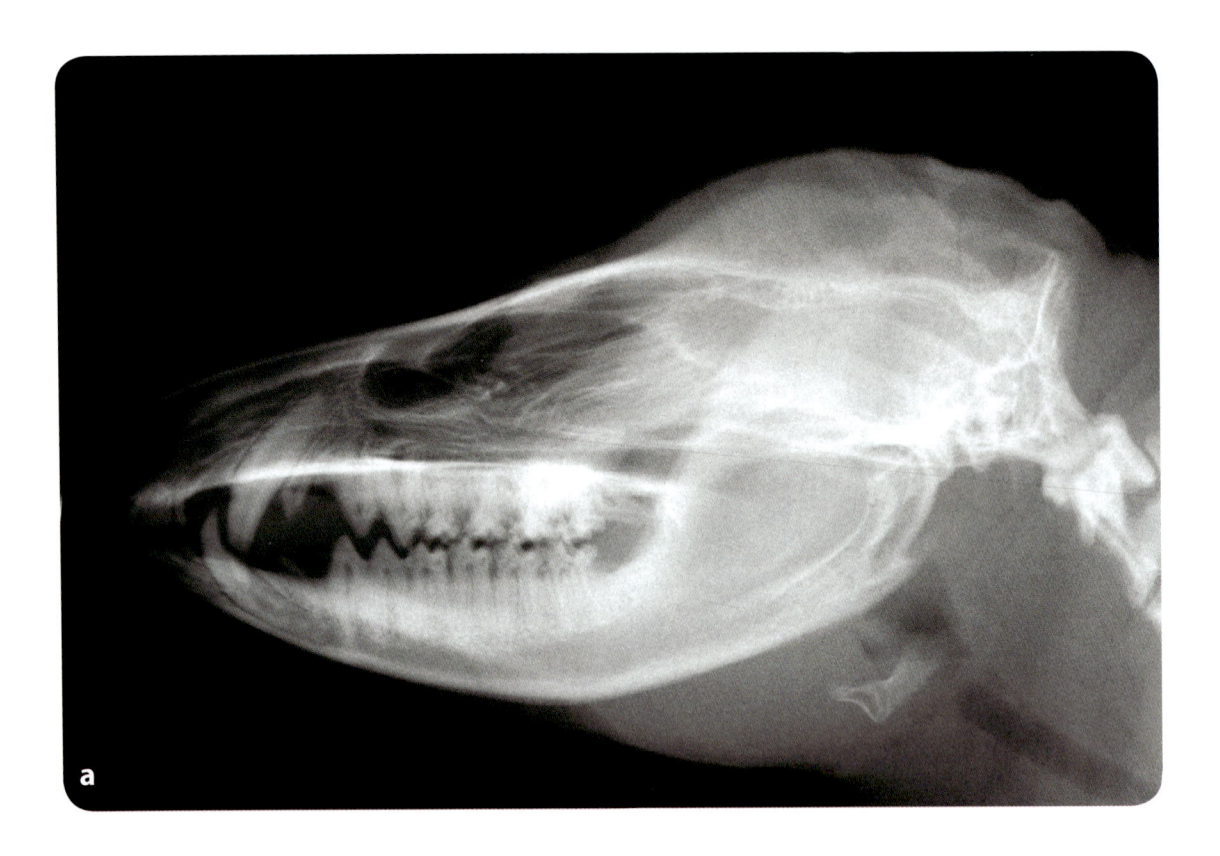

Figure 13.1a,b. Radiograph of the skull of a 3.5 kg adult female Virginia opossum, actual size.

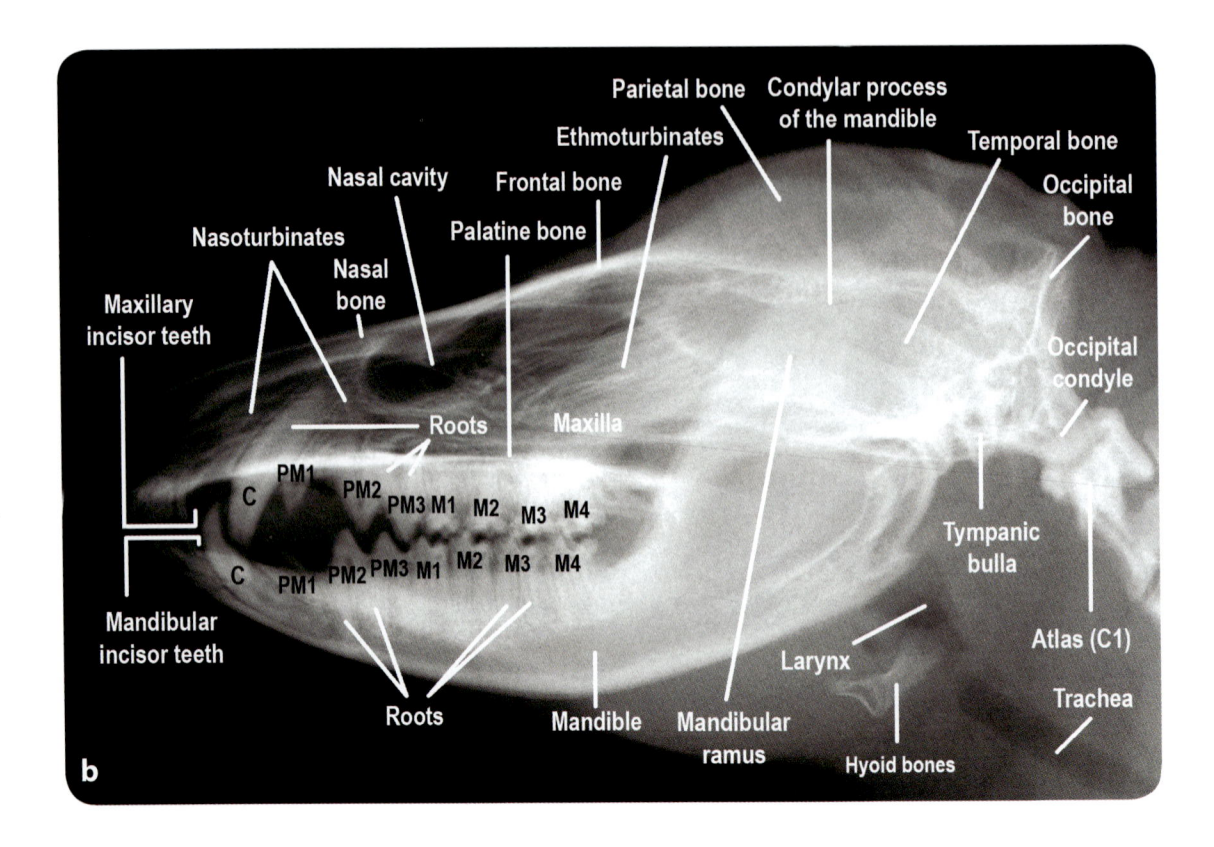

Ventrodorsal Projection

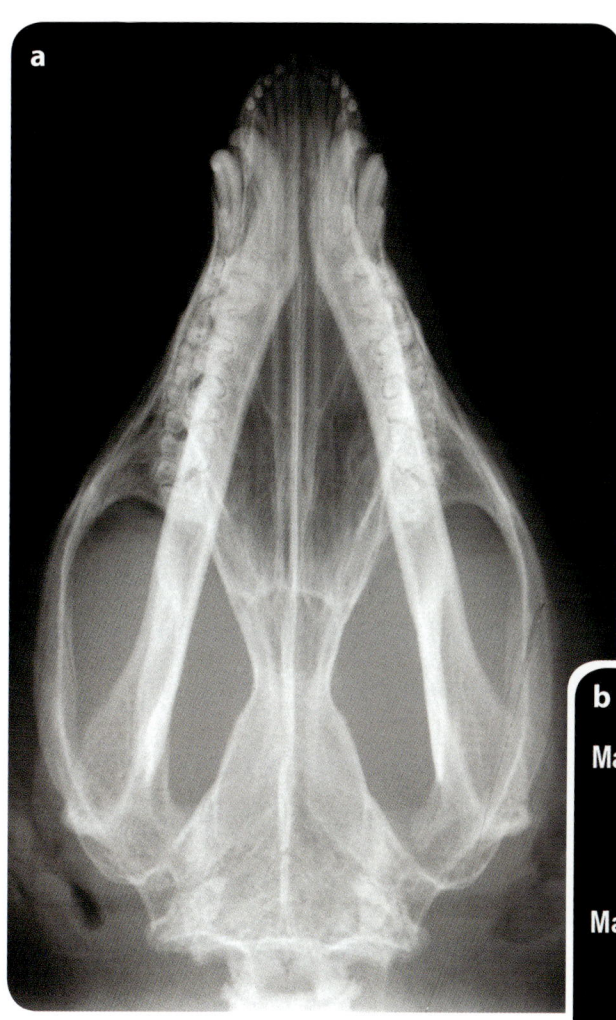

Figure 13.2a,b. Radiograph of the skull of a 3.5 kg adult female Virginia opossum, actual size.

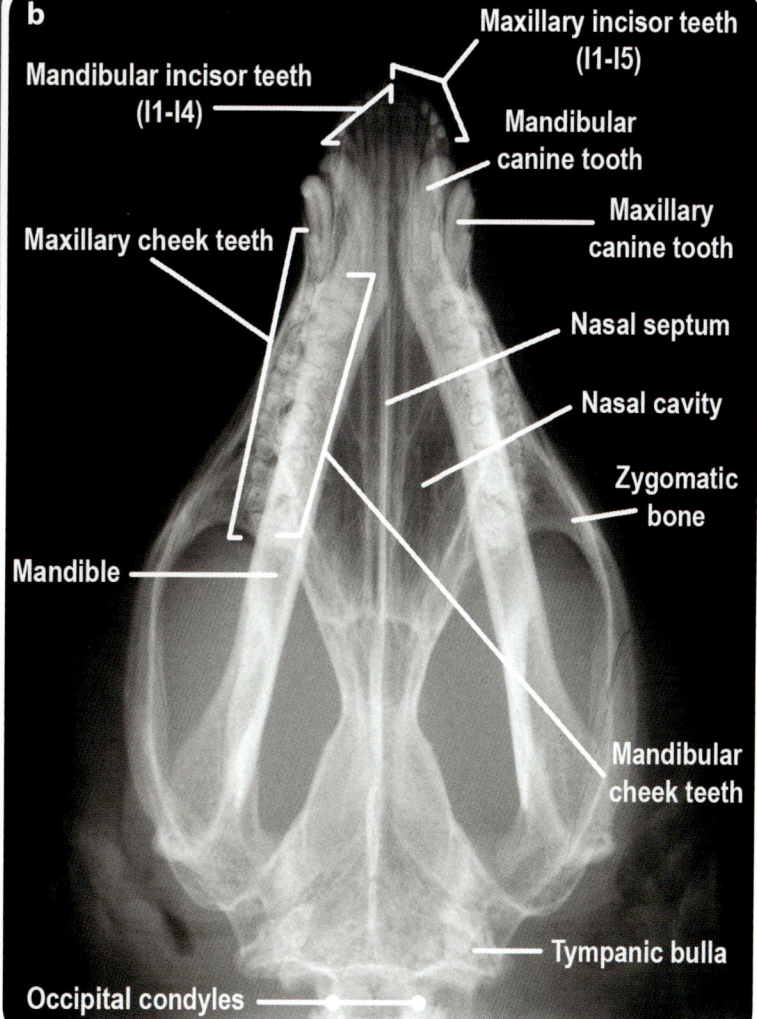

The NORMAL THORAX
The Cervical and Thoracic Vertebral Column

Lateral Projection

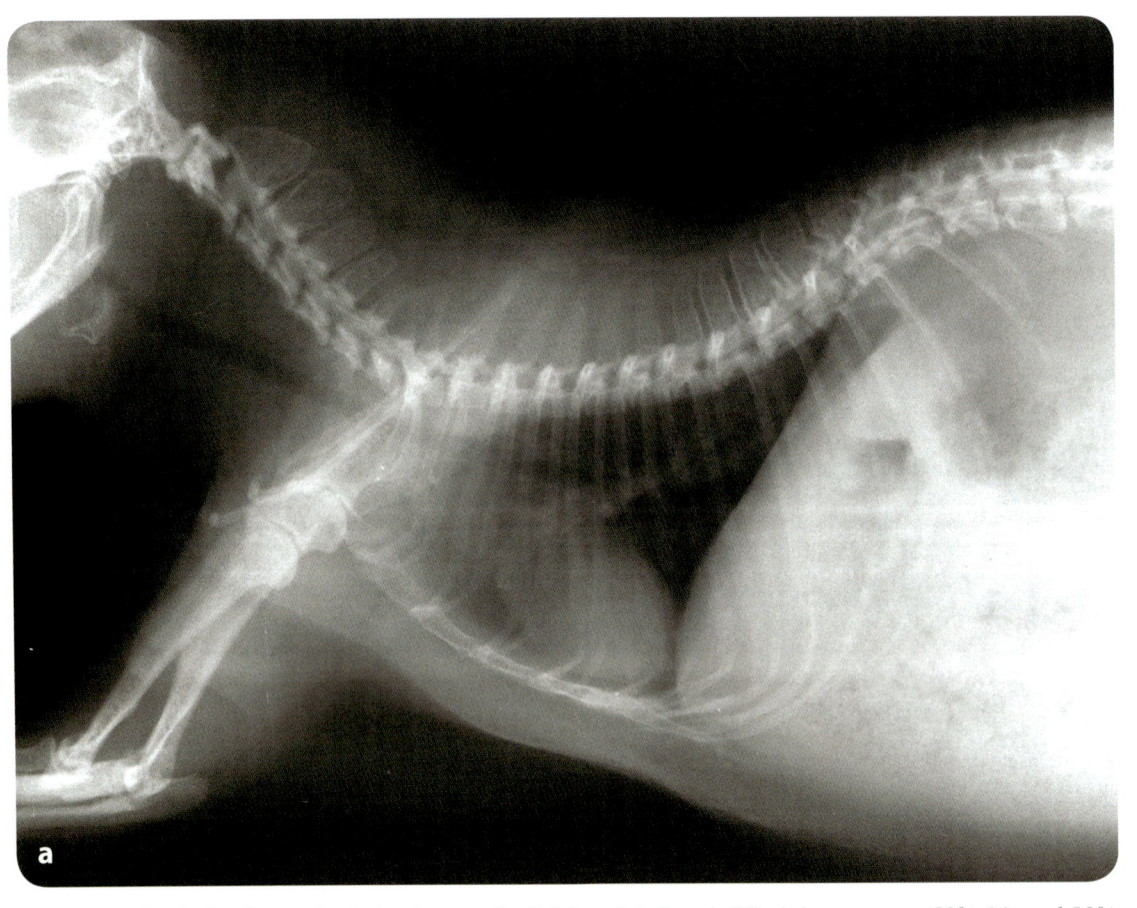

Figure 13.3a,b. Radiograph of the thorax of a 3.5 kg adult female Virginia opossum, 60% (a), and 50% (b) of actual size.

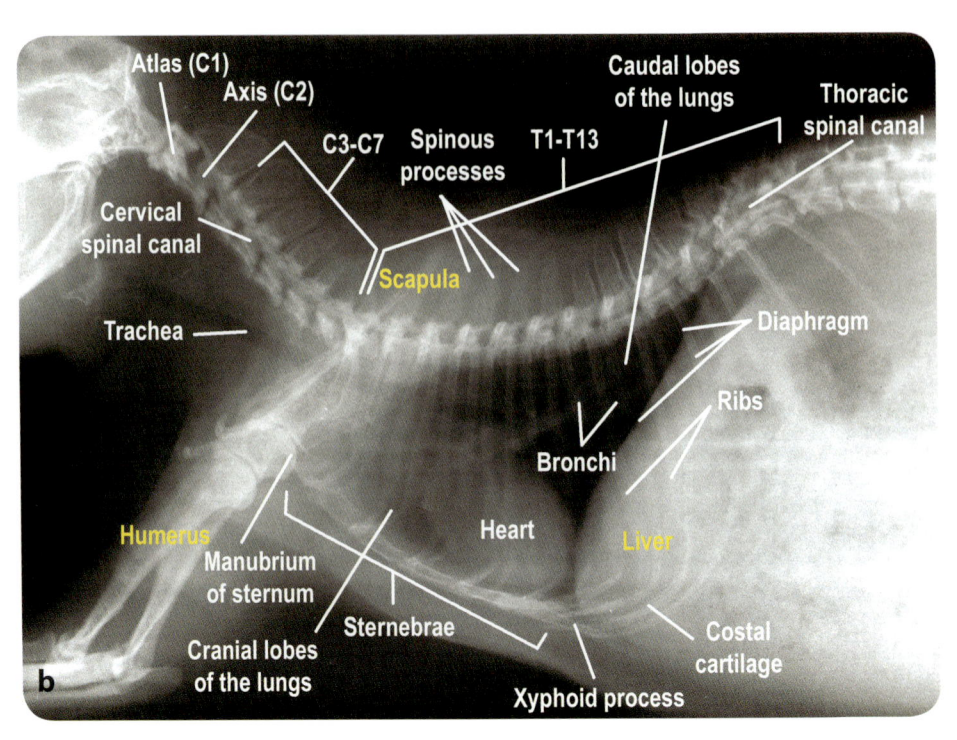

Ventrodorsal Projection

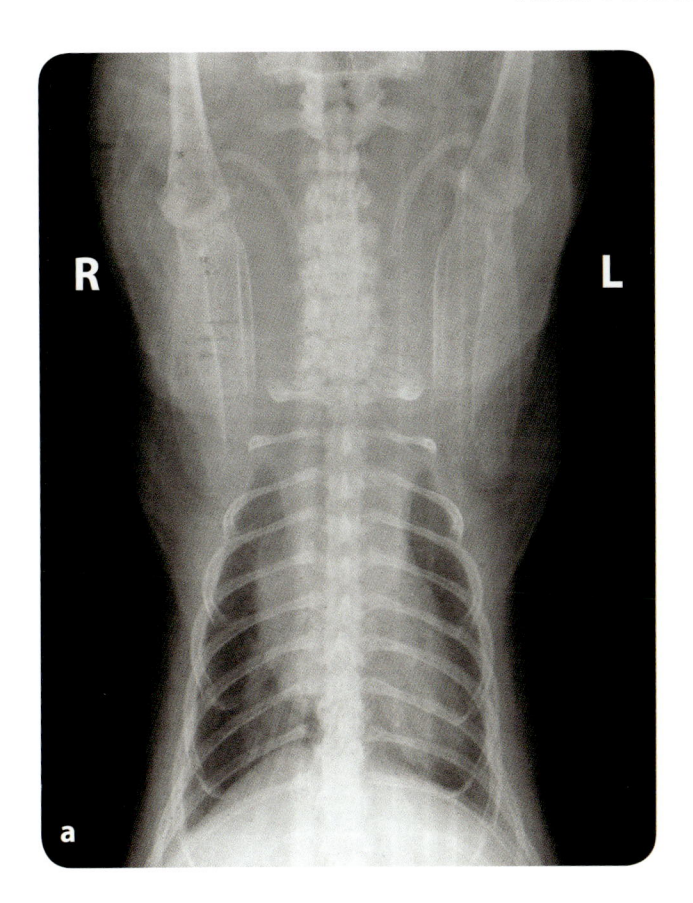

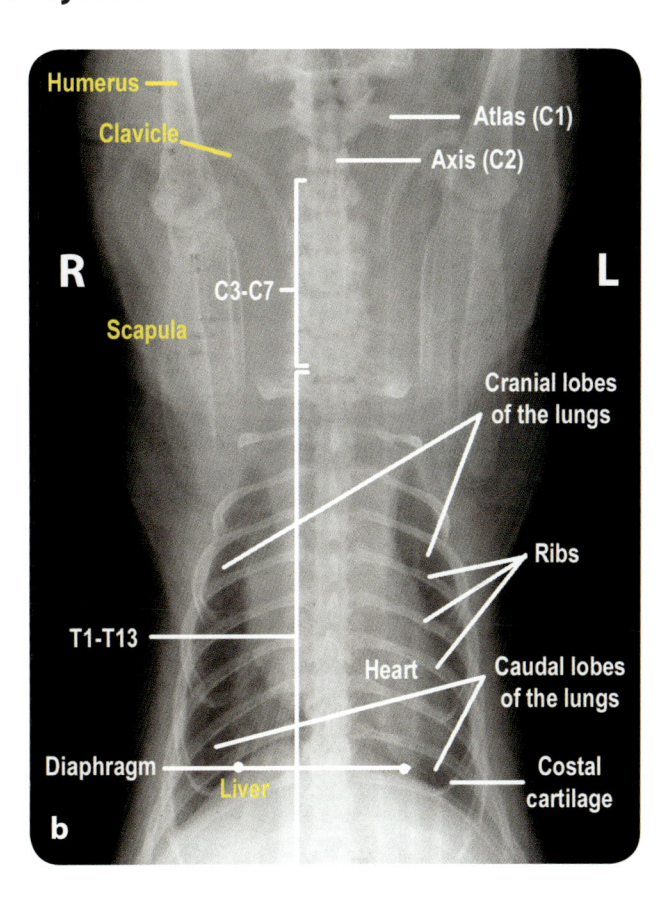

Figure 13.4a,b. Radiograph of the thorax of a 3.5 kg adult female Virginia opossum, 60% of actual size.

The NORMAL ABDOMEN
The Lumbar Vertebral Column

Lateral Projection

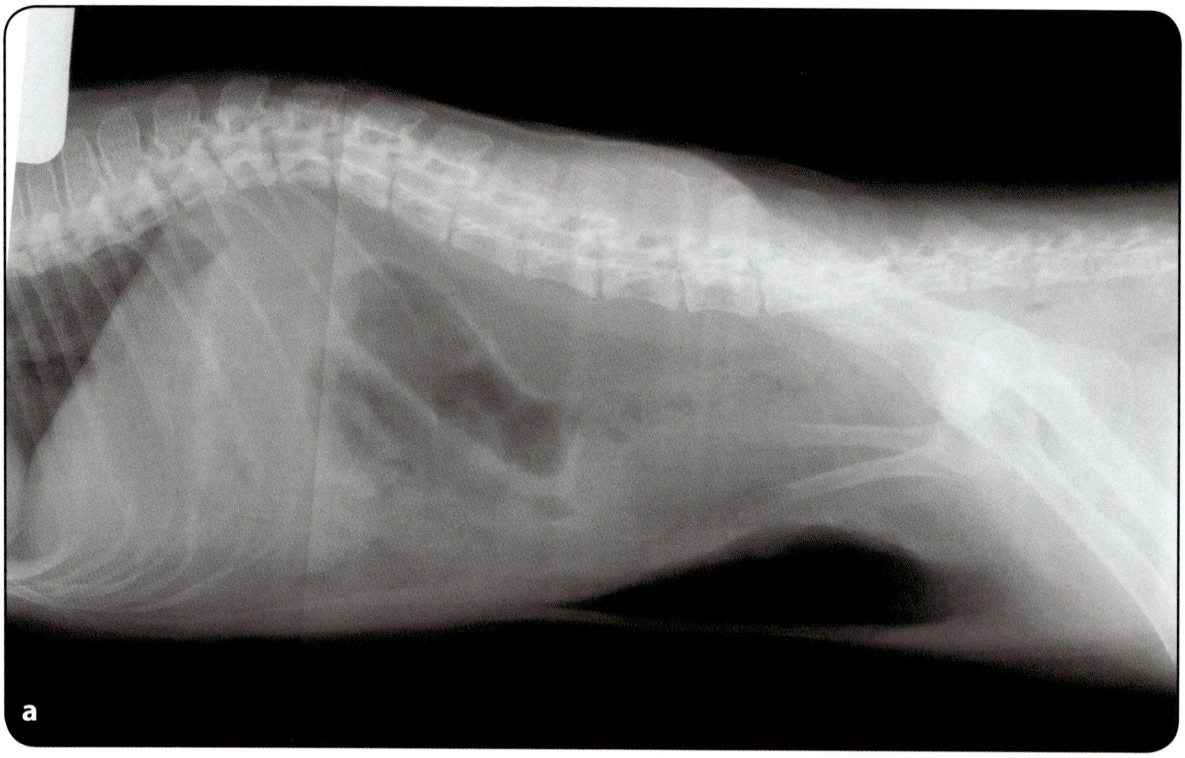
a

Figure 13.5a,b. Radiograph of the abdomen of a 3.5 kg adult female Virginia opossum, 60% of actual size. Note the ventral radiolucent area representing the marsupial pouch, and the presence of the eupubic bones.

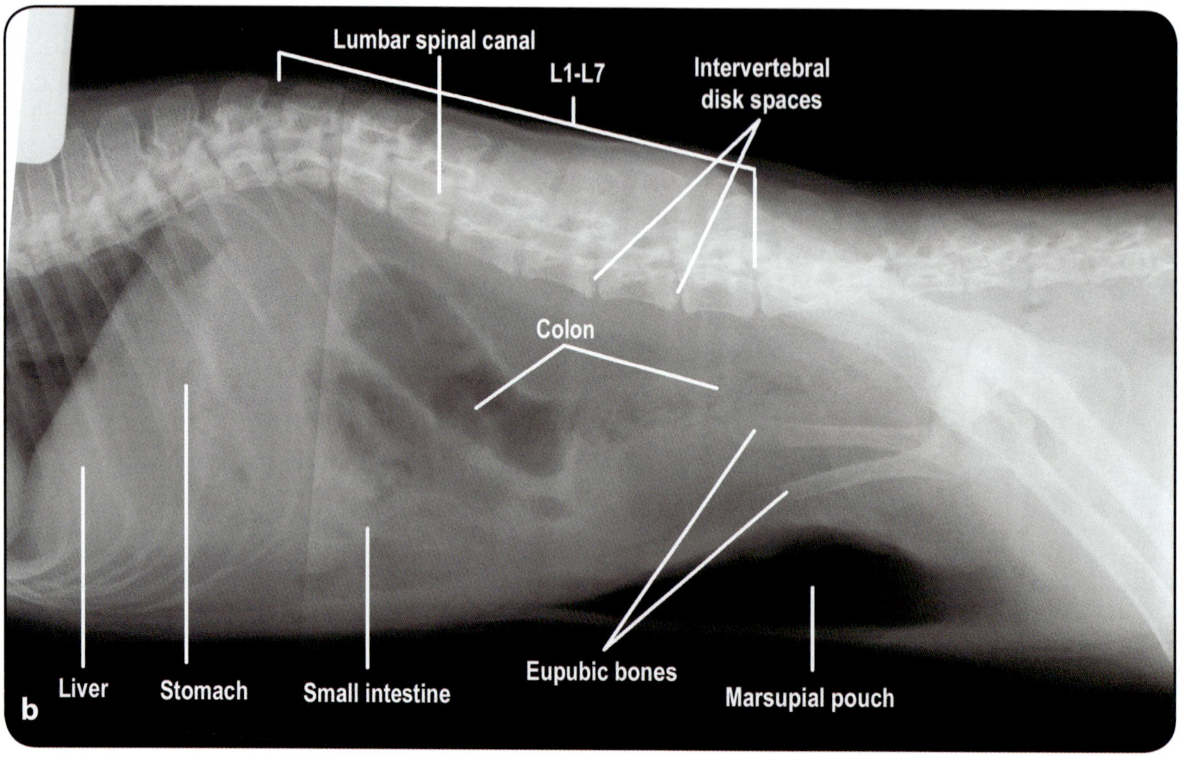
Lumbar spinal canal
L1-L7
Intervertebral disk spaces
Colon
Liver
Stomach
Small intestine
Eupubic bones
Marsupial pouch
b

Ventrodorsal Projection

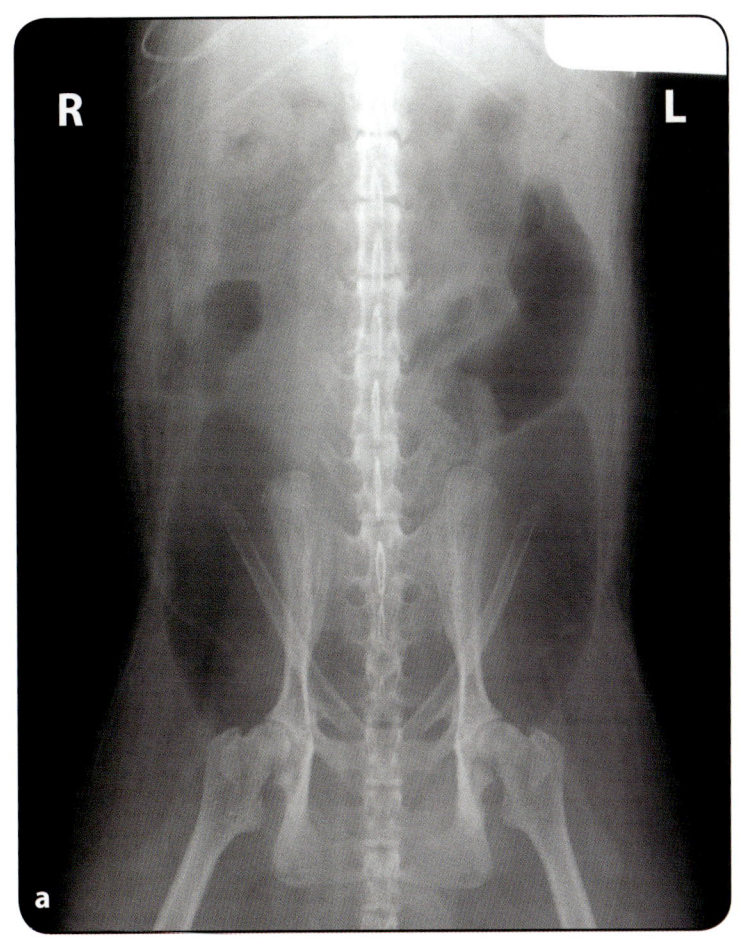

Figure 13.6a,b. Radiograph of the abdomen of a 3.5 kg adult female Virginia opossum, 60% (a) and 45% (b) of actual size. The bilaterally symmetrical radiolucent areas lateral to the abdomen represent air within the marsupial pouch. These opacities should not be confused with intestinal gas.

Note the presence of the marsupial bones *(ossa marsupialia)* that serve as attachment surfaces for several abdominal muscles. They rest on the pelvic and pubic bones, and articulate with them. They generally are boot-shaped, flattened, and vary in size in different species. They are considered comparable to abdominal ribs in reptiles, and are sometimes referred to as "eupubic bones."

They do not support the pouch, as they are also present in males (no pouch) and are atrophied or absent in marsupial moles and *Petaurus sp.* (gliders).

(Text courtesy of Cathy Johnson-Delaney, DVM).

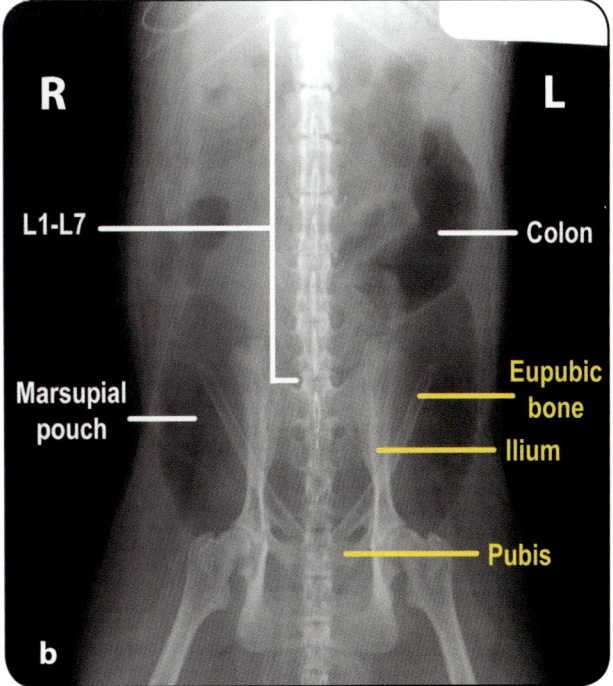

The NORMAL THORACIC LIMB
Lateral Projection

Due to varying soft tissue density, the superimposition of the thorax, and the relatively small size of the manus, it is impossible to obtain an optimal single radiograph of the entire forelimb. Different kVp settings are needed for the scapula and the humerus, and for the radioulnar segment and the manus.

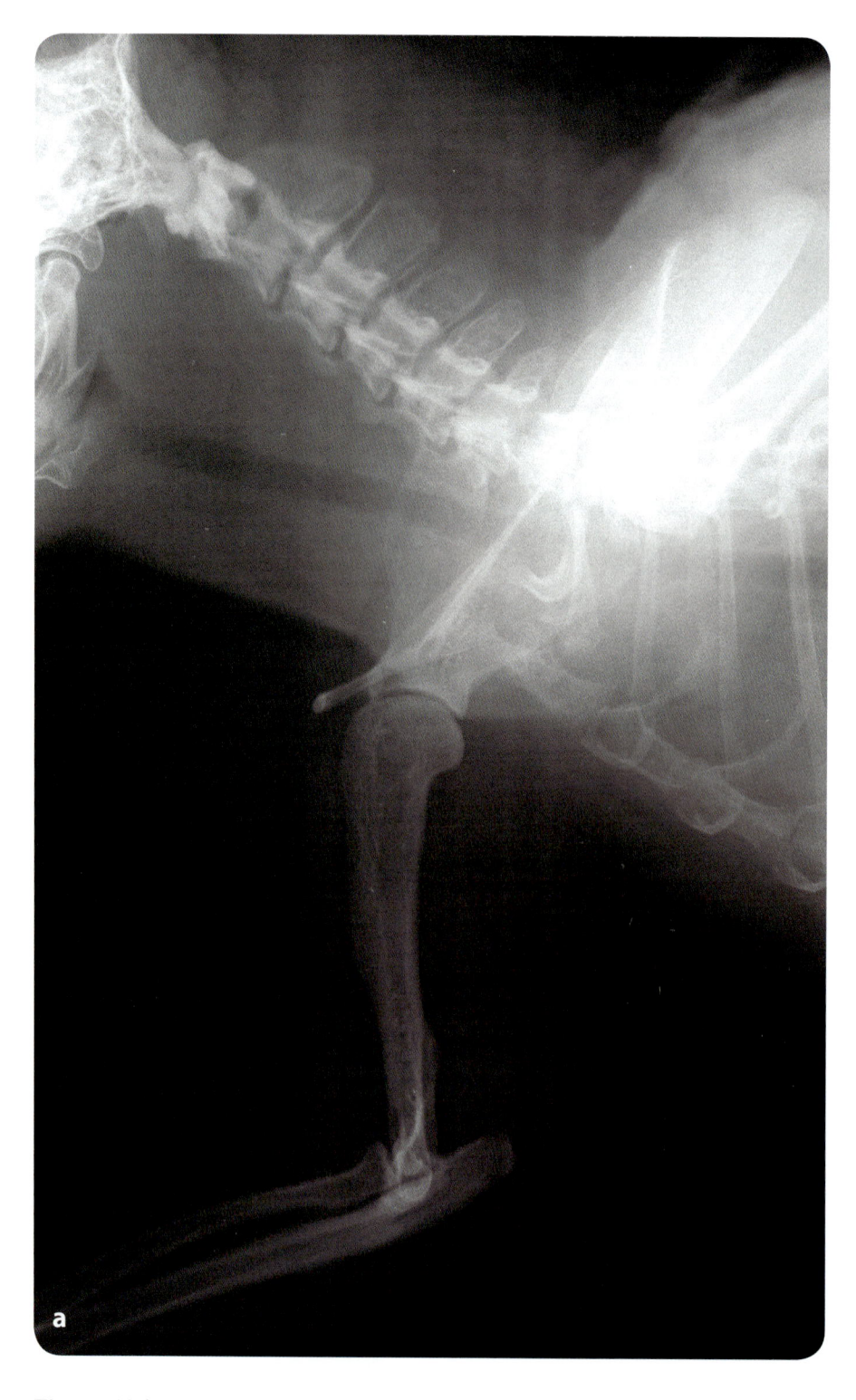

Figure 13.7a,b. Radiograph of the proximal thoracic limb of an adult female Virginia opossum, actual size.

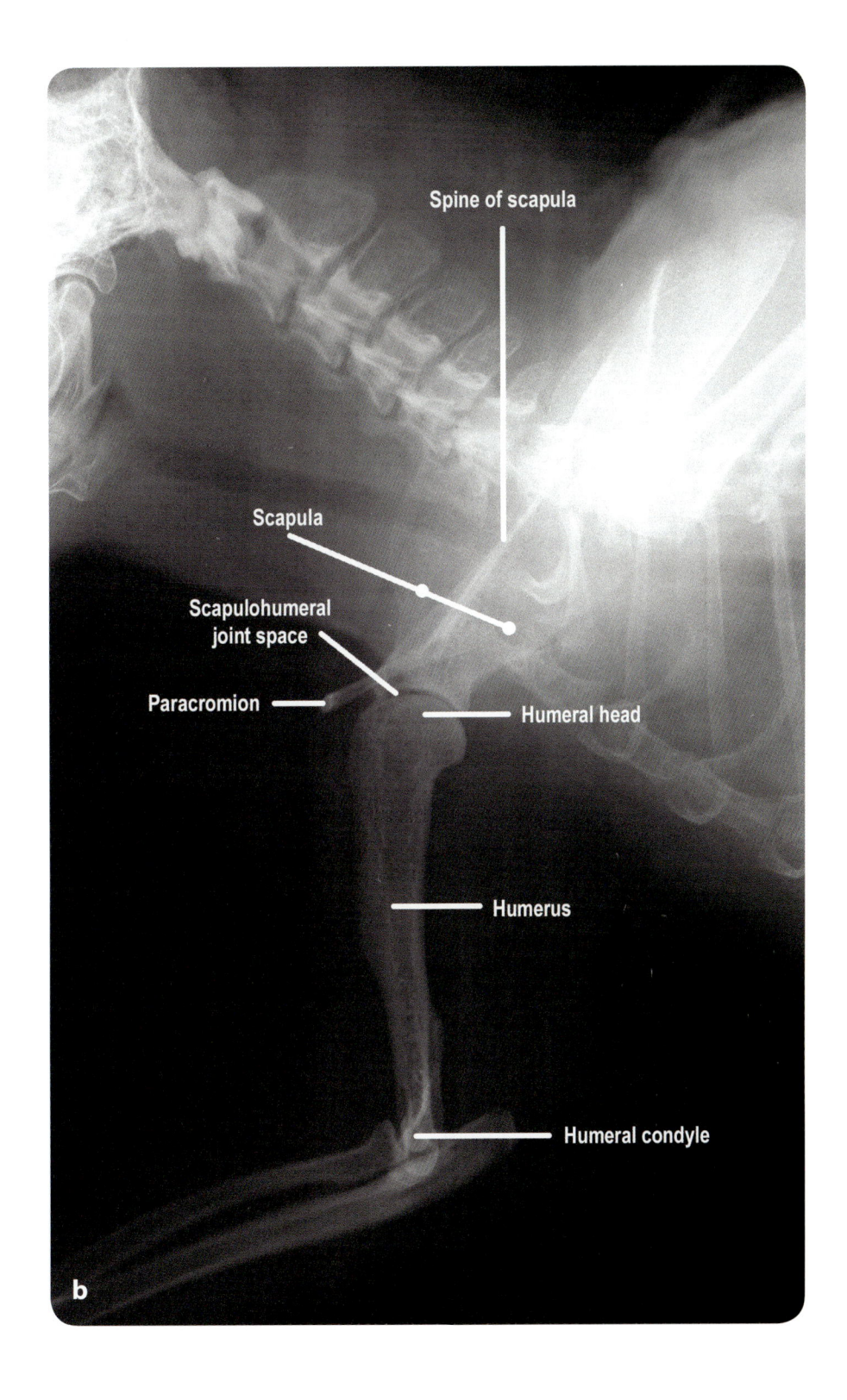

THORACIC LIMB 445

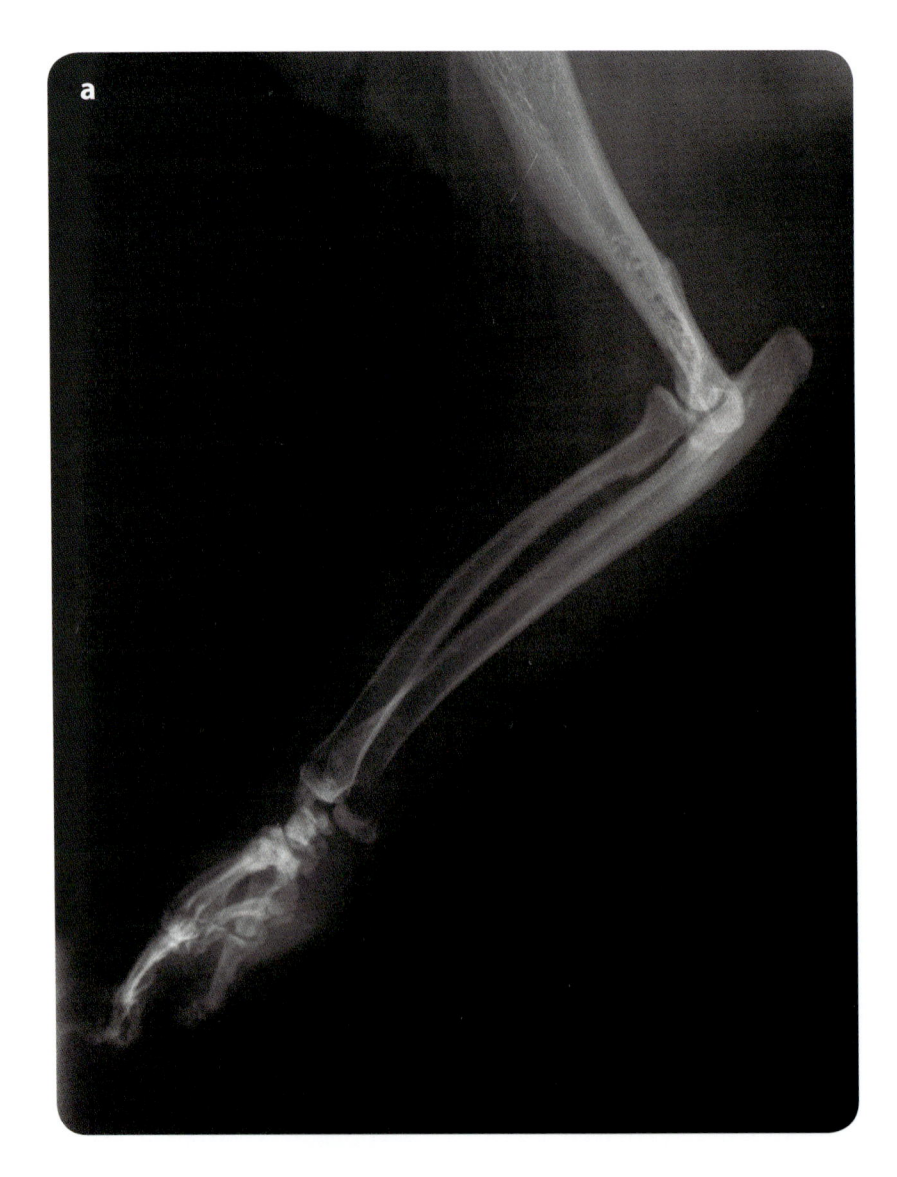

Figure 13.8a,b. Radiograph of the distal thoracic limb of a 3.5 kg adult female Virginia opossum, actual size.

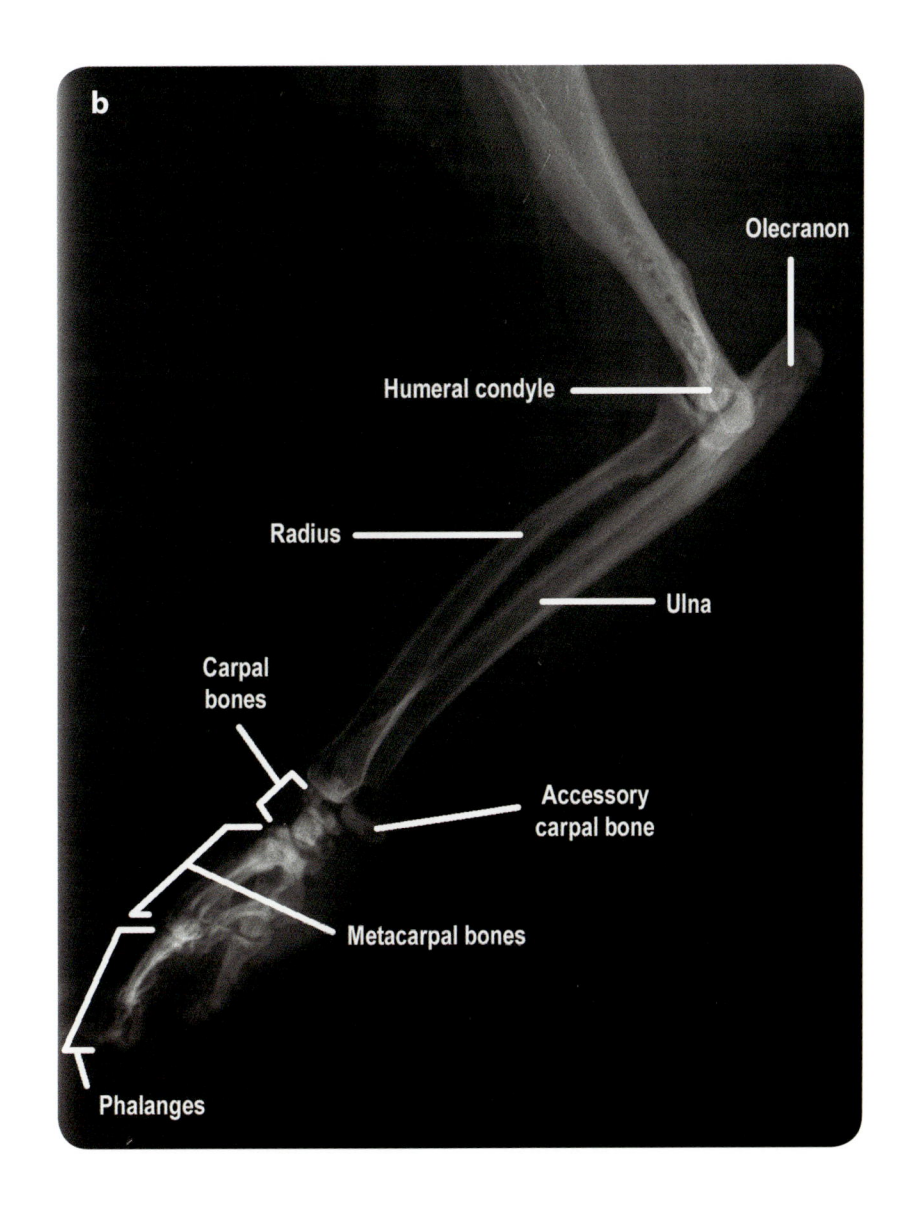

THORACIC LIMB 447

- Olecranon
- Humeral condyle
- Radius
- Ulna
- Carpal bones
- Accessory carpal bone
- Metacarpal bones
- Phalanges

Caudocranial Projection of the Proximal Thoracic Limb

Due to varying soft tissue thickness and the relatively small size of the manus, it is impossible to obtain an optimal single radiograph of the entire thoracic limb. Different kVp settings are needed for the scapula and the humerus, and for the radioulnar segment and the manus.
Note that in some brachymorphic species it is impossible to obtain a true caudocranial (or craniocaudal) view of every portion of the limb. In this radiographic position, all portions of the limb with the exception of the carpus will actually be oblique.

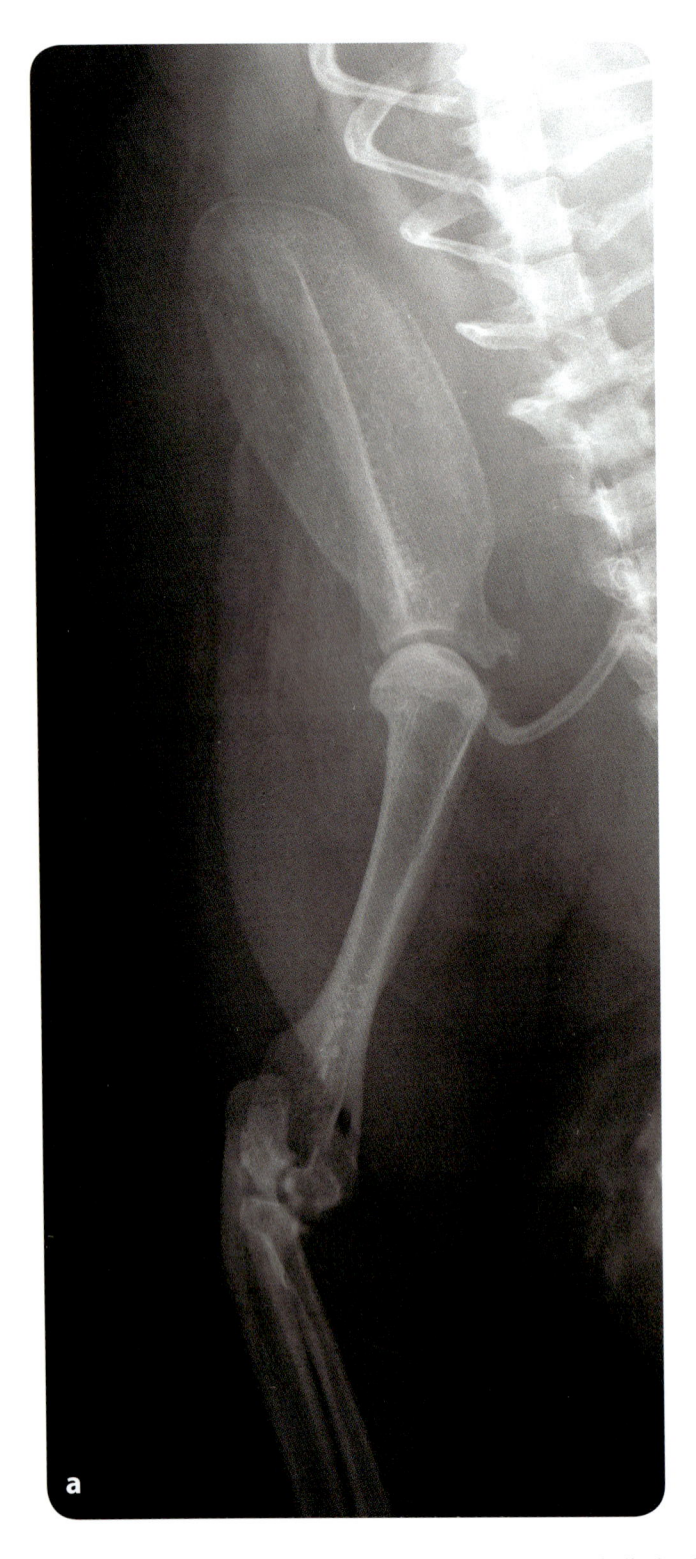

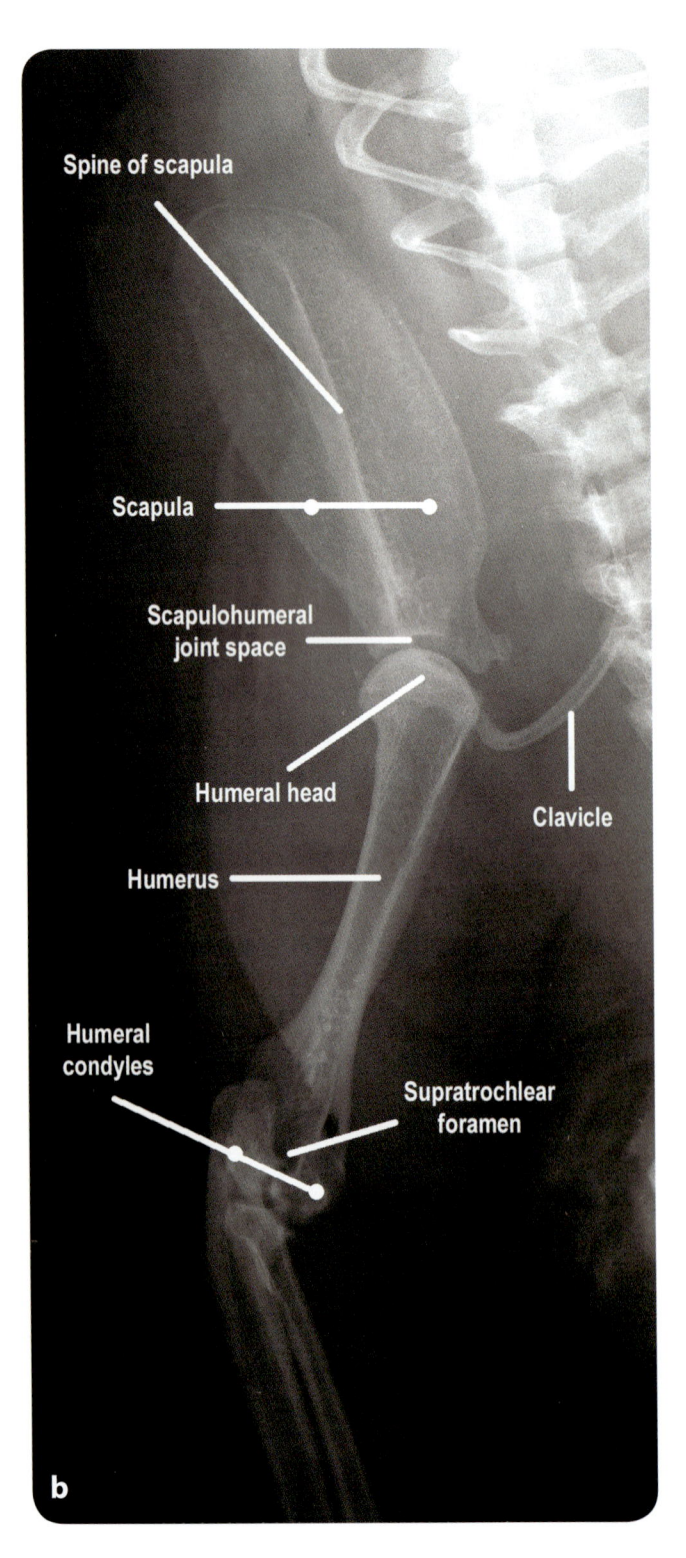

Figure 13.9a,b. Radiograph of the proximal thoracic limb of a 3.5 kg adult female Virginia opossum, actual size.

Craniocaudal Projection of the Distal Thoracic Limb

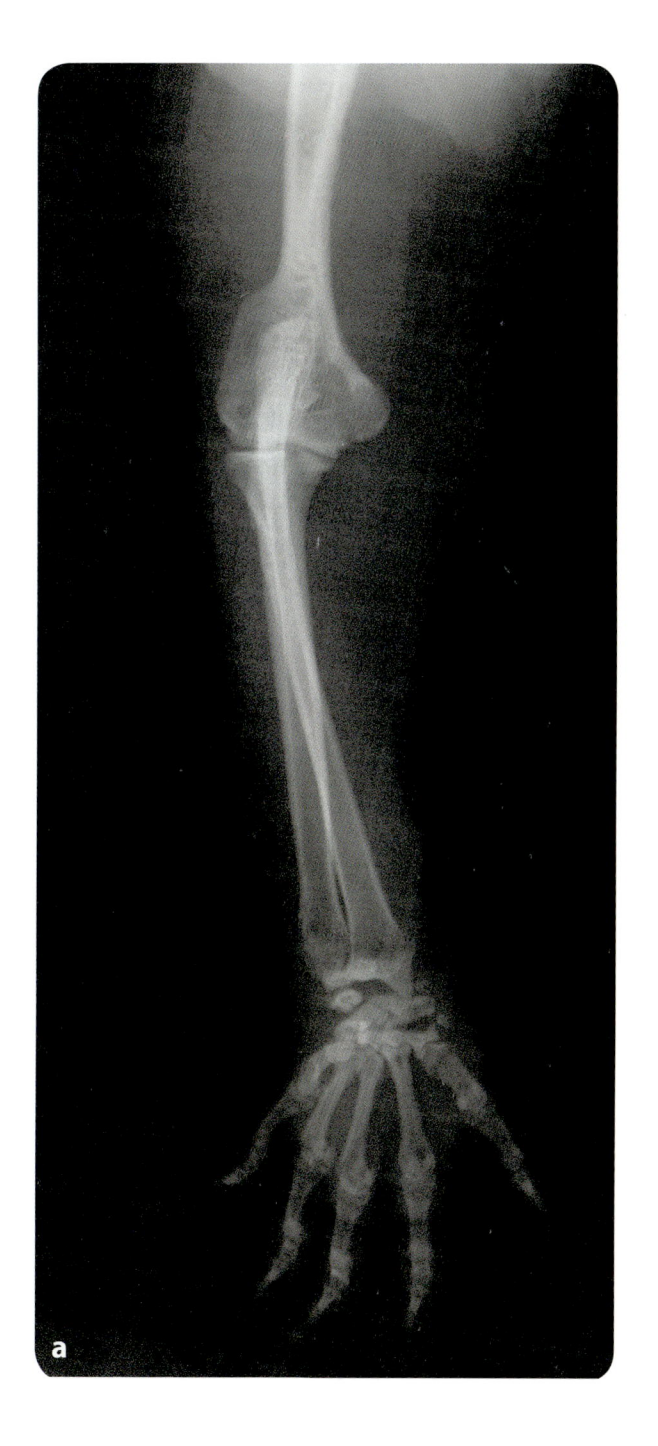

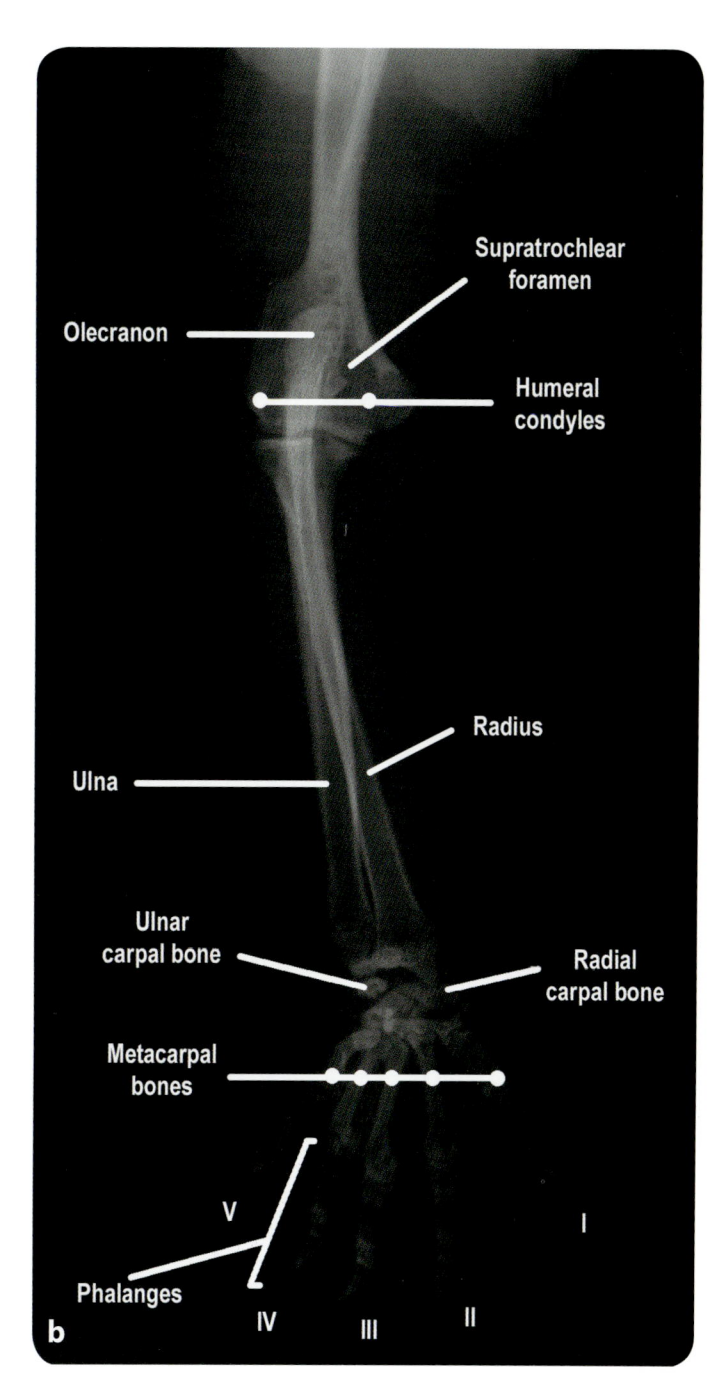

Figure 13.10a,b. Radiograph of the distal thoracic limb of a 3.5 kg adult female Virginia opossum, actual size.

The NORMAL PELVIC LIMB
Ventrodorsal Projection of the Pelvis
Craniocaudal Projection of the Proximal Pelvic Limb

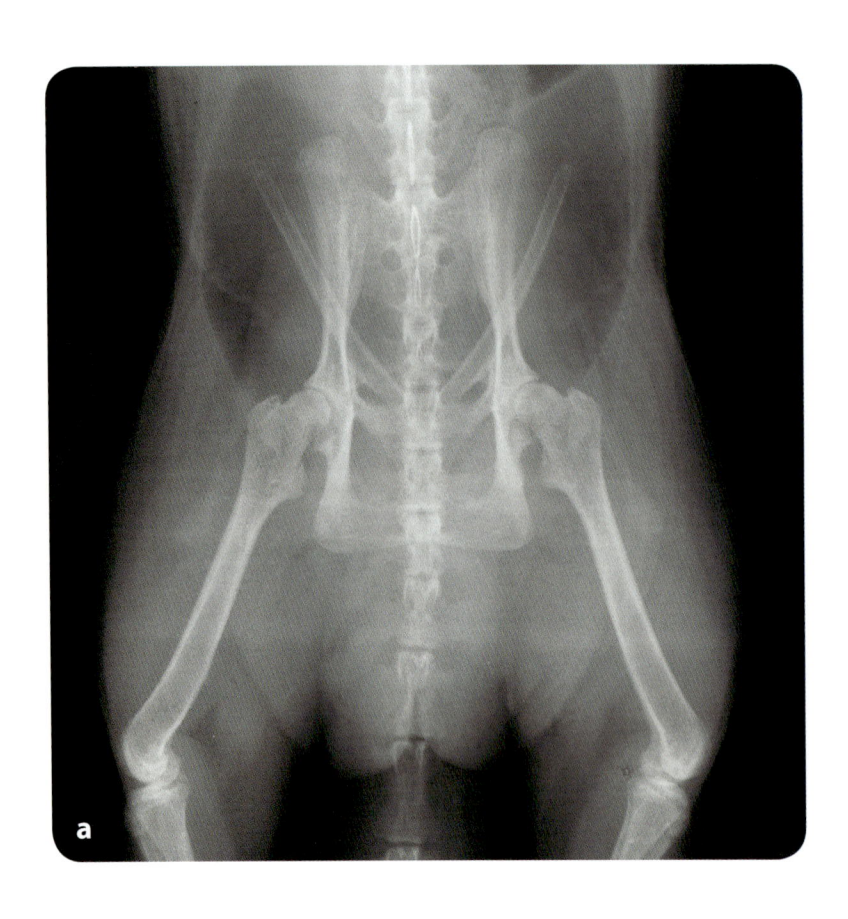

Figure 13.11a,b. Radiograph of the pelvis and femurs of a 3.5 kg adult female Virginia opossum, 60% of actual size.

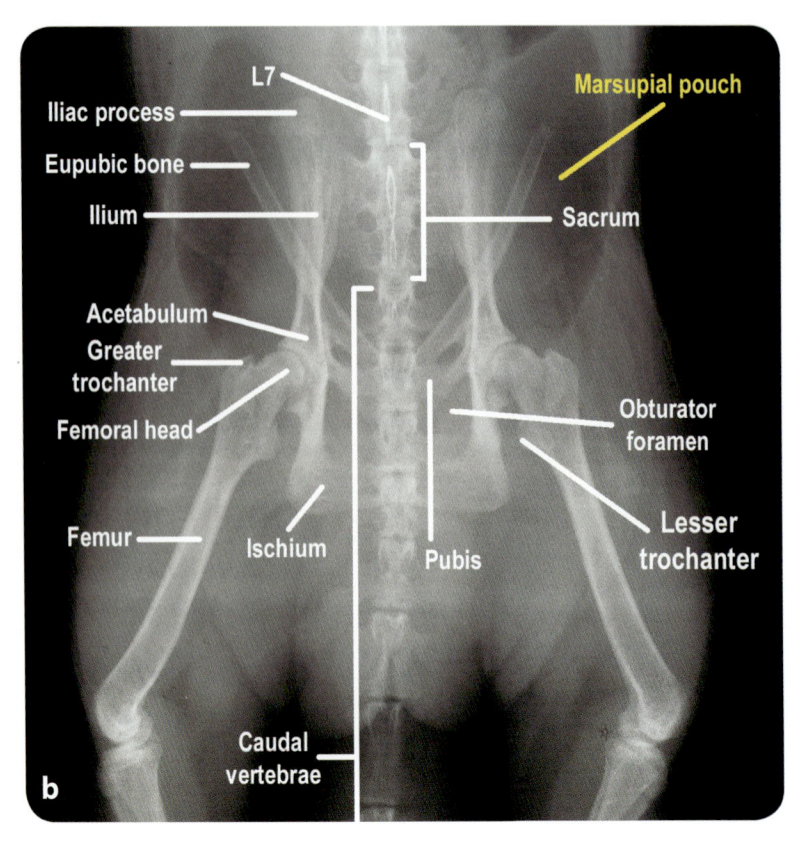

Labels (b):
- L7
- Iliac process
- Eupubic bone
- Ilium
- Marsupial pouch
- Sacrum
- Acetabulum
- Greater trochanter
- Femoral head
- Obturator foramen
- Lesser trochanter
- Femur
- Ischium
- Pubis
- Caudal vertebrae

Craniocaudal Projection of the Distal Pelvic Limb

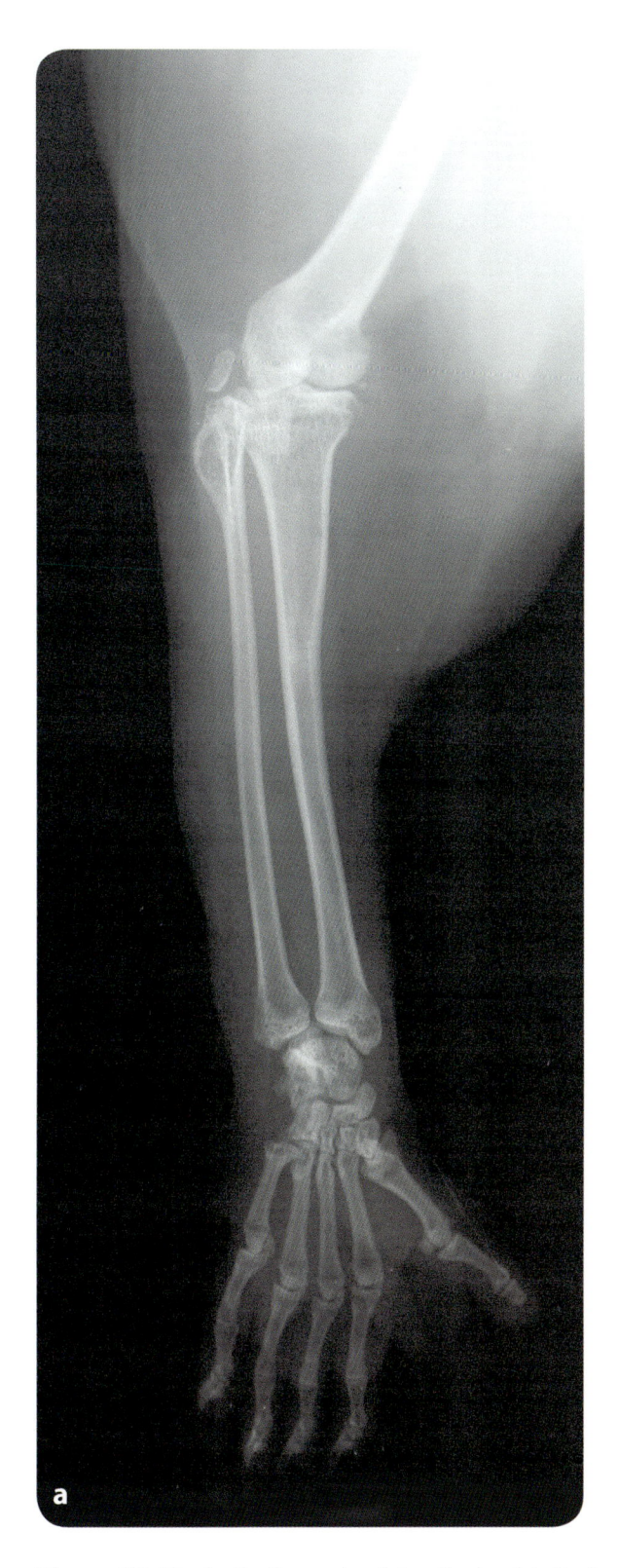

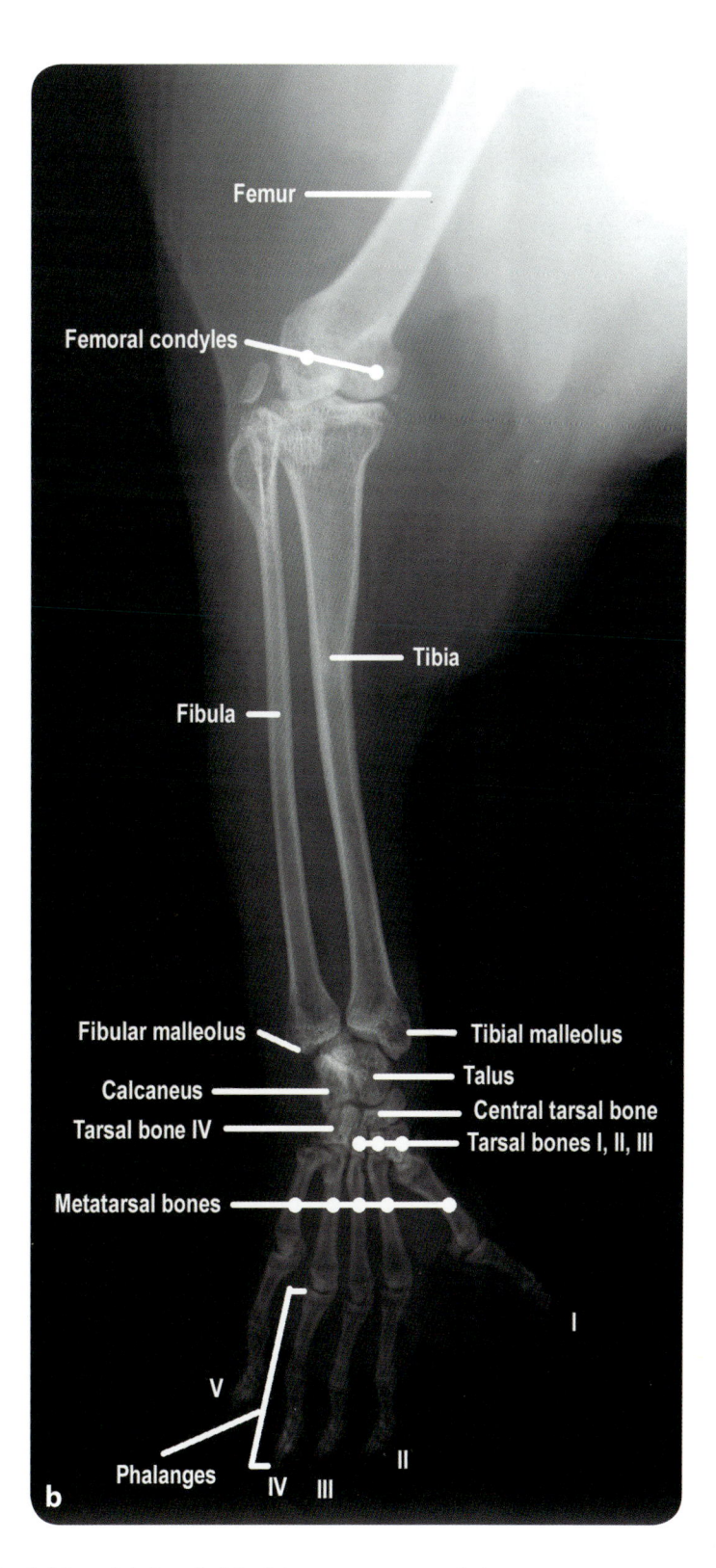

Figure 13.12a,b. Radiograph of the distal pelvic limb of a 3.5 kg adult female Virginia opossum, actual size.

Lateral Projection of the Distal Pelvic Limb

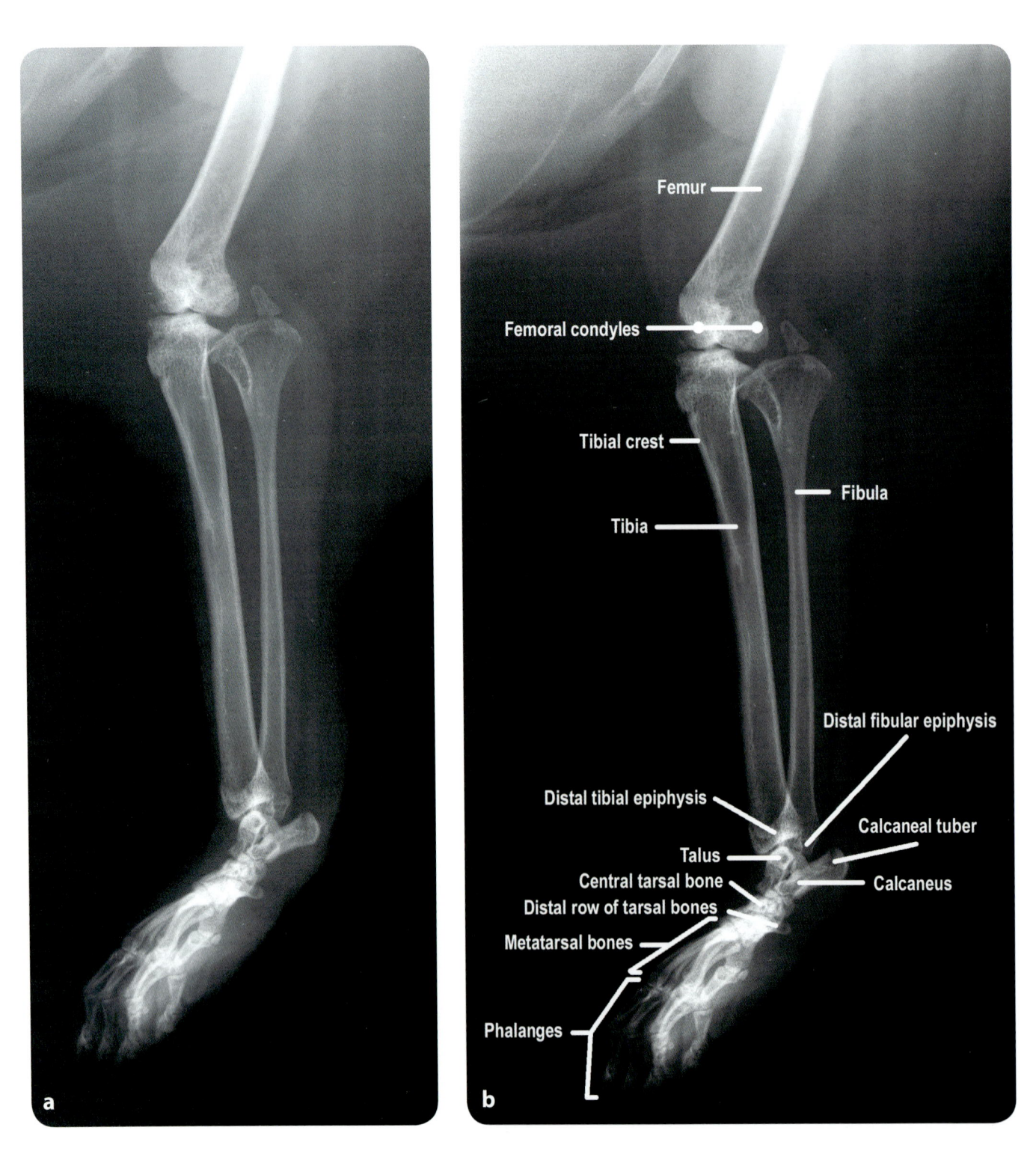

Figure 13.13a,b. Radiograph of the distal pelvic limb of a 3.5 kg adult female Virginia opossum, actual size.

MISCELLANEOUS ABNORMALITIES
Diseases of the Teeth

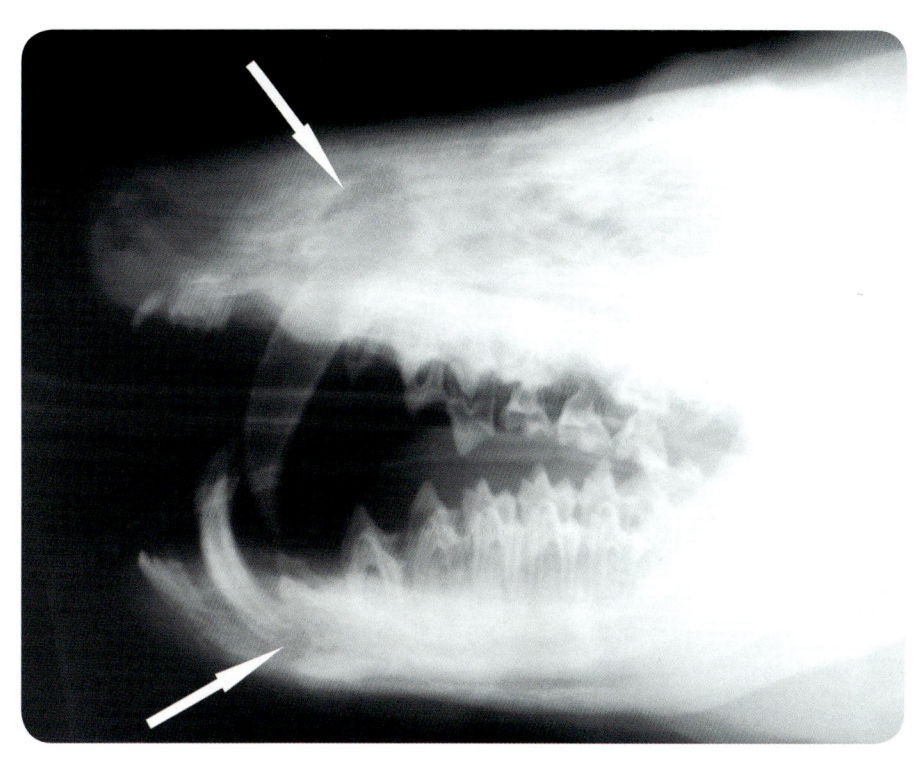

Figure 13.14. Lateral radiograph of the skull of a 2-year-old Virginia opossum with marked lysis of the rostral aspects of both the mandible and the maxilla (arrows). The right maxillary canine tooth is absent. *(Courtesy of Cathy Johnson-Delaney, DVM)*

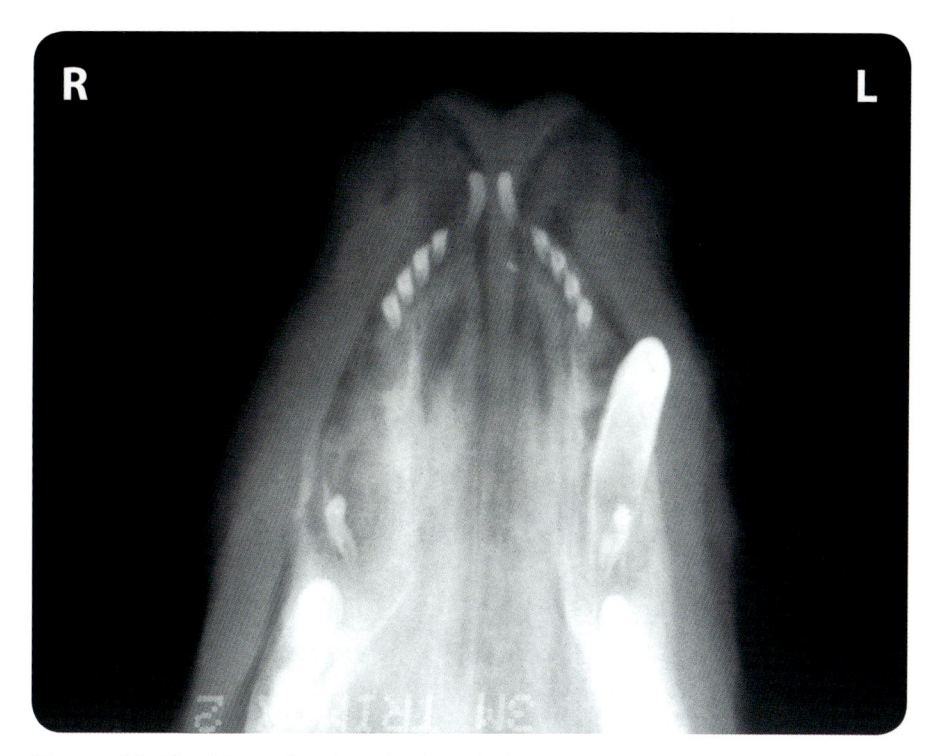

Figure 13.15. Ventrodorsal projection of the rostral portion of the maxilla in the same patient. Note marked lysis and the absence of the right maxillary canine tooth. *(Courtesy of Cathy Johnson-Delaney, DVM)*

Diseases of the Skeleton

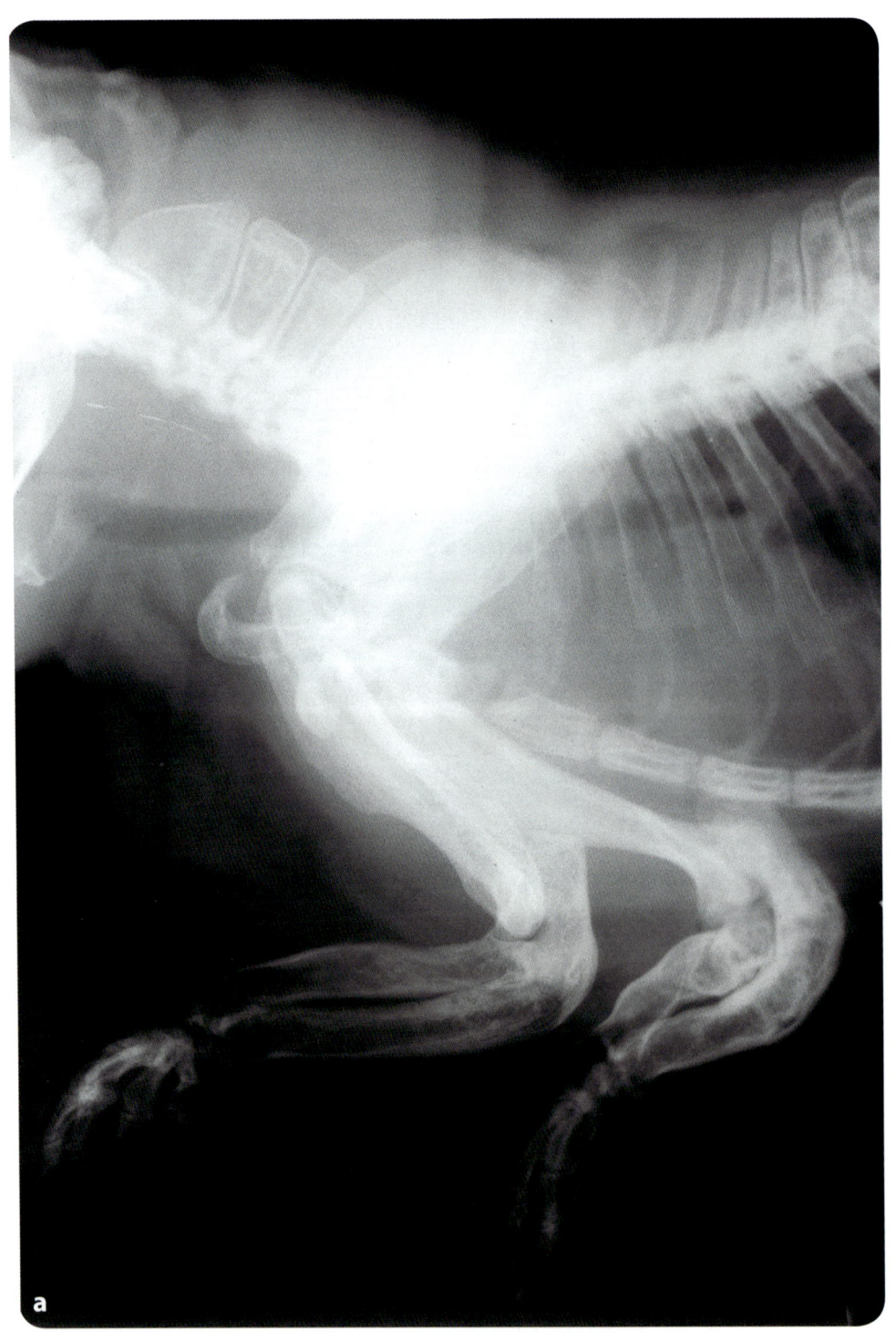

Figure 13.16a,b,c. Lateral (a,b) and ventrodorsal (c) radiographs of a 2-year-old male Virginia opossum with severe deformities of the long bones, secondary to previous metabolic bone disease. The ribs and the vertebrae also appear abnormal.
(Courtesy of Cathy Johnson-Delaney, DVM)

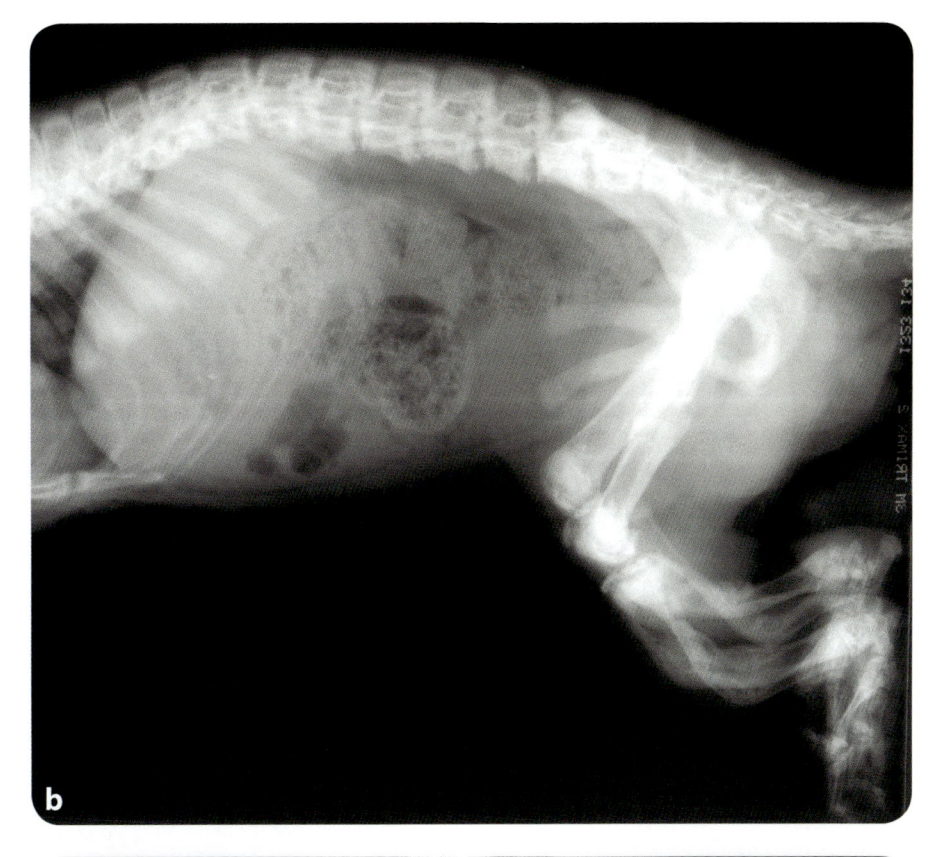

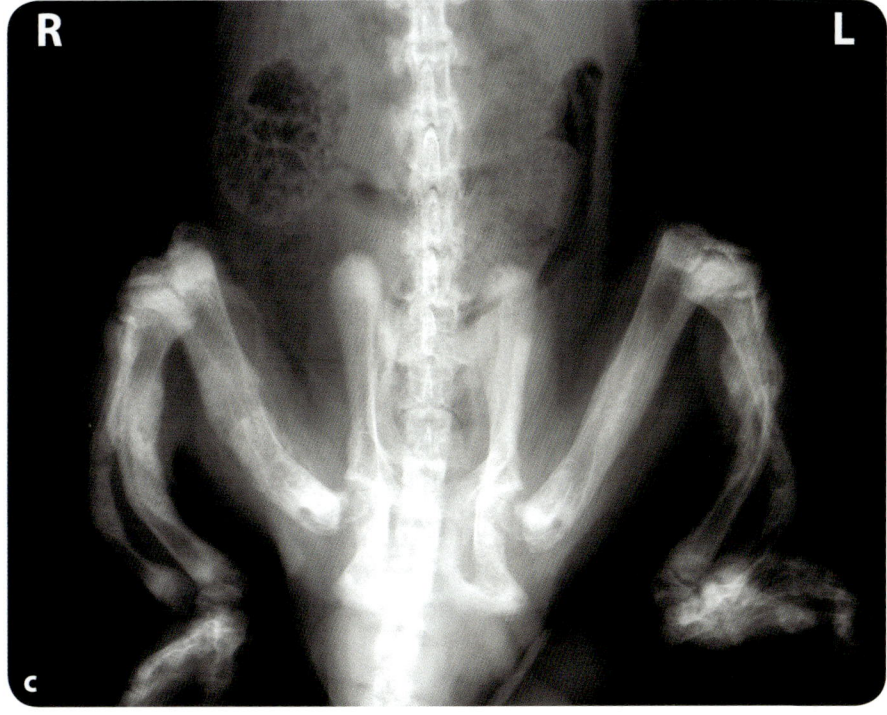

References:
 Johnson-Delaney CA: Marsupials. In: Exotic Companion Medicine Handbook. Lake Worth, FL: Zoological Education Network, 1996.
 Johnson-Delaney CA: What every veterinarian needs to know about Virginia opossums. Exotic DVM. 2005; 6(6):38-43.

POTBELLIED PIG

The NORMAL HEAD
Lateral Projection

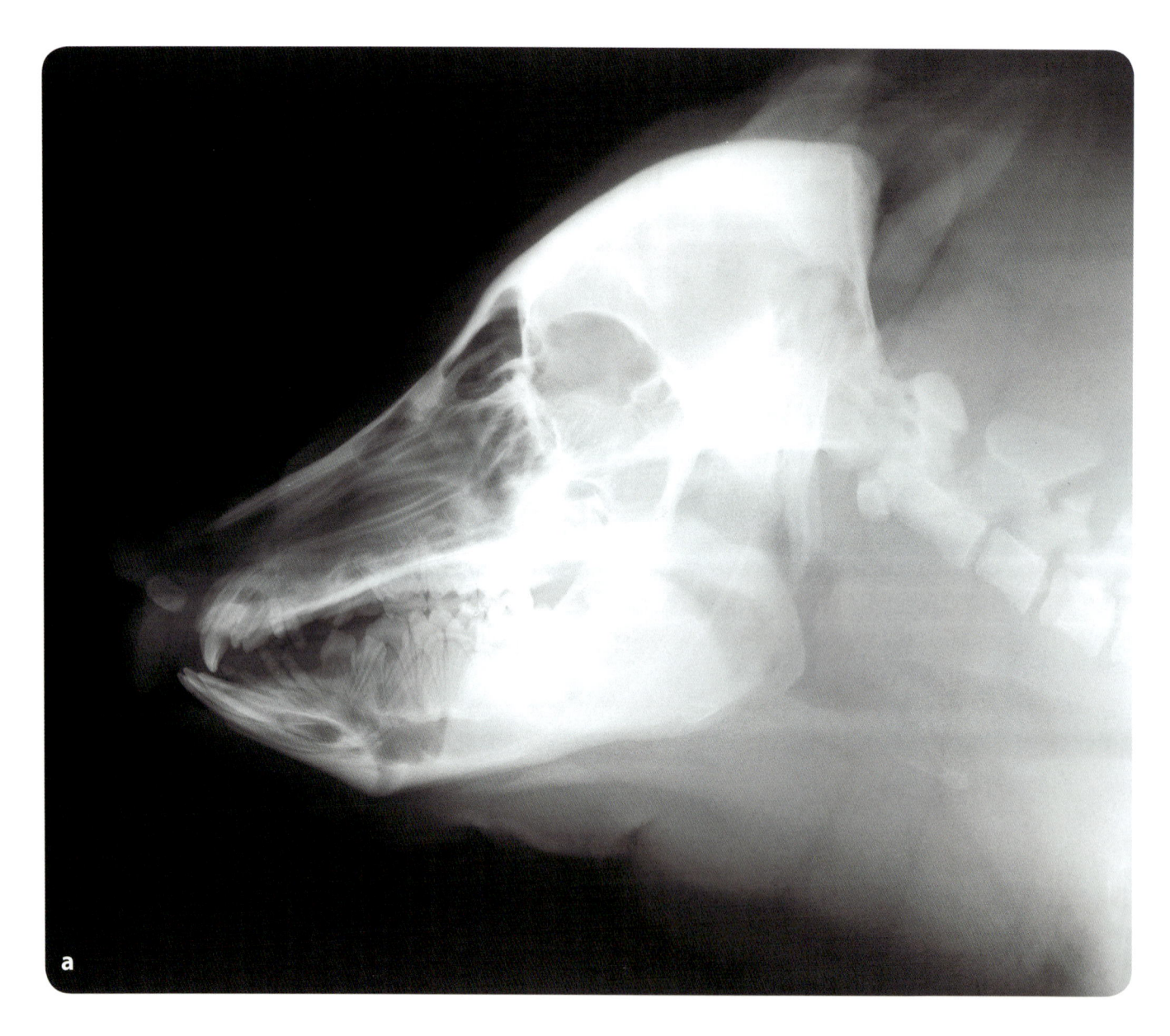

Figure 14.1a,b. Radiograph of the skull of a 15 kg, 6-month-old female pig, 75% of actual size.
Dentition in pigs can vary acccording to selected breeds. The deciduous dentition includes 3 incisor teeth, 1 small canine tooth, and 3 premolar teeth for each arcade (total: 28 teeth). The permanent dentition in standard breeds is I3 C1 PM4 M3 for each arcade (total 44 teeth). Eruption of permanent teeth varies as to tooth. Therefore, deciduous, permanent, and immature permanent teeth can be present in radiographs of the skull of immature pigs.
Note premolars and molar teeth (CT), incisors (I) and canine teeth (C) in Figure 14.2a,b.

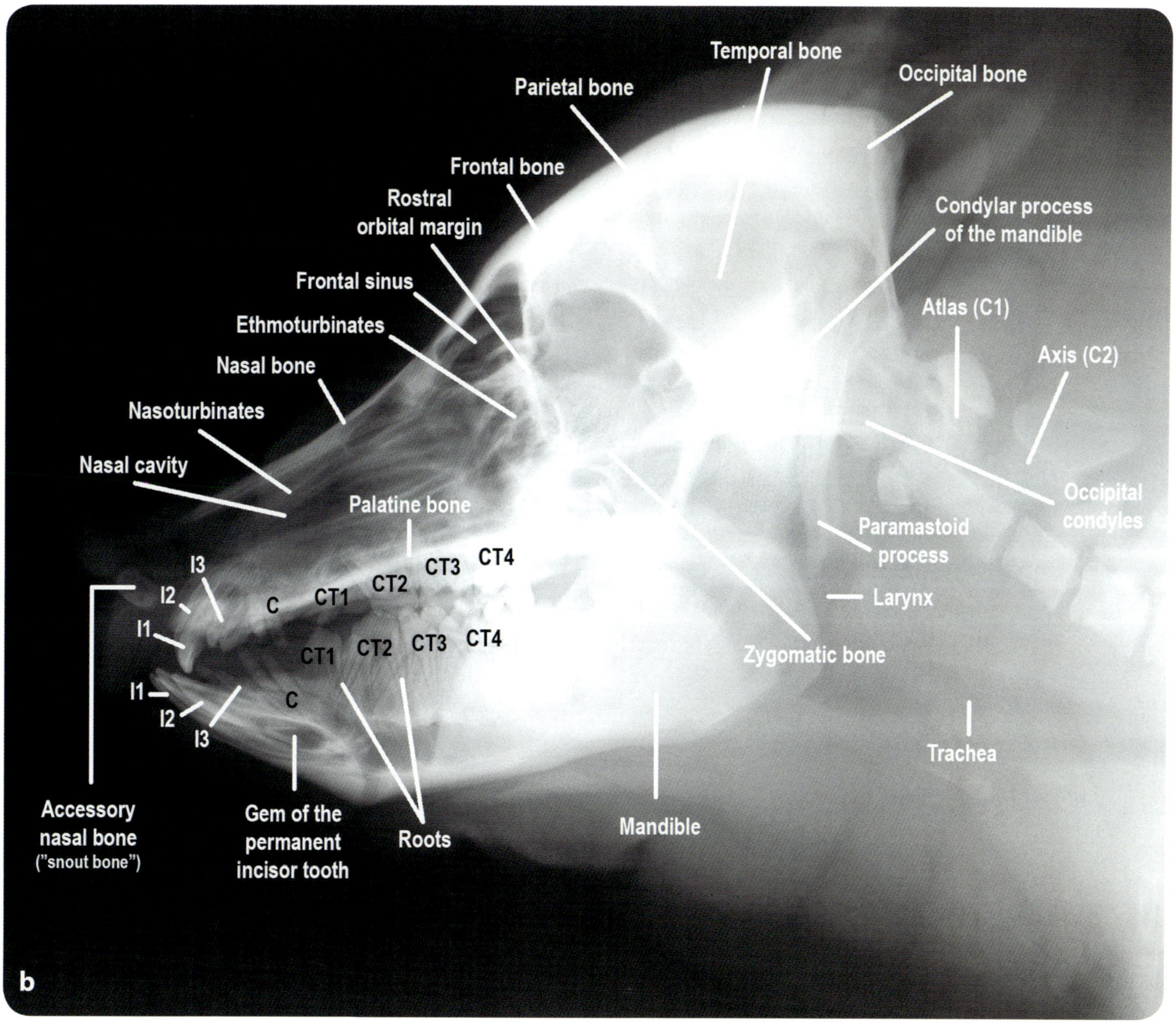
Temporal bone
Occipital bone
Parietal bone
Frontal bone
Rostral orbital margin
Condylar process of the mandible
Frontal sinus
Ethmoturbinates
Atlas (C1)
Nasal bone
Axis (C2)
Nasoturbinates
Nasal cavity
Occipital condyles
Palatine bone
Paramastoid process
I3
CT4
CT3
CT2
I2
C
CT1
Larynx
I1
CT1
CT2
CT3
CT4
Zygomatic bone
I1
C
I2
I3
Trachea
Accessory nasal bone ("snout bone")
Gem of the permanent incisor tooth
Roots
Mandible
b

Oblique Projection

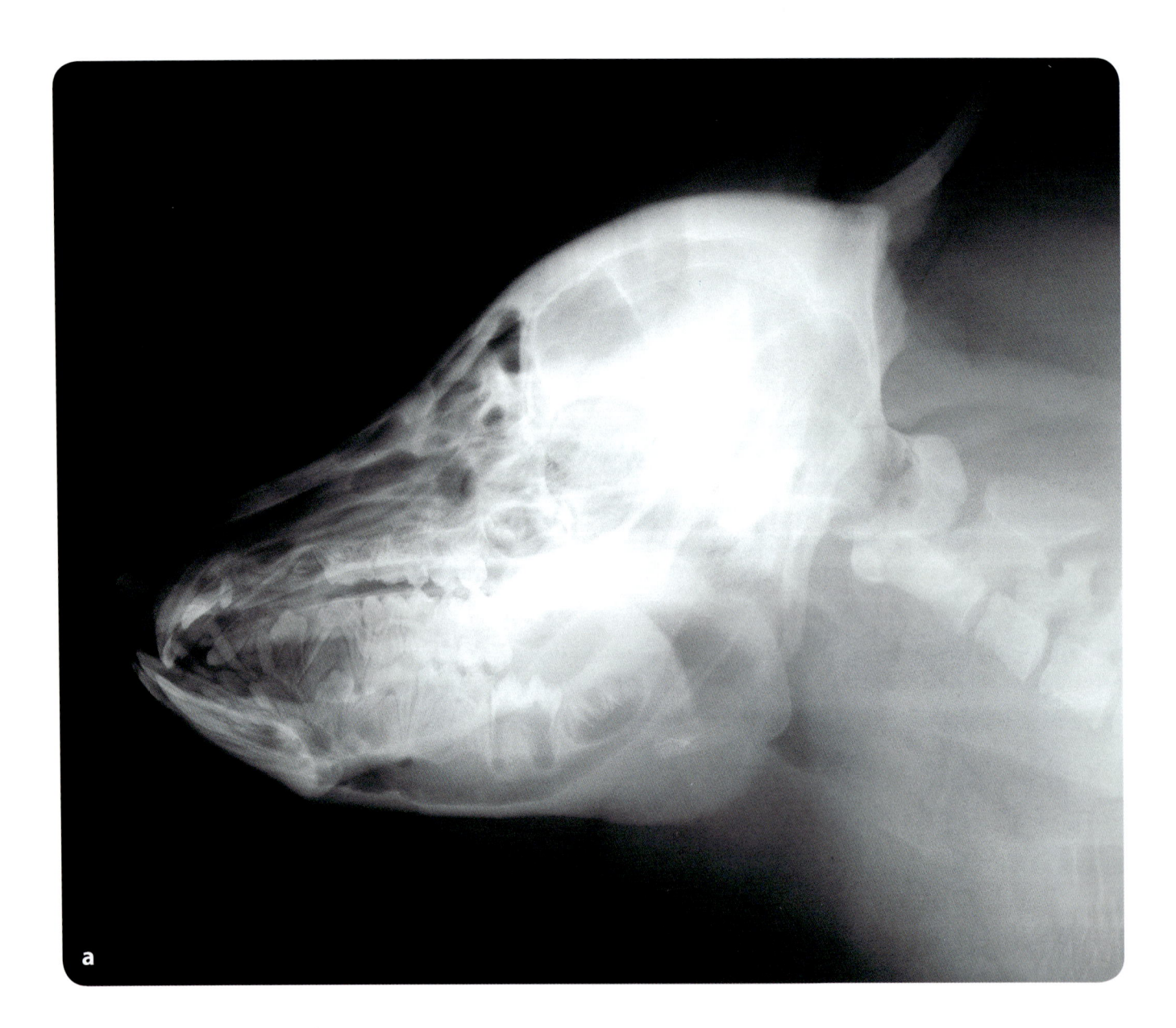

Figure 14.2a,b. Radiograph of the skull of a 15 kg, 6-month-old female pig, 75% of actual size. The oblique projection can be useful to avoid superimposition of the dental arcades, and to highlight the single roots. Intraoral films can be used in this species as well. Note progression of tooth roots from more mature (rostral) to less mature (caudal).

HEAD   461

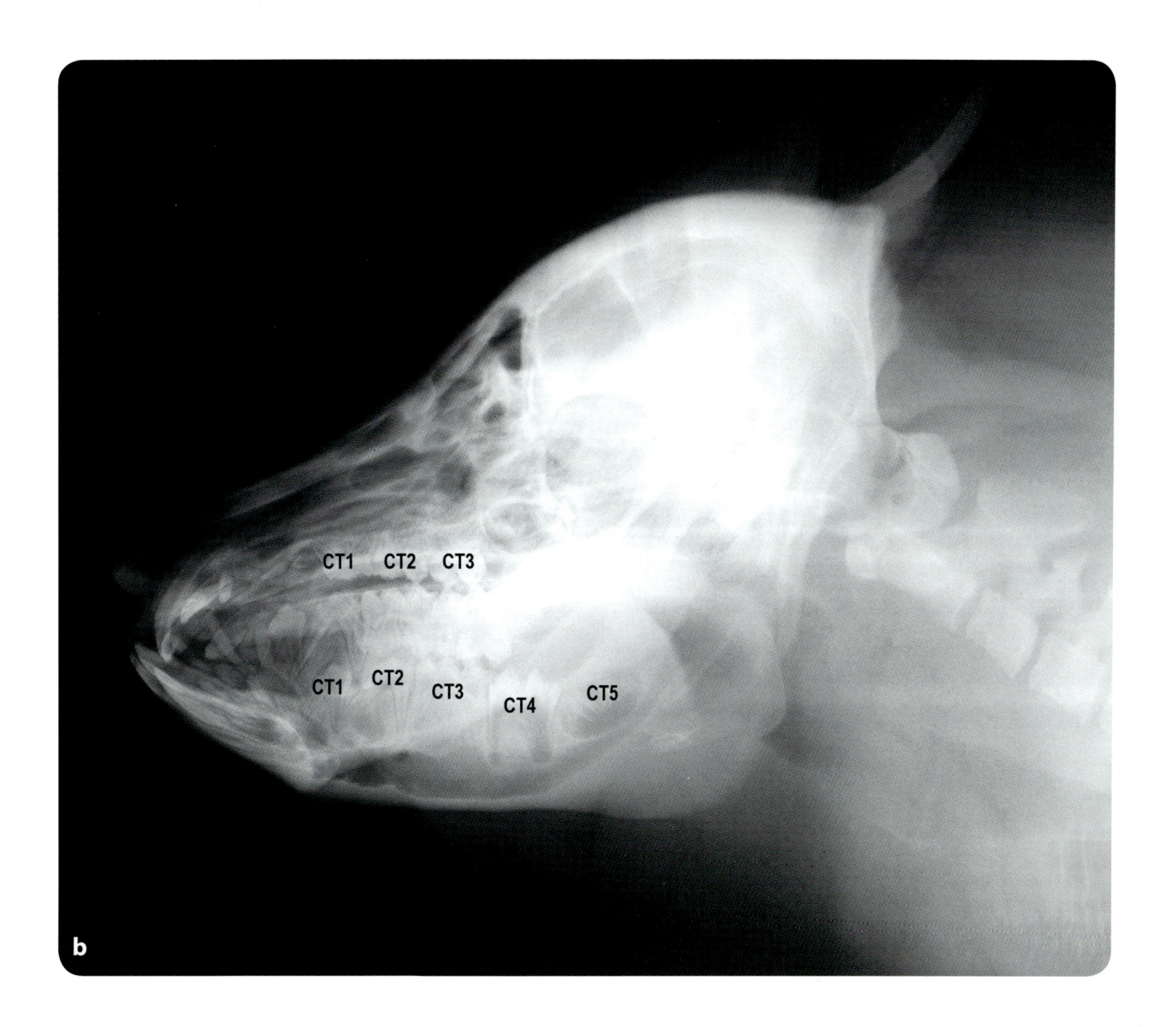

b

The NORMAL THORAX
The Cervical and Thoracic Vertebral Column
Lateral Projection

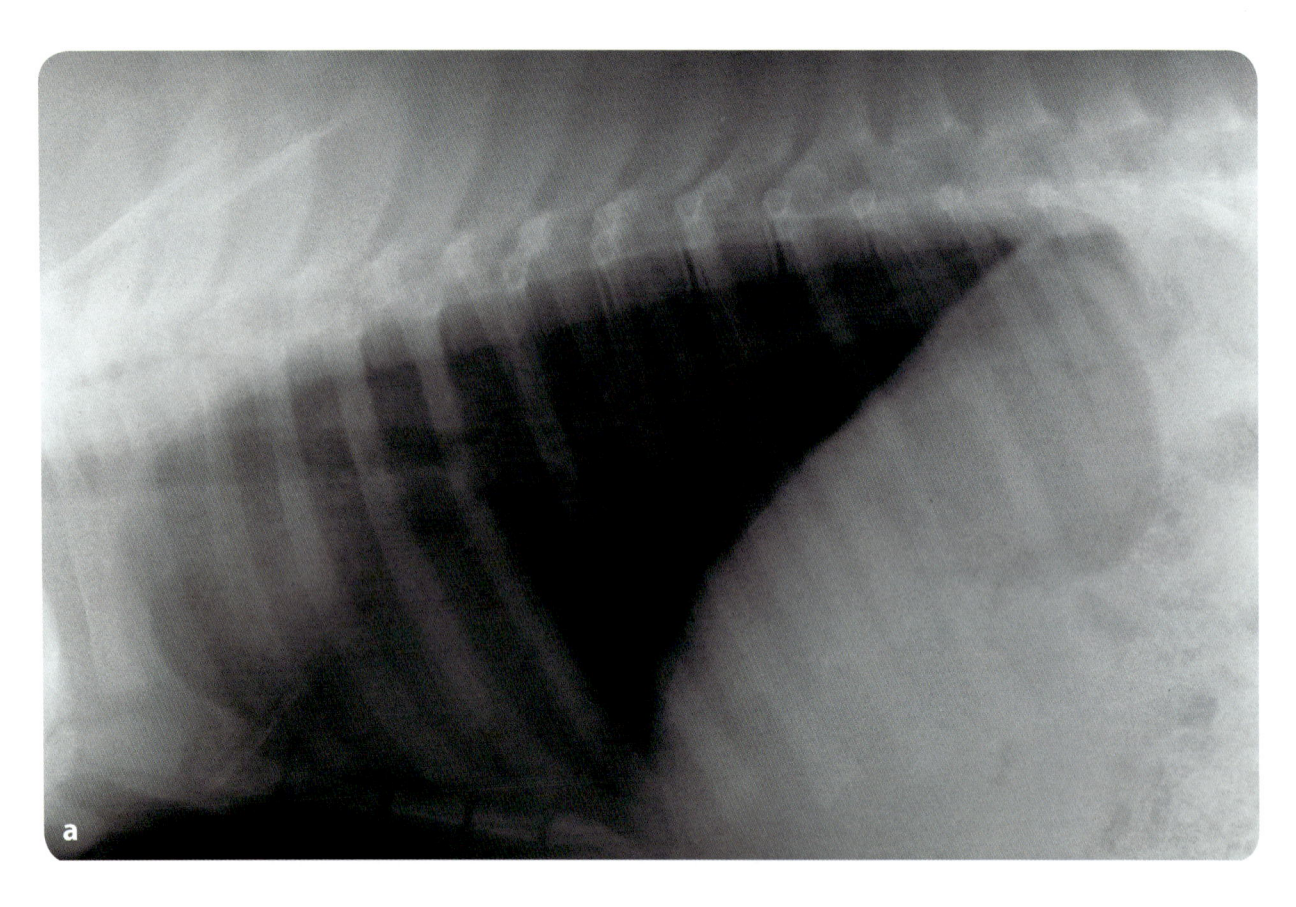

Figure 14.3a,b. Radiograph of the thorax of a 6 kg, 4-month-old female potbellied pig, 75% of actual size.

Spinous processes   T1-T15

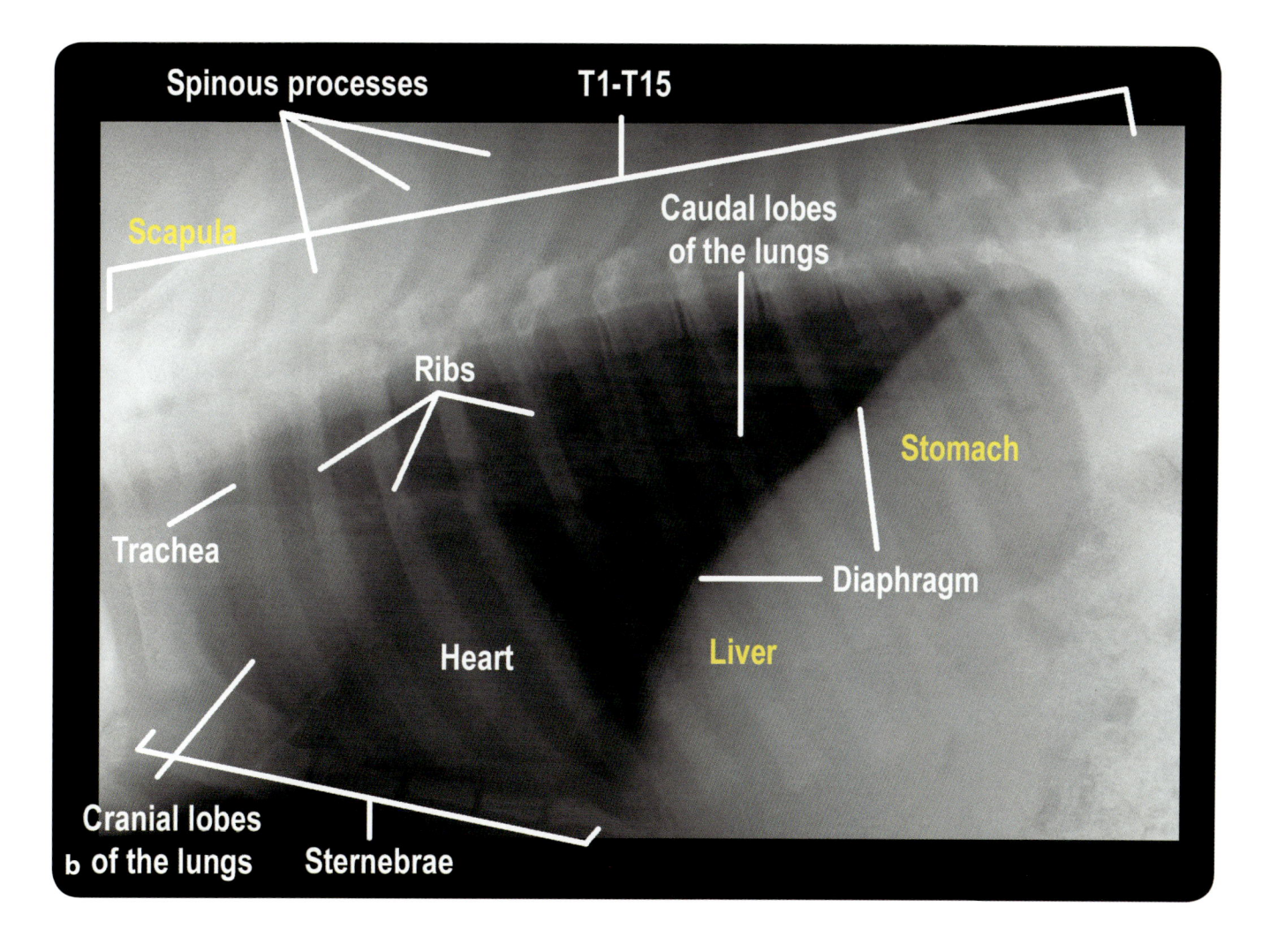

Scapula

Caudal lobes of the lungs

Ribs

Stomach

Trachea

Diaphragm

Heart   Liver

Cranial lobes
b of the lungs   Sternebrae

Ventrodorsal Projection

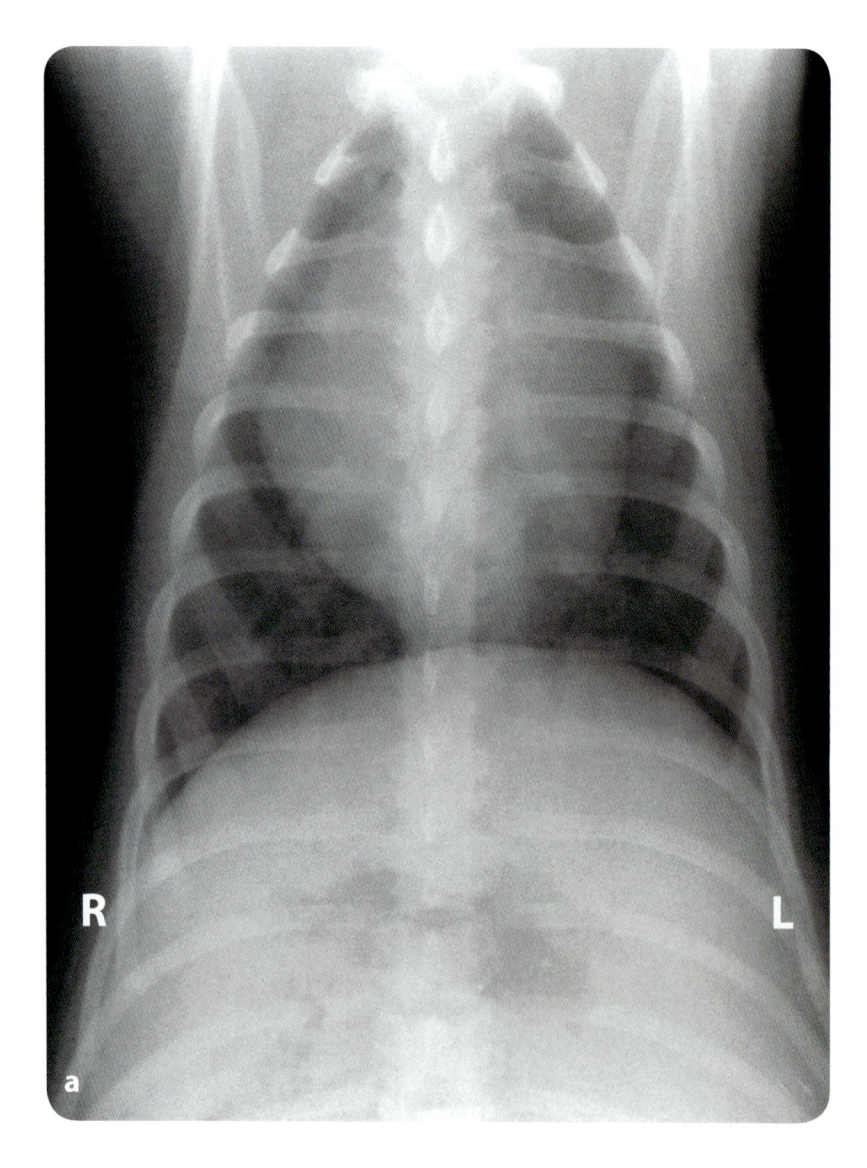

Figure 14.4a,b. Radiograph of the thorax of a 6 kg, 4-month-old female potbellied pig, 75% of actual size.

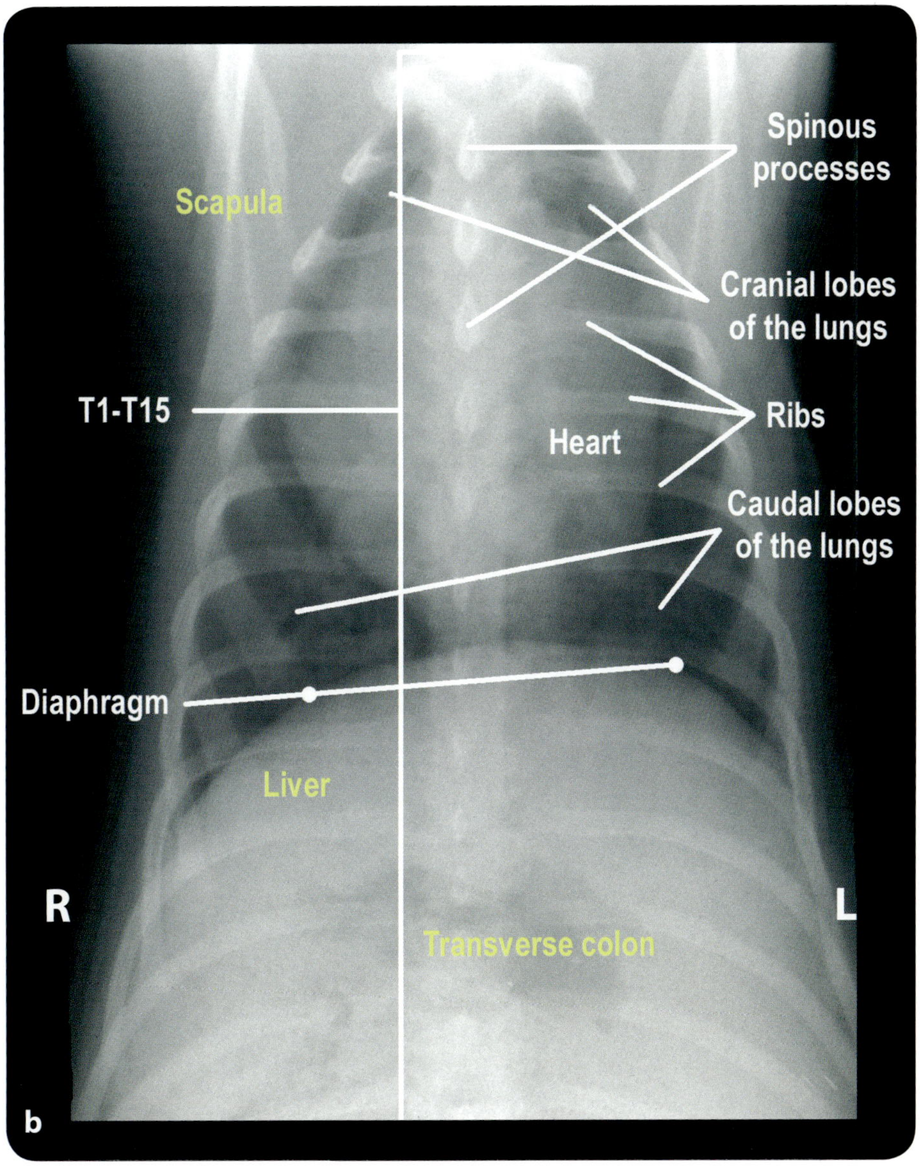
Spinous processes
Scapula
Cranial lobes of the lungs
T1-T15
Ribs
Heart
Caudal lobes of the lungs
Diaphragm
Liver
R
L
Transverse colon
b

The NORMAL ABDOMEN
The Lumbar Vertebral Column
Lateral Projection

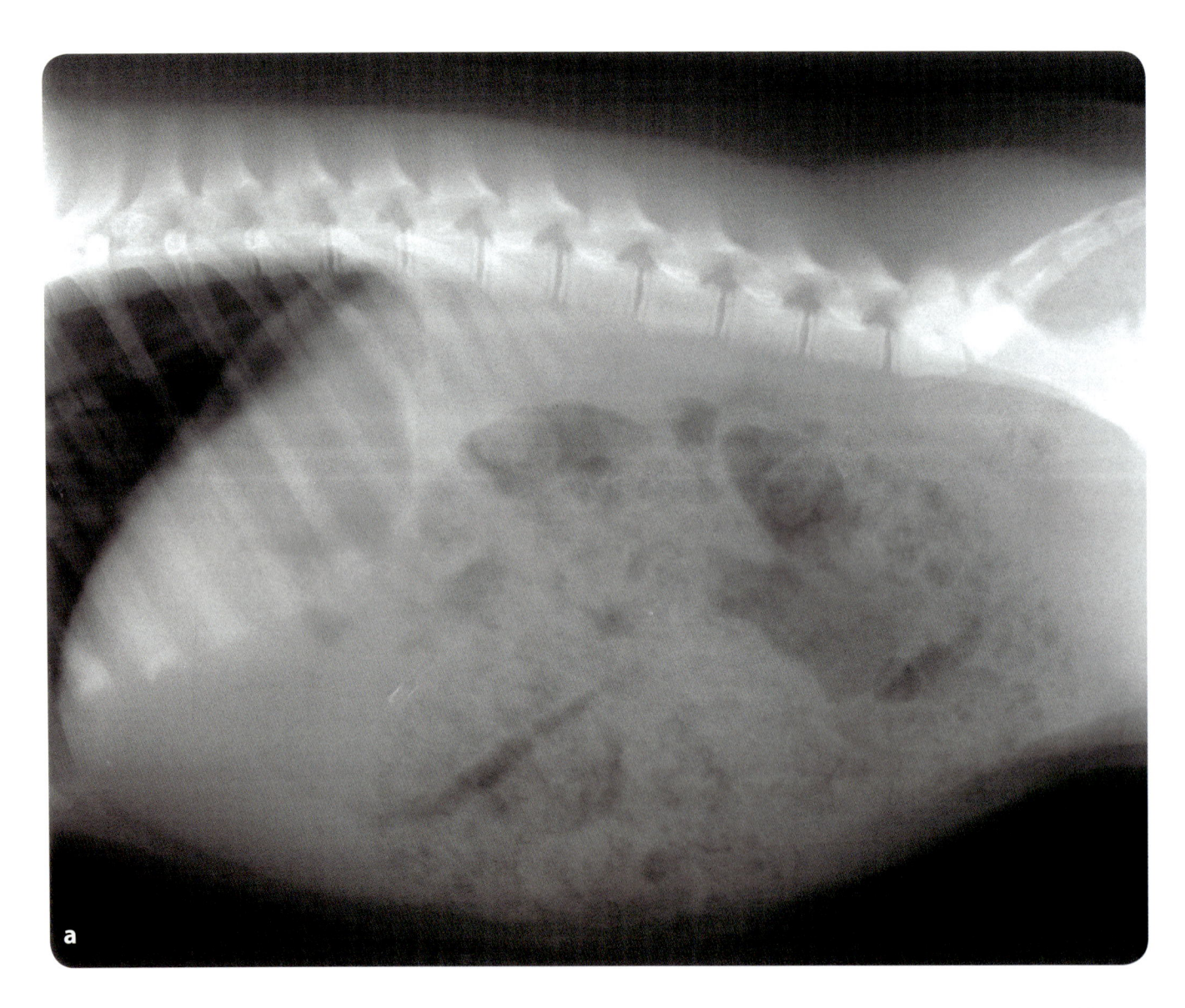

Figure 14.5a,b. Radiograph of the abdomen of a 6 kg, 4-month-old female potbellied pig, 75% of actual size. Normal ingesta can significantly interfere with identification of abdominal structures.

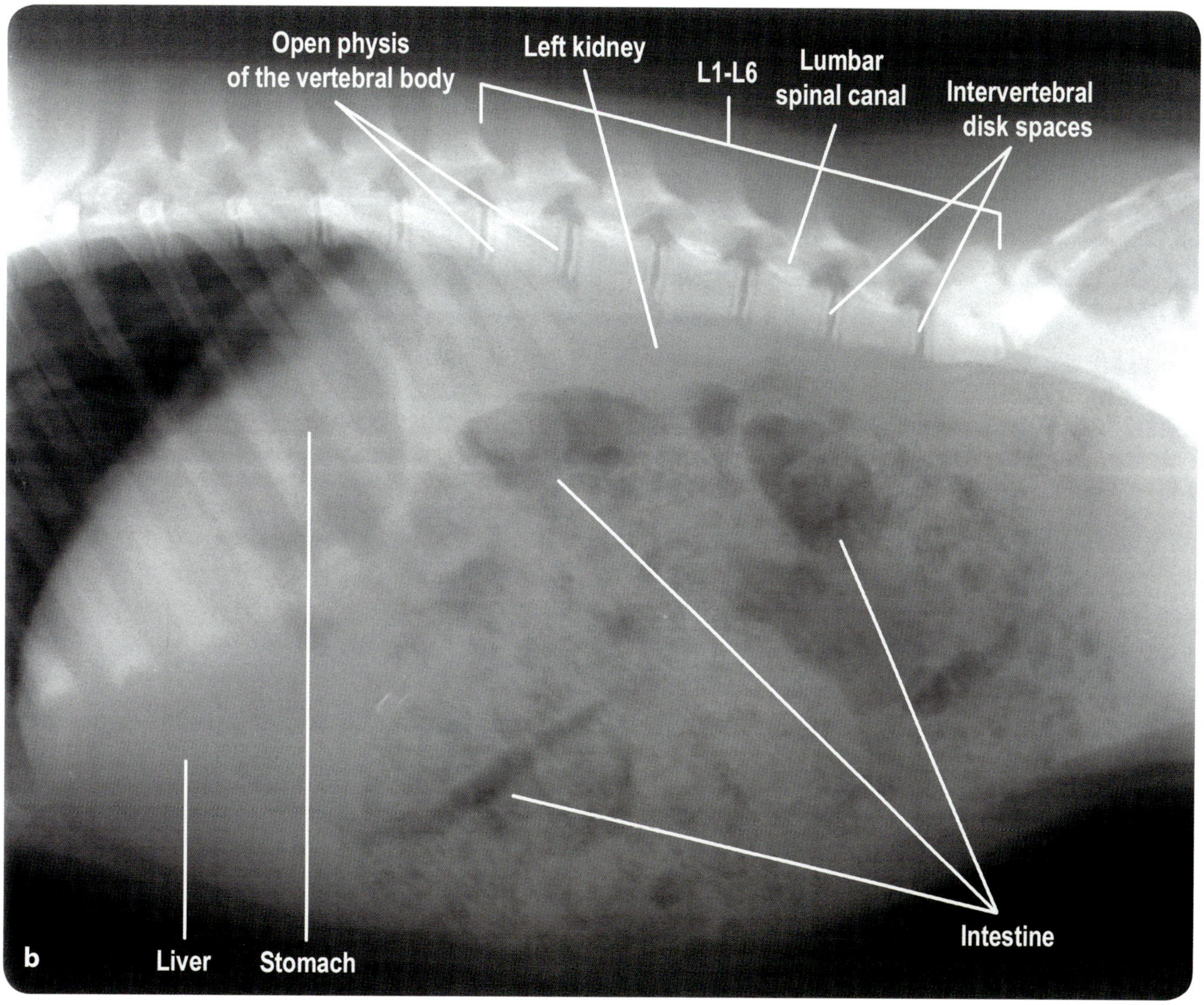
Open physis
of the vertebral body
Left kidney
L1-L6
Lumbar
spinal canal
Intervertebral
disk spaces
Intestine
b
Liver
Stomach

Ventrodorsal Projection

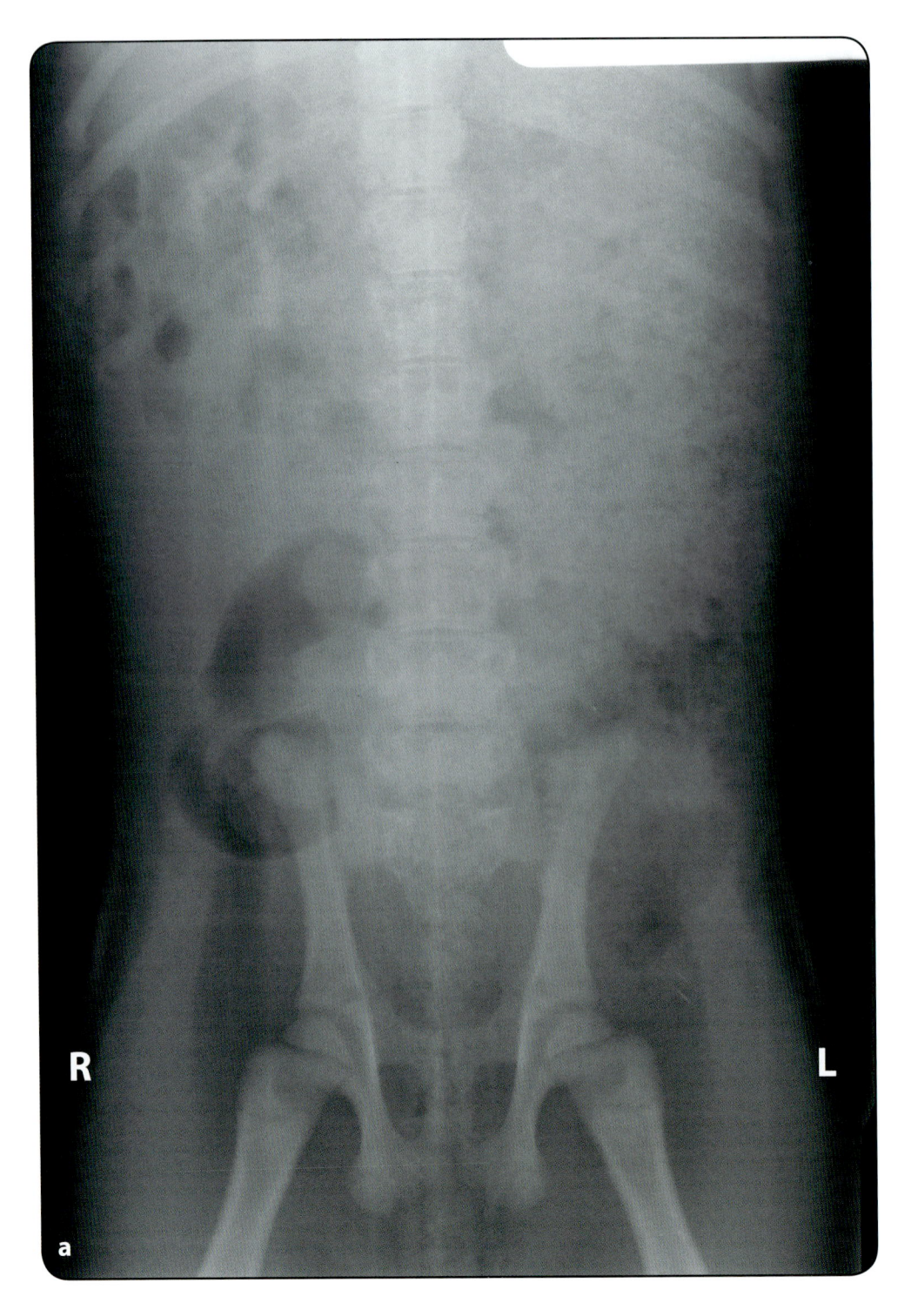

Figure 14.6a,b. Radiograph of the abdomen of a 6 kg, 4-month-old female potbellied pig, 75% of actual size.

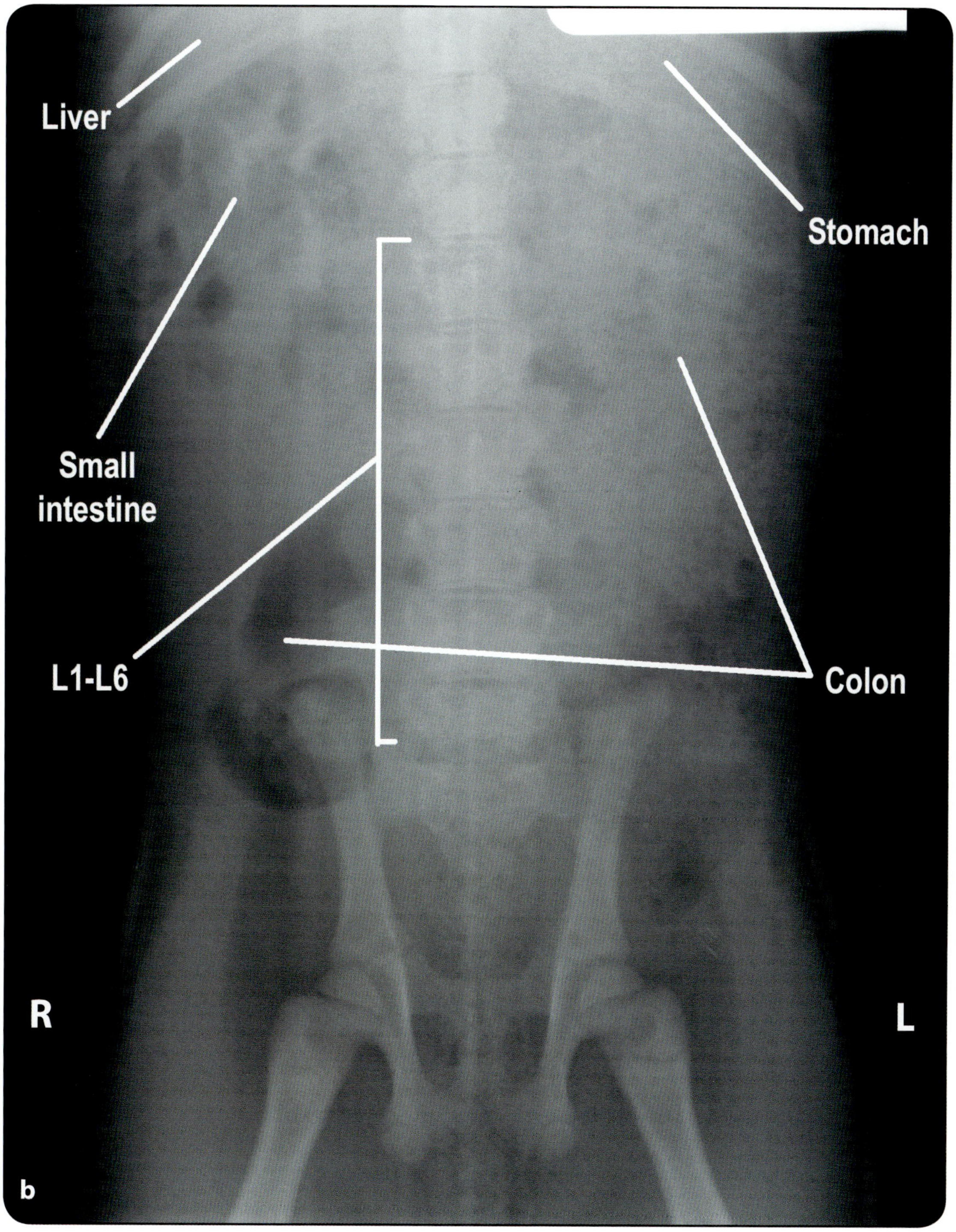
Liver
Stomach
Small intestine
L1-L6
Colon
R
L
b

The NORMAL THORACIC LIMB
Lateral Projection of the Distal Thoracic Limb

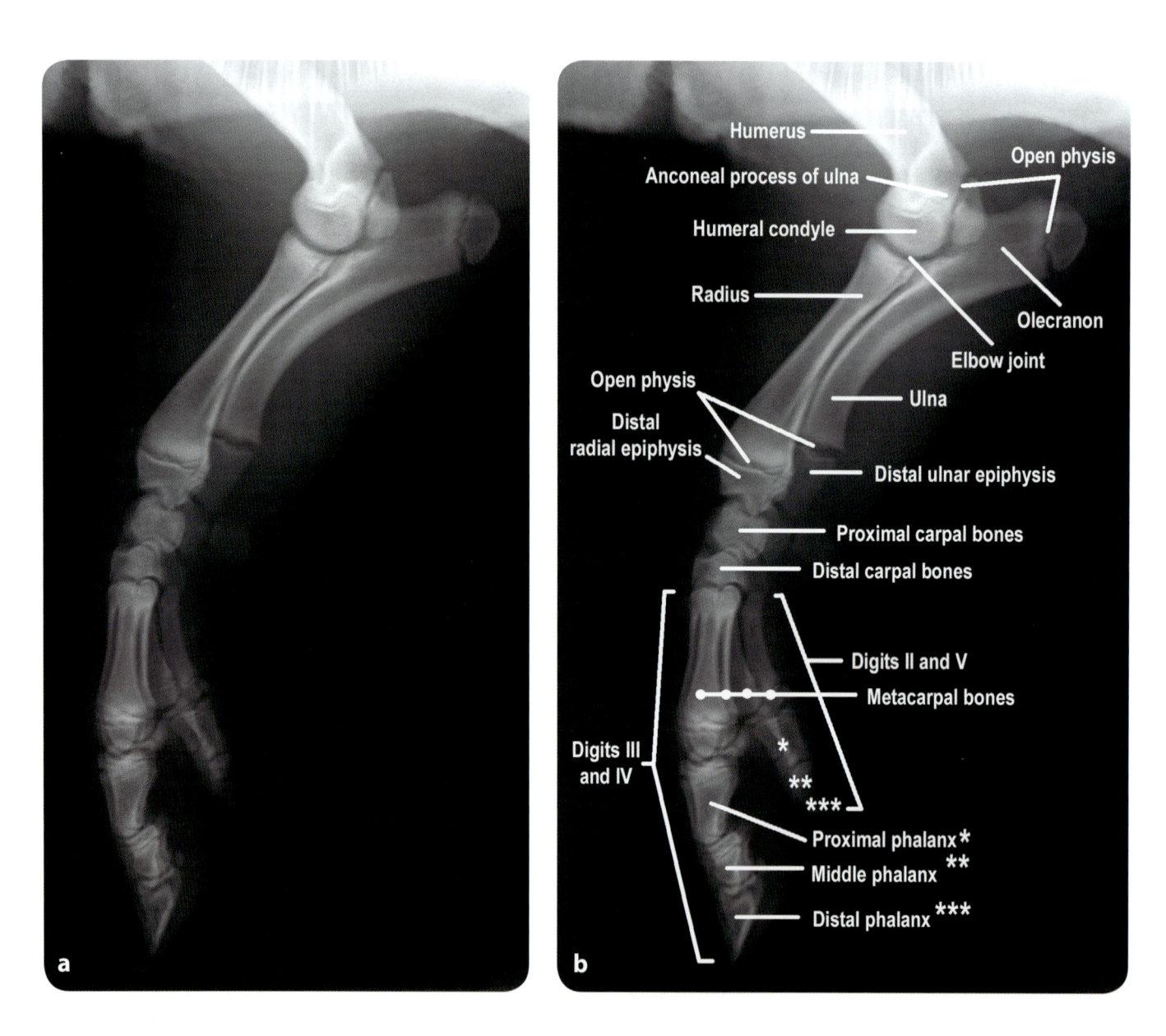

Figure 14.7a,b. Radiograph of the distal thoracic limb of a 15 kg, 6-month-old female pig, 75% of actual size.

Craniocaudal Projection of the Distal Thoracic Limb

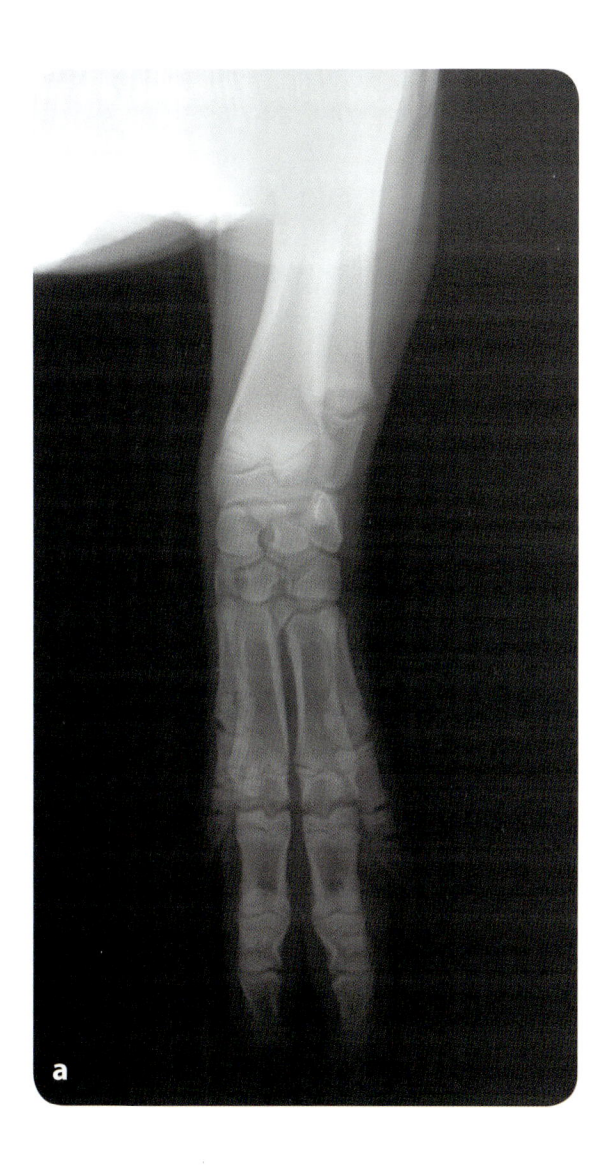

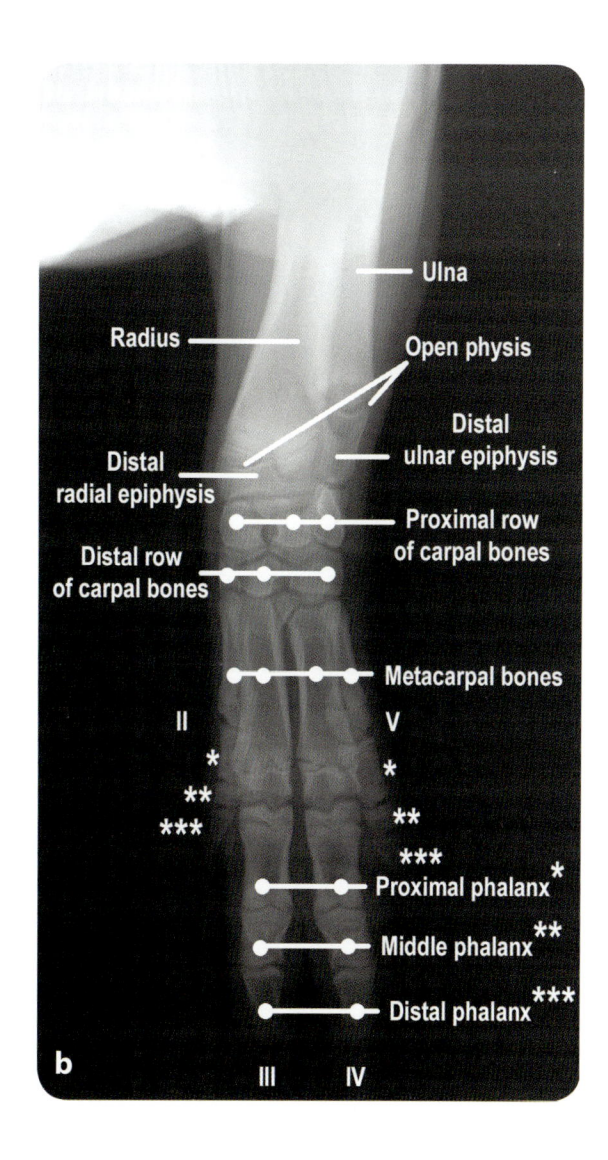

Figure 14.8a,b. Radiograph of the distal thoracic limb of a 15 kg 6-month-old female pig, 75% of actual size.

The NORMAL PELVIC LIMB
Lateral Projection of the Pelvis and of the Proximal Pelvic Limb

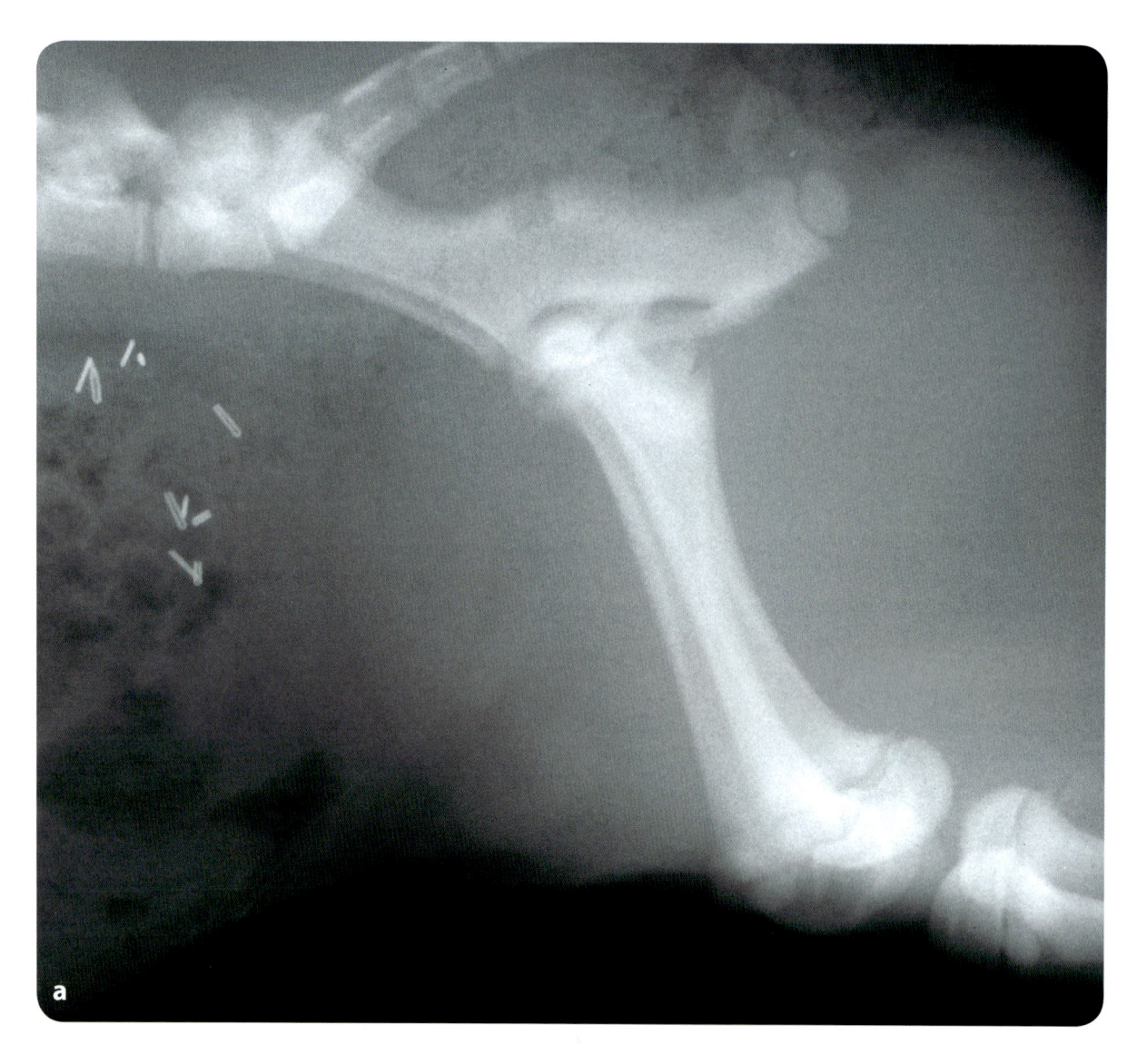

Figure 14.9a,b. Radiograph of the pelvis and the proximal pelvic limb of a 6 kg, 4-month-old female pig, actual size. Note metal hemostatic clips used for ovariohysterectomy.

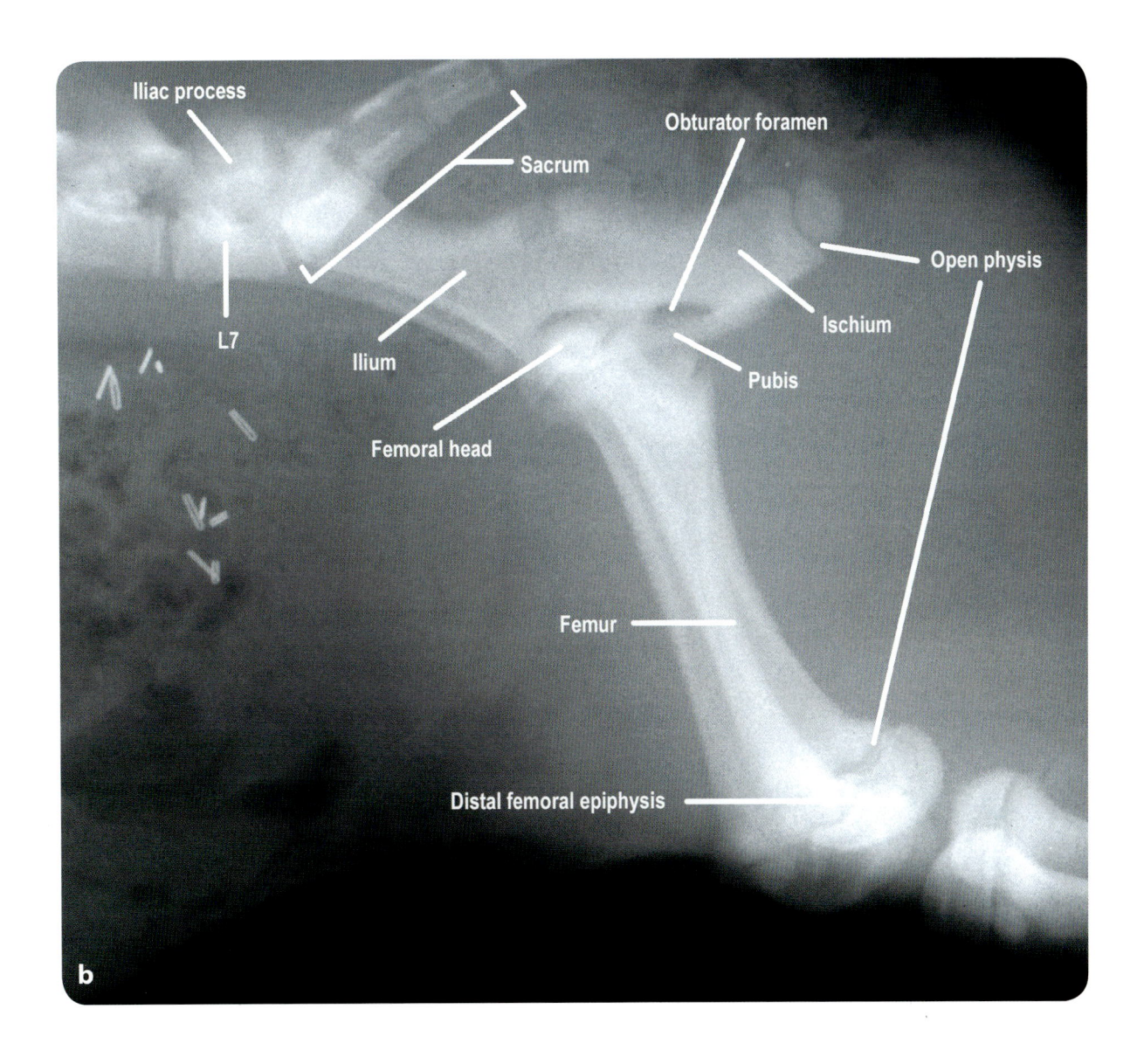

PELVIC LIMB

Ventrodorsal Projection of the Pelvis
Craniocaudal Projection of the Poximal Pelvic Limb

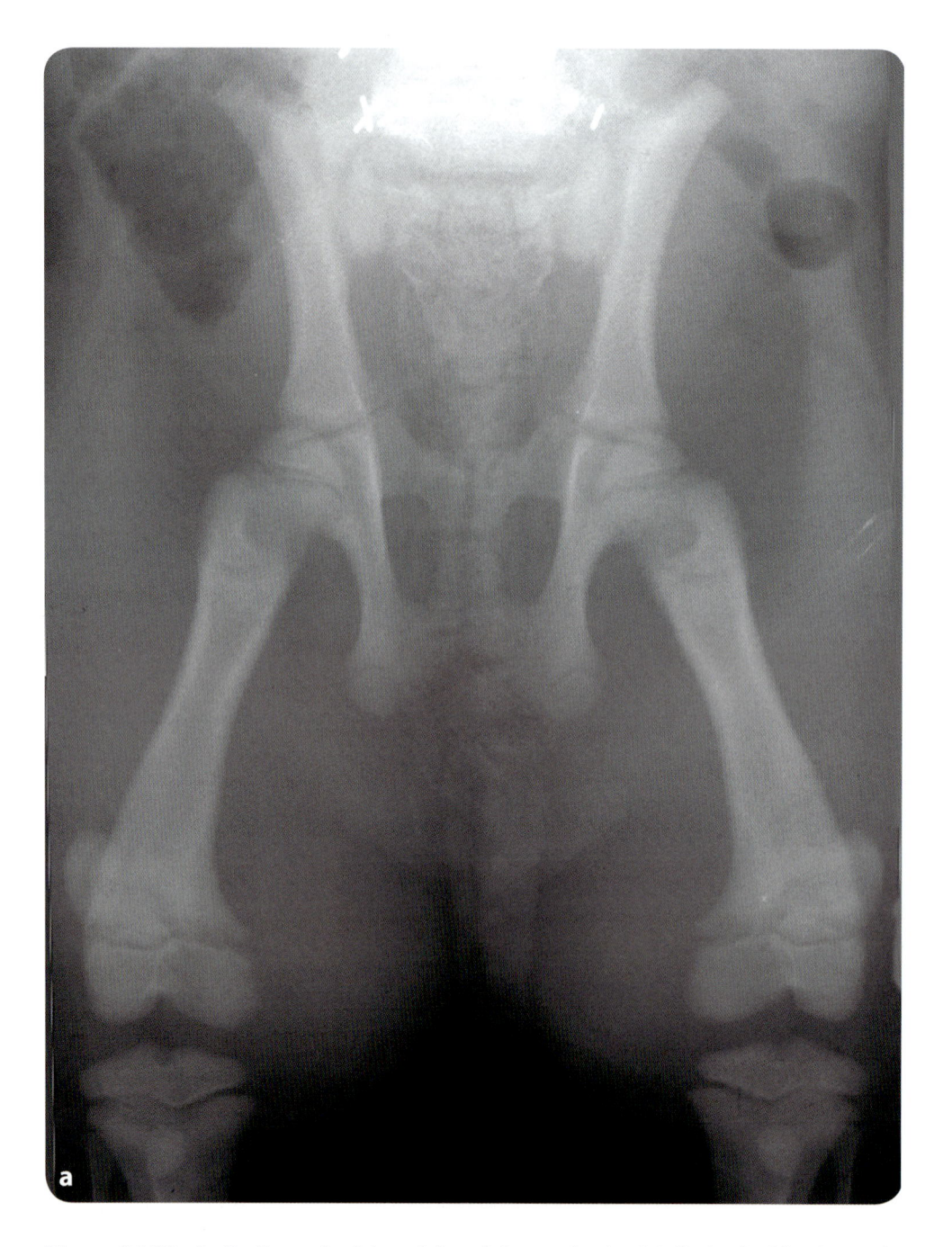

Figure 14.10a,b. Radiograph of the pelvis and the proximal pelvic limb of a 6 kg, 4-month-old female pig, actual size.

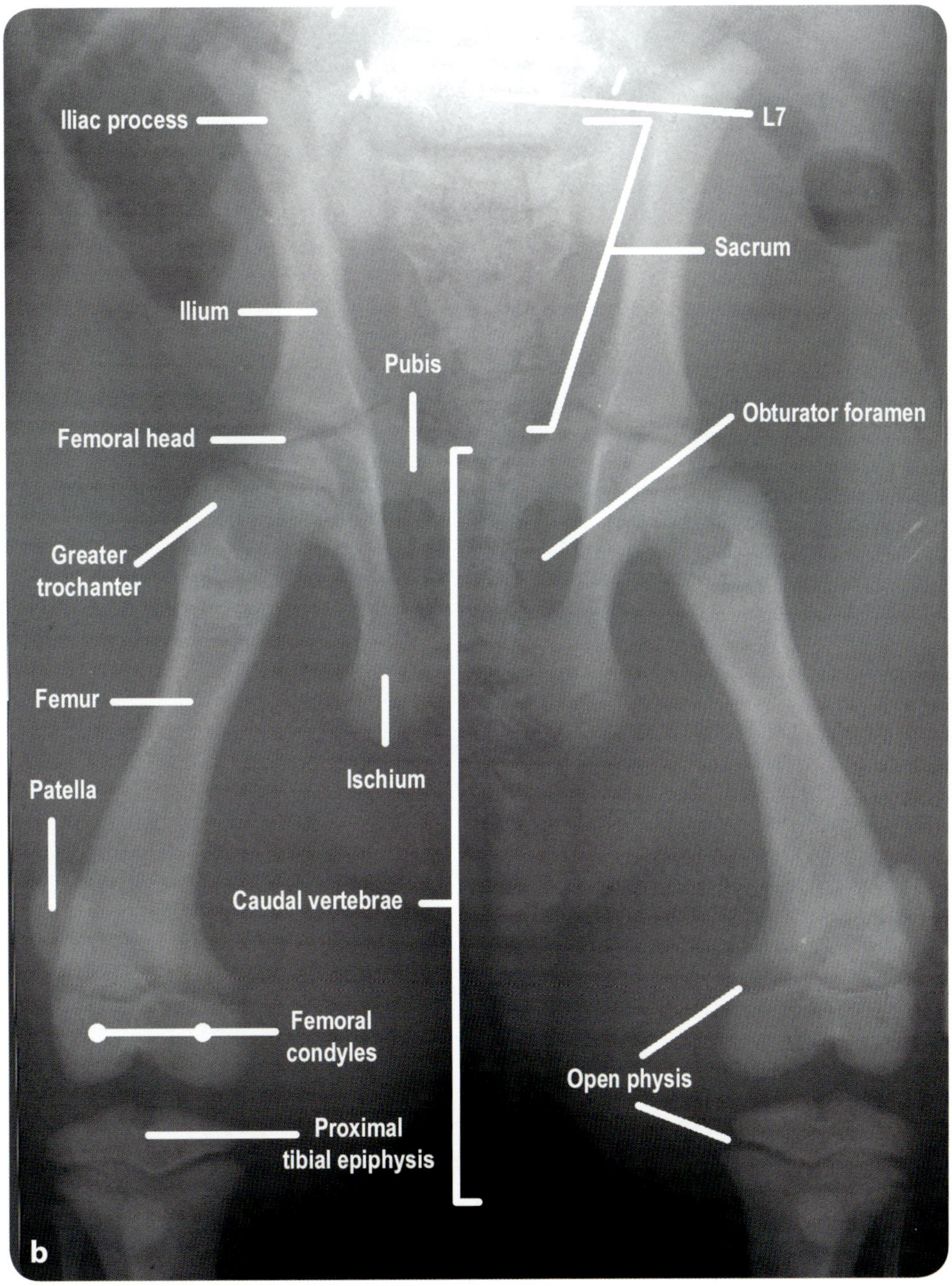
Iliac process
L7
Sacrum
Ilium
Pubis
Obturator foramen
Femoral head
Greater trochanter
Femur
Ischium
Patella
Caudal vertebrae
Femoral condyles
Open physis
Proximal tibial epiphysis
b

Lateral Projection of the Distal Pelvic Limb

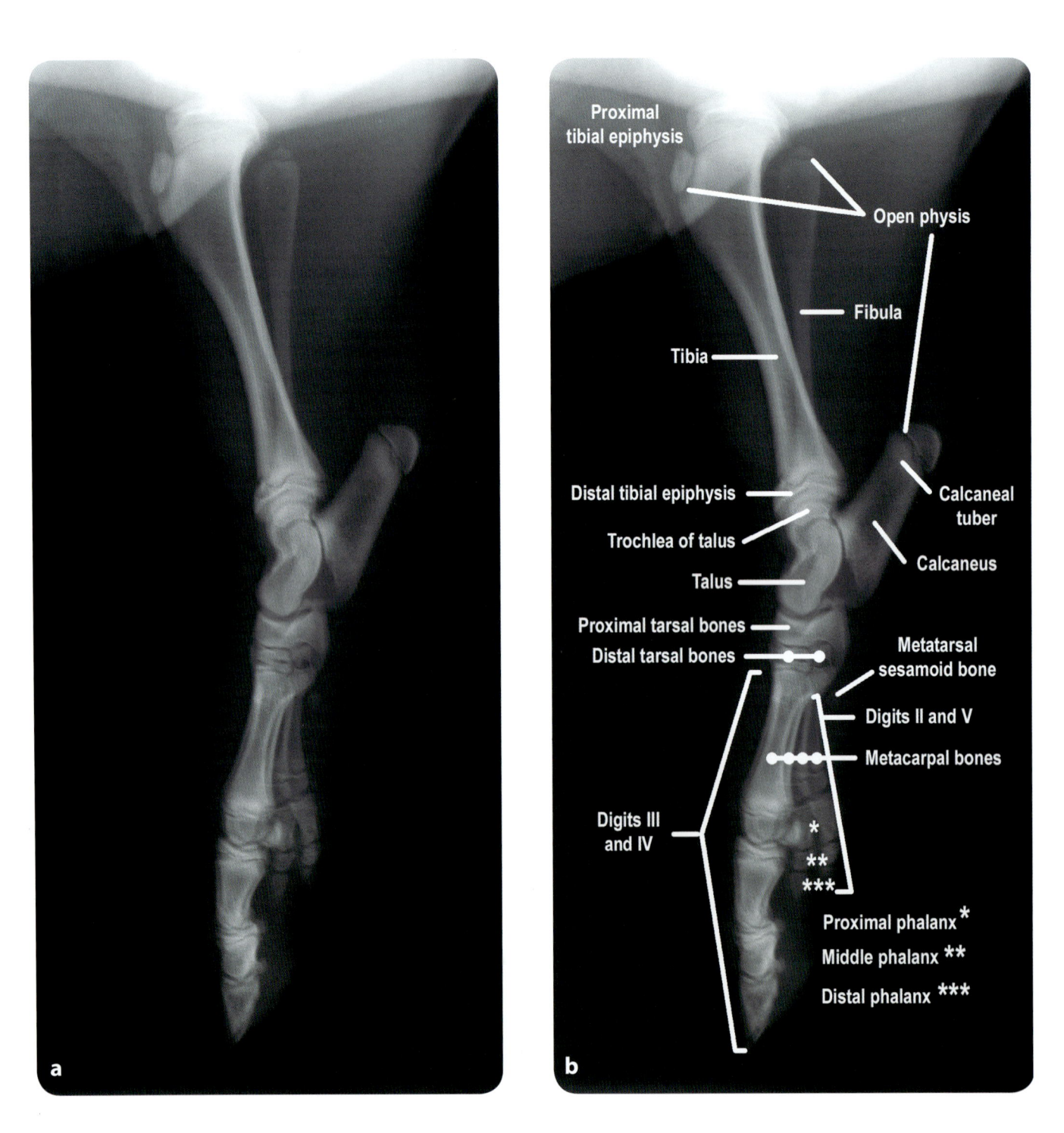

Figure 14.11a,b. Radiograph of the distal pelvic limb of a 15 kg 6-month female pig, 75% of the actual size.

Craniocaudal Projection of the Distal Pelvic Limb

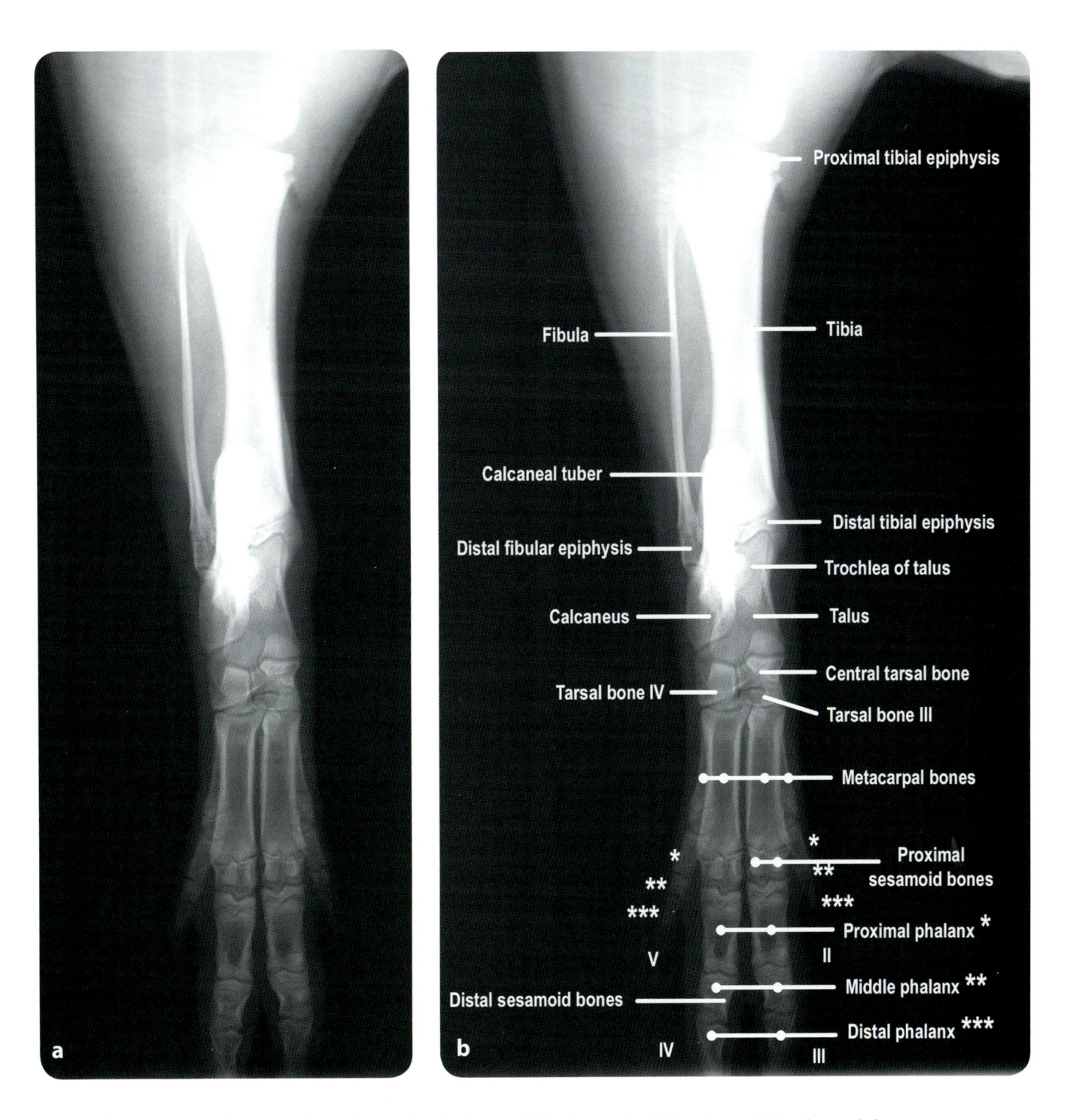

Figure 14.12a,b. Radiograph of the distal pelvic limb of a 15 kg 6-month-old female pig 75% of actual size.

ABNORMALITIES of the PELVIC LIMB

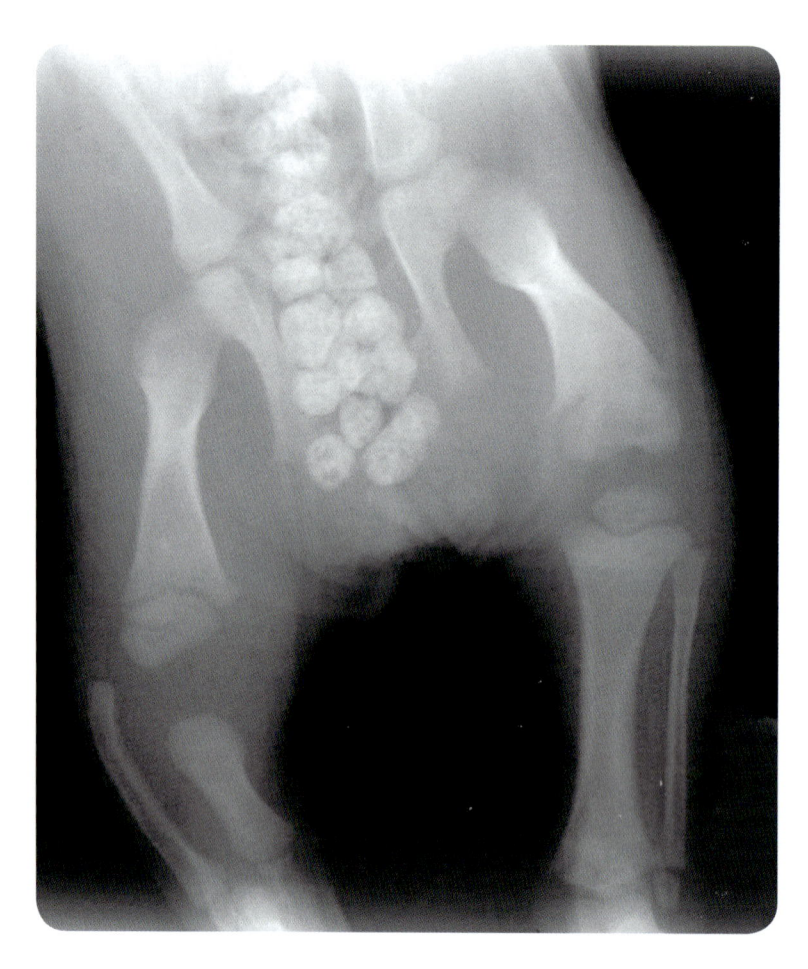

Figure 14.13. Congenital hypoplasia of the right tibia in a weanling potbellied pig. Note the presence of normal hard feces in the rectum. *(Courtesy of Purdue University School of Veterinary Medicine Large Animal Clinic, West Lafayette, Indiana).*

References:
Klaphake E: Tibial agenesis and patellar agenesis in two Vietnamese pot-bellied piglets. ExoticDVM. 2004; 6(1):5-7.
Swindle MM, Smith AC: Comparative anatomy and physiology of the pig. Scandinavian Journal of Laboratory Animal Science 1998; 25: 1-10.

African Pygmy
HEDGEHOG

The NORMAL WHOLE BODY SKELETON
Lateral Projection

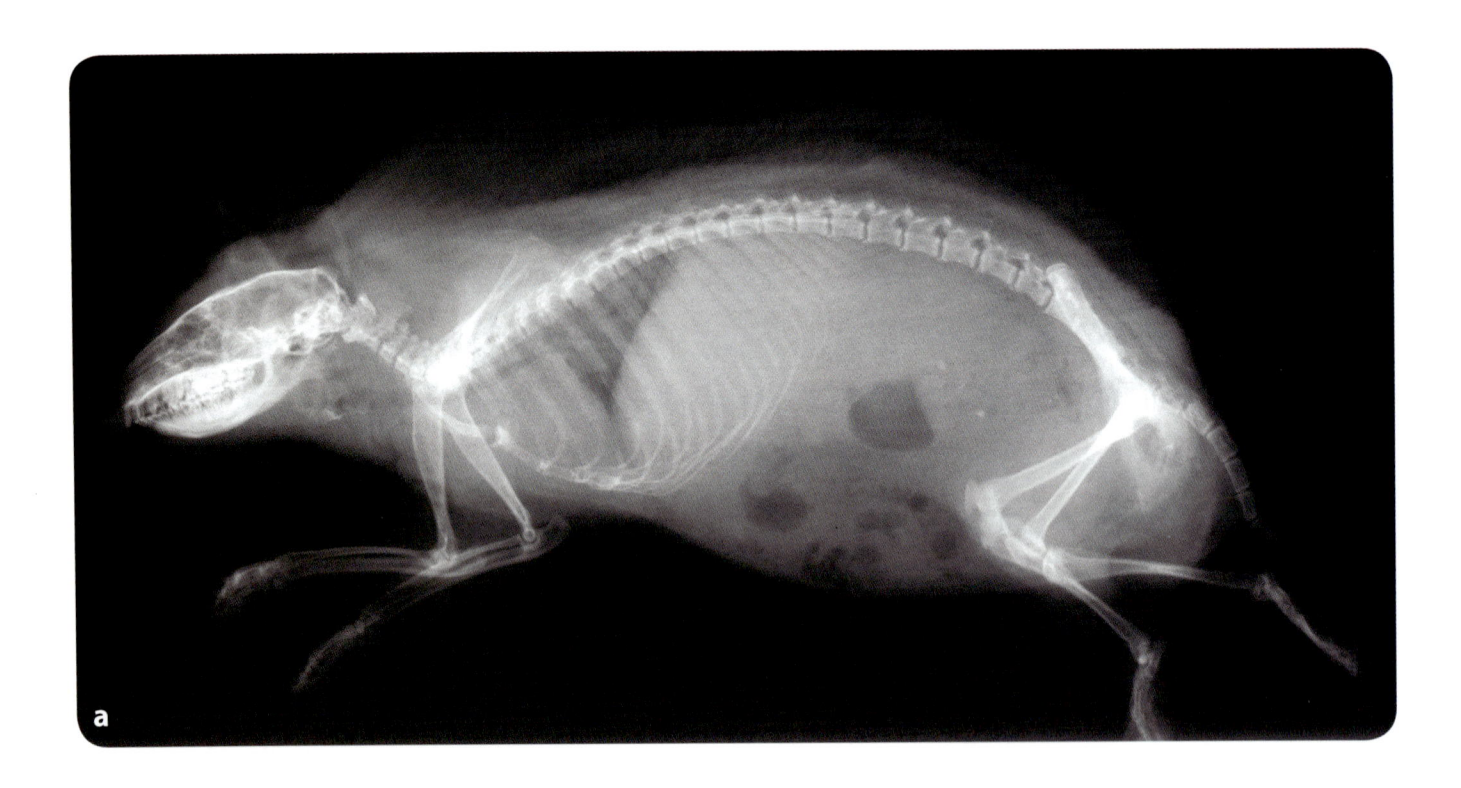

Figure 15.1a,b. Whole body skeleton of a 450 g female African pygmy hedgehog. 90% of actual size.

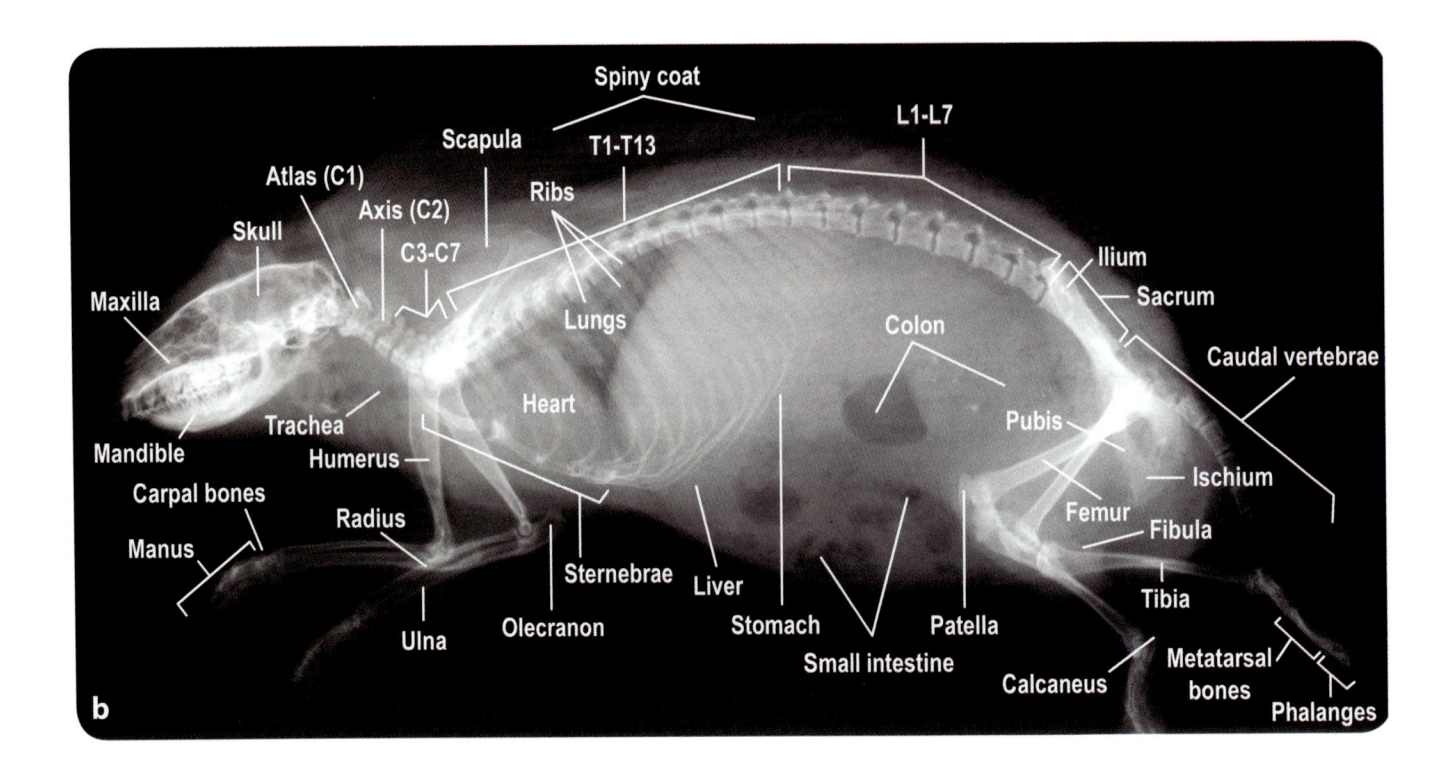

Ventrodorsal Projection

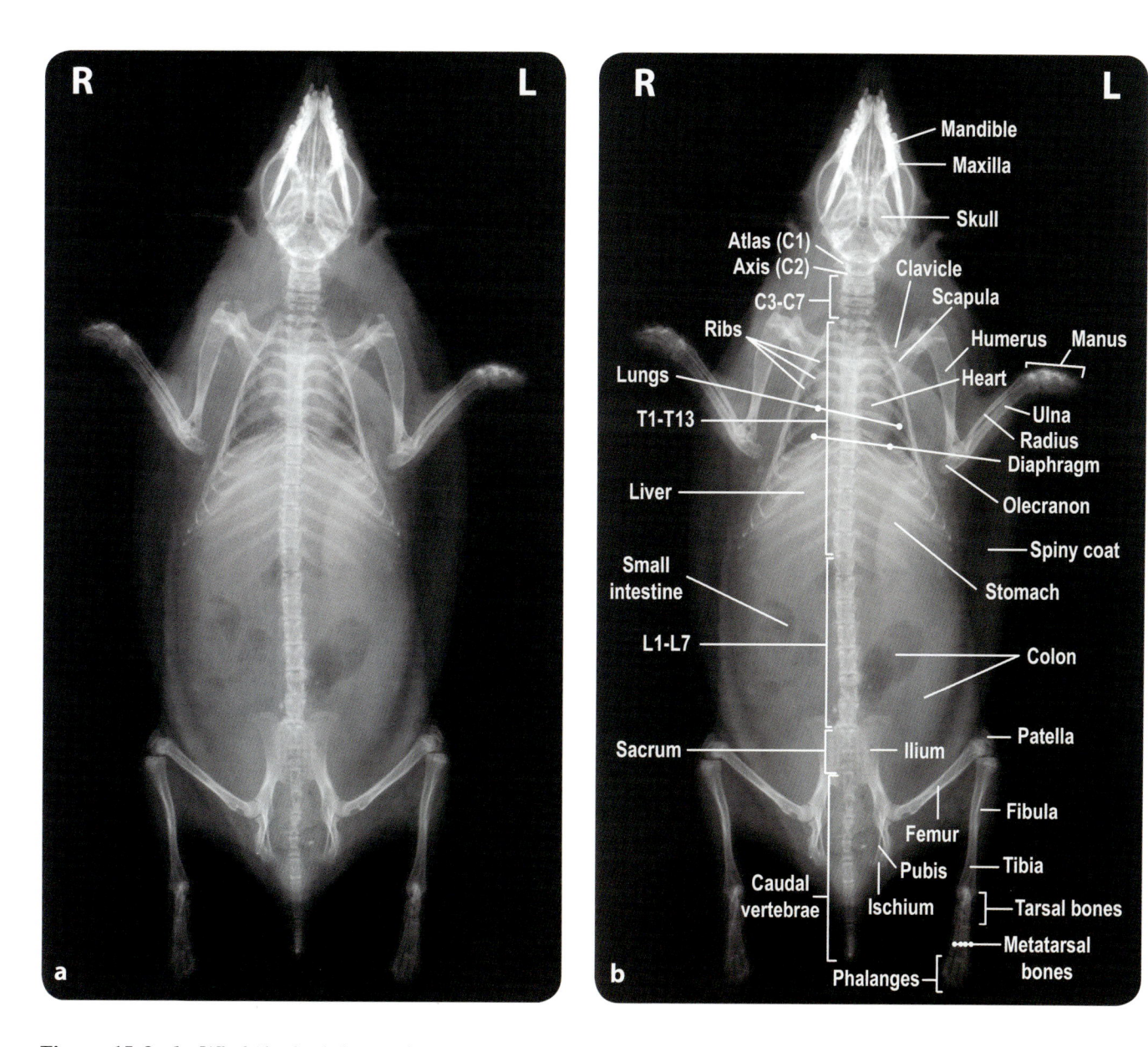

Figure 15.2a,b. Whole body skeleton of a 450 g female African pygmy hedgehog, 90% of actual size. Note diastasis of the ischiopubic symphysis in this non-pregnant animal with no history of parturition.

The NORMAL HEAD

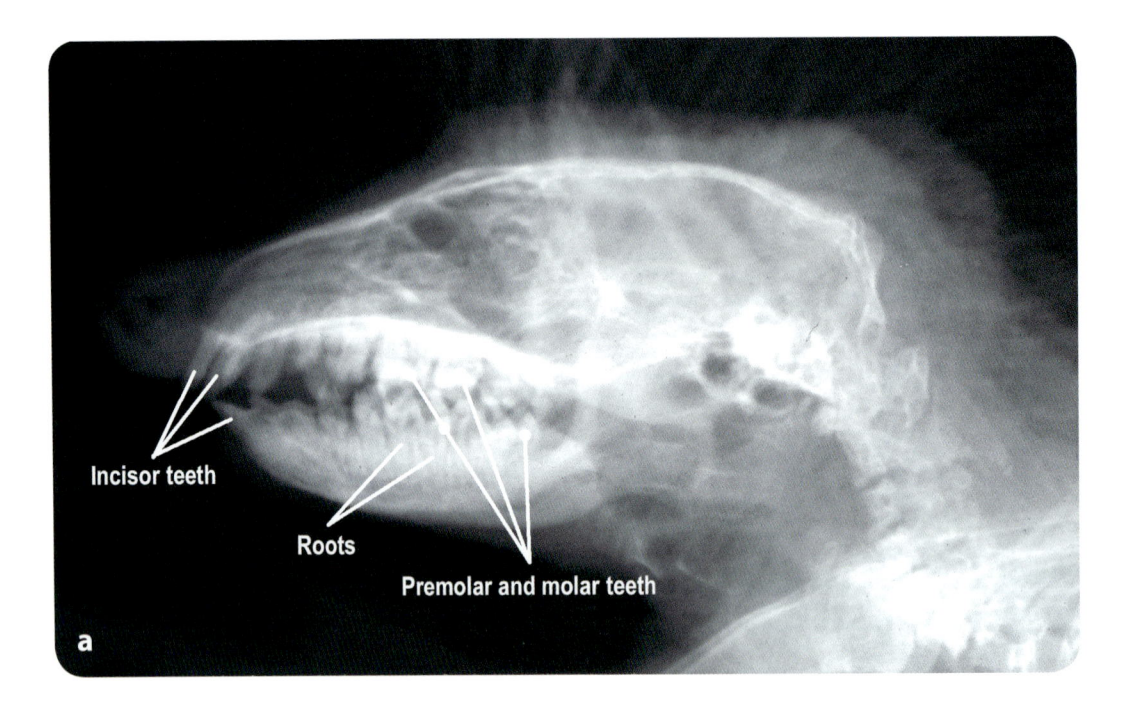

Figure 15.3a,b. Lateral (a) and ventrodorsal (b) projections of the skull. Dentition of this insectivore species is similar to that of carnivores and omnivores. Note forward projecting incisor teeth, and molar cusps of premolar and molars. Since dentition is elodont, true roots are also visible.

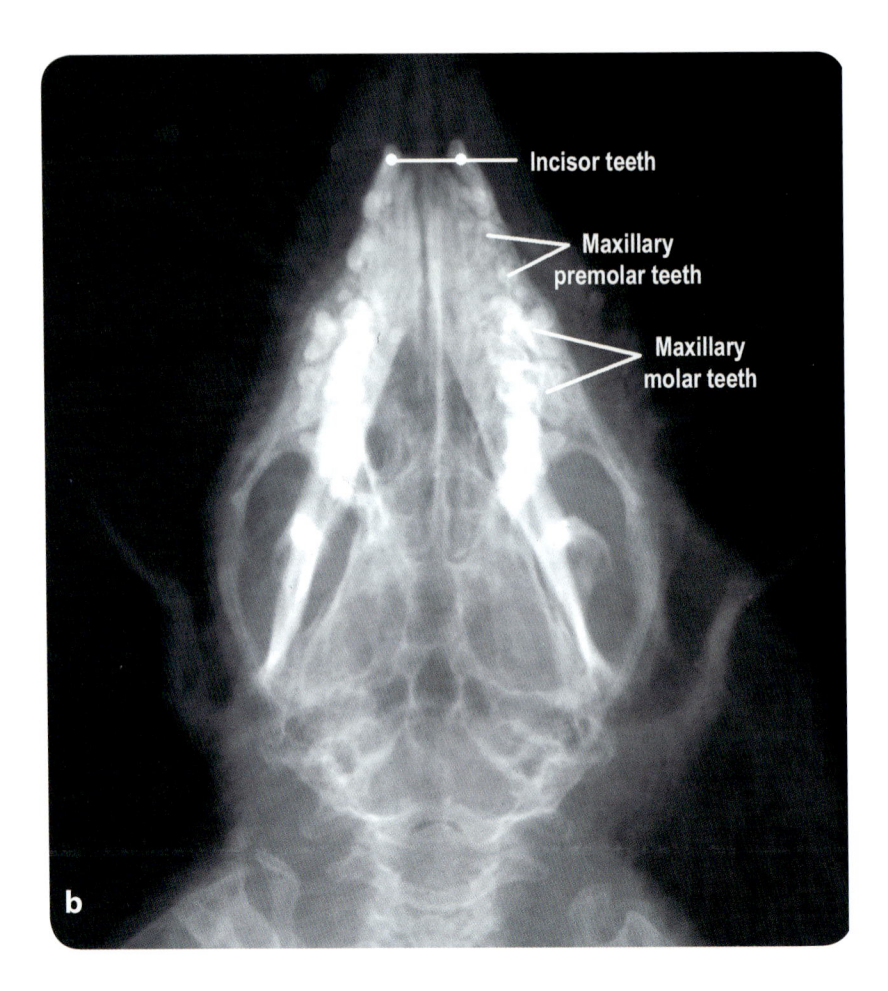

MISCELLANEOUS ABNORMALITIES
Diseases of the Thorax

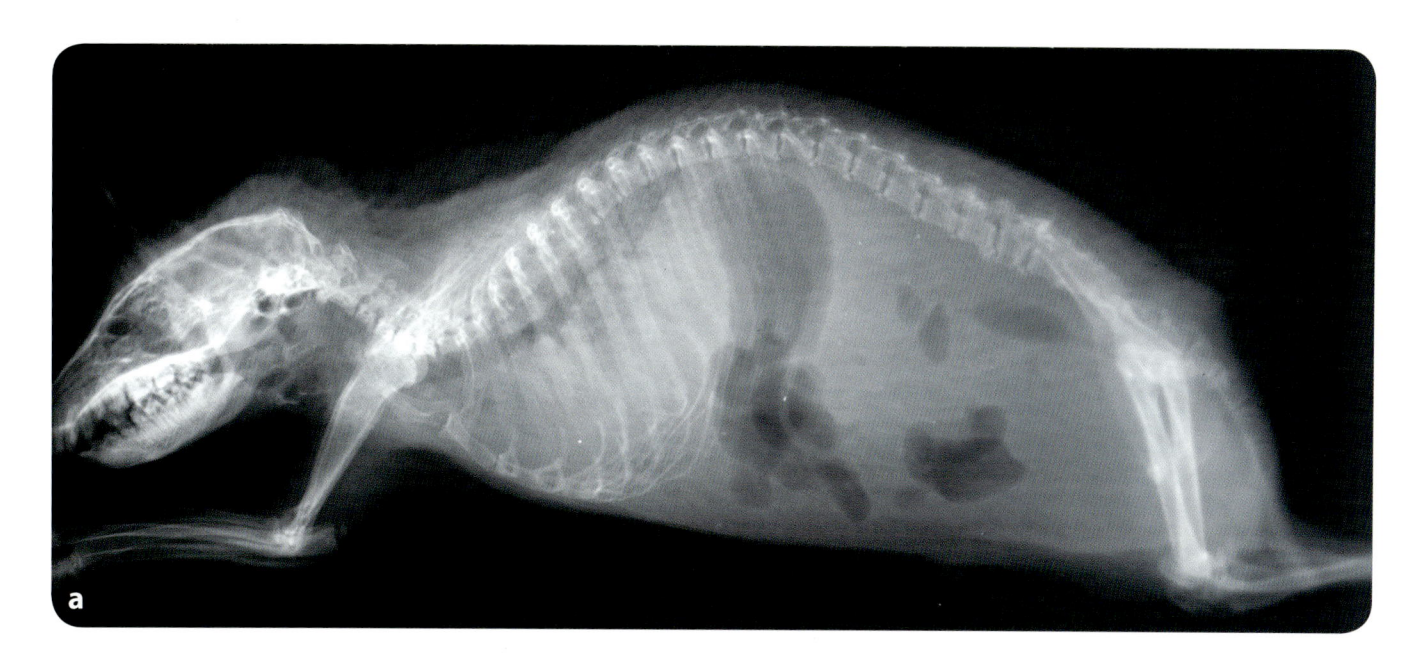

Figure 15.4a,b. Lateral (a) and ventrodorsal (b) projections of a hedgehog presenting with respiratory distress and depression. Note consolidation of both lung fields and obliteration of the cardiac silhouette. Loss of normal pulmonary opacity accentuates the air-filled trachea. Note mild aerophagia.

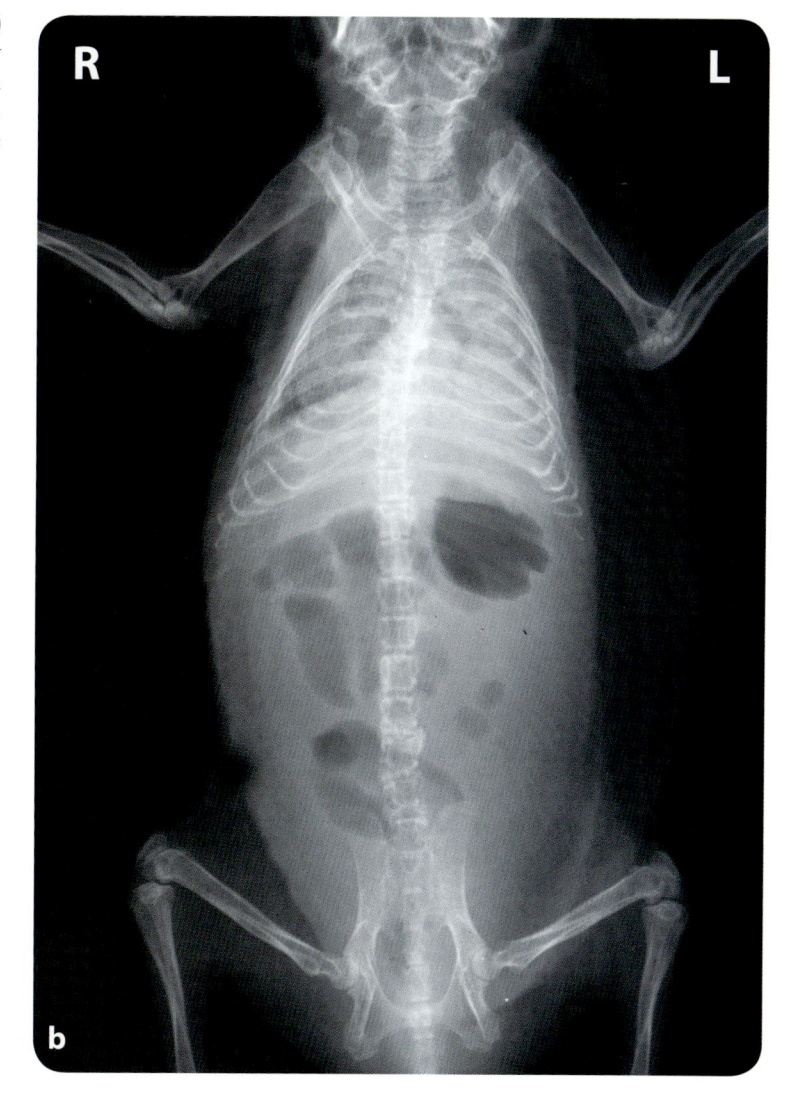

Diseases of the Abdomen

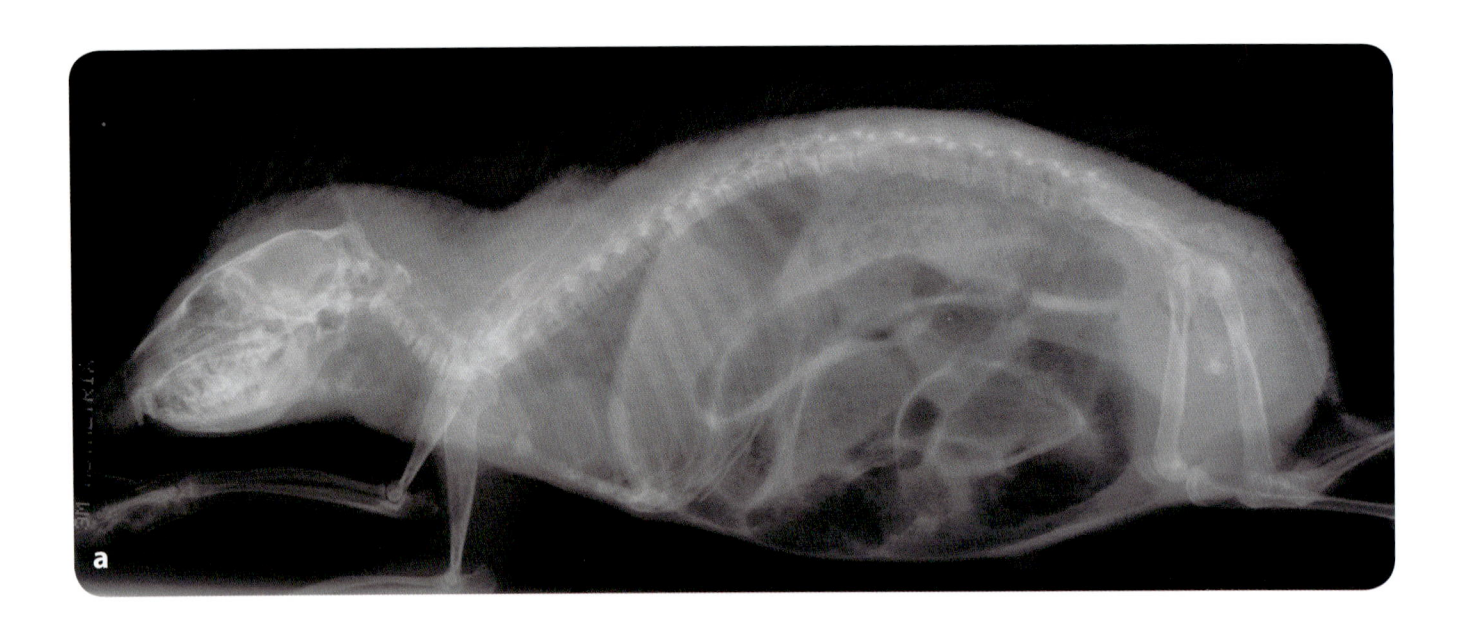

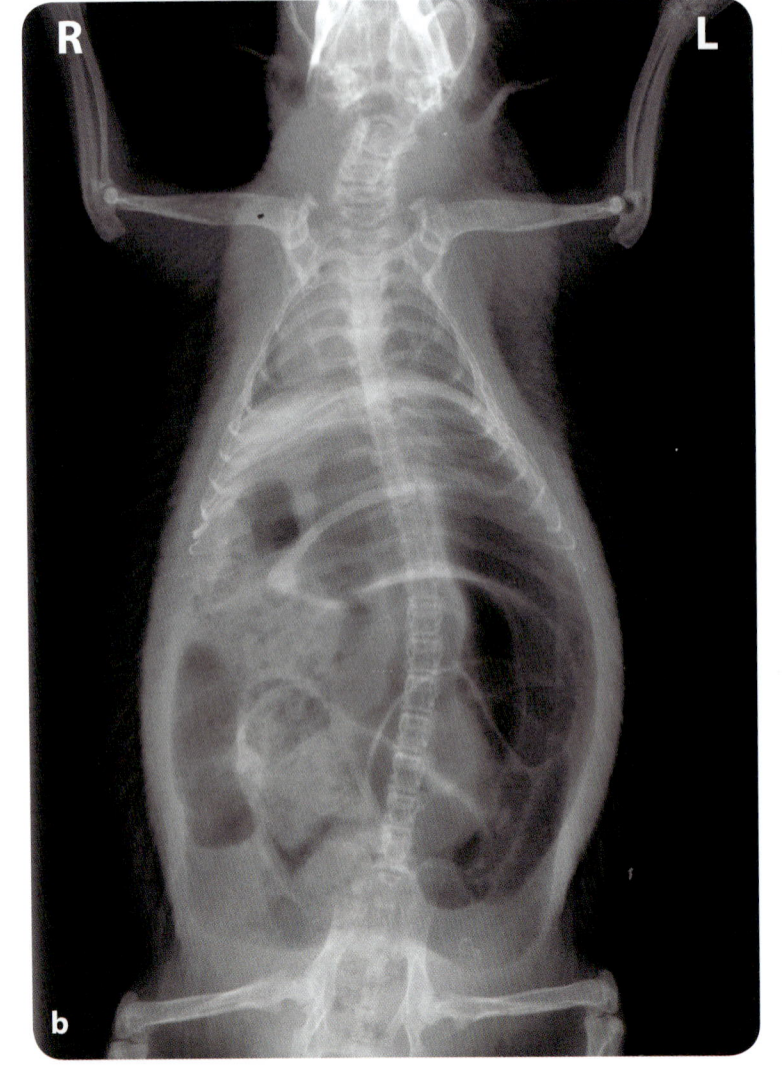

Figure 15.5a,b. Lateral (a) and ventrodorsal (b) projections of a hedgehog with depression, anorexia, and a distended abdomen. Note marked gas distention of the intestinal loops. This patient responded to supportive care and antibiotic therapy.

References:

Heatley JJ, Mauldin GE, Cho DY: A review of neoplasia in the captive African hedgehog *(Atelerix albiventrix)*. Sem Av Exotic Pet Med. 2005; 14(3):182-192.

Isenbugel E, Baumgartner RA: Diseases of the hedgehog. In: Fowler ME: Zoo & Wild Animal Medicine: Current Therapy, 3rd ed. Philadephia, PA: WB Saunders; 1993:294-302.

Ivey E, Carpenter JW: African hedgehogs. In: Quesenberry KE, Carpenter JW, eds. Ferrets, Rabbits and Rodents. Clinical Medicine and Surgery 2nd ed. Philadelphia, PA: Elsevier; 2004:339-353.

FINAL CUT

The authors enjoyed collecting and reviewing radiographs for this text. One of the hardest tasks, however, was deciding which exotic mammalian species to include and which to leave out. Some species and case selection was made simply by necessity, based on the radiographs we had on hand, and could coerce from our colleagues. While radiographs of rabbits and ferrets were plentiful, suitable images of smaller, less common species were considerably more difficult to find. A good number of interesting radiographs were deleted due to issues of less-than-publishable quality.

We would like to thank our clients who were willing and in some cases eager to allow us to radiograph their beloved pets in the interest of adding another normal representative species to our collection. It was especially interesting to find that at least two "normal" subjects were found to have significant radiographic abnormalities that were eventually included anyway!

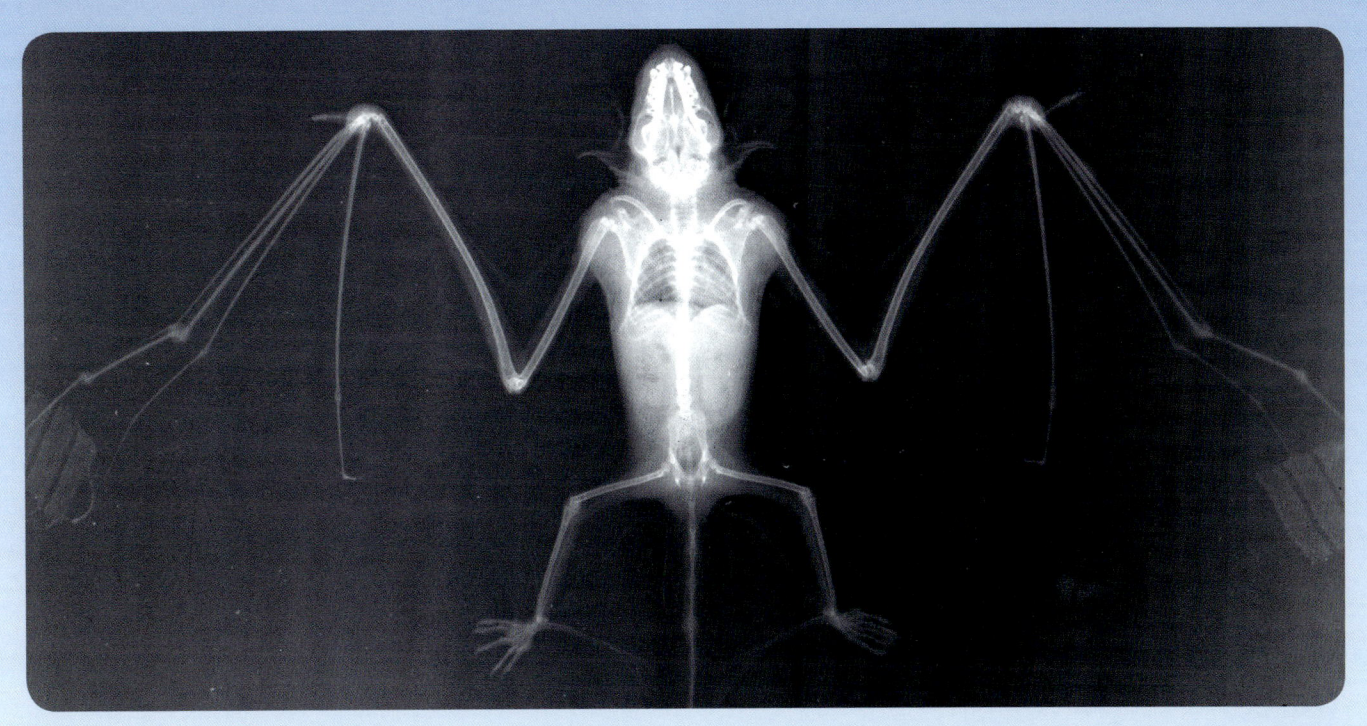

Figure 16.1a,b. Whole body radiograph of an Indiana brown bat, actual size. The thoracic limb of the bat is much longer than the pelvic limb. Most of the "wing" is composed of well-developed metacarpal bones and phalanges. These animals are occasionally presented to private practices by individuals or rehabilitators who encounter them in the wild. While occasionally a part of zoo collections, native bats are illegal to keep as pets.

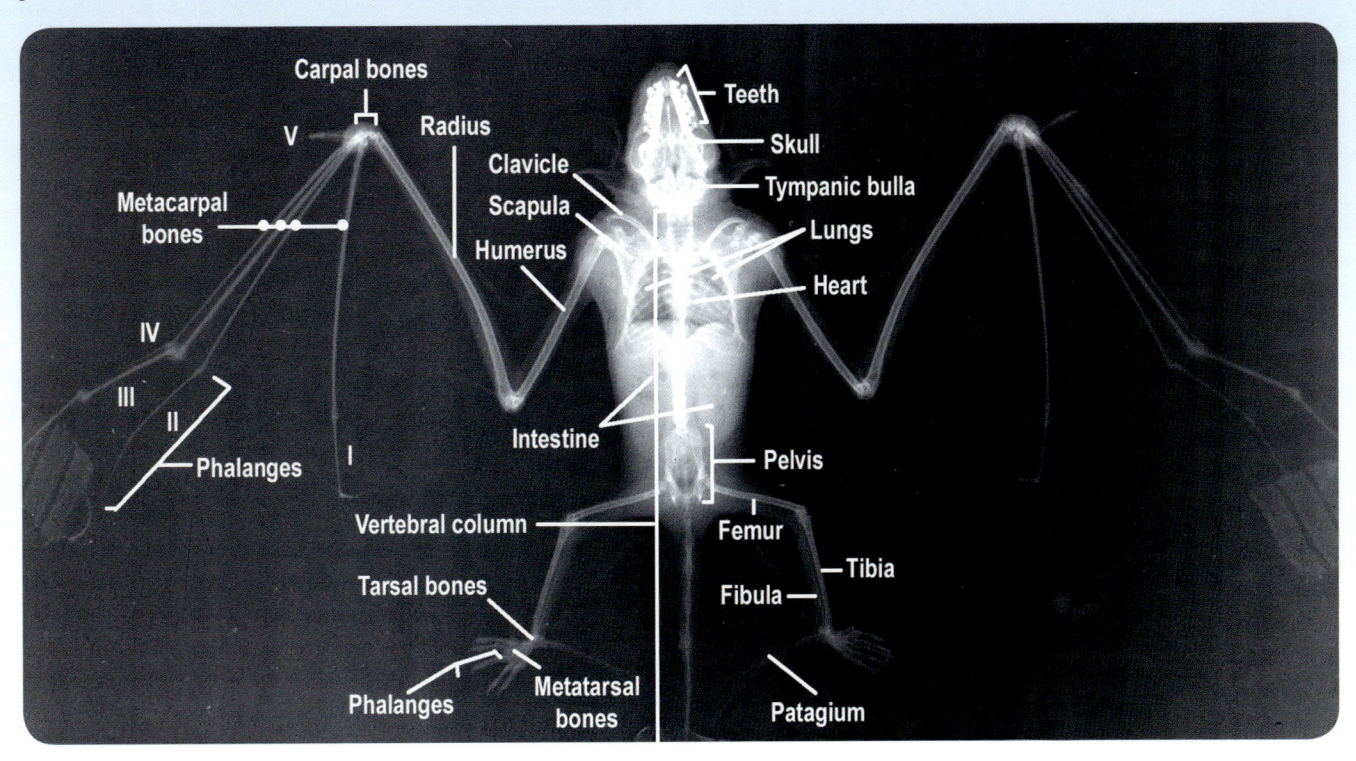

REFERENCES

1. Aiken S: Small mammal dentistry, part I: Surgical treatment of dental abscesses in rabbits. In: Quesenberry KE, Carpenter JW eds. Ferrets, Rabbits and Rodents. Clinical Medicine and Surgery 2nd ed. Philadelphia, PA: Elsevier; 2004:379-382.
2. Antinoff N, Hahn K: Ferret oncology: diseases, diagnostics, and therapeutics. Vet Clin N Am Exotic Anim Prac. 2004;7:579-625.
3. Barone R Pavaux C, Blin PC, Cuq P: Atlas d' Anatomie du Lapin, Masson et Cie;1973.
4. Bennett RA: Section 4. Small rodents. Soft tissue surgery. In: Quesenberry KE, Carpenter JW, eds. Ferrets, Rabbits and Rodents. Clinical Medicine and Surgery 2nd ed. Philadelphia, PA: Elsevier; 2004:316-328.
5. Bevilacqua L, Benato L: Uterine rupture with ectopic fetuses in a holland lop rabbit. Exotic DVM. 2006;8(2):3.
6. Booth R: Sugar gliders. Sem Av Exotic Pet Med. 2003; 12(4):228-231.
7. Brenner SZG, Hawkins MG, Tell LA, Hornof WJ, Plopper CG, Verstraete FJM: Clinical anatomy, radiography, and computed tomography of the chinchilla skull. Comp Cont Ed. 2005; 27:933-944.
8. Brown SA, Rosenthal LR: Question #30. In: Self Assessment Color Review of Small Mammals. London: Manson Publishing; 1997:27-28.
9. Brown SA, Rosenthal LR: Question #36. In: Self Assessment Color Review of Small Mammals. London: Manson Publishing; 1997:33-34.
10. Brown SA, Rosenthal LR: Question #45. In: Self Assessment Color review of Small Mammals. London: Manson Publishing;1997:41-42.
11. Brown SA, Rosenthal LR: Question #56. In: Self Assessment Color Review of Small Mammals. London: Manson Publishing; 1007:51-52.
12. Brown SA, Rosenthal LR: Question #66. In: Self Assessment Color Review of Small Mammals. London: Manson Publishing;1997:59-60.
13. Brown SA, Rosenthal LR: Question #77. In: Self Assessment Color Review of Small Mammals. London: Manson Publishing; 1997:39-40.
14. Brown SA, Rosenthal LR: Question #102. In: Self Assessment Color review of Small Mammals. London: Manson Publishing; 1997:89-90.
15. Brown SA, Rosenthal LR: Question #110. In: Self Assessment Color Review of Small Mammals. London: Manson Publishing; 1997:97-98.
16. Brown SA, Rosenthal LR: Question #169. In: Self Assessment Color Review of Small Mammals. London: Manson Publishing; 1997: 145-6.
17. Brown SA, Rosenthal LR: Question #209. In: Self Assessment Color Review of Small Mammals. London: Manson Publishing; 1997:181-2.
18. Capello V, Gracis M: Radiology of the skull and teeth. In: Lennox A, ed. Rabbit and Rodent Dentistry Handbook. Ames, IA: Blackwell Publishing, (Formerly Zoological Education Network, Lake Worth, FL); 2005:65-99.
19. Capello V, Gracis M: Dental diseases. In: Lennox A, ed. Rabbit and Rodent Dentistry Handbook. Ames, IA: Blackwell Publishing, (Formerly Zoological Education Network, Lake Worth, FL); 2005:113-164.
20. Capello V, Gracis M: Secondary diseases. In: Lennox A, ed. Rabbit and Rodent Dentistry Handbook. Ames, IA: Blackwell Publishing, (Formerly Zoological Education Network, Lake Worth, FL); 2005:165-186.
21. Capello V, Gracis M: Dental procedures. In: Lennox A, ed. Rabbit and Rodent Dentistry Handbook. Ames, IA: Blackwell Publishing, (Formerly Zoological Education Network, Lake Worth, FL); 2005:213-248.
22. Capello V, Gracis M: Surgical treatment of periapical abscessations. In: Lennox A, ed. Rabbit and Rodent Dentistry Handbook. Ames, IA: Blackwell Publishing, (Formerly Zoological Education Network, Lake Worth, FL); 2005:249-272.
23. Capello V: Diagnosis and treatment of urolithiasis in a pet rabbit. Exotic DVM 2004; 6(2):15-22.
24. Capello V: Surgical treatment of otitis externa and media in pet rabbits. Exotic DVM 2004; 6(3):15-21.
25. Capello V: Extraction of incisor teeth in pet rabbits. Exotic DVM 2004; 6(4):23-30.
26. Capello V: Extraction of cheek teeth and surgical treatment of periodontal abscessation in pet rabbits with acquired dental disease. Exotic DVM 2004; 6(4) 31-38.
27. Capello V: External fixation for fracture repair in small exotic mammals. Exotic DVM 2005; 7(6):21-37.
28. Capello V: Dental diseases and surgical treatment in pet rodents. Exotic DVM 2003;5(3):21-27.
29. Capello V: Pet hamster medicine and surgery part II: clinical evaluation and therapeutics. Exotic DVM 2001; 3(4):33-39.
30. Capello V: Pet hamster medicine and surgery part III: Infectious, parasitic and metabolis diseases. Exotic DVM 2002; 3(6):27-32.
31. Castanheira de Matos RE, Morrisey JK: Common procedures in the pet ferret. Vet Clin N Am Exotic Anim Prac. 2006;9:347-365.
32. Chambers JN, McBride MP, Hernandez-Divers SJ: Dynamic crossed-pin fixation of a distal femoral growth plate fracture in a domestic rabbit (Oryctolagus cuniculus). J Exotic Mam Med Surg. 2005:3(2): 4-5.
33. Chesney CJ: CT scanning in chinchillas. J Small Anim Pract. 1998; 39(11): 550.
34. Crossley DA, Jackson A, Yates J, et al. Use of computed tomography to investigate cheek tooth abnormalities in chinchillas (Chinchilla laniger). J Sm Anim Pract.1998;39:385-389.
35. Crossley DA: Oral biology and disorders of lagomorphs. Vet Clin N Am Exotic Anim Prac. 2003; 6:629-659.
36. Crossley DA: Clinical aspects of rodent dental anatomy. J Vet Dent. 1995;12(4):131-135.
37. Crossley DA: Rodent and rabbit radiology. In: DeForge DH, Colmery BH III,eds. An Atlas of Veterinary Dental Radiology. Ames: Iowa State University Press; 2000:247-260.
38. Crossley DA: Dental disease in chinchillas in the UK. J Small Anim Pract 2001; 42:12-19.
39. Crossley DA: Small mammal dentistry, part I: Dental anatomy and dental disease. In: Quesenberry KE, Carpenter JW, eds. Ferrets, Rabbits and Rodents. Clinical Medicine and Surgery 2nd ed. Philadelphia, PA: Elsevier; 2004:370-378.
40. Dal Pozzo G: Compendio di Tomografia Computerizzata e TC multi-strato. UTET, Torino, 2006.
41. Deeb B: Section 2. Rabbits. Respiratory disease and pasteurellosis. In: Quesenberry KE, Carpenter JW, eds. Ferrets, Rabbits and Rodents. Clinical Medicine and Surgery 2nd ed. Philadelphia, PA:Elsevier; 2004:172-182.
42. Deeb B: Section 2. Rabbits. Neurologic and muscoloskeletal diseases. In: Quesenberry KE, Carpenter JW, eds. Ferrets, Rabbits and Rodents. Clinical Medicine and Surgery 2nd ed. Philadelphia, PA:Elsevier; 2004:203-210.
43. Deeb B: The dyspneic rabbit. Exotic DVM 2005; 7(1):39-42.
44. Deeb B: Respiratory diseases in pet rats. Exotic DVM 2005; 7(1):31-33.
45. De Rycke LM, Gielen IM, Van Meervenne SA, Simoens PJ, Van Bree HJ: Computed tomography and cross-sectional anatomy of the brain in clinically normal dogs. Am J Vet Res. 2005; 66(10):1743-1756.
46. De Voe RS, Pack L, Greenacre CB: Radiographic and CT imaging of a skull associated osteoma in a ferret. Veterinary Radiology & Ultrasound 2002; 43(4): 346–348.
47. Divers SJ: Mandibular abscess treatment using antibiotic-impregnated beads. Exotic DVM 2000; 2(5):15-18.
48. Donnelly TM: Disease problems of small rodents. In: Quesenberry KE, Carpenter JW, eds. Ferrets, Rabbits and Rodents. Clinical Medicine and Surgery 2nd ed. Philadelphia, PA: Elsevier; 2004:299-315.
49. Eatwell K: Ovarian and uterine disease in guinea pigs: a review of 5 cases. Exotic DVM. 2003;5(5):37-39.
50. Fike JR, LeCouter RA, Cann CE: Anatomy of the canine orbital region. Multiplanar imaging by CT. Veterinary Radiology & Ultrasound 1984; 25(1): 32-36.
51. Fisher PG: Urethrostomy and penile amputation to treat urethral obstruction and preputial masses in male ferrets. ExoticDVM. 2002; 3(6):21-25.
52. Fisher PG: Exotic mammal renal disease: causes and clinical presentation. Vet Clin N Am Exot Anim Prac. 2006; 9:33-67 (2006).
53. Fisher PG: Exotic mammal renal disease: diagnosis and treatment. Vet Clin N Am Exotic Anim Prac. 2006; 9:69-96.
54. Garland MR, Lawler LP, Whitaker BR, et al. Modern CT applications in veterinary medicine. Radiographics 2002; 22(1):55-62.

REFERENCES

55. Funk RS: Medical management of prairie dogs. In: Quesenberry KE, Carpenter JW, eds. Ferrets, Rabbits and Rodents. Clinical Medicine and Surgery 2nd ed. Philadelphia, PA: Elsevier; 2004:266-273.
56. Girling SJ: Mammalian imaging and anatomy. In: Meredith A, Redrobe S, eds. BSAVA Manual of Exotic Pets 4th ed. Quedgeley, Glouchester: British Small Animal Veterinary Association; 2005:1-12.
57. Gobel T: Transurethral uroendoscopy in the female rabbit. Exotic DVM. 2002; 4(5) 23-27.
58. Greenacre CB: Spontaneous tumors of small mammals. Vet Clin Exot Anim. 2004; 7:627-651.
59. Guzman Sanchez-Migallon D, Mayer J, Gould J, Azuma C: Radiation therapy for the treatment of thymoma in rabbits *(Oryctolagus cuniculus)*. J Exotic Pet Med. 2006; 15(2):138-144.
60. Hanley CS, Wilson GH, Hernandez-Divers SJ: Secondary nutritional hyperparathyroidism associated with vitamin D deficiency in two domestic skunks *(Mephitis mephitis)*. Vet Rec. 2004; 115(8):233-7.
61. Harcourt-Brown FM: Dental disease. In: Textbook of Rabbit Medicine. Philadelphia, PA: Butterworth-Heinemann, imprint of Elsevier Science; 2002:165-205.
62. Harcourt-Brown FM: Abscesses In: Textbook of Rabbit Medicine. Philadelphia, PA: Butterworth-Heinemann, imprint of Elsevier Science; 2002:206-223.
63. Harcourt-Brown FM: Digestive disorders. In: Textbook of Rabbit Medicine. Philadelphia, PA: Butterworth-Heinemann, imprint of Elsevier Science; 2002:249-221.
64. Harcourt-Brown FM: Ophtalmic diseases. In: Textbook of Rabbit Medicine. Philadelphia, PA: Butterworth-Heinemann, imprint of Elsevier Science; 2002:229-306.
65. Harcourt-Brown FM: Neurological and locomotor disorders. In: Textbook of Rabbit Medicine. Philadelphia, PA: Butterworth-Heinemann, imprint of Elsevier Science; 2002:307-323.
66. Harcourt-Brown FM: Cardiorespiratory diseases. In: Textbook of Rabbit Medicine. Philadelphia, PA: Butterworth-Heinemann, imprint of Elsevier Science; 2002:324-333.
67. Harcourt-Brown FM: Urogenital diseases. In: Textbook of Rabbit Medicine. Philadelphia, PA: Butterworth-Heinemann, imprint of Elsevier Science; 2002:335-351.
68. Harcourt-Brown FM: Treatment of facial abscesses in rabbits. Exotic DVM. 1999; 1(3): 83-88.
69. Harcourt-Brown FM: Update on metabolic bone disease in rabbits. Exotic DVM. 2002; 4(3):43-46.
70. Harcourt-Brown FM: Dacryocystitis in rabbits. Exotic DVM. 2002; 4(3):47-49.
71. Harcourt-Brown FM: Intestinal obstruction in rabbits. Exotic DVM. 2002; 4(3):51-53.
72. Harcourt-Brown FM, Harcourt-Brown N: Surgical removal of a mediastinal mass in a rabbit. Exotic DVM. 2002; 4(3):59-60.
73. Harcourt-Brown FM: Radiology of rabbits: part 1. Soft tissue. Exotic DVM. 2004; 6(2):27-29.
74. Harcourt-Brown FM: Radiology of rabbits: part 2. Hard tissue. Exotic DVM. 2004; 6(2):30-32.
75. Harcourt-Brown N: Approach to selected orthopedic disorders in rabbits. Exotic DVM. 2004; 6(2) 33-36.
76. Hawkins MG: Diagnostic evaluation of urinary tract calculi in guinea pigs. Exotic DVM. 2006; 8(3):43-47.
77. Heatley JJ, Mauldin GE, Cho DY: A review of neoplasia in the captive African hedgehog *(Atelerix albiventrix)*. Sem Av Exotic Pet Med. 2005; 14(3):182-192.
78. Hernandez-Divers SJ: Molar disease and abscesses in rabbits. Exotic DVM. 2001; 3(3):65-69.
79. Hess L: Ferret lymphoma: the old and the new. Sem Av Exotic Pet Med. 2005;14(3):199-204.
80. Hoefer HL: Guinea pig urolithiasis. Exotic DVM. 2004;6(2):23-25.
81. Hoefer HL, Crossley DA: Chinchillas. In: Meredith A, Redrobe S, eds. BSAVA Manual of Exotic Pets 4th ed. Quedgeley, Glouchester: British Small Animal VeterinaryAssociation; 2005:65-75.
82. Hoefer HL: Section 1. Ferrets. Gastrointestinal diseases, part I. In: Quesenberry KE, Carpenter JW, eds. Ferrets, Rabbits and Rodents. Clinical Medicine and Surgery 2nd ed. Philadelphia, PA:Elsevier;2004:25-33.
83. Huston SM, Quesenberry KE: Section 2. Rabbits. Cardiovascular and lymphoproliferative diseases. In: Quesenberry KE, Carpenter JW, eds. Ferrets, Rabbits and Rodents. Clinical Medicine and Surgery 2nd ed. Philadelphia, PA:Elsevier; 2004:211-220.
84. Isenbugel E, Baumgartner RA: Diseases of the hedgehog. In: Fowler ME: Zoo & Wild Animal Medicine: Current Therapy, 3rd ed. Philadephia, PA: WB Saunders; 1993:294-302.
85. Ivey E, Carpenter JW: African hedgehogs. In: Quesenberry KE, Carpenter JW, eds. Ferrets, Rabbits and Rodents. Clinical Medicine and Surgery 2nd ed. Philadelphia, PA: Elsevier; 2004:339-353.
86. Jenkins JR: Section 2. Rabbits. Gastrointestinal diseases. In: Quesenberry KE, Carpenter JW, eds. Ferrets, Rabbits and Rodents. Clinical Medicine and Surgery 2nd ed. Philadelphia, PA: Elsevier; 2004:161-171.
87. Johnson-Delaney CA: Marsupials. In: Exotic Companion Medicine Handbook. Lake Worth, FL: Zoological Education Network, 1996.
88. Johnson-Delaney CA: What every veterinarian needs to know about Virginia opossums. Exotic DVM. 2005; 6(6):38-43.
89. Kalendar, WA: Computed Tomography. Fundamentals, System Technology, Image Quality, Applications. Publicis MCD Verlag: Werbeagentur GmbH, Munich, 2000.
90. Kapatkin A: Orthopedics in small mammals. In: Quesenberry KE, Carpenter JW, eds. Ferrets, Rabbits and Rodents. Clinical Medicine and Surgery 2nd ed. Philadelphia, PA:Elsevier; 2004:383-391.
91. Klaphake E: Tibial agenesis and patellar agenesis in two Vietnamese pot-bellied piglets. Exotic DVM 2004; 6(1):5-7.
92. Kottwitz J: Horizontal beam radiography in ferrets. Exotic DVM. 2004;6(1):37-41.
93. Kramer MH, Lennox AM: What veterinarians need to know about skunks. Exotic DVM. 2003; 5(1):36-39.
94. Legendre LFJ: Oral disorders of exotic rodents. Vet Clin N Am Exotic Anim Prac. 2003;6:601-628.
95. Mayer J, Azuma C: The use of radiation therapy in exotic cancer patients. Exotic DVM. 2006; 8(3):38-43.
96. McGavin MD, Zachary JF: Pathologic Basis of Veterinary Disease, 4th ed. Philadelphia, PA: Mosby Elsevier;2007.
97. Mehler SJ, Bennet RA: Surgical oncology of exotic animals. Vet Clin N Am Exot Anim Prac. 2004; 7:783-805.
98. Meredith A, Crossley DA: Rabbits. In: Meredith A, Redrobe S, eds. BSAVA Manual of Exotic Pets 4th ed. Quedgeley, Gloucester: British Small Animal Veterinary Association; 2005:79-92.
99. Morera N: Osteosarcoma in a Siberian chipmunk. Exotic DVM. 2004; 6(1):11-12.
100. Morgan JP., Silverman S.: Techniques of Veterinary Radiography, 4th ed. Davis, CA: Veterinary Radiology Associates;1984.
101. Morrisey JK, McEntee M: Therapeutic options for thymoma in the rabbit. Sem Avian Exotic Pet Med. 2005; 14(3):175-181.
102. Nemetz LP: Principles of high definition digital radiology for the avian patient. Proc Assoc Av Vet. 2006:39-45.
103. Ness RD, Booth R: Sugar gliders. In: Quesenberry KE, Carpenter JW, eds. Ferrets, Rabbits and Rodents. Clinical Medicine and Surgery 2nd ed. Philadelphia, PA: Elsevier; 2004:330-338.
104. Orcutt CJ: Ferret urogenital disease. Vet Clin N Am Exotic Anim Prac. 2003, 6:113-138.
105. Paré JA, Paul-Murphy J: Section 2. Rabbits. Disorders of the reproductive and urinary systems. In: Quesenberry KE, Carpenter JW, eds. Ferrets, Rabbits and Rodents. Clinical Medicine and Surgery 2nd ed. Philadelphia, PA:Elsevier; 2004:183-193.
106. Petrie JP: Section 1. Ferrets. Cardiovascular and other diseases, part I. In: Quesenberry KE, Carpenter JW, eds. Ferrets, Rabbits and Rodents. Clinical Medicine and Surgery 2nd ed. Philadelphia, PA:Elsevier;2004:58-66.
107. Pollock CG: Section 1. Ferrets. Urogenital diseases. In: Quesenberry KE, Carpenter JW, eds. Ferrets, Rabbits and Rodents. Clinical Medicine and Surgery 2nd ed. Philadelphia, PA:Elsevier;2004:41-49.
108. Popesko P, Rijtovà V, Horàk J: A Colour Atlas of Anatomy of Small Laboratory Animals. Vol. I: Rabbit, Guinea Pig. Bratislava: Príroda Publishing House; 1990.
109. Popesko P, Rijtovà V, Horàk J: A Colour Atlas of Anatomy of Small Laboratory Animals. Vol. II: Rat, Mouse, Hamster. Bratislava: Príroda Publishing House; 1990.
110. Porat-Mosenco Y, Schwarz T, Kass PH: Thick-section reformatting of thinly collimated computed tomography for reduction of skull-base-related artifacts in dogs and horses. Veterinary Radiology & Ultrasound 2004; 45(2): 131–135.
111. Powers LV: Pyothorax in a ferret: Exotic DVM, 1999;1(1):32.
112. Powers LV, Winkler K, Garner M, et al. Omentalization of prostatic abscesses and large cysts in ferrets *(Mustela putorius furo)*. J Exotic Pet Med. 2007;16(3):186-194.

113. Prokop M, Galanski M: Tomografia computerizzata spirale e multistrato. Masson, 2006.
114. Reusch B: Rabbit gastroenterology. Vet Clin N Am Exotic Pet Prac 2005; 8:351-375.
115. Rosenthal KL: Section 1. Ferrets. Respiratory diseases. In: Quesenberry KE, Carpenter JW, eds. Ferrets, Rabbits and Rodents. Clinical Medicine and Surgery 2nd ed. Philadelphia, PA:Elsevier;2004:72-78.
116. Ruelokke ML, Arnbjerg J: Retrobulbar abscess secondary to molar overgrowth in a guinea pig. Exotic DVM. 2003;5(2):10-16.
117. Ruelokke ML, Arnbjerg J, Martensen MR: Assessing gastrointestinal motility in guinea pigs using contrast radiography. Exotic DVM. 2003;5(2):10-16.
118. Schneider R. Hypocalcemia in a skunk. Exotic DVM. 2003; 5(1):5-6.
119. Silverman S, Tell LA: Radiology equipment and positioning techniques. In: Radiology of Rodents, Rabbits and Ferrets. An Atlas of Normal Anatomy and Positioning. Philadelphia, PA: Elsevier Saunders; 2005:1-8.
120. Silverman S, Tell LA: Domestic Rabbit *(Oryctolagus cuniculus)*. In: Radiology of Rodents, Rabbits and Ferrets. An Atlas of Normal Anatomy and Positioning. Philadelphia, PA:. Elsevier Saunders; 2005:159-230.
121. Silverman S, Tell LA: Domestic guinea pig *(Cavia porcellus)*. In: Radiology of Rodents, Rabbits and Ferrets. An Atlas of Normal Anatomy and Positioning. Philadelphia, PA: Elsevier Saunders; 2005:105-157.
122. Silverman S, Tell LA: Domestic chinchilla *(Chinchilla lanigera)*. In: Radiology of Rodents, Rabbits and Ferrets. An Atlas of Normal Anatomy and Positioning. Philadelphia, PA: Elsevier Saunders; 2005:67-104.
123. Silverman S, Tell LA: Norway rat *(Rattus norvegicus)*. In: Radiology of Rodents, Rabbits and Ferrets. An Atlas of Normal Anatomy and Positioning. Philadelphia, PA: Elsevier Saunders; 2005:19-43.
124. Silverman S, Tell LA: Laboratory mouse *(Mus musculus)*. In: Radiology of Rodents, Rabbits and Ferrets. An Atlas of Normal Anatomy and Positioning. Philadelphia, PA: Elsevier Saunders; 2005:9-17.
125. Silverman S, Tell LA: Syrian (Golden) Hamster *(Mesocricetus auratus)*. In: Radiology of Rodents, Rabbits and Ferrets. An Atlas of Normal Anatomy and Positioning. Philadelphia, PA: Elsevier Saunders; 2005: 45-65.
126. Silverman S, Tell LA: Domestic Ferret *(Mustela putorius)*. In: Radiology of Rodents, Rabbits and Ferrets. An Atlas of Normal Anatomy and Positioning. Philadelphia, PA: Elsevier Saunders; 2005:231-289.
127. Sjoberg JG: Hematuria in the rabbit. Exotic DVM. 2004; 6(4):23-30.
128. Smith AN, Burke H, Heatley JJ, Beard DM, Weiss RC, Blue JT: Chemotherapy for lymphosarcoma in a pet mouse. Exotic DVM. 2004; 6(5):5-8.
129. Stauber E, Finch N, Caplazi P: Diaphragmatic kidney herniation in a rabbit. Exotic DVM. 2005; 8(1):11-12.
130. Stefanacci JD, Hoefer HL: Radiology and ultrasound. In: Quesenberry KE, Carpenter JW, eds. Ferrets, Rabbits and Rodents. Clinical Medicine and Surgery 2nd ed. Philadelphia, PA: Elsevier; 2004:395-413.
131. Swindle MM, Smith AC: Comparative anatomy and physiology of the pig. Scandinavian Journal of Laboratory Animal Science 1998; 25: 1-10.
132. Tell LA, Silverman S, Wisner E: Imaging techniques for evaluating the head of birds, reptiles and small exotic mammals. Exotic DVM. 2003; 5.2:31-37.
133. Tell LA, Silverman S, Wisner E: Imaging techniques for evaluating the respiratory system of birds, reptiles and small exotic mammals: Exotic DVM. 2003; 5(2): 38-44.
134. Thrall D: Veterinary Diagnostic Radiology, 4th ed Phildelphia, PA: W.B. Saunders; 2002:4.
135. Ticer JW: Radiographic Technique in Veterinary Practice, 2nd ed, Philadelphia, PA:W.B. Saunders;1984.
136. Verstraete FJM: Advances in diagnosis and treatment of small exotic mammal dental disease. Seminars Avian Exotic Pet Med. 2003;12(1):37-48.
137. Wagner RA, Garman RH, Collins BM: Diagnosing odontomas in prairie dogs. Exotic DVM. 1999; 1(1):7-10.
138. Wagner R, Johnson D.: Rhinotomy for treatment of odontoma in prairie dogs. Exotic DVM. 2001; 3(5):29-34.
139. Widmer WR, Thrall DE, Shaw SM: Effects of low-level exposure to ionizing radiation: current concepts and concerns for veterinary workers. Vet Radiol and Ultrasound. 1996;37:227-239.
140. Wyre NR, Hess L: Clinical technique: ferret thoracocentesis. Sem Avian Exotic Pet Med 2005;14(1):22-25.
141. Yasutsugu M: Mandibulectomy for treatment of oral tumors (cementoma and chondrosarcoma) in two rabbits. Exotic DVM. 2006; 8(3):18-22.
142. The Fundamentals of Radiography, 12th ed. Rochester, NY: Eastman Kodak, Co;1980.

INDEX

A

Abdomen
  chinchilla, 242f–247f
  ferret, 374f–377f
  guinea pig, 192f–197f
  lateral projection, 29f
  opossum, 442f–443f
  potbellied pig, 466f–469f
  rabbit, 100f–105f
  radiographs, 28f, 29f
  skunk, 418f–419f
  ventrodorsal projection, 29f
Abdominal abnormalities
  chinchilla, 248f–252f
  ferret, 378f–389f
  guinea pig, 198f–209f
  hamsters, 312f–314f
  hedgehog, 486f
  prairie dogs, 344f–345f
  rabbit, 106f–129f
  rat, 285f–286f
Abdominal distention
  hamster, 312f
  hedgehog, 486f
  mouse, 294f
Abdominal hernia, ferret, 389f
Abdominal mass, ferret, 387f
Accessory nasal bone, potbellied pig, 459f
Acetabular fracture
  hamster, 319f
  rabbit, 152f
Acquired dental disease (ADD)
  chinchilla, CT, 236f, 237f
  rabbit, 62f, 63f, 64f, 65f, 66f, 72f, 74f
    CT of, 80f–87f
Acute respiratory distress, ferret, 370f
ADD. *See* Acquired dental disease
Adenocarcinoma metastases, rabbit, 95f
Adrenal gland diseases, ferret, 386f, 388f
African pygmy hedgehog. *See* Hedgehog
Agenesis, congenital, hamster, 316f
AIPMMA. *See* Antibiotic impregnated polymethylmethacrylate
Alveolar bulla
  guinea pig, 168f
  rabbit, 54f
  Alveolar margin
  chinchilla, 224f
  guinea pig, 168f
  hamster, 298f
  rabbit, 54f
  rat, 276f
Angular process of the mandible
  guinea pig, 170f
  hamster, 298f
  rabbit, 56f
Anode, 2f

Anorexia
  chinchilla, 250f, 251f
  ferret, 379f, 380f
  hedgehog, 486f
Antibiotic impregnated polymethylmethacrylate (AIPMMA), 70f
Apices
  chinchilla, 224f, 227f
  degu, 266f, 267f
  guinea pig, 168f, 169f
  oblique projection, 55f
  rabbit, 54f
Arachnoid, 40f
Atlas
  chinchilla, 224f, 226f
  degu, 266f, 268f
  ferret, 358f
  guinea pig, 168f
  hamster, 298f, 300f
  opossum, 438f
  potbellied pig, 459f
  prairie dog, 326f
  rabbit, 54f, 56f
  rat, 276f
  skunk, 412f
Axis
  potbellied pig, 459f
  skunk, 412f

B

Barium compounds, 40
Bat, total body projection, 15f, 489f
Beryllium window, 2f
Bezoar, guinea pig, 200f
Bitewing placement, intraoral projections, 58f
Bladder sludge
  guinea pig, 204f
  rabbit, 122f
Bone neoplasia, ferret, 407f
Bucky tray, 6
Bulla osteitis, rabbit, 75f

C

Calvarium, 40f
Cardiomyopathy, ferret, 369f
Carnassial tooth
  ferret, 358f
  skunk, 412f
Carpal diseases, rabbit, 147f
Cassette/film, for X-ray image recording, 3
Cathode, 2f
Cauda equina, 40f
Caudal view, 8f
Caudocranial/palmardorsal projection
  rabbit, 30f
  thoracic limb, 30f

Caudocranial/plantardorsal projection
  distal pelvic limb, 35f
  fibula, 35f
  metatarsus, 35f
  phalanges, 35f
  tarsus, 35f
  tibia, 35f
CCD detectors. *See* Charge coupled device detectors
Cecal gas
  chinchilla, 250f
  rabbit, 110f, 113f, 114f
Cecal impaction
  guinea pig, 201f
  prairie dogs, 344f, 345f
  rabbit, 115f
Centering, improper, of radiographic beam, 37f
Cerebral spinal fluid (CSF), 40
Cervical vertebral column, 26f, 27f
  chinchilla, 240f–241f
  opossum, 440f, 441f
  potbellied pig, 462f–465f
  prairie dogs, 336f–339f
Charge coupled device (CCD) detectors, 7
Cheek teeth. *See also* Mandibular cheek teeth; Maxillary cheek teeth
  chipmunk, 39f
Cheek teeth diseases
  chinchilla, 229f–231
  degu, 269f
  guinea pig, 174f–175f
  prairie dogs, 333f
  rabbit, 62f–66f
  rat, 279f
Cheek tooth elongation, rabbit, 65f
Cheek tooth malocclusion, chinchilla, 230f
Chest disease, guinea pig, 191f
Chinchilla
  abdomen, 242f–247f
  abdominal abnormalities, 248f–252f
  ADD, CT, 236f, 237f
  alveolar margin, 224f
  anorexia, 250f, 251f
  apices, 224f, 227f
  atlas, 224f, 226f
  cecal gas, 250f
  cervical vertebral column, 240f–241f
  cheek teeth diseases, 229f–231f
  cheek tooth malocclusion, 230f
  condylar process of the mandible, 224f, 225f, 227f
  crown elongation, 231f
  diastema, 224f
  ethmoturbinates, 224f
  female genital tract disease, 252f
  foramen magnum, 226f

Page numbers followed by f indicate figures; t, tables.

Chinchilla-*cont'd*
  frontal bone, 224f
  gastric gas, 250f
  gastrointestinal distention, 251f
  gastrointestinal impaction, 248f, 249f
  head, 224f–227f
  head abnormalities, 228f–231f
  incisive bone, 224f, 226f
  incisor tooth disease, 228f
  ischium fracture, 260f
  lamina dura, 225f
  lateral projection, 13f, 27f, 32f
  lumbar vertebral column, 242f–247f
  mandible, 224f, 226f
  mandibular abscess, 231f
  mandibular cheek teeth, 224f, 227f
  mandibular incisor tooth, 224f, 226f, 227f
  mandibular ramus, 224f
  mandibular symphysis, 227f
  maxilla, 226f
  maxillary cheek teeth, 224f, 227f
  maxillary incisor fracture, 228f
  maxillary incisor malocclusion, 228f
  maxillary incisor tooth, 224f, 226f, 227f
  nasal bone, 224f, 227f
  nasal cavity, 224f
  nasoturbinates, 224f
  occipital bone, 226f
  occlusal planes, 227f
  optic foramen, 224f
  palatine bone, 224f
  parietal bone, 224f, 227f
  pelvic limb, 256f–259f
  pelvic limb abnormalities, 260f–261f
  pregnancy, 246f, 247f
  pterygoid bone, 226f
  temporal bone, 224f, 227f
  temporomandibular joint, 227f
  thoracic limb, 254f–255f
  thoracic vertebral column, 240f–241f
  thorax, 240f–241f
  tibial fracture, 261f
  total body projection, 238f–239f
  tympanic bulla, 224f, 225f, 227f
  ventrodorsal projection, 13f
  vertebral column abnormalities, 253f
  vertebral fracture, 253f
  wave mouth, 229f
  zygomatic bone, 224f, 226f, 227f
Chipmunk
  cheek teeth, 39f
  maxillary incisor tooth fracture, 351f
  thoracic limb neoplasia, 354f
  total body projection, 350f–351f
Citellus
  maxillary incisor tooth fracture, 352f
  urethral obstruction, 355f
Collimation, of x-ray beam, 5f, 37f
Colon gas, rabbit, 111f
Compression fracture, rabbit, 130f, 131f, 132f
Computed radiography (CR), 7
Computed tomography (CT), 44–49
  rabbit head, 76f, 77f, 78, 79f
Computed tomography (CT) numbers, window of, 46f
Computed tomography (CT) scanners
  operation of, 46–47
  spiral, 44f

Computed tomography (CT) slices, symmetry, 44f
Condylar process of the mandible
  chinchilla, 224f, 225f, 227f
  degu, 266f, 267f
  ferret, 358f
  guinea pig, 169f, 171f
  hamster, 298f
  oblique projection, 55f
  opossum, 438f
  potbellied pig, 459f
  prairie dog, 326f, 328f
  rabbit, 54f
  rat, 276f, 278f
  rostrocaudal projection, 57f
  skunk, 412f
Congenital agenesis, hamster, 316f
Congenital hypoplasia, potbellied pig, 478f
Contrast, radiographic quality and, 4
Contrast media, 40
Contrast radiography, 40
Conus medullaris, 40f
Coronal fracture, guinea pig, 172f
Coronal reduction, rabbit, 64f
Coronoid process of the mandible
  ferret, 358f, 360f
  hamster, 298f
  skunk, 412f
CR. *See* Computed radiography
Cranial view, 8f
Craniocaudal/dorsopalmar projection
  rabbit, 30f, 31f
  thoracic limb, 30f
Craniocaudal/dorsoplantar projection
  distal pelvic limb, 35f
  fibula, 35f
  metatarsus, 35f
  pelvic limb, 36f
  phalanges, 35f
  tarsus, 35f
  tibia, 35f
Craniocaudal projection
  femur, 34f
  pelvic limb, 36f
Crown elongation, chinchilla, 231f
CSF. *See* Cerebral spinal fluid
CT. *See* Computed tomography
Cystic calculi, rabbit, 123f, 124f
Cystography, of urinary bladder, 40
Cystolithiasis, hamster, 314f
Cystoliths, guinea pig, 203f, 204f
Cystouroliths, rat, 286f

D
DDR. *See* Direct digital radiography
Degu
  apices, 266f, 267f
  atlas, 266f, 268f
  cheek teeth diseases, 269f
  condylar process of the mandible, 266f, 267f
  diastema, 266f
  ethmoturbinates, 266f
  frontal bone, 266f
  head, 266f–268f
  head abnormalities, 269f
  incisive bone, 266f, 268f
  mandible, 266f
  mandibular cheek teeth, 266f
  mandibular incisor tooth, 266f, 268f

  maxilla, 268f
  maxillary cheek teeth, 266f
  maxillary incisor tooth, 266f, 268f
  nasal bone, 266f
  nasal cavity, 266f
  parietal bone, 266f
  pregnancy, 272f
  temporal bone, 266f
  thoracic vertebral fracture, 273f
  total body projection, 270f–273f
  tympanic bulla, 266f, 267f
  zygomatic bone, 266f, 268f
Dehydration, ferret, 379f, 380f
Density, radiographic quality and, 4
Dental film, small, 3, 5f
Dental radiographic unit, 5f
Depression
  ferret, 380f
  hedgehog, 485f, 486f
Detail, radiographic quality and, 4
Diaphragmatic hernia, rabbit, 97f
Diastema
  chinchilla, 224f
  degu, 266f
  guinea pig, 168f
  rabbit, 54f
  rat, 276f
Diastemata, maxillary/mandibular, 62f
DICOM format, 7
Diffuse renal calcinosis, rabbit, 120f
Digital radiography, 7, 7f
Direct digital radiography (DDR), 7
Distal view, 8f
Distortion, radiographic quality and, 4
Distortion artifact, 38f
Dorsal recumbency, rostrocaudal projection, 24f
Dorsal-right ventral, oblique projection, 25f
Dorsopalmar projection, rabbit, 31f
Dorsoventral projection
  for dyspnea, 27f
  maxilla, 23f
  rabbit, 18f
  in ventral recumbency, 23f
Dura mater, 40f
Dura mater spinalis, 40f
Dwarf rabbit. *See* Rabbit
Dyspnea
  dorsoventral projection for, 27f
  rabbit, 95f, 97f, 98f
  skunk, 417f
Dysuria, rat, 286f

E
Ear canal, ventrodorsal projection, 56f
Elbow diseases, rabbit, 145f–146f
Elbow luxation, ferret, 396f
Electrons, 2f
Endometrial cystic hyperplasia, rabbit, 126f
Enteritis, hamster, 313f
Errors, in radiographic technique, 37–38
Ethmoturbinates
  chinchilla, 224f
  degu, 266f
  ferret, 358f
  guinea pig, 168f
  hamster, 298f
  opossum, 438f
  potbellied pig, 459f

prairie dog, 326f
rabbit, 54f
rat, 276f
skunk, 412f
Exposure time, X-ray machine, 3
Extrauterine pregnancy, rabbit, 128f

F

Facial tuber of maxilla, ventrodorsal projection, 56f
Female genital tract disease, chinchilla, 252f
Femoral diseases
   guinea pig, 218f–219f
   rabbit, 155f–156f
   Femoral fracture
     guinea pig, 218f
     hamster, 319f
     skunk, 425f
Femoral head ostectomy, guinea pig, 218f, 219f
Femoral neck fracture, ferret, 403f
Femoral shaft fracture, ferret, 404f, 405f
Femur
   craniocaudal projection, 34f
   lateral projection, 33f
   severe comminuted fracture, 37f
   superimposition prevention, lateral projection, 28f
Ferret
   abdomen, 374f–377f
   abdominal abnormalities, 378f–389f
   abdominal hernia, 389f
   abdominal mass, 387f
   acute respiratory distress, 370f
   adrenal gland diseases, 386f, 388f
   anorexia, 379f, 380f
   atlas, 358f
   bone neoplasia, 407f
   cardiomyopathy, 369f
   carnassial tooth, 358f
   chronic weight loss, 372f
   condylar process of the mandible, 358f
   coronoid process of the mandible, 358f, 360f
   dehydration, 379f, 380f
   depression, 380f
   elbow luxation, 396f
   ethmoturbinates, 358f
   femoral neck fracture, 403f
   femoral shaft fracture, 404f, 405f
   foreign body, 381f, 382f
   foreign body obstruction, 379f
   frontal bone, 361f
   gastric gas, 378f
   GDV, 379f
   head, 358f–361f
   head abnormalities, 362f–363f
   heart diseases, 369f
   heartworm disease, 370f
   hepatomegaly, 384f
   ilial fracture, 402f
   incisor teeth, 360f
   intestinal impaction, 383f
   lateral projection, 13f, 21f, 32f
   liver diseases, 384f
   lung diseases, 368f
   mandible, 358f, 360f
   mandibular canine tooth, 358f
   mandibular symphysis, 360f
   maxillary bone lysis, 363f
   maxillary canine tooth, 358f, 360f
   maxillary carnassial tooth, 360f
   maxillary incisor tooth, 358f
   mediastinal lymphoma, 371f
   mediastinum diseases, 370f–372f
   metabolic bone disease, 406f
   nasal bone, 358f
   nasal cavity, 358f, 360f, 361f
   nasal septum, 361f
   neck abscess, 362f
   occipital condyle, 358f, 360f
   occipital protuberance, 358f
   open mouth, lateral projection, 21f
   os penis fracture, 386f
   osteosarcoma, 362f
   osteosynthesis, 395f
   palatine bone, 361f
   palatine symphysis, 361f
   parietal bone, 358f
   pelvic limb, 398f–401f
   pelvic limb abnormalities, 402f–408f
   pneumothorax, 373f
   polydactylism, 408f
   prostatic cyst, 388f
   pterygoid bone, 360f
   radial fracture, 394f, 395f
   respiratory distress, 368f
   rib fracture, 373f
   sagittal crest, 360f
   scapular fracture, 392f, 393f
   spleen diseases, 385f
   splenic lymphoma, 385f
   temporomandibular joint, 358f, 360f, 361f
   thoracic abnormalities, 368f–373f
   thoracic limb, 390f–391f
   thoracic limb abnormalities, 392f–397f
   thoracic limb mass, 397f
   thorax, 366f–367f
   tibial fracture, 407f
   total body projection, 364f–365f
   tympanic bulla, 358f, 360f, 361f
   ulnar fracture, 394f, 395f
   ventrodorsal projection, 13f
   zygomatic bone, 360f, 361f
FFD. See Focal-film distance
Fibula
   caudocranial/plantardorsal projection, 35f
   craniocaudal/dorsoplantar projection, 35f
   lateral projection, 36f
Fibular fracture
   guinea pig, 220f
   hamster, 320f, 321f
   rabbit, 157f–162f
   skunk, 426f
Filament, 2f
Filum durae matris spinalis, 40f
Filum terminale, 40f
First maxillary incisor tooth
   rabbit, 54f
   ventrodorsal projection, 56f
Focal-film distance (FFD), 6
Focusing cup, 2f
Foramen magnum
   chinchilla, 226f
   guinea pig, 170f
   ventrodorsal projection, 56f
Foreign body
   ferret, 381f, 382f
   in guinea pig, 209f
   in rabbit, 75f
Foreign body obstruction, ferret, 379f
Frontal bone
   chinchilla, 224f
   degu, 266f
   ferret, 361f
   guinea pig, 168f
   opossum, 438f
   potbellied pig, 459f
   prairie dog, 326f
   rabbit, 54f
   rat, 276f
   skunk, 412f
Frontal sinus
   potbellied pig, 459f
   skunk, 412f

G

Gastric dilation, rabbit, 107f
Gastric dilation and volvulus (GDV), ferret, 379f
Gastric gas
   chinchilla, 250f
   ferret, 378f
   guinea pig, 198f, 199f
Gastric impaction, rabbit, 107f, 108f, 113f
Gastrointestinal distention, chinchilla, 251f
Gastrointestinal gas, rabbit, 112f
Gastrointestinal impaction
   chinchilla, 248f, 249f
   rabbit, 114f
GDV. See Gastric dilation and volvulus
Genital tract diseases, female guinea pig, 206f–209f
Genital tract enlargement, rabbit, 129f
Glass envelope, 2f
Golden hamster. See Hamster
Guinea pig
   abdomen, 192f–197f
   abdominal abnormalities, 198f–209f
   alveolar bulla, 168f
   alveolar margin, 168f
   angular process of the mandible, 170f
   apices, 168f, 169f
   atlas, 168f
   bezoar, 200f
   bladder sludge, 204f
   cecal impaction, 201f
   cheek teeth diseases, 174f–175f
   chest disease, 191f
   condylar process of the mandible, 169f, 171f
   coronal fracture, 172f
   cystoliths, 203f, 204f
   diastema, 168f
   ethmoturbinates, 168f
   female genital tract diseases, 206f–209f
   femoral diseases, 218f–219f
   femoral fracture, 218f
   femoral head ostectomy, 218f, 219f
   fibular fracture, 220f
   foramen magnum, 170f
   foreign body, 209f
   frontal bone, 168f
   gastric gas, 198f, 199f
   head, 168f–171f
     CT, 178f–181f
   head abnormalities, 172f–177f

Guinea pig-*cont'd*
  incisive bone, 168f, 170f
  incisor teeth diseases, 172f–173f
  kidney disease, 202f–203f
  lamina dura, 169f
  lateral projection, 29f
  lung diseases, 190f–191f
  lung neoplasia, 191f
  mandible, 168f, 170f
  mandibular arcade deformity, CT, 182f, 183f
  mandibular cheek teeth, 168f
  mandibular incisor alveolus fracture, 173f
  mandibular incisor tooth, 168f, 170f, 171f
  mandibular incisor tooth malocclusion, 172f, 173f
  mandibular ramus, 168f
  mandibular symphysis, 171f
  mandibular ventral cortex, 168f
  masseteric fossa, 168f
  maxilla, 170f
  maxillary cheek teeth, 168f, 171f
  maxillary incisor tooth, 168f, 170f, 171f
  nasal bone, 168f, 171f
  nasal cavity, 168f
  nasoturbinates, 168f
  nephrolithiasis, 202f
  occipital bone, 168f, 170f
  occipital condyle, 168f
  optic foramen, 168f
  osteodystrophy, 213f
  osteomyelitis, 176f–177f
    CT, 182f–183f
  ovarian cyst, 206f
  palatine bone, 168f
  parietal bone, 168f, 171f
  pelvic limb, 214f–217f
  pelvic limb abnormalities, 218f–221f
  periapical abscessation, CT, 182f–183f
  periapical infections, 176f–177f
  pregnancy, 196f, 197f
  pterygoid bone, 170f
  respiratory distress, 190f, 191f
  rostral orbital margin, 168f
  second maxillary incisor tooth, 168f
  stifle joint disease, 219f
  stomach diseases, 198f–200f
  temporal bone, 168f, 171f
  temporomandibular joint, 171f
  thoracic abnormalities, 190f–191f
  thoracic limb, 210f–212f
  thoracic limb abnormalities, 213f
  thorax, 186f–189f
  tibial fracture, 220f
  total body projection, 184f, 185f
  tympanic bulla, 168f, 169f
  tympanic cavity, 170f
  ureter disease, 202f–203f
  ureteroliths, 203f
  urethral calculi, 205f
  uterine horn enlargement, 208f
  ventrodorsal projection, 29f
  zygomatic bone, 168f, 170f, 171f

H

Hamster
  abdominal abnormalities, 312f–314f
  abdominal distention, 312f
  acetabular fracture, 319f
  alveolar margin, 298f
  angular process of the mandible, 298f
  atlas, 298f, 300f
  condylar process of the mandible, 298f
  congenital agenesis, 316f
  coronoid process of the mandible, 298f
  cystolithiasis, 314f
  enteritis, 313f
  ethmoturbinates, 298f
  femoral fracture, 319f
  fibular fracture, 320f, 321f
  forelimb neoplasia, 317f
  head, 298f–301f
  hepatic cysts, 313f
  incisive bone, 298f, 300f
  incisor teeth diseases, 302f
  intestinal impaction, 312f
  lateral projection, 29f
  mandible, 298f, 300f
  mandibular cheek teeth, 298f
  mandibular cheek tooth fracture, 303f
  mandibular incisor tooth, 298f, 300f, 301f
  mandibular symphysis, 301f
  maxillary cheek teeth, 298f
  maxillary incisor malocclusion, 302f
  maxillary incisor tooth, 298f, 300f, 301f
  maxillary incisor tooth fracture, 302f
  nasal bone, 298f
  nasal cavity, 298f
  occipital bone, 298f
  occipital condyle, 298f, 300f
  olecranon fracture, 316f
  osteomyelitis, 322f
  palatine bone, 298f
  pelvic limb abnormalities, 319f–323f, 323f
  pelvic mass, 322f, 323f
  pregnancy, 309f
  pterygoid bone, 300f
  pubic bone fracture, 319f
  tarsal fracture, 321f
  temporomandibular joint, 301f
  thoracic limb abnormalities, 316f–318f
  thoracic limb neoplasia, 317f, 318f
  tibial fracture, 320f, 321f
  total body projection, 14f, 304f–311f
  tympanic bulla, 298f, 299f, 300f
  ulnar fracture, 316f
  urogenital tract diseases, 314f
  uterine mass, 314f
  ventrodorsal projection, 14f, 29f
  vertebral column abnormalities, 315f
  vertebral fracture, 315f
  zygomatic bone, 298f, 300f, 301f
Head
  chinchilla, 224f–227f
    CT, 232f–237f
  degu, 266f–268f
  ferret, 358f–361f
  guinea pig, 168f–171f
    CT, 178f–181f
  hamster, 298f–301f
  hedgehog, 484f
  lateral oblique projection, 16f
  opossum, 438f–439f
  potbellied pig, 458f–461f
  prairie dog, 326f–329f
  rabbit, 15f–25f, 54f–59f
  rat, 276f–278f
  skunk, 412f
Head abnormalities
  chinchilla, 228f–231f
  degu, 269f
  ferret, 362f–363f
  guinea pig, 172f–177f
  rabbit, 60f–75f
  rat, 279f
  squirrel-like rodents, 351f–352f
Heart diseases, ferret, 369f
Heartworm disease, ferret, 370f
Hedgehog
  abdominal abnormalities, 486f
  abdominal distention, 486f
  anorexia, 486f
  depression, 485f, 486f
  head, 484f
  incisor teeth, 484f
  maxillary molar teeth, 484f
  maxillary premolar teeth, 484f
  molar teeth, 484f
  premolar teeth, 484f
  respiratory distress, 485f
  roots, 484f
  thoracic diseases, 485f
  total body projection, 482f–483f
Hemipelvis fracture, rabbit, 152f
Hepatic cysts, hamster, 313f
Hepatomegaly, ferret, 384f
High-resolution mammography x-ray films, 3f
Hip dysplasia, rabbit, 153f, 154f
Hip luxation, rabbit, 153f
Hip subluxation, rabbit, 154f
Humeral diseases, rabbit, 142f–143f
Hyperextended lateral projection, respiration impairment, 26f
Hyperextended neck, ventrodorsal projection with, 22f
Hyperflexion, technique, 42
Hypoplasia, congenital, potbellied pig, 478f

I

Ilial fracture, ferret, 402f
Image plate (IP), 7f
Image slice, 46f
Incisive bone
  chinchilla, 224f, 226f
  degu, 266f, 268f
  guinea pig, 168f, 170f
  hamster, 298f, 300f
  prairie dog, 326f, 328f
  rabbit, 54f
  rat, 276f, 278f
  ventrodorsal projection, 56f
Incisive bone fracture, rabbit, 73f
Incisor teeth
  ferret, 360f
  hedgehog, 484f
Incisor teeth diseases
  chinchilla, 228f
  guinea pig, 172f–173f
  hamster, 302f
  prairie dog, 330f–332f
  rabbit, 60f, 61f
  supernumerary first maxillary incisor tooth, 60f

Incisor tooth extraction, post-operative, 61f
Indian brown bat. *See* Bat
Intestinal diseases, rabbit, 110f–115f
Intestinal impaction
   ferret, 383f
   hamster, 312f
Intraoral projections
   bitewing placement, 58f
     mandibular incisor tooth, 58f
     mandibular left cheek teeth, 58f
     rabbit head, 58f
Intumescentia lumbalis, 40f
Iohexol, 40
   dosage, 42
Ionizing radiation, 50
IP. *See* Image plate
Ischium fracture, chinchilla, 260f

K
Kidney diseases
   guinea pig, 202f–203f
   rabbit, 118f–121f
Kilovoltage peak, X-ray machine, 3
Kyphosis, rabbit, 130f, 131f

L
Lamina dura
   chinchilla, 225f
   guinea pig, 169f
   oblique projection, 55f
Lateral dental radiograph, rabbit, 20f
Lateral oblique projection, head, 16f
Lateral projection
   abdomen, 29f
   chinchilla, 13f, 27f, 32f
   distal pelvic limb, 36f
   femur, 33f
   femur superimposition prevention and, 28f
   ferret, 13f, 21f, 32f
   fibula, 36f
   guinea pig, 29f
   hamster, 14f, 29f
   head, 15f
   metatarsus, 36f
   open mouth, ferret, 21f
   pelvis, 33f, 36f
   phalanges, 36f
   rabbit, 10f, 11f, 12f, 17f, 21f, 27f, 30f, 31f, 90f, 91f
   tarsus, 36f
   thoracic limb, 30f
   tibia, 36f
   uroliths and, 28f
Lead gloves, radiation safety in, 9f
Liver diseases
   ferret, 384f
   rabbit, 116f, 117f
Lumbar myelography
   positioning, 41f
   technique, 41f
Lumbar vertebral column, 28f, 29f
   chinchilla, 242f–247f
   potbellied pig, 466f–469f
   prairie dogs, 340f–343f
   rabbit, 100f–105f
Lung collapse, skunk, 414f
Lung diseases
   ferret, 368f
   guinea pig, 190f–191f
   rabbit, 94f–99f
Lung masses, 26f
Lung neoplasia, guinea pig, 191f
Lymphoma, rabbit, 98f

M
Malocclusion, severe, maxillary incisor tooth, 61f
Mammography film, 3
Mandible
   chinchilla, 224f, 226f
   degu, 266f
   ferret, 358f, 360f
   guinea pig, 168f, 170f
   hamster, 298f, 300f
   opossum, 438f, 439f
   potbellied pig, 459f
   prairie dog, 326f
   rabbit, 54f
     mineralized mass on, 74f
   skunk, 412f
   superimposition, 56f
   ventrodorsal projection, 22f, 56f
Mandible canine tooth, skunk, 412f
Mandible fracture, rabbit, 73f
Mandibular abscess, chinchilla, 231f
Mandibular arcade, overgrowth/malocclusion, rabbit, 64f
Mandibular arcade deformity, guinea pig, CT, 182f, 183f
Mandibular canine tooth
   ferret, 358f
   opossum, 439f
Mandibular cheek teeth. *See also* Cheek teeth; Maxillary cheek teeth
   artifact, 45f
   chinchilla, 224f, 227f
   degu, 266f
   guinea pig, 168f
   hamster, 298f
   intraoral projections, 58f
   opossum, 439f
   prairie dog, 326f, 328f
   rabbit, 54f
Mandibular CT1 fracture, rabbit, 65f
Mandibular CT2 sharp spur, rabbit, 64f
Mandibular incisor alveolus fracture, guinea pig, 173f
Mandibular incisor tooth
   chinchilla, 224f, 226f, 227f
   degu, 266f, 268f
   guinea pig, 168f, 170f, 171f
     malocclusion, 172f, 173f
   hamster, 298f, 300f, 301f
   intraoral projections, 58f
   opossum, 438f, 439f
   overgrown, 61f
   prairie dog, 326f, 328f
   rabbit, 20f, 54f
     fracture, 86f, 87f
   rat, 276f, 278f
   rostrocaudal projection, 57f
Mandibular ramus
   chinchilla, 224f
   guinea pig, 168f
   opossum, 438f
   prairie dog, 326f
   rabbit, 54f
Mandibular symphysis
   chinchilla, 227f
   ferret, 360f
   guinea pig, 171f
   hamster, 301f
   rostrocaudal projection, 57f
Mandibular ventral cortex
   guinea pig, 168f
   oblique projection, 55f
   rabbit, 54f
Manual restraint, for patient positioning, 9
Masses
   lung, 26f
   mediastinal, 26f
Masseteric fossa
   guinea pig, 168f
   rabbit, 54f
   rostrocaudal projection, 57f
Maxilla
   chinchilla, 226f
   degu, 268f
   dorsoventral projection, 23f
   guinea pig, 170f
   ventrodorsal projection, 22f, 56f
Maxillary bone lysis, ferret, 363f
Maxillary canine tooth
   ferret, 358f, 360f
   opossum, 439f
   skunk, 412f
Maxillary carnassial tooth, ferret, 360f
Maxillary cheek teeth. *See also* Cheek teeth; Mandibular cheek teeth
   chinchilla, 224f, 227f
   degu, 266f
   guinea pig, 168f, 171f
   hamster, 298f
   palatine bone, 298f
   prairie dog, 326f, 328f
   rabbit, 54f
   rostrocaudal projection, 57f
   ventrodorsal projection, 56f
Maxillary incisor fracture
   chinchilla, 228f
   hamster, 302f
Maxillary incisor malocclusion
   chinchilla, 228f
   hamster, 302f
Maxillary incisor tooth
   chinchilla, 224f, 226f, 227f
   chipmunk, fracture, 351f
   citellus, fracture, 352f
   degu, 266f, 268f
   ferret, 358f
   first
     rabbit, 54f
     ventrodorsal projection, 56f
   guinea pig, 168f, 170f, 171f
   hamster, 298f, 300f, 301f
   opossum, 438f, 439f
   prairie dog, 326f, 328f
     fracture, 330f
   rat, 276f, 278f
   rostrocaudal projection, 57f
   second
     guinea pig, 168f
     rabbit, 54f
Maxillary incisor tooth
   second-cont'd
     ventrodorsal projection, 56f
   severe malocclusion, 61f
   skunk, 412f
Maxillary/mandibular diastemata, 62f
Maxillary molar teeth, hedgehog, 484f

Maxillary premolar teeth, hedgehog, 484f
Maximum permissible dose (MPD), recommendation, 51
Mediastinal lymphoma, ferret, 371f
Mediastinal masses, rabbit, 26f
Mediastinum diseases, ferret, 370f–372f
Megaesophagus, skunk, 415f, 416f
Meninges, spinal cord nervous tissue and, 40f
Mesothelium, 40f
Metabolic bone disease
 ferret, 406f
 opossum, 454f–455f
 skunk, 421f
Metacarpal diseases, rabbit, 147f
Metatarsal fracture, rabbit, 163f
Metatarsus
 caudocranial/plantardorsal projection, 35f
 craniocaudal/dorsoplantar projection, 35f
 lateral projection, 36f
Methylprednisolone sodium succinate, 41
Milliamperage, 2
Molar teeth, hedgehog, 484f
Mouse
 abdominal distention, 294f
 total body projection, 292f–293f
MPD. *See* Maximum permissible dose
MPR. *See* Multiplanar reformation
Multiplanar reformation (MPR), two-dimensional reconstruction, 47f
Myelography, 40–42
 rabbit
  vertebral column, 134f–135f
  vertebral column abnormalities, 136f–137f

N
Nasal bone
 accessory, potbellied pig, 459f
 chinchilla, 224f, 227f
 degu, 266f
 ferret, 358f
 guinea pig, 168f, 171f
 hamster, 298f
 opossum, 438f
 potbellied pig, 459f
 prairie dog, 326f, 328f
 rabbit, 54f
 rat, 276f, 278f
 rostrocaudal projection, 57f
 skunk, 412f
Nasal cavity
 chinchilla, 224f
 degu, 266f
 ferret, 358f, 360f, 361f
 guinea pig, 168f
 hamster, 298f
 opossum, 438f, 439f
 potbellied pig, 459f
 prairie dog, 326f
 rabbit, 54f
 rat, 276f
 skunk, 412f
Nasal septum
 ferret, 361f
 opossum, 439f
Nasolacrimal duct, rabbit, 59f

Nasoturbinates
 chinchilla, 224f
 guinea pig, 168f
 opossum, 438f
 potbellied pig, 459f
 prairie dog, 326f
 rabbit, 54f
 rat, 276f
Neck abscess, ferret, 362f
Neoplasia, rat, 287f
Nephrolithiasis
 guinea pig, 202f
 rabbit, 118f, 119f
Nerve root tumors, 40
Nervous system tissue, 40f
Nomina Anatomica Veterinaria, 8, 8f
Non-occupational radiation exposure, 51

O
Oblique projection
 apices, 55f
 condylar process of the mandible, 55f
 dorsal-right ventral, 25f
 lamina dura, 55f
 mandibular ventral cortex, 55f
 pelvis, 33f
 periodontal space, 55f
 rabbit, 17f
 rabbit head, 55f
 right lateral, 25f
 tympanic bulla, 55f
Occasional radiation exposure, 51
Occipital bone
 chinchilla, 226f
 guinea pig, 168f, 170f
 hamster, 298f
 opossum, 438f
 potbellied pig, 459f
 prairie dog, 326f
 rabbit, 54f
 rat, 276f, 278f
Occipital condyle
 ferret, 358f, 360f
 guinea pig, 168f
 hamster, 298f, 300f
 opossum, 438f, 439f
 potbellied pig, 459f
 prairie dog, 326f, 328f
 rabbit, 54f
 rat, 276f, 278f
 skunk, 412f
 ventrodorsal projection, 56f
Occipital protuberance
 ferret, 358f
 skunk, 412f
Occlusal planes
 chinchilla, 227f
 rat, 278f
Occupational radiation exposure, 51
Olecranon fracture, hamster, 316f
Omnipaque, 40
Open mouth
 ferret, lateral projection, 21f
 rostrocaudal projection, 24f
Opossum
 abdomen, 442f–443f
 atlas, 438f
 cervical vertebral column, 440f, 441f
 condylar process of the mandible, 438f
 ethmoturbinates, 438f
 frontal bone, 438f

 head, 438f–439f
 mandible, 438f, 439f
 mandibular canine tooth, 439f
 mandibular cheek teeth, 439f
 mandibular incisor tooth, 438f, 439f
 mandibular ramus, 438f
 maxillary canine tooth, 439f
 maxillary incisor teeth, 438f, 439f
 metabolic bone disease, 454f–455f
 nasal bone, 438f
 nasal cavity, 438f, 439f
 nasal septum, 439f
 nasoturbinates, 438f
 occipital bone, 438f
 occipital condyle, 438f
 palatine bone, 438f
 parietal bone, 438f
 pelvic limb, 450f–452f
 skeletal diseases, 454f–455f
 teeth diseases, 453f
 temporal bone, 438f
 thoracic limb, 444f–449f
 thoracic vertebral column, 440f, 441f
 thorax, 440f–441f
 tympanic bulla, 438f, 439f
 zygomatic bone, 439f
Optic foramen
 chinchilla, 224f
 guinea pig, 168f
 prairie dog, 326f
 rabbit, 54f
 rat, 276f
Os penis fracture, ferret, 386f
Osteodystrophy, guinea pig, 213f
Osteomyelitis
 guinea pig, 176f–177f
  CT, 182f–183f
 hamster, 322f
 rabbit, 67f–72f
  CTs, 80f, 81f, 82f, 83f, 84f, 85f
Osteosarcoma, ferret, 362f
Osteosynthesis
 ferret, 395f
 skunk, 426f
Ovarian cyst
 guinea pig, 206f
 uterine neoplasia, 207f
Oxbow's critical care for, rabbit, 113f

P
Palatine bone
 chinchilla, 224f
 ferret, 361f
 guinea pig, 168f
 hamster, 298f
 opossum, 438f
 potbellied pig, 459f
 prairie dog, 326f
 rabbit, 54f
 scanning planes perpendicular to, 45f
Palatine symphysis, ferret, 361f
Palmar view, 8f
Paramastoid process, potbellied pig, 459f
Parietal bone
 chinchilla, 224f, 227f
 degu, 266f
 ferret, 358f
 guinea pig, 168f, 171f
 opossum, 438f
 potbellied pig, 459f
 prairie dog, 326f

rabbit, 54f
rat, 276f, 278f
rostrocaudal projection, 57f
skunk, 412f
Patient positioning, 9–10
  manual restraint for, 9
  pharmacologic restraint for, 9
  radiation safety in, 9, 9f
  sedation/anesthetic considerations in, 9
  techniques, 10
Pelvic fracture, rabbit, 133f
Pelvic limb
  caudocranial/plantardorsal projection, 35f
  chinchilla, 256f–259f
  craniocaudal/dorsopalmar projection, 35f
  craniocaudal/dorsoplantar projection, 35f, 36f
  craniocaudal projection, 36f
  ferret, 398f–401f
  guinea pig, 214f–217f
  lateral projection, 36f
  opossum, 450f–452f
  potbellied pig, 472f–477f
  prairie dogs, 348f–349f
  rabbit, 33f–36f, 148f–151f
  skunk, 422f–424f
Pelvic limb abnormalities
  chinchilla, 260f–261f
  ferret, 402f–408f
  guinea pig, 218f–221f
  hamster, 319f–323f
  potbellied pig, 478f
  rabbit, 152f–163f
  skunk, 425f–427f
Pelvic mass, hamster, 322f, 323f
Pelvis
  lateral projection, 33f, 36f
  oblique projection, 33f
  rabbit, 33f–36f
  ventrodorsal projection, 34f, 36f
  with caudal extension, 34f
  with internal rotation, 34f
Penumbra effect, 6
Periapical abscessation, guinea pig, CT, 182f–183f
Periapical infections
  guinea pig, 176f–177f
  rabbit, 67f–72f
    CTs, 80f, 82f, 84f, 85f
Perineal urine staining, rabbit, 122f
Periodontal space, oblique projection, 55f
Phalanges
  caudocranial/plantardorsal projection, 35f
  craniocaudal/dorsoplantar projection, 35f
  lateral projection, 36f
Pharmacologic restraint, for patient positioning, 9
Pia mater, 40f
Plantar view, 8f
Pneumothorax, ferret, 373f
Polydactylism, ferret, 408f
Potbellied pig
  abdomen, 466f–469f
  accessory nasal bone, 459f
  atlas, 459f

axis, 459f
cervical vertebral column, 462f–465f
condylar process of the mandible, 459f
congenital hypoplasia, 478f
ethmoturbinates, 459f
frontal bone, 459f
frontal sinus, 459f
head, 458f–461f
lumbar vertebral column, 466f–469f
mandible, 459f
nasal bone, 459f
nasal cavity, 459f
nasoturbinates, 459f
occipital bone, 459f
occipital condyles, 459f
palatine bone, 459f
paramastoid process, 459f
parietal bone, 459f
pelvic limb, 472f–477f
pelvic limb abnormalities, 478f
rostral orbital margin, 459f
snout bone, 459f
temporal bone, 459f
thoracic limb, 470f–471f
thoracic vertebral column, 462f–465f
thorax, 462f–465
zygomatic bone, 459f
Prairie dog
  abdominal abnormalities, 344f–345f
  atlas, 326f
  cecal impaction, 344f, 345f
  cervical vertebral column, 336f–339f
  cheek teeth diseases, 333f
  condylar process of the mandible, 326f, 328f
  ethmoturbinates, 326f
  frontal bone, 326f
  head, 326f–329f
  incisive bone, 326f, 328f
  incisor teeth diseases, 330f–332f
  lumbar vertebral column, 340f–343f
  mandible, 326f
  mandibular cheek teeth, 326f, 328f
  mandibular ramus, 326f
  maxillary cheek teeth, 326f, 328f
  maxillary incisor tooth, 326f, 328f
  maxillary incisor tooth fracture, 330f
  nasal bone, 326f, 328f
  nasal cavity, 326f
  nasoturbinates, 326f
  occipital bone, 326f
  occipital condyle, 326f, 328f
  optic foramen, 326f
  palatine bone, 326f
  parietal bone, 326f
  pelvic limb, 348f–349f
  pseudo-odontoma, 330f, 331f, 332f
  temporal bone, 326f, 328f
  thoracic limb, 346f–347f
  thoracic vertebral column, 336f–339f
  thorax, 336f–339f
  total body projection, 334f–335f
  tympanic bulla, 326f, 328f
  zygomatic bone, 326f, 328f
Pregnancy
  chinchilla, 246f, 247f
  degu, 272f
  guinea pig, 196f, 197f
  hamster, 309f
Premolar teeth, hedgehog, 484f
Primary X-ray beam, 2f

Prostatic cyst, ferret, 388f
Pseudo-odontoma, prairie dog, 330f, 331f, 332f
Pterygoid bone
  chinchilla, 226f
  ferret, 360f
  guinea pig, 170f
  hamster, 300f
  ventrodorsal projection, 56f
Pubic bone fracture
  hamster, 319f
  rabbit, 152f
Pulmonary nodules, rabbit, 96f

R
Rabbit
  abdomen, 100f–105f
  abdominal abnormalities, 106f–129f
  acetabular fracture, 152f
  ADD, 62f, 63f, 64f, 65f, 66f, 72f, 74f
    CT, 80f–87f
  adenocarcinoma metastases, 95f
  alveolar bulla, 54f
  alveolar margin, 54f
  angular process of the mandible, 56f
  atlas, 56f
  bladder sludge, 122f
  bulla osteitis, 75f
  carpal diseases, 147f
  caudocranial/palmardorsal projection, 30f
  cecal gas, 110f, 113f, 114f
  cecal impaction, 115f
  cheek tooth elongation, 65f
  colon gas, 111f
  compression fracture, 130f, 131f, 132f
  coronal reduction, 64f
  craniocaudal/dorsopalmar projection, 30f, 31f
  cystic calculi, 123f, 124f
  dental disease, 54f
  diaphragmatic hernia, 97f
  diffuse renal calcinosis, 120f
  dorsopalmar projection, 31f
  dorsoventral projection, 18f
  dyspnea, 95f, 97f, 98f
  elbow diseases, 145f–146f
  endometrial cystic hyperplasia, 126f
  extrauterine pregnancy, 128f
  femoral diseases, 155f–156f
  fibular diseases, 157f–162f
  fibular fracture, 157f–162f
  foreign body in, 75f
  gastric dilation, 107f
  gastric impaction, 107f, 108f, 113f
  gastrointestinal gas, 112f
  gastrointestinal impaction, 114f
  genital tract enlargement, 129f
  head, 15f–25f, 54f
    CT, 76f, 77f, 78, 79f
    intraoral projections, 58f
    nasolacrimal duct, 59f
    oblique projection, 55f
    rostrocaudal projection, 57f
    SSD, 79f
    ventrodorsal projection, 56f
  head abnormalities, 60–75
    cheek teeth diseases, 62f–66f
    incisor teeth diseases, 60f, 61f
  hemipelvis fracture, 152f
  hip dysplasia, 153f, 154f

INDEX

Rabbit-*cont'd*
  hip luxation, 153f
  hip subluxation, 154f
  humeral diseases, 142f–143f
  incisive bone fracture, 73f
  intestinal diseases, 110f–115f
  kidney diseases, 118f–121f
  kyphosis, 130f, 131f
  lateral dental radiograph, 20f
  lateral projection, 10f, 11f, 12f, 17f, 27f, 30f, 31f, 90f, 91f
  liver diseases, 116f, 117f
  lumbar vertebral column, 100f–105f
  lung diseases, 94f–99f
  lymphoma, 98f
  mandible, 54f
    mineralized mass, 74f
  mandible fracture, 73f
  mandibular arcade overgrowth/malocclusion, 64f
  mandibular cheek teeth, 54f
  mandibular CT1 fracture, 65f
  mandibular CT2 sharp spur, 64f
  mandibular incisor tooth, 20f, 54f
  mandibular incisor tooth fracture, CTs, 86f, 87f
  mandibular ramus, 54f
  mandibular ventral cortex, 54f
  masseteric fossa, 54f
  maxillary cheek teeth, 54f
  mediastinal masses, 26f
  metacarpal diseases, 147f
  metatarsal fracture, 163f
  myelography, of normal vertebral column, 134f–135f
  nasal bone, 54f
  nasal cavity, 54f
  nasoturbinates, 54f
  nephrolithiasis, 118f, 119f
  oblique projection, 17f
  occipital bone, 54f
  occipital condyle, 54f
  optic foramen, 54f
  osteomyelitis, 67f–72f
    CTs, 80f, 81f, 82f, 83f, 84f, 85f
  Oxbow's critical care for, 113f
  palatine bone, 54f
  parietal bone, 54f
  pelvic fracture, 133f
  pelvic limb, 33f–36f, 148f–151f
    abnormalities, 152f–163f
  pelvis, 33f–36f
  periapical infections, 67f–72f
    CTs, 80f, 82f, 84f, 85f
  perineal urine staining, 122f
  pubic bone fracture, 152f
  pulmonary nodules, 96f
  radial diseases, 143f–144f
  ramus fracture, 152f
  renal mineralization, 120f
  renal neoplasia, 121f
  renomegaly, 120f
  respiratory distress, 94f
  rostral orbital margin, 54f
  rostrocaudal projection, 19f
  sacral fracture, 133f
  sacral malalignment, 133f
  second maxillary incisor tooth, 54f
  small intestine gas, 111f
  spinal cord lesions, 130f
  step mouth, 63f
  stomach/cecum gas, 112f
  stomach diseases, 106f–109f
  tarsal diseases, 163f
  temporal bone, 54f
  thoracic abnormalities, 94f–99f
  thoracic limb, 30f, 31f, 32f, 138f–141f
  thoracic limb abnormalities, 142f–147f
  thoracic mass, 96f
  thoracotomy, 99f
  thorax, 26f, 27f, 90f–93f
  thymoma, 99f
  tibial fracture, 157f–162f
  total body projection, 10f–13f, 88f, 89f
  trichobezoars, 108f, 115f
  tympanic bulla, 54f
  ulnar diseases, 143f–144f
  ureter diseases, 118f–121f
  urethral diseases, 122f–125f
  urinary bladder diseases, 122f–125f
  urocystolith, 124f
  uroliths, 125f
  uterine adenocarcinoma, 127f
  uterine diseases, 126f–129f
  uterine neoplasia, 127f
  vaginal diseases, 126f–129f
  ventrodorsal projection, 10f, 11f, 12f, 18f, 27f
  vertebral column abnormalities, 130f–133f
  myelography, 136f–137f
  zygomatic bone, 54f, 57f
Radial diseases, rabbit, 143f–144f
Radial fracture, ferret, 394f, 395f
Radiation
  exposure
    non-occupational, 51
    occasional, 51
    occupational, 51
  ionizing, 50
  safety, 50
    lead gloves and, 9f
    in patient positioning, 9, 9f
  scattered, 6
  secondary, 2f
Radiographic beam
  improper centering, 37f
  improper collimation, 37f
Radiographic equipment, 4–6, 5f
Radiographic exposure, control of, 2
Radiographic quality
  contrast and, 4
  density and, 4
  detail and, 4
  distortion and, 4
  technique, 42
Radiographic technique, 35f
  common errors, 37f
Radiography
  contrast, 40
  digital, 7, 7f
  postioning for, 6
  storage of, 39–40
Radiology unit, 4f
Ramus fracture, rabbit, 152f
Rat
  abdominal abnormalities, 285f–286f
  alveolar margin, 276f
  atlas, 276f
  cheek teeth diseases, 279f
    condylar process of the mandible, 276f, 278f
  cystouroliths, 286f
  diastema, 276f
  dysuria, 286f
  ethmoturbinates, 276f
  frontal bone, 276f
  head, 276f–278f
  head abnormalities, 279f
  incisive bone, 276f, 278f
  mandibular incisor tooth, 276f, 278f
  maxillary incisor tooth, 276f, 278f
  nasal bone, 276f, 278f
  nasal cavity, 276f
  nasoturbinates, 276f
  neoplasia, 287f
  occipital bone, 276f, 278f
  occipital condyle, 276f, 278f
  occlusal planes, 278f
  optic foramen, 276f
  parietal bone, 276f, 278f
  respiratory distress, 284f
  temporal bone, 276f, 278f
  temporomandibular joint, 278f
  thoracic disease, 284f
  total body projection, 280f–283f
  tympanic bulla, 276f, 278f
  zygomatic bone, 276f, 278f
Renal mineralization, rabbit, 120f
Renal neoplasia, rabbit, 121f
Renomegaly, rabbit, 120f
Respiration impairment, hyperextended lateral projection, 26f
Respiratory distress
  acute, ferret, 370f
  ferret, 368f
  guinea pig, 190f, 191f
  hedgehog, 485f
  rabbit, 94f
  rat, 284f
Rib fracture, ferret, 373f
Roots, hedgehog, 484f
Rostral, 8f
Rostral orbital margin
  guinea pig, 168f
  potbellied pig, 459f
  rabbit, 54f
Rostrocaudal projection
  condylar process of the mandible, 57f
  dorsal recumbency, 24f
  mandibular cheek teeth, 57f
  mandibular symphysis, 57f
  masseteric fossa, 57f
  maxillary cheek teeth, 57f
  maxillary incisor tooth, 57f
  nasal bone, 57f
  open mouth, 24f
  parietal bone, 57f
  rabbit, 19f
  rabbit head, 57f
  temporal bone, 57f
  zygomatic bone, 57f
Russian hamster. *See* Hamster

S

Sacral fracture, rabbit, 133f
Sacral malalignment, rabbit, 133f
Sacrum, 40f
Sagittal crest, ferret, 360f
Scanner, for transparencies, 39f
Scanning plates, perpendicular to palatine bone, 45f
Scapular fracture, ferret, 392f, 393f

Scattered radiation, 6
  hazard of, 50f
Scout radiographic view, 45, 45f
Secondary radiation, 2f
Second maxillary incisor tooth
  guinea pig, 168f
  rabbit, 54f
  ventrodorsal projection, 56f
Sedation/anesthetic considerations, in patient positioning, 9
Shaded Surface Display (SSD), 49f
  rabbit head, 79f
Skeletal diseases, opossum, 454f–455f
Skunk
  abdomen, 418f–419
  atlas, 412f
  axis, 412f
  carnassial tooth, 412f
  condylar process of the mandible, 412f
  coronoid process of the mandible, 412f
  dyspnea, 417f
  ethmoturbinates, 412f
  femoral fracture, 425f
  fibular fracture, 426f
  frontal bone, 412f
  frontal sinus, 412f
  head, 412f
  lung collapse, 414f
  mandible, 412f
  mandible canine tooth, 412f
  maxillary canine tooth, 412f
  maxillary incisor tooth, 412f
  megaesophagus, 415f, 416f
  metabolic bone disease, 421f
  nasal bone, 412f
  nasal cavity, 412f
  occipital condyle, 412f
  occipital protuberance, 412f
  osteosynthesis, 426f
  parietal bone, 412f
  pelvic limb, 422f–424f
  pelvic limb abnormalities, 425f–427f
  temporal bone, 412f
  temporomandibular joint, 412f
  thoracic abnormalities, 414f–417f
  thoracic limb, 420f
  thoracic limb abnormalities, 421f
  thorax, 413f
  tibial fracture, 426f
  tympanic bulla, 412f
  vertebral fracture, 427f
  vertebral subluxation, 427f
Small dental film, 3, 5f
Small intestine gas, rabbit, 111f
Snout bone, potbellied pig, 459f
Spinal cord lesions, rabbit, 130f
Spinal cord nervous tissue, meninges and, 40f
Spinal needle, insertion of, 41f
Spiral CT scanners, 44f
Spleen diseases, ferret, 385f
Splenic lymphoma, ferret, 385f
Squirrel-like rodents, head abnormalities, 351f–352f
SSD. See Shaded Surface Display
Step mouth, rabbit, 63f
Stifle joint disease, guinea pig, 219f
Stomach/cecum gas, rabbit, 112f
Stomach diseases
  guinea pig, 198f–200f
  rabbit, 106f–109f
Storage, of radiographs, 39–40
Subarachnoid space, 40f
Sugar glider, total body projection, 430f–434f
Surface rendering, three-dimensional reconstructions, 49f
Symmetry, for CT slices, 44f

T

Tarsal diseases, rabbit, 163f
Tarsal fracture, hamster, 321f
Tarsus
  caudocranial/plantardorsal projection, 35f
  craniocaudal/dorsoplantar projection, 35f
  lateral projection, 36f
Teeth diseases, opossum, 453f
Temporal bone
  chinchilla, 224f, 227f
  degu, 266f
  guinea pig, 168f, 171f
  opossum, 438f
  potbellied pig, 459f
  prairie dog, 326f, 328f
  rabbit, 54f
  rat, 276f, 278f
  rostrocaudal projection, 57f
  skunk, 412f
Temporomandibular joint
  chinchilla, 227f
  ferret, 358f, 360f, 361f
  guinea pig, 171f
  hamster, 301f
  rat, 278f
  skunk, 412f
Thoracic abnormalities
  ferret, 368f–373f
  guinea pig, 190f–191f
  rabbit, 94f–99f
  skunk, 414f–417f
Thoracic cavity, 38f
Thoracic diseases
  hedgehog, 485f
  rat, 284f
Thoracic limb
  caudocranial/palmardorsal projection, 30f
  chinchilla, 254f–255f
  craniocaudal/dorsopalmar projection, 30f
  distal portion, 31f
  ferret, 390f–391f
  fully extended, 31f
  guinea pig, 210f–212f
  lateral projection, 30f
  neoplasia
    chipmunk, 354f
    hamster, 317f, 318f
  opossum, 444f–449f
  potbellied pig, 470f–471f
  prairie dogs, 346f–347f
  rabbit, 30f, 31f, 32f, 138f–141f
  skunk, 420f
Thoracic limb abnormalities
  ferret, 392f–397f
  guinea pig, 213f
  hamster, 316f–318f
  rabbit, 142f–147f
  skunk, 421f
Thoracic limb mass, ferret, 397f
Thoracic mass, rabbit, 96f
Thoracic vertebral column, 26f, 27f
  chinchilla, 240f–241f
  opossum, 440f, 441f
  potbellied pig, 462f–465f
  prairie dogs, 336f–339f
Thoracic vertebral fracture, degu, 273f
Thoracotomy, rabbit, 99f
Thorax
  chinchilla, 240f–241f
  ferret, 366f–367f
  guinea pig, 186f–189f
  opossum, 440f–441f
  potbellied pig, 462f–465f
  prairie dogs, 336f–339f
  rabbit, 26f, 27f, 90f–93f, 92f, 93f
  skunk, 413f
Three-dimensional reconstructions
  surface rendering, 49f
  volume rendering, 48f
Thymoma, rabbit, 99f
Tibia
  craniocaudal/dorsoplantar projection, 35f
  lateral projection, 36f
Tibial fracture
  chinchilla, 261f
  ferret, 407f
  guinea pig, 220f
  hamster, 320f, 321f
  rabbit, 157f–162f
  skunk, 426f
Total body projection
  bat, 15f, 489f
  chinchilla, 238f–239f
  chipmunk, 350f–351f
  degu, 270f–273f
  ferret, 364f–365f
  guinea pig, 184f, 185f
  hamster, 14f, 304f–311f
  hedgehog, 482f–483f
  mouse, 292f–293f
  prairie dogs, 334f–335f
  rabbit, 10f–13f, 88f, 89f
  rat, 280f–283f
  sugar glider, 430f–434f
Transparencies, scanner for, 39f
Trichobezoars, rabbit, 108f, 115f
Two-dimensional reconstruction, MPR, 47f
Tympanic bulla
  chinchilla, 224f, 225f, 227f
  degu, 266f, 267f
  ferret, 358f, 360f, 361f
  guinea pig, 168f, 169f
  hamster, 298f, 299f, 300f
  oblique projection, 55f
  opossum, 438f, 439f
  prairie dog, 326f, 328f
  rabbit, 54f
  rat, 276f, 278f
  skunk, 412f
  ventrodorsal projection, 56f
Tympanic cavity, guinea pig, 170f

U

Ulnar diseases, rabbit, 143f–144f
Ulnar fracture
  ferret, 394f, 395f
  hamster, 316f

Ureter diseases
  guinea pig, 202f–203f
  rabbit, 118f–121f
Ureteroliths, guinea pig, 203f
Urethral calculi, guinea pig, 205f
Urethral diseases, rabbit, 122f–125f
Urethral obstruction, citellus, 355f
Urinary bladder, cystography, 40
Urinary bladder diseases, rabbit, 122f–125f
Urocystolith, rabbit, 124f
Urogenital tract diseases, hamster, 314f
Uroliths
  lateral projection and, 28f
  rabbit, 125f
Uterine adenocarcinoma, rabbit, 127f
Uterine diseases, rabbit, 126f–129f
Uterine horn enlargement, guinea pig, 208f
Uterine mass, hamster, 314f
Uterine neoplasia
  ovarian cyst, 207f
  rabbit, 127f

V
Vaginal diseases, rabbit, 126f–129f
Ventral recumbency, dorsoventral projection in, 23f
Ventral view, 8f
Ventrodorsal projection
  abdomen, 29f
  angular process of the mandible, 56f
  atlas, 56f
  bat, 15f
  chinchilla, 13f
  ear canal, 56f
  facial tuber of maxilla, 56f
  ferret, 13f
  first maxillary incisor tooth, 56f
  foramen magnum, 56f
  golden hamster, 13f, 14f
  guinea pig, 29f
  hamster, 29f
  with hyperextended neck, 22f
  incisive bone, 56f
  mandible, 22f, 56f
  maxilla, 22f, 56f
  maxillary cheek teeth, 56f
  occipital condyle, 56f
  pelvis, 34f, 36f
    with caudal extension, 34f
    with internal rotation, 34f
  pterygoid bone, 56f
  rabbit, 10f, 11f, 12f, 18f, 27f
  rabbit head, 56f
  second maxillary incisor tooth, 56f
  symmetry of, 28f
  tympanic bulla, 56f
  zygomatic bone, 56f
Vertebral column, 38f
  diseases of, 40
  myelography, rabbit, 134f–135f
Vertebral column abnormalities
  chinchilla, 253f
  hamster, 315f
  rabbit, 130f–133f
    myelography, 136f–137f
Vertebral fracture
  chinchilla, 253f
  hamster, 315f
  skunk, 427f
Vertebral subluxation, skunk, 427f
Virginia opossum. *See* Opossum
Volume rendering, three-dimensional reconstructions, 48f
Voxel, 46f

W
Wave mouth, chinchilla, 229f
Weight loss, chronic, ferret, 372f
Window, of CT numbers, 46f

X
X-ray beam
  collimation, 5f, 37f
  magnification/distortion artifacts of, 6
X-ray film, 3, 5f
  processing, 39
  sizes, 5f
X-ray image formation, 2
X-ray image recording, cassette/film for, 3
X-ray machine
  adjusting settings on, 2–3
  control console, 4f
  exposure time, 3
  kilovoltage peak, 3
X-ray production, 2
X-ray projections, 8, 8f
X-ray tube, diagram, 2f

Z
Zygomatic bone
  chinchilla, 224f, 226f, 227f
  degu, 266f, 268f
  ferret, 360f, 361f
  guinea pig, 168f, 170f, 171f
  hamster, 298f, 300f, 301f
  opossum, 439f
  potbellied pig, 459f
  prairie dog, 326f, 328f
  rabbit, 54f, 57f
  rat, 276f, 278f
  ventrodorsal projection, 56f

Figures presented in this book have been provided by:

Dr. Claudio Maria Bussadori:
2.155, 2.161, 2.162.
10.24.

Dr. Vittorio Capello:
1.2, 1.3, 1.4, 1.5, 1.6, 1.9, 1.11, 1.12b,c, 1.13, 1.14, 1.15, 1.16, 1.17, 1.18, 1.19, 1.20, 1.21, 1.22, 1.23, 1.24, 1.29, 1.30, 1.31, 1.32, 1.33, 1.34, 1.35, 1.36, 1.39, 1.40. 1.41, 1.42, 1.43, 1.44, 1.45, 1.46, 1.47, 1.48, 1.49, 1.50, 1.51, 1.52, 1.53, 1.54, 1.55, 1.56, 1.57, 1.58, 1.59, 1.60, 1.61, 1.64, 1.65, 1.66, 1.67, 1.68, 1.69, 1.70, 1.71, 1.72, 1.73, 1.74, 1.75, 1.76, 1.77, 1.78, 1.79, 1.80, 1.81, 1.82, 1.83, 1.84, 1.85, 1.86, 1.87, 1.88, 1.89, 1.90, 1.91, 1.92, 1.93, 1.94.
2.1, 2.2, 2.3, 2.4, 2.5, 2.6, 2.7, 2.8, 2.9a, 2.13, 2.14, 2.15, 2.16, 2.17, 2.18, 2.19, 2.20, 2.21, 2.22, 2.23, 2.24, 2.25, 2.26, 2.27, 2.28, 2.29, 2.30, 2.31, 2.32, 2.33, 2.34, 2.35, 2.36, 2.37, 2.38, 2.39, 2.40, 2.41, 2.42, 2.43, 2.44, 2.45, 2.46, 2.47, 2.48, 2.49, 2.50, 2.51, 2.52, 2.53, 2.54, 2.55, 2.56, 2.57, 2.58, 2.60, 2.64, 2.65, 2.66, 2.67, 2.68, 2.69, 2.70, 2.71, 2.72, 2.75, 2.76, 2.77, 2.78, 2.79, 2.80a, 2.90, 2.91, 2.92, 2.93, 2.94, 2.95, 2.96, 2.97, 2.98, 2.99, 2.100, 2.101, 2.102, 2.103, 2.104, 2.105, 2.106, 2.107, 2.108, 2.109, 2.110, 2.111, 2.112, 2.113, 2.114, 2.115, 2.116, 2.117, 2.118, 2.119, 2.121, 2.122, 2.123, 2.124, 2.125, 2.126, 2.127, 2.128, 2.129, 2.130, 2.131, 2.132, 2.133, 2.134, 2.135, 2.136, 2.137, 2.138, 2.139, 2.140, 2.141, 2.142, 2.143, 2.144, 2.145, 2.146, 2.150, 2.153, 2.154, 2.156, 2.157, 2.158, 2.159, 2.160, 2.164, 2.165, 2.166, 2.167, 2.168, 2.169, 2.170, 2.171, 2.172, 2.173, 2.174, 2.175, 2.176, 2.177, 2.178, 2.179, 2.180, 2.181, 2.182, 2.183, 2.184, 2.185, 2.186, 2.187, 2.188, 2.189, 2.190, 2.191, 2.192, 2.193, 2.194, 2.204, 2.205, 2.206, 2.207, 2.208, 2.209, 2.210, 2.211, 2.212, 2.213, 2.214, 2.215, 2.216, 2.217, 2.218, 2.219, 2.220, 2.221, 2.222, 2.223, 2.224, 2.225, 2.226, 2.227, 2.228, 2.229, 2.230, 2.231, 2.232, 2.233, 2.234, 2.235, 2.236, 2.237, 2.238, 2.239, 2.240, 2.241, 2.242, 2.243, 2.244, 2.245, 2.246, 2.247, 2.248, 2.249, 2.250, 2.251, 2.252, 2.253, 2.254, 2.255, 2.256, 2.257, 2.258, 2.259, 2.260, 2.261, 2.262, 2.263, 2.264, 2.265, 2.266, 2.267, 2.268, 2.269, 2.270, 2.271, 2.272, 2.273, 2.274, 2.275, 2.276.
3.1, 3.2, 3.3, 3.4, 3.5, 3.6, 3.7, 3.8a, 3.9, 3.10, 3.11, 3.12, 3.13, 3.14, 3.15, 3.16, 3.17, 3.18, 3.19, 3.20, 3.21, 3.22, 3.24a, 3.28, 3.29, 3.30, 3.31, 3.32, 3.35, 3.36, 3.37, 3.38, 3.39, 3.40, 3.41, 3.42, 3.43, 3.44, 3.45, 3.46, 3.47, 3.48, 3.49, 3.50, 3.51, 3.52, 3.53, 3.54, 3.55, 3.56, 3.61, 3.62, 3.63, 3.64, 3.65, 3.66, 3.67, 3.68, 3.69, 3.70, 3.71, 3.72, 3.73, 3.74, 3.75, 3.76, 3.77, 3.78, 3.79, 3.83, 3.84, 3.85, 3.86, 3.89, 3.90.
4.1, 4.2, 4.3, 4.4, 4.5, 4.6a, 4.7, 4.8, 4.10, 4.11, 4.12, 4.13, 4.14, 4.15, 4.16, 4.17, 4.18, 4.19, 4.20, 4.21, 4.22a, 4.27, 4.28, 4.29, 4.30, 4.31, 4.32, 4.33, 4.34, 4.35, 4.36, 4.37, 4.38, 4.39, 4.40, 4.41, 4.42, 4.43, 4.44, 4.45, 4.46, 4.47, 4.48, 4.49, 4.50, 4.51, 4.52, 4.53, 4.54, 4.55, 4.56, 4.57, 4.58, 4.59, 4.60, 4.61, 4.62, 4.63, 4.64, 4.65, 4.66.
5.1, 5.2, 5.3, 5.4, 5.5, 5.6, 5.7, 5.8, 5.9, 5.10, 5.11, 5.12.
6.1, 6.2, 6.3, 6.4, 6.5, 6.6, 6.18, 6.19.
8.1, 8.2, 8.3, 8.4, 8.5, 8.6, 8.7, 8.8, 8.9, 8.10, 8.11, 8.12, 8.13, 8.14, 8.15, 8.16, 8.17, 8.19, 8.21, 8.22, 8.23, 8.24, 8.25, 8.26, 8.27, 8.28, 8.29, 8.30, 8.31, 8.32, 8.33, 8.34, 8.35, 8.36, 8.37, 8.38, 8.39, 8.40, 8.41, 8.42, 8.43, 8.44, 8.45, 8.46, 8.47, 8.48, 8.49, 8.50, 8.51, 8.52, 8.53, 8.54, 8.55, 8.56, 8.57, 8.58, 8.59, 8.60, 8.61, 8.62, 8.63, 8.64, 8.65.
9.1, 9.2, 9.3, 9.4, 9.5, 9.6, 9.7, 9.10, 9.11, 9.12, 9.13, 9.14, 9.15, 9.17, 9.18, 9.25, 9.26, 9.27, 9.32, 9.33, 9.34, 9.35, 9.36, 9.37, 9.38, 9.39, 9.40, 9.41, 9.42, 9.43, 9.44.
10.1, 10.2, 10.3, 10.4, 10.5, 10.6, 10.7, 10.8, 10.11, 10.12, 10.16, 10.17, 10.18, 10.19, 10.20, 10.21, 10.22, 10.23, 10.25, 10.26, 10.27, 10.28, 10.29, 10.31, 10.32, 10.33, 10.34, 10.35, 10.36, 10.37, 10.38, 10.39, 10.40, 10.41, 10.42, 10.43, 10.44, 10.45, 10.46, 10.47, 10.48, 10.49, 10.50, 10.51, 10.53, 10.54, 10.55, 10.56, 10.57, 10.58, 10.62, 10.63, 10.66, 10.67, 10.68, 10.69, 10.70, 10.71, 10.72, 10.73, 10.74, 10.76, 10.77, 10.78, 10.79, 10.80, 10.81, 10.85, 10.87, 10, 88, 10.89, 10.90, 10.91, 10.92, 10.93, 10.94, 10.95, 10.96, 10.97, 10.98, 10.99.
11.12, 11.18, 11.19, 11.20.
15.1, 15.2, 15.3.

Dr. Vittorio Capello and Dr. Alberto Cauduro:
1.99, 1.100, 1.101, 1.102a, 1.105, 1.106, 1.107, 1.108.
2.81, 2.83, 2.88.
3.25.
4.23, 4.24, 4.25.

Dr. Vittorio Capello and Dr. Margherita Gracis:
9.16.

Dr. Vittorio Capello and Dr. Angela Lennox:
1.25, 1.26, 1.27, 1.28, 1.62, 1.63.
2.59 2.61, 2.62, 2.63.
3.23, 3.87, 3.88.
4.9.
6.7, 6.8, 6.9, 6.10, 6.11, 6.12.
7.1, 7.2.
8.66.
9.19, 9.20, 9.21, 9.22, 9.23, 9.24, 9.28, 9.29, 9.30, 9.31.
11.3, 11.4, 11.5.
12.1, 12.2, 12.3, 12.4.
13.1, 13.2, 13.3, 13.4, 13.5, 13.6, 13.7, 13.8, 13.9, 13.10, 13.11, 13.12, 13.13.
14.1, 14.2, 14.7, 14.8, 14.11, 14.12.
16.1.

Dr. Alberto Cauduro:
1.102b.
2.9b, 2.80b,c,d,e,f,g,h,i,j,k,l, 2.82, 2.84, 2.87, 2.89.
3.8b, 3.24b,c,d,e,f,g,h,i,j,k,l.
4.6b, 4.22b,c,d,e,f,g,g,i,j,k.

Dr. Stefania Gianni and Dr. Vittorio Capello:
1.95, 1.96, 1.97, 1.98.
2.196, 2.197, 2.198, 2.199, 2.200, 2.201, 2.202, 2.203.

Dr. Margherita Gracis:
1.7, 1.8, 1.37, 1.38.
2.10, 2.11, 2.12.
9.8, 9.9.

Dr. Cathy Johnson-Delaney:
3.60.
7.3.
10.13, 10.14, 10.15, 10.55, 10.56, 10.57, 10.67, 10.68, 10.78, 10.86.
13.14, 13.15, 13.16.

Dr. Angela Lennox:
2.73, 2.120, 2.147, 2.148, 2.149, 2.151, 2.152, 2.163.
3.33, 3.34, 3.57, 3.58, 3.59, 3.80, 3.81, 3.82.
6.13, 6.14, 6.15, 6.16, 6.17.
8.18, 8.20
10.52, 10.62, 10.63, 10.64
11.1, 11.2, 11.6, 11.7, 11.8, 11.9, 11.10, 11.11, 11.13, 11.14, 11.15, 11.16, 11.17.
14.3, 14.4, 14.5, 14.6, 14.9, 14.10.
15.4, 15,5.

Dr. Angela Lennox and Dr. Vittorio Capello:
2.85, 2.86.
3.26, 3.27.

Dr. Joerg Mayer:
10.30.

Dr. Yasutsugu Miwa:
2.74.

Purdue University School of Veterinary Medicine, Large Animal Clinic:
14.13.

Dr. Giorgio Romanelli:
10.9, 10.10.

Dr. William R. Widmer:
1.1, 1.10, 1.12a, 1.103, 1.104, 1.109.